Recent Advances in Pharmaceutical Innovation and Research

Prati Pal Singh
Editor

Recent Advances in Pharmaceutical Innovation and Research

Springer

Editor
Prati Pal Singh
Department of Pharmacology and Toxicology
National Institute of Pharmaceutical Education and Research
Mohali, India

ISBN 978-981-99-2301-4 ISBN 978-981-99-2302-1 (eBook)
https://doi.org/10.1007/978-981-99-2302-1

This Springer imprint is published by the registered company Springer Nature Singapore Pte Ltd.
The registered company address is: 152 Beach Road, #21-01/04 Gateway East, Singapore 189721, Singapore

Foreword

The publication of the book *Recent Advances in Pharmaceutical Innovation and Research,* edited by Professor Prati Pal Singh, an internationally renowned pharmaceutical educationalist and scientist, is very much warranted and timely as the world is braving the ravages of the COVID-19 pandemic. The pandemic has greatly impacted the pharmaceutical industry and healthcare infrastructure in terms of science, technology, and economics. Therefore, these are times of both opportunities and challenges, and innovations and research in pharmaceuticals are expected to play defining roles more than ever before.

The editor Prof. Prati Pal Singh has more than 47 years of working experience in the pharmaceutical industry, and educational and research institutions. Prof. Singh has extensively contributed to the education and research of infectious diseases viz. malaria, tuberculosis, leishmaniasis, and amoebiasis. Prof. Singh has also contributed to the discovery of a new anti-amoebic drug satranidazole that is now on the market. He conceptualized and first reported that the opioid class of drugs can be used for the treatment of parasitic and infectious diseases. Additionally, Prof. Singh, as one of the principal investigators in the Indo-EU FP-7 project, immensely contributed to the discovery of new anti-tuberculosis molecules.

The chapters of the book have been authored by eminent experts and have been grouped under four heads viz. Pharmaceutical Innovation and Research, Drug Discovery and Development, Drug Delivery and Nanotechnology, and Pharmaceutical Microbiology. The contents of the book are very interesting, informative, and useful to students, innovators, researchers, and managers in the pharmaceutical academia and industry.

I congratulate Prof. Singh for his marvellous achievement of bringing out this important and timely compendium, and greatly appreciate the authors of the chapters.

Shashi Bala Singh

National Institute of Pharmaceutical
Education and Research
Hyderabad, India

National Institute of Pharmaceutical
Education and Research
Ahmedabad, India

Defence Research Development Organization
Delhi, India

Defence Institute of Physiology & Allied Sciences
New Delhi, India

Defence Institute of High Altitude Research
Leh, India
December 25, 2022

Preface

> *The process of turning an idea or invention into a product or service that adds value and/or that people would pay for is known as innovation. Focus on how the end-user customers perceive the impact of your innovation—rather than on how you, the innovators, perceive it.*
>
> (Thomas A. Edison)

The coronavirus disease 2019 (COVID-19) pandemic and the ensuing global healthcare-related problems, especially those centric on drugs and pharmaceuticals, and how the pandemic has impacted the global pharmaceutical industry, their business, and their innovations, have been the major driving forces behind the conceptualization of this book. On March 11, 2020, the World Health Organization declared COVID-19, caused by the coronavirus strain Severe Acute Respiratory Syndrome Coronavirus-2 (SARS-CoV-2), a pandemic. With an estimated 651,918,402 (651 million) confirmed cases and over 665,660 (6.6 million) deaths, globally, as of 4:54 pm CET, 23 December 2022, it is the most devastating global health problem the world has seen after the 1918 *Spanish flue* pandemic. The pandemic is not over yet, and the recent case surges in several countries have sent shock waves around. World over, during the pandemic, the healthcare infrastructure has been badly shaken up. There were several occasions when it was realized that our preparedness to handle such a calamity, in all its scale and dimensions, has been highly inadequate. Therefore, we require highly innovative scientific and technological solutions to meet future challenges of the entire healthcare sector; drugs and pharmaceuticals constitute one of its most important parts.

However, soon after the conceptualization of this book, I considered it expedient to temporize for a while thinking about its optics. The book has been entirely written and edited while the pandemic continues, and the ensuing scientific, social, and economic scars are still green. Therefore, the impending impact of the pandemic on pharmaceutical innovations, research, and developments, both in the short and long term, has never been subterranean to me. I have edited this book due to my deep interest, emotional connection, and over 40 years of educational, academic, research, and industrial experience in drugs, pharmaceuticals, and infectious diseases. Additionally, the grand success of my earlier four related books, viz. *Proceedings of International Conference on Biotechnological Approaches to Neuroimmunomodulation and Infectious Diseases* (2009), *Human Parasitic Infections of Pharmaceutical and National*

Health Importance (2009), *Water and Health* (2014), and *Infectious Diseases and Your Health* (2018), has provided me with the necessary enthusiasm and forte to edit this book.

I have, in this book, tried my best to get a reader to know the importance and significance of the recent advances in pharmaceutical innovation and research. With no intent to proselytize my readers, I must state that the ongoing business developments projected pharmaceutical sales, and the related dynamics strongly suggest that the global pharmaceutical industry is poised to grow tremendously. According to a recent estimate, the global pharmaceutical industry grew at a rapid pace, and by the end of 2021 touched a market value of approximately 1454.66 billion US dollars, and is projected to touch 1587.05 billion US dollars by 2022 (compound annual growth rate of 9.1%), and is further expected to grow to 1800 billion US dollars by 2026. Of this, the fast-growing global biopharmaceutical market valued at nearly 350.53 million US dollars is expected to approximately double in value by the end of this decade. The global pharmaceutical R&D spending, which is a key driver of innovation and research, of approximately 200 million US dollars in 2020 is also expected to grow to 230 million US dollars by 2026. It is quite apposite to mention here that new drug discovery is expensive (approx. 1.3 billion US dollars, the average cost of bringing a new drug into the market), labour intensive (involving nearly 10,000–15,000 people), long drawn (may take up to 15 years or even more), and a risky (nearly one in 10,000 chances, if one is a lottery winner) venture. The Indian pharmaceutical industry, on the other hand, is one of the major global players that stands 3rd by volume and 14th by value (42 billion US dollars domestic market in 2021) and is expected to touch 65 and 130 billion US dollars by 2024 and 2030, respectively. This is despite a statement in the UNDP report that nearly 65% of Indians have no access to even essential medicines. And while the entire global pharmaceutical industry was adversely impacted by the COVID-19 pandemic, the Indian pharmaceutical industry was resilient and continued to grow. It continues to supply hydroxychloroquine and other drugs to needy countries and has supplied COVID-19 vaccines to more than 100 countries. Therefore, given the expanding market potential, enormous costs, long and rigorous trials, and challenges and risks, drugs and pharmaceutical discovery and development are bound to be competitive and highly innovation- and research-centric. Nonetheless, it is one of the most assured and profitable markets, with earnings per share only second to the information technology sector.

The book consists of 31 chapters which have been grouped under four heads viz. Pharmaceutical Innovation and Research, Drug Discovery and Development, Drug Delivery and Nanotechnology, and Pharmaceutical Microbiology. Each chapter has been written by eminent experts, and there runs a deep current of innovation and research in each one of them, and then all these currents merge and form a confluence that takes the reader through the contents of the book.

It may not be possible for one to be innovative in isolation; one needs an innovative ecosystem along with an open competitive market/economy. As new drug discovery research and pharmaceutical innovations take quite a long time to accomplish, a researcher or an innovator has to bear in mind that the products obtained may see the market in the next 10–15 years or even longer. Therefore, they must have the foresight, to a fair measure of exactitude, as to what the market and its driving forces will be like, at that time. Additionally, the role(s) of technology push and demand pull has to be borne in mind. The advent of smart technologies like Artificial Intelligence (AI), Machine Learning (ML), and Natural Learning Processes (NLP) are also expected to greatly influence both the pharmaceutical industry and the market alike. In the present scenario, the green shoots are already visible.

I have tried to remain focused on drugs and pharmaceuticals and thus have deliberately avoided other related domains of healthcare which include medical devices, medical equipment, telemedicine, medical insurance, medical tourism, nursing homes, dispensaries, and hospitals. This has been primarily to ensure brevity and emphasis.

I thank Dr Vishwa Mohan Katoch, former Secretary, Department of Health Research, Govt. of India, and DG, ICMR; Dr Nitya Anand, former Director, Central Drug Research Institute (CDRI), Lucknow; late Prof. B. N. Dhawan, former Director, CDRI, Lucknow; late Prof. V. P. Sharma, former Director, Malaria Research Centre, Delhi, and ADG, ICMR; Dr K. Nagarajan, Director, Hindustan CIBA-GEIGY Research Centre, Mumbai; late Dr G. P. Dutta, former Scientist G, CDRI, Lucknow, and my Ph. D. supervisor; late Prof. R. C. Mahajan, Ex. HoD, PGI, Chandigarh; late Dr C. L. Kaul, Director, NIPER, Mohali; and Prof. A. P. Dash, Vice Chancellor, Central University, Puducherry, and former Director, National Malaria Research Institute, New Delhi, for their encouragement and support, and with whom, I have had intense and long discussions on various aspects of new drug research and infectious diseases. I am grateful to the late Prof. P. C. C. Garnham, CMG, FRS, for the encouragement and support during my early research years. My interactions with Prof. Wallace Peters, UK; Prof. Julius P. Kreier, USA; Prof. Richard F. Mortensen, USA; Prof. Paul F. Basch, USA; Prof. Robert M. Donahoe, USA; Prof. Howard E. Gendelman, USA; Prof. Trinidad Chakraborty, Germany; Prof. Dirk Schlüter, Germany; Prof. Bernhard Ryffel, France; and Prof. Donatella Taramelli, Italy, over several years, have always been stimulating and inspiring. I enjoyed intense intellectual discussions with some of my Indo-EU FP-7 tuberculosis research project collaborators Prof. Dr Piet Herdewyn, University of Leuven, Belgium; Prof. Dr Stefan Kaufmann, Max Planck Institute for Infection Biology, Berlin, Germany; Prof. Dr Marino Zerial, Max Planck Institute of Molecular Cell Biology and Genetics, Dresden, Germany; Dr Elaine Davis, National Institute for Medical Research, London, UK; Dr Matthias Wilmanns, European Molecular Biological Laboratory, Hamburg, Germany; and Dr Rajesh S. Gokhale, National Institute of Immunology, New Delhi, India (presently, Secretary, Department of Biotechnology, Govt. of India).

Mohali, Punjab, India Prati Pal Singh
December 30, 2022

Contents

Part I Pharmaceutical Innovation and Research

1 **Innovation in Stabilization of Biopharmaceuticals** 3
Shivcharan Prasad and Ipsita Roy

2 **Concepts in Pharmacogenomics: Tools and Applications** 41
Divya Vemula, Siva Singothu, and Vasundhra Bhandari

3 **Pharmacogenetics: A New Approach for the Selection of the Right Drug** 77
Hitesh Malhotra, Abhishek Dabra, Preeti, Peeyush Kaushik, Anjoo Kamboj, and Rupesh Gautam

4 **Meeting the Unmet Challenges of Pharmaceutical Research Using Artificial Intelligence** 99
Sonal Mehrotra and Kamla Pathak

5 **Pharmaceutical Innovation and Research in Ayurved** 117
Nilima Wadnerwar

6 **Ayurveda Intervention in Perimenopausal Syndrome** 137
Pratiksha P. Rathod

Part II Drug Discovery and Development

7 **Innovative Strategies in Drug Discovery and Pharmacoinformatics** 145
K. Nagarajan, Richa Goel, Roma Ghai, and Parul Grover

8 **Advances in Peptide Synthesis** 193
Gajanan K. Rathod, Rajkumar Misra, and Rahul Jain

9 **Peptidomimetics in Medicinal Chemistry: The Art of Transforming Peptides to Drugs** 215
K. Nagarajan and Garima Kapoor

10 **Diverse Pharmacological Activities of 4-Aminoquinoline and its Derivatives** 243
Deepak Kumar, Beena Negi, and Diwan S. Rawat

11 **Thiourea Catalysts for Synthesis of Active Pharmaceutical Ingredients** 293
Sharada Prasanna Swain, Kankanala Naveen Kumar, and V. Ravichandiran

12 **The Chemistry and Pharmacology of Mandarin Orange (*Citrus reticulata*)** 305
Inder Pal Singh, Dattatraya Dinkar Gore, Snehal Karkhele, and Charles Santhanaraju Vairappan

13 **Recent Progress in the Pharmacology of Phytoestrogens: Emerging Neuromodulators for Treating Anxiety and Depression** 321
Lovedeep Singh and Rajbir Bhatti

14 **Next-Generation Cell-Based Therapies to Combat Metastatic Brain Tumor** 341
Vignesh Balaji E and Karkala Sreedhara Ranganath Pai

15 **Pharmacokinetics and Pharmacodynamics: Fundamentals and Role(s) in Drug Discovery and Development** 357
Ashu Mittal, Roma Ghai, Alankar Srivastava, Deba Prasad Ghosh, and K. Nagarajan

16 **An Update on Pharmacokinetic Models** 395
Ravi Shankar and Kamla Pathak

Part III Drug Delivery and Nanotechnology

17 **Technological Innovations in Pharmaceutical Drug Nanocrystals** . . . 415
Sanika Jadhav, Amanpreet Kaur, and Arvind Kumar Bansal

18 **Fixed-Dose Combinations: Innovations and Research** 437
Sreedhar Dharmagadda

19 **Significance of Intranasal Drug Delivery Systems: Recent Trends and Clinical Investigations in Brain Disorders** 447
Deepshi Arora, Shailendra Bhatt, Manish Kumar, Rupesh K. Gautam, Yugam Taneja, and Muskan Chauhan

20 **Current Challenges and Nanotechnology-Based Pharmaceutical Approaches for the Treatment and Control of Visceral Leishmaniasis** 479
Ganesh Yadagiri and Shyam Lal Mudavath

21 Polymeric Nanoparticles-Based Drug Delivery Systems for Anticancer Therapy 499
Neeta Solanki and Harish Dureja

22 Carbon Nanotubes: A Review of Toxicity and Applicability in Biomedical Applications 517
Taranpreet Kaur and Raman Preet Singh

23 Recent Trends in Topical Herbal Nanotherapeutics for Psoriasis . . . 559
Deepti Katiyar, Abhishek Kumar, Priya Bansal, Surya Prakash, and N. G. Raghavendra Rao

Part IV Pharmaceutical Microbiology

24 Antibiotics: Past, Present, Future, and Clinical Pipeline 583
Raja Singh and Vibha Tandon

25 Innovative Strategies to Tackle Antimicrobial Resistance 621
Varsha Gupta and Nimitha K. Mohan

26 Molecular and Mechanistic Insights of Yeast Flocculation 633
Santhosh Kumar Sariki, Ramesh Kumawat, Ranu Singh, and Raghuvir Singh Tomar

27 Management of Hepatitis C Virus: Current and Future Treatments 653
Disha Arora, Smriti Parashar, and Rupesh K. Gautam

28 Tuberculosis: Experimental Models, Innovations, and Challenges . . . 669
Sarbjit Singh Jhamb, Raman Preet Singh, and Prati Pal Singh

29 Targeting Outer Membrane to Tackle Multidrug-Resistant Bacterial Pathogens 713
Grace Kaul, Manjulika Shukla, Isha Soni, Vaishnavi Veerareddy, Ravi Akunuri Kumar, Arunava Dasgupta, Nanduri Srinivas, and Sidharth Chopra

30 Diphtheria: A Paradigmatic Vaccine-Preventable Toxigenic Disease with Changing Epidemiology 749
Ivneet Kour, Lipika Singhal, and Varsha Gupta

31 Emerging and Re-Emerging Infections: An Overview 761
Varsha Gupta and Ritu Garg

Editor and Contributors

About the Editor

Prati Pal Singh, an internationally renowned educationist, scientist, and science writer/editor/communicator, has been working at the National Institute of Pharmaceutical Education and Research (NIPER), S. A. S Nagar, India, for the last 23 years and has now superannuated. For his Ph.D. degree (1980), he worked at the Central Drug Research Institute, Lucknow, and did postdoctoral research at Stanford University Medical Centre, CA, USA (1986), and the Ohio State University, OH, USA (1984–1985). Prof. Singh has been a visiting fellow at the Montreal General Hospital, Montreal, Canada (1985). He is an eminent medical microbiologist, immunologist, biotechnologist, and neuroimmunopharmacologist. Prof. Singh has worked extensively on the pre-clinical research and development of *satranidazole*, a new anti-amoebic drug now on the market. He first reported opiates as a new class of drugs for the treatment of parasitic/microbial infections and has made seminal contributions to the treatment of malaria, tuberculosis, leishmaniasis, amoebiasis, and trichomoniasis. Prof. Singh has edited/co-edited four books viz. *Infectious Diseases and Your Health* [2018, Springer Nature, ISBN: 978-981-13-1576-3 (Hardcover) 978-981-13-1577-0 (e-book); DOI 10.1007/978-981-13-1577-0], *Water and Health* [2014, Springer, ISBN 978-81-322-1028-3 (Hardcover) 978-81-322-1029-0 (e-book); DOI 10.1007/978-81-322-1029-0], *Human Parasitic Infections of Pharmaceutical and National Health Importance* (2009, ISSN 0369-8211), and *Biotechnical Approaches to Neuroimmunomodulation and Infectious Diseases*

(2009, ISBN 978-81-8465-013-6). He has over 200 research publications including one in *Nature Medicine*, five editorials, and nine patents including one US patent to his credit, and 81 Ph.D. and master's degrees have been awarded under his guidance. Prof. Singh has been the *Editor-in-Chief* for various journals and is on the editorial board of the US journal and several other international journals. Prof. Singh has been elected Fellow of *The National Academy of Sciences, India* (2004), and Fellow of the *Indian Academy of Microbiological Sciences* (2008), and was nominated for the fellowships *of* the *Indian National Science Academy* and the *Indian Science Academy*. He was elected *At-Large Councilor* (non-US), Society on Neuroimmune Pharmacology USA in 2010. Prof. Singh has been awarded the Bill and Melinda Gates Foundation Global Health Award in 2008, the National Institute on Drug Abuse, USA award in 2006, and the Tulasbai Somani Educational Trust Award of the Indian Academy of Neurosciences in 1992. Prof. Singh has been one of the Principal Investigators of the Indo-European Union 7th Framework Programme (2008) collaborative project "New Approaches to Target Tuberculosis". He has been invited to several lectures/official meetings both in India and in the USA, the UK, Canada, Switzerland, France, Germany, Austria, Italy, Belgium, and China. He has been the Nodal Officer of NIPER, Raebareli, and Member/Chairman of several high-powered Scientific Advisory Committees/sub-committees, various award committees, and selection committees of several institutes. Prof. Singh has been a Senator and Associate Dean (Academic Affairs) of NIPER. Presently, Prof. Singh is an Expert Specialist for various government research and funding agencies.

Contributors

Deepshi Arora M.M. College of Pharmacy, Maharishi Markandeshwar, Ambala, India
Guru Gobind Singh College of Pharmacy, Yamuna Nagar, India

Disha Arora Chandigarh College of Pharmacy, Mohali, Punjab, India

Arvind Kumar Bansal Solid State Pharmaceutics Lab, Department of Pharmaceutics, National Institute of Pharmaceutical Education and Research (NIPER), Mohali, Punjab, India

Priya Bansal Department of Pharmacology, KIET School of Pharmacy, KIET Group of Institutions, Ghaziabad, UP, India

Vasundhra Bhandari Department of Pharmacoinformatics, National Institute of Pharmaceutical Education and Research, Hyderabad, India

Shailendra Bhatt Department of Pharmaceutics, School of Medical and Allied Sciences, G.D. Goenka University, Gurugram, India

Rajbir Bhatti Department of Pharmaceutical Sciences, Guru Nanak Dev University, Amritsar, Punjab, India

Muskan Chauhan Guru Gobind Singh College of Pharmacy, Yamuna Nagar, India

Abhishek Dabra Guru Gobind Singh College of Pharmacy, Yamuna Nagar, Haryana, India

Harish Dureja Faculty of Pharmaceutical Sciences, Maharshi Dayanand University, Rohtak, India

Ritu Garg Department of Microbiology, Maharishi Markandeshwar Institute of Medical Sciences and Research, Mullana, India

Rupesh Gautam Department of Pharmacology, Indore Institute of Pharmacy, IIST Campus, Rau, Indore, Madhya Pradesh, India

Rupesh K. Gautam Department of Pharmacology, Indore Institute of Pharmacy, IIST Campus, Indore, MP, India

Roma Ghai KIET School of Pharmacy, KIET Group of Institutions, Ghaziabad, Uttar Pradesh, India

Deba Prasad Ghosh KIET School of Pharmacy, KIET Group of Institutions, Ghaziabad, India

Richa Goel KIET School of Pharmacy, KIET Group of Institutions, Ghaziabad, Uttar Pradesh, India

Dattatraya Dinkar Gore Department of Natural Products, National Institute of Pharmaceutical Education and Research, S.A.S. Nagar, India

Parul Grover KIET School of Pharmacy, KIET Group of Institutions, Ghaziabad, Uttar Pradesh, India

Varsha Gupta Department of Microbiology, Government Medical College and Hospital, Chandigarh, India

Sanika Jadhav Solid State Pharmaceutics Lab, Department of Pharmaceutics, National Institute of Pharmaceutical Education and Research (NIPER), Mohali, Punjab, India

Rahul Jain Department of Medicinal Chemistry, National Institute of Pharmaceutical Education and Research, Mohali, Punjab, India

Sarbjit Singh Jhamb Department of Pharmacology and Toxicology, National Institute of Pharmaceutical Education and Research, Mohali, India

Anjoo Kamboj Chandigarh College of Pharmacy, Mohali, Punjab, India

Garima Kapoor KIET School of Pharmacy, KIET Group of Institutions, Ghaziabad, India

Snehal Karkhele Department of Natural Products, National Institute of Pharmaceutical Education and Research, S.A.S. Nagar, India

Deepti Katiyar Department of Pharmacognosy, KIET School of Pharmacy, KIET Group of Institutions, Ghaziabad, UP, India

Amanpreet Kaur Solid State Pharmaceutics Lab, Department of Pharmaceutics, National Institute of Pharmaceutical Education and Research (NIPER), Mohali, Punjab, India

Taranpreet Kaur Department of Biotechnology, Government Mohindra College, Patiala, Punjab, India

Peeyush Kaushik Chandigarh College of Pharmacy, Mohali, Punjab, India

Ivneet Kour Department of Microbiology, Government Medical College and Hospital, Chandigarh, India

Abhishek Kumar Department of Pharmacology, KIET School of Pharmacy, KIET Group of Institutions, Ghaziabad, UP, India

Deepak Kumar Shaheed Rajguru College of Applied Sciences for Women, University of Delhi, Delhi, India

Girish Kumar School of Pharmaceutical Sciences, MVN University, Palwal, India

Kankanala Naveen Kumar National Institute of Pharmaceutical Education and Research, Kolkata, WB, India

Manish Kumar M.M. College of Pharmacy, Maharishi Markandeshwar, Ambala, India

Ramesh Kumawat Laboratory of Chromatin Biology, Department of Biological Sciences, Indian Institute of Science Education and Research, Bhopal, India

Hitesh Malhotra Guru Gobind Singh College of Pharmacy, Yamuna Nagar, Haryana, India

Sonal Mehrotra Manipal College of Pharmaceutical Sciences, Manipal, Karnataka, India

Rajkumar Misra Department of Medicinal Chemistry, National Institute of Pharmaceutical Education and Research, Mohali, Punjab, India

Ashu Mittal KIET School of Pharmacy, KIET Group of Institutions, Ghaziabad, India

Nimitha K. Mohan Aster Medcity, Kochi, India

Shyam Lal Mudavath Infectious Disease Biology Laboratory, Chemical Biology Unit, Institute of Nano Science and Technology, Mohali, Punjab, India

K. Nagarajan KIET School of Pharmacy, KIET Group of Institutions, Ghaziabad, Uttar Pradesh, India

Beena Negi Department of Chemistry, Gargi College, University of Delhi, Delhi, India

Karkala Sreedhara Ranganath Pai Department of Pharmacology, Manipal College of Pharmaceutical Sciences, Manipal Academy of Higher Education, Manipal, Karnataka, India

Smriti Parashar Vedic Institutes of Pharmaceutical Education and Research, Sagar, Madhya Pradesh, India

Kamla Pathak Faculty of Pharmacy, Uttar Pradesh University of Medical Sciences, Etawah, UP, India

Surya Prakash Department of Pharmaceutical Chemistry, KIET School of Pharmacy, KIET Group of Institutions, Ghaziabad, UP, India

Shivcharan Prasad Department of Biotechnology, National Institute of Pharmaceutical Education and Research, S.A.S. Nagar, Punjab, India

Preeti Guru Gobind Singh College of Pharmacy, Yamuna Nagar, Haryana, India

N. G. Raghavendra Rao Department of Pharmaceutics, KIET School of Pharmacy, KIET Group of Institutions, Ghaziabad, UP, India

Gajanan K. Rathod Department of Medicinal Chemistry, National Institute of Pharmaceutical Education and Research, Mohali, Punjab, India

Pratiksha P. Rathod Department of Prasutitantra-Strirog at MGACH, Datta Meghe Institute of higher education & research, Wardha, Maharashtra, India
Sanskar Kalash, Amravati, Maharashtra, India

V. Ravichandiran National Institute of Pharmaceutical Education and Research, Kolkata, WB, India

Diwan S. Rawat Department of Chemistry, University of Delhi, Delhi, India

Ipsita Roy Department of Biotechnology, National Institute of Pharmaceutical Education and Research, S.A.S. Nagar, Punjab, India

Santhosh Kumar Sariki Laboratory of Chromatin Biology, Department of Biological Sciences, Indian Institute of Science Education and Research, Bhopal, India
Molecular and Integrative Physiology Department, University of Michigan, Ann Arbor, MI, USA

Ravi Shankar Department of Pharmacy, Sagar Institute of Technology and Management, Barabanki, India

Inder Pal Singh Department of Natural Products, National Institute of Pharmaceutical Education and Research, S.A.S. Nagar, India

Lovedeep Singh Department of Pharmaceutical Sciences, Guru Nanak Dev University, Amritsar, Punjab, India
University Institute of Pharma Sciences, Chandigarh University, Mohali, Punjab, India

Prati Pal Singh Department of Pharmacology and Toxicology, National Institute of Pharmaceutical Education and Research, Mohali, India

Raja Singh Special Centre for Molecular Medicine, Jawaharlal Nehru University, New Delhi, India

Raman Preet Singh Department of Pharmacy, Government Polytechnic College, Bathinda, Punjab, India
Department of Pharmaceutical Sciences, Government Polytechnic College for Girls, Patiala, Punjab, India

Ranu Singh Laboratory of Chromatin Biology, Department of Biological Sciences, Indian Institute of Science Education and Research, Bhopal, India

Lipika Singhal Department of Microbiology, Government Medical College and Hospital, Chandigarh, India

Siva Singothu Department of Pharmacoinformatics, National Institute of Pharmaceutical Education and Research, Hyderabad, India

Neeta Solanki Faculty of Pharmaceutical Sciences, Maharshi Dayanand University, Rohtak, India

D. Sreedhar Department of Pharmaceutical Regulatory Affairs and Management, Manipal College of Pharmaceutical Sciences, Manipal Academy of Higher Education, Manipal, India

Alankar Srivastava KIET School of Pharmacy, KIET Group of Institutions, Ghaziabad, India

Sharada Prasanna Swain National Institute of Pharmaceutical Education and Research, Kolkata, WB, India

Vibha Tandon Special Centre for Molecular Medicine, Jawaharlal Nehru University, New Delhi, India

Yugam Taneja Zeon Lifesciences Pvt. Ltd, Paonta Sahib, Himachal Pradesh, India

Raghuvir Singh Tomar Laboratory of Chromatin Biology, Department of Biological Sciences, Indian Institute of Science Education and Research, Bhopal, India

Charles Santhanaraju Vairappan Institute for Tropical Biology and Conservation, Universiti Malaysia, Kota Kinabalu, Sabah, Malaysia

Divya Vemula Department of Pharmacoinformatics, National Institute of Pharmaceutical Education and Research, Hyderabad, India

E. Vignesh Balaji Department of Pharmacology, Manipal College of Pharmaceutical Sciences, Manipal Academy of Higher Education, Manipal, Karnataka, India

Reshu Virmani School of Pharmaceutical Sciences, MVN University, Palwal, India

Tarun Virmani School of Pharmaceutical Sciences, MVN University, Palwal, India

Nilima Wadnerwar Department of Agadtantra, Mahatma Gandhi Ayurved College Hospital and Research Centre, Salod (H), Datta Meghe Institute of Higher Education and Research (DU), Wardha, Maharashtra, India

Ganesh Yadagiri Infectious Disease Biology Laboratory, Chemical Biology Unit, Institute of Nano Science and Technology, Mohali, Punjab, India

Part I

Pharmaceutical Innovation and Research

Innovation in Stabilization of Biopharmaceuticals

1

Shivcharan Prasad and Ipsita Roy

Abstract

The applications of biopharmaceuticals, therapeutic molecules of biological origin, are increasing every day. The major reasons for this preference are target specificity and lower side effects. Production and downstream processing of biopharmaceuticals have their challenges that raise their final cost. The instability of proteins, during expression, purification, formulation, storage, and administration, is an issue that needs to be addressed if biopharmaceuticals are to achieve their full clinical potential. This requires a thorough understanding of the protein structure and changes that it undergoes when exposed to physical and chemical stressors. Changes in protein structure lead to a reduction in therapeutic efficacy but may introduce undesirable immunogenicity as well. A comprehensive appreciation of these changes will help devise rational usage of osmolytes as protein stabilizers. A novel direction in this is the use of nucleic aptamers that are capable of stabilizing proteins against a multitude of stress conditions. Changing/modifying pre-conceived notions of protein stabilizers is likely to yield long-lasting formulations.

Keywords

Aptamers · Biopharmaceuticals · Excipients · Formulations · Protein aggregation · Protein stabilization · Stabilizers

S. Prasad · I. Roy (✉)
Department of Biotechnology, National Institute of Pharmaceutical Education and Research, S.A.S. Nagar, Punjab, India
e-mail: ipsita@niper.ac.in

P. P. Singh (ed.), *Recent Advances in Pharmaceutical Innovation and Research*,
https://doi.org/10.1007/978-981-99-2302-1_1

1.1 Introduction

The expansion of various facets of life science has led to the opening up of new frontiers in the field of biopharmaceuticals. The aim is to develop new therapeutic strategies for disease conditions and the improvement of human health and quality of life. These therapeutic strategies rely on two types of drugs, viz., chemical and biological. The latter has some obvious, textbook advantages, the major one being their target specificity. This also results in lower side effects and toxicity (with optimization of delivery routes). Their importance also arises from the fact that they can target previously untreatable conditions. Biopharmaceuticals include proteins, antibodies in different formats, nucleic acids (antisense oligonucleotides, short interfering RNAs, aptamers), peptides, and vaccines. The global market for biopharmaceuticals was worth USD 401.32 billion (bn) in 2021 and is projected to reach ~USD 534.19 bn in 2027 (https://www.mordorintelligence.com/industry-reports/global-biopharmaceuticals-market-industry#:~:text=The%20Biopharmaceuticals%20Market%20was%20valued,impact%20on%20the%20biopharmaceutical%20industry, accessed on May 17, 2022). The relative contribution of different segments to global revenue in biopharmaceuticals is shown in Fig. 1.1a while the different disease areas where biologics find us are shown in Fig. 1.1b. Tremendous progress in recombinant DNA technology has allowed the large-scale production of biologics with an economical model. A classical example is the use of insulin for diabetics. In the early days, the only option available to diabetic patients was insulin isolated from pig and cow pancreas. This had associated disadvantages like immunogenicity and lack of starting material. The licensing of Humulin®, the first marketable product created via recombinant DNA technology, to Eli Lilly by Genentech in 1982 marked the advent of a new era in the biopharmaceutical industry (Kesik-Brodacka 2018).

Antibodies constitute a poster child of biopharmaceuticals (Table 1.1). The global market for monoclonal antibodies is pegged at USD 111.01 bn in 2021 and may reach USD 243.05 bn in 2028 (https://www.prnewswire.com/news-releases/

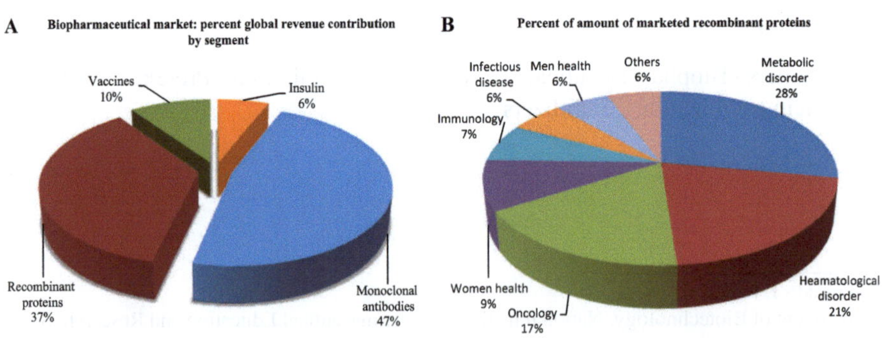

Fig. 1.1 (a) Global revenue contribution of recombinant protein in biopharmaceuticals. (b) Percent use of recombinant protein in the treatment of various sectors. (Adapted from Sanchez-Garcia et al. 2016)

Table 1.1 List of approved antibody therapeutics

Sr. no.	International non-proprietary name	Brand name	Target; format	Indication first approved	Year of first approval	Formulations (liquid); final concentration is per mL	Formulations (lyophilized); final concentration per mL after reconstitution
1.	Muromonab-CD3	Orthoclone Okt3	CD3; murine IgG2a	Reversal of kidney transplant rejection	1986	–	–
2.	Nivolumab	Opdivo	PD-1; human IgG4	Melanoma; non-small cell lung cancer	2014	Sodium citrate (5.88 mg)	–
3.	Pembrolizumab	Keytruda	PD-1; humanized IgG4	Melanoma	2014	L-histidine (1.55 mg); polysorbate 80 (0.2 mg); sucrose (70 mg)	L-histidine (1.55 mg); polysorbate 80 (0.2 mg); sucrose (70 mg)
4.	Blinatumomab	Blincyto	CD19 and CD3; murine bispecific tandem scFv	Acute lymphoblastic leukemia	2014	Citric acid monohydrate (5.25 mg); lysine hydrochloride (228.38 mg); polysorbate 80 (1 mg) as solution stabilizer	Citric acid monohydrate (1.11 mg); lysine hydrochloride (7.74 mg); polysorbate 80 (0.21 mg); trehalose dihydrate (31.83 mg)
5.	Alemtuzumab	Lemtrada; MabCampath; Campath-1H	CD52; humanized IgG1	Multiple sclerosis; chronic myeloid leukemia	2013; 2001	Sodium chloride (24.0 mg); dibasic sodium phosphate (3.5 mg); potassium chloride (0.6 mg); monobasic potassium phosphate (0.6 mg); polysorbate 80 (0.3 mg); disodium edentate (0.056 mg)	–
6.	Evolocumab	Repatha	PCSK9; human IgG2	High cholesterol	2015	Sodium acetate (1.2 mg); polysorbate 80 (0.1 mg); *proline* (25 mg)	–

(continued)

Table 1.1 (continued)

Sr. no.	International non-proprietary name	Brand name	Target; format	Indication first approved	Year of first approval	Formulations (liquid); final concentration is per mL	Formulations (lyophilized); final concentration per mL after reconstitution
7.	Idarucizumab	Praxbind	Dabigatran; humanized Fab	Reversal of dabigatran-induced anticoagulation	2015	Sorbitol (40.08 mg); polysorbate 20 (0.2 mg); sodium acetate (2.95 mg)	–
8.	Necitumumab	Portrazza	EGFR; human IgG1	Non-small cell lung cancer	2015	Mannitol (9.1 mg); polysorbate 80 (0.1 mg); sodium chloride (2.34 mg); sodium citrate (2.55 mg); citric acid (0.26 mg); glycine (9.98 mg)	–
9.	Dinutuximab	Unituxin	GD2; chimeric IgG1	Neuroblastoma	2015	Histidine (3.1 mg); polysorbate 20 (0.1 mg); sodium chloride (8.7 mg)	–
10.	Secukinumab	Cosentyx	IL-17a; human IgG1	Psoriasis	2015	L-histidine (3.103 mg); L-methionine (0.746 mg); polysorbate 80 (0.2 mg); trehalose dihydrate (75.67 mg)	L-histidine (4.656 mg); polysorbate 80 (0.6 mg); sucrose (92.43 mg)
11.	Mepolizumab	Nucala	IL-5; humanized IgG1	Severe eosinophilic asthma	2015	Citric acid monohydrate (0.95 mg); ethylenediaminetetraacetic acid (EDTA) disodium dihydrate (0.019 mg); polysorbate 80 (0.20 mg); sodium phosphate dibasic heptahydrate (4.16 mg); sucrose (120 mg)	Polysorbate 80 (0.67 mg); sodium phosphate dibasic heptahydrate (7.14 mg); sucrose (160 mg)

12.	Alirocumab	Praluent	PCSK9; human IgG1	High cholesterol	2015	Histidine (1.24 mg for 75 mg injection and 0.93 mg for 150 mg injection); polysorbate 20 (0.1 mg for both); sucrose (100 mg for both)	–
13.	Daratumumab	Darzalex	CD38; human IgG1	Multiple myeloma	2015	Glacial acetic acid (0.185 mg); mannitol (25.5 mg); polysorbate 20 (0.4 mg); sodium acetate trihydrate (2.93 mg); sodium chloride (3.5 mg)	–
14.	Elotuzumab	Empliciti	SLAMF7; humanized IgG1	Multiple myeloma	2015	–	Citric acid monohydrate (0.202 mg); polysorbate 80 (0.283 mg); sodium citrate (1.38 mg); sucrose (12.5 mg)
15.	Ixekizumab	Taltz	IL-17a; humanized IgG4	Psoriasis	2016	Citric acid anhydrous (0.51 mg); polysorbate 80 (0.3 mg); sodium chloride (11.69 mg); sodium citrate dihydrate (5.11 mg)	–
16.	Reslizumab	Cinqaero; Cinqair	IL-5; humanized IgG4	Asthma	2016	Glacial acetic acid (0.12 mg); sodium acetate trihydrate (2.45 mg); sucrose (70 mg)	–

(continued)

Table 1.1 (continued)

Sr. no.	International non-proprietary name	Brand name	Target; format	Indication first approved	Year of first approval	Formulations (liquid); final concentration is per mL	Formulations (lyophilized); final concentration per mL after reconstitution
17.	Olaratumab	Lartruvo	PDGFRα; human IgG1	Soft tissue sarcoma	2016	Glycine (7.5 mg); L-histidine (0.3 mg); L-histidine monohydrochloride (1.7 mg); mannitol (13.7 mg); polysorbate 20 (0.2 mg); sodium chloride (2.9 mg)	–
18.	Bezlotoxumab	Zinplava	*Clostridium difficile* enterotoxin B; human IgG1	Prevention of *Clostridium difficile* infection recurrence	2016	Citric acid monohydrate (0.8 mg); diethylene triamine pentaacetic acid (0.0078 mg); polysorbate 80 (0.25 mg); sodium chloride (8.77 mg); sodium citrate dihydrate (4.75 mg)	–
19.	Atezolizumab	Tecentriq	PD-L1; humanized IgG1	Bladder cancer	2016	Glacial acetic acid (16.5 mg); L-histidine (62 mg); sucrose (821.6 mg); polysorbate 20 (8 mg)	–
20.	Obiltoxaximab	Anthim	*Bacillus anthrasis* PA; chimeric IgG1	Prevention of inhalational anthrax	2016	Histidine (6.2 mg); polysorbate 80 (0.1 mg); sorbitol (36 mg)	–
21.	Brodalumab	Siliq; Lumicef	IL-17R; human IgG2	Plaque psoriasis	2017	Glutamate (4.33 mg); polysorbate 20 (0.1 mg); proline (24 mg)	–

22.	Dupilumab	Dupixent	IL-4R α; human IgG4	Atopic dermatitis	2017	L-arginine hydrochloride (5.25 mg); L-histidine (3.1 mg); polysorbate 80 (2 mg); sodium acetate (1 mg); sucrose (50 mg)	–
23.	Inotuzumab ozogamicin	Besponsa	CD22; humanized IgG4 ADC	Acute lymphoblastic leukemia	2017	–	Polysorbate 80 (0.09 mg); sodium chloride (0.54 mg); sucrose (45 mg); tromethamine (2.16 mg)
24.	Guselkumab	Tremfya	IL-23 p19; human IgG1	Plaque psoriasis	2017	L-histidine (0.6 mg); L-histidine monohydrochloride monohydrate (1.5 mg); polysorbate 80 (0.5 mg); sucrose (79 mg)	–
25.	Sarilumab	Kevzara	IL-6R; human IgG1	Rheumatoid arthritis	2017	Arginine (7.84 mg); histidine (3.25 mg); polysorbate 20 (2 mg); sucrose (50 mg)	–
26.	Avelumab	Bavencio	PD-L1; human IgG1	Merkel cell carcinoma	2017	D-mannitol (51 mg); glacial acetic acid (0.6 mg); polysorbate 20 (0.5 mg); sodium hydroxide (0.3 mg)	–
27.	Emicizumab	Hemlibra	Factor Ixa, X; humanized IgG4, bispecific	Hemophilia A	2017	L-arginine (26.1 mg); L-histidine (3.1 mg); poloxamer 188 (0.5 mg)	–
28.	Ocrelizumab	Ocrevus	CD20; humanized IgG1	Multiple sclerosis	2017	Glacial acetic acid (0.25 mg); polysorbate 20 (0.2 mg); sodium acetate trihydrate (2.14 mg); trehalose dihydrate (40 mg)	–

(continued)

Table 1.1 (continued)

Sr. no.	International non-proprietary name	Brand name	Target; format	Indication first approved	Year of first approval	Formulations (liquid); final concentration is per mL	Formulations (lyophilized); final concentration per mL after reconstitution
29.	Benralizumab	Fasenra	IL-5R α; humanized IgG1	Asthma	2017	L-histidine (1.4 mg); L-histidine hydrochloride monohydrate (2.3 mg); polysorbate 20 (0.06 mg); α,α-trehalose dihydrate (95 mg)	–
30.	Durvalumab	Imfinzi	PD-L1; human IgG1	Bladder cancer	2017	L-histidine (2 mg); L-histidine hydrochloride monohydrate (2.7 mg); α,α-trehalose dihydrate (104 mg); polysorbate 80 (0.2 mg)	–
31.	Gemtuzumab ozogamicin	Mylotarg	CD33; humanized IgG4 ADC	Acute myeloid leukemia	2017; 2000	Dextran 40 (8.2 mg); sodium chloride (5.22 mg); sodium phosphate dibasic anhydrous (0.54 mg); sodium phosphate monobasic monohydrate (0.09 mg); sucrose (13.96 mg)	–
32.	Erenumab	Aimovig	CGRP receptor; human IgG2	Migraine prevention	2018	Acetate (1.5 mg); polysorbate 80 (0.10 mg); sucrose (73 mg)	–
33.	Galcanezumab	Emgality	CGRP; humanized IgG4	Migraine prevention	2018	L-histidine (0.5 mg); L-histidine hydrochloride monohydrate (1.5 mg); polysorbate 80 (0.5 mg); sodium chloride (8.8 mg)	Galcanezumab

34.	Burosumab	Crysvita	FGF23; human IgG1	X-linked hypophosphatemia	2018	L-histidine (1.55 mg); L-methionine (1.49 mg); polysorbate 80 (0.5 mg); D-sorbitol (45.91 mg)	Burosumab
35.	Lanadelumab	Takhzyro	Plasma kallikrein; human IgG1	Hereditary angioedema attacks	2018	Citric acid monohydrate (4.1 mg); L-histidine (7.8 mg); polysorbate 80 (0.1 mg); sodium chloride (5.3 mg); sodium phosphate dibasic dihydrate (5.3 mg)	–
36.	Mogamulizumab	Poteligeo	CCR4; humanized IgG1	Mycosis fungoides or Sézary syndrome	2018	Citric acid monohydrate (0.44 mg); glycine (22.5 mg); polysorbate 80 (0.2 mg)	–
37.	Tildrakizumab	Ilumya	IL-23 p19; humanized IgG1	Plaque psoriasis	2018	L-histidine (0.495 mg); L-histidine hydrochloride monohydrate (1.42 mg); polysorbate 80 (0.5 mg); sucrose (70.0 mg)	–
38.	Fremanezumab	Ajovy	CGRP; humanized IgG2	Migraine prevention	2018	Disodium ethylenediaminetetraacetic acid (EDTA) dihydrate (0.136 mg); L-histidine (0.543 mg); L-histidine hydrochloride monohydrate (2.62 mg); polysorbate 80 (0.2 mg); sucrose (66 mg)	–

(continued)

Table 1.1 (continued)

Sr. no.	International non-proprietary name	Brand name	Target; format	Indication first approved	Year of first approval	Formulations (liquid); final concentration is per mL	Formulations (lyophilized); final concentration per mL after reconstitution
39.	Ravulizumab (ALXN1210)	Ultomiris	C5; humanized IgG2/4	Paroxysmal nocturnal hemoglobinuria	2018	Polysorbate 80 (0.2 mg); sodium chloride (8.77 mg); sodium phosphate dibasic (1.78 mg); sodium phosphate monobasic (0.46 mg)	–
40.	Cemiplimab	Libtayo	PD-1; human mAb	Cutaneous squamous cell carcinoma	2018	L-histidine (0.74 mg); L-histidine monohydrochloride monohydrate (1.1 mg); sucrose (50 mg); L-proline (15 mg); polysorbate 80 (2 mg)	–
45.	Ibalizumab; ibalizumab-uiyk	Trogarzo	CD4; humanized IgG4	HIV infection	2018	L-histidine (1.54 mg); polysorbate 80 (0.45 mg); sodium chloride (3 mg); sucrose (52 mg)	–
46.	Emapalumab; emapalumab-lzsg	Gamifant	IFNγ; human IgG1	Primary hemophagocytic lymphohistiocytosis	2018	L-histidine (1.55 mg); L-histidine monohydrochloride monohydrate (3.14 mg); polysorbate 80 (0.05 mg); sodium chloride (7.30 mg)	–
47.	Moxetumomab pasudotox	Lumoxiti	CD22; murine IgG1 dsFv immunotoxin	Hairy cell leukemia	2018	–	Glycine (80 mg); polysorbate 80 (0.2 mg); sodium phosphate monobasic monohydrate (3.4 mg); sucrose (40 mg)

48.	Caplacizumab	Cablivi	von Willebrand factor; humanized nanobody	Acquired thrombotic thrombocytopenic purpura	2019	–	Anhydrous citric acid (0.18 mg); polysorbate 80 (0.10 mg); sucrose (62 mg); trisodium citrate dihydrate (4.91 mg)
49.	Risankizumab	Skyrizi	IL-23 p19; humanized IgG1	Plaque psoriasis	2019	Disodium succinate hexahydrate (1.06 mg); polysorbate 20 (0.20 mg); sorbitol (40 mg); succinic acid (0.05 mg)	–
50.	Polatuzumab vedotin	Polivy	CD79b; humanized IgG1 ADC	Diffuse large B-cell lymphoma	2019	–	–
51.	Romosozumab	Evenity	Sclerostin; humanized IgG2	Osteoporosis in postmenopausal women at increased risk of fracture	2019	Acetate (3.24 mg); calcium (0.52 mg); polysorbate 20 (0.059 mg); sucrose (59.8 mg)	–
52.	Brolucizumab, brolucizumab-dbll	Beovu	VEGF-A; humanized scFv	Neovascular age-related macular degeneration	2019	Each 0.05 mL contains polysorbate 80 (0.02%), sodium citrate (10 mM), and sucrose (5.8%)	–
53.	Crizanlizumab	Adakveo	CD62 (aka P-selectin); humanized IgG2	Sickle cell disease	2019	Citric acid (0.54 mg); polysorbate 80 (0.2 mg); sodium citrate (5.05 mg); sucrose (75.33 mg)	–
54.	Enfortumab vedotin	Padcev	Nectin-4; human IgG1 ADC	Urothelial cancer	2019	Histidine (1.4 mg); histidine hydrochloride monohydrate (2.31 mg); polysorbate 20 (0.2 mg); trehalose dihydrate (55 mg)	–

(continued)

Table 1.1 (continued)

Sr. no.	International non-proprietary name	Brand name	Target; format	Indication first approved	Year of first approval	Formulations (liquid); final concentration is per mL	Formulations (lyophilized); final concentration per mL after reconstitution
55.	[Fam-] trastuzumab deruxtecan	Enhertu	HER2; humanized IgG1 ADC	HER2+ metastatic breast cancer	2019	–	L-histidine (0.89 mg); L-histidine hydrochloride monohydrate (4 mg); polysorbate 80 (0.3 mg); sucrose (90 mg)
56.	Isatuximab	Sarclisa	CD38; humanized IgG1	Multiple myeloma	2020	–	Citric acid monohydrate (0.20 mg); polysorbate 80 (0.28 mg); sodium citrate (1.38 mg); sucrose (42.5 mg)
57.	Teprotumumab	Tepezza	IGF-1R; human IgG1	Thyroid eye disease	2020	–	Contains inactive ingredients L-histidine, L-histidine hydrochloride, monohydrate, α,α-trehalose dehydrate, and polysorbate 20
58.	Inebilizumab	Uplizna	CD19; humanized IgG1	Neuromyelitis optica and neuromyelitis optica spectrum disorders	2020	–	–
59.	Eptinezumab	Vyepti	CGRP; humanized IgG1	Migraine prevention	2020	–	–
60.	Sacituzumab govitecan	Trodelvy	TROP-2; humanized IgG1	Triple-negative breast cancer	2020	–	–
61.	Satralizumab	Enspryng	IL-6R; humanized IgG2	Neuromyelitis optica spectrum disorder	2020	–	Satralizumab

62.	Tafasitamab	Monjuvi	CD19; humanized IgG1	Diffuse large B-cell lymphoma	2020	–	Tafasitamab
63.	Belantamab mafodotin	Blenrep	B-cell maturation antigen	Multiple myeloma	2020	–	Belantamab mafodotin
64.	Dostarlimab	Jemperli	PD-1; humanized IgG4	Endometrial cancer	2021	Dostarlimab-gxly (50 mg); citric acid monohydrate (0.48 mg); L-arginine hydrochloride (21.07 mg); polysorbate 80 (0.2 mg); sodium chloride (1.81 mg); trisodium citrate dihydrate (6.68 mg)	Dostarlimab
65.	Evinacumab	Evkeeza	Angiopoietin-like 3; human IgG4	Homozygous familial hypercholesterolemia	2021	Evinacumab-dgnb (150 mg); L-arginine hydrochloride (14.8 mg); L-histidine (0.74 mg); L-histidine monohydrochloride monohydrate (1.1 mg); L-proline (30 mg); polysorbate 80 (1 mg)	–
66.	Anifrolumab	Saphnelo	IFNAR1; human IgG1	Systemic lupus erythematosus	2021	Each vial contains 300 mg (150 mg/mL) of anifrolumab-fnia, L-histidine (3 mg), L-histidine hydrochloride monohydrate (6 mg), L-lysine hydrochloride (18 mg), polysorbate 80 (1 mg), and trehalose dihydrate (98 mg)	–

(continued)

Table 1.1 (continued)

Sr. no.	International non-proprietary name	Brand name	Target; format	Indication first approved	Year of first approval	Formulations (liquid); final concentration is per mL	Formulations (lyophilized); final concentration per mL after reconstitution
67.	Tebentafusp-tebn	Kimmtrak	gp100, CD3; bispecific immunoconjugate (TCR-scFv)	Metastatic uveal melanoma	2022	Each single-dose vial contains tebentafusp-tebn (100 µg), citric acid monohydrate (0.95 mg), disodium hydrogen phosphate (2.91 mg), mannitol (5 mg), polysorbate 20 (0.1 mg), and trehalose (25 mg)	–

ADC antibody–drug conjugate, *CD* cluster of differentiation, *CCR* CC chemokine receptor, *CGRP* calcitonin gene-related peptide, *FGF* fibroblast growth factor, *GD2* disialoganglioside, *HER* human epidermal growth factor receptor, *IFNAR* interferon alpha receptor, *IGF-1R* insulin-like growth factor 1 receptor, *IgG* immunoglobulin G, *EGFR* epidermal growth factor receptor, *HIV* human immunodeficiency virus, *IFNγ* interferon gamma, *IL* interleukin, *mAb* monoclonal antibody, *PA* protective antigen, *PCSK* proprotein convertase subtilisin/kexin, *PDGFR* platelet-derived growth factor receptor, *scFv* single-chain variable fragment, *SLAMF* signaling lymphocyte activation molecule family, *TCR* T cell receptor, *PD-1* programmed cell death protein 1, *PD-L1* programmed cell death ligand 1, *TROP* trophoblast cell surface antigen, *VEGF-A* vascular endothelial growth factor-A. Sources: https://www.antibodysociety.org/resources/approved-antibodies/, accessed on May 17, 2022; Kaplon and Reichert (2019), Walsh (2018)

Utmost care has been taken to prepare this table; however, please refer USP/BP/IP/EP for exact concentration of formulations

Formulation concentration has been obtained from the website of US FDA (www.fda.gov)

monoclonal-antibodies-market-size-worth-243bn-globally-by-2028-at-11-8-cagr%2D%2D-exclusive-report-by-the-insight-partners-01474558.html, accessed on May 17, 2022). The first commercial monoclonal antibody (mAb) was Muromonab-CD3 (Orthoclone OKT3, 1985; Janssen Biotech) for the prevention of glucocorticoid-resistant rejection of transplanted kidney, heart, and liver allografts. However, due to its short life and immunogenic nature, it was subsequently withdrawn by the U.S. Food and Drug Administration (FDA) (Sifniotis et al. 2019). The approval of Hemlibra (emicizumab-kxwh; Roche) by the US FDA to prevent or reduce the frequency of bleeding episodes in hemophilia A patients is a recent success story. This is a huge step in a disease where only a single prophylactic strategy existed to date (https://www.roche.com/media/releases/med-cor-2018-10-04c.htm, accessed on Jan 02, 2020).

The first biosimilar monoclonal antibody (mAb) therapy (infliximab-dyyb; Celltrion, Inc.) was approved by US FDA in 2016. Infliximab is a chimeric monoclonal IgG1 antibody composed of human and murine regions (Klotz et al. 2007). It binds to tumor necrosis factor-alpha (TNF-α) and inhibits its inflammatory activity. Recently, the fourth biosimilar (infliximab-axxq; Amgen) in this series was approved. The original molecule (tradename Remicade) was developed by Janssen Biotech Inc. and has proven to be a blockbuster molecule. This chapter discusses the various issues related to the (in)stability of biopharmaceuticals and what strategies are currently being adopted to overcome these roadblocks.

1.2 Stability Is the Major Concern for Biopharmaceuticals

Biopharmaceuticals are among the most sophisticated and elegant achievements of modern science. Production of biopharmaceuticals is a multistep production process and is influenced by various factors such as choice of a cell line, harvesting, purification, and further treatment of the finished product. Several factors can influence the stability and shelf-life of biopharmaceuticals. So, stringent measures need to be in place to maintain the integrity and efficacy of the product till it reaches the patient. Production of biopharmaceuticals (mainly proteins) follows a more arduous route than small molecules. Proteins are macromolecules consisting of a specified sequence of amino acids that determines their folding into a native three-dimensional structure to reach a functionally active conformation. This three-dimensional architecture is maintained by a fine balance of diverse covalent and non-covalent forces. The free energy of the unfolded state of the protein is low, typically 21–330 kJ/mol (Dagan et al. 2013). Hence, the folded state is only marginally stable as compared to the unfolded state, and any change in the physical and/or chemical parameters in its milieu can result in its denaturation or inactivation (Knotts et al. 2008; Pace et al. 2014). The long-term stability and safety profiles of proteins are major concerns. Therefore, pharmaceutical companies add excipients in the formulation to keep the properties of the active ingredient stable even after unexpected variations in storage temperature and other environmental factors. Biopharmaceuticals are very sensitive to changes in temperature and humidity, light and oxygen exposure, and shear forces, as compared to chemically synthesized drug. Pharmaceutical excipients are substances other than the pharmacologically active drug (i.e., non-reactive ingredients) added to the finished pharmaceutical dosage form during the manufacturing process. The market for pharmaceutical excipients was valued at USD 8.3 bn in 2021 and is projected to reach USD 10.6 bn by 2026 (https://www.bccresearch.com/market-research/pharmaceuticals/excipients-in-pharmaceuticals-global-markets.html, accessed on May 17, 2022). To reduce the cost of manufacturing and to speed up the drug development process, pharmaceutical companies are also developing new and novel excipients for formulation (Table 1.2). For example, the formulation of mini-tablets, used for pediatric and geriatric patients, is costly. The launch of a new excipient, SuperTab® 40LL, by DFE Pharma in 2017, which is highly flowable to allow for a uniform filling of the small die cavities and highly compactable to allow for fast(er) tableting, has proved helpful. Hence, therapeutic proteins need satisfactory stabilization; else they may end up being clinically inefficacious, with possible immunogenic reactions, with adverse effects on the patients' health. For example, the commercially available therapeutic antibody Humira® (adalimumab) was subjected to different stress. The monoclonal antibody was found to be unstable toward thermal and oxidative stress as monitored by size-exclusion chromatography and capillary electrophoresis (Katterle 2018). Another example is the aggregation of insulin during its storage and transportation in vials, infusion pumps, and controlled release devices. Agitation-induced stress led to the aggregation of the protein and a reduction in its biological availability (Brennan et al. 1985; Malik and Roy 2011). Shaking at the air–water interface leads to the unfolding of the protein and stabilization of the exposed hydrophobic interior by the hydrophobic air layer. This exposed core acts as the "seed" to facilitate the aggregation of the protein (Nielsen et al. 2001). Nowadays, aptamers have attracted the attention of many researchers as they are considered alternatives to antibodies along with having tremendous properties such as thermal stability, low cost, and unlimited applications. These have been discussed later.

Biopharmaceuticals with limited pharmacokinetic properties have been stabilized by attached polyethylene glycol (PEG) and polysialic acid (Swierczewska et al. 2015). First described in the 1970s (Abuchowski et al. 1977), pegylation refers to the covalent attachment of the polymer to the biologic. It may also refer to the encapsulation of the therapeutic molecule by PEG. Recently, a platform approach called "supramolecular modification" with PEG has been described, which can be extended to any protein–polymer pair (Webber et al. 2016). The ultimate result is enhanced solubility and attenuated immunogenicity of the biopharmaceutical, and increased systemic half-life (Harris and Chess 2003; Morpurgo and Veronese 2004; Turecek et al. 2016). As therapeutic proteins such as monoclonal antibodies and cytokines are eliminated by proteolysis and clearance in the kidney or liver, their bioavailability becomes limited. The first FDA-approved pegylated therapeutic proteins were Adagen® and Oncaspar®, which came into the market in 1990. Currently, more than 21 pegylated biologics have been approved by the US FDA (Kozma et al. 2020) and many more are in different stages of clinical trials (Table 1.3).

Table 1.2 List of top selling 15 drugs of 2018 and their formulations

Sl. no.	Product	Additives used	Company	Treatment
1	Humira®	Sodium chloride (6.16 mg), monobasic sodium phosphate dehydrate (0.86 mg), dibasic sodium phosphate dehydrate (1.52 mg), sodium citrate (0.3 mg), citric acid monohydrate (1.3 mg), mannitol (12 mg), and polysorbate 80 (1 mg)	AbbVie	Rheumatoid arthritis, psoriatic arthritis, ankylosing spondylitis, Crohn's disease, psoriasis, juvenile idiopathic arthritis, ulcerative colitis, hidradenitis suppurativa, and certain types of uveitis
2	Eliquis®	Tablet: anhydrous lactose, microcrystalline cellulose, croscarmellose sodium, sodium lauryl sulfate, and magnesium stearate	Bristol-Myers Squibb and Pfizer	Eliquis is a factor Xa inhibitor anticoagulant indicated to reduce the risk of stroke and systemic embolism in patients with non-valvular atrial fibrillation
3	Revlimid	Tablet: lactose anhydrous, microcrystalline cellulose, croscarmellose sodium, and magnesium stearate	Celgene	Multiple myeloma
4	Opdivo®	Liquid: mannitol (30 mg), pentetic acid (0.008 mg), polysorbate 80 (0.2 mg), sodium chloride (2.92 mg), and sodium citrate dihydrate (5.88 mg)	Bristol-Myers Squibb and Ono Pharmaceutical	Advanced skin cancer (melanoma), lung cancer, kidney cancer, and classical Hodgkin lymphoma
5	Keytruda®	Lyophilized powder: L-histidine (1.55 mg), polysorbate 80 (0.2 mg), and sucrose (70 mg)	Merck & Co.	Advanced skin cancer (melanoma or Merkel cell carcinoma), lung cancer, primary mediastinal large B-cell lymphoma, and advanced stomach cancer
6	Enbrel®	Liquid: 1% sucrose, 100 mM sodium chloride, 25 mM L-arginine hydrochloride, and 25 mM sodium phosphate	Amgen and Pfizer	Rheumatoid arthritis, psoriatic arthritis, or ankylosing spondylitis, and to prevent joint damage caused by these conditions
7	Herceptin®	α,α-trehalose dehydrate (400 mg), L-histidine hydrochloride (0.49 mg), L-histidine (0.32 mg), and polysorbate 20 (0.045 mg)	Roche (Genentech)	Certain types of breast cancer or stomach cancer
8	Avastin®	α,α-trehalose dehydrate (240 mg), sodium phosphate (monobasic, monohydrate) (5.8 mg), sodium phosphate (dibasic	Roche (Genentech)	Certain types of brain tumor, and certain types of cancers of the kidney, lung, colon, rectum, cervix, ovary, or fallopian tube

(continued)

Table 1.2 (continued)

Sl. no.	Product	Additives used	Company	Treatment
		anhydrous; 1.2 mg), and polysorbate 20 (0.4 mg)		
9	Rituxan® (also sold as MabThera)	Polysorbate 80 (0.7 mg), sodium citrate dihydrate (7.35 mg), and sodium chloride (9 mg)	Roche (Genentech) and Biogen	Non-Hodgkin's lymphoma or chronic lymphocytic leukemia in adults, and rheumatoid arthritis in adults
10	Xarelto®	Tablet: croscarmellose sodium, hypromellose, lactose monohydrate, magnesium stearate, microcrystalline cellulose, and sodium lauryl sulfate	Bayer and Johnson & Johnson	Deep vein thrombosis (DVT), and atrial fibrillation to lower the risk of stroke caused by a blood clot
11	Eylea®	40 mg in 10 mM sodium phosphate, 40 mM sodium chloride, 0.03% polysorbate 20, and 5% sucrose	Bayer and Regeneron Pharmaceuticals	Certain retinal diseases, such as wet age-related macular degeneration, diabetic macular edema, and macular edema following retinal vein occlusion
12	Remicade®	Sucrose (50 mg), polysorbate 80 (0.05 mg), monobasic sodium phosphate (0.22 mg), monohydrate, and 0.61 mg dibasic sodium phosphate, dihydrate	Johnson & Johnson and Merck & Co.	Crohn's disease, ulcerative colitis, rheumatoid arthritis, and ankylosing spondylitis
13	Prevnar 13®	Polysorbate 80 (0.1 mg), succinate buffer (0.59 mg), and aluminum as aluminum phosphate adjuvant (0.250 mg)	Prevenar 13® Pfizer	Pneumococcal pneumonia
14	Stelara	L-histidine and L-histidine monohydrochloride monohydrate (1 mg), polysorbate 80 (0.04 mg), and sucrose (76 mg)	Janssen Biotech (Johnson & Johnson)	Moderate or severe psoriasis
15	Lyrica®	Liquid: methylparaben, propylparaben, monobasic sodium phosphate anhydrous, dibasic sodium phosphate anhydrous, sucralose, and artificial strawberry #11545 Capsule: lactose monohydrate, cornstarch, and talc	Pfizer	This medication is used to treat pain—caused by nerve damage due to diabetes, shingles (herpes zoster) infection, or spinal cord injury—epilepsy, neuropathic pain, fibromyalgia, restless leg syndrome, and generalized anxiety disorder

Table 1.3 An illustrative list of marketed pegylated therapeutic proteins

Sr. no.	Trade name	Biomolecule	Condition	Company	Year of approval
1.	Esperoct	Recombinant factor VIII	Hemophilia A	Novo Nordisk	2019; launched in 2020
2.	Ziextenzo	G-CSF	Chemotherapy-associated infection	Sandoz	2019
3.	Udenyca	G-CSF	Chemotherapy-associated infection	Coherus Biosciences	2018
4.	Fulphila	G-CSF	Chemotherapy-associated infection	Mylan GmbH	2018
5.	Palynziq	Recombinant phenylalanine ammonia lyase	Phenylketonuria	BioMarin Pharmaceutical	2018
6.	Asparlas	L-Asparaginase	Leukemia	Servier Pharma	2018
7.	Revcovi	Recombinant adenosine deaminase	Severe combined immune deficiency	Leadiant Bioscience	2018
8.	Rebinyn	Recombinant factor IX	Hemophilia B	Novo Nordisk	2017
9.	Jivi	Recombinant factor VIII	Hemophilia A	Bayer Healthcare	2017
10.	Adynovate	Recombinant factor VIII	Hemophilia A	Baxalta	2015
11.	Plegridy	Interferon β-1a	Multiple sclerosis	Biogen	2014
12.	Omontys	Erythropoietin	Anemia	Takeda	2012
13.	Sylatron	Interferon-α-2b	Melanoma	Merck	2011
14.	Krystexxa	Recombinant uricase	Gout	Horizon Pharma	2010
15.	Cimzia	Certolizumab (tumor necrosis factor blocker)	Rheumatoid arthritis	UCB	2008
16.	Mircera	Erythropoietin	Anemia	Roche	2007
17.	Macugen	Aptamer	Age-related macular degeneration	Pfizer	2004
18.	Somavert	Human growth hormone	Acromegaly	Pfizer	2003
19.	Neulasta	G-CSF	Chemotherapy-associated infection	Amgen	2002
20.	Pegasys	Interferon-α-2a	Hepatitis B and C	Roche	2002
21.	Pegintron	Interferon-α-2b	Hepatitis C, melanoma	Schering	2001

(continued)

Table 1.3 (continued)

Sr. no.	Trade name	Biomolecule	Condition	Company	Year of approval
22.	Oncaspar	Asparaginase	Leukemia	Enzon	1994
23.	Adagen	Adenosine deaminase	Severe combined immune deficiency	Enzon	1990

G-CSF granulocyte colony-stimulating factor

An emerging concern is the presence of pre-existing anti-PEG antibodies that lead to adverse reactions in patients although the active ingredient is safe on its own (Kozma et al. 2020). This highlights the need to search for novel conjugation agents with better safety profiles.

Similar to other proteins, the stability of antibodies remains a major concern (Mahler and Allmendinger 2017). Because of their low bioavailability and consequent requirement of higher doses for therapeutic efficacy, antibodies are often formulated in concentrated forms. Apart from supporting aggregation, high concentration leads to high viscosity, which limits usage and causes injection site reactions. One of the ways of minimizing aggregation is to select the pH of the formulation correctly. A mild change in pH can affect the conformational stability of the protein and increase aggregation (Hartmann et al. 2004; Zhou and Pang 2018). The large size of antibodies also makes them prone to several chemical degradation mechanisms. These have been discussed later. Chemical reactions also depend on the solution pH, which determines the ionization of different amino acid residues of the protein molecule. Liquid formulations of biopharmaceuticals have been suggested to have mildly acidic pH to lower deamidation and isomerization (Cacia et al. 1996). For example, the stability of the monoclonal antibody against cluster of differentiation (CD)20 against oxidation could be improved by changing the pH from 7.5 to 5.0 (Johnson et al. 2004). The composition of the buffer is also of critical importance in formulations of monoclonal antibodies. Sodium phosphate and histidine are two commonly used buffers near the neutral pH (Warne 2011). Histidine has some distinct advantages for formulations of monoclonal antibodies at slightly acidic pH (Lee et al. 2011; Wörn and Plückthun 1998). Surfactants have been employed to minimize the lower agitation-induced aggregation of proteins. The preferential accumulation of surfactants at the interface interferes with and inhibits the proteins from being partitioned onto the interface, lowering their unfolding (Lee et al. 2011). The most frequently employed member of this class is polysorbates (polysorbate 20 and polysorbate 80; Jones et al. 2018; Kerwin 2008). Osmolytes like polyols can function as protein stabilizers and can protect biopharmaceutical formulations against a variety of stress conditions. Sugars like sucrose, mannose, and trehalose have been used as cryo- and/or lyo-protectant and as bulking agents (Chen et al. 2003; Jain and Roy 2009; Solanki et al. 2011) while mannitol and sorbitol used as a bulking agent during lyophilization. The addition of the disaccharide sucrose led to an enhancement of the melting temperature of RNase by 10°C (McIntosh et al.

1998). Mannitol and trehalose have been reported to improve the stability of recombinant human CD20 antibodies toward chemical denaturation by minimizing the number of negatively charged molecules (Lam and Rajaraman 2012).

Freeze-thawing is another factor for aggregation encountered during storage, transportation, and final use of therapeutic proteins (Carpenter and Crowe 1988; Vazquez-Rey and Lang 2011). Excessive freezing and thawing of therapeutic proteins produces several stresses in ice–liquid interface (Strambini and Gabellieri 1996) such as a change in pH of the buffer and freeze concentration of solutes, which is ultimately responsible for structural changes in protein structure (Izutsu 2018). Two commercially available formulations of recombinant human growth hormone, viz., Nordiflex® (Novo Nordisk) and Saizen® (Serono), were shown to form aggregates by freeze-thawing, which proved to be immunogenic in a mouse model (Fradkin et al. 2009). In such cases, excipients such as sugars, certain amino acids, polyols, salts, polymers, and surfactants have been shown to stabilize protein formulations (Arakawa et al. 2001; Arsiccio and Pisano 2017; Canchi and Garcia 2013; Izutsu 2005; Manning et al. 2010; Rayaprolu et al. 2018; Tonnis et al. 2015).

A recent study reports the higher post-reconstitution stability of inactivated polio vaccine formulation in the presence of a low concentration (0.4 M) of urea (Qi et al. 2018), which reduces the need for a cold chain. The requirement of excipients in liquid (Fig. 1.2) and lyophilized (Fig. 1.3) formulations of parenteral protein products may differ (Gervasi et al. 2018).

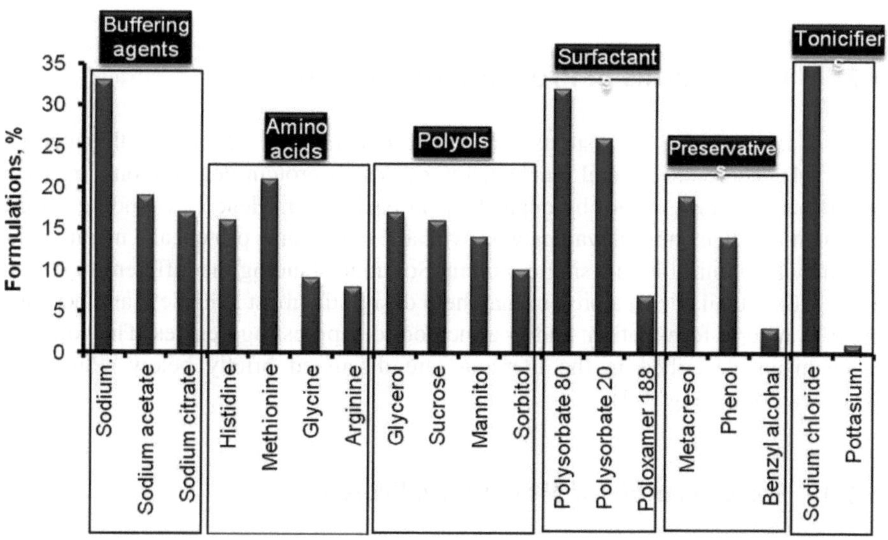

Fig. 1.2 The percentage of liquid parenteral protein products that contain excipients from each of the excipient categories: buffers, amino acids, polyols, surfactants, preservatives, and tonicifiers. (Reproduced from Gervasi et al. 2018, with permission from the publisher, https://doi.org/10.1016/j.ejpb.2018.07.011)

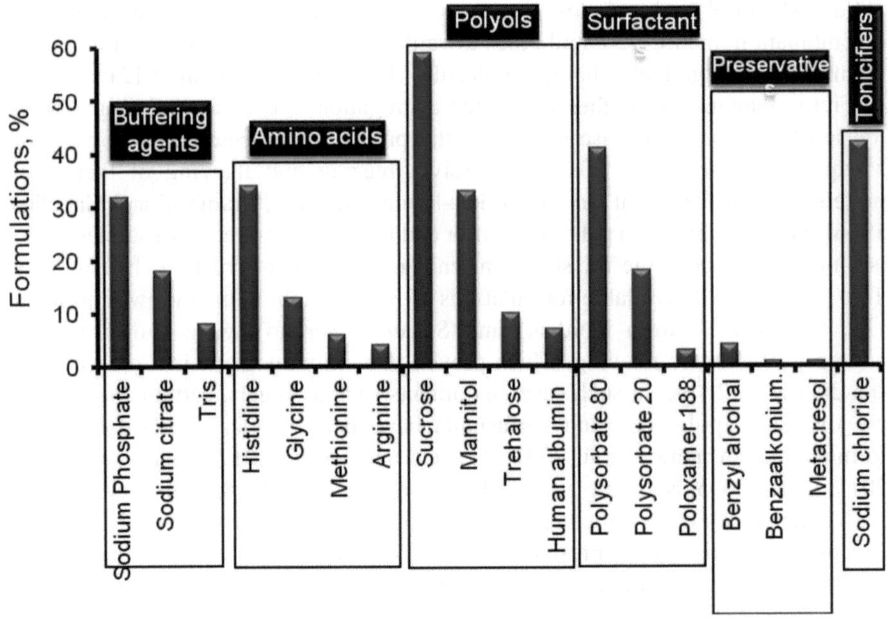

Fig. 1.3 The percentage of lyophilized parenteral protein products that contain excipients from each of the excipient categories: buffers, amino acids, polyols, surfactants, preservatives, and tonicifiers. (Reproduced from Gervasi et al. 2018, with permission from the publisher, https://doi.org/10.1016/j.ejpb.2018.07.011)

1.3 Mechanisms of Protein Degradation

Several factors may be responsible for the denaturation of protein therapeutics (Fig. 1.4). There are several mechanisms by which protein degradation can occur and these may or may not be dependent on each other. Hence, depending on the milieu, more than one pathway may be available to the same protein and no universal rule may be applicable to a single protein. So, understanding the different modes of aggregation available to a protein may help design the most suitable manufacturing process, a good formulation, and/or a method to suppress aggregates. The common mechanisms described in the literature are discussed briefly below (Philo and Arakawa 2009) (Fig. 1.5).

1.3.1 Self-Assembly of Monomeric Protein

This mechanism pertains to native proteins that have an inherent tendency to self-associate due to the presence of self-complementary surfaces. Their conformations favor interaction with similar surfaces, resulting in the formation of small reversible oligomers. With an increase in protein concentration, these small oligomers form

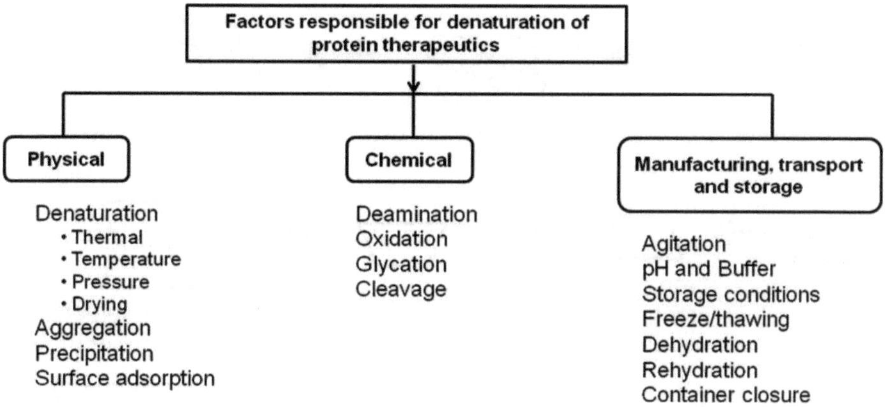

Fig. 1.4 Different factors responsible for protein denaturation

larger oligomers whereupon the process turns irreversible, transforming into protofibrils and mature fibrils. A classical example of this pathway is insulin, which associates readily to form reversible oligomers (Philo and Arakawa 2009). The interleukin-1 receptor antagonist too initially forms reversible dimers, which is followed by the formation of irreversible oligomers in a concentration-driven manner (Alford et al. 2008).

1.3.2 Aggregation of Conformationally Altered Monomers

In contrast to the above mechanism, in the current case, the native monomer does not self-associate readily. Any stress-induced conformational change in the monomer (a "non-native" state), however, may turn it susceptible to aggregation. Thus, this type of aggregation can be inhibited by adding excipients or altering conditions that stabilize the native conformation. For example, biologics like interferon-γ and granulocyte colony-stimulating factor (G-CSF) have been reported to undergo aggregation by this mechanism (Raso et al. 2005).

1.3.3 Aggregation of Chemically Modified Monomer

In this scenario, which is somewhat similar to the above mechanism, the "non-native" conformation results from chemical degradation during bioprocessing, e.g., oxidation of methionine (Met), hydrolysis, proteolysis, deamidation, isomerization, and glycation (Torosantucci et al. 2012). In some cases, the degree of chemical modification may also lead to alteration of properties of the biopharmaceutical, e.g., the extent of glycosylation has been seen to change the propensity of glycoproteins to aggregate. Covalent cross-linking may increase immunogenicity by causing

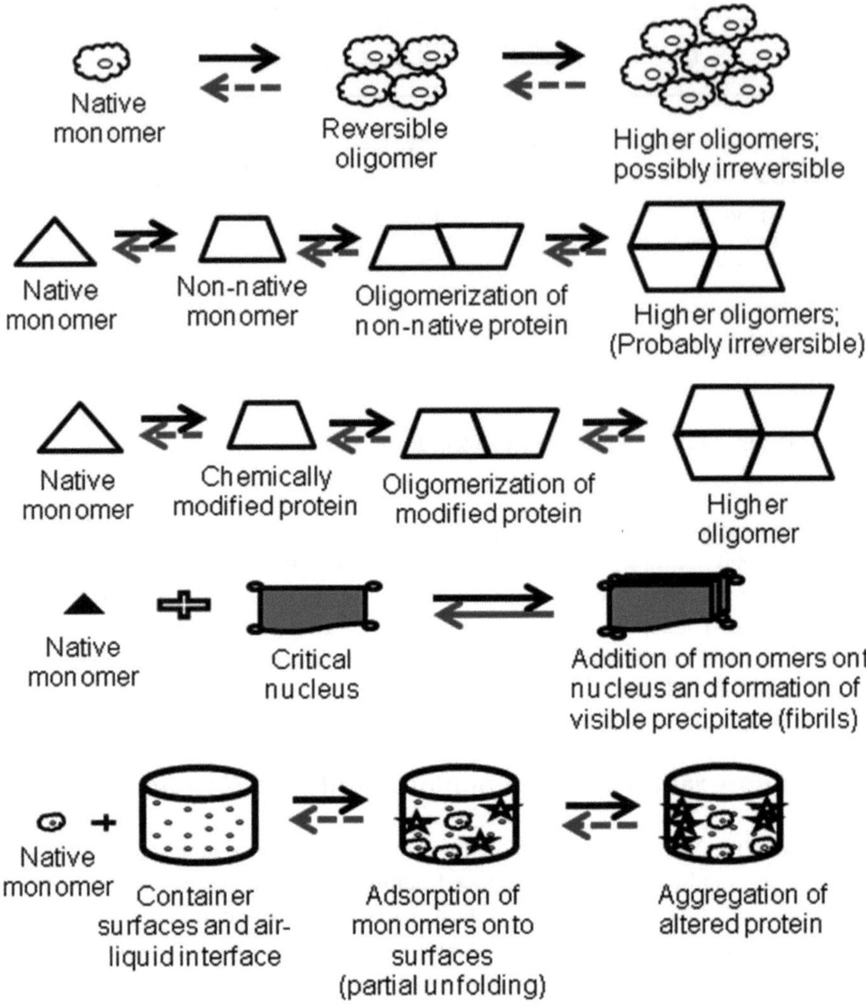

Fig. 1.5 Schematic diagram of different mechanisms of protein aggregation. (Adapted from Philo and Arakawa 2009)

aggregation; several instances have been cited where aggregation and immunogenicity have been correlated (Cleland et al. 1993; Rosenberg 2006).

1.3.4 Nucleation-Mediated Aggregation

This process occurs in three stages: the first stage is the lag phase, where the monomer forms reversible oligomers. The second stage is the growth/exponential phase in which the formation of the "critical nucleus" results in much larger species (fibrils) being formed. The third stage is known as the saturation phase in which equilibrium is reached between the formation and dissociation of terminal aggregates. This mode is also seen in the case of aggregation of several proteins associated with protein misfolding disorders.

1.3.5 Surface-Induced Aggregation

In this mechanism, the native monomer binds to the surface, because proteins tend to be adsorbed on a variety of surfaces of containers. Upon binding, the conformation of the monomer is altered, which may lead to its misfolding and further aggregation.

1.4 Instability of Protein Drugs

Protein therapeutics have many advantages over small-molecule drugs as they are highly specific to their targets. However, from manufacturing to patient use, protein therapeutics may be exposed to adverse environmental conditions leading to loss of structure of the protein and resulting in reduced activity and undesirable immunogenic properties. Generally, for a typical protein pharmaceutical, the shelf-life required for economic viability is 18–24 months. Keeping in view the inherent fragility of protein structure, maintaining the functional form of the protein becomes a challenge. Physical and chemical denaturation of proteins is one of the major stumbling blocks (Darkwah 2017; Manning et al. 2010; Tamizi and Jouyban 2016) as they cause changes in the activity of the protein as discussed above (Song et al. 2017).

The manufacturing process of biopharmaceuticals may expose them to different stress conditions that can cause protein destabilization, denaturation, and aggregation. A better understanding of the mechanisms of protein denaturation can provide suggestions to design successful quality control and their uses. Physical instability refers to structural or conformational alterations in the protein without any change in the chemical composition and results mostly from the realignment of non-covalent bonds holding the protein structure. Various facets of such modifications include denaturation, adsorption, precipitation, and aggregation (Akbarian et al. 2018; Laptos and Omersel 2018; Le Basle et al. 2020; Song et al. 2017; Zapadka et al. 2017). Chemical instability refers to changes in the protein structure due to chemical modifications of individual amino acid residues (Grassi and Cabrele 2019; Manning et al. 2010; Song et al. 2017). Cases have been reported where a change in the physical properties of the protein facilitates chemical denaturation and vice versa. An example is the non-enzymatic ribosylation of bovine serum albumin, which results in its misfolding and aggregation (Wei et al. 2009).

1.4.1 Denaturation

Protein molecules are biologically active in their native three-dimensional architecture, and any deviation from this state is referred to as denaturation (unfolding), which is frequently accompanied by loss of function (Angkawinitwong et al. 2015). The native conformation of a protein molecule is stabilized by several non-covalent interactions, such as hydrogen bonding, and hydrophobic and electrostatic interactions (Pace et al. 2014; Zhou and Pang 2018). However, when the protein is present in the solution state, these interactions are modified by several parameters, such as temperature, pH, chemicals (denaturants), and an organic solvent, which again promote denaturation (Kishore et al. 2012). Physical and chemical denaturation of proteins has been tried to be minimized by the addition of several excipients, such as sugars, surfactants, and amino acids (some of these have been discussed above; Bam et al. 1995, 1998), or by designing solid rather than liquid formulations wherever feasible (Kreilgaard et al. 1998; Salnikova et al. 2008).

1.4.2 Aggregation

Aggregation of therapeutic proteins is one of the most common problems encountered in almost all stages of biological drug development (Roberts 2014; Wang 2005, 2015). Different types of aggregates may be formed, such as amorphous (Trainor et al. 2017), oligomer, and amyloid fibrils (Riek and Eisenberg 2016), either in solution or on the surface due to adsorption (Kabiri and Unsworth 2014; Ozboyaci et al. 2016). Aggregates may be divided into covalent or non-covalent, visible or sub-visible, reversible or irreversible. The aggregated protein drug may show a loss of biological activity and may also induce toxicity and/or immunogenicity (Moussa et al. 2016; Rosenberg 2006; Wang et al. 2012). For example, interleukin-1 receptor antagonist forms aggregates in solution, and its activity is reduced by ~75% concerning the native form even without any major conformational change (Chang et al. 1996). Multiple freeze-thaw cycles can also induce protein aggregation. This has been shown in the case of tetanus toxoid where the adjuvant was shown to undergo agglomeration (Clausi et al. 2008). Inhibition of aggregation could be achieved by the use of a novel stabilizer, viz., RNA aptamers (Solanki et al. 2011). In most cases, temperature, pH, and agitation are generally responsible for protein aggregation. Thermal denaturation of proteins is generally irreversible, as the unfolded protein undergoes rapid irreversible aggregation (Manning et al. 2010; Vermeer et al. 1998). This is seen in the case of insulin, which undergoes aggregation due to agitation (during storage and transport), leading to an altered physiological release profile and failure to maintain normoglycemia (Mianowska et al. 2011; Radermecker et al. 2009). Hence, maintaining the physical and chemical integrity of insulin formulation is of prime concern. To prevent chemical and physical aggregation for insulin formulation, several research groups have tried different strategies, like protein encapsulation in lipid nanoparticle carriers (Angelov et al. 2012; Diwan and Park 2003; Meng et al. 2012). Stabilizers like Zn^{2+} (Gast et al. 2017) and surfactants (Shi et al. 2009) have also been used to avoid agitation-based aggregation. One novel approach has been to use RNA aptamers to prevent the aggregation of insulin (Malik and Roy 2013). The authors reported the selection of RNA aptamers against monomeric insulin that could bind specifically with the target with high affinity. This slowed down the association of monomers to form oligomers and further aggregates. The selected aptamer is a novel stabilizer of insulin, which decreases the rate of agitation-induced aggregation and helps insulin retain its physiological activity (Malik and Roy 2013).

1.4.3 Precipitation

Precipitation is another major problem faced during the manufacturing process of the protein and also due to its improper storage. The unfolding of the stress-exposed protein changes its hydrophobicity, altering its aqueous solubility and increasing its propensity to precipitate (Kiese et al. 2008). High concentration and subsequent precipitation also result in an increase in viscosity and turbidity of the protein formulation (Hong et al. 2018; Jezek et al. 2011). Similarly, solutions with increased viscosity may also result in sufficient shear stress to cause precipitation of the protein when passed through a syringe or narrow pump.

1.4.4 Chemical Instability

Deamination is a common chemical reaction that occurs in therapeutic proteins (Lipiainen et al. 2015; Wang 1999). In this, an amide group in the side chain of asparagine or glutamine is generally converted into an acidic group (Cleland et al. 1993). Deamination adds a negative charge to a protein molecule, which can alter physicochemical properties like a decrease in isoelectric point (pI) or structural changes in secondary/tertiary structure promoting aggregation, and ultimately reducing biological activity (Goolcharran et al. 2000; Pisal et al. 2010). This is also an example of chemical modification resulting in the physical denaturation of proteins. For example, the deamination of recombinant human deoxyribonuclease (rhDNase) led to a reduction in its activity by almost half as compared to their native activity (Shire 1996).

Oxidation is another route by which solid and liquid protein formulations can undergo chemical denaturation (Davies 2005, 2016; Li et al. 1995). Proteins containing sulfur or aromatic rings, such as Met, Cys, His, Trp, Tyr, and Phe, are more susceptible to oxidation (Davies and Truscott 2001). Oxidation in protein formulations mainly occurs due to the presence of trace amounts of transition metals, peroxides, light exposure, and atmospheric oxygen (Davies 2016; Luo et al. 2011). Hence, proper conditions need to be maintained during manufacturing, storage, and transportation as oxidized sidechains significantly affect the biological activity, immunogenicity, stability, and shelf-life of products (Davies 2005; Pan et al. 2009; Zhang and Kalonia 2007). Amino acids that are exposed to the solvent are amenable to oxidation as reactive oxygen species can interact with the side chain easily, while buried residues have lower avenues of interaction (Pan et al. 2009; Torosantucci et al. 2014). For example, it has been reported that in the case of recombinant human parathyroid hormone (rh-PTH), methionine (Met) sulfoxides are formed at Met18 and Met8 positions in the presence of H_2O_2, with Met18 being more vulnerable to oxidation (Nabuchi et al. 1995).

1.5 Novel Stabilizers

The approach of antibody–drug conjugates (ADCs) combines the specificity of antibodies and cytotoxicity of chemically conjugated agents to target tumor cells (Duerr and Friess 2019). Hartimath et al. (2019) have recently reported the synthesis of ADC of nimotuzumab (targeting anti-epidermal growth factor receptor I) with pegylated maytansine (a tubulin depolymerizer), and shown its successful application in a colorectal cancer model.

1.5.1 Immobilization of Enzymes

Immobilization of enzymes/proteins probably is the most likely approach to succeed in the continuing quest to make biocatalysis a routine tool for the industry. It arguably represents the most economically viable route of stabilizing enzymes for industrial use. Immobilization refers to the reduction in mobility of a protein or a cell, which decreases the flexibility of the native conformation, increases rigidity, and stabilizes the biomolecule by slowing down inactivation reactions (Cabral and Kennedy 1993; Guisan 2006; Katchalski-Katzir 1993; Sheldon and Brady 2018; Sheldon and van Pelt 2013). Immobilization offers advantages like high stability, resistance to shear stress, reusability of biocatalysts, and faster reaction rate due to high catalyst concentration (Cao 2005; Costa et al. 2005; Guisan 2006; Hernandez and Fernandez-Lafuente 2011; Roy and Gupta 2006). It also provides a platform for the facile separation of the product from the catalyst. Immobilization was able to solve several problems faced with the free enzyme, including difficulty in separating the product from the enzyme, recovery of the active enzyme, and a tedious purification protocol to separate the contaminants associated with the crude enzyme preparation. The Clinton Corn Processing Company has been using immobilized glucose isomerase for the production of high-fructose corn syrup. Because of the difference in size between the catalyst immobilized on an ion exchanger and the product (fructose), separation is easier. More stable enzyme preparations were later developed by industries like Novo Industri (now Novo Nordisk).

Immobilization can be matrix-based or matrix-less (Fig. 1.6). Immobilization methods can be categorized as follows: (a) binding to a prefabricated matrix, (b) entrapment in polymer gels like alginate and polyacrylamide, or encapsulation in hollow fiber matrices, and (c) cross-linking of enzyme molecules. For example, the commercially available immobilized *Candida antarctica* lipase B preparation

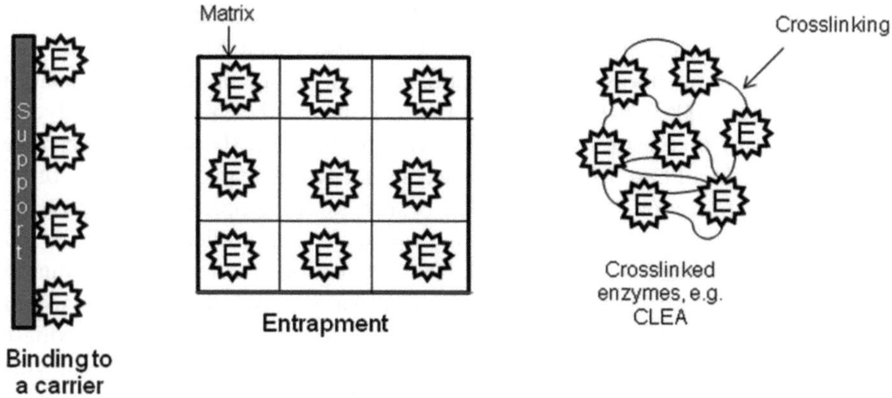

Fig. 1.6 Mechanisms of protein immobilization (*E* enzyme)

(Novozym® 435) is physically adsorbed on a macroporous acrylic resin and hence the enzyme is likely to leach out upon prolonged use. Covalent immobilization is the preferred route in such cases. Penicillin amidase immobilized on surface-activated acrylic resins, e.g., Eupergit® C, was able to retain more than half its original activity even after 800 cycles of use (Kallenberg et al. 2005). Although covalent binding is considered to be stable, inactivation of the enzyme may take place due to the formation of undesirable covalent interaction with the matrix.

Encapsulation may result in leakage of enzymes from the gel due to their smaller size (Smidsrod and Skjak-Braek 1990). Leaching can be avoided by increasing the size of the encapsulated species, for example, by inducing aggregation of the enzyme (>300 nm) (Mallardi et al. 2015). Thus, the ideal solution is to avoid the use of an external support system and still convert the enzyme into the solid form so that a heterogeneous reaction medium can facilitate solid–liquid separation. One option could be the use of stimuli-sensitive or "smart" polymers to immobilize enzymes. These preparations overcome diffusion limitations by allowing the reaction to be carried out in a homogeneous medium and precipitating the catalyst to allow separation of the substrate/product (Roy et al. 2004; Sheldon and van Pelt 2013). The bifunctional cross-linking reagent, glutaraldehyde, has been used quite commonly to circumvent the use of a matrix for immobilization (Lopez-Gallego et al. 2005; Richards and Knowles 1968). The reaction of glutaraldehyde with proteins, most commonly at the ε-amino group of lysine, results in the formation of the Schiff base. Uncontrolled modification and inactivation of the target enzyme can be avoided by incorporation of a feeder protein like bovine serum albumin (Shah et al. 2006). There is a possibility that glutaraldehyde may undergo self-polymerization by condensation to give rise to higher-ordered aggregates. Matrix-less immobilization of enzymes has been gaining increasing interest in different forms such as cross-linked enzyme crystals (CLECs; Jegan and Emilia 2004; Margolin and Navia 2001) and cross-linked enzyme aggregates (CLEAs; Mukherjee and Gupta 2016). One of the early commercially known examples of CLEC preparation is that of glucose isomerase (Visuri 1993). Altus Biologics, Inc. had conducted pre-clinical trials of TheraCLEC™ (also called Liprotamase) for the treatment of cystic fibrosis or pancreatitis by employing lipase as the catalyst (Borowitz et al. 2006). The formation of CLEC protected the enzyme in the adverse environment of the gastrointestinal tract. This formulation showed an acceptable safety profile. In a Phase III trial, the molecule showed increased coefficients of fat and nitrogen absorption (NCT00449904). A successful example of CLEC preparation is the synthesis of the dipeptide aspartame, an artificial sweetener by thermolysin, a protease (Persichetti et al. 1995). Retention of stereospecificity during synthesis cannot be ensured by chemical means and is accomplished by the use of the immobilized enzyme. CLEC of *Caldariomyces fumago* chloroperoxidase showed higher thermal stability and organic solvent resistance than the free enzyme (Ayala et al. 2002). The properties of the cross-linked preparation could be controlled by selecting homo- and hetero-bifunctional cross-linking reagents with variable lengths of spacer arms (Margolin and Navia 2001; Uy and Wold 1977). Immobilization of enzymes in the form of CLEA is relatively straightforward and involves either (a) aggregation due to extensive intermolecular cross-linking in the presence of a cross-linker, or (b) precipitation of the enzyme by the addition of an excipient followed by cross-linking (Schoevaart et al. 2004). It also overcomes the high cost of very pure enzymes required for CLECs. An example of CLEA is that of the alkaline protease, subtilisin Carlsberg, from *Bacillus licheniformis*, used in laundry detergents (de Beer et al. 2012). It is also widely used in organic synthesis. CLEAs can be highly efficient biocatalysts with enhanced thermal and environmental stability (Cabirol et al. 2008). The matrix-less approach offers distinct advantages like more concentrated biocatalytic activity (without the formation of multilayers as in carrier-based immobilization) and low process cost.

1.5.2 Aptamers

Aptamers are short, single-stranded oligonucleotides that can bind to their targets with high affinity and specificity due to their specific three-dimensional structures (Jain et al. 2020; Patel et al. 2017; Song et al. 2012; Zhang et al. 2019). They are of synthetic origin and hence cell-free (Cibiel et al. 2012; Yang et al. 2011). It is easy to scale up their production and modification. Their non-toxic and non-immunogenic nature makes them promising candidates for applications in the therapeutics domain, such as diagnostics, biosensors, and therapeutics molecules (Kaur and Roy 2008; Song et al. 2012; Zhang et al. 2019) (Table 1.4). Following the approval of pegaptanib (Macugen®) for the treatment of age-related macular degeneration, aptamers have attracted a lot of attention from the community of researchers. Aptamers are called alternates to antibodies and have several advantages such as thermal stability, low cost, and maneuverability, making them suitable for a variety of applications like therapy (Becker and Chan 2009), drug delivery systems (Min et al. 2011), stabilizer, and sensor (Song et al. 2012; Xu et al. 2005). A novel strategy has been tried to stabilize the monomeric structure of insulin in solution to preserve

Table 1.4 Applications of aptamers as therapeutics

Target	Aptamer	References
Vascular endothelial growth factor (VEGF-165)	SL (2)-B (DNA), RNV66 (DNA)	Gantenbein et al. (2015)
Nucleolin	FCL-II	Fan et al. (2017)
CXCL12 (stromal cell-derived factor-1)	NOX-A12	Zboralski et al. (2017)
Epidermal growth factor receptor (EGFR)	TuTu2231, KDI130	Wang et al. (2014)
Vimentin	NAS-24	Zamay et al. (2014)
E-selectin	ESTA	Morita et al. (2016)
Programmed cell death protein 1 (PD-1)	MP7	Prodeus et al. (2015)
Cytotoxic T-lymphocyte-associated protein 4 (CTLA-4)	AptCTLA-4	Huang et al. (2017)
C5a (complement component 5a)	AON-D21 I-aptamer	Ajona et al. (2017)
CD44/Epithelial cell adhesion molecule (EpCAM)	CD44-EpCAM aptamer	Zheng et al. (2017)
Thrombin	Anti-thrombin aptamer	Waters and Lillicrap (2009)

CD44 cluster of differentiation 44, *CXCL12* chemokine (CXC motif) ligand 12, *ESTA* E-selectin thioaptamer

the efficacy of the final formulation by RNA aptamer. This has been briefly discussed above. In this work, the authors have screened two RNA aptamers that can be used to stabilize insulin in solution against agitation-induced stress while retaining its biological activity. Similar kind of strategies has also been tried to stabilize moisture-induced aggregation of tetanus toxoid by using nucleic acid aptamer (Jetani et al. 2014). This approach also shows that since aptamers target the protein and not the solvent, they are successful in protecting the protein against a variety of stress conditions. This is not possible with traditional osmolytes where a different osmolyte needs to be employed to protect the protein against a different stress condition (Abbas et al. 2012).

1.6 Outlook

Stabilization of biopharmaceuticals is an integral part of the drug development process since biopharmaceuticals, unlike small molecules, are inherently fragile molecules. Any minor or insignificant change in the conformation may end up inactivating a protein or even introducing toxic/immunogenic properties. Thus, there is a continuous search for new molecules or novel mechanisms for existing molecules that can stabilize proteins. Osmolytes or small molecules have been traditionally employed for protein stabilization. Although relatively less expensive, they have several drawbacks, the major one being that one osmolyte is incapable of protecting a single protein against a variety of stress conditions. This means that if the same protein formulation is exposed to freezing or agitation or drying, the type of stabilizing osmolyte required will be different. On the other hand, aptamers, which are selected in vitro, bind with high affinity to the protein. Thus, they inhibit the initial step of aggregation, which is either misfolding of the protein or intermolecular interaction between monomers. Their activity is independent of the stress condition. This change from solvent-centric to protein-centric approach is a paradigm shift in the field of protein stabilization and reduces the components and hence the complexity of the protein formulation. Another novel stabilizer that is finding increasing mention in the literature is ionic liquids. As they are associated with several drawbacks, the development of this stabilizer will require support from chemical sciences. Thus, there is a stringent need to understand the structure and conformation of a protein before attempting to develop any successful stabilization strategy.

References

Abbas SA, Sharma VK, Patapoff TW et al (2012) Opposite effects of polyols on antibody aggregation: thermal versus mechanical stresses. Pharm Res 29:683–694

Abuchowski A, van Es T, Palczuk NC (1977) Alteration of immunological properties of bovine serum albumin by covalent attachment of polyethylene glycol. J Biol Chem 252:3578–3581

Ajona D, Ortiz-Espinosa S, Moreno H et al (2017) A combined Pd-1/C5a blockade synergistically protects against lung cancer growth and metastasis. Cancer Discov 7:694–703

Akbarian M, Ghasemi Y, Uversky V et al (2018) Chemical modifications of insulin: finding a compromise between stability and pharmaceutical performance. Int J Pharm 547:450–468

Alford JR, Kendrick BS, Carpenter JF et al (2008) High concentration formulations of recombinant human interleukin-1 receptor antagonist: II. Aggregation kinetics. J Pharm Sci 97:3005–3021

Angelov B, Angelova A, Papahadjopoulos-Sternberg B et al (2012) Protein-containing pegylated cubosomic particles: freeze-fracture electron microscopy and synchrotron radiation circular dichroism study. J Phys Chem B 116:7676–7686

Angkawinitwong U, Sharma G, Khaw PT et al (2015) Solid-state protein formulations. Ther Deliv 6:59–82

Arakawa T, Prestrelski SJ, Kenney WC et al (2001) Factors affecting short-term and long-term stabilities of proteins. Adv Drug Deliv Rev 46:307–326

Arsiccio A, Pisano R (2017) Stability of proteins in carbohydrates and other additives during freezing: the human growth hormone as a case study. J Phys Chem B 121:8652–8660

Ayala M, Horjales E, Pickard MA, Vazquez-Duhalt R (2002) Cross-linked crystals of chloroperoxidase. Biochem Biophys Res Commun 295(4):828–831. https://doi.org/10.1016/S0006-291X(02)00766-0

Bam NB, Randolph TW, Cleland JL (1995) Stability of protein formulations: investigation of surfactant effects by a novel EPR spectroscopic technique. Pharm Res 12:2–11

Bam NB, Cleland JL, Yang J et al (1998) Tween protects recombinant human growth hormone against agitation-induced damage via hydrophobic interactions. J Pharm Sci 87:1554–1559

Becker RC, Chan MY (2009) Reg-1, a regimen comprising RB-006, a factor IXa antagonist, and its oligonucleotide active control agent RB-007 for the potential treatment of arterial thrombosis. Curr Opin Mol Ther 11:707–715

Borowitz D, Goss CH, Stevens C et al (2006) Safety and preliminary clinical activity of a novel pancreatic enzyme preparation in pancreatic insufficient cystic fibrosis patients. Pancreas 32: 258–263

Brennan JR, Gebhart SS, Blackard WG (1985) Pump-induced insulin aggregation: a problem with the biostator. Diabetes 34:353–359
Cabirol FL, Lim AE, Hanefeld U et al (2008) Robust and efficient, yet uncatalyzed, synthesis of trialkylsilyl-protected cyanohydrins from ketones. J Org Chem 73:2446–2449
Cabral JMS, Kennedy JF (1993) Immobilisation techniques for altering thermal stability of enzymes. In: Gupta MN (ed) Thermostability of enzymes. Springer, Berlin, pp 163–179
Cacia J, Keck R, Presta LG et al (1996) Isomerization of an aspartic acid residue in the complementarity-determining regions of a recombinant antibody to human IgE: identification and effect on binding affinity. Biochemistry 35:1897–1903
Canchi DR, Garcia AE (2013) Cosolvent effects on protein stability. Annu Rev Phys Chem 64:273–293
Cao L (2005) Immobilised enzymes: science or art? Curr Opin Chem Biol 9:217–226
Carpenter JF, Crowe JH (1988) The mechanism of cryoprotection of proteins by solutes. Cryobiology 25:244–255
Chang BS, Beauvais RM, Arakawa T et al (1996) Formation of an active dimer during storage of interleukin-1 receptor antagonist in aqueous solution. Biophys J 71:3399–3406
Chen B, Bautista R, Yu K et al (2003) Influence of histidine on the stability and physical properties of a fully human antibody in aqueous and solid forms. Pharm Res 20:1952–1960
Cibiel A, Pestourie C, Duconge F (2012) In vivo uses of aptamers selected against cell surface biomarkers for therapy and molecular imaging. Biochimie 94:1595–1606
Clausi AL, Merkley SA, Carpenter JF (2008) Inhibition of aggregation of aluminum hydroxide adjuvant during freezing and drying. J Pharm Sci 97:2049–2061
Cleland JL, Powell MF, Shire SJ (1993) The development of stable protein formulations: a close look at protein aggregation, deamidation, and oxidation. Crit Rev Ther Drug Carrier Syst 10: 307–377
Costa SA, Azevedo HS, Reis RL (2005) Enzyme immobilization in biodegradable polymers for biomedical applications. In: Reis RL, Román JS (eds) Biodegradable systems in tissue engineering and regenerative medicine. CRC Press LLC, Boca Raton, pp 301–323
Dagan S, Hagai T, Gavrilov Y et al (2013) Stabilization of a protein conferred by an increase in folded state entropy. Proc Natl Acad Sci U S A 110:10628–10633
Darkwah J (2017) Protein stability: impact of formulation excipients and manufacturing processes in protein-based pharmaceuticals. PhD thesis. De Montfort University, Leicester. http://hdl.handle.net/2086/16284
Davies MJ (2005) The oxidative environment and protein damage. Biochim Biophys Acta 1703: 93–109
Davies MJ (2016) Protein oxidation and peroxidation. Biochem J 473:805–825
Davies MJ, Truscott RJ (2001) Photo-oxidation of proteins and its role in cataractogenesis. J Photochem Photobiol B Biol 63:114–125
de Beer RJ, Nuijens T, Wiermans L (2012) Improving the carboxyamidomethyl ester for subtilisin A-catalysed peptide synthesis. Org Biomol Chem 10:6767–6775
Diwan M, Park TG (2003) Stabilization of recombinant interferon-alpha by pegylation for encapsulation in PLGA microspheres. Int J Pharm 252:111–122
Duerr C, Friess W (2019) Antibody-drug conjugates- stability and formulation. Eur J Pharm Biopharm 139:168–176
Fan X, Sun L, Li K et al (2017) The bioactivity of d-/l-isonucleoside- and 2′-deoxyinosine-incorporated aptamer AS1411 including DNA replication/microRNA expression. Mol Ther Nucleic Acids 9:218–229
Fradkin AH, Carpenter JF, Randolph TW (2009) Immunogenicity of aggregates of recombinant human growth hormone in mouse models. J Pharm Sci 98:3247–3264
Gantenbein AR, Sarikaya H, Riederer F et al (2015) Postoperative hemicrania continua-like headache - a case series. J Headache Pain 16:526
Gast K, Schuler A, Wolff M et al (2017) Rapid-acting and human insulins: hexamer dissociation kinetics upon dilution of the pharmaceutical formulation. Pharm Res 34:2270–2286

Gervasi V, Dall Agnol R, Cullen S et al (2018) Parenteral protein formulations: an overview of approved products within the European Union. Eur J Pharm Biopharm 131:8–24
Goolcharran C, Cleland JL, Keck R et al (2000) Comparison of the rates of deamidation, diketopiperazine formation and oxidation in recombinant human vascular endothelial growth factor and model peptides. AAPS J 2:E5
Grassi L, Cabrele C (2019) Susceptibility of protein therapeutics to spontaneous chemical modifications by oxidation, cyclization, and elimination reactions. Amino Acids 51:1409–1431
Guisan JM (2006) Immobilization of enzymes and cells, 2nd edn. Humana Press, Totowa
Harris JM, Chess RB (2003) Effect of pegylation on pharmaceuticals. Nat Rev Drug Discov 2:214–221
Hartimath SV, El-Sayed A, Makhlouf A et al (2019) Therapeutic potential of nimotuzumab PEGylated-maytansine antibody drug conjugates against EGFR positive xenograft. Oncotarget 10:1031–1044
Hartmann WK, Saptharishi N, Yang XY et al (2004) Characterization and analysis of thermal denaturation of antibodies by size exclusion high-performance liquid chromatography with quadruple detection. Anal Biochem 325:227–239
Hernandez K, Fernandez-Lafuente R (2011) Control of protein immobilization: coupling immobilization and site-directed mutagenesis to improve biocatalyst or biosensor performance. Enzyme Microb Technol 48:107–122
Hong T, Iwashita K, Shiraki K (2018) Viscosity control of protein solution by small solutes: a review. Curr Protein Pept Sci 19:746–758
Huang BT, Lai WY, Chang YC et al (2017) A CTLA-4 antagonizing DNA aptamer with antitumor effect. Mol Ther Nucleic Acids 8:520–528
Izutsu K (2005) Stabilization of therapeutic proteins by chemical and physical methods. Methods Mol Biol 308:287–292
Izutsu KI (2018) Applications of freezing and freeze-drying in pharmaceutical formulations. Adv Exp Med Biol 1081:371–383
Jain NK, Roy I (2009) Effect of trehalose on protein structure. Protein Sci 18:24–36
Jain S, Kaur J, Prasad S et al (2020) Nucleic acid therapeutics: a focus on the development of aptamers. Expert Opin Drug Discov 16(3):255–274. https://doi.org/10.1080/17460441.2021.1829587
Jegan RJ, Emilia AT (2004) Strategies in making cross-linked enzyme crystals. Chem Rev 104:3705–3722
Jetani HC, Bhadra AK, Jain NK et al (2014) Nucleic acid aptamers stabilize proteins against different types of stress conditions. J Pharm Sci 103:100–106
Jezek J, Rides M, Derham B et al (2011) Viscosity of concentrated therapeutic protein compositions. Adv Drug Deliv Rev 63:1107–1117
Johnson RE, Qi H, Borgmeyer JR et al (2004) Stable pH optimized formulation of a modified antibody. WO Patent WO/2004/019861
Jones MT, Mahler H-C, Yadav S et al (2018) Considerations for the use of polysorbates in biopharmaceuticals. Pharm Res 35:148
Kabiri M, Unsworth LD (2014) Application of isothermal titration calorimetry for characterizing thermodynamic parameters of biomolecular interactions: peptide self-assembly and protein adsorption case studies. Biomacromolecules 15:3463–3473
Kallenberg A, van Rantwijk F, Sheldon R et al (2005) Immobilization of penicillin G acylase: the key to optimum performance. Adv Synth Catal. 347:905–926
Kaplon H, Reichert JM (2019) Antibodies to watch in 2019. MAbs 11:219–238
Katchalski-Katzir E (1993) Immobilized enzymes--learning from past successes and failures. Trends Biotechnol 11:471–478
Katterle B (2018) How stable are new biologics? Pharm Ind 80:1557–1563
Kaur G, Roy I (2008) Therapeutic applications of aptamers. Expert Opin Investig Drugs 17:43–60
Kerwin BA (2008) Polysorbates 20 and 80 used in the formulation of protein biotherapeutics: structure and degradation pathways. J Pharm Sci 97:2924–2935

Kesik-Brodacka M (2018) Progress in biopharmaceutical development. Biotechnol Appl Biochem 65:306–322

Kiese S, Papppenberger A, Friess W et al (2008) Shaken, not stirred: mechanical stress testing of an IgG1 antibody. J Pharm Sci 97:4347–4366

Kishore D, Kundu S, Kayastha AM (2012) Thermal, chemical and pH induced denaturation of a multimeric beta-galactosidase reveals multiple unfolding pathways. PLoS One 7:e50380

Klotz U, Teml A, Schwab M (2007) Clinical pharmacokinetics and use of infliximab. Clin Pharmacokinet 46:645–660

Knotts TA IV, Rathore N, de Pablo JJ (2008) An entropic perspective of protein stability on surfaces. Biophys J 94:4473–4483

Kozma GT, Shimizu T, Ishida T et al (2020) Anti-PEG antibodies: properties, formation, testing and role in adverse immune reactions to PEgylated nano-biopharmaceuticals. Adv Drug Deliv Rev 154–155:163–175. https://doi.org/10.1016/j.addr.2020.07.024

Kreilgaard L, Frokjaer S, Flink JM et al (1998) Effects of additives on the stability of recombinant human factor XIII during freeze-drying and storage in the dried solid. Arch Biochem Biophys 360:121–134

Lam KCL, Rajaraman G (2012) Assessment of P-glycoprotein substrate and inhibition potential of test compounds in MDR1-transfected MDCK cells. Curr Protoc Pharmacol 58:7.13.11–7.13.17

Laptos T, Omersel J (2018) The importance of handling high-value biologicals: physico-chemical instability and immunogenicity of monoclonal antibodies. Exp Ther Med 15:3161–3168

Le Basle Y, Chennell P, Tokhadze N et al (2020) Physicochemical stability of monoclonal antibodies: a review. J Pharm Sci 109:169–190

Lee HJ, McAuley A, Schilke KF et al (2011) Molecular origins of surfactant-mediated stabilization of protein drugs. Adv Drug Deliv Rev 63:1160–1171

Li S, Schöneich C, Borchardt RT (1995) Chemical instability of protein pharmaceuticals: mechanisms of oxidation and strategies for stabilization. Biotechnol Bioeng 48:490–500

Lipiainen T, Peltoniemi M, Sarkhel S et al (2015) Formulation and stability of cytokine therapeutics. J Pharm Sci 104:307–326

Lopez-Gallego F, Betancor L, Mateo C et al (2005) Enzyme stabilization by glutaraldehyde crosslinking of adsorbed proteins on aminated supports. J Biotechnol 119:70–75

Luo Q, Joubert MK, Stevenson R et al (2011) Chemical modifications in therapeutic protein aggregates generated under different stress conditions. J Biol Chem 286:25134–25144

Mahler HC, Allmendinger A (2017) Stability, formulation, and delivery of biopharmaceuticals. Protein Ther 2:469–491

Malik R, Roy I (2011) Probing the mechanism of insulin aggregation during agitation. Int J Pharm 413:73–80

Malik R, Roy I (2013) Stabilization of bovine insulin against agitation-induced aggregation using RNA aptamers. Int J Pharm 452:257–265

Mallardi A, Angarano V, Magliulo M et al (2015) General approach to the immobilization of glycoenzyme chains inside calcium alginate beads for bioassay. Anal Chem 87:11337–11344

Manning MC, Chou DK, Murphy BM et al (2010) Stability of protein pharmaceuticals: an update. Pharm Res 27:544–575

Margolin AL, Navia MA (2001) Protein crystals as novel catalytic materials. Angew Chem 40: 2204–2222

McIntosh KA, Charman WN, Charman SA (1998) The application of capillary electrophoresis for monitoring effects of excipients on protein conformation. J Pharm Biomed Anal 16:1097–1105

Meng W, Guo X, Qin M et al (2012) Mechanistic insights into the stabilization of srcSH3 by pegylation. Langmuir 28:16133–16140

Mianowska B, Szadkowska A, Pietrzak I et al (2011) Immunogenicity of different brands of human insulin and rapid-acting insulin analogs in insulin-naive children with Type 1 diabetes. Pediatr Diabetes 12:78–84

Min K, Jo H, Song K et al (2011) Dual-aptamer-based delivery vehicle of doxorubicin to both PSMA (+) and PSMA (-) prostate cancers. Biomaterials 32:2124–2132

Morita Y, Kamal M, Kang SA et al (2016) E-selectin targeting pegylated-thioaptamer prevents breast cancer metastases. Mol Ther Nucleic Acids 5:e399

Morpurgo M, Veronese FM (2004) Conjugates of peptides and proteins to polyethylene glycols. Methods Mol Biol 283:45–70

Moussa EM, Panchal JP, Moorthy BS et al (2016) Immunogenicity of therapeutic protein aggregates. J Pharm Sci 105:417–430

Mukherjee J, Gupta MN (2016) Lipase coated clusters of iron oxide nanoparticles for biodiesel synthesis in a solvent free medium. Bioresour Technol 209:166–171

Nabuchi Y, Fujiwara E, Ueno K et al (1995) Oxidation of recombinant human parathyroid hormone: effect of oxidized position on the biological activity. Pharm Res 12:2049–2052

Nielsen L, Khurana R, Coats A et al (2001) Effect of environmental factors on the kinetics of insulin fibril formation: elucidation of the molecular mechanism. Biochemistry 40:6036–6046

Ozboyaci M, Kokh DB, Corni S et al (2016) Modeling and simulation of protein-surface interactions: achievements and challenges. Q Rev Biophys 49:e4

Pace NC, Scholtz JM, Grimsley GR (2014) Forces stabilizing proteins. FEBS Lett 588:2177–2184

Pan H, Chen K, Chu L et al (2009) Methionine oxidation in human IgG2 Fc decreases binding affinities to Protein A and FcRn. Protein Sci 18:424–433

Patel KP, Sethi R, Dhara AR et al (2017) Challenges with osmolytes as inhibitors of protein aggregation: can nucleic acid aptamers provide an answer? Int J Biol Macromol 100:75–88

Persichetti RA, Clair NLS, Griffith JP et al (1995) Cross-linked enzyme crystals (CLECs) of thermolysin in the synthesis of peptides. J Am Chem Soc 117:2732–2737

Philo JS, Arakawa T (2009) Mechanisms of protein aggregation. Curr Pharm Biotechnol 10:348–351

Pisal DS, Kosloski MP, Balu-Iyer SV (2010) Delivery of therapeutic proteins. J Pharm Sci 99: 2557–2575

Prodeus A, Abdul-Wahid A, Fischer NW et al (2015) Targeting the PD-1/PD-L1 immune evasion axis with DNA aptamers as a novel therapeutic strategy for the treatment of disseminated cancers. Mol Ther Nucleic Acids 4:e237

Qi W, Orgel S, Francon A et al (2018) Urea improves stability of inactivated polio vaccine serotype 3 during lyophilization and storage in dried formulations. J Pharm Sci 107:2070–2078

Radermecker RP, Renard E, Scheen AJ (2009) Circulating insulin antibodies: influence of continuous subcutaneous or intraperitoneal insulin infusion, and impact on glucose control. Diabetes Metab Res Rev 25:491–501

Raso SW, Abel J, Barnes JM et al (2005) aggregation of granulocyte-colony stimulating factor in vitro involves a conformationally altered monomeric state. Protein Sci 14:2246–2257

Rayaprolu BM, Strawser JJ, Anyarambhatla G (2018) Excipients in parenteral formulations: selection considerations and effective utilization with small molecules and biologics. Drug Dev Ind Pharm 44:1565–1571

Richards FM, Knowles JR (1968) Glutaraldehyde as a protein cross-linkage reagent. J Mol Biol 37: 231–233

Riek R, Eisenberg DS (2016) The activities of amyloids from a structural perspective. Nature 539: 227–235

Roberts CJ (2014) Therapeutic protein aggregation: mechanisms, design, and control. Trends Biotechnol 32:372–380

Rosenberg AS (2006) Effects of protein aggregates: an immunologic perspective. AAPS J 8:E501–E507

Roy I, Gupta MN (2006) Design of smart biocatalysts: immobilization of enzymes on smart polymers. In: Guisan JM (ed) Immobilization of enzymes and cells. Humana Press, Totowa, pp 87–95

Roy I, Sharma S, Gupta MN (2004) Smart biocatalysts: design and applications. Adv Biochem Eng Biotechnol 86:159–189

Salnikova MS, Middaugh CR, Rytting JH (2008) Stability of lyophilized human growth hormone. Int J Pharm 358:108–113

Sanchez-Garcia L, Martin L, Mangues R (2016) Recombinant pharmaceuticals from microbial cells: a 2015 update. Microb Cell Factories 15:33

Schoevaart R, Wolbers MW, Golubovic M et al (2004) Preparation, optimization, and structures of cross-linked enzyme aggregates (CLEAs). Biotechnol Bioeng 87:754–762

Shah S, Sharma A, Gupta MN (2006) Preparation of cross-linked enzyme aggregates by using bovine serum albumin as a proteic feeder. Anal Biochem 351:207–213

Sheldon RA, Brady D (2018) The limits to biocatalysis: pushing the envelope. Chem Comm (Cambridge) 54:6088–6104

Sheldon RA, van Pelt S (2013) Enzyme immobilisation in biocatalysis: why, what and how. Chem Soc Rev 42:6223–6235

Shi K, Cui F, Yamamoto H et al (2009) Optimized formulation of high-payload PLGA nanoparticles containing insulin-lauryl sulfate complex. Drug Dev Ind Pharm 35:177–184

Shire SJ (1996) Stability characterization and formulation development of recombinant human deoxyribonuclease I [Pulmozyme, (Dornase Alpha)]. Pharm Biotechnol 9:393–426

Sifniotis V, Cruz E, Eroglu B et al (2019) Current advancements in addressing key challenges of therapeutic antibody design, manufacture, and formulation. Antibodies (Basel) 8:36

Smidsrod O, Skjak-Braek G (1990) Alginate as immobilization matrix for cells. Trends Biotechnol 8:71–78

Solanki VA, Jain NK, Roy I (2011) Stabilization of tetanus toxoid formulation containing aluminium hydroxide adjuvant against freeze-thawing. Int J Pharm 414:140–147

Song K-M, Lee S, Ban C (2012) Aptamers and their biological applications. Sensors (Basel) 12: 612–631

Song JG, Lee SH, Han H-K (2017) The stabilization of biopharmaceuticals: current understanding and future perspectives. J Pharm Investig 47:475–496

Strambini GB, Gabellieri E (1996) Proteins in frozen solutions: evidence of ice-induced partial unfolding. Biophys J 70:971–976

Swierczewska M, Lee KC, Lee S (2015) What is the future of pegylated therapies? Expert Opin Emerg Drugs 20:531–536

Tamizi E, Jouyban A (2016) Forced degradation studies of biopharmaceuticals: selection of stress conditions. Eur J Pharm Biopharm 98:26–46

Tonnis WF, Mensink MA, de Jager A et al (2015) Size and molecular flexibility of sugars determine the storage stability of freeze-dried proteins. Mol Pharm 12:684–694

Torosantucci R, Mozziconacci O, Sharov V et al (2012) Chemical modifications in aggregates of recombinant human insulin induced by metal-catalyzed oxidation: covalent cross-linking via Michael addition to tyrosine oxidation products. Pharm Res 29:2276–2293

Torosantucci R, Schoneich C, Jiskoot W (2014) Oxidation of therapeutic proteins and peptides: structural and biological consequences. Pharm Res 31:541–553

Trainor K, Broom A, Meiering EM (2017) Exploring the relationships between protein sequence, structure and solubility. Curr Opin Struct Biol 42:136–146

Turecek PL, Bossard MJ, Schoetens F et al (2016) PEgylation of biopharmaceuticals: a review of chemistry and nonclinical safety information of approved drugs. J Pharm Sci 105:460–475

Uy R, Wold F (1977) Introduction of artificial crosslinks into proteins. Adv Exp Med Biol 86A:169–186

Vazquez-Rey M, Lang DA (2011) Aggregates in monoclonal antibody manufacturing processes. Biotechnol Bioeng 108:1494–1508

Vermeer AW, Bremer MG, Norde W (1998) Structural changes of IgG induced by heat treatment and by adsorption onto a hydrophobic teflon surface studied by circular dichroism spectroscopy. Biochim Biophys Acta 1425:1–12

Visuri K (1993) Preparation of cross-linked glucose isomerase crystals. US Patent 5,437,993

Walsh G (2018) Biopharmaceutical benchmarks 2018. Nat Biotechnol 36:1136–1145

Wang W (1999) Instability, stabilization, and formulation of liquid protein pharmaceuticals. Int J Pharm 185:129–188

Wang W (2005) Protein aggregation and its inhibition in biopharmaceutics. Int J Pharm 289:1–30

Wang W (2015) Advanced protein formulations. Protein Sci 24:1031–1039
Wang W, Singh SK, Li N et al (2012) Immunogenicity of protein aggregates--concerns and realities. Int J Pharm 431:1–11
Wang Y, Luo Y, Bing T et al (2014) DNA aptamer evolved by cell-selex for recognition of prostate cancer. PLoS One 9:e100243
Warne NW (2011) Development of high concentration protein biopharmaceuticals: the use of platform approaches in formulation development. Eur J Pharm Biopharm 78:208–212
Waters B, Lillicrap D (2009) The molecular mechanisms of immunomodulation and tolerance induction to factor VIII. J Thromb Haemost 7:1446–1456
Webber MJ, Appel EA, Vinciguerra B et al (2016) Supramolecular pegylation of biopharmaceuticals. Proc Natl Acad Sci U S A 113:14189–14194
Wei Y, Chen L, Chen J et al (2009) Rapid glycation with D-ribose induces globular amyloid-like aggregations of BSA with high cytotoxicity to SH-SY5Y cells. BMC Cell Biol 10:10
Wörn A, Plückthun A (1998) Mutual stabilization of V_L and V_H in single-chain antibody fragments, investigated with mutants engineered for stability. Biochemistry 37:13120–13127
Xu D, Xu D, Yu X et al (2005) Label-free electrochemical detection for aptamer-based array electrodes. Anal Chem 77:5107–5113
Yang L, Zhang X, Ye M et al (2011) Aptamer-conjugated nanomaterials and their applications. Adv Drug Deliv Rev 63:1361–1370
Zamay TN, Kolovskaya OS, Glazyrin YE et al (2014) DNA-aptamer targeting vimentin for tumor therapy in vivo. Nucleic Acid Ther 24:160–170
Zapadka KL, Becher FJ, Gomes Dos Santos AL et al (2017) Factors affecting the physical stability (aggregation) of peptide therapeutics. Interface Focus 7:20170030
Zboralski D, Hoehlig K, Eulberg D et al (2017) Increasing tumor-infiltrating T cells through inhibition of CXCL12 with NOX-A12 synergizes with PD-1 blockade. Cancer Immunol Res 5:950–956
Zhang J, Kalonia DS (2007) The effect of neighboring amino acid residues and solution environment on the oxidative stability of tyrosine in small peptides. AAPS PharmSciTech 8:E102
Zhang Y, Lai BS, Juhas M (2019) Recent advances in aptamer discovery and applications. Molecules 24:941
Zheng J, Zhao S, Yu X et al (2017) Simultaneous targeting of CD44 and EpCAM with a bispecific aptamer effectively inhibits intraperitoneal ovarian cancer growth. Theranostics 7:1373–1388
Zhou H-X, Pang X (2018) Electrostatic interactions in protein structure, folding, binding, and condensation. Chem Rev 118:1691–1741

Concepts in Pharmacogenomics: Tools and Applications

2

Divya Vemula, Siva Singothu, and Vasundhra Bhandari

Abstract

Pharmacogenomics is the science of analyzing how an individual's genetic makeup influences his or her response to drugs. As the term indicates, it is a combination of pharmacology and genomics. It is the field concerned with the effect of genetic differences on drug responses in patients by correlating gene expression or single-nucleotide polymorphisms (SNPs) with a drug's effectiveness or toxicity. Thus, pharmacogenomics seeks to develop rational methods for optimizing drug therapy tailored to the genotype of the patient to ensure maximum efficacy with minimal side effects. It is a diverse field where various advanced cutting-edge biotechnological tools along with in silico approaches or computer-aided tools are utilized to find useful insights. This chapter will discuss the basics involved in pharmacogenomics, the tools used (both in vitro and in silico), various databases available, and its numerous applications.

Keywords

Database · Drug response · Genomics · Next-generation sequencing · SNPs

2.1 Background

Pharmacogenomics (PGx) is a pretty recent branch of science. It integrates pharmacology with genomics. Pharmacogenomics investigates the link between genetic differences and drug response in a patient's body. This genetic insight and

D. Vemula · S. Singothu · V. Bhandari (✉)
Department of Pharmacoinformatics, National Institute of Pharmaceutical Education and Research, Hyderabad, India
e-mail: vasundhra.b@niperhyd.ac.in

P. P. Singh (ed.), *Recent Advances in Pharmaceutical Innovation and Research*,
https://doi.org/10.1007/978-981-99-2302-1_2

information possesses the potential to enhance treatment effectiveness and reduce hazardous side effects. Our genes greatly influence how our bodies respond to medicine, which is poorly understood. Genes store the information for protein synthesis. These genes are many times expressed differently in different people. The inter-individual variability in the DNA sequence is linked to diseases, and the presence of specific proteins influences how medications work on a person's body.

PGx examines differences in these proteins or genes in response to drugs. The proteins, e.g., liver enzymes, alter medicines chemically. Chemical changes can sometimes make medications more or less active in the body. Even little variations in these liver enzyme genes can significantly impact a drug's safety or effectiveness. The human genome project's breakthrough technologies have transformed genetics (such as gene identification) and introduced genomics and pharmacogenomics. Pharmacogenomics, also referred to as pharmacogenetics, uses high-throughput DNA sequencing, gene mapping, and bioinformatics, resulting in a significant leap in the ability to uncover genes linked to physical characteristics, disease susceptibility, and treatment response (Bhagat 1999).

The rising interest in pharmacogenomics is due to two recent breakthroughs. The application of PGx in the systematic identification of genetic variance can open up the essential and feasible potential for generating novel therapeutic and diagnostic products based on genomics. The second is introducing a mechanism for discovering and analyzing genetic variation in human populations that can be used within the time restrictions and constraints of drug development (Housman and Ledley 1998).

2.2 Pharmacogenomics Concept

A pharmacogenomics study examines how the genome affects drug response. The combination of pharmacology and genomics is reflected in the term (pharmacy + genomics). The goal of pharmacogenomics is to establish reasonable methods for optimizing drug therapy according to the genotype of a patient in order to maximize effectiveness and minimize side effects (Becquemont 2009). It also aims to reduce the trial-and-error technique of prescription by allowing doctors to consider their patients' genes, how these genes function, and how this may affect the success of their present or future medicines (Sheffield and Phillimore 2009; Hauser et al. 2018). The application of pharmacogenomics may allow pharmaceutical drug therapies to move away from a "one-dose-fits-all" philosophy. As part of the study, gene expression or single-nucleotide polymorphisms (SNPs) are linked to pharmacokinetics (PK: the absorption, distribution, metabolism, and elimination of a drug) and pharmacodynamics (PD: the effects of the drug's biological targets) to examine the impact of inherited and acquired genetic variation on drug response (Johnson 2003; Gillis and Innocenti 2014; Tantisira and Weiss 2013).

2.3 History

Pythagoras noticed pharmacogenomics around 510 BC, where he linked the dangers of fava bean consumption to hemolytic anemia and oxidative stress. This identification was subsequently verified to G6PD (glucose-6-phosphate dehydrogenase) deficiency in the 1950s, giving rise to the term favism (Pirmohamed 2001; Prasad 2009). Despite the official start of this science in 1961 (Price and Clarke 1961), the roots of this science have long been documented. In 1956, the first case of prolonged paralysis after succinylcholine injection and fatal reactions was reported. It was diagnosed as pseudocholinesterase deficiency due to a genetic variation in a patient who lacked butyrylcholinesterase (pseudocholinesterase) (Johnson 2003; Kalow 2006). Friedrich Vogel of Heidelberg, Germany, coined pharmacogenetics in 1959, although some papers suggest it was 1957 or 1958 (Vogel 1959). In the late 1960s, twin studies confirmed the notion of genetic involvement in drug metabolism, with identical twins showing striking similarities in drug reaction when compared to fraternal twins (Motulsky and Qi 2006). Launching the Human Genome Project in 1990 marked a huge advance for personalized medicine. The complete human genome was sequenced as a result of a partnership between the U.S. National Institutes of Health (NIH) and the Office of Health and Environmental Research (OHER) of the Department of Energy. In 2003, the final findings were released and made available to the public. A number of pharmaceutical companies and organizations joined to create the SNP Consortium (TSC) in 1999 with the goal of building a public database of SNPs identified in the human genome. The discipline of pharmacogenetics expanded due to the growing accessibility of DNA technology and in vitro molecular assays. With the advent of the Human Genome Project and the growth of genome sciences in the 1990s, the word "pharmacogenomics" was first used (Motulsky and Qi 2006; Pirmohamed 2001). In 2005, the U.S. Food and Drug Administration (FDA) approved the first pharmacogenetic test (Squassina et al. 2010) for alleles in CYP2D6 and CYP2C19. Today, researchers use various high-throughput technologies to study drug response, such as DNA microarrays, RNA expression studies, protein arrays, DNA sequencing, and so on. To sequence the genomes of at least 1000 humans from various ethnic groups, the "1000 Genomes Project" was started in 2008. They published the sequencing findings for 1092 people in 2012. This effort made exome sequencing and whole-genome sequencing (WGS) more affordable and feasible for pharmacogenomic studies (Srivastava et al. 2018).

2.4 Inter-individual Differences in Therapeutic Response to Drugs

Two individuals may experience varying effects from the same drug dose; in other words, their drug pharmacokinetic profile differs since varying amounts of the drug reach specific receptors. Additionally, individual pharmacodynamics vary, such as the density of drug receptors or the physiological response to a particular drug concentration at the targeted site. While some variances are acquired with aging or illness, others are hereditary variations (https://uk.sagepub.com/sites/default/files/upmbinaries/65118_higgs_spotlight_2a.pdf). The other significant factors may be age (Albrecht et al. 1999), body weight, gender (Frezza et al. 1990), lifestyle, circadian and seasonal variation, exercise, comorbidities, and renal and hepatic function. Drug metabolism can be affected by genetic variations. Polymorphisms can influence enzymes involved in drug metabolism; it may cause minor changes in enzyme activity, or rarely the enzyme is inactivated (Rothman et al. 1997).

Genetic differences can cause diseases. There are several types of genetic variants. From among them, following are three main categories: (Alzu'bi et al. 2019):

1. **Single-Nucleotide Polymorphisms (SNPs)**
   A single base sequence change between individuals at a given site in the genome is known as a single-nucleotide polymorphism, or SNP (pronounced "snip"). Each individual has millions of SNPs, which can be utilized to distinguish individuals from one another. Most SNPs are biallelic (two alternative bases occur), which require a population frequency of at least 1%, and have the potential to alter the genomic sequence in the coding (exons), intergenic, or non-coding (introns) regions. The forensic DNA community has looked to SNPs as viable genetic markers because of their overall frequency, simplicity of analysis, affordable genotyping, and ability to conduct association studies using statistical and bioinformatics methods. A high-density SNP map of the human genome was created using the SNP Consortium (TSC) database (Vallejos-Vidal et al. 2020). SNP analysis helps understand the inter-individual variability of drug response. SNPs are useful in distinguishing drug responsive and unresponsive patients with the help of candidate-gene methods. Table 2.1 elaborately explains the role of specific SNPs in different drug responses. The classic example is of how SNP research on warfarin helped patients receive the correct dosage of the drug.
2. **Insertions and Deletions of Short Length (Indels)**
   A broad term used to describe the insertion, deletion, or combination of insertion and deletion of nucleotides in genomic DNA is referred to as "Indels." Indels are, by definition, less than 1 KB, and repeated DNA sequences frequently contain indels, making detection and annotation challenging. In clinical next-generation sequencing (NGS), indels play a crucial role because they are thought to be the primary cause of several physiological and oncologic disorders. Indels can affect gene expression in various ways, from transcription to translation to protein function. If an indel has not been previously reported and characterized in a clinical setting, it can be challenging to assess its functional impact. But numerous indels that cause human disease have been thoroughly studied (Sehn 2015). However, their correlation with drug responses is still underway.
3. **Copy Number Variants (CNVs)**
   CNVs are a class of structural DNA segment variations that affect a large portion of the human genome (Table 2.2). They include insertion, deletion, and

Table 2.1 List of SNPs involved in drug responses

S. no.	SNPs	Drugs	Drug response	Reference
1	VKORC1	Warfarin	Warfarin resistance	Soltani Banavandi and Satarzadeh (2020)
2	HLA-B*5701	Abacavir	Hypersensitivity reaction to antiretroviral abacavir	Ma et al. (2010)
3	SLCO1B1*5	Simvastatin	Reduces statin clearance, increasing the systemic exposure to the drug and the risk of unwanted effects	Li et al. (2015)
4	CYP2C19*17	Voriconazole	Low (sub-therapeutic) plasma levels in patients	Espinoza et al. (2019)
5	*DPYD*2A*, D949V	5-fluorouracil (5-FU)	Increased 5-FU toxicity	Lee et al. (2014)
6	*UGT1A1*28*, *UGT1A1*6*	Irinotecan	Serious side effects in patients receiving irinotecan-based chemotherapy	Takano and Sugiyama (2017)
7	TPMT *2, *3A, *3B, *3C, *4	Thiopurine drugs	Increased myelosuppression risk	Nguyen et al. (2011)
8	IL4 rs2070874T, IL4 rs2243250T, IL8RBrs1126580A, IL8RBrs2230054C/C	Antitumor necrosis factor drugs	Increased risk of developing rheumatoid arthritis	Canet et al. (2015)

duplication of DNA fragments ranging in size from one kilobase (kb) to several megabases (Mb) (Tuzun et al. 2005). Based on the length of the damaged sequence, researchers usually place CNVs into one of two basic types. Copy number polymorphisms (CNPs), which have an overall frequency of greater than 1% and are widespread in the general population, fall under the first group. CNPs are frequently enriched for genes that encode proteins vital to immunity and drug detoxification while being typically small (the majority are less than 10 kb in length). The second class of CNVs comprises of relatively infrequent variants that are significantly longer than CNPs, ranging in size from tens of thousands to over one million base pairs. These changes, which are sometimes referred to as microdeletions and microduplications, frequently arise more within a family (Zhang et al. 2009). Multiple hereditary disorders have CNVs as their genetic root (Kerkhof et al. 2017). Specific tools and methods are needed to detect them. This is primarily accomplished in genetic diagnostics utilizing either multiplex ligation-dependent probe amplification (MLPA) and array-based comparative genomic hybridization (aCGH) or software techniques that determine copy-number changes from NGS data produced during the diagnostics procedure (Eichler 2008).

Table 2.2 List of CNVs and its implications in diseases

S. no.	Name of CNVs	Diseases	Reference
1	Trinucleotide repeat (CAG in the HTT [Huntingtin] gene)	Huntington's disease	MacDonald et al. (1993)
2	A higher copy number of the CCL4/CCL4L genes	Psoriasis	Pedrosa et al. (2011)
3	Fewer copy number variants encoding CCL3L1 (Chemokine ligand 3-like 1)	HIV and acquired immunodeficiency syndrome (AIDS)	Gonzalez (2002)
4	High copies of ERBB2 (Erb-B2 Receptor Tyrosine Kinase 2)	Breast cancer	Peiro et al. (2004)
5	Deletion of SHPK (Sedoheptulokinase) gene	Cystinosis disorder	Kalatzis and Antignac (2002)
6	Deletion of HHAT (Hedgehog acyltransferase) (1q32.2), HLA-DPB1 (major histocompatibility complex, class II, DP beta 1) (6p21.3), PRKRA (protein activator of interferon-induced protein kinase) (2q31.2), and EEF1DP3 (eukaryotic translation elongation factor 1 delta pseudogene 3) (13q13.1) and (16p13.3)	Ankylosing spondylitis	Nijkamp et al. (2012)

2.5 Methods Used in Pharmacogenomics Research

Pharmacogenomics research aims to uncover the genes, variants, and pathways involved in drug activity in the human body. In pharmacogenomics research, these are categorized as (Mu and Zhang 2013):

1. Candidate-gene method
2. Whole-genome method
3. Cheminformatics method

1. **The Candidate-Gene Technique**
   The candidate-gene approach is the exploration of the genetic influences on a complex trait using the following techniques: constructing hypotheses about the pathogenesis of the disease and identifying potential candidate genes; locating polymorphisms in or close to those genes that may either modify the protein or its expression (Tabor et al. 2002; Kwon and Goate 2000). It is a traditional SNP-based (single-nucleotide polymorphism) method for discovering candidate regions of interest that uses genome-wide association studies (GWAS). In a GWAS investigation, a chip is coated with hundreds of thousands or millions of SNPs, indicating the regions of the genome with the largest inter-individual variation. Based on a 262-contingency table of alleles (or genotypes, depending on whether a dominant or recessive model is taken into consideration) and case/

control status for each SNP, a chi-squared test is used to assess the significance of the relationship between an SNP and a phenotype. In order to determine whether the two genotype groups have statistically different doses for a continuous independent variable, such as a medication dose, a likelihood-ratio test or a Wald test is used. Because each SNP is examined separately, significance (*p*-values) correction must be made using a Bonferroni correction or false discovery rate (FDR). Genes in or near the significant SNP, genes for which the SNP is an eQTL (an SNP related to the expression of another gene), and genes in the same pathway as these genes are all candidates for follow-up investigation after they acquire "genome-wide significance." The selection of representative genetic markers and phenotypically well-characterized patients are the two most significant aspects of the design of any pharmacogenomics study (including cases and controls). The first involves designing a suitable genotyping array that is technically simple and affordable. However, the actual design will rely on the desired balance of unbiased genome-wide studies and a focused SNP panel. The second consideration, case and control selection, characterization, and covariate identification, presents a considerable difficulty in any trait-association investigation. Because performing a million independent tests necessitates strict significance correction, finding an SNP with "genome-wide" significance necessitates many cases and controls, often in the thousands. When there is a strong signal from an SNP for the size of the study, SNP-based GWAS methods are useful (that is, when there is good separation between genotypes for the cases and controls). As a result, integrating data from several SNPs in a single gene can increase power while reducing the number of hypotheses for multiple hypothesis correction (Tatonetti et al. 2010). Alternatively, if we know the drug's mechanism of action, we can develop tailored SNP panels including just genes involved in the drug target's pathway, reducing the hypothesis space (Li et al. 2014). Many researchers are turning to exome or whole-genome sequencing to find genetic variables of medication response as the cost of high-throughput sequencing continues to decline. Such approaches have the benefit of maintaining impartiality in SNP discovery, finding rare (and even personal) mutations, and capturing larger-scale data, such as copy number variants (CNVs) and structural variants (SVs). The SNP platform (DNA microarray) utilized in significant association studies is frequently made up of SNPs that function as "tags" for a greater stretch of neighboring SNPs. Because of the presence of "linkage disequilibrium" in the genome, which is a phenomenon in which SNPs tend to be inherited together ("linked"), the structure of these "haploblocks" (where SNPs are frequently inherited together) is unique to each ethnic population. Optimized platforms for one population may not be the best solution for another since various populations have distinct linkage structures and a varied set of polymorphisms. Because of a hidden interaction with an alternate variation of another gene, the effect of a given SNP on drug response may be different (or even the reverse) due to underlying differences in genomes. Because many first genotyping systems were designed for Caucasian populations, studies on Africans or Asians will necessitate a different technique. Furthermore, the first SNP discovered is often an "associated" variant rather than a causative mutation. Further experimental or informatics investigation on genes and variations "related" to the associated variant is required to establish the exact proteins directly implicated in drug response. In association research, a replication strategy can be used to determine the prevalence of genes in specific groups of patients. It can also be used to confirm the accuracy of a marker over several samples. As a result, the replication scheme is flexible, combining several polymerase chain reaction (PCR)-based techniques such as PCR followed by restriction fragment length polymorphism (RFLP), traditional PCR plus sequencing, or reverse transcription (RT)-PCR, and in silico approaches (Kim et al. 2011; Ayuso et al. 2015; de Oliveira et al. 2017; Smith et al. 2014; Wadelius 2009). This technology is significantly cheaper and more accessible for study than next-generation sequencing, and it can be scaled and changed into an application with little effort (Van Nguyen et al. 2019; Pan et al. 2013).

2. **Whole-genome Method**

   DNA makes up an organism's genome or genetic material. Each organism has a DNA sequence of bases (A, T, C, and G). We can identify an organism's unique DNA fingerprint, or pattern, by knowing the sequence of its nucleotides. Sequencing is the process of determining the order of bases. Whole-genome sequencing is a laboratory approach for determining the order of bases in an organism's genome in a single step. In order to completely unravel the underlying mechanisms of the genetic contribution to drug response and to uncover new targets, it may be necessary to expand and supplement the candidate-gene strategy by using the whole-genome method to query the entire human genome, transcriptome, or epigenome. Technological advancements in genome-wide profiling technologies (e.g., microarray-based platforms) have made it possible to perform cost-effective whole-genome profiling of various molecular targets (e.g., mRNAs, SNP genotypes, microRNAs, DNA methylation) on a whole-genome scale, facilitating pharmacogenomic research and development. Other data sources, such as RNA expression data from microarrays or RNA-Seq experiments from drug-treated samples, can also be utilized to identify genes involved in drug response—for example, using expression profiles from individuals with a condition of interest to find genes implicated in disease pathogenesis and prospective therapeutic targets. Alternatively, expression profiles derived from a drug-treated sample (against a control) can be utilized to identify a drug's molecular response (Mardis 2008a, b). The sequencing-based whole-genome technique and associated data analysis are still in their infancy compared to array-based approaches. In an ideal situation, human pharmaceutical treatment research would produce an authentic in vivo physiological response. However, such studies are unethical for experimental (early phase) drugs, necessitate extensive regulatory approval, and are costly. As a result, established cell lines have given researchers a beneficial, low-cost resource for generating gene expression profiles. One such endeavor, the connection map (CMAP), has been used to examine expression profiles to find metabolite–protein interactions, small compounds with similar binding profiles, and metabolites that may mimic or inhibit diseases (Lamb et al. 2006).

Next-generation sequencing technology has given researchers unparalleled access to whole-genome analysis (Schuster 2008). Unlike traditional medium or high-throughput genetic screening technologies, such as microarray-based assays, this methodology promises to be highly effective in pharmacogenomics since it enables the acquisition of a complete picture of individual absorption, distribution, metabolism, elimination and toxicity (ADMET) gene variations. This is significant because each individual is likely to have unusual and/or novel functional variations in well-known pharmacogenetics, which can lead an individual/patient to be a poor or hyper metabolizer or non-responder to specific medications, which may go unnoticed by a genetic screening assay (Potamias et al. 2014). Whole-exome or whole-genome sequencing methods using NGS are becoming more popular. Whole-exome sequencing (WES) is a significant advancement for finding DNA variants, but it should only be used as a temporary solution because, even with complete technological precision, the exome only makes up around 1% of the complete genome.

The expense of sequencing technologies, the regulatory framework governing the use of (next-generation) sequencers as medical devices, the establishment of a centralized whole-genome sequencing facility, and educating physicians to interpret PGx data are currently the main obstacles to whole-genome sequencing becoming clinically viable (Katsila and Patrinos 2015).

3. **Cheminformatics Method**
While cheminformatics is not precisely a "pharmacogenomics" method, it has proven to be a valuable tool for researchers in the early phases of drug development (Table 2.3). Docking approaches, for example, predict the optimal fit of a molecule (or all molecules in a database like PubChem or ChEMBL) by reducing the conformation energy of the molecule–protein "fit" by combining knowledge about protein structure and small-molecule structure. Such approaches are computationally intensive because they examine a vast search space for each chemical and protein pair and employ molecular dynamics or genetic algorithms to find the best fit. To reduce the search space, molecule docking might be limited to the active site of a protein with a set of molecules. If a molecule has previously been shown to interact with a protein, molecular similarity metrics can be used to recommend related compounds as potential protein-binding possibilities. A search confined to ligands scoring above a similarity criterion to the known ligand would be substantially faster than a search across all of PubChem in this approach. While biochemistry (such as binding assays) must still be employed to confirm such predictions, these methods can narrow the hypothesis space for drug development by prioritizing the more expensive, lower-throughput biochemical assays. For a possible therapeutic target, cheminformatics tools can be used to find new "hits" or improve "leads" by recommending compounds that may disrupt the protein's function. Furthermore, approaches that combine the ligand's structure, and known interactions, can be used to find patterns of similar therapeutic targets (Keiser et al. 2007). The most important use of this technology is

Table 2.3 In silico tools used in pharmacogenomics research

S. no.	Name	Description	Reference
1	PharmGKB	PharmGKB is an NIH-funded resource that explains how human genetic variation influences pharmaceutical response; it is a knowledge repository that collects, curates, and disseminates therapeutically actionable gene–drug connections and genotype–phenotype relationships	https://www.pharmgkb.org/
2	Database of Genotypes and Phenotypes (dbGaP)	dbGaP was created to store and disseminate data and results from research that looked at how genotype and phenotype interact in humans	https://www.ncbi.nlm.nih.gov/gap/
3	Pharmacogenomics and Cell Database (PACdb)	PACdb is a database of connections between single-nucleotide polymorphisms, gene expression, and cellular susceptibility to various medications in cell-based models to aid in the identification of genetic variations associated with drug response	http://www.pacdb.org/
4	The Clinical Pharmacogenetics Implementation Consortium (CPIC)	In collaboration with the journal, the CPIC provides peer-reviewed guidelines; it also collaborates with Clinical Pharmacology and Therapeutics (www.nature.com/clpt) and PharmGKB; the CPIC guidelines are intended to assist doctors in determining how to effectively use available genetic test results to improve drug therapy	https://cpicpgx.org/
5	International HapMap Project	An international endeavor to find and document genetic variations in people that could affect drug response, disease propensity, and human health care	http://hapmap.ncbi.nlm.nih.gov/
6	Pan Asian SNP Database (PASNP)	It is a database containing over 50,000 SNPs and CNVs identified in 73 Asian communities; it is an effort to find biomarkers for the Asian population that are pharmacogenetically relevant	http://www4a.biotec.or.th/PASNP

(continued)

Table 2.3 (continued)

S. no.	Name	Description	Reference
7	e-PKGene	The e-PKGene database is a manually curated collection that offers access to quantitative data on drug exposure from the PGx literature as well as thorough examination of the effects of genetic variants of transporters and enzymes on drug and drug metabolite pharmacokinetic responses	http://www.druginteractioninfo.org/applications/pharmacogenetics-database
8	1000 Genomes Project	1000 Genomes is a global project that aims to sequence the entire genomes of 2500 people's DNA samples to find genetic variations with population frequencies of at least 1%	http://www.1000genomes.org/
9	FDA PGx Markers	A table of FDA-approved medications that ensures pharmacogenomic information in their labeling is available; some, but not all, of the medications' labels provide instructions for taking specific measures based on biomarker data	https://www.fda.gov/media/124784/download
10	Genetic Testing Registry (GTR)	The GTR is a central location where providers can submit information on genetic tests; its coverage includes the test's objective, methodology, validity, and utility evidence, as well as the identity and qualifications of the lab	www.ncbi.nlm.nih.gov/gtr
11	Indian Genome Variation Database (IGVdb)	It is a database of SNPs, CNVs, and repeats in over 1000 genes found in the genomes of people from India's four major linguistic groups, having applications in haplotype analysis, pharmacogenomics, and other areas	http://www.igvdb.res.in/
12	Search Tool for Interacting Chemicals (STITCH)	It is an online tool that uses public data and other sources to find interactions involving two substances, proteins, or molecules; it now contains interactions between 300,000	http://stitch.embl.de/

(continued)

Table 2.3 (continued)

S. no.	Name	Description	Reference
		compounds and 2.6 million proteins in 1133 organisms, which can be examined to discover new pharmacogenomic connections	
13	dbSNP	The National Center for Biotechnology Information (NCBI; www.ncbi.nlm.nih.gov) hosts the database of single-nucleotide polymorphism (dbSNP), a public-domain library holding a large collection (over 30 million) of simple genetic polymorphisms, or SNPs; it is the most extensive and actively accessed database of human DNA polymorphisms	www.ncbi.nlm.nih.gov/projects/SNP
14	Kyoto Encyclopedia of Genes and Genomes (KEGG)	By merging information on genes, proteins, and chemicals to create molecular interaction diagrams, this graphical database enables the systematic analysis of gene function	http://www.genome.jp/kegg/
15	Interpretome	It is a free, online interpretation tool that allows anyone to read their genomes, including discovering SNPs linked to diabetes, warfarin dose, and other biomarkers	http://esquilax.stanford.edu/
16	dbVar	dbVar is a public-domain database of genomic structural variation that comprises data and analysis from large-scale genomic variation investigations, as well as extra material such as a structural variation overview; ClinVar is connected to dbVar	www.ncbi.nlm.nih.gov/dbvar
17	ClinVar	ClinVar is a public-domain collection containing reports of links between medically significant variations and phenotypes; it is based on the phenotypic details recorded in MedGen (www.ncbi.nlm.nih.gov/medgen).	www.ncbi.nlm.nih.gov/clinvar

(continued)

Table 2.3 (continued)

S. no.	Name	Description	Reference
18	DGVa	The Database of Genomic Variations (DGVa) is a genomic variants database that catalogs, preserves, and freely disseminates copy number variants across many species, making it a useful resource for a vast community of researchers	www.ebi.ac.uk/dgva
19	HGVS	The Human Genome Variation Society (HGVS) promotes genomic variation discovery and characterization, as well as population distribution and phenotypic connections; the Society is a member of the Human Genome Organization (www.hugo-international.org/) and the International Federation of Human Genetics Societies (www.ifhgs.org/)	www.hgvs.org
20	NHGRI/GWAS	A genome-wide association studies (GWAS) catalogue, which is a publicly available, manually curated collection of published GWAS assaying at least 100,000 SNPs and all SNP-trait relationships, is kept up by the National Human Genome Research Institute (NHGRI); the catalog currently contains 1751 curated publications with 11,912 SNPs; along with the SNP-trait association data, the catalogue also publishes a quarterly graphic of all SNP-trait associations mapped to the SNPs' chromosomal positions	www.genome.gov
21	PheGenI	The Phenotype–Genotype Integrator (PheGenI) combines data from the NHGRI/GWAS catalog with data from several NCBI databases (dbGaP, Online Mendelian Inheritance in Man (OMIM), Genotype-Tissue Expression (GTEx), and dbSNP); it is intended for doctors and epidemiologists and can assist with biological	http://www.ncbi.nlm.nih.gov/gap/phegeni

(continued)

Table 2.3 (continued)

S. no.	Name	Description	Reference
		hypothesis creation, study design considerations, and variation prioritization; annotated tables of SNPs, genes, and association results, a dynamic genomic sequence viewer, and gene expression data are among the findings that users may read and download; users can also conduct searches by chromosomal position, gene, SNP, or phenotype	
22	FINDbase	A "one-stop shop" for pharmacogenomic marker allele frequency data across more than 100 populations and ethnic groups, provided by the Frequency of Inherited Disorders global database (FINDbase)	http://www.findbase.org
23	PGMD	The PharmacoGenomic Mutation Database (PGMD) is a database that lists all genetic variations that have been reported and have been shown to affect how patients respond to medications; the scientific literature contains every in vivo patient study that has discovered a strong correlation between genotype and therapeutic response; a number of access options are available for this database, including a user-friendly exploring interface and a data download for use with internal analysis procedures	http://www.biobase-international.com/product/pgmd
24	HGMD	The Human Gene Mutation Database (HGMD) collects information on known (published) gene mutations that cause inherited diseases in humans; it contains the first example of every human hereditary disease-causing or associated mutations, as well as disease-associated/functional polymorphisms	http://www.hgmd.org

(continued)

Table 2.3 (continued)

S. no.	Name	Description	Reference
		documented in the literature; HGMD now has over 150,000 polymorphisms in over 6100 genes, including several pharmacogenetics	
25	Pharmespresso	A full-text text-mining tool for extracting pharmacogenomic concepts and correlations	pharmspresso.stanford.edu
26	PGRN	A network of academic institutions known as the Pharmacogenomics Research Network (PGRN) aims to better understand how a person's genes influence how they react to medications; since 2000, it has been supported by funding from the U.S. National Institutes of Health (NIH) to advance genomics research and translational studies that will lead to safer and more efficient medication therapies	http://pgrn.org/display/pgrnwebsite/PGRN+Home
27	PMT database	The University of California, San Francisco, Pharmacogenetics of Membrane Transporters (PMT) database includes details on SNP locations and allele frequencies in significant racial and ethnic populations, as well as genetic variations in membrane transporter genes; variants that alter the protein sequences of the transporters are mapped to the secondary structure of the transporters, whereas variants that affect the gene structure of the transporters are linked to the gene structure; each transporter has a link to information in the NCBI databases	http://pharmacogenetics.ucsf.edu/
28	SNPedia and Promethease	SNPedia is a Wiki-based online database of genes and variations, as well as their medical, genealogical, phenotypic, and forensic correlations; it also includes the Promethease tool, which may be used for personal	http://www.snpedia.com/index.php/SNPedia

(continued)

Table 2.3 (continued)

S. no.	Name	Description	Reference
		genomic interpretation using SNPedia data	
29	Trait-O-Matic	It is an online tool for identifying genomic variations in personal geometries, which can be done by uploading entire genomes or downloading the tool for self-use	http://snp.med.harvard.edu/
30	ANNOtate VARiation (ANNOVAR)	It is a command-line-driven, open-source software program that may be used to annotate genetic variations in genomes of living species	http://www.openbioinformatics.org/annovar/
31	Pharmacogenetics for Every Nation Initiative (PGENI)	PGENI is a global initiative to help developed and developing countries integrate pharmacogenomics into their health care systems; it also conducts research by collecting blood samples from people of various ethnic backgrounds to uncover genetic indicators	http://www.pgeni.org/
32	PharmGenEd	It is a training program designed to educate health care professionals and patients on the concepts and implications of PGx; it has also formed partnerships with other datasets and resources to help with pharmacogenomics research	http://pharmacogenomics.ucsd.edu/about-us/pharmgened-objectives.aspx
33	Open Personal Genomics Consortium (OpenPGx)	Researchers from most South Asian nations are part of OpenPGx, an open access collaborative initiative to support PGx research; it also provides a database of genetic variations linked to pharmacological characteristics that have been carefully curated from the literature	http://www.openpgx.org/
34	Consortium on Breast Cancer Pharmacogenomics (COBRA)	It is a collaborative library of genetic variations that influence how tamoxifen and aromatase inhibitors work; COBRA is also involved in research to uncover new	http://medicine.iupui.edu/clinpharm/cobra/

(continued)

Table 2.3 (continued)

S. no.	Name	Description	Reference
		variations that may alter drug responsiveness in breast cancer treatment	
35	Population Reference Sample (POPRES)	It is a DNA database that was established by analyzing samples from over 6000 people to find potential genetic markers for usage in a range of applications, including pharmacogenomics	http://www.ncbi.nlm.nih.gov/projects/gap/cgi-bin/study.cgi?study_id=phs000145.v1.p1
36	Personal Genome Project (PGP)	It was formed as a follow-up to the Human Genome Project to collect DNA samples to better comprehend how genetic and environmental factors affect human phenotypes and how they might be used to treat and prevent disease	http://www.personalgenomes.org/
37	Koreangenome.org/ the Korean Genome Project (KPGP)	KPGP is a collection of personal genomes from people of Korean heritage	http://kpgp.kr/index.php/Main_Page
38	Genome Database of Latvian Population (LGDB)	It is a repository for Latvian genetic research advancing the country's genetic knowledge; eventually, it will be used in PGx, and 95 GWAS data of people of Latvian descent will be included in the future	http://www.p3gobservatory.org/catalogue.htm;jsessionid=ACE6E593F10B80573D64E965FA2DB3D8?measureId=18
39	Sri Lankan Genome Variation Database (SLGVD)	SLGVD is a database of SNPs found in the genomes of Sri Lanka's three major ethnic groups	http://www.p3gobservatory.org/catalogue.htm;jsessionid=ACE6E593F10B80573D64E965FA2DB3D8?measureId=18
40	1Malaysia Human Genome Variation Consortium (1MHGVC)	It is a database of genetic variants discovered in Malaysia's ethnically varied populations, to study migratory patterns, PGx uses, and other aspects of these variants about the Malaysian population	http://1mhgvc.kk.usm.my/
41	PharmGenEd	It is a training program designed to educate health care professionals and patients on the concepts and implications of pharmacogenomics; it has also partnered with other databases to help with pharmacogenomics research	http://pharmacogenomics.ucsd.edu/about-us/pharmgened-objectives.aspx

(continued)

Table 2.3 (continued)

S. no.	Name	Description	Reference
42	HIV-Pharmacogenomics.org	A database offering pharmacogenetic information on genes that influence drug responsiveness in HIV therapy	http://www.hiv-pharmacogenomics.org/
43	Comparative Toxicogenomics Database (CTD)	A database containing data about the effects of environmental factors on human health, not just pharmaceuticals, as well as gene, illness, and chemical databases	http://ctdbase.org/

the prediction of "polypharmacology," or the impact of a single drug on many targets, which can reveal new functions for currently available treatments, explained as "off-target" adverse outcomes (Keiser et al. 2009). These methods make use of small-molecule databases like PubChem and ChEMBL, which store small molecule and ligand structures and characteristics, as well as bioassay findings. In addition, a lot of scientific knowledge is available as free text in the biomedical literature (Garten et al. 2010). NGS investigations have the potential to uncover a huge number of PGx variations, the majority of which are novel, infrequent, and unsupported by biochemical or clinical evidence. It is not always possible to perform functional expression experiments for such large numbers of variations, which is why evaluating predictions produced from in silico methods is a viable option (Pandi et al. 2021).

2.6 OMICS Tools Used in Pharmacogenomics Research

2.6.1 Transcriptomics

Utilizing high-throughput technologies like microarray analysis, transcriptomics is the study of the transcriptome, which is the complete collection of RNA transcripts produced by the genome under specific circumstances or in a particular cell. Transcriptome comparison can be used to find genes that have variable levels of expression in various cell types or in response to various treatments. Transcriptomics has been utilized to find answers to specific gene expression-related problems. Consider the genes that are activated by a specific transcription factor. Comparing gene expression patterns in tissues where a transcription factor is active or inactive can help with this (https://www.nature.com/subjects/transcriptomics#:~:text=Transcriptomics%20is%20the%20study%20of,methods%2C%20such%20as%20microarray%20analysis). Transcriptome sequencing for disease genetics has been facilitated by community-wide efforts. The National Institute of Health-funded Pharmacogenomics Research Network (PGRN) leveraged the essential resources for a thorough transcriptome sequencing research, cataloging variance in genetic expression and splicing occurrences of 389 pharmacogenetics involved in drug disposal across liver, kidney, heart, adipose tissue, and lymphoblastoid cell lines (Chittani et al. 2015). In oncology, transcriptome profiling in tumor tissues has been successful in identifying treatment response predictors. In non-small-cell lung cancer (a group of lung cancers that behave similarly, such as squamous cell carcinoma and adenocarcinoma), co-expression of seven genes was linked to molecular subtype-specific vulnerabilities. The presence of this gene expression pattern suggests that a synthetic indolotriazine could be used as an ideal therapeutic. This study shows how gene expression analysis using RNA-Seq data can lead to a new potential for therapeutic advice (Kim et al. 2013). Following validation using cancer cell lines, small groups of coding and non-coding (lincRNAs) linked with trastuzumab resistance were identified using RNA-Seq for an integrated transcriptomic approach from trastuzumab-sensitive and trastuzumab-resistant HER2+ malignancies. They discovered that inhibiting S100P results in the reversal of trastuzumab resistance by functional analysis of top candidate genes (Merry et al. 2016). Transcriptomic research was carried out in lymphoblastoid cell lines taken from 150 people in the Cholesterol and Pharmacogenetics (CAP) simvastatin clinical trial, identifying genes involved in triglyceride statin response. Insulin-induced gene 1 (INSIG1) was the most significant biological candidate among the 23 genes connected with triglyceride statin response, according to extensive research linking this gene to cholesterol homeostasis regulation (Theusch et al. 2016). RNA-Seq was also used to explore a pharmacogenomic phenotype: glucocorticoid anti-inflammatory effects in the treatment of asthma. Dexamethasone was applied to airway smooth muscle cells for 18 h, then mRNA was extracted and high-throughput sequencing was performed. This method found 316 genes that had varied expression in response to dexamethasone therapy, highlighting CRISPLD2 as a possible asthma pharmacogenetics gene that modulates glucocorticoid anti-inflammatory effects (Himes et al. 2014).

The first transcriptome-wide analysis of gene expression impacting thiazide blood pressure (BP) response was made possible because of resources from the PGRN RNA-Seq group. Total 150 people from the Pharmacogenomics Evaluation of Antihypertensive Responses (PEAR) and PEAR-2 studies with extremes of BP response (responders and non-responders) to thiazide diuretics (50 whites from PEAR; 50 whites and 50 blacks from PEAR-2) were chosen for RNA-sequencing to find novel molecular markers of BP response to thiazide diuretics (McDonough et al. 2018). The use of RNA-Seq could lead to the identification of signature genes for drug response, as well as the identification of isoform diversity, cis/trans-acting regulatory variations, and gene expression networks that influence drug response heterogeneity (Sá et al. 2018).

2.6.2 Pharmacometabolomics

In order to anticipate how a person would respond to a treatment, pharmacometabolomics (PMx) research uses information from metabolic profiles (or metabolome) (Beger et al. 2020). Exogenous metabolites (those created by biotransformation of the pharmacological agent, i.e., the drug) and endogenous metabolites are the two primary subgroups of metabolites that can be employed for pharmacometabolomics-informed pharmacogenomic investigations. In other words, these two subgroups are the outcome of PK and PD variances, which may lead to individual differences in medication response (Neavin et al. 2016; Wilson 2009). Given the importance of the environment and gut microbiome in illness pathobiology, a "pharmacometabolomics-assisted pharmacogenomics" technique could be extremely useful, as pharmacogenomics alone does not address environmental impacts (Li and Jia 2013). The identification of an individual's metabolic state is referred to as "metabotype" (Kaddurah-Daouk et al. 2014; Abdin and Hamouda 2008; Ellero-Simatos et al. 2014; Li et al. 2008; Yapar et al. 2007). This could be used to create metabolic signatures and provide a better forecast of medication safety concerns (Schnackenberg et al. 2008; Ji et al. 2011; Abo et al. 2012). In the population-based cooperative health research of Augsburg (KORA), Wang-Sattler et al. (2012) identified 140 metabolites in 4297 serum samples. Even seven years before the disease started, glycine and lysophosphatidylcholine (18:2) were found to be strong predictors of glucose tolerance in their investigation. A panel of 196 metabolites was used in another investigation to identify a metabolic signature linked to the early signs of type 2 diabetes mellitus (T2DM), which was characterized by a large increase in glyoxylate (Padberg et al. 2014). Zhang et al. (2018) used case-control research to profile 180 metabolites in 189 people who had new-onset diabetes over a 12 years period and 189 propensity-matched controls from the same baseline test. These findings support the idea that branched chain amino acid (BCAA) metabolism plays a crucial role in adipose tissue, as evidenced by an increase in circulating BCAA in obese and prediabetic patients (Tobias et al. 2018; Hernández-Alvarez et al. 2017). The substantial association between insulin sensitivity and expression of BCAA metabolism genes in diabetic patients' adipose tissue (Roberts et al. 2014; Lynch and Adams 2014) emphasizes this point.

Recent research shows the importance of examining energy metabolism and the role it plays in brain function in this rush to understand the causes of neurodegenerative disorders (Jové et al. 2014; Magistretti and Allaman 2015; Everett 2013; Toledo et al. 2017). The human brain is well-known for being metabolically costly, consuming up to 20% of total body energy to function (Mink et al. 1981). The evolutionary results show a huge increase in genes involved in energy generation in the human cortex as compared to non-human primates, which corroborates this theory. Several metabolomics studies, ranging from Nuclear magnetic resonance (NMR) spectroscopy and Gas chromatography-mass spectrometry (GC-MS) to capillary electrophoresis-mass spectrometry (CE-MS), Direct-injection mass spectrometry (DIMS), and liquid chromatography-mass spectrometry (LC-MS), have investigated the metabolic features of Alzheimer's disease (AD), Parkinson's disease (PD), and Amyotrophic lateral sclerosis (ALS) in a variety of transgenic animal models and small human cohorts. Changes at the metabolite and pathway levels point to a disruption in mitochondrial energy production, as well as an increase in reactive oxygen species (ROS) production, which causes oxidative damage to DNA, electron transport chain proteins, and the plasma membrane redox system (due to the loss of function and structural integrity of amino acids, nucleic acids, and lipids as a result of oxidation) (Tönnies and Trushina 2017). These changes are expected to stimulate the formation and appearance of disease-specific components such as amyloid and p-tau proteins in Alzheimer's disease, synuclein in Parkinson's disease, huntingtin in Huntington's disease, and Cu/Zn-superoxide dismutase (SOD) in ALS, further initiating a neurodegenerative cascade of disrupted nutrition transport, neurotransmission, cell-to-cell communication, and synapse dysfunction and loss across the brain (Sun et al. 2004; Mattson 1998; Vickers 2017). Pharmacometabolomics offers the potential to increase our understanding of the mechanistic relationship between energy intake and brain function in relation with the aging process, oxidative stress, and neurodegeneration by providing a complete assessment of changes embedded in metabolic networks. This will be critical in identifying novel mechanisms and small-molecule targets for medicinal and diagnostic applications, as well as preventing pathogenesis before irreversible cell death.

Despite population-wide efforts in whole-genome sequencing, transcript profiling, and single-nucleotide polymorphism (SNP) characterization, it is evident that many diseases can be better understood when viewed from a metabolic as well as a genetic perspective. However, targetable disease genes are still lacking (Wishart 2016). Cancer is an excellent example of a rather complex disease whose understanding has progressed greatly since its metabolic component has been the subject of intense research. Recent metabolomics-assisted studies suggest that the majority of cancer-associated genes code for well-known metabolic enzymes and thus affect only a few pathways, such as aerobic glycolysis, glutaminolysis, and one-carbon metabolism (Yang et al. 2013), rather than challenging the exploration of millions of possible mutations. These oncopathways are involved in the creation of several oncometabolites, which are required for rapid tumor development and propagation. By changing signaling pathways and cell division processes, most oncometabolites play a crucial role in carcinogenesis. Most cancers rely on glucose, lactate, serine, and polyamines, with Myc-activated tumors relying on glutamine, breast cancer on glycine synthesis, acute lymphoblastic leukemia (ALL) on asparagine, and brain and prostate cancers on choline as examples of metabolic vulnerabilities. The ability to identify and measure a wide range of small-molecule metabolites in tumor tissues was made possible by advances in High-resolution mass spectrometry (HRMS)-based metabolomics, Mass spectrometry (MS)-based imaging, and magnetic resonance imaging (MRI). Overall, the combined use of various metabolomic and genomic techniques is paving the way for cellular metabolism manipulation and the development of effective anticancer medicines by targeting well-known metabolic enzymes (Ivanisevic and Thomas 2018). The application of metabolomics to systems biology is quickly increasing to provide a full view of molecular system dynamics (Bersanelli et al. 2016).

2.6.3 Proteomics

One significant drawback of genome-wide-related pharmacogenomics investigations is the lack of knowledge of the true biological mechanisms and full cellular pathways underlying the discovered genetic linkages. As a result, pharmacogenomics and pharmacoproteomics, a more modern area that combines proteome technologies for drug discovery and development, should be included in precision medicine (Jain 2004). Target enrichment, ionization, detection, and quantification are all used in proteomic research. Before analysis, complex samples obtained from tissue, plasma, serum, or urine are drained of complexity or enriched for a single analyte of interest (Silva et al. 2006; Torsetnes et al. 2014).

2.6.4 Genomics

The study of all the genes in an individual's genome, as well as how those genes interact with one another and with their environment, is known as genomics. A gene is typically the DNA unit that contains the instructions for producing a particular protein or set of proteins. The DNA of a cell may mutate, causing the production of an aberrant protein that can interfere with normal bodily functions and result in diseases like cancer. Identifying the precise sequence of the nucleotides in a strand of DNA is what is meant by sequencing. DNA sequencing can be used by researchers to look for genetic variants and/or mutations that may contribute to the onset or course of a disease (https://www.genome.gov/about-genomics/fact-sheets/A-Brief-Guide-to-Genomics).

Exome sequencing, whole-genome sequencing, and genome-wide association studies (GWAS) are a few genomics approaches that are used in pharmacogenomics research. In order to uncover genetic changes, two techniques—whole-exome sequencing and whole-genome sequencing—are being used more often in research and health care. Both techniques rely on modern technology that enables quick sequencing of vast amounts of DNA. These methods are referred to as "next-generation sequencing" (or next-gen sequencing) (https://medlineplus.gov/genetics/understanding/testing/sequencing/). Whole-genome sequencing (WGS) is a technology used in labs to determine nearly three billion nucleotides in a person's entire DNA sequence, including non-coding material, whereas exome sequencing is an efficient approach for sequencing the coding regions of the genome (https://www.cancer.gov/publications/dictionaries/genetics-dictionary/def/whole-genome-sequencing). The term "exome" refers to the collection of exons—transcribed regions of the human genome—that are translated into proteins (Burks et al. 2019). Exome sequencing has the potential to be a cutting-edge technique for identifying the genes causing complex diseases because it concentrates on exclusively sequencing the genomic DNA that codes for proteins (Gleason and Juul 2017). Future disease diagnosis will be aided by further investigation of exome and genome sequencing to see if novel genetic changes are linked to specific health issues. For instance, with WES, researchers recently discovered a loss-of-function mutation in tuberous sclerosis complex 1 (TSC1) in about 5% of advanced bladder tumors. This particular mutation was associated with everolimus tumor sensitivity, indicating that this subgroup of bladder cancer patients might benefit from everolimus therapy (Kreso et al. 2013).

Integrating genomics, transcriptomics, proteomics, and metabolomics explore an organism's response to xenobiotics in a systems biology approach, which gives a great opportunity to accelerate the understanding of biological pathway modifications associated with pathophysiological characteristics and the development of more classified biomarkers for individualized therapy (Pinu et al. 2019).

2.7 Applications

2.7.1 Precision Medicine

The provision of effective treatment while limiting or eliminating the risk of adverse drug reactions (ADRs)/events is included in the optimization of patient care related to medication therapy (Owen 2011). The development of pharmacogenomics as a source of information for clinical decision-making allows for the early implementation of optimal medication, avoiding "trial and error" medicines (Phapale et al. 2010). The realization that a person's response to medication is probably influenced by a number of genes gave rise to the more general term "pharmacogenomics." It helps to recognize the consequences related to pharmacokinetics (PK) and/or pharmacodynamics (PD).

Genome editing or engineering is currently a major focus for precision medicines, with particular emphasis on three genome-modifying methods that all use programmable nucleases, which can be thought of as "molecular tools." These include transcription activator-like effector nucleases (TALENs), zinc finger nucleases, and CRISPR-Cas9 (clustered, regularly interspaced, short palindromic repeats). To some extent, all of these nucleases have been applied to patient care. The initial patients of B-cell acute lymphoblastic leukemia (B-ALL) were treated with TALEN-engineered cells (Qasim et al. 2015). They generated universal CAR19 (UCART19) T-cells by lentiviral transduction of non-human leukocyte antigen-matched donor cells and simultaneous transcription activator-like effector nuclease (TALEN)-mediated gene editing of T-cell receptor α-chain and CD52 gene loci (Qasim et al. 2017). Current adult clinical studies using CRISPR-Cas9 gene editing technologies, however, are being conducted for advanced esophageal cancer, leukemia, lymphoma, transfusion-dependent beta-thalassemia, and relapsed refractory multiple myeloma, synovial sarcoma, and myxoid/round cell liposarcoma (Jing et al. 2018; Liu and Zhao 2018).

2.7.2 Drug Discovery and Development

Pharmacogenomics study tries to look into the genetic causes of individual variations in drug safety or efficacy. Pharmacogenomics is used to detect the genetic variants that affect diseases to estimate the chance of developing the condition and it has had an impact on drug therapy in recent years.

The utilization of pharmacogenomics has been expanded to include pharmacological efficacy and safety studies in drug discovery and development due to the rapid advancement of pharmacogenomics technology and methods. Pharmacogenomics can help with the two key factors that determine the success of drug discovery and development, namely safety and efficacy, which can be better predicted by identifying the susceptibility polymorphism of potential targets. This can increase the success rate of drug development by dividing subjects into different groups and modifying the dosage regimen in clinical trials (Xiang and Yuan 2020).

Pharmacogenomics has been extensively utilized in the treatment of several non-communicable and communicable diseases, which are discussed below.

1. **Non-communicable Diseases (NCDs)**
   The term "NCDs" refers to a set of disorders that are not primarily brought on by an acute infection, have an impact on long-term health, and frequently necessitate ongoing care and therapy. Cancers, cardiovascular diseases (CVDs), diabetes, and chronic respiratory problems are some of these conditions (https://www.paho.org/en/topics/noncommunicable-diseases). Non-communicable diseases (NCDs) account for 41 million annual deaths, or 71% of all fatalities worldwide. More than 15 million persons aged 30–69 die from an NCD every year; 85% of these "premature" deaths take place in low- and middle-income nations. Low- and middle-income nations account for 77% of all NCD fatalities. The majority of NCD deaths, or 17.9 million people per year, are caused by cardiovascular diseases, followed by malignancies (9.3 million), respiratory illnesses (4.1 million), and diabetes (1.5 million). Key elements of the approach to NCDs include palliative care, NCD detection, screening, and therapy (https://www.who.int/news-room/fact-sheets/detail/noncommunicable-diseases).
   (a) **Cancer**
   Globally, cancer was the primary cause of approximately ten million deaths in 2020, according to the World Health Organization (WHO). Breast, lung, colon, rectum, and prostate cancers are the most prevalent types of cancer. If diagnosed early and treated efficiently, many tumors are curable. For anti-cancer medications, toxicity is a significant issue. These drugs have a limited therapeutic index, with a tiny difference between the dose needed to have an antitumor effect and the dose that results in intolerable toxicity. According to the toxicity endpoint, their recommended doses are established.

   Alkylating compounds are often used as anticancer prodrugs that work by attaching an alkyl group to the DNA molecule's guanine base to cause cytotoxicity. Alkylating agents exhibit a wide spectrum of toxicities, including myelosuppression and renal and gastrointestinal toxicities. In addition to toxicity, investigations on the relationship between genetic variations and the pharmacokinetics of alkylating drugs are scarce, and few have specifically looked at this relationship. In the juvenile population of 49 B-cell non-Hodgkin lymphoma (NHL) patients, the impact of CYP2B6 on cyclophosphamide clearance was recently confirmed. Cyclophosphamide clearance was significantly reduced in patients with CYP2B6*6 (Veal et al. 2016).

Anthracycline-induced cardiotoxicity and the detoxifying enzyme UGT1A6*4 (rs17863783), as well as the genetic variants SLC28A3 (rs7853758 and rs885004), were previously linked (Visscher et al. 2013; Aminkeng et al. 2016). Additionally, genetic variants in CBR3, ABCC1 (rs3743527, rs246221), SLC28A3 (rs885004 and rs4877847), SLC22A17 (rs4982753, rs4149178), and sulfotransferase (SULT) SULT2B1 (rs12882406 and rs12896494) have been associated with anthracycline-induced cardiotoxicity (Blanco et al. 2012; Visscher et al. 2012; Conyers et al. 2018).

Busulfan, which is often used as part of conditioning regimens before hematopoietic stem cell transplantation, has a constrained therapeutic window and substantial interpatient pharmacokinetic heterogeneity. The medication is broken down in the liver by the isoenzyme glutathione *S*-transferase (GST). It is metabolized primarily by GSTA1, a GST isoenzyme. In a less significant way, GSTM1 and GSTT1 are implicated. AUC was shown to be lower in patients with the GSTA1*A/*A genotype (and higher in those with the GST*B/*B genotype), according to various research done in the past on both adult and pediatric patients (Myers et al. 2017).

(b) **Cardiovascular Diseases**

An estimated 17.9 million people die from cardiovascular diseases (CVDs), which are the main cause of death worldwide. The term "CVDs" refers to a group of conditions that affect the heart and blood arteries and includes cerebrovascular illness, rheumatic heart disease, and coronary heart disease. Heart attacks and strokes account for more than four out of every five CVD deaths, with premature deaths accounting for one-third of these deaths in those under the age of 70. Warfarin, a vitamin K antagonist, prevents coagulation by preventing the production of coagulation factors II, VII, IX, and X as well as proteins C and S (Fareed et al. 2012). The CYP2C9, VKORC1, and CYP4F2 genes are specifically targeted by the most recent evidence-based CPIC guidelines for warfarin dose. It is anticipated that patients who have the VKORC1-1639G>A variant will be more sensitive to the anticoagulant, requiring a lower dose. Both of these polymorphisms call for substantially lower warfarin dosages in affected individuals. The dose may also need to be increased by 5–10% for CYP4F2*3 carriers.

Beta-blockers compete with catecholamine to bind to β-1-adrenergic receptors (ADBR1). They are frequently prescribed for heart disease, angina, hypertension, and cardiac arrhythmia. CYP2D6, ADBR1, and ADBR2 are among the genes that have been linked to inter-individual variation in beta-blocker responses. Poor metabolizers of β-blockers like propranolol, timolol, and metoprolol are phenotypical traits of CYP2D6 loss-of-function variations. Five to ten percent of people have two or more CYP2D6 alleles that cause loss of function, which causes them to have higher levels of circulating drugs. It should be noted that not all β-blockers are metabolized by CYP2D6, such as atenolol and nadolol, and that decreased CYP2D6 metabolism is not necessarily associated with clinical consequences (e.g., carvedilol) (Bijl et al. 2009).

(c) **Diabetes Mellitus**

High blood glucose, commonly referred to as blood sugar, is the primary symptom of diabetes, a chronic metabolic disorder that over time can gravely injure the heart, blood vessels, eyes, kidneys, and nerves. Type 2 diabetes is the most common form and typically affects adults. It arises when the body produces insufficient insulin or becomes insulin resistant. Sulfonylureas are insulin secretagogues that bind the ABCC8-encoded SUR1 subunit, play a significant role in insulin secretion, and maybe a treatment option for type 2 diabetes (Weisnagel et al. 2001). The efficacy of sulfonylureas was linked to genetic variation at rs12255372 in a study conducted in India on a cohort of T2DM patients from Gujarat (Dhawan and Padh 2016). TCF7L2 genetic variations rs12255372 and rs7903146 are linked to poor treatment results while rs7903146 had the strongest connection with type 2 diabetes mellitus (T2DM) in people of different ethnic backgrounds (Holst et al. 2011; Holstein et al. 2011).

Glinides was taken into the blood binding to the KATP channel, which is made up of SUR1 (encoded by the ABCC8 gene) and Kir6.2 (encoded by the KCNJ11 gene). Patients with the ABCC8 rs1801261 SNP CT genotype have significantly lower FPG and HbA1c levels than patients with the CC genotype. In terms of repaglinide response in insulin sensitivity, the C/C homozygotes of the ABCC8 exon16- 3T/C variant performed better than the T/C and T/T genotypes of the KCNJ11 E23K variant (Zhou et al. 2019).

A nuclear hormone receptor called PPARγ is produced by the peroxisome proliferator-activated receptor γ gene (PPARG). The most prevalent mutation in the PPARG and a risk factor for T2DM is the polymorphism Pro12Ala (rs1801282) (Zeng et al. 2020). It has been repeatedly noted in a variety of groups, including those from South India, China, Iran, and menopausal women in Mexico; thus there may be a link between the Pro12Ala polymorphism and a stronger therapeutic response to pioglitazone (Hsieh et al. 2010; Namvaran et al. 2011; Priya et al. 2016; Ramírez-Salazar et al. 2008).

2. **Communicable Diseases**

Infectious diseases, often known as communicable diseases, are brought on by microorganisms including bacteria, viruses, parasites, and fungi that can transfer from one person to another either directly or indirectly. Others are brought on by consuming contaminated food or water, while some are spread by bug bites (https://www.afro.who.int/healthtopics/communicablediseases#:~:text=Tuberculosis,Overview,ingesting%20contaminated%20food%20or%20water).

(a) **Infectious Diseases**

Pathogenic microorganisms like bacteria, viruses, parasites, or fungus are the source of infectious diseases, which can be transmitted from one person to another either directly or indirectly. These diseases can be divided into three categories: those that result in high rates of mortality; those that exert a huge cost on populations in terms of impairment; and those that, due to the speed and unpredictability of their spread, may have serious global implications.

- **Viral Infections**

  SARS-CoV-2 virus-borne coronavirus disease-2019 (COVID-19) is an infectious disease. Many medications without a well-established track record of efficacy, safety, or data have been administered to patients in an effort to hastily alleviate this terrible tragedy (Khan et al. 2022). By choosing the best first-line therapies and the best dose of these medications in accordance with the individual's genetic makeup, the eradication of these medications' adverse effects can be improved by pharmacogenomics application. Additionally, by using pharmacogenomics, it is possible to identify the precise genetic markers that could improve the efficacy and lower the toxicity of COVID-19 drugs (Takahashi et al. 2020). In late March 2020, the FDA formally approved the first antiviral drugs, hydroxychloroquine (HCQ) and chloroquine (CQ), for use in treating COVID-19 in clinical settings (US Food and Drug Administration 2020). Human cytochrome P450 (CYP) enzymes CYP1A1, CYP2D6, CYP3A4, and CYP2C8 as well as the human liver microsomes (HLM) metabolize HCQ and CQ in the liver to produce one major metabolite, *N*-desethylchloroquine (DCQ) (Projean et al. 2003). Poor or intermediate CYP2D6 metabolizers (CYP2D6*4; CYP2D6*10) have been found to have higher HCQ concentrations than healthy individuals. According to a study, patients with the G/G genotype of the CYP2D6*10 (rs1065852) had a higher ratio of [DHCQ]:[HCQ] than those with the A/A genotype. In accordance with this, patients with the C/C genotype of CYP2D6*10 (rs1135840) had a higher ratio of [DHCQ]:[HCQ] than patients with the G/G genotype (Lee et al. 2016).

  Pharmacogenomics eventually takes on a significant role in foretelling negative effects brought on by antiretroviral medication therapy. The array of HIV treatment options has now been improved by highly active antiretroviral medication. Depending on the class of antiretroviral agents employed, antiretroviral medicines do, however, show specific adverse drug reactions (ADRs), which are typically characterized by short- and long-term toxicities (Lyimo et al. 2012). As an illustration, research by Mallal and colleagues demonstrated that the HLA-B5701 allele is suggestive of an abacavir hypersensitivity reaction (Mallal et al. 2008). Additionally, research by Young and colleagues has demonstrated that testing for the HLA-B5701 allele reduced the number of patients who experienced abacavir-related hypersensitivity to less than 1% from 4 to 8% in the absence of HLA testing (Young et al. 2008). In addition, the CYP2B6 gene's c.516G/T variant may serve as a pharmacogenetic marker for adverse drug reactions (ADRs) in individuals using efavirenz (Haas et al. 2004).

- **Bacterial Infections**

  Antimicrobial agent response, however, is frequently unpredictable and depends on the intricate interactions between the host, the bacterium, and the medication. Regarding the host, variations in pharmacokinetic profiles and pharmacodynamics, as well as predisposition to adverse effects, can have a significant impact on both the therapeutic response and the toxicity caused by drugs (Stocco et al. 2020). Erythromycin is a substrate for OATP1B1 (organic anion transporter polypeptide), which is expressed on the basolateral surface of hepatocytes and is encoded by the SLCO1B1 (solute carrier organic anion transporter family member 1B1) gene. Stably transfected Flp-In T-Rex293 cells producing OATP1B1*5 (V174A) showed a 50% reduction in erythromycin transport compared to OATP1B1*1A (wild type). In mice lacking the OATP1B2 transporter, there was a 52% drop in the rate of erythromycin metabolism, and in 91 cancer patients who underwent an erythromycin breath test, the c.521T>C mutation in SLCO1B1 (rs4149056), encoding OATP1B1*5, was linked to a decrease in erythromycin metabolism (Lancaster et al. 2012).

  A lipopeptide antibiotic called daptomycin is efficient against Gram-positive bacteria. It is also approved for the treatment of right-sided *Staphylococcus aureus* endocarditis and complex skin and soft tissue infections brought on by Gram-positive cocci. Three SNPs, 3435C>T (rs1045642), 1236C>T (rs1128503), and 2677G>T (rs2032582) in the ABCB1 gene, were examined in 23 Caucasian patients who received daptomycin treatment. In comparison to individuals who had the CT or CC genotype or the homozygous variant TT genotype, individuals with the homozygous variant TT genotype had a greater median dose-normalized AUC0-24. Daptomycin concentrations rose in individuals with the TT genotype due to reduced clearance of the drug. Even though a very small cohort of patients participated in the study, the findings imply that this mutation may be able to account for the significant level of pharmacokinetic heterogeneity seen in clinics (Baietto et al. 2015).

- **Fungal Infections**

  Azoles are a diverse class of antifungal medications. Lanosterol 14-demethylase, a member of the fungal cytochrome P450 family of enzymes, is the main target of azole class agents. A component of the fungal cell membrane, ergosterol, is disrupted as a result, altering the shape and functionality of the membrane and causing cell lysis and death (Nocua-Báez et al. 2020). Utilizing the biggest patient cohort available to date, the first GWAS assessing genetic vulnerability to candidiasis was carried out. This research found a substantial connection between new SNPs in the CD58, LCE4A-C1orf68, and TAGAP loci. Those who carried two or more risk alleles experienced a 19.4-fold increase in their candidemia risk (Kumar et al. 2014).

Although there is inter-individual variability in CYP3A4 activity, all triazoles are substrates of CYP3A4 to varying degrees, and although only a few common genetic variations have been discovered, currently, there is minimal data to justify dose modification for the bulk of medications metabolized by CYP3A4 (Amsden and Gubbins 2017; Saiz-Rodríguez et al. 2020). Isavuconazole is metabolized by CYP3A5, but to a smaller amount than by CYP3A4 (Darnaud et al. 2018; Townsend et al. 2017). Since of yet, interactions with CYP3A4/5 substrates have been shown to have the greatest impact on CYP3A4 and CYP3A5 genotypes' effects on azoles, as they can considerably raise azole concentrations in people who have the CYP3A5*3 variation, which is linked to a lack of CYP3A5 activity (Shirasaka et al. 2013). Fluconazole and voriconazole metabolism are mediated by two CYP2C subfamily enzymes, CYP2C19 and CYP2C9 (Niwa et al. 2005). The CYP2C9 gene is quite polymorphic, although CYP2C9*1, which is linked with normal enzyme activity, is the prevalent allele and is much less common than variants that result in impaired function (Scordo et al. 2001; García-García and Borobia 2021).

References

A Brief Guide to Genomics (n.d.). https://www.genome.gov/about-genomics/fact-sheets/A-Brief-Guide-to Genomics

Abdin AA, Hamouda HE (2008) Mechanism of the neuroprotective role of coenzyme Q10 with or without L-dopa in rotenone-induced parkinsonism. Neuropharmacology 55:1340–1346

Abo R, Hebbring S, Ji Y et al (2012) Merging pharmacometabolomics with pharmacogenomics using '1000 Genomes' single-nucleotide polymorphism imputation: selective serotonin reuptake inhibitor response pharmacogenomics. Pharmacogenet Genomics 22:247–253

Albrecht S, Ihmsen H, Hering W et al (1999) The effect of age on the pharmacokinetics and pharmacodynamics of midazolam. Clin Pharmacol Ther 65:630–639

Alzu'bi AA, Zhou L, Watzlaf VJ (2019) Genetic variations and precision medicine. Perspect Health Inf Manag 16:1a

Aminkeng F, Ross CJ, Rassekh SR et al (2016) Recommendations for genetic testing to reduce the incidence of anthracycline-induced cardiotoxicity. Br J Clin Pharmacol 82:683–695

Amsden JR, Gubbins PO (2017) Pharmacogenomics of triazole antifungal agents: implications for safety, tolerability and efficacy. Expert Opin Drug Metab Toxicol 13:1135–1146

Ayuso P, Plaza-Serón MDC, Blanca-López N (2015) Genetic variants in arachidonic acid pathway genes associated with NSAID-exacerbated respiratory disease. Pharmacogenomics J 16:825–839

Baietto L, D'Avolio A, Cusato J et al (2015) Effect of SNPs in human ABCB1 on daptomycin pharmacokinetics in Caucasian patients. J Antimicrob Chemother 70:307–308

Becquemont L (2009) Pharmacogenomics of adverse drug reactions: practical applications and perspectives. Pharmacogenomics J 10:961–969

Beger RD, Schmidt MA, Kaddurah-Daouk R (2020) Current concepts in pharmacometabolomics, biomarker discovery, and precision medicine. Metabolites 10(4):129

Bersanelli M, Mosca E, Remondini D (2016) Methods for the integration of multi-omics data: mathematical aspects. BMC Bioinform 17:167–177

Bhagat K (1999) Pharmacogenomics: today, tomorrow and beyond. Cent Afr J Med 45:335–337

Bijl MJ, Visser LE, Van Schaik RHN et al (2009) Genetic variation in the CYP2D6 gene is associated with a lower heart rate and blood pressure in β-blocker users. Clin Pharmacol Ther 85:45–50
Blanco JG, Sun CL, Landier W et al (2012) Anthracycline-related cardiomyopathy after childhood cancer: role of polymorphisms in carbonyl reductase genes—a report from the Children's Oncology Group. J Clin Oncol 30:1415
Burks AW, Holgate ST, O'Hehir RE et al (2019) Middleton's allergy E-Book: principles and practice. Elsevier, Amsterdam
Canet LM, Cáliz R, Lupiañez CB et al (2015) Genetic variants within immune-modulating genes influence the risk of developing rheumatoid arthritis and anti-TNF drug response: a two-stage case–control study. Pharmacogenet Genomics 25:432–443
Chittani M, Zaninello R, Lanzani C et al (2015) TET2 and CSMD1genes affect SBP response to hydrochlorothiazide in never-treated essential hypertensives. J Hypertens 33:1301
Conyers R, Devaraja S, Elliott D (2018) Systematic review of pharmacogenomics and adverse drug reactions in paediatric oncology patients. Pediatr Blood Cancer 65:26937
Darnaud L, Lamoureux F, Godet C et al (2018) Isavuconazole kinetic exploration for clinical practice. Drugs R D 18:317–321
de Oliveira Rodrigues R, Rabenhorst SHB, de Carvalho PG et al (2017) Association of IL10, IL4, IFNG and CTLA4 gene polymorphisms with efavirenz hypersensitivity reaction in patients infected with human immunodeficiency virus. Jpn J Infect Dis 70(4):430–436
Dhawan D, Padh H (2016) Genetic variations in TCF7L2 influence therapeutic response to sulfonylureas in Indian diabetics. Diabetes Res Clin Pract 121:35–40
Eichler EE (2008) Copy number variation and human disease. Nat Educ 1:1
Ellero-Simatos S, Lewis JP, Georgiades A et al (2014) Pharmacometabolomics reveals that serotonin is implicated in aspirin response variability. CPT Pharmacometrics Syst Pharmacol 3:1–9
Espinoza N, Galdames J, Navea D et al (2019) Frequency of the CYP2C19* 17 polymorphism in a Chilean population and its effect on voriconazole plasma concentration in immunocompromised children. Sci Rep 9:1–6
Everett JA (2013) The 12 item social and economic conservatism scale (SECS). PLoS One 8:82131
Fareed J, Thethi I, Hoppensteadt D (2012) Old versus new oral anticoagulants: focus on pharmacology. Annu Rev Pharmacol Toxicol 52:79–99
Frezza M, di Padova C, Pozzato G et al (1990) High blood alcohol levels in women. The role of decreased gastric alcohol dehydrogenase activity and first-pass metabolism. N Engl J Med 322: 95–99
García-García I, Borobia AM (2021) Current approaches and future strategies for the implementation of pharmacogenomics in the clinical use of azole antifungal drugs. Expert Opin Drug Metab Toxicol 17:509–514
Garten Y, Coulet A, Altman RB (2010) Recent progress in automatically extracting information from the pharmacogenomic literature. Pharmacogenomics J 11:1467–1489
Gillis NK, Innocenti F (2014) Evidence required to demonstrate clinical utility of pharmacogenetic testing: the debate continues. Clin Pharmacol Ther 96:655–657
Gleason CA, Juul SE (2017) Avery's diseases of the newborn e-book. Elsevier, Amsterdam
Gonzalez E (2002) Gene-containing CCL3L1 the influence of earth planet. Science 30:385
Haas DW, Ribaudo HJ, Kim RB et al (2004) Pharmacogenetics of efavirenz and central nervous system side effects: an Adult AIDS Clinical Trials Group study. AIDS 18:2391–2400
Hauser AS, Chavali S, Masuho I et al (2018) Pharmacogenomics of GPCR drug targets. Cell 172: 41–54
Hernández-Alvarez MI, Díaz-Ramos A, Berdasco M (2017) Early-onset and classical forms of type 2 diabetes show impaired expression of genes involved in muscle branched-chain amino acids metabolism. Sci Rep 7:1–12
Himes BE, Jiang X, Wagner P et al (2014) RNA-Seq transcriptome profiling identifies CRISPLD2 as a glucocorticoid responsive gene that modulates cytokine function in airway smooth muscle cells. PLoS One 9:99625

Holst JJ, Christensen M, Lund A et al (2011) Regulation of glucagon secretion by incretins. Diabetes Obes Metab 13:89–94

Holstein A, Hahn M, Körner A et al (2011) TCF7L2 and therapeutic response to sulfonylureas in patients with type 2 diabetes. BMC Med Genet 12:30

Housman D, Ledley FD (1998) Why pharmacogenomics? Why now? Nat Biotechnol 16:2–3

Hsieh MC, Lin KD, Tien KJ et al (2010) Common polymorphisms of the peroxisome proliferator-activated receptor–γ (Pro12Ala) and peroxisome proliferator-activated receptor–γ coactivator–1 (Gly482Ser) and the response to pioglitazone in Chinese patients with type 2 diabetes mellitus. Metabolism 59:1139–1144

Ivanisevic J, Thomas A (2018) Metabolomics as a tool to understand pathophysiological processes. Metab Clin Exp 1730:3–28

Jain KK (2004) Role of pharmacoproteomics in the development of personalized medicine. Pharmacogenomics 5:331–336

Ji Y, Hebbring S, Zhu H et al (2011) Glycine and a glycine dehydrogenase (GLDC) SNP as citalopram/escitalopram response biomarkers in depression: pharmacometabolomics-informed pharmacogenomics. Clin Pharmacol Ther 89:97–104

Jing Z, Zhang N, Ding L et al (2018) Safety and activity of programmed cell death-1 gene knockout engineered T cells in patients with previously treated advanced esophageal squamous cell carcinoma: an open-label, single-arm phase I study. J Clin Oncol 36:3054

Johnson JA (2003) Pharmacogenetics: potential for individualized drug therapy through genetics. Trends Genet 19:660–666

Jové M, Portero-Otín M, Naudí A et al (2014) Metabolomics of human brain aging and age-related neurodegenerative diseases. J Neuropathol Exp Neurol 73:640–657

Kaddurah-Daouk R, Weinshilboum RM, Pharmacometabolomics Research Network (2014) Pharmacometabolomics: implications for clinical pharmacology and systems pharmacology. Clin Pharmacol Ther 95:154–167

Kalatzis V, Antignac C (2002) Cystinosis: from gene to disease. Nephrol Dial Transplant 17:1883–1886

Kalow W (2006) Pharmacogenetics and pharmacogenomics: origin, status, and the hope for personalized medicine. Pharmacogenomics J 6:162–165

Katsila T, Patrinos GP (2015) Whole genome sequencing in pharmacogenomics. Front Pharmacol 6:61

Keiser MJ, Roth BL, Armbruster BN et al (2007) Relating protein pharmacology by ligand chemistry. Nat Biotechnol 25:197–206

Keiser MJ, Setola V, Irwin JJ et al (2009) Predicting new molecular targets for known drugs. Nature 462:175–181

Kerkhof J, Schenkel LC, Reilly J et al (2017) Clinical validation of copy number variant detection from targeted next-generation sequencing panels. J Mol Diagn 19:905–920

Khan GA, Elsayed YR, Said MO (2022) The pharmacogenomics aspects of drugs used in COVID-19 treatment. J Med 23:61–76

Kim SH, Lee KW, Song WJ et al (2011) Carbamazepine-induced severe cutaneous adverse reactions and HLA genotypes in Koreans. Epilepsy Res 97:190–197

Kim HS, Mendiratta S, Kim J et al (2013) Systematic identification of molecular subtype-selective vulnerabilities in non-small-cell lung cancer. Cell 155:552–566

Kreso A, O'Brien CA, Van Galen P et al (2013) Variable clonal repopulation dynamics influence chemotherapy response in colorectal cancer. Science 339:543–548

Kumar V, Cheng SC, Johnson MD et al (2014) Immunochip SNP array identifies novel genetic variants conferring susceptibility to candidaemia. Nat Commun 5:1–8

Kwon JM, Goate AM (2000) The candidate gene approach. Alcohol Res Health 24:164

Lamb J, Crawford ED, Peck D (2006) The Connectivity Map: using gene-expression signatures to connect small molecules, genes, and disease. Science 313:1929–1935

Lancaster CS, Bruun GH, Peer CJ et al (2012) OATP1B1 polymorphism as a determinant of erythromycin disposition. Clin Pharmacol Ther 92:642–650

Lee AM, Shi Q, Pavey E et al (2014) DPYD variants as predictors of 5-fluorouracil toxicity in adjuvant colon cancer treatment (NCCTG N0147). J Natl Cancer Inst 106:dju298

Lee JY, Vinayagamoorthy N, Han K et al (2016) Association of polymorphisms of cytochrome P450 2D6 with blood hydroxychloroquine levels in patients with systemic lupus erythematosus. Arthritis Rheumatol 68:184–190

Li H, Jia W (2013) Cometabolism of microbes and host: implications for drug metabolism and drug-induced toxicity. Clin Pharmacol Ther 94:574–581

Li M, Wang B, Zhang M (2008) Symbiotic gut microbes modulate human metabolic phenotypes. Proc Natl Acad Sci U S A 105:2117–2122

Li Y, Li W, Chen X et al (2014) Integrated analysis identifies interaction patterns between small molecules and pathways. Biomed Res Int 2:931825. https://doi.org/10.1155/2014/931825

Li JH, Suchindran S, Shah SH et al (2015) SLCO1B1 genetic variants, long-term low-density lipoprotein cholesterol levels and clinical events in patients following cardiac catheterization. Pharmacogenomics J 16:449–458

Liu X, Zhao Y (2018) CRISPR/Cas9 genome editing: fueling the revolution in cancer immunotherapy. Curr Res Transl Med 66:39–42

Lyimo RA, de Bruin M, van den Boogaard J et al (2012) Determinants of antiretroviral therapy adherence in northern Tanzania: a comprehensive picture from the patient perspective. BMC Public Health 12:1–8

Lynch CJ, Adams SH (2014) Branched-chain amino acids in metabolic signalling and insulin resistance. Nat Rev Endocrinol 10:723–736

Ma JD, Lee KC, Kuo GM (2010) HLA-B* 5701 testing to predict abacavir hypersensitivity. PLoS Curr 2:RRN1203

MacDonald ME, Ambrose CM, Duyao MP et al (1993) A novel gene containing a trinucleotide repeat that is expanded and unstable on Huntington's disease chromosomes. Cell 72:971–983

Magistretti PJ, Allaman I (2015) A cellular perspective on brain energy metabolism and functional imaging. Neuron 86:883–901

Mallal S, Phillips E, Carosi G et al (2008) HLA-B* 5701 screening for hypersensitivity to abacavir. N Engl J Med 358:568–579

Mardis ER (2008a) The impact of next-generation sequencing technology on genetics. Trends Genet 24:133–141

Mardis ER (2008b) Next-generation DNA sequencing methods. Annu Rev Genom Hum Genet 9: 387–402

Mattson MP (1998) Modification of ion homeostasis by lipid peroxidation: roles in neuronal degeneration and adaptive plasticity. Trends Neurosci 21:53–57

McDonough CW, Magvanjav O, Sá AC et al (2018) Genetic variants influencing plasma renin activity in hypertensive patients from the PEAR study (Pharmacogenomic Evaluation of Antihypertensive Responses). Circ Genom Precis Med 11:e001854

MedlinePlus (n.d.). https://medlineplus.gov/genetics/understanding/testing/sequencing/

Merry CR, McMahon S, Forrest ME et al (2016) Transcriptome-wide identification of mRNAs and lincRNAs associated with trastuzumab-resistance in HER2-positive breast cancer. Oncotarget 7: 53230

Mink JW, Blumenschine RJ, Adams DB (1981) Ratio of central nervous system to body metabolism in vertebrates: its constancy and functional basis. Am J Physiol Regul Integr Comp Physiol 241:R203–R212

Motulsky AG, Qi M (2006) Pharmacogenetics, pharmacogenomics and ecogenetics. J Zhejiang Univ Sci B 7:169–170

Mu W, Zhang W (2013) Molecular approaches, models and techniques in pharmacogenomic research and development. In: Pharmacogenomics. Academic, New York, pp 273–294

Myers AL, Kawedia JD, Champlin RE (2017) Clarifying busulfan metabolism and drug interactions to support new therapeutic drug monitoring strategies: a comprehensive review. Expert Opin Drug Metab Toxicol 13:901–923

Namvaran F, Azarpira N, Rahimi-Moghaddam P et al (2011) Polymorphism of peroxisome proliferator-activated receptor γ (PPARγ) Pro12Ala in the Iranian population: relation with insulin resistance and response to treatment with pioglitazone in type 2 diabetes. Eur J Pharmacol 671:1–6

National Cancer Institute (n.d.). https://www.cancer.gov/publications/dictionaries/genetics-dictionary/def/whole-genome-sequencing

Nature (n.d.). https://www.nature.com/subjects/transcriptomics#:~:text=Transcriptomics%20is%20the%20study%20of,methods%2C%20such%20as%20microarray%20analysis

Neavin D, Kaddurah-Daouk R, Weinshilboum R (2016) Pharmacometabolomics informs pharmacogenomics. Metabolomics 12:1–6

Nguyen CM, Mendes MA, Ma JD (2011) Thiopurine methyltransferase (TPMT) genotyping to predict myelosuppression risk. PLoS Curr 3:RRN1236

Nijkamp JF, van den Broek MA, Geertman JMA et al (2012) De novo detection of copy number variation by co-assembly. Bioinformatics 28:3195–3202

Niwa T, Shiraga T, Takagi A (2005) Effect of antifungal drugs on cytochrome P450 (CYP) 2C9, CYP2C19, and CYP3A4 activities in human liver microsomes. Biol Pharm Bull 28:1805–1808

Nocua-Báez LC, Uribe-Jerez P, Tarazona-Guaranga L et al (2020) Azoles of then and now: a review. Rev Chilena Infectol 37:219–230

Owen JA (2011) Integrating pharmacogenomics into pharmacy practice via medication therapy management: American Pharmacists Association. J Am Pharm Assoc 51:e64–e74

Padberg I, Peter E, Gonzalez-Maldonado S et al (2014) A new metabolomic signature in type-2 diabetes mellitus and its pathophysiology. PLoS One 9:85082

PAHO (n.d.). https://www.paho.org/en/topics/noncommunicable-diseases

Pan CW, Yu CH, Liao DL (2013) Carbamazepine-induced hypersensitivity syndrome in chronic schizophrenia. Gen Hosp Psychiatry 35:575–5e9

Pandi MT, Koromina M, Tsafaridis I et al (2021) A novel machine learning-based approach for the computational functional assessment of pharmacogenomic variants. Hum Genomics 15:1–13

Pedrosa E, Carretero-Iglesia L, Boada A et al (2011) CCL4L polymorphisms and CCL4/CCL4L serum levels are associated with psoriasis severity. J Invest Dermatol 131:1830–1837

Peiro G, Mayr D, Hillemanns P et al (2004) Analysis of HER-2/neu amplification in endometrial carcinoma by chromogenic in situ hybridization. Correlation with fluorescence in situ hybridization, HER-2/neu, p53 and Ki-67 protein expression and outcome. Mod Pathol 17: 277–287

Phapale PB, Kim SD, Lee HW et al (2010) An integrative approach for identifying a metabolic phenotype predictive of individualized pharmacokinetics of tacrolimus. Clin Pharmacol Ther 87:426–436

Pinu FR, Beale DJ, Paten AM et al (2019) Systems biology and multi-omics integration: viewpoints from the metabolomics research community. Metabolites 9:76

Pirmohamed M (2001) Pharmacogenetics and pharmacogenomics. Br J Clin Pharmacol 52:345

Potamias G, Lakiotaki K, Katsila T et al (2014) Deciphering next-generation pharmacogenomics: an information technology perspective. Open Biol 4:140071

Prasad K (2009) Role of regulatory agencies in translating pharmacogenetics to the clinics. Clin Cases Miner Bone Metab 6:29

Price DA, Clarke CA (1961) Pharmacogenetics. Br Med Bull 17:234–240

Priya SS, Sankaran R, Ramalingam S et al (2016) Genotype phenotype correlation of genetic polymorphism of PPAR gamma gene and therapeutic response to pioglitazone in type 2 diabetes mellitus—a pilot study. J Clin Diagn Res 10:11

Projean D, Baune B, Farinotti R et al (2003) In vitro metabolism of chloroquine: identification of CYP2C8, CYP3A4, and CYP2D6 as the main isoforms catalyzing N-desethylchloroquine formation. Drug Metab Dispos 31:748–754

Qasim W, Amrolia PJ, Samarasinghe S et al (2015) First clinical application of Talen Engineered Universal CAR19 T cells in B-ALL. Blood 126:2046

Qasim W, Zhan H, Samarasinghe S et al (2017) Molecular remission of infant B-ALL after infusion of universal TALEN gene-edited CAR T cells. Sci Transl Med 9:2013

Ramírez-Salazar M, Pérez-Luque E, Fajardo-Araujo M et al (2008) Effect of the Pro12Ala polymorphism of the PPARγ2 gene on response to pioglitazone treatment in menopausal women. Menopause 15:1151–1156

Roberts LD, Koulman A, Griffin JL (2014) Towards metabolic biomarkers of insulin resistance and type 2 diabetes: progress from the metabolome. Lancet Diabetes Endocrinol 2:65–75

Rothman N, Smith MT, Hayes RB et al (1997) Benzene poisoning, a risk factor for hematological malignancy, is associated with the NQO1 609C→ T mutation and rapid fractional excretion of chlorzoxazone. Cancer Res 57:2839–2842

Sá ACC, Sadee W, Johnson JA (2018) Whole transcriptome profiling: an RNA-Seq primer and implications for pharmacogenomics research. Clin Transl Sci 11:153

Saiz-Rodríguez M, Almenara S, Navares-Gómez M et al (2020) Effect of the most relevant CYP3A4 and CYP3A5 polymorphisms on the pharmacokinetic parameters of 10 CYP3A substrates. Biomedicines 8:94

Schnackenberg LK, Kaput J, Beger RD (2008) Metabolomics: a tool for personalizing medicine? Per Med 5:495–504

Schuster SC (2008) Next-generation sequencing transforms today's biology. Nat Methods 5:16–18

Scordo MG, Aklillu E, Yasar U et al (2001) Genetic polymorphism of cytochrome P450 2C9 in a Caucasian and a black African population. Br J Clin Pharmacol 52:447–450

Sehn JK (2015) Insertions and deletions (indels). In: Clinical genomics. Academic, New York, pp 129–150

Sheffield LJ, Phillimore HE (2009) Clinical use of pharmacogenomic tests in 2009. Clin Biochem Rev 30:55

Shirasaka Y, Chang SY, Grubb MF et al (2013) Effect of CYP3A5 expression on the inhibition of CYP3A-catalyzed drug metabolism: impact on modeling CYP3A-mediated drug-drug interactions. Drug Metab Dispos 41:1566–1574

Silva JC, Gorenstein MV, Li GZ et al (2006) Absolute quantification of proteins by LCMSE: a virtue of parallel MS acquisition. Mol Cell Proteomics 5:144–156

Smith RP, Eckalbar WL, Morrissey KM et al (2014) Genome-wide discovery of drug-dependent human liver regulatory elements. PLoS Genet 10:1004648

Soltani Banavandi MJ, Satarzadeh N (2020) Association between VKORC1 gene polymorphism and warfarin dose requirement and frequency of VKORC1 gene polymorphism in patients from Kerman province. Pharmacogenomics J 20:574–578

Spotlight 2A (n.d.). https://uk.sagepub.com/sites/default/files/upmbinaries/65118_higgs_spotlight_2a.pdf

Squassina A, Artac M, Manolopoulos VG et al (2010) Translation of genetic knowledge into clinical practice: the expectations and realities of pharmacogenomics and personalized medicine. Pharmacogenomics J 11:1149–1167

Srivastava A, Guin D, Kukreti R et al (2018) Pharmacogenomics: an evolution towards clinical practice. In: Pharmaceuticals medicine and translational clinical research. Academic, New York, pp 107–127

Stocco G, Lucafò M, Decorti G (2020) Pharmacogenomics of antibiotics. Int J Mol Sci 21:5975

Sun GY, Xu J, Jensen MD et al (2004) Phospholipase A2 in the central nervous system: implications for neurodegenerative diseases. J Lipid Res 45:205–213

Tabor HK, Risch NJ, Myers RM (2002) Candidate-gene approaches for studying complex genetic traits: practical considerations. Nat Rev Genet 3:391–397

Takahashi T, Luzum JA, Nicol MR et al (2020) Pharmacogenomics of COVID-19 therapies. NPJ Genom Med 5:35

Takano M, Sugiyama T (2017) UGT1A1 polymorphisms in cancer: impact on irinotecan treatment. Pharmgenomics Pers Med 10:61

Tantisira K, Weiss ST (2013) Overview of pharmacogenomics. UpToDate, Waltham, MA

Tatonetti NP, Dudley JT, Sagreiya H et al (2010) An integrative method for scoring candidate genes from association studies: application to warfarin dosing. BMC Bioinform BioMed Central 11:1–7. https://doi.org/10.1155/2014/931825

Theusch E, Kim K, Stevens K et al (2016) Erratum: Statin-induced expression change of INSIG1 in lymphoblastoid cell lines correlates with plasma triglyceride statin response in a sex-specific manner. Pharmacogenomics J 16:301

Tobias DK, Clish C, Mora S et al (2018) Dietary intakes and circulating concentrations of branched-chain amino acids in relation to incident type 2 diabetes risk among high-risk women with a history of gestational diabetes mellitus. Clin Chem 64:1203–1210

Toledo JB, Arnold M, Kastenmueller G et al (2017) Metabolic network failures in Alzheimer's disease: a biochemical road map. Alzheimers Dement 13:965–984

Tönnies E, Trushina E (2017) Oxidative stress, synaptic dysfunction, and Alzheimer's disease. J Alzheimers Dis 57:1105–1121

Torsetnes SB, Levernæs MS, Broughton MN (2014) Multiplexing determination of small cell lung cancer biomarkers and their isovariants in serum by immunocapture LC-MS/MS. Anal Chem 86:6983–6992

Townsend R, Dietz A, Hale C et al (2017) Pharmacokinetic evaluation of CYP3A4-mediated drug-drug interactions of isavuconazole with rifampin, ketoconazole, midazolam, and ethinyl estradiol/norethindrone in healthy adults. Clin Pharmacol Ther 6:44–53

Tuzun E, Sharp AJ, Bailey JA et al (2005) Fine-scale structural variation of the human genome. Nat Genet 37:727–732

US Food and Drug Administration (2020) Request for emergency use authorization for use of chloroquine phosphate or hydroxychloroquine sulfate supplied from the strategic national stockpile for treatment of 2019 coronavirus disease

Vallejos-Vidal E, Reyes-Cerpa S, Rivas-Pardo JA et al (2020) Single-nucleotide polymorphisms (SNP) mining and their effect on the tridimensional protein structure prediction in a set of immunity-related expressed sequence tags (EST) in Atlantic salmon (Salmo salar). Front Genet 10:1406

Van Nguyen D, Vidal C, Chu HC et al (2019) Developing pharmacogenetic screening methods for an emergent country: Vietnam. World Allergy Organ J 12:100037

Veal GJ, Cole M, Chinnaswamy G (2016) Cyclophosphamide pharmacokinetics and pharmacogenetics in children with B-cell non-Hodgkin's lymphoma. Eur J Cancer 55:56–64

Vickers NJ (2017) Animal communication: when I'm calling you, will you answer too? Curr Biol 27:R713–R715

Visscher H, Ross CJ, Rassekh S et al (2012) Pharmacogenomic prediction of anthracycline-induced cardiotoxicity in children. J Clin Oncol 30:1422–1428

Visscher H, Ross CJD, Rassekh SR et al (2013) Validation of variants in SLC28A3 and UGT1A6 as genetic markers predictive of anthracycline-induced cardiotoxicity in children. Pediatr Blood Cancer 60:1375–1381

Vogel F (1959) Moderne probleme der humangenetik [Modern human genetics problems]. In: Ergebnisse der Inneren Medizin und Kinderheilkunde [Results of internal medicine and pediatrics] (in German). Springer, Berlin, pp 52–125

Wadelius M (2009) Point: use of pharmacogenetics in guiding treatment with warfarin. Clin Chem 55:709–711

Wang-Sattler R, Yu Z, Herder C et al (2012) Novel biomarkers for pre-diabetes identified by metabolomics. Mol Syst Biol 8:615

Weisnagel SJ, Rankinen T, Nadeau A et al (2001) Decreased fasting and oral glucose stimulated C-peptide in nondiabetic subjects with sequence variants in the sulfonylurea receptor 1 gene. Diabetes 50:697–702

WHO (n.d.). https://www.who.int/news-room/fact-sheets/detail/noncommunicable-diseases

WHO Africa (n.d.). https://www.afro.who.int/healthtopics/communicablediseases#:~:text=Tuberculosis-,Overview,ingesting%20contaminated%20food%20or%20water

Wilson ID (2009) Drugs, bugs, and personalized medicine: pharmacometabonomics enters the ring. Proc Natl Acad Sci U S A 106:14187–14188

Wishart DS (2016) Emerging applications of metabolomics in drug discovery and precision medicine. Nat Rev Drug Discov 15:473–484

Xiang X, Yuan Y (2020) Application of pharmacogenomics in drug discovery and development. In: Pharmacogenomics in precision medicine. Springer, Singapore, pp 257–276

Yang M, Soga T, Pollard PJ (2013) Oncometabolites: linking altered metabolism with cancer. J Clin Invest 123:3652–3658

Yapar K, Kart A, Karapehlivan M et al (2007) Hepatoprotective effect of L-carnitine against acute acetaminophen toxicity in mice. Exp Toxicol Pathol 59:121–128

Young B, Squires K, Patel P et al (2008) First large, multicenter, open-label study utilizing HLA-B* 5701 screening for abacavir hypersensitivity in North America. AIDS 22:1673–1675

Zeng Z, Huang SY, Sun T (2020) Pharmacogenomic studies of current antidiabetic agents and potential new drug targets for precision medicine of diabetes. Diabetes Ther 11:2521–2538

Zhang F, Gu W, Hurles ME et al (2009) Copy number variation in human health, disease, and evolution. Annu Rev Genomics Hum Genet 10:451

Zhang Z-Y, Monleon D, Verhamme P, Staessen JA (2018) Branched-chain amino acids as critical switches in health and disease. Hypertension 72(5):1012–1022. https://doi.org/10.1161/HYPERTENSIONAHA.118.10919

Zhou X, Chen C, Yin D et al (2019) A variation in the ABCC8 gene is associated with type 2 diabetes mellitus and repaglinide efficacy in Chinese type 2 diabetes mellitus patients. Intern Med 58:2341–2347

Pharmacogenetics: A New Approach for the Selection of the Right Drug

3

Hitesh Malhotra, Abhishek Dabra, Preeti, Peeyush Kaushik, Anjoo Kamboj, and Rupesh Gautam

Abstract

Pharmacogenetics is the study of genes and the way an individual will respond to a particular drug. It is an advanced branch of medicine where the knowledge from pharmacology and genomics was utilized for the safe and effective prescription as per the individual genetic structure. The Human Genome Project was initiated to develop and learn the association of genes and their effect on physiology as well as drug response. Diversity in genetic makeup is responsible for alteration in therapeutic efficacy as well as the toxic profile of the drug. Besides, studying the genetic structure of the patient also enlightens about the future perspectives of drug response. Unfortunately, till now this novel approach has not been widely utilized in clinical practice. But due to the advancement in genetics as well as the introduction of pharmacogenetics drugs and gene-based diagnostic tests, it is crucial to introduce the concept of pharmacogenetics in therapeutics. Thus, in the coming future, pharmacogenomics will be used for the development of therapeutics for pandemic health problems such as neurodegenerative disorders, neoplasm, and cardiomyopathy.

H. Malhotra (✉) · A. Dabra · Preeti
Guru Gobind Singh College of Pharmacy, Yamuna Nagar, Haryana, India
e-mail: hitesh.3832@cgc.edu.in

P. Kaushik · A. Kamboj
Chandigarh College of Pharmacy, Mohali, Punjab, India

R. Gautam
Department of Pharmacology, Indore Institute of Pharmacy, IIST Campus, Rau, Indore, Madhya Pradesh, India

P. P. Singh (ed.), *Recent Advances in Pharmaceutical Innovation and Research*,
https://doi.org/10.1007/978-981-99-2302-1_3

Keywords

Allele · Carbamazepine · CYP · Drug allergy · HLA · Pharmacogenetics · Pharmacogenomics · Polymorphism

3.1 Introduction

In the past few decades, a significant advancement in the health care sector benefits the survival rate of patients. The rate of subsistence of patients and a substantial increase in quality-of-life span has been amplified because of the successful introduction of drugs with high selectivity, efficacy, and potency, and better patient care amenities (Arias 2010). The supreme challenge in the treatment and care of any patient is drug therapy. The prime level of treatment with the help of the best therapeutic agent necessitates the deep observance of associated pharmacokinetics, pharmacodynamics, adverse complications, and the cost of diverse agents.

Pharmacogenomics is termed as "the branch of medical science which is closely associated with the identification of the genomic characteristics of a person responsible for changeable reactions to drugs". Remarkably, the concept of pharmacogenomics has been developed with a motive to understand the configurations of inherited genetic variations in defined people like particular ethnicities, which are the possible triggers for the variation in pharmacological responses (Kolata 2012). The concept of pharmacogenomics is helpful to divulge the information that different individuals have discrepancies in responses to the same drug therapy. The key mechanistic pathways made possible to understand the alteration in drug responses are the study of a genome, base analysis of proteins, study of all biochemical processes which results in metabolites, and study of the transcriptome (A RNA transcripts set which are created by genome under definite conditions or in a particular cell). Every individual is blessed with a different genetic structure and this genomic makeup is responsible for the possibilities for the development of disease and variation in responses to an agent with therapeutic potential as well (Collins 1991).

The differences in the variation of drug responses are frequently higher if counted among members of a population as compared to the same person at various time intervals (Vesell 1989). The presence of massive population variations with small intra-patient variability is consistent with inheritance as a determinant of drug response. On estimation, it is determined that the role of genetics in a variation of drug displacement and efficacy can account for 20 to 95% (Kalow et al. 1998). There are plenty of nongenetic factors that alter the therapeutic effect of drugs. These triggering factors are age, functioning of organs, any parallel treatment, types of disease, and drug interactions. Several examples are there that confirm differences in drug response among the individuals are due to the variation in the genetic coding of drug-metabolizing enzymes, drug carriers protein, and the drug target protein (Evans and Relling 1999; Evans and Johnson 2001; McLeod and Evans 2001).

As pharmacogenomics is directly related to the pharmacodynamic and pharmacokinetic parameters, similarly severe drug reactions could be the result of

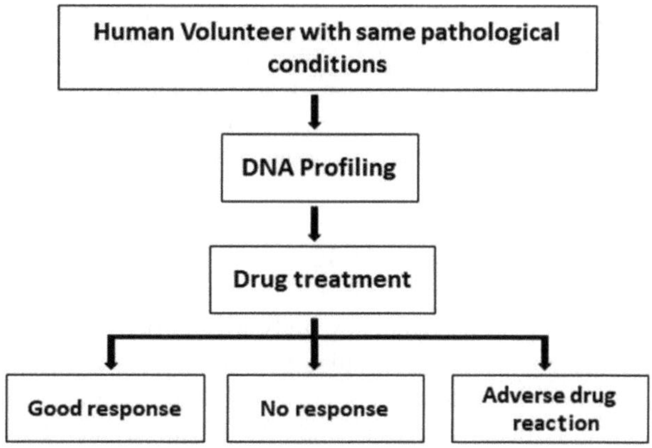

Fig. 3.1 Effect of pharmacogenomics on therapeutic strategies

pharmacokinetic parameters. Plenty of drug-related adverse drug effects and events have been reported in several publications which occurred due to genetic polymorphism (Spear et al. 2001; Lazarou et al. 1998). When a child suffered from leukemia has been treated by an oncologist with Mercaptopurine but it leads to unpredicted bone marrow toxicity followed by suppression of the immune system and serious fatal infections due to narrow therapeutic index with the dominance of white and red blood cell production (Fig. 3.1).

In the mid-1990s, scientists believed that maximum cases of life-threatening bone marrow toxicity can be described by genetics (Abbott 2003). Medicine history has plenty of medications that have been reported for unintended consequences. In the 1950s, an anesthesiologist used a drug named succinylcholine for its muscle relaxant property to carry out operations. A dreadful reaction (respiratory arrest) was observed by 1 in approximately 2500 persons. Later on, scientists disclosed that the individual who experienced respiratory arrest on treatment with succinylcholine has a defect in both genomic copies of enzyme cholinesterase. This enzyme ensures the metabolism of succinylcholine into its inactive form. Similarly, in 1980, an anti-anginal drug (Perhexiline) was reported to cause neurotoxicity and hepatotoxicity in a small population of patients. On investigation by the scientific team, it was found that the genetic polymorphism of an enzyme (CYP2D6, a drug-metabolizing enzyme) is responsible for such types of toxicities.

3.2 History of Pharmacogenomics

3.2.1 In the Early 1930s

Pharmacogenomics history can be mapped out from the time of Pythagoras who was a Greek mathematician and philosopher, who identified those individuals eating the bean *Vicia faba* suffered from severe hemolysis. In the Mediterranean countries, *V. faba* beans were a basic necessary food. The hemolysis occurred by *V. faba* bean was due to the deficiency of glucose-6-phosphate dehydrogenase deficiency. In initial of the 1930s, a scientist Sir Archibald Garrod published a book entitled "Inborn Factors of Disease" where he justified, "...a drug with therapeutic dose shows toxic effect in most of the individuals whereas other people show remarkable tolerability and termed as chemical sensitivity" (Katzung et al. 2012).

3.2.2 In the Late 1930s

The first pharmacogenetics study was carried out in the late 1930s. The bitter taste sensitivity of the participating individuals was tested. In the phenotype, major racial variances have been reported which produced data of non-taster frequency, that is, 9% of African Americans and 30% of Caucasians. The above-mentioned study was considered the first controlled study on humans indicating genetic polymorphism caused due to a chemical substance. The American Medical Association invited a professor of medical genetics at the University of Washington where he explained drug response and genetics of biochemical and enzymes. He justified "Variation in the therapeutic potential of the drug and adverse effects in the small populace could be the result of genetic inheritance" (Motulsky 1957). The term "pharmacogenetics" was introduced by Friedrich Vogel in the year 1959 and stated that it is the study of genetic inheritance that results in the variation in pharmacokinetic (what the body does to the drug) and pharmacodynamics (what the drug does to the body).

3.2.3 In the Period of the 1960s

The first monograph on pharmacogenetics was published by a scientist Kalow. New York was the center for the first international conference to be held, and a pharmacokinetic twin experiment was performed by a group headed by a clinical pharmacologist in Sweden. This group detailed the involvement of genetic variation in the maintenance of plasma concentration of antidepressants at a steady state. The Sixth IUPHAR congress was organized in Helsinki, Finland, where it has been demonstrated that the drugs sparteine and debrisoquine induced an adverse effect in people because of the inability to oxidize the active pharmaceutical ingredient. On continuous research on the concept, it has been revealed that this inability of oxidative metabolism was due to genetic polymorphism induced by CYP2D6. Several genetic polymorphisms of CYPs were recognized in the 1990s which had direct involvement in the pharmacokinetic parameters (CYP2C19, CYP2B6, and CYP2C9). Initially for such genetic polymorphisms the term pharmacogenetics was proposed and later pharmacogenomics was coined. (Meyer 1994; Lesko 2007).

3.2.4 The Present Scenario

At the current time, more precise data has been collected on how genes affect growth, development, health issues, and also drug metabolism (Wu et al. 2014). A project was launched with the title "Encyclopaedia of DNA Elements" (ENCODE) under the mentorship of the National Human Genome Research Institute (NHGRI) for human genome sequencing. The project started with the aim to identify all functional sequences of the human genome (Delves 2017). In 2009, it was launched with the title next-generation sequencing (NGS) and found of great significance for the recognition of variants that are associated with many inherited diseases. On the contrary, the identification of genomic variants in formerly known disease-associated genes is carried out through targeted sequencing (Lohmann and Klein 2014; Satoh et al. 2016). The understanding of molecular patterns responsible for various types of cancer was made possible with the help of The Cancer Genome Atlas (TCGA) program. TCGA proves its usefulness in the identification of molecular resemblances among several cancer patients and form of cancer as well as recording the shreds of evidence of the distinctiveness of each type of cancer (Tomczak et al. 2015). Several developed and underdeveloped organizations and institutions are going to develop their clinical services with their infrastructure to facilitate the testing of pharmacogenetics (Abul Husn et al. 2014; Johnson 2013).

3.3 Role of Pharmacogenomics in Pharmacokinetics and Pharmacodynamics

After the administration of a drug, the active pharmaceutical ingredient interacts with the physiological environment and triggers the cascade of events. The interaction of the drug with proteins (in its way) either acts as an agonist or antagonist on the receptor (could be made up of one or more proteins). At the cellular level, a communicative interaction takes place between metabolite and protein-active sites. As a result, propagation of biochemical cascade produces overall systemic physiological effects. All along the way, variances in the human genome can alter the way of action of drugs with these receptors which may either produce a therapeutic effect or cause potential adverse events (Katzung et al. 2012).

3.3.1 Pharmacokinetic Interactions

From the site of administration to the target site, and from the target site to the site of excretion, an active pharmaceutical ingredient (API) interacts with several proteins which increase support or hinder its progress. These interactions consist of the drug's pharmacokinetics such as absorption, dislocation, biotransformation (metabolism), and excretion processes. These parameters ensure not only the bioavailability but the duration of the action of the drug as well.

After administration, a drug gets absorbed by the body and is dislocated to different organs, tissues, and cells. Bioavailability is the important parameter that involves the amount of drug that reaches the systemic circulation, and this parameter is completely based on the mode of the administration of drugs. For example, a drug administered intravenously will have a 100% bioavailability as the drug has been directly injected into the bloodstream, whilst in the case of orally administered preparations such as tablets and capsules, several barriers are there which have to be crossed to enable the drug to reach the systemic circulation, and they hinder the absorption of the drug completely. In the case of orally administered drugs, bioavailability is altered by gastric transit-time, Gastrointestinal tract (GIT) enzyme action, absorption by GIT tract, and metabolism by liver secretions. After absorption, the drug is taken over by the portal vein to the liver. First, the pass metabolism of the absorbed drug is carried out in the liver

As the drug enters the systemic circulation, molecular transport components (proteins/ligands) lead to the alteration in the distribution of the drug. Modulation in the genome of a protein that is actively engaged in the coordination of these processes can affect the bioavailability (absorption) and dislocation (distribution) of certain drugs. For example, the class of ATP binding cassette (ABC) is the key transporter engaged in numerous transport processes in the circulation of active moiety and metabolites particularly in the GIT and through the blood-brain barrier (BBB). Alteration in the absorption and distribution of a few drugs is the result of the variation in these genes (Dietrich et al. 2003).

The metabolism of a drug in the body leads to the transformation of the precursor drug either into an active metabolite to produce a therapeutic effect or in an inactive form to proceed with the excretion. The most important protein involved in the biotransformation of drugs belongs to the member of the cytochrome P450 family, which is responsible for phase-I metabolism (oxidation, reduction, hydrolysis, etc.) of most known drugs (Evans and Relling 1999).

3.3.2 Pharmacodynamic Interaction

Pharmacodynamics deals with the pharmacological effect of the drug on its target site and the cascade of pathways associated with it. The interaction between the active moiety and target site may be of two types, one is "on-target" in which the interactions result in a potential therapeutic effect, whereas the second is "off-target" where an unwanted response could be the result of the interaction. Pharmacodynamics also engaged in the following:

1. How concentration of the drug influences the target
2. The desired concentration of drug used to induce maximum therapeutic effect and
3. Minimum quantity of drug used to arouse half of the maximum therapeutic effect.

In the majority of cases, the structurally similar molecules can complex with it and may alter the same region of the structure of the protein to lead to the induction of therapeutic response simultaneously. For example, both warfarin and vitamin K have the interaction selectivity for the vitamin K Epoxide Reductase Complex Subunit-1 (VKORC1). This enzyme is preferentially involved in the conversion of the inactive epoxidized form of vitamin K to the active form (Zimmermann and Matschiner 1974). The binding of warfarin to VKORC1, close to its catalytic site, causes the inhibition of the reduction reaction (Oldenburg et al. 2005). This ensures a decrease in the level of active vitamin K which subsequently catalyzes the anti-coagulant potential of warfarin (Owen et al. 2008). Genetic polymorphism of VKORC1 is strongly linked to the efficacy of warfarin as it alters the capacity of warfarin binding to vitamin K epoxide reductase complex subunit-1 (VKORC1) and leads to displacement of vitamin K. Because of the variation in the sensitivity to the warfarin expressively in individuals, it results in a 20-fold dose difference (Rost et al. 2004).

3.4 Investigation of Pharmacogenome

The domain of human genome sequencing has undergone a revolution in the last era (Yamey 2000). Presently, the much more global perspective has been adopted, which has impacted all areas of biomedical research, making them more attractive and dissolving the barriers which traditionally separated as distinct research disciplines. Advances in genomics-related technology, particularly in high throughput-next generation methods and its wide range of applications have enabled broadened perspectives in areas such as whole gene expression, mutation detection, genome-wide structural variation, complex human diseases, and the identification of inherited disorders, carrier screening, DNA library construction, and mitochondrial genome sequencing. Apart from advances in sequencing methodologies, the last 10 years will be regarded as that of the age of genome research. The fast sequencing of new genomes has been assisted by the development of potential techniques of HT-NGS. It has become feasible to assemble a vast genome from scratch and an excellent example is the current genome assembly, which uses only the brief readings delivered by future-generation sequencing technology (Li et al. 2010).

3.4.1 First-Generation Sequencers of DNA

Sanger in a 1975 Croonian speech proposed the DNA sequencing method and then subsequently published (Sanger 1975) a quick procedure to determine sequence data and the synthesis of data in DNA with the help of DNA polymerase. Two significant papers in DNA sequencing appeared in 1977: Frederick Sanger's enzyme technique-mediated dideoxy sequencing of DNA which depends upon chain termination reaction of dideoxynucleotide intermediates (Sanger et al. 1977) and the second one is Maxam and Gilbert's method of chemical degradation of DNA sequencing wherein marked DNA segments break from a particular base at the terminal (Maxam and Gilbert 1977) and isolated with the help of gel electrophoresis. Caltech was instrumental in developing the first DNA sequencing automatic technique (Smith et al. 1986). DNA sequencers, which Applied Biosystems later actively marketed, and the Pharmacia Amersham, as well as European Molecular Biology Laboratory (Ansorge et al. 1986, 1987) afterward General Electric healthcare. Then the sequencing method was refined and industrialized, resulting in widespread use within the scientific community. The hypoxanthine-guanine phosphoribosyltransferase gene was sequenced in its entirety for the first time by automatic fluorescent DNA sequencing equipment (Edwards et al. 1990), utilizing the paired-end sequencing approach. The first commercial DNA sequencer to use slab gel electrophoresis was the ABI Prism 310 and it was released in 1996 and 2003 that automated DNA sequencer was used to sequence the first human genome. That human genome project collaboration has worked for 13 years for $2.7 billion. A first tiny phage DNA, measuring 5386 nucleotides in length, was sequenced as well as approximately 3 billion bases of the human genome were sequenced in the subsequent years (Lander et al. 2001; Venter et al. 2001). The fact that certain improvements have been made utilizing procedures that are modifications of Sanger's basic "dideoxy" method released in 1977 is amazing.

3.4.2 The Beginning of High Throughput-Next-Generation Sequencing (HT-NGS)

Rothberg started more than 400 life sciences in 2000 that went on to produce the GS 20, its first commercially accessible NGS system. The GS instrument, manufactured with the help of 454 Life Sciences, was an NGS system to hit the industry in 2005. Combining the smPCR technique based on sequencing by synthesis principle with a shotgun process at 96% integration and accuracy of 99.96% in a single run successfully verified the new technique (Margulies et al. 2005). Four hundred fifty-four life sciences were purchased by Roche applied science in the following years as well as expanded to the next level of the 454 LS into GS FLX titanium. Separately, the Roche HT-NGS platform has created an smPCR technique in microcompartments containing emulsion of water-in-oil in the base (Tawfik and Griffiths 1998). The "sequencing by synthesis" approach underpins the pyro-sequencing process in general. A Swedish group in Stockholm collaborated to create the technology (Ronaghi et al. 1996). In 2007, the HT-NGS approaches were chosen as the methods of the year because they provide fresh possibilities and have significant studies in mammalian genomics (Schuster 2008).

3.4.3 HT-NGS Platforms of the Second Generation

In a single run, the second gen HT-NGS systems can output anything from 500 million to billions of raw bases (Roche). These new approaches are based on cyclic probing in multidirectional parallel. Three dominant second-generation HT-NGS systems are available currently for commercial use (Mardis 2008, 2009, 2010; Metzker 2010). A consortium by the NHGRI recently published the most complete and accurate blueprint of variation in human genetics through second-generation techniques of genome sequencing which sequentially characterize gene variations between 697 participants from 7 communities and 179 participants from

4 communities (Durbin et al. 2010), which was published in December 2010 and these studies lay the groundwork for investigating human genetic diversity, to create a complete, publicly accessible variation map which eventually underpins future genetics research.

3.4.4 HT-NGS Platforms of the Third Generation

The principle mentioned in second Gen HT-NGS systems was based on DNA amplification by emulsion PCR which enables the formation of intense signals from light signals for accurate identification of bases with imaging devices. This may create mistakes in the sequence of bases which only sends signals of the particular bases over the others in some cases, altering the frequency or availability of distinct DNA molecules that occurred before amplification. To circumvent this, the base sequence might be directly measured by a fragment of DNA, without the requirement of PCR amplification, allowing for eventual downsizing at the nanoscale or little biochemical consumption. This single-molecule DNA sequencing has currently been referred to as the technique of the third generation of HT-NGS (Schadt et al. 2010). Several businesses are now pursuing single-molecule sequencing which is described below (Fig. 3.2).

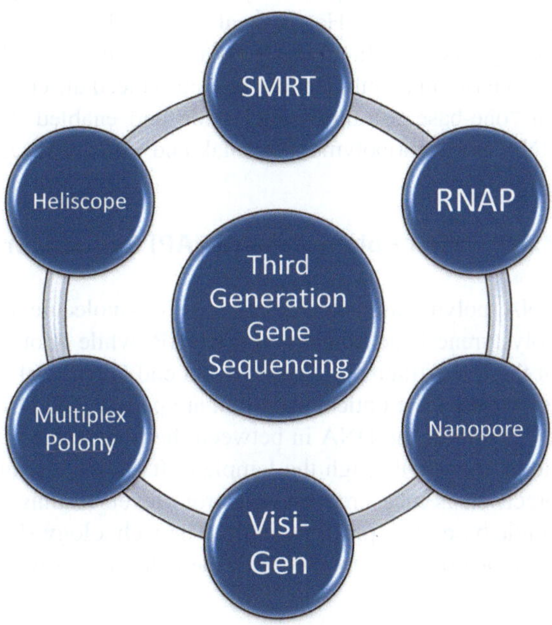

Fig. 3.2 Advanced technologies under third-generation gene sequencing

3.4.5 Sequencer for Single Molecules in Real-Time (SMRT™)

The SMRT is based on a synthesis process by the device of sequencing with several zero-mode waveguides. The sequencing method is carried out by the unique enzyme called DNA polymerase of a DNA segment that is connected from the bottom side of every zero-mode waveguide (Levene et al. 2003; Eid et al. 2009). As a result, every polymerase enzyme is located in the zero-mode detection zone. The SMRT analyzer, which the firm says can get long readings in 100 Gb in 1 h in a single run, was just released on the market. The platform, on the other hand, has much greater mistake rates of approximately 11- to 15%.

3.4.6 HeliScope™

Braslavsky et al. introduced procedures for sequencing out of a single DNA molecule in 2003. Helicos Biosciences gave license to the commercial device of DNA sequencing in 2007. HeliScope follows the idea of the "true single molecule sequencing" technique. The tSMS starts with both the production of a DNA library via DNA cuttings and the incorporation of polyadenylic acid tails towards the produced DNA molecules (Ozsolak et al. 2010). After that, DNA fragments are hybridized into poly-oligonucleotides that are coupled with flow cells and then sequenced through parallel processes. A Heliscope sequencer was then used to detect nucleotides (Harris et al. 2008). These sequencers can sequence approximately 28 GB within roughly 8 days in a single run. It may provide brief readings up to 55 bases in length. Helicos has announced the creation of a new series that is based on "one-base-at-a-time" nucleotide and enabled the precise direct sequencing of RNA and homopolymer (Ozsolak and Milos 2011a, b).

3.4.7 RNA Polymerase (RNAP) Sequencer

RNA polymerase (RNAP) is a single-molecule DNA sequencing method. One polystyrene bead contains the RNAP, while another bead contains the other side of the DNA fragment, that is, distal end (Greenleaf and Block 2006). The beads are levitated by the optical traps that are paired. The DNA fragment exposed to RNAP, the length of the DNA in between the two beads, is changed by RNAP's transcriptional motion through the template. It causes rearrangement of the two beads, and precision is measured by Angstrom-level, leading to the high resolution of every single base of a specific sample. The technology shows how DNA enzymes moved from one place to another with the help of the optical trap method.

3.4.8 DNA Sequencer Using Nanopores

Studies on the movement of DNA into different types of artificial nanopores led to the development of this technology. Nanopore sequencer does not require nucleotide tagging or detection when sequencing a DNA molecule. Sequencing of DNA by using nanopores is based upon the generation of electric signals when nucleotides pass via nanopore. The main binding site of nucleotide is a pore of alpha-hemolysin that is linked covalently to the molecule of cyclodextrin. This approach works by modulating the electric current into the pore as a DNA fragment passes through it and disclosing the molecule's features and attributes length, conformation, and diameter. The nucleotide blocks the flow or current of ions into the nanopore at the time of the sequencing process. Each base has its unique period which allows the proper determination of the given sequence of DNA (Astier et al. 2006; Rusk 2009). However, additional tweaks and refinements in the technique, for example, increasing the parameter or improving existing parameters in the instrument give the fastest and high-resolution technique to the industry.

3.4.9 VisiGen Biotechnologies

VisiGen biotechnologies are used for modified nucleotides that create a unique type of DNA polymerase enzyme which works like a "real-time sensor" containing a fluorescent dye. This dye is integrated adjacent to the catalytic site responsible for selecting the accurate nucleotide while synthesis. Each of the four nucleotides to be integrated was given a distinct acceptor dye. A suitable nucleotide is identified during synthesis, is picked out, and then inserted in the site of enzymatic action. Then the labeled fluorescent dye (donor dye) approaches the selected nucleotide and causes the transmission of energy that flows from the fluorescent dye to the selected nucleotide, resulting in a fluorescence resonance transfer of energy (Selvin 2000). Based on the marker in the nucleotides, the intensity of the signal changed. The company is now developing the instrument's initial iteration, which can create roughly 4 GB of data every day. In comparison to existing technologies, the single-molecule technique avoids a major portion of the cost by avoiding cloning and amplification. Furthermore, the instrument's read lengths are predicted to reach approximately 1 kb, which is significantly greater than any present platform.

3.4.10 Technology for Multiplex Polony

The multiplex polony technology was created and introduced through a private project by Prof. G Church named Personal Genome Project (PGP) (Mitra et al. 2003; Shendure et al. 2005). A thin layer of agar is used to deposit templates of DNA sequences that are a hundred in number and then determine the sequences simultaneously. The number of samples that may be studied simultaneously increases by several orders of magnitude with this method. It offers the advantage of reducing reaction volumes significantly, requiring lesser amounts of chemicals, thus leading to lower costs.

3.5 Clinical Application of Pharmacogenomics

The clinical purpose of pharmacogenetic testing is to identify patients with actionable genetic variations. There are three purposes for pharmacogenetic testing: testing to identify patients who are likely or unlikely to respond to an intended drug (i.e., indication); testing to identify patients who are likely to develop harm from a drug (i.e., contraindication/harm); and testing to guide drug dosing.

A polymorphism for certain proteins may alter due to population and clinical prevalence as well as its functional effects. Polymorphism is a key mechanism involved in the alteration of functional effects of proteins such as enzymes, drug transporters, and receptors. As a result, the activity of protein may enhance or reduce (Davaalkham et al. 2009). Furthermore, clinical prevalence can be determined by studying the relationship between drug, disease, and polymorphism. Here, polymorphism may affect the pharmacokinetic and pharmacodynamic profile of the drug as well as the diagnosis, pathology, and prognosis of the disease (Dendukuri et al. 2007).

3.5.1 Proton Pump Inhibitors (PPIs)

PPIs such as lansoprazole, pantoprazole, and rabeprazole are mainly indicated for the management of GIT ulcers like peptic ulcers, duodenal ulcers, and Zollinger–Ellison syndrome. PPIs are also indicated for *Helicobacter pylori*-induced gastric ulcers. PPIs retard the acid secretion from parietal cells by inhibiting the H^+/K^+ ATPase pump (Ogawa and Echizen 2010; Shi and Klotz 2008). The metabolism of all the PPIS is primarily done by CYP2C19 and CYP3A except rabeprazole. Rabeprazole is metabolized through a non-enzymatic reduction process (Kita et al. 2003; Vanden Branden et al. 1996). The gene for CYP2C19 is present on chromosome ten containing nine exons (Romkes et al. 1991). Around 27 variant alleles have been identified, with the most broadly studied being CYP2C19-1, CYP2C19-2 and CYP2C19-3. Variant CYP2C19-2 occurs due to a defect in exons-5 that results in early termination of protein synthesis (de Morais et al. 1994). Similarly, CYP2C19-3 is a condensed protein molecule that occurs due to polymorphism of stop codon and leads to negligible enzyme activity. Similarly, CYP2C19-4 to 8 have shown no enzymatic activity, while CYP2C19-9, 10, and 12 possess diminished activity. In contrast, CYP2C19-17 is associated with profound enzymatic activity (Blaisdell et al. 2012; Baldwin et al. 2008).

In Asians and Africans, the frequency of CYP2C19-2 is highest, that is, around 30%, while CYP2C19-3 frequency is 5% (Desta et al. 2002; Mizutani 2003; Roh et al. 1996). But the frequency of the CYP2C19-17 allele is approximately 18% in the Ethiopian population and 4% in the Asian population (Sim et al. 2006).

Therefore, the effect of the dose of drugs that are metabolized by CYP2C19 is markedly changed due to genotypic variation in CYP2C19 alleles. The fast metabolizers (FM) include individuals with two wild-type alleles of homozygous CYP2C19 and heterozygous CYP2C19 with one each of wild-type and null variant allele. However, individuals having heterozygous CYP2C19 with two null variants are poor metabolizers (PM). The population prevalence of CYP2C19 PM ranges from 4, 23, and 7% in Caucasians, Asians, and Africans, respectively (Kim et al. 2002).

Today, in clinical practices dosing strategies for PPIs for gastric ulcers do not base on the CYP2C19 genotype. The reason behind this is a large therapeutic window of PPIs and minimal adverse effects along with the availability of a multitude of therapy regimens. It has been suggested that while using high doses of PPIs in EMs, the genotype of CYP2C19 must be identified (Furuta et al. 1998). Furthermore, PPIs show characteristic alteration in pharmacokinetic and pharmacodynamic profiles due to the CYP2C19 genotype. For instance, omeprazole shows around ten times higher area under the curve as in PM as compared to EM (Furuta et al. 1999a, b; Sakai et al. 2001; Shirai et al. 2001). This may occur due to high gastric pH in individuals with CYP2C19 PM as compared to EM (Furuta et al. 1999a). A study was conducted on Japanese patients suffering from *H. pylori* infection where they were treated with omeprazole and amoxicillin. It has been observed that heterozygous PM shows a maximum cure rate as compared to homozygous FM. Studies were also conducted to verify this observation and consistent results were obtained with omeprazole chronic therapy (Aoyama et al. 1999; Sapone et al. 2003; Sheu et al. 2005). Furthermore, a clinical study was conducted where peptic ulcer patients were treated with a combination of amoxicillin, omeprazole, and clarithromycin for 7 weeks. The cure rates were found to be maximum for heterozygous PM (90%) and least for homozygous FM (76%) (Inaba et al. 2002).

3.5.2 Codeine

Codeine is an opiate agonist that acts on the central nervous system and is mainly employed for the relief of pain. Codeine is actively metabolized by CYP2D6 to morphine via o-demethylation. Besides, CYP2D6 is also responsible for the metabolism of neuroleptics, antidepressants, and beta-blockers through hydroxylation and demethylation (Cascorbi 2003). The gene for CYP2D6 is located on chromosome 22 and around 80 variants reported till now. The consequences of polymorphism of CYP2D6 are classified as PM, Moderate metabolizer (MM), FM, and Ultra-Fast metabolizer (UFM). Variant CYP2D6-3, 4, 5, 6, and 7 alleles have been responsible for decreased enzyme activity but decreased activity of CYP2D6 in MM contributes only to CYP2D6-9 and CYP2D6-10 alleles (Kirchheiner et al. 2007; Ingelman-Sundberg et al. 2007). In European Caucasians, Asians, and Africans, the frequency of CYP2D6-4, 5, and 10 alleles is highest. The most prevalent batch is CYP2D6-4 ranging from 12 to 21% followed by CYP2D6-3, that is, 7%. In Asians, allele CYP2D6-10 shows the highest frequency (70%) followed by CYP2D6-4 (1%) and

CYP2D6-5 (6%). While CYP2D6-4 allele has the lowest prevalence, that is, around 2% followed by CYP2D6-5 (4%) and CYP2D6-10 (6%) (Aynacioglu et al. 1999). CYP2D6 PM is seen maximum in Caucasians, that is, 10%, followed by Africans (3%). However, the frequency of CYP2D6 UFM is highest in Europeans and Saudi Arabians (10–25%) (Samer et al. 2010).

Till now no study has been reported that shows the deviation of codeine efficacy due to CYP2D6 polymorphism. A study conducted on the effect of CYP2D6 polymorphism on oxycodone demonstrates that FM has 2–20 times lower analgesic efficacy as compared to PM and for UFM the effect was enhanced by sixfold. While in the case of intravenous administration of oxycodone no change in analgesic activity was observed (Eckhardt et al. 1998). In PMs, the diminished analgesic effect of codeine accounts for impaired o-demethylation. But despite this the PMs experience codeine-related side effects such as sedation, dizziness, dry mouth and headache (Ciszkowski et al. 2009). On the contrary, UFM depicts greater analgesic effects with significant toxicity (Dalen et al. 1997; Koren et al. 2006). Furthermore, it has been reported that CYP2D6 polymorphism is responsible for toxic effects in breastfeeding neonates (Kurtz et al. 2010). A clinical report depicts an adverse drug event associated with intake of codeine 30 mg and paracetamol 600 mg by a milk-feeding mother for the management of episiotomy pain. Initially, she took 60 mg of codeine but on the second day she reduced the dose to half from the initial due to constipation and somnolence (http://www.clinicalpharmacologyip.com/Forms/Monograph/monograph.aspx?cpnum¼146&sec¼monpreg). Unfortunately, due to an overdose of morphine, her 13-day-old baby dies with a plasma concentration of 70 ng/mL. The concentration of morphine in breast milk appeared to be 87 ng/mL. After a complete investigation, it has been seen that the mother is UFM with CYP2D6-2 polymorphism and thus linked with increased codeine concentration and toxic effects in neonates. Afterward, FDA issues a warning memo regarding the addition of codeine in the prescription of breastfeeding mothers and states that the use of codeine is considered fatal for milk-feeding neonates. FDA also states that if possible don't include codeine in the prescription of such mothers and if the needed lowest possible dose should be added with regular examination of the mother as well as a neonate. Besides, the mother should be informed regarding the sign and symptoms of codeine toxicity in neonates which include sedation, drowsiness, respiratory depression, euphoria, mental confusion, and visual disturbances. Similarly, drugs that inhibit CYP2D6 are also responsible for poor analgesic activity similar to PMs. Hence, there should be strict monitoring of adverse drug interactions between CYP2D6 and codeine.

3.5.3 Carbamazepine

Carbamazepine is an anti-epileptic agent used for the treatment of generalized seizures, bipolar disorder, and trigeminal neuralgia by interfering with voltage-gated sodium channels in the brain. Carbamazepine is mainly metabolized through CYP3A4 to carbamazepine-10,11-epoxide (Kaniwa et al. 2010). The most problematic issue with the use of carbamazepine is its toxic allergic effects which may be fatal such as Stevens-Johnson syndrome and toxic epidermal necrolysis. The chances of developing such an allergic reaction are less than 2 patients per million with a mortality rate of around 30% (McCormack et al. 2011). It has been investigated that allergic reaction to carbamazepine is closely related to HLA-B 1502, 3101, and 1511 alleles (Ozeki et al. 2011; Edwards et al. 1999; Aihara 2011). Various studies and reports reveal that the development of hypersensitivity reactions such as Stevens-Johnson syndrome occurs mainly due to the involvement of HLA (Naisbitt et al. 2003). The development of Stevens-Johnson syndrome in patients receiving carbamazepine attributes to the HLA-B-1502 allele (Lee et al. 2010; Lonjou et al. 2006). A similar HLA allele is also responsible for the development of hypersensitivity reactions in subjects taking lamotrigine, oxcarbazepine, and phenytoin (Chung et al. 2004). The prime reason for such reaction includes the non-covalent association of HLA complex and carbamazepine that leads to cell death (Chen et al. 2011).

The incidence of carbamazepine-induced reaction is highest in the Asian population mainly in China, Indonesia, Thailand, Vietnam, and Taiwan, that is, 15%. However, only around 1% of the African, American, and European population suffers from such a reaction (Kuehn 2008). A study conducted on Chinese patients demonstrates that a 100% association exists between the HLA-B 1502 allele and carbamazepine-induced allergic reaction (Ikeda et al. 2010). A similar study was conducted where 59 patients from 60 reported Stevens-Johnson syndrome-like reaction with carbamazepine due HLA-B 1502 allele. Furthermore, a clinical study was conducted to confer the above results. The study concludes that the chances of development of allergic reaction in the HLA-B 1502 allele are 100% as compared to non-carriers, that is, 14.5% (Naisbitt et al. 2003). In a study conducted on 4877 Taiwanese patients who were already on carbamazepine therapy, 7.7% of patients showed significant positive results for HLA-B 1502-induced allergic reaction and thus shifted to some alternative therapy. While remaining patients do not show any allergic reaction due to HLA-B 1502 association with an allergic reaction (Kuehn 2008). Similarly, 81 patients were included in a clinical study already taking carbamazepine and phenytoin as anti-epileptic drugs. It had been seen that around 30 patients were suffering from Stevens-Johnson syndrome allergic reaction due to HLA linkages.

Besides HLA-B 1502, various other HLA alleles are also identified as key indicators for carbamazepine-induced allergic reactions. For instance, HLA-A 3101 is linked with carbamazepine-induced allergic reactions in European and Japanese populations. Also, HLA-B 1511 has shown to be a risk factor for Stevens-Johnson syndrome like an allergic reaction in carbamazepine therapy. HLA-B 1508, 1511, and 1521 are associated with such reactions in India and Thailand (Ikeda et al. 2010). Therefore, in 2007 FDA recommended that the prescriber must ensure the presence or absence of the HLA-B 1502 allele before initiating the carbamazepine therapy. Thus, patients who are at high risk for developing such reactions must be tested before initiating the therapy.

3.5.4 Allopurinol

Allopurinol is primarily used for the treatment of conditions like hyperuricemia such as gout but is also involved with the development of severe cutaneous adverse reactions (SCR). A clinical trial was conducted on 113 Taiwan subjects who are already on allopurinol therapy and gene analysis of these subjects reveals that 51 patients contain allopurinol linked SCR gene. Furthermore, they also found that the allopurinol-linked SCR gene is strongly associated with HLA-B 5801 similar to HLA-B 1502 linkage. The association was also observed in European and Japanese populations (Hung et al. 2005; Tassaneeyakul et al. 2009; Lonjou et al. 2008; Kaniwa et al. 2008). Unlike, in the Australian population the association of HLA-B 5801 with allopurinol-related gene is not so significant that it may produce some serious allergic reactions (Lee et al. 2012). Despite this, screening for HLA-B 5801 association with the allopurinol-related SCR gene is crucial to avoid allergic reactions.

3.5.5 Dapsone

Dapsone is an antimicrobial agent used for the cure and management of certain diseases like leprosy and pneumonia caused by *Pneumocystis jirovecii*. Besides, Dapsone is also used to treat dermatitis due to its anti-inflammatory activity. But Dapsone will produce certain hypersensitivity reactions along with eosinophilia and systemic manifestations like fever, hepatitis, and lymphadenopathy (Lorenz et al. 2012). A study conducted by Zhang et al. on 833 Chinese individuals taking dapsone for different reasons shows the strong association of hypersensitivity reaction with HLA-B 13:01 . Further based on screening it has been postulated that the chances of development of such allergic reactions are more in the Asian population as compared to the European population.

3.5.6 Asparaginase

Asparaginase is considered a primary drug for the treatment of acute lymphoblastic leukemia and is responsible for anaphylactic-like reactions. A study was conducted for the analysis of genome structure using 485 US children on asparaginase therapy due to leukemia. It has been observed that single nucleotide polymorphism exists in a gene located at 5q33 for ionotropic, glutamate, and AMPA receptors. This polymorphism was strongly associated with anaphylactic reaction due to asparaginase therapy (Rajic et al. 2015). Another study conducted in Hungary on 576 children also confirms the genetic association of anaphylactic reaction due to asparaginase therapy (Kutszegi et al. 2015). Furthermore, a study was carried out to elucidate the influence of HLA genes in patients suffering from asparaginase therapy using 1870 children with leukemia (Fernandez et al. 2014). They observed that the

HLA-DRB1- 07:01 allele is responsible for an asparaginase-induced anaphylactic reaction.

3.6 Conclusion

Since pharmacogenetics is not exclusively new, there has been a tremendous increase in the number of studies analyzing the associations with various reactions. To date, various studies have demonstrated that the genetic makeup of the individual is responsible for the immediate or delayed type of drug reactions. But in practice, no study is robust enough to make clinical references based on pharmacogenetic data alone. Furthermore, with advancements in genotype technology, new alleles discovered are clinically significant for drug allergic reactions. Thus, pharmacogenomics will play a significant role in a new era of making the prescription precise and accurate. Pharmacogenomics offers the optimism of tailoring to identify most susceptible patients, thereby recommending averting specific medicine for their interest.

References

Abbott A (2003) With your genes? Take one of these, three times a day. Nature 425:760 –762

Abul Husn NS, Owusu Obeng A, Sanderson SC et al (2014) Implementation and utilization of genetic testing in personalized medicine. Pharmgenomics Pers Med 7:227 –240

Aihara M (2011) Pharmacogenetics of cutaneous adverse drug reactions. J Dermatol 38:246 –254

Ansorge W, Sproat BS, Stegemann J et al (1986) A non-radioactive automated method for DNA sequence determination. J Biochem Biophys Methods 13:315 –323

Ansorge W, Sproat BS, Stegemann J et al (1987) Automated DNA sequencing: ultrasensitive detection of fluorescent bands during electrophoresis. Nucleic Acids Res 15:4593 –4602

Aoyama N, Tanigawara Y, Kita T et al (1999) Sufficient effect of 1-week omeprazole and amoxicillin dual treatment for Helicobacter pylori eradication in cytochrome P450 2C19 poor metabolizers. J Gastroenterol 34:80 –83

Arias E (2010) United States life tables (2006). Natl Vital Stat Rep 58:1 –40. http://www.cdc.gov/nchs/data/nvsr/nvsr58/nvsr58_21.pdf

Astier Y, Braha O, Bayley H (2006) Toward single molecule DNA sequencing: direct identification of ribonucleoside and deoxyribonucleoside 5′-monophosphates by using an engineered protein nanopore equipped with a molecular adapter. J Am Chem Soc 128:1705 –1710

Aynacioglu AS, Sachse C, Bozkurt A et al (1999) Low frequency of defective alleles of cytochrome P450 enzymes 2C19 and 2D6 in the Turkish population. Clin Pharmacol Ther 66:185 –192

Baldwin RM, Ohlsson S, Pedersen RS et al (2008) Increased omeprazole metabolism in carriers of the CYP2C19*17 allele; a pharmacokinetic study in healthy volunteers. Br J Clin Pharmacol 65:767 –774

Blaisdell J, Mohrenweiser H, Jackson J et al (2012) Identification and functional characterization of new potentially defective alleles of human CYP2C19. Pharmacogenetics 12:703 –711

Cascorbi I (2003) Pharmacogenetics of cytochrome p4502D6: genetic background and clinical implication. Eur J Clin Invest 33:17 –22

Chen P, Lin JJ, Lu CS et al (2011) Carbamazepine-induced toxic effects and HLA-B*1502 screening in Taiwan. N Engl J Med 364:1126 –1133

Chung WH, Hung SI, Hong HS et al (2004) Medical genetics: a marker for Stevens-Johnson syndrome. Nature 428:486
Ciszkowski C, Madadi P, Phillips MS et al (2009) Codeine, ultrarapid-metabolism genotype, and postoperative death. N Engl J Med 361:827 –828
Collins FS (1991) Of needles and haystacks: finding human disease genes by positional cloning. Clin Res 39:615 –623
Dalen P, Frengell C, Dahl ML et al (1997) Quick onset of severe abdominal pain after codeine in an ultrarapid metabolizer of debrisoquine. Ther Drug Monitor 19:543 –544
Davaalkham J, Hayashida T, Tsuchiya K et al (2009) Allele and genotype frequencies of cytochrome P450 2B6 gene in a Mongolian population. Drug Metab Dispos 37:1991 –1993
de Morais SM, Wilkinson GR, Blaisdell J et al (1994) The major genetic defect responsible for the polymorphism of S-mephenytoin metabolism in humans. J Biol Chem 269:15419 –15422
Delves PJ (2017) Human leukocyte antigen (HLA) system. In: Merck manual. Professional version. Merck Sharp & Dohme Corp, Rahway
Dendukuri N, Khetani K, McIsaac M et al (2007) Testing for HER2-positive breast cancer: a systematic review and cost effectiveness analysis. CMAJ 176:1429 –1434
Desta Z, Zhao X, Shin JG et al (2002) Clinical significance of the cytochrome P450 2C19 genetic polymorphism. Clin Pharmacokinet 41:913 –958
Dietrich CG, Geier A, Oude Elferink RPJ (2003) ABC of oral bioavailability: transporters as gatekeepers in the gut. Gut 52:1788 –1795
Durbin RM, Abecasis GR, Altshuler DL (2010) The 1000 Genomes Project Consortium. A map of human genome variation from population-scale sequencing. Nature 467:1061 –1073
Eckhardt K, Li S, Ammon S et al (1998) Same incidence of adverse drug events after codeine administration irrespective of the genetically determined differences in morphine formation. Pain 76:27 –33
Edwards A, Voss H, Rice P et al (1990) Automated DNA sequencing of the human HPRT locus. Genomics 6:593 –608
Edwards SG, Hubbard V, Aylett S et al (1999) Concordance of primary generalised epilepsy and carbamazepine hypersensitivity in monozygotic twins. Postgrad Med J 75:680 –681
Eid J, Fehr A, Gray J et al (2009) Real-time DNA sequencing from single polymerase molecules. Science 323:133 –138
Evans WE, Johnson JA (2001) Pharmacogenomics: the inherited basis for interindividual differences in drug response. Annu Rev Genomics Hum Genet 2:9 –39
Evans WE, Relling MV (1999) Pharmacogenomics: translating functional genomics into rational therapeutics. Science 286:487 –491
Fernandez CA, Smith C, Yang W et al (2014) HLADRB1* 07:01 is associated with a higher risk of asparaginase allergies. Blood 124:1266 –1276
Furuta T, Ohashi K, Kamata T et al (1998) Effect of genetic differences in omeprazole metabolism on cure rates for Helicobacter pylori infection and peptic ulcer. Ann Intern Med 129:1027 – 1030
Furuta T, Ohashi K, Kobayashi K et al (1999a) Effects of clarithromycin on the metabolism of omeprazole in relation to CYP2C19 genotype status in humans. Clin Pharmacol Ther 66:265 – 274
Furuta T, Ohashi K, Kosuge K et al (1999b) CYP2C19 genotype status and effect of omeprazole on intragastric pH in humans. Clin Pharmacol Ther 65:552 –561
Greenleaf WJ, Block SM (2006) Single-molecule, motion-based DNA sequencing using RNA polymerase. Science 313:801
Harris TD, Buzby PR, Babcock H et al (2008) Single-molecule DNA sequencing of a viral genome. Science 320:106 –109
Hung SI, Chung WH, Liou LB et al (2005) HLA-B*5801 allele as a genetic marker for severe cutaneous adverse reactions caused by allopurinol. Proc Natl Acad Sci U S A 102:4134 –4139
Ikeda H, Takahashi Y, Yamazaki E et al (2010) HLA class I markers in Japanese patients with carbamazepine-induced cutaneous adverse reactions. Epilepsia 51:297 –300

Inaba T, Mizuno M, Kawai K et al (2002) Randomized open trial for comparison of proton pump inhibitors in triple therapy for Helicobacter pylori infection in relation to CYP2C19 genotype. J Gastroenterol Hepatol 17:748 –775

Ingelman-Sundberg M, Sim SC, Gomez A et al (2007) Influence of cytochrome P450 polymorphisms on drug therapies: pharmacogenetic, pharmacoepigenetic and clinical aspects. Pharmacol Ther 116:496 –526

Johnson JA (2013) Pharmacogenetics in clinical practice: how far have we come and where are we going? Pharmacogenomics 14:835 –843

Kalow W, Tang BK, Endrenyi I (1998) Hypothesis: comparisons of inter- and intra-individual variations can substitute for twin studies in drug research. Pharmacogenetics 8:283 –928

Kaniwa N, Saito Y, Aihara M et al (2008) HLA-B locus in Japanese patients with anti-epileptics and allopurinol-related Stevens-Johnson syndrome and toxic epidermal necrolysis. Pharmacogenomics 9:1617 –1622

Kaniwa N, Saito Y, Aihara M et al (2010) HLA-B*1511 is a risk factor for carbamazepine-induced Stevens-Johnson syndrome and toxic epidermal necrolysis in Japanese patients. Epilepsia 51: 2461 –2465

Katzung BG, Masters SB, Trevor AJ (2012) Basic & clinical pharmacology. McGraw-Hill Medical, New York

Kim MJ, Bertino JS Jr, Gaedigk A et al (2002) Effect of sex and menstrual cycle phase on cytochrome P450 2C19 activity with omeprazole used as a biomarker. Clin Pharmacol Ther 72:192 –199

Kirchheiner J, Schmidt H, Tzvetkov M et al (2007) Pharmacokinetics of codeine and its metabolite morphine in ultra-rapid metabolizers due to CYP2D6 duplication. Pharmacogenomics J 7:257 – 265

Kita T, Sakaeda T, Baba T et al (2003) Different contribution of CYP2C19 in the in vitro metabolism of three proton pump inhibitors. Biol Pharm Bull 26:386 –390

Kolata G (2012) In treatment of leukemia, glimpses of the future. New York Times, 7 Jul

Koren G, Cairns J, Chitayat D et al (2006) Pharmacogenetics of morphine poisoning in a breastfed neonate of a codeine-prescribed mother. Lancet 368(9536):704

Kuehn BM (2008) FDA: epilepsy drugs may carry skin risks for Asians. JAMA 300:2845

Kurtz M, Black Golde P, Berlinger N (2010) Ethical considerations in CYP2D6 genotype testing for codeine-prescribed breastfeeding mothers. Clin Pharmacol Ther 88:760 –762

Kutszegi N, Semsei AF, Gezsi A et al (2015) Subgroups of paediatric acute lymphoblastic leukaemia might differ significantly in genetic predisposition to asparaginase hypersensitivity. PLoS One 10:e0140136

Lander ES, Linton LM, Birren B et al (2001) Initial sequencing and analysis of the human genome. Nature 409:860 –921

Lazarou J, Pomeranz BH, Corey PN (1998) Incidence of adverse drug reactions in hospitalised patients. JAMA 279:1200 –1205

Lee MT, Hung SI, Wei CY et al (2010) Pharmacogenetics of toxic epidermal necrolysis. Expert Opin Pharmacother 11:2153 –2162

Lee MH, Stocker SL, Anderson J et al (2012) Initiating allopurinol therapy: do we need to know the patient's human leucocyte antigen status? Intern Med J 42:411 –416

Lesko LJ (2007) Paving the critical path: how can clinical pharmacology help achieve the vision. Clin Pharmacol Ther 81:170 –177

Levene MJ, Korlach J, Turner SW et al (2003) Zero-mode wave guides for single-molecule analysis at high concentrations. Science 299:682 –686

Li R, Fan W, Tian G et al (2010) The sequence and *de novo* assembly of the giant panda genome. Nature 463:311 –317

Lohmann K, Klein C (2014) Next generation sequencing and the future of genetic diagnosis. Neurotherapeutics 11:699 –707

Lonjou C, Thomas L, Borot N et al (2006) A marker for Stevens-Johnson syndrome: ethnicity matters. Pharmacogenomics J 6:265 –268

Lonjou C, Borot N, Sekula P et al (2008) A European study of HLA-B in Stevens-Johnson syndrome and toxic epidermal necrolysis related to five high-risk drugs. Pharmacogenet Genomics 18:99 –107

Lorenz M, Wozel G, Schmitt J (2012) Hypersensitivity reactions to dapsone: a systematic review. Acta Derm Venereol 92:194 –199

Mardis ER (2008) Next-generation DNA sequencing methods. Annu Rev Genomics Hum Genet 9: 387 –402

Mardis ER (2009) New strategies and emerging technologies for massively parallel sequencing: applications in medical research. Genome Med 1:40

Mardis ER (2010) The $1,000 genome, the $100,000 analysis? Genome Med 2:84

Margulies M, Egholm M, Altman WE et al (2005) Genome sequencing in microfabricated high-density picolitre reactors. Nature 437:376 –380

Maxam AM, Gilbert W (1977) A new method for sequencing DNA. Proc Natl Acad Sci U S A 74: 560 –564

McCormack M, Alfirevic A, Bourgeois S et al (2011) HLA-A*3101 and carbamazepine-induced hypersensitivity reactions in Europeans. N Engl J Med 364:1134 –1143

McLeod HL, Evans WE (2001) Pharmacogenomics: unlocking the human genome for better drug therapy. Annu Rev Pharmacol Toxicol 41:101 –121

Metzker ML (2010) Sequencing technologies - the next generation. Nat Rev Genet 1:31 –46

Meyer UA (1994) Pharmacogenetics: the slow, the rapid, and the ultrarapid. Proc Natl Acad Sci U S A 91:1983 –1984

Mitra RD, Butty VL, Shendure J et al (2003) Digital genotyping and haplotyping with polymerase colonies. Proc Natl Acad Sci U S A 100:5926 –5931

Mizutani T (2003) PM frequencies of major CYPs in Asians and Caucasians. Drug Metab Rev 35: 99 –106

Motulsky AG (1957) Drug reactions, enzymes and biochemical genetics. JAMA 165:835 –837

Naisbitt DJ, Britschgi M, Wong G et al (2003) Hypersensitivity reactions to carbamazepine: characterization of the specificity, phenotype, and cytokine profile of drug-specific T cell clones. Mol Pharmacol 6:732 –741

Ogawa R, Echizen H (2010) Drug–drug interaction profiles of proton pump inhibitors. Clin Pharmacokinet 49:509 –533

Oldenburg J, Watzka M, Rost S et al (2005) VKORC1: molecular target of coumarins. J Thromb Haemost 5:1 –6

Owen RP, Altman RB, Klein TE (2008) PharmGKB and the International Warfarin Pharmacogenetics Consortium: the changing role for pharmacogenomic databases and single-drug pharmacogenetics. Hum Mutat 29:456 –460

Ozeki T, Mushiroda T, Yowang A et al (2011) Genome-wide association study identifies HLA-A*3101 allele as a genetic risk factor for carbamazepine-induced cutaneous adverse drug reactions in Japanese population. Hum Mol Genet 20:1034 –1041

Ozsolak F, Milos PM (2011a) RNA sequencing: advances, challenges and opportunities. Nat Rev Genet 12:87 –98

Ozsolak F, Milos PM (2011b) Transcriptome profiling using single-molecule direct RNA sequencing. Methods Mol Biol 733:51 –61

Ozsolak F, Kapranov P, Foissac S et al (2010) Comprehensive polyadenylation site maps in yeast and human reveal pervasive alternative polyadenylation. Cell 143:1018 –1029

Rajic V, Debeljak M, Goricar K et al (2015) Polymorphisms in GRIA1 gene are a risk factor for asparaginase hypersensitivity during the treatment of childhood acute lymphoblastic leukemia. Leuk Lymphoma 56:3103 –3108

Roh HK, Dahl ML, Tybring G et al (1996) CYP2C19 genotype and phenotype determined by omeprazole in a Korean population. Pharmacogenetics 6:547 –551

Romkes M, Faletto MB, Blaisdell JA et al (1991) Cloning and expression of complementary DNAs for multiple members of the human cytochrome P450IIC subfamily. Biochemistry 30:3247 –3255

Ronaghi M, Karamohamed S, Pettersson B et al (1996) Real-time DNA sequencing using detection of pyrophosphate release. Anal Biochem 242:84 –89

Rost S, Fregin A, Ivaskevicius V et al (2004) Mutations in VKORC1 cause warfarin resistance and multiple coagulation factor deficiency type 2. Nature 427:537 –541

Rusk N (2009) Cheap third-generation sequencing. Nat Methods 6:244 –245

Sakai T, Aoyama N, Kita T et al (2001) CYP2C19 genotype and pharmacokinetics of three proton pump inhibitors in healthy subjects. Pharm Res 18:721 –727

Samer CF, Daali Y, Wagner M et al (2010) The effects of CYP2D6 and CYP3A activities on the pharmacokinetics of immediate release oxycodone. Br J Pharmacol 160:907 –918

Sanger F (1975) The Croonian Lecture, 1975: nucleotide sequences in DNA. Proc R Soc Lond B Biol Sci 191(1104):317–333

Sanger F, Nicklen S, Coulson AR (1977) DNA sequencing with chain-terminating inhibitors. Proc Natl Acad Sci U S A 74:5463 –5467

Sapone A, Vaira D, Trespidi S et al (2003) The clinical role of cytochrome p450 genotypes in Helicobacter pylori management. Am J Gastroenterol 98:1010 –1015

Satoh JI, Yanaizu M, Tosaki Y et al (2016) Targeted sequencing approach to identify genetic mutations in Nasu-Hakola disease. Intractable Rare Dis Res 5:269 –274

Schadt EE, Turner S, Kasarskis A (2010) A window into third-generation sequencing. Hum Mol Genet 19:R227–R240

Schuster SC (2008) Method of the year, next-generation DNA sequencing: functional genomics and medical applications. Nat Methods 5:11 –21

Selvin PR (2000) The renaissance of fluorescence resonance energy transfer. Nat Struct Biol 7: 730 –734

Shendure J, Porreca GJ, Reppas NB et al (2005) Accurate multiplex polony sequencing of an evolved bacterial genome. Science 309:1728 –1732

Sheu BS, Kao AW, Cheng HC et al (2005) Esomeprazole 40 mg twice daily in triple therapy and the efficacy of Helicobacter pylori eradication related to CYP2C19 metabolism. Aliment Pharmacol Ther 21:283 –288

Shi S, Klotz U (2008) Proton pump inhibitors: an update of their clinical use and pharmacokinetics. Eur J Clin Pharmacol 64:935 –951

Shirai N, Furuta T, Moriyama Y et al (2001) Effects of CYP2C19 genotypic differences in the metabolism of omeprazole and rabeprazole on intragastric pH. Aliment Pharmacol Ther 15: 1929 –1937

Sim SC, Risinger C, Dahl ML et al (2006) A common novel CYP2C19 gene variant causes ultrarapid drug metabolism relevant for the drug response to proton pump inhibitors and antidepressants. Clin Pharmacol Ther 79:103 –113

Smith LM, Sanders JZ, Kaiser RJ et al (1986) Fluorescence detection in automated DNA sequence analysis. Nature 321:674 –679

Spear B, Heath-Chiozzi M, Huff J (2001) Clinical applications of pharmacogenetics. Trends Mol Med 7:201 –204

Tassaneeyakul W, Jantararoungtong T, Chen P et al (2009) Strong association between HLA-B*5801 and allopurinol-induced Stevens-Johnson syndrome and toxic epidermal necrolysis in a Thai population. Pharmacogenet Genomics 19:704 –709

Tawfik DS, Griffiths AD (1998) Man-made cell-like compartments for molecular evolution. Nat Biotechnol 16:652 –656

Tomczak K, Czerwinska P, Wiznerowicz M (2015) The Cancer Genome Atlas (TCGA): an immeasurable source of knowledge. Contemp Oncol (Pozn) 19:A68–A77

Vanden Branden M, Ring BJ, Binkley SN et al (1996) Interaction of human liver cytochromes P450 in vitro with LY307640, a gastric proton pump inhibitor. Pharmacogenetics 6:81 –91

Venter JC, Adams MD, Myers EW et al (2001) The sequence of the human genome. Science 291: 1304 –1351

Vesell ES (1989) Pharmacogenetic perspectives gained from twin and family studies. Pharmacol Ther 41:535 –552

Wu J, Li Y, Jiang R (2014) Integrating multiple genomic data to predict disease-causing nonsynonymous single nucleotide variants in exome sequencing studies. PLoS Genet 10: e1004237

Yamey G (2000) Scientists unveil first draft of human genome. BMJ 321:7

Zimmermann A, Matschiner JT (1974) Biochemical basis of hereditary resistance to warfarin in the rat. Biochem Pharmacol 23:1033 –1040

Meeting the Unmet Challenges of Pharmaceutical Research Using Artificial Intelligence

4

Sonal Mehrotra and Kamla Pathak

Abstract

Over the last decade, artificial intelligence (AI) has piqued attention in the various areas of robotics, healthcare, E-commerce, and so on. But the pharmaceutical sector has geared up its applications recently as it lessens the time and millions of dollars in research. Artificial intelligence (AI) is the design and application of algorithms for the analysis of learning and interpretation of data. This chapter presents a review that highlights the impressive use of AI in the drug discovery and development process. Some of the methods like artificial neural networks, fuzzy logic, and so on are discussed for modeling and optimization purposes to reduce the human workload as well as achieve targets in a short period. AI is the next big thing, and pharmaceutical corporations that are more adaptable and quicker to incorporate AI will undoubtedly gain a strategic advantage. In reality, AI will be needed to compete in the sector soon.

Keywords

Artificial intelligence · Artificial neural network · Fuzzy logic · Machine learning

S. Mehrotra
Manipal College of Pharmaceutical Sciences, Manipal, Karnataka, India

K. Pathak (✉)
Faculty of Pharmacy, Uttar Pradesh University of Medical Sciences, Etawah, UP, India

P. P. Singh (ed.), *Recent Advances in Pharmaceutical Innovation and Research*,
https://doi.org/10.1007/978-981-99-2302-1_4

4.1 Introduction

Artificial intelligence (AI) is a discipline of computer science that encompasses problem solving using symbolic programming. Its enormous growth as a problem-solving science has found application in healthcare, business, and engineering (Krishnaveni et al. 2019). In a span of more than a decade, AI is being used for analyzing and interpreting biological or genetic data, and for expedited drug discovery (Hassanzadeh et al. 2019) via a myriad of advanced tools and networks that can simulate human intelligence (Yang and Siar 2018). The development of AI can be divided into two groups, first is a method and system, like expert systems, that models the human brain and draws conclusions from a set of rules. The second group comprises of systems that simulate the brain functions (Agatonovic-Kustrin and Beresford 2000).

A variety of branches of statistical and machine learning, clustering, pattern recognition, and similarity-based approaches are included in AI (Duch et al. 2007). A defined algorithm is used to further classify data using the process of machine learning to find patterns in the data. Deep learning (DL) forms a subcategory of machine learning which utilizes artificial neural networks. These are a group of interconnected, powerful computer units that mimic the electrical impulse transmission in the human brain by using "perceptrons" that are analogous to biological human neurons (Beneke and Mackenrodt 2019). Multilayer perceptron (MLP) networks, convolutional neural networks (CNNs), and recurrent neural networks (RNNs) are examples of ANNs that make use of either supervised or unsupervised training processes (Bielecki 2019; Kalyane et al. 2020). These networks have served as the basis for numerous technologies that form the core of AI architecture. Figure 4.1 summarizes several examples of AI approach domains.

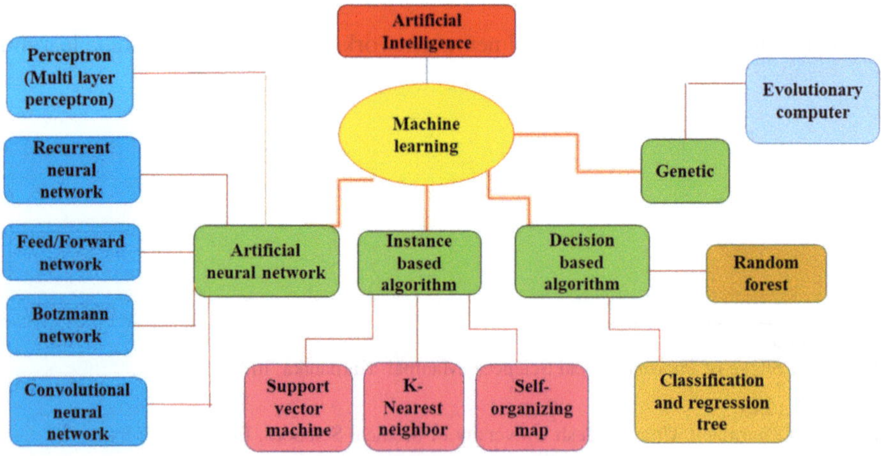

Fig. 4.1 Several approaches of AI

4.2 AI in the Lifecycle of Pharmaceutical Product

AI helps in rationale drug disovery, identifies the best course of treatment and level of care for a patient, and faciliates supervision of clinical trials for future drug development, from the bench to the bedside. AI is likely to be used in the creation of pharmaceuticals (Duch et al. 2007; Blasiak et al. 2019). The text below identifies the pharmaceutical areas wherein AI plays significant roles.

4.2.1 In Drug Discovery

The huge chemical space, which contains about $>10^{65}$ compounds, encourages the discovery of many medicinal compounds (Mak and Pichika 2019). The testing of drugs against samples of diseased cells is time- and labor-consuming operation which might get tackled by AI (Vyas et al. 2018). The ability of AI to distinguish between hit and lead compounds enables quicker medicinal target assessment and refinement over the design of structure (Mak and Pichika 2019; Sellwood et al. 2018). AI faces significant data challenges, including the quantity, development, diversity, and complexity of the data. Pharmaceutical companies may have millions of compounds in their research database, making it difficult for standard machine learning algorithms to manage them. A computer model derived on quantitative structure activity relationship (QSAR) can swiftly predict huge numbers of compounds or basic physicochemical characteristics like log *P* or log *D*. QSAR, however, confronts issues such as less information of training sets, inaccuracy in training data sets, and an inadequate experimental validation. When compared to traditional ML techniques, recently developing AI technologies, such as DL and related modeling research, display significant predictability (Zhu 2020; Ciallella and Zhu 2019). Structure and ligand-based approaches, as well as a variety of in silico techniques for screening of molecules from digital chemical regions, enable more efficient profile analysis, quicker elimination of non-lead compounds, and selection of therapeutic molecules at reduced costs (Mak and Pichika 2019). Decision trees, support vector machines, random forests, and other types of QSAR modeling tools have been used to find novel drug candidates and have developed into AI-based QSAR techniques that might be utilized to accelerate QSAR analysis (Zhang et al. 2017; Jain et al. 2015; Wang et al. 2015). Figure 4.2 shows a variety of AI uses in drug discovery.

4.2.2 Forecasting of Physicochemical Properties

When creating a novel molecule, it is important to consider how a drug's physicochemical characteristics, such as solubility, partition coefficient (log *P*), degree of ionization, and intrinsic permeability, may affect its pharmacokinetic properties and target receptor family. Numerous AI-based techniques can be used to predict

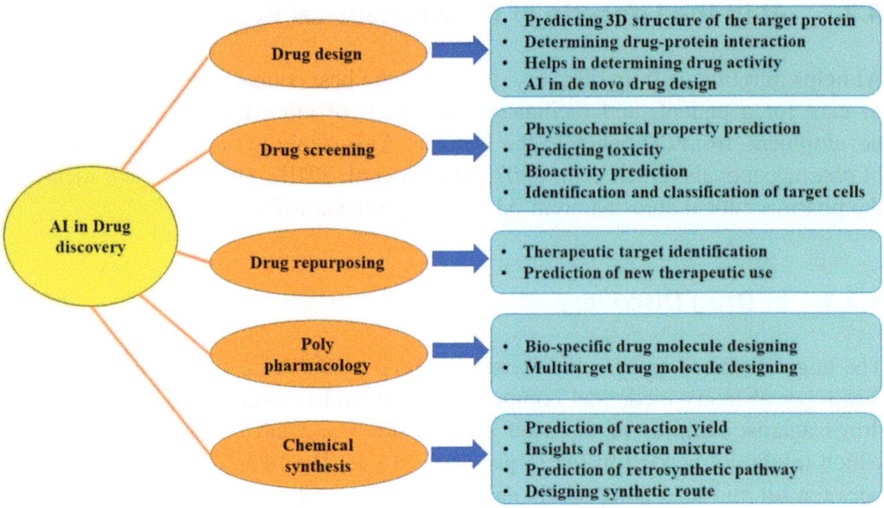

Fig. 4.2 AI in drug discovery

physical and chemical properties. For instance, ML makes use of enormous data sets produced via earlier compound optimization. Some cell lines studies including Madin-Darby canine kidney cells (MDCK) and human colon adenocarcinoma (Caco-2) cells have also been used to create cellular permeability data for a variety of compounds (Yang et al. 2019). In a similar vein, in silico models, based on RF and DNN, have been built to predict intestinal absorption in human for a range of biological/chemical substances (Thafar et al. 2019).

4.2.3 Forecasting of Pharmacological Activity

The potency of drugs depends on how well they bind to the target molecule or receptor. In order to forecast drug-target interactions, DTBA (drug target binding affinity) is essential (Ozturk et al. 2018). By examining the features or similarities between the drug and its target, AI-based techniques can evaluate the drug's binding affinity. KronRLS, SimBoost, DeepDTA, and PADME are some of the machine learning and deep learning algorithms that occupied to determine DTBA. Kronecker-regularized least squares (KronRLS) analyze the resemblance among drug and protein molecules in order to estimate DTBA. SimBoost, on the other hand, utilized regression trees to forecast DTBA and considered interactions based on both similarity and features (Lounkine et al. 2012).

4.2.4 In Pharmaceutical Product Development

The subsequent incorporation of a novel therapeutic chemical into an appropriate dosage form with desired delivery qualities is known as drug development. In this situation, AI can replace the traditional trial-and-error approach (Guo et al. 2018). Various computational tools can be used to tackle challenges in formulation design.

4.3 Artificial Neural Network

The organization and operation of the human nervous system have long sparked interest. Both medical and computer scientists have investigated the principles of brain activities such as learning and memory. Machine learning algorithms are based on the rules that govern human thinking and reasoning. ANNs are among various approaches based on neural computing which have been created and implemented in general science. Many textbooks have gone through their benefits and drawbacks in great detail (Freeman and Skapura 1991; Veelenturf 1995; Krose and Van der Smagt 1996; Gurney and Gurney 1997; Kasabov 1998; Haykinm 1999; Gupta et al. 2003; Dreyfus 2005; Rabunal and Dorado 2006; Taylor 2006). An artificial neural network (ANN) is a training network built on computer technology that can replicate the neurological and cognitive processes of the human brain (Achanta et al. 1995). The connectivity of neurons in a network is what gives brain calculations their power (Agatonovic-Kustrin and Beresford 2000). By establishing correlations between input and output factors, they may answer "what if" queries. ANNs is widely used for the prediction of nonlinear relationship between input factors and the output response (Djuris et al. 2013).

4.3.1 ANN vs. Human Brain

ANN is a biologically inspired system that mimics the human brain's neurological processing capabilities. The human brain has around 100 billion neurons and 100,000 times number of neuron connections. The cell body of a neuron has a soma as its center, dendrites as inputs, and axons as outputs. Interconnections between the neuron are called as synapse. Neurons are always connected in a network and operate as messengers, receiving and transmitting messages. As a result, a smart brain capable of learning, prediction, and recognition has been created. ANN mimics the function of human brain. The ANN does not learn (or are trained) through programming, but rather via experience (Simpson 1990; Moudgil et al. 2002).

4.3.2 Theory of ANN

Multiple layers of processing units known as neurons make up ANNs (nodes). Processing units known as artificial neurons are built on the same principles as biological neurons. This means that the artificial neuron takes data, processes it, and then sends it on to other processing units. A systematic representation of artificial neuron is shown in Fig. 4.3. Weights are connections between neurons that correspond to physiological synapses and are particularly crucial for learning and other adaptive processes in neural networks. According to biological principles, a neuron in an ANN receives one or more inputs and transmits one output, which is duplicated and relayed to additional units. An ANN is trained in a similar way to how the human brain learns, where the network generates predictions about future events based on past performance (Takayama et al. 2003).

4.3.3 The Architecture of ANNs

The ANNs (units) are structured into layers: An input, hidden, and output layer of neurons make up a simple ANN. One input layer and one output layer are often employed. Individual neurons (nodes) in each layer are infinite and typically equal to the number of inputs (independent variables) and outputs (dependent variables) in each layer (Erb 1993).

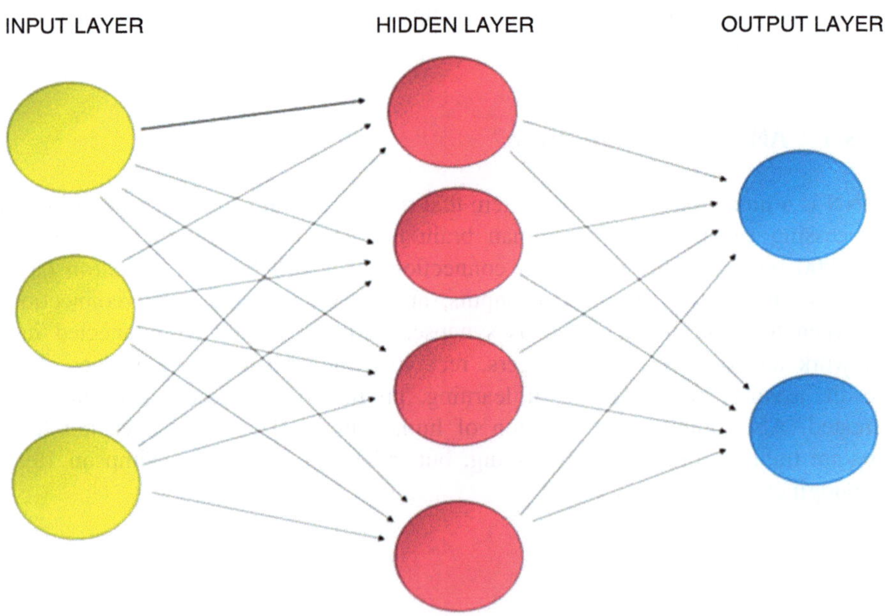

Fig. 4.3 Systematic view of ANNs

The number of hidden layers and nodes in each layer, on the other hand, varies depending on the type of input and the network's function. To generate the best predictions with the least amount of memorization, the number of hidden nodes should be optimized. In contrast to the hidden layers' units, which apply functions and carry out computations on the given data, the network's input and output layers reflect the data that is fed into the network and the data that is returned as a network result, respectively. It has already been noted that ANNs are non-linear computing techniques. Input and hidden layer signals are transported to hidden and output layers, which are then positioned. The transmitted signal, that neurons in the hidden layer receive is activated by a specific activation function, which is most typically a sigmoid function. There are numerous activation functions that can be used, including logistic sigmoid functions, step functions, Gaussian functions, exponential functions, and more (Agatonovic-Kustrin and Beresford 2000). Equation (4.1) shows how each neuron's activation (transformation) function multiplies input signals by their corresponding weights and converts them to output:

$$y_q = \sum{}^{z} w_{pq} x_p \tag{4.1}$$

Where w_{pq} is the weight of the link from the current layer's neuron (unit) q to the previous layer's unit p, and x_p is the previous layer's output value (it is the input value for the neuron in the current layer).

Carpenter and Hoffman (1995) proposed Eq. (4.2) for associating the number of units in the input, hidden, and output layers:

$$n_{\mathrm{s}} = \beta[n_{\mathrm{h}}(n_{\mathrm{i}} + 1) + n_{\mathrm{o}}(n_{\mathrm{h}} + 1)] \tag{4.2}$$

The total number of units is n_{h} for the hidden units, n_{i} for the input units, n_{o} for the output units, and n_{s} for the training data pairs. The constant () is the parameter that relates to the level of overdetermination. The equation was used by Jadid and Fairbairn (1996) to suggest a maximum number of hidden units:

$$n_{\mathrm{h}} = \frac{n_{\mathrm{s}}}{[R + (n_{\mathrm{i}} + n_{\mathrm{o}})]} \tag{4.3}$$

where R is a constant that has a range of values from 5 to 10.

Network connections in a typical ANN are feedforward, meaning that signals are not passed back to neurons in the previous layer once they have been sent to the next layer of neurons in the network(s). A little more challenging is the feedback type of network connection, which describes the creation of cycles in connections where data is transmitted back and developed over time. Figure 4.4 depicts the organization of a typical neural network. A network's neurons can also be fully or partially coupled. When every neuron in the layer above is coupled to every neuron in the layer below, this is known as a full connection. The fact that a neuron has partial connections means that it is not connected to every neuron in the layer above.

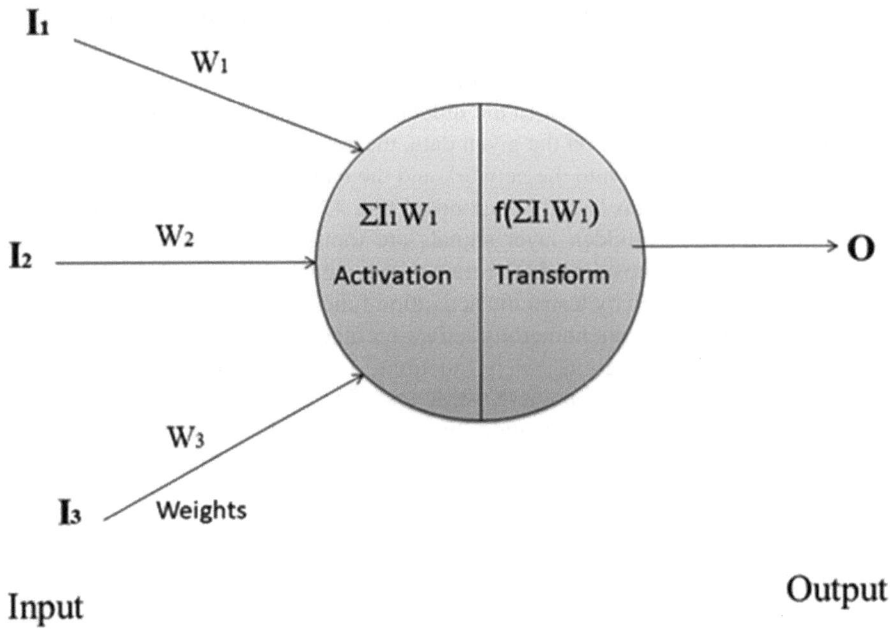

Fig. 4.4 Organization of ANN

4.3.4 Training

In order to achieve the best non-linear relationship between the parameters used as the network's inputs and outputs, an ANN must be trained by changing weight values. In actuality, network training is a "learning process." It could be separated into two halves.

1. Supervised learning: when the input and output pairs are shared and the network weights are modified depending on the known data. Supervised learning is the most used method of ANN training.
2. Unsupervised learning: It is a challenge that arises while attempting to uncover a hidden structure in untagged data. The examples provided to the students are not labeled; therefore, there is no indication of a mistake or incentive for considering a potential solution.

Data is separated into two subsets at the start of the training process: training and test. The network's prediction power is tested internally using test data, while training data is used to find the ideal weight values (Basheer and Hajmeer 2000). The following equation can be used to compute the sum of squared errors (SSE) for the training and test subsets.

$$\text{SSE} = \frac{1}{N} \sum_{p=1}^{N} \sum_{i=1}^{M} \left(t_{pi} = O_{pi}\right)^2 \tag{4.4}$$

where t_{pi} is the training (actual) output of the ith output node from the pth sample and o_{pi} is the output predicted by the network of the ith output node from the pth sample; The number of output nodes is M, while the number of training samples is N.

During the network training process, both the training and test data sets' progress in terms of change in error are simultaneously assessed. In a training data set, SSE continually decreases as the quantity of hidden nodes or training iterations rises (epochs). Initially, learning causes SSE for the training set to reduce quickly, while memory or overfitting causes it to diminish more slowly after that whereas SSE drops initially but increases later in the test data set due to memorizing and overfitting of the ANN model. Therefore, it is preferable to halt training once the test error starts to increase and select the amount of hidden nodes while the test error is minimal (Sun et al. 2003).

The training process can be set to last until the predetermined error value is found, or it can be configured to last for a specific amount of time (Reis et al. 2004). One of the most popular training techniques is the back- propagation (BP) algorithm (i.e., weight adjustment method). In addition to the BP algorithm, there are several other methods for modifying weights during the training phase (i.e., error minimization), including the Gauss-Newton method, the steepest descent method, conjugate gradient method, simulated annealing methodology, and others (Peh et al. 2000).

4.3.5 Working

The first step in this process determines the amount of neurons in the input, output, and hidden layers as well as the number of hidden layers. The researched problem determines the number of neurons in the input and output layers, and it has previously been stated that the number of neurons in the input layer should be small. Choosing the hidden layers and the quantity of neurons in each hidden layer is the most difficult assignment at this point. The trial-and-error method, as well as numerous optimization strategies, is still widely utilized to solve this problem. Once the network architecture (topology) has been defined, the type of transfer (activation) function, learning rate, smooth factor, and other parameters must typically be chosen in this stage. Some of these characteristics are predefined in the software, while others require adjusting. The network's real training is the next phase. Data is usually divided into three categories: training, testing, and validation. During the training process, the network is provided with both training and test data sets. Test data is saved and used to evaluate the network's predictive ability, while training data is used to establish and improve neuron weights in the network.

As previously stated, if the test error stops dropping or starts to increase, the network is overfitting, and the training process should be terminated. If network overfitting is expected, reducing the number of hidden layers or neurons in those hidden layers is recommended. Validation data is not used during training; rather, it is used immediately once training is complete. Validation data includes samples that were not supplied to network during training, but they must be contained within the data range specified by the training data (Djuris et al. 2013; Sun et al. 2003). In the literature, there are numerous examples of ANN use in pharmacological research, product development, and process improvement. Some of them are shown Table 4.1 with a brief description of the approach employed.

4.4 Fuzzy Logic

Fuzzy logic, a subfield of logic which deals with human inference, reasoning, and reasoning, acknowledges and makes use of the fact that everything in life is a matter of degree (Sproule et al. 2002). A fuzzy set theory is the foundation of fuzzy logic. It is based on specific aspects, such as height, that do not have simply two values (low or high, short or tall); rather, there are several levels to express it more precisely. If binary logic is considered as a red- and-yellow approach, then fuzzy logic comes from shades of orange.

4.4.1 Theory of Fuzzy Logic

Fuzzy logic is a multi-valued logic with truth values ranging from 0 (totally false) to 1 (totally true) (Klenner et al. 2010). As a result, binary logic elements can only exist in two states: 0 (false) and 1 (true) (true). The presence of a feature as well as its quantification is possible with fuzzy logic. When applying the fuzzy methodology, word changes can be progressive, and the binary all-or-nothing choice is transformed into the extreme of a continuum (Sproule et al. 2002). Neuro-fuzzy systems are predictive data mining tools that provide sets of language rules, that make the processes under study simpler to understand.

Samples of data sets can have variable degrees of membership, and are the foundations of understanding and modeling fuzzy system. In a binary system, a sample's degree of membership is either 1 or 0, while in a fuzzy system, a sample may belong to several classes with different levels of membership (that all add up to 1). Fuzzy units are typically trapezoidal in shape in the graphic depiction, but they can also have irregular shapes like bells, triangles, sigmoid shapes, or triangular shapes (Cox 1994). Fuzzy units include type 2 fuzzy units, L-fuzzy units, bipolar fuzzy units, and intuitive fuzzy units (Zadeh 2007).

In the form of ***if... then...*** statements, fuzzy systems are frequently expressed as rules of language that are simple to comprehend. Various system states can be expressed using this way (like with the decrement in particle size the dissolution rate is increased). The technique of defining if... then... rules is known as fuzzy modelling. To set rules, experts use their skills, experience, and experimental data. As with any other modeling, the first steps in fuzzy modeling are to define and gather inputs and outputs. Fuzzy C-means, Fuzzy first-class clustering, Fuzzy LVQ, Fuzzy

Table 4.1 Pharmaceutical applications of ANN

Aim	Input/output study	Network type	Reference
Drug release control and system understanding of sucrose ester matrix tablets using ANN	HLB values of sucrose esters (SEs), tablet volume, tablet porosity, SEs concentration and tablet tensile strength were used as network inputs. The outputs were in vitro dissolution profiles and characteristics suggestive of burst release, such as release exponent and mean dissolution time.	MLP	Chansanroj et al. (2011)
Use of DoE and multilayer perceptron neural network for the optimization of release of diclofenac sodium from Carbopol 71G matrix tablets	The quantity of polymer used and the crushing strength of the MUPS tablets had been community inputs. Outputs of the release profiles had been network outputs.	MLP	Ivic et al. (2010)
Design of release of diclofenac sodium from polyethylene oxide matrix tablets	Polymer weight ratio and compression pressure have been used as inputs, while in vitro dissolution profiles have been used as networks outputs. Dissolution profiles have been dealt with as time collection the usage of dynamic neural networks.	MLP, GMDNN, OLDNN	Petrovic et al. (2009)
Evaluation of salbutamol's relative lung bioavailability and clinical effect in healthy volunteers and asthmatic patients using dry powder inhalers	In vitro aerodynamic capabilities of the system and demographic facts of volunteers/sufferers have been used to train the ANN network, even as in vivo data (urinary excretion of the drug and its metabolite) have been used as network outputs.	MLP	De Matas et al. (2008)
ANN-based optimization of benznidazole chitosan microparticles	The polymer concentration, NaOH concentration, stirring rate, and spraying rate had been all used as inputs for the network training (microparticles had been organized via way of means of the coacervation method). Microparticle size, encapsulation efficiency, yield, and dissolution rates had been the network training's outputs.	MLP	Leonardi et al. (2009)
Analyzing the diffusion coefficient using ANN from	The sodium dodecyl sulphate dose, flow rate, vaginal fluid simulant pH, rotation speed,	MLP	Lee et al. (2008)

(continued)

Table 4.1 (continued)

Aim	Input/output study	Network type	Reference
mucoadhesive vaginal controlled drug delivery system	delivery device weight was all used as inputs for the network training. The network training outputs were chosen to be release profiles (diffusion coefficients).		
Determining the pellet properties by using of artificial neural networks	The network was trained using parameters defining pellet quantitative and qualitative composition, as well as preparation method, as inputs, and mean dissolving time (MDT)and pellet aspect ratio (AR) as outputs.	MLP	Mendyk et al. (2010)
Prediction of the release kinetics of doxorubicin from sulfopropyl dextran ion exchange microspheres	Three independent variables, drug loading level, NaCl and $CaCl_2$ concentration in the release medium were used as ANN inputs and fractional release of doxorubicin at four different times as outputs.	MLP, HNN	Li et al. (2005)
Prediction of drug release profiles in transdermal iontophoresis.	The inputs from the neural networks were the process conditions of pH, ionic strength and current as well as the point in time. The output was the predicted rate of permeability of the drug (diclofenac sodium).	RBFNN	Lim et al. (2003)

MLP multilayered perceptron, *RBFNN* radial basis function neural network, *GMDNN* gamma memory dynamic neural network, *OLDNN* one-layer dynamic neural network

adaptive resonance theory, Fuzzy C-regression version clustering, and other approaches are used to cluster the outputs of the fuzzy system (Bezdek 1981; Emami et al. 1998; Baraldi and Blonda 1999; Kim et al. 1997).

4.4.2 The Architecture of Fuzzy Logic

The architecture has been grouped into four sections. Rules come first, followed by fuzzifier, defuzzifier, and intelligence. Rules comprise of all the rules and *if... then...* criteria offered by experts for controlling the decision-making system. Recent advances in fuzzy theory have resulted in a variety of practical strategies for designing and tuning fuzzy controllers. They will also lessen the number of fuzzy

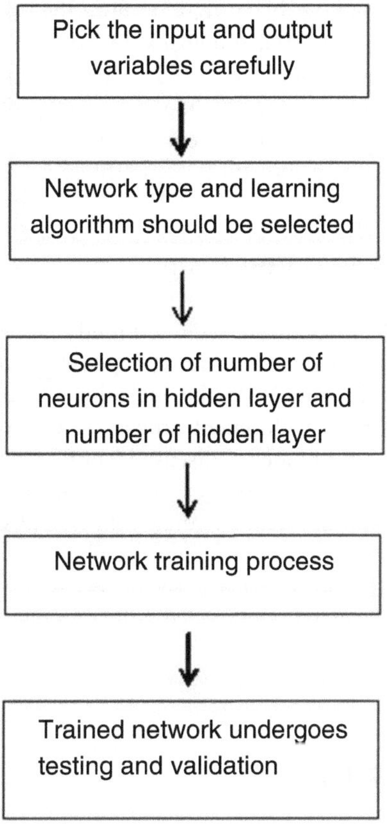

Fig. 4.5 Architecture of fuzzy logic

rules (Sproule et al. 2002). A simple architecture of fuzzy logic has been shown in Fig. 4.5.

4.4.3 Membership Functions

Within a set, the element will have a degree of membership that may be described by a membership function (Zadeh 1965). For example, a set of granules with a size range of 200–400 mm is considered classic or "crisp," but a set of granules with a size range of "about 300" is considered fuzzy. An element in a crisp or Boolean set is either a complete member of a given set, as indicated by a membership value of one on the *y*-axis, or it is not a member, as represented by a membership value of zero. Items in a fuzzy set can have degrees of membership on the actual continuous interval, with the end points of interval (zero and one) signifying no and complete membership, respectively (Pannier et al. 2003).

While more sophisticated tools that automatically create the rule base (based on inputs and outputs) are still not widely available, most software solutions for applying fuzzy logic require the user to construct fuzzy sets and rules. This is the fundamental drawback of using fuzzy logic to solve real-world situations. Fuzzy logic should be used in combination with other systems theory, like ANNs, to address some of the shortcomings. The adaptive learning capabilities of ANNs and the generality of representation of fuzzy logic are combined in a hybrid technology called neuro-fuzzy logic (Shao et al. 2007).

4.5 Conclusion

The healthcare sector currently faces a number of complex challenges, such as rising medical and therapy expenses, and society needs significant advancements in this area. With the inclusion of artificial intelligence, both the time and cost in drug development process can be reduced. AI technologies offer the tremendous opportunities in disease identification, radiology, radiotherapy, clinical trial research, drug discovery, personalized medicine, rare disease identification, treatment robotic pharmacy, and medical robots in pharma and healthcare. AI can also aid in decision-making and the implementation will help of the researched medication in the appropriate dose form, which will speed up the production of higher-quality items. Artificial intelligence will soon play a significant role in the healthcare industry.

Acknowledgment All the authors declare no funding support and no conflict of interest.

References

Achanta AS, Kowalski JG, Rhodes CT (1995) Artificial neural networks: implications for pharmaceutical sciences. Drug Dev Ind Pharm 21:119–155

Agatonovic-Kustrin S, Beresford R (2000) Basic concepts of artificial neural network (ANN) modelling and its application in pharmaceutical research. J Pharm Biomed 22:717–727

Baraldi A, Blonda P (1999) A survey of fuzzy clustering algorithms for pattern recognition II. IEEE Trans Syst Man Cybern Syst 29:786–801

Basheer IA, Hajmeer M (2000) Artificial neural networks: fundamentals, computing, design and application. J Microbiol Methods 43:3–31

Beneke F, Mackenrodt MO (2019) Artificial intelligence and collusion. IIC Int Rev Ind Prop Copyr Law 50:109–134

Bezdek JC (1981) Pattern recognition with fuzzy objective function algorithm. Plenum Press, New York

Bielecki A (2019) Foundation of artificial neural networks. In: Kacprzyk J (ed) Model of neurons & perceptrons: selected problems & challenges. Springer, Berlin, pp 15–28

Blasiak A, Khong J, Kee T (2019) CURATE. AI: optimizing personalized medicines with artificial intelligence. SLAS Technol 25:95–105

Carpenter WC, Hoffman ME (1995) Understanding neural network approximations and polynomial approximations help neural network performance. Artif Intell:31–33

Chansanroj K, Petrovic J, Ibric S et al (2011) Drug release control and system understanding of sucrose esters matrix tablets by artificial neural networks. Eur J Pharm Sci 44:321–334

Ciallella HL, Zhu H (2019) Advancing computing toxicology in the big data era by artificial intelligence: data-driven and mechanism-driven modeling for chemical toxicology. Chem Res Toxicol 32:536–547

Cox E (1994) Using and maintaining fuzzy systems. In: The fuzzy systems handbook. A practitioner's guide to building. Academic Press, Cambridge, MA; Book and Disk

De Matas M, Shao Q, Richardson CH et al (2008) Evaluation of in vitro in vivo correlations for dry powder inhaler delivery using artificial neural network. Eur J Pharm Sci 33:80–90

Djuris J, Ibric S, Djuric Z (2013) Neural computing in pharmaceutical products and process development. In: Djuris J (ed) Computer-aided applications in pharmaceutical technology. Woodhead Publishing, Sawston, pp 92–93

Dreyfus G (2005) Neural networks: methodology and applications. Springer, Berlin

Duch W, Swaminathan K, Meller J (2007) Artificial intelligence approaches for rational drug design and discovery. Curr Pharm Des 13:1497–1508

Emami MR, Turksen IV, Goldenberg AA (1998) Development of systematic methodology of fuzzy logic modeling. IEEE Trans Fuzzy Syst 6:346–361

Erb RJ (1993) Introduction to back propagation neural network computation. Pharm Res 10:165–170

Freeman JA, Skapura DM (1991) Neural networks: algorithms, applications, and programming techniques. Addison-Wesley, Boston, MA

Guo KY, Fokoue A, Luo H et al (2018) Interpretable drug target prediction using deep neural representation. In: Lang J (Ed) Proceeding of twenty-seventh international joint conferences on artificial intelligence organization. IJCAI, pp 3371–3377

Gupta MM, Jin L, Homma N (2003) Static and dynamic neural networks: from fundamentals to advanced theory. Wiley-Interscience IEEE Press, Hoboken, NJ

Gurney K, Gurney KN (1997) An introduction to neural networks. UCL Press, London

Hassanzadeh P, Atyabi F, Dinarvand R (2019) The significance of artificial intelligence in drug delivery system design. Adv Drug Deliv Rev 151–152:169–190

Haykinm S (1999) Neural networks: a comprehensive foundation. Prentice-Hall, Hertfordshire

Ivic B, Ibric S, Cvetkovic N et al (2010) Application of design of experiment and multilayer perceptrons neural network in the optimization of diclofenac sodium extended release tablets with Carbopol® 71G. Chem Pharm Bull 58:947–949

Jadid MN, Fairbairn DR (1996) Predicting moment-curvature parameters from experimental data. Eng Appl Artif Intell 9:303–319

Jain N, Gupta S, Sapre N et al (2015) *In silico de novo* design of novel NNRTIs: a bio molecular modelling approach. RCS Adv 5:14814–14827

Kalyane D, Sanap G, Paul D et al (2020) Artificial intelligence in pharmaceutical sector: current scene and future prospect. In: Tekade RK (ed) The future of pharmaceutical product development and research. Elsevier, Amsterdam, pp 73–107

Kasabov NK (1998) Foundations of neural networks, fuzzy systems, and knowledge engineering. A Bradford Book, 2nd edn. The MIT Press, Cambridge, MA

Kim E, Park M, Ji S et al (1997) A new approach to fuzzy modelling. IEEE Trans Fuzzy Syst 5: 328–337

Klenner A, Hartenfeller M, Schneider P et al (2010) 'Fuzziness' in pharmacophore-based virtual screening and *de novo* design. Drug Discov Today Technol 7:e237–e244

Krishnaveni C, Arvapalli S, Sharma JVC et al (2019) Artificial intelligence in pharma industry—a review. Int J Innov Pharm Sci Res 7:37–50

Krose B, Van der Smagt P (1996) An introduction to neural networks. The University of Amsterdam, Amsterdam

Lee Y, Khemka A, Yoo JW et al (2008) Assessment of diffusion coefficient from mucoadhesive barrier devices using artificial neural networks. Int J Pharm 351:119–126

Leonardi D, Salomon CJ, Lamas MC et al (2009) Development of novel formulations for Chagas' disease: optimization of benznidazole chitosan microparticles based on artificial neural networks. Int J Pharm 367:140–147

Li Y, Rauth AM, Wu XY (2005) Prediction of kinetics of doxorubicin release from sulfopropyl dextran ion- exchange microspheres using artificial neural networks. Eur J Pharm Res 24:401–410

Lim CP, Quek SS, Peh KK (2003) Prediction of drug release profiles using an intelligent learning system: an experimental study in transdermal iontophoresis. J Pharm Biomed 31:159–168

Lounkine E, Keiser MJ, Whitebread S et al (2012) Large-scale prediction and testing of drug activity on side-effect targets. Nature 486:361–367

Mak KK, Pichika MR (2019) Artificial intelligence in drug development: present status and future prospect. Drug Discov Today 24:773–780

Mendyk A, Kleinebudde P, Thommes M et al (2010) Analysis of pellet properties with use of artificial neural networks. Eur J Pharm Sci 41:421–429

Moudgil P, Tiwary AK, Gupta S (2002) Artificial intelligence in pharmaceutical research: theory and applications. Indian J Pharm Sci 64(6):509–514

Ozturk H, Ozgur A, Ozkirimli E (2018) Deep DTA: deep drug target binding affinity prediction. Bioinformatics 34(17):i821–i829

Pannier AK, Brand RM, Jones DD (2003) Fuzzy modelling of skin permeability coefficients. Pharm Res 20:143–148

Peh KK, Lim CP, Quek SS et al (2000) Use of artificial neural networks to predict drug dissolution profiles and evaluation of network performance using similarity factor. Pharm Res 17:1384–1388

Petrovic J, Ibric S, Betz G et al (2009) Application of dynamic neural networks in modeling of drug release from polyethylene oxide matrix tablets. Eur J Pharm Sci 38:172–180

Rabunal JR, Dorado J (2006) Artificial neural networks in real-life applications. Idea Group Publishing, Hershey, PA

Reis MAA, Sinisterra RD, Belchior JC (2004) An alternative approach based on artificial neural networks to study controlled drug release. J Pharm Sci 93:418–430

Sellwood MA, Ahmed M, Segler M et al (2018) Artificial intelligence in drug discovery. Future Sci 10:2025–2028

Shao Q, Rowe RC, York P (2007) Comparison of neuro-fuzzy logic and decision trees in discovering knowledge from experimental data of an immediate release tablet formulation. Eur J Pharm Sci 31:129–136

Simpson PK (1990) Artificial neural systems. Pergamon Press, New York, p 79

Sproule BA, Naranjo CA, Turksen IB (2002) Fuzzy pharmacology; theory and applications. Trends Pharmacol Sci 23:412–417

Sun Y, Peng Y, Chen Y et al (2003) Application of artificial neural networks in the design of controlled release drug delivery systems. Adv Drug Deliv Rev 55:1201–1215

Takayama K, Fujikawa M, Obata Y et al (2003) Neural network based optimization of drug formulations. Adv Drug Deliv Rev 55:1217–1231

Taylor BJ (ed) (2006) Methods and procedures for the verification and validation of artificial neural networks. Springer Science, New York

Thafar M, Raies AB, Albaradei S et al (2019) Comparison study of computational prediction tools for drug-target binding affinities. Front Chem 7:1–19

Veelenturf LPJ (1995) Analysis and applications of artificial neural networks. Prentice Hall, Hertfordshire

Vyas M, Thakur S, Riyaz B et al (2018) Artificial intelligence: the beginning of a new era in pharmacy profession. Asian J Pharm Sci 12:72–76

Wang Y, Guo Y, Kuang Q et al (2015) A comparative study of family-specific protein-ligand complex affinity prediction based on random forest approach. J Comput Aided Mol Des 29: 349–360

Yang Y, Siar K (2018) Qualitative research on marketing & sales in artificial intelligence age. In: MWAIS

Yang X, Wang Y, Byrne R et al (2019) Concept of artificial intelligence for computer-assisted drug discovery. Chem Rev 119:10520–10594

Zadeh LA (1965) Fuzzy sets. Inf Control 8:338–353
Zadeh LA (2007) Fuzzy logic. Scholarpedia 3:1766
Zhang L, Tang J, Han D et al (2017) From machine learning to deep learning: progress in machine learning for rational drug discovery. Drug Discov Today 22:1680–1685
Zhu H (2020) Big data and artificial intelligence modelling for the drug discovery. Annu Rev Pharmacol Toxicol 60:573–589

Pharmaceutical Innovation and Research in Ayurved

5

Nilima Wadnerwar

Abstract

Ayurvedic drugs are used effectively for treatment but the question arises for the validation of their efficacy and safety. To validate the safety and efficacy of various formulations in Ayurved, some experimental studies have been carried out on the basis of reverse pharmacology. It was observed that the study drugs undertaken are safe and efficacious. In some studies it was observed that the pharmaceutical procedures like *shodhana* and *marana* are beneficial when tested using modern techniques and parameters. Their limitations are also ruled out. Even the use of poisonous drugs and heavy metals is also justified. Apart from routine treatment, some effective remedies have also been developed.

Keywords

Agadtantra · Ashtang Ayurved · Bhaishajya Kalpana · Dravyaguna · *Marana* · Rasamanikya · Rasashastra · Shodhana · Shwitra · Tribhuvankirti rasa

5.1 Pharmaceutical Research in Ayurved

5.1.1 Introduction

Though Ashtang Ayurved (eight faculties of Ayurved) contributes to Ayurvedic therapeutics, Dravyaguna (Ayurvedic pharmacology), Agadtantra (Ayurvedic toxicology), and Rasashastra and Bhaishajya Kalpana (Ayurvedic pharmaceutics) are an

N. Wadnerwar (✉)
Department of Agadtantra, Mahatma Gandhi Ayurved College Hospital and Research Centre, Salod (H), Datta Meghe Institute of Higher Education and Research (DU), Wardha, Maharashtra, India
e-mail: nilima.wadnerwar@dmiher.edu.in

P. P. Singh (ed.), *Recent Advances in Pharmaceutical Innovation and Research*,
https://doi.org/10.1007/978-981-99-2302-1_5

integral part of Ayurved. All types of treatment include the use of medicine, and Ayurved is glorified with the use of several kinds of herbal, mineral, and herbomineral medicines.

Dravyaguna is a science that includes the pharmacology of various herbal drugs, minerals, and food materials. Agadtantra deals with the study of herbal, mineral, animal, and artificial poisons, their sources, properties, manifestations, diagnosis, and treatment. Rasashastra is the science that deals with the study of minerals and metals, their properties, purification, incineration, and therapeutic effects whereas Bhaishajya Kalpana includes the principles and techniques of preparations of various herbal, mineral, and herbomineral formulations along with different dosage forms, their standardization, and quality control.

5.1.1.1 Use of Heavy Metals in Ayurveda

The formulations described in Ayurveda are used according to their *Rasa, Guna, Virya Vipaka, Prabhava,* etc. from ancient times. Traditional medicine was used in ample amounts by the sages almost for all diseases. This period was followed by the trend to use minerals to increase the potency of the formulation. Knowing the potency of minerals, they were used as main ingredients which were known as *Rasakalpa.* The minerals also include some heavy metals which possess toxic properties. But the solution was also described by the sages in the form of some purification procedures called as *Shodhana.* Sushruta has described as inanimate poisons which include poisonous plants and *Dhatuvisha* (toxic mineral). These *Visha Dravya* are used after proper *Shodhana* methods to prepare formulation for the treatment. The formulations containing heavy metals are known as *Rasakalpa* and are popularly used in the treatment of Ayurved.

5.1.1.2 Pharmaceutical Procedures in Ayurved

In the classical texts of Ayurved, specific techniques and procedures are described for the preparation of formulations and different dosage forms. The ancient methods and techniques were very slow. Earthen and/or iron containers and instruments were used along with all naturally available organic materials. The parameters for standardization and quality control are also mentioned wherever necessary. For the use of toxic drugs in formulations, specific *shodhana* (purification) and *marana* (incineration) procedures were described. These procedures involved use of such drugs with the help of which untoward effects of the toxic drug can be avoided or overcome.

5.1.1.3 Strength of the Formulations in Ayurved

The formulations thus prepared according to the classical references with specified techniques are efficacious in different acute and chronic cases. They should be used considering the constitution of the body (*prakriti*), strength, age, time, place, etc. If such formulations are prepared scientifically and used judiciously, they may prove more beneficial without untoward effects.

5.1.1.4 Research Gap

In the classical text, methods of purification, incineration, and formulation are described in different places. There are more than two methods of purification, incineration, and formulation of most of the drugs. Hence, there may be a dilemma that which method of purification, incineration, and formulation should be used for a particular patient or particular disease. However, every time it is not possible to prepare different formulations for different patients.

Moreover, the dose and duration of administration of the drug are mentioned in the context of some drugs but it is not described for all formulations. In some places, common doses, duration, and common methods of purification and incineration are described. In all such conditions, the physician has to decide which method of purification, incineration, and formulation should be used on the basis of efficacy, safety, cost-effectiveness, and convenience. The dose and duration of administration of the drug may be decided on the basis of common principles of treatment. For instance, *Arogyavardhini vati* is prescribed at most for a period of 21 days for skin disorders. It is described for the formulation containing minerals and metals that they should be used only for 7–14 days and if required for more periods, they should be used after a washout period of 7–14 days.

To fulfil this gap, scientific research is essential in the field of pharmaceutical sciences in Ayurved. Hence, most of the research is conducted to assess the effect of purification and incineration on the drug, its constituents, their nature, and its safety and efficacy. It involves all the types of pharmaceutical studies, standardization of drugs, and their preclinical and clinical studies.

5.1.1.5 Need of Research in Ayurved

In ancient times, physicians used to prepare and dispense medications for their patients. In the present era, only 10% of physicians are preparing and dispensing medicine as the process is very complex, time-consuming, and costly to prepare formulations on a small scale. Due to commercialization, many pharmaceutical companies are established and they are manufacturing ancient formulations on a large scale by adopting modern methods and techniques. In this scenario, all the traditional and ancient techniques of formulation have been modified and standardized to maintain the quality control of the drugs. Hence in this modernization, earthen and iron vessels are replaced by plastic and other metal containers. Small-sized pieces of equipment are replaced by modern, huge pieces of machinery. The ancient techniques are replaced by contemporary techniques along with other factors involving maintenance. But they have also maintained certain standards of quality control.

But in recent years, some reports are published in articles and in newspapers which have raised questions regarding the safety and efficacy of Ayurvedic drugs, especially the medicines containing metals and minerals (Saper et al. 2004). Hence, it has become very necessary to re-examine and revalidate the formulations of Ayurved. Along with this, it is essential to revalidate ancient techniques and compare them with modern techniques based on safety, efficacy, and quality control. In this context, many more studies have been conducted in the last 20 years by various government and private research institutes in the form of postgraduate and doctoral research projects. The Ministry of Health and Family Welfare in collaboration with the Central Council for Research in Ayurved, Siddha, and Unani has conducted various research projects and published the data as standard in the Ayurvedic Pharmacopoeia of India and Ayurvedic Formulary of India. The author intends to share her experience in this field.

5.2 Study 1: Chronic Toxicity Study of *Rasamanikya* (an Arsenical Formulation)

5.2.1 Introduction

Arsenic is a heavy metal. Chronic Arsenic poisoning is a major concern in contemporary toxicology. *Rasamanikya* (Bhatt 2000) is an Ayurvedic formulation prepared from processed orpiment which is arsenic trisulphide. Though *Rasamanikya* is a medicine, it can produce toxic effects when administered in a higher dose. Chronic poisoning may occur due to its long-term use and/or without prescription. So to avoid acute, sub-acute, and chronic toxicity of *Rasamanikya*, it is necessary to limit the doses and duration of administration. A number of research are carried out with self-prepared *Rasamanikya* for its standardization, to evaluate its therapeutic efficacy, antimicrobial activity, and acute, sub-acute, and chronic toxicity in animals. In clinical studies, no adverse drug reactions were found. But in some animal studies, the self-prepared *Rasamanikya* was found to be toxic on chronic use. No research regarding safety and efficacy was conducted with the market preparations of *Rasamanikya*. In the present scenario, many pharmaceutical companies are manufacturing Ayurvedic formulations. Hence they are easily available and maximum people are using them. Hence, we have selected market samples of *Rasamanikya* for further study. Whether finished market preparations of *Rasamanikya* are of the same structure and have the same content of metallic arsenic and are within permissible limit was a research question before us as on this issue no previous research is available. Hence, this research was undertaken to study the physicochemical nature of the finished market preparations and their safety in animals so that they can be safely used in humans.

5.2.2 Methodology

In this study three market preparations of *Rasamanikya* were procured from different pharmaceutical companies. They were coded and subjected for different physicochemical parameters which are nowadays essential for standardization and quality assurance of the finished product. They were also tested for the levels of heavy metals. The formulation containing highest amount of arsenic was subjected to acute and chronic toxicity study.

5.2.3 Results

In a qualitative analysis of all the three market samples of *Rasamanikya*, 21 elements (Ag, Al, As, B, Ba, Ca, Cl, Cr, Cu, Fe, Hg, K, Mg, Mn, Na, S, Sb, Si, Sr, Ti, and Zn) were found in each sample. Quantitative analysis suggests that sample C (67.18%) contains more arsenic in comparison to sample A (58.29%) and sample B (64.42%).

In XRD of sample A, two compounds Orpiment (As_2S_3/*Haratala*) and Quartz (O_2 Si_1) were detected. A total of 33 peaks were observed. The average crystal size of sample A is 110.308 nm and the average lattice strain is 0.0028.

In XRD of sample B, only one compound Arsenolite (As_2O_3) was detected. A total of 17 peaks were observed. The average crystal size of sample B is 167.884 nm and the average lattice strain is 0.0046. In the XRD of sample C, the pattern list was not found. Only two peaks were observed. The average crystal size of sample C is 296.505 nm and the average lattice strain is 0.0391. Sample C on XRD was observed to be amorphous. Hence only two peaks have been detected. The amorphous state is achieved when sulphur or arsenic or arsenic trisulphide is extremely heated or cooled immediately after heating. It causes the presence of moisture which makes the drug amorphous. It also causes the change in colour to dark red. Sample C is also dark reddish yellow with a black tint. The method of preparation of sample C may be faulty. The procedure of preparation of sample C may have been done by heating *Haratala* extremely to prepare *Rasamanikya* or may have cooled the finished product immediately after heating, producing moisture in it which causes an amorphous state. The colour of sample C (dark reddish yellow with a blackish tint) is different from sample A (golden yellow) and sample B (yellow). Overheating causes *Khara paka* (exceeded rough product) which is unacceptable according to Ayurved. Due to overheating, the molecules went away from each other. The particle size is the lattice strain. The average crystal size is finer in sample A (110.308 nm) in comparison to sample B (167.884 nm) and sample C (593.01 nm). It is maximum in sample C. Lattice strain is more in sample C (0.0391) than in sample A (0.0028) and sample B (0.0046). Lattice strain denotes the bonding or strength between the crystals. It is minimum in sample A, more in sample B, and maximum in sample C. The nature of sample C is different from the expected nature of *Rasamanikya*. Hence it is difficult to absorb and may have potential toxic properties. All three samples after the analytical study show huge variations in all respects. Sample C shows a higher concentration of metallic arsenic in comparison to samples A and B. Crystal size in all the three samples did not match each other.

Acute oral toxicity of *Rasamanikya* sample C performed at a limit dose of 2000 mg/kg by OECD 423 was found to be safe. No mortality was observed during the course of the study (14 days). The gross behaviour of all the animals was found to be normal during the period of study. From this, it can be inferred that the LD50 value of *Rasamanikya* is higher than 2000 mg/kg body weight in the rats. The safety of *Rasamanikya* prepared by *Abhraka patra* method after *Shodhana* in *Kushmanda Swarasa* was reported previously (Chaudhary et al. 2018). The marketed *Rasamanikya* sample is also prepared by the *Abhraka Patra* method according to the reference of Bhaishajya Ratnawali (Shastri and Shastri 2008).

In the chronic toxicity study, no behavioural changes were found in all the experimental groups during 90 days and no mortality was observed in any of the groups at TED, 5 TED, and 10 TED levels. Faecal and urine output remained unaffected in all four groups. There was no significant change in body weight during the 90 days of study in all experimental groups in comparison with control group. There was no significant difference observed in haematological and biochemical parameters at the interval of 30 days and 60 days after administration of *Rasamanikya* to Wistar rats when compared among all experimental groups. But the significant decrease in white blood cells, haemoglobin, haematocrit value, and platelet count was observed only at 90 days intervals which were pathologically insignificant. This indicates that *Rasamanikya* has no toxic effect on haematological parameters up to 60 days after administration of *Rasamanikya* with threefold dose levels in Wistar rats. Hence it can be inferred from these results that *Rasamanikya* may not produce a toxic effect on haematological and biochemical parameters with TED, 5 TED, and 10 TED up to 15 days in human beings.

Relative organ weights of all the studied organs are not significant except that of the lungs which is statistically significant but pathologically insignificant when compared among all experimental groups.

In arsenic poisoning brain, heart, lungs, stomach, intestines, liver, kidneys, skin, and nerves are the targets for toxicity. In the present study, minimal to mild changes in the cytoarchitecture of the liver and kidney were observed which are pathologically insignificant. However, the biochemical parameters of the liver and kidney are not significantly increased in all the experimental groups. This indicates that the changes that occurred in the organs are mild and not severe to cause impaired liver and kidney functions.

To rule out heavy metal toxicity due to mineral-containing formulations in Ayurved was the main concern in the present study. It indicates that *Rasamanikya* is safe for 15 days duration in the therapeutic dose in human beings. Hence, if *Rasamanikya* is needed to be used for a longer period, it may be repeated after a washout period of 15 days in human beings as a preventive or safety measure. Chronic arsenic toxicity due to *Rasamanikya* is not possible if it is consumed in a therapeutic dose for short duration, i.e. 10–15 days (Wadnerwar and Patkar 2020).

5.3 Study 2: Efficacy of Gunja (*Abrus precatorius* Linn) Seeds in Inflammatory Conditions of Arthritis

5.3.1 Study Design (Table 5.1)

Table 5.1 Study design

Type of research	Study design	Study drug	Study population	Intervention	Duration	Assessment parameters
Interventional study	RCT	Gunja (*Abrus precatorius*)	Patients ages 21–70 years, suffering from transient mono or bi-arthropathies of the knee, i.e. having a history of pain and swelling of either one or both joints with an unattended short duration of history	Local application	3, 5, or 7 days	DAS 28

5.3.2 Results

The trial drug was found efficacious in reducing pain and swelling in arthritis when it was applied for 7 days. But there were some local untoward effects like rash, itching, and redness which were less complicated and manageable. Hence it was indicated that the application should be retained on the body only for 10 min.

As *Gunja* has been proved efficacious for external application in reducing pain and inflammation, the liniments or spray with the ingredients of *Gunja* may open new dimensions in sports medicine to relieve the pain and swelling instantly.

5.4 Study 3: Role of *Shwetra lepa* in *Shwetra* (Vitiligo)

5.4.1 Study Design (Table 5.2)

5.4.2 Methodology

This study involves preparation of *Shwetra lepa* with its dosage modification and its pilot study in patients of *Shwetra* (Vitiligo). *Shwetra lepa* contains fine powders of purified *Vatsanabha* (*Aconitum ferox*), *Gunja* (*Abrus orecatorius*), *Bhallataka* (*Semicarpus anacardium*), *Chitraka* (*Plumbago zeylanica*), and *Nimba* fruits (*Azadirecta indica*) to be applied by mixing with lemon juice (Mishra 2011).

The application of *lepa* in powder form along with *nimbu swarasa* may have problems such as waste of powder, an uneven quantity of fresh *nimbu swarasa* daily, and differences in consistency of *lepa*. Considering these difficulties, it was decided to modify the powder form used for *lepa* into *lepavati* for clinical use (Fig. 5.1a–c).

Number of patches, region of patches, discoloration and gradation of patch (by VETI score), and photographs of patches were assessed before and after treatment. The participants were assessed for presence/absence of *daha* (burning sensation) and *kandu* (itching) in the patches (Fig. 5.2a–c).

5.4.3 Results

The modified form of *shwetra lepa* by maintaining the pH was very convenient but the duration of application was found insufficient. As *shwetra* is itself chronic in nature, rapidly spreading disease and takes more time for treatment, the drug should be used for 2–6 months (Deshmukh et al. 2020).

Table 5.2 Study design

Type of research	Study design	Study drug	Study population	Intervention	Duration	Assessment parameters
Interventional study	Experimental study	*Shwetra lepa*	Patients with age group of 20–60 years with newly diagnosed white, thin patch or patches on extremities and trunk and having no discoloration of hair were selected for the study	Local application	1 month	VETI score

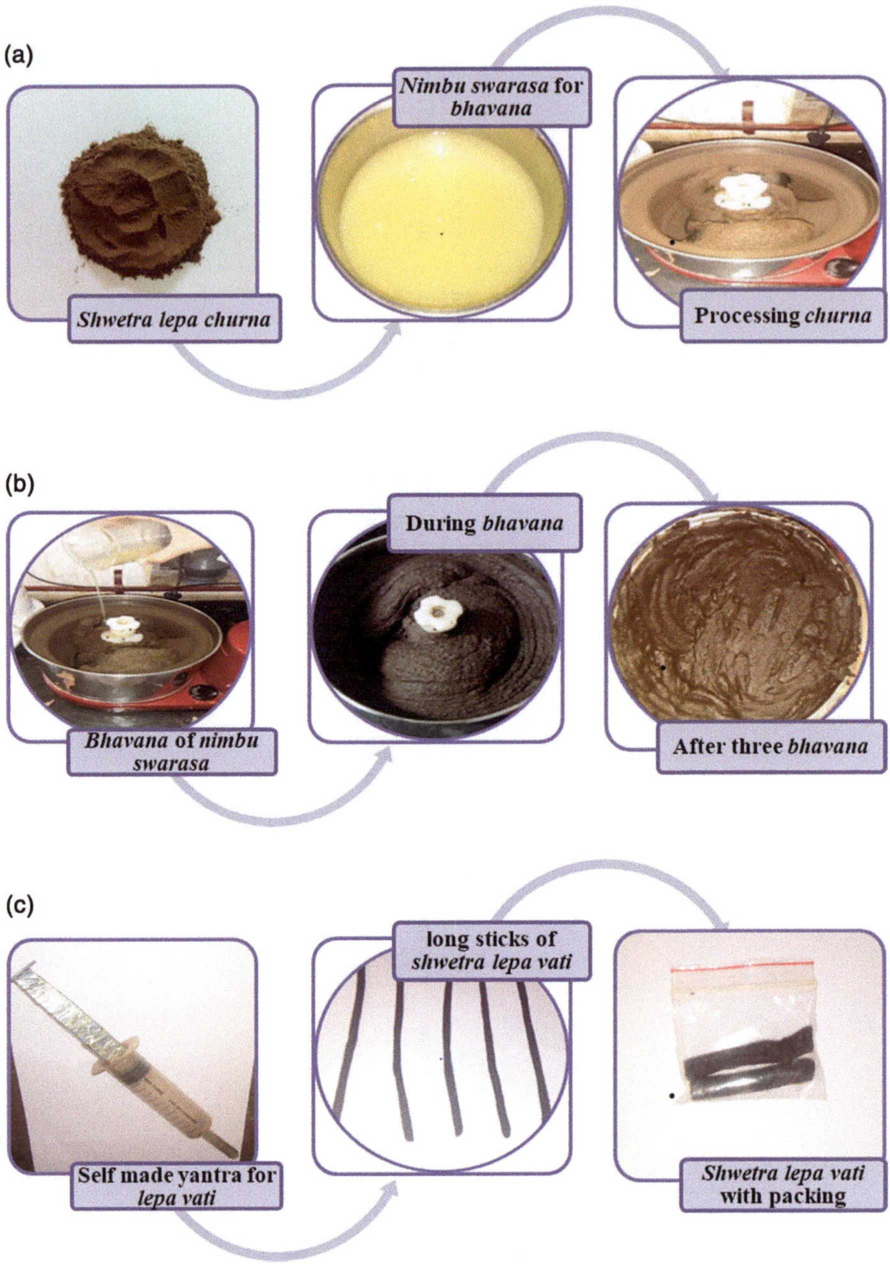

Fig. 5.1 (**a**–**c**) Preparation of *Shwetra lepa*

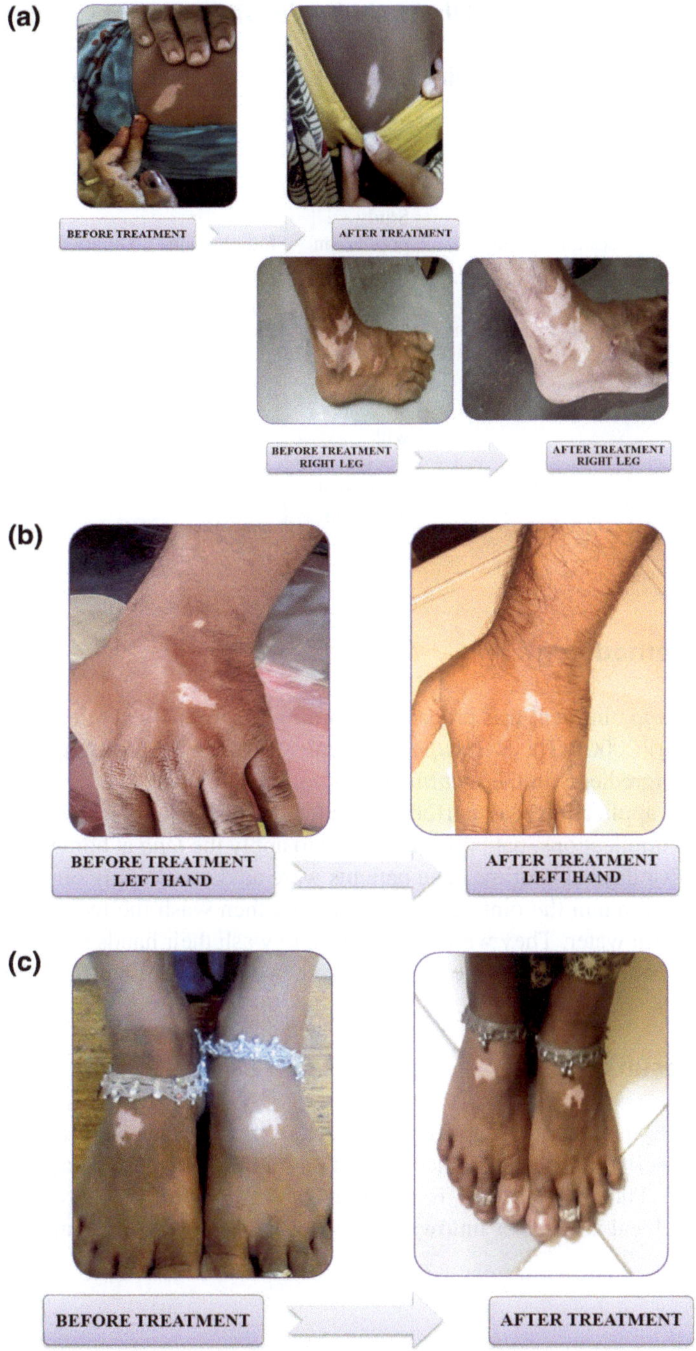

Fig. 5.2 (**a**–**c**) Images of patients showing effect of *Shwetra Lepa* therapy

5.5 Study 4: Role of *Dhattura beeja* Ointment in Cracked Heels

5.5.1 Study Design (Table 5.3)

Table 5.3 Study design

Type of research	Study design	Study drug	Study population	Intervention	Duration	Assessment parameters
Interventional study	RCT	Dhattura seeds ointment	Patients of *Vipadika* having age between 20 and 60 years with cracks associated with pain	Local application	7 days	Number of cracks per foot, and the number of cracks associated with pain per foot and photographs

5.5.2 Methodology

Dhattura beeja ointment was prepared according to the classical reference (Dwivedi 1997; Shastri 2000). The *Sarshapa* ointment was prepared for comparison as it is one of the ingredients in the *Dhattura beeja* ointment.

The first application was performed under the physician's supervision and for subsequent applications, patients were asked to apply the *lepa* at home every day for 7 consecutive days at bedtime. The patients were advised to apply ointment with an applicator and retain the ointment overnight and then wash the feet in the morning with lukewarm water. They were also advised to wash their hands immediately after application and not to touch their eyes and other external orifices (Fig. 5.3a–c).

5.5.3 Results

Dhattura beeja ointment was found efficacious in reducing pain and number of cracks on heel when it was applied once a day for 7 consecutive days with proper precautions. Though *Dhattura* is a cerebral delirium poison, there was not any systemic or local adverse or untoward effect of the drug (Meena et al. 2020).

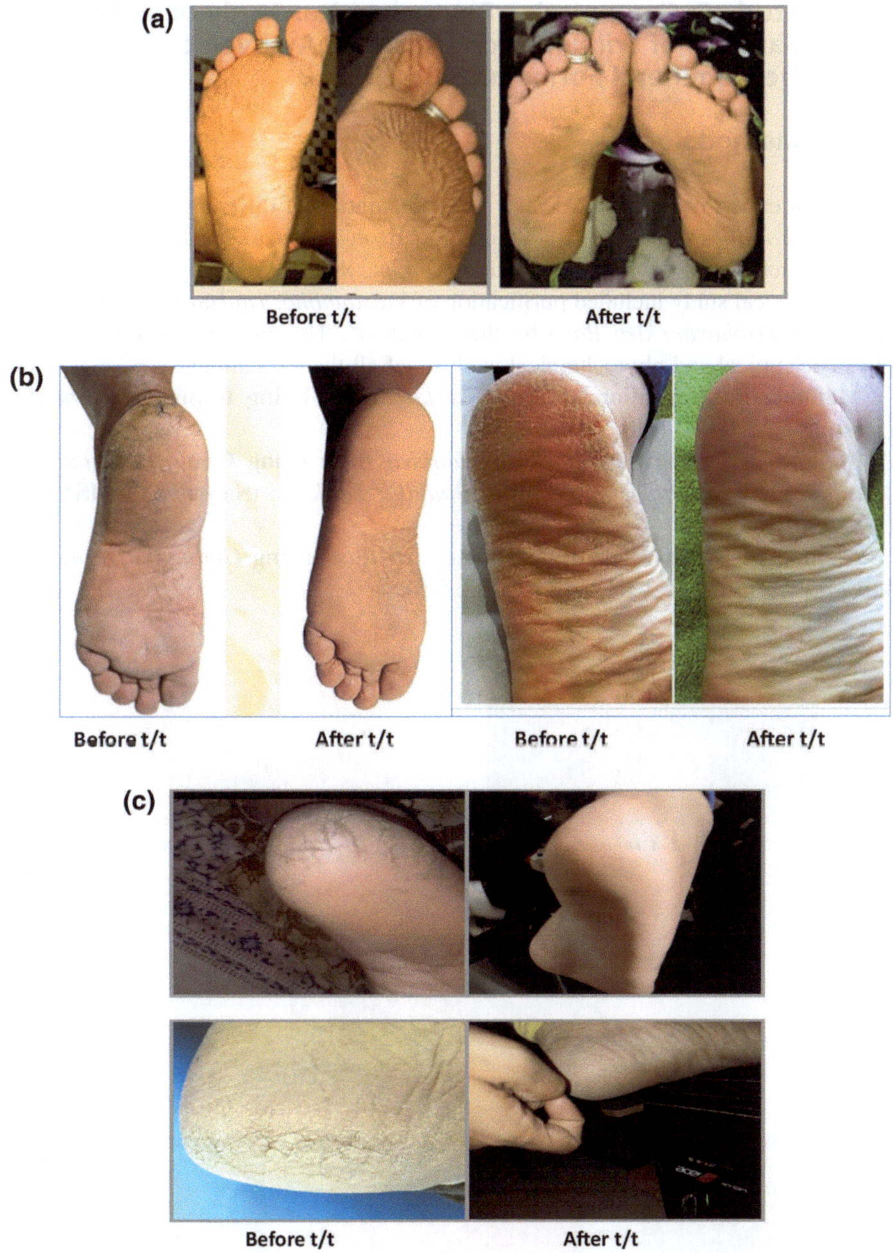

Fig. 5.3 (**a**–**c**) Images of patients showing effect of *Dhattura beeja* ointment therapy

5.6 Study 5: Comparative Phytochemical Study of Different Samples of *Tribhuvankirti Rasa* (an Ayurvedic Herbomineral Formulation)

5.6.1 Methodology

This is an experimental study design with preparation of different samples of self-prepared *Tribhuvankirti Rasa* and their comparison with market samples. The study was conducted in two phases: (1) Pharmaceutical study, (2) Analytical study. The pharmaceutical study included purification of *Vatsanabha, Tankana, Hingula*, preparation of *Tribhuvanakirti Rasa* by three methods. The analytical study included physicochemical and phytochemical analysis of all the six samples.

Method 1: Preparation of *Tribhuvankirti Rasa* Using Impure *Vatsanabha* (Sample 1 ASVTK) (Fig. 5.4).

Method 2: Preparation of *Tribhuvankirti Rasa* using Gomutra *Shoddhita Vatsanabha, Shuddha Hingula,* and *Shuddha Tankana* (Sample 2 GMSVTK) (Fig. 5.5).

Method 3: Preparation of *Tribhuvankirti Rasa* using *Godugdha shoddhita Vatsanabha* (Sample 3 GDSVTK) (Fig. 5.6).

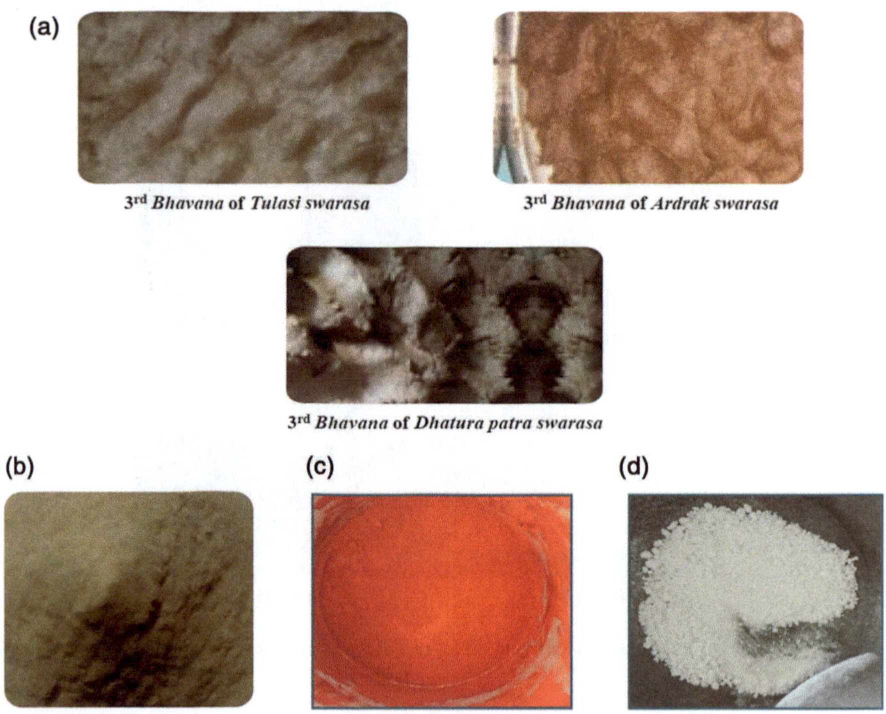

Fig. 5.4 (**a**–**d**) Preparation of *Tribhuvankirti Rasa* using impure *Vatsanabha* (**Sample 1 ASVTK**). (**b**) *Shuddha Vatsanabha*. (**c**) *Shuddha Hingula*. (**d**) *Shuddha Tankana*

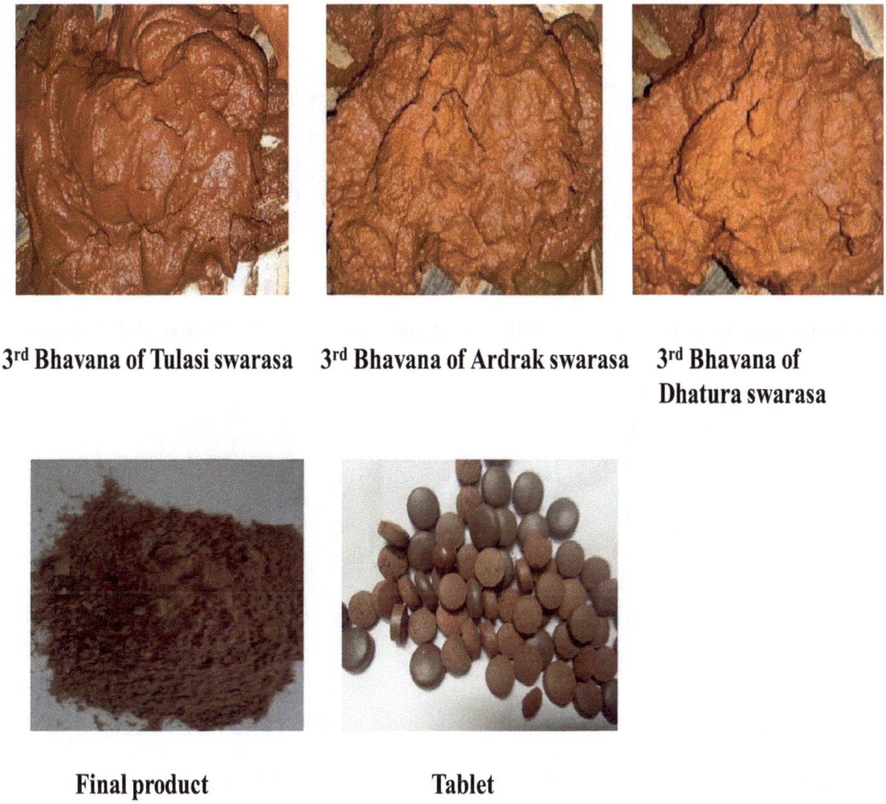

Fig. 5.5 Preparation of *Tribhuvankirti Rasa* using *Gomutra Shoddhita Vatsanabha, Shuddha Hingula,* and *Shuddha Tankana* (Sample 2 GMSVTK)

5.6.2 Results

HPLC graphs indicate that the numbers of peaks are increased in the sample GDSTK and all the market samples in comparison with *Ashodhita* and *Gomutrashodhita* samples of *Tribhuvankirti Rasa* (GMSVTK). This may be due to the effect of the *shodhana* procedure, *shodhana* media, and drugs used for *bhavana* (trituration). The drugs used for *shodhana* and *bhavana* may exert an antidotal effect on the toxic principle in *Vatsanabha* (Aconitine) and mercury. From the phytochemical analysis of all the samples, it can be said that the market samples of *Tribhuvankirti Rasa* may be from *Godugdha shodhita Vatsanabha* (Gaikwad and Wadnerwar 2021) (Figs. 5.7, 5.8, 5.9, 5.10, 5.11, and 5.12).

*Godugdha Shoddhita*Vatsanabha

3rd *Bhavana* of *Tulasi swarasa*

3rd *Bhavana* of *Ardrak swarasa*

3rd *Bhavana* of *Dhatura swarasav*

Final product

Fig. 5.6 Preparation of *Tribhuvankirti Rasa* using *Godugdha Shodhita Vatsanabha* (Sample 3 GDSVTK)

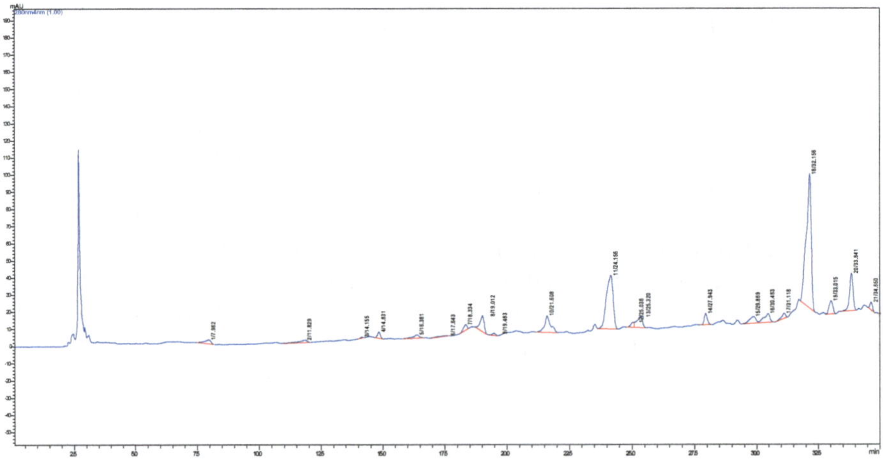

Fig. 5.7 HPLC of ASVTK (*Tribhuvankirti Rasa* prepared from *Ashodhita Vatsanabha*)

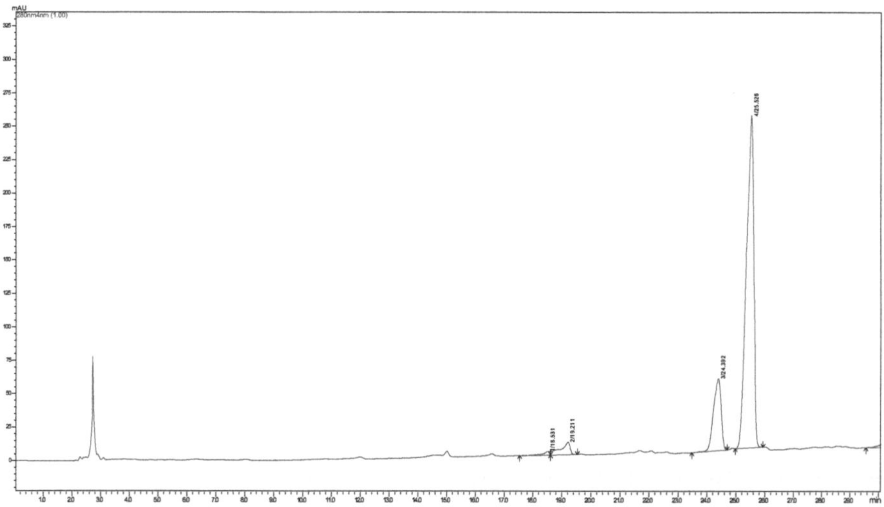

Fig. 5.8 GMSVTK (*Tribhuvankirti Rasa* prepared from *Gomutra Shoddhita Vatsanabha*)

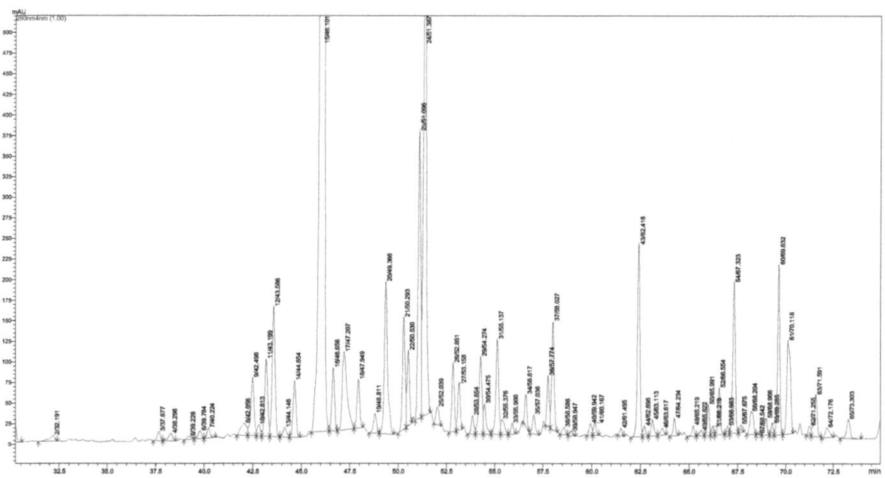

Fig. 5.9 HPLC of GDSVTK (*Tribhuvankirti Rasa* prepared from *Godugdha Shodhita Vatsanabha*)

In the references for the preparation of *Tribhuvankirti Rasa,* it is advised to use *Shuddha Vatsanabha* but media *Shodhana* of *Vatsanabha* is not mentioned in the preparation. The HPLC graphs of *Tribhuvankirti Rasa* prepared with *Godugdha shodhita Vatsanabha* are nearly similar to the HPLC graphs of a market sample of *Tribhuvankirti Rasa*; hence it can be said that the market samples might be prepared from *Godugdha shodhita Vatsanabha.*

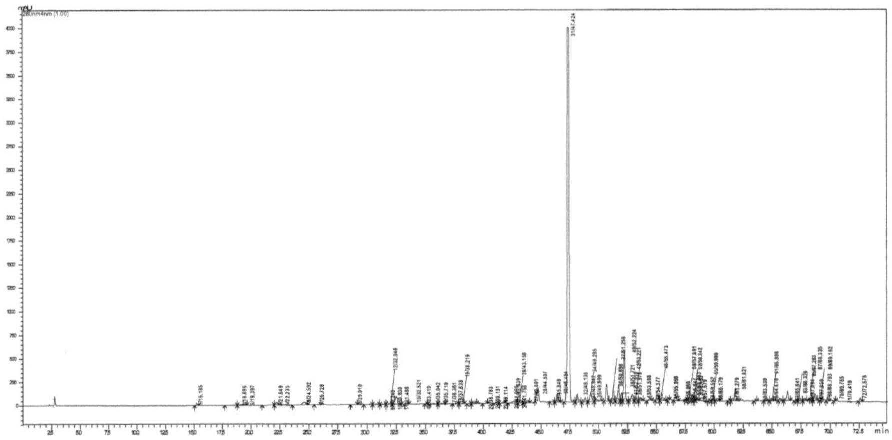

Fig. 5.10 HPLC of MRSTK-A (market sample of *Tribhuvankirti Rasa*)

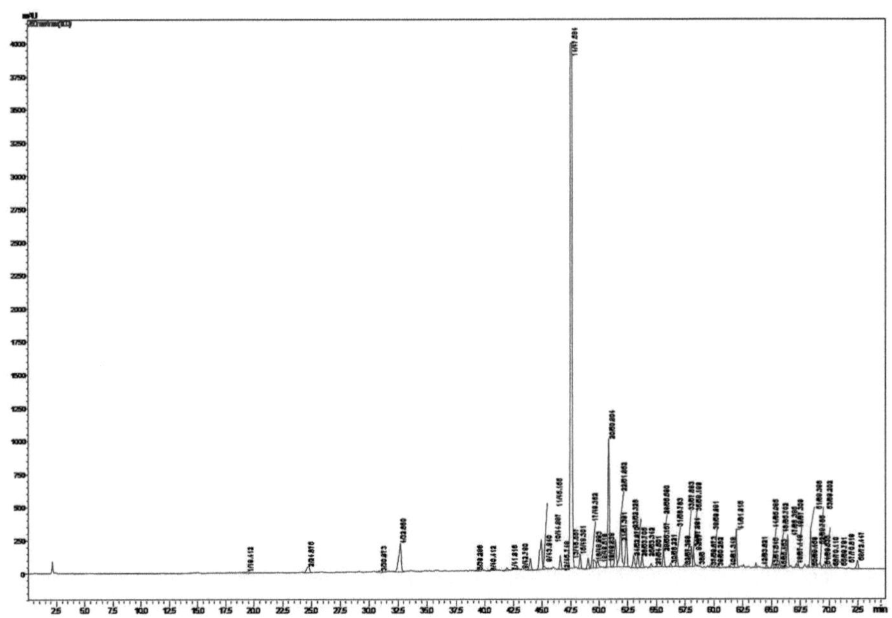

Fig. 5.11 HPLC of MRSTK-B (market sample of *Tribhuvankirti Rasa*)

HPLC has separated the phytoconstituents very well. With the help of HPLC, it is concluded some constituents are decreased and some are added with the effect of *shodhana* and *shodhana* media. In addition, flavonoids, glycosides, and antioxidants are found which are beneficial for health. Thus, HPLC is a sophisticated method of

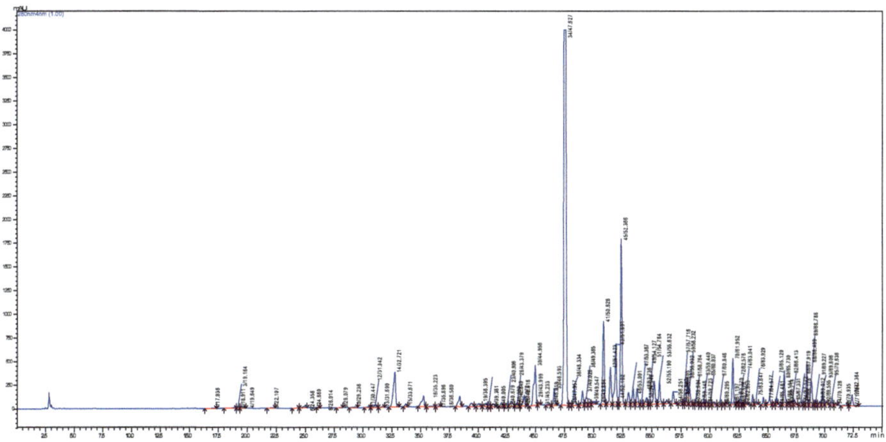

Fig. 5.12 HPLC of MRSTK-C (market sample of *Tribhuvankirti Rasa*)

identifying, separating, and testing the phytoconstituents in the formulation from an analytical perspective.

Acknowledgments I acknowledge the support received from Mahatma Gandhi Ayurved College, Hospital and Research Centre, Salod (Hirapur) as well as Datta Meghe Institute of Higher Education & Research (DU), Wardha, Maharashtra, India.

References

Bhatt K (2000) Siddha Bheshaja Manimala. Chaukhamba Krishnadas Academy, Varanasi, pp 159–160

Chaudhary S, Biradar S, Nariya M et al (2018) Acute and chronic toxicity of *Rasamanikya*, an ayurvedic formulation in rats. Indian J Pharm Sci 80:325–333

Deshmukh A, Wadnerwar N, Gaikwad A (2020) A pilot study on efficacy of Vishakalpa (Shwetra lepa) in Shwetra (Vitiligo). Int J Ayurvedic Med 11(4):747–753

Dwivedi R (1997) Chakrapanidatta. Chaukhambha Sanskrit Sansthana, Varanasi

Gaikwad A, Wadnerwar N (2021) Comparative HPLC analysis of different samples of Tribhuvankirti Rasa. Int J Ayurvedic Med 12(1):84–89

Meena R, Wadnerwar N, Jagtap N (2020) Comparative study of efficacy of Dhattura Beeja Ointment & Sarshapa Ointment in the management of Vipadika (cracked heel). Int J Ayurvedic Med 11(4):716–722

Mishra S (2011) Rasaratna Samucchaya by Vagbhata. Chaukhamba Oriantalia, Varanasi, p 661

Saper R, Kales S, Paquin J et al (2004) Heavy metal content of Ayurvedic herbal medicine products. J Am Med Assoc 292:2868–2873

Shastri K (2000) Rasatarangini. Motilal Banarasidas, Prakashan, New Delhi, p 711

Shastri A, Shastri R (2008) Bhaishajya Ratnawali by Govindadas. Choukhambha Prakashan, Varanasi, pp 895–896

Wadnerwar N, Patkar A (2020) Chronic toxicity evaluation of Ayurvedic arsenical formulation Rasamanikya with special reference to heavy metal toxicity. Int J Pharm Res 12(1):889–898

Ayurveda Intervention in Perimenopausal Syndrome

6

Pratiksha P. Rathod

Abstract

Menopausal transition or perimenopause is the period characterized by irregularities of menses, lengthy and heavy menses with episodes of amenorrhea, sub-fertility, hot flushes which falls under vasomotor symptoms, and psychological symptoms like insomnia. Currently, the menopausal women's count is about 43 million and is estimated to be 103 million in 2026. Western medicine offers Hormone replacement therapy (HRT) which has its limitations. Ayurveda offers efficient, economic, natural, and safer formulations to manage perimenopausal symptoms. *Ashwagandha Ksheer Basti* is the administration of a medication regime per rectum. Ashwagandha has shown anti-stress, anti-inflammatory, immunity-enhancing, and sleep-inducing effects and mood-stabilizing activity in clinical anxiety and depression. A woman suffering from on and off hot flush, insomnia, and mood swings for the last 10 years was administered *Ashwagandha ksheer basti* for 8 days. Within a month the patient experienced adequate sleep, fewer mood swings, and negligible episodes of a hot flush, and serum estradiol levels were found to be elevated post-treatment. Ayurveda considers the base of the reproductive system to be in the bone marrow and female reproductive components are generated and derived from the bone marrow. Recent studies show bone marrow transplantation restores follicular maturation. The intervention category of drug, *Tikta* (bitter) herb, *ksheer* (milk), *Ghee* (clarified butter), and route of administration (*Basti*) is an exclusive treatment for bone marrow-related disorders as per Ayurveda. Signs and symptoms of perimenopause fall

P. P. Rathod (✉)
Department of Prasutitantra-Strirog at MGACH, Datta Meghe Institute of higher education & research, Wardha, Maharashtra, India

Sanskar Kalash, Amravati, Maharashtra, India
e-mail: Pratiksha.rathod@dmiher.edu.in

P. P. Singh (ed.), *Recent Advances in Pharmaceutical Innovation and Research*,
https://doi.org/10.1007/978-981-99-2302-1_6

within the criteria of *Vata & Pitta* disorders and *Ksheer Basti* is a mild *Vata-Pitta* alleviating *Niruha Basti*.

Keywords

Ashwagandha · Hot flush · *Ksheer basti* · Menopause · Perimenopause

6.1 Introduction

Menopausal transition, or "perimenopause," is a defined period beginning with the onset of irregular menstrual cycles until the last menstrual period and is marked by fluctuations in reproductive hormones (Soares and Taylor 2007). This period is characterized by irregularities of menses, lengthy and heavy menses with episodes of amenorrhea, sub-fertility, hot flushes which fall under vasomotor symptoms, and psychological symptoms like insomnia. Four years is an average span during which these symptoms appear before menopause (Baram 1997). The primary source of sex hormones like estrogen, progesterone, and androgens is the ovary. Menopause resulting in infertility secondary to oocyte depletion is the major event. According to a study, perimenopausal women showed greater symptoms as compared to menopausal women. In another study (Taylor and Fonseka 1999). The perimenopausal period has been considered a peak phase for menopausal symptoms. Also, the period of postmenopause showed declining symptoms in some women (Gunaratna et al. 1999). An Indian study found that the perimenopausal group included more prevalent vasomotor symptoms while the postmenopausal age group was suffering from musculoskeletal symptoms in the majority. According to a study (Lagos et al. 1998), a greater prevalence of biological and psychological symptoms has been recorded in women from a low socioeconomic group. Women all over the world now have to spend almost 1/3rd of their lives in menopause years because average life expectancy is increasing. Western medicine offers hormone replacement therapy (HRT) which has its limitations. The Collaborative Group on Hormonal Factors in Breast Cancer showed a significant increase in breast cancer risk in 52,000 women using HRT-Estrogen replacement therapy for >5 years (Bluming 1993). The WHI findings have drawn attention to non-hormonal treatments of hot flushes and other menopausal symptoms. The use of HRT, bisphosphonates, SERMs, and anabolic steroids is not devoid of adverse effects (Freedman et al. 1990).

Menopause has not been reflected as an abnormality in classical Ayurveda texts. The uniform 50 years of age of menopause is mentioned by almost all Ayurveda scholars without any difference of opinion (Sushrut 2006; Vagbhat 2005, 2007; Bhavamishra 2005).

Currently, menopausal women's count is about 43 million and is estimated to be 103 million in 2026 (Bavadam 2012). Menopausal health needs to be prioritized. It is the need of the hour to carry out research for finding efficient, economic, natural, and safer formulations to manage perimenopausal symptoms.

Ashwagandha Ksheer Basti is the administration of a medication regime per rectum. The anti-inflammatory activity of Ashwagandha (*Withania somnifera*) by affecting the central nervous system and the immune system, Ashwagandha is useful in reducing hot flushes. Hot flush involves the inflammation caused by prostaglandins (PG). Enzymes such as cyclooxygenase-2 (COX-2) like enzymes are generated by the PG. A study showed that Ashwagandha inhibited the enzyme activity of the COX-2 enzyme showing its anti-inflammatory activity (Millar 2012). A potent vasodilator, interleukin-8 (IL 8) is produced by macrophages under stressful situations. In a study of 179 women having hot flashes, IL-8 was found elevated suggesting the decline in estrogen is sensed by macrophages. Ashwagandha was found to decrease the genetic expression of IL-8 in the study (Millar 2012). The anti-stress activity was shown by a withanolide-free aqueous fraction which was isolated from Ashwagandha roots (Khare 2007), in a dose-dependent manner, in mice (Singh et al. 2003). The forced swim-induced "behavioral despair" and "learned helplessness" tests were used to compare the antidepressant effects of Ashwagandha and imipramine. Ashwagandha was found to be a mood stabilizer in clinical anxiety and depression (Bhattacharya et al. 2000).

6.2 Case Report

6.2.1 Patient's Information

Age—49 years
Gender—Female
Ethnicity—Indian
Occupation—Nurse
Medical history—K/C/O Migraine and she consumes the western medicines as per need for the same.
Personal history—The patient is having stressful life including night shifts.
Menstrual history—The patient was observing absent menses for 12 months. The period is considered recent menopause but the patient is still under perimenopause conditions.
Clinical examination—When the patient visited the OPD her vitals were stable. Systemic and pelvic examination was performed.
Timeline—The patient experienced on and off hot flush, insomnia, and mood swings for the last 10 years and visited OPD for the same.
Diagnostic assessment—Athens insomnia scale, hot flush scale, panic scale, and serum estradiol levels estimation were performed before and after the intervention of the treatment (Table 6.1).
Screening—Screening for blood sugar levels and hemoglobin was done and found normal.
Therapeutic intervention—Per rectal medicines (*Basti*) containing *Ashwagandhadi* drugs were instilled for 8 days.

Table 6.1 Diagnostic assessment of the patient

Parameters (standard scales used)	Scores before intervention	Scores after intervention
Hot flush	Elevated	Reduced
Insomnia	Elevated	Reduced
Panic	Elevated	Reduced
Laboratory parameter	Date—28.11.2019	Date—6.12.2019
Serum estradiol	75.7 pg/mL	184.4 pg/dL

Follow-up and outcome—within a month a patient experienced adequate sleep, fewer mood swings, and negligible episodes of a hot flush. Serum estradiol levels were found elevated post-treatment.

6.3 Discussion

Ashwagandha ksheer Basti has shown significant improvement in the patient. There is a possibility of insomnia associated with hot flush and occurred as a secondary symptom to it. But the patient reported mixed timings of hot flush and not only at night. Insomnia might be a dominant symptom acting as a stressor for other vasomotor and neurological symptoms.

Ashwagandha is a sleep-inducing herb (Kuboyama et al. 2014). There can be an improvement in estrogen receptor activity or normalization of FSH or direct secretions of beta-estradiol resulting in improvement of symptoms considered as a manifestation of sub-clinical estrogen inadequacy. Ayurveda states that the base of the reproductive system is in the bone marrow and female reproductive components are generated and derived from the bone marrow (Sharma 2017). The intervention category of drug, *Tikta* (bitter) herb, *ksheer* (milk), *Ghee* (clarified butter), and route of administration (*Basti*) is an exclusive treatment for bone marrow-related disorders as per Ayurveda texts (Charak Samhita 2007). It possibly enhances the production of follicles called folliculogenesis and thereby estrogen secretion. Intervention with *Ashwagandha ksheer basti* possibly increases the rate of folliculogenesis or might increase the receptivity of follicles for FSH to enhance folliculogenesis and thereby more estrogen secretion. To observe the estrogen fluctuations in serum by recording two samples, before and post-treatment, this could see the relationship of the drug, its route, and serum estradiol levels, hypothesizing the intervention affects through bone marrow and folliculogenesis for enhancing or at least fluctuating serum estradiol levels.

Signs and symptoms of perimenopause (Santoro et al. 2015) fall within the criteria of *Vata and Pitta* (Neurological and Metabolic) disorders and hence abnormal perimenopause can be seen as an abnormal *Vata-Pitta* condition. The *Basti* (per rectal instillation of drugs) as the route itself is effective and indicated in *Vata* disorders and drug medium *ksheer* (milk) is indicated in *Pitta* disorders. As per Ayurveda scholars, *Vata* resides in bone (*Asthi*) (Ashtanga Hridayam 2011). Also, *Ksheer Basti* is a mild *Vata-Pitta* alleviating *Niruha Basti*. According to Ayurveda commentator Arunadatta, the combination of *Snigdha* (soft), *Shoshana* (absorbing), and *Khara* (rough) properties together will improve the *Asthi Dhatu* (bone system). It can be achieved through *Tikta* (bitter) herbs medicated *with Ksheer Basti.* Usually, *Tikta* substances cause *Vata* aggravation but when it is processed with *Ksheer* (milk) it appropriates the *Vata. Tikta* (bitter) herbs medicated *with Ksheer Basti* are considered the best treatment for joint disorders (Arunadatta 2011). Vitiation of *Vata* leads to not only degeneration of bones like osteoporosis but also reproductive dysfunctions, neurological, vasomotor, and endocrine abnormality as its residence is in the bone as per Ayurveda (Acharya 2003). *Basti* works in the bone, bone marrow (*Majja*) looking after neurological functions. As of now, one must be able to consider menopause syndrome or disorders closely related to bone health.

A hot flush is closely related to Pitta abnormality. Also as it is a vasomotor symptom and is equally controlled by *Vata.* The type of *Pitta* called "*Sadhak Pitta*" is said to maintain psychological health (Sushrut Samhita 1980). The symptoms of menopause like anxiety, panicking, insomnia, and mood swings is a manifestation of abnormal *Sadhak Pitta.* Correction of *Pitta* in the above treatment involves the correction of *Sadhak Pitta.* The principal treatment for bone-related disorders is *Tikta-ksheer basti* meaning the rectal medication containing milk medicated with bitter herbs. Ashwagandha falls under the bitter herb category. Ashwagandha is Balya (strengthening herb). It is rejuvenating and anti-aging (Shah 2005). It is proven for promotive action, sleep induction, and neurodegenerative disorders (Kuboyama et al. 2014). These properties of this herb justify the selection of Ashwagandha as *the* main drug in this case. Recent studies show bone marrow transplantation restores follicular maturation and steroid production in mouse models for primary ovarian failure (Ghadami et al. 2012). Follicular recruitment occurs in the bone marrow before the ovaries. Female-reproductive component (HPO) is a product of bone marrow. There is an inter-relationship of bone—bone marrow and ovaries/follicles. The span of the treatment was of 8 days. On the 8th day, *basti* reaches *the* bone system as per Scholar Text of Sushrut.

Patient Perspective—The patient got satisfied with the treatment outcome. She also experienced that while taking this treatment, she didn't get any gastric-related symptoms that she used to get with other medicines.

References

Acharya JT (2003) Susrutha samhita with Nibandhasangraha commentary of Dalhan, reprint edn. Chaukambha Sanskrit Sansthan, Varanasi, p 101

Arunadatta (2011) Sarvangasundara commentery, ashtanga hridayam, sutrasthana, 11/31 Chaukhamba Orientalia, Varanasi, p 187

Ashtanga Hridayam (2011) Ashtanga hridayam of vagbhat, Vidyotini Hindi Commentary by Kaviraja Atrideva Gupta, Vidyalankara, Bhisagratns, edited by Vaidya Yadunandana Upadhyaya, Chaukhambha Prakashan, edition: reprint 2011, A.Ha.Su11A.hr.su.11/28

Baram D (1997) Physiology and symptoms of menopause. In: Steward DE, Robinson GE (eds) A clinician's guide to menopause. Health Press International, Washington, DC, pp 9–28

Bavadam L (2012) HRT and older women in India. HAI News, 108. http://www.haiweb.org/pubs/hainews/aug1999.html. Accessed 19 Mar 1999

Bhattacharya SK, Bhattacharya A, Sairam K et al (2000) Anxiolytic-antidepressant activity of Withania somnifera glycowithanolides: an experimental study. Phytomedicine 7(6):463–469. https://doi.org/10.1016/S0944-7113(00)80030-6. PMID: 11194174

Bhavamishra (2005) Bhavaprakasha Purva Khanda 3/1. Pandit Shree B. S. Mishra, 9th edn. Chaukhamba Samskrit Samsthana, Varanasi, p 204

Bluming AZ (1993) Hormone replacement therapy: benefits and risks for the general postmenopausal female population and women with a history of previously treated breast cancer. Semin Oncol 20:662–674

Freedman RR, Woodward S, Sabharwal SC (1990) Alpha 2-adrenergic mechanism in menopausal hot flushes. Obstet Gynecol 76:573–578

Ghadami M et al (2012) Bone marrow transplantation restores follicular maturation and steroid hormone production in a mouse model for primary ovarian failure. PLoS One 7(3):e32462. https://doi.org/10.1371/journal.pone.0032462. Epub 2012 Mar 7. PMID: 22412875; PMCID: PMC3296713

Gunaratna C, Fonseka P, Wijeywardene K (1999) Perimenopausal symptoms in Sri Lankan women. Ceylon Med J 44:63–69

Khare CP (2007) Indian medicinal plants: an illustrated dictionary. Springer Science + Business Media, New York, p 719

Kuboyama T, Tohda C, Komatsu K (2014) Effects of ashwagandha (roots of withania somnifera) on neurodegenerative diseases. Biol Pharm Bull 37:892–897. https://doi.org/10.1248/bpb.b14-00022

Lagos X, Navarro N, Illanes E et al (1998) Prevalence of biological and psychological symptoms in perimenopausal women from different socioeconomic levels in the city of Temuco. Rev Med Chil 126:1189–1194

Millar S (2012) Ashwagandha and hot flashes. [Article available on internet] [Last updated Jul 28]. http://www.livestrong.com/article/187966-ashwagandha-hotflashes

Santoro N, Epperson CN, Mathews SB (2015) Menopausal symptoms and their management. Endocrinol Metab Clin North Am 44:497–515. https://doi.org/10.1016/j.ecl.2015.05.001

Shah NC (2005) Bharat Bhaishajya Ratnakara, Part 1, B. Jain Publishers, New Delhi, p 27

Sharma P (2017) Charak samhita of agnivesh, vol II, Commentator—Acharya Vidyadhar Shukla, Acharya Ravidatta Tripathi, Chowkhamba Sanskrit Pratishthan, Delhi, reprinted, Cha. Chi. 15/16, p 361

Singh B, Chandan BK, Gupta DK (2003) Adaptogenic activity of a novel withanolide-free aqueous fraction from the roots of Withania somnifera Dun. (Part II). Phytother Res 17(51):531–536

Soares CN, Taylor V (2007) Effects and management of the menopausal transition in women with depression and bipolar disorder. J Clin Psychiatry 6(Suppl 9):16–21

Sushrut (1980) Sushrut samhita of sushruta, edited by Yadavaji, 4th edn. Chaukhambha Orientalia, Varanasi. Uttartantra, shoshapratishedha, 41/41, p 715

Sushrut (2006) Sushruta Samhita, Sutra Sthan, Shonitvarnaniya Adhyay 14/6, Ambika Dutta Shastri, 'Ayurveda-Tattva-Samdipika' Vyakhya, reprinted 2nd edn. Chaukhamba Samskrit Samsthan, Varanasi, p 48

Taylor DL, Fonseka P (1999) Symptom experience of Filipino American midlife women. Menopause 6:105–114

Tripathi B (2007) Charak samhita of charak. Chaukhambha Sanskrit Pratishthan, Delhi, reprinted 2007, Cha. Su. 28/27, p 550

Vagbhat (2005) 1. Ashtanga Samgraha, Sharira Sthana 1/11, Commentary by Kaviraja Atrideva Gupta, reprinted. Chaukhamba Krishnadas Acedemy, Varanasi

Vagbhat (2007) Asthanga Hridaya, Sharira Sthana, 1/7, Commentary by Kaviraj Atridev Gupta, reprinted. Chaukhamba Surbharti Prakashan, Varanasi, p 170

Part II

Drug Discovery and Development

Innovative Strategies in Drug Discovery and Pharmacoinformatics

7

K. Nagarajan, Richa Goel, Roma Ghai, and Parul Grover

Abstract

Drug discovery comprises all the activities involved in transforming a compound from a drug candidate to a product approved for marketable form and gaining regulatory permission to market it for use in the target indication(s). Bioinformatics, as related to genetics and genomics, is a scientific subdiscipline that involves using software tools to collect, store, compare, analyze and understand biological data and information, such as DNA, RNA, or protein and amino acid sequences or annotations about those sequences. The various innovative approaches in drug discovery include gene sequencing strategies, the role of bioinformatics, tissue expression patterns for target validation, ultra-high-throughput screening for lead identification, biology and chemistry approaches for developability screens, and finally the selection of a candidate molecule for clinical development. The above methodologies are incorporated in this chapter in detail. The overall detail explicates the progression of molecular targets to novel therapeutics under a new paradigm for drug discovery. These details will be useful for researchers working in the domain of drug discovery and pharmacoinformatics and eventually may help bring more effective drugs to patients.

Keywords

Bioinformatics · Clinical development · Drug discovery · Gene sequencing · Target validation · Ultra-high-throughput screening

K. Nagarajan · R. Goel (✉) · R. Ghai · P. Grover
KIET School of Pharmacy, KIET Group of Institutions, Ghaziabad, Uttar Pradesh, India
e-mail: k.nagarajan@kiet.edu; richa.goel@kiet.edu; roma.ghai@kiet.edu; parul.grover@kiet.edu

P. P. Singh (ed.), *Recent Advances in Pharmaceutical Innovation and Research*, https://doi.org/10.1007/978-981-99-2302-1_7

7.1 Gene Sequencing Strategies and Methods

7.1.1 DNA Structure and Significance

All the species including humans have DNA, or deoxyribonucleic acid, as their hereditary material. The DNA molecule is packaged into thread-like structures called chromosomes. The DNA of nearly every cell in a person's body is identical. The majority of DNA is contained in the cell nucleus (also known as nuclear DNA), although a tiny quantity is also present in the mitochondria (where it is called mitochondrial DNA or mtDNA).

Human DNA is made up of around 3 billion bases, which make up the human genome with over 99% of those bases being identical in all humans. Similar to how letters of the alphabet appear in a specific order to form words and sentences, the arrangement, or sequence, of these bases dictates the information accessible for creating and maintaining an organism.

This information in the genome is stored as a code consisting of four chemical bases that provide the instructions for the creation and maintenance of a human being. These four chemical bases are adenine (A), guanine (G), cytosine (C), and thymine (T). They form base pairs of double helixes of DNA when they link with their specific partner (base pair). Adenine (A) and thymine (T) pair together, while cytosine (C) and guanine (G) pair together (G).

Each base has a sugar molecule and a phosphate molecule connected to it and together make a nucleotide. The same framework but without a phosphate group is called a nucleoside. Nucleotides are structured in a spiral known as a double helix, which is made up of two long strands. The structure of the double helix is similar to that of a ladder, with the base pairs forming the rungs and the sugar and phosphate molecules serving as the ladder's vertical side pieces, as depicted in Fig. 7.1.

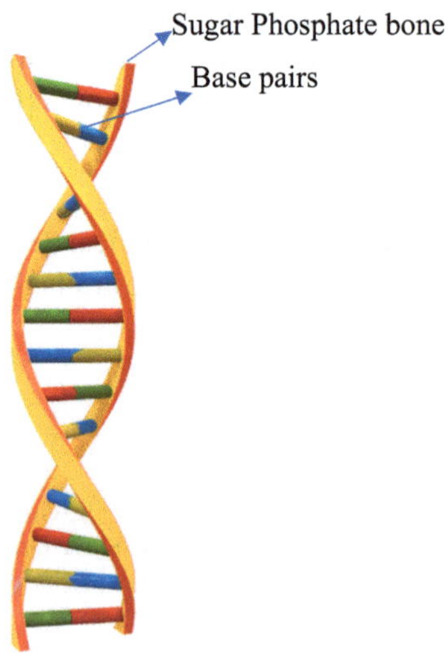

Fig. 7.1 Structure of DNA

For thousands of years, humans have been selecting species with the qualities we want by modifying their gene sequences. Modern genetic manipulation entails scientific processes that either add new DNA to an organism's gene sequences or silence them.

7.1.2 Chemical Synthesis of DNA

DNA is a massive molecule made up of millions of nucleotides. **Oligonucleotides (oligo means few)** are single-stranded or double-stranded DNA or RNA segments that are small in length. Oligonucleotides, unlike DNA, normally include 13–25 nucleotides, but they can also be greater. They rarely reach 200 nucleotides, though.

The ability to create and manipulate DNA sequences has revolutionized the creation of diagnostic tools, medicines, and vaccines for diseases such as COVID-19 that are currently afflicting our civilization.

They are most commonly used as primers for DNA sequencing and amplification, probes for detecting complementary DNA or RNA via molecular hybridization, tools for the targeted introduction of mutations and restriction sites, and the synthesis of artificial genes. Another application is the production of antisense oligonucleotides (ASOs). These are short synthetic nucleotide chains that have been chemically changed and can target any gene product of interest. An ASO is a single-stranded sequence that is complementary to the target gene's transcribed messenger RNA (mRNA) sequence within a cell, thus known to inhibit gene expression through the recognition of cellular RNAs (Rinaldi and Wood 2018; Bennet 2019). The ASO attaches to the target mRNA or pre-mRNA once within the cell, causing it to degrade and preventing it from being translated into a harmful protein product. So, they are a viable therapy option for diseases caused by hereditary factors (Khvorova and Watts 2017).

7.1.3 Methods of DNA Synthesis

DNA synthesizers are specialized machines that have made the synthesis of oligonucleotides comparatively easier. These synthesizers consist of a set of valves and pumps which can be programmed to introduce specific nucleotides in the correct order and also incorporate reagents simultaneously for coupling the nucleotides and developing specific oligonucleotides.

The chemical DNA synthesis can be done by phosphoramidite chemistry which involves the addition of phosphoramidites. Using solid phase synthesis, which involves attaching the growing DNA strand to a solid support, allows all of the reactions to be carried out in a single reaction vessel, the reagents from one reaction step to be easily removed before the reagents for the next step are added, and the reagents to be used in excess to move the reactions as close to completion as possible (Glick et al. 2010). The complete synthesis is as follows:

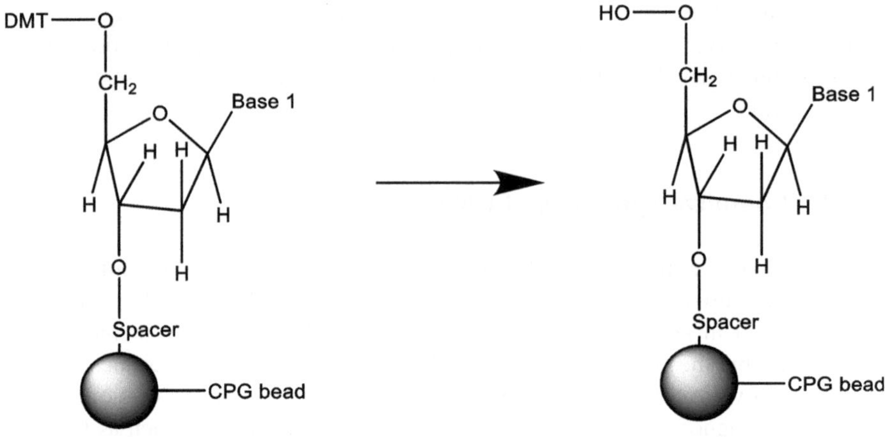

Fig. 7.2 Detritylation to yield reactive OH groups

The first nucleoside (i.e., base and sugar) is attached to inert solid support like controlled pore glass beads via a spacer.

This nucleoside shall be the 3′terminal nucleotide of the synthesized strand. It has a 5-DMT (dimethoxytrityl) group bound to 5′terminal to prevent the 5'-OH group from reacting non-specifically before the addition of the second nucleotide.

Every nucleotide being added to the growing chain is a PHOSHORADIMITE (i.e., having a 5'DMT group as well as a diisopropylamine attached to a methylated 3'phosphite group.

The whole procedure can be represented in chronological order as:

1. Attachment of the first nucleoside to CPG beads via a spacer in the reaction column.
2. Washing of column with an anhydrous agent like acetonitrile.
3. Flushing of the column with Argon to remove the agent.
4. Removal of DMT group, i.e., DETRITYLATION by treatment with TCA to yield reactive 5′-OH groups. This is represented in Fig. 7.2.
5. Washing with the anhydrous agent to remove TCA (trichloroacetic acid).
6. Flushing of the column with Argon to remove acetonitrile.
7. Addition of next phosphoramidite and Tetrazole simultaneously for the coupling reaction.

Tetrazole helps in activation of next phosphoramidite and thus **formation of a covalent bond between 3′phosphite group of phosphoramidite and 5′hydroxyl group of initial nucleosides** (INTER-NUCLEOTIDE LINKAGE). This is depicted in Fig. 7.3.

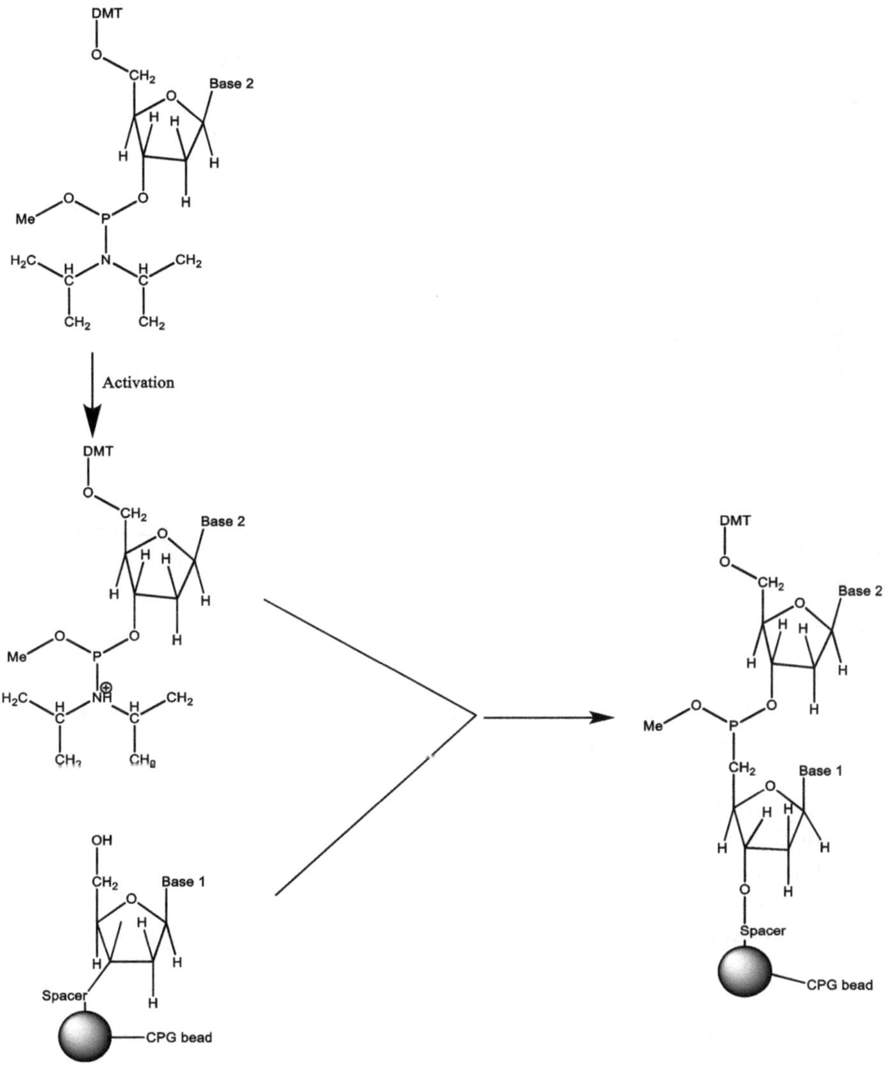

Fig. 7.3 Activation followed by coupling

7.1.3.1 Capping Process

Acetic anhydride and Dimethyl aminopyridine are added to acetylate the unreacted 5′ hydroxyl groups. If capping is not conducted, then the resultant growing chain may differ in both length and nucleotide sequence. Fig. 7.4 shows capping.

7.1.3.2 Oxidation Process

Initially, the internucleotide linkage (which is in the form of a phosphite triester bond) is unstable and is liable to break, so the Oxidation step helps in the formation

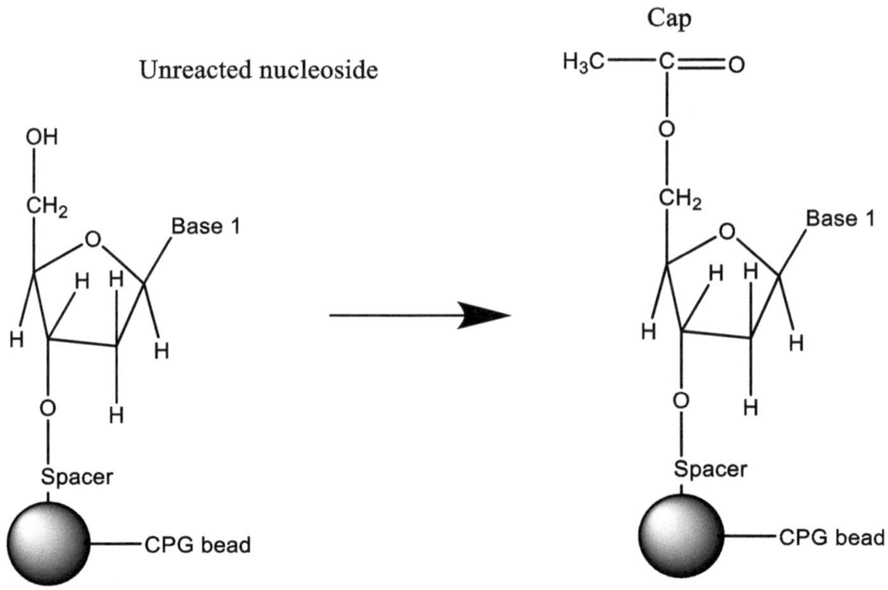

Fig. 7.4 Capping to acetylate the unreacted 5' hydroxyl groups

of a more stable linkage. The phosphite triester is oxidized with an iodine mixture to form a more stable pentavalent phosphate triester. This is represented in Fig. 7.5.

1. Washing of the column.
2. Steps 2–9 are repeated, i.e., washing, flushing, detritylation, phosphoramidite addition and activation, coupling reaction, capping, and oxidation are repeated in chronological order.
3. After n coupling reactions or cycles, a DNA strand of $n + 1$ oligonucleotides is produced.

This newly synthesized DNA strand is bound to the inert solid support, i.e., CPG beads.

Its phosphate triester contains a methyl group, bases like guanine, cytosine, and adenine carries amino-protecting groups like benzoyl, and isobutyl groups and 5'terminus carries a DMT group.

The methyl groups are removed by chemical treatment in the reaction column. DNA strands are cleaved from the spacer.

The oligonucleotide is removed from the column. The benzoyl and isobutyl groups are also removed and DNA is deacetylated (removal of DMT).

4. The 5'terminus of the DNA strand is phosphorylated. This can be done by the addition of the enzyme T4 polynucleotide kinase and ATP or by any chemical method.

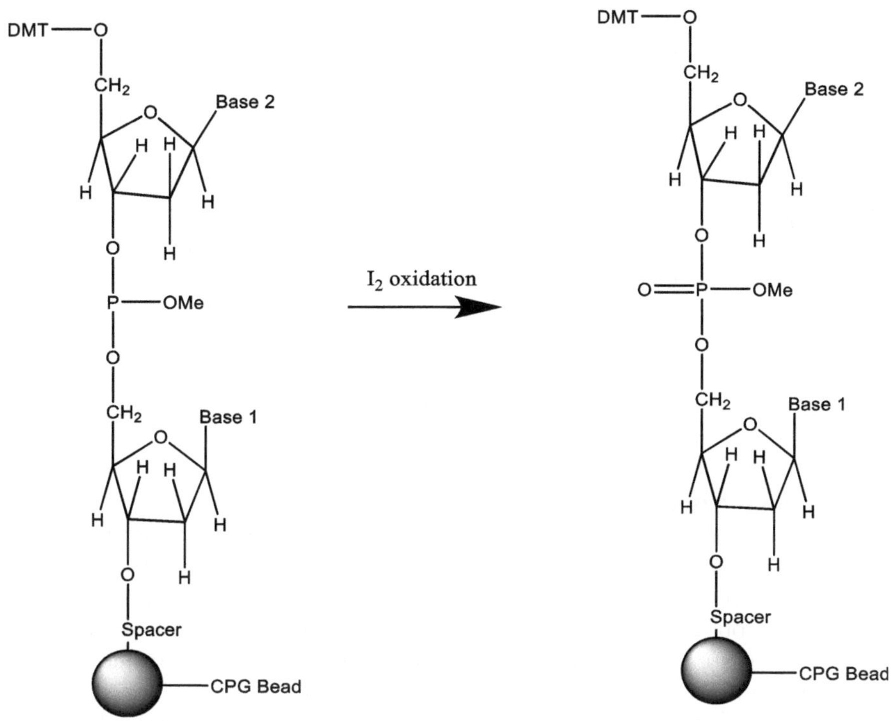

Fig. 7.5 Oxidation of phosphite triester using iodine to form pentavalent moiety

The length of the oligonucleotide is usually denoted by "-mer" (from Greek meros, "part"). For example, an oligonucleotide of six nucleotides (nt) is a hexamer, while one of 25 nucleotides will be called "25-mer".

7.1.4 DNA Sequencing Technique

Nucleotide sequences can frequently be used to infer a gene's function, for instance, by comparing the inferred sequence to sequences of genes with established functions.

The process of determining a genome's complete nucleotide sequence is known as DNA sequencing. In other words, it's a procedure of determining the order of the 4 nucleotide bases, i.e., adenine, thymine, cytosine, and guanine that form the deoxyribonucleic acid which contains significant genetic data.

Complementary base pairing is the basis for the process by which DNA molecules are copied, transcribed, and translated as well as the majority of DNA sequencing techniques.

A DNA segment, a complete genome, or a complex microbiome can be sequenced to reveal the genetic data hidden inside it. Using DNA sequence data, scientists can infer which genes and regulatory instructions are present in a DNA molecule. The DNA sequence can be analyzed to look for gene-specific elements like CG islands and open reading frames (ORFs). Homologous DNA sequences from various organisms can be compared for evolutionary research between species or populations.

DNA sequencing can show alterations in a gene that could cause a disease. So, DNA sequencing has been utilized in medicine for a variety of purposes, including disease diagnosis and therapy, as well as epidemiology investigations. Sequencing has the potential to transform public health and reduce disease breakout risks.

There are many methods of performing DNA sequencing.

1. Fred Sanger method.
2. Using Bacteriophage M13 to obtain single-stranded DNA.
3. Primer Walking method.

7.1.4.1 Fred Sanger Method

The technique of Frederick Sanger is regarded as the preferred approach. Fred Sanger's method uses dideoxynucleotides for sequencing DNA. Unlike deoxynucleotide which has a 3'hydroxyl group on the sugar moiety, dideoxynucleotide does not have a hydroxyl group on 3'C as well as 2'C of the sugar unit. It is a human-made molecule that, if added at the end of the growing chain, stops the DNA synthesis as a phosphodiester bond cannot be formed with the next incoming nucleotide. In this technique, an oligonucleotide primer is fused to a single-stranded DNA template and expanded by DNA polymerase in the basic dideoxy sequencing reaction. This is done in the presence of 4 deoxyribonucleoside triphosphates (dNTPs), one being 35S-labeled. One of the four ddNTPs (dideoxynucleotide triphosphates) which stop elongation when integrated into the expanding DNA chain, is also added in the process. It could be either ddATP, ddCTP, ddGTP, or ddTTP which can be added. The results of the sequencing processes are electrophoresed on a high-resolution denaturing polyacrylamide gel when they have been completed, and the DNA sequence is then visible using autoradiography.

7.1.5 Procedure/Process

The oligonucleotide serves as a primer sequence by supplying a 3'-OH group to start DNA synthesis.

Four different reaction tubes are divided up to hold the primed DNA sample. Four deoxyribonucleotides (one is radiolabeled) and one of the four dideoxynucleotides are added to each tube.

Dideoxynucleotide concentration in each reaction tube is meticulously controlled to assure that it is incorporated into the mixture of developing chains at every conceivable site and not simply at the first appearance of the template's complementary nucleotide (Slatko et al. 2001).

Each reaction tube meets this requirement for the experiment.

As a result, following enzyme-assisted DNA synthesis using DNA polymerase, each reaction tube will contain a different collection of oligonucleotides, each of which comprises a primer sequence.

The addition of formamide stops the DNA synthesis reaction as the chemical substance prevents DNA strands from the base pairing.

Finally, DNA molecules are separated by polyacrylamide gel electrophoresis. Fragments of DNA that differ in size by just one nucleotide can be separated using this method.

Each lane of the gel is loaded with the contents of four reaction tubes that contain one of the dideoxyribonucleotide (ddATP, ddCTP, ddGTP, or ddTTP).

Only the radiolabeled DNA fragments created during the enzymatic DNA synthesis step are visible on an autoradiograph of the gel.

The order of the bands on the autoradiograph, from bottom to top, can be used to determine the sequencing of a segment of a cloned fragment of DNA (Slatko et al. 2001), as depicted in Fig. 7.6.

7.1.6 Using Bacteriophage M13 to Obtain Single Stranded DNA

This technique was an earlier approach that used *Escherichia coli* bacteriophage M13 as a vector. DNA sequencing and cloning have both been accomplished using the bacteriophage M13 vector system. The addition of cloned DNA was made possible by eliminating a non-essential section of the M13 DNA without reducing the infectiousness of fresh virus particles. The M13 system has two benefits: the single-stranded phage DNA may be utilized as a template for DNA sequencing and the double replicative form can be extracted and handled as a plasmid.

The single-stranded M13 cloned DNA construct is combined with a primer oligonucleotide that hybridizes to the vector DNA close to the insertion site, the dideoxynucleotide reactions are carried out, and the sequence of the insert is read from the autoradiographs. The M13 sequencing system is not feasible for pieces of DNA that are $>>$ 5000 bp. It also becomes a time-consuming process.

7.1.7 Primer Walking

Long DNA lengths can be quickly and easily sequenced using primer walking, as depicted in Fig. 7.7.

7.1.8 Pyro-Sequencing

When a nucleotide is integrated into a developing DNA strand, pyrophosphate is released as a result. The pyrophosphate released can be identified using a coupled enzyme reaction in which ATP sulfurylase changes Adenosine 5 phosphosulfate

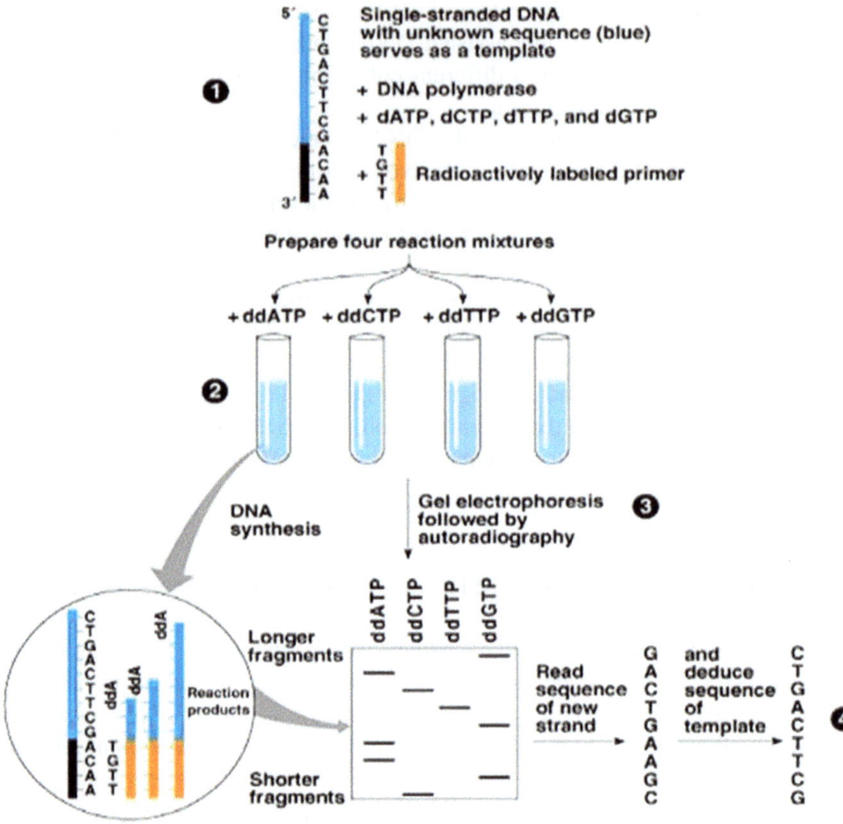

Fig. 7.6 Procedure of conducting Sanger's method

(APS) into ATP, allowing the second enzyme, luciferase, to react with its substrate (luciferin), resulting in visible light that can be detected using a detector. The correct dNTP must be added to the DNA synthesis step for pyrophosphate to be produced. The figure representing the reaction is given below in Fig. 7.8.

7.2 Role of Bioinformatics in Drug Discovery

Bioinformatics has evolved as a combination of Biology and Information Technology. Bioinformatics, as related to genetics and genomics, is a scientific subdiscipline that involves using software tools to collect, store, compare, analyze, and understand biological data and information, such as DNA, RNA, or protein and amino acid sequences or annotations about those sequences. Scientists and clinicians use databases that organize and index such biological information to increase our understanding of evolutionary aspects of molecular biology, and more generally

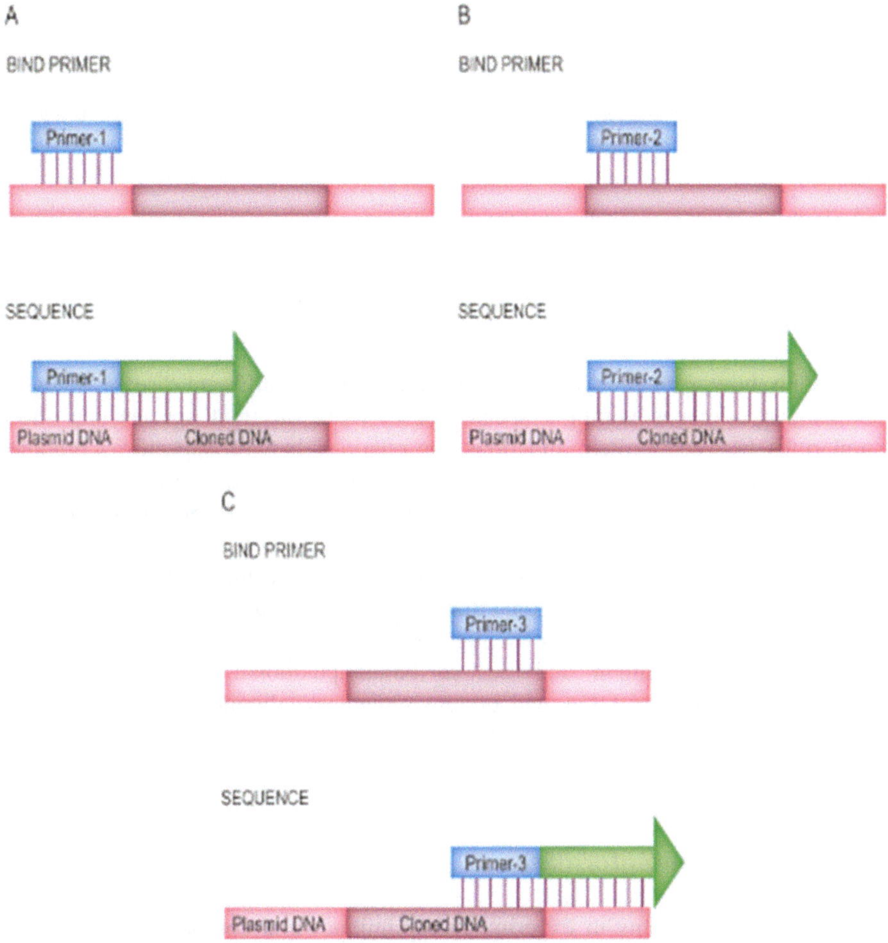

Fig. 7.7 Procedure of Conducting Primer Walking. (**a**) A primer-binding site inside the vector is used to sequence the cloned DNA as far as possible. A second primer can be created using this sequence data and the primer binds close to the far end of the known sequence. This leads to the sequencing of more parts of cloned DNA as depicted in (**b**). The process of developing primers and binding continues till the whole cloned DNA is completely sequenced (**c**). (Source: David and Nanette 2013)

the understanding of health and disease. Moreover, clinical bioinformatics provides benefits of improving healthcare, disease prevention, and health maintenance as we move toward the era of personalized medicine (Teufel et al. 2006).

Bioinformatics is a field of computational science that is particularly useful in comparing genes and other sequences in proteins and other sequences within an organism or between organisms, looking at evolutionary relationships between organisms, and using the patterns that exist across DNA and protein sequences to figure out what their function is. Moreover, genome sequencing and annotation help

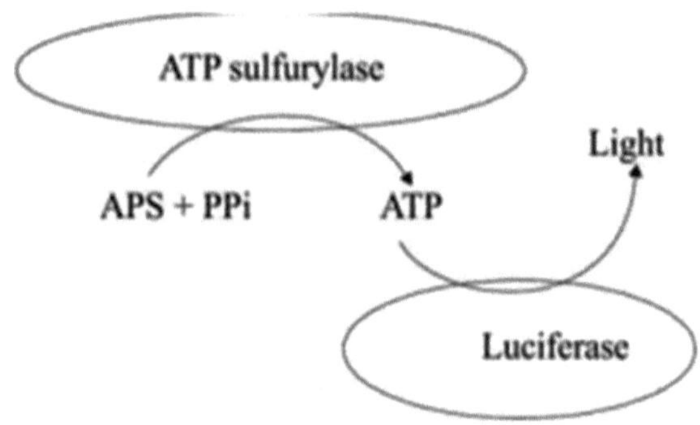

Fig. 7.8 Principle showing pyro-sequencing. (Source: Nagarajan and Bodla 2018)

us to study how normal cellular activities are altered in different disease states. A comparison of genes within a species or between different species can show similarities between protein functions, or relations between species. Computer programs such as BLAST are used routinely to search sequences—as of 2008, from more than 260,000 organisms, containing over 190 billion nucleotides (Rankovic and Morphy 2010; Patrick 2001).

Bioinformatics now entails the creation and advancement of rich sequence databases, algorithms, computational and statistical techniques, and theory to solve formal and practical problems arising from the management and analysis of biological data. Common activities in biocomputational approaches include mapping and analyzing DNA and protein sequences, aligning DNA and protein sequences to compare them, and creating and viewing 3-D models of protein structures (Nagarajan and Bodla 2018).

7.2.1 Protein Bioinformatics

Enormous developments in experimental techniques and computing power have created an unprecedented amount of biological data at the genomic and proteomic levels. Protein bioinformatics refers to the application of bioinformatics techniques and methodologies to the analysis of protein sequences, structures, and functional hypotheses for all known proteins. It helps in detecting the existence of an evolutionary relationship between proteins and exploring how proteins interact with each other and with ligands, both physically and logically. Also, how the knowledge of the three-dimensional structures of proteins can be experimentally determined or inferred, and then exploited to understand the role of a protein. It helps uncover the possibility of designing completely new proteins tailored to specific tasks and the ways to modify the functional properties of proteins (Kalimuthu et al. 2017).

7.2.2 Software and Tools

Software tools for bioinformatics range from simple command-line tools, to more complex graphical programs and standalone web services available from various bioinformatics companies or public institutions.

7.2.3 Computer Representations of Bioinformatics Models (Prejzendanc et al. 2016)

There are 3 classes of formats for representing the Bioinformatics model.

1. Formats with hidden structure.
2. Specialized programming languages.
3. Controlled natural languages.

7.2.3.1 Formats with Hidden Structure

The logical representation of data can be made with the design and analysis of biological models can be done using high-level computer software as follows:

1. E.g., Copasi ML with suitable GUI (Graphical user interface).
2. BioPAX.
3. SBML.

7.2.3.2 Specialized Programming Languages

1. E.g., Jarnae
2. Antimony.
3. MML (Mathematical modeling language).
4. Logo.
5. RePAST.
6. Kappa.
7. Matlab.
8. Octave.

7.2.3.3 Controlled Natural Languages

E.g., ACE,

1. PENG,
2. CPL,
3. Mode Lang.

A brief comparison of various formats is summarized in Table 7.1.

With the completion of the human genome sequence and the advent of high-throughput genomics-based technologies, it is now possible to study the entire human genome and epigenome.

Table 7.1 Brief comparison of various formats

Parameter	Formats with hidden structure	Specialized programming languages	Controlled natural languages (CNLs)
Extensibility	Extensible	Constrained by language designer's concept.	Extensible, constraints required.
Self-documenting	Incapable	Depends on the language	Completely
Popularity	Very popular	Average popularity	Unpopular
Way of use	Database for specialized software	Harder manual development, easier parsing	Easier manual development, harder parsing
Version control systems integration	Usually difficult, but depends on the structure	Easy	Easy
Models scale	Small to large	Small to medium	Small to medium

7.2.4 The Current Bioinformatics Approaches and Tools to Rank Human Disease Candidate Genes Are: (Kubinyi et al. 1998)

7.2.4.1 Approaches Based on Disease Gene Properties

1. Data Type Used:
   (a) Sequence,
   (b) Expression;
2. Training Sets:
   (a) Not applicable.

7.2.4.2 Approaches Using Links between Genes and Phenotypes

1. Data Type Used.
   (a) Sequence.
   (b) Go.
   (c) Literature mining,
   (d) Expression,
   (e) Phenotype.
2. Training Sets:
   (a) Phenotype.
   (b) Go terms.
   (c) Known genes.
   (d) Concept.
   (e) Disease.
   (f) Known genes.
   (g) Disease loci.

7.2.4.3 Approaches Using Functional Relatedness between Candidate Genes

1. Data Type Used:
   (a) Phenotype,
   (b) Go,
   (c) Sequence,
   (d) Protein interactions,
   (e) Expression,
   (f) Pathways,
   (g) Literature mining,
   (h) Mouse phenotype.
2. Training Sets:
   (a) Known genes and disease loci.
   (b) Disease loci.
   (c) Known genes.

Few examples of tools to rank human disease candidate genes are mentioned in Table 7.2.

7.2.5 Software's Used in Bioinformatics (Blass 2015; Larson 2006; Martin 2010)

1. **Grand Unified Schema (GUS) Project:** (http://www.gusdb.org): The GUS platform integrates the genome, transcriptome, and proteome of 1 or more organisms, gene regulation and networks, gene expression, and inter-organism comparisons (Computational Biology and Informatics Laboratory (n.d.)).

Table 7.2 Few examples of tools to rank human disease candidate genes

S. No.	Disease	Disease genes	Drug	Known target genes
1.	Alzheimer's (43 genes)	A2M, ACE, CASP3	Donepezil Galantamine Memantine Rivastigmine Tacrine	ACHE ACHE GRIN 3A ACHE ACHE
2.	Parkinson's (34 genes)	ATP13A2, CYP2D6, CYP2E1 GSTA4 UCHL1 PARK10, PARK12 STAP1	Amantadine Apomorphine Entacapone Tolcapone Memantine Selegiline	DRD1 DRD1, D2, D3, D4, D5 COMT COMT GRIN 3A MAOB
3.	Schizophrenia (70 genes)	BTN 3A1, BTN 3A2, DISC2	Aripiprazole Entacapone	DRD2 COMT

2. **Seq Hound:** This is a freely accessible bioinformatics database warehouse made available to the public through a preliminary web interface but mainly using a remote application programming interface (API).
3. **Cancer Bioinformatics Infrastructure Objects (CaBIo):** (http://ncicb.nci.nih.gov/) CaBIo is the primary architecture for data integration at the National Cancer Institute (NCI). They consist of a variety of NIH datasets, including genomic, expression, pathway, and clinical trial data (National Cancer Institute: Centre for Biomedical Informatics & Information Technology (n.d.))
4. **Atlas:** It is a biological data warehouse that locally stores and integrates biological sequences, molecular interactions, homology information, functional annotations of genes, and biological ontologies.
5. **SRS:** (Sequence Retrieval System): It is an indexing system for flat-file libraries that remains freely available for academics.
6. **K2:** It is a distributed query system that relies on a set of data drivers each of which handles the low-level details of communicating with a single class of underlying data sources such as the BLAST family of similarity search programs.
7. **Discovery Link:** This uses database middleware technology to provide integrated access to biological data sources.
8. **BioMart:** It is a system focused on a large-scale aspect of distributed data integration.
9. **The Bio* Family:** Bio Perl, Bio Java, Bio Python, Bio Ruby, Bio Corba, and Bio Pipe are the family of packages that provide open-source modules and scripts for life science research in the major programming languages of bioinformatics.
10. **GCG:** (Genetics Computer Group): The GCG Wisconsin package is an integrated package of nucleotide and protein manipulation and analysis tools.
11. **EMBOSS:** (European Molecular Biology open software suite): **EMBOSS** handles protein sequence alignment, searching databases with a sequence pattern, protein motif identification, nucleotide sequence pattern analysis, codon usage analysis, identification of sequence patterns in large sequence sets, and presentation tools for publication.
12. **D3C:** (A classification method for Microarrays based on diversity): D3C is a novel ensemble method which is accurately predicting gene classification based on K-means clustering and the dynamic selection and circulating combination.
13. **SVMs:** (Support Vector Machines) The application of support vector machines is used to accurately discriminate between active and decoy compounds. The SVM performance outperforms other well-known methodologies like AutoDOCK Vina.
14. **SOMS:** (Self-organizing Maps): A solution for two protein fold classification is provided by using Kohonen's self-organizing maps. Additionally, they compare their proposal against other machine learning techniques like Fisher's Linear Discriminant Analysis (FLD), K-Nearest Neighbor (KNN), Support Vector Machine (SVM), and Multilayer Perceptron (MLP). Among all other

techniques, SOM obtains better performances than other classification approaches when classifying protein fold with six attributes.

15. **Micro Array and Gene Expression (MAGE):** Standard data model to capture, represent and exchange information from microarray data, allows simultaneous measurement of thousands of genes.
16. **CASPAR:** Exploratory program designed to study the genetics of complex diseases.
17. **Steered Molecular Dynamics (SMD)**: Steered molecular dynamics are used to investigate binding of ligand to biomolecule surfaces. The main idea of using SMD to screen out leads is based on the hypothesis that the larger is the force needed to unbind a ligand from a receptor, the higher is its binding affinity. Thus, instead of binding free energy, the rupture force defined as the maximum on the force-time/displacement profile, is used as a score function.
    SMD is as accurate as the Molecular Mechanics-Poisson-Boltzmann Surface area method in predicting ligand binding affinity, but the former is computationally much more efficient. The high correlation level between theoretically determined rupture forces and experimental data on binding energies implies that SMD is a promising tool for drug design. This may attract attention to recent studies on inhibitors of influenza viruses.
18. **Markov-Randic Indices (or) The Randic-Index:** The Randic index is a well-known topological index (TI) used in QSAR/QSPR studies to quantify the molecular structure represented by a graph.

7.2.6 List of Bioinformatics Software Commonly Used for Data Mining and Analysis in Ruminant Research

1. DAVID.
2. Genespring Gx http://www.genomics.agilent.com/ (Agilent (n.d.))
3. IPA http://ingenuity.com/
4. Genesis.
5. KEGG (General metabolism) http://www.genome.jp/kegg/ (Kyoto Enclyclopedia of Genes and Genomes (n.d.))
6. DIA Dynamic Impact Approach.
7. Metacore.
8. GOSeq http://www.bioconductor.org/packages/2.9/bioc/html/goseq.html (Bioconductor (n.d.))

The softwares used in de novo drug design are mentioned in Table 7.3.

7.2.7 Selected Software Tools for Fragment Based de novo drug design (Anderson 2003)

1. **SPROUT:** 2 steps.

Table 7.3 Softwares used in de novo drug design

S.No.	Software	Application
1	LUDI	Docks and scores fragments
2	GRID	Calculates binding energies for functional groups
3	MCSS	Exhaustive search of binding site for functional group minima
4	CONCERTS	Fills active site with molecular fragments, links fragments
5	Legend	Grows molecule atom by atom
6	Grow Mol	Builds ligands from a library of atom types
7	GROW	Constructs a peptide by residue addition
8	SPROUT	Generates skeletons that fit site
9	CAVEAT	Searches database of small molecules to connect fragments
10	HOOK	Searches database of molecular skeletons for fit to binding site

Step (1): Builds up molecular skeletons that satisfy steric constraints (1° constraints).

(a) Skeletons are constructed by a stepwise assembly of template fragments. Each template fragment represents a set of molecular fragments as it consists of dummy atoms that only have hybridization states assigned but no element types.

(b) Step (2): Element types are assigned to fulfil electrostatic and hydrophobic requirements (2° constraints).

(c) 1° and 2° constraints are derived from a 3D structure of a ligand binding site. SPROUT has been extended to incorporate additional design for synthetic feasibility via virtual chemical synthesis.

2. **TOPAS/Flux:** It is an example of a software tool for ligand-based de novo design.
3. **Skelgen:** Ligand based design in case a 3D model for receptor is not available.
4. **BREED:** Breed overlays known ligands in their receptor-bound state and recombines parts of them to novel potential ligands.
5. **Fragment shuffling:** The idea is to change a part of a known ligand (e.g., a lead compound) by replacing it with a bioisosteric and chemically compatible fragment from other ligands.
6. **SQUIRREL novo:** A strategy to incorporate human expertise as part of the process of computer aided fragment assembly. Most of the de novo design methods deliver proposals for complete ligands. Squirrel novo further evaluates them after a design run with regard to desired **ADMET properties** and **Synthetic feasibility.** SQUIRREL novo is suitable for bioisosteric replacement of side chains and scaffold-hopping.
7. **NP chemical space (Natural Product Chemical Space):** Chemical space describes the set of all organic molecules that are chemically feasible. NP chemical space is the subspace containing all NPs. It is a small but particular fraction of chemical space. In 2008, the dictionary of Natural products lists 215,000 NPs and analogues. Upon continuous discovery of new NPs, we estimate that the NP chemical space to contain in the range of 10^6–10^7 compounds, a small but particular fraction of chemical space, Website: http://molsoft.com/mprop.

8. **Chemicalize.org:** A free web-based drug-likeness tool from ChemAxon. It has tools namely calculation, structure search, document search, and web viewer. The query structures can be submitted using MarvinSketch Java applet. The molecular properties such as polanzability, log P. log D and pKa can be calculated. Website: http://chemicalize.org Chemicalize (n.d.).
9. **PASS:** Prediction of activity spectra for substance: It is a software application (available online and download) to predict the biological activity spectra of drug-like molecules. It estimates the biological activity profile of virtual molecules based on their structure. It also predicts carcinogenicity, mutagenicity, teratogenicity, and embrotoxicity information.
10. **PreADMET:** A web-tool for drug-likeness analysis and absorption, distribution, metabolism, and excretion (ADME) predictions. It also supports in toxicity predictions and molecular visualization. A commercial tool (PreADMET 2.0) is also available.
11. **ALOGPS2.1:** Virtual Computational Chemistry Laboratory (VCCLAB) features the ALOGPS2.1 (interactive on-line predictor) for predicting molecular water solubility, log P. pKa, Log D. Log Wand Log S values. It is developed based on the associative neural network (ASNN). Website: www.vcclab.org/lab/alogps/ (Virtual Computational Chemistry Laboratory (n.d.)).
12. **Commercial Software Tools**: The software tools used commercially for ADME prediction, metabolism prediction, and toxicity prediction are mentioned in Tables 7.4, 7.5, 7.6 respectively.

7.2.7.1 ADME Prediction Tools

Table 7.4 Software tools for the ADME analysis

Software tool (Company)	Features
Carlus2-ADME (Accelrys, Inc., UK)	Predicts solubility, permeability, distribution, and toxicity data
ACDM-Lab (Advanced Chemistry Development, Inc., USA)	LogP. pKa and logD predictions
DEAM (Lion Bioscience AG, Germany)	Predicts intestinal absorption, absorption rate, and first pass metabolism effects
Predictive ADME simulation system Yonsei University, Seoul, Republic of Korea)	Predicts permeability of the compounds and human intestinal absorption (HIA) percentage
OikProp (Shrodinger, Inc.)	Predicts drug-like properties (log S, % HOA, PSA, SASA) including CNS activity (log BB)
GastroPlus (Simulation Plus, Inc., Lancaster, CA)	Predicts molecular properties (solubility. Permeability, absorption, bioavailability, and distribution). It simulates dissolution trans and pharmacodynamics
ADMET predictor (Simulation Plus, Inc., Lancaster, CA)	ADMET property predictor and QSAR model building application
VoiSurf (Molecular Discovery, UK)	Predicts absorption, distribution, metabolism, and excretion (ADME) data

7.2.7.2 Metabolism Prediction Tools

Table 7.5 Software tools for the molecular metabolism predictions

Software tool (Developer)	Features
Metabol expert (CompuDRUG International, Inc., USA)	Predicts the metabolic sites on molecules, site of metabolism and possible metabolites
META Ultra (MultiCASE, Inc., USA)	Predicts the sites of metabolic enzymes attack and metabolites
METEOR nexus (Lhasa Ltd., UK)	Predicts the metabolic fate of drugs
Meta site (Molecular Discovery, UK)	Predicts metabolic transformations related to cytochrome and flavin containing monooxygenase mediated phase-l reactions

7.2.7.3 Toxicity Prediction Tools

Table 7.6 Software tools for the molecular toxicity predictions

Software tool (Developer)	Features
DEREK nexus; deductive estimation of risk from existing knowledge (Lhosa Ltd., UK)	Predicts genotoxicity. Acute neurotoxicity carcinogenicity, hepatotoxicity, toxicity and mutagenicity, teratogenicity, skin sensitization
DS-TOPKAT (Accelrys, Inc., UK)	Predicts carcinogenicity, mutagenicity, LD adverse effects and skin sensitization effects in rodent models
Hazard expert (Compu DRUG International, Inc., USA)	Provides carcinogenicity teratogenicity, indication, mutagenicity, skin sensitization, immunotoxicity, and neurotoxicity data
VITAL (V Life Technologies, Pune, India)	Provides mutagenicity carcinogenicity, and genotoxicity data

7.3 Tissue Expression Patterns for Target Validation

Tissue specificity is an important aspect of many diseases that reflects the potentially different roles of proteins and pathways in diverse cell lineages. Therefore, it is commonly assumed that drug targets are expressed in tissues relevant to their indicated diseases, even under normal conditions (Kumar et al. 2016).

The target identification is performed by mathematical modeling and predictive bio simulation tools. There are two hypotheses involved in mathematical modeling: Top-down hypothesis and bottom-up hypothesis. When a stimulus is presented short and clarity is uncertain that gives a vague stimulus, the perception becomes top-down hypothesis, whereas, the approach wherein there is a progression from the individual elements to the whole is called as bottom-up hypothesis. Top-down mathematical approach covers the whole genome and considered as a "potentially complete" approach in that it deals with all genome wide transcriptomic information. A step-wise summary of the top-down mathematical approach is described below in Fig. 7.9 (Leon and Markel 2006).

7.3.1 Target Validation for Tissue Expression

A crucial step to validate the most promising molecular targets for drug development is to determine their biological relevance for a given disease. The possible strategies that are usually used to understand the function of a specific gene in a cell are: either based on techniques that impair the expression of the candidate target gene, or alternatively rely on the use of products that act by specifically interfering or inhibiting the function, but not the expression, of the final product. In both cases, the resulting phenotype turns out as a powerful source of information on the function of the target protein (Totowa 2007). Schematic representation of the systematic evolution of ligands by exponential enrichment (SELEX) process is given in Fig. 7.10.

7.3.2 Microarray Technology

Microarray technology is expanding rapidly providing an extensive as well as promising source of data for better addressing complex questions involving biological processes. It is an innovative platform to analyze thousands of genes at once. Microarray-based studies have provided the essential impetus for biomedical experiments, such as identification of disease-causing genes in malignancies and regulatory genes in the cell cycle mechanism. Microarrays can identify genes for new and unique potential drug targets, predict drug responsiveness for individual patients and, finally, initiate gene therapy and prevention strategies (Krogsgaard-Larsen et al. 2004). An overview of Microarray technology is represented in Fig. 7.11.

7.3.2.1 Methods Available to Perform High Throughout miRNA Expression Profiling

1. Bead-based methods using flow cytometry.
2. q RT-PCR.
3. Hybridization platforms.
4. Oligonucleotide microarrays.
5. Multi spectral imaging.

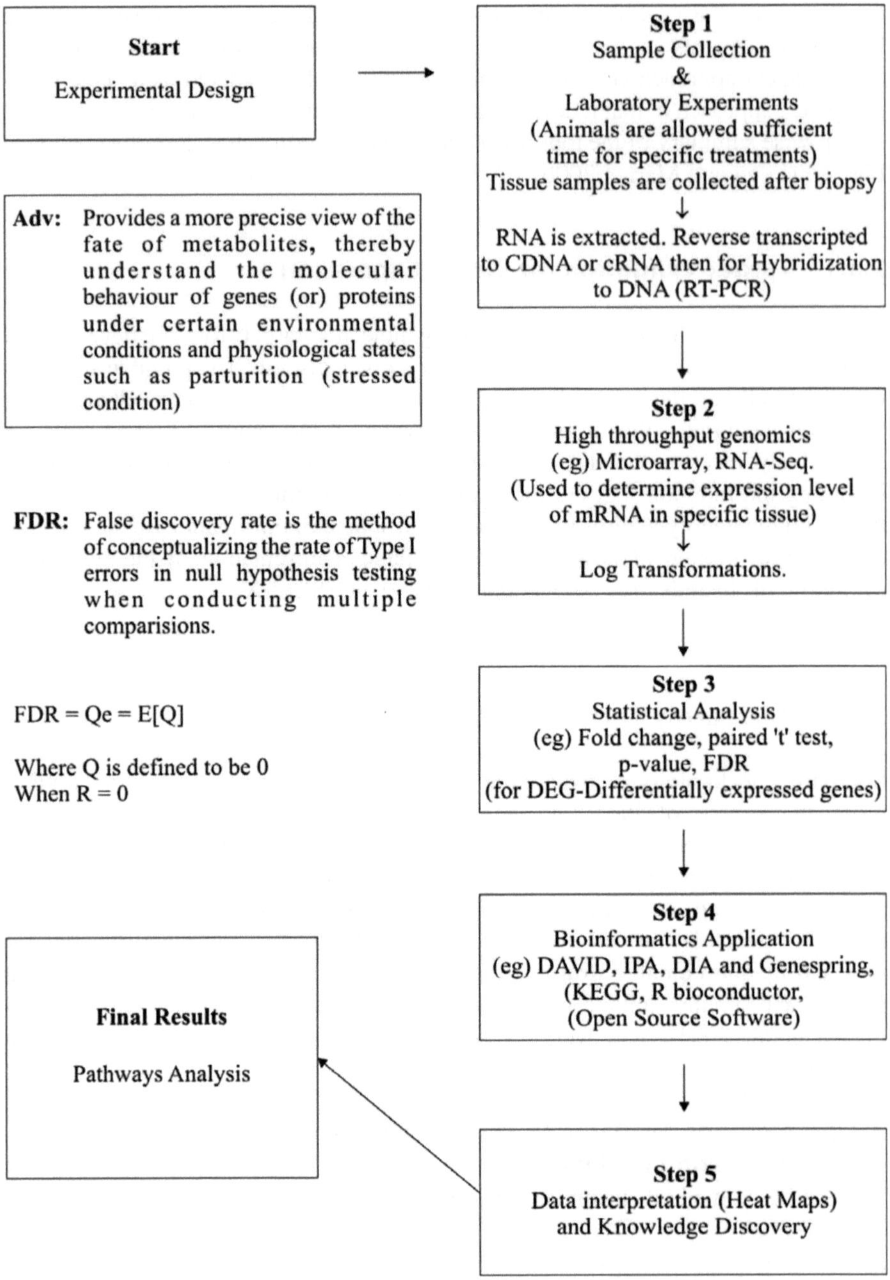

Fig. 7.9 Top-down mathematical approach, e.g., DNA microarrays

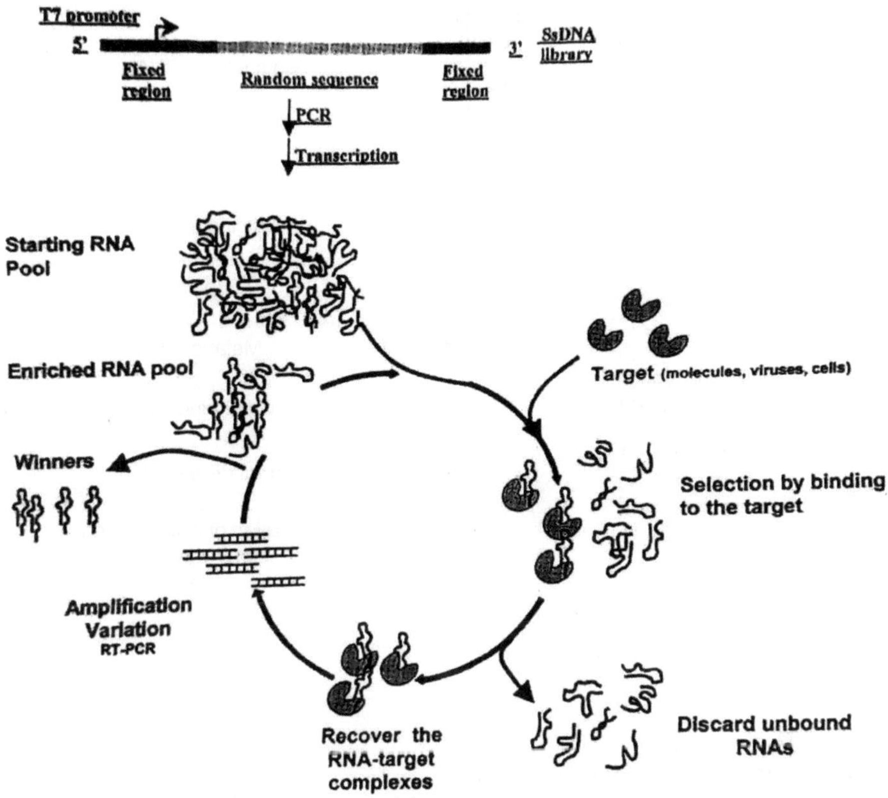

Fig. 7.10 Schematic representation of the SELEX process. The single-stranded (ss) DNA library is amplified by PCR to generate the double-stranded DNA pool that will be transcribed by T7RNA polymerase. The pool of RNA molecules with different conformations will be used for the selection process (Source: Totowa 2007)

7.3.2.2 Microarray Hybridization Platform (Agilent)

Step 1: RNA isolated from the tissue of choice is first labelled and purified.
Step 2: Then hybridized with microarray slides.
Step 3: Scan data and feature extractions are used to measure miRNA expression.

7.3.2.3 Procedure

1. **Isolate.**
2. **Labeling of RNA is done in three steps:**
   (a) Dephosphorylation.
   (b) Denaturation.
   (c) Ligation.
3. **Ligation master mix:**
   (a) 10 × T_4 RNA ligase buffer (2 μL)
   (b) RNase free water (2 μL).

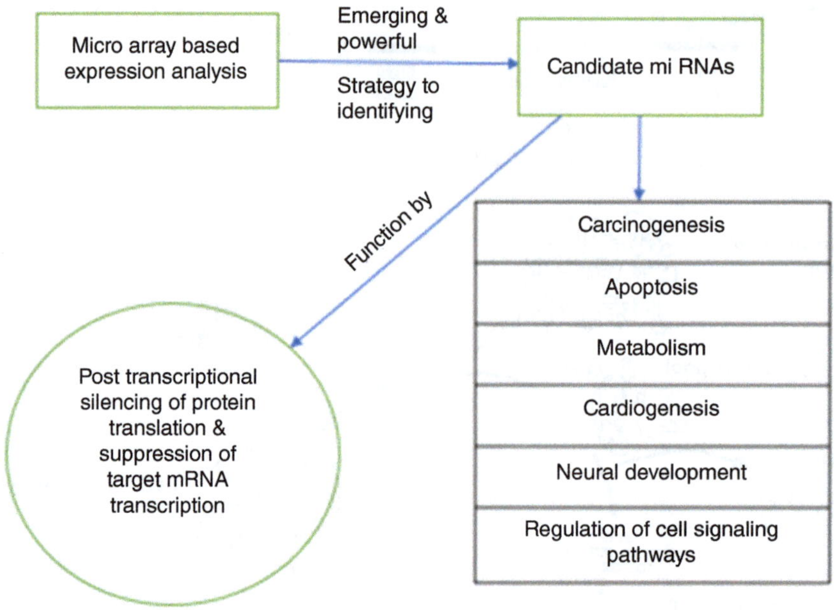

Fig. 7.11 Overview of microarray technology

   (c) pC_p-Cy_3 (3 μL).
   (d) T_4 RNA ligase (20 U/μL): 1 μL.
4. **Purification of Labeled RNA:** Labeled RNA is purified by Micro Bio-spin Six Columns (Gel bed).
5. **Hybridization:** It is performed in two steps:
   (a) Preparation of Hybridization Sample.
   Sample + GE blocking agent + Hi-RPM hybridization buffer + Incubate at 100°C to 5 mt. To hybridize the sample, insert in Agilent microarray Hybridization chamber at 55°C for 20 h at 20 rpm in a rotating hybridization oven.
   (b) Preparation of Hybridization Assembly.
   Place the hybridization chamber cover, rotate, place the assembled slide chamber in Hybridization oven set to 55°C at 20 rpm for 20 h.
6. **Microarray Wash** (Table 7.7):

Table 7.7 Wash condition for washing the microarray slides after hybridization

Dish No.	Procedure	Wash buffer	Time (min)	Temperature
1.	Disassembly	Buffer 1	–	Room temp
2.	First wash	Buffer 1	5	Room temp
3.	Second wash	Buffer 2	5	37 °C

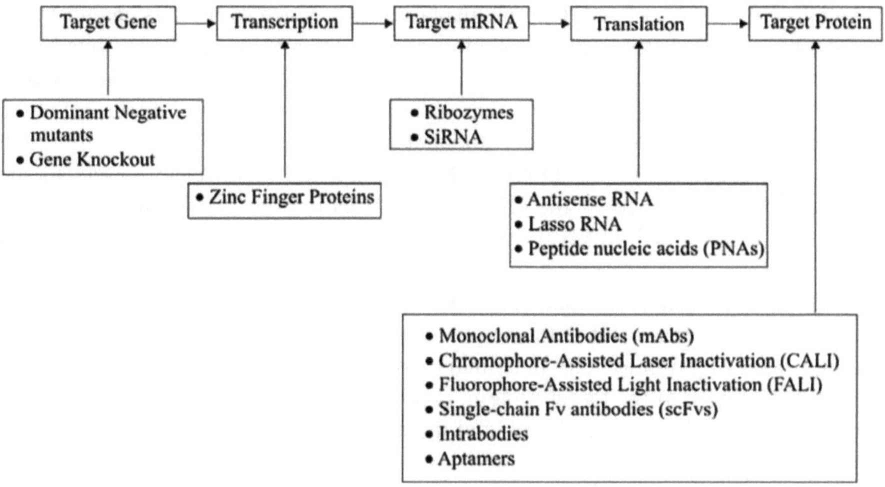

Fig. 7.12 Genomic methods of target validation

7. **Scanning Slides and Using the Feature Extraction Software**
   Scan the slides on scanner settings for miRNA scans. Micro array data analysis begins with feature extraction software and add images to be extracted.
8. **Data Analysis**
   The data files in the .txt format are analyzed using Genespring Gx software.
   Steps of Genespring Gx:
   (a) Summary report as Box whisker plot.
   (b) Experimental grouping.
   (c) Quality control on samples—to evaluate reproducibility data and reliability of micro array data.
   (d) "Filter probesets"—Filtered based on their flag values. (Present, absent, marginal).
   (e) Calculates "Significance analysis" ("t" test/ANOVA).
   (f) Entries satisfying significance analysis are passed on for the "Fold Change Analysis," which helps to identify genes with expression ratios.
9. **Gene ontology (Go) classification scheme:** This allows quick categorization of genes by biological process, molecular function, cellular component. A statistical test performed again and "*p*" value is assigned to each category (Shoaib et al. 2021). An overview of genomic methods of target validation is summarized in Fig. 7.12.

7.3.3 Antisense Oligonucleotides (ASO)

The basis of antisense oligonucleotide (ASO) based therapy is to interrupt the flow of genetic information from a *gene* to a *protein* by using synthetic oligonucleotides targeted for specific mRNA sequences (or) the "Sense" Sequences, by the recognition of Watson–Crick Complementary bases. Since ribosomes cannot translate double stranded RNA, the translation of a given mRNA can be inhibited by a segment of its complementary sequence, the so-called antisense RNA. This result in blocking the translation of the RNA message to generate a specific protein and in the degradation of the mRNA strand by ribonuclease H (RNAase H). "Sense" refers to the original sequence of the DNA or RNA molecule. "Antisense" refers to the complementary sequence of the DNA or RNA molecules. An example of sense and antisense RNA is 5′ACGU3′ mRNA and 3′UGCA5′ respectively (Rinaldi and Wood 2018).

Ideally, an antisense therapeutic is expected to fulfil all of the following criteria: nuclease resistance; lack of cytosine-guanine motifs; acceptable pharmacokinetics and bioavailability; nontoxicity in animal models; transport through cell membrane; maximal inhibition of the target protein; preclinical efficacy; cost-effective scale-up and production; clinical tolerability; and therapeutic benefit over existing treatment modalities.

The criteria of a clinically and commercially compatible ASO delivery system include nanoparticle size (0.1 μm in diameter), biodegradability, high encapsulation/entrapment efficiency, ASO stability and favorable pharmacokinetics, tumor-specific drug delivery, easy formulation, scalability, and safety profiles in animals and humans. Such an optimal in vivo delivery system offers merits that may complement not only the chemically modified ASO but also natural ASO (Michne 2010).

The therapeutic objective of antisense technology is to block the production of disease technology through blockade of the production of disease-causing proteins. Since over expression (or) mutation of specific genes (oncogenes) causes cancer, downregulation of their expression offers the possibility of a selective tumor ablation. To reach this goal, it's necessary for the oligonucleotides not only to have a high and selective affinity toward the target mRNA sequence but also to elude the action of nucleases in cells and body fluids, which rapidly degrade native oligonucleotides, to reach the target cells and to be retained in to them (Johanna and Stefano 2011).

Clinically useful oligonucleotides: The first-generation antisense nucleotides in clinical use are characterized by having one of the phosphate nonbridging oxygen's of the phosphodiester linkages reduced by sulfur. These phosphorothioate linkages lead to resistance to RNAase H, which slows the degradation of antisense oligonucleotide in the cells.

7.3.4 Inserting Antisense into Cells

(a) **Endocytosis:** One of the simplest methods to get nucleotide in the cell, it relies on the cells natural process of receptor mediated endocytosis. The drawbacks to this method are the long amount of time for any accumulation to occur, the unreliable result, and the inefficiency.

(b) **Micro-Infection:** As the name implies, the antisense molecule would be injected into the cell. The yield of this method is very high, but because of the precision needed to inject a very small cell with smaller molecules only about 100 cells can be injected per day.
(c) **Liposome–Encapsulation:** This is the most effective method, but also a very expensive one. Liposome encapsulation can be achieved by using products such as lipofect ACE to create a cationic phospholipids bilayer that will surround the nucleotide sequence. The resulting liposome can merge with the cell membrane allowing the antisense to enter the cell.
(d) **Electroporation:** The conventional method of adding a nucleotide sequence to a cell can also be used. The antisense molecule should transverse the cell membrane offers a shock is applied to the cells.
(e) **Antisense PG gene:** The PG enzyme is responsible for the breakdown pectin. Pectin is a building block in cell walls, and is what gives tomatoes their firmness. In an attempt to slow the softening process, the Flavr Savr employs antisense technology to block PG enzyme production.

Type of Antisense Oligonucleotides: Antisense oligonucleotides can comprise of any type of nucleotide subunit, e.g., they can be DNA, RNA, analogues such as peptide nucleic acids (PNA), or mixtures of the preceding. RNA oligonucleotides form a more stable duplex with a target mRNA molecule, but the unhybridized oligonucleotides are less stable intracellularly than other types of oligonucleotides and oligonucleotide analogues. Different types of antisense oligonucleotides are summarized in Table 7.8.

7.3.5 Identification and Binding of Antisense Oligonucleotide

Algorithms for identifying oligonucleotide sequences with the highest predicted binding affinity for their target mRNA, based on a thermodynamic cycle that accounts for the energy of structural alterations in both the target mRNA and the oligonucleotide, are available. Binding of an antisense oligonucleotide (antisense oligonucleotide analogue) to a target mRNA molecule can lead to the enzymatic cleavage of the hybrid by intracellular RNaseH. In certain cases, formation of an antisense RNA–mRNA hybrid can interfere with correct splicing. In both cases, the number of intact, functional target mRNAs, suitable for translation, is reduced or eliminated.

Antisense oligonucleotides can be expressed from a nucleic acid construct administered to a cell or tissue. Optionally, expression of the antisense sequences is controlled by an inducible promoter, such that expression of antisense sequences can be switched on and off in a cell or tissue. Alternatively, antisense oligonucleotides can be chemically synthesized and administered directly to a cell or tissue, as part of a pharmaceutical composition.

SiRNA: A method of inhibition of a lysyl oxidase type enzyme is RNA interference (RNAi), an approach which utilized double-stranded small interfering RNA

Table 7.8 Various types of antisense oligonucleotide agents

Type	Target	Antisense oligonucleotide agents	Uses
Serine Thereonine Kinases	PKCs	ISIS - 3521 (CGP - 64128) ISIS - 5132	They are the two phosphorothioate antisense oligonucleotides, which hybridize to the PKC mRNA and are in clinical trials with locally advanced (or) metastatic colorectal cancer
Ras-Raf 1 MEK pathway	Ras Raf	ISIS - 2503 ISIS - 5132	Inhibitors of Ras protein expression targeted at H-Ras mRNA. They contain 20 nucleotides for pancreatic carcinoma in combination with gemcitabine, used for colorectal cancer
Family of 25 apoptotic and anti-apoptotic proteins	BCL-2 genes (regulate the release of cytochrome 'c' from mitochondria, which is promoted by pro-apoptotic BCL-2 proteins	Oblimersen sodium	These 18-mer antisense oligonucleotides reduces the expression of anti-apoptotic BCL-2 genes used in combination with dacarbazine to treat malignant melanoma
hTR	RNA subunit of telomerase	GRN-163 L	They are 13-mer oligonucleotide belongs to thiophosphoramidate family targeting the RNA subunit of telomerase called hTR, since hTR is not an mRNA and will not be translated into a protein, these compounds will not have to compete with the ribosomal machinery, its toxicity is low and good selectivity is achieved in chronic lymphocytic leukemia and solid tumors
Delivery system of antisense oligodeoxy nucleotides	HeLA cells	(ASoD ns)-PEI-MWCNT (carbon nanotubes)	ASoDN interacts with positively charged amine groups on PEI -MWCNT

(SiRNA) molecules that are homologous to a target mRNA and lead to its degradation. Qualifying target sequences are selected as templates for SiRNA synthesis. Selected sequences can include those with low G/C content as these have been shown to be more effective in mediating gene-silencing, compared to those with G/C content higher than 55%. Several target sites can be selected along the length of the target gene for evaluation. For better evaluation of the selected SiRNAs, a negative control is used in conjunction, negative control SiRNA can include a sequence with the same nucleotide composition as a test SiRNA, but lacks significant homology to the genome.

The SiRNA molecules can be transcribed from expression vectors which can facilitate stable expression of the SiRNA transcripts once introduced in to a host cell. These vectors are engineered to express small hairpin RNAs (ShRNAs), which are processed in vivo in to SiRNA molecules capable of carrying out gene-specific silencing. Small hairpin RNAs are single-stranded polynucleotides that form a double stranded, hairpin loop structure. SiRNAs, Sh RNAs, and vectors encoding them can be introduced in to cells by a variety of methods, e.g., lipofection. Vector-mediated methods have been developed. For example, SiRNA molecules can be delivered in to cell using retroviruses. Delivery of SiRNA using retroviruses can provide advantages in certain situations, since retroviral delivery can be efficient, uniform, and immediately selects for stable "Knock-down" cells (Nagarajan and Bodla 2018).

7.3.5.1 Automated High-Throughput Target Validation by Using siRNAs

Among multiple approaches, target validation often involves modulation of gene expression. Analysis of changes in gene expression could be achieved by using various functional genomic tools, such as antisense, siRNA, and various viral vectors. Changes in mRNA levels are relatively easy to follow with reverse transcription RT-PCR and microarray assays.

7.3.5.2 Delivery of Biomolecules, Gene Transfection, and as Biosensors

The lipophilic nature of biological membranes is the major checkpoint to the direct intracellular delivery of potential drugs and molecular probes. Small molecular weight nanotubes (SWNTs) are the materials of interest as carriers of biologically active molecules such as small interfering RNAs (SiRNAs), due to their ability to cross the cell membranes. The chemical functionalization of SWNTs with hexamethylene diamine (HMDA) and poly diallyldimethyl ammonium chloride (PDDA) has been performed to obtain a material that can bind to negatively charged SiRNA by electrostatic interactions. PDDA-HMDA-SWNTs loaded with extracellular signal-regulated kinase (ERK) SiRNA were able to cross the cell membrane and suppressed 75% expression of the ERK target proteins in primary cardiomyocytes (Dale and Schantz 2007).

PDDA-functionalized SWNTs have been used as an effective carrier system for application in SiRNA-mediated gene silencing (Table 7.9).

The new wave of interest in the antisense field arises from the discovery that dsRNAs can induce a potent targeted degradation of complementary RNA sequences, a process referred to RNAi, and that the effectors components of the RNAi pathway can be chemically synthesized or expressed from plasmid/viral vectors, similarly to ribozymes.

Table 7.9 PDDA-functionalized SWNTs as an effective carrier system

Delivery system	Biological system employed	Results
SiRNA-PDDA-HMDA-SWCNT	Isolated rat heart cells	PDDA-HMDA-SWCNT bound negatively charged SiRNA by electrostatic interactions
SiRNA-PL-PEG-SWCNT	Human T-cells and primary cells	CNT were capable of SiRNA delivery to human T-cells and PBMCs and caused RNAi of CXCR4 and CD4 receptors
SiRNA/DNA-PL-PEG-SWCNT	HeLA cells	Amine or maleimide terminal of PL-PEG-SWCNT could bind to various biomolecules
TERT SiRNA-SWCNT-CoNH $(CH_4)_6$ $NH_3^+Cl^-$	HeLa cells	TERT SiRNA specifically targeted TERT expression and led to growth arrest of tumor cells

7.4 Ultrahigh Throughput Screening for Lead Identification

7.4.1 Cheminformatics

Chemoinformatics/cheminformatics can be broadly defined as the field of solving chemical problems with computers and pharmacology, where chemical processes are involved or studied. In the chemical industry, computational methods are frequently used, particularly in biotechnology and pharmaceutical research. Additionally, the terms "computer-aided molecular design" and "drug design" are based on these techniques, where one frequently searches for molecules with biological properties. Computational chemistry and chemoinformatics are widely used in fields other than pharmaceutical research, such as the creation of novel biotechnological methods and products.

7.4.2 Developability Screens Predictive Toxicology

The term "developability" encompasses the feasibility of molecules to successfully progress from discovery to development via evaluation of their physicochemical properties. These properties include the tendency for thermal stability, self-interaction and aggregation, colloidal stability and optimization of their properties through sequence engineering. A successful and efficient CMC phase is made possible by choosing the best antibody molecule-based on biological function, efficacy, safety, and developability.

7.5 Biology and Clinical Approaches for Developability Screens

7.5.1 Target Validation

The process of verifying a small molecule's predicted molecular target, such as a protein or nucleic acid, is known as target validation. It can include determining the SAR of analogs of the small molecule; making a drug-resistant mutant of the presumed target; of the presumed target; and monitoring the known signaling pathways that lead to the suspected target (Bentley et al. 2008).

Lack of efficacy and/or toxicity, both of which are frequently attributed to insufficient pre-clinical target validation, are the main reasons why drugs in development fail in the clinic.

The first step in developing a new drug is target validation, which typically takes 2–6 months. To show that drug effects on the target can provide a therapeutic benefit with a reasonable safety window, a variety of techniques are applied during the process. Early, detailed target validation improves understanding of the relationship between target manipulation and disease efficacy, increasing the likelihood of clinical success. The project enters the hit identification phase once a target has attained an acceptable level of validation and disease linkage (Tanner et al. 2016).

Below is a summary of the spectrum of different validation techniques (Fig. 7.13):

Functional Analysis

- Value-adding in vitro assays to characterize pharmacology, measure biological activity of the target, and evaluate the effects of modulating function.
- To demonstrate desired in vitro biological effect, use of "tool" molecules.

Expression Profile

- To ascertain target expression and function in both healthy and disease states, analyze the distribution of mRNA and proteins.
- Using disease-relevant cells or tissue, correlating expression with disease progression or exacerbation.

Cell-Based Models

- Co-culture models and 3D cultures, human stem cells (iPSC), including access to the disease cells.

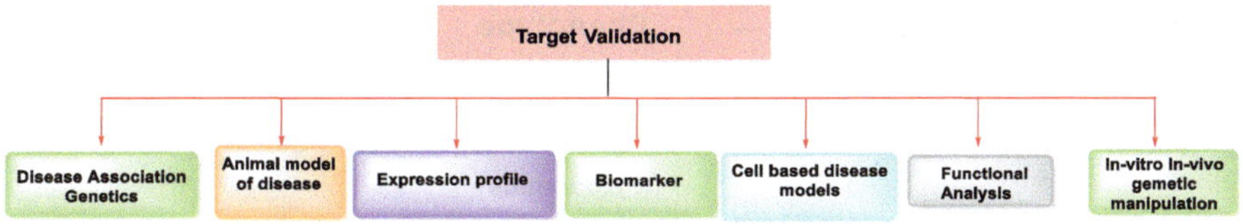

Fig. 7.13 Spectrum of different validation techniques

Biomarkers

- Identification and verification of biomarkers using a variety of methods, such as flow cytometry, protein analyte detection with luminex, and transcriptomics (qPCR platforms).

7.5.2 Functional Genomics

The study of how genes and intergenic regions of the genome contribute to various biological processes is known as functional genomics. In this field, a researcher typically examines genes or regions "genome-wide" (i.e., all or several genes/ regions at once), with the goal of reducing them to a list of candidate genes or regions to further examine.

Understanding how the various parts of a biological system interact to produce a specific phenotype is the aim of functional genomics. Functional genomics focuses on the dynamic expression of gene products in a particular context, such as during a disease or a particular stage of development. In functional genomics, an attempt to build a model connecting genotype to phenotype using current understanding of gene function is made (Franks et al. 2014).

The various specific functional genomics approaches are (Fig. 7.14):

- Genomics and epigenomics (DNA level).
- Transcriptomics (RNA level).
- Proteomics (protein level).
- Metabolomics (metabolite level).

Transcriptomics, proteomics, and metabolomics collectively describe the transcripts, proteins, and metabolites of a biological system, and their integration is anticipated to result in a comprehensive model of the biological system under investigation.

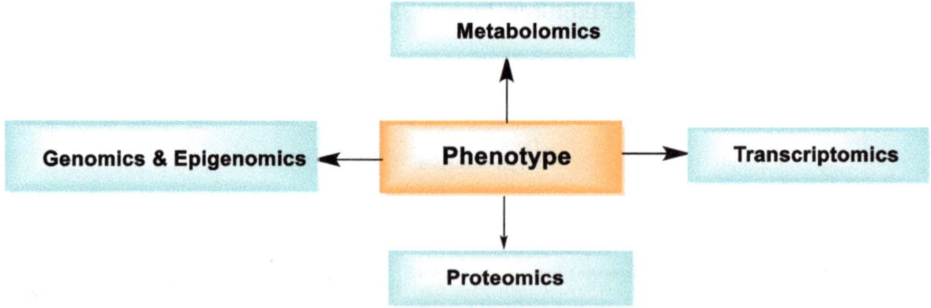

Fig. 7.14 Functional genomic approaches

7.5.3 Common Study Types in Functional Genomics

7.5.3.1 Transcription Profiling

This study type, also known as "expression profiling," is one of the most common. It entails measuring the transcription (RNA) level gene expression of numerous genes in cell or tissue samples. In order to create "snapshots" of expression patterns, quantification can be done by gathering biological samples and extracting RNA (in most cases, total RNA) after a treatment or at fixed time-points in a time-series.

A researcher can decide to concentrate on quantifying transcription of all or a subset of transcripts, genes, coding exons, non-coding RNA, and other biological components for common reference genomes with well-annotated transcripts and genes (such as the human genome) (Ashburn and Thor 2004).

7.5.3.2 Genotyping

Studies known as genotyping are those that reveal variations in a sample's DNA sequence (i.e., genotype). In order to identify differences in the genotype that may account for the difference in phenotype, genomic DNA samples are frequently taken from two contrasting groups of samples, such as drought-resistant rice cultivars and their counterparts that are drought-sensitive.

At three different levels, DNA sequence differences can be detected in genotyping studies:

- **Single nucleotide polymorphisms** (SNPs): SNP analysis focuses on differences in the DNA sequence at the single nucleotide level.
- **Copy number variations** (CNVs): CNVs refer to an increase or decrease in the number of copies of a segment of DNA (e.g., a gene, or a locus-specific DNA repeat element). Each "copy" can be as short as 50 bases or up to 100 kilobases.
- **Structural variations**: they are an order of magnitude larger than CNVs and often cover megabases of DNA, and can be caused by chromosomal rearrangement events.

Genome-wide association studies are one frequent extension of genotyping studies in people (GWAS). The genotyping of samples from cases (such as rheumatoid arthritis patients) and controls (such as healthy people) is followed by statistical analysis to identify SNPs that are significantly more common in one group (e.g., the disease cases). The presence of such SNPs may then indicate a link between the SNPs and disease susceptibility (Marcotte et al. 2016).

7.5.3.3 Epigenetic Profiling

The study of how biochemical modifications or physical interaction of DNA/chromatin affect gene regulation in a cell is called epigenetics. In this, modifications/interactions are not related to changes in the underlying DNA sequence.

At the DNA level, methylation of CpG dinucleotides (commonly found near gene promoters) can be detected by first converting unmethylated cytosines to uracil with bisulfite, allowing methylated and unmethylated cytosines to be distinguished.

At the chromatin level, modifications of the tails of histone proteins (e.g., methylation, acetylation) can be mapped by "immunoprecipitation," where chromatin and proteins are chemically cross-linked reversibly. Then, using particular antibodies raised against the modification or protein of interest, the genomic DNA associated with it is "pulled-down" (precipitated). Following precipitation, the cross-linking is undone to liberate the genomic DNA for additional investigation (Grant et al. 2007).

7.5.4 DNA/RNA–Protein Interactions

Transcription factors, ribosomes, and other DNA/RNA-binding proteins can bind to nucleic acid sequences and influence the transcription and translation of genes. The immunoprecipitation technique has also been applied to study protein binding sites on RNA.

7.5.5 Meta-Analysis

Meta-analysis is a branch of functional genomics in which data from pre-existing experiments is combined to create statistically more powerful models of a biological process. This type of analysis has become popular as it allows the identification of subtle events that could not be detected in smaller studies.

Functional genomics databases such as Array Express and Expression Atlas play an important role in these studies as reliable, well-annotated sources of functional genomics data (Shendure and Lieberman 2012).

7.5.6 Imaging Technologies

A "non-invasive visualisation, characterization, and quantification of molecular and biochemical events that occur at the cellular or subcellular level within intact living organisms" is what is meant by the term "molecular imaging." An understanding of integrative biology, disease characterization, earlier detection, and treatment evaluation are all made possible by this method, which typically makes use of specific molecular probes as well as intrinsic tissue characteristics from the source of image contrast. Additionally, it provides the opportunity for repeated, uniform, non-invasive, and comparatively automated studies of living subjects using the same or different imaging assays at various time points, as well as the attachment of statistics in longitudinal studies, both of which are necessary to reduce the number of animals used in experiments and the associated costs. Conventional imaging is the term used when imaging for in vivo methods depends on gross anatomy. It is now

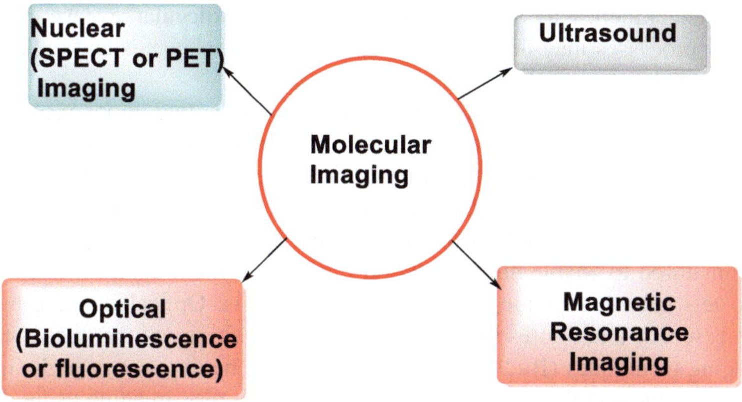

Fig. 7.15 Different molecular imaging modalities

possible to image physiological parameters in living subjects thanks to imaging agents, a process known as functional imaging. Molecular imaging can be used to see drug targets and track the kinetics of medicines that have been taken. The more recent molecular imaging tools need to have high sensitivity and high spatiotemporal resolution to be effective. More target-specific molecular probes must be created to improve the value of molecular imaging for drug discovery and development because there are currently only a few target-specific molecular probes available.

Molecular imaging techniques have an advantage over traditional anatomical imaging in that they can study biological properties in vivo with sufficient temporal and spatial resolution without being invasive. Different diagnostic imaging methods are common in clinical radiology, and they are equally important in the context of experimental research.

There are benefits and drawbacks to each imaging technique, including optical, nuclear, ultrasound, and magnetic resonance imaging (MRI; Fig. 7.15). Imaging techniques such as MRI, ultrasound imaging, single-photon emission computed tomography (SPECT), optical imaging, nuclear imaging, and single-photon emission tomography (PET) can all be used to perform molecular imaging.

To image the body noninvasively, imaging technologies use the interaction of various forms of energy in tissues. While some imaging techniques, like computed tomography (CT) and MRI, rely solely on energy–tissue interactions, others, like PET or optical imaging, call for the administration of optical or radioactive probes. Because different imaging techniques are typically complementary rather than competitive, choosing an imaging modality based primarily on the questions to be addressed for drug development may be a good option. Due to the excellent soft tissue contrast properties of MRI, it is frequently used in pharmaceutical research. It also provides useful physiological data, although in much smaller amounts than nuclear or optical imaging. The traditional anatomical imaging technique, CT, is ideal for morphology-based studies. The SPECT and PET nuclear imaging techniques provide the extremely high sensitivity needed to assess drug distribution and pharmacokinetics as well as image particular molecular events. Numerous molecular processes may be evaluated depending on the ligands and radionuclides used (Kalimuthu et al. 2017).

Fluorescence and Bioluminescence Imaging (BLI), two more recent optical imaging techniques, are particularly useful for mapping specific molecular events and noninvasively tracking cells in living mice. They don't need radionuclides and are also quick, affordable, and cheap. Numerous noninvasive technologies have been created and are currently being used in clinical settings, but recently they have been miniatured to enable high-resolution imaging of small animals, which can be used to test new therapeutics in small animal models. Optical imaging based on bioluminescence and fluorescence has the highest sensitivity of the various molecular imaging techniques (Rudin, 2009). In addition to having a very high signal-to-noise ratio, optical molecular imaging also offers multiplex imaging using different probes with different optical spectra and requires the least amount of money to install the instrument. These advantages make optical imaging the most widely used preclinical study technique. The absorption and scattering of light signals on their way to the detector system, which makes it challenging to see an animal's internal organs and prevents further clinical applications, is a significant drawback of this technique. To get around these restrictions, researchers have worked to create reporters that emit photons at a longer wavelength, like red light, which is more effective at penetrating tissues (Massoud and Gambhir 2003; Ottobrini et al. 2006; Gross and Piwnica-Worms 2006; Weissleder and Pittet 2008).

Additionally, compared to its planar counterpart, a quantitative three-dimensional (3D) image of optical imaging signals offers more accurate biological information. Advanced mathematical algorithms that localize the source position and resolve photon scattering deep within the tissue are used to generate the 3D image. Nuclear imaging methods, like PET and SPECT (with nanomolar blood concentrations of administered radiotracers), provide the necessary 3D distribution of the administered tracer and have high sensitivity, resolution, and tissue penetration depth. They may be able to identify cellular and molecular alterations brought on by diseases. These benefits enable the use of these imaging techniques in clinical and experimental settings. Quantitative nuclear imaging makes it possible to evaluate treatment effectiveness, and kinetic modeling makes it possible to measure the metabolic rates of both diseased and healthy organs (Jaffer et al. 2009; Lu et al. 2011; Pichler et al. 2008).

A magnetic field and radio waves are used in the MRI technique to produce fine-grained images of the organs and tissues inside a body. Without posing a radiation risk, MRI simultaneously conveys molecular and anatomical information. Diffusion-weighted imaging (DWI) uses Brownian motion's variability to its advantage and sheds light on how water molecules diffuse through tissues (Massoud and Gambhir 2003; Lecchi et al. 2007). The primary use of perfusion-weighted imaging includes measurements of cerebral hemodynamics at the microcirculation level and the assessment of ischemic conditions. With the help of magnetic resonance (MR) spectroscopy, tissues can be examined to determine the concentration and presence of different metabolites like choline, creatine, and *N*-acetyl aspartate (Rossi et al. 2010). High-resolution anatomical data as well as a lot of information on physiological, biological, and metabolic structures have been made available by these functional MR techniques. MRI is several orders of magnitude less sensitive than nuclear or optical imaging at obtaining molecular information, despite having high tissue contrast and anatomical resolution. The concentrations of the lesion-targeting molecular probes must be higher than the tracer levels due to this lower sensitivity. The development of methods to increase the sensitivity of MRI, such as specialized coils, stronger magnetic fields, conditional MRI contrast agents, or activatable probes, requires more effort (Velde et al. 2009; Hartman et al. 2008; Cosentino et al. 2009).

Pharmaceutical research might find solutions using the molecular imaging method that has high specificity, sensitivity, and temporal and spatial resolution. One molecular imaging technology cannot be the best for all potential applications because each modality has specific advantages and disadvantages. To overcome the limitations of single-modality imaging, particularly in vivo, multimodal imaging tools combine technologies like CT, PET, and MRI. Since the introduction of hybrid PET/CT and PET/MR in the late 1990s, the latter's success has made the use of the PET system on its own impossible. The hybrid PET/MR approach, which has the advantages of no additional radiation and high tissue contrast compared to the combined PET/CT method, rectifies the shortcomings of standalone PET and MRI and is used in clinical practice and preclinical research. Because hybrid technologies can incorporate traditional MRI or functional MR spectroscopy, it is anticipated that translational research and preclinical drug discovery and development will benefit greatly from and see an increase in the use of PET/MR (Boss et al. 2010; Zaidi and Prasad 2009).

7.5.7 Proteomics

The study of the proteome is known as proteomics, and it makes use of technology to recognize and count the various proteins, as well as the interactions between proteins and nucleic acids that occur within the proteome, as well as the post-translational alterations that influence the activity of proteins. Recently, computational proteomic technologies have advanced over many other complementary techniques. This makes it possible for researchers to examine many proteins in clinically distinct samples, which aids in the discovery of disease biomarkers, the identification, and validation of drug targets, the design of more potent medications, the evaluation of drug efficacy and patient response-basically, in interfering with nearly every step in the current drug discovery process. Finding an unstable protein that is having an undesirable effect and using a molecule to change its effect are both steps in the proteomic approach to drug discovery. Aspects of biology, chemistry, engineering, and information science are combined by proteomics and applied to all areas of drug discovery (Burbaum and Tobal, 2002). Since proteomics technologies face numerous obstacles, their development will necessitate the simultaneous development of several techniques. Proteomics relies heavily on two-dimensional gel electrophoresis and mass spectrometry, but other technologies are also available and required.

7.5.8 Use of Proteomics to Identify Biomarkers

Biomarkers are biological parameters that can be measured and quantified, such as pathogenic processes, environmental exposure, or pharmacologic responses to a drug. Small circulating proteins or peptides from degraded molecules in various disease states are the signatures of disease, and proteomes are the end product of interactions between genetic background and environmental factors. RNA-based expression studies may not be able to detect changes in post-translational modifications, cellular trafficking, or even total expression levels, but proteomics. Mass spectrometry and two-dimensional gel electrophoresis both have important roles in proteomics, but they are not the only tools available or required. Some of the many excellent results of biomarker discovery are listed below and condensed in Fig. 7.16. It can offer specific details regarding the presence of a disease and/or the stage of a disease that facilitates early disease diagnosis. During preclinical or clinical development, proteome analysis may enable the identification of potential markers for the efficacy prediction of drugs (Zhou et al. 2020).

7.5.9 Metabolomics

An approach for the identification and semi-quantification of all known and unknown metabolites (small molecules with a weight below 1500 Da) within a biological system or systems (i.e., a cell, tissue, organ, biological fluid, or organism) is known as metabolomics, and it is an interdisciplinary field (Viant 2007). The small metabolites found in biological fluids, tissues, and environmental samples may

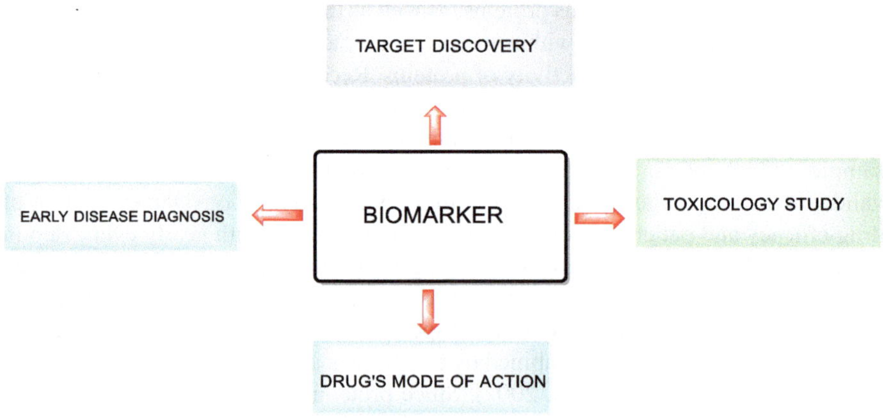

Fig. 7.16 Potential use of biomarkers

occasionally come from an exogenous source (e.g., food additives, plasticizers, cosmetics, and personal care products). The term "xenometabolome" refers to the combination of the metabolome, xenobiotics, and their end products from phases I and II metabolism (Al-Salhi et al. 2012).

Currently, high-resolution mass spectrometry (HRMS) and nuclear magnetic resonance (NMR) spectroscopy are the two primary analytical methods used in metabolomics. HRMS is frequently used in conjunction with contemporary separation techniques like liquid chromatography (LC), gas chromatography (GC), or capillary electrophoresis (CE). The most common method among all of these is probably LC-HRMS, which typically uses an electrospray ionization (ESI) source. It's increased sensitivity when compared to NMR analysis and the ESI's soft ionization process, which enables structural elucidation, are two of the primary causes.

A literature survey recently conducted by Miggiels et al. (2019) shows the increasing adaptation of metabolomics, a trend that was previously observed by Kuehnbaum and Britz-McKibbin (2013). The trends visualized show that LC-MS is recently a dominant technique, while CE-MS is rather a niche technique. Untargeted and targeted experiments in metabolomics can typically be separated into two groups. Untargeted metabolomics seeks to unbiasedly identify and characterize as much of the metabolome as possible (Patti et al. 2012). A specific group of compounds is being quantified using real standards in targeted studies (Soule et al. 2015). Untargeted metabolomics can measure thousands of features at once, whereas targeted approaches only cover a small subset of metabolites (Naz et al. 2014). Consequently, this method offers a wealth of information and the chance to discover new biomarkers.

A mass spectrometry (MS)-based approach's development and use present a few difficulties: The choice of a thorough extraction method can be difficult because (I) metabolites can differ greatly in their chemical properties and concentrations; (II) chromatography's ability to resolve many molecules at once makes high-throughput analyses difficult; and (III) the choice of ionization coverage and efficiency may also present difficulties. For analytes with specific physical and chemical properties, commonly used techniques like electron ionization (EI), atmospheric pressure chemical ionization (APCI), and ESI work well. For instance, ESI is more appropriate for polar metabolites, whereas EI (without sample derivatization) is more appropriate for nonpolar compounds. In addition, the existence of substances with identical masses, such as isomers and isobars, makes confident metabolite identification challenging.

7.6 Clinical Development

Clinical development is a way to transform scientific acumen into medicine. It is the stage at which all the information collected from the fundamental research and preclinical animal studies is applied to clinical practice to examine if the scientific data can be translated into a useful new treatment for patients. The clinical development program's primary goal is to generate clinical data to support product labeling, which ultimately instructs the physicians as well as the patients on using the medicine safely and effectively.

Medication is a drug used to diagnose, cure, treat, or prevent disease. Various aspects of a drug need to be studied before it comes into the market, viz. ADME (absorption, distribution, metabolism, and excretion), dosage, dosing regimen, indication, mechanism of action, drug interactions, warnings/ precautions, contraindications, effectiveness, and adverse effects. All of these data must be derived via a development effort aimed at elucidating these special attributes.

The clinical development program is carried out not only for a drug or medicine but is also required for any intervention like a diagnostic product, vaccines, medical devices, and any intervention that could be used to modify health-related processes. Clinical development should be conducted in such a manner meeting the requirements of the regulatory bodies that shall be approved for the marketing of a product.

Bringing novel pharmaceuticals to market is not only difficult, but also expensive, time-consuming, and resource-intensive. Between 2009 and 2018, data on novel pharmacological agents authorized by the US Food and Drug Administration (USFDA) were studied to estimate the amount of money spent on research and development to bring a new drug to market. Along with public statistics on clinical trial success rates, data were obtained from the US Securities and Exchange Commission, FDA database, and ClinicalTrials.gov. The average cost of producing a new medication has been a source of contention, with estimates varying from $314 million to $2.8 billion in recent years (Wouters et al. 2020). The number of new medications licensed for sale climbed by 60% between 2010 and 2019, reaching a high of 59 in 2018. Emerging biopharmaceutical businesses account for 80% of the whole research pipeline and generated 14% of revenue in 2018. Many early-stage trials from small and medium-sized companies have been added to the vaccine pipeline over the last two decades during the COVID-19 pandemic period (Hwang and Kesselheim 2016).

7.6.1 Phases of Clinical Development

Clinical development program consists of various trials that are conducted in human beings for investigating efficacy and safety of an intervention.

Trials are conducted in phases with specific objectives. Each phase is treated as a separate clinical trial.

Typically, these phases have been carried out in a step-by-step fashion, with choices to go on to the next stage once the previous one had been finished successfully.

These phases are:

1. Phase I:
2. Phase II.

3. Phase III.
4. Phase IV.

Depending on the nature of the medicine and its therapeutic use, the scope and duration of these studies will vary significantly.

Clinical trials in different countries are approved and monitored by the following different regulatory bodies:

(a) India: DCGI office under CDSCO.
(b) UK: MHRA advised by CSM.
(c) USA: FDA.

A clinical development plan is used to chart the development of a therapeutic substance from its early Phase I research through its commercialization. It contains a summary of pre-clinical findings and a summary of market research done for the drug. The clinical development plan includes mainly the target dates of the studies, the objective of each study, and the proposed study design for achieving the objective. If any drug successfully passes through Phase I followed by Phase II and then Phase III, it will usually be approved by the national regulatory authority for use in the general population. Phase IV is the post-approval studies.

Any company must do significant pre-clinical investigations before beginning clinical trials on the medications. This research includes both in vitro (test tube) experiments and animal species testing. To acquire preliminary effectiveness, toxicity, and pharmacokinetic information, a wide range of dosages of the test medication is given to animal subjects or an in vitro substrate, as well as to aid pharmaceutical manufacturers in selecting if it is beneficial to proceed with additional research.

After completion of pre-clinical studies, IND needs to be filed for taking approval to conduct clinical studies. The first step is the submission of [CFR 21 Part 312] (IND) application (Investigational New Drug) to the regulatory authorities like FDA requesting permission to move a clinical candidate into human study. IND provides regulatory agencies with detailed preclinical data describing: –

Animal pharmacology and toxicology studies.

Chemical manufacturing information includes formulation, stability studies and Quality Control measures.

Detailed clinical protocols that describe how the clinical compounds will be studied in human populations if the studies are approved.

7.6.2 Phase I: First in Human Trials

The first step in drug research is to determine whether or not a medicine is safe for people. The initial stage of testing in healthy human volunteers is known as Phase I trials.

Normally, just a limited number of healthy volunteers, such as 20 to 80, are chosen. This phase includes tests to determine a drug's safety, tolerability, and pharmacokinetics as well as an assessment of the maximum tolerated dose. The maximum tolerated dosage (MTD) is the level at which unbearable side effects and/or unacceptable safety findings occur. The MTD must be determined in order to calculate the therapeutic window, which is the dosage range within which the medicine is effective while being safe and well tolerated.

These studies are frequently carried out at an in-patient facility where the participant may be monitored by full-time professionals. The individual who receives the medicine is frequently kept under observation until many half-lives have passed. Dosage-ranging studies, also known as dose escalation studies, are also included in Phase I trials to determine the optimal dose for therapeutic usage (Casazza et al. 2022). Healthy volunteers are frequently included in Phase I studies. They include young men and women (mainly men) of age 18–45 years.

In cancer and HIV treatment studies, patients are also used. Real patients are employed in some situations, such as when patients are at the end of their lives and have no other therapeutic alternatives. All research participants are compensated for their time spent at the trial facility with an inconvenience fee.

There are several types of Phase I trials.

(a) **SAD (Single ascending dose):** In this study, a single dosage of the medication is given to a small sample of participants who are then followed and assessed for a length of time. The dose is increased and a new set of participants is given a larger dose if there are no serious side effects and the pharmacokinetic data are approximately in line with projected safe values (Shafiq et al. 2014). This process is repeated until predetermined pharmacokinetic safety levels are attained or severe adverse effects appear (at which point the drug is said to have reached the maximum tolerated dose).
Each group normally has 6–8 people in it. The greatest dosage at which no adverse effects were observed in the most sensitive species evaluated in toxicological studies should be used to determine the doses to be used in SAD trials. The beginning dosage should be 10 times lower than the NOAEL (no observed adverse effect level).
(b) **MAD (Multiple ascending doses):** These are conducted to better understand the pharmacokinetics and pharmacodynamics of multiple doses of the drug. ADME (i.e., absorption, distribution, metabolism, and excretion) data can be obtained from the collected blood samples from a group of patients who are receiving multiple low doses of the medicines. Samples of blood and other fluids are collected at various time points and then analyzed how the drug is processed within the body. This dose is subsequently increased for further groups up to a predetermined level.
(c) **Food Effect:** A short study to see if there are any changes in drug absorption by the body, caused by eating before the drug is given. These trials are typically conducted as a crossover study, with participants receiving two equal doses of the medicine on two separate occasions: one while fasting and the other after eating (Hoover et al. 2016).

Phase I should address questions like how fast the drug is absorbed, in which tissues drug gets distributed, which organ/s involved in metabolism and how quickly the drug is eliminated from the body.

Regulatory bodies can impose a clinical hold for reasons of safety or because of sponsor failure to accurately disclose the risk of study to the investigators.

7.6.3 Phase II Trials (First in Patients)

Phase II trials are conducted on a larger number of patients after Phase I trials have validated the study medicine's initial safety. They are aimed to determine how effectively the drug works. Adaptive designs are proposed in Phase II studies in order to assess earlier determination of failure thereby help to decrease the development costs (Van Norman 2019).

Phase II studies are designed to establish clinical efficacy, quantify the occurrence of adverse events, define the optimal therapeutic dosage, and give complete pharmacokinetic and pharmacological data to support an adequate drug trial. There are two types of Phase II trials: Phase IIA and Phase IIB. Phase IIA is dedicated to determining dosage needs (how much medicine should be given) and Phase IIB is dedicated to determining efficacy (how well the drug works at the prescribed dose).

Phase II trials are designed as randomized clinical trials where some patients receive the test drug/intervention and other group receives a placebo/standard therapy. Placebo controlled groups are not always used. For example, it would be unethical to use a placebo in a cancer study. In such cases, an active control should be used (i.e., a drug which shows therapeutic activity) in the disease/condition.

Double blinded trials are desirable in Phase II studies as they minimize bias when results are analyzed. In double blinded studies, the identity of the intervention is hidden from the patient as well as the investigator, i.e., both are blinded toward the treatment.

7.6.4 Phase III CT

Phase III studies are huge patient groups randomized controlled multicenter trials (typically 300–3000 or more depending on the disease/medical condition investigated) aimed at determining the drug's efficacy in comparison to current "gold standard" treatment (Mahan 2014). Phase III trials are the most expensive, time intensive, and challenging to design and operate, especially in therapy for chronic medical diseases, due to their size and comparatively long duration.

The prime objective of Phase III trials is to assess the overall benefit/risk relationship, assessment of adverse reactions in a large group of patients over a longer period of exposure, assessing the dosage regimen and to take decision if the drug should be allowed to be marketed.

Certain Phase III trials are frequently continued while regulatory submissions are pending with the appropriate regulatory agency. This permits patients to continue

Table 7.10 Summary of characteristics of the different phases of clinical trials required before new drug is marketed

Name of the phase	No. of participants	Nature of study & type of population	Tentative time period	Objective	Tentative cost
Phase I (first in human) trials	10–100	Open label, usually healthy volunteers	Months to 1 year	Safety and tolerability	$10 million
Phase II (first in patients)	100–500	Randomized and controlled, may be blinded	1 to 2 years	Efficacy and dose ranging	$20 million
Phase III (multisite trials)	Few hundred to thousands	Randomized and controlled or uncontrolled (usually double blinded)	3 to 5 years	Confirmation of efficacy in larger population and comparison with gold standard therapy	$50–100 million

receiving potentially life-saving medications until the drug can be purchased. Other reasons for conducting trials at this point include the sponsor's desire to "label expand" (demonstrate the medicine works for different categories of patients/diseases beyond the original use for which the drug was approved for marketing), collect further safety data, or back up marketing claims. Some companies classify studies in this phase as PHASE IIIB investigations.

After Phase III trials have shown that a medicine is safe and effective, the results are usually compiled into a substantial document that includes a detailed description of the methodology and outcomes of human and animal research, manufacturing techniques, formulation information, and shelf life. This material is assembled into a "regulatory submission" that is sent to the proper regulatory authorities in various countries for assessment. They'll look through the submission and, hopefully, give the sponsor permission to market the medicine.

A summary of Characteristics of the Different Phases of Clinical Trials required before new drug is marketed is represented in Table 7.10.

After the initial Phase III trials are completed, the sponsor (typically a pharmaceutical corporation) submits an application to the FDA for authorization to market the drug, which is known as a New Drug Application (NDA) or a Biological Licence Application (BLA). These applications include detailed information, such as individual case report forms from the thousands of individuals who got the medicine during Phase III testing. In complicated circumstances, FDA may request the assistance of panels of external experts to assess applications (Goodman et al. 2006).

A new drug application submitted to appropriate regulatory body should contain:

- All details of animal and human studies.
- All safety findings (adverse effects of the intervention).
- Manufacturing procedures including method of analysis.

- Detailed formulation information for all dosing methods studied and storage conditions.

Regulatory reviews – can lead to requests for additional information regarding the submission, or even additional clinical trials to further establish either safety or efficacy. (This leads to regulatory approval, including labeling and approval to market the new drug).

Before a medicine can be sold, the manufacturer and the FDA must agree on the content of the label (package insert), which is the official prescription information. This label contains clinical pharmacological information, such as dose, adverse reactions, and particular warnings and precautions, as well as the drug's approved indications for usage. Pharmaceutical businesses' promotional materials must adhere to the facts stated in the package insert.

Phase IV is Post marketing surveillance where the patients are in treatment with the approved drug and the drug is prescribed by the physician. As the patient knows the identity of the drugs, so Phase IV studies are open label. The drugs are being monitored for their adverse events, rare adverse effects, compliance, and drug–drug interactions in a larger population in real life settings. Phase IV studies delineate additional information regarding new indications, risks, optimal doses, and schedules.

The outcomes of Phase IV studies may result in labeling changes based on safety information, a warning against using a new treatment in combination with another drug, or even the revocation of marketing approval if the results are particularly concerning.

Example 1:NSAID Cox-2 inhibitor Rofecoxib, for instance, was taken off the market due to a higher risk of ischemic events in patients undergoing Phase IV treatment.

Example 2: Cerivastatin, an inhibitor of HMG CoA reductase (high cholesterol and CV disease), was discontinued after reports of fatal rhabdomyolysis.

Skeletal muscle damage causes rhabdomyolysis, which causes the muscle to degrade rapidly. The patient's urine is brown tea-colored because of myoglobin, which damages the kidneys and causes kidney failure (due to very high creatine kinase levels).

Apart from Phase I, II, and III trials, early-stage trial also called Phase O is suggested to speed up the development of promising new drugs. It involves micro-dosing in small number of people to treat, a particular condition. The dose administered to the volunteer is a sub-therapeutic dose just to understand the behavior of molecule in the body (Kummar et al. 2008).

Acknowledgments All the authors are highly grateful to the Director Dr. (Col) A. Garg and Joint Director Dr. Manoj Goel from the KIET Group of Institutions for their continuous inspiration and support.

References

Agilent (n.d.). http://www.genomics.agilent.com/. Assessed 22 July 2022

Al-Salhi R, Abdul-Sada A, Lange A et al (2012) The xenometabolome and novel contaminant markers in fish exposed to a wastewater treatment works effluent. Environ Sci Technol 46: 9080–9088

Anderson AC (2003) The process of structure-based drug design. Chem Biol 10:787–797

Ashburn TT, Thor KB (2004) Drug repositioning: identifying and developing new uses for existing drugs. Nat Rev Drug Discov 3:673–683

Bennett CF (2019) Therapeutic antisense oligonucleotides are coming of age. Ann Rev Med 70: 307–321

Bentley DR, Balasubramanian S, Swerdlow HP et al (2008) Accurate whole human genome sequencing using reversible terminator chemistry. Nature 456:53–59

Bioconductor (n.d.) Open-Source Software for Bioinformatics. http://www.bioconductor.org/packages/2.9/bioc/html/goseq.html. Assessed 22 July 2022.

Blass BE (2015) Basic principles of drug discovery and development. Academic, New York, NY

Boss A, Kolb A, Hofmann M et al (2010) Diffusion tensor imaging in a human PET/MR hybrid system. Investig Radiol 45:270–274

Burbaum J, Tobal GM (2002) Proteomics in drug discovery. Curr Opin Chem Biol 6:427–433

Casazza JP, Cale EM, Narpala S (2022) Safety and tolerability of AAV8 delivery of a broadly neutralizing antibody in adults living with HIV: a phase 1, dose-escalation trial. Nat Med 28: 1022–1030. https://doi.org/10.1038/s41591-022-01762-x

Chemicalize(n.d.). http://chemicalize.org. Assessed 22 July 2022

Computational Biology and Informatics Laboratory (n.d.). http://www.gusdb.org. Assessed 22 July 2022

Cosentino U, Pitea D, Moro G (2009) Conformational behaviour determines the low-relaxivity state of a conditional MRI contrast agent. Phys Chem Chem Phy 11:3943–3950

Dale JW, Schantz MV (2007) From genes to genomes. Concepts and applications of DNA technology, 2nd edn. Wiley, London

David PC, Nanette JP (2013) DNA sequencing. In: Clark DP, Pazdernik NJ (eds) Molecular biology, 2nd edn. Academic, New York, NY, pp 227–247. ISBN 9780123785947.

Franks SJ, Weber JJ, Aitken SN (2014) Evolutionary and plastic responses to climate change in terrestrial plant populations. Evol Appl 7:123–139

Glick BR, Pasternak JJ, Patten CL (2010) Molecular biotechnology: principles and applications of recombinant DNA, 4th edn. ASM Press, Washington, DC

Goodman LS, Gilman A, Brunton LL et al (2006) Goodman & Gilman's the pharmacological basis of therapeutics. McGraw-Hill, New York, NY

Grant GR, Manduchi E, Stoeckert CJ Jr (2007) Analysis and management of microarray gene expression data. Curr Protoc Mol Biol 77:19–26

Gross S, Piwnica-Worms D (2006) Molecular imaging strategies for drug discovery and development. Curr Opin Chem Biol 10:33–342

Hartman KB, Laus S, Bolskar RD et al (2008) Gadonanotubes as ultrasensitive pH-smart probes for magnetic resonance imaging. Nano Lett 8:415–419

Hoover R, Hunt T, Benedict M et al (2016) Single and multiple ascending-dose studies of oral delafloxacin: effects of food, sex, and age. Clin Ther 38:39–52. https://doi.org/10.1016/j.clinthera.2015.10.016

Hwang TJ, Kesselheim AS (2016) Vaccine pipeline has grown during the past two decades with more early-stage trials from small and medium-size companies. Health Aff 35:219–226. https://doi.org/10.1377/hlthaff.2015.1073

Jaffer FA, Libby P, Weissleder R (2009) Optical and multimodality molecular imaging: insights into atherosclerosis. Arterioscler Thromb Vasc Biol 29:1017–1024

Johanna K, Stefano D (2011) Disease gene identification: methods and protocols. Springer, New York, NY

Kalimuthu S, Jeong JH, Oh JM (2017) Drug discovery by molecular imaging and monitoring therapy response in lymphoma. Int J Mol Sci 18:1639

Khvorova A, Watts JK (2017) The chemical evolution of oligonucleotide therapies of clinical utility. Nat Biotechnol 35:238–248

Krogsgaard-Larsen P, Liljefors T, Madsen U (2004) Text book of drug design and discovery, 3rd edn. Taylor & Francis, London

Kubinyi H, Folkers G, Martin YC (1998) 3D QSAR in drug design: recent advances, vol 3. Kluwer Academic Publishers, Norwell, MA

Kuehnbaum NL, Britz-McKibbin P (2013) New advances in separation science for metabolomics: resolving chemical diversity in a post-genomic era. Chem Rev 113:2437–2468

Kumar V, Sanseau P, Simola DF et al (2016) Systematic analysis of drug targets confirms expression in disease-relevant tissues. Sci Rep 6:36205

Kummar S, Rubinstein L, Kinders R et al (2008) Phase 0 clinical trials: conceptions and misconceptions. Cancer J 14:133–137. https://doi.org/10.1097/PPO.0b013e318172d6f3

Kyoto Enclyclopedia of Genes and Genomes (n.d.). http://www.genome.jp/kegg/. Assessed 22 July 2022

Larson RS (2006) Bioinformatics and drug discovery. Humana Press, Totowa, NJ

Lecchi M, Ottobrini L, Martelli C et al (2007) Instrumentation and probes for molecular and cellular imaging. Q J Nucl Med Mol Imaging 51:111

Leon D, Markel S (2006) In silico technologies in drug target identification and validation. Taylor & Francis, London

Lu Y, Machado HB, Bao Q (2011) In vivo mouse bioluminescence tomography with radionuclide-based imaging validation. Mol Imaging Biol 13:53–58

Mahan VL (2014) Clinical trial phases. Int J Clin Med 5:1374–1383

Marcotte R, Sayad A, Brown KR et al (2016) Functional genomic landscape of human breast cancer drivers, vulnerabilities, and resistance. Cell 164:293–309

Martin YC (2010) Quantitative drug design: a critical introduction, 2nd edn. CRC, Boca Raton, FL

Massoud TF, Gambhir SS (2003) Molecular imaging in living subjects: seeing fundamental biological processes in a new light. Genes Dev 17:545–580

Michne WF (2010) Lead discovery: the process. In: Hit-to-lead process—lead compound discovery, milestone to new small molecule drug discovery. Wiley, New York, pp 1–19

Miggiels P, Wouters B, van Westen GJ (2019) Novel technologies for metabolomics: more for less. TRAC Trend Anal Chem 120:115323

Nagarajan K, Bodla RB (2018) Fundamentals principles of drug discovery. LAP Lambert Academic Publishing, Saarbruecken. ISBN: 9786139981441.

National Cancer Institute: Centre for Biomedical Informatics & Information Technology (n.d.). http://ncicb.nci.nih.gov/. Assessed 22 July 2022

Naz S, Vallejo M, García A et al (2014) Method validation strategies involved in non-targeted metabolomics. J Chromatogr 1353:99–105

Ottobrini L, Ciana P, Biserni A (2006) Molecular imaging: a new way to study molecular processes in vivo. Mol Cell Endocrinol 246:69–75

Patrick GC (2001) An introduction to medicinal chemistry, 2nd edn. Oxford University Press, New Delhi

Patti GJ, Yanes O, Siuzdak G (2012) Metabolomics: the apogee of the omics trilogy. Nat Rev Mol Cell Biol 13:263–269

Pichler BJ, Wehrl HF, Judenhofer MS (2008) Latest advances in molecular imaging instrumentation. J Nucl Med 49(Suppl 2):5S–23S

Prejzendanc T, Wasik S, Blazewicz J (2016) Computer representations of bioinformatics models. Curr Bioinforma 11:551–560

Rankovic Z, Morphy R (2010) Lead generation approaches in drug discovery. Willey, Hoboken, NJ

Rinaldi C, Wood MJA (2018) Antisense oligonucleotides: the next frontier for treatment of neurological disorders. Nat Rev Neurol 14:9–21

Rossi A, Gandolfo C, Morana G (2010) New MR sequences (diffusion, perfusion, spectroscopy) in brain tumours. Pediatr Radiol 40:999–1009

Rudin M (2009) Noninvasive structural, functional, and molecular imaging in drug development. Curr Opin Chem Biol 13:360–371

Shafiq N, Rajagopalan S, Kushwaha HN et al (2014) Single ascending dose safety and pharmacokinetics of CDRI-97/78: first-in-human study of a novel antimalarial drug. Malar Res Treat 2014:372521. https://doi.org/10.1155/2014/372521

Shendure J, Lieberman AE (2012) The expanding scope of DNA sequencing. Nat Biotechnol 30: 1084–1094

Shoaib M, Singh A, Gulati S et al (2021) Chapter 8. In: Mapping genomes by using bioinformatics data and tools, chemoinformatics and bioinformatics in the pharmaceutical sciences, 1st edn. Academic, New York, NY, pp 245–278, ISBN 9780128217481.

Slatko BE, Albright LM, Tabor S et al (2001) DNA sequencing by the dideoxy method. Curr Protoc Mol Biol 7:Unit7.4A. https://doi.org/10.1002/0471142727.mb0704as47

Soule MC, Longnecker K, Johnson WM et al (2015) Environmental metabolomics: analytical strategies. Mar Chem 177:374–387

Tanner GJ, Blundell MJ, Colgrave ML et al (2016) Creation of the first ultra-low gluten barley (*Hordeum vulgare* L.) for celiac and gluten-intolerant populations. Plant Biotechnol J 14:1139–1150

Teufel A, Krupp M, Weinmann A et al (2006) Current bioinformatics tools in genomic biomedical research (review). Int J Mol Med 17:967–973

Totowa H (2007) Target discovery and validation reviews and protocols: emerging molecular targets and treatment options, vol 2. Humana Press, Totowa, NJ, p XIV, 345

Van Norman GA (2019) Phase II trials in drug development and adaptive trial design. JACC Basic Transl Sci 4:428–437. https://doi.org/10.1016/j.jacbts.2019.02.005

Velde GV, Baekelandt V, Dresselaers TE et al (2009) Magnetic resonance imaging and spectroscopy methods for molecular imaging. Q J Nucl Med Mol Imaging 53:565

Viant MR (2007) Metabolomics of aquatic organisms: the new 'omics' on the block. Mar Ecol Prog Ser 332:301–306

Virtual Computational Chemistry Laboratory (n.d.). www.vcclab.org/lab/alogps/. Assessed 22 July 2022

Weissleder R, Pittet MJ (2008) Imaging in the era of molecular oncology. Nature 452:580–589

Wouters OJ, McKee M, Luyten J (2020) Estimated research and development investment needed to bring a new medicine to market, 2009-2018. JAMA 323:844–853. https://doi.org/10.1001/jama.2020.1166

Zaidi H, Prasad R (2009) Advances in multimodality molecular imaging. J Med Phys 34:122

Zhou B, Zhou Z, Chen Y et al (2020) Plasma proteomics-based identification of novel biomarkers in early gastric cancer. Clin Biochem 76:5–10

Advances in Peptide Synthesis

8

Gajanan K. Rathod, Rajkumar Misra, and Rahul Jain

Abstract

Peptide drug development has made significant progress over the last century. The discovery of solid-phase peptide synthesis has enabled chemists to synthesize various peptides with divergent sequence patterns. However, due to the increased demand for various peptide sequences in the modern pharmaceutical industry, there is always room for new methods to modify the existing methods to improve yield, purity, and synthesis time. The current century has witnessed a lot of progress in the field of peptide synthesis, including developments in new synthetic strategies, suitable selection of protecting groups, and introduction of efficient coupling reagents, as well as the development of automated peptide synthesizers. This chapter will give a summary of the recent reports on the most significant breakthroughs in peptide chemical synthesis in current years.

Keywords

Amino acids · Microwave · Peptides · Solid phase · Solution phase

8.1 Introduction

Proteins and peptides are ubiquitous in every living organism and play a crucial role in the development of life. They are responsible for a variety of physiological and biological activities that occur in the living system, including the transport of diverse ingredients across membranes and intercellular communication, and serve as the

G. K. Rathod · R. Misra · R. Jain (✉)
Department of Medicinal Chemistry, National Institute of Pharmaceutical Education and Research, Mohali, Punjab, India
e-mail: rahuljain@niper.ac.in

P. P. Singh (ed.), *Recent Advances in Pharmaceutical Innovation and Research*,
https://doi.org/10.1007/978-981-99-2302-1_8

primary component of antibodies that protect the organisms. In addition, due to their inherent biocompatibility as well as high specificity, polypeptides and proteins are considered active pharmaceutical ingredients (API) in most of the modern pharmaceutical industries (Liu et al. 2012; Domalaon et al. 2016; Våbenø et al. 2015; Uhlig et al. 2014; Fosgerau and Hoffmann 2015). Polypeptides, which are composed of 20 natural amino acids have exceptional therapeutic and biological properties, including antioxidant, antibacterial, anti-inflammatory, anticancer, and anti-HIV activity. Due to the high biological potential of the peptides, more than 150 peptides have been approved as drugs and some of them are in clinical trials (Wang et al. 2022). Currently, the peptide therapeutics market is valued at approximately USD 28 billion and is expected to increase by more than 10% over the coming years (Henninot et al. 2018). Even though peptides and proteins have great therapeutic potential; however, poor bioavailability and proteolytic degradation by endogenous enzymes limit their in vivo application (Diao and Meibohm 2013). These shortcomings have been overcome by the incorporation of non-natural amino acids (Werner et al. 2016; Cheloha et al. 2016) or the introduction of other functionalities such as glycosylation, and PEGylation (Harris and Chess 2003), lipidation, or *N*-methylation (Chatterjee et al. 2013). Polypeptides and proteins are synthesized either chemically or biologically. The chemical synthesis of peptides can be accomplished by either solution phase or solid-phase synthesis and each methodology has its own set of benefits and drawbacks. While short peptides (≤10–15 amino acids) can be chemically synthesized using the solution-phase fragmentation condensation strategy, the larger or longer peptides have been synthesized employing solid-phase methods. The foundation work by Bruce Merrifield, the development of solid-phase synthesis revolutionized peptide synthesis by simplifying the purification step involved with solution-phase peptide synthesis (Merrifield 1963). SPPS is now a popular choice for the production of longer and difficult peptide sequences; however, solution-phase synthesis can still be effective for large-scale production of a given peptide. The constant high demand for peptides in various disciplines stimulated the advancement of peptide synthesis to achieve high yield, high purity, and a shorter synthesis time. Over the last 20 years, several approaches have emerged for the development of peptide synthesis both in solid and solution phase methods. In this chapter, we'll cover all of the new advanced approaches.

8.2 Classical Approaches of Chemical Peptide Synthesis

Prior to Merrifield's discovery of solid-phase peptide synthesis, solution-phase synthesis was the only way to make the peptides. In this traditional approach, peptide synthesis is manifested using the fragment condensation method. Suitably protected amino acids are synthesized or commercially supplied and coupled together using coupling reagents (Tsuda and Okada 2010; Anderson 1960). Typically, the N-terminal group is protected by Boc, and the C-terminal group is converted to ester. The amide bond formation is accomplished with carbodiimide-based coupling reagents such as DCC or EDC in DMF and HOBt used as the racemization suppressor. Finally, the N-terminal Boc and C-terminal ester are de-protected by strong acids like TFA and strong bases such as NaOH. The methodology can be very helpful for synthesizing short and ultra-short peptides but is not applicable for longer and complex peptide sequences (Verlander 2007).

8.3 Solid-Phase Peptide Synthesis

Solid-phase synthesis is one of the most successful methods for synthesizing a wide range of peptides, from short to long sequences, and has gained widespread acceptance in the peptide industry. The method was first introduced by R. Bruce Merrifield in 1963 and 20 years later he got a Nobel Prize (Merrifield 1985).

The SPPS method revolutionizes peptide synthesis by simplifying the extensive and complicated purification steps involved with the solution phase synthesis. The use of reactive solid support allows for the sequential coupling of individual amino acids. After the amino acids have been attached, the unreacted amino acids and coupling reagents are removed by filtration and washing. Once the anchoring of individual amino acids of the desired peptide sequence is completed, the peptide is released from the solid support, and simultaneously the temporary side chain protective groups of amino acids are also removed under the same conditions. The principle of SPPS is shown in Scheme 8.1. Typically, the two most common N-terminal protected amino acids are used one is acid-labile Boc (Merrifield 1986) and the other is base-labile Fmoc (Fig. 8.1) (Carpino and Han 1970). The peptide synthesized by Boc-protected amino acid requires the use of extremely hazardous and toxic HF to cleave the peptide from the solid support (Muttenthaler et al. 2015) and thus is less often used than the Fmoc-strategy.

The Fmoc- strategy has been extensively applied to the synthesis of diverse peptide sequences since this protective group can be easily removed under mild conditions with secondary amines, usually 20% piperidine–DMF. Crosslinked polystyrene (PS)-based resins (Merrifield 1985) or copolymer of polyethylene glycol and polystyrene (PEG-PS)-based resins are routinely used in Fmoc SPPS (Barany et al. 1992; Rapp et al. 1988; Carpino et al. 1994). The swelling of resin is carried out in presence of solvents such as DMF and DCM which expose the reactive functional groups and make them available for amino acid coupling.

The side chains of the amino acids like Cys, Asn, Gln, His Asp, Glu, Lys, Ser, Tyr, and Arg need to be protected during the FMOC SPPS as they can react during the peptide synthesis. The commonly used protecting group for Glu, Asp, Ser, Thr, and Tyr is *tert*-butyl (*t*-Bu); 2,2,4,6,7-pentamethyl-dihydrobenzofuran-5-sulfonyl (Pbf) for Arg; and trityl (Trt) for Cys, Asn, Gln, and His. The coupling reaction between the free α-amino group and the Fmoc-protected α-carboxylic acid of the next amino acid is mediated by various coupling reagents. Carbodiimide-based coupling reagents such as DCC were used earlier in SPPS and later it is replaced by DIC as their urea is more soluble in organic solvent and can be removed simply by washing the resin (Fig. 8.2) (Carpino 1993).

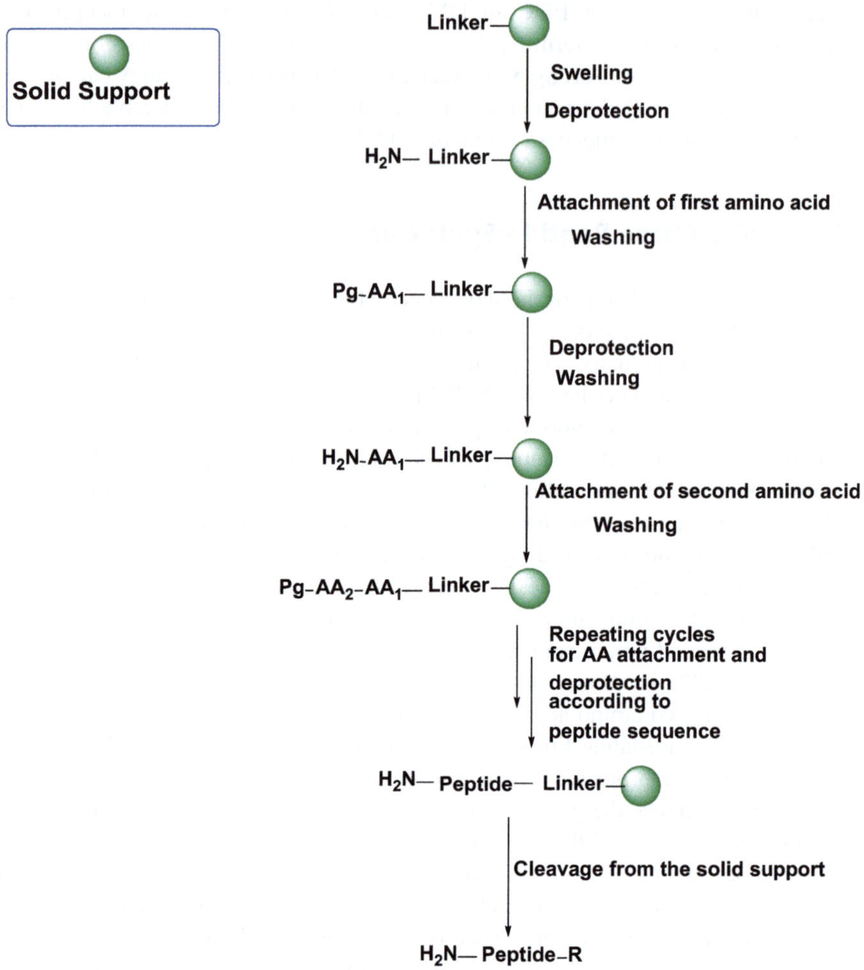

Scheme 8.1 Principle of Solid-Phase Peptide Synthesis

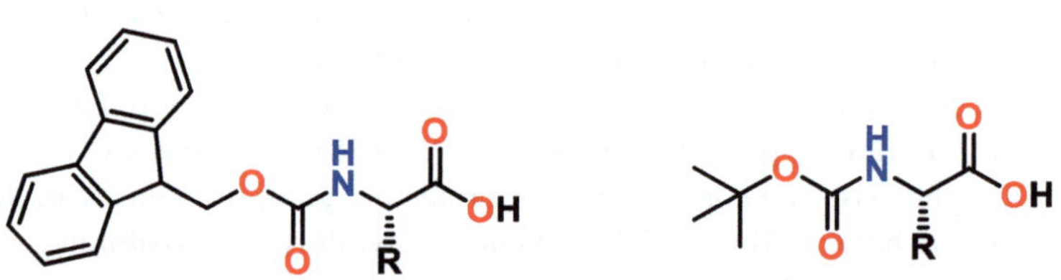

Fig. 8.1 Commonly used protected amino acid in solid-phase peptide synthesis

Benzotriazole derivatives such as HOBt and HOAt are generally utilized as additives to activate the C-terminal amino acids and as racemization suppressors during the coupling reaction (Fig. 8.3) (Carpino 1993).

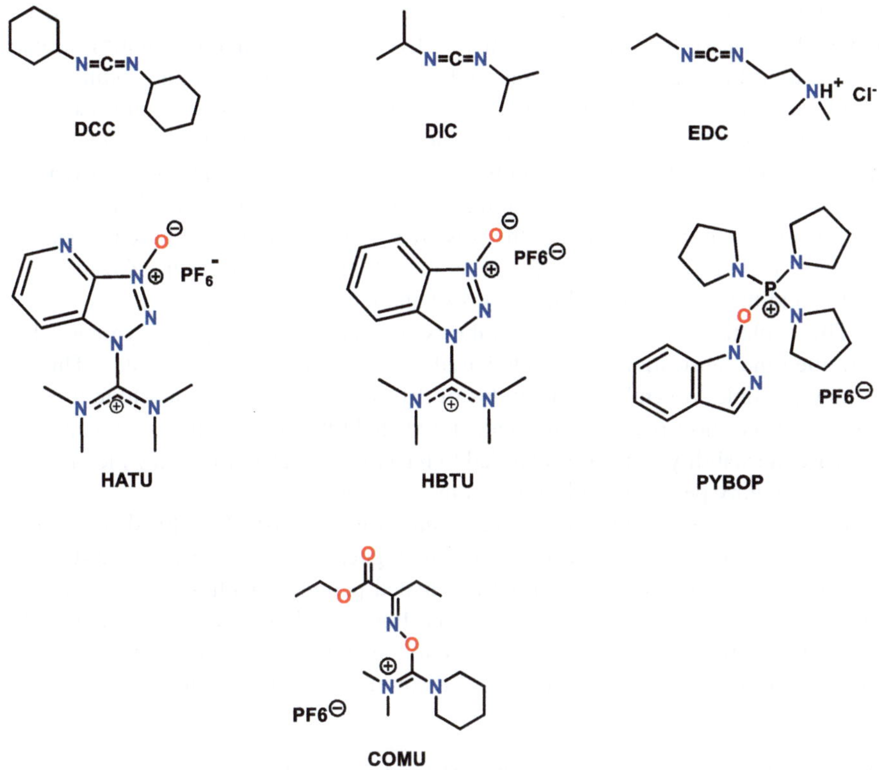

Fig. 8.2 Commonly used coupling reagents in solid and solution-phase peptide synthesis

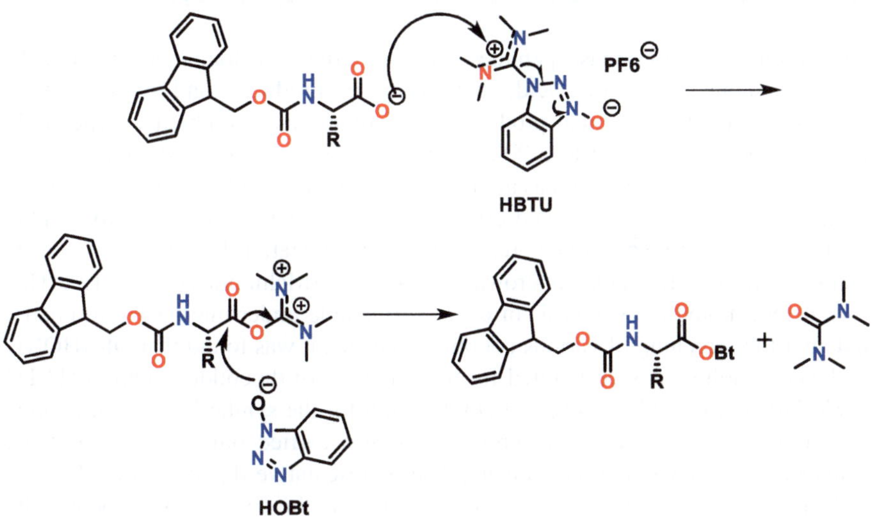

Fig. 8.3 Coupling reaction mediated by HBTU and HOBt

However, because of thermal instability and considering class 1 explosive materials limited their use. Recently Ethyl 2-cyano-2-(hydroxyimino) acetate (OxymaPure) appeared to be a safer and more efficient carbodiimide additive than HOBt and HOAt and demonstrated exceptional racemization suppressor and activating abilities in both manual and automated synthesis (Subirós-Funosas et al. 2009). In addition to additives, significant advancements in the coupling reagent have been achieved to decrease coupling time and minimize epimerization. The most significant breakthrough since the discovery of carbodiimide-based coupling reagents includes HBTU, HATU (Carpino et al. 2002), PyBOP (Coste et al. 1990), and the new COMU (Fig. 8.2) (El-Faham et al. 2009).

The evolution of solid-phase synthesis into large-scale peptide synthesis has permitted the fabrication of peptide-based "active medicinal ingredients." Thus, it is now possible to synthesize peptides of 30–50 amino acids on a kilogram scale within a very short period. This synthetic availability along with the strategies to increase the stability of peptides has led to the commercialization of a wide range of short to lengthy peptides, with more on the horizon.

Over the years solid-phase synthesis has been extensively refined to achieve superior outcomes, such as faster synthesis, higher yields, and greater purity. The developments of microwave-assisted automated peptides synthesizer, implementation of native chemical ligation (NCL), and flow-based approach revolutionize the SPPS. In addition, efforts have been made to make peptide synthesis more environmentally friendly. We will discuss all of them in the coming sections.

8.4 Microwave-Assisted Peptide Synthesis

8.4.1 Microwave-Assisted Solution-Phase Peptide Synthesis

Conventional solution-phase peptide synthesis requires a long reaction time for coupling, along with low yield, and purity of products, and often leads to enantiomerization of the final product. To overcome these problems earlier Babu and co-workers (Babu and Rao 2005) reported microwave-assisted solution-phase peptide synthesis; however, it has not been explored very well. This prompted many researchers to explore robust and efficient protocols with wide scope for peptide synthesis. Jain and co-workers reported the racemization-free protocol for the synthesis of peptides under microwave irradiation (Mahindra et al. 2012). They have demonstrated the synthesis of di- and tripeptide under microwave irradiation and after the screening of various coupling reagents, it was found that DIC/HONB-mediated coupling was best suited for the synthesis of dipeptide, whereas HATU/HOAt/DIEA was the best coupling combination for the synthesis of the tripeptide. The conventional synthesis of peptides was also carried out and as expected it produced a lower yield comparatively. The representative dipeptide Boc-Trp-His-OMe as shown in Scheme 8.2 was synthesized using the developed protocol with a 70% yield.

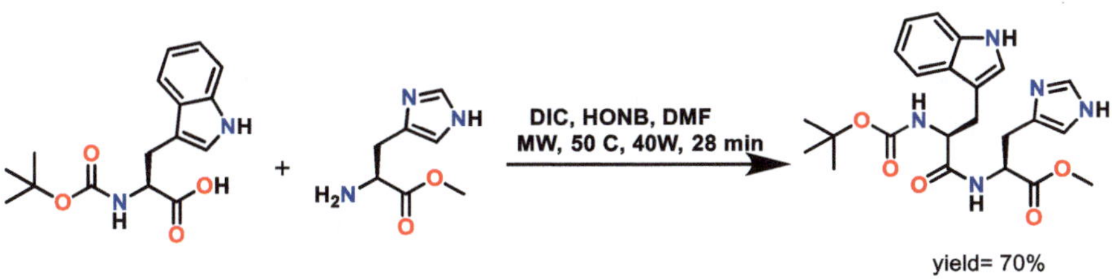

Scheme 8.2 Synthesis of Boc-Trp-His-OMe

In the quest to find an environmentally benign method for peptide synthesis, Jain and co-workers next reported microwave-assisted solution-phase peptide synthesis in neat water as a reaction medium (Mahindra et al. 2013). This was the first report of solution-phase peptide synthesis in water as a medium. The use of water as a reaction medium offers a great advantage as it is considered a greener solvent that is non-toxic, non-flammable, and reduces the generation of waste. Along with the use of a greener solvent, the use of this protocol produces the racemization-free peptide in a short reaction time and both protecting group BOC and Fmoc can be used. While optimizing reaction protocol various parameters were screened such as reaction time and temperature, solvent, and coupling reagents. Optimization studies lead to the use of a coupling combination of TBTU/HOBt in water and DIC/HONB in *tert*-Butyl methyl ether (TBME), DIEA as base produced the highest yield up to 90% at 60 °C in 30 min of microwave irradiation. Later the developed protocol was applied for the synthesis of pentapeptide pentagastrin in water as shown in Scheme 8.3. It is a synthetic analogue of endogenous gastrin and is used in diagnostic aid. The pentagastrin peptide was efficiently synthesized with high yield and purity.

8.4.2 Microwave-Assisted Solid-Phase Peptide Synthesis

In the conventional SPPS method, it is believed that the elongation of the peptide chain leads to an increase in steric hindrance and aggregation. Moreover, coupling of secondary amino acids such as proline or other sterically hindered amino acids decreases the reaction rates and leads to incomplete reaction. As a result, such synthesis required excess reagents and solvents *N, N*-dimethylformamide (DMF), and *N*-methyl-2-pyrrolidinone (NMP) for coupling and washing to remove impurities in addition to long reaction times that reduce the efficiency of SPPS. Moreover, an incomplete reaction could result in the formation of deleted sequence, which can be difficult to remove from the desired product (Jad et al. 2019; Collins et al. 2014). So, to improve the efficiency of the SPPS protocol researchers started using different energy sources one of them being microwave irradiation or employing automated peptide synthesizers or heating. In the microwave, heating energy is directly transferred to a solvent via the interaction of molecules in a solvent with an electromagnetic field. Based on data it is suggested that SPPS microwave

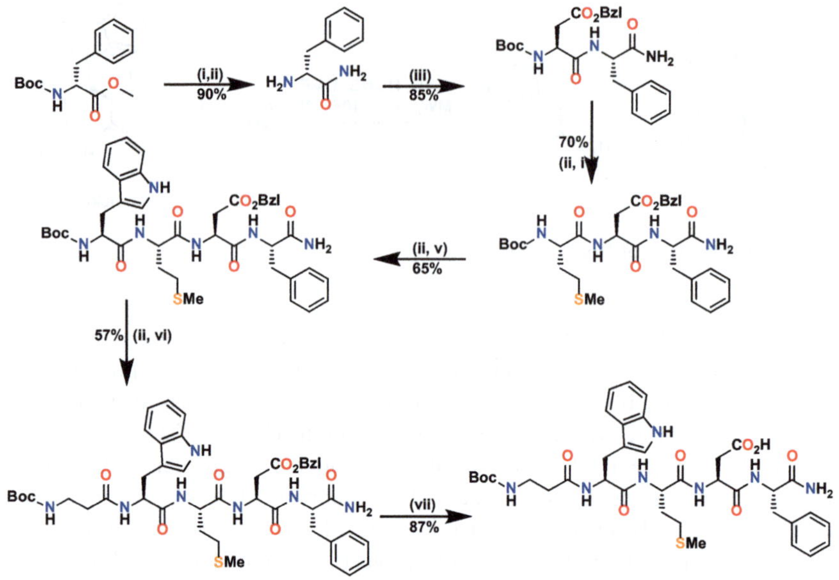

Reagents and conditions: (i) $HCONH_2$, NaOMe, 100 °C; (ii) 6N HCl, 15 min, 25 °C; (iii) Boc-Asp(Bzl)-OH, TBTU, HOBt, DIEA, H_2O, MW, 60 °C; (iv) Boc-Met-OH, TBTU, HOBt, DIEA, H_2O, MW, 60 °C; (v) Boc-Trp-OH, TBTU, HOBt, DIEA, H_2O, MW, 60 °C; (vi) Boc-β-Ala-OH, TBTU, HOBt, DIEA, H_2O, MW, 60 °C; (vii) 10% Pd/C, $HCOONH_4$, CH_3OH, reflux

Scheme 8.3 Synthesis of Pentagastrin

heating can improve the reaction time and yield of the synthesized products (Pedersen et al. 2012).

Gogoll and co-workers (Erdelyi and Gogoll 2002) reported the first-time use of microwave irradiation in SPPS. It has been known that the use of microwaves in organic synthesis improves the reaction yield and time. Gogoll et al. demonstrated the synthesis of sterically hindered peptides which have been difficult to access via SPPS at room temperature. The desired peptides can be obtained in a short reaction time of 1.5–20 min without racemization.

As we know SPPS involves excess use of highly toxic organic solvents for coupling, washing, and deprotection therefore to further improve the SPPS technique by using an environmentally benign solvent such as water is of high interest. Grøtli and co-workers (Galanis et al. 2009) reported the solid phase peptide synthesis using water as a solvent for coupling and deprotection. It involves the use of Boc-protected amino acid along with the combination of EDC/HONB as coupling reagents under microwave irradiation. This is the first report in which whole peptide synthesis has been carried out in the water. Moreover, peptides are obtained in high yield and purity without racemization (Scheme 8.4).

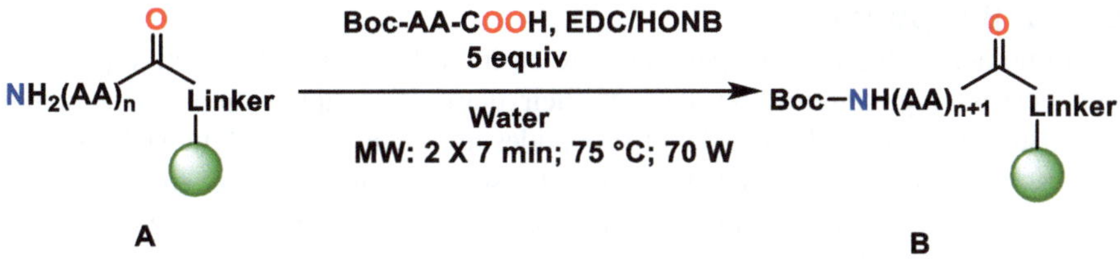

Scheme 8.4 Microwave-Assisted Solid-Phase Synthesis of Peptide in Water

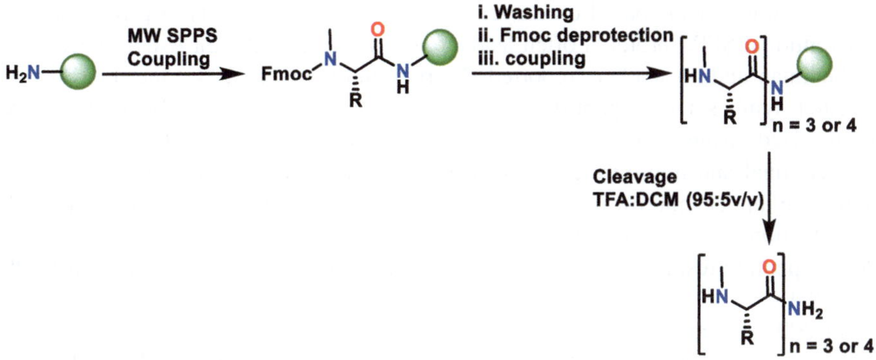

Scheme 8.5 Microwave-Assisted Synthesis of *N*-Methylated Peptides

8.4.3 Synthesis of Difficult Peptides Using Microwave-Assisted SPPS

8.4.3.1 N^{α}-Methylated Peptides

Peptides are highly susceptible to proteolytic degradation; however, several approaches have been developed to improve their proteolytic stability. Among recent advances, the synthesis of N^{α}-methylated peptides has emerged as an effective strategy for improving their short half-life (Chatterjee et al. 2008). Angell and co-workers (Angell et al. 1994) reported the coupling of sterically hindered *N*-methylated amino acids under SPPS protocol using HOAt as coupling reagents. However, the coupling of *N*-methyl amino acids often results in low yields and requires excess coupling reagents along with a long reaction time. Microwave heating can be used to overcome these problems therefore Albericio and co-workers (Rodríguez et al. 2010) reported the efficient coupling of Fmoc-*N*-methyl amino acid under microwave irradiation in SPPS conditions using rink amide MBHA resin. The *N*-methyl-rich peptides were obtained in high yield and purity using a developed protocol. The short *N*-methylated peptides were synthesized using microwave irradiation at 35 °C for 20 min using DIC/HOAt as a coupling combination in DCM as solvent (Scheme 8.5).

8.4.3.2 β-Peptides

β-Amino acids oligomers (β-peptides) are an important class of bioactive peptides that possess high interest because of their demanding application in medicinal chemistry. The major drawbacks of α-peptides are the low proteolytic stability and possess high conformational freedom. These problems can be overcome by incorporating β-amino acids into bioactive peptides as they are proven to be effective in modulating structures and physicochemical properties (Cabrele et al. 2014). There are two types of β-amino acids often used as building blocks for bioactive peptides. Homologues of α-amino acids consist of an extra methylene group and the second one is β-amino acids based on a cycloalkane ring. The synthesis of β-peptides via conventional SPPS protocol often leads to low yield and difficult *N*-deprotection.

Gellman and co-workers reported (Murray and Gellman 2005) the microwave-assisted synthesis of β-peptides. The 14-helical hexa and deca-β-peptides were synthesized using *trans*-2-aminocyclohexane carboxylic acid (ACHC) via both conventional and microwave SPPS. It was identified that for the synthesis of this β-peptide under microwave irradiation, coupling at 60 °C for 2 min and *N*-deprotection at 50 °C, 4 min results in higher purity of 80% and 57% of hexa and deca-β-peptide respectively, which is much superior relative to conventional SPPS (Fig. 8.4).

In 2003 the completely automated microwave-assisted peptide synthesizer was introduced by CEM Corporation. Ultrafast automation of microwave-based Liberty Blue synthesizer can be used in the synthesis of longer peptides within a short reaction time with more than 90% reduction in solvent wastage. The single-mode microwave reactor Discover has been employed widely in the Liberty system. The system uses the amino acid stock solution and consists of reagent ports for washing, deprotection, activation, and cleavage (Collins and Collins 2003). It is also connected with a nitrogen cylinder, which maintains the inert atmosphere while performing coupling, deprotection, and washing steps. All procedures in liberty are

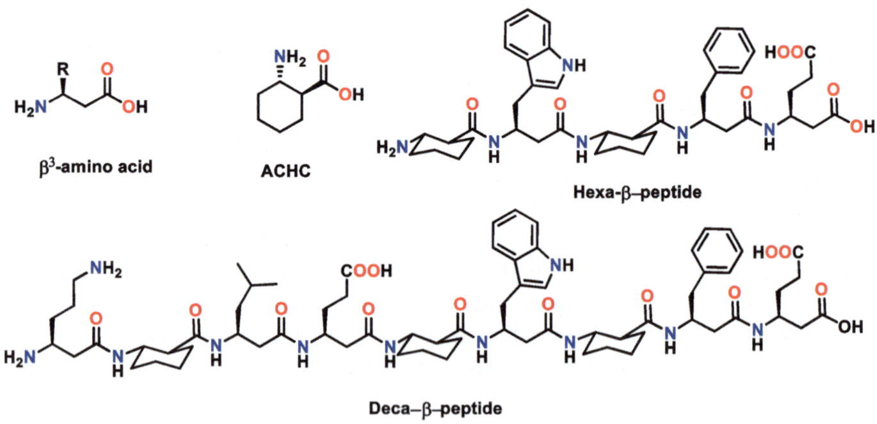

Fig. 8.4 β-Amino acids along with hexa and deca-beta-peptide

controlled by an external computer, the complete manual is available by Discover SPS.

8.5 Peptide Synthesis Using Ball Mill

Despite well-established protocols for the production of peptides, they are associated with major problems that involve the use of a huge amount of highly volatile and toxic solvents mainly in SPPS. Therefore, exploration of a convenient and environmentally benign protocol is highly desirable. One of the attempts to avoid the use of toxic solvents is the use of a ball mill for the synthesis of peptides (Sato et al. 2006).

Lamaty and co-workers (Declerck et al. 2009) reported the solvent-free synthesis of peptides without epimerization using ball milling. The developed protocol involves the coupling of urethane-protected α-amino acid *N*-carboxyanhydride (UNCA) with amino acid ester in presence of base $NaHCO_3$. It was confirmed that for coupling reaction all the solids are necessary and the absence of one of the reagents results in failure of reaction. They have also demonstrated for the first-time synthesis of artificial sweetener aspartame in the absence of solvent using ball milling technology. Aspartame is a dipeptide composed of α-L-aspartyl-L-phenylalanine and its sweetening ability is 150 times greater than sucrose (Scheme 8.6).

Lamaty and co-workers (Bonnamour et al. 2013) in their further study have demonstrated the environmentally benign protocol liquid-assisted ball milling for the synthesis of short and long peptides such as tetra and pentapeptide on a gram scale with high yield. It involves the coupling of Boc-protected α-amino acid *N*-carboxyanhydrides or *N*-hydroxysuccinimide esters (Boc-AA-OSu) (Boc-AA-NCA or OSu) with amino acid ester in presence of $NaHCO_3$ and a small amount of ethyl acetate under ball milling. The developed protocol was used for the synthesis of pentapeptide leu-enkephalin in nine steps with a 46% of yield (Fig. 8.5).

8.6 Native Chemical Ligation Strategy for Peptide Synthesis

The native chemical ligation method is one of the most widely employed for the synthesis of larger polypeptides and proteins. Conventional SPPS is ineffective for the synthesis of larger polypeptides because of the increased aggregation and steric crowding caused by elongation of the peptide chain, which resulted in undesired side

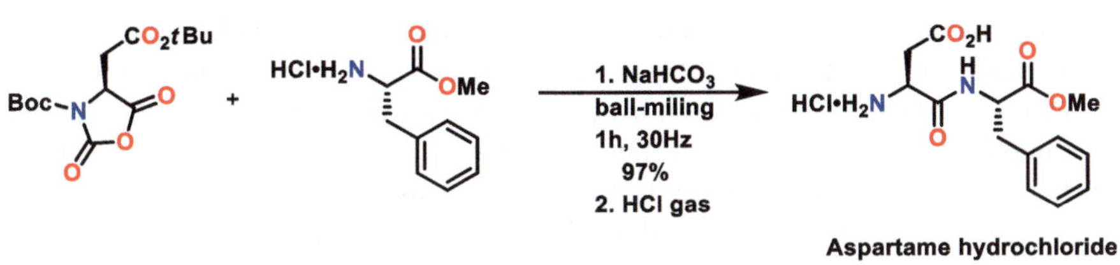

Scheme 8.6 Synthesis of Aspartame Using Ball Milling Technology

Fig. 8.5 Synthesis of leu-enkephalin using ball mill

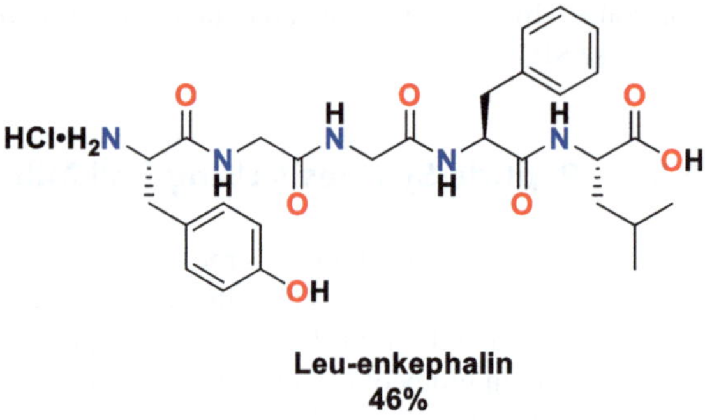

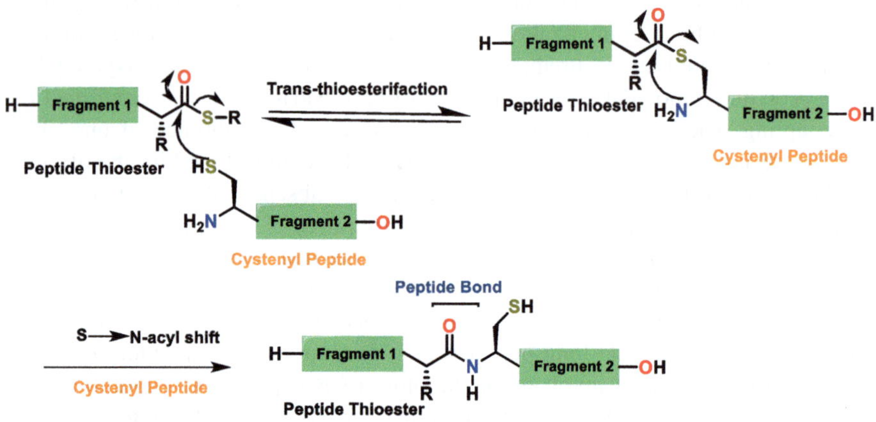

Scheme 8.7 Mechanism of Native Chemical Ligation (NCL). Initially peptide thioester reacts with cysteinyl peptide fragments via an initial trans-thioesterification followed by an S-N acyl shift to produce a native peptide (amide) bond

products and epimerization. The invention of the native chemical ligation (NCL) technique has overcome these flaws, and using this strategy hundreds of protein targets have been synthesized to date. The principle of the strategy was first introduced by Kent and co-workers in the early 1990s (Dawson et al. 1994) which is further extrapolated and modified by Dawson (Dirksen and Dawson 2008) and others (Haase and Seitz 2008; Macmillan 2006; Bondalapati et al. 2016). The reaction involves the condensation of two unprotected peptide segments in aqueous conditions at neutral pH to form a native amide bond in a chemoselective and high-yielding manner. In general, C-terminal peptide thioesters react with another peptide containing an N-terminal Cys residue, and following the reversible transthioesterification reaction, an irreversible S-to-N acyl shift occurs through the five-membered ring intermediate that joins the two peptides via an amide bond (Scheme 8.7). The important key feature of the method is that the condensation of the two peptide fragments is carried out in aqueous conditions and physiological pH which helps the solubility of unprotected peptide and protein fragments. The success of the strategy is highlighted by the synthesis of HIV1 protease (Torbeev and Kent

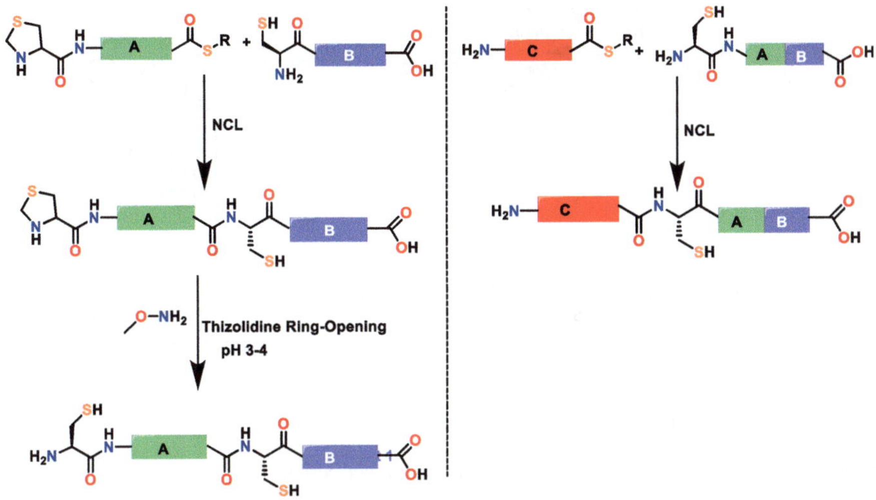

Scheme 8.8 Coupling of more than two peptide fragments through ligation chemistry: The N-terminal Cys of one of the peptide fragments is protected as a thiazolidine (Thz) moiety which after the ligation is removed and coupled with another peptide thioester through native chemical ligation (NCL)

2007), D-Dpo4 enzyme (Jiang et al. 2017), and different head-to-tail cyclic peptides (Clark and Craik 2010).

The major drawback of the methodology is the requirement of a Cys residue which is not frequently found in protein. To address the issue, Dawson and co-workers (Yan and Dawson 2001) introduced β- or γ-thiol-substituted amino acids in the N-terminal position.

Similar to Cys, these amino acids have thiol functionality at the β- or γ-position, which is involved in the ligation akin to the original strategy and provides the peptide with an additional thiol group that is subsequently desulfurized to get the target protein. However, desulfurization is not chemoselective in the presence of other unprotected cysteines in the sequences. This limitation can be handled by the introduction of selenocysteine as well as selenoamino acids (Hondal et al. 2001; Quaderer et al. 2001; Gieselman et al. 2001). The main advantage of using selenoamino acids instead of thio amino acids for ligation chemistry is that chemoselective deselenization can be done under mild conditions without affecting exposed Cys residues.

Further, to couple more than two peptide segments, the N-terminal Cys residue of one peptide segment is protected by thiazolidine (Thz) (Bang and Kent 2004). Following successful ligation of the two peptide segments, the N-terminal Thz moiety was removed to generate a native Cys residue peptide, which was then used in another NCL to obtain the desired protein (Scheme 8.8).

The peptide thioester is an important reactive intermediate in native chemical ligation, and a substantial amount of work has been done on the development of an efficient synthetic pathway for peptide thioesters. Although (Boc)-SPPS has been

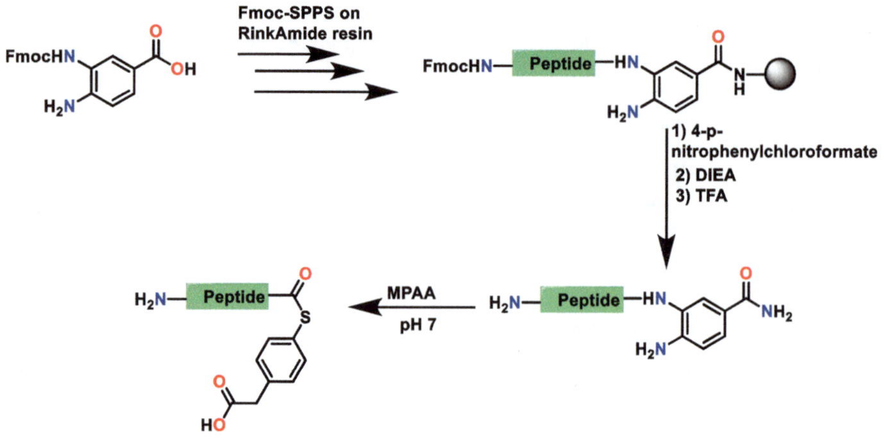

Scheme 8.9 Synthesis of Peptide Thioester through Fmoc-SPPS

used in the past (Schnölzer et al. 1992), however, in order to make it widely employed for protein synthesis, it must also be possible to prepare peptides using fluronylmethoxy carbonyl (Fmoc)-SPPS. This unmet requirement prompted the development of new synthetic ways for preparing the crucial intermediate using Fmoc strategies, which have shown to be extremely useful in protein synthesis (Zheng et al. 2013). Dawson and co-workers disclosed a strategy for preparing the peptide thioester in Fmoc SPPS. In this approach, 3,4-diaminobenzoic acid is initially attached to Rink amide resin (Blanco-Canosa and Dawson 2008). The first amino acid is then coupled into one of the amino groups of 3,4-diaminobenzoic acid, and the peptide chain is elongated. After completion of the whole sequence, the resin-bound peptide is treated with 4-nitro chloroformate under basic conditions to form *N*-acyl-benzimidazolinone (Nbz). Finally, the peptide is deprotected from the resin and reacted with aryl thiol to afford the corresponding thioester (Scheme 8.9). Interestingly, this method is applicable to all essential amino acids; however, the branching of Dbz functionality has been observed when a Gly residue is present at the C-terminus, which can be prevented by allyloxycarbonyl (alloc) protection of the linker's free amine (Mahto et al. 2011).

The native chemical ligation (NCL) has proven to be a highly successful method for the ligation of peptide segments. In combination with SPPS and NCL, it is now possible to synthesize larger proteins, which was very difficult in earlier days. In the future, it is expected that the native chemical ligation method will be intended to enable the synthesis of diverse complex protein targets for a variety of biological applications.

8.7 Automated Flow-Based Approach for Peptide Synthesis

The continuous shifting of the drug development from small molecules to peptides fuelled the demand for efficient automated peptide synthesizers which can synthesize peptides with high yield and purity within a very limited amount of time. Although numerous automated systems are available for Fmoc/Boc solid-phase peptide synthesis (SPPS), the complex setup, microwave heating and the requirement of a significant amount of excess reagent, and optimization of microwave heating limited their frequent use. In this context, the flow-based approach to peptide synthesis gains much attention in recent years (Simon et al. 2014; Mong et al. 2014). In the flow-based approach, the pre-heated reagents and solvents deliver to the solid support with a very high flow rate which increases the efficiency and speed of coupling cycles thereby reducing the time required for peptide synthesis and minimizing the requirement for excess reagents (Fig. 8.6). In addition, due to the very high coupling efficiency, it reduces the amount of waste generated during the synthesis.

Pentelute and co-workers made a significant breakthrough in the flow-based approach to peptide synthesis (Mijalis et al. 2017). They developed a fully automated, flow-based solid-phase polypeptide synthesis technology that allows for amide bond formation in 7 s, and the complete cycle for each amino acid is finished in 40 s. In addition, the presence of a UV–visible spectrometer enables the monitoring of Fmoc deprotection and gives information about the synthetic yields of each step. Further, the automated replacement of disposable reactors facilitates the quick shift from one peptide synthesis to the next. The efficiency of the technology is validated by synthesizing difficult sequences and longer peptides. For instance, the insulin B chain is synthesized in high purity and with greater yield compared to the synthesized by other existing methods. In traditional microwave-assisted peptide synthesis, epimerization of Cys and His residues is always a possibility; however, AFPS minimizes the chances of epimerization by increasing flow rate and thus shortening the residence time at high temperature for activated His and Cys monomers.

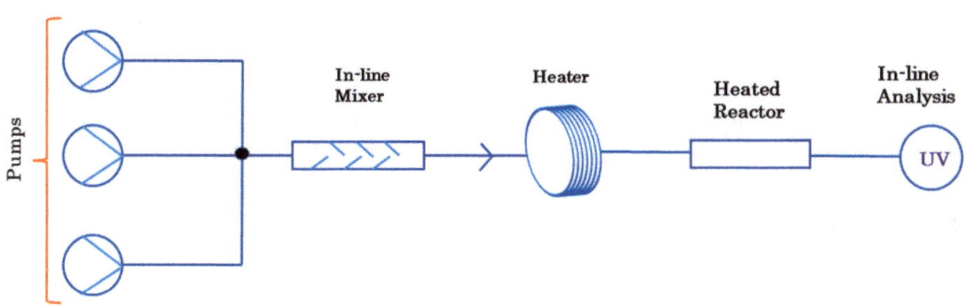

Fig. 8.6 Flow diagram of automated fast-flow peptide synthesizer described by Pentelute and Co-workers

In another interesting study (Hartrampf et al. 2020), Pentelute and co-workers described the proficiency of an automated fast-flow peptide synthesizer by synthesizing various protein chains that represent enzymes within hours. More intriguingly, compared to other conventional methods, AFPS produces better outcomes in terms of synthesis time, purity, and yields of the crude protein. Surprisingly, the physicochemical and enzymatic properties of the synthesized proteins are similar to those of the biologically expressed proteins. It is believed that shortly AFPS will show great potential for large-scale synthesis of peptides and proteins for therapeutic applications.

8.8 Enzymatic Approach for Peptide Synthesis

The enzymatic approach to peptide synthesis is one of the greener approaches for peptide synthesis. The chemical synthesis of peptides involves the use of vast amounts of toxic organic solvents which can be avoided using the chemoenzymatic method (Guzmán et al. 2007). In addition, there are no requirements for protected monomers and peptide synthesis can be carried out in a very selective and stereo-specific manner with high yield (Kumar and Bhalla 2005).

Proteolytic enzymes play a very important role in various biological systems and are ubiquitous in living organisms. It has a significant contribution to cell division, tissue dissociation, immune function, and protein recycling, among other important processes (Mótyán et al. 2013). Although the primary role of the proteolytic enzymes is the hydrolyzing proteins; however, they can also function in the reverse direction, thereby facilitating peptide bond formation.

Halling and co-workers reported protease catalyzed amide bond formation on the solid support (Ulijn et al. 2002). They explored PEGA1900 as a solid substrate for the attachment of amino acids. After the coupling of the first amino acid via Wang linker and followed by deprotection, the coupling next amino acid was performed in the presence of thermolysin. Finally, the peptide was cleaved from the resin and purified through HPLC (Fig. 8.7). The efficiency of the reaction was demonstrated by the formation of high yield and purity of the various dipeptides. Further, this methodology was explored for the formation of longer peptides where the yield varied from 3–55% for sequences containing up to six amino acids (Ulijn et al. 2003).

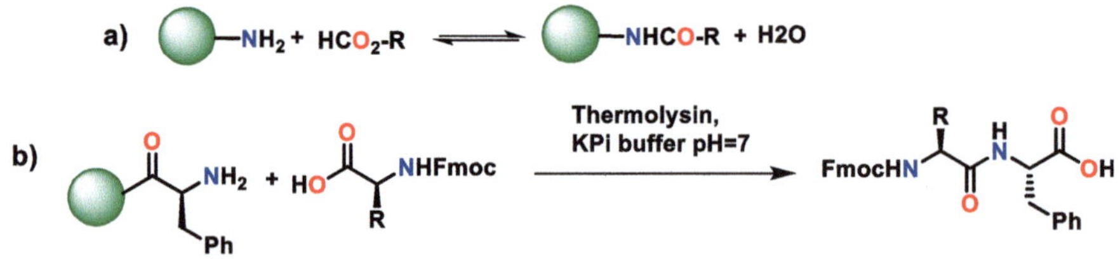

Fig. 8.7 (**a**) Amide bond formation/hydrolysis on solid substrate, (**b**) thermolysin catalyzed synthesis of dipeptides on PEGA190

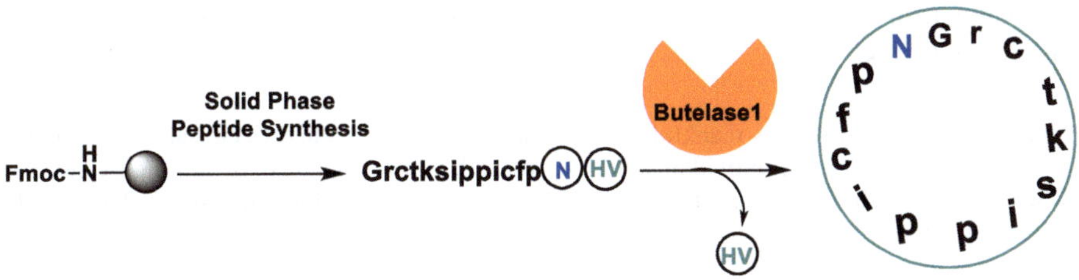

Fig. 8.8 Butelase catalyzed synthesis of cyclic dipeptide comprised of D-amino-acid

The cyclic peptide has great potential in peptide therapeutics due to its high biostability and bioavailability (Gibson and Lecci 2006). Inspired by the therapeutic import of the cyclic peptide, Guo and co-workers explored SrtA a class of transpeptidases found in gram-positive bacteria (Mazmanian et al. 1999) to form head-to-tail cyclization of a 19 residues peptide having a sorting signal of SrtA and a diglycine motif at its C- and N-termini, respectively (Wu et al. 2011). It is found that at 0.5 mM peptide and 40 mM SrtA concentration direct head-to-tail cyclization product is observed with an overall yield of 79%.

Tam and co-workers explored for the first time the Butelase-1 enzyme to catalyze cyclization of linear peptides composed of D-amino acid and Asn as well as Histidine and Valine (Fig. 8.8). The C-terminal Asn was used for the ligating side for butelase 1 and the dipeptide segment His-Val was used for substrate recognition. It has been found that butelase 1 efficiently catalyzes the cyclization of a variety of D-amino acid-containing linear peptide sequences (Nguyen et al. 2016).

Chemoenzymatic synthesis of peptides can be exploited as a surrogate for the traditional synthesis of the peptides due to their high yield and lesser use of hazardous chemicals. However, there are some limitations such as the difficulty in controlling the sequences and difficulty in synthesizing longer peptides which must be circumvented by engineering the protease and optimizing the reaction conditions.

8.9 Conclusion

As demonstrated in the chapter, recent advances in peptide synthesis enabled the synthesis of long and complex peptide sequences. In light of increasing interest in peptide therapeutics, the above-mentioned approaches and technologies will be of great assistance in the quick access to complex bioactive peptides with high yield and purity. The emergence of automated microwave peptide synthesizers and automated flow-based technology will play a central role in the future of synthesizing longer and more complex peptides. Peptide chemists in the twenty-first century are emphasizing the use of green, non-hazardous solvents and the improvement of atom economy in the protection/deprotection methods to make peptide synthesis more environmentally friendly. Moreover, the greener approaches must be widely used and applied to the various chemical alterations of peptides, such as cyclization and *N*-methylation, before they can be of real worth for commercial uses.

Acknowledgment G.K.R. is thankful to the National Institute of Pharmaceutical Education and Research, S. A. S. Nagar for providing a Senior Research Fellowship. R.M. thanks the Department of Science and Technology, New Delhi for DST Faculty Inspire Fellowship (No. DST/INSPIRE/04/2020/002499).

References

Anderson GW (1960) New approaches to peptide synthesis. Ann N Y Acad Sci 88:676–688

Angell YM, García-Echeverría C, Rich DH (1994) Comparative studies of the coupling of N-methylated, sterically hindered amino acids during solid-phase peptide synthesis. Tetrahedron Lett 35:5981–5984

Babu VVS, Rao RVR (2005) Microwave irradiated high-speed solution synthesis of peptide acids employing Fmoc-amino acid pentafluorophenyl esters as coupling agents. Indian J Chem 44B:2328–2332

Bang D, Kent SB (2004) A one-pot total synthesis of crambin. Angew Chem Int Ed 43:2534–2538

Barany G, Albricio F, Biancalana S et al (1992) Biopolymer syntheses on novel polyethylene glycol-polystyrene (PEG-PS) graft supports. Minnesota University Minneapolis Department of Chemistry, Minneapolis, MN

Blanco-Canosa JB, Dawson PE (2008) An efficient Fmoc-SPPS approach for the generation of thioester peptide precursors for use in native chemical ligation. Angew Chem Int Ed 120:6957–6961

Bondalapati S, Jbara M, Brik A (2016) Expanding the chemical toolbox for the synthesis of large and uniquely modified proteins. Nat Chem 8:407–418

Bonnamour J, Métro TX, Martinez J et al (2013) Environmentally benign peptide synthesis using liquid-assisted ball-milling: application to the synthesis of Leu-enkephalin. Green Chem 15: 1116–1120

Cabrele C, Martinek TA, Reiser O et al (2014) Peptides containing β-amino acid patterns: challenges and successes in medicinal chemistry. J Med Chem 57:9718–9739

Carpino LA (1993) 1-Hydroxy-7-azabenzotriazole. An efficient peptide coupling additive. J Am Chem Soc 115:4397–4398

Carpino LA, Han GY (1970) 9-Fluorenylmethoxycarbonyl function, a new base-sensitive amino-protecting group. J Am Chem Soc 92:5748–5749

Carpino LA, El-Faham A, Minor CA (1994) Advantageous applications of azabenzotriazole (triazolopyridine)-based coupling reagents to solid-phase peptide synthesis. J Chem Soc Chem Comm 2:201–203

Carpino LA, Imazumi H, El-Faham A et al (2002) The uronium/guanidinium peptide coupling reagents: finally, the true uronium salts. Angew Chem Int Ed 41:441–445

Chatterjee J, Gilon C, Hoffman A et al (2008) N-methylation of peptides: a new perspective in medicinal chemistry. Acc Chem Res 41:1331–1342

Chatterjee J, Rechenmacher F, Kessler H (2013) N-methylation of peptides and proteins: an important element for modulating biological functions. Angew Chem Int Ed 52:254–269

Cheloha RW, Watanabe T, Dean T (2016) Backbone modification of a parathyroid hormone receptor-1 antagonist/inverse agonist. ACS Chem Biol 1:2752–2762

Clark RJ, Craik DJ (2010) Invited review native chemical ligation applied to the synthesis and bioengineering of circular peptides and proteins. Peptide Sci 94:414–422

Collins JM, Collins MJ (2003) Biopolymers 71:267

Collins JM, Porter KA, Singh SK et al (2014) High-efficiency solid phase peptide synthesis (HE-SPPS). Org Lett 16:940–943

Coste J, Le-Nguyen D, Castro B (1990) PyBOP®: a new peptide coupling reagent devoid of toxic by-product. Tetrahedron Lett 31:205–208

Dawson PE, Muir TW, Clark-Lewis I et al (1994) Synthesis of proteins by native chemical ligation. Science 266:776–779

Declerck V, Nun P, Martinez J et al (2009) Solvent-free synthesis of peptides. Angew Chem Int Ed 48:9318–9321

Diao L, Meibohm B (2013) Pharmacokinetics and pharmacokinetic–pharmacodynamic correlations of therapeutic peptides. Clin Pharmacokinet 52:855–868

Dirksen A, Dawson PE (2008) Expanding the scope of chemoselective peptide ligations in chemical biology. Curr Opin Chem Biol 12:760–766

Domalaon R, Zhanel GG, Schweizer F (2016) Curr Top Med Chem 16:141–155

El-Faham A, Funosas RS, Prohens R et al (2009) COMU: a safer and more effective replacement for benzotriazole-based uronium coupling reagents. Chem Eur J 15:9404–9416

Erdelyi M, Gogoll A (2002) Rapid microwave-assisted solid phase peptide synthesis. Synthesis 11: 1592–1596

Fosgerau K, Hoffmann T (2015) Peptide therapeutics: current status and future directions. Drug Discov Today 20:122–128

Galanis AS, Albericio F, Grøtli M (2009) Solid-phase peptide synthesis in water using microwave-assisted heating. Org Lett 11:4488–4491

Gibson SE, Lecci C (2006) Amino acid derived macrocycles-an area driven by synthesis or application? Angew Chem Int Ed 45:1364–1377

Gieselman MD, Xie L, Van Der Donk WA (2001) Synthesis of a selenocysteine-containing peptide by native chemical ligation. Org Lett 3:1331–1334

Guzmán F, Barberis S, Illanes A (2007) Peptide synthesis: chemical or enzymatic. Electron J Biotechnol 10:279–314

Haase C, Seitz O (2008) Extending the scope of native chemical peptide coupling. Angew Chem Int Ed 47:1553–1556

Harris JM, Chess RB (2003) Effect of pegylation on pharmaceuticals. Nature Rev Drug Disc 2:214–221

Hartrampf N, Saebi A, Poskus M et al (2020) Synthesis of proteins by automated flow chemistry. Science 368:980–987

Henninot A, Collins JC, Nuss JM (2018) The current state of peptide drug discovery: back to the future? J Med Chem 61:1382–1414

Hondal RJ, Nilsson BL, Raines RT (2001) Selenocysteine in native chemical ligation and expressed protein ligation. J Am Chem Soc 123:5140–5141

Jad YE, Kumar A, El-Faham A et al (2019) Green transformation of solid-phase peptide synthesis. ACS Sustain Chem Eng 7:3671–3683

Jiang W, Zhang B, Fan C et al (2017) Mirror-image polymerase chain reaction. Cell Discov 3:1–7

Kumar D, Bhalla TC (2005) Microbial proteases in peptide synthesis: approaches and applications. Appl Microbiol Biotechnol 68:726–736

Liu Y, Hu Y, Liu T (2012) Recent advances in non-peptidomimetic dipeptidyl peptidase 4 inhibitors: medicinal chemistry and preclinical aspects. Curr Med Chem 19:3982–3999

Macmillan D (2006) Evolving strategies for protein synthesis converge on native chemical ligation. Angew Chem Int Ed 45:7668–7672

Mahindra A, Sharma KK, Jain R (2012) Rapid microwave-assisted solution-phase peptide synthesis. Tetrahedron Lett 53:6931–6935

Mahindra A, Nooney K, Uraon S et al (2013) Microwave-assisted solution phase peptide synthesis in neat water. RSC Adv 3:16810–16816

Mahto SK, Howard CJ, Shimko JC (2011) A reversible protection strategy to improve Fmoc-SPPS of peptide thioesters by the N-acylurea approach. Chembiochem 12:2488–2494

Mazmanian SK, Liu G, Ton-That H et al (1999) Staphylococcus aureus sortase, an enzyme that anchors surface proteins to the cell wall. Science 285:760–763

Merrifield RB (1963) Solid phase peptide synthesis I. the synthesis of a tetrapeptide. J Am Chem Soc 85:2149–2154

Merrifield RB (1985) Solid phase synthesis (Nobel lecture). Angew Chem Int Ed 24:799–810

Merrifield RB (1986) Solid phase synthesis. Science 232:341–347

Mijalis AJ, Thomas DA, Simon MD et al (2017) A fully automated flow-based approach for accelerated peptide synthesis. Nat Chem Biol 13:464–466

Mong SK, Vinogradov AA, Simon MD et al (2014) Rapid total synthesis of DARPin pE59 and barnase. Chembiochem 15:721–733

Mótyán JA, Tóth F, Tőzsér J (2013) Research applications of proteolytic enzymes in molecular biology. Biomol Ther 3:923–942

Murray JK, Gellman SH (2005) Application of microwave irradiation to the synthesis of 14-helical β-peptides. Org Lett 7:1517–1520

Muttenthaler M, Albericio F, Dawson PE (2015) Methods, setup and safe handling for anhydrous hydrogen fluoride cleavage in Boc solid-phase peptide synthesis. Nat Protoc 7:1067–1083

Nguyen GK, Hemu X, Quek JP (2016) Butelase-mediated macrocyclization of d-amino-acid-containing peptides. Angew Chem Int Ed 128:12994–12998

Pedersen SL, Tofteng AP, Malik L (2012) Microwave heating in solid-phase peptide synthesis. Chem Soc Rev 41:1826–1844

Quaderer R, Sewing A, Hilvert D (2001) Selenocysteine-mediated native chemical ligation. Helv Chim Acta 84:1197–1206

Rapp W, Zhang L, Habich R et al (1988) Polystyrene-polyoxyethylene graft copolymers for high-speed peptide synthesis, pp 199–201

Rodríguez H, Suarez M, Albericio F (2010) A convenient microwave-enhanced solid-phase synthesis of short chain N-methyl-rich peptides. J Pept Sci 16:136–140

Sato AK, Viswanathan M, Kent RB et al (2006) Therapeutic peptides: technological advances driving peptides into development. Curr Opin Biotechnol 17:638–642

Schnölzer M, Alewood P, Jones A et al (1992) In situ neutralization in Boc-chemistry solid phase peptide synthesis: rapid, high yield assembly of difficult sequences. Int J Pept Protein Res 40: 180–193

Simon MD, Heider PL, Adamo A et al (2014) Rapid flow-based peptide synthesis. Chembiochem 15:713–720

Subirós-Funosas R, Prohens R, Barbas R et al (2009) Oxyma: an efficient additive for peptide synthesis to replace the benzotriazole-based HOBt and HOAt with a lower risk of explosion. Chem Eur J 15:9394–9403

Torbeev VY, Kent SB (2007) Convergent chemical synthesis and crystal structure of a 203 amino acid "covalent dimer" HIV-1 protease enzyme molecule. Angew Chem Int Ed 46:1667–1670

Tsuda Y, Okada Y (2010) Amino acids, peptides and proteins in organic chemistry. Wiley, Hoboken, NJ, pp 201–251

Uhlig T, Kyprianou T, Martinelli FG (2014) The emergence of peptides in the pharmaceutical business: from exploration to exploitation. EuPA Open Proteom 4:58–69

Ulijn RV, Baragaña B, Halling PJ et al (2002) Protease-catalyzed peptide synthesis on solid support. J Am Chem Soc 124:10988–10989

Ulijn RV, Bisek N, Halling PJ et al (2003) Understanding protease catalysed solid phase peptide synthesis. Org Biomol Chem 1:1277–1281

Våbenø J, Haug BE, Rosenkilde MM (2015) Progress toward rationally designed small-molecule peptide and peptidomimetic CXCR4 antagonists. Future Med Chem 7:1261–1283

Verlander M (2007) Industrial applications of solid-phase peptide synthesis–a status report. Int J Pept Res Ther 13:75–82

Wang L, Wang N, Zhang W et al (2022) Therapeutic peptides: current applications and future directions. Signal Transduct Target Ther 7:48

Werner HM, Cabalteja CC, Horne WS (2016) Peptide backbone composition and protease susceptibility: impact of modification type, position, and tandem substitution. Chembiochem 17:712–718

Wu Z, Guo X, Guo Z (2011) Sortase A-catalyzed peptide cyclization for the synthesis of macrocyclic peptides and glycopeptides. Chem Commun 47:9218–9220

Yan LZ, Dawson PE (2001) Synthesis of peptides and proteins without cysteine residues by native chemical ligation combined with desulfurization. J Am Chem Soc 123:526–533

Zheng JS, Tang S, Huang YC et al (2013) Development of new thioester equivalents for protein chemical synthesis. Acc Chem Res 46:2475–2484

Peptidomimetics in Medicinal Chemistry: The Art of Transforming Peptides to Drugs

9

K. Nagarajan and Garima Kapoor

Abstract

A peptidomimetic is a small protein-like chain designed to mimic a peptide with adjusted molecular properties such as enhanced stability or biological activity. Small-molecule medicines acting as receptor ligands or enzyme inhibitors are produced using this method. The principles of synthetic methods behind the construction of peptidomimetic bioactive substances are described in this chapter on peptidomimetics. Topics include peptide scaffold-based peptidomimetics, which focuses on design and synthetic considerations, and the introduction and scope of peptidomimetics in chemistry in the synthetic pathway of amino acids. Medicinal chemistry peptidomimetic therapeutic uses and delivery methods are covered in detail. In this chapter, case studies of the design, synthesis and amino acid composition of fullerene-based peptidomimetics as anticancer and hepatoprotective medicines, as well as the analysis of fullerene-based smaller chain peptidomimetics targeting tuberculosis, are concentrated. The entire data analysis focuses on the development of ligands from hit to lead, demonstrating the effective use of peptidomimetics in drug discovery.

Keywords

Bioactive · Fullerene · Glycopeptides · Hepatoprotective · Peptides · PeptiHdomimetics

K. Nagarajan · G. Kapoor (✉)
KIET School of Pharmacy, KIET Group of Institutions, Ghaziabad, India
e-mail: k.nagarajan@kiet.edu; garima.kapoor@kiet.edu

P. P. Singh (ed.), *Recent Advances in Pharmaceutical Innovation and Research*,
https://doi.org/10.1007/978-981-99-2302-1_9

9.1 Introduction and Scope of Peptidomimetics

9.1.1 Introduction

Numerous peptides, including hormones, neuroactive peptides and vasoactive peptides, have been found; these peptides play important roles in pharmacy because of their interactions with membrane-bound receptors. Additionally, there are significant limitations that restrict their use, such as limited oral absorption, low diffusion in the CNS and rapid elimination (Giannis and Thomas 1993). Despite all of these drawbacks, biomedical research is constantly aimed at developing novel therapies based on peptides and proteins while keeping the characteristics that give these substances their biological activity. This includes both structural and functional modifications. The creation of peptidomimetics meets each of these requirements (Gante 1994; Adessi and Soto 2002). In this approach, peptides and proteins are considered tools for the discovery of other classes of compounds.

9.1.2 Definition and Classes

A compound having a secondary structure like the original peptide and binds to enzymes or receptors more potently than the peptide from which it is derives id called as peptidomimetic. Overall, the natural peptide actions are either augmented (antagonist) or suppressed (inhibitor) (agonist). The use of peptidomimetics in both organic and medicinal chemistry has shown significant promise since they were first proposed as a novel approach to the development of therapeutic candidates (Liskamp 1994). Peptidomimetics exhibit higher oral bioavailability than native peptides and have longer biological activity due to less enzymatic breakdown, in addition to being significantly more efficient and selective than native peptides, which results in less adverse effects (Olson et al. 1993) Knowledge of the electrical and conformational properties of the original peptide and its receptor, or active site, is the main focus of peptidomimetics creation. Therefore, several fundamental criteria must be considered while creating peptidomimetics as molecules with potential biological activity (Farmer 1980), including the following:

- A non-peptide framework is used in place of the peptide backbone: Peptide bonds may be eliminated from the template if the presence of an amide bond has no impact on biological function or if the amide bond is not accessible to the active site.
- Preservation of biologically active side chains, which are essential to the pharmacophore. The development of second-generation mimetics may involve several alterations to enhance biological activity, such as changing the chain length, adding restrictions, replacing the cyclopeptide link with a covalent one, and introducing isosteric replacements (Marshall 1993).

According to their structural and functional traits, peptidomimetics can be grouped into three classes (Ripka and Rich 1998):

- Type I, sometimes referred to as structural mimetics, are equipped with all the features required to interact with an enzyme or receptor in a specific spatial orientation. They show similarities between the local topography and the natural substrate.
- Type II, or functional mimics, lack obvious structural similarities and instead base their resemblance to the genuine chemical on how they interact with the target receptor or enzyme.
- Functional-structural mimetics, often referred to as Type III mimetics, are frequently believed to have a scaffold with a structure different from that of the substrate, in which all the functional groups necessary for biological interactions are mounted in a well-defined manner. Type II mimics, sometimes referred to as functional mimics, differ from type I mimics in that they rely on their interactions with target receptors or enzymes. The literature contains numerous examples of situations when an artificial framework takes the role of the peptide backbone and carries a peptide.

9.2 Peptidomimetic Design and Synthetic Considerations

The nature of the individual amino acid side chains and the polyamide peptide backbone itself governs the physical and chemical properties of peptides and proteins. Fig. 9.1 provides the structures of the 20 basic amino acids. Hydrophobic residues and hydrophilic residues are two categories of amino acids. Both those with aliphatic side chains (Ala, Val, Ile, Leu and Met) and those with aromatic side chains make up the first category (Phe, Tyr, Trp). The hydrophilic group of amino acids consists of those with neutral, polar (Ser, Thr, Asn, Gln), acidic (Asp and Glu) or basic side chains (Lys, Arg and His). Cys and Pro, two different amino acids, have unique characteristics that make them distinct. A Cys residue has a thiol group that can be coupled oxidatively to another Cys residue to form a disulphide bond that stabilises secondary and/or tertiary structure or holds two distinct peptide chains together. On the other hand, some proteins include free thiols, which operate as nucleophiles in proteolytic enzymes, ligands for metal chelation, or as carboxyl activators in acyl transferases. Pro is a cyclic residue that modifies the shape of the peptide or protein backbone in a specific way. Because there are numerous conformations that are energetically feasible for each residue in small peptides, they frequently exhibit significant structural flexibility. Three torsional angles can be used to define the peptide backbone conformation (Fig. 9.2).

The rationale behind the bioisosteric alteration of lead compounds can be found in an analysis of the similarities between numerous physicochemical properties of atoms, groups, radicals and molecules (Langmuir 1919). The term 'bioisosterism' was coined as a result of the widespread application of isosterism to modify

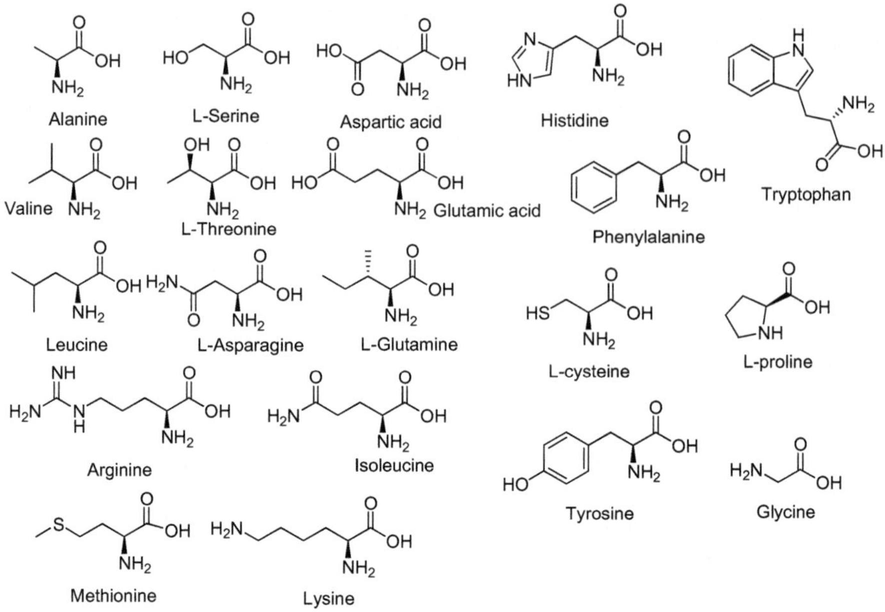

Fig. 9.1 Twenty primary amino acids

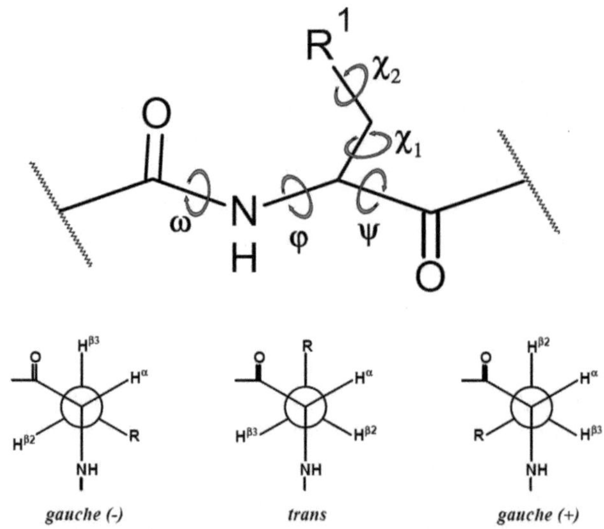

Fig. 9.2 Three staggered rotamers n L-amino acids with backbone, side chain torsional angles and Newman projection

biological function. Bioisosteres are all atoms and molecules that meet the broadest definition of isosteres and exhibit a similar type of biological activity.

In order to increase the metabolic resistance of the peptide, peptide bonds are typically replaced with suitable moieties in peptidomimetic chemistry. This modification of the peptide backbone typically involves the introduction of new fragments

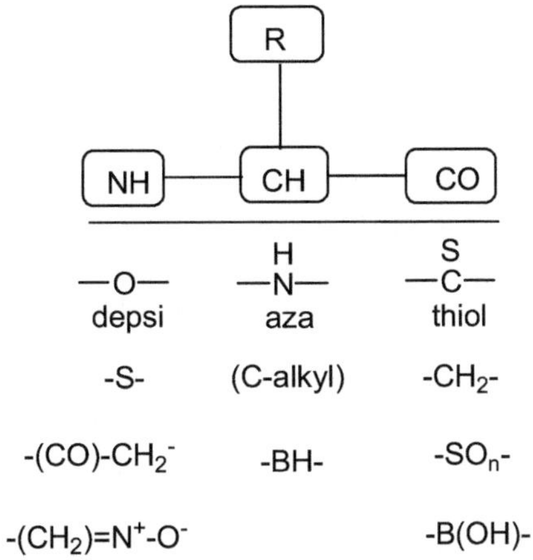

Fig. 9.3 Amino acid structure isosterically alteration

as well as isosteric or isoelectronic exchanges between units in the peptide chain. A peptide's conformational effects may be modified by the addition of amide bond isosteres since the stereoelectronic properties and hydrogen-bonding abilities of various amide bond substitutes vary significantly (Choudhary and Raines 2011). There have been developed several potential bioisosteric amide bond substitutes (Fig. 9.3). The ability of these groups to precisely replicate the function of the parent amide bond in the setting of the pharmacophore determines their eligibility as bioisosteres.

A hierarchical approach that includes several structural alterations to the original bioactive peptide, such as size reduction, alanine scanning, d-amino acid scanning, N-methylation and the introduction of local and global constraints to define the bioactive conformation, is used to introduce chemical modifications and comprehend structure-activity data.

In order to identify the crucial peptide sequence directly interacting with the active site of the target enzyme or receptor, the size of the original peptide sequence is reduced. The outcome of this process is crucial for creating small molecule peptidomimetic derivatives, and this step is approached by testing a variety of peptides made by methodically removing amino acids from either the N-termini or the C-termini, which leads to the minimal peptide sequence carrier of bioactivity. This is the hit peptide sequence that includes the pharmacophore and is considered for further peptide structure adjustments.

The process of synthesizing and biologically analysing a variety of peptides that have an alanine residue in place of a different amino acid from the original bioactive peptide sequence is known as alanine scanning. Since there are no side chains engaging with the target receptor, the bioactivity of the peptides that substitute an alanine residue for essential amino acids in the original sequence is lost. This information is crucial for determining which amino acids in the reference peptide actively interact with the target enzyme or receptor.

A strategy related to alanine scanning is the systematic introduction of d-amino acids into the parent peptide sequence to identify the important residues that contain bioactivity. Furthermore, this method provides insight into the structural organization of the bioactive conformation since the change in configuration leads to a distinct arrangement of the side chains.

Another tried-and-true method for determining the function of each amino acid that makes up a bioactive peptide is the production and testing of N-methylated peptides. One method for determining which amino acids function as hydrogen-bonding donors in the interaction with the target enzyme or receptor is the alkylation of amide bonds. In addition, the N-methylation produces a tertiary amide bond that is likely to achieve cis/trans equilibrium at the amide bond, helping to shed light on the relationship between the peptide's preferred shape and its bioactivity.

In order to create a compound with reduced peptide character and potentially improve both the intrinsic binding affinity to the target receptor or enzyme and the pharmacokinetics/pharmacodynamics (PK/PD) profile from a pharmacological standpoint, the minimal peptide sequence carrier of bioactivity and the role of each amino acid constituting the parent peptide are identified.

To increase the peptide's resistance, a great deal of work has been put into replacing peptide bonds with suitable moieties; this type of peptide backbone modification typically entails the introduction of new fragments as well as the isosteric or isoelectronic exchange of certain peptide chain units. Usually, single amino acid changes and the addition of dipeptide analogues are used to accomplish this purpose. Both strategies are intended to either limit the side chains' rotational freedom or enable the topological orientation needed by the pharmacophore, which would provide an orderly configuration that might also mimic the secondary structure, such as helices or turns.

A single amino acid has been found to undergo modifications at every point (Haubner et al. 1997). In particular, (1) the amino group can be substituted with isosteric atoms or groups, such as oxygen, keto-methylene, or N-hydroxyl moieties; (2) the alpha carbon can be substituted with nitrogen atoms, boron atoms or C-alkyl to produce quaternary amino acids; and (3) the carbonyl group can be substituted with thiol, methylene, or phosphinic (Fig. 9.3). It has also been suggested to create peptides known as 'retro-inverso peptides', which include an amino acid moiety with the original amino and carboxylic groups' relative positions reversed.

A few amide bond isosteres that resemble the structural features of the peptide bond and, in some circumstances, alter the conformational profile and hydrogen-bonding ability have been proposed (Fig. 9.4). The introduction of amide bond substitutes is the main strategy for backbone alteration with the aim of enhancing the peptide stability in vivo (Burger 1995).

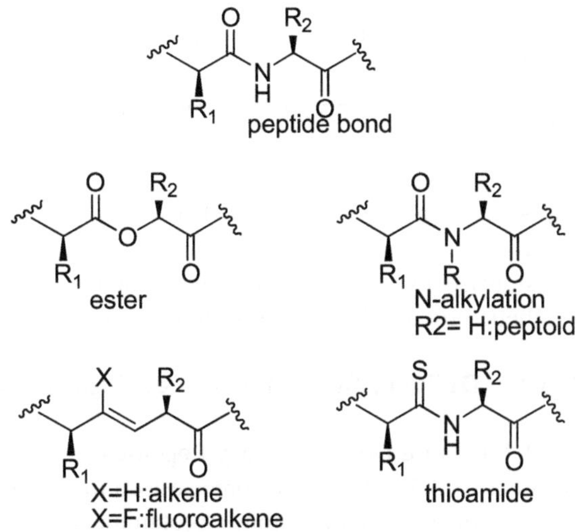

Fig. 9.4 Common peptide bond isosteres

9.3 Applications and Delivery Strategies of Peptidomimetic Therapeutics in Medicinal Chemistry

9.3.1 Applications

N. Andre Sasaki developed a unique synthetic methodology to produce enantiopure 3-, 4- and substituted prolines. In this, 3-, 4- and 5-monosubstituted prolines can be used as probes in the search for powerful neuroexcitatory compounds or the creation of conformationally restricted peptides. These prolines are stereochemically well-defined. Understanding the structural prerequisites for these amino acids' binding to the proteins involved in neurotransmission may be improved by simulating various conformations of excitatory amino acids (Bridges et al. 1991). Syntheses of conformationally constrained peptides are now showing promise for the creation of pharmaceuticals derived from peptides. The process for making 3-, 4- and 5-monosubstituted prolines begins from a single chiral derivative produced from L-serine (Sasaki 1999).

On a wide scale, it is simple to manufacture the chiral educts from N-Boc-L-serine methyl ester, and both (R)- and (S)-2,3-O-isopropylidene glycerol and glyceraldehyde are readily accessible at a fair price. In order to provide any one of the four stereoisomers of 3-, 4- and 5-substituted prolines in enantiomerically pure form, the above-described synthetic approach provides a practical option (Humphrey et al. 1994). But the current approach makes it simple to obtain cis- and trans-5-substituted prolines in enantiopure form. Only a few effective synthetic methods have been described for the asymmetric synthesis of 4-abistoured prolines, even though the widely accessible trans-4-hydroxy-L-proline has been used in the synthesis of numerous 4-substituted proline derivatives. The hydroxyethyl group will be

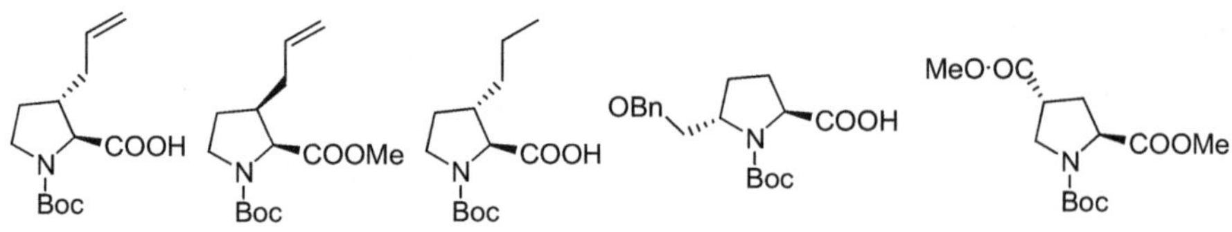

Fig. 9.5 Example of enantiopure 3-, 4- and substituted prolines

transformed by a variety of 4-substituted enantiopure prolines using the current synthetic process (Fig. 9.5) (Hashimoto et al. 1994).

9.3.1.1 Delivery Strategies of Peptidomimetic

9.3.1.1.1 Gene Delivery Using Peptides

In order to deliver genes to precise locations, positively charged macromolecules such as poly (L-lysine), histones, protamine, or poly (L-ornithine) may be linked to a cell-specific ligand and electrostatically bonded to plasmids. The resulting complexes maintain their capacity to precisely interact with target cell receptors, which causes the complex to enter the cells by receptor-mediated internalization. Glycoproteins, transferrin, polymeric immunoglobulin, insulin, epidermal growth factor (EGF), lectins, folate, malaria circumsporozoite protein, a2-macroglobulin, CD3-T cell, sugars, integrins, thrombomodulin, surfactant protein A and B, mucin and the c-kit receptor are some of the receptor ligands that are currently being researched.

Site-specific gene delivery and expression are influenced by factors such as the degree of DNA condensation, the method of complexation, the molecular weights of the plasmid and polycations, and the number of ligand residues bound per polycation molecule. Synthetic peptides rich in lysine have been used to avoid the cytotoxicity and molecular heterogeneity of poly (L-high lysine) for the goal of gene transfer. It is common knowledge that the active regions of enzymes, receptor ligands and antibodies are typically composed of 5–20 amino acids. It should therefore be able to produce short synthetic peptides that mimic the active areas of viral proteins and produce peptide/DNA complexes that are just as efficient as viruses but without their downsides. A DNA binding peptide to condense the plasmid; a targeting peptide to impart cell and/or tissue specificity; a lytic peptide to stimulate the endosomal release of the plasmid; and a NLS peptide to facilitate nuclear entrance of the plasmid are the main elements of a peptide-based delivery system.

Synthetic peptide-based systems featuring an endosomolytic motif and a lysine-rich DNA binding motif have been developed for in vivo gene delivery. One such gene delivery system consists of a galactosylated peptide that condenses the plasmid into monodisperse nanoparticles of about 100 nm diameter and enables specific recognition and binding to asialoglycoprotein receptors, as well as an amphipathic, pH-selective peptide that enables the plasmid to exit the endosomes before their fusion with lysosomes and entry into the cytoplasm.

Tryptophan is an amino acid that improves DNA condensation and binding, which suggests that it may be advantageous to include hydrophobic interactions in DNA complexes. The stability of peptide-based DNA delivery systems would be increased by peptides containing moieties that allow cooperative hydrophobic behaviour of alkyl chains of cationic lipids. Two general classes of lipopeptide analogues of Tyr-Lys-Ala-Lys_n-Trp-Lys peptides have been prepared by including a hydrophobic anchor. The general structures are N, N-dialkyl-Gly-Tyr-Lys-Ala-Lys_n-Trp-Lys and N^{α},N-diacyl-Lys-Lys_n-Trp-Lys. These peptides distinguish themselves from their source structures by forming micelles through self-association in aqueous solutions. Dialkyl or diacyl chains increase the cationic peptides' capacity to bind DNA and lessen the complexes' tendency to aggregate in ionic environments.

According to recent studies, short synthetic peptides with various analogues of the first 19–23 amino acid residues of influenza hemagglutinin protein (HA) terminus may be appealing due to their pH-dependent lytic properties, with little activity at pH 7 but an approximately 100-fold increase in transfection efficiency at pH 5. The aspartyl and glutamyl side chain carboxyl groups are protonated, allowing the peptides to take an a-helical shape that may be incorporated into the membrane bilayer, which reveals the lytic properties.

JTS-1, Gly-Leu-phe-glu-le-leu-glu-le-leu-glu-llu-leu-trp-glu-le-Leu-Leu-Glu-Ala, an amphipathic membrane associating peptide, was also recently created. Only strongly apolar amino acids are present in the hydrophobic face, while Leu-Leu-Glu-Ala has also recently been created. Only strongly apolar amino acids are present on the hydrophobic face, whereas negatively charged glutamic acid residues predominate on the hydrophilic face at normal pH. The hydrophobic face of JTS-1 seems to promote self-association, hole creation in one side of the endosomal membrane, membrane destabilization and membrane rupture. The condensing reaction gives cationic DNA complex with peptide Tyr-Lys-Ala-Lyss-Trp-Lys having negatively charged JTS-1, which spontaneously incorporates through electrostatic interactions to form the ternary complex. At a given charge ratio of condensing peptide to the plasmid, the transfection efficiency has been shown to be proportional to the concentration of the endosomolytic peptide added to the complex. In vitro transfection efficiency up to 10,000-fold higher than that of DNA/Tyr-Lys-Ala-Lys-Trp-Lys_8-Trp-Lys complex alone has been reported (Hillery et al. 2001).

9.3.1.2 Delivery of Antisense Drugs to the Brain

Novel antisense oligonucleotides called peptide nucleic acids (PNA) have a polypeptide backbone. Their transport to the CNS can be aided by receptor-mediated transcytosis. For instance, it has been demonstrated that PNAs can be attached to anti-transferrin (OX26) receptor antibodies to improve brain absorption of the PNAs while maintaining their ability to hybridise to target mRNA. Antisense drugs, however, need to be released from endosomes into the cytosol in order to have pharmacological effects in living organisms after being delivered to cells via receptor-mediated endocytosis processes. Therefore, the endosomal release must be supported by the design and optimization of current antisense drug delivery systems (Boado 1995).

9.4 Fullerene-Based Peptidomimetics: Design, Preparation and Amino Acid Composition as Anticancer Agents

A method for creating shorter chain peptides using liquid phase peptide synthesis and testing them on ovarian cancer and leukaemia cell lines for in vitro anticancer activity was developed and studied. Four peptide scaffolds were chosen and manufactured utilizing Lipinski rule, Bowman index and HDAC docking as the basis for rational drug design., viz., peptidomimetics Fullerene C60-Arg-Lys (RK), Fullerene C60-Lys-Hist (KH), Fullerene C60-Arg-Arg-Hist (RRH) and Fullerene C60- Arg-Arg-Lys (RRK) (Lipinski et al. 1997). Using the MTT assay, the anticancer activity of various drug concentrations was measured. The test compounds' IC50 values were then compared to those of reference medicines as shown in Table 9.1. Peptides were combined with buckminsterfullerene (fullerene C60) and provided stability and better bioavailability (Sato et al. 2013). Peptidomimetics Fullerene C60- Arg-Arg-Tys (RRK) (Fig. 9.6) was found to be the best anticancer agent against leukaemia cell line with an IC50 value of 8 µg/mL whereas test drug 2 showed the maximum activity against the ovarian cancer cell line with an IC50 value of 7.1 µg/mL. Fullerene C60-Arg-Arg-Hist showed the maximum activity against the ovarian cancer cell line with % inhibition of 17.1%, which clearly indicates that these peptidomimetics are potent anticancer agents for leukaemia and ovarian cancer cell lines (Smakhtin et al. 2002). The results show that in recent years drugs against cancer have rapidly increased. Peptide anticancer drugs have shown properties of LTX-315, a novel synthetic anticancer peptide, against syngeneic B16 melanomas (Ding et al. 2013). Niclosamide has been identified to potently inhibit the activation, nuclear translocation and transactivation of STAT3 (Camilio and Rekdal 2014). Recently inhibitors of the 20S proteasome, a validated target in the treatment of multiple myeloma has been of great significance. The synthesized compounds showed a good inhibitory profile against the ChT-L activity of the 20S proteasome (Li et al. 2008). Compounds bearing a ß-alanine residue at the P2 position were the most active (Chen et al. 2013). Peptide drugs have more bioavailability by using fullerene C_{60} as nanotubes and estimating its anticancer properties (Veber et al. 2002). On comparing all the four test drugs for their in vitro MTT assay

Table 9.1 MTT assay by HL-60 cell line

S. No.	Compound	% cell viability				IC_{90} (µg/mL)
		0	10	30	100	
1	C_{60}-RK	100	50	20	20	10
2	C_{60}-KH	100	78	23	20	17.5
3	(C_{60}-RRH)	100	104	28	20	25
4	(C_{60}-RRK)	100	34	25	20	8.0

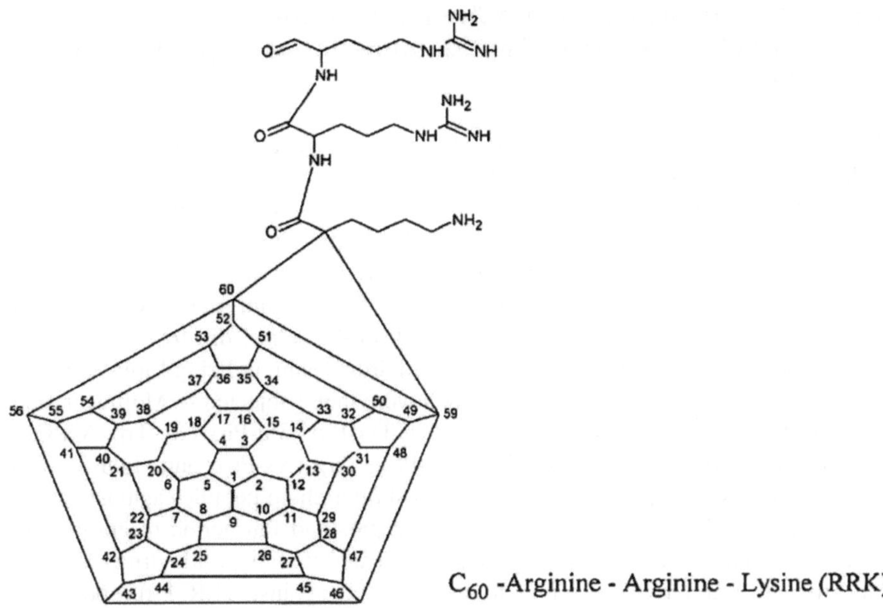

Fig. 9.6 Fullerene C60-RRK

Table 9.2 MTT assay of ovarian cell line

S. No.	Concentration (μM)	% inhibition			
		C_{60}-RK	C_{60}-KH	(C_{60}-RRH)	(C_{60}-RRK)
1	0.156	2.56	3.51	−5.43	6.97
2	0.313	0.71	1.98	3.07	5.8
3	0.625	2.63	−1.09	−6.28	7.86
4	1.25	−2.86	1.57	−5.39	0.88
5	2.5	−3.42	28.26	−3.23	1.02
6	5.0	0.46	44.85	−7.58	23.06
7	10.0	57.72	58.23	32.8	50.62
8	20.0	77.93	65.7	57.12	63.64
IC_{50} value		**9.3**	**7.1**	**17.1**	**10**

by using two different cell lines (HL-60 cell line and ovarian cell line), it was found that the test drug Fullerene C60-RRK had very high anticancer potency and primary lead for research shown in Table 9.2 (Hasan et al. 2006). As the drugs are actively binding to HDAC enzyme thus they can control abnormal regulation of apoptosis, or programmed cell death which is main implication in many diseases, including cancer, stroke, myocardial infarction, viral infections and several other diseases (Das and Khanna 1997). Apoptotic failures can lead to cancer resistance towards chemo- or radiotherapy. Therefore, molecules and pathways that govern the PCD process have become an attractive target for the development of novel anticancer strategies (Goldberg et al. 1997).

9.5 Design, Synthesis and Amino Acid Composition of Fullerene-Based Peptidomimetics as Hepatoprotective Agents

Because peptide and protein-protein interactions have a dominating position in molecular recognition and signalling in living systems, the design and manufacturing of peptidomimetics are extremely significant (Chafee and Greisheimer 1974). One of the amino acids in shorter chain di-, tri-, tetra-, penta- and hexapeptides was selected together with the remaining amino acids using a hit-and-trial strategy based on a literature review of a four-point search (Chazov et al. 1987). Two hundred fifty shorter chain peptides (50 each of di-, tri-, tetra-, penta- and hexapeptides) were treated to ligand-based drug design utilizing Lipinski's Rule of Five. The best result was 38 leads, which were then subjected to Molinspiration. This resulted in ten peptides, namely, Ala-Tyr, Tyr-Pro, Glu-Pro, Gly-Trp, Ala-Glu-Pro, Ile-Asp-Pro, Lys-Glu-Leu, Lys-Phe-Pro, Arg-Lys-Pro and Leu-Ser-Pro. Docking studies were performed with the ten shorter chain peptides against various targets such as CYP 3A4 enzyme, TLR-4 receptor and Purinergic receptor (Rink et al. 2010). The results of docking clearly indicated that dipeptide drugs Val-Tyr, Ala-Trp and Val-Trp were found to be effective against TLR-4 receptors and dipeptide drugs Val-Tyr, Gly-Trp and Phe-Pro were found to be effective against Purinergic receptors (Grosdidica et al. 2011a, b). In addition, tripeptide drugs Arg-Lys-Pro, Lys-Glu-Leu and Ala-Glu-Pro were found to be effective against the CYP 3A4 enzyme target (Sapakal et al. 2008; Saraswat and Churchard 2013). The five best leads of ligand-based drug design outcomes, namely, Val-Tyr, Gly-Trp, Ala-Trp, Lys-Glu-Leu and Arg-Lys-Pro, were synthesized and made into peptide scaffolds using fullerene soot and were tested for hepatoprotective activity initially by in vitro testing followed by in vivo animal testing by selecting the right dose of the drugs with acute toxicity testing using OECD 423 guidelines (Haura et al. 2011). All five peptidomimetics test compounds were put through in vitro testing using the DPPH technique, reducing power method, serum glutamate oxaloacetate transaminase (SGOT) method, and serum glutamate pyruvate transaminase (SGPT) method (Kafsara et al. 2006). According to the DPPH assay results, the best leads were found to be FS-GW and FS-KEL with IC50 values of 0.5 and 4 g/mL, respectively, in comparison to conventional ascorbic acid's 34.5 g/mL. The SGOT results with the average values of all test chemical concentrations (1 g/mL, 2 g/mL, 4 g/mL, 8 g/mL, 16 g/mL and 32 g/mL) revealed that the tripeptides FS-KEL (−0.09 U/L) and FS-RKP (−0.18 U/L) were more effective than the ascorbic acid standard (−0.18 U/L) (Jayaprakash et al. 2001). Similarly, the results of SGPT with average readings of all concentrations of the test compound (1 μg/mL, 2 μg/mL, 4 μg/mL, 8 μg/mL, 16 μg/mL and 32 μg/mL) showed that test drugs FS-VY (−0.06 U/L) and FS-KEL (−0.15 U/L) were most potent compounds in comparison with the standard ascorbic acid (−0.09 U/L) (Khafagy and Morishita 2012). In addition, five methods—SGOT, SGPT, alkaline phosphatase (ALP), thio-barbituric acid reacting substances (TBARS), and reduced glutathione (GSH)—were used to investigate the peptidomimetics' in vivo hepatoprotective effect in rats (Perrisaud and Testa 1982).

Table 9.3 DPPH scavenging assay IC50 values for peptidomimetics and ascorbic acid

S. No	Sample	IC50 (μg/mL)
1	Standard	34.5
2	VY	8
3	AW	10
4	GW	0.5
5	KEL	4
6	RKP	12.5

As a result of the tests, FS-KEL was found to be more effective than normal rats in both SGOT and SGPT methods, with values of 5.038 U/L and 5.922 U/L, respectively, as opposed to their average readings of 2.828 U/L (SGOT) and 7.428 U/L (SGPT), respectively. Similar to this, the substances FS-GW and FS-KEL demonstrated improved results in the alkaline phosphatase assay with values of 1.152 and 3.363 Kat/L, respectively, as compared to the average reading of 1.427 Kat/L for normal rats. In addition, the test compounds FS-KEL, FS-VY, FS-GW and FS-RKP have shown promising results in TBARS assay (in vivo) with corresponding IC_{50} values as 1.2 μg/mL, 5.4 μg/mL, 5.5 μg/mL and 9.2 μg/mL in comparison with the standard ascorbic acid (IC50 = 23.8 μg/mL). Moreover, the test compounds FS-KEL, FS-GW, FS-VY and FS-AW have shown promising results in in vivo reduced GSH assay with corresponding IC50 values as 5.8 μg/mL, 10.4 μg/mL, 11 μg/mL and 15.8 μg/mL in comparison with the standard ascorbic acid (IC50 = 18.8 μg/mL) (Ellmann 1959). Using the student's t-test, all test chemicals were statistically significant at the 1% and 5% levels. Following in vivo testing, the liver portion underwent histology. Hepatic cells in the control group of rats had well-preserved cytoplasm, nuclei and central veins, while the normal architecture of the carbon tetrachloride-treated rats had vacuolated hepatocytes and degraded nuclei (Curevac 2015). Along with normal 5-fluoro uracil, the livers of rats treated with FS-KEL, FS-RKP and FS-AW showed a promising recovery, with the best hepatoprotective activity. The tripeptide scaffold peptidomimetic drug fullerene has definite evidence that it is hepatoprotective in vivo based on study investigations based on drug design, in vivo hepatoprotective potency and histological findings soot-Lys-Glu-Leu (FS-KEL) showing best activity against CYP 3A4 enzyme target (full fitness energy = −621643.22 kcal/mol & ΔG value = −2123.03 kcal/mol) with promising hepatoprotective action, thereby having a significant attenuation from CCl_4 induced liver damage (Table 9.3) (Edward et al. 1999). With peptide-based biofriendly pharmacological therapy, FS-KEL is a potent pharmaceutical increasing the function of the liver and its cells at a lower cost for the benefit of society (Ertl et al. 2000).

The results of the DPPH assay have shown that FS-GW and FS-KEL were identified as best leads with IC_{50} values as 0.5 μg/mL and 4 μg/mL, respectively, as compared with the standard ascorbic acid (IC_{50} = 34.5 μg/mL, Fig. 9.7) Grosdidica et al. 2011a, b.

According to the results of the reducing power assay, the best leads were found to be FS-VY, FS-KEL and FS-RKP, with respective IC50 values of 2, 2.5 and 2.5 g/mL

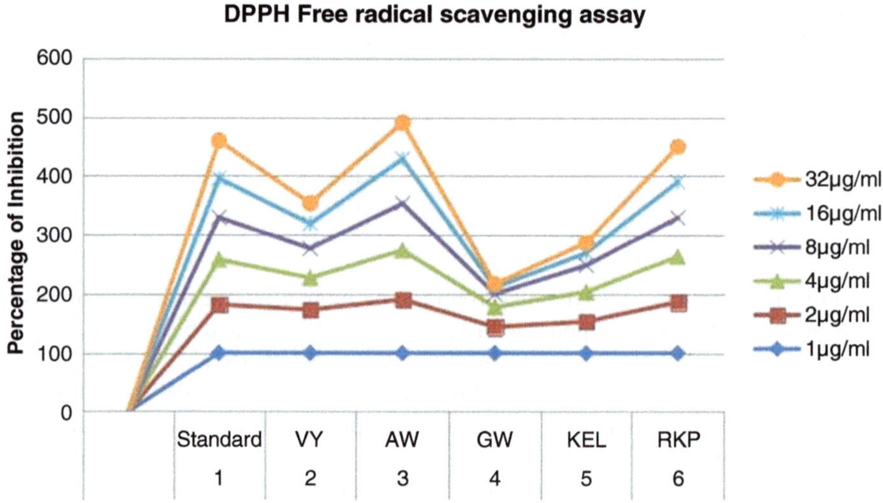

Fig. 9.7 In vitro DPPH Free radical scavenging assay

Table 9.4 IC50 values of peptidomimetics and ascorbic acid by reducing power method (RP)

S.No	Compound	IC50 values (μg/mL)
1	Standard	9.8
2	VY	2
3	GW	18.75
4	KEL	2.5
5	RKP	2.5

when compared to the ascorbic acid standard (IC50 = 9.8 g/mL), as shown in Table 9.4 and Fig. 9.8 (Gupta 2011).

9.5.1 Histopathological Investigation of the Isolated Liver

Histological analysis supported the hepatoprotective effects of FS-AW, FS-GW, FS-VY, FS-RKP and FS-KEL (Panagiota and Yiannis 2004). Hepatic cells with well-preserved cytoplasm, a conspicuous nucleus and a central vein were seen in the liver slices of healthy control animals (Fig. 9.9, More et al. 2017).

Figure 9.10 shows the appearance of vacuolated hepatocytes and degenerated nuclei, vacuolization, fatty metamorphosis in the adjacent hepatocytes, cell infiltration of lymphocytes and Kupffer cells and severe necrosis of hepatocytes in the centrilobular region in carbon tetrachloride-treated rats. These changes were also seen in areas other than the centrilobular regions (Nagarajan et al. 2008).

Figures 9.11 and 9.12, which depict the livers of rats given FS-KEL and regular 5-Fluorouracil treatment, show that the liver damage caused by CCl_4 has been significantly attenuated. This is shown by the presence of normal hepatocytes with

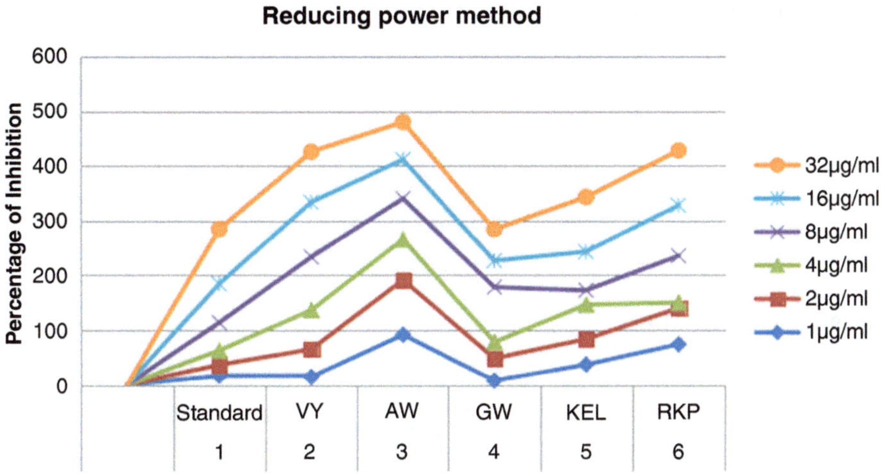

Fig. 9.8 Percent inhibition of peptidomimetics and ascorbic acid by in vitro reducing power method (RP Method)

Fig. 9.9 Histological view of normal rat's liver

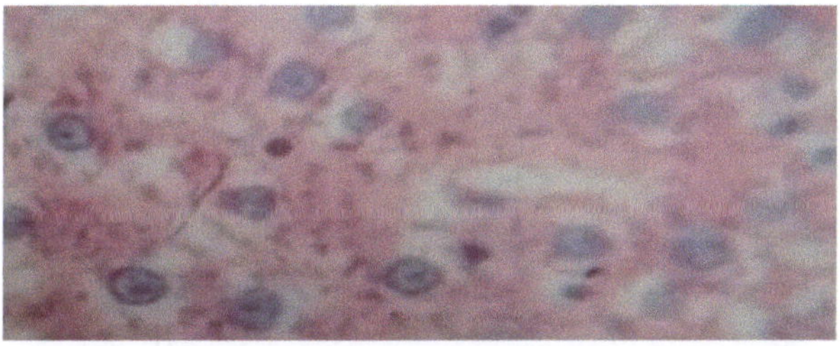

Fig. 9.10 Histological view of CCl_4 -induced hepatotoxicity in the rat's liver

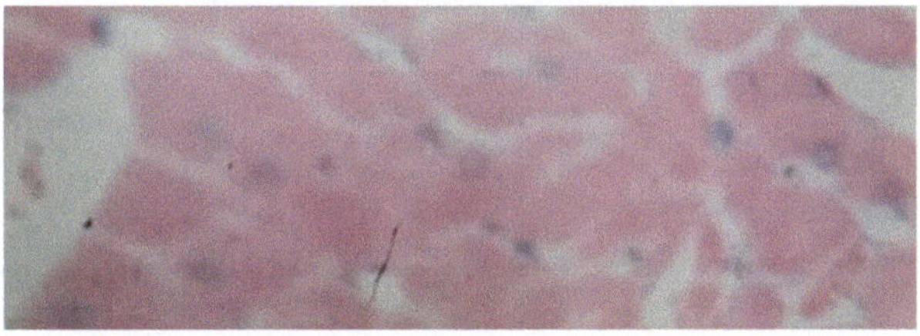

Fig. 9.11 Histological view of FS-KEL treated rat's liver

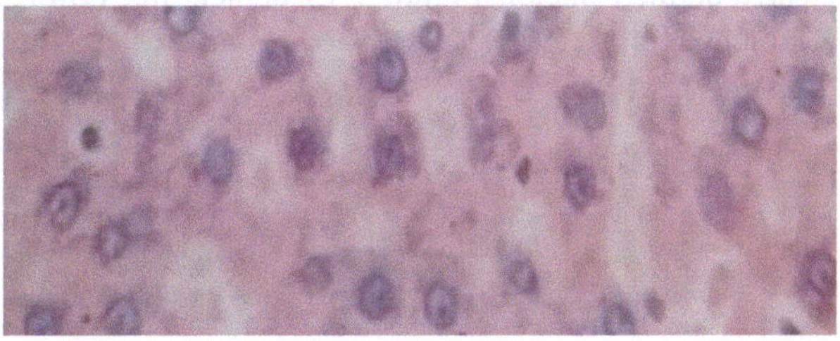

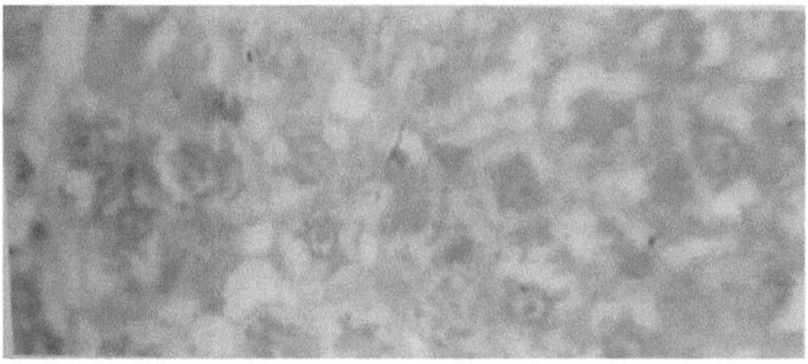

Fig. 9.12 Histological view of 5-flurouracil-treated rat's liver

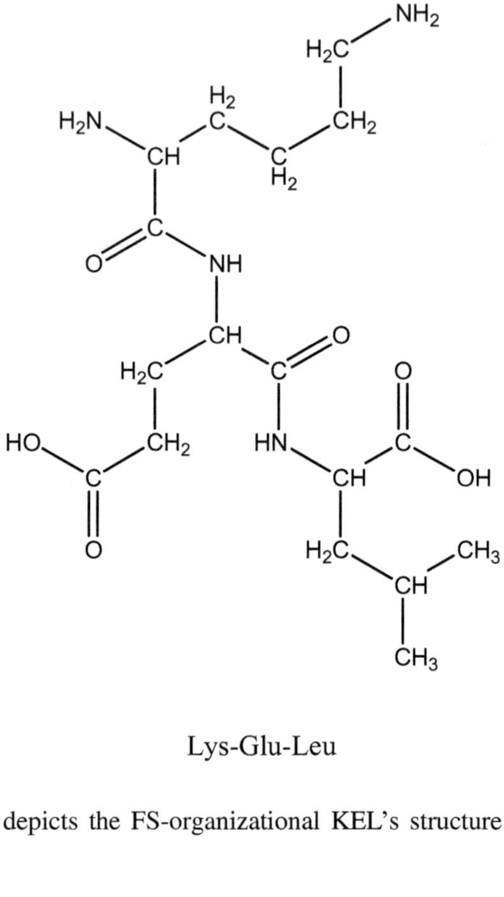

Fig. 9.13 Structure of KEL (Lys-Glu-Leu)

clearly defined nuclei. Figure 9.13 depicts the FS-organizational KEL's structure (Ottolenghi et al. 1979).

9.6 Analysis of Smaller Chain Peptidomimetics Based on Fullerenes that Target Tuberculosis and Creation of Glycopeptides Containing N- and O-Linked Smaller Chains

9.6.1 Analysis of Smaller Chain Peptidomimetics Based on Fullerene that Fight TB

There are diverse strains of tuberculosis which are found to be resistant to the known drugs, responsible for the threat to the population. There is an urgent requirement for the development of new therapeutic agents for this cause (Khatun et al. 2019). Rifampin (2 μg/mL) as reference for testing Glutamic acid – Tryptophan – Lysine (EWK), Threonine- Phenylalanine- Arginine (TFR), Phenylalanine – Arginine – Asparagine (FRN), Glutamic acid –Histidine – Arginine (EHR) and Threonine – Arginine – Asparagine (TRN) by in vitro BACTEC radiometric assay against *Mycobacterium tuberculosis* H37 Ra has been performed (Srivastava et al. 2021). The value of MIC of the tested tripeptides was found to be 50 μM. Two compounds that showed good activity are Phe-Arg-Asn (FRN) and Thr- Arg – Asn (TRN), which may act as initial lead for tuberculosis therapy.

The tripeptides of 63 templates have Boman index of the good range which is satisfactory in interaction with proteins (Prabowo et al. 2019). Among them, 5 template analogues were selected for their antitubercular potency by rational drug design approach based on their high Boman index value, viz., EWK = 3.34 kcal/mol; TFR = 4.83 kcal/mol; FRN = 6.19 kcal/mol; EHR = 8.79 kcal/mol; and TRN = 8.04 kcal/mol. The results of these selected analogs including their hydrophobic ratio and total net charge are given in Table 9.1. Practical bioactivity of these compounds is to be confirmed upon further synthesis and testing by BACTEC radiometric assay (Babu and Rangappa 2008).

9.6.2 In Vitro BACTEC Radiometric Antitubercular Assay of Tripeptides

The efficacy of shorter chain synthetic tripeptides, namely, EWK, TFR, FRN, EHR and TRN were evaluated for their antitubercular activity against *Mycobacterium tuberculosis* H_{37} Ra strain using BACTEC 460 TB system and the results are summarized in Table 9.5 (Reis et al. 2004). Due to the lead's propensity to interact

Table 9.5 Activity of test compounds against *Mycobacterium tuberculosis* H37Ra by BACTEC 460 TB system

S. No.	Test compound	Result
1.	EWK	Inactive at 50 μM
2.	TFR	Inactive at 50 μM
3.	FRN	Inactive at 50 μM
4.	EHR	Inactive at 50 μM
5.	TRN	Inactive at 50 μM

with a wide variety of proteins, the higher Boman index value clearly indicates that the lead will play a variety of roles within the cell. Consequently, from the outcomes (Boman 2003), we discovered that tripeptides EHR and TRN will likely display more (B.I. = 8.79 kcal/mol and 8.04 kcal/mol, respectively) pronounced antitubercular potency. On the contrary, the test results of the BACTEC radiometric assay showed FRN and TRN as biologically active antitubercular compounds, which indicates that, other than the Boman index, hydrophobic ratio (%) and total net charge also play a crucial role and have to be considered upon in deciding a proper target for MTB cell membrane disruption in tuberculosis. This was exemplified by the obtained results, which showed the best tripeptide lead FRN has a 33% hydrophobic ratio, +1 total net charge and a Boman index of 6.19 K. Cal/mol. The tripeptides (EWK, TFR, FRN, EHR and TRN) have MIC of 50 μM (Yan and Adams 1998). Because of this, all 5 tripeptide medication combinations had improved antitubercular activity above this threshold. In comparing the antitubercular potency among the five tripeptides, test compounds Phe-Arg-Asn (FRN) and Thr-Arg-Asn (TRN) showed moderate antitubercular potency with a lesser growth index of 413 and 431, respectively, on fourth day, whereas the standard Rifampin showed a much lesser growth index of 6 on fourth day (Martin and Portaels 2007). These results given in Table 9.5 clearly indicated that our test drugs FRN and TRN showed moderately potent antitubercular action as compared with the standard. Nurbo et al. in 2007 made a statistical molecular design approach based on the heptapeptide Ac-Glu-Asp-Asp-Trp-Asp-Phe-OH which was synthesized and evaluated against *Mycobacterium tuberculosis* ribonucleotide reductase, which showed that Trp 5 and Phe 7 were important for inhibitory potency. Recent studies made by Khara et al. in 2014 also showed that a series of synthetic cationic α-helical antimicrobial peptides (AMPs) modified with different hydrophobic amino acids increased the efficacy of the primary peptide against all strains tested, including clinically isolated multi-drug resistant mycobacteria. The best antitubercular lead was Phe-Arg- Asn (FRN). Figure 9.14 also contains large size Phe-residue, which is hydrophobic (33%) and ensured an efficient disruption of the membrane integrity, supporting the higher activity of Phe-containing peptides (Stahl 1969).

In comparing the two potent antimycobacterial tripeptides Phe-Arg- Asn (FRN) and Thr-Arg-Asn (TRN), it may be assumed that the selective antitubercular activity against MTB of the above peptides can be attributed to the presence of Arg-residue, which ensures a strong electrostatic interaction between the peptides and the negatively charged bacterial surface possessing high content of mycolic acid (Nurbo et al. 2007).

Investigations into the structure-activity relationship of synthetic peptide analogues would benefit from the information from the current effort. In conclusion, shorter chain tripeptides are potent as novel templates and developing therapy for the treatment of tuberculosis as these have small size, easy of synthesize, may have low toxicity (Khara et al. 2014).

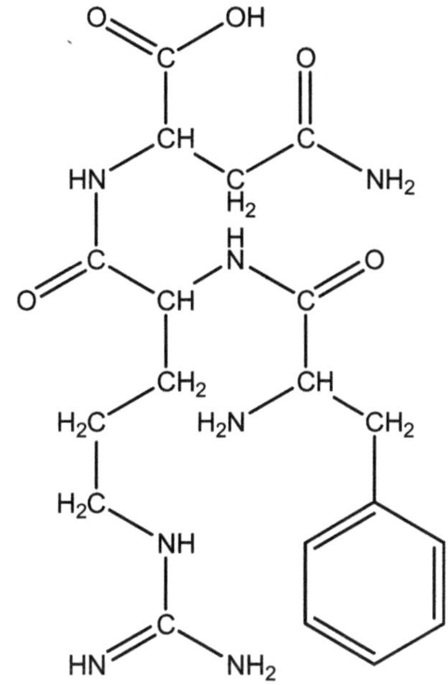

Fig. 9.14 Structure of Phe-Arg- Asn (FRN)

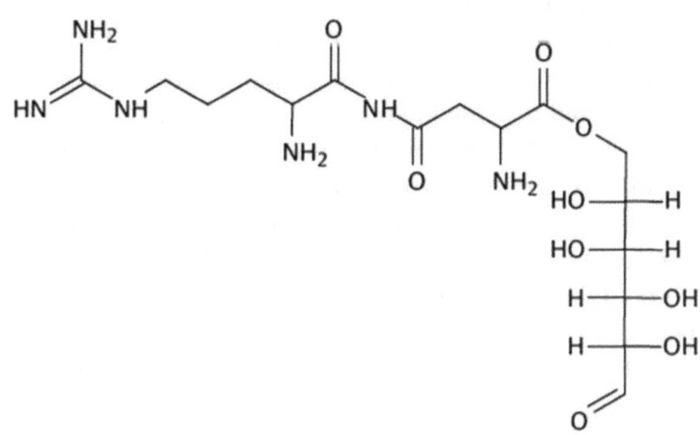

Fig. 9.15 Structure of arginine-asparagine-mannose (N-linked glycopeptide)

9.6.3 N-Linked and O-Linked Smaller Chain Glycopeptides Preparation as Antibacterial Activity

Mono-, di- and polysaccharides are used to synthesize the shorter chain N-linked and O-linked glycopeptides, with the main goal being the identification of antibacterial substances (Cappuccino and Sherman 1995). Seven glycopeptides, specifically Arg-Asn-MannoseMono, di and polysaccharides, are used to synthesize the shorter chain N-linked and O-linked glycopeptides, with the main goal being the identification of antibacterial substances (Cappuccino and Sherman 1995). Seven glycopeptides, namely, Arg-Asn-Mannose (Fig. 9.15), Arg-Asn-Lactose, His-Asn-

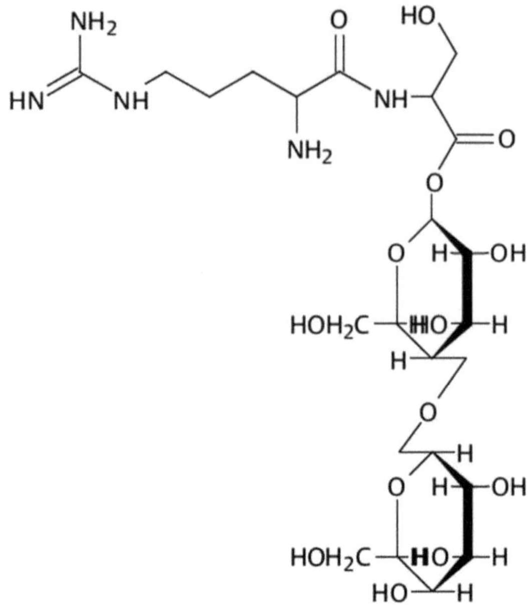

Fig. 9.16 Structure of arginine-serine-lactose (O-linked glycopeptide)

Mannose, His-Asn-Lactose (N-glycopeptides), Arg-Ser-Lactose (Fig. 9.16), Arg-Thr-Lactose and Arg-Thr-Starch (O-glycopeptides), were prepared by dicyclohexyl carbodiimide (DCC) coupling for amino acids using microwave oven (50 W power; 15 min) and activated and coupled with respective sugar moieties using microwave oven at 120 W for 20–25 min (Talbot et al. 2003). With 3-g positive *S. aureus*, *B. subtilis* and *S. caprae* and 3-g negative *E. coli*, *P. aeruginosa* and *S. sonnei* strains, the column-eluted compounds were examined for disc diffusion assay at various doses predicted by pH and inhibitory concentrations (Kobayashi and Pillai 2003) (Cirioni et al. 2003). Against all the examined microbiological strains, His-Asn-Lactose, one of the test glycopeptides and three additional test compounds, viz., His-Asn- Mannose, Arg-Thr-Lactose and Arg-Thr-Starch, also demonstrated effectiveness against tested strains that were 2 g positive and 2 g negative (Chanda and Parekh 2006). Maximum activity was observed at a concentration of 450 μg/mL (747.51 μM) for the N-glycopeptide His-Asn-Lactose with the corresponding zonal inhibition diameters (15 mm, 19 mm, 14 mm, 18 mm, 16 mm, 17 mm) against *S. aureus*, *B. subtilis*, *S. caprae*, *E. coli*, *P. aeruginosa* and *S. sonnei* (Biavasco et al. 1999) (Bambeke 2006). This is the first scientifically supported report demonstrating the antibacterial activity of our N-glycopeptide, His-Asn-Lactose, against both

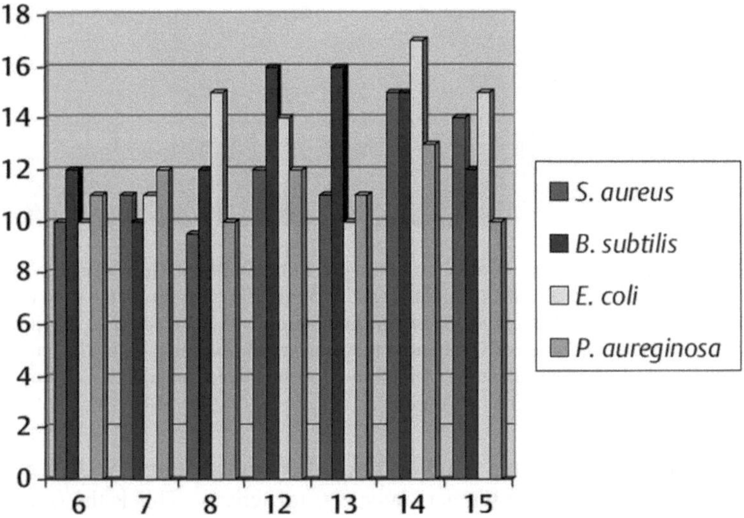

Fig. 9.17 Antimicrobial activity X-axis representation: 6 = SAL; 7 = MPA; 8 = PAL; 12 = ATL; 13 = HAM; 14 = LHA; 15 = ATS (Nagarajan et al. 2014)

gram-positive and gram-negative microorganisms (Ruzin et al. 2004; Hoffmann et al. 1999). Our potent small chain N-glycopeptide will undergo further in vivo testing and clinical investigations to be included in the arsenal of novel broad-range anti-gram-positive and gram negative medications (Goldstein et al. 1995).

Modern anti-infective chemotherapy is not complete without the glycopeptide antibiotics vancomycin and teicoplanin, which have exceptional action against gram-positive organisms from the genera *Streptococcus, Staphylococcus* and *Enterococcus* (Cooper et al. 1996; Lin et al. 1999). Results with (Arg-Asn-Mannose) and (Arg-Ser-Lactose) demonstrated very strong antibacterial effectiveness against VRSA at 450 g/mL of the concentration (997.78 μM; 747.51 μM; 500.56 μM) (Pootoolal et al. 2002; Judice and Pace 2003). This shows that gram-positive *Staph. aureus*, the most prevalent cause of biomaterial infections, including the most lethal vascular prosthetic graft infection, is very susceptible to our smaller chain glycopeptides (Grappel et al. 1985). The narrow spectrum of activity that glycopeptides display mostly covers gram-positive bacteria and a few anaerobic pathogens, including *Clostridium difficile*. In addition, they have bacteriostatic or slowl bactericidal effects on certain infections (Jorgensen et al. 1989). Our research revealed that the N-linked glycopeptide His-Asn-Lactose, which is essentially a gramme positive and two gramme negative bacterial strains, was found to be particularly effective against both types of bacteria (Figs. 9.17 and 9.18) (Haiyin

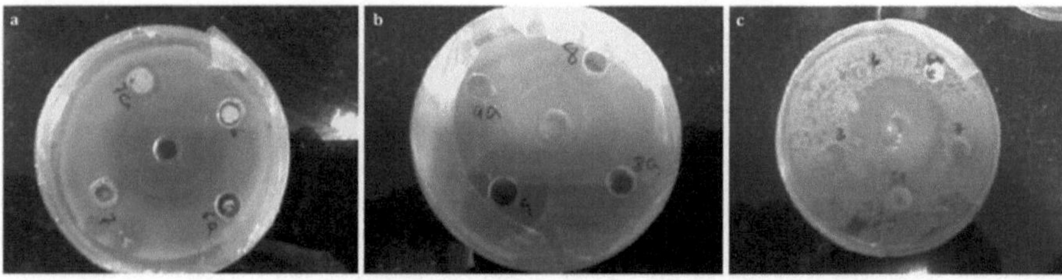

Fig. 9.18 Glycopeptides Antimicrobial Action against the Different Gram-Positive and Gram–Negative Bacteria (Bauer et al. 1996; Higgins et al. 2005). (**a**) The disk shows the zone of inhibition of the two Compounds I and V against *B. subtilis* comparing with tetracycline (± 25 mm). (**b**) The disk showing a zone of inhibition of Compound II against *B. subtilis* comparing with tetracycline (± 25 mm). (**c**) The disk showing the zone of inhibition of the three Compounds I, II and V against *E. coli* comparing with tetracycline (± 20 mm)

et al. 2002). This disproves past studies that had previously employed glycopeptides as test drugs but had only shown narrow-spectrum activity. This is the first scientifically validated assertion that a small chain N-linked glycopeptide exhibited robust broad-spectrum antibacterial activity (Shao et al. 2011). Most glycopeptides just need to block the initial stages of peptidoglycan production to have an antibacterial impact (Nagarajan et al. 2012). Additional hypothesized modes of action for several of the glycopeptides include changes in membrane integrity, disruptions of fatty acid synthesis, or direct reduction of the activity of transglycosylases, an enzyme involved in peptidoglycan formation (Arimoto et al. 1999). His-Asn-Lactose, a powerful test N-glycopeptide that we identified, can be tested further for in vivo and clinical research shown in Tables 9.6 and 9.7, which will be surely helpful to position these smaller chain glycopeptides in the arsenal of new broad-spectrum anti-gram positive and anti-gram-negative agent (Ge et al. 1999; Hossain et al. 2004).

Table 9.6 Antibacterial activity (Candiani et al. 1999)

S. No.	Organism	Strain No.	Test (μg/mL)	(MPA-7) Test Comp I	(PAL-8) Test comp II	(HAM-13) Test comp III	Diameter of zone of inhibition (mm) (LHA-14) Test comp IV	Diameter of zone of inhibition (mm) (SAL-6) Test comp V	(ATL-12) Test comp VI	(ATS-15) Test comp VII	Standard drug (tetracycline (50 μg/disc)
1	*S. aureus*	NCIM 2079	150	11	9.5	11	15	10	12	14	
			300	13	10	12	15	11	12	14	20
			450	15	10	13	15	11	13	15	
2	*B. subtilis*	NCIM 2063	150	10	12	14	15	12	16	12	
			300	10	13	16	16	13	17	12	25
			450	11	13	17	19	14	17	14	
3	*S. caprae*	ATCC 35538	150	10	9	11	10	9.5	10	13	
			300	12	9	12	12	10	11	13	19
			450	13	9.5	14	14	10	11	13	

NCIM National collection of industrial microorganisms, India, *ATCC* American type culture collection, *MTCC* Microbial type culture collection, *MPA7* arg-asn-mannose, *PAL8* arg-asn-lac, *HAM13* his-asn-mannose, *LHA14* his-asn-lac, *SAL6* arg-ser-lac, *ATL12* arg-thr-lac, *ATS15* arg-thr-starch.

Table 9.7 N-linked glycopeptides and O-linked glycopeptides antimicrobial action (Kahne et al. 2005)

S. No.	Organism	Strain No.	Test (μg/mL)	(MPA-7) Test comp I	(PAL-8) Test comp II	(HAM-13) Test comp III	Diameter of zone of inhibition (mm) (LHA-14) Test comp IV	Diameter of zone of inhibition (mm) (SAL-6) Test comp V	(ATL-12) Test comp VI	(ATS-15) Test comp VII	Standard drug (tetracycline (50 μg/disc)
1	*E. Coli*	NCIM 2118	150	11	15	10	17	10	14	15	
			300	13	16	13	17	11	14	15	20
			450	15	17	15	18	12	16	18	
2	*P. aeruginosa*	NCIM 2036	150	12	10	11	13	11	12	10	
			300	13	10	13	15	12	13	13	30
			450	13	12	15	16	14	15	15	
3	*S. sonnei*	MTCC 2957	150	13	12	13	15	10	14	10	
			300	13	13	14	16	12	15	10	23
			450	11	14	14	17	12	15	13	

NCIM National collection of industrial microorganisms, India, *ATCC* American type culture collection, *MTCC* Microbial type culture collection, *MPA7* arg-asn-mannose, *PAL8* arg-asn-lac, *HAM13* his-asn-mannose, *LHA14* his-asn-lac, *SAL6* arg-ser-lac, *ATL12* arg-thr-lac, *ATS15* arg-thr-starch

References

Adessi C, Soto C (2002) Converting a peptide into a drug: strategies to improve stability and bioavailability. Curr Med Chem 9:963–978. https://doi.org/10.2174/0929867024606731. PMID: 11966456

Arimoto H, Nishimura K, Hayakawa I et al (1999) Multi-valent polymer of vancomycin: enhanced antibacterial activity against VRE. Chem Commun:1361–1362

Babu SNN, Rangappa KS (2008) Design, synthesis, antibacterial and antitubercular activity of cationic antimicrobial peptide, ovine Bactenecin5. Ind J Chem 47B:297–304

Bambeke FV (2006) Glycopeptides and glycodepsipeptides in clinical development: a comparative review of their antibacterial spectrum, pharmacokinetics and clinical efficacy. Curr Opin Investig Drug 7:740–749

Bauer AW, Kirby WMM, Sherris JC et al (1996) Antibiotic susceptibility testing by standardised single disk method. Am J Clin Pathol 45:493–496

Biavasco F, Vignaroli C, Lupidi R et al (1999) In-vitro antibacterial activity of LY 333328, a new semisynthetic glycopeptide. Antimicrob Agents Chemother 41:2165–2172

Boado RJ (1995) Antisense drug delivery through the blood-brain barrier. Adv Drug Deliv Rev 15: 73–107. PMID: 35524391

Boman HG (2003) Antibacterial peptides: basic facts and emerging concepts. J Intern Med 254(3): 197–215. https://doi.org/10.1046/j.1365-2796.2003.01228.xH

Bridges RJ, Stanley MS, Anderson MW et al (1991) Conformationally defined neurotransmitter analogues. Selective inhibition of glutamate uptake by one pyrrolidine-2,4-dicarboxylate diastereomer. J Med Chem 34:717–725

Burger (1995) Burger's medicinal chemistry and drug discovery volume 1, principles and practice. In: Wolff, Manfred E (eds) , vol 72, 5th edn. Wiley, New York, pp 803–861. https://doi.org/10.1021/ed072pa170.3

Camilio KA, Rekdal O (2014) Short synthetic anticancer peptide andnovel immunotherapeutic agent. Onco Targets Ther 4:315

Candiani G, Abbondi M, Borgonovi M et al (1999) In-vitro and in-vivo antibacterial activity of BI 397, a new semisynthetic glycopeptide antibiotic. J Antimicrob Chemother 44:179–192

Cappuccino JG, Sherman N (1995) Microbiology-a laboratory manual, 4th edn. The Benjamin Cummings Publishing Co. Inc, San Francisco, CA, pp 21–23

Chafee EE, Greisheimer EM (1974) Basic physiology and anatomy, 3rd edn. J B Lippincott Company, Philadelphia, PA, p 505

Chanda S, Parekh J (2006) In-vitro antimicrobial activity of some Indian folklore medicinal plants. J Cell Tissue Res 6:577–580

Chazov EI, Samirnav VN, Vinegradov VA, et al. (1987) U S Patent US 4698419A

Chen H, Yang Z, Ding C, Chu L et al (2013) Discovery of O-alkylamino-tethered niclosamide derivatives as potent and orally bioavailable anticancer agents. ACS Med Chem Lett 4:180–185

Choudhary A, Raines RT (2011) An evaluation of peptide-bond Isosteres. Chembiochem 12:1801–1807. https://doi.org/10.1002/cbic.201100272

Cirioni O, Giacometti A, Ghiselli R et al (2003) Propylactic efficacy of topical temporin a and RNA III-inhibiting peptide in a subcutaneous rat pouch model of graft infection attributable to staphylococci with intermediate resistance to glycopeptides. Circulation 108:767–771

Cooper RD, Snyder NJ, Zweifel MJ et al (1996) Reductive alkylation of glycopeptideantibiotics: synthesis and antibacterial activity. J Antibiot (Tokyo) 49:575–581

Curevac AG (2015) Complexes of RNA and cationic peptide for transfaction and for immunostimulation. Korean Patent PCT/EP2007/007702, application Number: 1020107007226, Publication number 2015.04.28

Das M, Khanna SK (1997) Clinic-epidemiological, toxicological, and safety evaluation studies on argemone oil. Crit Rev Toxicol 5:273–297

Ding PM, Kumar N, Miller CM, Loren M (2013) Structure-based optimization of angiostatic agent 6DBF7, an allosteric antagonist of galectin-1. J Pharmocol Exp Ther 344:3589–3599

Edward CM, Cohen MA, Bloom SR (1999) Peptides as drug. QJM 2(92):1–4
Ellmann GL (1959) Tissue sulphydryl group. Arch Biochem Biophys 82:70–77
Ertl P, Rohde B, Selzer P (2000) Fast calculation of molecules polar surface area as sum of fragment-based contribution and its prediction to the drug transport properties. J Med Chem 43:3714–3717
Farmer PS (1980) In: Ariens EJ (ed) Drug design. Academic, New York, pp 119–143
Gante J (1994) Peptidomimetics—tailored enzyme inhibitors. Angew Chem 33:1699–1720. https://doi.org/10.1002/anie.199416991
Ge M, Chen Z, Onishi HR et al (1999) Vancomycin derivatives that inhibit peptidoglycan biosynthesis without binding D-ala-D-ala. Science 284:507–511
Giannis A, Thomas K (1993) Peptidomimetics for receptor ligands—discovery, development, and medical perspectives. Angew Chem 32:1244–1267. https://doi.org/10.1002/anie.199312441
Goldberg AL, Akopian TN, Kisselev AF et al (1997) New insights into marine collagen peptides protect against early alcoholic liver injury in rats. Br J Nutr 107:1160–1166
Goldstein BP, Candiani G, Arain TM et al (1995) Antimicrobial activity of MDL 63,246, a new semisynthetic glycopeptide antibiotic. Antimicrob Agents Chemother 39:1580–1588
Grappel SF, Giovenella AJ, Philips L et al (1985) Antimicrobial activity of aridicins, novel glycopeptide antibiotics with high and prolonged levels in blood. Antimicrob Agents Chemother 28:660–662
Grosdidica A, Zoete V, Micheilin O (2011a) Fast docking using the CHARMM force field with E a dock DSS. J Comput Chem 32:2149–2159
Grosdidica A, Zoete V, Micheilin O (2011b) Swiss dock a protein small, molecular docking web service based on E a dock DSS. Nucleic Acids Res 39:270–277
Gupta SP (2011) Statistical method, 40th edn. Sultan Chand & Sons, New Delhi, pp 910–929
Haiyin HE, Williamson RT, Shen BO et al (2002) Mannopeptimycins, novel antibacterial glycopeptides from Streptomyces hygroscopicus, LL-AC98. J Am Chem Soc 124:9729–9736
Hasan S, Daugelat S, Rao PS et al (2006) Prioritizing genomic drug targets in pathogens: application to mycobacterium tuberculosis. PLOS Comput Biol 2:539–550
Hashimoto K, Yamamoto O, Horikawa M et al (1994) Synthesis and neurobiological actions of pyrrolidine-2,3-dicarboxylic acids (PRDA) Conformationally restricted analogues of L-aspartate. Bioorg Med Chem Lett 4:1851–1854
Haubner R, Finsinger D, Kessler H (1997) Stereoisomeric peptide libraries and Peptidomimetics for designing selective inhibitors of the αvβ3 integrin for a new. Cancer Ther 36:1374–1389. https://doi.org/10.1002/anie.199713741
Haura R, Turcic P, Gabricevic M et al (2011) Interaction of α-Melanocortin and its pentapeptide antisense RVKAT: effects on hepatoprotection in male CBA mice. Molecules 16:7331–7343
Higgins DL, Chang R, Debabov DV et al (2005) Telavancin, a multifunctional lipoglycopeptide, disrupts both cell wall synthesis and cell membrane integrity in methicillin-resistant Staphylococcus aureus. Antimicrob Agents Chemother 49:1127–1134
Hillery AM, Lloyd AW, Swarbrick J (2001) Drug delivery and targeting: for pharmacists and pharmaceutical scientists, 1st edn. CRC, Boca Raton, FL. https://doi.org/10.1201/b12801
Hoffmann R, Bulet P, Urge L et al (1999) Range of activity and metabolic stability of synthetic antibacterial glycopeptides from insects. Biochim Biophys Acta 1426:459–467
Hossain SM, Easmin S, Islam SM et al (2004) Novel thiocyanato complexes with potent cytotoxic and antimicrobial properties. J Pharm Pharmacol 56:1519–1520
Humphrey JM, Bridges RJ, Hart JA et al (1994) 2.3- Pyrrolidinedicarboxyles as neurotransmitter conformer mimics: enantioselective synthesis via chelation-controlled enolate alkylation. J Org Chem 59:2467–2472
Jayaprakash GK, Singh RP, Sakrish KK (2001) Antioxidant activity of grape seed extract on peroxidation model in vitro. J Agric Food Chem 11:1018–1022
Jorgensen JH, Redding JS, Maher LA (1989) Antibacterial activity of the new glycopeptide antibiotic SKF 104662. Antimicrob Agents Chemother 33:560–561

Judice JK, Pace JL (2003) Semi-synthetic glycopeptide antibacterials. Biorg Med Chem Lett 13: 4165–4168
Kafsara M, Tselios T, Deraos S et al (2006) Round and round we go: cyclic peptides in disease. Curr Med Chem 4:2221–2232
Kahne D, Leimkuhler C, Lu W et al (2005) Glycopeptide and lipoglycopeptide antibiotics. Chem Rev 105:425–448
Khafagy, Morishita M (2012) Oral biodrug using cell penetrating peptides. Adv Drug Deliv Rev 2: 531–539
Khara JS, Wang Y, Ke XY et al (2014) Anti-mycobacterial activities of synthetic cationic α-helical peptides and their synergism with rifampicin. Biomaterials 35:2032–2038. https://doi.org/10.1016/j.biomaterials.2013.11.035
Khatun MS, Hasan MM, Kurata H (2019) Efficient computational model for identification of anti-tubercular peptides by integrating amino acid patterns and properties. FEBS Lett 1873–3468: 3029. https://doi.org/10.1002/1873-3468.13536
Kobayashi K, Pillai SK (2003) Applied statistics in toxicology and pharmacology. Oxford and IBH Publishing Co. Pvt. Ltd, New Delhi
Langmuir I (1919) Isomorphism, isosterism and covalence. J Am Chem Soc 41(10):1543–1559. https://doi.org/10.1021/ja02231a009
Li XL, Liu JY, Lu R et al (2008) Evaluation of the therapeutic efficacy of tripeptide tyroserleutide (YSL) for human hepatocarcinoma by in-vivo hollow fibre assay. Invest New Drugs 26:525–529
Lin J, Opuku AR, Geheeb-keller M et al (1999) Preliminary screening of some traditional zulu medicinal plants for anti-inflammatory and antimicrobial activities. J Ethnopharmacol 68:267–274
Lipinski CA, Lomabando DBW, Feeney PJ (1997) Experimental and computational approaches to estimate solubility and permeability in drug discovery and development setting. Adv Drug Deliv Rev 23:4–25
Liskamp RMJ (1994) Conformationally restricted amino acids and dipeptides, (non)-peptidomimetics and secondary structure mimetics. Recl Trav Chim Pays-Bas 113(1):1–19. https://doi.org/10.1002/recl.19941130102
Marshall GR (1993) A hierarchical approach to peptidomimetic design. Tetrahedron 49(17): 3547–3558. https://doi.org/10.1016/s0040-4020(01)90214-5
Martin A and Portaels F (2007) Drug resistance and drug resistance detection, tuberculosis—from basic science to patient care, 1st edn. Brazil, p 643
More SS, Nugent J, Vartak AP et al (2017) Hepatoprotective effect of Ψ-Gluthazione in a murine model of acetaminophen induced liver toxicity. Chem Res Toxicol 30:777–784
Nagarajan K, Mazumdar A, Ghosh LK (2008) In vitro antioxidant activity alcoholic extract of wrightia tomentosa. Pharmacologyonline 3:196–203
Nagarajan K, Singh S, Taleuzzaman M et al (2012) Assessment of antioxidant potency of small chain glycopeptides using DPPH free radical scavenging assay. J Pharm Nutr Sci 2:178–190
Nagarajan K, Taleuzzamman M, Kumar V et al (2014) Antimicrobial glycopeptides: synthesis and antibacterial activity of N-linked and O-linked smaller chain glycopeptides. Drug Res (Stuttg) 64:186–194. https://doi.org/10.1055/s-0033-1354413
Nurbo J, Roos AK, Muthas D et al (2007) Design, synthesis and evaluation of peptide inhibitors of mycobacterium tuberculosis ribonucleotide reductase. J Pept Sci 13:822–832. https://doi.org/10.1002/psc.906
Olson GL, Bolin DR, Bonner MP et al (1993) Concepts and progress in the development of peptide mimetics. J Med Chem 15(36):3039–3049. https://doi.org/10.1021/jm00073a001. PMID: 8230089
Ottolenghi H, Onishi N, Yogi K (1979) Interaction of ascorbica acid and mitochondrial lipids. Arch Biochem Biophys 1:355

Panagiota S, Yiannis E (2004) Synthesise of a proline rich [60] fullerene peptide with potential bioactivity. Tetrahedron 60:2823–2822

Perrisaud D, Testa B (1982) Hepatic pharmacology: mechanism of action and classification of antinecrotic hepatoprotective agents. Trends Pharmacol Sci 3:365–367

Pootoolal J, Thomas MG, Marshall CG et al (2002) Assembling the glycopeptide antibiotic scaff old: the biosynthesis of A47394 from Streptomyces toyocaensis NRRL 15009. Proc Natl Acad Sci 99:8962–8967

Prabowo SA, Painter H, Zelmer A et al (2019) RUTI vaccination enhances inhibition of mycobacterial growth ex vivo and induces a shift of monocyte phenotype in mice. Front Immunol 10: 894. https://doi.org/10.3389/fimmu.2019.00894

Reis RS, Neves IJ, Lourenco SL et al (2004) Comparison of flow cytometric and Alamar blue tests with the proportional method for testing susceptibility of mycobacterium tuberculosis to rifampin and isoniazid. J Clin Microbiol 42:2247–2248

Rink R, Arkema A, Baudoin I et al (2010) To protect pharmaceuticals against peptidases. J Pharmacol Toxicol Methods 1:210–218

Ripka AS, Rich DH (1998) Peptidomimetic design. Curr Opin Chem Biol 2:441–452. https://doi.org/10.1016/s1367-5931(98)80119-1. PMID: 9736916

Ruzin A, Singh G, Severin A et al (2004) Mechanism of action of the mannopeptimycins, a novel class of glycopeptide antibiotics active against vancomycin-resistant gram-positive bacteria. Antimicrob Agents Chemother 48:728–738

Sapakal VD, Shikalgar TS, Ghadge RV et al (2008) In vivo screening of antioxidant profile: a review. J Herb Med Toxicol 2:1–8

Saraswat P, Churchard P (2013) Biochemical and histological study of rat liver and kidney injury induced by cisplatin. J Toxicol Pathol 26:293–299

Sasaki NA (1999) A novel synthetic protocol for the preparation of enantiopure 3-, 4-, and 5-substituted prolines. Methods Mol Med 23:489–512. https://doi.org/10.1385/0-89603-517-4:489. PMID: 21380915

Sato R, Egashira Y, Ono S et al (2013) Identification of hepatoprotective peptide in a wheat gluten hydrolysate against galactosamine induced hepatitis in rats. J Agric Food Chem 61(26): 6304–6310

Shao C, Zhou W, Zhang S et al (2011) Synthesis and antibacterial activity of N 4 -mono alkyl derivatives of novel glycopeptide LYV07ww01. Biorg Med Chem Lett 21:6732–6738

Smakhtin MYU, Severyanova LA, Konoplya AI et al (2002) Tripeptide Gly-his-Lys is a hepatotropic immunosupressor. Bull Exp Biol Med 133:586–587

Srivastava S, Chapagain M, van Zyl J, Deshpande D, Gumbo T (2021) Potency of vancomycin against mycobacterium tuberculosis in the hollow fiber system model. J Glob Antimicrob Resist 24:403–410. https://doi.org/10.1016/j.jgar.2021.01.005

Stahl E (1969) Thin layer chromatography, 2nd edn. Springer, Berlin, p 21

Talbot P, Shur BD, Myles DG (2003) Cell adhesion and fertilization: steps in oocyte transport, spermzona pellucida interactions, and sperm egg fusion. Biol Reprod 68(1):1

Veber DF, Johnson SR, Cheng HY et al (2002) Molecular properties that influence the oral bioavailability of drug candidates. J Med Chem 45:2615–2623

Yan L, Adams ME (1998) Lycotoxins, antimicrobial peptides from venom of the wolf SpiderLycosa carolinensis. J Biol Chem 273:2059–2066. https://doi.org/10.1074/jbc.273.4.2059

Diverse Pharmacological Activities of 4-Aminoquinoline and its Derivatives

10

Deepak Kumar, Beena Negi, and Diwan S. Rawat

Abstract

Compounds containing 4-aminoquinoline moiety are known to exhibit various pharmacological properties such as antimalarial, anticancer, antitubercular and antivirus. Chloroquine, hydroxyquinoline and amodiaquine are a few well-known 4-aminoquinoline-based medications. Many new derivatives of this compound have also been synthesized and developed as lead molecules in the drug development process. The already existing 4-aminoquinoline-based drugs are also explored for the treatment of more recent diseases like coronavirus disease 2019 (COVID-19). Drug resistance also gives scope for testing newer molecules against various other pathogenic infections. This molecule has always drawn the attention of medicinal chemists for the synthesis of new derivatives. This article gives an overview of the recent development of 4-aminoquinoline derivatives and their pharmacological activities.

Keywords

Anticancer · Antimalarial · Antitubercular · Coronavirus

D. Kumar
Shaheed Rajguru College of Applied Sciences for Women, University of Delhi, Delhi, India

B. Negi (✉)
Department of Chemistry, Gargi College, University of Delhi, Delhi, India
e-mail: beenanegi@gargi.du.ac.in

D. S. Rawat (✉)
Department of Chemistry, University of Delhi, Delhi, India
e-mail: dsrawat@chemistry.du.ac.in

P. P. Singh (ed.), *Recent Advances in Pharmaceutical Innovation and Research*,
https://doi.org/10.1007/978-981-99-2302-1_10

10.1 Introduction

Nitrogen-containing heterocyclic compounds are present in several biologically active natural products which are being used in traditional medicines or approved prescribed drugs. Many of their synthetic counterparts are also available in the clinical market for the treatment of various diseases (Kerru et al. 2020). Some of the notable synthetic nitrogen heterocyclic-based drugs are chlorpromazine, diazepam, metronidazole, azidothymidine, isoniazid, barbituric acid, chloroquine, captopril etc. (Parashar and Negi 2015). In addition to serving as the foundation of medicinal chemistry, the nitrogen heterocyclic compounds are also a crucial component of the majority of vitamins, enzymes, coenzymes, hormones, nucleic acids and alkaloids (Al-Ghorbani et al. 2015; Heravi and Zadsirjan 2020). Due to these reasons, nitrogen heterocycles have always been attractive targets to synthetic organic and medicinal chemists.

Among various nitrogen-containing heterocyclic compounds, 4-aminoquinoline pharmacophores have gained a lot of interest from researchers to design and develop biologically important molecules (Matada et al. 2021; Yadav and Shah 2021). In 1810, Gomes et al. isolated quinine from Cinchona bark and it was the first 4-aminoquinoline-based drug that was used for the treatment of malaria (Meshnick and Dobson 2001). In 1856, William Henry Perkins attempted to synthesize quinine but he isolated the first synthetic textile dye called 'mauveine'. However, later in 1944, the first total synthesis of quinine was accomplished (Woodward and Doering 1944). This sparked the interest of medicinal chemists in the 4-aminoquinoline pharmacophore that resulted in many drugs including chloroquine (Deshpande and Kuppast 2016; Manohar et al. 2014). Further research demonstrated that the 4-aminoquinoline class of compounds exhibits a wide range of biological activities such as antimalarial (Hu et al. 2017; Narula et al. 2019; Raj et al. 2015; Vandekerckhove and D'hooghe 2015), anticancer (Musiol 2017), antimicrobial (Musiol et al. 2017), antifungal (Musiol et al. 2010), antiviral (Kaur and Kumar 2021), antitubercular (Rangappa and Siddappa 2014; Rawat and Beena 2013; Singh et al. 2015), anti-inflammatory (Ambatkar and Khedekar 2019; Mukherjee and Pal 2013) and anti-HIV (Chokkar et al. 2019). The main objective of this article is to update the literature regarding the recent development of 4-aminoquinoline-based compounds and their biological activities.

10.2 Antimalarial Activity

Among all biological activities, the antimalarial properties of 4-aminoquinolines have received the greatest attention. Chloroquine, a 4-aminoquinoline-based drug, was first synthesized by Hans Andersag in 1934 (Krafts et al. 2012). Initially, it was considered to be toxic, but subsequent studies demonstrated its potential antimalarial properties, making it one of the most popular medications for the treatment of malaria. Later on, a number of analogues, namely amodiaquine, isoquine, sontoquine, piperaquine and primaquine were discovered (Fig. 10.1). Although the

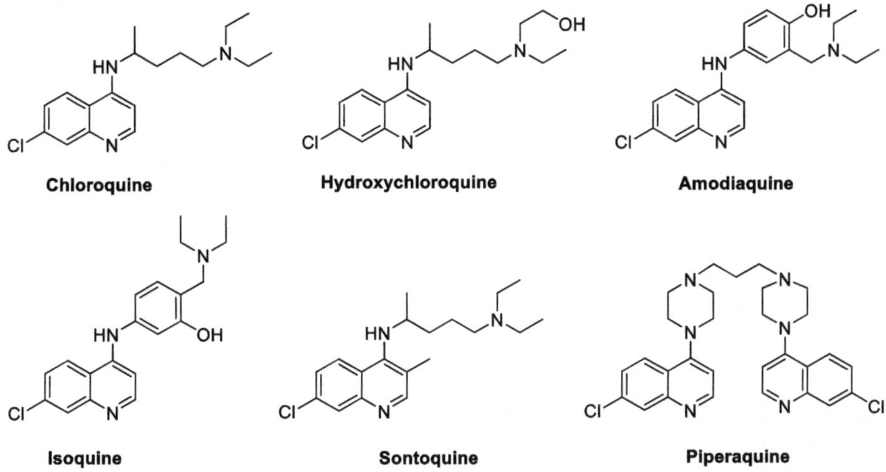

Fig. 10.1 4-Aminoquinoline-based antimalarial drugs

prevalence of malarial infection was reduced by these medications, the development of drug resistance, particularly against *Plasmodium falciparum*, prompted medicinal chemists to look for other safer medications. This led to the discovery of artemisinin (Wang et al. 2019). Currently, the first-line treatment for malarial infection caused by *Plasmodium falciparum* is artemisinin-based combination therapy (ACT), wherein the artemisinin and its derivatives are co-administered with some drug from a different class of antimalarials such as amodiaquine, mefloquine and piperaquine (Kremsner and Krishna 2004). However, in 2008, the first case of resistance in *P. falciparum* against artemisinin and its derivatives was observed in Western Cambodia (Dondorp et al. 2009; Noedl et al. 2008). The emergence of multidrug resistance and the parasite's increasing capacity to tolerate the artemisinin-based combination therapy urges researchers to develop novel treatments and antimalarial medications to further reduce malaria occurrences and fatalities.

To tackle drug-resistant issues, the concept of hybrid molecules has been the most fascinating topic in medicinal chemistry, where two or more pharmacophores are connected covalently to create one single molecule (Meunier 2008; Muregi and Ishih 2010). The two units of these compounds may exert dual pharmacological action by acting on different targets, or one molecule may be able to balance off the side effects caused by the other molecule. With this aim, many research groups have prepared a large variety of aminoquinoline derivatives by modifying the side chain of chloroquine or combining chloroquine with new pharmacophores acting on different targets.

A series of 4-aminoquinoline-linked Mannich bases (**1a**-**1i** and **2a**-**2c**) were synthesized and tested for in vitro antimalarial activity against *P. falciparum* chloroquine-sensitive strain (3D7) (Singh et al. 2021). All of the compounds demonstrated mild activity, with MIC values ranging from 15.6 to 125 μg/mL.

Compounds **1b** and **1c** were found to be the most active. Both compounds had MIC values of 15.6 μg/mL when compared to chloroquine (MIC = 0.4 μg/mL).

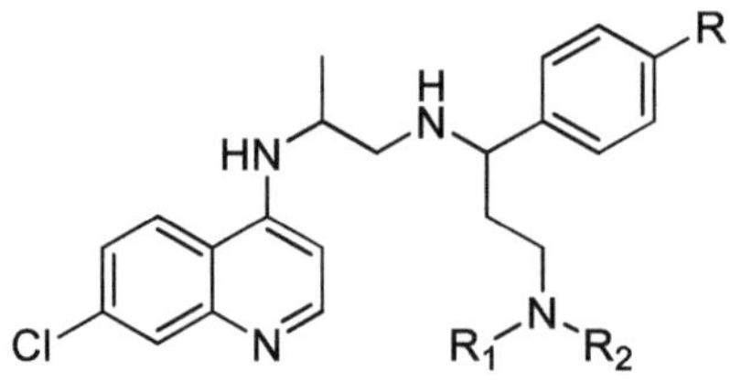

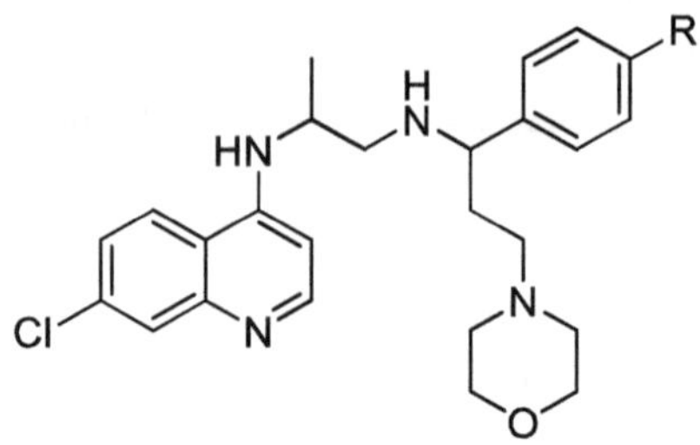

1a; R = H, $R_1 = CH_2C_6H_5$, $R_2 = CH_2C_6H_5$
1b; R = CH_3, $R_1 = CH_2C_6H_5$, $R_2 = CH_2C_6H_5$
1c; R = OCH_3, $R_1 = CH_2C_6H_5$, $R_2 = CH_2C_6H_5$
1d; R = H, $R_1 = CH_2CH_3$, $R_2 = CH_2CH_3$
1e; R = CH_3, $R_1 = CH_2CH_3$, $R_2 = CH_2CH_3$
1f; R = OCH_3, $R_1 = CH_2CH_3$, $R_2 = CH_2CH_3$
1g; R = H, $R_1 = CH_3$, $R_2 = CH_3$
1h; R = CH_3, $R_1 = CH_3$, $R_2 = CH_3$
1i; R = OCH_3, $R_1 = CH_3$, $R_2 = CH_3$

2a; R = H
2b; R = CH_3
2c; R = OCH_3

The hybrid molecules **3a-3e** were synthesized by conjugating 4-aminoquinoline with pyrano[2,3-c]pyrazole and tested against *P. falciparum* chloroquine susceptible (3D7) and chloroquine-resistant (K1) strains (Shamsuddin et al. 2021). Compounds **3a**, **3b**, **3c** and **3e** displayed very good antimalarial activity against the sensitive strain 3D7, with EC_{50} values of 0.19 μM, 0.0130 μM, 0.113 μM and 0.026 μM, respectively. Aside from their potent antimalarial activity against the chloroquine susceptible strain, compounds **3a**, **3b** and **3d** also had potent activity against the resistant strain K1, with EC_{50} values of 0.25 μM, 0.02 μM and 0.30 μM, respectively. When the cytotoxicity of compounds was evaluated against the Vero mammalian cell line, it was discovered that they exhibited modest cytotoxic activities and high selectivity index values ranging from 355 to 1354.

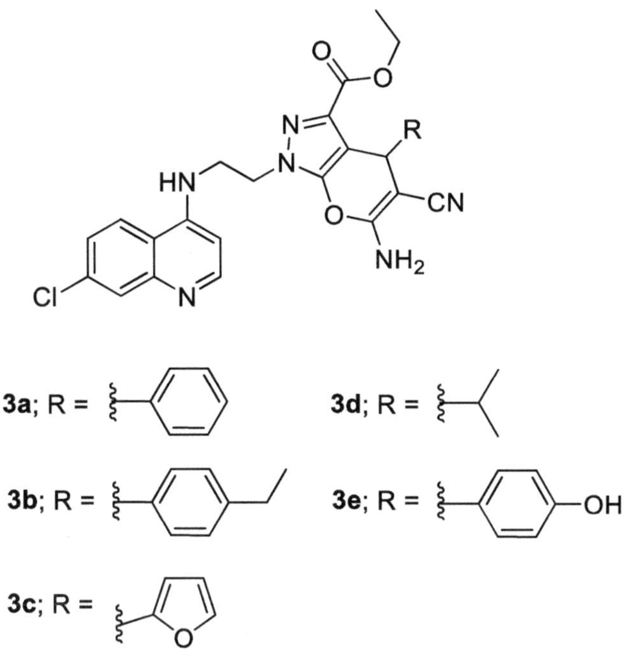

1,2,3,4-Tetrahydro-β-carbolines (THβC) are present in a variety of pharmacologically active substances with putative antimalarial properties. Keeping this in mind, a series of hybrid molecules **4a-4b**, **5a-5d**, **6a-6h** and **7a-7b** were synthesized by coupling this moiety with 4-aminoquinoline where the two pharmacophoric groups were linked with either a triazole or an acyl hydrazide linker (Sharma et al. 2020). The antimalarial activity of these compounds was tested against the chloroquine-resistant W2 strain of *P. falciparum*. In general, the acyl hydrazide hybrids **6** and **7** were found to be less active than the triazole-linked hybrids **4** and **5**. The insertion of a flexible alkyl chain between the triazole and quinoline moiety in the 1*H*-1,2,3-triazole tethered THβC-4-aminoquinoline conjugates significantly improved the antimalarial activity. Compounds **5a-5d** were more active than **4a-4b** ($IC_{50} = 4.02–9.28$ μM), having an IC_{50} value of 0.49–1.37 μM. The antimalarial activity of the aliphatic acyl hydrazide linked hybrids **6a-6h** was improved by lengthening the alkyl chain; however, the kind of substituent at the C-1 position of THβC did not appear to affect the antimalarial activity. Cytotoxic activity of the most active compounds **5a**, **5c** (triazole as linker) and **7a**, **7b** (acyl-hydrazide as linker) was assessed on mammalian Vero cell lines. As anticipated, compounds **7a** and **7b** were mildly cytotoxic, whereas **5a** and **5c** were found to be noncytotoxic, with $SI > 300$.

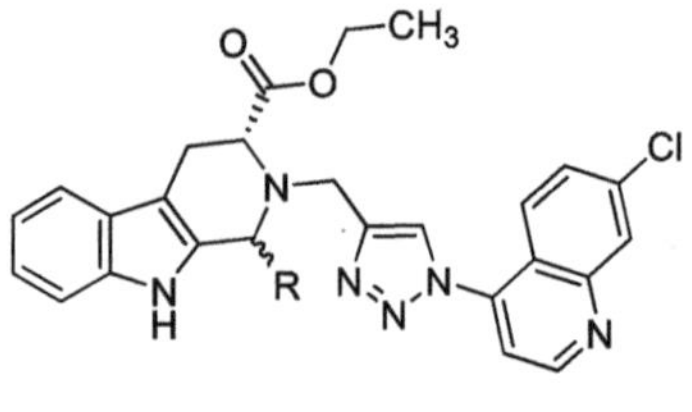

4a; R = H
4b; R = Cis-CH_3

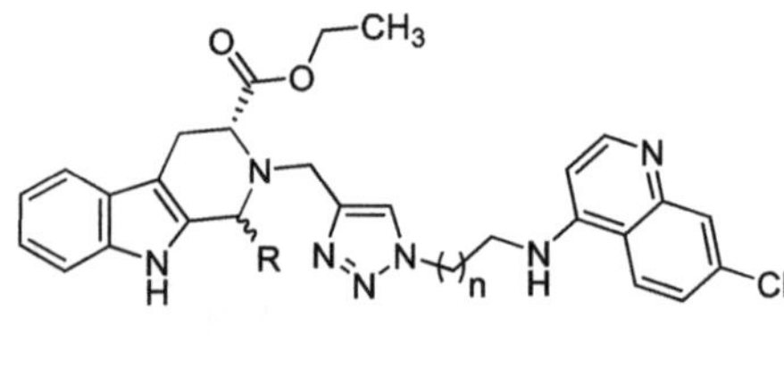

5a; R = H
5b; R = Cis-CH_3
5c; R = H
5d; R = Cis-CH_3

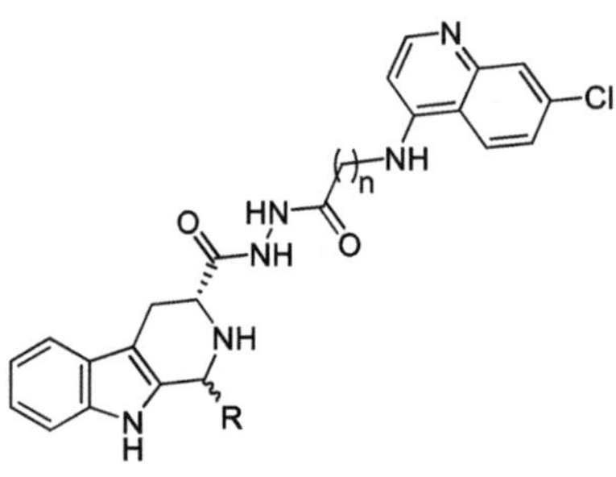

6a; R = H, n = 1
6b; R = Cis-CH_3, n = 1
6c; R = H, n = 2
6d; R = Cis-CH_3, n = 2
6e; R = H, n = 3
6f; R = Cis-CH_3, n = 3
6g; R = H, n = 5
6h; R = Cis-CH_3, n = 5

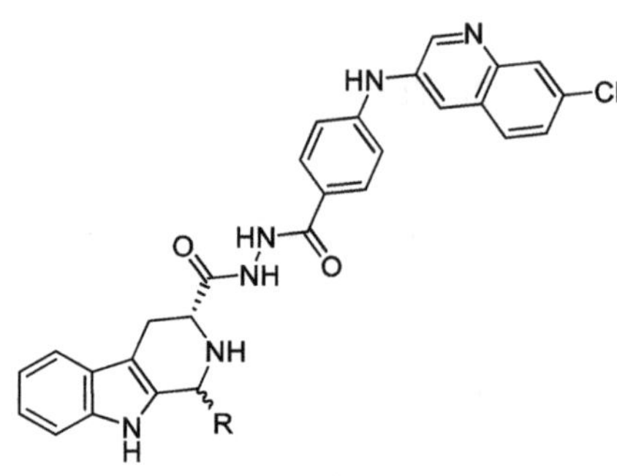

7a; R = H
7b; R = Cis-CH_3

The artemisinin-quinoline hybrids **8–20** were synthesized and tested for their antimalarial efficacy against chloroquine susceptible strain (3D7) and chloroquine-resistant strains (Dd2 and K1) of *P. falciparum* (Çapci et al. 2019). All the compounds displayed extremely interesting activity with EC_{50} values from 0.78 to 27.5 nM and were discovered to be more effective against the resistant strains. Compounds **8–11**, in which the carbon chain length between artesunic acid and quinoline moieties was varied from one to four carbon atoms, demonstrated better antimalarial activity than the standard drug chloroquine (EC_{50} = 12.7 nM), with EC_{50} values in the nanomolar range (5.3, 8.0, 5.4 and 6.2 nM, respectively). The compound with the shortest linker demonstrated the highest antimalarial activity against all the strains. In the case of the resistant strains, hybrids **8–11** were found to be 34–39 times more potent (EC_{50} = 4.2–4.9 nM) against the Dd2 strain and 116–168 times more potent (EC_{50} = 1.8–2.6 nM) against K1 strain than chloroquine. Furthermore, artemisinin-quinoline hybrids **12–16** demonstrated excellent activity with EC_{50} values between 2.0 nM and 27.5 nM. These compounds were more effective against resistant strains than sensitive strains. When tested towards three distinct strains, compounds **12, 14, 15** and **16** showed nearly identical potency.

These compounds had EC_{50} values of 9.0 nM, 8.8 nM, 9.4 nM and 8.4 nM towards 3D7 strain; 5.9 nM, 4.5 nM, 5.1 nM and 6.1 nM towards Dd2 strain; and 4.3 nM, 2.0 nM, 3.8 nM and 3.7 nM against K1 strain. The EC_{50} values for the antimalarial activity of compounds **17–19** ranged from 0.78 nM to 14.8 nM. Compound **18** with NH at C4 of the 7-chloroquinoline unit was found to be more active against all three strains than compound **17** with O-atom at C4 of the 7-chloroquinoline unit. Compound **19**, obtained by replacing the second oxygen atom in the linker with a nitrogen atom, demonstrated higher activity than compounds **17** and **18** with EC_{50} values of 1.0 nM, 0.78 nM and 2.7 nM towards Dd2, K1 and 3D7 strains, respectively. Two compounds **16** and **18** were examined for in vivo activity in mice infected with *P. berghei.* Unfortunately, compound **16** failed to suppress parasitaemia when given subcutaneously for 4 days at a concentration of 105 μmol/kg and hence did not receive attention for additional studies. However, compound **18** reduced parasitaemia by more than 99% at a concentration of 105 μmol/kg for 4 days, while artesunate was unable to treat any animal at this concentration. To investigate the mechanism of action, compound **18** was tested in vitro and in vivo for its capability to inhibit the formation of *β*-hematin. The investigations demonstrated a 71.9% reduction in β-hematin formation in vivo and the IC_{50} value of 0.56 mM was observed for in vitro inhibition of the formation of β-hematin. The in vitro inhibition was comparable to chloroquine ($IC_{50} = 0.89$ mM), although artesunate did not show any significant activity.

Eighteen 4-aminoquinoline derivatives **21a-21f**, **22a-22f** and **23a-23f** were synthesized and evaluated against chloroquine-sensitive 3D7 strain and chloroquine resistant K1 strain of *P. falciparum* (Kondaparla et al. 2017). Four of the 18 compounds examined (**21b**, **22b**, **22c** and **22e**), with corresponding IC_{50} values of 28.49, 22.21, 27.16 and 73.05 nM, demonstrated extremely good activity against the 3D7 strain. Eleven compounds were discovered to be more effective (IC_{50} = 31.19–252.28 nM) than chloroquine (IC_{50} = 255 nM) against the chloroquine-resistant K1 strain. The cytotoxic activity of the compounds was determined using the MTT assay against the Vero cell line, and the majority of the compounds demonstrated a modest cytotoxic effect with a high selectivity index. Furthermore, some of the compounds (**21b-21f**, **22b-22f** and **23f**) were tested for in vivo activity against chloroquine-resistant *Plasmodium yoelii* (N-67 strain) in Albino mice of the Swiss strain, and all the tested compounds completely suppressed parasitaemia on day 4 when administered orally at a dose level of 100 mg/kg.

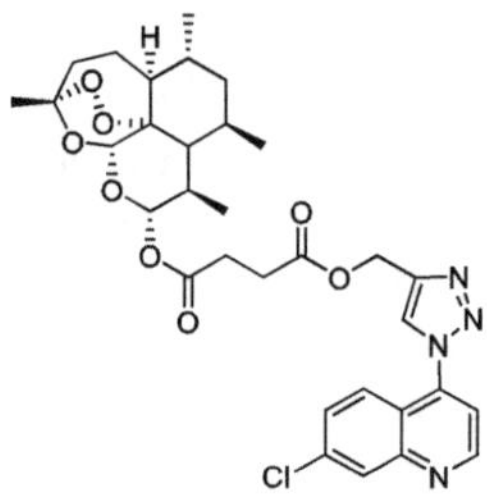

8

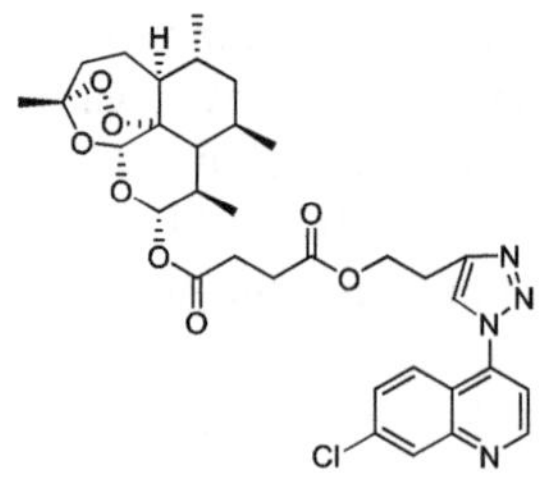

9

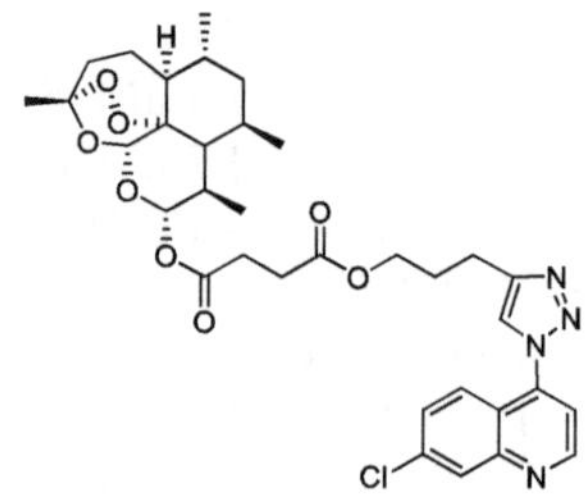

10

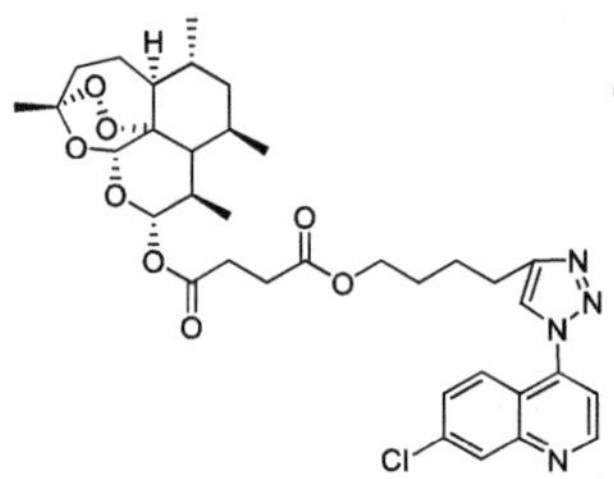

11

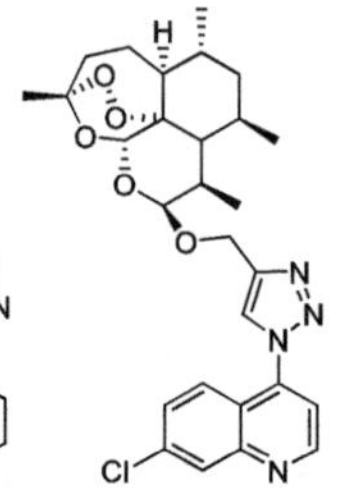

12

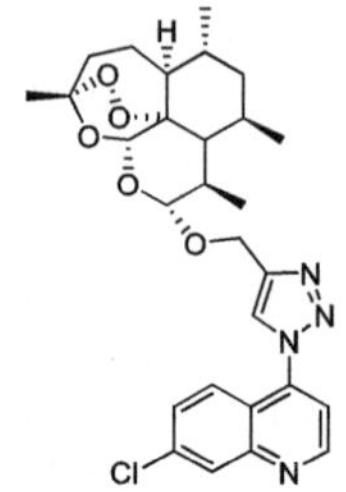

13

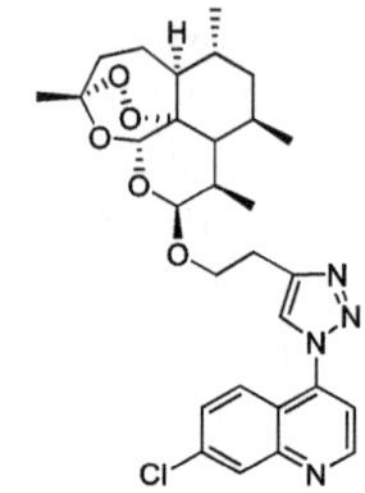

14

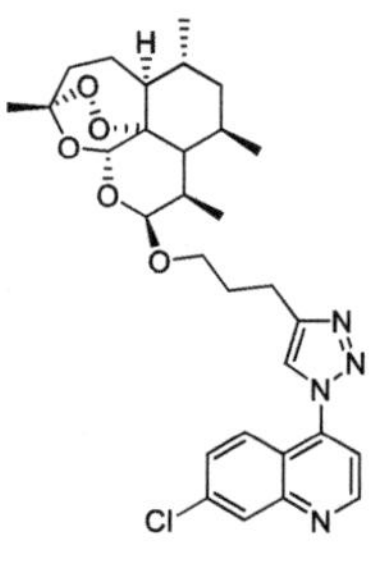

15

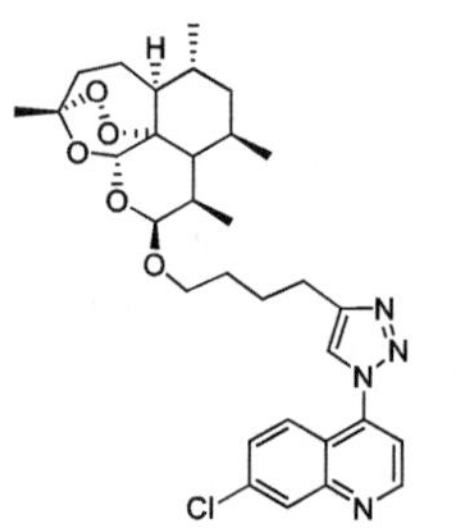

16

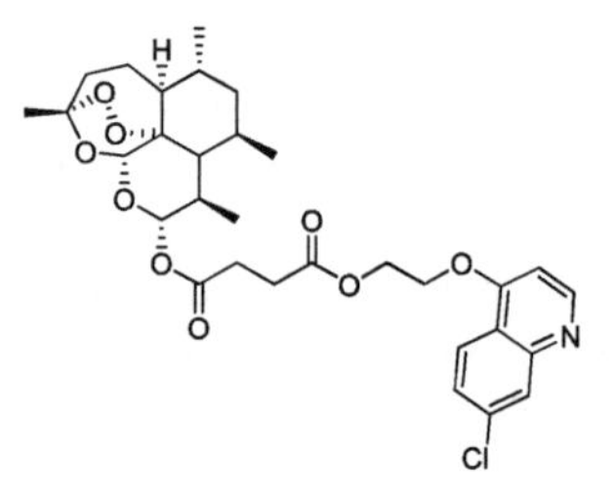

17

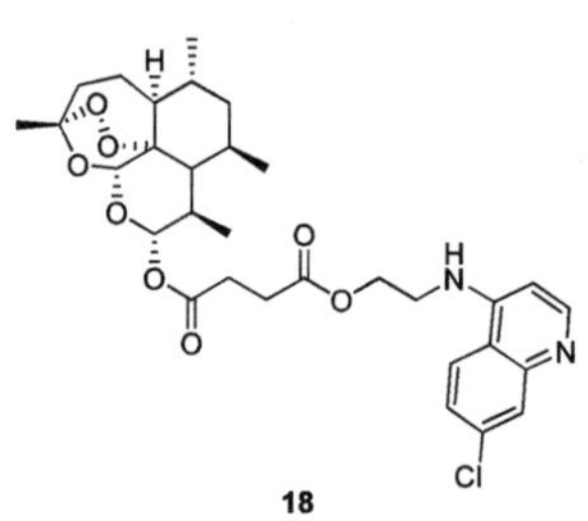

18

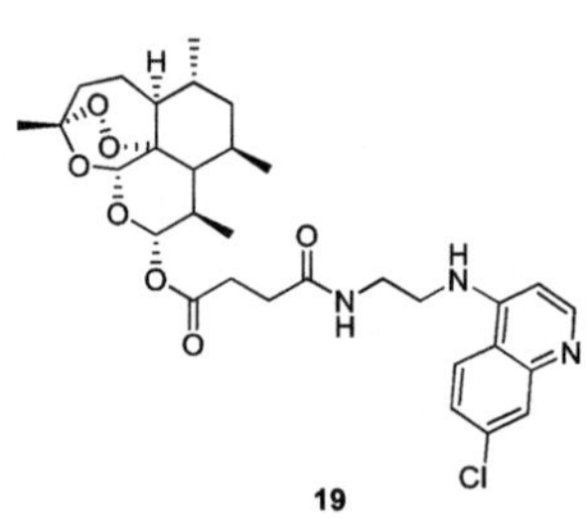

19

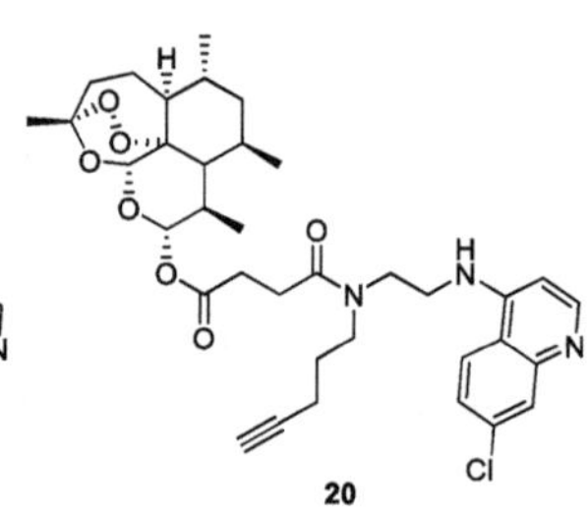

20

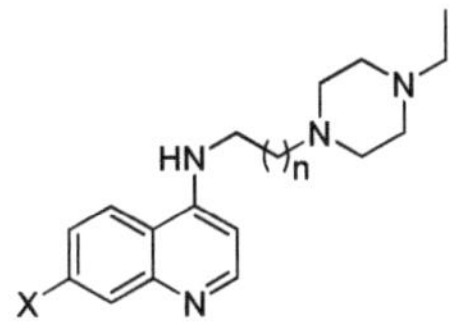

21a; X = Cl, n = 1
21b; X = CF_3, n = 1
21c; X = Cl, n = 2
21d; X = CF_3, n = 2
21e; X = Cl, n = 3
21f; X = CF_3, n = 3

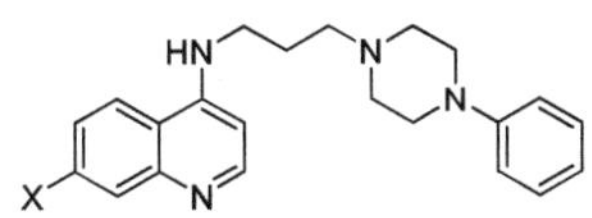

22a; X = Cl, n = 1
22b; X = CF_3, n = 1
22c; X = Cl, n = 2
22d; X = CF_3, n = 2
22e; X = Cl, n = 3
22f; X = CF_3, n = 3

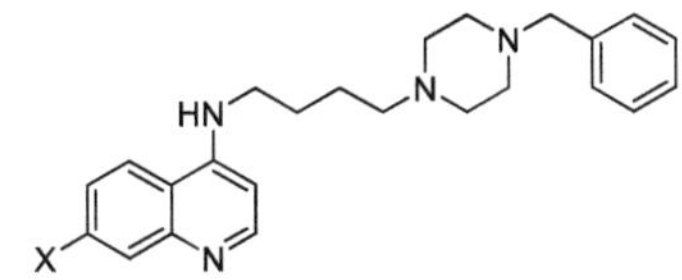

23a; X = Cl, n = 1
23b; X = CF_3, n = 1
23c; X = Cl, n = 2
23d; X = CF_3, n = 2
23e; X = Cl, n = 3
23f; X = CF_3, n = 3

The 4-aminoquinoline-imidazole hybrids **24a-24r** were synthesized and tested for in vitro antimalarial efficacy against both chloroquine-sensitive 3D7 and chloroquine-resistant K1 strain of *P. falciparum* (Kondaparla et al. 2018). All of the substances showed moderate to high activity, with IC_{50} values against the sensitive strain 3D7 ranging from 0.079 to 5 μM and against the resistant strain K1 from 0.291 to 5 μM. Compounds (**24a**, **24b**, **24c**, **24 g** and **24j**) with a heterocyclic moiety attached to an imidazole ring demonstrated promising activity against the K1 strain of *P. falciparum*, with IC_{50} values of 1.15, 0.46, 1.70, 1.33 and 0.97 μM, respectively. Compounds with bulky aromatic substituents on the imidazole moiety (compounds **24h** and **24p**) were found to be inactive against both *P. falciparum* strains (3D7 and K1). Compound **24m** with cyclopentyl substitution demonstrated stronger activity against chloroquine-resistant K1 strain (IC_{50} = 0.34 μM) than compound **24q** with cyclohexyl substitution (IC_{50} = 0.52 μM). Furthermore, among the various substitutions at the phenyl ring attached to *N*-(2-(1H-imidazole-1-yl)ethyl)-7-chloroquine in 4-amine moiety, isopropyl group containing compound **24e** was shown to be the most effective in the series against the chloroquine-resistant strain with IC_{50} value of 0.29 μM (IC_{50} of chloroquine = 0.255 μM). Compound **24i** with *tert*-butyl group substitution also demonstrated substantial efficacy against the resistant strain (IC_{50} = 0.50 μM). Overall, the author concluded that alkyl groups and alkyl-substituted phenyl rings are the best substitutions which result in compounds with increased antimalarial activity against both strains. Furthermore, no compounds significantly displayed cytotoxicity against the mammalian Vero cell line. To ascertain the most likely mode of action, heme binding assays were also conducted. All of the compounds showed a strong complex formation with hematin and inhibited the formation of β-hematin, implying the same mode of action as that of chloroquine.

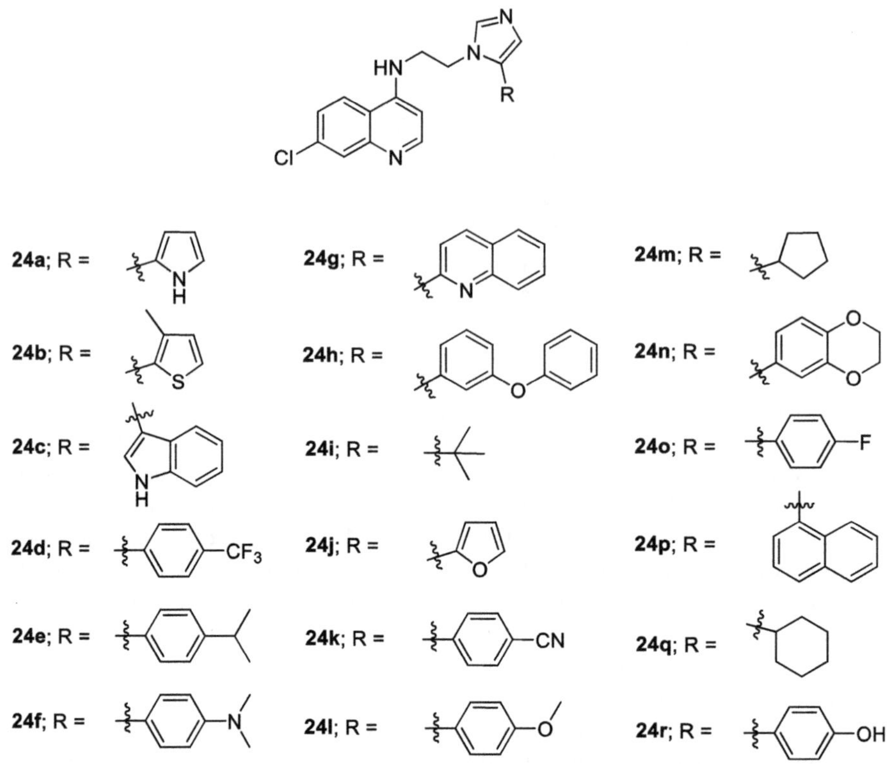

Due to the shown potential of pyrimidine-based substances, such as pyrimethamine and trimethoprim, to block dihydrofolate reductase (DHFR), an enzyme involved in the folate pathway of the *Plasmodium* parasite, pyrimidine is an essential component in designing of antimalarial drugs (Singh and Kaur 2016). As a result, the synthesis of quinoline-pyrimidine hybrids is an appealing strategy to find antimalarial agents.

Compounds **25a-25 l** and **26a-26f**, the triazole-tethered hybrids of 7-chloroquinoline and pyrimidine-5-carboxylate, were synthesized and evaluated for antimalarial activity against the NF54 strain (chloroquine sensitive) of *P. falciparum* (Chopra et al. 2018). Compounds **25a-25 l**, having an alkyl chain between triazole and quinoline rings, were found more potent than hybrids **26a-26f** having triazole moiety directly linked to the quinoline ring. The compounds **25a-25l**, which were found to be the most effective against NF54, were also investigated against the Dd2 strain (chloroquine resistant) of *P. falciparum*. With IC_{50} values of less than 1 μM, 9 of the 12 analogues (**25a-25 l**) demonstrated very promising antimalarial activity. In general, compounds with a propylene linker were more effective than those with an ethylene linkage against both strains. Compound **25d**, which was found to be the most effective among all hybrids against the sensitive strain (NF54), proved to be less effective against the resistant strain (Dd2), whereas compound **25c** was proven to be the most potent against the resistant strain.

Furthermore, there were no cytotoxic effects observed against the mammalian Vero cell line. The most active hybrid **25d** against the sensitive strain NF54 had a high value (317.50) of the selectivity index. Heme binding and DNA binding studies were conducted to determine the likely mode of action of these hybrids. The complex formation of the compound **25d** with heme was observed with binding constant values log $K = 4.679$ (monomeric) and log $K = 5.434$ (μ-oxo dimeric), implying the same mechanism of action as that observed with chloroquine. The high values of the binding constant, log K = 3.74 and 3.92 for CT DNA and pUC 18, respectively, further suggest that the antimalarial activity of the synthesized compounds may also be mediated by interaction with parasite DNA in addition to the inhibition of heme polymerization.

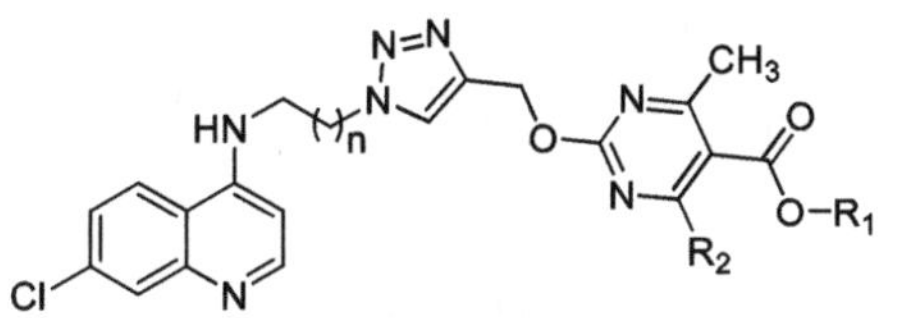

25a; $R_1 = CH_3$, $R_2 = C_6H_5$, n = 2
25b; $R_1 = C_2H_5$, $R_2 = C_6H_5$, n = 2
25c; $R_1 = (CH_3)_2CH$, $R_2 = C_6H_5$, n = 2
25d; $R_1 = (CH_3)_2CH$, $R_2 = 4\text{-}NO_2\text{-}C_6H_5$, n = 2
25e; $R_1 = (CH_3)_2CH$, $R_2 = 2\text{-}NO_2\text{-}C_6H_5$, n = 2
25f; $R_1 = C_2H_5$, $R_2 = CH_3$, n = 2
25g; $R_1 = CH_3$, $R_2 = C_6H_5$, n = 1
25h; $R_1 = C_2H_5$, $R_2 = C_6H_5$, n = 1
25i; $R_1 = (CH_3)_2CH$, $R_2 = C_6H_5$, n = 1
25j; $R_1 = (CH_3)_2CH$, $R_2 = 4\text{-}NO_2\text{-}C_6H_5$, n = 1
25k; $R_1 = (CH_3)_2CH$, $R_2 = 2\text{-}NO_2\text{-}C_6H_5$, n = 1
25l; $R_1 = C_2H_5$, $R_2 = CH_3$, n = 1

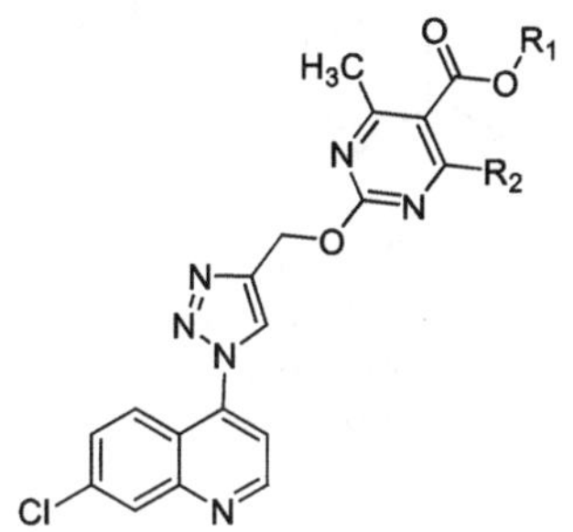

26a; $R_1 = CH_3$, $R_2 = C_6H_5$
26b; $R_1 = C_2H_5$, $R_2 = C_6H_5$
26c; $R_1 = (CH_3)_2CH$, $R_2 = C_6H_5$
26d; $R_1 = (CH_3)_2CH$, $R_2 = 4\text{-}NO_2\text{-}C_6H_5$
26e; $R_1 = (CH_3)_2CH$, $R_2 = 2\text{-}NO_2\text{-}C_6H_5$
26f; $R_1 = C_2H_5$, $R_2 = CH_3$

The 4-aminoquinoline-pyrimidine hybrids **27a-27d, 28a-28d and 29a-29n** were synthesized and investigated for in vitro antimalarial efficacy against chloroquine susceptible (D6) and chloroquine-resistant (W2) strains of *P. falciparum* (Manohar et al. 2012). Out of 22 compounds, 11 compounds (**29b**, **29c**, **29e**, **29f** and **29h-29n**) showed superior antimalarial activity (IC_{50} = 0.005–0.03 μM) than chloroquine against the sensitive strain, and 12 compounds (**29b-29f** and **29h-29n**) showed better activity (IC_{50} = 0.01–0.21 μM) than chloroquine against the resistant strain of *P. falciparum*. Four compounds (**29i**, **29j**, **29l** and **29m**) showed better effectiveness as compared to pyrimethamine against both strains (D6 and W2). According to the SAR analyses, no clear pattern of activity, with changing the length of the carbon chain, was seen for a specific amino-substituted 4-aminoquinoline-pyrimidine hybrid. However, switching the amino groups for a specific carbon chain length from piperidine to morpholine to 4-methyl piperazine to 4-ethyl piperazine, the antimalarial activity significantly increased. Most of the compounds were found to be nontoxic towards Vero, HepG2 and LLC-PK_{11} cells up to a concentration of 60 μM, while a few of them showed minor toxicity but had a very high selectivity index. The most effective compounds, **29i** and **29m**, were chosen for in vivo testing and demonstrated exceptional effectiveness in a *P. berghei* mice model without any toxic effects. Compound **29i** displayed better activity as compared to **29m** and chloroquine. Compound **29i** provided almost complete parasitaemia suppression and cured 80% of the treated mice at three doses of 30 mg/kg, whereas **29m** only cured 20% of the mice, and chloroquine had no effect.

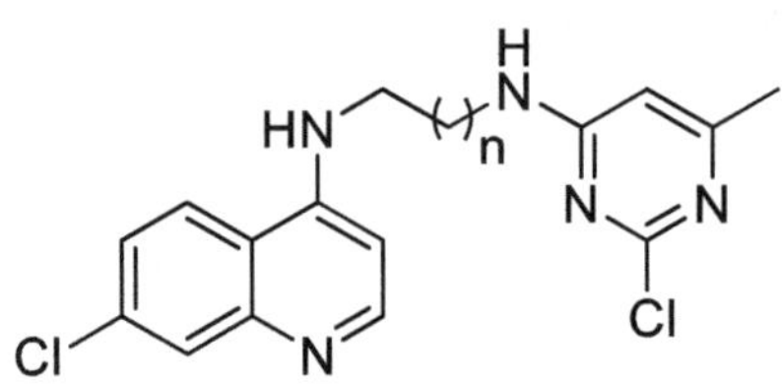

27a; n = 1
27b; n = 2
27c; n = 3
27d; n = 5

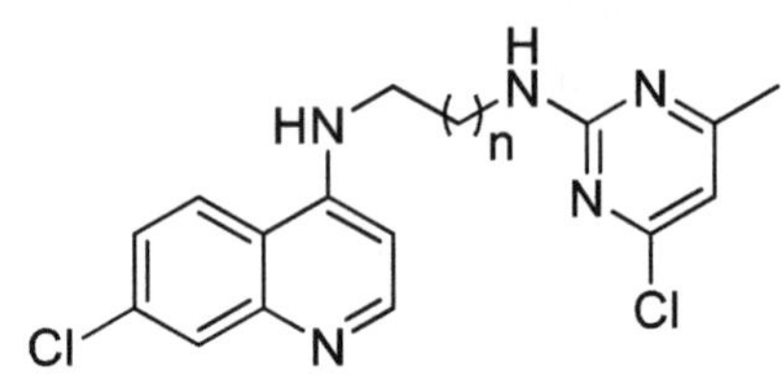

28a; n = 1
28b; n = 2
28c; n = 3
28d; n = 5

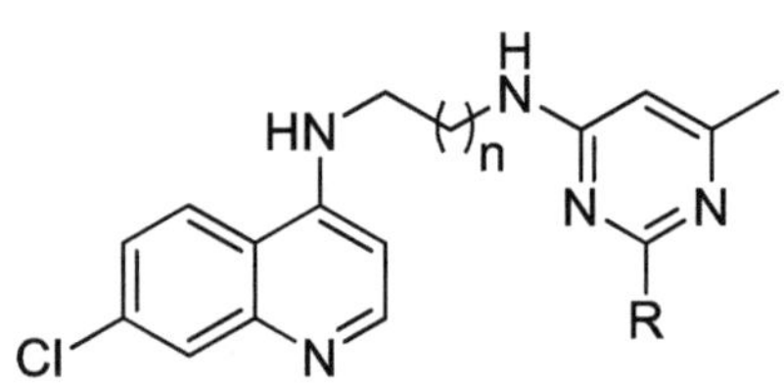

29a; n = 1, R = piperidine
29b; n = 2, R = piperidine
29c; n = 3, R = piperidine
29d; n = 5, R = piperidine
29e; n = 1 R = morpholine
29f; n = 2, R = morpholine
29g; n = 3, R = morpholine
29h; n = 5, R = morpholine
29i; n = 1, R = N-methylpiperazine
29j; n = 2, R = N-methylpiperazine
29k; n = 3, R = N-methylpiperazine
29l; n = 5, R = N-methylpiperazine
29m; n = 2, R = N-ethylpiperazine
29n; n = 3, R = N-ethylpiperazine

Another series of 4-aminoquinoline-pyrimidine hybrids **30a-30d**, **31a-31d**, **32a-32l** and **33a-33l** was synthesized and tested for antimalarial efficacy against chloroquine susceptible (NF54) and resistant (Dd2) strains *P. falciparum* by the same author (Manohar et al. (2015). Out of 32 compounds, 19 (IC_{50} = 0.003–0.198 μM) showed superior activity to chloroquine (IC_{50} = 0.222 μM) against the resistant strain (Dd2). When tested against the chloroquine susceptible strain, six compounds (IC_{50} = 0.009–0.025 μM) displayed better activity than chloroquine (IC_{50} = 0.027 μM). Compound **32f** was discovered to be the most potent, with an IC_{50} value of 0.003 μM against the resistant strain (Dd2) and 0.028 μM against the sensitive strain (NF54). The cytotoxicity of the selected active compounds against CHO (Chinese Hamster Ovary) cell lines was also tested, and they were found to be nontoxic. One compound, **32e**, was tested for in vivo antimalarial activity against mice infected with *P. berghei* and it was found that a dose of 30 mg/kg on day 4 suppresses parasites by 93.9%, with a mean survival time of 11 days after infection. Chloroquine, on the other hand, demonstrated 90.3% suppression at a dosage of 15 mg/kg. Furthermore, the complex formation between heme and compound **32f** suggested that the heme is the most likely mechanism of action for these hybrids.

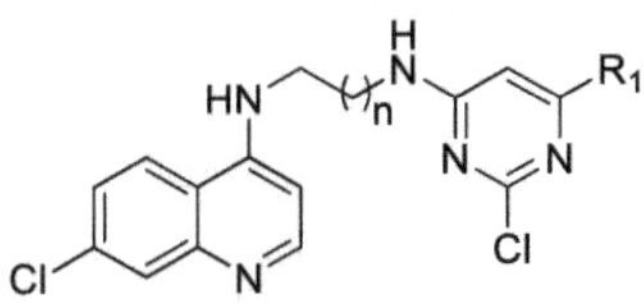

30a; n = 1, R_1 = H
30b; n = 1, R_1 = Cl
30c; n = 2, R_1 = H
30d; n = 2, R_1 = Cl

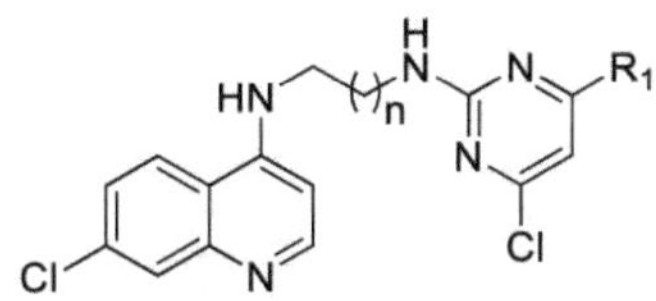

31a; n = 1, R_1 = H
31b; n = 1, R_1 = Cl
31c; n = 2, R_1 = H
31d; n = 2, R_1 = Cl

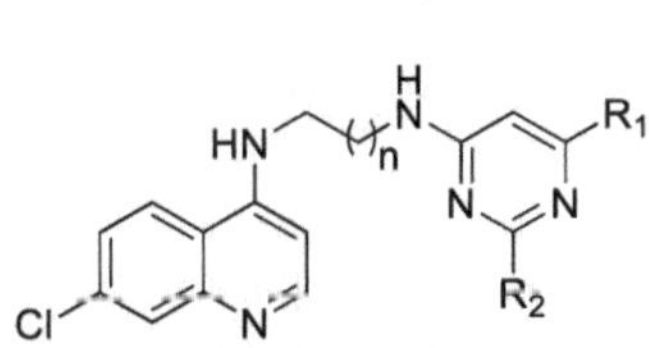

32a; n = 1, R_1 = H, R_2 = morpholine
32b; n = 1, R_1 = H, R_2 = piperidine
32c; n = 1, R_1 = Cl, R_2 = morpholine
32d; n = 1, R_1 = Cl, R_2 = piperidine
32e; n = 1, R_1 = Cl, R_2 = N-methylpiperazine
32f; n = 1, R_1 = Cl, R_2 = N-ethylpiperazine
32g; n = 2, R_1 = H, R_2 = N-methylpiperazine
32h; n = 2, R_1 = H, R_2 = N-ethylpiperazine
32i; n = 2, R_1 = Cl, R_2 = morpholine
32j; n = 2, R_1 = Cl, R_2 = piperidine
32k; n = 2, R_1 = Cl, R_2 = N-methylpiperazine
32l; n = 2, R_1 = Cl, R_2 = N-ethylpiperazine

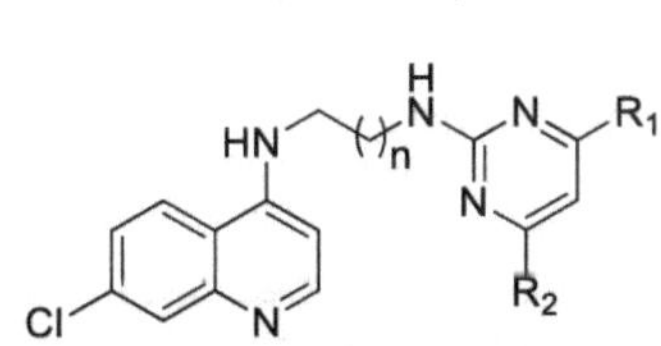

33a; n = 1, R_1 = H, R_2 = morpholine
33b; n = 1, R_1 = H, R_2 = piperidine
33c; n = 1, R_1 = Cl, R_2 = morpholine
33d; n = 1, R_1 = Cl, R_2 = piperidine
33e; n = 1, R_1 = Cl, R_2 = N-methylpiperazine
33f; n = 1, R_1 = Cl, R_2 = N-ethylpiperazine
33g; n = 2, R_1 = H, R_2 = N-methylpiperazine
33h; n = 2, R_1 = H, R_2 = N-ethylpiperazine
33i; n = 2, R_1 = Cl, R_2 = morpholine
33j; n = 2, R_1 = Cl, R_2 = piperidine
33k; n = 2, R_1 = Cl, R_2 = N-methylpiperazine
33l; n = 2, R_1 = Cl, R_2 = N-ethylpiperazine

A novel series of *N*-substituted 4-aminoquinoline-pyrimidine hybrids **34a-34j** and **35a-35j** was synthesized (Maurya et al. 2017). To study the SAR, structural variations were made at two points. The first variation was made in the group that connects the aminoquinoline and pyrimidine moieties, and the second was made in the fourth position of the pyrimidine ring. The majority of the compounds demonstrated potent antimalarial efficacy against chloroquine-sensitive (D6) and resistant (W2) strains of *P. falciparum*. Ten compounds (IC_{50} = 0.038–0.044 μM) were shown to be almost as strong as artemisinin (IC_{50} = 0.045 μM) against the D6 strain, whereas 22 compounds exhibited better activity (IC_{50} = 0.039–0.257 μM) than chloroquine (IC_{50} = 0.317 μM) against W2 strain of *P. falciparum*. In the SAR study, there was no clear pattern of activity observed in altering the carbon chain length or between the two sets of pyrimidine regio-isomers (**34a-34j** and **35a-35j**) but varying the amino substituent for a specific carbon chain length changed the activity. It was discovered that compounds with morpholine, piperidine, or pyrrolidine moieties at the pyrimidine nucleus were less active than those with methyl or ethyl piperazine moieties. Furthermore, none of the compounds were found to be cytotoxic to Vero cell lines. The most active compound, **34b**, was studied for heme binding, and good interaction with monomeric heme and μ-oxodimeric heme was observed, implying that heme could be a target of these hybrids.

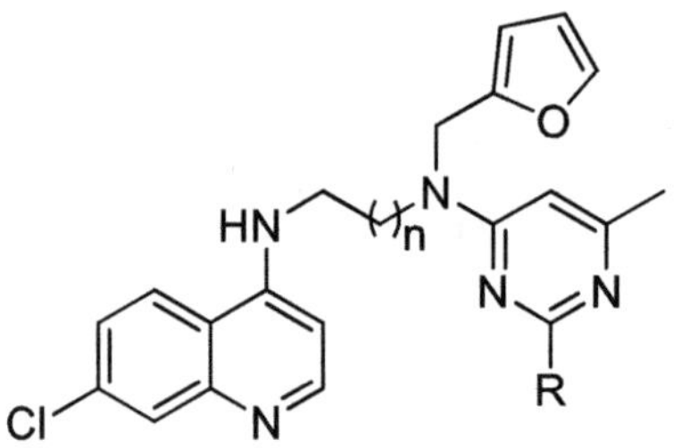

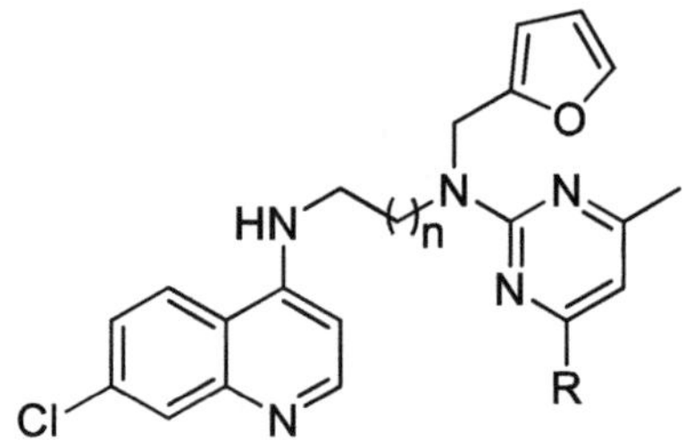

34a; R = N-Me piperazine, n = 1
34b; R = N-Et piperazine, n = 1
34c; R = pyrrolidine, n = 1
34d; R = piperidine, n = 1
34e; R = morpholine, n = 1
34f; R = piperidine, n = 2
34g; R = morpholine, n = 2
34h; R = N-Me piperazine, n = 2
34i; R = pyrrolidine, n = 2
34j; R = N-Et piperazine, n = 3

35a; R = N-Me piperazine, n = 1
35b; R = N-Et piperazine, n = 1
35c; R = pyrrolidine, n = 1
35d; R = piperidine, n = 1
35e; R = morpholine, n = 1
35f; R = piperidine, n = 2
35g; R = morpholine, n = 2
35h; R = N-Me piperazine, n = 2
35i; R = pyrrolidine, n = 2
35j; R = N-Et piperazine, n = 3

Further, a new series of 4-aminoquinoline-pyrimidine hybrids **36a-36j**, **37a-37j**, **38a-38e** and **39a-39e** were developed and tested for their ability to treat malaria in both chloroquine-sensitive (D6) and resistant (W2) strains of *P. falciparum* (Maurya et al. 2019). The majority of the compounds were strongly active, with IC_{50} values ranging from 0.027 to 0.661 μM against the D6 strain and 0.0189 to 1.443 μM against the W2 strain of *P. falciparum*. Twenty-three compounds were shown to be superior to chloroquine in combating the chloroquine-resistant strain (W2). Among all, compound **36d** was discovered to be the most potent compound against the D6 strain (IC_{50} = 0.027 μM), whereas **39e** was the most efficient compound against the W2 strain (IC_{50} = 0.0189 μM). Regarding the SAR, there was no discernible effect of pyrimidine regio-isomers, alkyl linker chain length, or NH substitution on activity. The most active compound, **36d**, was examined for heme binding and found to have good interactions with both monomeric heme and μ-oxodimeric heme, indicating that heme could be a target of these hybrids.

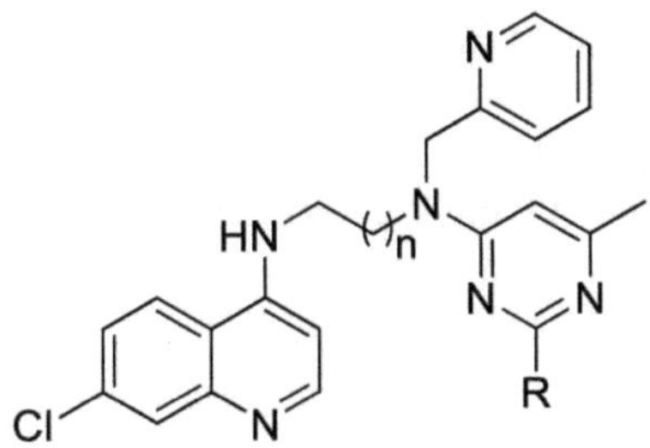

36a; R = piperidine, n = 1
36b; R = pyrrolidine, n = 1
36c; R = morpholine, n = 1
36d; R = N-Me piperazine, n = 1
36e; R = N-Et piperazine, n = 1
36f; R = piperidine, n = 2
36g; R = pyrrolidine, n = 2
36h; R = morpholine, n = 2
36i; R = N-Me piperazine, n = 2
36j; R = N-Et piperazine, n = 3

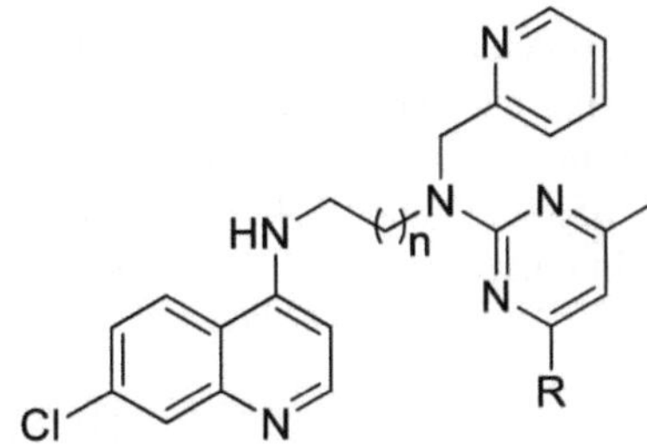

37a; R = piperidine, n = 1
37b; R = pyrrolidine, n = 1
37c; R = morpholine, n = 1
37d; R = N-Me piperazine, n = 1
37e; R = N-Et piperazine, n = 1
37f; R = piperidine, n = 2
37g; R = pyrrolidine, n = 2
37h; R = morpholine, n = 2
37i; R = N-Me piperazine, n = 2

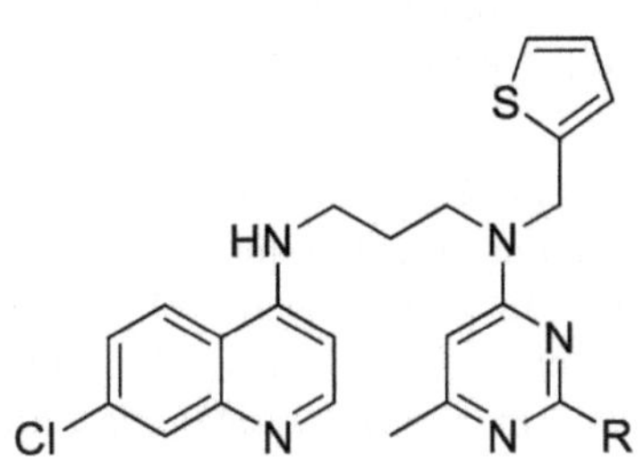

38a; R = piperidine, n = 1
38b; R = pyrrolidine, n = 1
38c; R = morpholine, n = 1
38d; R = N-Me piperazine, n = 1
38e; R = N-Et piperazine, n = 1

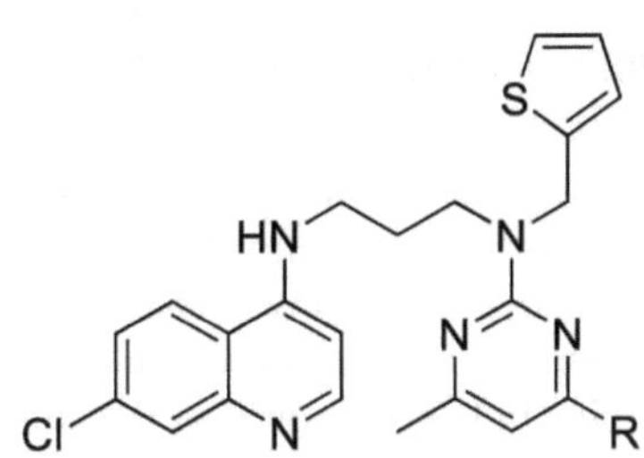

39a; R = piperidine, n = 1
39b; R = pyrrolidine, n = 1
39c; R = morpholine, n = 1
39d; R = N-Me piperazine, n = 1
39e; R = N-Et piperazine, n = 1

The 4-aminoquinoline and pyrimidine hybrids **40a-40j**, **41a-41j** and **42a-42j** were synthesized and investigated for their antimalarial efficacy against *P. falciparum* chloroquine-sensitive (D6) and chloroquine-resistant (W2) strains (Kumar et al. 2015a, b). Eight compounds (**40a**, **40f**, **40g**, **41d**, **41f**, **41g**, **42b** and **42d**) showed antimalarial activity with $IC_{50} < 0.05$ M, and compound **41f** was shown to be just as effective as chloroquine against the D6 strain. Except for two,

40e and **42c**, all the compounds demonstrated superior efficacy to chloroquine against the resistant W2 strain. From the activity data, it was shown that compounds that performed better against chloroquine-sensitive strains also performed well against chloroquine-resistant strains. There was no discernible trend of activity for a specific 4-aminoquinoline pyrimidine conjugate with increasing or decreasing carbon chain length. Compound **40f** with significant in vitro activity was tested for in vivo antimalarial activity in mice infected with *P. berghei* via the oral route of administration, and it was found that the compound **40f** suppresses parasites by 17.85, 37.62 and 96.42% at doses of 11.1, 33.3 and 100 mg/kg, respectively, on day 5, in contrast to the 100% suppression shown by chloroquine.

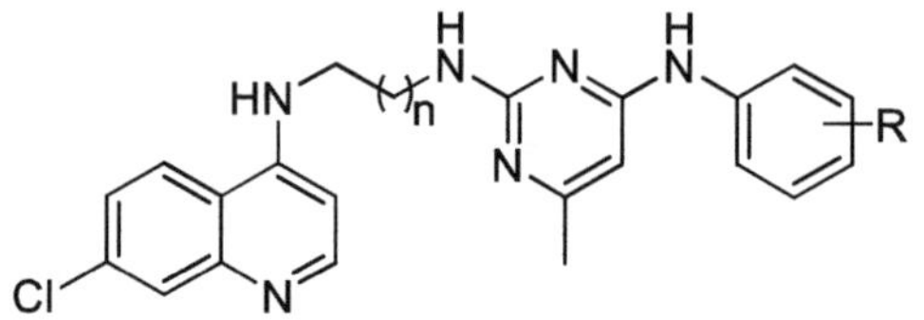

40a; R = H, n = 1	**41a**; R = H, n = 2	**42a**; R = H, n = 3
40b; R = 4-F, n = 1	**41b**; R = 4-F, n = 2	**42b**; R = 4-F, n = 3
40c; R = 4-Cl, n = 1	**41c**; R = 4-Cl, n = 2	**42c**; R = 4-Cl, n = 3
40d; R = 4-Br, n = 1	**41d**; R = 4-Br, n = 2	**42d**; R = 4-Br, n = 3
40e; R = 4-CH_3, n = 1	**41e**; R = 4-CH_3, n = 2	**42e**; R = 4-CH_3, n = 3
40f; R = 4-OCH_3, n = 1	**41f**; R = 4-OCH_3, n = 2	**42f**; R = 4-OCH_3, n = 3
40g; R = 3,5-OCH_3, n = 1	**41g**; R = 3,5-OCH_3, n = 2	**42g**; R = 3,5-OCH_3, n = 3

A novel series of 4-aminoquinoline-purine hybrids **43a-43t** were synthesized and tested for in vitro antimalarial activity against *P. falciparum* chloroquine-sensitive (D6) and resistant (W2) strains (Reddy et al. 2017). In designing these hybrids, structural variations were made at two points. The first variation was made in the diamine linker that connects the aminoquinoline and purine moieties, and the second was made in the secondary amine substitution at the sixth position of the purine ring. Eighteen compounds demonstrated antimalarial activity with IC_{50} values less than 1 μM (0.061–0.833 μM) against the sensitive strain (D6), while eight compounds (**43a**, **43f**, **43h-43j**, **43r-43t**) showed superior activity as compared to chloroquine against the resistant strain (W2) of *P. falciparum*. The strongest compound in the series was found to be compound **43i**, with IC_{50} values of 0.061 μM (D6 strain) and 0.080 μM (W2 strain). Except for two compounds, all of the compounds were shown to be nontoxic towards mammalian (Vero) cells up to the highest tested concentration of 11.86 μM. Furthermore, strong binding with heme revealed that heme may be the primary target of these hybrids for antimalarial activity.

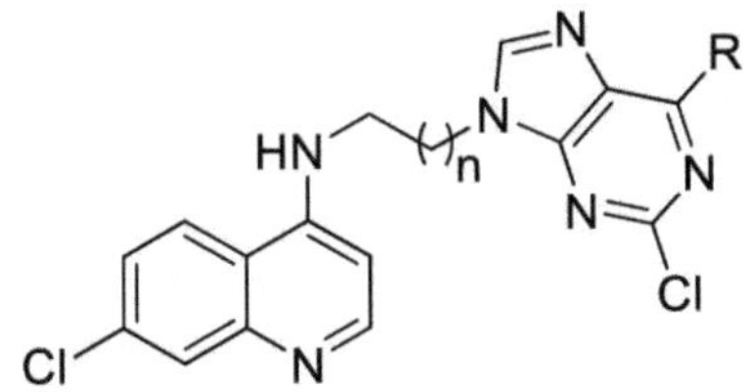

43a; n = 1, R = dimethylamino
43b; n = 1, R = diethylamino
43c; n = 1, R = pyrrolidin-1-yl
43d; n = 1, R = piperidin-1-yl
43e; n = 1, R = azepan-1-yl
43f; n = 1, R = morpholin-1-yl
43g; n = 1, R = thiomorpholin-1-yl
43h; n = 1, R = 4-methylpiperazin-1-yl
43i; n = 1, R = 4-ethylpiperazin-1-yl
43j; n = 1, R = 4-phenylpiperazin-1-yl
43k; n = 2, R = dimethylamino
43l; n = 2, R = diethylamino
43m; n = 2, R = pyrrolidin-1-yl
43n; n = 2, R = piperidin-1-yl
43o; n = 2, R = azepan-1-yl
43p; n = 2, R = morpholin-1-yl
43q; n = 2, R = thiomorpholin-1-yl
43r; n = 2, R = 4-methylpiperazin-1-yl
43s; n = 2, R = 4-ethylpiperazin-1-yl
43t; n = 2, R = 4-phenylpiperazin-1-yl

A library of 4-aminoquinoline-guanylthiourea derivatives was designed and based on molecular docking investigations, and nine compounds were chosen for synthesis and tested for in vitro antimalarial efficacy against chloroquine-sensitive strain (D6) and resistant strain (W2) of *P. falciparum* (Bhagat et al. 2019). Eight of the nine compounds displayed activity against the D6 strain in the range of 0.61–7.55 μM and the W2 strain in the range of 0.43–8.04 μM. Compound **44c** was found to be most effective with MIC values of 0.6 μM and 0.4 μM against D6 and W2 strains, respectively. Furthermore, none of the compounds were shown to be cytotoxic at the highest concentrations tested against Vero cell lines.

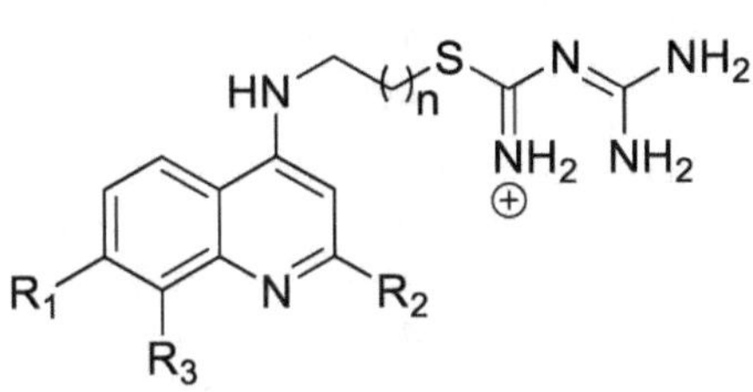

44a; R_1 = Cl, R_2 = H, R_3 = H, n = 1
44b; R_1 = Cl, R_2 = H, R_3 = H, n = 2
44c; R_1 = Cl, R_2 = H, R_3 = H, n = 3
44d; R_1 = CF_3, R_2 = H, R_3 = H, n = 1
44e; R_1 = CF_3, R_2 = H, R_3 = H, n = 2
44f; R_1 = CF_3, R_2 = H, R_3 = H, n = 3
44g; R_1 = H, R_2 = CF_3, R_3 = CF_3, n = 1
44h; R_1 = H, R_2 = CF_3, R_3 = CF_3, n = 2
44i; R_1 = H, R_2 = CF_3, R_3 = CF_3, n = 3

A series of 4-aminoquinoline-phthalimide hybrids were synthesized and tested for antimalarial activity against the chloroquine-resistant and mefloquine-sensitive strain (W2) of *P. falciparum* (Rani et al. 2018). To investigate SAR, the length of the alkyl chain linker between the phthalimide and 4-aminoquinoline, as well as the nature of substituents at the C3 or C4 position of phthalimide, was varied. And it was established that on increasing the number of carbon atoms in the linker, antimalarial activity increases. The optimum activity was achieved for the 6 and 8 carbons chain length. The compound **45v** having a hexyl chain as a linker and a 3,4,5,6-tetrabromo substituent on the phthalimide ring was the most effective compound in the series, with an IC_{50} value of 0.10 μM. Some of the active compounds were tested for cytotoxicity using J774 murine macrophage cells and found to be noncytotoxic, with selectivity indexes ranging from 126 to 291. However, these analogues were not as effective as chloroquine (IC_{50} = 77 nM), prompting the research team to continue optimizing these compounds.

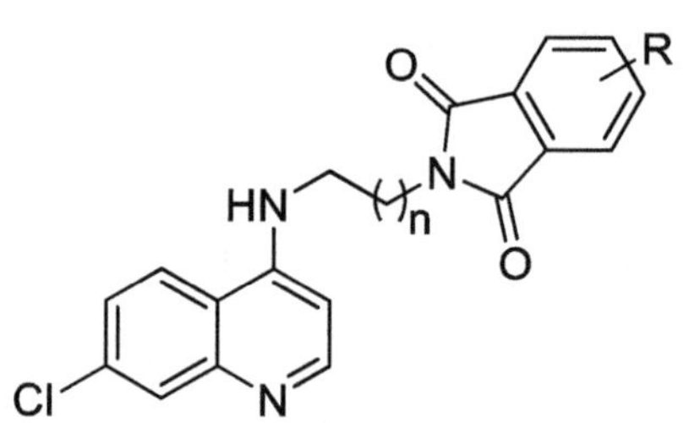

45a; R = H, n = 2
45b; R = H, n = 3
45c; R = H, n = 5
45d; R = H, n = 7
45e; R = 3-F, n = 1
45f; R = 3-F, n = 2
45g; R = 3-F, n = 3
45h; R = 3-F, n = 5
45i; R = 3-F, n = 7
45j; R = 4-F, n = 1
45k; R = 4-F, n = 2
45l; R = 4-F, n = 3
45m; R = 4-F, n = 5
45n; R = 4-F, n = 7
45o; R = 3,4,5,6-Cl, n = 1
45p; R = 3,4,5,6-Cl, n = 2
45q; R = 3,4,5,6-Cl, n = 3
45r; R = 3,4,5,6-Cl, n = 5
45s; R = 3,4,5,6-Br, n = 1
45t; R = 3,4,5,6-Br, n = 2
45u; R = 3,4,5,6-Br, n = 3
45v; R = 3,4,5,6-Br, n = 5
45w; R = 3-NO_2, n = 1
45x; R = 4-NO_2, n = 5

In continuation of previous research, a variable chain length amide linker was introduced between phthalimide and 4-aminoquinoline rings (Rani et al. 2019). Additionally, several cyclic tertiary amines were introduced at the fifth position of the phthalimide ring. All of the compounds were tested towards the chloroquine-resistant and mefloquine-sensitive strain (W2) of *P. falciparum*. Seven of the 36 compounds displayed better activity than chloroquine (IC_{50} = 0.23 μM). The

SAR study revealed that compounds having $m = 2$ had improved antimalarial activity when compared to compounds with $m = 1$ or 3. In addition, compounds with tertiary amine substituents at the fifth position of the phthalimide ring demonstrated higher activity than their unsubstituted or 5-F-substituted analogues. Furthermore, it was found that none of the compounds were harmful to mammalian Vero cells. Among all compounds, **47j** was shown to be the most active compound with an IC_{50} value of 0.097 μM.

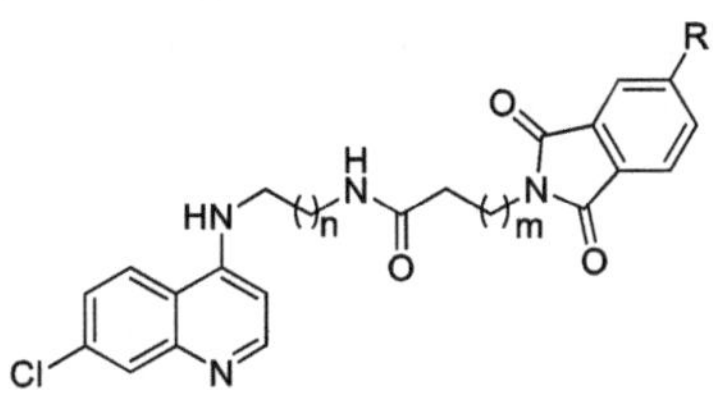

46a; R = H, n = 1, m = 0
46b; R = H, n = 3, m = 0
46c; R = H, n = 5, m = 0
46d; R = H, n = 7, m = 0
46e; R = H, n = 1, m = 1
46f; R = H, n = 3, m = 1
46g; R = H, n = 5, m = 1
46h; R = H, n = 7, m = 1
46i; R = H, n = 1, m = 2
46j; R = H, n = 3, m = 2
46k; R = H, n = 5, m = 2
46l; R = H, n = 7, m = 2
46m; R = 5-F, n = 1, m = 0
46n; R = 5-F, n = 3, m = 0
46o; R = 5-F, n = 5, m = 0
46p; R = 5-F, n = 1, m = 1
46q; R = 5-F, n = 3, m = 1
46r; R = 5-F, n = 5, m = 1

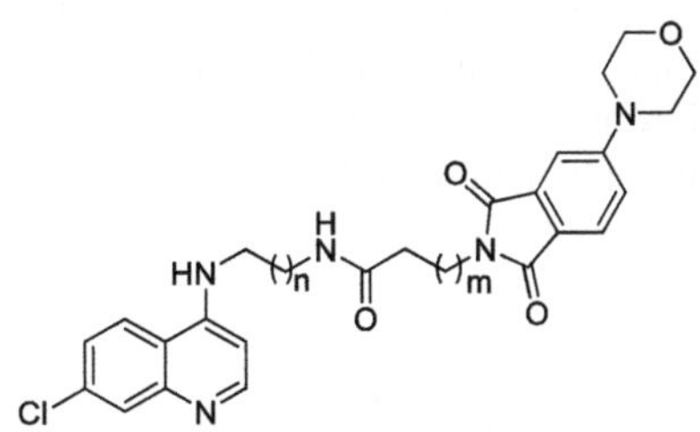

47a; n = 1, m = 0
47b; n = 3, m = 0
47c; n = 5, m = 0
47d; n = 1, m = 1
47e; n = 3, m = 1
47f; n = 5, m = 1

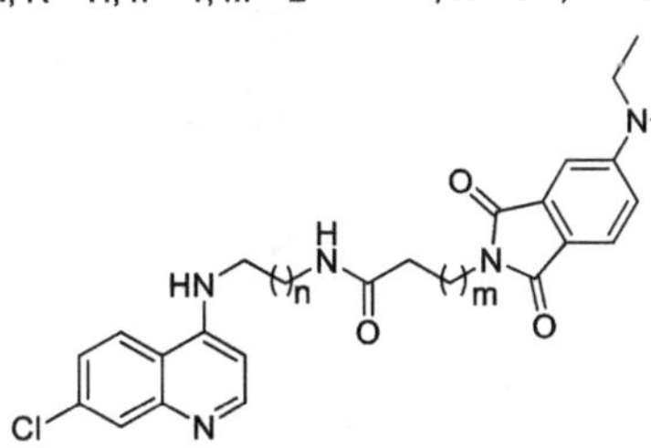

47g; n = 1, m = 0
47h; n = 3, m = 0
47i; n = 5, m = 0
47j; n = 1, m = 1
47k; n = 3, m = 1
47l; n = 5, m = 1

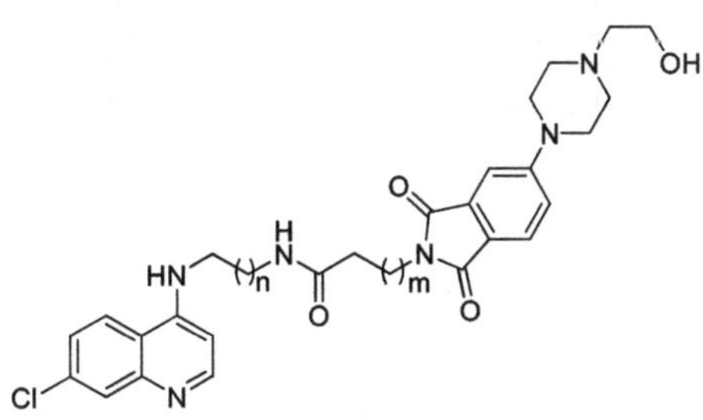

47m; n = 1, m = 0
47n; n = 3, m = 0
47o; n = 5, m = 0
47p; n = 1, m = 1
47q; n = 3, m = 1
47r; n = 5, m = 1

Further, a new series of phthalimide-4-aminoquinoline hybrids (**48a-48 t**) with a variable length alkyl chain and a cyclic tertiary amine linked at the C4 or C5 position of the phthalimide ring was synthesized (Rani et al. 2020). The in vitro antimalarial activity of the synthesized compounds was tested against the chloroquine-resistant strain (W2) of *P. falciparum*, and most of them were found to be more effective than chloroquine ($IC_{50} = 0.079$ M). It was found that the type of the substituent at the C4 or C5 positions of the isoindoline-1,3-dione as well as the length of the alkyl chain linker had an impact on the antimalarial activity of the compounds. The author found that switching the substitution on the phthalimide ring from piperidine to morpholine and then to hydroxyethyl piperazine increased the antimalarial activity. The activity was also enhanced by moving the substituent from C4 to the C5 position. Furthermore, increasing the chain length of the linker resulted in improved antimalarial activity, except for the hydroxyethyl piperazine substituent, for which all the related compounds demonstrated good activities regardless of chain length. With an IC_{50} of 0.006 μM, compound **48r** having hydroxyethyl-piperazine at the C5 position and a propyl chain as a linker was the most active compound among all. Heme binding studies revealed that heme could be the principal mechanism of action of these compounds. The heme binding constant of **48r** was found to be comparable to that of chloroquine. Furthermore, none of the compounds showed cytotoxic effects on the mammalian Vero cells.

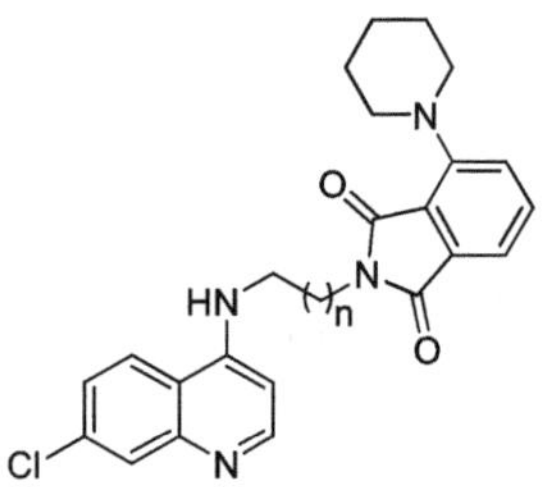

48a; n = 1
48b; n = 2
48c; n = 3
48d; n = 5

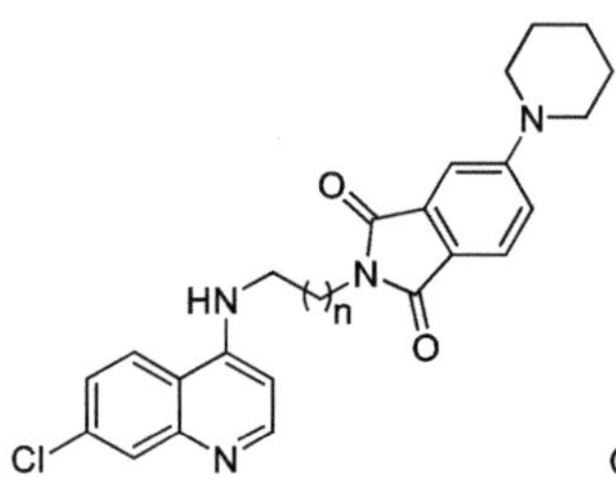

48e; n = 1
48f; n = 2
48g; n = 3
48h; n = 5

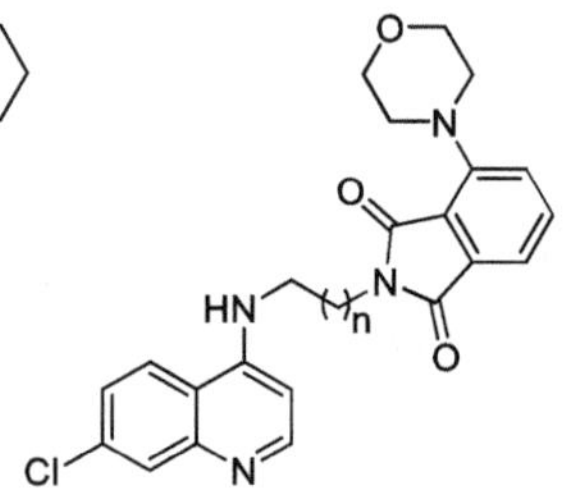

48i; n = 1
48j; n = 2
48k; n = 3
48l; n = 5

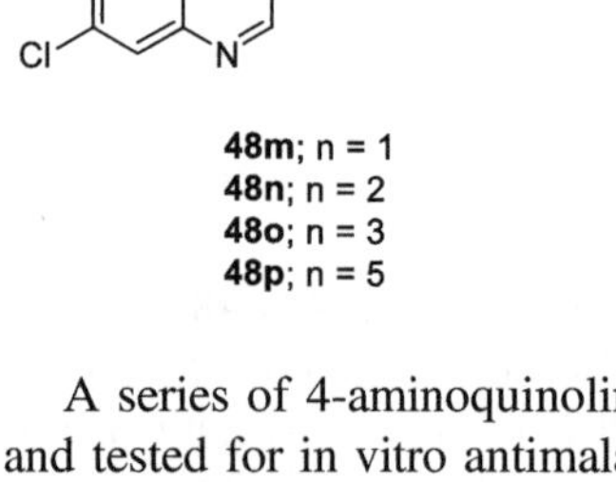

48m; n = 1
48n; n = 2
48o; n = 3
48p; n = 5

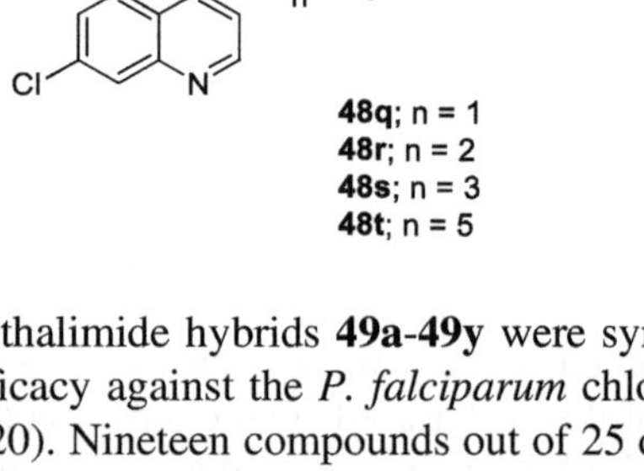

48q; n = 1
48r; n = 2
48s; n = 3
48t; n = 5

A series of 4-aminoquinoline-naphthalimide hybrids **49a-49y** were synthesized and tested for in vitro antimalarial efficacy against the *P. falciparum* chloroquine-resistant strain (W2) (Shalini et al. 2020). Nineteen compounds out of 25 displayed better activity than chloroquine (IC_{50} = 232.7 nM). In general, compounds with shorter alkyl chain linkers outperformed compounds with longer alkyl chain linkers. The substitution by morpholine or piperidine on the naphthalimide ring displayed no noticeable impact on the antimalarial activity, whereas hydroxyethyl piperazine substituted analogues significantly increased antimalarial activity. Compound **49v**, having propylene linker and hydroxyethyl piperazine substituent at the C-4 position of the naphthalimide ring, displayed the strongest activity (IC_{50} = 15.4 nM) and it was ~15 times more potent than chloroquine (IC_{50} = 232.6 nM). Furthermore, it was found that none of the compounds were cytotoxic to the mammalian Vero cell lines.

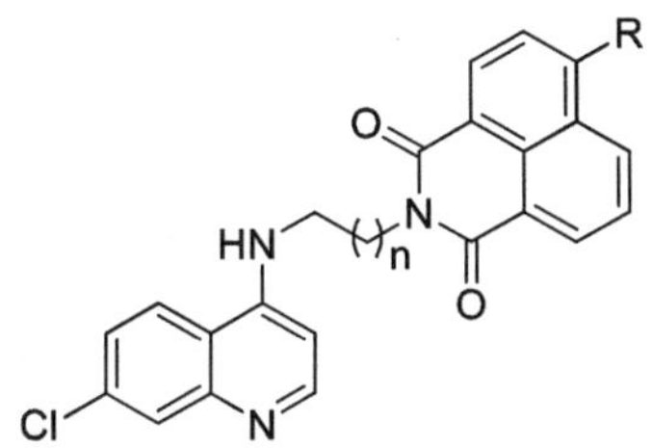

49a; R = H, n = 1	**49i**; R = Br, n = 5	**49q**; R = morpholine, n = 2
49b; R = H, n = 2	**49j**; R = Br, n = 7	**49r**; R = morpholine, n = 3
49c; R = H, n = 3	**49k**; R = Piperidine, n = 1	**49s**; R = morpholine, n = 5
49d; R = H, n = 5	**49l**; R = Piperidine, n = 2	**49t**; R = morpholine, n = 7
49e; R = H, n = 7	**49m**; R = Piperidine, n = 3	**49u**; R = Hydroxyethyl piperazine, n = 1
49f; R = Br, n = 1	**49n**; R = Piperidine, n = 5	**49v**; R = Hydroxyethyl piperazine, n = 2
49g; R = Br, n = 2	**49o**; R = Piperidine, n = 7	**49w**; R = Hydroxyethyl piperazine, n = 3
49h; R = Br, n = 3	**49p**; R = morpholine, n = 1	**49x**; R = Hydroxyethyl piperazine, n = 5
		49y; R = Hydroxyethyl piperazine, n = 7

A series of 4-aminoquinoline-benzimidazole hybrids (**50a-50e**) and their ferrocenyl analogues (**51a-51e**) were synthesized and tested for in vitro antimalarial efficacy towards chloroquine-sensitive strain (NF54) and multidrug resistant strain (K1) of *P. falciparum* (Baartzes et al. 2019). The majority of the compounds were more effective against the K1 strain as compared to the NF54 strain of *P. falciparum*. Compounds **50c** and **50e** were discovered to be more active than chloroquine (IC_{50} = 0.205 μM) against the resistant strain, with IC_{50} values of 0.151 μM and 0.179 μM, respectively. The cytotoxic activities of the compounds were investigated against the non-tumorigenic Chinese hamster ovarian (CHO) cell line, and the majority of compounds proved to be nontoxic. The antimalarial effectiveness of the most potent phenyl (**50c**) and ferrocenyl (**51c**) aminoquinoline benzimidazole hybrids was also examined for in vivo activity in mice infected with *P. berghei*. At a dose level of 4 × 50 mg/kg, a considerable suppression of parasitaemia (58% for compound **50c** and 92% for compound **51b**) was observed. The β-hematin inhibitory activity of both compounds was also investigated which revealed more than four times higher potency of both compounds (IC_{50} = 16 μM) as compared to chloroquine (IC_{50} = 74 μM) to inhibit the formation of hemozoin.

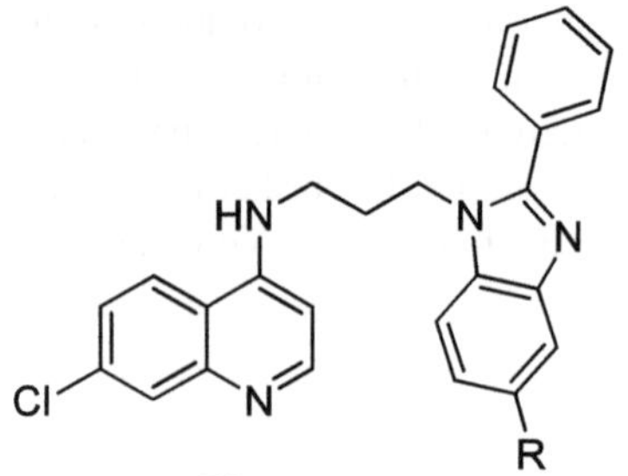

50a; R = H
50b; R = CH_3
50c; R = CF_3
50d; R = SO_2CH_3
50e; R = CN

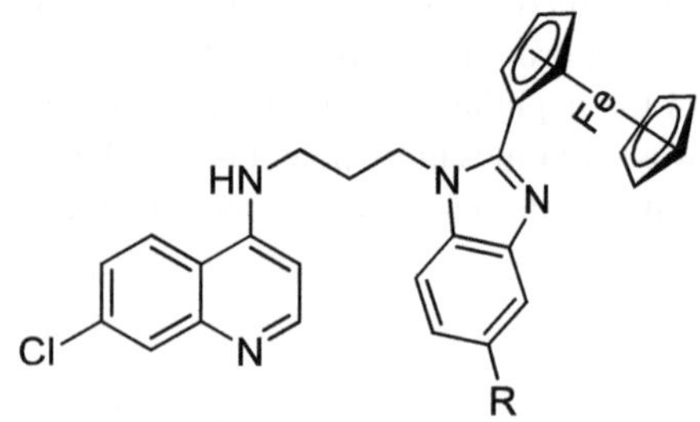

51a; R = H
51b; R = CH_3
51c; R = CF_3
51d; R = SO_2CH_3
51e; R = CN

A series of 4-aminoquinoline-ferrocene hybrids (**52a-52g**) with various linkers between the two pharmacophores were synthesized and tested for in vitro antimalarial activity towards chloroquine-sensitive (D10) and resistant (Dd2) strains of *P. falciparum* (N'Da and Smith 2014). Compounds having rigid linkers were shown to be ineffective, whereas those with flexible linkers were found to be active against both D10 and Dd2 strains of *P. falciparum*. Compound **52g** with a 3-aminopropyl methylamine linker was the most active compound. Its antimalarial activity was found to be comparable to chloroquine against the susceptible strain D10 but significantly more active than chloroquine and the equimolar combination of chloroquine and ferrocene against the resistant strain D2d.

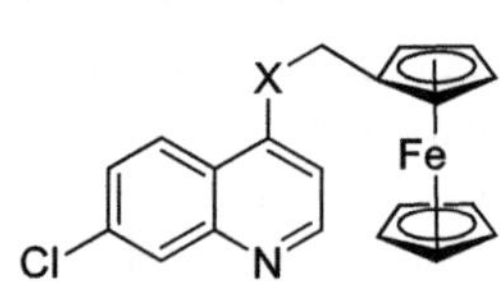

52a; R = $-NHCH_2CH_2CH_2NH-$

52b; X = $-NHCH_2CH_2CH_2CH_2NH-$

52c; X = 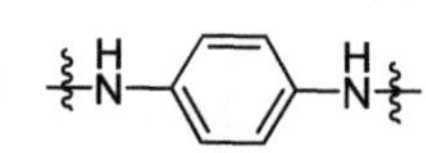

52d; X = 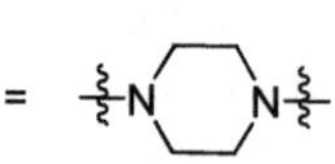

52e; X = 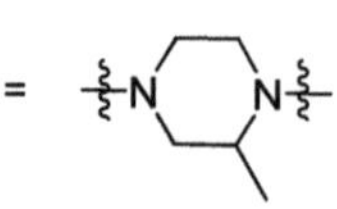

52f; X = 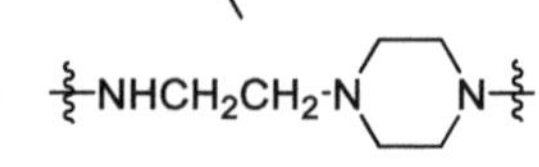

52g; X = 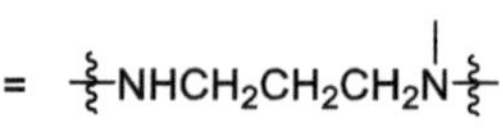

The chloroquine-quinoxaline 1,4-di-*N*-oxide hybrids were synthesized and tested for antimalarial efficacy against *P. falciparum* chloroquine-sensitive (3D7) and multidrug-resistant (FCR-3) strains (Bonilla-Ramirez et al. 2018). Furthermore, they tested in vitro and in vivo liver stage efficacy as well as the capacity to prevent transmission of the active compounds in the mosquito stage. Except for **53c**, all of the compounds demonstrated antimalarial efficacy with IC_{50} values less than 1 μM towards the chloroquine susceptible strain; however, only two compounds displayed the same range of IC_{50} towards the multidrug-resistant strain. Compounds **53b** and **53e** were shown to be the most effective against both sensitive and resistant strains, with IC_{50} values ranging from 0.40 to 0.90 μM. Interestingly, compound **53b** demonstrated moderate efficacy towards three liver stage parasites, most notably *P. falciparum* liver stage with an IC_{50} value of 6.39 μM.

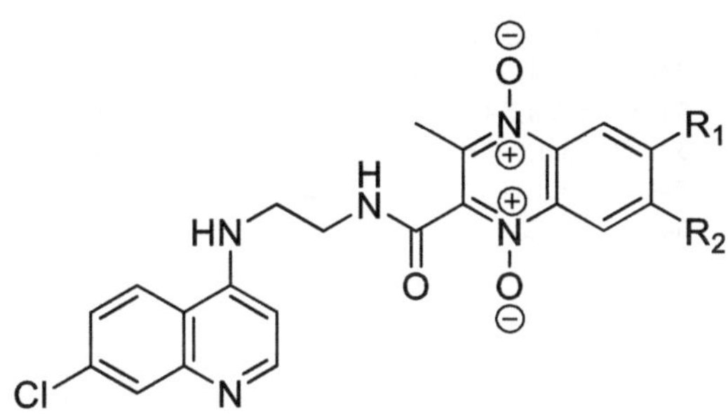

53a; R_1 = H, R_2 = H
53b; R_1 = H, R_2 = Cl
53c; R_1 = H, R_2 = OCH_3
53d; R_1 = H, R_2 = CH_3
53e; R_1 = CH_3, R_2 = CH_3

Several quinoline-chalcone (**54a-54k**, **55a-55k**) and quinoline-pyrazoline (**56a-56g**, **57a-57i**) hybrids were synthesized and evaluated for in vitro antimalarial efficacy (Charris et al. 2019). As a preliminary screening for antimalarial activity, the in vitro *β*-hematin formation inhibitory activity was investigated. Ten compounds were found to inhibit more than 80% of heme crystallization, with chloroquine inhibiting 98.52%. These compounds were then chosen for in vivo testing in mice infected with *P. berghei.* The antimalarial effectiveness of the compounds was assessed by their ability to reduce parasitaemia and increase survival on the fourth day after infection as compared to the untreated control group. Mice were given either compound (20 mg/kg) or chloroquine (20 mg/kg) intraperitoneally once daily for 4 days. On the fourth day, the mean survival time and parasitaemia percentage were compared to control mice given only saline. Compounds **54e** and **54h** were found to be the most active as they extended the infected mice's average survival period to 22.44 and 24.60 days, respectively, which is comparable to chloroquine (24.99 days) and threefold better than the control (6.8 days). They were, however, unable to reduce or postpone the progression of malaria in the same way that chloroquine did (1.40% and 1.81% parasitaemia for compounds **54e** and **54h**, respectively on day 4, compared to 0.32% for chloroquine).

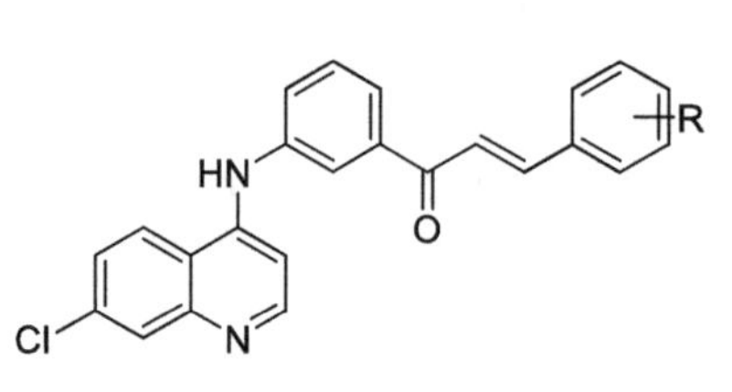

54a; R = 4-CH_3
54b; R = 2,3-OCH_3
54c; R = 2,4-OCH_3
54d; R = 2,5-OCH_3
54e; R = 3,4-OCH_3
54f; R = 3,5-OCH_3
54g; R = 2,4,5-OCH_3
54h; R = 3,4,5-OCH_3
54i; R = 3,4,5-CH_3
54j; R = -OCH_2O-
54k; R = 4-Cl

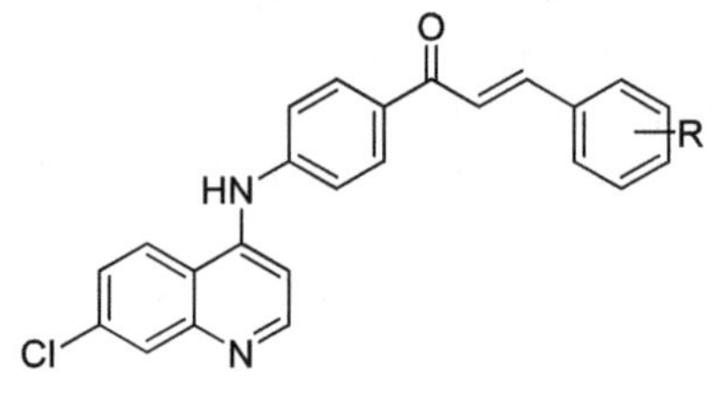

55a; R = 4-CH_3
55b; R = 2,3-OCH_3
55c; R = 2,4-OCH_3
55d; R = 2,5-OCH_3
55e; R = 3,4-OCH_3
55f; R = 3,5-OCH_3
55g; R = 2,4,5-OCH_3
55h; R = 3,4,5-OCH_3
55i; R = 3,4,5-CH_3
55j; R = -OCH_2O-
55k; R = 4-Cl

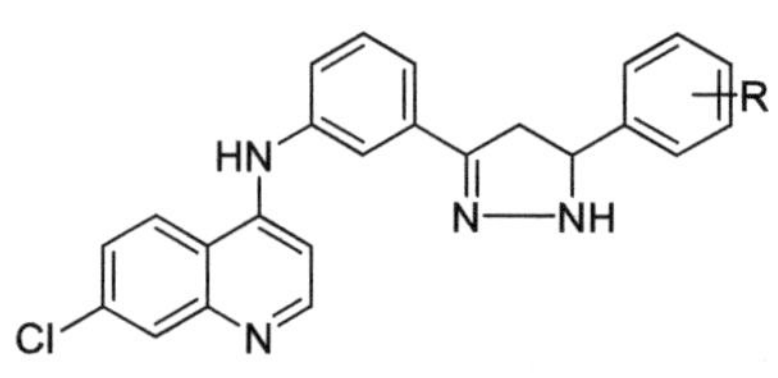

56a; R = 3,4-OCH_3
56b; R = 3,5-OCH_3
56c; R = 2,4,5-OCH_3
56d; R = 3,4,5-OCH_3
56e; R = 3,4,5-CH_3
56f; R = -OCH_2O-
56g; R = 4-Cl

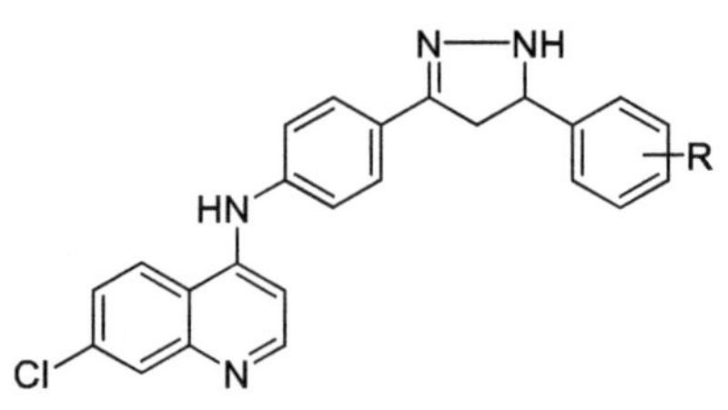

57a; R = 2,4-OCH_3
57b; R = 2,5-OCH_3
57c; R = 3,4-OCH_3
57d; R = 3,5-OCH_3
57e; R = 2,4,5-OCH_3
57f; R = 3,4,5-OCH_3
57g; R = 3,4,5-CH_3
57h; R = -OCH_2O-
57i; R = 4-Cl

10.3 Anticancer Activity

Cancer is a leading cause of death worldwide. The number of new cancer cases is increasing daily even though better medications and targeted cancer therapies are readily available. No single treatment of cancer is completely successful and patients frequently receive a combination of therapies such as surgery, radiation therapy, immunotherapy, chemotherapy, gene therapy, or hormonal therapy, depending on the type and stage of cancer, patient's health status, age and personal characteristics (http://www.cancer.gov/) (Cancer. gov n.d.). Due to the genotoxicity, resistance and toxic effects, anticancer drugs have limited therapeutic applicability (Aydemir and Bilaloglu 2003). Thus, there is always a need for comparatively more effective and safer compounds for the treatment of cancer. Recently, quinoline has attracted much attention from researchers across the globe due to its presence in well-known anticancer drugs such as camptothecin and cabozantinib (Srivastava et al. 2005). The number of quinoline compounds having anticancer action has grown over the past few years, according to scientific research papers (Musiol 2017; Afzal et al. 2015; Marganakop et al. 2014; Aly 2010; Tseng et al. 2013; El-Gamal et al. 2014; Solomona et al. 2019; Ahadi and Emami 2020; Akkachairin et al. 2020).

A series of *N′*-substituted methylene-4-(quinoline-4-amino) benzoyl hydrazides **58a-58p** were synthesized and tested against HepG2 cells (Li et al. 2020). Compounds **58h** and **58j** were found to have significant antiproliferative activity, with IC_{50} values of 12.6 μM and 27.3 μM, respectively. With IC_{50} values of 9.6 μM and 6.3 μM, respectively, against SMMC-7721 and Huh7 cells (human hepatic cancer cell lines), compound **58h** also demonstrated strong cytotoxicity. Inspiringly, both **58h** and **58j** were less cytotoxic in healthy cells as compared to hepatocellular cancer cells. Additional bioassays revealed that **58h** and **58j** can reduce the mRNA level and expression level of c-Myc. Furthermore, both compounds were found to induce the pro-apoptotic protein Bax to be up-regulated and the anti-apoptotic protein Bcl-2 to be down-regulated. The anticancer activity of **58h** and **58j** in HepG2 cells was discovered to be due to their anti-survival effects, induction of apoptotic cell death, cell cycle arrest and inhibition of cell migration. According to SAR studies, alkyl substitutions (**58a** and **58b**) at the nitrogen atom (*N′*) of *N*-acylhydrazone do not affect the cytotoxic activity significantly. On the other hand, different substituents on the benzene ring attached to the *N′*-methylene group showed a substantial correlation with cytotoxic activity. Compounds **58d**, **58e** and **58k** having carboxyl (COOH) or nitro (NO_2) groups on the benzene ring were found to be inactive.

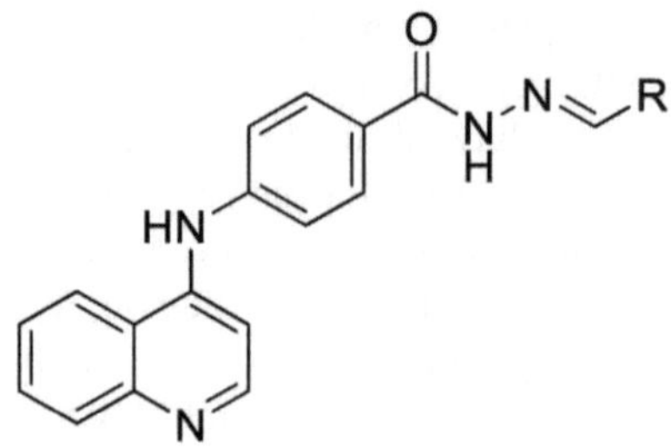

58a; R = methyl
58b; R = n-propyl
58c; R = 2-(trifluoromethyl)phenyl
58d; R = 2-benzoic acid-1-yl
58e; R = 3-benzoic acid 1-yl
58f; R = 4-phenyl acetate 1-yl
58g; R = 2,6-dimethoxyphenyl
58h; R = 3,4-dimethoxyphenyl
58i; R = 3,4-dihydroxyphenyl
58j; R = 2,5-dihydroxyphenyl
58k; R = 4-nitrophenol-2-yl
58l; R = 3,5-dibromo-4-hydroxyphenyl
58m; R = 1*H*-imidazole-2-yl
58n; R = 1*H*-pyrazole-5-yl
58o; R = 4-methylthiazole-5-yl
58p; R = 1*H*-benzo[*d*]imidazole-2-yl

The hydrazone derivatives of quinoline were synthesized and their anticancer activity was determined towards the full NCI 60 human cancer cell line panel (Katariya et al. 2020). Nine compounds (**59c, 59d, 59e, 60a, 60b, 60c, 60d, 60e** and **60f**) exhibited significant anticancer activity at 10 μM concentration and were subsequently examined at ten-fold dilutions of five different concentrations (0.01, 0.1, 1, 10 and 100 μM). All of the compounds displayed antiproliferative action with GI_{50} values in the range of 0.33–4.87 μM and LC_{50} values in the range of 4.67–100 μM. Compound **60e** had the highest activity of all, with GI_{50} values from 0.33 to 2.05 μM and LC_{50} values from 5.15 to 24.4 μM. Furthermore, a molecular docking analysis showed that compounds exhibited strong interaction with DNA and GTP nucleotides and also in the active site of the topoisomerase enzyme via hydrogen bonding.

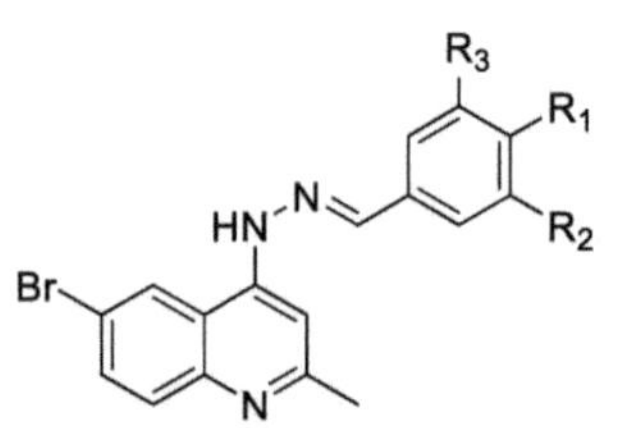

59a; $R_1 = OC_2H_5$, R_2 = H, R_3 = H
59b; $R_1 = OC_3H_7$, R_2 = H, R_3 = H
59c; $R_1 = OC_4H_9$, R_2 = H, R_3 = H
59d; $R_1 = OC_5H_{11}$, R_2 = H, R_3 = H
59e; $R_1 = OC_6H_{13}$, R_2 = H, R_3 = H
59f; $R_1 = OC_8H_{17}$, R_2 = H, R_3 = H
59g; $R_1 = OC_8H_{17}$, R_2 = H, $R_3 = OC_8H_{17}$
59h; $R_1 = OC_8H_{17}$, $R_2 = OC_8H_{17}$, $R_3 = OC_8H_{17}$

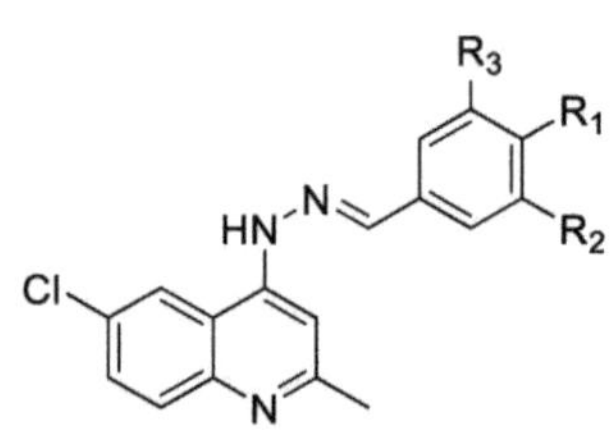

60a; $R_1 = OC_2H_5$, R_2 = H, R_3 = H
60b; $R_1 = OC_3H_7$, R_2 = H, R_3 = H
60c; $R_1 = OC_4H_9$, R_2 = H, R_3 = H
60d; $R_1 = OC_5H_{11}$, R_2 = H, R_3 = H
60e; $R_1 = OC_6H_{13}$, R_2 = H, R_3 = H
60f; $R_1 = OC_8H_{17}$, R_2 = H, R_3 = H
60g; $R_1 = OC_8H_{17}$, R_2 = H, $R_3 = OC_8H_{17}$
60h; $R_1 = OC_8H_{17}$, $R_2 = OC_8H_{17}$, $R_3 = OC_8H_{17}$

Twenty-six new 4-aminoquinoline derivatives were synthesized and evaluated for their ability to inhibit cell proliferation in A549, A2780/T, A2780S, HCT116, HCT-8, HCT-8/T, HCT-8/V, MCF-7, MCF-7/ADR and MDA-MB-231 cells (Zhou et al. 2017). Among all, compound **63h** demonstrated the most effective cytotoxic activity with IC_{50} values in the range of 1.5–3.9 nM towards all the examined cancer cell lines and displayed significant activity against the multidrug-resistant cancer cells also. The immunofluorescence tests, EBI examination, microtubule dynamics test and molecular modelling investigations identified compound **63h** as a novel depolymerization agent binding to the colchicine active site. Moreover, compound **63h** displayed significant in vivo anticancer activity in the HCT116 xenograft model without appreciable body weight loss.

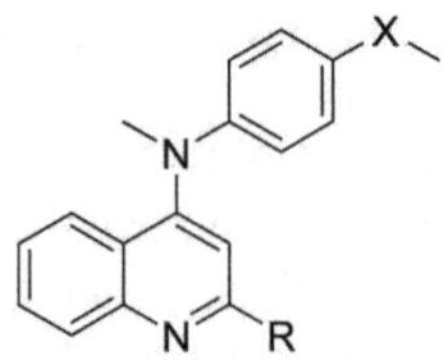

61a; R = Cl, X = O
61b; R = CH_3, X = O
61c; R = H, X = O
61d; R = CH_3, X = S

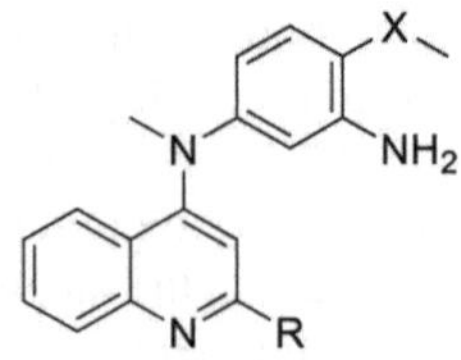

62a; R = Cl, X = O
62b; R = CH_3, X = O
62c; R = H, X = O
62d; R = CH_3, X = S

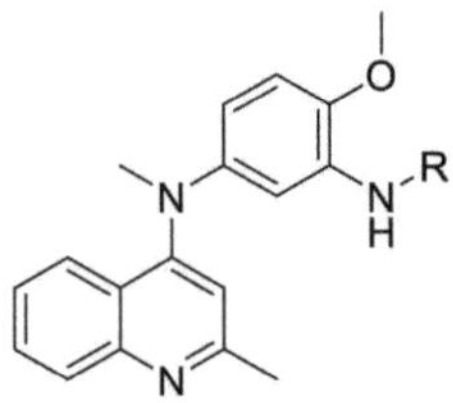

63a-m

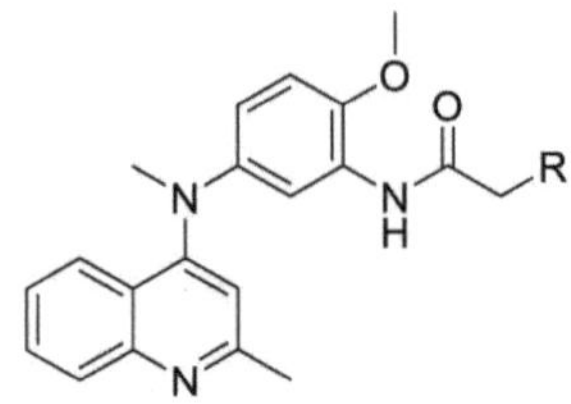

64a-e

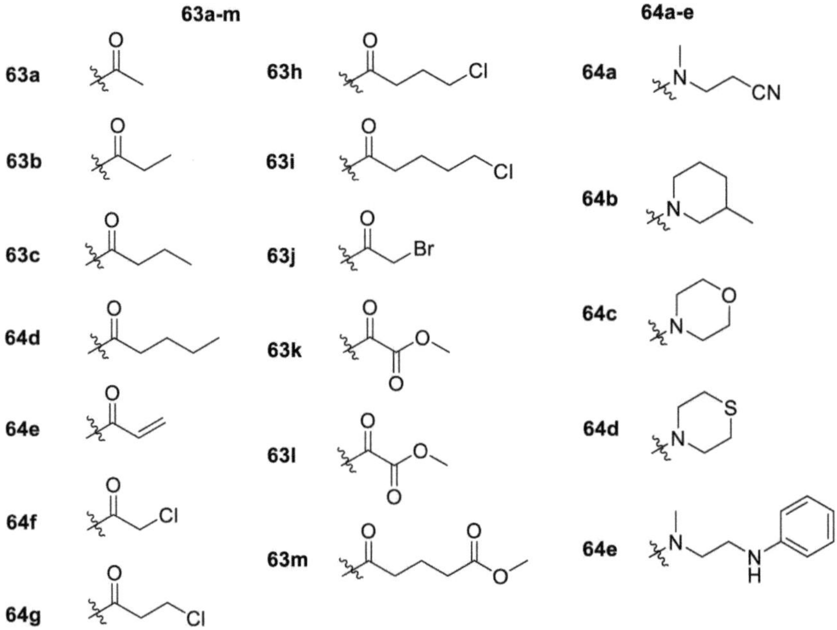

A series of new 4-aminoquinoline derivatives **65a-65s** and **66a-66y** were synthesized and evaluated for their anticancer action against cervical carcinoma (Hela), gastric carcinoma (BGC-823), human colorectal carcinoma (HCT-116 and RKO), non-small-cell lung carcinoma (A549 and NCI-H1650) and ovarian carcinoma (A2780) cells (Su et al. 2019). At a concentration of 10 μM, compounds **65a-65s** exhibited no cytotoxic activity against any of the cancer cell lines. Compounds

66a-66n and **66w** demonstrated anticancer effects selectively towards Hela cell lines with IC_{50} values <10 μM. Compounds **66b**, **66d-66j**, **66l-66n** and **66w** demonstrated selectivity for A2780 cancer cell lines, with IC_{50} values smaller than 10 μM. Furthermore, compounds **66a**, **66e**, **66f**, **66h**, **66j**, **66m** and **66n** demonstrated antiproliferative action specifically against BGC-823 cell lines with IC_{50} values less than 10 μM. Surprisingly, five compounds, **66p**, **66s**, **66v**, **66x** and **66y**, demonstrated strong anticancer activity against all the examined cancer cell lines with IC_{50} values <10 μM. Compound **66x** was discovered to be the most effective compound against HCT-116, Hela, A2780 and RKO cell lines, with IC_{50} values of 2.56, 2.71, 3.46 and 3.67 μM, respectively. SAR analysis revealed that the presence of a bulky group at the seventh position and 4-methoxyaniline at the fourth position of the quinoline nucleus is favourable for antiproliferative activity. The in vivo anticancer activity of the representative compound **66x** was also determined, and the results revealed that compound **66x** significantly slowed down tumour growth and reduced tumour size in animal models. According to mechanistic studies, compound **66x** suppresses colorectal cancer growth through an ATG5-dependent autophagy mechanism.

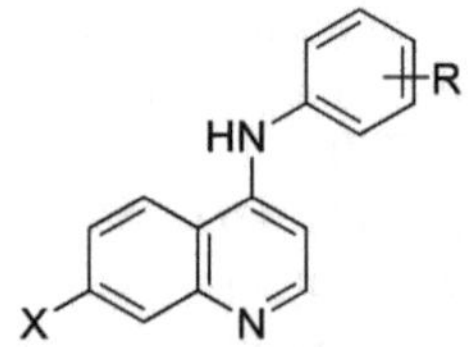

65a; X = F, R = 3-F C_6H_5
65b; X = F, R = 3-Cl C_6H_5
65c; X = F, R = 3-Br C_6H_5
65d; X = F, R = 3-NO_2 C_6H_5
65e; X = F, R = 3-OCH_3 C_6H_5
65f; X = Cl, R = 3-F C_6H_5
65g; X = Cl, R = 3-NO_2 C_6H_5
65h; X = Cl, R = 3-OCH_3 C_6H_5
65i; X = Br, R = 3-F C_6H_5
65j; X = Br, R = 3-Cl C_6H_5
65k; X = Br, R = 3-Br C_6H_5
65l; X = Br, R = 3-NO_2 C_6H_5
65m; X = Br, R = 3-OCH_3 C_6H_5
65n; X = OCH_3, R = 3-F C_6H_5
65o; X = OCH_3, R = 3-Cl C_6H_5
65p; X = OCH_3, R = 3-Br C_6H_5
65q; X = OCH_3, R = 3-NO_2 C_6H_5
65r; X = OCH_3, R = 3-OCH_3 C_6H_5
65s; X = OCH_3, R = 3,4-Cl C_6H_5

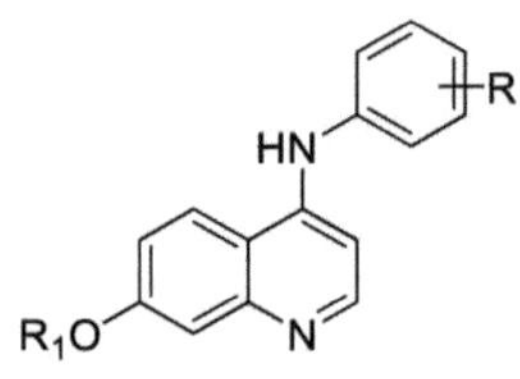

66a; R_1 = C_2H_5; R = 3-F C_6H_5
66b; R_1 = C_2H_5; R = 3-OCH_3 C_6H_5
66c; R_1 = C_2H_5; R = 3-$NO_2C_6H_5$
66d; R_1 = $CH(CH_3)_2$; R = 3-F C_6H_5
66e; R_1 = $CH(CH_3)_2$; R = 3-OCH_3 C_6H_5
66f; R_1 = $CH(CH_3)_2$; R = 3-NO_2 C_6H_5
66g; R_1 = $CH_2CH{=}CH_2$; R = 3-F C_6H_5
66h; R_1 = $CH_2CH{=}CH_2$; R = 3-OCH_3 C_6H_5
66i; R_1 = $CH_2CH{=}CH_2$; R = 3-NO_2 C_6H_5
66j; R_1 = n-C_4H_9; R = 3-F C_6H_5
66k; R_1 = n-C_4H_9; R = 3-Br C_6H_5
66l; R_1 = n-C_4H_9; R = 3-OCH_3 C_6H_5
66m; R_1 = n-C_4H_9; R = 3-NO_2 C_6H_5
66n; R_1 = C_8H_{17}; R = 3-OCH_3 C_6H_5
66o; R_1 = C_8H_{17}; R = 3-NO_2 C_6H_5
66p; R_1 = C_8H_{17}; R = 3-F C_6H_5
66q; R_1 = CH_2Ph; R = 3-F C_6H_5
66r; R_1 = CH_2Ph; R = 3-OCH_3 C_6H_5
66s; R_1 = CH_2Ph; R = 3-NO_2 C_6H_5
66t; R_1 = $CH_2(4\text{-F})C_6H_5$; R = 3-F C_6H_5
66u; R_1 = $CH_2(4\text{-F})C_6H_5$; R = 3-OCH_3 C_6H_5
66v; R_1 = $CH_2(4\text{-F})C_6H_5$; R = 3-NO_2 C_6H_5
66w; R_1 = $CH_2CH_2CH_2C_6H_5$; R = 3-F C_6H_5
66x; R_1 = $CH_2CH_2CH_2C_6H_5$; R = 3-OCH_3 C_6H_5
66y; R_1 = $CH_2CH_2CH_2C_6H_5$; R = 3-NO_2 C_6H_5

A series of quinoline-based chalcones (**67a-67f**) and pyrazoles (**68a-68f**, **69a-69f**, **70a-70f**, **71a-71f** and **72a-72f**) were synthesized (Ramírez-Prada et al. 2017). Fourteen compounds (**67a**, **67c**, **67d**, **68a**, **68c**, **69a**, **69c**, **69d**, **70a**, **70c**, **71a**, **71c**, **71d** and **72a**) were selected by NCI for a single dose (10 μM) testing against 60 different human cancer cell lines. Out of all, four compounds, **67c**, **67d** (chalcones) and **69a**, **69c** (pyrazolines), demonstrated significant activity (GI_{50} = 0.28–11.7 μM and LC_{50} = 2.6–100 μM) at 10 μM concentration and were chosen for five dose screening. Compounds **67c**, **67d** and **69a**, **69c** demonstrated high GI_{50} values in five-dose screening against a variety of cell lines, and some of the values were less than 1 μM. The GI_{50} and LC_{50} values of compound **67c** ranged from 0.49 to 3.85 μM and 23.7 to >100 μM, respectively. The best cytostatic effect of **67c** was demonstrated in the SR leukaemia cell line (GI_{50} = 0.49 μM), and the best cytotoxic effect was demonstrated in the U251 CNS cancer cell line (LC_{50} = 23.7 μM), both of which were superior to the reference drug adriamycin. Compound **67d** was discovered to be the most efficient for which the GI_{50} values ranged from 0.31 to 2.49 μM and LC_{50} values from 2.6 to >100 μM. The best cytostatic effect of **67d** was demonstrated towards the leukaemia cell line RPMI-8226 (GI_{50} = 0.31 μM), and the best cytotoxic effect was demonstrated towards the ovarian cancer cell line SK-OV-3 (LC_{50} = 2.6 μM), both of which were superior to adriamycin. The GI_{50} and LC_{50} values of compound **69a** ranged from 0.28 to 9.42 μM and 5.1 to >100 μM, respectively. Compound **69a** had the best cytostatic effect on the breast cancer cell line MDA-MB-468 (GI_{50} = 0.28 μM) and the greatest cytotoxic effect on the melanoma cell line SK-MEL-5 (LC_{50} = 5.1 μM). The GI_{50} and LC_{50} values of compound **69c** ranged from 0.37 to 11.7 μM and 17.4 to >100 μM, respectively. The strongest cytostatic effect was determined towards the melanoma cell line MDA-MB-435 (GI_{50} = 0.37 μM) and the greatest cytotoxic effect towards the melanoma cell line SK-MEL-5 (LC_{50} = 17.4 μM).

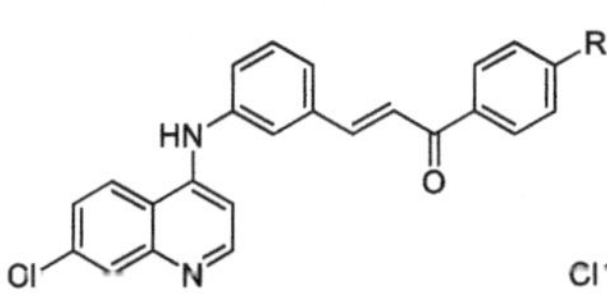

67a; R = Cl
67b; R = Br
67c; R = OCH_3
67d; R = 3,4,5-OCH_3
67e; R = CH_3
67f; R = H

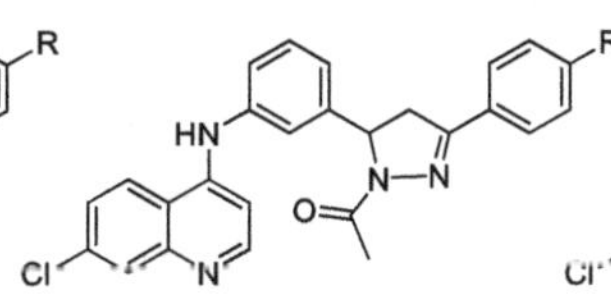

68a; R = Cl
68b; R = Br
68c; R = OCH_3
68d; R = 3,4,5-OCH_3
68e; R = CH_3
68f; R = H

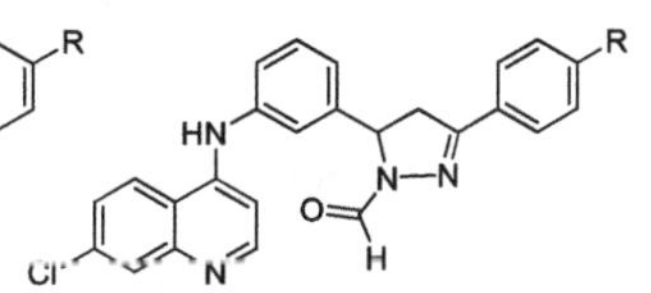

69a; R = Cl
69b; R = Br
69c; R = OCH_3
69d; R = 3,4,5-OCH_3
69e; R = CH_3
69f; R = H

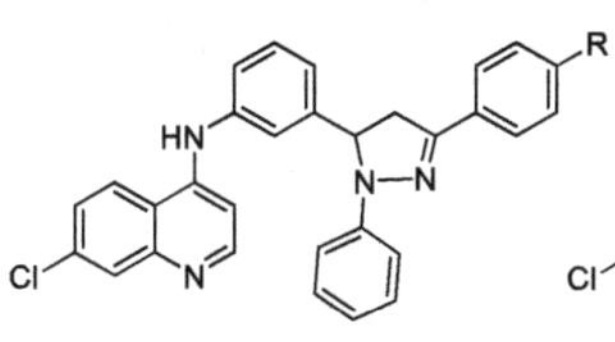

70a; R = Cl
70b; R = Br
70c; R = OCH_3
70d; R = 3,4,5-OCH_3
70e; R = CH_3
70f; R = H

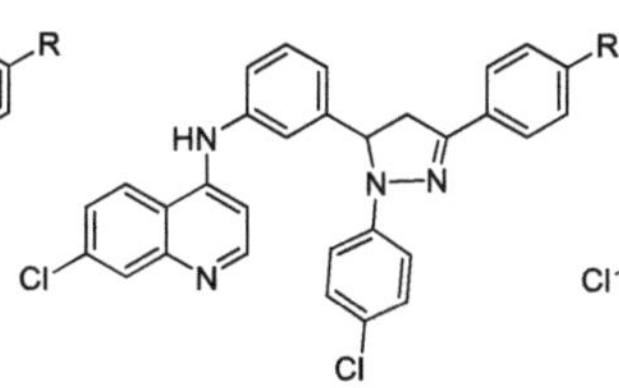

71a; R = Cl
71b; R = Br
71c; R = OCH_3
71d; R = 3,4,5-OCH_3
71e; R = CH_3
71f; R = H

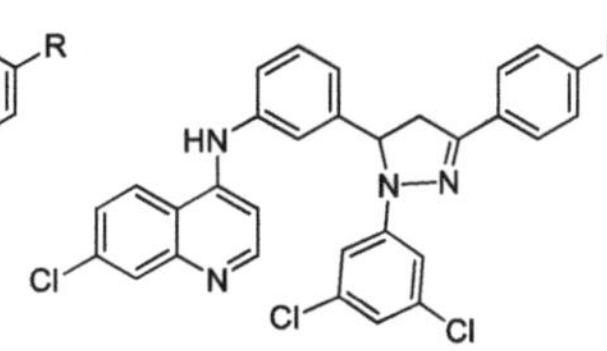

72a; R = Cl
72b; R = Br
72c; R = OCH_3
72d; R = 3,4,5-OCH_3
72e; R = CH_3
72f; R = H

Four different series of aminoquinoline derivatives, **73a-73f**, **74a-74f**, **75a-75f** and **76a-76f**, were synthesized and evaluated for their antiproliferative activity against MCF-7, H460, HCT 116 and SW620 cancer cell lines (Zorc et al. 2019). Compounds **73a-73f** and **74a-74f** demonstrated low to moderate activity against all four cancer cell lines. Compounds **75a-75f** and **76a-76f**, on the other hand, demonstrated good antiproliferative activity in micromolar concentrations. Compound **75f** demonstrated potent anticancer activity in MCF-7 and HCT 116 cells, with IC_{50} values of 0.4 μmol/L and 0.3 μmol/L, respectively. Compounds **76d** and **76f** also demonstrated strong activity towards MCF-7 cell lines with IC_{50} values of 0.4 and 0.3 μmol/L, respectively.

73a

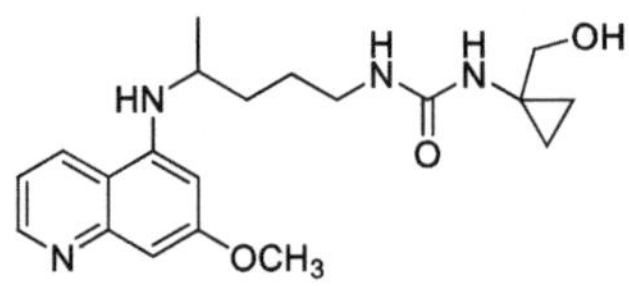

73b

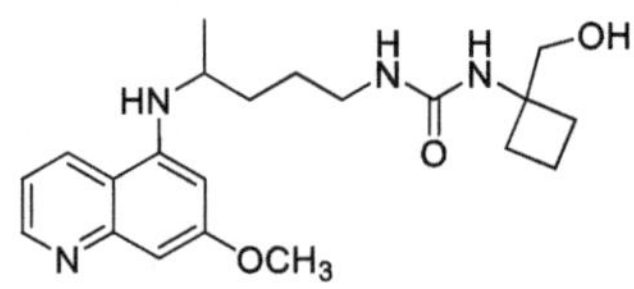

73c

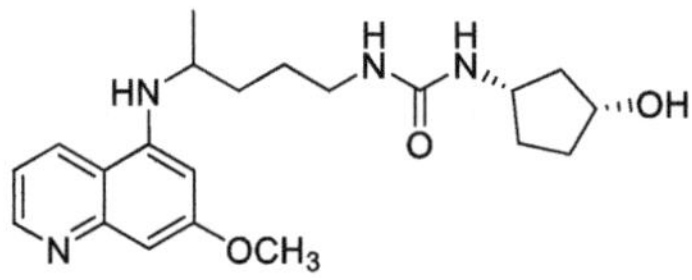

73d

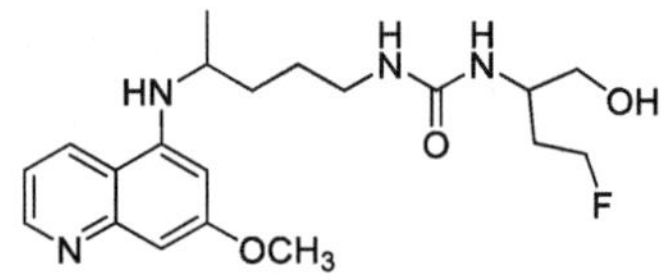

73e

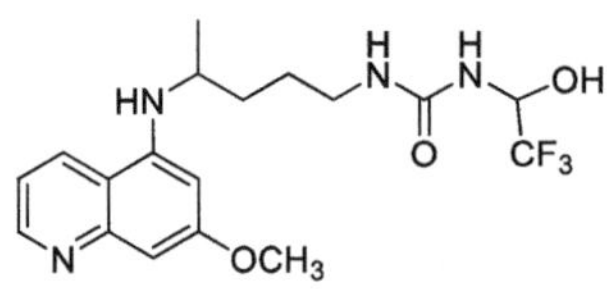

73f

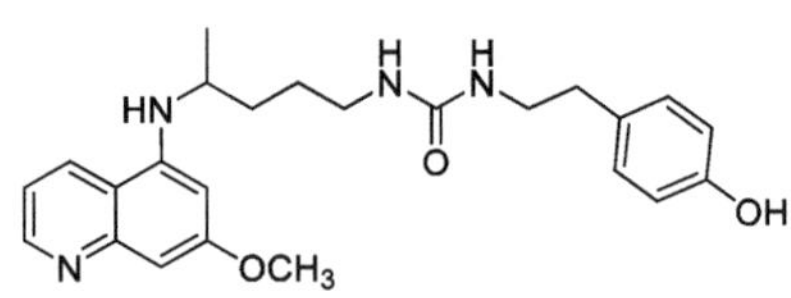

74a

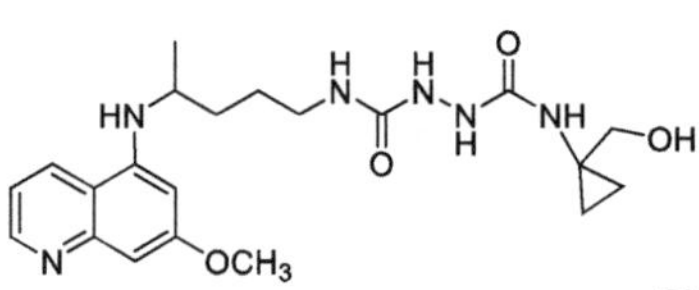

74b

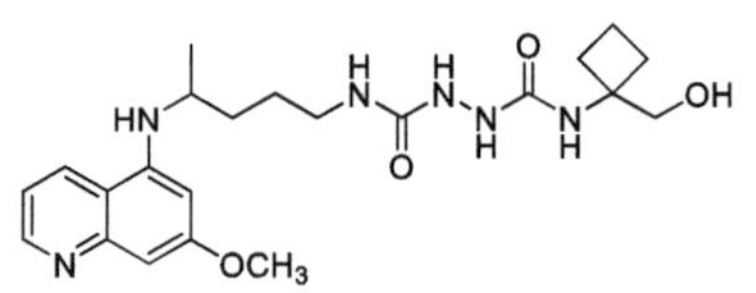

74c

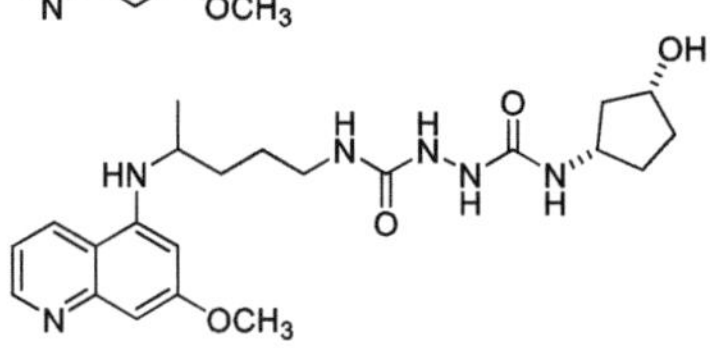

74d

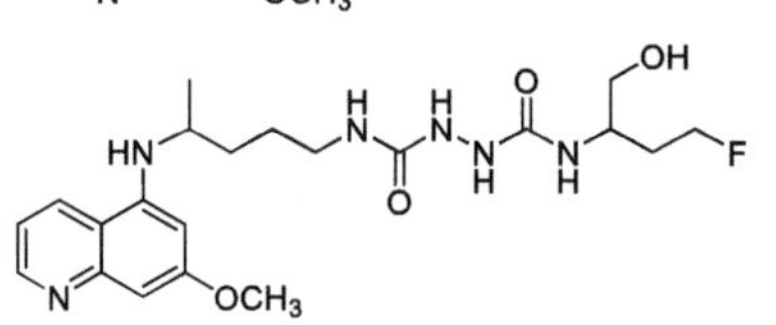

74e

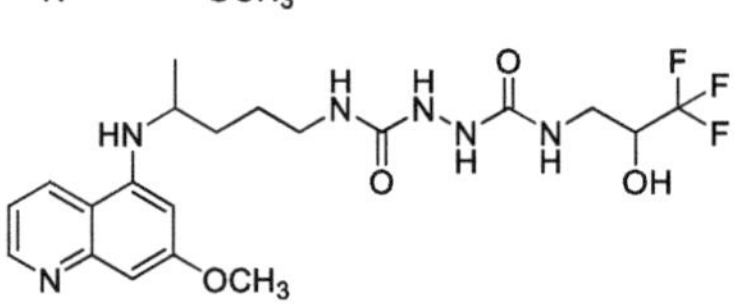

74f

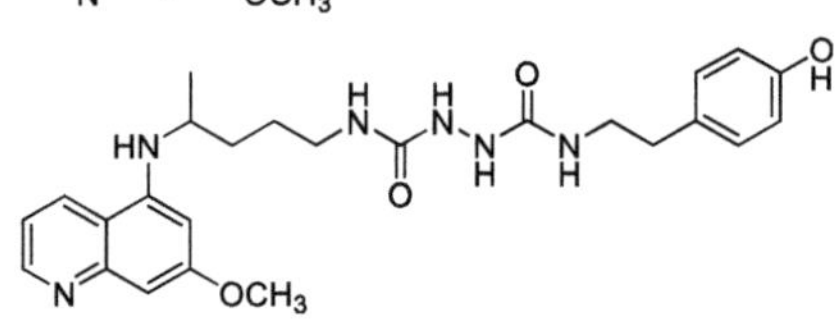

75a

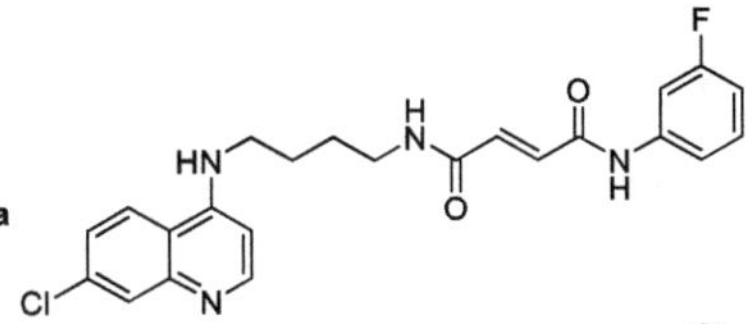

75b

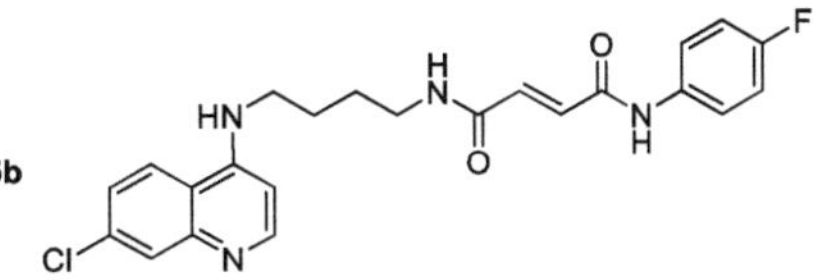

75c

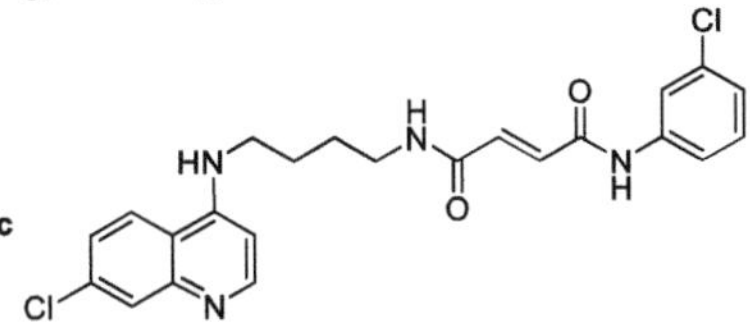

75d

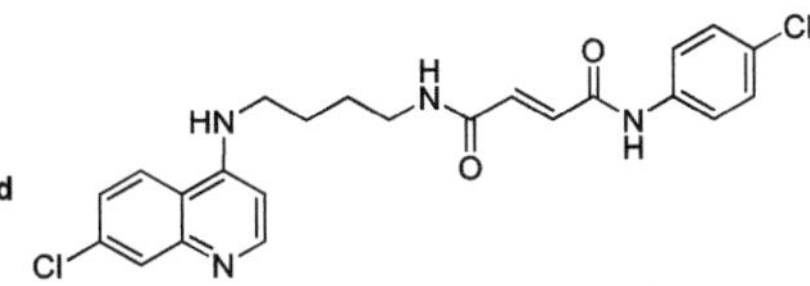

75e

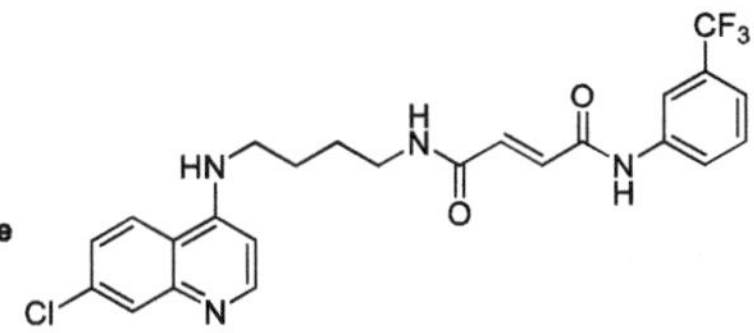

75f

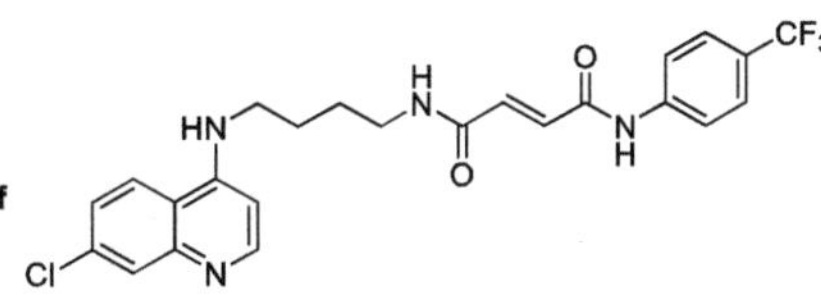

76a

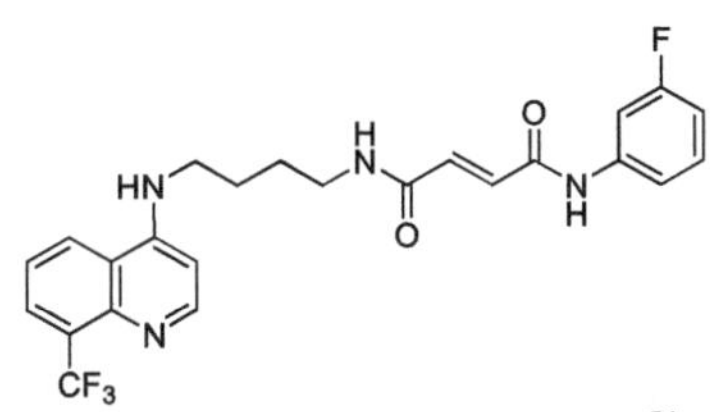

76b

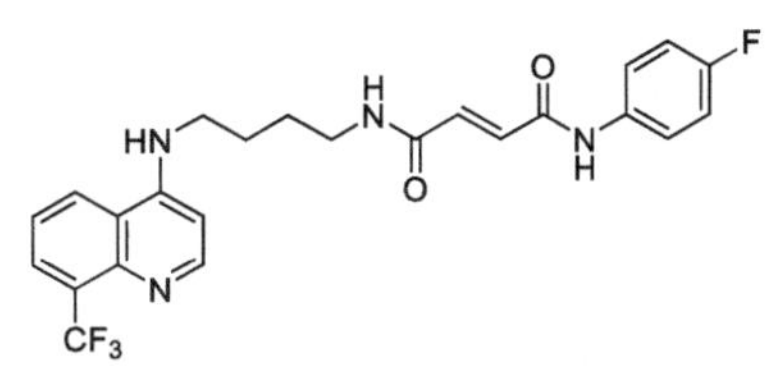

76c

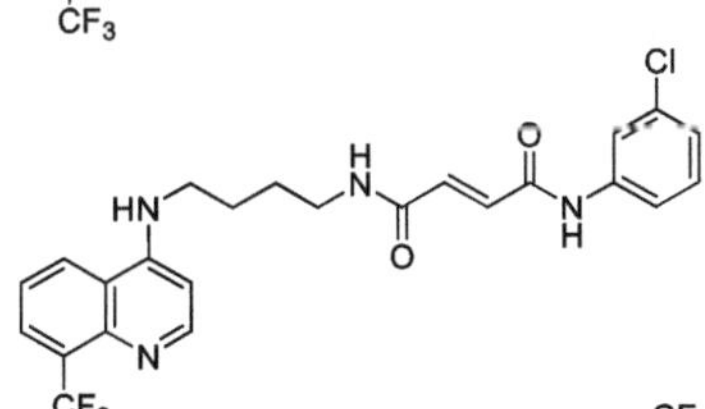

76d

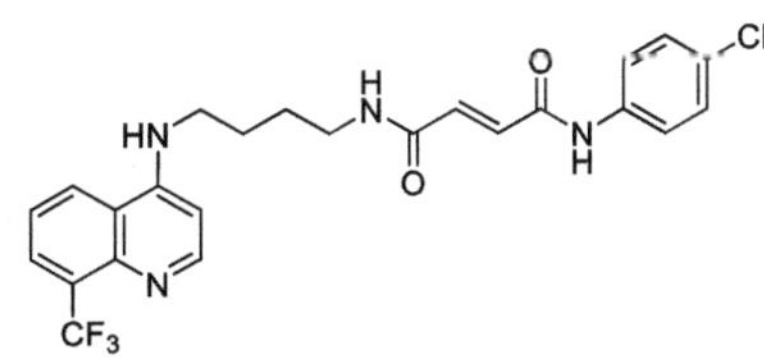

76e

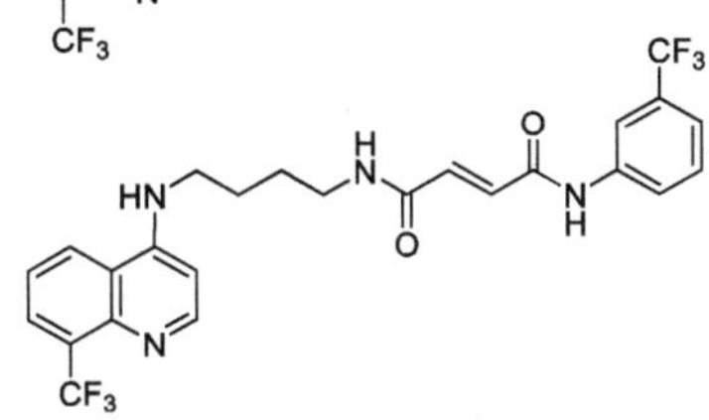

76f

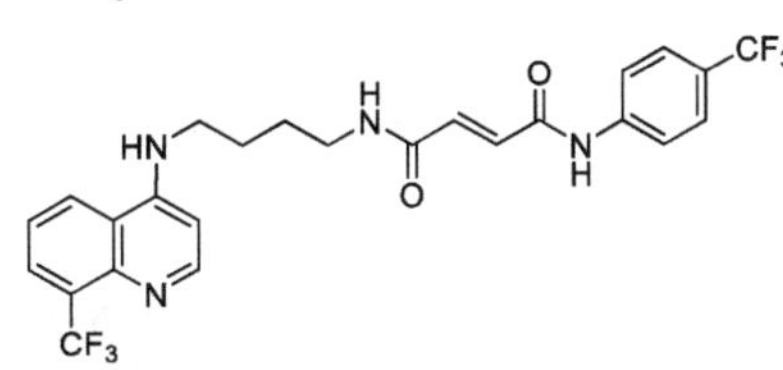

4-Aminoquinoline and 1,3,5-triazine-based hybrid molecules were synthesized and evaluated for their anticancer activity against the four cancer cell lines, HeLa, MCF-7, HL-60 and HepG2, as well as MCF 12A (normal epithelial cell) (Bhat et al. 2020). Compound **77e** was shown to be the most efficient compound against all the cell lines tested. When chloro (**77e**) was replaced with fluoro (**77i**), the activity against all cell lines was significantly reduced. The replacement of fluoro (**77i**) with nitro (**77c**) results in a further decrease in activity. Interestingly, none of the compounds displayed toxicity towards the normal cell lines MCF-12A.

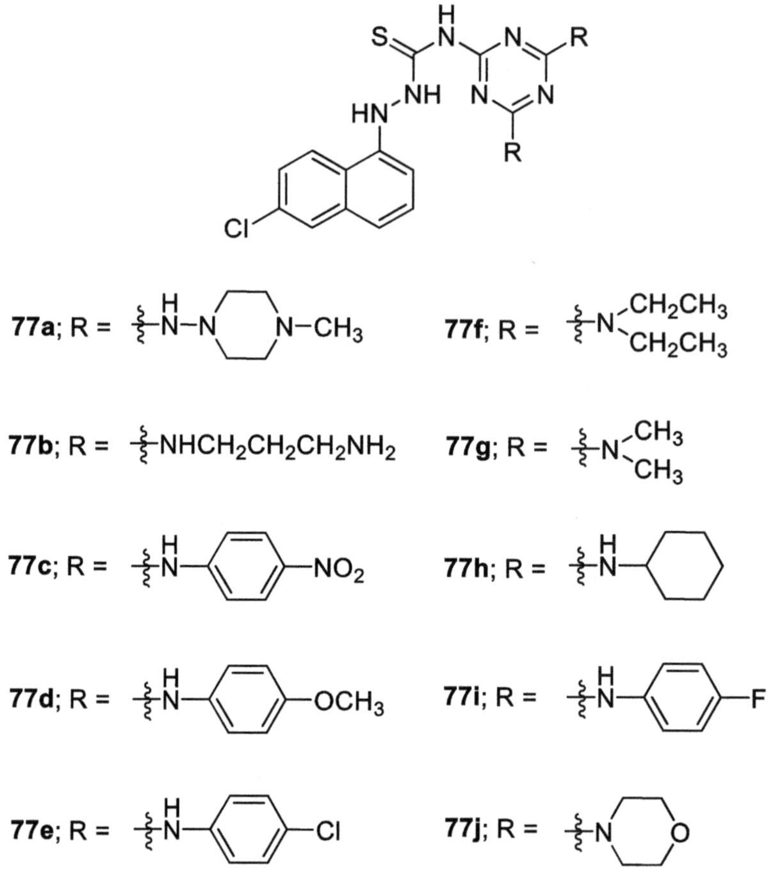

A series of molecular hybrids were synthesized by conjugating C5-curcuminoids and 4-aminoquinoline moieties through the 1,2,3-triazole ring (Kandi et al. 2015). Some of the molecules were selected by NCI for a single dose (10 μM) testing against 60 different human cancer cell lines. Only two compounds **78d** and **78f** demonstrated significant inhibition of several cell lines and were chosen for five-dose screening. Compound **78d** was shown to be more efficient ($GI_{50} = 0.189–2.17$ μM and $LC_{50} = 0.77–100$ μM) than **78f** against all the cancer cell lines. The strongest cytostatic effect of **78d** was demonstrated against the melanoma cancer cell line LOX IMVI ($GI_{50} = 0.189$ μM), and the strongest cytotoxic effect was demonstrated against the colon cancer cell line HCT-116 ($LC_{50} = 0.77$ μM). Mechanistic studies like Annexin V screening, DNA fragmentation and caspase activation showed apoptotic induction as a cause for their anticancer activity.

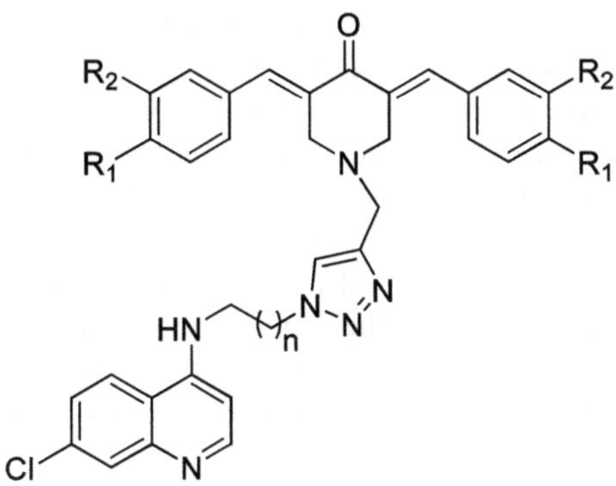

78a; $R_1 = CH_3$, $R_2 = H$, n = 1
78b; $R_1 = CH_2CH_3$, $R_2 = H$, n = 1
78c; $R_1 = F$, $R_2 = H$, n = 1
78d; $R_1 = Cl$, $R_2 = H$, n = 1
78e; $R_1 = Br$, $R_2 = H$, n = 1
78f; $R_1 = OCH_3$, $R_2 = OCH_3$, n = 1
78g; $R_1 = H$, $R_2 = Br$, n = 1
78h; $R_1 = CH_3$, $R_2 = H$, n = 2
78i; $R_1 = CH_2CH_3$, $R_2 = H$, n = 2
78j; $R_1 = F$, $R_2 = H$, n = 2
78k; $R_1 = Cl$, $R_2 = H$, n = 2
78l; $R_1 = Br$, $R_2 = H$, n = 2
78m; $R_1 = OCH_3$, $R_2 = OCH_3$, n = 2
78n; $R_1 = OCH_3$, $R_2 = H$, n = 2

10.4 Antitubercular Activity

According to WHO, approximately ten million new cases and 1.5 million deaths were reported in 2020 due to tuberculosis (https://www.who.int/news-room/fact-sheets/detail/tuberculosis WHO n.d.-b; Kumar et al. 2015a, b). The drug-sensitive variant of tuberculosis can be treated with a conventional 6-month course of four antimycobacterial drugs (isoniazid, ethambutol, pyrazinamide and rifampicin). However, the treatment of multidrug-resistant (MDR) and extensively drug-resistant (XDR) tuberculosis remains a big challenge worldwide (Bahuguna and Rawat 2020). There is an urgent need to find new and better therapeutic options for the therapy due to the resistance developed by *M. tuberculosis* against several first-line medications. Quinoline has always drawn the attention of researchers due to the presence of this nucleus in various already established antimicrobial drugs such as ciprofloxacin, moxifloxacin mefloquine and gatifloxacin. Bedaquiline, a new quinoline-based compound, received FDA approval in 2012 for its use in the treatment of MDR-TB. Unfortunately, resistance to bedaquiline was also observed shortly after it was used in clinical settings (Veziris et al. 2017). Therefore, it is necessary to continuously discover antitubercular agents with new mechanisms of action.

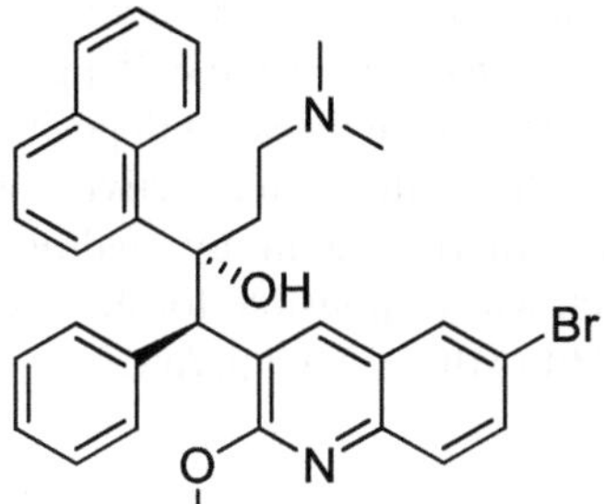

Bedaquiline

4-Quinolinyl hydrazone derivatives **79a-79u** were synthesized and tested for in vitro antimycobacterial activity against *M. tuberculosis* H37Rv (Candéa et al. 2009). Some of the compounds showed reasonably good activity (MIC = 2.5–6.25-μg/mL). Three compounds **79f**, **79i** and **79o** exhibited very good activity with a MIC value of 2.5 μg/mL. The activity of these three compounds was found to be better than the standard antitubercular drugs ethambutol (MIC = 3.12 μg/mL). According to the SAR analyses, compounds containing chloro, fluoro, hydroxy and methoxy groups showed good activity, whereas bromo, cyano and nitro group-containing compounds were shown to be less effective. All of the compounds were also examined for their cytotoxic activities using the BCG-infected and non-infected macrophage cell line J774s and the majority of the compounds were found to be noncytotoxic.

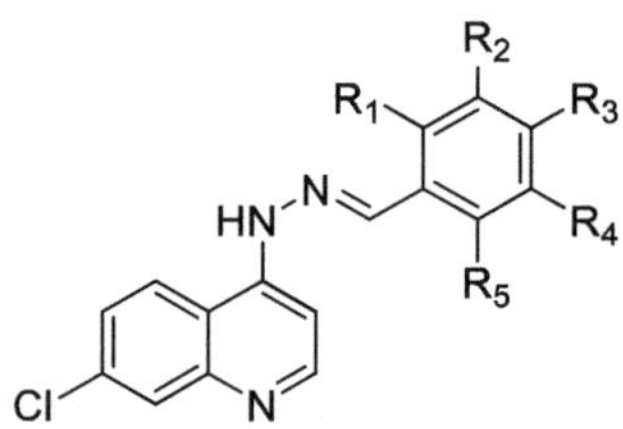

79a; R_1 = Cl, $R_2 = R_3 = R_4 = R_5$ = H
79b; R_1 = H, R_2 = Cl, $R_3 = R_4 = R_5$ = H
79c; R_1 = H, R_2 = H, R_3 = Cl, $R_4 = R_5$ = H
79d; R_1 = Br, $R_2 = R_3 = R_4 = R_5$ = H
79e; R_1 = H, R_2 = Br, $R_3 = R_4 = R_5$ = H
79f; R_1 = H, R_2 = H, R_3 = Br, $R_4 = R_5$ = H
79g; R_1 = F, $R_2 = R_3 = R_4 = R_5$ = H
79h; R_1 = H, R_2 = F, $R_3 = R_4 = R_5$ = H
79i; R_1 = H, R_2 = H, R_3 = F, $R_4 = R_5$ = H
79j; R_1 = OH, $R_2 = R_3 = R_4 = R_5$ = H
79k; R_1 = H, R_2 = OH, $R_3 = R_4 = R_5$ = H
79l; R_1 = H, R_2 = H, R_3 = OH, $R_4 = R_5$ = H
79m; R_1 = OMe, $R_2 = R_3 = R_4 = R_5$ = H
79n; R_1 = H, R_2 = OMe, $R_3 = R_4 = R_5$ = H
79o; R_1 = H, R_2 = H, R_3 = OMe, $R_4 = R_5$ = H
79p; R_1 = NO2, $R_2 = R_3 = R_4 = R_5$ = H
79q; R_1 = H, R_2 = NO_2, $R_3 = R_4 = R_5$ = H
79r; R_1 = H, R_2 = H, R_3 = NO_2, $R_4 = R_5$ = H
79s; R_1 = H, R_2 = CN, $R_3 = R_4 = R_5$ = H
79t; R_1 = H, R_2 = H, R_3 = CN, $R_4 = R_5$ = H
79u; $R_1 = R_2 = R_3 = R_4 = R_5$ = H

4-Aminoquinoline derivatives were synthesized and assessed for in vitro antimycobacterial efficacy against H37Rv strain of *M. tuberculosis* (De Souza et al. 2009). The compounds containing 2–4 carbon atoms in the diamine part displayed weak activity (MIC = 100 μg/mL), whereas compounds with 6, 8 and 10 carbon atoms demonstrated significantly increased activity. For example, in the series (**81a-81f**), the MIC value varies from 100 μg/mL (**81a**, **81b**, **81c**, where $n = 2$, 3, 4, respectively) to 25 μg/mL (**81d**, $n = 6$) to 6.25 μg/mL (**81e**, $n = 8$) to 3.12 μg/mL (**81f**, $n = 10$). The chlorine atom at the seventh position in the quinoline nucleus was found to be essential for the antitubercular activity, which can be demonstrated by the substitution of the chlorine atom by hydrogen in the compound **82f** (MIC = 12.5 μg/mL), which was responsible for the decrease in the antitubercular activity of the compound **83d** (MIC = 50.0 μg/mL).

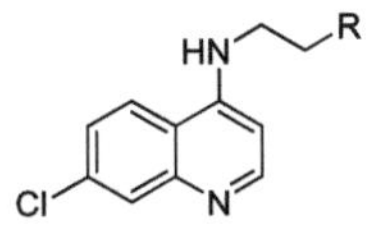

80a; R = N_3
80b; R = $NHCH_2CH_2OH$
80c; R = $NHCH_3$
80d; R = $NHCH_2CH_2CH_3$
80e; R = $NHCH(CH_3)_2$
80f; R = $NHCH_2CH_2CH_2CH_3$
80g; R = NHCyclohexyl
80h; R = $NHCH_2C_6H_5$

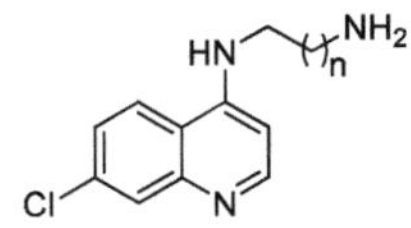

81a; n = 1
81b; n = 2
81c; n = 3
81d; n = 5
81e; n = 7
81f; n = 9

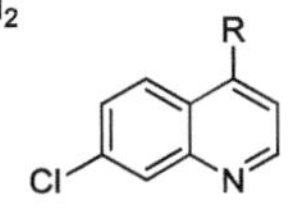

82a; R = N_3
82b; R = NH_2
82c; R = $NHCH_3$
82d; R = $NHCH_2CH_2CH_3$
82e; R = $NHCH(CH_3)_2$
82f; R = $NHCH_2CH_2CH_2CH_3$
82g; R = $NHCH_2CH_2CH_2OH$
82h; R = (R)-$NHCH(CH_2CH_3)CH_2OH$
82i; R = (S)-$NHCH(CH_2CH_3)CH_2OH$
82j; R = (R,S)-$NHCH(CH_2CH_3)CH_2OH$

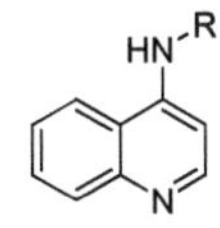

83a; R = CH_2CH_2OH
83b; R = CH_2CH_2Cl
83c; R = H
83d; R = $CH_2CH_2CH_2CH_3$

A series of novel 4-aminoquinoline-chalcone and 4-aminoquinoline-pyrimidine conjugates were synthesized and examined for in vitro antitubercular activity against H37R_V *M. tuberculosis* (Sharma et al. 2009). Quinolinyl-chalcones (**84a-84p**) outperformed quinolinyl-pyrimidine hybrids in terms of antitubercular activity (**85a-85p**). With MIC values of 3.12 μg/mL and being noncytotoxic to Vero and MBMDM cell lines, the 2,3-dimethoxyphenyl (**84g**) and 2,5-dimethoxyphenyl (**84h**) substituted derivatives of chalcones were found to be the most effective against *M. tuberculosis*, while their pyrimidine analogues (**85g, 85h**) were discovered to be ineffective against the same strain.

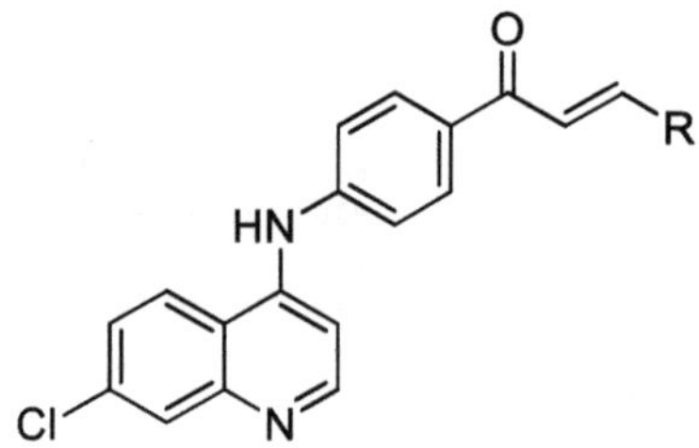

84a-84p

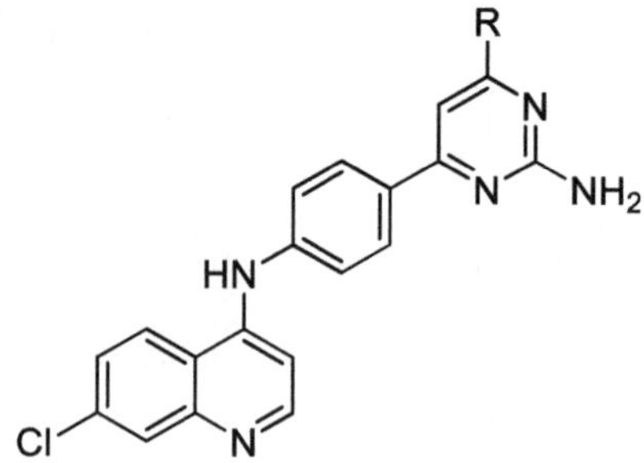

85a-85p

84a, 85a; R = 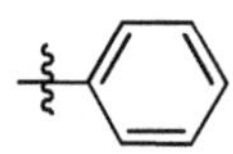

84b, 85b; R = 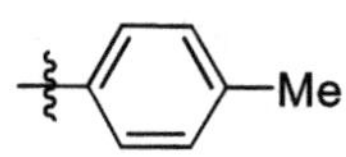

84c, 85c; R = 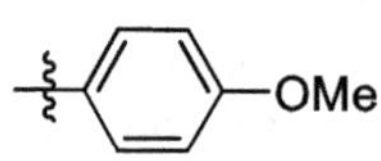

84d, 85d; R = 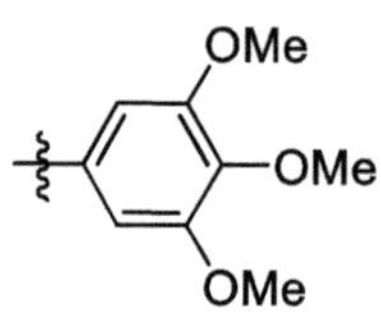

84e, 85e; R = 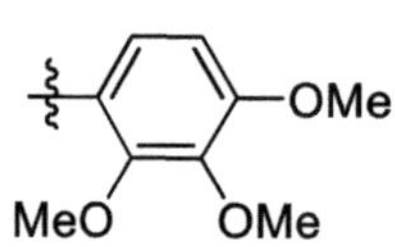

84f, 85f; R = 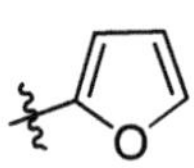

84g, 85g; R = 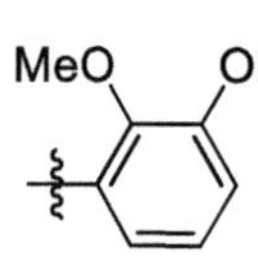

84h, 85h; R = 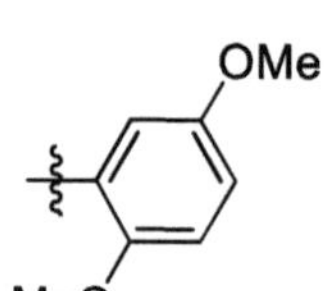

84i, 85i; R = 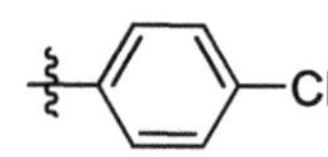

84j, 85j; R = 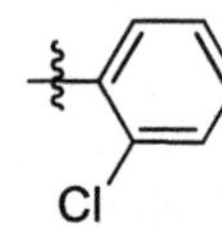

84k, 85k; R = 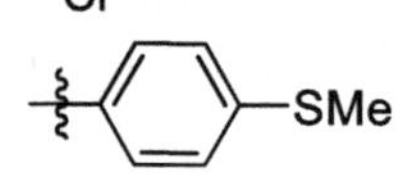

84l, 85l; R = 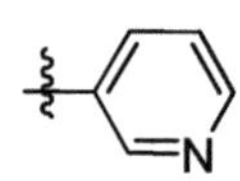

84m, 85m; R = 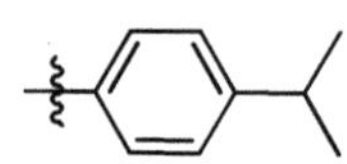

84n, 85n; R = 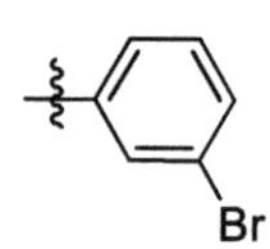

84o, 85o; R = 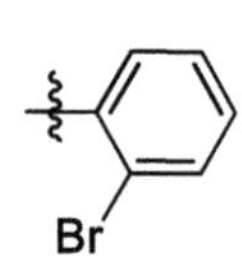

84p, 85p; R = 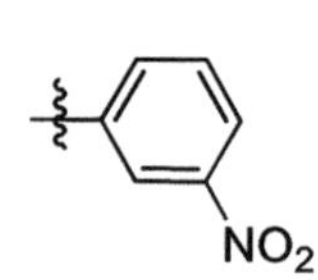

Four different series of 4-aminoquinolines derivatives **86a-86 t**, **87a-87e**, **88a-88c** and **89a-89d** were synthesized (Eswaran et al. 2010). Except for **88a-88c**, all the tested compounds exhibited good to moderate inhibitory activity against *M. tuberculosis* and multidrug-resistant tuberculosis (MDR-TB). Compounds **86k**,

86s and **89a** demonstrated very good activity and they exhibited 99% growth inhibition of *M. tuberculosis* at 3.12 μg/mL concentration, while compounds **86d**, **86g**, **86i**, **86l**, **86q**, **87b**, **89b** and **89d** showed significant activity and inhibited 95% growth of *M. tuberculosis* at 6.25 μg/mL concentration. Additionally, all the compounds were tested against MDR-TB (*M. tuberculosis* is resistant to three antitubercular drugs, namely, isoniazid, rifampicin and ethambutol). The majority of compounds exhibited decent activity (MIC = 6.25–25 μg/mL) against MDR-TB strains, and many of them were found to be more effective than isoniazid and rifampicin. Interestingly, nine compounds, namely, **86d**, **86g**, **86i**, **86k**, **86l**, **86s**, **87d**, **89a** and **89b**, were discovered to be more efficient than isoniazid (MIC = 12.5 μg/mL) with a MIC value of 6.25 μg/mL, whereas compounds **86c**, **86h**, **86m**, **86o**, **86q**, **87a**, **87c**, **89c** and **89d** were discovered to be more efficient than rifampicin (MIC = 25 μg/mL) with MIC value of 12.5 μg/mL. Further, compounds **86b**, **86n**, **86p** and **88b** were shown to be as potent as rifampicin. Surprisingly, compound **89a** unexpectedly exhibited strong antimycobacterial action against sensitive and MDR strains of tuberculosis with MIC values of 3.12 and 6.25 μg/mL, respectively.

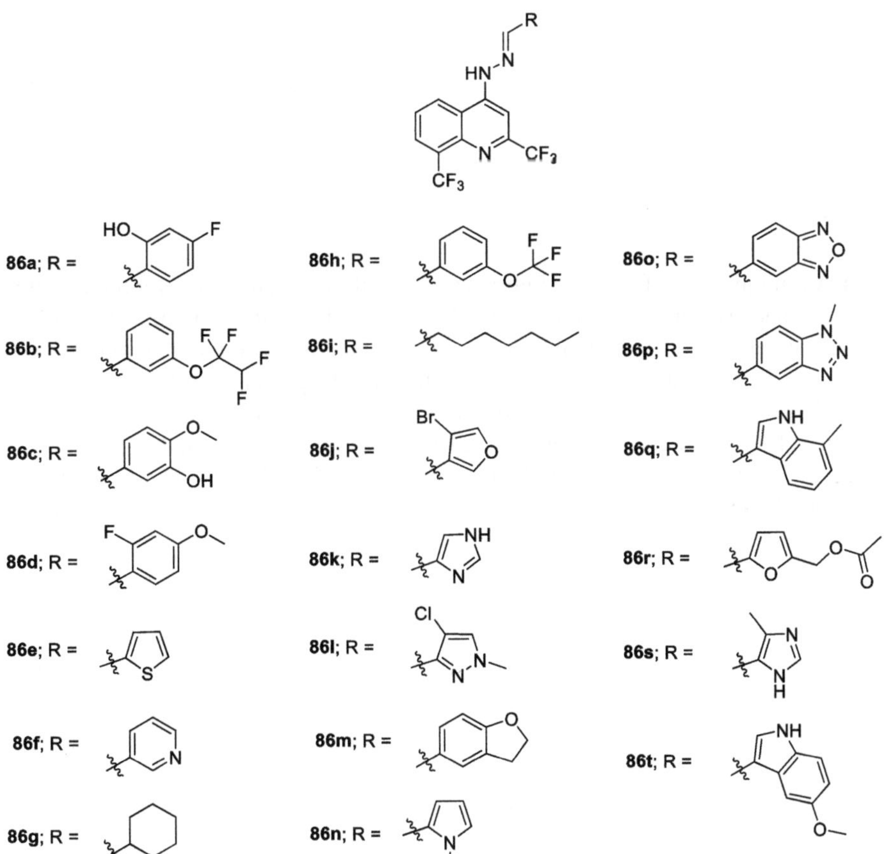

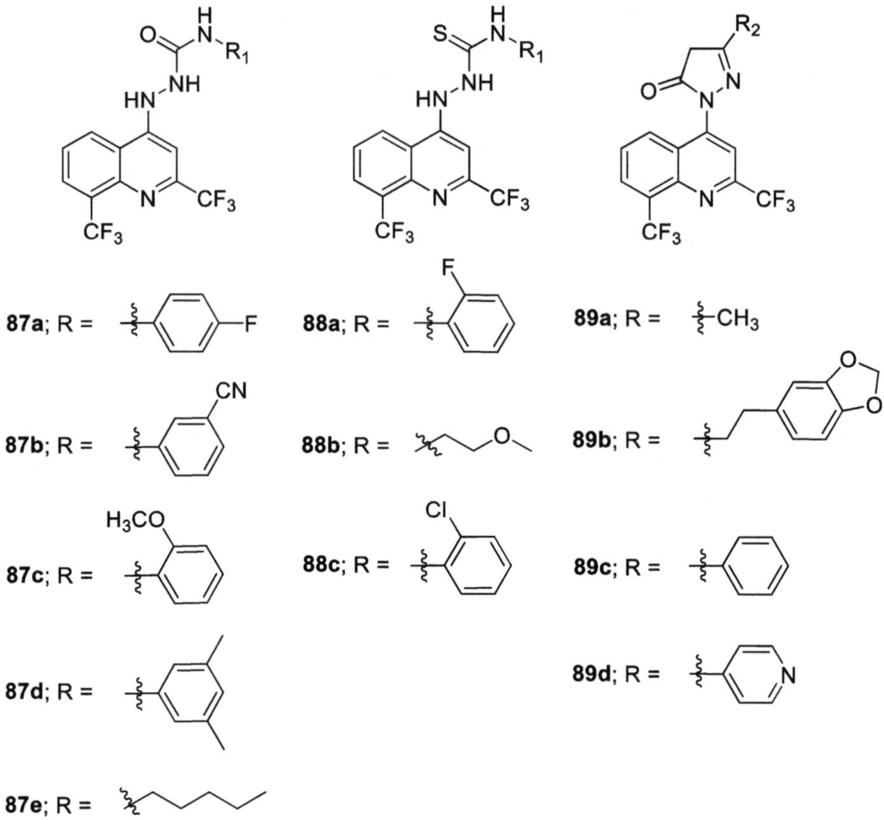

4-Aminoquinoline-ferrocene hybrids were synthesized and tested for their in vitro antimycobacterial activity against the Mtb mc^2 7000 strain of *M. tuberculosis* (Mahajan et al. 2011). Only one compound **90a** exhibited good activity (MIC = 2.5–5 μg/mL) comparable to the standard drug ethambutol (MIC = 1–2.5- μg/mL), while all other compounds demonstrated no activity even at very high dosages (100 μg/mL). Unexpectedly, ferroquine (FQ) also exhibited potent activity (MIC = 10–15 μg/mL), although its activity was found to be less than that of compound **90a**.

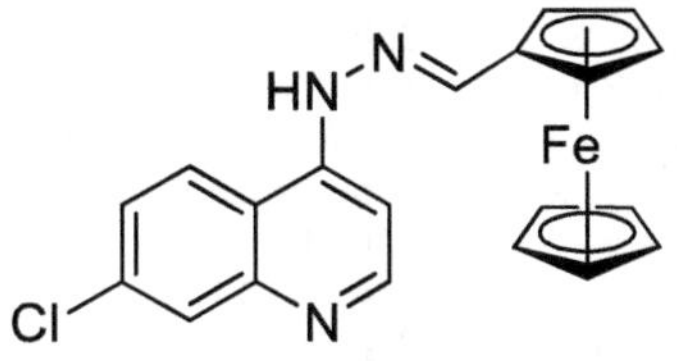

90a

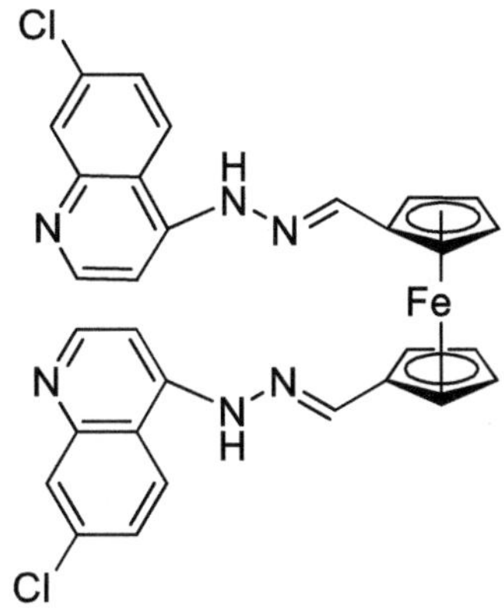

90b

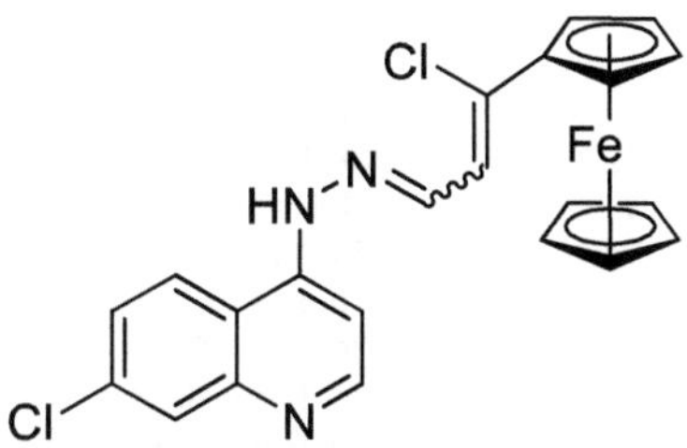

90c

A series of novel 4-aminoquinoline derivatives comprising semicarbazone and thiosemicarbazone were synthesized and screened for their in vitro antimycobacterial activity against H37Rv *M. tuberculosis* (Alegaon et al. 2020). All of the compounds showed significant antitubercular activity (MIC = 1.56–50-μM). The majority of the compounds were discovered to be more effective than pyrazinamide (MIC = 50 μM). Compounds, **91l** and **91p** were the most potent of all the compounds examined, with a MIC value of 1.56 μM, and were equally effective as the standard antitubercular drug isoniazid. Three compounds **91j**, **91r** and **92j** exhibited superior activity than ciprofloxacin (MIC = 12.5 μM) and pyrazinamide (MIC = 50 μM) with a MIC value of 6.25 μM. Nine compounds **91a**, **91b**, **91d**, **91e**, **91m**, **91n**, **92a**, **92b**, **92k** and **92l** were shown to be just as effective as ciprofloxacin (12.5 μM). Cytotoxicity of the compounds having MIC values less than 12.5 μM was determined against normal HDF cell line using MTT cell proliferation assay. None of the active compounds showed toxic effects on HDF cell lines.

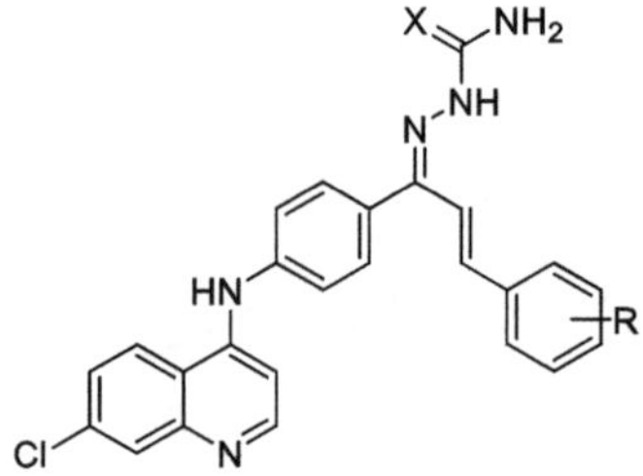

91a; R = 4-CH_3, X = S
91b; R = 4-OCH_3, X = S
91c; R = 3,4-OCH_3, X = S
91d; R= 3,4,5-OCH_3, X = S
91e; R = 4-Cl, X = S
91f; R = 4-F, X = S
91g; R = 4-Br, X = S
91h; R = 4-NO_2, X = S
91i; R = 4-OH, X = S

91j; R = 4-CH_3, X = O
91k; R = 4-OCH_3, X = O
91l; R = 3,4-OCH_3, X = O
91m; R= 3,4,5-OCH_3, X = O
91n; R = 4-Cl, X = O
91o; R = 4-F, X = O
91p; R = 4-Br, X = O
91q; R = 4-NO_2, X = O
91r; R = 4-OH, X = O

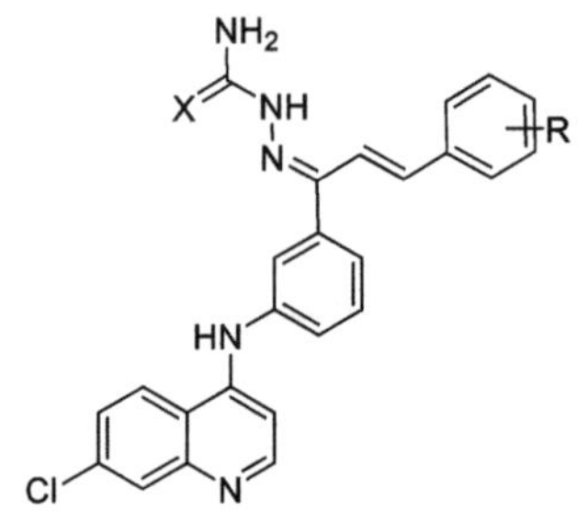

92a; R = 4-CH_3, X = S
92b; R = 4-OCH_3, X = S
92c; R = 3,4-OCH_3, X = S
92d; R= 3,4,5-OCH_3, X = S
92e; R = 4-Cl, X = S
92f; R = 4-F, X = S
92g; R = 4-Br, X = S
92h; R = 4-NO_2, X = S
92i; R = 4-OH, X = S

92j; R = 4-CH_3, X = O
92k; R = 4-OCH_3, X = O
92l; R = 3,4-OCH_3, X = O
92m; R= 3,4,5-OCH_3, X = O
92n; R = 4-Cl, X = O
92o; R = 4-F, X = O
92p, R = 4-Br, X = O
92q, R = 4-NO_2, X = O
92r, R = 4-OH, X = O

10.5 Anticoronavirus Activity

Coronavirus disease (COVID-19) is a viral infection caused by Severe Acute Respiratory Syndrome Coronavirus 2 (SARS-CoV-2). The first coronavirus case was discovered in December 2019 in the Chinese city of Wuhan, and it quickly spread to almost every country in the world (Gorbalenya et al. 2020). As a result, in March 2020, WHO declared this a worldwide emergency. It has infected over 500 million people and killed over six million people worldwide, as well as causing long-term health problems in those who survived (https://covid19.who.int/) (WHO n.d.-a). COVID-19 causes mild to severe respiratory diseases (Clark et al. 2020). So far, seven human coronaviruses (HCoV) have been discovered (Wang et al. 2020a, b). SARS-CoV-1, Middle East respiratory syndrome coronavirus (MERS-CoV), and SARS-CoV-2, belonging to the beta coronavirus family, are extremely deadly viruses. The HCoV-229E and HCoV-NL63 (both alpha coronaviruses) and HCoV-OC43 and HCoV-HKU1 (both beta coronaviruses), on the other hand, cause minor upper respiratory tract infections. Treatment of COVID-19 depends upon the infection's severity. Proper rest and antipyretic medication are often enough for milder symptoms. More severe cases may require hospitalization, supplementary oxygen, ventilation and other supportive measures. Even though several vaccines have already been developed, no efficient antiviral treatment for COVID-19 has been established. As a result, it is highly desirable to discover new and effective drugs for the treatment of SARS-CoV-2. The most common approach to providing new drugs, especially in a short period, is the repurposing of drugs which have already received approval and license for use in treating other diseases (Bazotte et al. 2021; Pushpakom et al. 2019).

A commonly used antiviral drug called remdesivir was repurposed and discovered to exhibit potential action against SARS-CoV-2 in vitro as well as in preclinical research and human clinical trials (Wang et al. 2020a, b). It was the first SARS-CoV-2 treatment that the FDA had approved. Despite initial activity data that looked promising, there were not enough clinical investigations, and recently, due to limited efficacy, its usage for treating COVID-19 is no longer advised (Young et al. 2021; Jiang et al. 2021; Kokic et al. 2021). The other two antimalarial drugs chloroquine and hydroxychloroquine have also been reported to be effective against SARS-CoV-2 (Cortegiani et al. 2020). Although both drugs were highly effective against SARS-CoV-2, neither in vitro testing nor clinical trial studies produced findings that were persuasive enough to be approved for either single use or use in combination with other antiviral medications (Giacomelli et al. 2021). Due to the promising activity of quinoline-based compounds against SARS-CoV-2, researchers have been working on this moiety to generate more effective therapeutic molecules.

The anticoronavirus activity of ten quinoline analogues was investigated against HCoV-229E (alpha coronavirus), HCoVOC43, SARS-CoV-1 and SARS-CoV-2 (all are beta coronaviruses) (Persoons et al. 2021). Except for piperaquine, all analogues demonstrated antiviral activity against one or more coronaviruses tested. The most potent analogues against the various coronaviruses were chloroquine and hydroxychloroquine, with EC_{50} values in the range of 0.12–12 μM. With an EC_{50} value of 0.12 μM and a selectivity index of 165, chloroquine has particularly remarkable action against HCoV-OC43 in HEL cells. Overall, the reduced cytotoxicity of hydroxychloroquine in the different cell lines, which led to more favourable selectivity indices, made the profile of the drug more appealing. Although mefloquine demonstrated very good antiviral action against the various coronaviruses, it was associated with significant toxicity in Huh7, HEL and Vero E6 cell lines. In addition to showing strong antiviral activity, amodiaquine and ferroquine also had lower cytotoxicity and higher selectivity indices. Furthermore, primaquine, quinidine, quinine and tafenoquine exhibited antiviral action against a particular coronavirus only at higher concentrations ($EC_{50} = >13$ μM), which suggests a lack of selective antiviral action.

The 4-aminoquinoline derivatives were synthesized and tested for antiviral activity against SARS-CoV-2 (Herrmann et al. 2022). All the tested compounds had anti-SARS-CoV-2 activity comparable to or greater than the reference compound chloroquine. The EC_{50} values in Caco-2 cells ranged from 5.9 μM to 18.9 μM (EC_{50} of chloroquine = 12.7 μM) and from 1.5 μM to 2.9 μM in Vero76 cells (EC_{50} of chloroquine = 3.1 μM). Compound **93** was shown to be the most efficient against Vero 76 cells, with an EC_{50} value of 1.5 μM, while compound **94** was the most effective against Caco-2 cells, with an EC_{50} value of 5.9 μM.

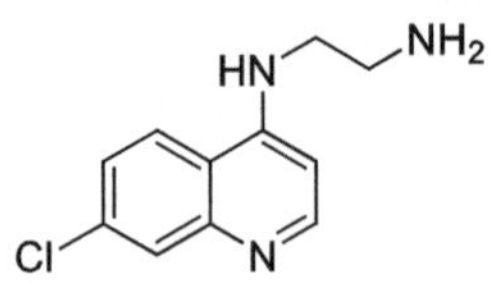

93

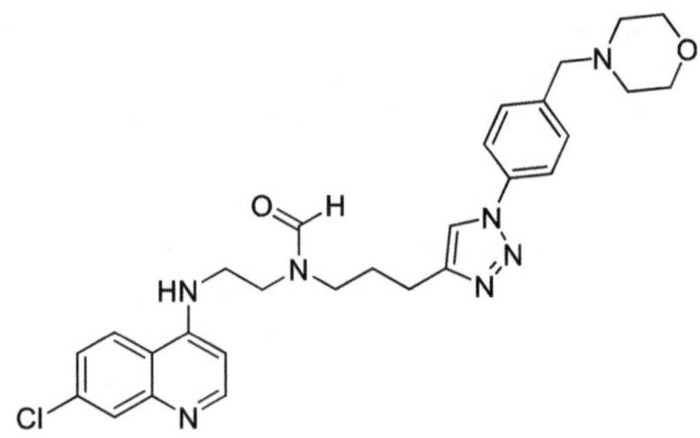

95

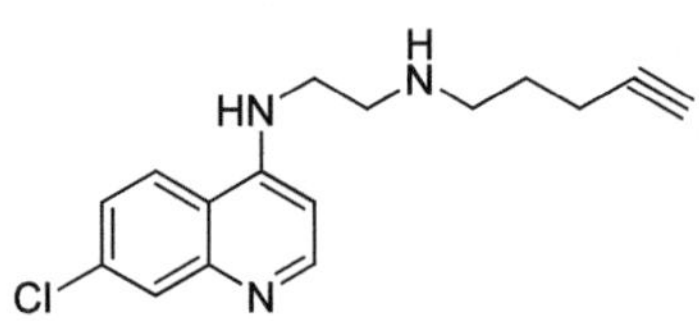

94

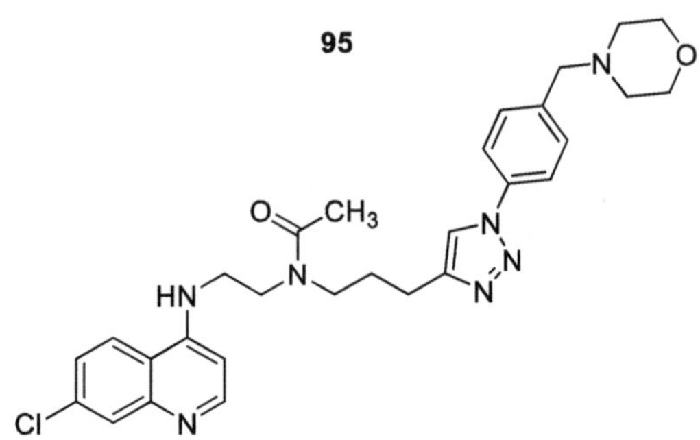

96

Fifteen hybrid compounds containing artemisinin and 4-aminoquinoline were synthesized and tested for their antiviral activity against SARS-CoV-2. All hybrids containing artesunic acid (**97**–**103**) demonstrated very good anti-SARS-CoV-2 activity, with EC_{50} values in the range of 7.8–46 μM. With an EC_{50} value of 7.8 μM, compound **103** having a morpholine ring was discovered to be the most active. It was also substantially more active than its parent compound, artesunic acid ($EC_{50} > 50$ μM). From the quinoline-based hybrids, compound **111** containing adamantane ring ($EC_{50} = 1.5$ μM) was the most effective, outperforming both the reference compounds chloroquine ($EC_{50} = 3.8$ μM) and remdesivir ($EC_{50} = 4.0$ μM). Notably, synthetic peroxide-based hybrids (**104**–**109**) were also extremely effective against SARS-CoV-2, with the compound **106** being the most active ($EC_{50} = 11.5$ μM). Remarkably, none of the hybrid molecules had cytotoxic effects on Vero E6 cells.

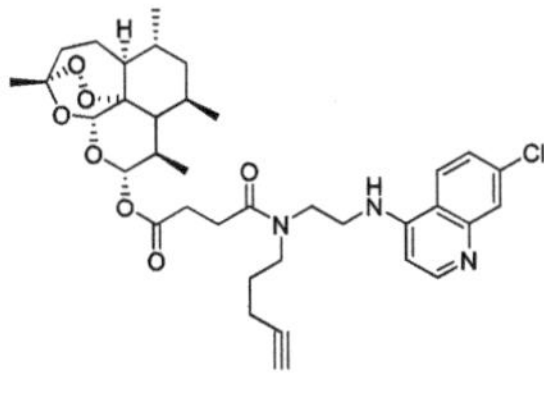

97

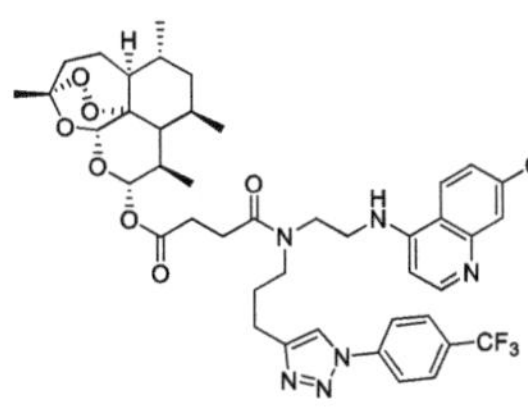

98

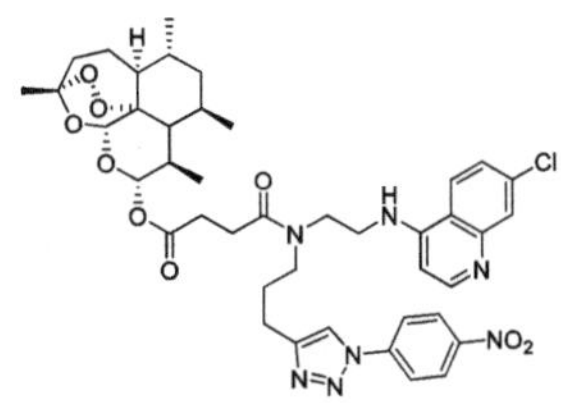

99

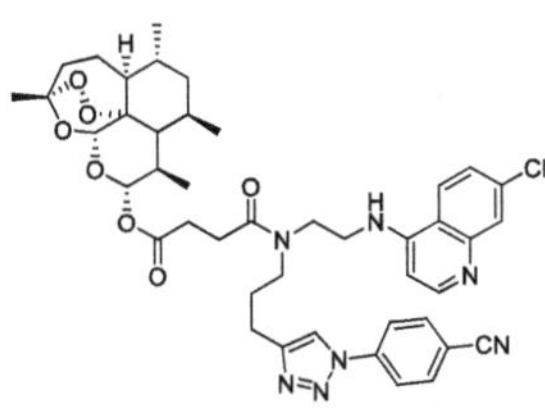

100

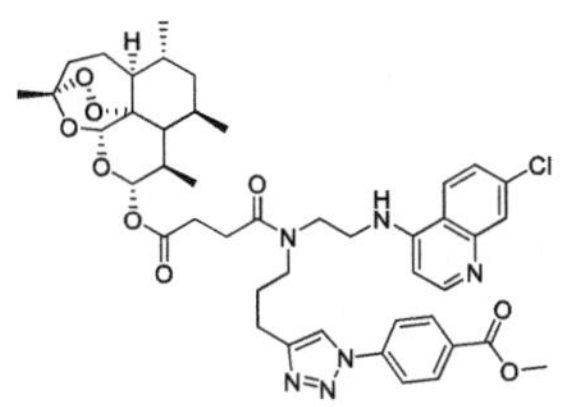

101

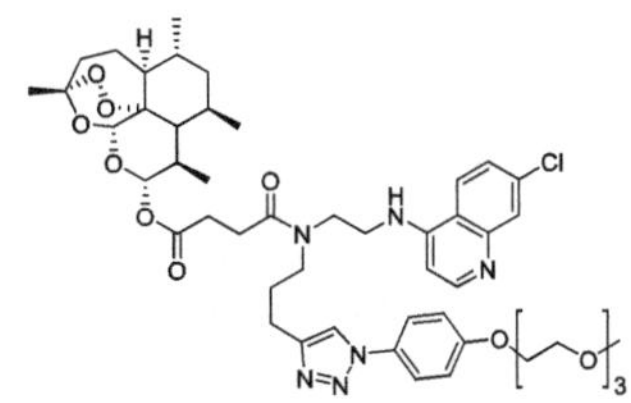

102

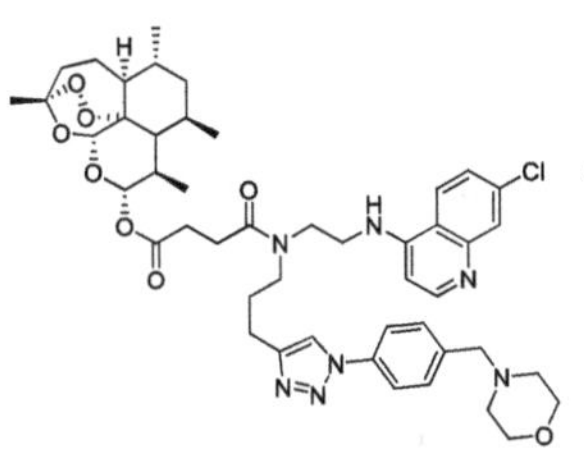

103

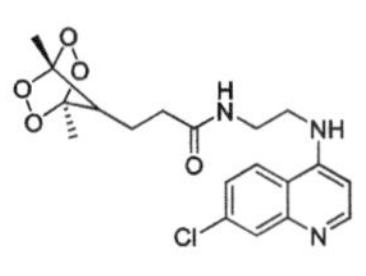

104

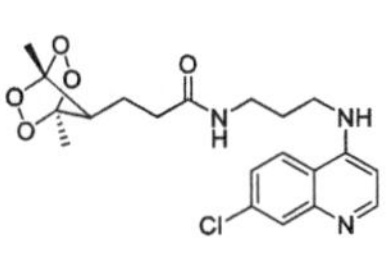

105

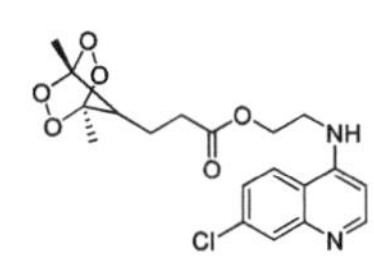

106

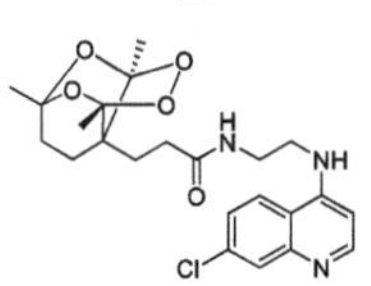

107

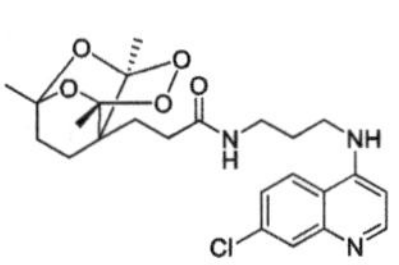

108

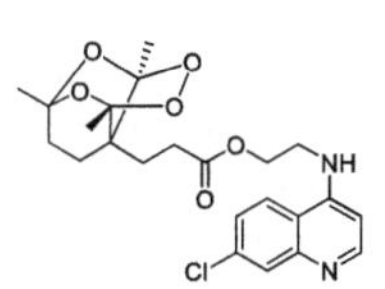

109

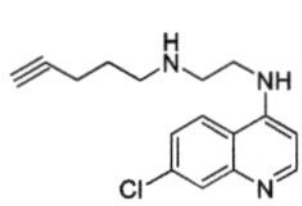

110

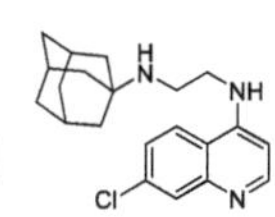

111

10.6 Conclusion

The synthesis of new derivatives of 4-aminoquinolines has proven to be an effective chemotherapeutic approach. In this article, the effectiveness of compounds, having an aminoquinoline scaffold, against malaria parasites, tuberculosis bacteria, cancer-causing agents, and coronavirus has been discussed. Some of the compounds discussed in this article may serve as a starting lead for the development of novel medications to treat a variety of diseases brought on by parasites and other pathogens.

References

Afzal O, Kumar S, Haider MR et al (2015) A review on anticancer potential of bioactive heterocycles quinoline. Eur J Med Chem 97:871–910

Ahadi H, Emami S (2020) modification of 7-piperazinylquinolone antibacterials to promising anticancer lead compounds: synthesis and in vitro studies. Eur J Med Chem 187:11

Akkachairin B, Rodphon W, Reamtong O et al (2020) Synthesis of neocryptolepines and carbocycle fused quinolines and evaluation of their anticancer and antiplasmodial activities. Bioorg Chem 98:103732

Alegaon S, Kashniyal K, Kuncolienkar S et al (2020) Synthesis and biological evaluation of some 4-aminoquinoline derivatives as potential antitubercular agents. Futur J Pharm Sc 6:2

Al-Ghorbani M, Begum B, Zabiulla MS et al (2015) Piperazine and morpholine: synthetic preview and pharmaceutical application. J Chem Pharm Res 7:281–301

Aly EI (2010) Design, synthesis and in vitro cytotoxic activity of new 4-anilino-7-chloro quinoline derivatives targeting EGFR tyrosine kinase. J Am Sci 6:73–83

Ambatkar MP, Khedekar PB (2019) Quinoline as TRPV1 antagonists: a new approach against inflammation. J Drug Deliv Ther 9:782–788

Aydemir N, Bilaloglu R (2003) Genotoxicity of two anticancer drugs, gemcitabine and topotecan, in mouse bone marrow in vivo. Mutat Res 537:43–51

Baartzes N, Stringer T, Seldon R et al (2019) Bioisosteric ferrocenyl aminoquinoline-benzimidazole hybrids: antimicrobial evaluation and mechanistic insights. Eur J Med Chem 180:121–133

Bahuguna A, Rawat DS (2020) Recent trends and strategies for the anti-tubercular drug development. Med Res Rev 40:263–292

Bazotte RB, Hirabara SM, Serdan TAD et al (2021) 4-Aminoquinoline compounds from the Spanish flu to COVID-19. Biomed Pharmacother 135:111138

Bhagat S, Arfeen M, Das G et al (2019) Design, synthesis and biological evaluation of 4-aminoquinolineguanylthiourea derivatives as antimalarial agents. Bioorg Chem 91:103094

Bhat HR, Masih A, Shakya A et al (2020) Design, synthesis, anticancer, antibacterial, and antifungal evaluation of 4-aminoquinoline-1,3,5-triazine derivatives. J Heterocycl Chem 57:390–399

Bonilla-Ramirez L, Rios A, Quiliano M et al (2018) Novel antimalarial chloroquine- and primaquine-quinoxaline 1,4-di-N-oxide hybrids: design, synthesis, plasmodium life cycle stage profile, and preliminary toxicity studies. Eur J Med Chem 158:68–81

Cancer.gov (n.d.). http://www.cancer.gov/. Accessed 10 July 2022

Candéa ALP, Ferreira ML, Pais KC et al (2009) Synthesis and antitubercular activity of 7-chloro-4 quinolinylhydrazones derivatives. Bioorg Med Chem Lett 19:6272–6274

Çapci A, Lorion MM, Wang H et al (2019) Artemisinine(iso)quinoline hybrids by CH activation and click chemistry: combating multidrug-resistant malaria. Angew Chem Int Ed 58:13066–13079

Charris JE, Monasterios MC, Acosta ME et al (2019) Antimalarial, antiproliferative, and apoptotic activity of quinoline-chalcone and quinolinepyrazoline hybrids. A dual action. Med Chem Res 28:2050–2066

Chokkar N, Kalra S, Chauhan M et al (2019) A review on quinoline derived scaffolds as anti-HIV agent. Mini Rev Med Chem 19:510–526

Chopra R, Chibale K, Singh K (2018) Pyrimidine-chloroquinoline hybrids: synthesis and antiplasmodial activity. Eur J Med Chem 148:39–53

Clark A, Jit M, Warren-Gash C et al (2020) Global, regional, and national estimates of the population at increased risk of severe COVID-19 due to underlying health conditions in 2020: a modelling study. Lancet Glob Health 8:1003–1017

Cortegiani A, Ingoglia G, Ippolito M et al (2020) A systematic review on the efficacy and safety of chloroquine for the treatment of COVID-19. J Crit Care 57:279–283

de Souza MVN, Pais KC, Kaiser CR et al (2009) Synthesis and in vitro antitubercular activity of a series of quinoline derivatives. Bioorg Med Chem 17:1474–1480

Deshpande S, Kuppast B (2016) 4-aminoquinolines: an overview of antimalarial chemotherapy. Med Chem 6:1–11

Dondorp AM, Nosten F, Yi P et al (2009) Artemisinin resistance in Plasmodium falciparum malaria. N Engl J Med 361:455–467

El-Gamal MI, Khan MA, Abdel-Maksoud MS et al (2014) A new series of diarylamides possessing quinoline nucleus: synthesis, in vitro anticancer activities, and kinase inhibitory effect. Eur J Med Chem 87:484–492

Eswaran S, Adhikari AV, Chowdhury IH et al (2010) New quinoline derivatives: synthesis and investigation of antibacterial and antituberculosis properties. Eur J Med Chem 45:3374–3383

Giacomelli A, Pagani G, Ridolfo AL et al (2021) Early administration of lopinavir/ritonavir plus hydroxychloroquine does not alter the clinical course of SARS-CoV-2 infection: a retrospective cohort study. J Med Virol 93:1421–1427

Gorbalenya AE, Baker SC, Baric RS et al (2020) The species severe acute respiratory syndrome-related coronavirus: classifying 2019-nCoV and naming it SARS-CoV-2. Nat Microbiol 5:536–544

Heravi MM, Zadsirjan V (2020) Prescribed drugs containing nitrogen heterocycles: an overview. RSC Adv 10:44247–44311

Herrmann L, Hahn F, Wangen C et al (2022) Anti-SARS-CoV-2 inhibitory profile of new quinoline compounds in cell culture-based infection models. Chem Eur J 28:e202103861

Hu YQ, Gao C, Zhang S et al (2017) Quinoline hybrids and their antiplasmodial and antimalarial activities. Eur J Med Chem 139:22–47

Jiang Y, Chen D, Cai D et al (2021) Effectiveness of remdesivir for the treatment of hospitalized COVID-19 persons: a network meta-analysis. J Med Virol 93:1171–1174

Kandi SK, Manohar S, Ve'lez Gerena CE et al (2015) C5-curcuminoid-4-aminoquinoline based molecular hybrids: design, synthesis and mechanistic investigation of anticancer activity. New J Chem 39:224–234

Katariya KD, Shah SR, Reddy D (2020) Anticancer, antimicrobial activities of quinoline based hydrazone analogues: synthesis, characterization and molecular docking. Bioorg Chem 94: 103406

Kaur R, Kumar K (2021) Synthetic and medicinal perspective of quinolines as antiviral agents. Eur J Med Chem 215:113220

Kerru N, Gummidi L, Maddila S et al (2020) A review on recent advances in nitrogen-containing molecules and their biological applications. Molecules 25:1909

Kokic G, Hillen HS, Tegunov D et al (2021) Mechanism of SARS-CoV-2 polymerase stalling by remdesivir. Nat Commun 12:279

Kondaparla S, Soni A, Manhas A et al (2017) Antimalarial activity of novel 4-aminoquinolines active against drug resistant strains. Bioorg Chem 70:74–85

Kondaparla S, Manhas A, Dola VR et al (2018) Design, synthesis and antiplasmodial activity of novel imidazole derivatives based on 7-chloro-4-aminoquinoline. Bioorg Chem 80:204–211

Krafts K, Hempelmann E, Skorska-Stania A (2012) From methylene blue to chloroquine: a brief review of the development of an antimalarial therapy. Parasitol Res 11:1–6

Kremsner PG, Krishna S (2004) Antimalarial combinations. Lancet 364:285–294

Kumar D, Khan SI, Tekwani BL et al (2015a) 4-Aminoquinoline pyrimidine hybrids: synthesis, antimalarial activity, heme binding and docking studies. Eur J Med Chem 89:490–502

Kumar D, Negi B, Rawat DS (2015b) The anti-tuberculosis agents under development and the challenges ahead. Future Med Chem 7:1981–2003

Li B, Zhu F, He F et al (2020) Synthesis and biological evaluations of N′-substituted methylene-4-(quinoline-4-amino) benzoylhydrazides as potential anti-hepatoma agents. Bioorg Chem 96: 103592

Mahajan A, Kremer L, Louw S et al (2011) Synthesis and in vitro antitubercular activity of ferrocene-based hydrazons. Bioorg Med Chem Lett 21:2866–2868

Manohar S, Rajesh UC, Khan SI et al (2012) Novel 4-aminoquinoline-pyrimidine based hybrids with improved in vitro and in vivo antimalarial activity. ACS Med Chem Lett 3:555–559

Manohar S, Tripathi M, Rawat DS (2014) 4-Aminoquinoline based molecular hybrids as antimalarials: an overview. Curr Top Med Chem 14:1706–1733

Manohar S, Pavan VS, Taylor D et al (2015) Highly active 4-aminoquinoline–pyrimidine based molecular hybrids as potential next generation antialarial agents. RSC Adv 5:28171

Marganakop SB, Kamble RR, Hoskeri J et al (2014) Facile synthesis of novel quinoline derivatives as anticancer agents. Med Chem Res 23(6):2727–2735

Matada BS, Pattanashettar R, Yernale NG (2021) A comprehensive review on the biological interest of quinoline and its derivatives. Bioorg Med Chem 32:115973

Maurya SS, Khan SI, Bahuguna A et al (2017) Eur J Med Chem 129:175–185

Maurya SS, Bahuguna A, Khan SI et al (2019) *N*-substituted aminoquinoline-pyrimidine hybrids: synthesis, in vitro antimalarial activity evaluation and docking studies. Eur J Med Chem 162: 277–289

Meshnick S, Dobson M (2001) The history of antimalarial drugs. In: Rosenthal P (ed) Antimalarial chemotherapy. Humana Press, Totowa, NJ, pp 15–25

Meunier B (2008) Hybrid molecules with a dual mode of action: dream or reality. Acc Chem Res 41:69–77

Mukherjee S, Pal M (2013) Quinolines: a new hope against inflammation. Drug Discov Today 18: 389–398

Muregi FW, Ishih A (2010) Next-generation antimalarial drugs: hybrid molecules as a new strategy in drug design. Drug Dev Res 71:20–32

Musiol R (2017) An overview of quinoline as a privileged scaffold in cancer drug discovery. Expert Opin Drug Discov 12:583–597

Musiol R, Serda M, Hensel-Bielowka S et al (2010) Quinoline-based antifungals. Curr Med Chem 17:1960–1973

Musiol R, Malarz K, Mularski J (2017) Quinoline alkaloids against neglected tropical diseases. Curr Org Chem 21:1896–1906

N'Da DD, Smith PJ (2014) Synthesis, in vitro antiplasmodial and antiproliferative activities of a series of quinoline-ferrocene hybrids. Med Chem Res 23:1214–1224

Narula AK, Azad CS, Nainwal LM (2019) New dimensions in the field of antimalarial research against malaria resurgence. Eur J Med Chem 181:111353

Noedl H, Se Y, Schaecher K et al (2008) Evidence of artemisinin-resistant malaria in western Cambodia. N Engl J Med 359:2619–2620

Parashar RK, Negi B (2015) Chemistry of heterocyclic compounds. Ane Books Pvt Ltd, New Delhi. ISBN 978–93–8546-292-4

Persoons L, Vanderlinden E, Vangeel L et al (2021) Broad spectrum anti-coronavirus activity of a series of anti-malaria quinoline analogues. Antivir Res 193:105127

Pushpakom S, Iorio F, Eyers PA et al (2019) Drug repurposing: Progress, challenges and recommendations. Nat Rev Drug Discov 18:41–58

Raj R, Land KM, Kumar V (2015) 4-Aminoquinoline-hybridization en route towards the development of rationally designed antimalarial agents. RSC Adv 5:82676

Ramírez-Prada J, Robledo SM, V'elez ID et al (2017) Synthesis of novel quinoline–based 4,5-dihydro–1H–pyrazoles as potential anticancer, antifungal, antibacterial and antiprotozoal agents. Eur J Med Chem 131:237–254

Rangappa SK, Siddappa AP (2014) Quinoline: a promising antitubercular target. Biomed Pharmacother 68:1161–1175

Rani A, Singh A, Gut J et al (2018) Microwave-promoted facile access to 4-aminoquinoline-phthalimides: synthesis and anti-plasmodial evaluation. Eur J Med Chem 143:150–156

Rani A, Legac J, Rosenthal PJ et al (2019) Substituted 1,3-dioxoisoindoline-4-aminoquinolines coupled via amide linkers: synthesis, antiplasmodial and cytotoxic evaluation. Bioorg Chem 88: 102912

Rani A, Kumar S, Legac J et al (2020) Design, synthesis, heme binding and density functional theory studies of isoindoline-dione-4-aminoquinolines as potential antiplasmodials. Future Med Chem 12:193–205

Rawat DS, Beena (2013) Antituberculosis drug research: a critical overview. Med Res Rev 33:693–764

Reddy PL, Khan SI, Ponnan P et al (2017) Design, synthesis and evaluation of 4-aminoquinoline-purine hybrids as potential antiplasmodial agents. Eur J Med Chem 126:675–686

Shalini J, Legac AA, Adeniyi P et al (2020) Functionalized naphthalimide-4-aminoquinoline conjugates as promising antiplasmodials, with mechanistic insights. ACS Med Chem Lett 11: 154–161

Shamsuddin MA, Ali AH, Zakaria NH et al (2021) Synthesis, molecular docking, and antimalarial activity of hybrid 4-aminoquinoline-pyrano[2,3-c]pyrazole derivatives. Pharmaceuticals 14: 1174

Sharma M, Chaturvedi V, Manju YK et al (2009) Substituted quinolinylchalcones and quinolinylpyrimidines as a new class of antiinfective agents. Eur J Med Chem 44:2081–2091

Sharma B, Kaur S, Legac J et al (2020) Synthesis, antiplasmodial and cytotoxic evaluation of 1H-1,2,3-triazole/acyl hydrazide integrated tetrahydro-b-carboline-4-aminoquinoline conjugates. Bioorg Med Chem Lett 30:126810

Singh K, Kaur T (2016) Pyrimidine-based antimalarials: design strategies and antiplasmodial effects. Med Chem Commun 7:749–768

Singh S, Kaur G, Mangla V et al (2015) Quinoline and quinolones: promising scaffolds for future antimycobacterial agents. J Enzyme Inhib Med Chem 30:492–504

Singh B, Chetia D, Kumawat MK (2021) Synthesis and in vitro antimalarial activity evaluation of some new 1,2-diaminopropane side-chain-modified 4-aminoquinoline Mannich bases. Pharm Chem J 55:7

Solomona VR, Pundira S, Lee H (2019) Design and synthesis of 4-piperazinyl quinoline derived urea/thioureas for anti-breast cancer activity by a hybrid pharmacophore approach. J Enzyme Inhib Med Chem 34(1):620–630

Srivastava V, Negi AS, Kumar JK et al (2005) Plant-based anticancer molecules: a chemical and biological profile of some important leads. Bioorg Med Chem 13:5892–5908

Su T, Zhu J, Sun R et al (2019) Design, synthesis and biological evaluation of new quinoline derivatives as potential antitumor agents. Eur J Med Chem 178:154–167

Tseng CH, Chen YL, Hsu CY et al (2013) Synthesis and antiproliferative evaluation of 3-phenylquinolinylchalcone derivatives against non-small cell lung cancers and breast cancers. Eur J Med Chem 59:274–282

Vandekerckhove S, D'hooghe M (2015) Quinoline-based antimalarial hybrid compounds. Bioorg Med Chem 23:5098–5119

Veziris N, Bernard C, Guglielmetti L et al (2017) Rapid emergence of mycobacterium tuberculosis bedaquiline resistance: lessons to avoid repeating past errors. Eur Respir J 49:1601719

Wang J, Xu C, Wong YK et al (2019) Artemisinin, the magic drug discovered from traditional Chinese medicine. Engineering 5:9–32

Wang M, Cao R, Zhang L et al (2020a) Remdesivir and chloroquine effectively inhibit the recently emerged novel coronavirus (2019-nCoV) in vitro. Cell Res 30:269–271

Wang Y, Grunewald M, Perlman S (2020b) Coronaviruses: an updated overview of their replication and pathogenesis. Methods Mol Biol 2203:1–29

WHO (n.d.-a). https://covid19.who.int/. Accessed 20 July 2022

WHO (n.d.-b). https://www.who.int/news-room/fact-sheets/detail/tuberculosis. Accessed 15 July 2022

Woodward R, Doering W (1944) The total synthesis of quinine. J Am Chem Soc 66:849

Yadav P, Shah K (2021) Quinolines, a perpetual, multipurpose scaffold in medicinal chemistry. Bioorg Chem 109:104639

Young B, Tan TT, Leo YS (2021) The place for remdesivir in COVID-19 treatment. Lancet Infect Dis 21:20–21

Zhou Y, Yan W, Cao D et al (2017) Design, synthesis and biological evaluation of 4-anilinoquinoline derivatives as novel potent tubulin depolymerization agents. Eur J Med Chem 138:1114–1125

Zorc B, Rajic Z, Perkovic I (2019) Antiproliferative evaluation of various aminoquinoline derivatives. Acta Pharma 69:661–672

Thiourea Catalysts for Synthesis of Active Pharmaceutical Ingredients

11

Sharada Prasanna Swain, Kankanala Naveen Kumar, and V. Ravichandiran

Abstract

Thiourea is an important building block found in several drug molecules such as thioacetazone, enzalutamide, and thiocarlide. Thiourea derivatives have been used for the activation of carbonyl and imine compounds to facilitate Michael addition reactions, as oxyanion stabilizer for [Ir] catalyzed amination of alcohols without using any base or acid. Chiral bifunctional thiourea catalysts have been successfully applied for the asymmetric synthesis of several drug molecules.

Keywords

Baclofen · Laninamivir · Sitagliptin · Thiourea · Zanamavir

11.1 Introduction

Thiourea was synthesized by the Polish chemist Marceli Nencki in the year 1873, as the first urea analogue (Ronchetti et al. 2021). Thiourea moiety is present in a variety of drugs and bioactive compounds such as anti-viral, anti-convulsant (Siddiqui et al. 2011), anti-inflammatory (Keche et al. 2012), antimicrobial (Hashem et al. 2020), and anti-tumor agents (Abbas et al. 2020), (Fig. 11.1).

Thiourea catalysts act as hydrogen bond donor catalysts and activate electrophiles such as carbonyl compounds, imines, and nitro functional groups to react with nucleophiles (Wende and Schreiner 2012). *N,N′*-bis[3,5-bis(trifluoromethyl)-phenyl]thiourea known as Schreiner thiourea acts as a hydrogen bond donor catalyst,

S. P. Swain (✉) · K. N. Kumar · V. Ravichandiran
Department of Medicinal Chemistry, National Institute of Pharmaceutical Education and Research, Kolkata, WB, India
e-mail: spswain@niperkolkata.edu.in

P. P. Singh (ed.), *Recent Advances in Pharmaceutical Innovation and Research*,
https://doi.org/10.1007/978-981-99-2302-1_11

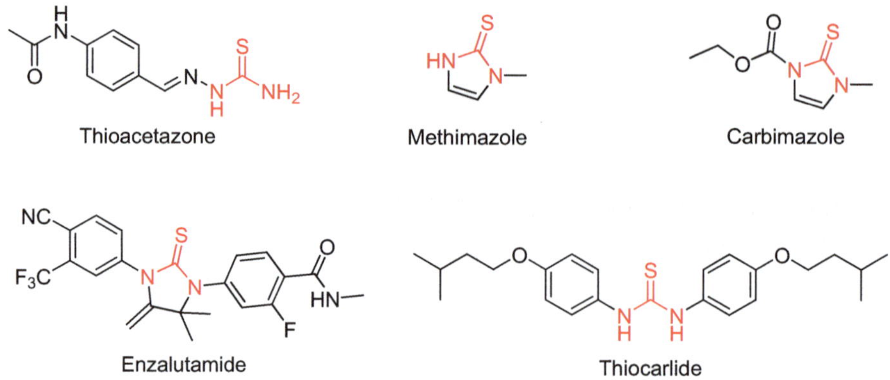

Fig. 11.1 Thiourea-containing drug molecules

which has been widely used for several nucleophilic additions and other reactions (Zhang et al. 2014). Loh and co-workers have reported thiourea as a Brønsted acid for glycosylation reaction (Xu and Loh 2018). Thiourea can also act as a Lewis base for oxidation of alcohols as described by Mukherjee and co-workers (Tripathi and Mukherjee 2012). Schreiner (Kotke and Schreiner 2007) and Kass (Beletskiy et al. 2012) proved that the oxyanion stabilization by thiourea as a double hydrogen bond donor for the addition of alcohols to tetrahydropyran is feasible, and thioureas can also catalyze acetalization of carbonyl compounds by oxyanion stabilization (Kotke and Schreiner 2006). Many bifunctional thioureas such as Takemoto (Parvin et al. 2020), Jacobsen (Doyle and Jacobsen 2007), and other heterocyclic-containing thiourea catalysts have been synthesized, which utilize both hydrogen-bonding interactions and enamine formation (Taylor and Jacobsen 2006).

The pharmaceutical industries are keen to develop green processes for the synthesis of active pharmaceutical ingredients (API) and heterocyclic compounds (Cue and Zhang 2009). They are spending on research and development for generating environment-friendly friendly processes and eliminating hazardous chemicals, and thiourea catalysts have been utilized by chemists for the green synthesis of heterocyclic compounds (Takemoto 2010).

11.2 Applications of Thiourea Catalysts for Synthesis of Active Pharmaceutical Ingredients (API)

11.2.1 Asymmetric Aza-Michael Addition: Synthesis of Sitagliptin

Aza-Michael addition reaction is a classical method for synthesizing β-amino acids. Takemoto and co-workers explored the possibility of intermolecular aza-Michael addition of benzylhydroxylamine to carboxylic acids **2** using the hybrid thiourea catalyst **1** (Takemoto 2021). The reaction was carried out in carbon tetrachloride in the presence of a 10 mol% catalyst and 4 Å molecular sieves (MS). The chiral

Scheme 11.1 Thiourea catalyzed asymmetric Michael addition reaction

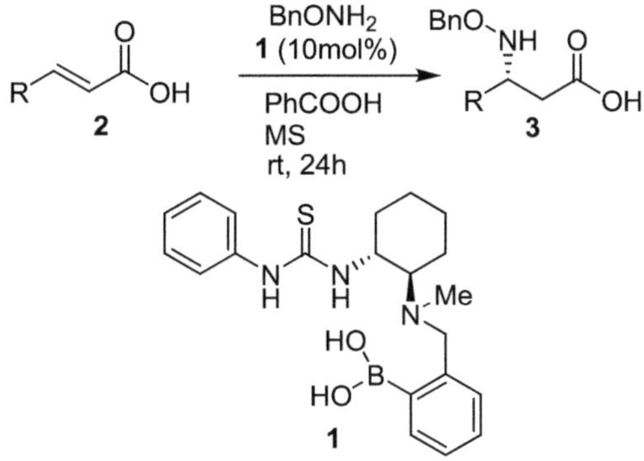

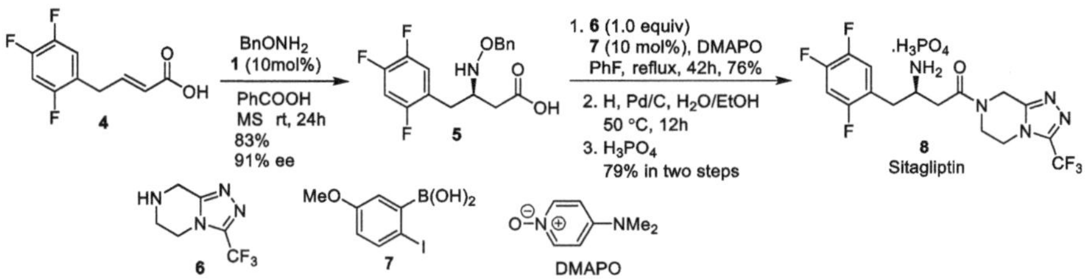

Scheme 11.2 Synthesis of Sitagliptin catalyzed by chiral thiourea catalyst

thiourea hybrid catalyst also contains a boronic acid moiety (Scheme 11.1). The 4 Å MS facilitated the catalytic reaction efficiently. The methodology was successfully applied for the formal asymmetric synthesis of sitagliptin (antidiabetic drug). Sitagliptin synthesis was accomplished in four steps, starting from the, β-unsaturated carboxylic acid **4** without using any chiral auxiliary or protecting groups (Scheme 11.2). The thiourea catalyzed reaction provided the β-amino acid in 83% yield and good enantioselectivity (91%ee). Then it was coupled with amine 6, followed by deprotection of -OBn group by hydrogenation. Finally, the phosphoric acid salt was made to get the desired sitagliptin (**8**).

11.2.2 Enantioselective Aza-Henry Reaction: Synthesis of Anti-HIV Drug DPC 083

Dihydroquinazolinones are an important class of heterocyclic compounds that possess broad biological activities such as antiviral, and antiobesity, used for the treatment of cardiovascular diseases, inflammation, and pain. A drug candidate called DPC 083 developed by Dupont, bearing a dihydroquinazolinone ring along with a chiral trifluoromethyl moiety is a potent inhibitor of HIV-1 nonnucleoside reverse transcriptase.

Wang and co-workers developed chorial thiourea **9**-catalyzed aza-Henry reaction for the synthesis of trifluoromethylquinazolin-2(1H)-ones in high yields and good-

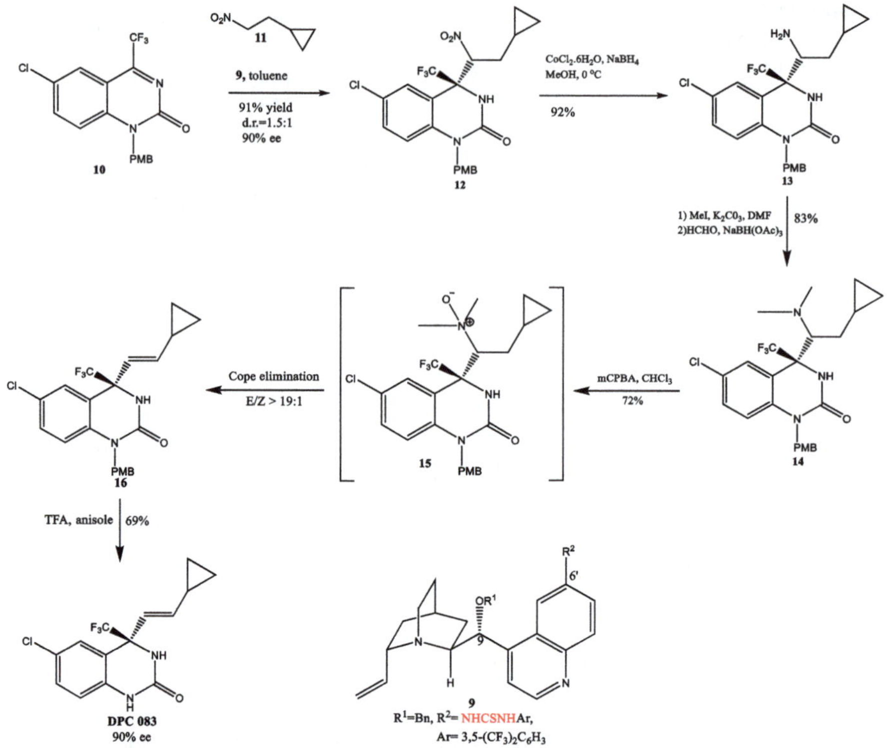

Scheme 11.3 Generation of quaternary chiral carbon center and synthesis of DPC 083

to-excellent enantioselectivities (Xie et al. 2011). A variety of nitroalkanes 2 were also investigated. This asymmetric aza-Henry reaction was applied for the synthesis of the anti-HIV drug DCP 083 (Scheme 11.3). The thiourea-catalyzed aza-Henry reaction between dihydroquinazolinone **10** and nitrocyclopropylalkane **11** generated a quaternary stereogenic center in 91% yield, with diastereomeric ratio 1.5:1 and 90% enantiomeric excess. The diastereomers were easily separated by silica gel column chromatography. The synthesis of the target molecule DPC 083 was accomplished in five steps. Reduction in the nitro group in compound **12** in presence of $CoCl_2{\cdot}6H_2O$ and $NaBH_4$ in MeOH produced amine compound **13** in 92% yield. N-dimethylation of the amino group using an excess of iodomethane (>3 equiv) resulted in a mixture of mono- and bis-products. Hence, monomethylation with MeI in the presence of K_2CO_3 and followed by bismethylation by reductive amination with formaldehyde and $NaBH(OAc)_3$ generated compound **14** in 83% yield. N-oxide and spontaneous Cope elimination reactions afforded olefin compound **16** in 72% over the two steps (E/Z > 19:1). Finally, the deprotection of the PMB group was performed by treating with TFA in the presence of anisole to provide the target DPC 083 in 69% and no racemization of the quaternary stereogenic center was observed as analyzed by chiral HPLC.

11.2.3 Michael Addition of Acetone with Tert-Butyl (2-Nitrovinyl) Carbamate

Zanamivir, laninamivir, and CS-8958 are neuraminidase inhibitors used for the treatment of influenza virus infection. Ma and co-workers (Tian et al. 2014) studied asymmetric Michael addition reaction of acetone with less reactive tert-butyl (2-nitrovinyl)carbamate **17** (Scheme 11.4). The use of chiral bifunctional primary amine–thiourea catalyst 20 (20 mol%) provides desired Michael addition product 19 in 96% yield with 82%ee. The reduction of catalyst to 5 mol% did not reduce enantioselectivity, but yield was reduced to 82%.

The core moiety of zanamavir and laninamivir **22** was synthesized by asymmetric Henry reaction using chiral proline catalyst **23** (Scheme 11.5).

Reduction in nitro group in presence of Zn and followed by acetylation provided desired compound **25a** in 77% yield and **25b** in 72% yield (Scheme 11.6). Oxidation of methyl group with SeO_2 and further oxidation with sodium chlorite produced acid compounds **26a** and **26b** in 60% yield. Deprotection of amine (Boc protection) by treating with HCl provided free amine compound **27a** and **27b** in 100% yield. Reaction of amine compound **27a** and **27b** with **29** generated zanamivir (**28a**) in 88% yield and laninamivir (**28b**) in 85% yield, respectively. The esterification of **28b** with n-$C_7H_{15}C(OMe)_3$ in methanolic HCl gave CS-8958 (**30**) in 92% yield.

Scheme 11.4 Thiourea-catalyzed asymmetric Michael addition reaction of acetone with nitro vinyl compound

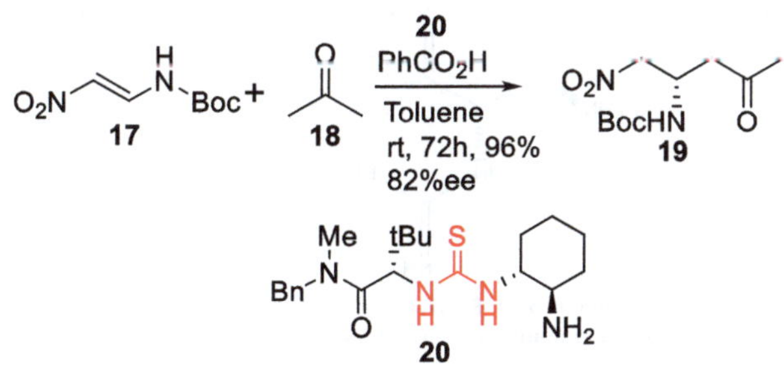

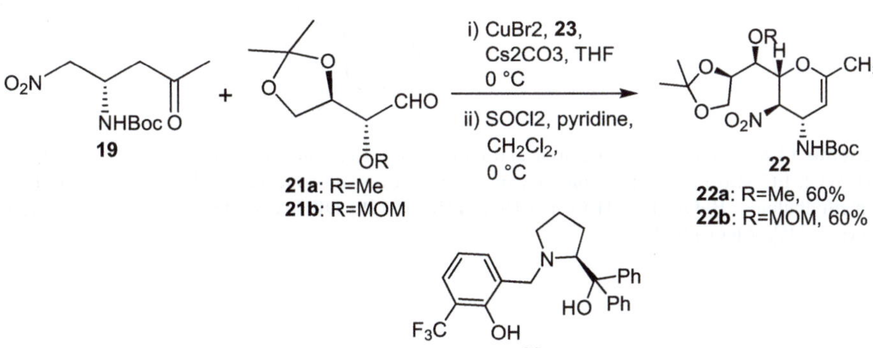

Scheme 11.5 Chiral synthesis of zanamivir and laninamivir core moiety

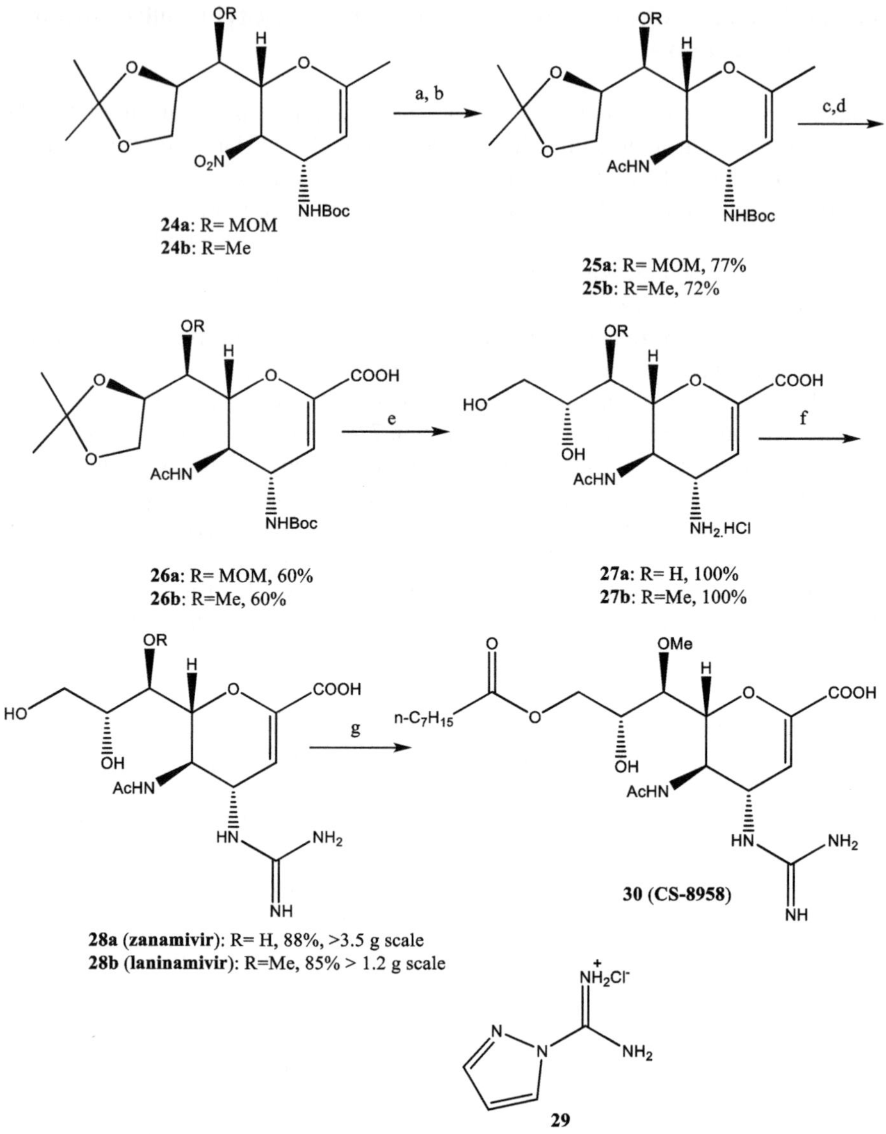

Scheme 11.6 Synthesis of zanamivir and laninamivir. Reagents and conditions: (**a**) Zn, HOAc; (**b**) AcCl, Et3N; (**c**) SeO_2, pyridine, 4A molecular sieves, dioxane/THF; (**d**) $NaClO_2$, NaH_2PO_4, 2-methylbutene, $^{t}BuOH/THF/H_2O$; (**e**) HCl, THF; (**f**) **29**, DIPEA, DMF, 50 °C; (**g**) n-$C_7H_{15}C(OMe)_3$, HCl, MeOH,92%

11.2.4 Enantioselective Michael Addition of 1,3-Dicarbonyl Compounds to Nitroolefins

The Michael addition reaction of β-nitrostyrene compound **32** with diethyl malonate in toluene and 10 mol% of Takemoto chiral thiourea catalyst **31** (Scheme 11.7) gave the compound **33** in 80% yield and 99% ee (Takemoto and Miyabe 2007). Both rigidity of the chiral diamine scaffold and cooperative function of two N–H bonds and the tertiary amino group in the catalyst were critical for the enantioselectivity reaction. Reduction in the nitro group of compound 33 in presence of $NiCl_2$, $NaBH_4$; hydrolysis of ester in basic medium, and finally decarboxylation of one carboxylic group in acidic medium generated baclofen. Baclofen is a derivative of GABA (γ-aminobutyric acid), and is prescribed as an antispastic agent in racemic form.

(−)-Epibatidine is a potent nicotinic acetylcholine receptor agonist. The reaction of γ,δ-unsaturated β-ketoester **36** with nitroalkene **35** in presence of chiral thiourea catalyst **31** gave 3,4-anti-4,5-syn cycloadduct **38** (Scheme 11.8) via Michael adduct intermediate **37** (Takemoto and Miyabe 2007) with excellent diastereoselectivity and moderate enantioselectivity (75%ee). This is the first reported asymmetric synthesis of three contiguous stereogenic centers by the tandem Michael addition reaction with nitroalkenes. Then, the total synthesis of (−)-epibatidine was accomplished in six steps from the Michael adduct 36 (Scheme 11.8).

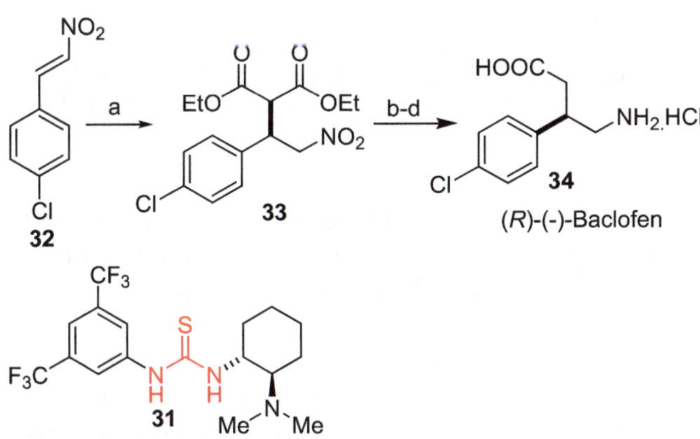

Scheme 11.7 Synthesis of (−)-Baclofen. Reagents and Conditions: (**a**) Diethyl malonate, **31**, toluene, rt., 24 h, 80% (>99% ee after single recrystallization from hexane/AcOEt); (**b**) $NiCl_2.6H_2O$, $NaBH_4$, MeOH, rt., 7.5 h, 94%; (**c**) NaOH, EtOH, rt., 45 h; then toluene, reflux, 6.5 h, 84%; (**d**) 6 N HCl, reflux, 24 h, 94%

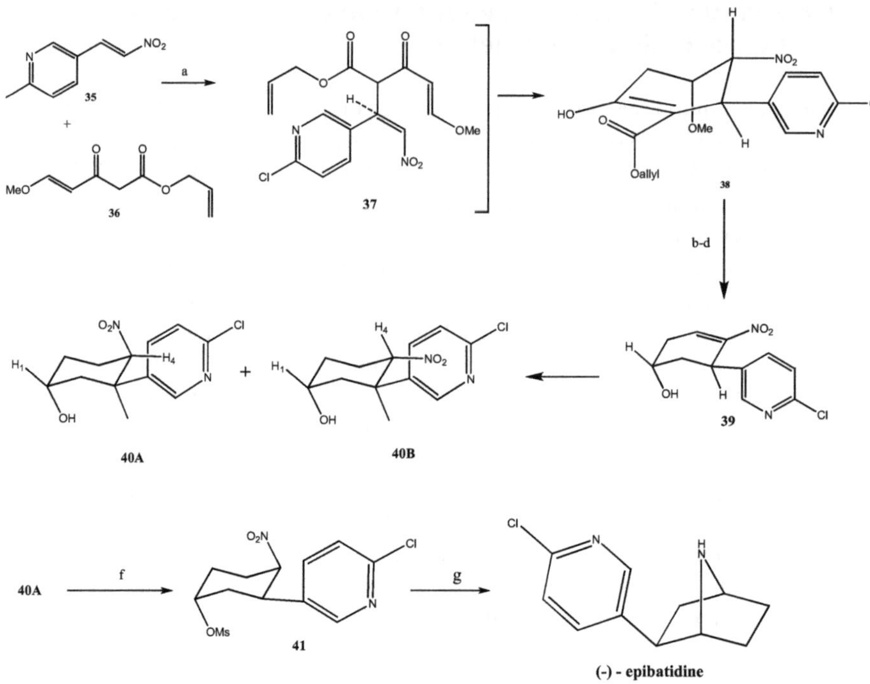

Scheme 11.8 Synthesis of (−)-epibatidine. Reagents and conditions: **31**, toluene, 0 °C; KOH, EtOH, 0 °C,77% (75% ee); (**b**) $Pd(OAc)_2$, PPh_3, HCOOH, Et_3N, THF, rt., 99%; (**c**) L-Selectride, THF, −78 °C, 71%; (**d**) NaOMe, ^tBuOH, 71%; (**e**) $NaBH_3CN$, AcOH, MeOH, −20 °C, 87% (**40A/40B** = 9/1); (**f**) MsCl, ET_3N, DMAP, CH_2Cl_2, 0 °C, 91%; (**g**) Zn, AcOH, THF, rt., $CHCl_3$, 60 °C, 85%

11.2.5 Michael-Michael Addition Reactions Promoted by Secondary Amine-Thiourea: Stereocontrolled Construction of Barbiturate-Fused Tetrahydropyrano

The barbituric acid scaffold is an important class of building block found in the pharmaceutical products. Drug molecules containing barbituric acid moiety are widely used for treatment of many diseases. Those are applied as sedatives (Johns 1975), anticancer agents (Singh et al. 2009), and anticonvulsants (Smith and Riskin 1991). The tetrahydropyrano scaffold is a structural motif which attracts intensive attention (Larghi and Kaufman 2006). Thus, as a combination of these two scaffolds, barbiturate-fused tetrahydropyrano compounds have a great importance for several biological activities, such as antimicrobial (Singh et al. 2014), antiproliferative (Venkatesham et al. 2012), and antituberculosis activities (Kalaria et al. 2014) (Fig. 11.2).

Bifunctional secondary amine-thiourea organocatalyst **46** was applied for asymmetric catalysis in the chiral synthesis of barbiturate-fused tetrahydropyrano moieties **49** and **50** (Zhang et al. 2017). The catalyst **46** possessed excellent catalytic

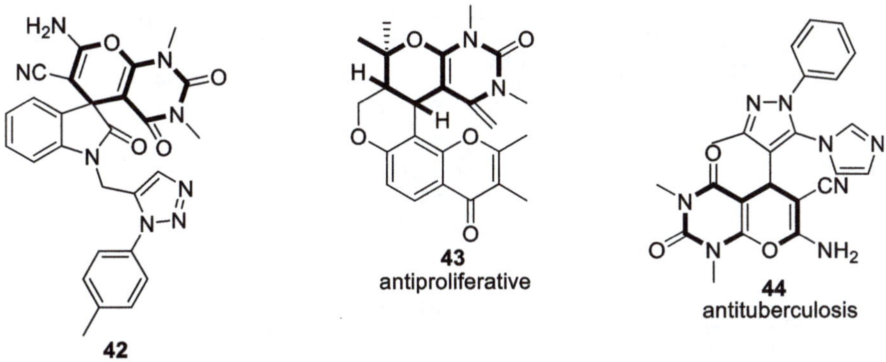

Fig. 11.2 Drug molecules containing barbiturate-fused tetrahydropyrano moiety

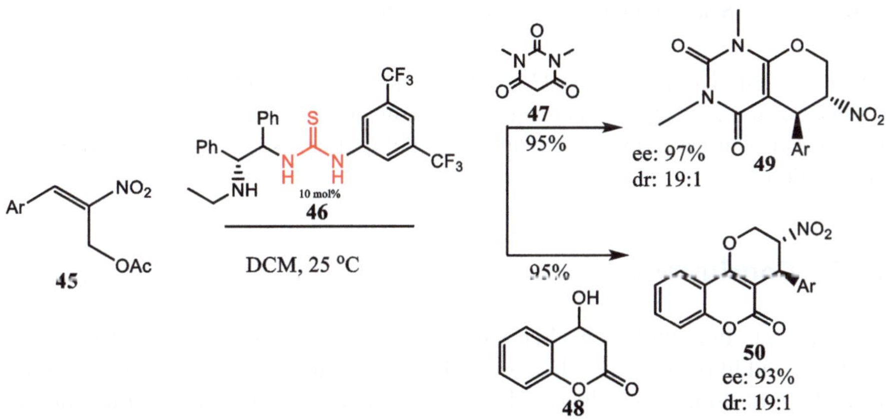

Scheme 11.9 Thiourea catalyzed asymmetric synthesis of barbiturate-fused tetrahydropyrano moieties

activity in the domino Michael-Michael reaction between N, N′-dimethylbarbituric acid **48**, and Morita-Baylis-Hillman acetates of nitroalkenes to get pharmaceuticals in good yields with excellent enantioselectivities (97% ee), as described in Scheme 11.9. This catalytic reaction could also be applied to prepare active pyranocoumarin compounds.

11.2.6 Thiourea as Oxyanion Stabilizer for Amination of Alcohols: Synthesis of Cardiovascular Drug Ticlopidine

Schreiner's thiourea can act as an oxyanion stabilizer and facilitate [Ir]-catalyzed amination of alcohols **51** without any base or strong acid (Scheme 11.10). Our laboratory has developed $[Cp^*IrCl_2]_2$ and thiourea catalyzed amination of alcohol in

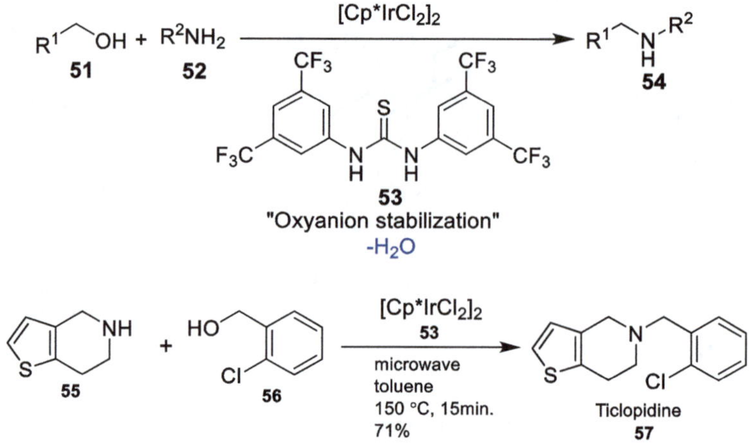

Scheme 11.10 Synthesis of ticlopidine catalyzed by thiourea

60–82% yield (Kumar et al. 2022). The method was successfully applied for the applied for the synthesis of cardiovascular drug ticlopidine in 71% yield (Scheme 11.10). This a green process for synthesis of ticlopidine, as H2O is the by-product of the reaction.

11.3 Conclusions

Thiourea was first applied by Schriener for activation of carbonyl, and imine for Michael's addition reaction. Jacobsen, Takemoto, and other chemists developed several chiral thiourea catalysts, which were applied for the asymmetric synthesis of several drug molecules. These thiourea catalysts were used for the synthesis of sitagliptin, baclofen, zanamivir, laninamivir, ticlopidine, etc.

Acknowledgements The authors acknowledge the financial support from the National Institute of Pharmaceutical Education and Research-Kolkata, Ministry of Chemicals and Fertilizers, Govt. of India.

References

Abbas SY, Al-Harbi RAK, Sh El-Sharief MAM (2020) Synthesis and anticancer activity of thiourea derivatives bearing a benzodioxole moiety with EGFR inhibitory activity, apoptosis assay and molecular docking study. Eur J Med Chem 198:112363

Beletskiy EV, Schmidt J, Wang XB et al (2012) Three hydrogen bond donor catalysts: oxyanion hole mimics and transition state analogues. J Am Chem Soc 134:18534–18537

Cue BW, Zhang J (2009) Green process chemistry in the pharmaceutical industry. Green Chem Lett Rev 2:193–211

Doyle AG, Jacobsen EN (2007) Small-molecule H-bond donors in asymmetric catalysis. Chem Rev 107:5713–5743

Hashem HE, Amr AEE, Nossier ES et al (2020) Synthesis, antimicrobial activity and molecular docking of novel thiourea derivatives tagged with thiadiazole, imidazole and triazine moieties as potential DNA gyrase and topoisomerase IV inhibitors. Molecules 25:2766

Johns MW (1975) Sleep and hypnotic dugs. Drugs 9:448–478

Kalaria PN, Satasia SP, Raval DK et al (2014) Synthesis, characterization and biological screening of novel 5-imidazopyrazole incorporated fused pyran motifs under microwave irradiation. New J Chem 38:1512–1521

Keche AP, Hatnapure GD, Tale RH et al (2012) A novel pyrimidine derivatives with aryl urea, thiourea and sulfonamide moieties: synthesis, anti-inflammatory and antimicrobial evaluation. Bioorg Med Chem Lett 22:3445–3448

Kotke M, Schreiner PR (2006) Acid-free, organocatalytic acetalization. Tetrahedron 62:434–439

Kotke M, Schreiner PR (2007) Generally applicable organocatalytic tetrahydropyranylation of hydroxy functionalities with very low catalyst loading. Synthesis 5:779–790

Kumar KN, Reddy MM, Panchami H et al (2022) Thiourea as oxyanion stabilizer for iridium catalysed, base free green synthesis of amines: synthesis of cardiovascular drug ticlopidine. Mol Catal 524:112324

Larghi EL, Kaufman TS (2006) The Oxa-Pictet-Spengler cyclization: synthesis of Isochromans and related pyran-type heterocycles. Synthesis 2:187–220

Parvin T, Yadav R, Choudhury LH (2020) Recent application of thiourea-based organocatalysts in asymmetric multicomponent reactions (AMCRs). Org Biomol Chem 18:5513–5532

Ronchetti R, Moroni G, Carotti A et al (2021) Recent advances in urea- and thiourea-containing compounds: focus on innovative approaches in medicinal chemistry and organic synthesis. RSC Med Chem 12:1046–1064

Siddiqui N, Alam MS, Stables JP (2011) Synthesis and anticonvulsant properties of 1-(amino-*N*-arylmethanethio)-3-(1-substituted benzyl-2, 3-dioxoindolin-5-yl) urea derivatives. Eur J Med Chem 46:2236–2242

Singh P, Kaur M, Verma P (2009) Design, synthesis and anticancer activities of hybrids of indole and barbituric acids—identification of highly promising leads. Bioorg Med Chem Lett 19:3054–3058

Singh H, Sindhu J, Khurana JM et al (2014) Ultrasound promoted one pot synthesis of novel fluorescent triazolyl spirocyclic oxindoles using DBU based task specific ionic liquids and their antimicrobial activity. Eur J Med Chem 77:145–154

Smith MC, Riskin BJ (1991) The clinical use of barbiturates in neurological disorders. Drugs 42: 365–378

Takemoto Y (2010) Development of chiral thiourea catalysts and its application to asymmetric catalytic reactions. Chem Pharm Bull 58:593–601

Takemoto Y (2021) Molecular transformation based on an innovative catalytic system. Chem Pharm Bull 69:819–831

Takemoto Y, Miyabe H (2007) The amino thiourea-catalyzed asymmetric nucleophilic reactions. Chimia 61:269–275

Taylor MS, Jacobsen EN (2006) Asymmetric catalysis by chiral hydrogen-bond donors. Angew Chem Int Ed 45:1520–1543

Tian J, Zhong J, Li Y et al (2014) Organocatalytic and scalable synthesis of the anti-influenza drugs zanamivir, laninamivir, and CS-8958. Angew Chem Int Ed 53:1–5

Tripathi CB, Mukherjee S (2012) Lewis base catalysis by thiourea: *n*-bromosuccinimide-mediated oxidation of alcohols. J Org Chem 77:1592–1598

Venkatesham A, Rao RS, Nagaiah K et al (2012) Synthesis of new chromeno-annulated *cis*-fused pyrano[3,4-*c*]pyran derivatives *via* domino Knoevenagel–hetero-Diels–Alder reactions and their biological evaluation towards antiproliferative activity. Med Chem Commun 3:652–658

Wende RC, Schreiner PR (2012) Evolution of asymmetric organocatalysis: multi- and retrocatalysis. Green Chem 14:1821–1849

Xie H, Zhang Y, Zhang S et al (2011) Bifunctional cinchona alkaloid thiourea catalyzed highly efficient, enantioselective aza-henry reaction of cyclic trifluoromethyl ketimines: synthesis of anti-HIV drug DPC 083. Angew Chem Int Ed 50:11773–11776

Xu C, Loh CCJ (2018) An ultra-low thiourea catalyzed strain-release glycosylation and a multicatalytic diversification strategy. Nat Commun 9:4057

Zhang Z, Bao Z, Xing H (2014) N, N′-Bis[3,5-bis(trifluoromethyl)phenyl]thiourea: a privileged motif for catalyst development. Org Biomol Chem 12:3151–3162

Zhang J, Yin G, Du Y et al (2017) Michael-Michael addition reactions promoted by secondary amine-thiourea: stereocontrolled construction of barbiturate-fused tetrahydropyrano scaffolds and pyranocoumarins. J Org Chem 82:13594–13601

The Chemistry and Pharmacology of Mandarin Orange (*Citrus reticulata*) 12

Inder Pal Singh, Dattatraya Dinkar Gore, Snehal Karkhele, and Charles Santhanaraju Vairappan

Abstract

Citrus fruits and their juices are among the most widely consumed fruit products as a regular source of nutrition, flavor, and health remedy. In particular, mandarin orange (*Citrus reticulata* Blanco) is often known as an excellent source of vitamin C and is well utilized in culinary and important nutraceuticals. Almost every component of mandarin orange, peels, pulps, leaves, barks, and roots, has been reported to contain a diverse array of phytochemicals. Essential oils that are derived from peels, fruits, and leaves contain 85–99% of terpenes and non-terpene hydrocarbons. Flavonoids that are represented in the forms of flavones, flavanones, flavonols, anthocyanidins, flavonols, and isoflavones are also the other major chemical components. Some *C. reticulata* populations have been shown to contain coumarins and furanocoumarins. Other minor chemicals consist of benzoic acid derivatives and steroidal glycosides. The relative composition of its chemical constituents depends on their geographical location and variety. Phytoconstituents of *C. reticulata* exhibit potent antimicrobial, antioxidant, cytolytic, and anticancer potentials. Due to its wide geographical distribution and ability to synthesize a wide array of important chemicals, mandarin oranges could be used as an important source of nutraceuticals for the growing global population.

I. P. Singh (✉) · D. D. Gore · S. Karkhele
Department of Natural Products, National Institute of Pharmaceutical Education and Research, S.A. S. Nagar, India
e-mail: ipsingh@niper.ac.in

C. S. Vairappan
Institute for Tropical Biology and Conservation, Universiti Malaysia, Kota Kinabalu, Sabah, Malaysia

P. P. Singh (ed.), *Recent Advances in Pharmaceutical Innovation and Research*,
https://doi.org/10.1007/978-981-99-2302-1_12

Keywords

Citrus reticulata · Coumarins · Essential oil · Flavonoids

12.1 Introduction

Citrus fruits are popular worldwide due to their enormous health benefits. These genus 'Citrus' fruits belong to the family Rutaceae and encompass various species including *C. limon* (lemon), *C. medica* (citron), *C. aurantium* (sour orange), *C. paradisi* (grapefruit), *C. reticulata* (mandarin, tangerine), *C. clementina* (clementine), and *C. sinensis* (sweet orange). There are four ancestral taxa, namely, *C. medica C. reticulata* Blanco, *C. maxima* (Burm.), and *C. micrantha* (Wester). The other cultivated species have derived mainly from the above (Mabberley 1997). The mandarin orange (*C. reticulata*, order: Sapindales, family: Rutaceae), is commonly known as the mandarin or tangerine; Santara in Hindi, Narangi in Marathi, Aravta in Sanskrit, Kamala lebu in Bengali, and Kamalapplam, Naarinija in Tamil. *C. reticulata* is a widely cultivated extensively grown fruit tree in many warm regions with temperate to tropical climate. This evergreen tree is small and grows to about 8 m tall. It has axillary thorns and the leaves are dark green, slender, long, and narrow. The tree produces flowers that are star-shaped, borne singly, and white. Mandarin fruits, resembling tangerines, are oblate and reddish-orange. Flowers are white and scented and flowering occurs in mid to late autumn. Leaves are dark green, lance-shaped on long stalks with small wings. Peels are abundant in essential oils which have been historically employed as antiseptic, antispasmodic, stomachic, sedative, diuretic, as well as to enhance circulation. The essential oils have been also commonly used in perfumes, cosmetics, and flavoring food and drinks. Due to the presence of phenolic compounds, flavonoids and flavonoid glycosides, coumarins, and other important secondary metabolites, *C. reticulata* is considered useful in treating constipation, cramps, colic, diarrhea, bronchitis, tuberculosis, cough, cold, obesity, menstrual disorders, angina, hypertension, skin disorders, thirst, hiccup, earache, nausea, indigestion, and vomiting. *C. reticulata* is renowned for being a prominent reservoir of health-beneficial phytoconstituents that include phenolic compounds (primarily flavonoids) and other nutrients and non-nutrients (vitamins, minerals, dietary fiber, essential oils, limonoids, and carotenoids). *C. reticulata* is consumed globally as a superb source of vitamin C, which is a powerful natural antioxidant that builds the body's immune system. Peel extracts, essential oil, and leaf extracts of *C. reticulata* have been reported for several pharmacological activities, that include antioxidant, immunomodulatory, anti-inflammatory, anti-cancer, anti-diabetes, and antiobesity (Tripoli et al. 2007; Barreca et al. 2020) (Fig. 12.1).

Numerous chemical constituents/phytoconstituents are primarily produced by the plants for the purpose of defense and are called secondary metabolites. *C. reticulata* is abundant in these secondary metabolites which play a significant role in the health benefits attributed to this plant. Various types of pharmacologically active

Fig. 12.1 Fruits of *Citrus reticulata*

compounds have been identified in the fruits, peel, leaves, and roots of *C. reticulata*. This chapter presents the chemical composition and explores the biological activities associated with *C. reticulata*.

12.2 Essential Oil

The essential oils (EO) of citrus plants are mainly obtained from their peel (rind) or fruit or leaves. These essential oils are generally extracted by steam distillation or by cold pressing. The volatiles can also be extracted using non-polar solvents like pentane or hexane to yield organic extracts. The essential oil fraction particularly obtained from the fruit peels of *C. reticulata* is rich in many flavoring compounds. Volatile and semi-volatile compounds represent 85–99% of the entire oil fraction with hydrocarbons, monoterpenes, and sesquiterpenes being the major constituents (González-Mas et al. 2019). Aliphatic and olefinic C6–C12 non-terpene aldehydes, alcohols, ketones, esters, and acids constitute the remainder volatile fraction. The volatile constituents of essential oils are mainly analyzed using Gas Chromatography-Mass spectrometry. The oils are normally diluted with hexane or pentane for GC-MS analysis. Approximately 400 compounds are reported from different citrus species with their content depending on specific cultivars. Monoterpene hydrocarbons are the most abundant chemical constituents of *C. reticulata* EO, with limonene comprising about 60–95%, γ-terpinene up to 15%, followed by β-myrcene, α-ανδ β-pinene, linalool, and β-citronellal. Sesquiterpenes, namely α-sinensal, β-copaene, and non-terpenes octanal and decanal are also present in small amounts. The other compounds are usually less than 0.1%. The EO of *C. reticulata* mainly differs qualitatively from other citrus species in the presence of non-terpenoid aldehydes including *E,E*-2,4-heptadienal or (Z)-2-dodecanal. The volatiles identified in peel EO have also been reported from the leaf EO. Limonene is found to be the most abundant volatile constituting up to 95% of EO (González-Mas et al. 2019). Cold-pressed oils of mandarin and tangerine were compared with other citrus oils qualitatively and quantitatively using GC and IR spectroscopy (Ashoor and Bernhard 1967), and the composition of oils has been found to be similar in various oils. In a different study, limonene was found to be above 80% of EO in a cold-pressed oil of *C. reticulata* Blanco (Ponkan), a popular commercial cultivar of mandarin in China and Japan. The EO was separated into hydrocarbon fraction and oxygenated fraction by chromatography on silica gel where the oxygenated fraction was found to contain 40 compounds (Sawamura et al. 2004).

In a study on essential oil obtained by hydro-distillation of peels from *C. reticulata* Blanco from Assam, India, 37 compounds were identified, the major volatiles being limonene (46.7%), geranial (19%), neral (14.5%). The other main components were geranyl acetate (3.9%), geraniol (3.5%), b-caryophyllene (2.6%), nerol (2.3%), and neryl acetate (1.1%). The essential oil has been evaluated for antifungal activity against five common plant pathogenic fungi, *Alternaria alternata, Rhizoctonia salani, Curvularia lunata, Fusarium oxysporum* and *Helminthosporium oryzae* and showed good activity against the first three mentioned fungi. Fungal sporulation was inhibited in *A. alternata, R. salani,* and *F. oxysporum* (Chutia et al. 2009) (Fig.12.2).

In another study, essential oil from peels (limonene content 50.4%) showed antioxidant activity in the DPPH assay (Goyal and Kaushal, 2018). Essential oil from mandarin from Spain (Limonene content 74.4%) showed strong antimicrobial activity against *Enterococcus faecium, Staphylococcus aureus, Pseudomonas aeruginosa, Salmonella enterica,* and *Escherichia coli*. The EO at low concentration (2 μg/mL) combined with mild heat treatment (54 °C/10 min) has also shown synergistic lethal effects. Thirty volatile constituents were identified by GC-MS (Espina et al. 2011). In general, the essential oils, as well as isolated volatiles, are well reported for antifungal and antimicrobial activity.

The major component of Citrus essential oils is d-limonene (*R*-(+) enantiomer). It is extensively used as a flavor and fragrance compound in cosmetics, foods, beverages, and pharmaceuticals. d-Limonene has shown various pharmacological effects including antioxidant, anti-inflammatory, anti-cancer, anti-diabetic, antimicrobial, and several other activities (Erasto and Viljoen 2008; Crowell et al. 1994; Vieira et al. 2018). The anti-diabetic effect of d-limonene was evaluated in streptozotocin-induced diabetic male albino Wistar rats. A single intraperitoneal injection of 40 mg/kg body weight (bw) streptozotocin was used to induce diabetes. Oral administration of d-limonene at a dose of 100 mg/kg bw was found to gradually decrease plasma glucose level in 45 days. d-Limonene is also reported to decrease

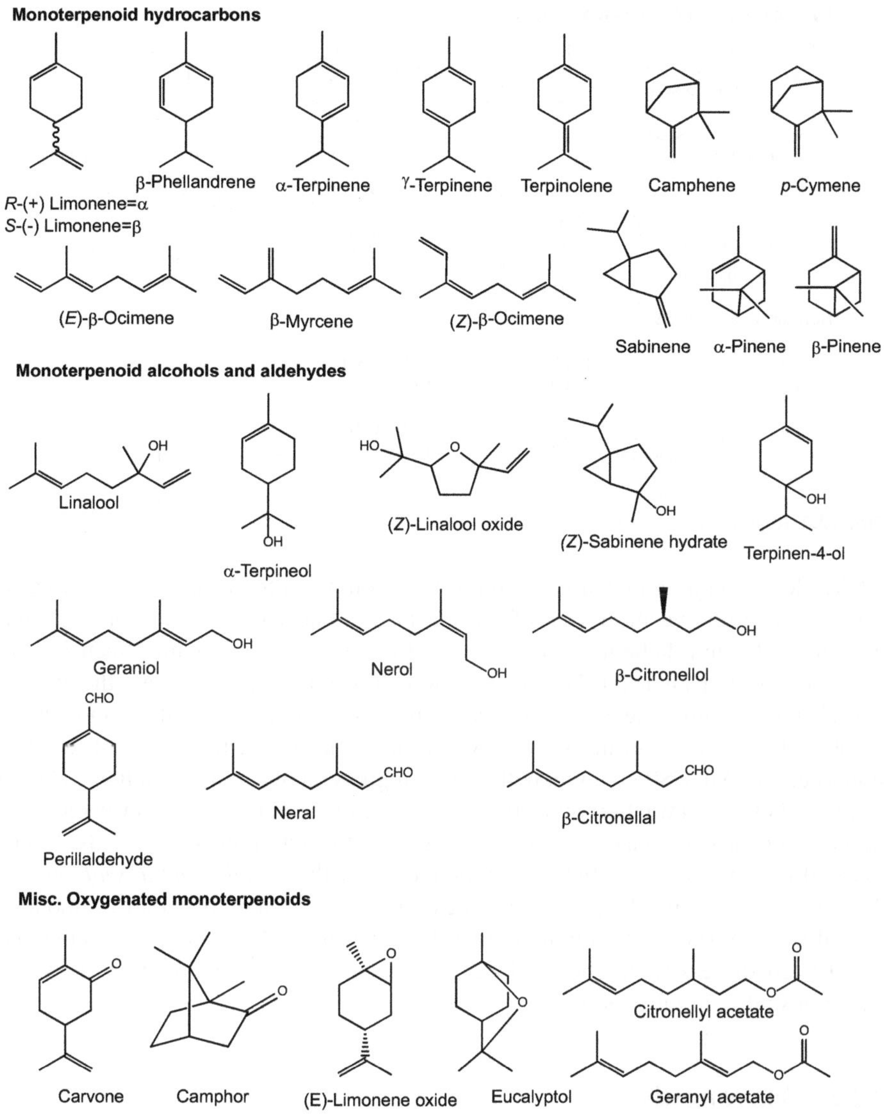

Fig. 12.2 Chemical constituents in *C. reticulata* essential oil

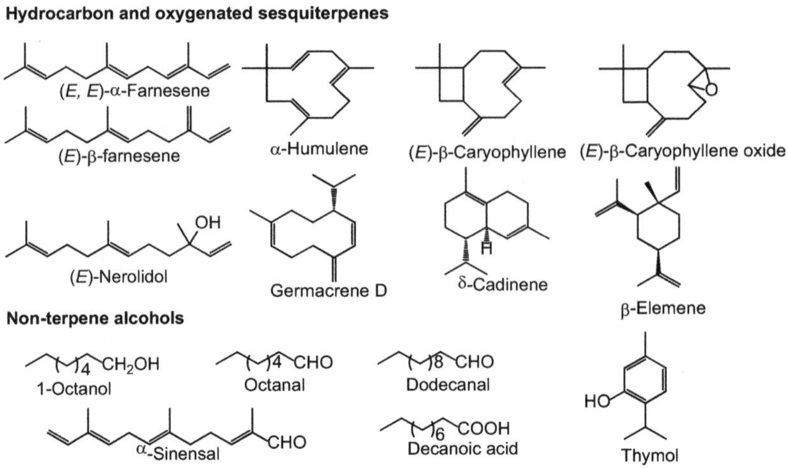

Fig. 12.2 (continued)

HbA1c levels with an increase in Hb levels as compared to normal rats. It was also observed to improve the levels of plasma insulin and carbohydrate metabolizing enzymes. The anti-diabetic effect was found to be comparable with glibenclamide (Murali and Saravanan 2012). d-Limonene is also reported as a chemotherapeutic and chemopreventive agent in chemically induced tumors in rodents. Limonene in combination with berberine showed synergistic anticancer activity in gastric carcinoma cell line MGC803 with more than 80% regression of carcinomas. It was shown to affect both the initiation and promotion stages of carcinogenesis in chemically induced rodent mammary, skin, liver, lung, and stomach tumors in rodents. Limonene also showed up to 99% inhibition of growth of *Salmonella senftenberg, E. coli, S, aureus,* and *Pseudomonas* species at a concentration of 1 ml/L. Limonene has also been shown to improve transdermal drug delivery by enhancing the penetration of several lipophilic drugs. Limonene is pharmacologically active in numerous other in vitro assays.

12.3 Flavonoids

Flavonoids are a significant group of plant phytochemicals, which are hydroxylated polyphenol compounds. They are abundantly found in fruits and vegetables and constitute an important group of dietary flavonoids. Flavonoids are divided into six classes based on differences in chemical structures and include flavones, flavanones, flavonols, anthocyanidins, flavonols, and isoflavones. Flavonoids are secondary plant metabolites containing two aromatic rings (ring A and ring B) which are linked through a pyrone or a dihydropyrone ring (ring C). Flavonoids are derived from the shikimate and phenylpropanoid metabolic pathways and contain a 15-carbon, C6-C3-C6 skeleton. The importance of flavonoids in the diet comes from their physiological and pharmacological role and their health benefits, particularly for their anti-inflammatory and antioxidant properties. Besides, citrus flavonoids are also reported to be useful in cardiovascular diseases and cancer. Citrus fruits including oranges, mandarins, grapefruits, and lemons are considered one of the main sources of dietary flavonoids, that are present as their glycosides or as aglycones (flavanones and flavones). A large number of flavonoids in citrus juices result from many different combinations possible between variously hydroxylated aglycones and a limited number of sugar molecules. The most common sugar moieties incorporated in flavonoid glycosides include D-glucose and L-rhamnose. Most of the citrus glycosides have a sugar moiety joined by a glycosidic linkage to the aglycone hydroxyl group at C-7 or C-3. In addition, a few C-glycosides have also been isolated from citrus fruits or juices. These flavonoid compounds are present in seeds, leaves, juices, stems, and peels, majorly in the fruit and peels of citrus species. Citrus peels, especially from orange and tangerine, are food industry by-products, but have been used in traditional medicine mainly for stomach problems and cough. Another large group of flavonoid compounds occurring in different citrus fruits is the polymethoxyflavones (PMFs). Citrus PMFs have exhibited various biological activities including anti-inflammation, anti-cancer, anti-microbial, and anti-atherosclerosis. PMFs containing up to seven methoxy groups have been reported from *C. reticulata* (Fig. 12.3).

Gattuso et al. have compiled the quantitative data reported in the literature on various flavonoids in juices of different citrus species. Both mandarin (*C. reticulata*) and sweet orange (*C. sinensis*) juice are reported to have similar flavonoid profiles. Mandarin juice contains hesperidin (24.3 mg/100 mL) as the major flavanone glycoside followed by narirutin (3.92 mg/100 mL), didymin (1.44 mg/100 mL), and eriocitrin (0.31 mg/100 mL) together with minor amounts of neoeriocitrin. The major polymethoxyflavone aglycones are sinensetin (1.05 mg/100 mL), tangeretin (0.26 mg/100 mL), and nobiletin (0.23 mg/100 mL) together with minor amounts of heptamethoxyflavone and quercetagetin. The aglycone acacetin has also been reported in small amounts in mandarin juice (Gattuso et al. 2007).

In another study on *C. reticulata* from Northern Thailand, Stuetz et al. compared the amounts of PMFs, flavanone glycosides, and carotenoids in hand-pressed juice and the peeled fruit from organic farms. Three flavanone glycosides, namely hesperidin, narirutin, and didymin, and three PMFs tangeretin, nobiletin, and sinensetin were identified in pulp samples. The PMFs were identified in high amounts in hand-

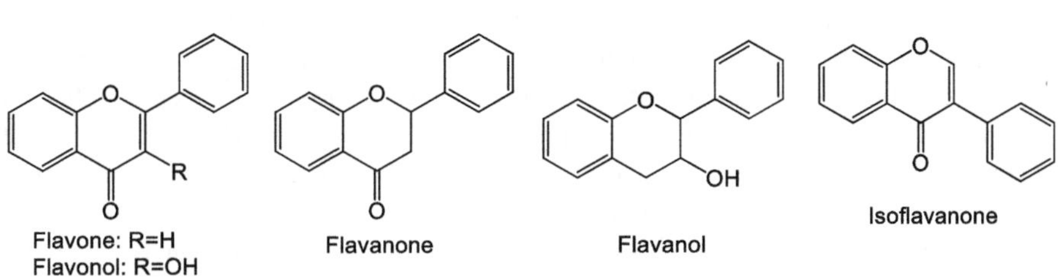

Fig. 12.3 Flavonoid skeletons sound in *C. reticulata*

pressed juice, whereas these were low in peeled fruit. It was reasoned that PMFs are located in the oil glands of peel, while flavonoid glycosides are present in fruit sacs. Several carotenoids including β-cryptoxanthin, zeaxanthin, lutein, lycopene, and β-carotene along with ascorbic acid and tocopherol were also identified and quantified (Stuetz et al. 2010). Similar results were shown in a study on ponkan fruits (*C. reticulata*) from Japan. Major PMFs identified in peels of ponkan cultivar were isosinensetin, sinensetin, tetramethyl-O-isoscutellarein, and nobiletin. The isolated compounds showed antiproliferative activity against human ovarian, breast, myeloid leukemia, and lung adenocarcinoma epithelial cancer cell lines (Du and Chen 2010). Mizuno et al. have isolated two new PMFs (tetra- and penta-methoxyflavone) together with eight known polyoxygenated flavones from the dried leaves of *C. reticulata* (Mizuno et al. 1987) (Fig. 12.4).

Liu et al. developed a rapid resolution liquid chromatography coupled with a triple quadrupole electrospray tandem mass spectrometry method for simultaneous determination of six bioactive flavonoids in the dried ripe pericarp of *C. reticulata* from six different regions of China. The identified compounds included naringin, hesperidin, nobiletin, tangeretin, and two PMFs, 3,5,6,7,8,3′,4′--heptamethoxyflavone, and 5-hydroxy-6,7,8,3′4’-pentamethoxyflavone (Liu et al. 2013). In a later study, the same group isolated eight different PMFs from peels of *C. reticulata* and investigated their biological activities. 3,5,6,7,8,3′,4’--Heptamethoxyflavone showed potent sterol regulatory element-binding proteins inhibition activity, while 5-hydroxy-6,7,8,3′4’-pentamethoxyflavone exhibited strong antiproliferative activity against MCF-7, A549, and HepG2 human cancer cell lines. The PMFs also decreased NO production; it was shown that PMFs with more methoxy groups showed higher anti-inflammatory activity (Duan et al. 2017).

Saito et al. demonstrated that major PMFs especially nobiletin potentiated the cytolytic activity of KHYG-1 natural killer leukemia cells via enhancing the expression of granzyme B, which is a serine protease with a crucial role in the cytolytic activity. Nobiletin is also reported to increase the levels of phosphorylated CREB, ERK1/2, and p38 MAPK (Saito et al. 2015).

Xu et al. observed antiviral activity of *C. reticulata* pericarp supercritical fluid extract against the human respiratory syncytial virus (RSV) A2, Long, and B strains with low IC_{50} values pointing towards PMFs as active constituents. Five PMFs were isolated and evaluated in vitro for activity; tangeretin and nobiletin showed activity better than the positive control ribavirin. Out of all five compounds, only these two had fully methoxylated ring A, underlining the importance of methoxy groups (accompanying hydrophobicity) for the PMFs to show activity. Further, tangeretin was shown to inhibit plaque formation in Hep-2 cells in a dose- dependent manner. Tangeretin did not inhibit the attachment of RSV absorption to Hep-2 cells but inhibited the penetration suggesting a possible mode of action, which was different from ribavirin. Tangeretin also decreased the RSV P protein expression in a dose-dependent manner (Xu et al. 2015).

In a comprehensive study, Wang et al. studied the antioxidant capacity, anticancer ability, and flavonoid composition of 35 varieties of *C. reticulata*. The authors identified a total of 39 flavonoids including 4 flavones, 9 flavanones, and 26 PMFs.

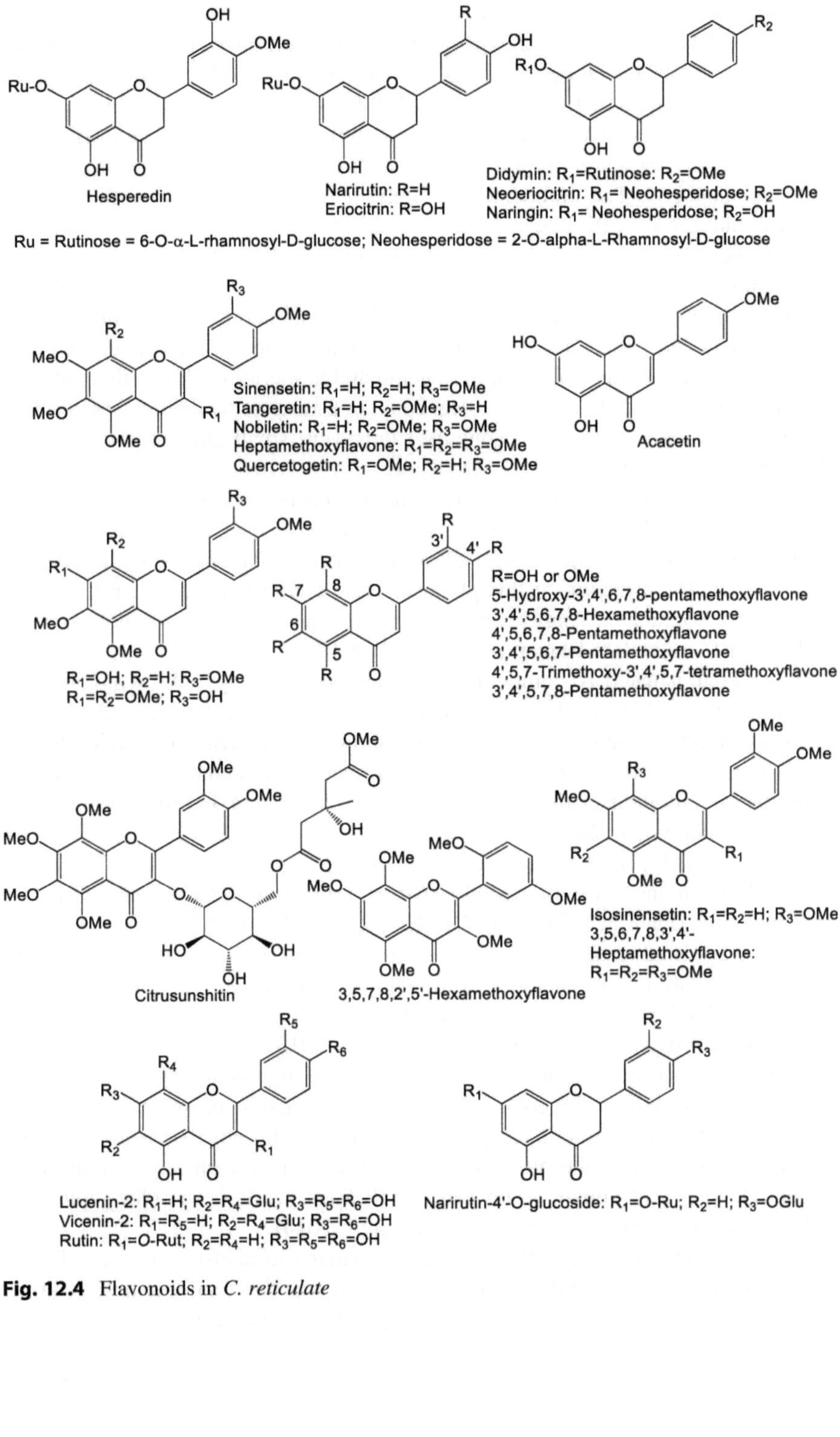

Fig. 12.4 Flavonoids in *C. reticulate*

The correlation analysis suggested phenolics to be responsible for antioxidant activity and PMFs for activity against gastric cancer cell lines (Wang et al. 2017).

Xing et al. 2017 developed a fast and efficient UPLC-QTOF-MS method for the identification and quantitative analysis of 42 PMFs with a run time of 12 min. *Citrus reticulata* peels were shown to contain more than 20 PMFs. The PMFs were identified based on mass fragmentation patterns and also by comparison with authentic standards. The number of methoxy or hydroxyl groups could be determined by adding $n \times 16$ or $n \times 30$ to the molecular weight (222 Da) of the core flavone skeleton, for example, a peak at m/z of 403 (M + 1) indicated a hexamethoxyflavone (222 + (30 × 6) (Xing et al. 2017). In another study, five flavonoids including naringin, hesperidin, didymin, tangeretin, and nobiletin were determined in the peel, pith, endocarp, pulp, and the seed of *C. reticulata.* Pith contained the maximum amount of naringenin and hesperidin, while pulp contained the maximum amount of hesperidin. Peel contained maximum amounts of tangeretin and nobiletin, while other parts contained only very small amounts of these two PMFs (Sun et al. 2010).

Zhong et al. isolated a new glycoside named citrusunshitin together with five other PMFs including nobiletin, sinensetin, isosinensetin, 3,5,6,7,8,3′4′--heptamethoxyflavone, and 3,5,6,7,8,2′,5′-hexamethoxyflavone from the peels of *C. reticulata.* The structure of this glycoside was established by spectral data and X-ray crystallography. These compounds were proposed as chemotaxonomic markers for *C. reticulata* Blanco (Zhong et al. 2016).

Uckoo et al. isolated 10 compounds including several polymethoxyflavones and limonin from the dried peels of Miaray mandarin and evaluated these for biofilm inhibitory activity. The compounds did not show any bactericidal activity against *Vibrio harveyi* at 50 μM concentration. However, hexamethoxyflavone showed inhibition of autoinducer-mediated cell-cell signaling and biofilm formation in *V. harvey* (Uckoo et al. 2015).

In another study, Barreca et al. isolated several polymethoxy, C- and O-glycosyl flavonoids from the juice of tangelo (*Citrus reticulata* × *Citrus paradisi*) and evaluated antioxidant activity using DPPH, $O2^{\bullet -}$, $OH^{\bullet}$, and ABTS radical assays. Didymin was found to be the most abundant compound followed by narirutin. The crude juice showed strong activity in the DPPH assay, ferric reducing antioxidant assay, and ABTS assay. A fraction containing all detected flavonoids showed weak antimicrobial activity at 1.0 mg/mL (Barreca et al. 2013).

Citrus PMFs show inhibition of carcinogenesis through various mechanisms like cell cycle arrest and angiogenesis. Mechanisms of action of several PMFs and flavonoids including nobiletin, tangeretin, and hesperidin are reviewed. Nobiletin and tangeretin inhibited the progression of cell cycle in G1 phase for both breast cancer cell lines and human colon cancer cell lines, while hesperidin suppressed proliferation by promoting apoptosis (Wang et al. 2014).

Gao et al. reviewed the activities and molecular mechanisms of citrus PMFs. Sources of 78 PMFs are compiled in the review. It was observed that free hydroxyl groups in polyphenolics restrict their absorption and accelerate metabolism and excretion. Therefore, less polar PMFs show high permeability and are easily absorbed, while those with a higher number of methoxy groups show higher biological activities due to their hydrophobic nature. Nobiletin and tangeretin are the most studied PMFs for their biological activities. Both these compounds led to a decrease in serum LDL cholesterol, and serum and liver triglycerides in the hamster model. The hypolipidemic activity was also observed in different animal models. Tangeretin has also shown anti-diabetic activity in several animal models. It stimulates glucose uptake. In HFD-induced diabetic mice, tangeretin at a dose of 200 mg/kg activated AMPK resulting in glucose uptake and improved obesity-induced glucose intolerance. Oral intake of tangeretin also led to a decrease in plasma glucose and glycosylated hemoglobin in streptozotocin-induced diabetic rats. Nobiletin also improved hyperglycemia and insulin resistance in diabetic *ab/ab* mice. Several PMFs have also been shown to prevent diet-induced obesity. Nobiletin was also shown to interfere with LPS and cyclo-oxygenase-induced prostaglandin E2 production. Both nobiletin and tangeretin have the potential to be developed as nutraceuticals. More striking is the role of PMFs to exhibit anti-cancer activity. PMFs can absorb UV light and protect against DNA damage. Tangeretin has exhibited anti-proliferative activities against several cancer cell lines that include breast, colon, lung, gastric and human non-small-cell lung cancer cells. Nobiletin has also shown inhibition of breast cancer and hepatocellular carcinoma cell lines. Tangeretin and nobiletin are reported to block cell cycle progression at mitosis interphase. Several PMFs have also shown anti-angiogenesis activity in the zebrafish model. Nobiletin has also shown growth inhibition of metastasis of human gastric adenocarcinoma and nasopharyngeal carcinoma cell lines. Abundant literature is available on the mechanism of anti-cancer activity, which is not discussed in this chapter. Additionally, PMFs are also known to possess antibacterial, antifungal, and antiviral activities (Gao et al. 2018).

Besides PMFs nobiletin and tangeretin, the major flavonoid hesperidin is also reported for several activities. Hesperidin showed anti-inflammatory activity comparable to indomethacin in vivo, though the doses were different. Pretreatment of male or female albino Wistar rats at a dose of 50 and 100 mg/kg, s.c. reduced carrageenan-induced paw oedema by 47% and 63%. Analgesic activity similar to indomethacin was also observed with hesperidin in the abdominal constriction test, but no analgesic activity was observed in the tail-flick assay. Hesperidin administration did not produce any gastric mucosal injury whereas indomethacin caused severe ulceration of gastric mucosa (Emim et al. 1994). Hesperidin showed an anti-atherosclerotic effect in high-fat diet-fed LDLr$^{-/-}$ mice mediated through improving insulin resistance, antioxidative effects, lipid profiles, and anti-inflammatory action. Hesperidin ameliorated HFD-induced metabolic syndrome, and hepatic steatosis inhibited atherosclerotic lesion formation, inhibited macrophage foam cell formation, reduced oxidative stress, and reduced inflammation in HFD-fed LDLr$^{-/-}$ mice (Sun et al. 2017). It showed in vitro antifungal activity against *Botrytis cinerea, Trichoderma glaucum* and *Aspergillus fumigatus* at a dose of 1–10 mg. Hesperidin showed anti-viral activity against virus-vesicular somatitis, influenza virus, herpes type-I, polio virus type-I, HSV,-1, RSV, and rotavirus. Hesperidin alone or in combination with diosmin has shown chemotherapeutic as well as chemopreventive activity in chemically induced tumors in animal models. Hesperidin has also shown inhibition of platelet and cell aggregation. In a study on horses, hesperidin reduced the aggregation of erythrocytes. In addition, several other biological activities of hesperidin have been reviewed (Garg et al. 2001).

The variations in levels of six flavonoids including PMFs in Pericarpium *Citri reticulate* (PCR) were investigated during a three year storage period at 25 °C by HPLC-PDA. The results showed a decrease in content of hesperidin, while the contents of all five PMFs increased with time. This observation suggested that longer storage durations improve quality as the amounts of active constituents increase with time (Fu et al. 2017).

12.4 Coumarins and Furanocoumarins

Coumarins (benzo-α-pyrones) and furanocoumarins in citrus fruits are associated with plant defense against pathogens. The pathogens in citrus fruits easily proliferate due to high nutrient content, humidity, and low pH. In general, plants produce antimicrobial secondary metabolites (phytoalexins) to protect themselves from microbial damage. In a study on citrus fruits grown in Colombia, Ramirez-Pelayo et al. identified two furanocoumarins, 5-methoxypsoralen (bergapten) and 5,8-dimethoxypsoralen (isopimpinellin) in the methanolic extract of peels of *C. reticulata*. The two furanocoumarins were quantified in the peels. The isolated compounds from different citrus species were evaluated for antifungal activity against *Colletotrichum* sp. and it was found that furanocoumarins showed better activity than coumarins. Besides, some other coumarins were tentatively identified. Furanocoumarins are known to be phototoxic compounds and can cause dermatitis after skin contact or ingestion (Ramírez-Pelayo et al. 2019). Phetkul et al. isolated a few more alkylated and functionalized coumarins from bark, peels, and leaves. 7-Hydroxy-5-prenyloxy coumarin, crenulatin, 8(3′-b-glucosyl-2-'hydroxy-3′--methyl)butyl-7-methoxycoumarin, and scopoletin were isolated and characterized (Phetkul et al. 2014). Two coumarin compounds, scoparone and xanthyletin, were isolated from the *n*-hexane extract of stem bark that showed cytotoxicity against three cancer cell lines. Scoparone showed low IC50 values against three tested cell lines A549, MCF7, and PC3 (Tahsin et al. 2017). Five Geranyloxycoumarin together with two known coumarins xanthotoxin and 7-geranyloxycoumarin were isolated from the peels of *C. reticulata* procured from Karachi (Saleem et al. 2005) (Fig. 12.5).

12.5 Miscellaneous Compounds

Several other secondary metabolites have also been reported from different parts of *C. reticulata*. Phetkul et al. isolated a new compound named depcitrus A from high polarity fractions from dried branch bark after their methylation. Several other

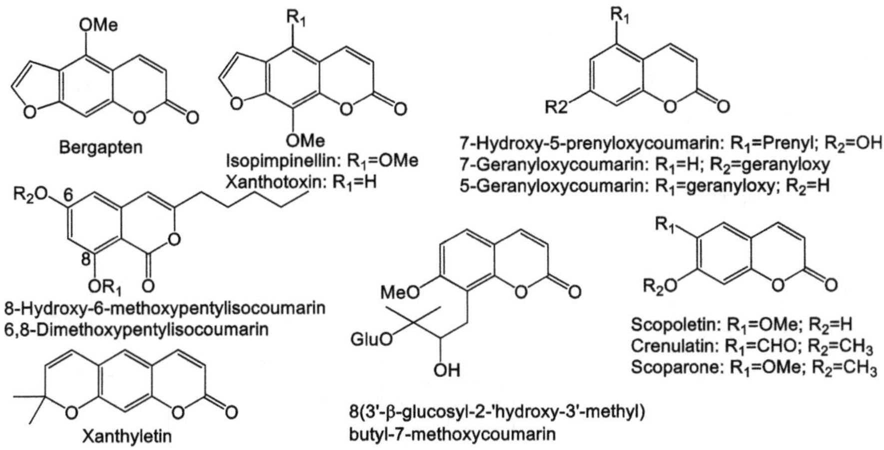

Fig. 12.5 Coumarins and furanocoumarins in *C. reticulata*

known compounds were isolated from the methylene chloride and acetone extracts of peels, leaves, and branch barks. A total of 32 compounds were isolated and several of these were evaluated for anti-oxidant, antimicrobial, and cytotoxic activities (Phetkul et al. 2014). A simple benzoic acid derivative named reticulatal was isolated and characterized (Saleem et al. 2005). Three new steroidal glycosides, namely reticulate ursenoside, citrusteryl arachidate, and citrulanosteroside were isolated and characterized from the peels of *C. reticulata* (Khan et al. 2010). This chapter lists some of the important compounds from *C. reticulata*; several other compounds are also reported from various parts (Fig. 12.6).

12.6 Conclusions

Importance of citrus in global economy could not be over emphasized as the average citrus production between 2007 and 2017 was approximately ten million tons (Raspo et al. 2020). Mandarin orange (*Citrus reticulata*) in particular is an important source of vitamin C, essential oil, and secondary metabolites with nutraceutical relevance. The review presented in this chapter has provided over-whelming information to substantiate the importance of *C. reticulata* as a great source of diverse chemicals and as a raw material for nutraceutical product development. Emerging reports on the biological importance of its chemicals will further expand the industrial utilization of this species. Due to its versatile importance, post-harvest handling, and quality control could not be over-looked, and more research is expected to be conducted and reported. Wide global distribution and abundance of supply further strengthen its potential utilization as an ideal source of nutraceutical and industrial commodity in product development.

Acknowledgements The authors are thankful to Director, NIPER-SAS Nagar for support.

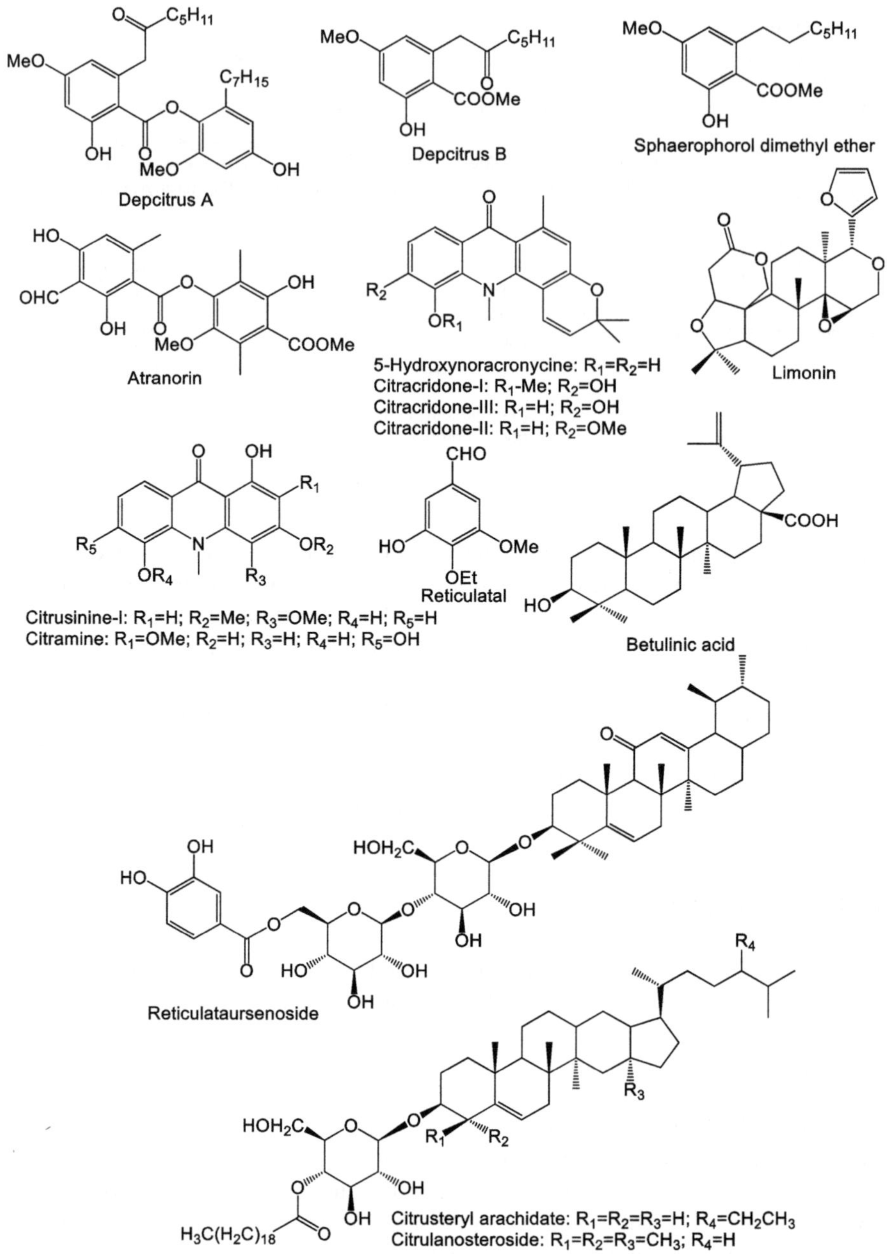

Fig. 12.6 Miscellaneous compounds in *C. reticulata*

References

Ashoor SH, Bernhard RA (1967) Isolation and characterization of terpenes from *Citrus reticulata* and their comparative distribution among other citrus species. J Agric Food Chem 15:1044–1047

Barreca D, Bisignano C, Ginestra G et al (2013) Polymethoxylated, C-and O-glycosyl flavonoids in tangelo (*Citrus reticulata* × *Citrus paradisi*) juice and their influence on antioxidant properties. Food Chem 141:1481–1488

Barreca D, Mandalari G, Calderaro A et al (2020) Citrus flavones: an update on sources, biological functions, and health promoting properties. Plan Theory 9:288

Chutia M, Bhuyan PD, Pathak MG et al (2009) Antifungal activity and chemical composition of *Citrus reticulata* Blanco essential oil against phytopathogens from north East India. LWT-Food Sci Technol 42:777–780

Crowell PL, Elson CE, Bailey HH et al (1994) Human metabolism of the experimental cancer therapeutic agent d-limonene. Cancer Chemother Pharmacol 35:31–37

Du Q, Chen H (2010) The methoxyflavones in *Citrus reticulata* Blanco cv. Ponkan and their antiproliferative activity against cancer cells. Food Chem 119:567–572

Duan L, Dou LL, Yu KY et al (2017) Polymethoxyflavones in peel of *Citrus reticulata* 'Chachi' and their biological activities. Food Chem 234:254–261

Emim JADS, Oliveira AB, Lapa AJ (1994) Pharmacological evaluation of the anti-inflammatory activity of a citrus bioflavonoid, hesperidin, and the isoflavonoids, duartin and claussequinone, in rats and mice. J Pharm Pharmacol 46:118–122

Erasto P, Viljoen AM (2008) Limonene-a review: biosynthetic, ecological and pharmacological relevance. Nat Prod Commun 3:1193–1202

Espina L, Somolinos M, Lorán et al (2011) Chemical composition of commercial citrus fruit essential oils and evaluation of their antimicrobial activity acting alone or in combined processes. Food Control 22:896–902

Fu M, Xu Y, Chen Y et al (2017) Evaluation of bioactive flavonoids and antioxidant activity in pericarpium Citri Reticulatae (*Citrus reticulata* 'Chachi') during storage. Food Chem 230:649–656

Gao Z, Gao W, Zeng SL et al (2018) Chemical structures, bioactivities and molecular mechanisms of citrus polymethoxyflavones. J Funct Foods 40:498–509

Garg A, Garg S, Zaneveld JD et al (2001) Chemistry and pharmacology of the citrus bioflavonoid hesperidin. Phytother Res 15:655–669

Gattuso G, Barreca D, Gargiulli C et al (2007) Flavonoid composition of citrus juices. Molecules 12:1641–1673

González-Mas MC, Rambla JL, López-Gresa MP et al (2019) Volatile compounds in citrus essential oils: a comprehensive review. Front Plant Sci 10:12

Goyal L, Kaushal S (2018) Evaluation of chemical composition and antioxidant potential of essential oil from *Citrus reticulata* fruit peels. Adv Res 15:1–9. https://doi.org/10.9734/AIR/2018/41981

Khan MA, Ali M, Alam P (2010) Phytochemical investigation of the fruit peels of *Citrus reticulata* Blanco. Nat Prod Res 24:610–620

Liu EH, Zhao P, Duan L et al (2013) Simultaneous determination of six bioactive flavonoids in *Citri reticulatae* pericarpium by rapid resolution liquid chromatography coupled with triple quadrupole electrospray tandem mass spectrometry. Food Chem 14:3977–3983

Mabberley DJ (1997) A classification for edible citrus (Rutaceae). Telopea 7:167–172

Mizuno M, Matoba Y, Tanaka T et al (1987) Two new flavones in *Citrus reticulata*. J Nat Prod 50: 751–753

Murali R, Saravanan R (2012) Antidiabetic effect of d-limonene, a monoterpene in streptozotocin-induced diabetic rats. Biomed Prev Nutr 2:269–275

Phetkul U, Phongpaichit S, Watanapokasin R et al (2014) New depside from *Citrus reticulata* Blanco. Nat Prod Res 28:945–951

Ramírez-Pelayo C, Martínez-Quiñones J, Gil J et al (2019) Coumarins from the peel of citrus grown in Colombia: composition, elicitation and antifungal activity. Heliyon 5:e01937. https://doi.org/10.1016/j.heliyon.2019.e01937

Raspo MA, Vignola MB, Andreatta AE et al (2020) Antioxidant and antimicrobial activities of citrus essential oils from Argentina and the United States. Food Biosci 36:100651

Saito T, Abe D, Nogata Y (2015) Polymethoxylated flavones potentiate the cytolytic activity of NK leukemia cell line KHYG-1 via enhanced expression of granzyme B. Biochem Biophys Res Commun 456:799–803

Saleem M, Afza N, Aijaz Anwar M et al (2005) Aromatic constituents from fruit peels of *Citrus reticulata*. Nat Prod Res 19:633–638

Sawamura M, Thi Minh TN, Onishi Y et al (2004) Characteristic odor components of *Citrus reticulata* Blanco (Ponkan) cold-pressed oil. Biosci Biotechnol Biochem 68:1690–1697

Stuetz W, Prapamontol T, Hongsibsong S et al (2010) Polymethoxylated flavones, flavanone glycosides, carotenoids, and antioxidants in different cultivation types of tangerines (*Citrus reticulata* Blanco cv. Sainampueng) from Northern Thailand. J Agric Food Chem 58:6069–6074

Sun Y, Wang J, Gu S et al (2010) Simultaneous determination of flavonoids in different parts of *Citrus reticulata* 'Chachi' fruit by high performance liquid chromatography—photodiode array detection. Molecules 15:5378–5388

Sun YZ, Chen JF, Shen LM et al (2017) Anti-atherosclerotic effect of hesperidin in LDLr−/− mice and its possible mechanism. Eur J Pharmacol 815:109–117

Tahsin T, Wansi JD, Al-Groshi A et al (2017) Cytotoxic properties of the stem bark of *Citrus reticulata* Blanco (Rutaceae). Phytother Res 31:1215–1219

Tripoli E, La Guardia M, Giammanco S et al (2007) Citrus flavonoids: molecular structure, biological activity and nutritional properties: a review. Food Chem 104:466–479

Uckoo RM, Jayaprakasha GK, Vikram A et al (2015) Polymethoxyflavones isolated from the peel of Miaray mandarin (*citrus miaray*) have biofilm inhibitory activity in *Vibrio harveyi*. J Agric Food Chem 63:7180–7189

Vieira AJ, Beserra FP, Souza MC et al (2018) Limonene: aroma of innovation in health and disease. Chem Biol Interact 283:97–106

Wang L, Wang J, Fang L et al (2014) Anticancer activities of citrus peel polymethoxyflavones related to angiogenesis and others. Biomed Res Int 2014:1. https://doi.org/10.1155/2014/453972

Wang Y, Qian J, Cao J et al (2017) Antioxidant capacity, anticancer ability and flavonoids composition of 35 citrus (*Citrus reticulata* Blanco) varieties. Molecules 22:1114

Xing TT, Zhao XJ, Zhang YD et al (2017) Fast separation and sensitive quantitation of polymethoxylated flavonoids in the peels of citrus using UPLC-Q-TOF-MS. J Agric Food Chem 65:2615–2627

Xu JJ, Liu Z, Tang W et al (2015) Tangeretin from *Citrus reticulata* inhibits respiratory syncytial virus replication and associated inflammation in vivo. J Agric Food Chem 63:9520–9527

Zhong WJ, Luo YJ, Li J et al (2016) Polymethoxylated flavonoids from *Citrus reticulata* Blanco. Biochem Syst Ecol 68:11–14

Recent Progress in the Pharmacology of Phytoestrogens: Emerging Neuromodulators for Treating Anxiety and Depression

13

Lovedeep Singh and Rajbir Bhatti

Abstract

Phytoestrogens are plant-derived non-steroidal xenoestrogens that possess an ability to bind with endogenous estrogen receptors and mildly mimic the estrogenic activity of endogenous estrogens. They are commonly found in soy, beans, red clover, sprouting plants, etc. Several preclinical studies have documented that phytoestrogen improves health and brain functions. As per the WHO, depression and anxiety are significant contributors to the global burden of non-fatal diseases. Depression is ranked as one of the largest contributors to a disability, whereas anxiety disorders are ranked sixth among the global burden of non-fatal diseases. Till now there have been several synthetic treatments for depression and anxiety. However, they present a lot of problems that affect the daily lives of the patients. Therefore, scientists are now focusing on developing plant-derived drugs to exclude these problems. Over the last decade, phytoestrogens have gained a lot of attention due to the lower prevalence of chronic diseases in the Japanese and Chinese populations taking soy isoflavones in their diet. The promising benefits of various phytoestrogens and the role of different mediators in mediating the antidepressant and anxiolytic effects of phytoestrogens have been widely explored. Inhibition of monoamine oxidase-A (MAO-A), increased serotonergic transmission, activation of the gamma-aminobutyric acid-A receptor ($GABA_A$ receptor), oxidative stress reduction, glucocorticoid secretion, and inhibition of

L. Singh (✉)
Department of Pharmaceutical Sciences, Guru Nanak Dev University, Amritsar, Punjab, India

University Institute of Pharma Sciences, Chandigarh University, Mohali, Punjab, India
e-mail: lovedeep.e13279@cumail.in

R. Bhatti
Department of Pharmaceutical Sciences, Guru Nanak Dev University, Amritsar, Punjab, India
e-mail: rajbir.pharma@gndu.ac.in

P. P. Singh (ed.), *Recent Advances in Pharmaceutical Innovation and Research*,
https://doi.org/10.1007/978-981-99-2302-1_13

cytokines release, etc. are the major modulations produced by the phytoestrogens in eliciting antidepressant and anxiolytic effect. This chapter is designed to discuss the role and possible pathways leading to the antidepressant and anxiolytic effects of best-researched phytoestrogens.

Keywords

Anxiety · Biochanin-A · Coumestrol · Daidzein · Depression · Genistein · Kaempferol

13.1 Introduction

Phytoestrogens are plant-derived nonsteroidal compounds having structural similarity with the endocrine hormone 17-β-estradiol. These are xenoestrogens as they possess an ability to bind with the same receptor with which endogenous estradiol does (Yildiz 2005). They have been classified into three types, that is, isoflavones (genistein, daidzein, biochanin-A), coumestans (Coumestrol, repensol), and lignans (enterodiol, enterolactone). Isoflavones and coumestans are the most active and best-researched groups amongst all phytoestrogens. Isoflavones are commonly found in soy, beans, and red clover, whereas coumestans are found in sprouting plants (Price and Fenwick 1985; Glazier and Bowman 2001). Over the last decade, phytoestrogens have gained much attention and are also recognized as food ingredients associated with the increased life expectancy of the Japanese population consuming soy isoflavones among other lifestyle factors (Tsugane 2021). Epidemiological studies had elucidated that Japanese migrants in the USA had a greater incidence of developing cancers. The deficiency/reduced phytoestrogens-containing diet was considered a confounding factor in the development of cancers of the breast, prostate as well as other western diseases (Kolonel et al. 1986; Barnes et al. 1995; Glazier and Bowman 2001). This potential benefit was considered a fundamental reason for the development of various isoflavone-containing drinks and supplements (Glazier and Bowman 2001; Rietjens et al. 2017). Various health-related benefits of phytoestrogens have been reported in the literature, including reduced risk of menopausal symptoms (hot flushes and osteoporosis), heart disease, diabetes, obesity, brain function disorders, and cancers (Eden 2012; Lethaby et al. 2013; Karahalil 2006; Cederroth and Nef 2009; Patisaul and Jefferson 2010).

Depression and anxiety are amongst the most prevalent illnesses affecting the brain globally (Tiller 2013; WHO 2017). According to the WHO, more than 300 million people are suffering from depression. An 18% increase in patients suffering from depression has been observed over the past 10 years. Further, these disorders were found to be more prevalent in women than men. The suicide rate is very high in depressed patients and approximately 8 lakh suicidal deaths are reported every year (WHO 2017). It has been estimated that about 3.6% of the global population is suffering from various types of anxiety disorders which are also more common among females compared to males, that is, 4.6% of females to 2.6% of males (WHO 2017). Newer generation antidepressants are widely used as the first-line treatment for depressive disorders. However, they also present a lot of adverse events, including cardiovascular disturbances, genitourinary symptoms, sexual dysfunction, osteoporosis, hyponatremia, hepatotoxicity, lowering of seizure threshold, cognitive troubles, hyperprolactinemia, gastrointestinal symptoms such as nausea, diarrhea, and dyspepsia (Carvalho et al. 2016).

Pre-clinical studies have shown the potential benefits of various phytoestrogens in anxiety and depression (Kageyama et al. 2010; Kalandakanond-Thongsong et al. 2007; Wu et al. 2017). Several plants, such as *Opuntia ficus indica var.* saboten, *Apocynum venetum*, and *Macrotyloma uniflorum*, are well-documented sources of different phytoestrogens (Park et al. 2010; Yan et al. 2015). Antidepressant and anti-anxiety effects of different phytoestrogens have been explored by several investigators (Sloley et al. 2000; Wang et al. 2008; Phadnis et al. 2018). However, the underlying mechanism for the antidepressant and anxiolytic effects of different phytoestrogens has not been explored yet. There is a need to pay considerable attention to exploring the various underlying mechanisms involved in the actions of different phytoestrogens in depression and anxiety. Therefore, this chapter focuses on exploring the possible pathways leading to the antidepressant and anxiolytic effects of undertaking phytoestrogens.

13.2 Possible Mediators and Their Interplay Leading to the Anti-Depressant Effect of Different Phytoestrogens

13.2.1 Genistein

Genistein ($C_5H_{10}O_5$) is a naturally occurring 4, 5, 6-trihydroxy isoflavone, which closely resembles 17- β-estradiol, structurally (Jaiswal et al. 2019). Genistein has shown promising results in combating various ailments. Genistein has been documented to reduce the risk of breast cancer, prostate cancer, the incidence of type 2 diabetes, high blood pressure, and osteoporosis in women (Dong and Qin 2011; Ding et al. 2016; Taku et al. 2010; van Die et al. 2014). Several preclinical studies have also explored the antidepressant potential of genistein (Atteritano et al. 2014; Kageyama et al. 2010; Shen et al. 2018). Atteritano et al. conducted a double-blind, randomized study in osteopenic postmenopausal women. Out of the total, 139 received genistein tablets (54 mg/day), and 123 received a placebo tablet. A short questionnaire and Zung self-rating scale were used to study the mental, physical, emotional, vitality, and pain problems after 1 year and after completing the treatment period at the end of the intervention period which was 2 years. A significant improvement was observed in the genistein-treated group in comparison to the placebo group of postmenopausal women at the end of the intervention period, thereby indicating the role of genistein in improving quality of life after menopause and of sex hormones in maintaining health integrity (Atteritano et al. 2014). A pre-clinical study revealed that genistein regulates the serotonergic pathway. A forced swim test (FST) trial was used to induce exertion, which significantly reduced the hippocampal serotonin level as indicated by the enhanced 5-hydroxy-indole acetic acid 5-hydroxy-tryptamine ratio (5-HIAA/5-HT) in ovariectomized rats. Treatment for 14 days with genistein (10 mg/kg) significantly downregulated the 5-HIAA/5-HT ratio and immobility duration in FST, suggesting antidepressant action of genistein via modulating the serotonin level (Kageyama et al. 2010). Support for the study above came from another report, which also showed the anti-depressant effect of genistein (at doses of 5, 15, and 45 mg/kg) on its chronic administration to mice for 3 weeks. A significant reduction in the immobility time was observed in the FST and tail suspension test (TST) in comparison to the vehicle-treated group. It has been delineated that higher dose (45 mg/kg) results were approximately similar to positive control imipramine (15 mg/kg). Furthermore, an increased level of serotonin (5-HT) and noradrenaline (NA) in the hippocampus and frontal cortex was observed in genistein-treated mice in a dose-dependent manner. A concomitant suppression of MAO activity was also evidenced with genistein treatment (Hu et al. 2017). Repeated post-treatment with 5-HT depleter, para-chlorophenylalanine (PCPA) for 5 days after 21 days of genistein pretreatment showed a significant reversal in the antidepressant effect of genistein. However, NA depleter, N-(2-chloroethyl)-N-ethyl-2-bromobenzylamine (DSP-4) did not affect the antidepressant effect of genistein, suggesting that 5-HT but not NA has a role in a genistein-induced anti-depressant effect. Pretreatment with WAY-100635, a 5-HT_{1A} receptor antagonist, significantly abolished the genistein-induced anti-depressant effect. In contrast, ritanserin and ondansetron (5-$HT_{2A/C}$ and 5-HT_3 receptor antagonists, respectively) did not affect the anti-depressant action of genistein, further indicating the implications of 5-HT_{1A} receptor in genistein-induced anti-depressant effect (Hu et al. 2017). Furthermore, one report demonstrated the MAO-A and MAO-B inhibiting potential of genistein on recombinant human isoenzyme MAO-A and MAO-B. Genistein exhibited an inhibitor constant *(Ki)* value of 1.45 μM for human MAO-B (hMAO-B) and 4.31 μM for human MAO-A (hMAO-B). However, genistein was found to be more potent in inhibiting hMAO-A as compared to the standard deprenyl. It has also been reported that genistein inhibited the hMAO-B tyramine oxidation more than the hMAO-A, leading to the reduced generation of toxic H_2O_2 (Zarmouh et al. 2016).

Shen et al. (2018) also demonstrated genistein's role in alleviating major depression in unpredictable chronic mild stress mouse models (UCMS). Genistein treatment significantly increased nest-building score, frequency, and grooming time in nest building and splash tests, respectively, compared to vehicle-treated UCMS mice. Genistein also decreased and increased the mice's expression of miR-221/222 targets and connexin-43 (Cx43) in the prefrontal cortex, respectively. In support of the report above, a study on the human breast tumor cell line delineated that genistein treatment significantly enhanced the level of Cx-43 (Conklin et al. 2007). Various reports have demonstrated that Cx-43 is pathologically implicated in depression (Miguel-Hidalgo et al. 2014; Nagy et al. 2016). The reduced level of Cx-43 in the orbitofrontal cortex of depressed patients and post-mortem studies on suicides due to depression has also been postulated (Miguel-Hidalgo et al. 2014; Nagy et al. 2016). Moreover, various research groups have also investigated the effect of antidepressants on the levels of Cx-43 (Fatemi et al. 2008; Sun et al. 2012;

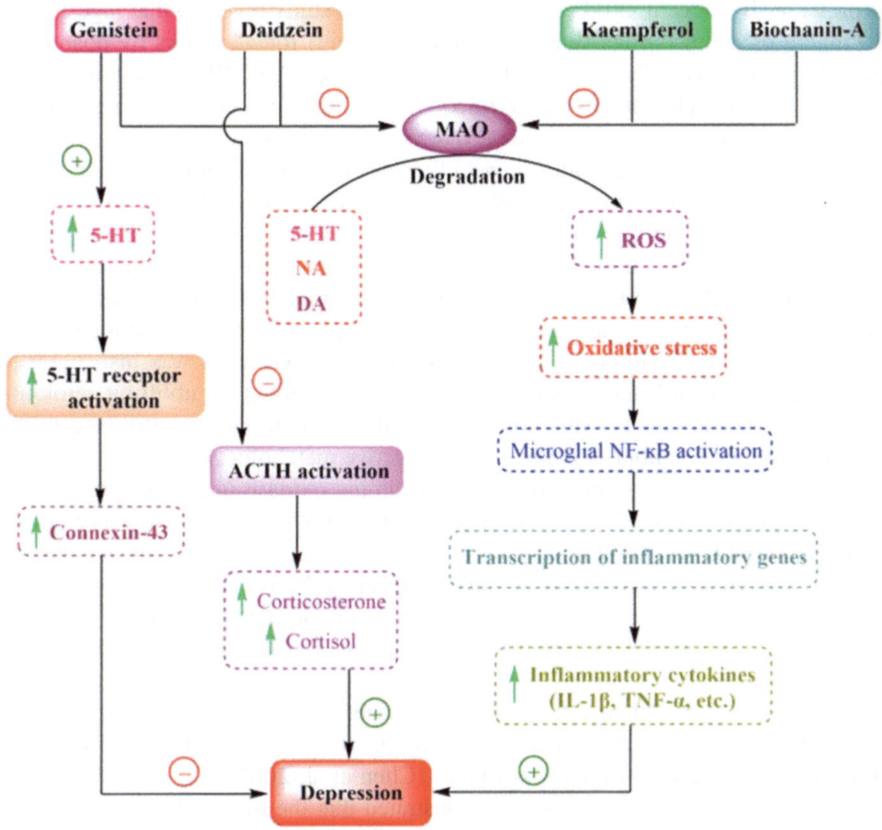

Fig. 13.1 The various modulations produced by different phytoestrogens leading to an anti-depressant effect: implication of 5-HT receptor, ACTH, microglial NF-Kb, and inflammatory cytokines

Mostafavi et al. 2014). An increased Cx-43 protein and mRNA expression have been detected in the prefrontal cortex of rats chronically treated with duloxetine (Fatemi et al. 2008; Sun et al. 2012). It has been reported that fluoxetine treatment also up-regulated the level of Cx-43 in the prefrontal cortex of rat brain (Sun et al. 2012; Fatemi et al. 2008) and human astrocytoma cell culture (Mostafavi et al. 2014). These findings are summarized in Fig. 13.1.

13.2.1.1 Daidzein and Its Metabolite

Daidzein is a 4′,7-dihydroxyisoflavone exclusively found in the roots of kudzu wine, soybeans, fava beans, and other legumes (Veitch 2007; Harsha Pedapati et al. 2018). Daidzein has been reported to have weak estrogenic and anti-estrogenic activity. It also possesses an antioxidant, anticarcinogenic, and antiatherogenic activity along with some protection against osteoporosis (Aldred 2008). Studies have also explored the antidepressant potential of daidzein; and reported promising results in abrogating depression (Blake et al. 2011; Aso 2010). A clinical study revealed that ingestion of 30 mg/kg of equol (an active metabolite of daidzein) for 12 weeks by postmenopausal non-producers showed an anti-depressant effect (Aso 2010). The participants who have excreted less than 10 μg/L of equol within 24 h via urine were categorized as non-producers (Aso 2010). It has been documented that equol treatment decreases the immobility time in FST and increases the level of 5-HT in natural ovarian failure (NOF) in female Long-Evans rats (Blake et al. 2011). Blake et al. demonstrated that all female rats at the age of around 295 days experienced NOF. At the age of 353 days (approx.), animals taking a low phytoestrogen-containing diet were exposed to equol (5 mg/kg/day) and continued till animals attained the age of 365. A significant reduction in immobility time in FST was observed after the seventh day of equol treatment. Moreover, an increase in the 5-HT level was also evident, suggesting that equol may produce its antidepressant effect by uplifting serotonin levels (Blake et al. 2011). This upregulation of serotonin may be attributed to the MAO inhibitory potential of daidzein and its active metabolite, equol. Some reports have shown the in-vitro mitochondrial monoamine oxidase inhibitory potential of daidzein (Rooke et al. 2000; Gao et al. 2001). Another study also reported the monoamine oxidase inhibitory potential of equol, a metabolite of daidzein isolated from bovine wine, showed an IC_{50} value of 158 μM. Moreover, its methylated and carbazole derivatives have shown superior monoamine oxidase inhibitory activity with an IC_{50} value of 28 and 16 μM as compared to 158 μM of equol (Dewar et al. 1986).

Literature reports delineated that Th-2 cytokines upregulate the expression of the MAO-A enzyme (Chaitidis et al. 2004, 2005; Sturza et al. 2019) and daidzein downregulates the level of Th-2 cytokines (Chen et al. 2002; Wei et al. 2012; Sun et al. 2016). Chaitidis et al. (2005) demonstrated that the transfection of peripheral human monocytes with IL-4 significantly increases MAO-A expression. In their latest study, they reported the upregulation of MAO-A with Th-2 cytokines treatment (IL-4 and IL-13). No effect was observed with Th-2 cytokines on MAO-B expression (Chaitidis et al. 2004). In another report, it has been noticed that the IL-6 stimulation of mesentery artery branches extracted from patients gone through abdominal surgery increases the expression of MAO-A mRNA (Sturza et al. 2019). Studies conducted on different cell lines showed a significant decrease in the release of IL-6 from human osteoblast cell lines (hFOB1.19 and hFOB/ER9) with daidzein treatment (Chen et al. 2002; Sun et al. 2016). Other than IL-6, the effect of daidzein on the level of other Th-2 cytokines has also been observed. Tanaka et al. (2014) delineated that daidzein treatment suppresses the release of LPS-induced IL-6, IL-12, and TNF-α from the human monocyte THP-1 cells. From these reports, one of the evident hypotheses arises that daidzein may induce an antidepressant effect by increasing the serotonin level through MAO-A inhibition. The reduced level of Th-2 cytokines might downregulate the expression of the MAO-A enzyme, which could be the reason for the upregulation of the serotonin level by daidzein.

Literature reports suggest that an increased level of glucocorticoids also plays a pivotal role in the progression of depressive disorders (de Kloet et al. 2007; Mazarati et al. 2009; Shishkina and Dygalo 2017). A rat model of depression has been developed using exogenous corticosterone administration, which has been employed to explore the antidepressant potential of various pharmacological agents (Johnson et al. 2006; Zhao et al. 2008). Zhang et al. demonstrated the role of the HPA-axis in the daidzein-induced anti-depressant effect in rats exposed to chronic unpredicted mild stress (CUMS) for 21 days. Daidzein (30 mg/kg/day) and duloxetine (10 mg/kg/day) showed improved novelty-suppressed feeding test, rearing, and crossings in the open-field test, in comparison to CUMS group. The results in both groups, that is, daidzein and duloxetine were comparable to each other. Daidzein and duloxetine treatments significantly abrogated the CUMS-induced increase in ACTH and corticosterone levels, suggesting that the downregulation of ACTH and corticosterone hormone may be the underlying antidepressant mechanism of daidzein (Zhang et al. 2015). Some other reports on cell lines have also demonstrated that daidzein suppresses the ACTH-stimulated glucocorticoid hormone release (Mesiano et al. 1999; Kaminska et al. 2013). In the cultured human postnatal and fetal adrenal cortical cells, daidzein (0.4–40 μM/L) reduced the ACTH-stimulated cortisol production to a normal level (Mesiano et al. 1999). Kaminska et al. documented that a 10 μM concentration of daidzein suppresses the in vitro secretion of ACTH-stimulated cortisol and corticosterone (Kaminska et al. 2013). Thus, these literature findings were also delineating the role of HPA-axis modulation in the antidepressant effect of daidzein (Fig. 13.1).

13.2.2 Coumestrol

Coumestrol is a dihydroxy substituted member of the coumestan family (Song et al. 2019). It is abundantly found in various plants such as alfalfa sprouts, soybeans, sunflower seeds, clover, and legumes (Barakat et al. 2018). Coumestrol possesses 30 to 100 times higher anti-estrogenic activity than isoflavones at ER-α and ER-β receptors (Barakat et al. 2018). It has been widely explored in various ailments and reported to have effects such as antioxidative, antimicrobial, antitubercular, antineoplastic, and anti-anxiety (Luo et al. 2019). The modulation of ER- β receptors plays a crucial role in the pathophysiology of depression (Ryan and Ancelin 2012). Walf et al. (2004) tested the antidepressant potential of the estrogen receptor modulator in FST. It has been demonstrated that ER-β (estrogen β receptor) receptor modulators, that is, coumestrol, and diaryl-propionitrile (DPN), significantly reduced immobility and increased swimming and struggling in FST. On the other hand, no effect was observed with ER-α (estrogen α receptor) receptor modulators, suggesting that ER-β receptors play a pivotal role in mediating the anti-depressant effect of coumestrol and DPN. Another study also investigated the role of ER-β receptors in modulating the antidepressant effect of coumestrol in rats. This study examined the regional difference of ER-β receptors in eliciting the antidepressant effect. Vehicle, coumestrol (more affinity toward ER-β receptors), 17-β-estradiol (equal affinity toward ER-α and ER-β receptors), and 17-α-estradiol (more affinity toward ER-α receptors) were administered directly to the hippocampus and ventral tegmental area of rat brains in specific treatment groups. Coumestrol and 17- β E(2) in the hippocampus showed an antidepressant effect as indicated by positive results in the open field and forced swim test. However, the animals treated with coumestrol and 17-β E (2) in the ventral tegmental area and 17-α E(2) in both regions had no difference in both tasks than vehicle controls. This suggests that only hippocampal ER-β receptors are involved in modulating coumestrol's antidepressant effect (Walf and Frye 2007).

13.2.3 Kaempferol

Kaempferol, 3,4′,5,7-tetrahydroxyflavone, is a potent chemoprotective agent, widely found in vegetables and fruits such as *broccoli, ginkgo biloba* leaves, grapes, and tomatoes (Hu et al. 2020). Besides the anti-cancer effect, it also possesses various other activities, such as anti-inflammatory, antioxidant, *anti-diabetic*, anti-microbial, and anti-Alzheimer (Shields 2017; Beg et al. 2018). Literature reports are available in which the role of kaempferol in depression has been explored (Hosseinzadeh et al. 2007; Park et al. 2010; Yan et al. 2015). Administration of 100 mg/kg and 200 mg/kg of kaempferol to mice and 50 mg/kg to rats significantly decreased the immobility in FST (Hosseinzadeh et al. 2007). Park et al. isolated the kaempferol and quercitirin from *Opuntia ficus indica var. saboten.* Kaempferol and quercitirin treatment significantly reduced the immobility time in FST and tail suspension test (TST) in chronic restraint immobilized stressed mice. Moreover, the extract and fractions also increased the plasma β-endorphin level (Park et al. 2010). Yan et al. also explored the antidepressant activity of kaempferol, quercetin, and kaempferol 3-O- β-D-glucose in mice. It was found that kaempferol significantly decreased immobility in FST and TST. A decrease in 5-HIAA metabolite and an increase in norepinephrine, dopamine, and serotonin levels were observed with kaempferol treatment in mice. This delineated that kaempferol produces an antidepressant effect by modulating different neurotransmitter levels (Yan et al. 2015). Sloley et al. (2000) evaluated that pure kaempferol possesses in-vitro MAO inhibitory potential and inhibited MAO-A enzyme more pronouncedly than MAO-B, with an IC-50 value of 7×10^{-7} for MAO-A enzyme. A report in 2016 suggested that kaempferol inhibits the in-vitro hMAO-A enzyme selectively. However, no effect was observed on hMAO-B at maximal 100 μM tested kaempferol concentration (Gidaro et al. 2016).

Various reports delineate that an enhanced oxidative insult and inflammation are associated with depression (Andreazza et al. 2008; Bajpai et al. 2014; Grases et al. 2014). The metabolic degradation of biogenic amines, like serotonin by MAO-A, generates reactive oxygen species (ROS) which initiates lipid peroxidation and inflammation (Bianchi et al. 2005; Peña-Silva et al. 2009; Chelombitko 2018). There have been reports suggesting that kaempferol plays a crucial role in attenuating oxidative stress and inflammation (Kowalski et al. 2005; Kong et al. 2013; Guo and Feng 2017; Gao et al. 2019). Gao et al. explored the antioxidant profile of kaempferol along with its anti-inflammatory profile in depressed mice. It has been demonstrated that kaempferol (20 mg/kg) increases the activity of antioxidant enzymes (SOD, CAT, GPx, GST) and reduces the level of malondialdehyde (MDA), interleukin-1beta (IL-1β), and tumor necrosis factor (TNF-α) as compared to chronic social defeat stress (CSDS) group. Moreover, kaempferol (10 mg/kg) treatment also enhances the social interaction time and percentage of sucrose preference. In the tail suspension test, immobility time was significantly reduced by kaempferol (20 mg/kg), suggesting that kaempferol might exert its antidepressant effect by improving oxidant and inflammatory derangements (Gao et al. 2019). Furthermore, there have been reports which also explored the antioxidant and anti-inflammatory activity of kaempferol in models other than depression (Kowalski et al. 2005; Kong et al. 2013; Xiao et al. 2016; Guo and Feng 2017; Cheng et al. 2018).

Reports delineate that oxidative stress promotes the activation of NF-κB, which further induces the expression of chemokines, pro-inflammatory cytokines, etc. (Dornas et al. 2017; Liu et al. 2017). Furthermore, there has been literature evidence that showed that kaempferol inhibits the activation of NF-κB (Kim et al. 2010; Luo et al. 2012) and microglia cells (Lee et al. 2015; Yang et al. 2019). Lee et al. reported the neuroglia inhibition in the cerebral cortex region of a mouse brain with kaempferol (Lee et al. 2015). In contrast, Yang et al. observed the microglial inhibition in mice brain's striatal region with kaempferol treatment (Yang et al. 2019). Evidence suggests that early inflammation in the brain activates microglia, and this neuroglia secrete pro-inflammatory cytokines, such as TNF- α and IL-1β. (Smith et al. 2012; Lee et al. 2015). Cumulatively, an evident hypothesis arises that kaempferol may inhibit the MAO-A enzyme activation, leading to reduced depletion of serotonin and generation of ROS. This reduced generation of ROS may inhibit microglial cells' activation through inhibition of NF-KB activation, leading to the decreased secretion of pro-inflammatory cytokines (Fig. 13.1).

13.2.3.1 Biochanin-A

Biochanin-A is a 5,7-dihydroxy-4-methoxy-isoflavone often found in red clover, cabbage, alfalfa, chickpea, peanuts, and also in various herbal products (Yu et al. 2019). It possesses multiple activities, including anti-cancer, anti-diabetic, anti-inflammatory, neuroprotective, antimicrobial, and hepatoprotective. Biochanin-A has been reported to exert its neuroprotective effect by inhibiting microglial activation and neuronal apoptosis (Sarfraz et al. 2020). The role of MAO-A and NF-κB in depression has been well established. Increased MAO-A activity and activation of NF-κB play a pathological role in depression (Walia 2016; Wu et al. 2018). Studies have shown the in-vitro MAO and in-vivo NF-κB inhibitory potential of biochanin-A (Zarmouh et al. 2017; Kole et al. 2011; Qiu et al. 2012), delineating that biochanin-A may be a useful molecule to be explored in depression. Zarmouh et al. demonstrated that biochanin-A inhibited both recombinant hMAO-A and hMAO-B. However, biochanin-A exhibited a lower inhibitory constant of 3.8 nM for hMAO-B than 99.3 nM for hMAO-A. This was further strengthened by the docking study suggesting that the H-bonding and hydrophobic interaction at the active of hMAO-A and hMAO-B for its selectivity and reversibility (Zarmouh et al. 2017).

Furthermore, reports have delineated that enhanced oxidative insult and inflammation are associated with depression (Andreazza et al. 2008; Bajpai et al. 2014; Grases et al. 2014). It has been reported that monoamine degradation, oxidative stress, and inflammation are interlinked with each other (Delcambre et al. 2016; Hermida-Ameijeiras et al. 2004; Biswas 2016). As stated earlier, increased depletion of biogenic amine by MAO increases oxidative stress and inflammation (Hermida-Ameijeiras et al. 2004; Bianchi et al. 2005; Delcambre et al. 2016). NF-κB is a protein complex that controls cytokine production and responds to stressful stimuli such as free radicals (ROS) (Gilmore 2006; Brasier 2006; Dornas et al. 2017). To date, there has been no study that explored the effect of biochanin-A on NF-KB in depression. However, there have been reports other than depression, in which biochanin-A is reported to inhibit the expression of NF-κB (Berghe et al. 2006; Kole et al. 2011; Qiu et al. 2012). A report investigating the molecular depth of various isoflavones in inhibiting the NF-κB expression reported that Biochanin-A inhibits the NF-κB expression induced IL-6 release (Berghe et al. 2006). Various other studies have also reported that biochanin-A inhibits the NF-κB activation in LPS-induced RAW 264.7 cells (Kole et al. 2011; Qiu et al. 2012). Moreover, biochanin-A also abrogated the LPS-induced increased secretion of TNF-α and IL-6 (Qiu et al. 2012). Thus, these findings conclude that biochanin-A might be a potential candidate to be explored in depression, in which MAO-A and NF-KB could be a site to be focused.

13.3 Possible Mediators and Their Interplay Leading to the Anti-Anxiety Effect of Different Phytoestrogens

13.3.1 Genistein

Kalandakanond-Thongsong et al. explored the anxiolytic potential of genistein. Genistein (0.5 mg/kg) treatment significantly lowered the latency of avoidance in elevated T maze (ETM) (Kalandakanond-Thongsong et al. 2007). Studies have demonstrated that ovariectomized rats significantly develop significant anxiogenic behavior and are considered a model to study anxiety (Walf and Frye 2010). The reduction in ovarian hormones prompts the chances of physical and mood-related problems (Rodríguez-Landa et al. 2009). Rodriguez-Landa et al. evaluated the anxiolytic effect of genistein (0.25, 0.5, 1 mg/kg) in the 12 weeks of post-ovariectomized. Wistar rats using a black and white compartment model. Genistein (0.5 and 1 mg/kg) significantly enhanced the white zone frequency and exploration time in addition to increased rearing as well as grooming behavior in the genistein-treated group (Rodríguez-Landa et al. 2009). Furthermore, Rodriguez-Landa et al. also explored the role of ER-β receptors in the anxiolytic effect of genistein at a dose of 1 mg/kg in ovariectomized Wistar rats. Genistein treatment significantly reduced the latency and increased the time spent in a light section of the dark/light test apparatus. Pretreatment with tamoxifen (ER-β receptors antagonist) (5 mg/kg) for 6 days significantly abrogated the effect of genistein, suggesting that ER-β receptor involvement in mediating the anxiolytic effect of genistein (Rodríguez-Landa et al. 2012). The anxiolytic property of genistein was further explored in the mice's offspring by Rodriguez-Gomez and his coworkers. Daily treatment of female parent mice, from gestation period (day 11th) to post-partum (day 8th) to genistein (100 μg/g w/w), resulting a male offspring has shown significant anxiolytic behavior in EPM in comparison to genistein (5 μg/g w/w) and control group (Rodriguez-Gomez et al. 2014). Furthermore, in another study, Rodriguez –Landa et al. compared the anxiolytic potential of genistein and 17β-estradiol in ovariectomized rats at the same dose level. Treatment with 0.09 mg/kg and 0.18 mg/kg of genistein and 17β-estradiol for 7 days post 12 weeks after ovariectomy. Both compounds increased the open arm entries in EPM and also increased the grooming and rearing time. The anxiolytic effect was abolished by tamoxifen (ER-β receptor antagonist), suggesting that genistein and 17β-estradiol have equal anxiolytic potential, and this effect was mediated through the activation of the ER-β receptor (Rodríguez-Landa et al. 2017).

Studies have also revealed that genistein treatment modulates obsessive-compulsive disorder (OCD) in mice (Phadnis et al. 2018). OCD is a chronic anxiety disorder characterized by uncontrollable thoughts, urges, and repetitive behavioral activities (Stein et al. 2010). Serotonergic transmission plays a pivotal role in modulating anxiogenic behavior (Barr et al. 1992; Ressler and Nemeroff 2000; Lissemore et al. 2018). It has been reported that serotonin synthesizing capacity is dysregulated in OCD (Barr et al. 1992; Lissemore et al. 2018). Genistein at a dose of 10 mg/kg administered twice daily for 14 days significantly reduced the obsessive-compulsive behavior in the marble digging test and decreased the 5-HIAA/5-HT ratio in the cortex region of treated mice brain compared to untreated controls (Phadnis et al. 2018). Some indirect reports also suggested that genistein enhances the level of 5-HT (Kageyama et al. 2010; Hu et al. 2017). The 5-HT1$_A$ receptor agonists are used to treat certain anxiety disorders (Baldwin and Rudge 1995). Hu et al. also demonstrated the role of 5-HT$_{1A}$ receptors in a genistein-induced anti-depressant effect. Pretreatment with WAY-100635 (5-HT$_{1A}$ receptor antagonist) significantly abrogated the anti-depressant effect of genistein (Hu et al. 2017). However, this hypothesis lacks enough literature evidence, but this can give researchers a window to further explore the involvement of 5-HT$_{1A}$ receptors in the genistein-induced anxiolytic effect.

The anxiolytic effect of genistein has also been explored in rats with post-traumatic stress disorder (PTSD) (Wu et al. 2017). Genistein (2–8 mg/kg) treatment significantly reduced the freezing behavior and increased the open arm entries in the EPM test in a dose-dependent manner and also increased the serotonin level, phosphorylation of CaMKII/cAMP response element-binding protein (CREB) in the amygdala thereby delineating the downstream role of serotonin and p-CaMKII/p-CREB signaling in alleviating anxiety-like behavior in PTSD (Wu et al. 2017). Furthermore, many other studies have also shown that genistein up-regulates the CREB phosphorylation (Jiang et al. 2017; Sárvári et al. 2009; Si et al. 2012). The support for the above contention about the role of CREB signaling in anxiety came from the report of Wand, which delineated that the decreased phosphorylation of CREB in the amygdala is associated with increased anxiety-like behavior (Wand

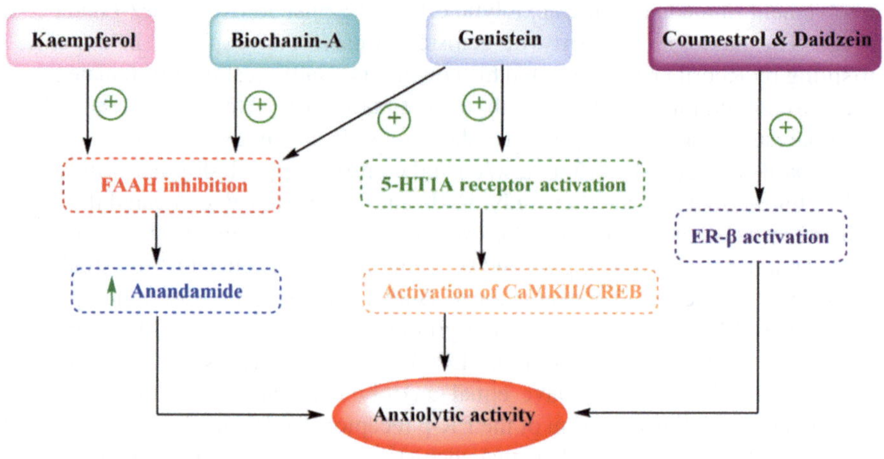

Fig. 13.2 The possible modulations of FAAH, CaMKII/CREB, ER-B by different phytoestrogens leading to anxiolytic effect

2005). Serotonin is a major neurotransmitter that is involved in up-regulating CREB phosphorylation (Mahgoub et al. 2006). These findings lead to a contention that genistein might produce its anti-anxiety effect through activation of CaMKII/cAMP through enhanced serotonergic transmission (Fig. 13.2).

Other than the reduced serotonergic transmission, the fatty acid amide hydrolase (FAAH) enzyme is also implicated in anxiogenic behavior modulation (Kathuria et al. 2003). FAAH is an enzyme that is responsible for the hydrolysis of anandamide. FAAH inhibition significantly enhances the level of anandamide, which is negatively correlated with anxiolytic behavior (Kathuria et al. 2003). FAAH also plays a pivotal role in anandamide uptake and inhibition of the same is associated with reduced uptake. The reduced uptake is allied with limited degradation, as anandamide uptake is primarily followed by its metabolism (Di Marzo et al. 1994). Thor et al. observed the FAAH inhibitory potential of genistein. Genistein significantly inhibited FAAH with a constant inhibitory value of 8 μM and reduced the anandamide concentration available for uptake by RBL2H3 cells (Thors et al. 2007a, b). These findings corroborate the hypothesis that FAAH may also be a site to be targeted with genistein to alleviate anxiety (Fig. 13.2).

13.3.1.1 Coumestrol and Daidzein

ER-β receptor activation is associated with the anxiolytic effect. The ER-β receptor agonists were reported to reduce the anxiogenic behavior in rats and mice (Lund et al. 2005; Walf et al. 2008; Weiser et al. 2009). Walf and Frye reported the anti-anxiety effect of Coumestrol and the potential role of ER-β receptors in modulating anxiogenic behavior in rats. In coumestrol and diaryl-propionitrile (DPN) (higher affinity for ER-β receptors) treated ovariectomized rats, a significant anti-anxiety effect was observed when compared to the vehicle-treated group. However, no change was seen in the propyl pyrazole triol and 17α-E2 (selective ER- α SERMs) treated groups as compared to the vehicle control group, suggesting the effectiveness of coumestrol and DPN and the role of ER-β receptors activation in the modulation of anti-anxiety behavior (Walf and Frye 2005). Zeng et al. demonstrated that treatment of male mice with daidzein for 30 days significantly increased the open arm entries in the elevated plus-maze (EPM) as well as increased the locomotor activity as evidenced by improved transitions in the open field test (OFT) and elevated plus maze (EPM) test (Zeng et al. 2010). Daidzein (0.5 mg/kg) treatment has also been shown to lower the latency of avoidance in elevated T maze (ETM) (Kalandakanond-Thongsong et al. 2007). As stated earlier, the role of the ER-β receptor in anxiety has been well established. Several studies have delineated that daidzein predominantly activates the ER-β receptor compared to the ER-α receptor (An et al. 2001; Ramasamy et al. 2017), suggesting that daidzein may exert its anti-anxiety effect through the activation of ER-β receptors. However, this hypothesis needs stronger evidence or in-vivo studies to demonstrate the role of ER-β in the daidzein-induced anxiolytic effect (Fig. 13.2).

13.3.2 Kaempferol

Several studies have explored the potential of kaempferol in anxiety with different routes of administration (Grundmann et al. 2009; Aguirre-Hernández et al. 2010; Vissiennon et al. 2012). Aguirre-Hernandez et al. have demonstrated that kaempferol treatment (10 mg/kg, i.p) significantly increased the time spent in the open arm in the elevated plus-maze as compared to normal vehicle control (Aguirre-Hernández et al. 2010). Vissiennan et al. demonstrated no effect on intraperitoneal administration of kaempferol (0.1–2 mg/kg). However, oral administration of kaempferol significantly enhanced the open arm entries in the EPM test (Vissiennon et al. 2012). It has also been documented that kaempferol acts as a pro-drug and requires activation by intestinal microflora. This hypothesis was confirmed by antibiotic treatment and the anxiolytic potential of active metabolites. Enrofloxacin (7.5 mg/kg/day) treatment for 4 days significantly abrogated the effect of kaempferol, and its metabolites (p-HPAA and DOPAC) showed a similar anxiolytic effect as that of oral kaempferol on intraperitoneal administration (Vissiennon et al. 2012). However, this hypothesis related to activation by intestinal microflora for kaempferol action lacks much literature support and therefore needs futuristic insight. Moreover, it has been also demonstrated that pretreatment with flumazenil (selective $GABA_A$ antagonist) significantly reverted the effect of kaempferol (0.08 mg/kg, p.o). However, no effect was seen with Way-100,635 ($5\text{-}HT_{1A}$ receptor antagonist) on kaempferol-induced anxiolytic effect, delineating the probable role of $GABA_A$ receptors (Grundmann et al. 2009). Moreover, similar to genistein, Thors et al. also observed the FAAH inhibiting potential of kaempferol. It has been reported that kaempferol inhibits anandamide hydrolysis in rat brains and intact RBL2H3 cells. Kaempferol inhibits the FAAH with a constant inhibitory value of 5 μM. Also, it inhibited the FAAH-dependent uptake of anandamide (Thors et al.

2008), suggesting that FAAH may also be a site to be targeted with kaempferol to alleviate anxiety (Fig. 13.2).

13.4 Conclusion

In this chapter, several reports on the anti-depressant and anti-anxiety effects of phytoestrogens have been discussed, and their possible underlying pathways have been summarized. The antidepressant effect of different phytoestrogens may be attributed to the inhibition of MAO, upregulation of serotonergic pathways, and modulation of the HPA axis. Genistein has been observed to exert an antidepressant effect via inhibiting the MAO enzyme and increasing the 5-HT_{1A} receptor activation leading to an increase in gap junction turnover by stimulating the MAPK and Akt cascade of Cx-43 phosphorylation. Moreover, different phytoestrogens have been observed to exert an anti-anxiety effect through the modulation of FAAH, CREB, GABA receptors, and ER- β receptors.

References

Aguirre-Hernández E, González-Trujano ME, Martínez AL et al (2010) HPLC/MS analysis and anxiolytic-like effect of quercetin and kaempferol flavonoids from Tilia americana var. mexicana. J Ethnopharmacol 127:91–97

Aldred EM (2008) Pharmacology e-book: a handbook for complementary healthcare professionals. Elsevier, London

An J, Tzagarakis-Foster C, Scharschmidt TC et al (2001) Estrogen receptor β-selective transcriptional activity and recruitment of coregulators by phytoestrogens. J Biol Chem 276:17808–17814

Andreazza AC, Kauer-Sant'Anna M, Frey BN et al (2008) Oxidative stress markers in bipolar disorder: a meta-analysis. J Affect Disord 111:135–144

Aso T (2010) Equol improves menopausal symptoms in Japanese women. J Nutr 140:1386–1389

Atteritano M, Mazzaferro S, Bitto A et al (2014) Genistein effects on quality of life and depression symptoms in osteopenic postmenopausal women: a 2-year randomized, double-blind, controlled study. Osteoporos Int 25:1123–1129

Bajpai A, Verma AK, Srivastava M et al (2014) Oxidative stress and major depression. J Clin Diagn Res 8:CC04–CC07

Baldwin D, Rudge S (1995) The role of serotonin in depression and anxiety. Int Clin Psychopharmacol 4:41–45

Barakat R, Park CJ, Perez PA et al (2018) Female antiestrogens. In: Encyclopedia of reproduction. Elsevier, Oxford

Barnes S, Peterson TG, Coward L (1995) Rationale for the use of genistein-containing soy matrices in chemoprevention trials for breast and prostate cancer. J Cell Biochem 59:181–187

Barr LC, Goodman WK, Price LH (1992) The serotonin hypothesis of obsessive compulsive disorder: implications of pharmacologic challenge studies. J Clin Psychiatry 53:17–28

Beg T, Jyoti S, Naz F et al (2018) Protective effect of kaempferol on the transgenic drosophila model of Alzheimer's disease. CNS Neurol Disord Drug Targets 17:421–429

Berghe WV, Dijsselbloem N, Vermeulen L et al (2006) Attenuation of mitogen-and stress-activated protein Kinase-1–driven nuclear factor-κB gene expression by soy Isoflavones does not require estrogenic activity. Cancer Res 66:4852–4862

Bianchi P, Kunduzova O, Masini E et al (2005) Oxidative stress by monoamine oxidase mediates receptor-independent cardiomyocyte apoptosis by serotonin and postischemic myocardial injury. Circulation 112:3297–3305

Biswas SK (2016) Does the interdependence between oxidative stress and inflammation explain the antioxidant paradox? Oxid Med Cell Longev 2016:5698931

Blake C, Fabick KM, Setchell KD (2011) Neuromodulation by soy diets or equol: anti-depressive & anti-obesity-like influences, age-& hormone-dependent effects. BMC Neurosci 12:28

Brasier AR (2006) The NF-κB regulatory network. Cardiovasc Toxicol 6:111–130

Carvalho AF, Sharma MS, Brunoni AR et al (2016) The safety, tolerability and risks associated with the use of newer generation antidepressant drugs: a critical review of the literature. Psychother Psychosom 85:270–288

Cederroth CR, Nef S (2009) Soy, phytoestrogens and metabolism: a review. Mol Cell Endocrinol 304:30–42

Chaitidis P, Billett EE, O'Donnell VB et al (2004) Th2 response of human peripheral monocytes involves isoform-specific induction of monoamine oxidase-a. J Immunol 173:4821–4827

Chaitidis P, Billett E, Kuban RJ et al (2005) Expression regulation of MAO isoforms in monocytic cells in response to Th2cytokines. Med Sci Monit 11:259–265

Chelombitko MA (2018) Role of reactive oxygen species in inflammation: a Minireview. Moscow Univ Biol Sci Bull 73:199–202

Chen XW, Garner SC, Anderson JJ (2002) Isoflavones regulate interleukin-6 and osteoprotegerin synthesis during osteoblast cell differentiation via an estrogen-receptor-dependent pathway. Biochem Biophys 295:417–422

Cheng X, Yang YL, Yang H et al (2018) Kaempferol alleviates LPS-induced neuroinflammation and BBB dysfunction in mice via inhibiting HMGB1 release and down-regulating TLR4/MyD88 pathway. Int Immunopharmacol 56:29–35

Conklin CM, Bechberger JF, MacFabe D et al (2007) Genistein and quercetin increase connexin43 and suppress growth of breast cancer cells. Carcinogenesis 28:93–100

de Kloet ER, Derijk RH, Meijer OC (2007) Therapy insight: is there an imbalanced response of mineralocorticoid and glucocorticoid receptors in depression. Nat Clin Pract Endocrinol Metab 3:168–179

Delcambre S, Nonnenmacher Y, Hiller K (2016) Dopamine metabolism and reactive oxygen species production. In: Mitochondrial mechanisms of degeneration and repair in Parkinson's disease. Springer, Cham

Dewar D, Glover V, Elsworth J et al (1986) Equol and other compounds from bovine urine as monoamine oxidase inhibitors. J Neural Transm Suppl 65:147–150

Di Marzo V, Fontana A, Cadas H et al (1994) Formation and inactivation of endogenous cannabinoid anandamide in central neurons. Nature 372:686–691

Ding M, Pan A, Ding M et al (2016) Consumption of soy foods and isoflavones and risk of type 2 diabetes: a pooled analysis of three US cohorts. Eur J Clin Nutr 70:1381–1387

Dong JY, Qin LQ (2011) Soy isoflavones consumption and risk of breast cancer incidence or recurrence: a meta-analysis of prospective studies. Breast Cancer Res Treat 125:315–323

Dornas WC, Cardoso LM, Silva M et al (2017) Oxidative stress causes hypertension and activation of nuclear factor-κB after high-fructose and salt treatments. Sci Rep 7:46051

Eden JA (2012) Phytoestrogens formenopausal symptoms: a review. Maturitas 72:157–159

Fatemi SH, Folsom TD, Reutiman TJ et al (2008) Expression of astrocytic markers aquaporin 4 and connexin 43 is altered in brains of subjects with autism. Synapse 62:501–507

Gao GY, Li DJ, Keung WM (2001) Synthesis of potential antidipsotropic isoflavones: inhibitors of the mitochondiral monoamine oxidase-aldehyde dehydrogenase pathway. J Med Chem 44: 3320–3328

Gao W, Wang W, Peng Y et al (2019) Antidepressive effects of kaempferol mediated by reduction of oxidative stress, proinflammatory cytokines and up-regulation of AKT/β-catenin cascade. Metab Brain Dis 34:485–494

Gidaro MC, Astorino C, Petzer A et al (2016) Kaempferol as selective human MAO-A inhibitor: analytical detection in calabrian red wines, biological and molecular modeling studies. J Agric Food Chem 64:1394–1400

Gilmore TD (2006) Introduction to NF-κB: players, pathways, perspectives. Oncogene 25:6680

Glazier MG, Bowman MA (2001) A review of the evidence for the use of phytoestrogens as a replacement for traditional estrogen replacement therapy. Arch Intern Med 161:1161–1172

Grases G, Colom MA, Fernandez RA et al (2014) Evidence of higher oxidative status in depression and anxiety. Oxid Med Cell Longev 2014:1. https://doi.org/10.1155/2014/430216

Grundmann O, Nakajima JI, Kamata K et al (2009) Kaempferol from the leaves of Apocynum venetum possesses anxiolytic activities in the elevated plus maze test in mice. Phytomedicine 16:295–302

Guo P, Feng YY (2017) Anti-inflammatory effects of kaempferol, myricetin, fisetin and ibuprofen in neonatal rats. Trop J Pharm Res 16:1819–1826

Harsha Pedapati S, Wahab RA, Garcia Aloy M et al (2018) Biomarkers of legume intake in human intervention and observational studies: a systematic review. Genes Nutr 13:25

Hermida-Ameijeiras Á, Méndez-Álvarez E, Sánchez-Iglesias S et al (2004) Autoxidation and MAO-mediated metabolism of dopamine as a potential cause of oxidative stress: role of ferrous and ferric ions. Neurochem Int 45:103–116

Hosseinzadeh H, Motamedshariaty V, Hadizadeh F (2007) Antidepressant effect of kaempferol, a constituent of saffron (crocus sativus) petal, in mice and rats. Pharmacologyonline 2:367–370

Hu P, Ma L, Wang YG, Ye F et al (2017) Genistein, a dietary soy isoflavone, exerts antidepressant-like effects in mice: involvement of serotonergic system. Neurochem Int 108:426–435

Hu WH, Wang HY, Xia YT et al (2020) Kaempferol, a major flavonoid in ginkgo folium, potentiates Angiogenic functions in cultured endothelial cells by binding to vascular endothelial growth factor. Front Pharmacol 11:526

Jaiswal N, Akhtar J, Singh SP et al (2019) An overview on Genistein and its various formulations. Drug Res (Stuttg) 69:305–313

Jiang T, Wang XQ, Ding C et al (2017) Genistein attenuates isoflurane-induced neurotoxicity and improves impaired spatial learning and memory by regulating cAMP/CREB and BDNF-TrkB-PI3K/Akt signaling. Korean J Physiol Pharmacol 21:579–589

Johnson SA, Fournier NM, Kalynchuk LE (2006) Effect of different doses of corticosterone on depression-like behavior and HPA axis responses to a novel stressor. Behav Brain Res 168:280–288

Kageyama A, Sakakibara H, Zhou W et al (2010) Genistein regulated serotonergic activity in the hippocampus of ovariectomized rats under forced swimming stress. Biosci Biotechnol Biochem 74:2005–2010

Kalandakanond-Thongsong S, Daendee S, Poonyachoti S (2007) P29 the effect of Genistein and Daidzein on anxiety levels: in comparison to estrogen in Ovariectomized rat. In: Proceedings Chula Univ Vet Sci Ann Con, p 26–27

Kaminska B, Ciereszko R, Kiezun M et al (2013) In vitro effects of genistein and daidzein on the activity of adrenocortical steroidogenic enzymes in mature female pigs. J Physiol Pharmacol 64:103–108

Karahalil B (2006) Benefits and risks of phytoestrogens. Phytoestrogens in functional foods. Taylor and Francis, New York, NY

Kathuria S, Gaetani S, Fegley D et al (2003) Modulation of anxiety through blockade of anandamide hydrolysis. Nat Med 9:76

Kim JM, Lee EK, Kim DH et al (2010) Kaempferol modulates pro-inflammatory NF-κB activation by suppressing advanced glycation endproducts-induced NADPH oxidase. Age 32:197–208

Kole L, Giri B, Manna SK et al (2011) Biochanin-a, an isoflavon, showed anti-proliferative and anti-inflammatory activities through the inhibition of iNOS expression, p38-MAPK and ATF-2 phosphorylation and blocking NFκB nuclear translocation. Eur J Pharmacol 653:8–15

Kolonel LN, Hankin JH, Nomura AM (1986) Multiethnic studies of diet, nutrition and cancer in Hawaii. Diet Nutr Cancer 16:29–40

Kong L, Luo C, Li X et al (2013) The anti-inflammatory effect of kaempferol on early atherosclerosis in high cholesterol fed rabbits. Lipids Health Dis 12:115

Kowalski J, Samojedny A, Paul M et al (2005) Effect of apigenin, kaempferol and resveratrol on the expression of interleukin-1beta and tumor necrosis factor-alpha genes in J774. 2 macrophages. Pharmacol Rep 57:390–394

Lee HG, Kim DH, Kim YS et al (2015) Effects of Kaempferol on Lippolysaccharide-induced inflammation in mouse brain. Korea J Herbology 30:77–84

Lethaby A, Marjoribanks J, Kronenberg F et al (2013) Phytoestrogens for menopausal vasomotor symptoms. Cochrane Database Syst Rev 12:Cd001395

Lissemore JI, Sookman D, Gravel P et al (2018) Brain serotonin synthesis capacity in obsessive-compulsive disorder: effects of cognitive behavioral therapy and sertraline. Transl Psychiatry 28:82

Liu T, Zhang L, Joo D et al (2017) NF-κB signaling in inflammation. Signal Transduct Target Ther 2:17023

Lund TD, Rovis T, Chung WCJ et al (2005) Novel actions of estrogen receptor-β on anxiety-related behaviors. Endocrinology 146:797–807

Luo H, Rankin GO, Juliano N et al (2012) Kaempferol inhibits VEGF expression and in vitro angiogenesis through a novel ERK-NFκB-cMyc-p21 pathway. Food Chem 130:321–328

Luo G, Tang Z, Li X et al (2019) 3, 9-di-O-substituted coumestrols incorporating basic amine side chains act as novel apoptosis inducers with improved pharmacological selectivity. Bioorg Chem 85:140–151

Mahgoub MA, Sara Y, Kavalali ET et al (2006) Reciprocal interaction of serotonin and neuronal activity in regulation of cAMP-responsive element-dependent gene expression. J Pharmacol Exp Ther 317:88–96

Mazarati AM, Shin D, Kwon YS et al (2009) Elevated plasma corticosterone level and depressive behavior in experimental temporal lobe epilepsy. Neurobiol Dis 34:457–461

Mesiano S, Katz SL, Lee JY et al (1999) Phytoestrogens alter adrenocortical function: genistein and daidzein suppress glucocorticoid and stimulate androgen production by cultured adrenal cortical cells. J Clin Endocrinol Metab 84:2443–2448

Miguel-Hidalgo JJ, Wilson BA, Hussain S et al (2014) Reduced connexin 43 immunolabeling in the orbitofrontal cortex in alcohol dependence and depression. J Psychiatr Res 55:101–109

Mostafavi H, Khaksarian M, Joghataei MT et al (2014) Fluoxetin upregulates connexin 43 expression in astrocyte. Basic Clin Neurosci 5:74

Nagy C, Torres-Platas SG, Mechawar N et al (2016) Repression of astrocytic connexins in cortical and subcortical brain regions and prefrontal enrichment of H3K9me3 in depression and suicide. Int J Neuropsychopharmacol 20:50–57

Park SH, Sim YB, Han PL (2010) Antidepressant-like effect of kaempferol and quercitirin, isolated from Opuntia ficus-indica var. saboten. Exp Neurobiol 19:30–38

Patisaul HB, Jefferson W (2010) The pros and cons of phytoestrogens. Front Neuroendocrinol 31: 400–419

Peña-Silva RA, Miller JD, Chu Y (2009) Serotonin produces monoamine oxidase-dependent oxidative stress in human heart valves. Am J Physiol Heart Circ Physiol 297:1354–1360

Phadnis P, Sarkar PD, Rajput MS (2018) Improved serotonergic neurotransmission by genistein pretreatment regulates symptoms of obsessive-compulsive disorder in streptozotocin-induced diabetic mice. J Basic Clin Physiol Pharmacol 29:421–425

Price KR, Fenwick GR (1985) Naturally occurring oestrogens in foods- a review. Food Addit Contam 2:73–106

Qiu L, Lin B, Lin Z et al (2012) Biochanin a ameliorates the cytokine secretion profile of lipopolysaccharide-stimulated macrophages by a PPARγ-dependent pathway. Mol Med Rep 5:217–222

Ramasamy K, Samayoa C, Krishnegowda N et al (2017) Therapeutic use of estrogen receptor β agonists in prevention and treatment of endocrine therapy resistant breast cancers: observations from preclinical models. Prog Mol Biol Transl Sci 151:177–194

Ressler KJ, Nemeroff CB (2000) Role of serotonergic and noradrenergic systems in the pathophysiology of depression and anxiety disorders. Depress Anxiety 12:2–19

Rietjens IM, Louisse J, Beekmann K (2017) The potential health effects of dietary phytoestrogens. Br J Pharmacol 174:1263–1280

Rodriguez-Gomez A, Filice F, Gotti S et al (2014) Perinatal exposure to genistein affects the normal development of anxiety and aggressive behaviors and nitric oxide system in CD1 male mice. Physiol Behav 133:107–114

Rodríguez-Landa JF, Hernández-Figueroa JD, del Carmen H-CB et al (2009) Anxiolytic-like effect of phytoestrogen genistein in rats with long-term absence of ovarian hormones in the black and white model. Prog Neuropsychopharmacol Biol Psychiatry 33:367–372

Rodríguez-Landa JF, Hernández-López F, Saavedra M (2012) Involvement of estrogen receptors in the anxiolytic-like effect of phytoestrogen genistein in rats with 12-weeks postovariectomy. Pharmacol Pharm 3:439

Rodríguez-Landa JF, Cueto-Escobedo J, Puga-Olguín A et al (2017) The phytoestrogen genistein produces similar effects as 17β-estradiol on anxiety-like behavior in rats at 12 weeks after ovariectomy. Biomed Res Int 2017:9073816

Rooke N, Li DJ, Li J et al (2000) The mitochondrial monoamine oxidase-aldehyde dehydrogenase pathway: a potential site of action of diadzin. J Med Chem 43:4169–4179

Ryan J, Ancelin ML (2012) Polymorphisms of estrogen receptors and risk of depression. Drugs 72: 1725–1738

Sarfraz A, Javeed M, Shah MA et al (2020) Biochanin A: a novel bioactive multifunctional compound from nature. Sci Total Environ 12:137907

Sárvári M, Szegő ÉM, Barabás K et al (2009) Genistein induces phosphorylation of cAMP response element-binding protein in neonatal hypothalamus in vivo. J Neuroendocrinol 21:1024–1028

Shen F, Huang WL, Xing BP et al (2018) Genistein improves the major depression through suppressing the expression of miR-221/222 by targeting Connexin 43. Psychiatry Investig 15: 919

Shields M (2017) Chemotherapeutics. Elsevier, In Pharmacognosy

Shishkina GT, Dygalo NN (2017) The glucocorticoid hypothesis of depression: history and prospects. Russ J Genet Appl Res 7:128–133

Si H, Yu J, Jiang H et al (2012) Phytoestrogen genistein up-regulates endothelial nitric oxide synthase expression via activation of cAMP response element-binding protein in human aortic endothelial cells. Endocrinology 153:3190–3198

Sloley BD, Urichuk LJ, Morley P et al (2000) Identification of kaempferol as a monoamine oxidase inhibitor and potential neuroprotectant in extracts of Ginkgo biloba leaves. J Pharm Pharmacol 52:451–459

Smith JA, Das A, Ray SK et al (2012) Role of pro-inflammatory cytokines released from microglia in neurodegenerative diseases. Brain Res Bull 87:10–20

Song X, Luo X, Sheng J et al (2019) Copper-catalyzed intramolecular cross dehydrogenative coupling approach to coumestans from 2′-hydroxyl-3-arylcoumarins. RSC Adv 9:17391–17398

Stein DJ, Fineberg NA, Bienvenu OJ et al (2010) Should OCD be classified as an anxiety disorder in DSM-V. Depress Anxiety 27:495–506

Sturza A, Popoiu CM, Ionică M et al (2019) Monoamine oxidase-related vascular oxidative stress in diseases associated with inflammatory burden. Oxid Med Cell Longev 3:e000713, 1

Sun JD, Liu Y, Yuan YH et al (2012) Gap junction dysfunction in the prefrontal cortex induces depressive-like behaviors in rats. Neuropsychopharmacology 37:1305–1320

Sun J, Sun WJ, Li ZY et al (2016) Daidzein increases OPG/RANKL ratio and suppresses IL-6 in MG-63 osteoblast cells. Int Immunopharmacol 40:32–40

Taku K, Lin N, Taku K et al (2010) Effects of soy isoflavone extract supplements on blood pressure in adult humans: systematic review and meta-analysis of randomized placebo-controlled trials. J Hypertens 28:1971–1982

Tanaka K, Ohgo Y, Katayanagi Y et al (2014) Anti-inflammatory effects of green soybean extract irradiated with visible light. Sci Rep 4:4732

Thors L, Alajakku K, Fowler CJ et al (2007a) The 'specific' tyrosine kinase inhibitor genistein inhibits the enzymic hydrolysis of anandamide: implications for anandamide uptake. Br J Pharmacol 150:951–960

Thors L, Eriksson J, Fowler CJ (2007b) Inhibition of the cellular uptake of anandamide by genistein and its analogue daidzein in cells with different levels of fatty acid amide hydrolase-driven uptake. Br J Pharmacol 152:744–750

Thors L, Belghiti M, Fowler CJ (2008) Inhibition of fatty acid amide hydrolase by kaempferol and related naturally occurring flavonoids. Br J Pharmacol 155:244–252

Tiller JW (2013) Depression and anxiety. Med J Aust 199:28–31

Tsugane S (2021) Why has Japan become the world's most long-lived country: insights from a food and nutrition perspective. Eur J Clin Nutr 75:921–928

van Die MD, Bone KM, Williams SG et al (2014) Soy and soy isoflavones in prostate cancer: a systematic review and meta-analysis of randomized controlled trials. BJU Int 113:119–130

Veitch NC (2007) Isoflavonoids of the Leguminosae. Nat Prod Rep 24:417–464

Vissiennon C, Nieber K, Kelber O et al (2012) Route of administration determines the anxiolytic activity of the flavonols kaempferol, quercetin and myricetin-are they prodrugs. J Nutr Biochem 23:733–740

Walf AA, Frye CA (2005) ER β-selective estrogen receptor modulators produce antianxiety behavior when administered systemically to ovariectomized rats. Neuropsychopharmacology 30:1598–1609

Walf AA, Frye CA (2007) Administration of estrogen receptor beta-specific selective estrogen receptor modulators to the hippocampus decrease anxiety and depressive behavior of ovariectomized rats. Pharmacol Biochem Behav 86:407–414

Walf AA, Frye CA (2010) Raloxifene and/or estradiol decrease anxiety-like and depressive-like behavior, whereas only estradiol increases carcinogen-induced tumorigenesis and uterine proliferation among ovariectomized rats. Behav Pharmacol 21:231

Walf AA, Rhodes ME, Frye CA (2004) Antidepressant effects of ERβ-selective estrogen receptor modulators in the forced swim test. Pharmacol Biochem Behav 78:523–529

Walf AA, Koonce CJ, Frye CA (2008) Estradiol or diarylpropionitrile decrease anxiety-like behavior of wildtype, but not estrogen receptor beta knockout, mice. Behav Neurosci 122: 974–981

Walia V (2016) Role of enzymes in the pathogenesis of depression. J Crit Rev 3:11–16

Wand G (2005) The anxious amygdala: CREB signaling and predisposition to anxiety and alcoholism. J Clin Investig 115:2697–2699

Wang H, Li H, Moore LB et al (2008) The phytoestrogen coumestrol is a naturally occurring antagonist of the human pregnane X receptor. Mol Endocrinol 22:838–857

Wei J, Bhatt S, Chang LM et al (2012) Isoflavones, genistein and daidzein, regulate mucosal immune response by suppressing dendritic cell function. PloS One 7:e47979

Weiser MJ, Wu TJ, Handa RJ (2009) Estrogen receptor-beta agonist diarylpropionitrile: biological activities of R- and S- enantiomers on behavior and hormonal response to stress. Endocrinology 150:1817–1825

WHO (2017) Depression and other common mental disorders: global health estimates. World Health Organization, Geneva

Wu ZM, Ni GL, Shao AM (2017) Genistein alleviates anxiety-like behaviors in post-traumatic stress disorder model through enhancing serotonergic transmission in the amygdala. Psychiatry Res 255:287–291

Wu Y, Wang L, Hu K et al (2018) Mechanisms and therapeutic targets of depression after intracerebral hemorrhage. Front Psych 9:682

Xiao HB, Lu XY, Liu ZK et al (2016) Kaempferol inhibits the production of ROS to modulate OPN–α v β 3 integrin pathway in HUVECs. J Physiol Biochem 72:303–313

Yan SX, Lang JL, Song YY et al (2015) Studies on anti-depressant activity of four flavonoids isolated from Apocynum venetum Linn (Apocynaceae) leaf in mice. Trop J Pharm Res 14: 2269–2277

Yang YL, Cheng X, Li WH (2019) Kaempferol attenuates LPS-induced striatum injury in mice involving anti-neuroinflammation, maintaining BBB integrity, and down-regulating the HMGB1/TLR4 pathway. Int J Mol Sci 20:491

Yildiz F (2005) Phytoestrogens in functional foods. Taylor & Francis, Boca Raton, FL

Yu C, Zhang P, Lou L et al (2019) Perspectives on the role of Biochanin a in human. Front Pharmacol 10:793

Zarmouh NO, Messeha SS, Elshami FM et al (2016) Evaluation of the isoflavone genistein as reversible human monoamine oxidase-A and-B inhibitor. Evid Based Complement Alternat Med 2016:1. https://doi.org/10.1155/2016/1423052

Zarmouh NO, Eyunni SK, Soliman KF (2017) The Benzopyrone Biochanin-a as a reversible, competitive, and selective monoamine oxidase B inhibitor. BMC Complement Altern Med 17: 34

Zeng S, Tai F, Zhai P et al (2010) Effect of daidzein on anxiety, social behavior and spatial learning in male Balb/cJ mice. Pharmacol Biochem Behav 96:16–23

Zhang JR, Sun DL, Shi JJ et al (2015) Antidepressive-like effect of daidzein in rats and its mechanism. Chin J New Drugs 24:1531–1536

Zhao Y, Ma R, Shen J et al (2008) A mouse model of depression induced by repeated corticosterone injections. Eur J Pharmacol 581:113–120

Next-Generation Cell-Based Therapies to Combat Metastatic Brain Tumor

14

Vignesh Balaji E and Karkala Sreedhara Ranganath Pai

Abstract

Glioblastoma (GBM) is a grade IV aggressive type brain tumor that causes rapid proliferation and growth rate. The average survival period is 14 to 16 months after diagnosis and the progression-free survival rate (PFS) is 7 to 10 months. The existing therapies such as radiation, surgical resection, and chemotherapy are not showing curable results due to the development of resistance and the plethora of side effects in the treatment of an advanced brain tumor. To destroy the formidable glioma cells and bring the patient's outcome, we need to look for a novel therapy that is more advanced and effective against metastatic brain tumor cells. Several cellular-based therapeutic approaches significantly identified the tumor cells and destroyed them by stimulating the host's innate and adaptive immune system and repressing oncogenic intracellular pathways. Monitoring the therapeutic molecule's action and patient's outcome can be done with the help of artificial intelligence (AI). AI is a growing technique used to predict the tumor stage and the therapeutic benefits associated with the treatment. Treating advanced metastatic brain tumors with this cellular therapy and analyzing the patient's therapeutic outcome using AI might be a promising approach to curing the heterogeneous brain tumor.

Keywords

Anti-tumor · Artificial intelligence · Brain tumor · Genetic mutation · Metastasis

Vignesh Balaji E · K. S. R. Pai (✉)
Department of Pharmacology, Manipal College of Pharmaceutical Sciences, Manipal Academy of Higher Education, Manipal, Karnataka, India
e-mail: ksr.pai@manipal.edu

P. P. Singh (ed.), *Recent Advances in Pharmaceutical Innovation and Research*,
https://doi.org/10.1007/978-981-99-2302-1_14

14.1 Introduction

Brain cancer is the foremost cause of death worldwide. Nearly, 18 million new cases have been identified every year and causing around 10 million deaths worldwide reported by International Agency for Research on Cancer (IARC). Most of the patients with advanced brain tumor-like glioblastoma (GBM) have less survival period. Conventional therapies like chemotherapy, surgery, and radiotherapy are becoming resistant to destroying tumor growth (Hemminki et al. 2020). Moreover, tumor cell metastasis to the brain predominantly occurs via breast carcinoma, lung adenocarcinoma, and melanoma. Development of the metastatic brain tumor occurs when cancer cells from a primary tumor anywhere in the systemic circulation reach the brain via the bloodstream and begin to multiply and grow. Brain cancer metastasis is a multistep process that contains primary tumor invasion into the surrounding tissues, intravasation, and dissemination of the tumor cells into the systemic circulation, and colonization of the tumor cell formation gives rise to metastasis (Hosonaga et al. 2020).

In the need for effective treatment, cell-based therapies have received more attention from researchers in many parts of the world. Oncolytic virotherapy (OVT) is one of the possible novel therapeutic molecules to destroy tumor cells effectively without harming healthy cells. OVT helps to accelerate the suppressed immune system in the tumor microenvironment and facilitates a strong immune cell response against glioma cells (Patel and Kratzke 2013). Likewise, the intervention of immunotherapy offered an excellent option to cure the advanced stage of brain tumors (Smith et al. 2020). Furthermore, targeting cancer cell surface receptors is considered a potential approach to understanding the molecular mechanisms and destroying the intracellular oncogenic tumor proteins. Targeting the receptor with a suitable ligand inhibits the intracellular signaling pathways which are responsible for tumor growth, proliferation, and survival. Likewise, genetically engineered stem cells (SCs) have a tumor-homing capacity to allow the delivery of anti-cancer molecules into the tumor sites and reduce tumor growth. A key regulator in the central dogma such as miRNA is a non-coding RNA that regulates gene expression by controlling the translation process via complementary binding to the messenger RNA (mRNA). Inhibition of the tumor cells through microRNA (miRNA) is an upcoming possible therapeutic approach to halt and destroy cancer cells (Srinivasan et al. 2011). In that manner, genome editing with clustered regularly interspaced short palindromic repeats-associated protein 9 (CRISPR Cas9) brought breakthrough advancement in the field of cancer therapy. Insertion or deletion of the single gene at the targeted site has increased the therapeutic profile and destroyed the cancer cells (Meca-cortés et al. 2017). CRISPR-Cas9 technology has gained more therapeutic advantages in the alleviation of GBM pathogenic progression (Daisy et al. 2021).

Furthermore, ongoing studies are working in a full-fledged manner to discover the potential therapy to destroy the formidable brain tumor. In this article, we took a broad vision of the various gene and cell-based therapies which are currently employed to treat a metastatic brain tumor. Also, the advancement in the prediction of tumor grades and their migration potential using artificial intelligence (AI) has been explained in a futuristic manner.

14.2 Receptor-Based Therapy to Destroy GBM Cells

Several receptors have been identified to target and destroy GBM cells by inducing caspase-mediated apoptotic pathways and preventing drug resistance. In that order, the death receptor (DR) is a member of the tumor necrosis factor receptor that initiates cytotoxic signals in the tumor cells to induce apoptotic-induced cell death. Studies found that tumor necrosis factor-related apoptosis-induced ligand (TRAIL) is an excellent activator for death receptor 5 (DR5) which is widely present in advanced glioma cells like U87, and LN229 (Bagci-Onder et al. 2012). Triggering DR5 through TRAIL stimulates the apoptosis signaling pathway as well as the mitochondria-mediated pathway including activation of caspase 3/8 resulting in the promotion of tumor cell death with high efficacy and low toxicity (Cha et al. 2019). The aberrant epidermal growth factor receptor (EGFR) signaling pathway is widely associated with various types of cancers including GBM. EGFR amplification has been found in more than 90% of GBM patients (Nicholas et al. 2006). Studies reported that EGFR-targeted nanobody and TRAIL combination blocks the EGFR signaling pathway as well as induced the DR5 clustering at the cell plasma membrane resulting in to activate the caspase-mediated apoptotic tumor cell death in LN229 and GBM6 cells (Zhu et al. 2017).

Brain metastasis occurs mainly through mutation in the EGFR and anaplastic lymphoma kinase (ALK) translocation in non-small cell lung cancer (NSCLC), BRAF mutation in melanoma, and HER2 mutation in breast cancer (Dodson et al. 2020). Studies found that approximately 40% of metastatic cancer patients develop a brain tumor. Mutations in the EGFR frequently develop NSCLC-caused brain metastasis and gefitinib and erlotinib are first-line treatments to inhibit tyrosine kinase that target EGFR mutation. First-generation ALK inhibitors such as crizotinib inhibit tyrosine kinase to prevent NSCLC-induced brain metastasis (Sun et al. 2018). Likewise, breast cancer is the second leading cancer that causes brain metastasis. Triple-negative breast cancer patients have a high chance to cause brain metastasis and frequently induce cranial diseases. Approximately, 50% of HER2-positive breast cancer patients develop brain metastasis and it might be prevented via anti-HER2 agents such as trastuzumab. Dual inhibition of EGFR and HER2 using lapatinib and neratinib may improve breast cancer brain metastasis patients' outcomes (Venur and Ahluwalia 2016). Nearly 30% of NSCLC patients have an ALK mutation that causes brain metastasis. Tyrosine kinase inhibitors such as crizotinib repress the brain metastasis from NSCLC via targeting ALK gene rearrangement along with radiation therapy (Fig. 14.1) (Wang et al. 2019).

Chemotherapeutic agents ultimately inhibit cancer cell mobility from the primary tumor site to the secondary tumor sites. These chemotherapeutic agents act on the intracellular signaling pathway in various cancers including lung and breast to prevent brain tumor metastasis.

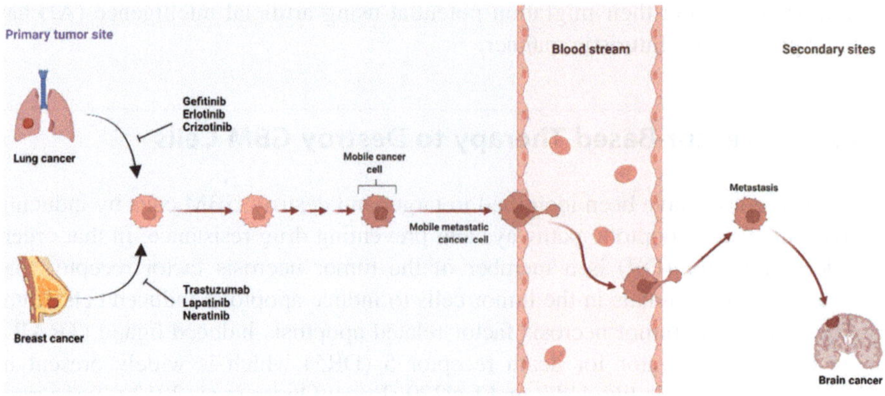

Fig. 14.1 Chemotherapeutic agents inhibit tumor metastasis

14.3 Potential Role of Stem Cells Therapy in the GBM Treatment

Stem cells (SC) are specialized cells that have the potential to develop various types of cells in the body such as muscle cells, blood cells, and brain cells. In that order, mesenchymal stem cells (MSC) originate from various parts of our body mainly from adipose tissues, peripheral blood, umbilical cord, and bone marrow. It has self-renewal and differentiation capability with minimal immunogenic capacity (Shah 2016). MSC inhibits tumor cell growth by releasing TRAIL, arresting the cell cycle, and activating the caspase-mediated apoptotic pathway to induce cell death (Parker Kerrigan et al. 2017). Administration of MSCs can be carried out through various routes such as intratumoral, intracranial, intra-arterial, and intravascular routes to kill the tumor cells (Parker Kerrigan et al. 2017). MSCs can boost the host immune cells including interleukins-2 (IL-2), IL-18, IL-12, and T-cells in the tumor microenvironment to support and destroy the cancer cells (Zhang et al. 2018).

In that order, neural stem cells (NSC) are also called neural precursor cells isolated mainly from the subventricular and periventricular zone in the brain (Matarredona and Pastor 2019). Therapeutic molecules can be delivered through the NSC to alleviate tumor growth. NSC can support chemotherapy like irinotecan, and 5-fluorocytosine (5-FC) via inhibiting topoisomerase-1 and arresting DNA replication in the tumor cells to inhibit the migration and act against the glioma cells (Kim 2011). HSCs are isolated from the bone marrow and used to support the formation of blood cells (Adair et al. 2012). HSC can migrate and release certain antibodies and tumor-specific lymphocytes via chemokine receptor activation at the tumor site to destroy the tumor cells and enhance the survival period of the tumor-bearing mice. Likewise, iPSC is isolated from the inner mass of the blastocyst to produce various types of cells except for placenta cells. Induced pluripotent stem cells (iPSCs) is a vital source of immune cells and prevents tumor invasion by releasing cytotoxic cytokines at the tumor site (Chu et al. 2020). iPSC is a potential

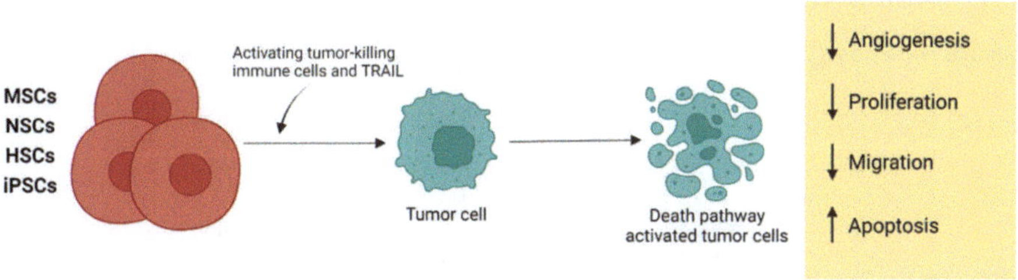

Fig. 14.2 Anti-tumor action of various stem cells. Stem cells (SC) have an anti-tumor activity via triggering the anti-tumor immune response and secreting TRAIL. SC enhances the host immune response in the tumor microenvironment. SC-released ligands such as TRAIL have a potent anti-cancer action against GBM cells. SC represses various growth factors involved in tumor progression and activates caspase-mediated cell death. *MSCs* mesenchymal stem cells, *NSCs* neural stem cells, *HSCs* hematopoietic stem cells; *iPSCs* induced pluripotent stem cells, *TRAIL* TNF-related apoptosis-inducing ligands

delivery vehicle for CD8+ cytotoxic T lymphocytes and proapoptotic genes such as caspase to predict the immune surveillance and study the multiple stages of the intracranial tumor to promote the anti-tumor effect in the glioma cells (Fernandez et al. 2013).

Melanoma is a complicated form of skin cancer that causes brain metastasis to increase morbidity and mortality. Stem cell therapy might be a potential treatment for metastasis melanoma in the brain. Blood–brain barrier (BBB) is composed of microvascular endothelium connected by gap junction, neurons, astrocytes, and basement membrane. Breakdown and leaky BBB cause brain metastasis of melanoma and worsen the metastasis of brain cancer patients. Stem cell therapy repairs the damaged BBB by promoting angiogenesis and vasculogenesis and suppressing inflammation in the tumor microenvironment to prevent metastatic brain tumors (Kaneko et al. 2015). Breast cancer promotes brain metastasis through upregulating EGFR and DR4/5. Simultaneous targeting of DR4/5 and EGFR in breast cancer through engineered allogeneic mesenchymal stem cells and neural stem cells delivered anti-EGFR and potential ligand of DR4/5 tremendously reduced the tumor growth and metastasis (Kitamura et al. 2021) (Fig. 14.2).

14.4 Molecular Mechanism of miRNA and Its Role in the Treatment of GBM

MicroRNA (miR) are small endogenous non-coding RNA that contains approximately 22–24 nucleotides which repress the gene expression through binding into the 3′-untranslated region (3’UTR) and inhibit or degrade the specific messenger RNA (mRNA) strand and causes inhibition of the translation process (Srinivasan et al. 2011). miR-21 is an anti-apoptotic gene that increased in the glioma cells and suppressed apoptotic-induced cell death. Knockdown of miR-21 reversed the temozolomide (TMZ) resistance by upregulating the apoptosis-inducing genes

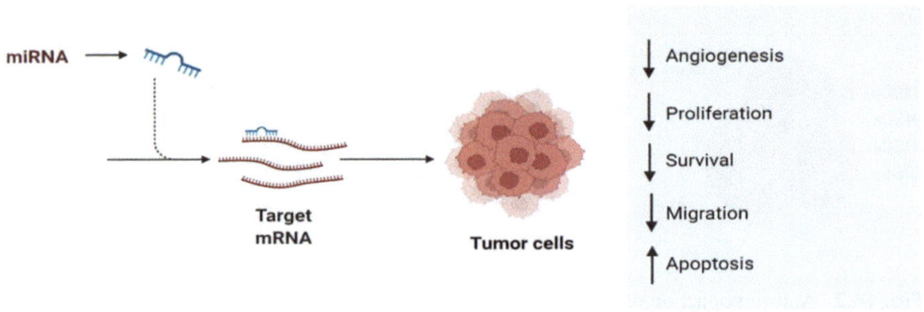

Fig. 14.3 Inhibition of the tumor cell progression by Mirna

such as Bax and phosphatase and tensin homolog (PTEN) protein through repressing the anti-apoptotic genes such as Bcl-2 results in the induction of the cell death through apoptosis in the chemotherapy-resistant glioma cells (Bhere et al. 2020). Neuroblastoma can be prevented through the binding of N-MYC mRNA by miR-204 which results in the suppression of NB tumor progression. Neuroblastoma cells' proliferation and migration would be repressed through the binding of miR-203 into the Sam68 mRNA strand (Balacescu et al. 2019).

Various miRNAs are aberrantly expressed in metastatic brain tumors that promote tumor formation via upregulating oncogenic intracellular pathways and protein secretion which is mainly responsible for tumor cell proliferation and growth. Microarray analysis plays a vital role in the miRNA expression analysis in tumor cells and tissues (Balachandran et al. 2020). miR-141 knockdown inhibited the metastasis potential of triple-negative breast cancer cells to the brain and prevented the formation of colonization in the brain (Kanchan et al. 2020). Studies found that miR-146a is downregulated in brain metastasis compared to the non-metastasis brain tumor considered a significant prognostic factor for metastatic brain tumor (Alsidawi et al. 2014). Colorectal cancer (CRC) is a common digestive tract malignancy caused by changes in diet and lifestyle. Over the past decade, many therapeutic agents such as oxaliplatin, irinotecan, and cetuximab provided potential benefits to patients. Approximately, 1 to 4% of CRC causes brain metastasis in that order, and various miRNAs such as miR-145, miR-143, miR-22, miR-125, and miR-31 act as a biomarker to predict CRC with brain metastasis (Li et al. 2012) (Fig. 14.3).

Tumor suppressor miRs are employed to repress the cancer cell growth by acting on the particular target. miRs directly enter the cell and break down the mRNA strand in the cytoplasmic region to prevent the translation and further oncoprotein synthesis. It leads to the inhibition of the tumor progression and induction of apoptotic cell death.

14.5 A Novel Oncolytic Virotherapy to Act against GBM

Oncolytic virus (OV) is a genetically engineered non-virulent virus that replicates and kills cancer cells without harming healthy tissues. AV can infect the cancer cells through receptor-mediated entry such as CD60 or CD80 which is widely present on the surface of advanced brain tumors to destroy the GBM cells (Vähä-Koskela et al. 2007). AV stimulates the immunological signals including pathogen-associated molecular pattern (PAMP), damage-associated molecular pattern (DAMP), IL-6, tumor necrosis factor- α (TNF-α), and cytotoxic T-lymphocytes to target and kill the tumor cells (Chiocca and Rabkin 2014). AV Delta-24-RGD is considered an anti-glioma agent by reducing tumor volume and enhancing the survival rate in the tumor-bearing xenograft mice model (Zhang and Liu 2020).

Genetically engineered herpes simplex virus 1 (HSV-1) derived thymidine kinase inhibitors convert pro-drugs into cytotoxic agents to destroy the GBM cells like U87MG through bystander effect (de Melo et al. 2015). Nearly 50% of melanoma patients develop brain metastasis and there is no treatment available to patients. More than 90% of melanoma brain metastasis patients die and the average survival time is 17–22 weeks after diagnosis. Recently, the Food and Drug Administration (FDA) approved the oncolytic virus such as herpes simplex virus (HSV) for advanced melanoma brain metastasis patients. Intracarotid administration of MSC-armed HSV (MSC-HSV) effectively tracks the metastatic potential of melanoma brain metastasis patients derived M12 cells and prolongs the survival period of tumor-bearing mice. The addition of MSC-HSV with Programmed Cell Death Ligand 1 (PD-L1) blocker enhances the Interferon γ (IFN γ) produced tumor-infiltrating lymphocytes (TILs) such as $CD8^+$ shows anti-tumor efficacy against melanoma brain metastasis (Du et al. 2017).

Furthermore, the new castle disease virus (NDV) induces autophagy signals in glioma cells including GL261, and U87MG, and causes apoptosis-mediated cell death. NDV arrests the oncogenic signaling pathway like PI3k/Akt and restores the tumor suppressor gene to kill the glioma cell growth (Kazimirsky et al. 2016). Measles virus (MV) can interact with the CD46 receptor in the GBM cells to promote syncytia formation leading to caspase-mediated apoptotic cell death. Likewise, poliovirus (PV) enters through the CD155 glioma surface receptor and it effectively destroys the tumor cells resulting in an increased patient survival rate. Likewise, reovirus (RV) and varicella-zoster virus (VZV) showed excellent oncolytic properties against GBM cells including U373, U87MG through repressing the tumor cell signaling pathway, and inhibiting anti-apoptotic genes (Fig. 14.4) (Muik and Von Laer 2011).

14.6 Targeting GBM Through Checkpoint Inhibitors

Over the last decade, immunotherapy has potentially been considered an emerging therapeutic approach to treat advanced cancer. Immunotherapy can be used as an alternative therapy to prevent TMZ-induced chemo-resistance and destroy cancer

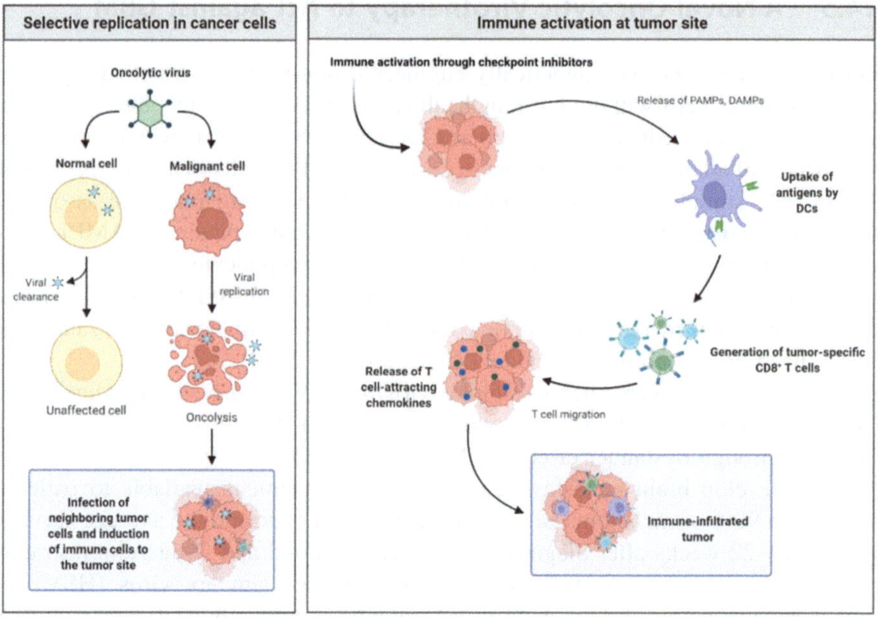

Fig. 14.4 Role of oncolytic virus and immune checkpoint inhibitors in cancer treatment. Oncolytic virus (OVs) selectively replicates in the tumor cells and induces cell death. Also, OVs can migrate nearby tumor cells and cause tumor cell death. Immune checkpoint inhibitors activate various immune cells and their mediators to infiltrate the tumor-killing immune cells to inhibit the tumor cells. *PAMP* pathogen-associated molecular patterns, *DAMP* damage-associated molecular patterns, *DCs* dendritic cells

cells (Wang et al. 2020). Administration of dendritic cells (DC)-based vaccines to GBM patients caused increasing T-cell infiltration and showed an excellent anti-cancer effect. Likewise, the injection of DC enhanced the cytokine release and increased the survival rate of tumor-bearing patients. Studies revealed that PD-L1 expression was observed in 94 GBM patients and found that 61% have PD-L1 positive cells. The efficacy of the PD-L1 inhibitors was increased with the combination of radiation therapy in the glioma mice model (Huang et al. 2017).

Brain metastasis is a series of complications that spread across the central nervous system (CNS) and cause high morbidity and mortality. The increasing prevalence of brain metastasis limited the therapy's efficacy in fighting against tumor cells. Various monoclonal antibodies like pembrolizumab and nivolumab target checkpoint inhibitors such as CTLA-4 and PD-L1 have been developed to treat brain metastasis (Nieblas-Bedolla et al. 2021). Phase II clinical study was performed to evaluate the therapeutic efficacy of anti-CTLA-4 such as ipilimumab and fotemustine in advanced melanoma brain metastasis (Berghoff et al. 2021). Clinical studies found that pembrolizumab and atezolizumab (PD-L1 inhibitors) received melanoma or NSCLC brain metastasis patients improved their overall survival rate by 33% through activating TILs such as CD8+ and CD4+ T-lymphocytes (Amin et al. 2020; Vilariño et al. 2020). Renal cell carcinoma (RCC) brain metastasis patients show poor prognosis and OS of 6 to 10 months. Studies revealed that checkpoint inhibitors such as nivolumab and ipilimumab are safe and effective to treat RCC brain metastasis patients (Fallah and Ahluwalia 2019). Intravenous infusion of CAR-T cells EGFRvIII repressed the growth factor protein expression and reduced the tumor volume. A combination of EGFR-based targeted vaccines and chemotherapy like TMZ potentiated the host immune response and extended the OS and PFS in glioma patients (Fig. 14.1) (Khamis et al. 2020).

14.7 CRISPR-Cas 9 Technology to Repress GBM Cell Growth

In 2002, CRISPR-Cas9 was first developed using *Escherichia coli*. Alteration of DNA or RNA through gene therapy is a form of treatment that remedies the malfunctioning genes. Among all the available tools, a recent breakthrough in CRISPR-Cas9 gene-editing technology has revolutionized the field of gene therapy. It is a simple, high-accuracy, strong specificity, cost-efficient, and time-saving eukaryotic cell gene-editing method (You et al. 2019). Studies found that CRISPR-Cas9 mediated aryl hydrocarbon receptor (AhR) knockdown in GBM cells enhanced tumor growth, invasion, migration, and survival. Epigenetic alterations and genetic mutations are considered the primary cause of tumor formation (Ratan et al. 2018). Scientists have continually worked to modify the CRISPR-Cas9 to use as a delivery vehicle to increase the efficacy of therapeutic molecules and to reduce the off-target effect (Chen et al. 2016).

In cancer research, CRISPR-Cas9 has been developed to explore tumor occurrence and metastasis. Genetic manipulation can be done through the DNA knockout results in a transcriptional alteration to treat disease. Moreover, CRISPR-Cas9 is used to build a preclinical tumor model by causing gene mutations including Kras, liver kinase b1 (Lkb1), and p53 in mice (Jiang et al. 2019). CRISPR-Cas9 can identify the genes which are resistant to other therapies and also can be used to knock out the mutated genes to promote the treatment to destroy the cancer cells. CRISPR-Cas9 showed a potential therapeutic effect in immunotherapy through universal CAR T-cells creation deleting PD-1 and CTLA-4 genes to stimulate the advanced solid tumor cells destruction (Martinez-lage et al. 2018). Using genome-editing CRISPR-Cas9 technology, the genes responsible for glioma stem cell (GSCs) growth, proliferation, and survival can be identified to support the treatment (MacLeod et al. 2018). Downregulation of the MGMT andAPOBEC3B gene has been achieved through CRISPR-Cas9 results in an enhancement of temozolomide sensitivity and reduction the chemoresistance in the GBM cells including LN229, A172 (Fig. 14.5) (Yang et al. 2021).

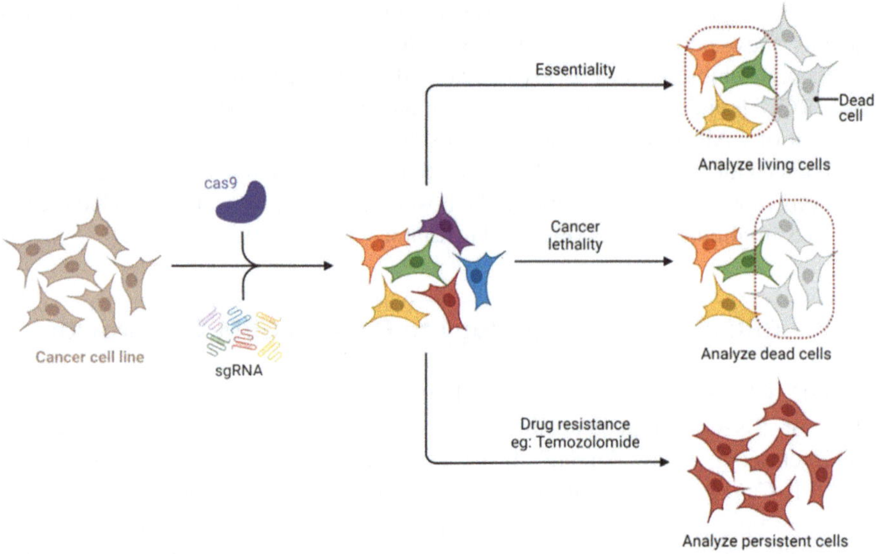

Fig. 14.5 Anti-cancer action of CRISPR- Cas9. CRISPR-Cas9 ultimately inhibits the tumor progression by knock outing the mutated gene and reversing the drug resistance. *CRISPR* clustered regularly interspaced short palindromic repeats, *Cas9* CRISPR-associated protein 9, *sgRNA* single guide RNA

14.8 Artificial Intelligence in GBM Therapy

Artificial intelligence (AI) is a branch of computer science that deals with analyzing medical data, diagnosis, and treatment outcomes. AI is generally classified into generalized, super, and narrow. AI gained more attention in the scientific world because of its problem-solving and disease-diagnosing ability (Iqbal et al. 2021). In AI, machine learning (ML) is one of the advanced techniques used to predict brain-related diseases such as memory impairment, multiple sclerosis, epilepsy, cerebral artery, schizophrenia, and cancer. ML mainly employed to detect the brain structure and its pathological modifications including brain tumor segmentation. Magnetic resonance imaging (MRI) plays a vital role in the molecular and pathological stage of advanced brain tumors. MRI helps to identify the brain tumor grades faster and more accurately without performing a patient's biopsy test. Also, ML tends to detect the tumor tissue deformation so that the tissue contains future risk also can be recognized (Segato et al. 2020). Another novel technique such as the stationary wavelet-based radiomics approach performed to evaluate the glioma grades like low-grade glioma (LGG) to high-grade glioma (HGG) patients more accurately in a non-invasive manner (Kumar et al. 2020). Usually, LGG did not show any contrast enhancement or MR sequences compared to HGG with nodular enhancement, increased perfusion, and central necrosis (Sotoudeh et al. 2019). ML shows promising results in the automated interpretation of brain tumor imaging and histopathology to identify the tumor burden and treatment effect with the help of AI.

AI algorithms significantly identify the aberrant genetic mutation protein expression at the very initial stage. ML and deep learning (DL) can identify genetic mutations such as iso-citrate dehydrogenase 1/2 (IDH1/2), and O-methylguanine-DNA methyltransferase (MGMT) with a high accuracy rate of 95% which is responsible for the development of tumor formation. The researchers developed novel deep learning (DL) CNN method to estimate the mutated gene in the brain tumor such as codeletion of chromosome arms 1p/19q with an accuracy rate of 92% (Khalsa et al. 2020). DL improved to diagnose and predict various health complications such as brain tumors. DL can also identify the brain tumor segmentation using MRI scan. MRI contains multiple sequencing techniques such as T1-weighed and T2-weighed fluid attenuation inversion recovery (FLAIR) to assess the grades of advanced metastatic brain tumors. Convolutional Neural Network (CNN) is a potential image recognition technique that tends to predict the survival rate of brain tumor patients. 3D-CNN plays an essential role in the classical regression methods to predict HGG patient's survival time with better accuracy including tumor shape, location, and the metastatic range (Nadeem et al. 2020). Radiomics analysis in radiotherapy tends to helps the neuroimaging in patients with glioma and brain metastasis with accurate segmentation. Radiomics with conventional MRI combined to predict the tumor grading and the metastatic potential of the brain tumor. Also, ML and radiomics are used the identify the progression-free survival (PFS) and overall survival (OS) rate of brain tumor patients (Kocher et al. 2020).

14.9 Conclusion and Its Future Perspectives

In the past decades, treating GBM and destroying its growth including GSC is the biggest hurdle to treat an advanced metastatic brain tumor. Due to the high resistance and drug tolerance in the existing therapies failed to cure the disease and did not improve the survival rate of the patients. Tumor formation in the glial cells not only affects the infected area rather it disturbs the normal physiological function and also causes psychological disorders including memory impairment. Moreover, brain tumor occurs through metastasis breast, lung, melanoma, and other types of cancer including renal, and colorectal cancers. In that order, some novel receptors have been discovered like DR4/5 is considered as an effective target to induce apoptosis in the GBM cells. SCs and miRNA-based therapies are directly entering into the cancer cells and targeting the messenger RNA (mRNA) or other cellular materials which are responsible for the induction of cancer cell formation to destroy without inducing any undesirable side effects or resistance. Moreover, OVs and checkpoint inhibitors are attractive immunotherapeutic treatment options to induce the host immune system in the tumor microenvironment and to repress the GBM cell growth. Likewise, the CRISPR-Cas9 gene-editing tool is also an emerging technique to identify and remove the aberrantly expressed gene and prevent the disease. Moreover, monitoring tumor growth and the treatment regimen is a significant area to develop that helps the researchers and physicians to identify the tumor volume and the effectiveness of the following treatment. By this, AI is a growing technique in the medical field to predict the tumor grades and their pathological pathways help to promote the therapy option to cure diseases. So, the mentioned novel therapies might be a potential alternative way to destroy the cancer cells and many of these therapies have been moved forward to clinical studies. In the future, we can expect these therapies widely in the field of brain tumor treatment.

Acknowledgement All authors acknowledge the Manipal College of Pharmaceutical Sciences, MAHE, Manipal for supporting this work.

References

Adair JE, Beard BC, Trobridge GD et al (2012) Extended survival of glioblastoma patients after chemoprotective HSC gene therapy. Sci Transl Med 4:1–12. https://doi.org/10.1126/scitranslmed.3003425

Alsidawi S, Malek E, Driscoll JJ (2014) Micrornas in brain metastases: potential role as diagnostics and therapeutics. Int J Mol Sci 15:10508–10526. https://doi.org/10.3390/ijms150610508

Amin S, Baine MJ, Meza JL et al (2020) Association of immunotherapy with survival among patients with brain metastases whose cancer was managed with definitive surgery of the primary tumor. JAMA Netw Open 3:1–12. https://doi.org/10.1001/jamanetworkopen.2020.15444

Bagci-Onder T, Agarwal A, Flusberg D et al (2012) Real-time imaging of the dynamics of death receptors and therapeutics that overcome TRAIL resistance in tumors. Oncogene 32:1–10. https://doi.org/10.1038/onc.2012.304

Balacescu O, Visan S, Baldasici O et al (2019) MiRNA-based therapeutics in oncology, realities, and challenges. Antisense Ther:1–27. https://doi.org/10.5772/intechopen.81847

Balachandran AA, Larcher LM, Chen S et al (2020) Therapeutically significant MicroRNAs in primary and metastatic brain malignancies. Cancers (Basel) 12:1–29. https://doi.org/10.3390/cancers12092534

Berghoff AS, Venur VA, Preusser M et al (2021) Immune checkpoint inhibitors in brain metastases: from biology to treatment. Am Soc Clin Oncol Educ Book 35:116–122

Bhere D, Arghiani N, Lechtich ER et al (2020) Simultaneous downregulation of miR-21 and upregulation of miR-7 has anti-tumor efficacy. Sci Rep 10:1–10. https://doi.org/10.1038/s41598-020-58072-w

Cha Z, Cheng J, Xiang H et al (2019) Celastrol enhances TRAIL-induced apoptosis in human glioblastoma via the death receptor pathway. Cancer Chemother Pharmacol 84:719–728. https://doi.org/10.1007/s00280-019-03900-8

Chen S, Sun H, Miao K et al (2016) CRISPR-Cas9: from genome editing to cancer. Int J Biol Sci 12:1427–1436. https://doi.org/10.7150/ijbs.17421

Chiocca E, Rabkin S (2014) Oncolytic viruses and their application to cancer immunotherapy. Cancer Immunol Res 2:295–300. https://doi.org/10.1158/2326-6066.CIR-14-0015.Oncolytic

Chu DT, Nguyen TT, Tien NLB et al (2020) Recent progress of stem cell therapy in cancer treatment: molecular mechanisms and potential applications. Cell 9:563. https://doi.org/10.3390/cells9030563

Daisy PS, Shreyas KS, Anitha TS (2021) Will CRISPR—Cas9 have cards to play against cancer? An update on its applications. Mol Biotechnol 63:93–108. https://doi.org/10.1007/s12033-020-00289-1

de Melo SM, Bittencourt S, Ferrazoli EG et al (2015) The anti-tumor effects of adipose tissue mesenchymal stem cell transduced with HSV-Tk gene on U-87-driven brain tumor. PloS One 10:1–13. https://doi.org/10.1371/journal.pone.0128922

Dodson C, Richards TJ, Smith DA et al (2020) Tyrosine kinase inhibitor therapy for brain metastases in non-small-cell lung cancer: a primer for radiologists. AJNR Am J Neuroradiol 41:738–750. https://doi.org/10.3174/ajnr.A6477

Du W, Seah I, Bougazzoul O et al (2017) Stem cell-released oncolytic herpes simplex virus has therapeutic efficacy in brain metastatic melanomas. Proc Natl Acad Sci U S A 114:E6157–E6165. https://doi.org/10.1073/pnas.1700363114

Fallah J, Ahluwalia MS (2019) The role of immunotherapy in the management of patients with renal cell carcinoma and brain metastases. Ann Transl Med 7:8–12. https://doi.org/10.21037/atm.2019.10.31

Fernandez TDS, De Souza FC, Mencalha AL (2013) Human induced pluripotent stem cells from basic research to potential clinical applications in cancer. Biomed Res Int 2013:1–12. https://doi.org/10.1155/2013/430290

Hemminki O, Manuel J, Hemminki A (2020) Oncolytic viruses for cancer immunotherapy. J Hematol Oncol 13:1–15. https://doi.org/10.1186/s13045-020-00922-1

Hosonaga M, Saya H, Arima Y (2020) Molecular and cellular mechanisms underlying brain metastasis of breast cancer. Cancer Metastasis Rev 39:711–720. https://doi.org/10.1007/s10555-020-09881-y

Huang B, Zhang H, Gu L et al (2017) Advances in immunotherapy for glioblastoma multiforme. J Immunol Res 2017:1–12. https://doi.org/10.1155/2017/3597613

Iqbal MJ, Javed Z, Sadia H et al (2021) Clinical applications of artificial intelligence and machine learning in cancer diagnosis: looking into the future. Cancer Cell Int 21:1–11. https://doi.org/10.1186/s12935-021-01981-1

Jiang C, Meng L, Yang B et al (2019) Application of CRISPR/Cas9 gene editing technique in the study of cancer treatment. Clin Genet 97:1–16. https://doi.org/10.1111/cge.13589

Kanchan RK, Siddiqui JA, Mahapatra S et al (2020) microRNAs orchestrate pathophysiology of breast cancer brain metastasis: advances in therapy. Mol Cancer 19:1–16. https://doi.org/10.1186/s12943-020-1140-x

Kaneko Y, Tajiri N, Staples M et al (2015) Bone marrow-derived stem cell therapy for metastatic brain cancers. Cell Transplant 24:625–630. https://doi.org/10.3727/096368914X685096

Kazimirsky G, Jiang W, Slavin S et al (2016) Mesenchymal stem cells enhance the oncolytic effect of newcastle disease virus in glioma cells and glioma stem cells via the secretion of TRAIL. Stem Cell Res Ther 7:1–10. https://doi.org/10.1186/s13287-016-0414-0

Khalsa SSS, Hollon TC, Adapa AR et al (2020) Automated histologic diagnosis of CNS tumors with machine learning. CNS Oncol 9:1–10. https://doi.org/10.2217/cns-2020-0003

Khamis ZI, Al-akkary N, Hua T et al (2020) Clinical investigations of immunotherapy for human primary brain tumors. Neuroimmunol Neuroinflamm 7:1–20. https://doi.org/10.20517/2347-8659.2020.43

Kim SU (2011) Neural stem cell-based gene therapy for brain tumors. Stem Cell Rev Reports 7:130–140. https://doi.org/10.1007/s12015-010-9154-1

Kitamura Y, Kanaya N, Moleirinho S et al (2021) Anti-EGFR VHH-armed death receptor ligand–engineered allogeneic stem cells have therapeutic efficacy in diverse brain metastatic breast cancers. Sci Adv 7:1–18. https://doi.org/10.1126/sciadv.abe8671

Kocher M, Ruge MI, Galldiks N, Lohmann P (2020) Applications of radiomics and machine learning for radiotherapy of malignant brain tumors. Strahlenther Onkol 196:856–867. https://doi.org/10.1007/s00066-020-01626-8

Kumar R, Gupta A, Raman B (2020) CGHF: a computational decision support system for glioma classification using hybrid Radiomics- and stationary wavelet-based features. IEEE Access 8:79440–79458. https://doi.org/10.1109/ACCESS.2020.2989193

Li Z, Gu X, Fang Y et al (2012) microRNA expression profiles in human colorectal cancers with brain metastases. Oncol Lett 3:346–350. https://doi.org/10.3892/ol.2011.497

MacLeod G, Bozek DA, Rajakulendran N et al (2018) The functional genomic circuitry of human glioblastoma stem cells. bioRxiv:1–42. https://doi.org/10.1101/358432

Martinez-lage M, Puig-serra P, Menendez P et al (2018) CRISPR/Cas9 for cancer therapy: hopes and challenges. Biomedicine 6:1–13. https://doi.org/10.3390/biomedicines6040105

Matarredona ER, Pastor AM (2019) Neural stem cells of the subventricular zone as the origin of human glioblastoma stem cells. Therapeutic implications. Front Oncol 9:1–12. https://doi.org/10.3389/fonc.2019.00779

Meca-cortés O, Guerra-rebollo M, Garrido C et al (2017) CRISPR/Cas9-mediated knockin application in cell therapy: a non-viral procedure for bystander treatment of glioma in mice. Mol Ther Nucleic Acids 8:395–403. https://doi.org/10.1016/j.omtn.2017.07.012

Muik A, Von Laer D (2011) Oncolytic virotherapy of glioma: what does it need to make it work? Future Virol 6:1289–1297. https://doi.org/10.2217/fvl.11.111

Nadeem M, Al GMA, Hussain M et al (2020) Brain tumor analysis empowered with deep learning: a review, taxonomy, and future challenges Muhammad Waqas. Brain Sci 10:1–33. https://doi.org/10.3390/brainsci10020118

Nicholas MK, Lukas RV, Jafri NF et al (2006) Epidermal growth factor receptor- mediated signal transduction in the development and therapy of gliomas. Clin Cancer Res 12:7261–7271. https://doi.org/10.1158/1078-0432.CCR-06-0874

Nieblas-Bedolla E, Nayyar N, Singh M et al (2021) Emerging immunotherapies in the treatment of brain metastases. Oncologist 26:231–241. https://doi.org/10.1002/onco.13575

Parker Kerrigan BC, Shimizu Y, Andreeff M et al (2017) Mesenchymal stromal cells for the delivery of oncolytic viruses in gliomas. Cytotherapy 19:445–457. https://doi.org/10.1016/j.jcyt.2017.02.002

Patel MR, Kratzke RA (2013) Oncolytic virus therapy for cancer: the first wave of translational clinical trials. Transl Res 161:355–364. https://doi.org/10.1016/j.trsl.2012.12.010

Ratan ZA, Son YJ, Haidere MF et al (2018) CRISPR-Cas9: a promising genetic engineering approach in cancer research. Ther Adv Med Oncol 10:1–15. https://doi.org/10.1177/1758834018755089

Segato A, Marzullo A, Calimeri F et al (2020) Artificial intelligence for brain diseases: a systematic review. APL Bioeng 4:1–36. https://doi.org/10.1063/5.0011697

Shah K (2016) Stem cell-based therapies for tumors in the brain: are we there yet? Neuro Oncol 18: 1066–1078. https://doi.org/10.1093/neuonc/now096

Smith C, Lineburg KE, Martins JP et al (2020) Autologous CMV-specific T cells are a safe adjuvant immunotherapy for primary glioblastoma multiforme. J Clin Invest 130:6041–6053. https://doi.org/10.1172/JCI138649

Sotoudeh H, Shafaat O, Bernstock JD et al (2019) Artificial intelligence in the management of Glioma: era of personalized medicine. Front Oncol 9:1–11. https://doi.org/10.3389/fonc.2019.00768

Srinivasan S, Rosita I, Patric P et al (2011) A ten-microRNA expression signature predicts survival in glioblastoma. PloS One 6:1–7. https://doi.org/10.1371/journal.pone.0017438

Sun Y, Xu J, Zhou J et al (2018) Targeted drugs for systemic therapy of lung cancer with brain metastases. Oncotarget 9:5459–5472

Vähä-Koskela MJV, Heikkilä JE, Hinkkanen AE (2007) Oncolytic viruses in cancer therapy. Cancer Lett 254:178–216. https://doi.org/10.1016/j.canlet.2007.02.002

Venur VA, Ahluwalia MS (2016) Targeted therapy in brain metastases: ready for primetime? Am Soc Clin Oncol Educ Book 36:e123–e130. https://doi.org/10.14694/edbk_100006

Vilariño N, Bruna J, Valiente M et al (2020) Immunotherapy in NSCLC patients with brain metastases. Understanding brain tumor microenvironment and dissecting outcomes from immune checkpoint blockade in the clinic. Cancer Treat Rev 89:102067. https://doi.org/10.1016/j.ctrv.2020.102067

Wang W, Sun X, Hui Z (2019) Treatment optimization for brain metastasis from anaplastic lymphoma kinase rearrangement non-small-cell lung cancer. Oncol Res Treat 2019:42599–42606. https://doi.org/10.1159/000502755

Wang J, Shen F, Yao Y et al (2020) Adoptive cell therapy: a novel and potential immunotherapy for glioblastoma. Front Oncol 10:1–13. https://doi.org/10.3389/fonc.2020.00059

Yang Q, Zhou Y, Chen J et al (2021) Gene therapy for drug-resistant glioblastoma via lipid-polymer hybrid nanoparticles combined with focused ultrasound. Int J Nanomedicine 16:185–199. https://doi.org/10.2147/IJN.S286221

You L, Tong R, Li M et al (2019) Advancements and obstacles of CRISPR-Cas9 technology in translational research. Mol Ther Methods Clin Dev 13:359–370. https://doi.org/10.1016/j.omtm.2019.02.008

Zhang Q, Liu F (2020) Advances and potential pitfalls of oncolytic viruses expressing immunomodulatory transgene therapy for malignant gliomas. Cell Death Dis 11:1–11. https://doi.org/10.1038/s41419-020-2696-5

Zhang Q, Xiang W, Yi DY et al (2018) Current status and potential challenges of mesenchymal stem cell-based therapy for malignant gliomas. Stem Cell Res Ther 9:1–9. https://doi.org/10.1186/s13287-018-0977-z

Zhu Y, Bassoff N, Reinshagen C et al (2017) Bi-specific molecule against EGFR and death receptors simultaneously targets proliferation and death pathways in tumors. Sci Rep 7:1–11. https://doi.org/10.1038/s41598-017-02483-9

Pharmacokinetics and Pharmacodynamics: Fundamentals and Role(s) in Drug Discovery and Development

15

Ashu Mittal, Roma Ghai, Alankar Srivastava, Deba Prasad Ghosh, and K. Nagarajan

Abstract

The process of finding new drugs begins with the identification of a target and its validation, then hits are found, leads are generated, and they are optimized. Eventually, a suitable and appealing candidate is chosen, who will then be sent for further development. This chapter's primary goal is to provide an overview of the fundamental concepts behind pharmacokinetics (PK) and pharmacodynamics (PD), as well as how PK works. In the many stages of the drug discovery and development procedures, pharmacodynamics is used. Pharmacokinetic parameters from PK experimental research are used to comprehend and anticipate the drug response, such as clearance, apparent Vd and elimination half-life, and bioavailability. This chapter also discusses the different types of compartmental models with their implications for drug development. Principles from pharmacodynamics are used to enable the safe and effective therapeutic management of patients; hence, various receptor and non-receptor-mediated mechanisms of action are included in this chapter. To assure the accuracy of the PK data generated in the drug development process, it is crucial to establish and validate an appropriate bioanalytical procedure; the application of LC-MS methods in bioanalysis has been discussed in this chapter.

Keywords

Bio-analysis · Drug development · Drug discovery · Pharmacodynamics · Pharmacokinetics

A. Mittal (✉) · R. Ghai · D. P. Ghosh · K. Nagarajan
KIET School of Pharmacy, KIET Group of Institutions, Ghaziabad, India
e-mail: ashu.mittal@kiet.edu; roma.ghai@kiet.edu; daba.prasad@kiet.edu; k.nagarajan@kiet.edu

A. Srivastava
Shri Rawatpura Sarkar Institute of Pharmacy, Kumhari Durg, Chhattisgarh, India

P. P. Singh (ed.), *Recent Advances in Pharmaceutical Innovation and Research*,
https://doi.org/10.1007/978-981-99-2302-1_15

15.1 Introduction

Novel active pharmaceutical ingredients are always in constant demand in the pharmaceutical and healthcare industry to counter the challenges that are arising from unfulfilled medical requirements across a wide variety of therapeutic domains. Thus, the primary endeavor of any pharmaceutical industry involved in the research and development of novel drugs is to introduce new active pharmaceutical ingredients into the market through the complicated, tedious, laborious, and expensive activity of drug discovery and development. Identification of a target and its validation, then hit identification, followed by lead creation and optimization, and eventually, the identification of a suitable and desirable candidate which can then be ultimately sent for further development all make up this drug discovery process. The research on drug development, on the contrary, consists of the optimization of the chemical synthesis process as well as the formulation, conducting toxicity studies in animals, doing clinical trials, and eventually acquiring all the necessary regulatory approvals (David 2015). The primary objective of this chapter is to provide an overview of the general principles of pharmacokinetics and pharmacodynamics and how pharmacokinetics, commonly known as PK, and pharmacodynamics (PD) are applied in the various phases of drug discovery and drug development processes. Pharmacokinetics is the mathematical study of the movement of a drug and/or its metabolites inside the body. Some of the paramount pharmacokinetic properties that are required to be understood during any drug discovery and development phase include the absorption and the disposition of a drug. These are the vital pharmacokinetic parameters for understanding and predicting the action of a drug. Pharmacological activity relies on the requisite drug levels at the intended site of action. Pharmacokinetics (PK) studies' primary function in the development of new medications is to aid in the ADME (absorption, distribution, metabolism, and excretion) properties optimization of lead compounds, to achieve the main purpose of creating a suitable therapeutic candidate with a sufficient concentration vs. time profile inside the body for the desired clinical action. Data generated from pharmacokinetic studies also aids in the optimization of the drugs in humans by comparing the properties involved in the gastrointestinal absorption of an extravascularly administered drug. This includes quick and inexpensive determination of the distribution, clearance, elimination, and adverse effect potential of that drug. Therefore, the goal of the drug discovery and development process is to provide both effective and safe drugs for human consumption (Patrick and Vladimir 2019). Data obtained from pharmacokinetics and pharmacodynamics studies and PKPD studies together play a very important role in determining the safety and efficacy of any drug. Another very important application of PKPD studies is to develop drug product labels before launching new drugs or novel or modified formulations. This product label, often known as the package insert, is the "manual" for using a drug safely and effectively. It is necessary to list down the crucial bio-pharmaceutical, pharmacological, and pharmacokinetic characteristics of a drug product on the label in particular sections, as per the requirement of the CFR (Code of Federal Regulations), and also in a language and format as recommended by the United States Food and

Drug Administration (USFDA) in several of its guidance documents for the industry. During drug discovery and development, experiments collect bio-pharmaceutical, pharmacological, and pharmacokinetic properties of the drug. These properties optimize the lead compound by evaluating factors like gastrointestinal absorption (for oral medications), distribution, metabolism, clearance, and elimination. The aim is to efficiently and affordably achieve these optimizations, developing a clinical candidate with the desired concentration vs. time profile for the intended therapeutic effect (Patrick and Jogarao 2002).

15.2 Drug Development and Drug Discovery: An Overview

Drug discovery is a process of locating chemical entities with the prospective to turn into medicinal agents. The primary objective of drug development programs is to identify new molecular entities which may become useful in the management of ailments that have already been qualified as unmet medical needs as well as better alternatives compared to an already existing medication. The unmet medical needs stand for those disorders which practically have no definite treatments and are either potentially or currently (Kiriiri et al. 2020). Currently, the commercialization of only a few medication target categories has happened. These include G-protein coupled receptors, nuclear (hormone) receptors, and ion channels. But these are the targets for somewhat less than half of all the medications that are currently available in the market. A significant portion of marketed drugs are designed to target enzymes. However, it may be necessary to explore newer types of drug targets to address therapeutic limitations. Selecting a target that holds potential value becomes crucial, especially when venturing into less-investigated types of drug targets, despite the lack of substantial existing knowledge (Alexander et al. 2017). The traditional pharmaceutical research and development process have considerably high rate of attrition. From tens of thousands of chemicals, researches first narrow down to a few therapeutic leads after going through over 100 screens; this laborious and tedious work is necessary to buy a new medicine into the market. Thus, the discovery of a lead compound is both an expensive and also time-consuming process. According to some estimates, the discovery of a lead compound requires more than $200 million and up to 5 years, again not to mention the even longer time and costs associated with drug development (Atanas et al. 2015). Regardless of finding a promising lead compound, a variety of unsuspected reasons at the time of the lead discovery phase can lead to its failure at the time of the drug development phase. Undesirable toxicity, insufficient levels of efficacy in the in-vivo models mimicking the disease of particular interest, issues regarding its attractiveness in the market, and meager biopharmaceutical properties are all can be the potential reasons behind the failure. Synthetic intricacy, low potency, and confusing toxicological findings can all slow down development significantly (Richard and Nigel 2017).

Drug discovery tactics must be carefully considered and implemented, especially when setting the foot in the realm of a new disease or novel drug target. The four critical aspects in drug discovery include the determination of pharmacological targets, developing diversification of chemicals as a source of new molecular entities, use of screening methodologies to define lead structures, and certifying the lead molecules for preliminary trials (Hughes et al. 2011). Performing DMPK (Drug Metabolism and Pharmacokinetics) research, also called the ADMET (Absorption, Distribution, Metabolism, Elimination, and Toxicity) investigations, is an integral part of the drug discovery and development process. These experiments help in deciding the feasibility of drug candidate by answering the following crucial questions:

- Absorption—How much and how swiftly is the medication getting absorbed? (bioavailability).
- Distribution—Where does the drug become distributed within the body? What is the rate of distribution and size of distribution?
- Metabolism—How rapidly does the drug break down? What is the mechanism behind its breakdown? What type of metabolite is getting generated, and is it harmful or useful?
- Elimination—When and how will the drug excrete out?
- Toxicity—Is this drug poisonous to any of the body's organs or systems?

As per estimation, nearly half of all possible drug candidates fall short due to inadequate efficacy, and the failure rate because of previously detected toxicity has been up to 40% of all potential drug candidates. Due to the toxicity or drug-drug interactions, medications like soruvidine, mibefradil, and phenylpropanolamine hydrochloride have been withdrawn from the market. Realizing this, the pharmaceutical companies and the regulators both have reached the conclusion that, in addition to pharmacological properties, investigations of ADME/Tox parameters are vital for the success of a drug candidate in the market (Gail 2019). In vitro and in vivo investigations are quite useful in assisting researchers involved in the drug development process to determine whether a new chemical entity should be selected as a possible drug candidate and sent forward onto late-stage preclinical and clinical trials or not. ADME properties are needed for approval by the regulatory authorities and to help drug developers determine the safety and efficacy of a new therapeutic entity. While each medicine is unique, FDA guidance documents describe distinct models and related assays that aid the scientists to ascertain which ADME aspects should be examined. In vitro ADME investigations usually use liver microsomes and whole hepatocyte models, both of which contain metabolic enzymes such as CYP450 and UDP-glucuronosyltransferase (UGT). These in vitro models can be used in CYP inhibition and induction assays. In vitro tests such as CACO-2 or MDCK cell-based research are used to assess intestinal permeability. In vivo investigations are undertaken to examine pharmacokinetic (PK) qualities during drug discovery, as well as in late-stage preclinical and non-clinical trials. Nonhuman primates are used to generate PK data to evaluate properties such as drug clearance, bioavailability, exposure, half-life, and distribution volume (Steinmetz and Spack 2009).

15.2.1 Drug Discovery and Research: Role of Molecular Bio-pharmaceutics Research

Molecular biopharmaceutics is a scientific domain that is both complex and still developing. It extensively encompasses the indispensable reciprocity between the membrane transporters and the chemistry of drug molecules. This also helps to direct investigators in constructing experiments that may aid in understanding the mechanisms and kinetics involved in transport-drug absorption and delivery. Molecular biopharmaceutics considers the following aspects of understanding bioavailability by (1) physicochemical characterization, (2) membrane transport, and (3) biosimulation methods. These issues are of increasing importance in drug development. For drug research and development, biopharmaceutics will be a valuable resource. It covers the essentials of the interaction between the membrane transporters and the chemistry of pharmacological molecules in a complicated and rapidly evolving scientific field. It also intends to assist drug developers and researchers in designing studies that will aid in the study of the kinetics and mechanisms involved in transport-drug absorption and delivery. To understand drug bioavailability, it considers physicochemical characterization, membrane transport, and biosimulation approaches. In drug development, these strategies are becoming increasingly important (Alavijeh et al. 2012).

15.2.2 Pharmacokinetics: Principles in Drug Discovery and Development

Pharmacokinetics plays a pivotal role in the field of drug discovery and development. It encompasses the principles that govern the fate of a drug within the body, focusing on its absorption, distribution, metabolism, and elimination. By studying pharmacokinetics, researchers aim to understand how a drug is absorbed, distributed to its target site, metabolized, and ultimately eliminated from the body. These insights help in optimizing drug formulations, dosage regimens, and therapeutic strategies. In drug discovery, pharmacokinetics guides the selection and optimization of lead compounds by assessing their bioavailability and pharmacokinetic profiles. It helps identify compounds with favorable absorption characteristics and suitable distribution patterns to reach the intended target site. Additionally, pharmacokinetic considerations aid in predicting drug-drug interactions, evaluating the potential for adverse effects, and estimating the appropriate dosage for subsequent studies. During drug development, pharmacokinetics is instrumental in determining the appropriate dosage form and administration route for a drug. It assists in establishing the relationship between the administered dose and the resulting plasma concentration, helping define dosing intervals and therapeutic ranges. Moreover, pharmacokinetic studies provide crucial information for regulatory submissions, guiding the design of clinical trials and ensuring the safety and efficacy of the drug candidate. Overall, understanding the principles of pharmacokinetics is vital for making informed decisions throughout the drug discovery and development process, ultimately leading to the successful translation of potential therapeutic agents into clinical practice. Pharmacokinetics is a quantitative study that examines the rate processes of drug absorption and drug disposition (distribution, metabolism, and excretion), focusing on changes in plasma drug concentration. The plasma concentration of a drug rises and falls based on the rates of three primary processes: absorption, distribution, and elimination. First, drug absorption involves the movement of a drug from its administration site (e.g., gastrointestinal tract) into the bloodstream. The rate of absorption depends on various factors such as drug permeability, absorption characteristics, affinity of the drug for tissues, and formulation. Second, distribution occurs when the drug reversibly exits the bloodstream, distributing into extravascular tissues including interstitial and intracellular fluids Nishant et al. (2011). Third, the drug's elimination from the blood rests on two processes, that is,

1. metabolism of the drug into one or more metabolites (Biotransformation) which takes place mainly in the liver; and,
2. the excretion of the parent drug (unchanged drug) or its metabolite(s), mainly through the kidneys.

The characteristics of the above-mentioned processes can be further expressed by their pharmacokinetic and pharmacodynamic parameters. To interpret and comprehend the mystery behind the ways by which a medication interacts with the body, research utilizes the pharmacokinetic (PK) parameters. The PK parameters are immensely useful for a drug discovery scientist. The PK has been proven to be quite an effective tool in the process of drug development both in the cases of therapeutics and also in defining the disposition characteristics of the drug.

It deduces

1. How a medication is absorbed after being taken.
2. How the body distributes the medicine among the many organs, tissues, or physiological compartments,
3. How the body metabolizes or breaks down the medication,
4. The body's mechanisms for excreting or eliminating the drug, or how it is being eliminated by the body.

At the time of drug discovery and before any maiden human dose of a novel drug candidate under investigation, the preclinical data is being used by the researchers to predict the pharmacokinetic (PK) performance of the novel drug in humans. This becomes quite helpful in informing the feasibility of attaining therapeutic exposures in initial clinical trials. Once clinical data are made available, the observed PK parameters in humans are compared with the previous predictions. This ultimately gives a great opportunity to assess and further refine or fine-tune the prediction methods. In the recent age of artificial intelligence and advances in data computing, the probability of successful PK predictions has been maximized due to the implementation of best practices in experimental data generation and predictive methodologies. The focus on understanding a candidate drug's disposition properties prior to its nomination for clinical development in a focused and robust mechanistic manner has also contributed immensely to PK predictions (Gupta and Sahu 2021). Thus, it can be said with great certainty that drug developers can learn a lot from the predictions of PK parameters. For each new drug candidate, there are crucial pharmacokinetic parameters that should be determined, both in test animal species and in humans (Ana et al. 2008). These 11 critical pharmacokinetic parameters have been listed below in order of their importance. These parameters are as follows:

1. Clearance,
2. Apparent volume of distribution,
3. Elimination half-life,
4. Elimination rate constant,
5. Absorption rate constant,
6. Concentration at a steady state,
7. Effective drug concentrations (target concentration),
8. Extent of absorption (area under the curve),
9. Fraction of the available dose excreted unchanged,
10. Toxic concentrations,
11. Protein binding.

Some of these important parameters have been discussed in the next segment.

15.3 Important PK Parameters in Drug Design and Development

1. **Clearance:** Because of its implications for both dose level and frequency, clearance rate is one of the most critical pharmacokinetic parameters to be considered while designing new drug candidates. A proper drug metabolic and pharmacokinetic (DMPK) profile remains a significant barrier for lowering the risk and increasing the productivity in pharmaceutical R & D. This has been found to be responsible for roughly 40% of all failures amongst the new drug candidates. For orally administered medicines, poor intestinal absorption and/or a high degree of clearance usually result in poor and variable bioavailability, which can be considered as the most common causes of this failure). However, the presence of active metabolite(s) and drug-drug interactions has been found to be the two further reasons for failure. It can be very grueling and difficult to obtain the required dosage profile for the desired therapeutic efficacy when the drug candidate has a poor pharmacokinetic profile. The key purpose of DMPK studies in drug development is to ascertain drug metabolism and pharmacokinetics in humans. With successful predictions of DMPK, the rate of attrition during the drug research and development phase can be reduced (Zhoupeng and Wei 2018). As a consequence, the importance and the necessity of conducting DMPK in the

drug discovery process have been identified and acknowledged. This, in association with the demand to screen an ever-increasing number of molecules has resulted in significant improvements in both the process of drug discovery and the technologies involved. The term "clearance" refers to the apparent volume of plasma cleared of drug per unit of time via metabolic or excretion processes. It is generally expressed in proportion to body surface area or weight. As it is a proportionality factor, it relates the threat of elimination of the drug from the body with the measured plasma drug concentration (Smith et al. 2019). Renal excretion and other extrarenal pathways (non-renal clearance) mainly hepatic metabolism results in drug clearance from the body. Total clearance is the parameter that indicates drug elimination from the central compartment but without any reference to the mechanism of the actual process. Clearance is constant for drugs that get eliminated by first-order kinetics. The term "renal Clearance" denotes the clearance by the kidneys, whereas clearance by all other organs is referred to as non-renal clearance. Drugs and their metabolites are excreted in urine by three mechanisms:

(a) Glomerular filtration.
(b) Active secretion by renal tubules, and.
(c) Passive reabsorption by tubules.

The GFR and fractional plasma protein binding affect the glomerular filtration of tiny molecules like those found in anesthetic medications. In case of the heavily protein-bound drugs, glomerular filtration will be inefficient.

In the distal and proximal renal tubules, back diffusion passively reabsorbs non-ionized acidic and basic substances. Ionized versions of these weak acids and bases, on the other hand, are retained within renal tubules, resulting in enhanced renal elimination by urine acidification or alkalinization. Drug clearance is an extremely important PK parameter in drug design and development. Understanding the concept of clearance is important is essential. It describes how efficiently or rapidly a drug is eliminated from the body.

Its clinical relevance is to understand how much drug is to be administered, how frequently to dose a patient, and how two interacting drugs will affect a patient. Administration of the same dose of a drug to a group of individuals who have different CLt results in different rates of decline of the blood drug concentration if Vd is similar in these individuals. Individuals with higher CLt eliminate the drug at a faster rate (Larger k and shorter t1/2) and will have longer AUC. The prediction of clearance (CL) remains a notable challenge in drug discovery, particularly when complex processes such as drug transporters are involved. In drug discovery, it is of critical importance to accurately predict human clearance. Thus, to obtain a better understanding of the clearance mechanisms and human translation, in vitro—in vivo extrapolation (IVIVE) of hepatic clearance has been established using large sets of compounds for four preclinical species (mouse, rat, dog, and non-human primate) (David 2015). An Extended Clearance Classification System (ECCS) could be advantageous for optimal compound characterization. HepatoPac and plated hepatocytes can be satisfactory instruments for the assessment of metabolic and active uptake clearance, respectively, for a substantial number of marketed drugs. These tools can support a broad strategy to select optimal in vitro tools and to attain an Extended Clearance Classification System-dependent in vitro to in vivo extrapolation for the prediction of human clearance (David et al. 2022).

2. **Apparent volume of distribution:** After entering the systemic circulation, a drug undergoes distribution throughout the body's tissues. This process is typically uneven due to variations in blood perfusion, tissue binding (such as lipid content), regional pH, and the permeability of cell membranes. The concept of apparent volume of distribution relates to the theoretical fluid volume needed to dilute the total administered drug concentration to the same level as in the plasma. Diffusion across cell membranes then becomes the step that determines the rate at which distribution occurs. After attaining the equilibrium, the concentrations of the drug in the tissues and in the extracellular fluids can be represented by the plasma concentration. The metabolism and excretion of the drug occur simultaneously alongside the distribution, which make the process complex and dynamic. After a drug has entered the tissues, drug distribution into the interstitial fluid is primarily determined by perfusion. For poorly perfused tissues (e.g., muscle and fat), the rate of distribution can become very slow, especially if the tissue possesses a high affinity for the drug. Many acidic drugs (e.g., warfarin, aspirin) can be highly protein-bound in nature and thus can have a small apparent volume of distribution. Many basic drugs (e.g., amphetamine, meperidine) are taken up by tissues in a considerable amount and thus can have an apparent volume of distribution that can be even larger than the volume of the entire body. Thus, in the early phase of drug discovery and molecule development, the determination of the volume of distribution of a drug becomes very crucial and is even considered to be one of the most important pharmacokinetic properties of a drug candidate. It plays a crucial role not only as a major determinant for the half-life of the drug but also as an important parameter for calculating the dosing frequency of the drug. Although having an almost close to even a similar log P, a basic molecule will prefer to showcase a relatively higher volume of distribution than that can be shown by a neutral molecule. Acidic drugs often display a low volume of distribution. Although strategizing a design against the volume of distribution can be very beneficial in obtaining a desirable dosing regimen, it has to be well-directed in order to avoid undesirable effects on the other important parameters. Several strategies can be adopted to increase the volume of distribution, including the addition of lipophilicity and the introduction of basic functional groups. But this has to be done in a way that does not increase or alter the metabolic clearance. Administration of the same dose of the same drug to a group of individuals who have different Vd results in lower initial drug concentration CP_0 in the individuals with larger Vd. The difference in the rate of elimination of the drug (K and half-life) in these individuals will depend on whether CLt is similar in these individuals. This is because k and t1/2 depend on both Clt and Vd. (Patrick and Vladimir 2019).

Elimination half-life: The half-life t1/2 can be defined as the time required for the amount of the drug in the body or the drug's blood concentration to reduce by one-half (50%). The half-life of a drug has important implications for calculating the dosing regimen and to determine the peak-to-trough ratio at the steady state. Generally, a half-life of 12–48 h is optimal for once-daily dosing of orally administered drugs. If the half-life of any drug is too short, more frequent dosing may be required in order to maintain the desired plasma drug concentration for the desired exposures in the form of required pharmacological or therapeutic activity and to avoid any unnecessary high peak concentrations or plasma drug concentrations beyond minimum toxic dosing.

This may present many challenges to achieving optimal efficacy, safety, and patient compliance. For any drug with a much longer half-life, the time for which that drug will get accumulated inside the body and then the subsequent occurrence of its elimination from the body may get prolonged. This may create many challenges in the management of adverse effects and in the designing of efficient clinical trials. The half-life is considered a key parameter for optimization in the research and development of new therapeutic entities. The preferred means of modulating the half-life include structural modification which literally affects the clearance of the drug, and to some lesser extent the volume of distribution. A practical method for the half-life optimization calls for an apprehension of the many drawbacks associated with its interpretation and estimation (Birkett 1996).

3. **Bioavailability:** The term "bioavailability" generally refers to the quantity and rate of absorption of an active drug component or an active moiety from a drug product at the intended site of the medication's action. The bioavailability of medications has recently been an interesting topic in drug research as well as in the early stages of drug discovery. This is a technique for determining the actual causes of a new prospective medicinal entity's failure and according to several of these studies, drug researchers have largely reached the opinion that the majority of candidate medications that failed in clinical trials did so due to toxicity and absorption, distribution, metabolism, and excretion issues rather than a lack of therapeutic efficacy, that is, ADME. The pharmaceutical industry is working extremely hard to increase the success rates of new prospective drug candidates by taking toxicology and ADME considerations into account during the drug discovery process. Therefore, bioavailability studies are to be viewed as an essential component of drug disposition studies and ought to be given considerable weight across all stages of the drug-development process, from preclinical research through the various IND clinical phases (phases I–III). One of the most important factors in determining the systemic exposure profile and dosage guidelines for orally delivered medication candidates is oral bioavailability (F). The selection of a promising drug candidate is the aim of drug discovery; hence, predicting a drug's bioavailability potential appears adequate. The further optimization of a therapeutic candidate, though, necessitates the use of increasingly precise and adaptable techniques. Thus, for the correct and safe dosage during the first human trials with new drug candidates as well as for effective drug development and pharmacokinetic (PK) optimization, an accurate prediction of this bioavailability parameter becomes extremely important. If the bioavailability (F) of a drug is overestimated, it can lead to inadequate and uneven exposure of

the drug in the systemic circulation. Consequently, such miscalculations can result in the failure of the drug candidate during the clinical phase. On the other hand, a significant underestimation of bioavailability during first-in-man studies can have detrimental effects, causing undesirable side effects in the first dose panels. (Shein-Chung 2014). The oral bioavailability (F) can be determined by several underlying factors including

(a) The intestinal permeability (passive and active) and gastrointestinal solubility/dissolution, which control the amount of absorption (fabs).
(b) The metabolism of the gut wall.
(c) Intrinsic hepatic metabolic clearance (CLint),
(d) Bile extraction,
(e) Binding to blood components (such as plasma proteins and red blood cells) and,
(f) The flow characteristics in the intestines and in the liver.

4. Protein binding: Although it is common knowledge that drugs bind to plasma proteins, the complexity of drug action in vivo has only lately been fully comprehended. Plasma proteins, which are highly concentrated, control the amounts of free medicines in plasma and compartments that are in equilibrium with plasma, effectively lowering the potency of medications in vivo. Plasma protein binding (PPB) elevations can lower clearance (CL) by decreasing hepatic and renal access to unbound medicines, but they can also result in a low, stable distribution volume that is equal to the distribution volume of albumin. An increase in PPB can reduce clearance by limiting the number of unbound drugs that can reach the liver and kidneys, but they can also cause a low, stable distribution volume that is comparable to albumin's distribution volume; thus, the effective half-life can be increased. Modulation of plasma protein binding can be part of an effective drug discovery technique to maximize the effective half-life of drug candidates in humans, provided that the pharmacologic efficacy is high enough. The optimization of pharmacokinetics characteristics of acidic drugs is particularly difficult, but it is necessary for once- or twice-daily therapeutic dosage. Unless active hepatic uptake is a deciding factor, the majority of acidic medicines have low steady-state distribution volumes (Vss) in the range of 0.1–0.3 L/kg as a result of a combination of strong plasma protein binding (PPB) and low tissue distribution. Therefore, to obtain a reasonable effective half-life, drug clearance must likewise be very low. Even though PPB shouldn't often be a factor in drug discovery initiatives, in the specific instance of acidic compounds, this can be a viable tactic. The distribution volume of albumin is roughly 0.1 L/kg, and increasing the PPB can cause Vss to approach this lower limiting value. As a result, further PPB increases cannot lower Vss below this value, but they can reduce clearance (CL) by limiting unbound drug access to the hepatocytes. If the goal is to keep free drug concentrations above a set minimum effective drug concentration, then increasing half-life and decreasing Cmax (and consequently dose) through an impact on CL but not Vss will achieve the desired result. If there is excellent pharmacological potency that enables effective free blood levels to be maintained at target receptors, this method for optimizing acidic medicines may

work. In light of these ideas, it is intriguing to note that in a previous study of the pertinent literature, it was discovered that seven of the nine examples of commercially available oral acidic medicines with half-lives of more than 8 h had plasma protein binding of less than 99% (naproxen, piroxicam, atovaquone, diflunisal, cetirizine, warfarin, and oxaprozin). PPB is important for effective modulation over the past few decades; drug development organizations have engaged in much discussion and debate over pharmaceutical concentration on pharmacological target sites. Pharmacokinetics pharmacodynamics drug development models have been used extensively to explain correlations in pharmacological target receptors in tissues. Any abrupt increase in a drug's free concentration has the potential to be hazardous and may call for dose modifications. Free drug concentration is beneficial for determining the effective dosage of medications and may be able to eliminate interactions with other drugs that affect metabolism (or transporter). Animal disease models are thoroughly evaluated before a drug enters development. Under the direction of defining safety margins and the selection of first in human (FIH) dose and human effective dose, unbound drug concentration, and therefore PPB information in species, is particularly instructive (Trainor 2007; Philip et al. 2019).

15.4 Pharmacokinetic Modeling in Drug Discovery and Development

To optimize the absorption, distribution, metabolism, and excretion (ADME) properties of lead compounds, pharmacokinetics (PK) models are used in drug discovery. The end goal is to develop a clinical candidate that has a concentration-time profile in the body that is sufficient for the desired efficacy and safety profile. An early generation of PK/PD correlations combining in vitro potency and target exposure/engagement with an expression of pharmacological activity (mode-of-action) and efficacy in animal investigations is part of a full assessment of the lead compounds attempting to identify the inherent PK liabilities. PK models can be as simple as one compartment or as complicated as models with hundreds of compartments. From the most basic, one-compartment model to more complicated models, each type has its own uses and can be employed to extract useful information from clinical or non-clinical data. A model-based technique for estimating PK parameters is compartmental modeling. The body is divided into fictitious sections to use this strategy. Often, the containers employed in compartmental modeling serve as a surrogate for PK parameters rather than representing genuine physiological tissues in the body. For example, the rates of absorption and/or clearance between compartments can be changed in compartmental modeling to simulate the effects of such alterations, which represent a disease state or the activation of metabolism.

A model with at least one compartment is referred to generally as "compartmental modeling." The number of compartments in a typical compartmental model ranges from one to three, although depending on the application, more compartments may be included. For instance, there may be one compartment for each body organ in whole-body physiologically based pharmacokinetic (PBPK) modeling, a sort of compartmental modeling (Andreas and Philip 2016).

15.4.1 Compartmental Model vs. Non-compartmental Analysis

Compartmental models can provide information about a drug's underlying effects that a non-compartmental analysis (NCA) may not always be able to provide. While both a one-compartment model and a non-compartmental analysis presume that the entire body functions as a single container, the difference between the two is that a one-compartment model is a model-based approach; thus parameters, such as clearance, can be altered to investigate the effects of particular disease states, like renal impairment, or the stimulation of metabolism on the PK properties of a drug. With an NCA, PK parameters are directly inferred from the observed data without the need for prior knowledge of the PK properties of the drug in the body (i.e., absorption rate or rate of elimination). The body does not need to be divided into compartments for an NCA. An NCA has the advantage of not requiring drug-specific properties, which makes it less complicated and more affordable than compartmental modeling while still providing important data on a drug's PK characteristics. Compartmental models have an advantage over NCAs in that they can be applied to a greater variety of situations. In a compartmental model, for instance, the dosing schedule can be changed to generate simulated concentration-time profiles that can be used to comprehend the effects of changing the dose. An NCA estimates the PK parameters directly from the data; however, because this is a model-independent approach, modifications to the protocol (such as an alternative dosing regimen) can only be extrapolated using basic superposition, and the range of dosing regimens that can be employed is constrained. When completing a population pharmacokinetic analysis, compartmental models can be utilized to address inter-subject variability, such as how patient-specific variables affect drug disposition, in addition to being used to estimate PK parameters (James et al. 2012).

15.4.2 One-Compartment Model

The central compartment is the only compartment that a one-compartment model considers to be home to all of the body's tissues. This model's simplicity makes it easy to build, has a clear interpretation, and is reasonably priced. This type of model has the disadvantage of assuming an unequal distribution of the drug throughout the body, which is rarely the case.

Furthermore, clearance is expected to proceed linearly because there isn't a distribution parameter. This kind of model may be the best choice for pharmaceuticals that do not diffuse throughout the body due to their simplicity. A more complicated model may be required for medications that are widely dispersed throughout the body (Subhashis and Harish 2020).

15.4.3 Two-Compartment Model

Two-compartment models address a distribution parameter that is not captured by one-compartment models by dividing the single compartment into two separate containers: the "central" and "peripheral" compartments. The central compartment represents plasma and highly perfused tissues such as the kidneys and liver. In contrast, the peripheral compartment represents tissues with lower blood flow, such as muscle.

In this approach, the drug is dispersed between these two distinct compartments rather than being thought to be spread evenly throughout the body. It is possible to more correctly describe factors like drug distribution by grouping tissues with comparable blood flow rates. However, a two-compartment model might not be enough when the medication has three separate elimination phases (Subhashis and Harish 2020).

15.4.4 Three-Compartment Model

In addition to the two compartments mentioned above, three-compartment variants also feature a peripheral compartment. By using this additional peripheral compartment, the plasma and highly perfused tissue are separated. The central compartment of the three-compartment concept only represents plasma. The highly perfused tissues are represented by the first peripheral compartment, while the sparsely perfused tissues are represented by the second peripheral compartment.

This kind of model combines all the advantages of the two-compartment model with the potential to more accurately depict a medication with three phases of elimination. This sort of model's (and all of the compartment models listed above) disadvantage is that distribution to a particular tissue cannot be investigated, which could be useful if medication concentration in the liver, for example, is particularly important (Sara et al. 2013; Subhashis and Harish 2020).

15.4.5 Whole-Body PBPK Model

Another example of compartmental modeling is whole-body PBPK modeling, in which each tissue in the body is represented by a separate compartment. A PBPK model's compartments, in contrast to models with one, two, or three compartments, are biologically-based and physiologically based. This model can take into account organ size, tissue partitioning coefficients, and blood flow rates between certain organs.

A whole-body PBPK model becomes significantly more adaptable with the addition of this physiology. From drug-drug interaction (DDI) research to first-in-human (FIH) dose prediction, PBPK models have been applied in a variety of contexts. But the extracomplexity makes building the model more expensive and time-consuming. In addition, compared to smaller compartmental the model, this form of model requires a lot more data to be created.

Drug developers can use compartmental modeling to better comprehend how medications affect the body. This may entail several things, such as PK parameter estimation, PK characterization across many studies, comprehension of observed patient variability, and much more. There are various kinds of compartmental models, including whole-body PBPK models and models with one, two, or three compartments. Each sort of model has unique benefits and drawbacks (Edginton et al. 2008).

The variety of applications for a compartmental model is its main benefit. An essential point is that using numerous compartments in a model may help it better describe the observed data. A compartmental model can be employed at many different stages of drug development.

15.5 Role of Pharmacodynamics in Drug Discovery

Establishing that a molecule is safe and effective is a crucial component of any drug development approach. Pharmacodynamics (PD) is a broad division of pharmacology that provide a better understanding of the drug's efficacy and safety. The term pharmacodynamics is formed from two words: Pharmakon meaning drug, and dynamics meaning activity.

Pharmacodynamics refers to the study of a drug's biochemical and physiological effects, as well as its mode of action. It focuses on understanding the relationship between a drug's plasma concentration, the response it elicits, and the duration of that response. In simpler terms, pharmacodynamics explores "what a drug does to the body." The specific effect produced by a drug is often referred to as its pharmacological effect. For instance, the dilation of pupils, known as mydriasis, can be produced by the drug atropine, which acts as an anticholinergic agent, in eye investigations.

Pharmacokinetics as well as pharmacodynamics screening is extremely essential for the drug development process as they both are necessary preconditions for the therapeutic use of medications.

Regulatory agencies would not permit the use of medicine without first investigating its pharmacokinetics and pharmacodynamics. Therefore, PK and PD play a significant part in drug discovery and determining whether or not a medicine is approved, checking with FDA-approved standards. Additionally, in order to estimate key parameters, PK and PD both employ complex data processing strategies.

Principles from pharmacodynamics and pharmacokinetics are both used to enable the safe and effective therapeutic management of patients. Understanding the relevance of PK/PD is crucial for patients with chronic diseases since it will determine whether or not the prescribed pharmacological treatment is effective.

Drug exposure is described via pharmacokinetic and pharmacodynamic analyses.

A drug's exposure is explained by pharmacokinetics, whereas its reaction in terms of biochemical interactions is shown by pharmacodynamics. Understanding how exposure and response are related is essential for the development and approval of every medicine.

The optimum and safest manner to use a medicine in a clinic must be determined by quantifying the extent of drug exposure. Researchers can determine dose requirements beforehand in the medication development process with the aid of pharmacokinetics and pharmacodynamics analyses. Therapeutic formulation changes are relatively frequent in the initial stages of drug development. Furthermore, early on in the drug development process, it's critical to anticipate the biological impacts of tiny dosage modifications. Pharmacokinetics and pharmacodynamics concepts assist researchers in understanding how rapidly a person absorbs and eliminates medications enabling them to take decisions regarding drug formulation and dosage needs. Pharmacodynamics also aids to understand concentration-effect interactions: Because the concentration-effect relationship forms the cornerstone of pharmacodynamics (PD), it is essential to identify the variables affecting it in order to support an effective pharmaceutical development program.

Pharmacokinetic and pharmacodynamic modeling assist to compute safety margins, that is, the analysis of dosing thresholds to assess the drug's safety and establish the maximum tolerated dosage. Additionally, optimizing the dose, providing individualized care, and preventing side effects all depend heavily on PK/PD, which in turn can affect the clinical result (Hassan et al. 2011).

Drugs are known to exert their action either at extracellular, cellular, or intracellular sites levels.

15.5.1 Extracellular Site

Some examples of drug acting on extracellular site are:

(a) Antacids neutralize the gastric acidity.
(b) Chelators form complexes with mercury, lead, chromium, cadmium, and arsenic (heavy metals).
(c) Magnesium sulphate retains fluid inside the intestinal lumen and increases fecal bulk.

15.5.2 Cellular Site of Action

Drugs can penetrate the plasma membrane by a variety of processes, including the passive diffusion, facilitated diffusion, and active transport systems. The size, lipid solubility, and charge of the drug molecule all affect passive diffusion of the drug through the bilayer lipid structure of the plasma membrane (Ratain and Plunkett Jr 2003).

Some examples of drugs acting on cellular sites are:

(a) Acetylcholine binds to nicotinic receptors on the motor end plate thereby producing contraction.
(b) Cardiac glycoside inhibits membrane-bound Na + K+ ATPase pump in cardiac muscles thereby inhibiting ATP-dependent sodium-potassium exchange.
(c) Adrenaline acts on adrenergic receptors in the heart to increase heart rate.

15.5.3 Intracellular Site of Action

Many anticancer medications need to be activated before they can have a cytotoxic effect. In either healthy or malignant tissues, the activation process may involve chemical or enzymatic processes.

Ara-C, fludarabine, and gemcitabine are a few examples of nucleoside analogues that must be phosphorylated into active nucleotide triphosphate forms and incorporated into deoxyribonucleic acid (DNA) in order to have a cytotoxic effect.

Another example of a drug acting on an intracellular site is trimethoprim which hinders the synthesis of folic acid, an intracellular component (Sharma and Sharma 2017).

15.5.4 Principles and Mechanism of Action of Drugs

Drugs do not provide novel functions to any cell/tissue/organ, they just modulate the action of a specific activity. Drugs can act either by receptor-mediated mechanisms or by non-receptor mediated mechanism or by targeting particular genetic changes.

Non-receptor mediated mechanism of action: This refers to drugs that do not act by binding to specific regulatory macromolecules.

Non-receptor mediated mechanisms are as follows:

1. Stimulation
2. Depression
3. Irritation
4. Replacement
5. Cytotoxic
6. Action through enzymes
7. Action through ion channels
8. Action through transporters

15.5.4.1 Stimulation

It means an increase in the level of specialized cell activity, for example, pilocarpine stimulates human salivary glands; adrenaline stimulates the heart; picrotoxin (high dose) produces central nervous system (CNS) stimulant function.

15.5.4.2 Depression

It means a reduction in the level of specialized cell activity, for examples, barbiturates depress heart function; acetylcholine depresses the sinoatrial node in the heart.

15.5.4.3 Irritation

It has often a toxic impact on the specialized cells/tissue. Strong irritation leads to inflammation followed by necrosis.

15.5.4.4 Replacement

It means the usage of hormones, natural metabolites, and derivatives in situations of insufficiency. For examples, levodopa is used in Parkinson characterized by dopamine deficiency; diabetes mellitus is treated with Insulin; iron is utilized in anemia to replenish hemoglobin.

15.5.4.5 Cytotoxic Action

It means a damaging effect on cancer cells or developing parasites without having a substantial negative impact on host cells. Anticancer drugs work by cytotoxic action.

15.5.4.6 Action Through enzymes

In particular biological reactions, enzymes, which are complex structural proteins produced by cells, operate as a catalyst.

Enzyme induction can lead to an increase in enzyme activity, whereas enzyme inhibition results in a decrease of enzyme activity. Drugs can either speed up or slow down reactions that are mediated by enzymes. As an illustration, physostigmine and acetylcholine compete for the enzyme cholinesterase, while adrenaline stimulates the hepatic glycogen phosphorylases.

15.5.4.7 Through Ion Channels

Drugs can modulate the opening and closing of the ion channel. Drugs act on ion channel present on the cell membrane and can directly bind to them and can modulate them. There are primarily two types of ion channels: ligand-gated ion channel and voltage-gated ion channels.

Voltage-gated sodium channels are essential for the generation and transmission of pain-signaling action potentials in sensory nerves. Lidocaine, novocaine, and other locally applied non-subtype-selective sodium channel blockers completely eliminate pain by blocking conduction (Kaczorowski et al. 2008).

A few more Illustrations of drugs acting via ion channels are as follows:

(a) Antiarrhythmic drug, quinidine, acts by blocking myocardial Na + ion channel;
(b) Antiepileptic drug, ethosuximide, acts by inhibiting T- type Ca2 + ion-channel.
(c) Anti-epileptic phenytoin acts by modulating Na + ion channel.
(d) Anti-anginal nicorandil opens ATP-sensitive K+ channel and thus allows potassium ions to move out of the plasma membrane leading to a hyperpolarization state (Andersson 1992).

15.5.4.8 Through Transporters

Transporters are membrane-bound proteins that carry substances both within and outside of cells. Many drugs interact with transporters, bind to them, and exhibit their distinct activities. They go by the name "carriers" as well. Examples include:

(a) The vesicular uptake of nor-epinephrine is blocked by reserpine.
(b) The ascending loop of Henle's Na + Ka + 2Cl-cotransporter is inhibited by furosemide.
(c) In renal tubules, probenecid limits the active transport of penicillin and uric acid.
(d) The $Na^{+}Cl^{-}$ symporter in the distal tubule is constrained by hydrochlorothiazide.

15.5.4.9 Through Receptors

Receptors are protein-based cellular structures that are hypothesized to exist in order to act as a bridge between a chemical agent that acts on cells and elicits a physiological response. They are macromolecules with an elaborate structure. They serve as the binding sites within or on the surface of the effector cell that start a response when a specific drug binds to it. The following receptors are found on the plasma membrane of the cell.

(a) G-protein-coupled receptors (GPCR).
(b) Ionotropic receptors.
(c) Kinase linked receptors.
(d) Cytosolic receptors.

15.5.4.9.1 G-Protein-Coupled Receptors

The majority of cellular responses to hormones and neurotransmitters are mediated by the G protein-coupled receptor (GPCR) family of membrane proteins, which is also responsible for the physiology of taste, smell, and vision. The presence of seven membrane-spanning -helical segments, spaced by alternate intracellular and extracellular loop sections, is the fundamental characteristic shared by all GPCRs. Rhodopsin, secretin, glutamate, adhesion, and frizzled/taste are the five families into which GPCRs in vertebrates are typically subdivided based on their structural and sequence similarity. By far the largest and most diversified of these groups is the rhodopsin family (Rosenbaum et al. 2009).

Unique combinations of signal-transduction activities involving several G-protein subtypes, as well as G-protein-independent signaling pathways and intricate regulation mechanisms, can be found in each particular GPCR.

GPCRs are one of crucial communication centers that connect cells' internal and exterior environments.

Heterotrimeric G proteins (G, G/G subunits) are one of the most important aspects of the cell signaling cascade. A conformational change occurs in the GPCR, upon ligand activation, which then activates the G proteins by accelerating the exchange of GDP/GTP linked to the G subunit. As a result, the $G_{\beta}/G\gamma$ dimer separates from $G\alpha$. Once both of these moieties are liberated, they can each operate on their respective downstream effectors and start their own individual intracellular signaling reactions. After the signal has been transmitted, the GTP in Gα-GTP is hydrolyzed to GDP and Gα becomes inactive (Gα-GDP), which causes it to reassociate with the $G_\beta/G\gamma$ dimer to create the inactive heterotrimeric complexes as depicted in Fig. 15.1 (Tuteja 2009).

The GPCR is also capable of signal transduction via a G protein-free pathway. GPCRs control cell cycle progression as well. There are thousands of GPCRs known from the animal world that have little similarities in common, but only one GPCR has been found in the plant system.

The function of GPCRs is to connect the binding of agonists to the activation of specific heterotrimeric G proteins, which then regulate downstream effector proteins. In cardiac myocytes, the β2 adrenergic receptor couples with both the stimulatory subunit Gαs and the inhibitory subunit Gαi. Additionally, it can initiate signaling through MAP kinase pathways independently of G proteins by utilizing arrestin. Arrestin is a signaling and regulating protein that promotes the activation of extra-cellular signal-regulated kinases (ERK), inhibits the activation of G proteins, and facilitates the internalization of the receptor through clathrin-coated pits (Shenoy et al. 2006).

There are two major signal transduction pathways involving G-protein coupled receptors:

1. Cyclic adenosine monophosphate (cAMP) signal pathway and
2. Phosphatidylinositol Signal Pathway

15.5.4.9.1.1 cAMP Signal Pathway

The process through which glycogen is broken down to glucose by adrenaline can be used to show the cAMP pathway. Adrenaline interacts with a GPCR in the skeletal muscles to act as a stimulator molecule. This active receptor then activates a G-protein (**Gs**)whose alpha subunit in turn activates adenylate cyclase leading to the formation of cAMP from ATP. The tiny intracellular signaling molecule cAMP is a second messenger and is increased or produced in response to an external signal. cAMP activates and phosphorylates protein kinase A (PKA), which further stimulates another enzyme known as phosphorylase kinase. The glycogen phosphorylase is then activated by this kinase, allowing glycogen to be broken down. The enzyme phosphodiesterase is responsible to degrade cAMP to AMP to terminate the signaling.

Other receptors that function through the cAMP pathway are the histamine receptor, serotonin, and dopamine (D1) receptors.

Drugs that activate **Gi** protein will inhibit the adenyl cyclase and thus prevents the formation of cAMP. Activation of M2 receptors in cardiac muscle by acetylcholine will inhibit SA Node thus causing a decrease in heart rate.

Other receptors that function by inhibiting the cAMP pathway are the adrenergic (α_2) receptor, opioid, and dopamine (D2) receptors.

It is noted that increases in cAMP can trigger the transcription of specific target genes in many animal cells that include a regulatory region known as the cAMP

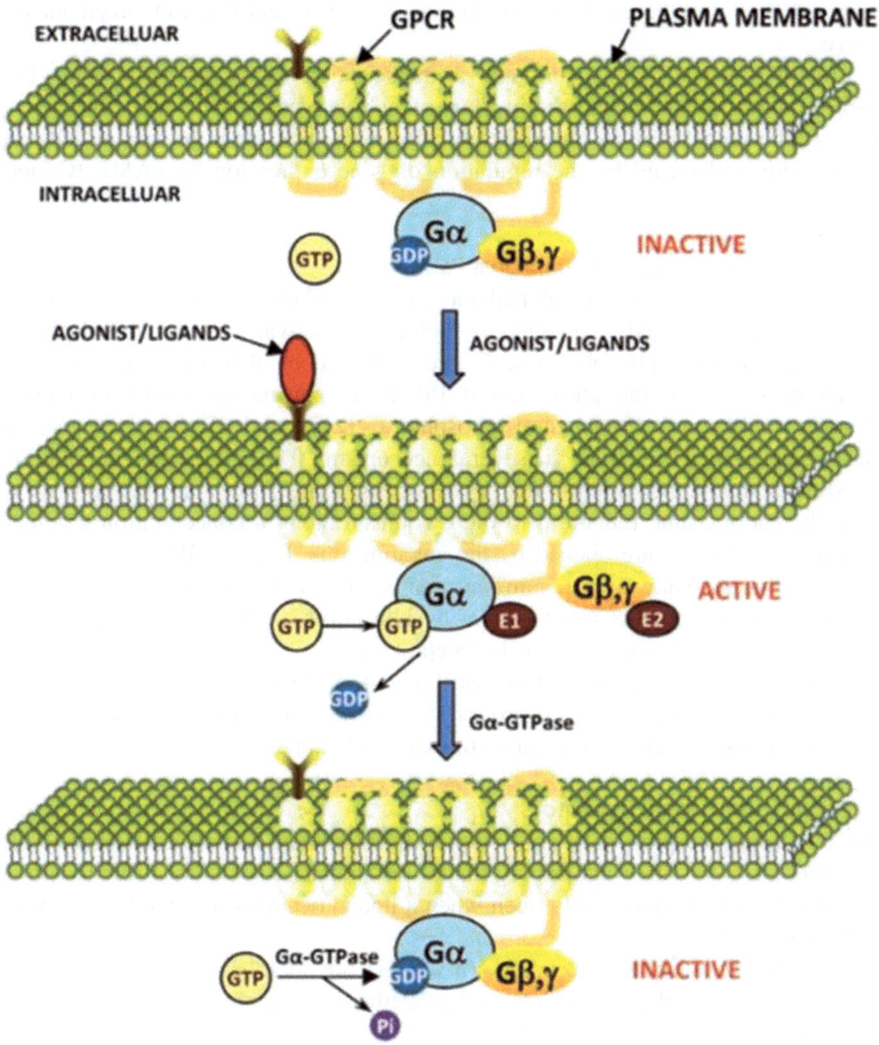

Fig. 15.1 The active and inactive states of GPCR. (Source: Rehman S, Dimri M. Biochemistry, G Protein Coupled Receptors. [Updated 2021 Jul 22]. In: StatPearls [Internet]. Treasure Island (FL): StatPearls Publishing; 2022 Jan. Available from: https://www.ncbi.nlm.nih.gov/books/NBK518966/)

response element, abbreviated as CRE. The catalytic subunit of protein kinase A, which can enter the nucleus after being released from the regulatory subunit, transports the signal from the cytoplasm to the nucleus.

Protein kinase A has two regulatory (R) and two catalytic (C) subunits in its inactive state. When cAMP binds to the regulatory subunits, a conformational shift occurs that causes the catalytic subunits to dissociate and become enzymatically active.

Protein kinase A phosphorylates a transcription factor termed CREB (for CRE-binding protein) inside the nucleus, which activates cAMP-inducible genes. The proliferation, survival, and differentiation of a wide range of animal cells are significantly influenced by the regulation of gene expression by cAMP (Cooper 2000).

15.5.4.9.1.2 Phosphatidylinositol Signal Pathway

The phosphatidylinositol signal pathway starts similarly to the cAMP pathway. When a GPCR is bound by a signaling molecule, it activates a particular G-protein (Gq). Subsequently, phospholipase C gets triggered by this activated G-protein, which breaks down the phospholipid PIPin the plasma membrane into DAG (diacylglycerol) and IP (inositol phosphate). DAG and IP both are second messengers as they further act via several mechanisms. The endoplasmic reticulum's gated calcium channel when bound by IP opens it and allows calcium ions to go from the endoplasmic reticulum to the cytoplasm along a concentration gradient. Additional pathways are opened by the calcium ions. Because IP was the second messenger in this pathway, calcium ions can be viewed as the third messenger.

Examples of receptors that function by phospholipase pathway are adrenergic (α_1), muscarinic M1, and angiotensin receptor (AT_1).

G-proteins and GPCR are widely distributed in the human body and are involved in a variety of physiological processes. Several disorders, such as retinitis pigmentosa and cholera, can result from mutations in these proteins.

15.5.4.9.2 Ionotropic Receptors

The ionotropic receptors are found on the cell membrane and are connected to ion channels directly. These agonist-regulated ion channels, sometimes referred to as ligand-gated ion channels, only open when a ligand occupies a site at the receptor.

Examples are:

Nicotinic acetylcholine receptors (nAChRs) are members of the "Cys-loop" superfamily of ligand-gated ion channels, together with serotonin (5-HT3), glycine, and $GABA_A$ receptors (Dani 2015).

A nicotinic cholinergic receptor comprises five protein subunits ($2\alpha + \beta + \Upsilon + \delta$) surrounding a central pore. When a molecule of acetylcholine binds to each α subunit, the ligand-gated ion channel opens and the subsequent flow of ions (in this case Sodium ions) produces a response. Figure 15.2 shows the opening of the channel when the ligand/neurotransmitter binds to it.

15.5.4.9.3 Kinase Linked Receptors

Numerous members of this family of cell surface receptors have been identified as important regulators of vital cellular processes, including cell migration, cell cycle control, cell survival and metabolism, proliferation, and differentiation.

With a ligand-binding area in the extracellular domain, a single transmembrane helix, and a cytoplasmic region containing the protein tyrosine kinase (TK) domain

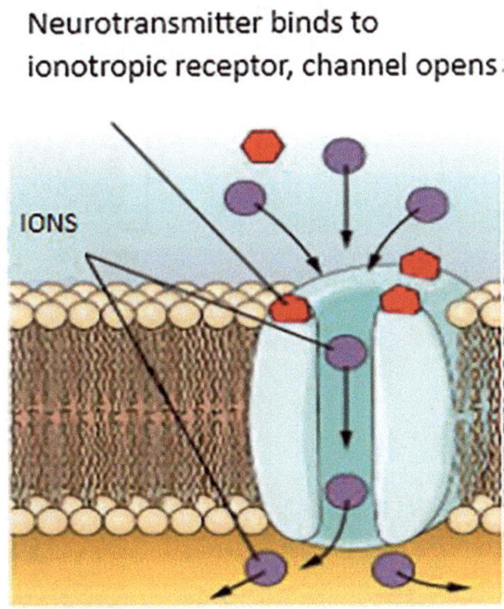

Fig. 15.2 Opening of channel and movement of ions through ionotropic receptor

along with additional carboxy (C-) terminal and juxtamembrane regulatory domains, all human receptor tyrosine kinases share a similar molecular architecture (Lemmon and Schlessinger 2010).

Ligands like growth factor, insulin receptors, or IGF-1 can bind to sites in the extracellular domain which produces a conformational change that leads to dimerization of the receptor. The intracellular tyrosine kinase domain is activated as a result of ligand-induced dimerization of the extracellular domain. This is followed by the autophosphorylation of tyrosine residues. These phosphorylated tyrosine residues then couple with the -SH2 domain of the Grb2 protein and initiate cellular events. The steps of functioning of tyrosine kinase receptors are shown in Fig. 15.3.

A few hormones like ANP (atrial natriuretic peptide) have the structure that fits into such types of receptors but the intracellular domain is guanylyl cyclase instead of tyrosine kinase.

JAK/STAT Pathway: A much more direct link between protein-tyrosine kinases and transcription factors is provided by the JAK/STAT Pathway. Tyrosine phosphorylation of proteins directly impacts the positioning and operation of transcription factors in this pathway.

The STAT proteins (signal transducers and activators of transcription), which were first discovered in studies of cytokine receptor signaling, are the essential components of this pathway. SH2 domains are found in the STAT protein family of transcription factors. They are confined to the cytoplasm of unstimulated cells, where they are dormant. Recruitment of STAT proteins occurs as a result of cytokine receptor stimulation. STAT proteins bind to phosphotyrosine-containing regions in the cytoplasmic domains of receptor polypeptides via their SH2 domains. The JAK

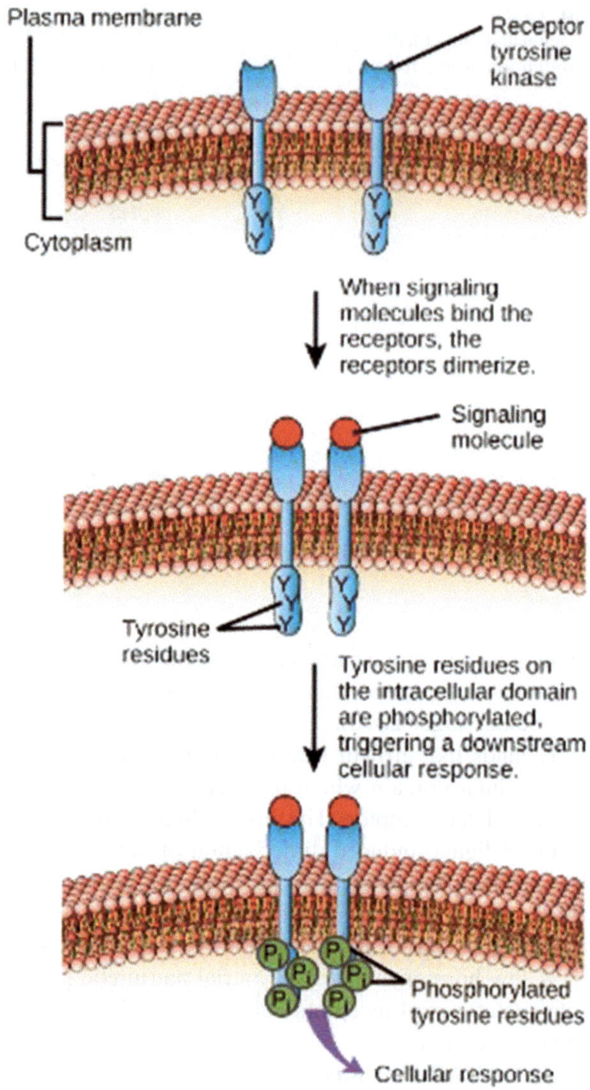

Fig. 15.3 Functioning of tyrosine kinase receptors

family of nonreceptor protein-tyrosine kinases phosphorylates the STAT proteins. Tyrosine phosphorylation helps STAT proteins dimerize so they can move to the nucleus and activate the transcription of their target genes.

15.5.4.9.4 Intracellular Receptors

These are also called cytosolic receptors. Intracellular receptors are receptor proteins found on the inside of the cell, typically in the cytoplasm or nucleus. Intracellular receptors are receptor proteins located inside the cells, usually in the cytoplasm or nucleus. In most cases, the ligands of intracellular receptors are small, hydrophobic (water-hating) molecules, since they must be able to traverse the cell membrane to reach their receptors.

Corticosteroids, Vit A, Vit D, sex hormones, and thyroid hormones produce their actions via these receptors and can alter gene expression.

By attaching to heat shock protein 90 (HSP90), which masks the receptor's nuclear localization signal, the receptor is retained in the cytoplasm in the absence of a ligand. After ligand interaction, HSP90 is released, allowing the receptor to move quickly to the nucleus. As transcription factors, steroid receptors bind to particular steroid response regions on the DNA once they enter the nucleus. These sequences are located in the regulatory region of genes, and steroid receptor interaction can change the gene's transcriptional activity (Cato et al. 2002; Lounsbury 2009).

15.6 Understanding Drug Effect in Pharmacodynamics

15.6.1 Drug Binds to Receptor to Produce Action

The binding of the drug to the receptor is usually specific and reversible as there is the formation of hydrogen bonds, Vander Waals forces or electrostatic bonds. However, in some cases, drugs do bind irreversibly when there is the formation of a covalent bond.

Binding may also be stereoselective, that is, if a drug is having two optical isomers, levo or dextro, only one form is found to be active, while the other form is not active; for example, l-epinephrine and d-amphetamine are known to be active forms.

The following two elements are responsible for the overall pharmacological effect:

1. Affinity
2. Intrinsic activity.

Affinity: The capacity of a drug to interact with a receptor to form a drug-receptor complex is referred to as affinity. A drug won't bind to a receptor if it has no affinity for it. All drugs that work by binding to receptors have some affinity for those receptors. Drugs with high affinity towards their receptor can be used in low concentrations.

Intrinsic Activity: The capacity to activate the receptor following binding to it is known as its intrinsic activity. It ranges from -1 through 0 to $+1$.

Based on their intrinsic activity (IA), drugs/ligands can be classified into four categories.

1. **Agonist:** It has high affinity and high intrinsic activity. Agonists on binding to the receptor can activate it maximally and thus trigger a maximum biological response i. e. IA is +1.
2. **Antagonist:** It has only affinity but no intrinsic activity, that is, IA = 0. These drugs can bind to the receptor but cannot activate it. It can interfere with the binding of an agonist. It will prevent the agonist to bind at a particular site as that site is occupied by an antagonist.
3. **Partial Agonist**: Pentazocine is a partial agonist at the μ receptor type of opioid receptor, which means it exhibits full affinity for the receptor but has low intrinsic activity. It activates the receptor sub-maximally, with an intrinsic activity ranging between 0 and +1. While it can produce a similar effect in the absence of an agonist, it diminishes the effect of a pure agonist. When agonists such as adrenaline and nor-adrenaline are present, pentazocine exerts an antagonistic effect.
4. **Inverse Agonist**: These types of drugs bind to the receptor and produce opposite effects (IA is negative). They have full affinity but intrinsic activity ranges between 0 and − 1, for example, β carboline is an inverse agonist at BZD receptors.

There are different types of **antagonists.** These may be physical, chemical, physiological, or pharmacological.

(a) **Physical antagonist** binds to a particular drug and thus prevents its absorption, for example, charcoal binds to the alkaloids and prevents their absorption.
(b) **Chemical antagonist** combines with a substance chemically like chelating agents bind with the metals and form complexes.
(c) **Physiological antagonist** produces an action opposite to a substance but by binding to the different receptors. Adrenaline for example is a physiological antagonist of histamine as adrenaline delates the bronchioles by binding to the β_2 receptors, which is opposite to the bronchoconstriction produced by histamine through H1 receptors.
(d) **Pharmacological antagonists** produce opposite actions by binding to the same receptor. For example, beta-blockers block the action of beta-agonists by occupying the receptor site.

15.7 Measurement of Drug Effects

The effects of a drug can be assessed after administration, for quantitative evaluation of its safety and effectiveness.

15.7.1 Graded Dose-Response Curve

The dose-response curve can be established by administering increasing doses of a drug to an isolated tissue. When the dose is enhanced, the magnitude of its effect also rises until a condition comes when a further rise in the dose produces no further increase in effect. Such response at this stage is referred to as the "Maximal response" or the "Ceiling response" while that corresponding dose is referred to as the "Maximal dose" or the Ceiling dose.

Such a simple graded dose-response curve is graphically represented by a hyperbola as represented in Fig. 15.4.

The initial portion is so steep that it is virtually impossible to gauge the magnitude of an increase in the response corresponding to a small increase in dose. It is difficult to compare drugs using hyperbola curves.

Since curved lines cannot give good mathematical comparisons, so the dose is converted to log dose to form log DRC, which gives a sigmoid-shaped curve (Fig. 15.5).

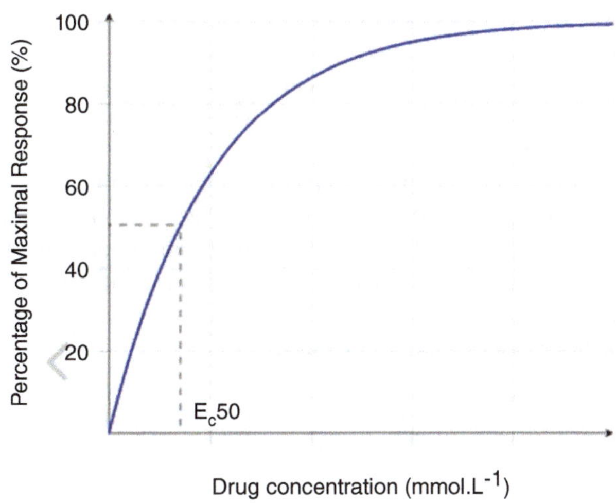

Fig. 15.4 Drug concentration vs. response curve. (Source: https://partone.litfl.com/dose-response_curves.html)

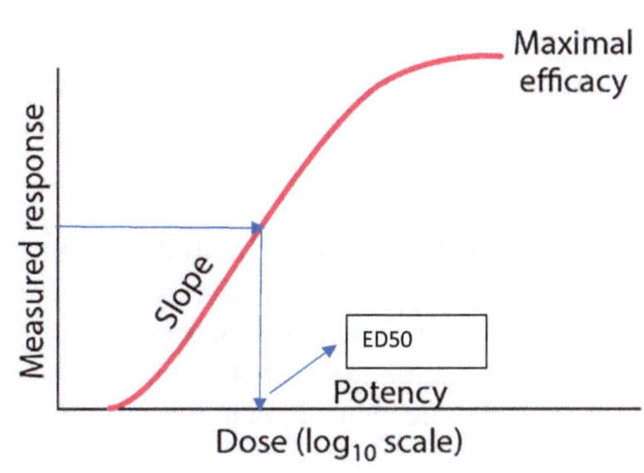

Fig. 15.5 Log dose-response curve

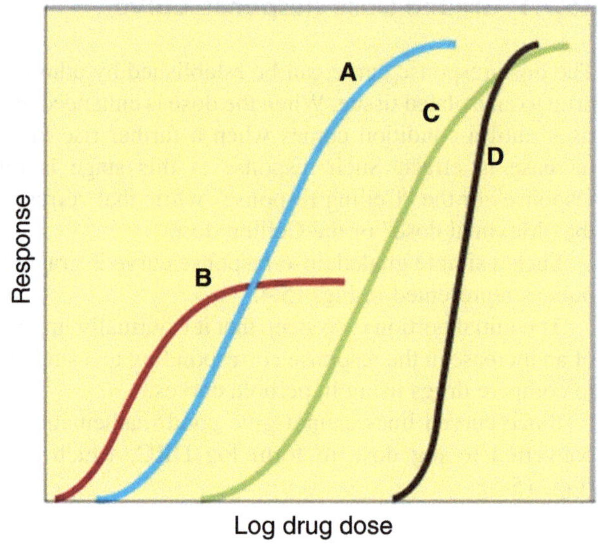

Fig. 15.6 Graded dose response curves for four drugs

In the log dose-response curve, the middle part of the curve is a straight line. From the middle portion of the curve (i.e., linear segment), the ED_{50} of a drug can be determined.

ED50 refers to the median effective dose which can provide 50% of the maximal response. Lesser the value of ED50, the more potent, that is, stronger the drug.

Another advantage of converting it into logarithmic form is that wide differences in doses can be depicted on the same curve after converting it to logarithmic form.

DRC can be used to calculate three crucial characteristics, including potency, efficacy, and slope of the curve.

1. Potency: The amount of a pharmaceutical product that must be taken to get the desired result. The closer the log dose-response (LDR) curve towards the ordinate, the lesser the dose needed to generate the given effect and therefore higher is the potency.

A less potent drug's LDR curve would come on the right side of the dosage axis.

As seen from Fig. 15.6, due to the relative placements of their dose-response curves along the dose axis in the above picture, medicines A and B are considered to be more potent than drugs C and D.

1. Potency is the amount of a drug that must be taken to achieve 50% of the drug's maximum effect (EC50 or ED50). Therefore, drug A in Fig. 15.6 has a lower pharmacologic potency than drug B. The potency of a drug depends in part on the affinity (Kd) of receptors for binding the drug and in part on the efficiency with which drug-receptor interaction is coupled to response.

2. Efficacy: the maximal response as reflected by the height of the LDR curve on its ordinate (i.e., The response axis).
   Drugs A, C, and D in the above figure have equal maximal efficacy, and all have greater maximal efficacy than drug B.
3. Slope: A steeper DRC indicates that a small dosage increase will result in a big rise in response. Therefore, drugs with steeper DRC have narrower therapeutic indices than those with less steep curves (like barbiturates) (e.g., benzodiazepines).
   Additionally, DRC can be used to determine whether a drug is a competitive or non-competitive inhibitor.

15.7.2 Quantal Dose Response Curve

Numerous pharmacological effects can't be assessed using a continuous scale with a graded response. For instance, the criteria for a response might be either the death of an experimental animal or no death if one wanted to test the acute toxicity of medicine.

In that case, the only way to measure the response is on an ALL OR NONE basis.

Shaded boxes and bell-shaped curves depicted in Fig. 15.7 indicate the frequency distribution of doses of the drug required to produce a specified effect.

The quantal dose-effect curve is often characterized by stating the median effective dose (ED50), which is the dose at which 50% of individuals exhibit the specified quantal effect.

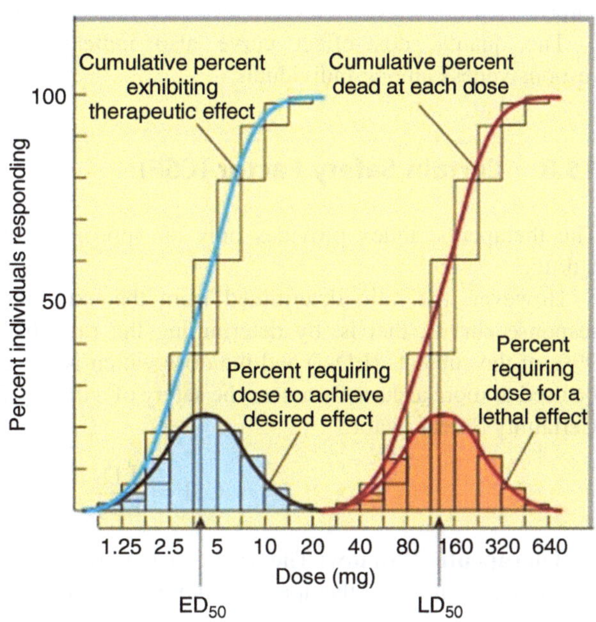

Fig. 15.7 Quantal dose-response plot. (Source: Bertram G. Katzung: Basic & Clinical Pharmacology, Fourteenth Edition)

Similarly, the dose required to produce a particularly toxic effect in 50% of animals is called the median toxic dose (TD50). It is possible to experimentally quantify a median lethal dosage (LD50) if the toxic result is animal mortality. Such values offer an easy way to compare the potencies of medications in both clinical and experimental settings: As a result, the first drug can be considered to be 100 times more potent than the second if the ED_{50} of two pharmaceuticals for causing a defined quantal effect are 5 and 500 mg, respectively. Similar to this, comparing a drug's ED50s for two distinct quantal effects in a population can yield a useful indicator of how selectively it acts (e.g., cough suppression vs. sedation for opioid drugs).

Quantal dose-effect curves can also be used to generate data about the level of safety that can be anticipated when a specific pharmaceutical is used to achieve a particular effect.

The therapeutic index is one metric that compares the dosage of a drug needed to achieve the desired outcome against the dosage needed to get an undesirable outcome.

Therapeutic Index: It is the ratio of the median lethal dose to the median effective dose of the drug.

The median lethal dose (LD_{50}) is the dose that is lethal to 50% of the subjects and the median effective dose (ED_{50}) is the dose of the drug that provides a therapeutic effect in 50% of the subjects.

$$\text{Therapeutic Index} = \frac{LD_{50}}{ED_{50}}$$

For a safe drug, the therapeutic index should be at least more than 1, and hence a drug having a larger value of LD_{50}, but a smaller value of ED_{50} is considered to be safer.

The quantal dose-effect curve also indicates the potential variability of responsiveness among individuals.

15.8 Certain Safety Factor (CSF)

The therapeutic index provides only an approximation of the relative safety of a drug.

However, the ratio determined from the extremes of the respective quantal response curves, that is, by determining the ratio between the dose effective in 99% of the subjects (ED_{99}) and the dose which is lethal to 1% of the subjects, can be used to more accurately assess the safety of a drug (LD_1). This ratio is known as a Certainity Safety factor.

$$\text{CSF} = \frac{LD_1}{ED_{99}}$$

Therapeutic Window: The drug concentration range that produces the intended effect is known as the therapeutic window. It is the range of drug dosages that can

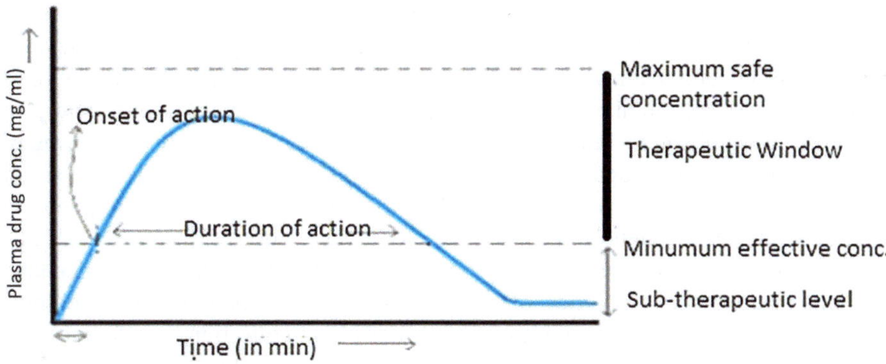

Fig. 15.8 Plasma drug concentration (mcg/mL) vs. Time (min) Graph

treat a disease effectively while staying within the safety range. In other words, it is the dosage of medication between the amount that gives an effect (effective dose) and the amount that gives more adverse effects than desired effects.

All parameters like MSC, MIC, therapeutic window, duration of action and sub-therapeutic level can be calculated from the graph plotted between plasma drug concentration vs. time as shown in Fig. 15.8.

15.9 LC MS: Key for Analysis in Drug Discovery

The revelation and improvement of effective and safe new drugs is a long and complex procedure. Drug organizations commonly contribute 9–15 years of examination and a huge number of dollars into this work; a low pace of progress has generally been accomplished. Interest in inventive and exceptionally good drugs will build because of higher way-of-life assumptions and changing demographic profiles (Meibohm and Derendorf 2002). Drug development depends fundamentally on procuring information on pharmacokinetics for new drug entities. For instance, if another drug is adequately powerful; however, it doesn't arrive at the objective site in the right focus for a predefined time frame, not entirely set in stone by the dosing routine, it will be of little worth to the patient (Schoenwald 2002).

The initial years of the twenty-first century were a difficult period for drug manufacturers to zero in on research of drugs. A few changes in innovation and market elements have happened over the past years and they unavoidably reformed the drug discovery and development (DDD) process. The gigantic advancement in biomedical knowledge and the fulfillment of the project of human genome sequencing have sped up the revelation of the sub-atomic premise of diseases of humans and have permitted the distinguishing proof of significant new organic remedial targets. Furthermore, the synergistic connection between judicious medication plan, recombinant biotechnology and combinatorial science incited a quick augmentation of the quantity of new chemical entities (NCEs) with molecular characteristic hopeful to interface with explicit biological target. Albeit most of the drugs available commercially are small molecule drugs, headways in molecular biology, as well as the need to foster designated treatments for a few disease areas, for example, oncology have prompted an expansion in biologics development. Biologics are large molecules, for example, recombinant therapeutic proteins and antibodies commonly comprising of proteins, nucleic acids, sugars, and so on that are gotten from biological processes or living cells while small molecule drugs like ibuprofen and dextromethorphan are normally inferred through synthesis chemically in laboratories (Thakur et al. 2021).

Animal experiments are frequently performed in the initial stages to understand the pharmacokinetic properties of newly developed compounds, and the most important part is the investigation of blood. For blood collection, a wide range of techniques are used and some of them impact the pharmacokinetic assessments of the examined compounds. There is a wide range of techniques for blood assortment, and some of them impact the pharmacokinetic assessments of the examined compounds. Determination of a suitable testing strategy is hence significant; in any case, until now, no normalized exists for reference. The decision depends exclusively on the analyst's insight, testing volume, and the exploratory species utilized. Albeit a couple of studies have investigated that the plasma drug concentrations are reliant upon the sampling site utilized, there is no reasonable component presently to recognize which variables will cause such variety (Chen et al. 2019).

The information gathered through bioanalytical method development and validation (BMDV) is crucial for sponsors of different drug approval processes like veterinary drug approval, ANDAs, INDs, and NDAs. BMDV can be performed with different approval boundaries by utilizing LC-MS/MS and other logical methods. Additionally, there are different dependability rules and methodologies were set which are helpful for bioanalysis (Deshpande et al. 2019) (Fig. 15.9).

In the publication of James CA et al., the term "bioanalysis "was described as "a term generally used to describe the quantitative measurement of a compound (drug) or their metabolite in biological fluids, primarily blood, plasma, serum, urine or tissue extracts" (James et al. 2004).

The term "pharmacokinetic" was coined by Dost in 1953 (Gladtke 1988). Pharmacokinetics refers to the study of how drugs behave in biological systems over time, as well as the mathematical relationships needed to create models that accurately represent the time-dependent changes in drug concentrations or amounts within those systems. The application of pharmacokinetics in clinical pharmacology has rapidly advanced in the past 20 years, largely due to advancements in bioanalytical technologies. The ultimate goal of pharmacokinetics is to understand the relationship between the toxic and pharmacological effects of drugs and their concentrations in body fluids (Peng and Chiou 1990).

LC-MS is a strong logical strategy utilized for the identification, separation, and quantification of both known and unknown components in the sample. Moreover, this can also be utilized to understand the chemical and structural properties of diverse types of samples. It is extremely valuable for small molecules, offering higher selectivity and sensitivity in the trace examination of complex samples

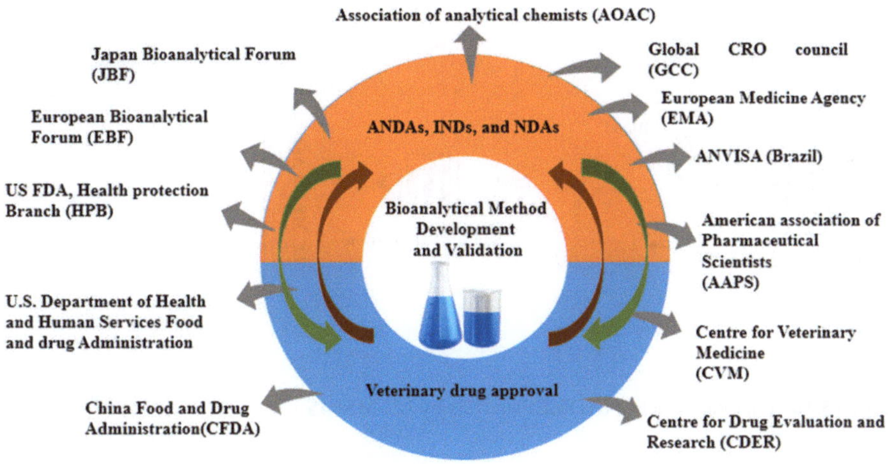

Fig. 15.9 The importance of data of BMDV

(Mukherjee 2019). Coupling of LC with MS detectors has forever been alluring because of the delicate and profoundly specific nature of MS contrasted with other chromatographic detectors. The GC-MS coupling was accomplished during the 1950s with commercial instruments accessible from the 1970s. After the availability of GC-MS systems, the coupling of MS with LC (LC-MS) was an undeniable expansion yet progress in this space was restricted for a longer period because of the general contrariness of existing MS particle sources with a constant fluid stream. However, several interfaces were tried till take-up by clinical labs was extremely restricted due to non-reliability and complexity. A few points of interaction were grown; however, they were lumbering to utilize and questionable, so take-up by clinical labs was extremely restricted. This present circumstance changed with the improvement of the electrospray ion source by Fenn during the 1980s. Makers quickly created instruments furnished with electrospray sources, which incredibly affected protein and peptide natural chemistry. Fenn, along with Koichi Tanaka, was awarded Nobel Prize in 2002 for the creation of a grid that helped laser desorption ionization, one more very valuable MS ionization method for the examination of biological molecules. The cost and execution of LC-MS instruments had improved to the degree that clinical biochemistry labs in the mid-1900s, had the option to exploit the new innovation. Biochemical hereditary qualities were quite possibly the earliest region to do as such, and the examination of neonatal dried blood spot tests for a scope of inalienable errors of digestion was a significant early application. There are various other clinical utilizations of LC-MS, and the procedure is more commonly material than GC-MS attributable to the more extensive scope of biological molecules that can be analyzed and the more prominent utilization of LC separations in clinical research facilities. The purposes behind picking LC-MS over LC with other conventional detectors are basically equivalent to GC-MS, in

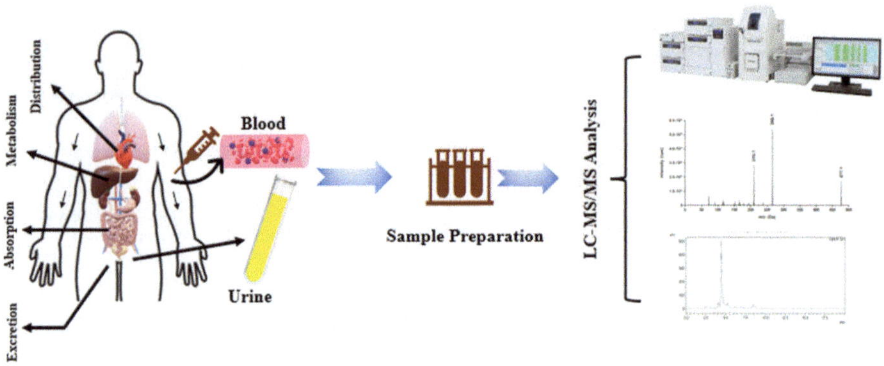

Fig. 15.10 Pharmacokinetics (ADME) analysis by LC-MS/MS

particular high explicitness and the capacity to deal with complex combinations (Pitt 2009).

The science of bioanalytical analysis first grew gradually, and afterward rapidly, resembling the improvement of analytical bioassay strategies that currently can evaluate a drug's plasma concentration having high distribution volume (V_d), arriving at the low limit of quantification (LLOQ) with values going from pg·mL^{-1} to ng·mL^{-1}. This is currently attainable with chromatographic techniques, primarily with pair mass spectrometry (LC-MS) that guarantees the most noteworthy sensitivity, specificity, and short chromatographic runs (Marzo et al. 2016) (Fig. 15.10).

The two primary parts of a bioanalytical method are the preparation of the sample and the detection of the compound. The most ordinarily involved sample preparation strategies for PK studies are solid-phase extraction (SPE), plasma protein precipitation, and liquid-liquid extraction. Because of the straightforward and quick tidy-up of tests, protein precipitation is the most utilized strategy for the extraction of mixtures from biological matrixes. However, for the specific removal of any analyte SPE approaches is mostly used. Another option and strong methodology of bioanalytical studies for online extraction is turbulent flow chromatography. This innovation further reduces the time in sample preparation and improves sensitivity. Presently, because of its selectivity and sensitivity, LC-MS is the guideline strategy utilized in triple quadrupole mass spectrometers for quantitative bioanalysis and is the instrument of decision for PK studies. Be that as it may, the most recent age of high-resolution mass spectrometer systems is advancing to offer a productive option for the quantitative examination of biofluids. LC-MS has likewise arisen as a promising other option or a correlative innovation to make up for the absence of specificity of customary methodologies embraced in macromolecule drug bioanalysis of, for example, ligand-binding assays (Veneziano 2021).

The bioanalytical method is always aimed to address challenges and specific issues associated with the development of each moiety as a drug during the drug development phase. The important consideration of BMDV is the identity of analytes, analytical methodology, and timing of method development. The bioanalytical methods support oral /inhaled biotherapeutic drugs and bioconjugates can be extremely wide, biomarker discovery, metabolite identification, pharmacokinetic assessment, immunogenicity assessment, and quantification, co-medication quantification and tissue biodistribution, and among others (Mu et al. 2022).

References

Alavijeh MO, Brown K, Holder J et al (2012) The importance of biopharmaceutical properties in successful drug design. Drugs Future 37:879–883

Alexander SH, Misty MA, Mathias RA et al (2017) Trends in GPCR drug discovery: new agents, targets and indications. Nat Rev Drug Discov 16:829–842

Ana RG, Marival B, Aaron M et al (2008) Pharmacokinetics in drug discovery. J Pharm Sci 97: 654–690

Andersson KE (1992) Clinical pharmacology of potassium channel openers. Pharmacol Toxicol 70: 244–254

Andreas R, Philip L (2016) Pharmacokinetics in drug discovery: an exposure-centered approach to optimising and predicting drug efficacy and safety. Handb Exp Pharmacol 232:235–260

Atanas GA, Birgit W, MPW E et al (2015) Discovery and resupply of pharmacologically active plant-derived natural products: a review. Biotechnol Adv 33:1582–1614

Birkett D (1996) Pharmacokinetics made easy 11 designing dose regimens. Aust Prescr 19:76–78

Cato AC, Nestl A, Mink S (2002) Rapid actions of steroid receptors in cellular signalling pathways. Sci STKE 2002(138):re9

Chen WC, Huang PW, Yang WL et al (2019) Fundamentals of pharmacokinetics to assess the correlation between plasma drug concentrations and different blood sampling methods. Pharm Res 36:32

Cooper GM (2000) The cell: a molecular approach, 2nd edn. Sinauer Associates, New York, NY

Dani JA (2015) Neuronal nicotinic acetylcholine receptor structure and function and response to nicotine. Int Rev Neurobiol 124:3–19

David Taylor (2015) Book series issues in environmental science and technology: the pharmaceutical industry and the future of drug development. Royal Society of Chemistry

David AT, Sangwoo R, Di L (2022) In vitro—in vivo extrapolation of hepatic clearance in preclinical species. Pharm Res 39:1615–1632

Deshpande MM, Kasture VS, Chavan MMMJ (2019) Bioanalytical method development and validation: a review in recent advances in analytical chemistry. IntechOpen

Edginton A, Theil FP et al (2008) Whole body physiologically based pharmacokinetic models: their use in clincal drug development. Expert Opin Drug Metab Toxicol 4:1143–1152

Gail AVN (2019) Phase II trials in drug development and adaptive trial design. JACC Basic Transl Sci 4:428–437

Gladtke E (1988) History of pharmacokinetics. In: Pecile A, Rescigno A (eds) Pharmacokinetics. Plenum PresFortunaAs, New York, NY, pp 1–10

Gupta RS, Sahu DM (2021) Artificial intelligence to deep learning: machine intelligence approach for drug discovery. Mol Divers 25:1315–1360

Hassan M, Sallam H, Hassan Z (2011) The role of pharmacokinetics and pharmacodynamics in early drug development with reference to the cyclin-dependent kinase (Cdk) inhibitor—Roscovitine. Sultan Qaboos Univ Med J [SQUMJ]11:165–178

Hughes JP, Rees S, Kalindjian BS et al (2011) Principles of early drug discovery. Br J Pharmacol 162:1239–1249

James CA, Breda M, Barattè S et al (2004) Analysis of drug and metabolites in tissues and other solid matrices. Chromatographia 59:149–156

James BB, Erik B, Jardine B, Gary MR (2012) Compartmental modeling in the analysis of biological systems. Methods Mol Biol 929:391–438

Kaczorowski GJ, McManus OB, Priest BT et al (2008) Ion channels as drug targets: the next GPCRs. J Gen Physiol 131:399–405

Kiriiri GK, Njogu PM, Mwangi AN (2020) Exploring different approaches to improve the success of drug discovery and development projects: a review. Future J Pharm Sci 6:27

Lemmon MA, Schlessinger J (2010) Cell signaling by receptor tyrosine kinases. Cell 141:1117–1134

Lounsbury K (2009) Signal transduction and second messengers. In: Pharmacology principles and practice. Academic Press, London

Marzo M, Ciccarelli R, Di Iorio P et al (2016) Synergic development of pharmacokinetics and bioanalytical methods as support of pharmaceutical research. Int J Immunopathol Pharmacol 29: 168–179

Meibohm B, Derendorf H (2002) Pharmacokinetic/pharmacodynamic studies in drug product development. J Pharm Sci 91:18–31

Mu R, Yuan J, Huang Y et al (2022) Bioanalytical methods and strategic perspectives addressing the rising complexity of novel bioconjugates and delivery routes for biotherapeutics. BioDrugs 36:181–196

Mukherjee PK (2019) LC–MS: a rapid technique for understanding the plant metabolite analysis. In: Mukherjee PK (ed) Quality control and evaluation of herbal drugs. Elsevier

Nishant T, Sathish KD, Kumar A et al (2011) Role of pharmacokienetic studies in drug discovery. J Bioequiv 3:263–267

Patrick JM, Jogarao G (2002) The product label: how pharmacokinetics and pharmacodynamics reach the prescriber. Clin Pharmacokinet 41:161–169

Patrick MG, Vladimir RM (2019) Pharmacokinetic and pharmacodynamic properties of drug delivery systems. J Pharmacol Exp Ther 370:570–580

Peng GW, Chiou WL (1990) Analysis of drugs and other toxic substances in biological samples for pharmacokinetic studies. J Chromatogr B Biomed Appl 531:3–50

Philip G, Rhona JC, Ken G (2019) Plasma protein binding as an optimizable parameter for acidic drugs. Drug Metab Dispos 47:865–873

Pitt JJ (2009) Principles and applications of liquid chromatography-mass spectrometry in clinical biochemistry. Clin Biochem Rev 30:19–34

Ratain MJ, Plunkett WK Jr (2003) General mechanisms of drug action. In: Kufe DW, Pollock RE, Weichselbaum RR et al (eds) Holland-Frei cancer medicine, 6th edn. Decker, Hamilton, Ont

Richard CM, Nigel HG (2017) Drug discovery and development: role of basic biological research. Alzheimers Dement (N Y) 3:651–657

Rosenbaum DM, Rasmussen SG, Kobilka BK (2009) The structure and function of G-protein-coupled receptors. Nature 459:356–363

Sara C, Gaetano L, Giuseppe T et al (2013) Pharmacokinetics of remifentanil: a three-compartmental modeling approach. Transl Med UniSa 7:18–22

Schoenwald RD (2002) Basic principles. In: Schoenwald RD (ed) Pharmacokinetics in drug discovery and development. CRC Press, Boca Raton, FL

Sharma HL, Sharma KK (2017) Principles of pharmacology. Paras Medical Publishers, Hyderabad

Shein-Chung C (2014) Bioavailability and bioequivalence in drug development. Wiley Interdiscip Rev Comput Stat 6:304–312

Shenoy SK, Drake MT, Nelson CD et al (2006) Beta-arrestin-dependent, G protein-independent ERK1/2 activation by the beta2 adrenergic receptor. J Biol Chem 281:1261–1273

Smith DA, Beaumont K, Maurer TS et al (2019) Clearance in drug design. J Med Chem 62:2245–2255

Steinmetz KL, Spack EG (2009) The basics of preclinical drug development for neurodegenerative disease indications. BMC Neurol 9:S2

Subhashis D, Harish KTH (2020) An overview on pharmacokinetics and pharmacokinetic modeling. Asian J Res Pharm Sci 10:124–130

Thakur A, Tan Z, Kameyama T (2021) Bioanalytical strategies in drug discovery and development. Drug Metab Rev 53:434–458
Trainor G (2007) The importance of plasma protein in drug discovery. Expert Opin Drug Discovery 2:51–64
Tuteja N (2009) Signaling through G protein coupled receptors. Plant Signal Behav 4:942–947
Veneziano M (2021) Bioanalysis and pharmacokinetic studies. Int J Pharm 5:1
Zhoupeng Z, Wei T (2018) Drug metabolism in drug discovery and development. Acta Pharm Sin B 8:721–732

An Update on Pharmacokinetic Models 16

Ravi Shankar and Kamla Pathak

Abstract

The process and kinetics involved in drug distribution and disposition are complex, and drug events often happen simultaneously. The process is governed by a variety of factors that must be properly defined and quantified for designing optimum drug therapy regimens through pharmacokinetic models. A pharmacokinetic model is a hypothesis using mathematical terms to describe quantitative relationships and is efficient in describing the time course of the drug throughout the body and is helpful in computing and calculating desired pharmacokinetic parameters which are needed for achieving the overall objective of drug therapy. The predictive capability of a model lies in the proper selection and development of mathematical function(s) that parameterize the essential factors governing the kinetic process. Such mathematical models can be devised to simulate the rate processes of drug absorption, distribution and elimination to describe and predict drug concentrations in the body as a function of time. The field is under constant upgradation to match with recent and novel drug delivery systems and therapeutic approaches for achieving the overall objective of the therapeutic regimen.

Keywords

Conventional models · Pharmacokinetic models · Novel models

R. Shankar
Department of Pharmacy, Sagar Institute of Technology and Management, Barabanki, India

K. Pathak (✉)
Faculty of Pharmacy, Uttar Pradesh University of Medical Sciences, Etawah, India

P. P. Singh (ed.), *Recent Advances in Pharmaceutical Innovation and Research*,
https://doi.org/10.1007/978-981-99-2302-1_16

16.1 Introduction

The movement of drugs inside the body is a very complex and continuous process starting from the blood to extracellular fluid to intracellular compartments and from this state to metabolism and finally excretion. The biological nature of drug distribution and disposition is complex, and drug events often happen simultaneously. The process is governed by a variety of factors, including the properties of drug molecules, blood flow rate and permeability across different membranes, and affinity between drugs and different tissue components. So, it is of utmost importance to consider these important factors when designing drug therapy regimens.

The process of designing an effective dosage regimen consists of determining the dose of the drug and time interval which directly depends upon the state and rate of different on-going processes.

A pharmacokinetic model is a hypothesis using mathematical terms to describe quantitative relationships and is efficient in describing the time course of the drug throughout the body and is helpful in computing and calculating desired pharmacokinetic parameters which are needed for achieving the overall objective of drug therapy. The predictive capability of a model lies in the proper selection and development of mathematical function(s) that parameterize the essential factors governing the kinetic process. A pharmacokinetic function relates an independent variable to a dependent variable, often through the use of parameters. Such mathematical models can be devised to simulate the rate processes of drug absorption, distribution and elimination to describe and predict drug concentrations in the body as a function of time (Jones and Rowland-Yeo 2013; Shargel et al. 2016). Pharmacokinetic models find their applications in:

1. Prediction of plasma, tissue and urine drug levels with any dosage regimen and relates it to the optimum therapeutic response or dosage regimen.
2. Calculation of the optimum dosage regimen for each patient individually called individualization of dosage regimen.
3. Estimation of the possible accumulation of drugs and/or metabolites.
4. Correlation of drug concentrations with pharmacologic or toxicology activity of the drug.
5. Evaluation of differences in the rate and extent of availability between formulations (bioequivalence).
6. Describing the effect of physiological alterations on the pharmacokinetic parameters finally relating it to pharmacodynamic parameters and the development of an effective therapeutic regimen. There should be insightful studies and caution should be there in ensuring that a suitable model would be chosen to fit the experimental data so that the correct pharmacokinetic parameter could be derived (Brahmankar and Jaiswal 2015).

16.2 Types of Pharmacokinetic Models

There are various well-established pharmacokinetic models (Fig. 16.1) that have been consistently used by academicians and researchers to understand the kinetic behaviour of drug molecules in the body. The subsequent text describes the traditional models used to elucidate the pharmacokinetic behaviour of drugs. These can be classified broadly as compartment models, non-compartment models and physiological models.

16.2.1 Compartment Models

Compartment modelling is the simplest and most commonly used model approach for the determination of pharmacokinetic parameters. This approach is based on the simple determination of plasma concentration with time data and its interpolation of the data to calculate various parameters. A compartment is not a real physiologic or anatomic region but is considered as tissue or group of tissues that have similar blood flow and drug affinity. Within each compartment, the drug is considered to be uniformly distributed. Mixing of the drug within a compartment is rapid and homogeneous and is considered to be 'well stirred', so that the drug concentration represents an average concentration, and each drug molecule has an equal probability of leaving the compartment. Rate constants are used to represent the overall rate processes of drug entry into and exit from the compartment. The model is an *open system* because the drug enters the system and also can be eliminated from the system simultaneously. Compartment models are based on linear assumptions using

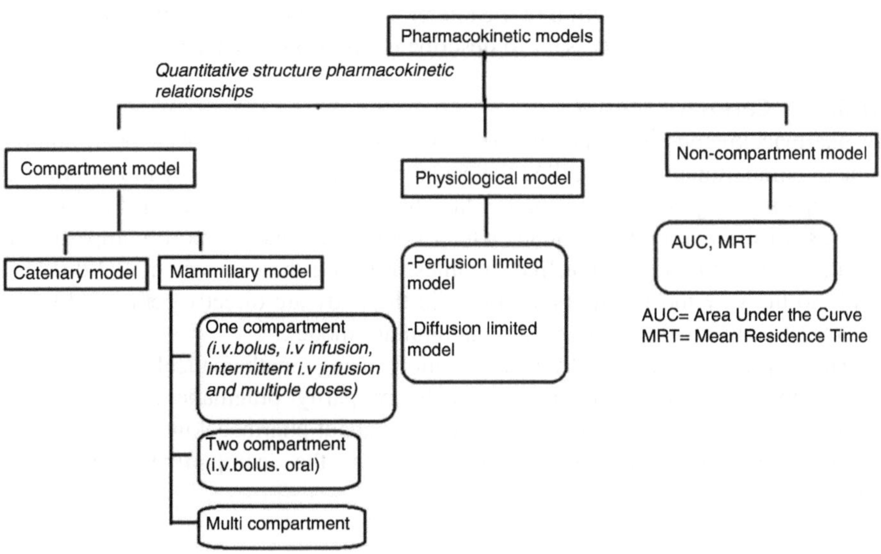

Fig. 16.1 Classification of conventional pharmacokinetic models

linear differential equations. Depending upon the arrangement of compartments, compartment modelling is further classified into (1) mammillary model and (2) catenary model (Notari 2013).

16.2.1.1 Mammillary Model

The mammillary model is the most common compartment model used in pharmacokinetics. The mammillary model is a strongly connected system because one can estimate the amount of drug in any compartment of the system after a drug is introduced into a given compartment. The central compartment is assigned to represent plasma and highly perfused tissues that rapidly equilibrate with a drug. When an intravenous dose of a drug is administered, the drug enters directly into the central compartment. Elimination of drugs occurs from the central compartment because the organs involved in drug elimination, primarily the kidney and liver, are well-perfused tissues.

In a two-compartment model, the drug can move between the central or plasma compartment to and from the tissue compartment. Although the tissue compartment does not represent a specific tissue, the mass balance accounts for the drug present in all the tissues. In this model, the total amount of drugs in the body is simply the sum of drugs present in the central compartment plus the drug present in the tissue compartment. The compartmental models are particularly useful when little information is known about the tissues.

Several types of compartment models are described in Fig. 16.2. The pharmacokinetic rate constants are represented by the letter *k*. Compartment 1 represents the plasma or central compartment, and compartment 2 represents the tissue compartment. The drawing of models has three functions. The model (1) enables writing differential equations to describe drug concentration changes in each compartment, (2) gives a visual representation of the rate processes and (3) shows the pharmacokinetic constants that are necessary to describe the process adequately.

16.2.1.2 Catenary Model

In pharmacokinetics, the mammillary model must be distinguished from another type of compartment model called the catenary model. The catenary model consists of compartments joined to one another like the compartments of a train (Fig. 16.3) in contrast to the mammillary model which consists of one or more compartments around a central compartment like satellites. Because the catenary model does not apply to the way most functional organs in the body are directly connected to the plasma, it is not used as often as the mammillary model.

The representation of drug distribution using mathematical compartment modelling has various limitations, which are partly pragmatic and partly the consequences of various assumptions we make about pharmacokinetics.

The common limitations associated with the compartment modelling approach are:

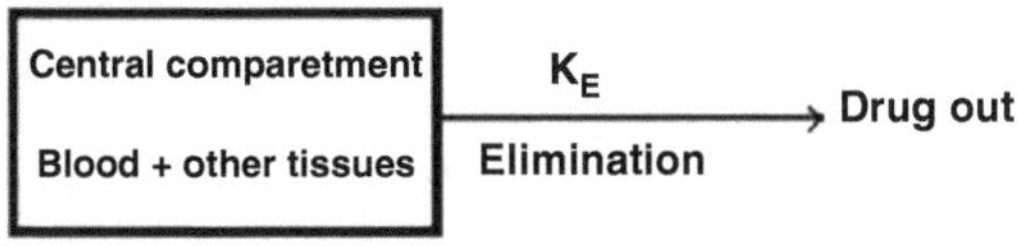

Model 1: One comparetment open model, intravenous administration

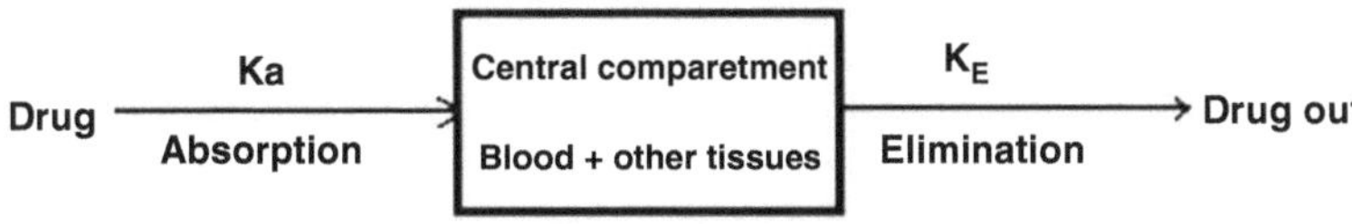

Model 2: One comparetment open model, extravascular administration

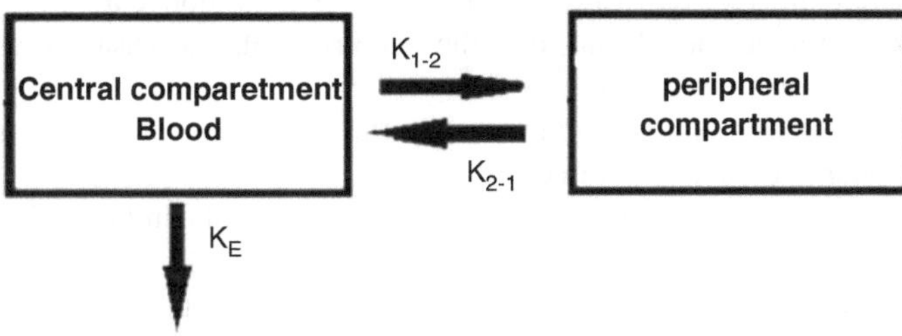

Model 3: Two comparetment open model, Intravenous administration

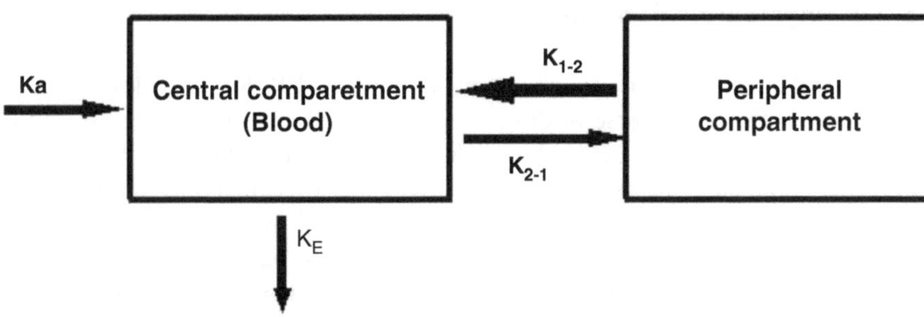

Model 4: Two comparetment open model, extravascular administration

Fig. 16.2 Types of compartment models

- The compartments and parameters are hypothetical and bear no relationship with actual anatomy and physiology that finally needs complex data interpretation for developing information,
- That the central compartment is the only compartment from which the drug is eliminated. In the case of cisatracurium where the drug degrades spontaneously

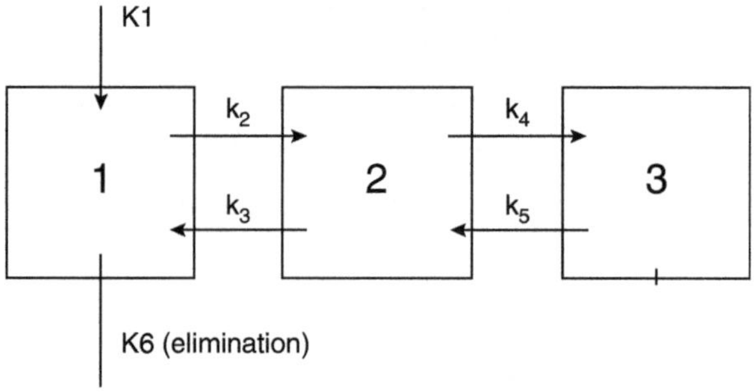

Fig. 16.3 Diagram for catenary model

no matter where it is in the body, i.e., elimination takes place in numerous compartments simultaneously.

- That the multicompartment model is more accurate the more compartments it has. This is not the fact and often the plasma concentration data derived from a two-compartment model matches the empiric data at least as well as the multicompartment prediction.
- The model utilizes complex multiexponential mathematical equations for determining pharmacokinetic parameters.
- The selection of models may vary with the drug, route of administration and study population.
- Interpretation and extrapolation of animal data to human data are quite a complex task (Notari 2013).

16.2.2 Physiological Model

The physiologically based pharmacokinetic model (PBPK) provides an exact description of the time course of drug concentration in any organ or tissue and is, therefore, able to provide greater insight into drug distribution in the body. Also, since the parameters of these models correspond to actual physiological and anatomical measures, such as organ blood flows and volumes, changes in the disposition kinetics of drug because of physiological or pathological alterations in body function may be predicted by perturbation of the appropriate parameter (s) (Aarons 2005). Finally, these models introduce the possibility of animal scale-up which would provide a rational basis for the correlation of drug data among animal species. A physiological pharmacokinetic model is composed of a series of lumped compartments (body regions) representing organs or tissue spaces whose drug concentrations are assumed to be uniform. The compartments are arranged in a flow diagram as illustrated in Fig. 16.4.

Drug concentrations in the various tissues are predicted by organ tissue size, blood flow and experimentally determined drug tissue–blood ratios. Second, blood flow, tissue size and the drug tissue-blood ratios may vary due to certain pathophysiologic conditions. Thus, the effect of these variations on drug distribution must be taken into account in physiologic pharmacokinetic models. The model shows that

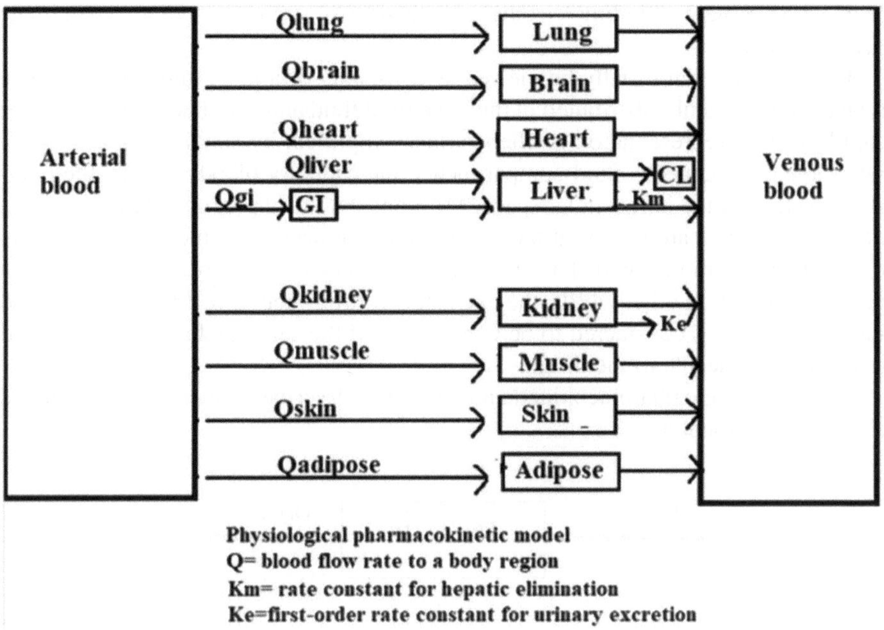

Fig. 16.4 Schematic representation of physiological model

adipose tissue accumulates drugs slowly because of low blood supply and is classified as (slowly equilibrating tissues). In contrast, vascular tissues, like the lung, equilibrate rapidly (rapidly equilibrating tissues) with the blood and start to decline as soon as the drug level in the blood starts to fall. The physiologic pharmacokinetic model provides a realistic means of modelling tissue drug levels. The real significance of the physiologically based model is the potential application of this model in the prediction of human pharmacokinetics from animal data. Unfortunately, the simulated tissue levels cannot be verified in humans because drug levels in tissues are not available. The common representation of a physiological model is shown in Fig. 16.4. Major tissues/organs are represented by compartments from which blood flows carry a drug into and out of tissue/organ. The rate of drug presentation to a particular tissue or organ is dependent upon the rate of perfusion of blood to the tissue or organ and the permeability of the drug is dependent on its partition coefficient between blood and tissue components. Thirteen compartments such as lungs, liver and kidney get maximum blood inflow so are present at the top, followed by other highly perfused organs termed as (HPT) followed by organs which are poorly perfused by blood (PPT) (Gerlowski and Jain 1983; Winter 2004).

Once the selection has been made, the kinds of information required by the model can be classified as (1) anatomical (e.g. organ and tissue volumes), (2) physiological (e.g. blood flow rates and enzyme reaction parameters), (3) thermodynamic (e.g. drug-protein binding isotherms) and (4) transport (e.g. membrane permeabilities).

Body regions can usually be viewed as consisting of a large number of a single type of cell randomly distributed in the interstitial fluid and supplied with blood by a capillary. This representation is often further simplified, by subdividing the region into three homogeneous fluid compartments: the capillary blood volume, the interstitial water and the intracellular space. Most physiological pharmacokinetic models developed to date are based on the assumption that drug movement within a body region is much more rapid than the rate of delivery of drug to the region by the perfusing blood. In other words, the exchange of drugs between capillary blood and interstitial water is considered to be very rapid and the cell membrane is considered to be very permeable to the drug.

The physiologic pharmacokinetic model can be described for a single well-mixed tissue compartment as

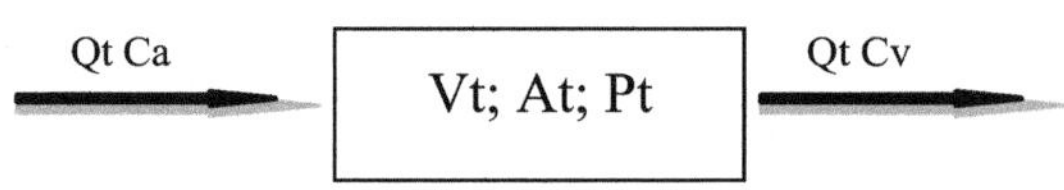

$$\text{Mass} - \text{balance equation}: \frac{dA}{dT} = V\frac{dC}{dt} = QtCa - QtCv \tag{16.1}$$

Where Qt = tissue blood flow, Cv = venous blood concentration, Ca = arterial blood concentration, Pt = tissue/blood partition coefficient, Vt = volume of tissue and At = amount of chemical in tissue.

The physiology-based pharmacokinetic models (PBPK) models are complex and depend upon various parameters. Generally, these parameters represent the combined effects of not only the drug that is administered but also the subject to which the drug is administered. The simple PBPK models have various limitations which are quite a challenging activity for a variety of reasons including:

- Lack of complete knowledge of dynamics, whose features change with time, and are differentially expressed by various species, and within single species by the different individuals.
- Lack of specific and detailed mathematical formulation to acknowledge the experimental data.
- An increasing number of parameters grows with the equation complexity and this may lead to unmanageable formulations.
- Furthermore, several parameters introduce specificity and this reduces the model's flexibility to describe different systems.

This led to the development of pharmacokinetic models with the lowest complexity and at the same time capable of describing the concerned process and representing it with the help of easy mathematical formulas.

16.2.3 Non-compartment Model

The non-compartmental method is based on statistical moment theory and is not dependent on assumptions of a specific compartment model for either drug or metabolite. In fact, this method can be applied to any compartmental model, provided that alinear pharmacokinetics is assumed. The time course of drug concentration in plasma can usually be regarded as a statistical distribution curve.

The route of drug administration does not affect the process and the first three (zero to second) statistical moments are defined as follows:

$$\text{Zero moment AUC} = \text{AUC} = \int_0^\infty \text{Cdt} \tag{16.2}$$

$$\text{First moment MRT} = \frac{\text{A} \cup \text{MC}}{\text{A} \cup \text{C}} = \frac{\int_0^\propto \text{TCdt}}{\int_0^\infty \text{Cdt}} \tag{16.3}$$

$$\text{Second moment}: \quad \text{VRT} = \frac{\int_0^\propto \text{T2Cdt}}{\int_0^\infty \text{Cdt}} \tag{16.4}$$

where MRT stands for mean resident time and VRT stands for the variance of the mean resident time of a drug in the body. AUC, MRT and VRT are termed as the zero, first and second moment, respectively, of the drug concentration-time-curve. The area under the curve of a plot of the product of concentration and time versus time from zero to infinity is often referred to as the area under the (first) moment curve, AUMC.

The first moment of the blood level-time curve, mean residence time, is the statistical moment analogue to half-life ($t_{1/2}$). It is defined as the average amount of time spent by the drug in the body before being eliminated. The MRT represents the time for 63.2% of the administered dose to be eliminated statistically.

It is evident that statistical moment theory permits a wide range of analyses that, in most instances, will be adequate to characterize the pharmacokinetics of the drug including bioavailability, clearance and apparent volume of distribution etc. It is also useful in determining half-life, rate of absorption and absorption rate constant without the complex procedures of compartment modelling irrespective of the number of compartments. Certain problems are not addressed by this theory; non-linear events are not adequately treated by the statistical moment theory. Statistical moments provide only limited information regarding the time course of drug concentrations; for the most part, we deal with averages.

16.3 Novel Pharmacokinetic Models

One of the major issues of complex PBPK models is to derive certain mathematical parameters to represent and express the process of absorption, distribution, metabolism and excretion in the system. There are certain updated models which inculcate new scientific technologies and methodologies to better describe the characteristics of the drugs in the body.

16.3.1 Lumped and Flexible PBPK Model

There have been certain advantages and limitations of both compartmental and PBPK models. Most of the cases utilize classical compartmental PK or the physiologically-based PK approach independently or in parallel, with little to no overlap or cross-fertilization.

The lumped flexible PBPK model is an approach that establishes the link between mechanistic PBPK models and classical compartmental models. The proposed method has several advantages over existing methods: Perfusion and permeability rate-limited models can be lumped; the lumped model allows for predicting the original organ concentrations, and the volume of distribution at a steady state is preserved by the lumping method as seen in Fig. 16.5 (Nestorov et al. 1998).

The compartments (rectangles) are organs or lumped organ tissues which have similar pharmacokinetic characteristics. In this approach, the researchers have the choice of exclusively selecting the specific organs depending upon their objective of

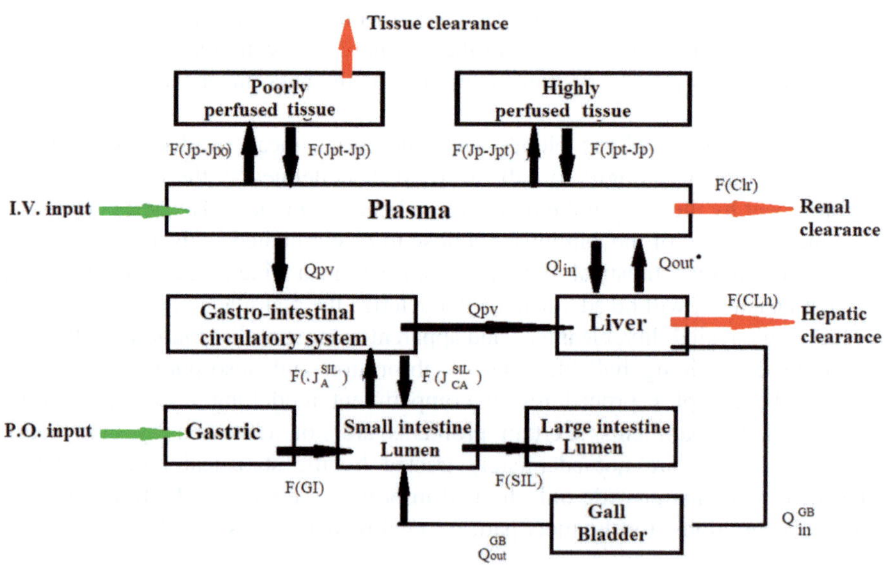

Fig. 16.5 Schematic representation of lumped physiological pharmacokinetic model

study. In most cases, the organs that have a primary role in the ADME paths are considered, while those that are merely distribution sites are lumped into a single compartment. A case where highly perfused organs (HO) and the poorly perfused tissues (PT) compartments, which gather the organs and tissues that are not explicitly represented by a dedicated compartment. The difference between HO and PT compartments is linked to the blood vessel perfusion of the single organs/tissues based on the assumption that the higher the perfusion, the easier the drug transport. This model, contrary to most of the PBPK literature models, considers only the liquid fraction of blood, i.e., plasma. This is convenient because the plasma consists mostly of water that acts as a solvent and suspending medium, and transports the drug to different organs and tissues (Brochot et al. 2005; Di Muria et al. 2010).

The gastrointestinal circulatory system (GICS) compartment lumps several vessels from the gastrointestinal tract that transport nutrients and possibly drugs to the liver via portal vein. In addition, GICS allows considering the so-called first-pass effect of orally administered drugs. This is related to hepatic metabolic activity. The portal vein (belonging to the GICS compartment) conveys drugs right after intestinal absorption to the liver where, depending on the specific active principle undergoes a metabolic action, which produces a partial loss of the administered dose. The liver compartment is considered individually as it has both anatomical and physiological relevance. At the anatomical level, it receives a large amount of the administered drug from both the systemic circulation (via the hepatic artery) and the intestinal region (via the portal vein). At the physiological level, it plays an important role in the drug metabolism using the hepatocyte's action. The gastrointestinal region is schematized into three compartments: the gastric (GL), small intestinal (SIL) and large intestinal (LIL) lumina, and characterizes the entire drug absorption process in case of oral administration, starting with the ingestion into the GL, up to faecal excretion from LIL. Finally, an additional compartment, the gall bladder, allows modelling the bile enterohepatic circulation process, which assists the digestion of lipids.

This periodic process of accumulation and release produces a characteristic effect of multiple drug concentration peaks in the PK profile of some drugs, e.g., sorafenib, erythromycin, ampicillin and phenolphthalein (Shiffman et al. 1990; Roberts et al. 2002).

The lumped PBPK model introduces and considers the issue of the binding process between drug molecules and plasma proteins (i.e. albumin, lipoproteins and globulins). Every drug has its tendency to bind specific proteins, thus reaching a dynamic equilibrium. When a drug is bound to a protein, it is confined within the plasma, as the drug-protein ensemble cannot diffuse through the endothelium of blood vessels. A specific parameter (R) accounts for a kind of drug-protein passive nature in the blood compartment.

In summary, the proposed lumping scheme comprised the following steps:

1. Simulate the whole-body PBPK model to predict the concentrations C_{tis} in all organs and tissues.

2. Plot the normalized concentrations and identify the groups of organs/tissues with similar normalized concentration-time profiles.
3. For each group of organs/tissues L, determine the lumped volume, blood flow and partition coefficient.
4. The process of simulation is applied to the lumped model and the lumped concentration (CL) for all groups of organs/tissues is determined.
5. The original tissue concentration (C_{tis}) is determined from CL for each organ/tissue group.

16.3.2 Pharmacokinetic Models for Optimizing Nanomedicines

Nanomedicines have been developed for more than four decades to optimize the pharmacokinetics of drugs, especially absorption, distribution and stability in vivo. Unfortunately, only a few drug products have reached the market. One reason among others is the lack of proper PK modelling and evaluation, which impedes the optimization of these promising drug delivery systems. So, it is extremely necessary to determine the biodistribution of nanomedicines in the body. The physical characteristics of nanomedicines are quite different from simple drug molecules due to a variety of reasons including a very high surface area to mass ratio, surface charge and release of activity from the nanosystem. The simple pharmacokinetic models do not take into account the specificity of these nano-drug delivery systems (Alexis et al. 2008).

The parameters that have to be included in the pharmacokinetic model for describing the complete profile of nanomedicines are described in Table 16.1 and shown underneath in Fig. 16.6:

16.3.2.1 Absorption

- In case of absorption, the model should consider the stability of the nanocarrier in GIT and should be suitably included.
- Mucus penetration of NPs is a limiting factor in the case of absorption of a nanomedicine. So, it should be included in the PK model.
- Absorption mechanism and nanoparticle integrity is considered an important parameter to be considered for inclusion in the PK model (Groo et al. 2015).

16.3.2.2 Distribution

- The complete distribution and interaction of NPs with the body tissues must be considered.
- The factors playing a crucial role in distribution parameters must be studied which include size, shape and charge.
- Volume of distribution (V_d) could not explain the distribution pattern of NPs.

The most important information that needs to be utilized for developing an accurate PK model is to consider the dynamic equilibrium that exists between the

Table 16.1 A comparison of ADME Properties and PK models of small molecule drugs and nanoparticles

Subject	Small molecules	Nanoparticles
Molecules to be modelled	Active ingredient and/or Metabolite(s)	• Nanoparticle • Nanoparticle-associated Drug and released drug
Major application	• Prediction of rate of absorption and rate of excretion Food-drug and drug -drug interaction studies affecting ADME • Special population PK Prediction (paediatrics, renal/ hepatic impairment, pharmacogenetics, sex, race, pregnancy, obesity)	• PK prediction and risk assessment • Optimization of formulation. Physicochemical property–ADME relationships • Prediction of in vivo drug release from nanoparticles and correlation with in vivo activity
Transport mechanism incorporated into models	• Absorption and distribution: diffusion and/or active transport by transporters • Metabolism: CYP enzymes and non-CYP enzymes; • Excretion: renal and biliary excretion	• *Absorption*: paracellular Transport, transcytosis and M cell uptake (oral), Macrophage uptake and diffusion (s.c., i.m. or inhalation), lymphatic uptake • *Distribution:*opsonization, MPS uptake, target-mediated disposition, EPR effect, lymphatic transport • *Metabolism*: extracellular Degradation, endocytosis and phagocytosis • *Excretion*: renal and biliary Excretion
Challenges	• Prediction of active transport and non-CYP metabolism • Prediction of patients suffering with renal or hepatic disease with altered pharmacokinetics	• Limited understanding on the transport mechanisms and ADME • Immunogenicity and nanotoxicity affect ADME of nanoparticles • Un equal distribution in tissues • Variability in MPS effect • Insufficient initial model parameters

encapsulated drug, free drug and the free NP. Regular PK models do not consider this approach (Longmire et al. 2008).

16.3.3 Pharmacokinetic Models for Describing Direct Delivery to the Lungs

There are certain conditions and disorders where direct targeting of drugs to the desired tissue is the preferred method and the best example is the delivery of drugs to the lungs. The novel drug delivery system via inhalation is the preferred method for delivering therapeutic aerosols to the respiratory tract. The efficacy of an inhaled

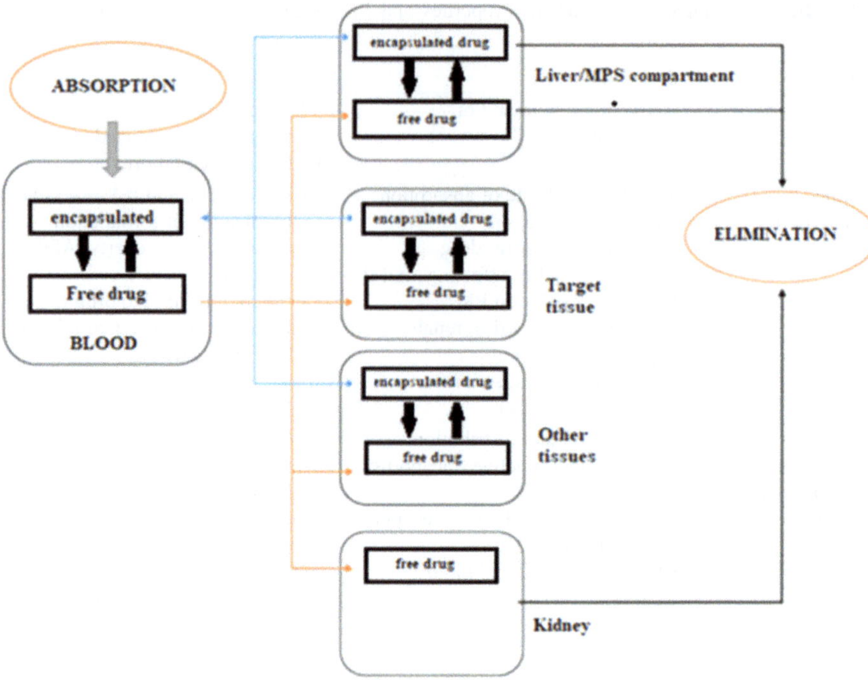

Fig. 16.6 Schematic proposed model adapted for describing pharmacokinetics of nanoparticles and their bio-distribution

therapy depends primarily on the quantity of drug deposited in the lung which in turn depends upon certain factors including formulation characteristics, device characteristics and patient characteristics (Chapman et al. 2011).

The normal physiological models could not assess the chain of clinically relevant aspects which are present in the delivery to the lungs and could correlate and characterize the performance of the device including flow characteristics of the inhaler mouthpiece, powder emptying (i.e. the efficiency by which the powder is released from the capsule after inhalation) and detachment (i.e. the detachment of the active substance particles from the surface of the carrier particles), and physiological parameters such as inhalation rate and airway anatomy.

The Biophysical model is an approach to predict inhaled drug deposition in patients with respiratory diseases and quantitatively investigate sources of variability in the delivery of inhaled drug with the device utilized formulation type, and patient variability. The model uses certain drugs, e.g., Indacaterol and glycopyrronium; and a novel drug delivery system called 'Breezhaler' (DPI), Novartis, Bael, Switzerland. The 'Biophysical model' is a complex combination of integrated computational fluid dynamics, in vitro experiments, and in vivo lung measurements (Fig. 16.7). The model was utilized to assess certain important parameters, namely flow characteristics of inhaler, powder emptying, detachment of active from carrier particles, and the effect of physiological variables including inhalation rate, airway

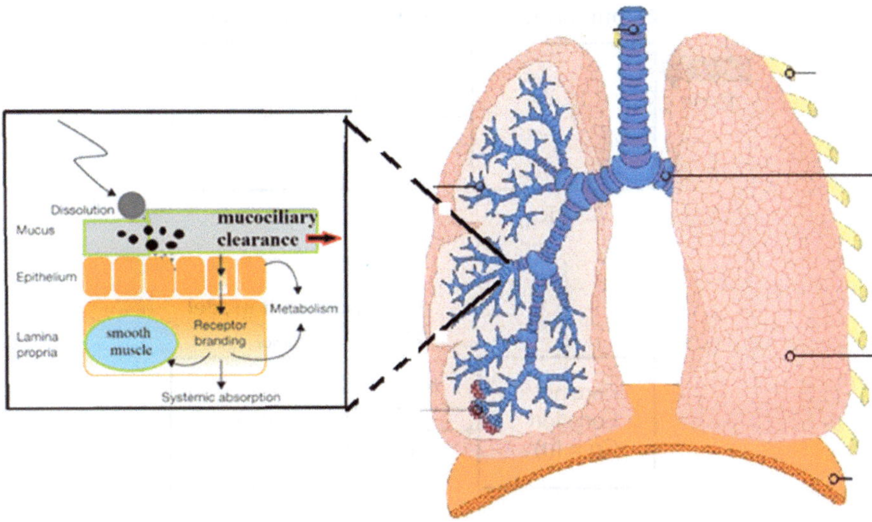

Fig. 16.7 Diagrammatic presentation of biophysical model for direct delivery to lungs

anatomy, inclination angle etc. To compute the aerodynamic particle size distribution and the drug dose delivered into the patient's lungs, 3D computational models of the human oropharynx were used with the inhaler mouthpieces attached. The anatomical model utilized for the process is the 'Alberta mouth throat model'. The biophysical lung model utilizes integrated computational fluid dynamics in combination with in vitro aerosol and in vivo lung measurements (Grgic et al. 2004; Dolovich et al. 2019).

16.3.4 Hybrid Pharmacokinetic Model for Characterization of PK Parameters in Tumours

The current era is mainly utilizing and trying to formulate molecular and targeted anticancer therapeutics. It is very essential to understand drug dynamics in the tumour. It is very advantageous to be able to relate drug concentrations in tumours to corresponding biological endpoints. A novel physiologically based hybrid pharmacokinetic model is developed to predict human tumour drug concentrations. Such models consist of a forcing function, describing the plasma drug concentration-time profile, which is linked to a model describing drug disposition in tumours. The hybrid models are originally derived from preclinical data and then scaled to humans. Integral to the scale-up procedure is the ability to derive human forcing functions directly from clinical pharmacokinetic data (Dogra et al. 2020).

Translation of these preclinical hybrid models to humans used a Monte Carlo simulation technique that accounted for intra-subject and inter-subject variability. Different pharmacokinetic endpoints, such as the AUC tumour, are extracted from

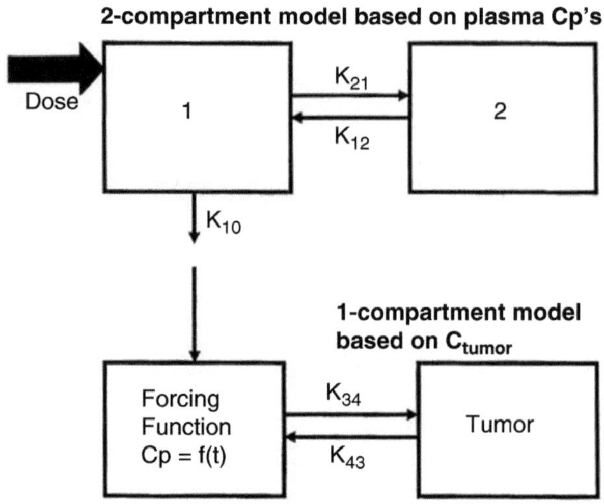

Fig. 16.8 Schematic representation of hybrid pharmacokinetic model having two-compartment plasma disposition characteristics and one compartment tumour model

the simulated human tumour drug concentrations to show how the predicted drug concentrations might be used to select drug-dosing regimens. It is believed that this modelling strategy can be used as an aid in the drug development process by providing key insights into drug disposition in tumours and by offering a foundation to optimize drug regimen design. The schematic representation of the hybrid model is shown in Figs. 16.8 and 16.9 (Gallo et al. 2004).

16.3.5 Multiorgan-on-a-Chip: A Systemic Approach to Model Inter-Tissue/Organ Transfer

The delivery of drugs follows a specific sequence and consequences in the body which is characterized by the body itself. The small intestine absorbs the (digested) substances, the liver metabolizes them, they are then delivered to target organs via the blood circulation and the kidney excretes corresponding waste products. This complex process of absorption/distribution/metabolism/excretion/toxicity (ADMET; affects the fate, distribution, efficacy and possible toxicity of exogenous substances). There is complex communication between organs and systems at the chemical and neuroglia levels. Together, this systemic and cross-organ communication is the key to deciphering and emulating the temporal processes involved in physiological functions.

But in vivo models suffer from numerous limitations: high experimental costs, limited throughput, ethical concerns and differences in genetic background. More importantly, they exhibit large physiological differences in terms of drug effects and/or disease phenotypes compared with humans, which explain the frequent

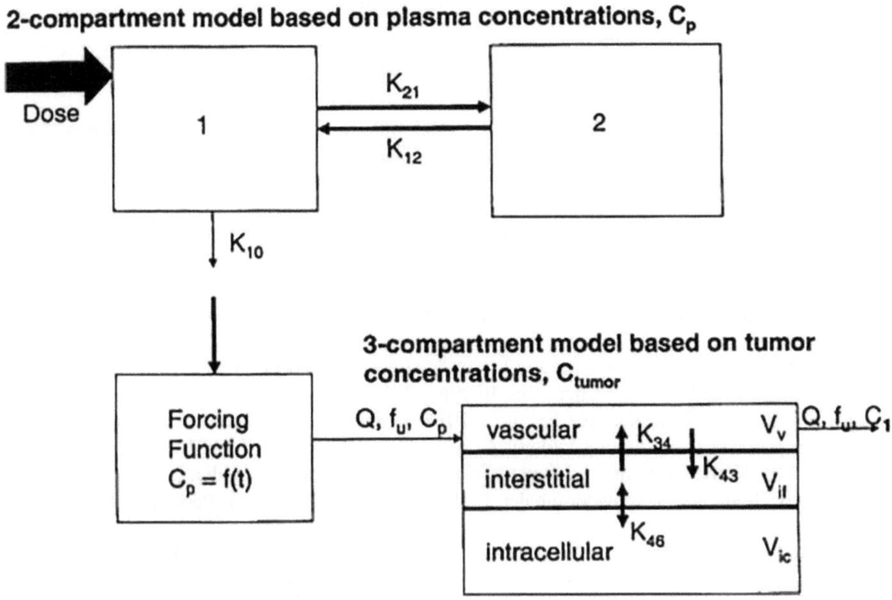

Fig. 16.9 Schematic representation of hybrid pharmacokinetic model having two-compartment plasma disposition model and three-compartment tumour model

failure of clinical trials. Overall, animals do not allow an analysis of inter-organ crosstalk, determination of quantitative pharmacokinetics (PK) or prediction of ADMET parameters, as recently highlighted. Therefore, advanced in vitro approaches incorporating a systemic dimension and multiple organs must be developed to faithfully emulate human health and pathophysiology.

16.4 Conclusion

In drug discovery and preclinical development, pharmacokinetic modelling is frequently used to optimize the selection of drug candidates based on in silico and in vitro data that are likely to have the desired in vivo pharmacokinetic properties. After entry into clinical development, there is a shift from mechanistic to empirical modelling and the clinical drug development does not directly benefit from the effort that has been made to develop a PBPK model in the preclinical stage. Thus, approaches that can establish links between the mechanistic PK models and the empirical models are highly desirable to bridge the gap between preclinical and clinical model development.

References

Aarons L (2005) Physiologically based pharmacokinetic modelling: a sound mechanistic basis is needed. Br J Clin Pharmacol 60:581–583

Alexis F, Pridgen E, Molnar LK, Farokhzad OC (2008) Factors affecting the clearance and biodistribution of polymeric nanoparticles. Mol Pharm 5:505–515

Brahmankar DM, Jaiswal SB (2015) Biopharmaceutics and pharmacokinetics - a treatise. Vallabh Prakashan, India

Brochot C, Tóth J, Bois FY (2005) Lumping in pharmacokinetics. J Pharmacokinet Pharmacodyn 32:719–736

Chapman KR, Fogarty CM, Peckitt C et al (2011) Delivery characteristics and patients' handling of two single-dose dry-powder inhalers used in COPD. Int J Chron Obstruct Pulmon Dis 6:353–363

Di Muria M, Lamberti G, Titomanlio G (2010) Physiologically based pharmacokinetics: a simple, all-purpose model. Ind Eng Chem Res 49:2969–2978

Dogra P, Butner JD, Ruiz Ramírez J et al (2020) A mathematical model to predict nanomedicine pharmacokinetics and tumour delivery. Comput Struct Biotechnol J 18:518–531

Dolovich MB, Kuttler A, Dimke TJ et al (2019) Biophysical model to predict lung delivery from a dual bronchodilator dry-powder inhaler. Int J Pharm X 1:100018. https://doi.org/10.1016/j.ijpx.2019.100018

Gallo JM, Vicini P, Orlansky A et al (2004) Pharmacokinetic model-predicted anticancer drug concentrations in human tumors. Clin Cancer Res 10:8048–8058

Gerlowski LE, Jain RK (1983) Physiologically based pharmacokinetic modelling: principles and applications. J Pharm Sci 72:1103–1127

Grgic B, Finlay WH, Heenan AF (2004) Regional aerosol deposition and flow measurements in an idealized mouth and throat. J Aerosol Sci 35:21–32

Groo AC, Bossiere M, Trichard L et al (2015) In vivo evaluation of paclitaxel-loaded lipid nanocapsules after intravenous and oral administration on resistant tumor. Nanomedicine (Lond) 10:589–601

Jones HM, Rowland-Yeo K (2013) Basic concepts in physiologically based pharmacokinetic modelling in drug discovery and development. CPT Pharmacometrics Syst Pharmacol 2:1–12

Longmire M, Choyke PL, Kobayashi H (2008) Clearance properties of nano-sized particles and molecules as imaging agents: considerations and caveats. Nanomedicine (Lond) 3:703–717

Nestorov IA, Aarons LJ, Arundel PA (1998) Lumping of whole-body physiologically based pharmacokinetic models. J Pharmacokinet Biopharm 26:21–46

Notari RE (2013) Biopharmaceutics and clinical pharmacokinetics- an introduction, 4th edn. New York, NY, Marcel Dekker

Roberts MS, Magnusson BM, Burczynski FJ et al (2002) Enterohepatic circulation: physiological, pharmacokinetic and clinical implications. Clin Pharmacokinet 41:751–790

Shargel L, Wu-Pong S, Yu ABC (2016) Applied biopharmaceutics & pharmacokinetics, 6e. Mc GrawHill, New York

Shiffman ML, Sugerman HJ, Moore EW (1990) Human gallbladder mucosal function. Effect of concentration and acidification of bile on cholesterol and calcium solubility. Gastroenterology 99:1452–1459

Winter ME (2004) Basic clinical pharmacokinetics, 4th edn. Lippincott Williams & Wilkins, Philadelphia, PA

Part III

Drug Delivery and Nanotechnology

Technological Innovations in Pharmaceutical Drug Nanocrystals

17

Sanika Jadhav, Amanpreet Kaur, and Arvind Kumar Bansal

Abstract

Nanocrystals are innovative drug delivery systems to improve solubility, dissolution rate, and bioavailability of poorly water-soluble drugs. In this review, the techniques used for the preparation of drug nanocrystals are presented along with their technological limitations. Numerous advancements in manufacturing and equipment have been introduced in case of both top-down and bottom-up approaches. Furthermore, combination methods have been introduced to overcome challenges of individual technologies. The technological gaps can be bridged through further innovation in production methods and techniques for the generation of nanocrystals. Moreover, with the better understanding of technological aspects, the success of the production method can be ensured. In this review, the strategies used for mapping of innovation have also been presented. A novel spray drying based technique, NanoCrySP, developed in our lab at NIPER, SAS Nagar has been discussed. This is an example of incremental innovation in the domain of generation of drug nanocrystals. In comparison to the exiting proprietary techniques, NanoCrySP is a one-step process to produce drug nanocrystals in powder form. This significantly overcomes the challenges associated with downstream processing of drug nanosuspensions produced using existing approaches. A case study of application of innovation radar tool in comparing process efficiency of a novel and existing technique has also been presented. Thus, NanoCrySP was proved to offer an innovative solution to minimize downstream processing and ease production of nanocrystal based finished drug products.

S. Jadhav · A. Kaur · A. K. Bansal (✉)
Solid State Pharmaceutics Lab, Department of Pharmaceutics, National Institute of Pharmaceutical Education and Research (NIPER), Mohali, Punjab, India
e-mail: akbansal@niper.ac.in

P. P. Singh (ed.), *Recent Advances in Pharmaceutical Innovation and Research*, https://doi.org/10.1007/978-981-99-2302-1_17

Keywords

Dissolution rate · Drug delivery · Drug nanocrystals · Oral bioavailability · Particle size reduction · Solubility

17.1 Introduction

Approximately 40% of new chemical entities (NCEs) are reported to be poorly water-soluble which inturn negatively impact their oral bioavailability. Numerous molecules are withdrawn from development pipeline due to their poor aqueous solubility. Development of such molecules into bioavailable formulations is a major challenge and hence, requires enabling technologies to improve aqueous solubility and oral bioavailability.

A plethora of techniques such as particle size reduction, salt formation, inclusion complexation, and modification of crystallinity are employed to improve aqueous solubility. Particle size reduction, using micronization and nanonization, is one of the most promising approaches to improve solubility and dissolution rate of poorly soluble drugs.

Drug nanocrystals are produced by "top-down" or "bottom-up" technologies. Numerous products in the category of oral, parenteral, and topical formulations have been commercialized using drug nanocrystals. Despite their advantages, current technologies for generation of nanocrystals suffer from certain disadvantages such as contamination from milling media, drug degradation, challenges in solvent removal, downstream processing and most importantly in controlling the crystal size in nano range. There has been an unmet need to develop newer technologies for generation of drug nanocrystals. Development of NanoCrySP technology is an outcome of the extensive research to bridge the gap in current technologies for nanocrystals. NanoCrySP is a novel, efficient "bottom-up" spray drying based technique for generation of drug nanocrystals. Development of oral dosage forms using NanoCrySP technology requires reduced downstream processing and offers opportunities for commercial exploitation.

The current review discusses proprietary techniques currently available for the generation of drug nanocrystals. The review also details technological aspects of a spray drying based technology developed in NIPER, SAS Nagar for one-step production of drug nanocrystals in powder form, which can be readily formulated into orally compliable finished dosage forms such as tablets or capsules.

17.2 Pharmaceutical Innovation

Innovation is a stepwise process for continuous improvement. It starts with problems and loopholes in the existing products or technologies followed by determination to find solution and produce sustainable results. Innovation is the key for market differentiation and competitive advantage (Burnett 2011; Gamal et al. 2011). Among the different ways to analyze innovation, incremental and radical innovation are the most common classifications used in pharmaceutical industry (Cardinal 2001; Norman and Verganti 2014). The main criterion for distinguishing between radical and incremental innovation is whether the innovation is considered as a continuous modification of preexisting technology/product or whether it is novel, unique, and discontinuous (Norman and Verganti 2014).

Incremental innovations consist of advancements of existing technology/product using well-established concepts, ideas, and knowledge. The purpose of these innovations is cost reduction, repositioning, introduction into new markets or adaptation to new circumstances, such as new regulatory laws and standards. Strategies such as new combinations of existing drugs, new dosage forms, new indications, and formula changes are considered incremental innovations (Cardinal 2001; Robbins and O'Gorman 2015). The risks with incremental innovation is that huge marketing expenses are necessary in mature markets to become a competitive player in the market (Norman and Verganti 2014). Also, competition in research and development resources limits advancement of new products/technologies. Another challenge related to incremental innovation is that the market tends to change with the introduction of disruptive innovations. In such a case, just relying on incremental innovation may not be sufficient to sustain in the market (Kylliäinen 2019).

In contrast, radical innovations involve development of novel technologies or products with generation of new knowledge domains and new paradigms. New chemical entities (NCEs), i.e., new drugs with significant therapeutic advances and novel technologies are considered radical innovations. Radical innovations are associated with substantial technical risk as there is no preexisting knowledge space available unlike incremental innovations. Also, the time and cost required for the development of new techniques is significantly higher in case of radical innovations as compared to incremental innovations (Cardinal 2001; Norman and Verganti 2014). Radical innovation is often called as disruptive or competence destroying, or as breakthrough, because they all use the same concept, i.e., radical innovation implies a discontinuity with the previously developed technologies (Norman and Verganti 2014).

Many successful products/technologies follow incremental innovation for the development of profitable and efficient outputs. Pharmaceutical company should decide the type of innovation depending on overall goals of the organization. A company can create sustainable competition in the market with good profit margin, by following incremental innovation as it involves low technical risk, low cost, and comparatively higher commercial viability. On the other side, if company is willing to enter into new, undiscovered field with novel products and technologies, radical innovation is the suitable choice for the organization (BMILab 2017).

17.3 Mapping of Innovation

Measuring innovation is important to understand the extent of innovativeness of the product/process in a definite way. It helps to analyze different technical, commercial aspects associated with innovation. Mapping of innovation is essential for business growth and profitability as several market variables depend on innovation (Burnett 2011; Gamal et al. 2011). Sustainable innovation results in continuous revenue growth which further accelerates more extensive innovation (Gamal et al. 2011). Also, mapping of innovation helps to identify areas of strength to capitalize on and provide opportunities for increasing innovation. Such examination assists to focus on weaker aspects to the enhance innovation process and spread awareness about innovation concepts (Gamal et al. 2011).

Two broad ways are used to measure innovation in the pharmaceutical industry. The most common tool measures innovation through innovation inputs, such as research and development (R&D) expenditure and innovation outputs (like patents and patent-related index). The understanding of the cost effectiveness of a pharmaceutical process is crucial while mapping innovative product/process (Ketokivi and Ali-Yrkkö 2010; van der Aalst et al. 2006). It is important to determine the commercial success of the process. However, these approaches capture narrow aspects of innovative processes neglecting different technological aspects. Such mapping of innovation is unclear and improper concerning the growth and scope of the process (Ketokivi and Ali-Yrkkö 2010; Dodgson and Hinze 2000; Carroll et al. 2017).

Innovation is a complex and multidimensional activity that cannot be measured directly or with a single indicator and hence a composite measure of innovative process is required. Therefore, along with R&D cost, consideration of different formulation and process parameters for measuring innovation of pharmaceutical process is essential for a broader perspective (Ketokivi and Ali-Yrkkö 2010). Various frameworks such as innovation funnel, diamond model, and innovation radar are utilized to map innovativeness of a novel process or a product. Most of these models capture various dimensions involved in the innovation, but they differ in terms of perspectives of focus, and visualization of the innovation process (Gamal et al. 2011; Burnett 2011; Longdon 2008).

17.3.1 Innovation Funnel

A pictorial representation of innovation funnel is depicted in Fig. 17.1. In this model, ideas flow through a "funnel" and are analyzed considering various criteria of innovation such as business strategy, implementation, commercial feasibility, and market potential. Several experts evaluate ideas at every stage of the innovation. Every innovation is passed through funnel entrance and come out with a particular outcome. This model connects "people with problems" to "people with solutions" (Burnett 2011; Gamal et al. 2011).

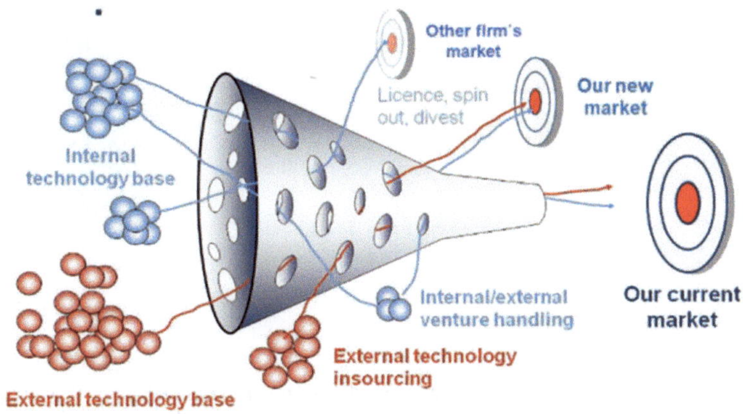

Fig. 17.1 Representation of innovation funnel. (Reproduced with permission from Yun JJ, Yun JJ. Concept, Structures, and Decision Factors of Open Innovation. Business Model Design Compass: Open Innovation Funnel to Schumpeterian New Combination Business Model Developing Circle. 2017:63–85)

17.3.2 Diamond Model

Innovation diamond model considers following five dimensions for mapping of innovation (Gamal et al. 2011).

- Strategy: Systematic planning of innovation is a key aspect while mapping innovation. Well-managed and analyzed strategy outlook helps company to effectively implement the innovation.
- Process: This parameter takes care of the robustness and flexibility of the organization's new product development (NPD) process. This dimension also considers various internal processes involved in the innovative process.
- Organization: This dimension mainly considers whether the organizational structure encourages coordination within the firm and employees to come up with novel ideas.
- Linkages: Networking with external entities such as suppliers, customers, the academe, other industries, specialist individuals, and competitors is important for advancement of the innovation.
- Learning: This dimension captures learning parameters in terms of training of employees and analyzing knowledge/information.

Figure 17.2 shows the results of assessment based on diamond models for two different organizations.

17.3.3 Innovation Radar

This tool was developed to provide comprehensive summary of the multidimensional nature of innovation and to measure its innovativeness (Sawhney et al. 2006).

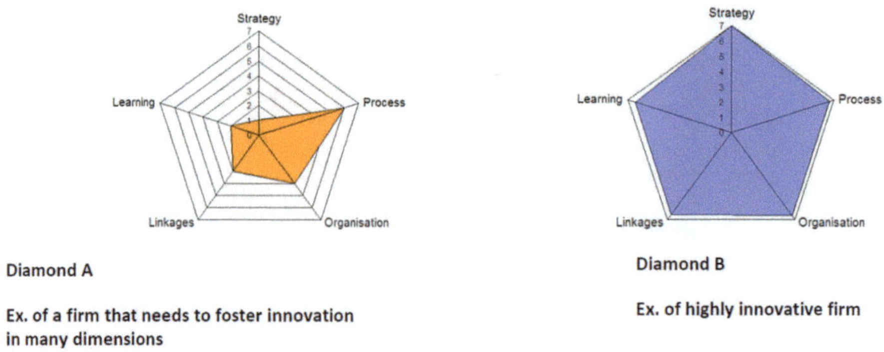

Fig. 17.2 Schematic illustrations of diamond models of two different firms

It distinguishes 12 different ways in which a firm can seek innovation, i.e., offerings, platforms, solutions, customers, customer experience, value capture, processes, organization, supply chain, presence, networking, and brand. Our group had modified this model for parameters to be considered to measure the innovativeness of a novel technology, NanoCrySP, for generation of drug nanocrystals in comparison to an existing technology (Jadhav et al. 2020). The different aspects used are neither mutually exclusive nor wholly independent of one another. However, they are distinct and unique in their emphasis (Sawhney et al. 2006). Innovation radar gives a pictorial representation in the form of the radar of process efficiency. This makes it easier for researchers to interpret innovation capabilities concerning each aspect and necessary strategies to be enforced for further development (Sawhney et al. 2006).

17.4 Nanocrystals: Proprietary Techniques and Gaps

In the past two decades, drug nanocrystals are the most extensively investigated and developed drug delivery system (Shegokar and Müller 2010; Möschwitzer 2013). According to the US FDA, nanocrystal is defined as a material or end product which should have at least one dimension in the nanoscale range (approximately 1 nanometer (nm) to 100 nm) and/or material or end product is engineered to exhibit properties or phenomena, including physical or chemical properties or biological effects, that are attributable to its dimension(s), even if these dimensions fall outside the nanoscale range, up to 1 μm (1000 nm). Their size ranges from 10 to 1000 nm, which offers drug loading as high as 100% (Möschwitzer 2013). Simplicity of nanocrystal formulations and enabling development from a milligram scale to production level makes nanocrystals the most studied and conspicuous discipline in pharmaceutical sciences in recent years (Möschwitzer 2013; Gao et al. 2008). Nanocrystals can be formulated as liquid drug nanosuspensions or solid drug nanoparticles. Nanosuspension is usually formulated into solid dosage forms to overcome its instability and crystal growth related challenges (Mohana Raghava Srivalli and Mishra 2015; Chen et al. 2011; Möschwitzer 2013).

A large number of products based on nanocrystals have been commercialized due to multiple benefits associated with drug nanocrystals (Shegokar and Müller 2010). Nanocrystals show an increment in the dissolution rate of a drug owing to large particle surface area. Drug nanocrystals eventually exhibit enhanced in vivo bioavailability (Gao et al. 2013; Möschwitzer 2013; Junghanns and Müller 2008).

This section focuses on various technologies employed to produce drug nanocrystals considering their industrial applicability and scope. These methods are mainly classified into three groups, i.e., top-down, bottom-up, and combination technologies. In the case of top-down approach, the micronized drug is suspended in an aqueous media and size is reduced to the nano range by applying high energy. The bottom-up technologies generally involve controlled precipitation of drug nanocrystals from a drug solution. Combination technologies include a pre-treatment step followed by a reduction of particle size using high energy (Shegokar and Müller 2010; Möschwitzer 2013).

17.4.1 Top-Down Technologies

17.4.1.1 Bead/Pearl Milling

Wet media milling (bead/pearl milling) technique is the most widely used technique for the production of nanosuspension, i.e., liquid dispersion of drug nanocrystals. Liversidge and his colleagues developed NanoCrystal® technology based on bead/pearl mill to achieve particle size reduction of crystalline drugs (Liversidge et al. 1992).

The wet media milling technique can be employed on lab as well as industrial level. For the lab scale, the low energy wet media milling (LE-WBM) process is commonly used wherein a vial is filled with milling media, dispersion medium containing one or more stabilizers, and micronized drug. With the aid of a magnetic stirrer, drug particles interact with milling media and thus shear forces generated by the movement of the beads lead to particle size diminution (Möschwitzer 2013). Different types of milling media are used such as smaller or larger coated milling pearls of ceramics (cerium or yttrium stabilized zirconium dioxide), stainless steel, glass, or highly cross-linked polystyrene resin-coated beads.

In the pharmaceutical industry, NanoCrystal® technology is based on high energy wet ball milling process (HE-WBM) (Merisko-Liversidge and Liversidge 2008). In recent times, agitated ball mills in re-circulation mode are used for large-scale production of nanosuspension (Liversidge et al. 1992; Kwade 1999).

A common problem associated with this technology is erosion from the milling media during the milling process. Therefore, contamination of drug nanosuspension from milling media typically from glass or zirconium oxide beads due to abrasion of either of the milling beads or parts of the milling chamber poses a major challenge (Juhnke et al. 2012; Hennart et al. 2010). However, the use of highly cross-linked polystyrene beads as a milling material minimizes the contamination of nanosuspension.

The milling time required for the generation of nanosuspension can vary from about 30 min to hours or several days. The various parameters such as drug amount, the hardness of the drug, viscosity, temperature, energy input, number of milling pearls, size of the milling media and surfactant/stabilizer concentration used, milling speed, and milling time should be optimized to obtain nanosuspension with a narrow particle size distribution and a very little batch-to-batch variation (Liversidge et al. 1992).

The first commercial product based on this technique is Rapammune®, an immunosuppressant agent, marketed by Wyeth Research Laboratories in 2002. The particle size reported for nanosuspensions prepared by wet ball milling are in the range of 100 and 300 nm, irrespectively whether LE-WBM or HE-WBM was used. Also, NanoCrystal® technology has expanded its application for the generation of nanosuspension for oral, inhalation, intravenous, subcutaneous, intramuscular, and ocular delivery (Merisko-Liversidge and Liversidge 2008).

17.4.1.2 High-Pressure Homogenization

High-pressure homogenization (HPH) is another commonly used technique for generation of drug nanosuspension. First, the coarse micronized drug and dispersion medium containing stabilizers are mixed using high-speed stirrers. Subsequently, the coarse suspension (also refer as macrosuspension) is pumped several times through the gap having an opening of only a few micrometers, of high-pressure homogenizer. The pressure is built up gradually to 1000–2000 bar to avoid clogging of the narrow homogenization gap (Möschwitzer 2013). The high energy produced during this process due to cavitation forces, shear forces, and particle collision against each other results in particle size reduction (Gao et al. 2008; Muller and Keck 2004; Liedtke et al. 2000).

The desired nanosuspension formulation is produced by maintaining a balance between primarily three parameters, i.e., applied pressure, number of homogenization cycles, and temperature (Gao et al. 2008). Mostly, the maximum pressure of 1500 bar is essential for the generation of drug nanocrystals. For the temperature sensitive drugs, regulation of temperature while homogenization is needed (Müller and Peters 1998).

There are three technologies based on HPH- Microfluidizer technology, Dissocubes® technology, and Nanopure® technology (Shegokar and Müller 2010). The Microfluidizer technology is based on the jet stream principle and SkyePharma Canada Inc. employs this principle for their Insoluble Drug Delivery-Particles (IDD-PTM) technology to produce submicron particles of poorly soluble drugs (Keck and Müller 2006; Haynes 1992; Möschwitzer 2013). The commercial product Triglide® uses this technology to produce drug nanocrystals of fenofibrate.

Muller and his co-workers developed piston-gap homogenizers based technology, which is named Dissocubes™ considering the cubic shape of the drug nanocrystals produced with this process (Muller et al. 1999). First, a micronized drug is mixed with an aqueous dispersion medium containing surfactant/stabilizers. Subsequently, the coarse suspension is forced under pressure (up to 4000 bar, typically 1500–2000 bar) through a tiny gap (e.g., 5–20 μm). The drug nanocrystals are generated due to the high power of the shockwaves caused by cavitation (Muller et al. 2000). Nanopure® technology is another method based on piston-gap homogenizer (prev. PharmaSol GmbH, now Abbott). This technology uses non-aqueous liquids such as oils, liquid, and solid (melted) PEG, or water reduced media (hydro-alcoholic systems) as a primary dispersion medium. The homogenization is carried out at low temperatures which facilitates the production of nanosuspension of thermo labile drugs (Gao et al. 2008).

HPH is preferred over WBM as there are fewer chances of contamination of nanosuspension due to abrasion and wearing of the milling equipment (Möschwitzer 2013).

17.4.1.3 Dry Co-Grinding

SoluMatrix Fine Particle Technology™ (Trademark of iCeutica Inc.) is one such dry milling method to produce nanocrystals of pharmaceutical compounds (Cammarano et al. 2014). This technology is licensed to Iroko Pharmaceuticals, LLC. In this method, a drug is grinded in the presence of a mill able grinding compound in a dry mill which results in solid dispersion of drug nanocrystals in the millable grinding compound. Using this technique, it is possible to produce drug nanocrystals in the size range of 200–1000 μm (Desjardins et al. 2015). Nanocrystals of indomethacin, meloxicam, diclofenac acid are already commercialized in the US market by Iroko Pharmaceuticals employing SoluMatrix™ technology. These products showed optimal dissolution behavior and oral bioavailability along with rapid onset of action. The technology helped in the production of low dose combinations of existing products available in the market (Desjardins et al. 2015).

Another method for generation of drug nanocrystals using a dry grinding process along with polyvinylpyrrolidone (PVP) and sodium dodecylsulfate (SDS) has been reported for several poorly water-soluble drugs: griseofulvin, glibenclamide, and nifedipine (Itoh et al. 2003; Wongmekiat et al. 2002). The addition of PVP and SDS inhibits aggregation of nanocrystals and provide stability to drug nanocrystal formulation. (Chingunpituk 2007; Wongmekiat et al. 2002; Mura et al. 2002; Otsuka and Matsuda 1995).

Dry grinding is considered an easy and economical process due to the absence of organic solvents for industrial production of nanocrystals. However, the prevention of partial amorphization of a crystalline drug during grinding eventually affecting physicochemical properties of a compound, poses a major challenge (Wongmekiat et al. 2002; Itoh et al. 2003).

17.4.2 Bottom-up: Precipitation Methods

Before top-down techniques were developed for the production of nanosuspension, drug nanoparticles were generated using simple, low cost bottom-up techniques; commonly known as precipitation techniques (Gao et al. 2008; Sinha et al. 2013). First, a drug is dissolved in a suitable solvent system to form a drug solution. Consequently, rapid addition of antisolvent leads to the super saturation of the system followed by precipitation of fine crystalline or amorphous solids. For the successful generation of drug nanocrystals using this approach, the drug has to soluble in one solvent that is miscible with non-solvent (antisolvent) (Chingunpituk 2007; Kipp et al. 2003b). However, it is challenging to restrict drug particle size in the nano range and control crystal growth due to Ostwald ripening. The addition of stabilizers such as HPMC, PVP, Tween 80, Poloxamer 188, or lecithin is the most commonly utilized approach to avoid spontaneous aggregation of particles (Gassmann et al. 1994; Van Eerdenbrugh et al. 2008; Chen et al. 2011). The drug nanoparticles with uniform and narrow particle size distribution can be generated by controlling the following parameters: stirring rate, the volume ratio of antisolvent to solvent, drug content, and temperature (Douroumis and Fahr 2006; Zhang et al. 2006; Gao et al. 2008). Precipitation techniques did not attain popularity in the pharmaceutical industry due to difficulty in controlling the crystal growth process (Möschwitzer 2013).

17.4.2.1 Hydrosols

This bottom-up technique was developed by Sucker around 1980; the intellectual property of was later acquired by Sandoz, Novartis (List and Sucker 1988; Sucker and Gassmann 1994). It is a classical precipitation method wherein the drug is dissolved in a solvent and subsequent addition of non-solvent leads to precipitation of fine particles (Shegokar and Müller 2010). No pharmaceutical product is commercialized yet based on this technology due to complexities in the manufacturing process.

17.4.2.2 NanoMorph®

This is another precipitation process developed by Auweter and Horn. NanoMorph® is the trade name of Soliqs® (Abbott GmbH & Co. KG, Ludwigshafen) previously Knoll/BASF (Auweter et al. 1998). This technique yields spherical amorphous drug nanoparticles (Auweter et al. 2002). However, there are high chances of partial recrystallization of amorphous form affecting the overall bioavailability of the product. This technology is applied to develop a commercial product in the food sector, e.g., Lucarotin® as partial recrystallization is not a major concern in food products.

17.4.2.3 Precipitation Using Spray Freezing into Liquid

De Waard et al. reported a modified freeze-drying method to produce drug nanocrystals. The research group modified the mixing step by using a 3-way nozzle, i.e., for solvent, antisolvent, and atomizing air (de Waard et al. 2008, 2009). Atomizing air generates the fine droplets of solvent and antisolvent as they both get mixed immediately on passing through the nozzle. Subsequently, the mixture enters the liquid nitrogen which leads to freezing and lyophilization of the mixture (Hu et al. 2002; Yang et al. 2008; Möschwitzer 2013). The nanocrystals produced by this method are relatively smaller in size and porous as compared to other precipitation methods. This is advantageous to obtain a higher dissolution rate of nanocrystals (Hu et al. 2003; Sinha et al. 2013).

17.4.2.4 Supercritical Fluid Methods

Different techniques such as supercritical antisolvent process, rapid expansion of supercritical solution (RESS) process, and precipitation with compressed antisolvent (PCA) process are used to produce nanometer-sized drug crystals (York 1999; Fages et al. 2004; Möschwitzer 2013). In the case of the supercritical antisolvent process, the supercritical fluid acts as an antisolvent for the drug, whereas in RESS technique, a drug is solubilized in supercritical fluids such as supercritical carbon dioxide (Matson et al. 1987; Türk 2009). In the case of PCA method, a drug solution is atomized into the carbon dioxide compressed chamber resulting in solvent evaporation. The drug solution gets supersaturated and drug nanoparticles are generated by the precipitation process. The use of hazardous solvents is a major disadvantage of the above-mentioned techniques (Sinha et al. 2013).

17.4.2.5 Sonoprecipitation

Ultrasonic sound is used to produce drug nanocrystals using the precipitation technique (Li et al. 2003; Kumar et al. 2009). Generally, an antisolvent is mixed with the drug solution kept under stirring. The mixing is enhanced by ultrasonic energy by introduction of a probe sonicator (Dhumal et al. 2008; Beck et al. 2010). The uncontrolled particle growth and aggregation after the precipitation of drug particles can be reduced by increasing the micromixing of the supersaturated solution by sonication (Xia et al. 2010; Sinha et al. 2013). The particle size of drug nanoparticles can be controlled by optimizing sonication duration and intensity, horn length, depth of horn immersion, and cavitation (Sinha et al. 2013).

17.4.3 Combination Technologies

These techniques are the combination of bottom-up and top-down methods (Shegokar and Müller 2010). Crystal growth during industrial production poses a major challenge in case of bottom-up approaches. On the other hand, longer production time in case of sole use of WBM or HPH and clogging of the equipment caused by the use of coarse drug suspension are major concerns for widely applied top-down approaches (Möschwitzer 2013). Thus, the high energy particle diminution process after the precipitation method reduces the chances of crystal growth of drug particles. Also, the processing time of the top-down process can be significantly reduced when the drug is pre-treated.

17.4.3.1 NANOEDGE™ Technology

This technology was developed by Baxter which employs a precipitation step followed by the annealing process by applying high energy, e.g., high-pressure homogenization (Kipp et al. 2003a; Kipp 2004; Rabinow 2004). HPH process acts as an annealing step wherein the growth of precipitated nanocrystals is prevented. The thermodynamically unstable form converts into a stable form by repeated application of energy followed by thermal relaxation which eventually leads to a more ordered lattice structure (Hancock and Zografi 1997).

17.4.3.2 smartCrystal® Technology

Abbott owns smartCrystal® technology and is marketed by its drug delivery company Soliqs. It involves a variety of combination processes, e.g. H42, H69, and H96, customized according to a specific application and drug properties (Shegokar and Müller 2010). These processes generate nanocrystals below 100 nm which is not possible to produce using HPH alone (Keck et al. 2008).

H42 involves spray drying followed by HPH wherein few homogenization cycles lead to the generation of fine nanosuspension (Möschwitzer and Müller 2006; Salazar et al. 2013). H69 is a combination of precipitation and HPH, whereas in the case of H96 process lyophilization is carried out first followed by HPH (Müller and Möschwitzer 2015; Möschwitzer and Lemke 2006). H69 and H96 processes are used to generate nanocrystals of amphotericin B with a size below 50 nm (Shegokar and Müller 2010). During large-scale production, lower homogenization pressure around 300 bars is sufficient to produce the desired size of drug nanocrystals. Thus, this technology is useful for industrial application as it is fast, minimizes wearing of the machine, and more economical (Möschwitzer 2013).

17.4.4 Unmet Needs

The outcome of technology used for the production of drug nanocrystals is influenced by different drug and material related factors and process attributes. Drug-related properties like physicochemical properties of the drug such as solubility and crystal hardness are primary deciding factors for the selection of nanocrystal preparation method (Shete and Bansal 2016). The critical material and process attributes associated with the technology, industrial scalability of the method, the time required, %yield, reproducibility, environmental concern are other important parameters to be considered while choosing a particular method for the generation of drug nanocrystals.

Top-down methods suffer from drawbacks such as the use of a high proportion of surfactants, long processing times, difficulty in achieving a uniform size distribution, low yield, high energy input, and possible contamination from the milling media. A solid-state transformation and chemical degradation of the drug can occur during the top-down process due to the application of high energy (Shegokar and Müller 2010; Müller et al. 2011; Sinha et al. 2013).

The uncontrolled particle growth and aggregation followed by the precipitation process is a major challenge possessed by bottom-up technologies. This disadvantage has restricted the industrial viability of these methods. Also, bottom-up technologies suffer from other pitfalls such as possible decomposition of the drug, contamination from a toxic organic solvent, the limited solubility of the drug in the solvent, and long processing times (Junghanns and Müller 2008; de Waard et al. 2011; Sinha et al. 2013; Shete and Bansal 2016).

Moreover, these processes generate drug nanocrystals in the form of liquid dispersion, i.e., nanosuspension. Therefore, post-processing of nanosuspension is required to commercialize drug nanocrystals in physically stable and patient compliant oral solid dosage forms. Hence, research is going on to generate drug nanocrystals as a solid dispersion for ease of the development of various oral solid and injectable dosage forms. The production of solid nanoparticles will also be advantageous considering industrial scalability, stability, and time and cost required for post-processing (Jacobs et al. 2000; Shegokar and Müller 2010; Shete and Bansal 2016).

17.5 NanoCrySP Technology

NanoCrySP technology is a novel, innovative bottom-up spray drying based process developed in our lab at NIPER, S.A.S Nagar to generate drug nanocrystals (Bansal et al. 2017). This method produces spray dried powder product known as "nanocrystalline solid dispersion" (NCSD). It consists of primary spray dried particles of size 2–50 μm wherein drug nanocrystals of size ranging from 10 to 1000 nm are dispersed in the matrix of non-polymeric crystallization-inducing agent. Drug along with crystallization-inducing agent is first solubilized in a suitable solvent or solvent system and this feed solution is then spray dried to produce NCSD (Shete et al. 2015a, b; Shete and Bansal 2016). This technology has been discussed in proceeding sections in detail by considering technical, biopharmaceutical, translational, efficiency, and innovation related aspects.

17.5.1 Technical Aspects

Evaporation of droplet containing drug and crystallization-inducing agent during spray drying, results in the formation of the amorphous phase of the drug. This amorphous phase of the drug later crystallizes into a more stable crystalline form during the drying phase and thus produces drug nanocrystals in the range of 10–1000 nm (Fig. 17.3) (Shete and Bansal 2016; Bansal et al. 2017). Crystallization-inducing excipients are selected from the category of generally regarded as safe (GRAS) excipients, comprising mainly small molecule sugars (glucose, dextrose, and maltose), polyols (xylitol, sorbitol, and mannitol), amino acids (glycine, alanine, and leucine), and dicarboxylic acids (adipic acid, citric acid, fumaric acid, malic acid, and tartaric acid) (Bansal et al. 2017). Possibility of formation of aggregates in NCSD during storage can be overcome by the addition of a different type of stabilizers. They provide electrostatic repulsion and/or steric hindrance which might help in minimizing aggregation of nanocrystals. The most commonly employed stabilizers are either polymer (synthetic and semisynthetic) or surfactants (ionic and non-ionic) (Shete et al. 2016).

Various drug and excipient-related factors and process parameters are crucial for optimization of the spray drying process used in NanoCrySP technology. Excipient-related parameters include the crystallization-inducing effect of excipients through plasticization, primary heterogeneous nucleation, and creating a physical barrier to crystal growth (Bhatt et al. 2015; Shete and Bansal 2016; Bansal et al. 2017; Thakur

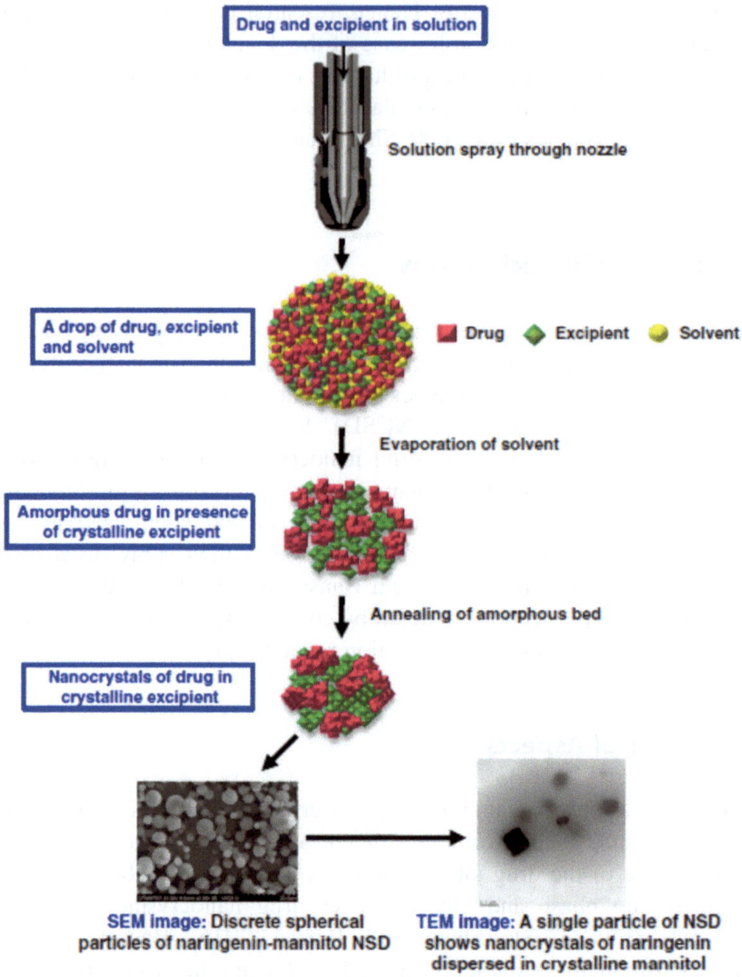

Fig. 17.3 Schematic representation of generation of NCSD using NanoCrySP technology (Reproduced with permission from Shete et al. 2016)

et al. 2019). Drug nanocrystals can be formed in the presence of crystallization-inducing excipient by employing a combination of the above-mentioned mechanisms. Molecular level miscibility between hesperetin and mannitol (1:1) leads to the formation of hesperetin nanocrystals due to plasticization (Shete et al. 2015b). However, plasticization cannot be achieved in the case of celecoxib and mannitol (1:1) due to their immiscibility (Bhatt et al. 2015). Thus, it is essential to have a molecular level miscibility between drug and excipient to achieve plasticization. Drug nanocrystals can be produced in the presence of crystallization-inducing excipient with fast crystallization kinetics through primary heterogeneous nucleation. Celecoxib nanocrystals were produced through this mechanism owing to the fast crystallization kinetics of mannitol (Bhatt et al. 2015). In the drug-excipient matrix, the excipient provides a physical barrier to the growth of drug crystals and limit their size in the nano range. Aspirin-mannitol combination produced aspirin nanocrystals when spray dried together whereas micron sized aspirin crystals are generated when it was spray dried alone (Bansal et al. 2017). The optimization of spray drying process parameters, e.g., the temperature of drying air, drug supersaturation, atomization pressure, and the flow rate is essential to produce uniform and narrow particle size distribution of drug nanocrystals embedded in the matrix of NCSD (Bansal et al. 2017).

The determination of nanocrystals' size embedded in the matrix of NCSD is challenging. Size determination of nanocrystals on an industrial scale is often carried out using Zetasizer® by following standard sample preparation protocol and testing procedure. Employment of additional techniques such as Morphologi G3-ID, scanning electron microscopy (SEM), and transmission electron microscopy (TEM) during the development stage gives further information about nanocrystal size and identity (Kaur et al. 2019).

17.5.2 Biopharmaceutical Aspects

Drug nanocrystals produced using NanoCrySP technology have shown improved dissolution, permeability, oral bioavailability, and efficacy. For example, hesperetin nanocrystals embedded in NCSD showed enhanced release rate and oral bioavailability of hesperetin (Shete et al. 2015b). Also, better mucoadhesion provided by drug nanocrystals embedded in the matrix showed local supersaturation and concentration flux facilitating the absorption of the Dipyridamole (Girdhar et al. 2018).

NCSD improves biopharmaceutical performance due to the following reasons: (1) increased dissolution rate, (2) increased apparent solubility, and (3) increased penetration and adhesiveness to the gastric mucosa, thus opening new avenues for its industrial sustainability and applicability (Gao et al. 2008; Müller et al. 2011).

17.5.3 Translational Aspects

NanoCrySP technology is protected through patents in India, the USA, and the European Union (Bansal et al. 2017). NanoCrySP technology can be applied for (1) the pharmaceutical development of new chemical entities, (2) differentiated products of existing molecules, and (3) generic drug products (Shete and Bansal 2016).

This technology has been utilized till now to develop nanocrystals of different drugs, e.g., celecoxib, aspirin, aceclofenac, diclofenac, fenofibrate, hesperetin, and naringenin (Bansal et al. 2017). Due to its commercial feasibility, extensive work is going on to formulate different dosage forms of drug nanocrystals to be delivered through various routes of administration such as oral, ophthalmic, parenteral, and topical (Shete et al. 2015b; Jadhav et al. 2020). Also, NanoCrySP technology has been employed to successfully generate nanocrystals of a fixed dose combination of drugs. Nanocrystals of simvastatin and ezetimibe were produced in the presence of mannitol with an improved optimal dissolution profile as compared to micronized drug formulations (Nandi et al. 2020).

17.5.4 Efficiency and Innovation Related Aspects

New technologies with improved performance can be created efficiently by the strategic application of innovative and novel methods (Burnett 2011; Gamal et al. 2011). The development of NanoCrySP technology involved many steps, ranging from determining disadvantages of existing methods, application of creative and viable solutions, and their successful implementation (Bansal et al. 2017). NanoCrySP technology is unique in terms of the "intermediate product state" in which drug nanocrystals are obtained. The previously explained approaches produce liquid dispersion of drug nanocrystals whereas NanoCrySP technology generates drug nanocrystals directly in powder form (Shete and Bansal 2016). Thus, NCSD does not suffer from various physical and chemical stability issues, unlike nanosuspension (Chin et al. 2014; Malamatari et al. 2016). Also, downstream processing of NCSD to formulate different solid dosage form is simple as compared to nanosuspension (Dolenc et al. 2009; Jadhav et al. 2020). Moreover, NanoCrySP technology differs from other solvent removal processes, as spray drying is usually carried out in the temperature range of 40–150 °C, while spray freezing and freeze drying are carried out at sub-ambient temperatures (Shete and Bansal 2016).

The determination of different aspects related to its industrial scalability, critical quality parameters, and process efficiency are essential to establish proof of concept for this technology (Ketokivi and Ali-Yrkkö 2010; Lugovoi et al. 2018). It is equally important to measure the innovativeness of NanoCrySP technology to understand multidimensional attributes about the formulation, process, and scale up approaches (Longdon 2008; Burnett 2011).

Thus in our recently published paper, NanoCrySP technology was compared with commonly used process, i.e., wet media milling followed by wet granulation using fluidized bed processor, to formulate oral solid dosage form of diclofenac acid nanocrystals (Jadhav et al. 2020). Such comparison is essential to ascertain the advantages of NanoCrySP technology over established processes. We have analyzed and compared various attributes of NanoCrySP technology such as costing, formulation, and process parameters (Jadhav et al. 2020). We used the "innovation radar" tool to compare the process efficiency of NanoCrySP with the conventional approach. This tool maps process efficiency as it compares multidimensional aspects at a time facilitating a comparison of different processes.

It is important to determine cost effectiveness of a pharmaceutical process for the commercial success of the process (van der Aalst et al. 2006; Ketokivi and Ali-Yrkkö 2010). The costing parameters were calculated in two phases: (1) phase 1: cost required to convert raw materials into nanocrystal and (2) phase 2: cost needed for conversion of nanocrystals into final dosage form, i.e., for downstream processing. The technology reduced Phase 1 cost of nanocrystals by increasing total solid content, drug loading in NCSD, and by improving the yield obtained during the spray drying process (Jadhav et al. 2020). It was evident that the downstream processing cost for NanoCrySP technology was 80% lower than that of the conventional process, proving NanoCrySP as a more cost efficient process (Jadhav et al. 2020).

The next step for this comparison was the quantification of all qualitative parameters using uniform rating, weightage, and scoring scales (Srnka and Koeszegi 2007; Longdon 2008). Formulation parameters such as intermediate and final product nature and stability, nanocrystal size in intermediate and final dosage form, number of total unit processes required, % drug dissolved with respect to time were considered for comparison of process efficiency. Process parameters include number of optimized parameters, % yield, the time required to formulate nanocrystals to dosage form, and environmental concern due to organic solvents (Jadhav et al. 2020). The phase 2 cost, i.e., downstream processing cost, was considered as a major contributing factor for comparison of process efficiency (Carroll et al. 2017).

The comparison of radars (Fig. 17.4) made it evident that NanoCrySP technology is more uniformly distributed toward the circumference of the circle and thus is the more efficient process as compared to the conventional approach (Jadhav et al. 2020). This significant finding has opened new translational paths to employ innovative NanoCrySP technology to commercialize different drug nanocrystals through different dosage forms and routes of administration. Comparison of process efficiency using innovation radar tool also gave insights regarding facets of both NanoCrySP technology and conventional process to further enhance their biopharmaceutical and commercial viability (Jadhav et al. 2020). Such mapping of innovation and comparison of process efficiency should be done more extensively in the case of pharmaceutical processes to focus on the few important aspects and

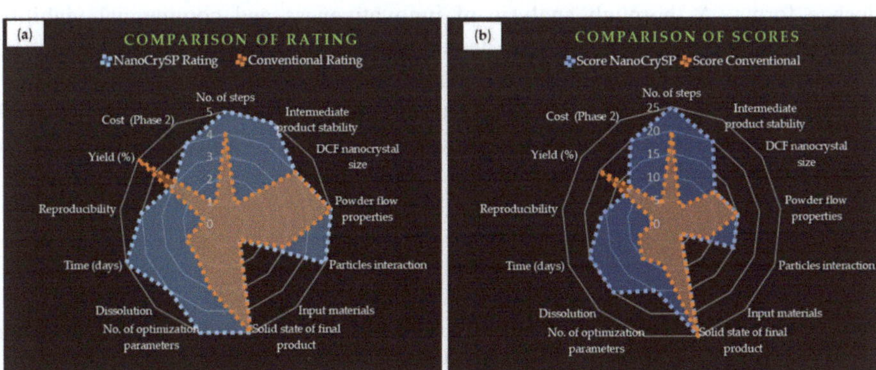

Fig. 17.4 Comparison of (**a**) rating and (**b**) scores for NanoCrySP and conventional approach showing the area covered by each process (Reproduced from Jadhav et al. 2020)

eliminate many trivial parameters to develop quick, efficient, and profitable techniques (Sawhney et al. 2006; Gamal et al. 2011).

17.6 Summary and Future Perspectives

The nanocrystalline drug delivery system is a well-established and substantially applied formulation strategy for poorly soluble drugs. Top-down and bottom-up approaches are extensively explored to produce drug nanocrystals considering ease of scalability and commercial feasibility. Combination techniques are also gaining attention in the past few years due to the production of narrow and uniform particle size distribution of drug nanocrystals. Over the past two decades, several products based on drug nanocrystals have been commercialized making nanonizing a widely utilized technique in the pharmaceutical industry. Various pitfalls of existing technologies encourage the development of novel innovative platform technologies to manufacture efficient formulations of poorly soluble drugs.

A novel NanoCrySP technology produces drug nanocrystals directly in powder form and thus minimizes downstream processing to formulate the finished dosage forms. This technology has been implied to produce nanocrystals of various drugs and to be administered through different routes. The mapping of innovation of NanoCrySP technology assisted in scientific understanding of the technology and different attributes to work on to improve the acceptability and applicability of the process.

It is now evident that nanonization is a most promising approach to commercialize products with improved oral bioavailability of poorly soluble drugs. In the future, more efficient technologies to generate drug nanocrystals will be developed considering multidimensional technological and industrial aspects. Drug nanocrystals can show improvement in the areas of more rationale selection of excipients, targeted drug delivery using specialized ligands, and post-processing to formulate final dosage forms. A thorough analysis of innovativeness and commercial viability will increase the success rate for the development of novel technologies. Moreover, this shall open up new avenues for the advancement of efficient drug delivery systems based on drug nanocrystals. This progress in formulation strategies will eventually optimize the desired pharmacodynamics effects of drugs resulting in an increment in overall therapeutic efficacy.

References

Auweter H, André V, Horn D et al (1998) The function of gelatin in controlled precipitation processes of nanosize particles. J Dispers Sci Technol 19:163–184

Auweter H, Bohn H, Heger R, et al (2002) Precipitated water-insoluble colorants in colloid disperse form US patent 6494924 B1

Bansal AK, Dantuluri AKR, Bhaskarao SG et al (2017) Nanocrystalline solid dispersion compositions and process of preparation thereof US patent 9801855 B2

Beck C, Dalvi SV, Dave RN (2010) Controlled liquid antisolvent precipitation using a rapid mixing device. Chem Eng Sci 65:5669–5675

Bhatt V, Shete G, Bansal AK (2015) Mechanism of generation of drug nanocrystals in celecoxib: mannitol nanocrystalline solid dispersion. Int J Pharm 495:132–139

BMILab (2017) Two strategic approaches to innovation: incremental vs radical. https://bmilab.com/blog/2017/8/3/two-strategic-approaches-to-innovation-incremental-vs-radical

Burnett M (2011) Measuring innovation-sustaining competitive advantage by turning ideas into value. Bearing point–management and technology consultants; www.bearingpoint.com/files/Innovation_High_Res.pdf

Cammarano R, Meiser F, Postma A et al (2014) Methods for the preparation of biologically active compounds in nanoparticulate form US patent 8808751 B2

Cardinal LB (2001) Technological innovation in the pharmaceutical industry: the use of organizational control in managing research and development. Org Sci 12:19–36

Carroll GP, Srivastava S, Volini AS et al (2017) Measuring the effectiveness and impact of an open innovation platform. Drug Discov Today 22:776–785

Chen H, Khemtong C, Yang X et al (2011) Nanonization strategies for poorly water-soluble drugs. Drug Discov Today 16:354–360

Chin WWL, Parmentier J, Widzinski M et al (2014) A brief literature and patent review of nanosuspensions to a final drug product. J Pharm Sci 103:2980–2999

Chingunpituk J (2007) Nanosuspension technology for drug delivery. Walailak J Sci Technol 4: 139–153

De Waard H, Hinrichs W, Frijlink H (2008) A novel bottom–up process to produce drug nanocrystals: controlled crystallization during freeze-drying. J Control Release 128:179–183

De Waard H, Grasmeijer N, Pfaffenbach PP, Frijlink HW et al (2009) Preparation of drug nanocrystals by controlled crystallization: application of a 3-way nozzle to prevent premature crystallization for large scale production. Eur J Pharm Sci 38:224–229

De Waard H, Frijlink HW, Hinrichs WL (2011) Bottom-up preparation techniques for nanocrystals of lipophilic drugs. Pharm Res 28:1220–1223

Desjardins PJ, Olugemo K, Solorio D et al (2015) Pharmacokinetic properties and tolerability of low-dose solu matrix diclofenac. Clin Ther 37:448–461

Dhumal RS, Biradar SV, Yamamura S et al (2008) Preparation of amorphous cefuroxime axetil nanoparticles by sonoprecipitation for enhancement of bioavailability. Eur J Pharm Biopharm 70:109–115

Dodgson M, Hinze S (2000) Indicators used to measure the innovation process: defects and possible remedies. Res Eval 9:101–114

Dolenc A, Kristl J, Baumgartner S et al (2009) Advantages of celecoxib nanosuspension formulation and transformation into tablets. Int J Pharm 376:204–212

Douroumis D, Fahr A (2006) Nano-and micro-particulate formulations of poorly water-soluble drugs by using a novel optimized technique. Eur J Pharm Biopharm 63:173–175

Fages J, Lochard H, Letourneau J-J et al (2004) Particle generation for pharmaceutical applications using supercritical fluid technology. Powder Technol 141:219–226

Gamal D, Salah ET, Elrayyes EN (2011) How to measure organization innovativeness. Technology Innovation and Entrepreneurship Center

Gao L, Zhang D, Chen M (2008) Drug nanocrystals for the formulation of poorly soluble drugs and its application as a potential drug delivery system. J Nanopart Res 10:845–862

Gao L, Liu G, Ma J et al (2013) Application of drug nanocrystal technologies on oral drug delivery of poorly soluble drugs. Pharm Res 30:307–324

Gassmann P, List M, Schweitzer A et al (1994) Hydrosols: alternatives for the parenteral application of poorly water soluble drugs. Eur J Pharm Biopharm 40:64–72

Girdhar A, Thakur PS, Sheokand S et al (2018) Permeability behavior of nanocrystalline solid dispersion of dipyridamole generated using NanoCrySP technology. Pharmceutics 10:160

Hancock BC, Zografi G (1997) Characteristics and significance of the amorphous state in pharmaceutical systems. J Pharm Sci 86:1–12

Haynes DH (1992) Phospholipid-coated microcrystals: injectable formulations of water-insoluble drugs US patent 5091188

Hennart S, Domingues M, Wildeboer W et al (2010) Study of the process of stirred ball milling of poorly water soluble organic products using factorial design. Powder Technol 198:56–60

Hu J, Rogers TL, Brown J et al (2002) Improvement of dissolution rates of poorly water soluble APIs using novel spray freezing into liquid technology. Pharm Res 19:1278–1284
Hu J, Johnston KP, Williams Iii RO (2003) Spray freezing into liquid (SFL) particle engineering technology to enhance dissolution of poorly water soluble drugs: organic solvent versus organic/aqueous co-solvent systems. Eur J Pharm Sci 20:295–303
Itoh K, Pongpeerapat A, Tozuka Y et al (2003) Nanoparticle formation of poorly water-soluble drugs from ternary ground mixtures with PVP and SDS. Chem Pharm Bull 51:171–174
Jacobs C, Kayser O, Müller R (2000) Nanosuspensions as a new approach for the formulation for the poorly soluble drug tarazepide. Int J Pharm 196:161–164
Jadhav S, Kaur A, Bansal AK (2020) Comparison of downstream processing of nanocrystalline solid dispersion and nanosuspension of diclofenac acid to develop solid oral dosage form. Pharmaceutics 12:1015
Juhnke M, Märtin D, John E (2012) Generation of wear during the production of drug nanosuspensions by wet media milling. Eur J Pharm Biopharm 81:214–222
Junghanns JUA, Müller RH (2008) Nanocrystal technology, drug delivery and clinical applications. Int J Nanomedicine 3:295
Kaur A, Parmar PK, Bansal AK (2019) Evaluation of different techniques for size determination of drug nanocrystals: a case study of celecoxib nanocrystalline solid dispersion. Pharmaceutics 11: 516
Keck CM, Müller RH (2006) Drug nanocrystals of poorly soluble drugs produced by high pressure homogenisation. Eur J Pharm Biopharm 62:3–16
Keck C, Kobierski S, Mauludin R et al (2008) Second generation of drug nanocrystals for delivery of poorly soluble drugs: smartCrystals technology. Dosis 24:124–128
Ketokivi M, Ali-Yrkkö J (2010) Innovation does not equal R&D: strategic innovation profiles and firm growth. ETLA Discussion Papers
Kipp J (2004) The role of solid nanoparticle technology in the parenteral delivery of poorly water-soluble drugs. Int J Pharm 284:109–122
Kipp J, Wong J, Doty M et al (2003a) Microprecipitation method for preparing submicron suspensions. US patent 6607784
Kipp JE, Wong JCT, Doty MJ et al (2003b) Microprecipitation method for preparing submicron suspensions. US patent 6607784 B2
Kumar V, Wang L, Riebe M et al (2009) Formulation and stability of itraconazole and odanacatib nanoparticles: governing physical parameters. Mol Pharm 6:1118–1124
Kwade A (1999) Wet comminution in stirred media mills—research and its practical application. Powder Technol 105:14–20
Kylliäinen J (2019) Types of innovation–the ultimate guide with definitions and examples; https://www.viima.com/blog/types-of-innovation
Li H, Wang J, Bao Y et al (2003) Rapid sonocrystallization in the salting-out process. J Cryst Growth 247:192–198
Liedtke S, Wissing S, Müller R et al (2000) Influence of high pressure homogenisation equipment on nanodispersions characteristics. Int J Pharm 196:183–185
List M, Sucker H (1988) Pharmaceutical colloidal hydrosols for injection. GB patent. 2200048
Liversidge GG, Cundy KC, Bishop JF et al (1992) Surface modified drug nanoparticles US patent 5145684
Longdon M (2008) Innovation metrics: the innovation process and how to measure it. Innovation Labs White Paper, Innovation Labs
Lugovoi I, Andritsos D, Senot C (2018) Process innovation in the pharmaceutical industry. HEC Paris Research Paper No. MOSI-2018-1314
Malamatari M, Somavarapu S, Taylor KM et al (2016) Solidification of nanosuspensions for the production of solid oral dosage forms and inhalable dry powders. Expert Opin Drug Deliv 13: 435–450
Matson DW, Fulton JL, Petersen RC et al (1987) Rapid expansion of supercritical fluid solutions: solute formation of powders, thin films, and fibers. Ind Eng Chem Res 26:2298–2306
Merisko-Liversidge EM, Liversidge GG (2008) Drug nanoparticles: formulating poorly water-soluble compounds. Toxicol Pathol 36:43–48

Mohana Raghava Srivalli K, Mishra B (2015) Drug nanocrystals: four basic prerequisites for formulation development and scale-up. Curr Drug Targets 16:136–147
Möschwitzer JP (2013) Drug nanocrystals in the commercial pharmaceutical development process. Int J Pharm 453:142–156
Möschwitzer J, Lemke A (2006) Method for carefully producing ultrafine particle suspensions and ultrafine particles and use thereof. EP000001868574 B1
Möschwitzer J, Müller RH (2006) New method for the effective production of ultrafine drug nanocrystals. J Nanosci Nanotechnol 6:3145–3153
Muller RH, Keck CM (2004) Challenges and solutions for the delivery of biotech drugs–a review of drug nanocrystal technology and lipid nanoparticles. J Biotechnol 113:151–170
Müller RH, Möschwitzer J (2015) Method and device for producing very fine particles and coating such particles. US patent 9168498 B2
Müller RH, Peters K (1998) Nanosuspensions for the formulation of poorly soluble drugs: I. preparation by a size-reduction technique. Int J Pharm 160:229–237
Muller RH, Becker R, Kruss B et al (1999) Pharmaceutical nanosuspensions for medicament administration as systems with increased saturation solubility and rate of solution US patent 5858410
Muller R, Bohm B, Grau J (2000) Nanosuspensions: a formulation approach for poorly soluble and poorly bioavailable drugs. In: Wise D (ed) Handbook of pharmaceutical controlled release technology. Marcel Dekker, NewYork, NY, pp 345–357
Müller RH, Gohla S, Keck CM (2011) State of the art of nanocrystals–special features, production, nanotoxicology aspects and intracellular delivery. Eur J Pharm Biopharm 78:1–9
Mura P, Cirri M, Faucci M et al (2002) Investigation of the effects of grinding and co-grinding on physicochemical properties of glisentide. J Pharm Biomed Anal 30:227–237
Nandi S, Kaur A, Bansal AK (2020) Dual drug nanocrystals loaded microparticles for fixed dose combination of simvastatin and ezetimibe. Pharm Dev Technol 25:40–53
Norman DA, Verganti R (2014) Incremental and radical innovation: design research vs. technology and meaning change. Des Issues 30:78–96
Otsuka M, Matsuda Y (1995) Effect of cogrinding with various kinds of surfactants on the dissolution behavior of phenytoin. J Pharm Sci 84:1434–1437
Rabinow BE (2004) Nanosuspensions in drug delivery. Nat Rev Drug Discov 3:785–796
Robbins P, O'gorman C (2015) Innovating the innovation process: an organisational experiment in global pharma pursuing radical innovation. R D Manag 45:76–93
Salazar J, Müller RH, Möschwitzer JP (2013) Performance comparison of two novel combinative particle-size-reduction technologies. J Pharm Sci 102:1636–1649
Sawhney M, Wolcott RC, Arroniz I (2006) The 12 different ways for companies to innovate. MIT Sloan Manag Rev 47:75
Shegokar R, Müller RH (2010) Nanocrystals: industrially feasible multifunctional formulation technology for poorly soluble actives. Int J Pharm 399:129–139
Shete G, Bansal AK (2016) NanoCrySP technology for generation of drug nanocrystals: translational aspects and business potential. Drug Deliv Transl Res 6:392–398
Shete G, Modi SR, Bansal AK (2015a) Effect of mannitol on nucleation and crystal growth of amorphous flavonoids: implications on the formation of nanocrystalline solid dispersion. J Pharm Sci 104:3789–3797
Shete G, Pawar YB, Thanki K et al (2015b) Oral bioavailability and pharmacodynamic activity of hesperetin nanocrystals generated using a novel bottom-up technology. Mol Pharm 12:1158–1170
Shete G, Jain H, Punj D et al (2016) Stabilizers used in nano-crystal based drug delivery systems. J Excip Food Chem 5:941
Sinha B, Müller RH, Möschwitzer JP (2013) Bottom-up approaches for preparing drug nanocrystals: formulations and factors affecting particle size. Int J Pharm 453:126–141
Srnka KJ, Koeszegi ST (2007) From words to numbers: how to transform qualitative data into meaningful quantitative results. Schmalenbach Business Rev 59:29–57
Sucker H, Gassmann P (1994) Improvements in pharmaceutical compositions. GB patent 2269536A

Thakur PS, Sheokand S, Bansal AK (2019) Factors affecting crystallization kinetics of fenofibrate and its implications for the generation of nanocrystalline solid dispersions via spray drying. Cryst Growth Des 19:4417–4428

Türk M (2009) Manufacture of submicron drug particles with enhanced dissolution behaviour by rapid expansion processes. J Supercrit Fluids 47:537–545

Van Der Aalst WM, De Medeiros AA et al (2006) Process equivalence: comparing two process models based on observed behavior. In: International conference on business process management. Springer, Berlin, Heidelberg, pp 129–144

Van Eerdenbrugh B, Van Den Mooter G, Augustijns P (2008) Top-down production of drug nanocrystals: nanosuspension stabilization, miniaturization and transformation into solid products. Int J Pharm 364:64–75

Wongmekiat A, Tozuka Y, Oguchi T et al (2002) Formation of fine drug particles by cogrinding with cyclodextrins. I. the use of β-cyclodextrin anhydrate and hydrate. Pharm Res 19:1867–1872

Xia D, Quan P, Piao H et al (2010) Preparation of stable nitrendipine nanosuspensions using the precipitation–ultrasonication method for enhancement of dissolution and oral bioavailability. Eur J Pharm Sci 40:325–334

Yang W, Tam J, Miller DA et al (2008) High bioavailability from nebulized itraconazole nanoparticle dispersions with biocompatible stabilizers. Int J Pharm 361:177–188

York P (1999) Strategies for particle design using supercritical fluid technologies. Pharm Sci Technol Today 2:430–440

Zhang JY, Shen ZG, Zhong J et al (2006) Preparation of amorphous cefuroxime axetil nanoparticles by controlled nanoprecipitation method without surfactants. Int J Pharm 323:153–160

Fixed-Dose Combinations: Innovations and Research 18

Sreedhar Dharmagadda

Abstract

Innovation is a key element that drives the pharmaceutical industry. Pharmaceutical companies strive hard to find answers to the disease through new drugs. Most companies are involved in innovations and the majority of which are incremental. Pharmaceutical incremental innovations may include new indications for the current drugs, new dosage forms, and new doses in a certain formulation with increased bioavailability. A fixed-dose combination (FDC) product, a kind of incremental innovation, is a combination of two or more active pharmaceutical agents in a single dosage form. FDCs provide the advantage of combined and complementary action of combined therapy while reducing the number of pills and prescriptions. However, their real contributions have been muddled by disadvantages and stayed away from primary therapy. The chapter consists of a brief description of the ongoing debate of rational and irrational combinations, the reason behind many pharmaceutical companies veering toward incremental innovations. It gives an overview of the regulatory and patent scenario of FDCs. Statistics related to FDC approvals, approvals given by US Food and Drug Administration, and their description have been dealt with. Overall, fixed-dose combinations are a boon and have benefits to offer to their stakeholders.

Keywords

Fixed-dose combinations · Innovation · Patent · Regulatory and research

S. Dharmagadda (✉)
Department of Pharmaceutical Regulatory Affairs and Management, Manipal College of Pharmaceutical Sciences, Manipal Academy of Higher Education, Manipal, India
e-mail: d.sreedhar@manipal.edu

P. P. Singh (ed.), *Recent Advances in Pharmaceutical Innovation and Research*,
https://doi.org/10.1007/978-981-99-2302-1_18

18.1 Introduction

Most pharmaceutical innovations are incremental. Pharmaceutical incremental innovations may include new indications for the current drugs, new dosage forms, and new doses in a certain formulation with increased bioavailability. The combination of two or more active agents having added benefits represents a type of incremental innovation. This gave rise to combination products, or fixed-dose combinations (FDC). FDC is a combination of two or more active pharmaceutical agents in a single dosage form (Wertheimer et al. 2001).

The most accepted definition of FDCs is given by the World Health Organization (WHO), "a combination of two or more active ingredients in a fixed ratio of doses". Typically, the active ingredients used to combine in an FDC are already approved and marketed. The term, FDC, is generically used for a particular combination of active ingredients. It may be administered as a single entity made up of two or more active ingredients given concurrently or as finished pharmaceutical products (WHO 2005).

Poly pill, co-packaged formulation, and combination product are a few widely used terminologies in various journals, periodicals, and informal discussions. There exists confusion about these terminologies and one can be wrongly inferred for that of another. Poly pill, nearly similar to fixed-dose combination, refers to FDC containing three or more drugs in a single pill; here pill restricts to tablet formulation whereas fixed-dose combination may be in a tablet or capsule formulation. Poly pill came into popularity after an article published in British Medical Journal on June 28, 2003, by Wald and Law (2003). They proposed a poly pill consisting of six active ingredients. On the other hand, co-packaged formulation refers to two separate pharmaceutical products in their final dosage form that are packaged together. A detailed explanation of combination products is given by the Office of Combination Products, US Food and Drug Administration (USFDA) in 21 CFR § 3.2(e) 2.

Combination product includes the following:

(a) A product comprised of two or more regulated components, i.e., drug/device, biologic/device, drug/biologic, or drug/device/biologic, that are physically, chemically, or otherwise combined or mixed and produced as a single entity;
(b) Two or more separate products packaged together in a single package or as a unit and comprised of drug and device products, device and biological products, or biological and drug products;
(c) A drug, device, or biological product packaged separately that according to its investigational plan or proposed labeling is intended for use only with an approved individually specified drug, device, or biological product where both are required to achieve the intended use, indication, or effect and whereupon approval of the proposed product the labeling of the approved product would need to be changed, e.g., to reflect a change in intended use, dosage form, strength, route of administration, or significant change in dose; or
(d) Any investigational drug, device, or biological product packaged separately that according to its proposed labeling is for use only with another individually

specified investigational drug, device, or biological product where both are required to achieve the intended use, indication, or effect.

Looking finitely into the above-given definitions, although fixed-dose combination, poly pill, co-packaged formulation, and combination product give an impression that they are synonyms, they are essentially different.

FDCs provide the advantage of combined and complementary action of combined therapy while reducing the number of pills and prescriptions. However, their real contributions have been muddled by disadvantages and stayed away from primary therapy (Table 18.1) (Sarwar and Hossain 2012; Bailey and Day 2009). Previous research identified the synergistic and additive effects of combined drugs in some ailments, such as AIDS, diabetes, hypertension, and tuberculosis. Especially in AIDS and tuberculosis, FDCs are endorsed for first-line treatment. FDCs offer simplified treatment with a reduction in the number of tablets per day. Benefits are amplified when there is complementary action of two or more agents given concomitantly.

Sarwar and Hossein identified four factors that are essential for the success of an FDC formulation (Sarwar and Hossain 2012). These include the following:

1. Formulation development challenges: Incompatibility, release profile differences, particle size, delivery challenges, and regulatory requirements.
2. Patent feasibility: The feasibility of grant of a patent for FDCs should be established. The "obviousness" (one of the three criteria for getting patents—novelty, utility, and non obviousness) barrier is getting tougher each year.
3. Pricing and reimbursement issues: Reimbursement at premium pricing is feasible only if a clear clinical advantage for FDCs exists.
4. Physician considerations: The issues include relative dosing of combination components on an individual patient, the need of titrating the dose of a medication may lead to complications in case of FDC, difficulty to identify the source of side effects, and physicians may perceive patients receiving drugs that they do not require.

18.2 The Rationale for Fixed-Dose Combinations

The debate on the rationality of fixed-dose combinations is a never-ending saga as there are innumerable examples of both rational and irrational FDCs. Rational FDCs are backed by scientific evidence while the irrational FDCs have taken shelter under less defined regulations. Profiteering is the motto behind irrational combinations. However, regulatory authorities have banned quite a few irrational combinations and have put forward guidelines to curb them.

Leaving irrational combinations aside, the reason to combine two or more products is based on the complementary modes of action. Angiotensin converting enzyme (ACE)-inhibitor and calcium channel blocker are good examples of additive effects in treating hypertension. Thiazide diuretic and ACE-inhibitor nullify the

Table 18.1 Potential advantages and disadvantages of FDCs

Advantages	Disadvantages
• Simpler dosage schedule, patient compliance, and improved treatment outcomes	• Sometimes expensive when compared to separate tablets
• Greater efficacy than any of the active ingredients given alone at the same dose	• If a patient is allergic or has a side effect to one drug of FDC, it is difficult to withdraw the suspected drug alone
• Potential to reduce inadvertent medication errors	• Dosing cannot be easily adjusted according to the needs of the patient as each patient has unique characteristics, like weight, age, pharmacokinetics, and comorbidity, that may alter drug metabolism and effect
• Potential to reduce drug shortages by simplifying drug storage and handling, and thus lower risk of being "out of stock"	• Drug interactions may lead to alteration of the therapeutic effect
• Only one expiry date simplifies dosing (single products may have different expiry dates)	• The use of FDCs can lead to polypharmacy
• Procurement, management, and handling of drugs are simplified	• The product (tablets or capsules) is so large that patients find it difficult to swallow
• Potential for lower production, packing, and shipping costs	• Incompatible pharmacokinetics is irrational because of different elimination half-lives of individual components
• The incidence of adverse reactions in response to treatment with the FDC may be lower than that of the component active ingredients given alone as a result of a lower dose of one component or a protective effect of one component particularly having less adverse effects	• Profit motive with least/without any therapeutic benefit. A pharmaceutical company may develop FDC to extend the market potential of a drug product. Since FDCs may be protected by patents, a company can prolong sales even though the individual active ingredients may be off-patent and available as less expensive generics
• The potential for drug abuse can be minimized	• There is potential for high incidence or greater severity of adverse reactions to the FDC compared to individual ingredients or there may be adverse reactions not captured in response to treatment with any of the individual active ingredients
• FDC may reduce errors by physicians and other health care professionals	• FDCs may reduce the range of treatment options available as manufacturers often stop selling less frequently used dose strength and produce a limited range of treatments

hypokalemic effect of a thiazide diuretic with the hyperkalemic effect of an ACE-inhibitor. There are several such examples from different therapeutic classes (Table 18.2).

Table 18.2 Some combination products and the benefits they offer

Disease	Drugs	Specific benefits	Common benefits
Cardiovascular disease and hypertension	• Lisinopril + amlodipine • Lisinopril + hydrochlorothiazide • Telmisartan + amlodipine • Telmisartan + hydrochlorothiazide	• Improved effectiveness with different mechanisms of action • One component offset the side effect of other	• Ease of administration • Improved adherence • Minimal adverse effects
Diabetes	• Metformin + sulfonyl urea • Metformin + dipeptidyl peptidase-4 • Metformin + thiazolidinedione • Metformin + sodium-glucose transport protein 2 inhibitor • Alogliptin + pioglitazone	• Improved disease control • Reduced complications	
HIV	• Abacavir + lamivudine • Dolutegravir + lamivudine + tenofovir • Efavirenz + emtricitabine + tenofovir • Efavirenz + lamivudine + tenofovir • Lamivudine + nevirapine + zidovudine	• Improved treatment outcomes with different mechanisms of action • Synergism • Prevent resistant strains • Simplified treatment regimens	
Pain	• Acetaminophen + hydrocodone • Acetaminophen + tramadol • Paracetamol + NSAIDS • NSAIDs + gastro-protective drugs (proton pump inhibitor or H2 receptor antagonists)	• Improved pain management • Improved effectiveness with different mechanisms of actions	
Respiratory disease	• Budesonide + formoterol • Fluticasone + salmeterol • Budesonide + formoterol	• Improved acceptance of inhalers • Reduced doses	
Tuberculosis (TB)	• Isoniazid + pyridoxine + sulfamethoxazole + trimethoprim, • Ethambutol + isoniazid + pyrazinamide + rifampicin	• Prevent of the emergence of resistant strains including multidrug-resistant tuberculosis • Increased effectiveness • Dispersible dosage form for children	

18.3 Innovations and Research

Early humans might have learned to treat diseases by trial and error. Effects on plants, animals, and other materials were observed. This keen observation translated to the use or discard of processed or semi-processed products. As time passed, the observations were recorded for the benefit of future generations (Jee 1998). Then came the contributions written in volumes by Charaka and Sushruta. You can find discoveries, inventions, and innovations in these volumes which were based on original observations (Jee 1998; Sharma 2003). Gradually there was improvement through technological advances, innovations, and inventions. A paradigm shift happened from experience-based medicine to evidence-based medicine.

Discovery and research of new drugs should target unmet needs. However, it is now primarily driven by commercial purposes. The cost of drug development and the expense of failure may account for a minimum of US$1 billion and may go up to US$3 billion based on the therapeutic area at which the new drug is targeted (Jonathan 2020).

Medicines go through stringent paths before they are approved for marketing. They have to prove three parameters for approval, namely, effectiveness, safety, and quality. They will undergo testing in animals (preclinical) and humans (clinical). Randomized multicentric clinical trials are mandatory for the approval of new medicines. It takes roughly 12 years for research and only 1 in 12 research molecules reach the market, yet there is a greater chance of failure due to some unexpected adverse effect or anticipated efficacy (Dimasi 2001). Overall drug discovery development is not expensive but requires considerable time and is also risky.

18.4 Incremental Innovation

Innovation is categorized into "radical" and "incremental." A new product, process, or system that results from a technological breakthrough or the application of technology having a far-reaching impact is Radical innovation and Incremental innovation includes technical modifications of any existing product, process, or system that results in improvement thereto. Despite their classification, radical and incremental innovations are often interrelated and dependent on each other.

From a pharmaceutical innovation point of view, radical innovation includes the discovery of new molecules that result in new drug classes based on their mechanism of action. On the other hand, incremental innovations involve modifications resulting in a greater number of drugs in a given drug class. Penicillin is a traditional example of radical innovation which kills bacteria effectively. Ampicillin, which is modified penicillin, offered a broad spectrum of activity and was effective against penicillin-resistant bacteria. Incremental innovations, as new dosing formulations (once a day formulations or new drug delivery systems), have played a pivotal role in improving treatments of several diseases. Unfortunately, many incremental innovations are considered "me-too" drugs that provide little benefit over existing drugs. However, 60% of the drugs on WHO's Model List of Essential Medicines reflect incremental innovations, and the criteria for such inclusion include evidence of efficacy and comparative cost-effectiveness. Between 1989 and 2000, 65% of all drugs approved by the USFDA were based on new forms of known substances. Examples of incremental innovations include, but are not limited to, controlled release drug delivery systems, alternative salt form, FDCs, and modification of existing drugs' life and diabetic agents, antibiotics, and H2 antagonists.

18.5 Regulatory Scenario

Normally a new drug approval demands a series of tests, from preclinical through clinical. It mainly includes various safety tests, proof of concept, and quality tests. Though the procedures and processes are developed and understood, there are some unanticipated challenges that require a constant exchange of communication between the company and the regulatory body. The scenario is slightly different for FDCs. Whereas the same steps are somewhat similar, approval of FDCs requires some additional steps like toxicology profile, interactions profile, and efficacy and safety tests of combined actives. Moreover, the dose adjustments of both drugs should determine the best possible ratio of the actives.

The company usually looks at the data of individual drugs that are previously approved. So all the company needs to look at is how they behave when combined. Moreover, the two activities might have been used concomitantly. Efficacy and safety can be determined by specified tests but the challenges are mostly related to the formulation, dose adjustments, and drug-drug interactions. It is observed that most FDC products approved by USFDA include at least one drug approved previously. However, there are certain instances where there are some pressing clinical needs that require a new drug combined with an already approved drug.

18.6 Patent Scenario

A patent is granted only if the invention satisfies the patentability requirements. Though FDC is considered a new drug, it cannot be considered an invention and cannot be patented. Exceptions being that the two actives are new to the market and are not patented. Moreover, an FDC cannot be patented if it includes a drug that has been patented by a different company.

There are instances where a company has been granted a patent, despite the individual drugs were already patented, for example, an anti-retroviral FDC containing stavudine, lamivudine, and nevirapine. The three drugs were patented by Bristol Mayer Squibb, GlaxoSmithKline, and Boehringer Ingelheim, respectively. However, the patent regime was different back then. There is a possibility of patenting FDCs through a patent pool. In a patent pool, owners of the patents would allow a manufacture against agreed terms and royalty payments. Thus, patent pool paves the way to manufacture FDCs despite the actives being already patented (Satyanarayana and Srivastava 2010). So far patent pool has been effectively applied only for ant-retroviral drugs. However, it can essentially be extended to other therapeutic areas such as diabetes, pain management, and hypertension.

18.7 FDC Approvals

The USFDA approved a total of 131 FDCs between 1990 and 2012. In terms of average, 5.5 FDCs were approved by the USFDA per year. There was only one FDC which was approved in the year 1991, and there were a total of 12 approvals in the year 2012. Only two active ingredients were present in 117 FDCs, whereas three actives and four actives were present in 12 and 2 FDCs, respectively. A total of 13 out of 131 FDCs were discontinued, mostly due to poor commercial performance. Though most of the 131 drugs have been approved by other regulatory authorities in their respective country, 41 FDCs are approved only in the USA and Canada (Kararli et al. 2014b).

It is noted that most of the FDCs contain actives that had been approved by USFDA previously. The approval process for a new chemical entity is quite challenging and getting approval for two or more unapproved NCEs would be more daunting. However, one FDC contained two new actives of 131 FDCs, and 21 has at least one active which is first approval and the rest two or more actives were already approved individually previously (Kararli et al. 2014a). Of the 131 approved FDCs, the majority were developed either alone or in partnership with big pharma companies. Among the leading pharmaceutical companies, Merck, Novartis, and GlaxoSmithKline have 10, 9, and 8 approved FDCs, respectively (Table 18.3).

Most of the FDCs have been approved for cardiovascular diseases (a total of 35 FDCs, out of which 28 alone for hypertension). For endocrinology-related disorders and infectious diseases, 28 and 17 FDCs were approved, respectively. In the case of central nervous system disorders, 11 FDCs were approved, and the majority of them targeted pain management (Table 18.4).

Table 18.3 Sponsor and their FDC approvals

Sponsor	# Approved FDCs
Merck	10
Novartis	9
GlaxoSmithKline	8
Abbvie/Abbott	6
Pfizer	6
Bayer	5
Boehringer Ingelheim	5
AstraZeneca	4
Bristol-Myers Squibb	4
Gilead	4
Takeda	4
All others (less than 4 FDCs each)	66

Table 18.4 Therapeutic category-wise approval of FDCs

Therapeutic category	Approved FDCs
Cardiovascular	35
Endocrinology	28
Infectious disease	17
Central nervous system	11
Respiratory	9
Allergy	7
Ophthalmology	6
Dermatology	5
All others (less than five approvals only)	13

A total of 17 FDCs were approved by the Central Drug Standard Control Organization of India. Of these, four belong to the cardiovascular therapeutic category (three of them for hypertension) followed by three for kidney disease (Fixed-dose combinations 2021).

18.8 Conclusion

FDCs are a boon and provide benefits to all the stakeholders. For a pharmaceutical company, an FDC can improve sales and can extend market exclusivity. It may also increase profit due to price increase, especially with advanced formulations. For a patient, a combination product may reduce the extra cost and the number of drugs to be consumed. In a way, FDC is also a satisfying experience for the regulatory body because of the added therapeutic benefit, reduced side effects, and better acceptance and compliance.

Acknowledgments The author acknowledges the support and facilities provided by Manipal College of Pharmaceutical Sciences, Manipal Academy of Higher Education, Manipal, Karnataka, India for conducting this study.

References

Bailey CJ, Day C (2009) Fixed-dose single tablet anti-diabetic combinations. Diabetes Obes Metab 11:527 –533

Dimasi JA (2001) New drug development in the United States from 1963 to 1999. Clin Pharmacol Ther 69:286 –296

Fixed-Dose Combinations (2021). Fixed-dose combinations approved by DCG(I) From 1st January 2021 to 30th June 2021. https://cdsco.gov.in/opencms/opencms/system/modules/CDSCO.WEB/elements/download_file_division.jsp?num_id=NzM3Mg==. Accessed 30 May 2022

Jee BS (1998) History of Hindoo medical science. Logos Press, New Delhi

Jonathan G (2020) New estimate puts the cost to develop a new drug at $1B, adding to long-running debate. https://www.biopharmadive.com/news/new-drug-cost-research-development-market-jama-study/573381/. Accessed 30 May 2022

Kararli TT, Sedo K, Bossart J (2014a) Fixed-dose combinations—fixed-dose combination products—a review (part 1—introduction). https://drug-devcom/fixed-dose-combinations-fixed-dose-combination-products-a-review-part-1-introduction. Accessed 30 May 2022

Kararli TT, Sedo K, Bossart J (2014b) Fixed-dose combinations—fixed-dose combination products—a review (part 2—analysis). https://drug-dev.com/fixed-dose-combinations-fixed-dose-combination-products-a-review-part-2-analysis/. Accessed 30 May 2022

Sarwar S, Hossain D (2012) Fixed dose combination and disease management. Int Res J Pharm 3: 17 –21

Satyanarayana K, Srivastava S (2010) Patent pooling for promoting access to antiretroviral drugs (ARVs)—a strategic option for India. Open AIDS J 4:41 –53

Sharma PV (2003) Caraka samhita. Chaukhamba Orientalia, Varanasi

Wald NJ, Law MR (2003) A strategy to reduce cardiovascular disease by more than 80%. Br Med J 326:1419

Wertheimer A, Levy R, O'Connor T (2001) Too many drugs? The clinical and economic value of incremental innovations. In: Farquhar I, Summers K, Sorkin A (eds) Investing in health: the social and economic benefits of health care innovation (research in human capital and development, Vol. 14). Emerald Group Publishing Limited, Bingley, pp 77–118. https://doi.org/10.1016/S0194-3960(01)14005-9

WHO Expert Committee on Specifications for Pharmaceutical Preparations (2005) WHO technical report series 929. Twenty Ninth Report, Geneva

Significance of Intranasal Drug Delivery Systems: Recent Trends and Clinical Investigations in Brain Disorders

19

Deepshi Arora, Shailendra Bhatt, Manish Kumar, Rupesh K. Gautam, Yugam Taneja, and Muskan Chauhan

Abstract

Intranasal drug administration has enormous research potential, including both drugs designed specifically for intranasal use and off-label applications of commercially available generic medications. The olfactory canal architecture has a well-vascularized mucosal framework, which permits the drug molecule to be transported fast through a single epithelium layer and into the systemic bloodstream without having to go through hepatic and intestinal processing first. The majority of CNS therapy treatments now available rely on systemic medication delivery to the brain via the oral route and intravenous injection. The current CNS therapy treatments mostly rely on systemic medication delivery to the brain via the oral route and intravenous injection. The primary drawback of these approaches is the limited pharmacological access of therapeutic compounds in

D. Arora
M.M. College of Pharmacy, Maharishi Markandeshwar, Ambala, India

Guru Gobind Singh College of Pharmacy, Yamuna Nagar, India

S. Bhatt
Department of Pharmaceutics, School of Medical and Allied Sciences, G.D. Goenka University, Gurugram, India

M. Kumar
School of Pharmaceutical Sciences, CT University, Ludhiana, Punjab, India

R. K. Gautam (✉)
Department of Pharmacology, Indore Institute of Pharmacy, IIST Campus, Indore, MP, India
e-mail: rupesh.gautam@indoreinstitute.com

Y. Taneja
Nutra Science, Haridwar, Uttarakhand, India

M. Chauhan
Guru Gobind Singh College of Pharmacy, Yamuna Nagar, India

P. P. Singh (ed.), *Recent Advances in Pharmaceutical Innovation and Research*,
https://doi.org/10.1007/978-981-99-2302-1_19

the blood supply to the brain, which is linked with decreased therapeutic efficacy and, eventually, increased peripheral adverse effects. The current study highlights the clinical status of numerous medications employed in intranasal drug delivery systems invented by diverse scientists for various brain illnesses. The paper also focuses on current advancements in intranasal medication delivery systems modeled for brain transport.

Keywords

CNS · Blood–brain barrier · Intranasal · Non-invasive route

19.1 Introduction

19.1.1 Central Nervous System Overview

The central nervous system (CNS), one of the two basic divisions of the nervous system, is made up of the brain and spinal cord. The other division is the peripheral nervous system (PNS), which is found outside of the brain and spinal cord. The CNS's primary role is to consolidate incoming information, coordinate, and impact the functioning of all bodily components. CNS and brain ailments impact more than a hundred million individuals globally, according to a World Health Organization (WHO) estimate, and require long-term therapy (World Health Organization 2006). There are various illnesses affecting the brain and CNS that impair a person's motor functional activity. It has proven difficult to get drugs to the brain to treat neurological illnesses (Barchet and Amiji 2009; MacDonald et al. 2000). The most effective route to transport medications to any part of the body is orally, intranasally, or intravenously. There are several obstacles to these medication delivery pathways for brain targets (Bozdağ Pehlivan 2013).

19.2 Barriers to Brain Drug Delivery

Physiological barriers that prevent medications from accessing the brain commonly hamper the treatment of CNS diseases. Despite the availability of a number of brain targeting methods, none have proved successful in producing therapeutic concentrations in the brain (Lesniak and Brem 2004). Three primary barriers separate blood and the CNS: the BBB, the blood–cerebrospinal fluid barrier (BCSFB), and the blood–tumor barrier. Aside from these obstacles, efflux transporter proteins such as P-glycoprotein (P-gp) and others operate as gatekeepers, preventing drug molecules from reaching the brain (Ricci et al. 2006).

19.2.1 Blood–Brain Barrier

The blood–brain barrier (BBB) is a capillary-based endothelial barrier that runs throughout the brain parenchyma. Endothelial tight junctions contain intracellular junction proteins termed occludin and claudin. These structures provide a barrier to the flow of chemicals between the bloodstream and the brain (Sanchez-Covarrubias et al. 2014; Chen and Liu 2012).

The existence of tight junctions between cells, which inhibits paracellular transport owing to high electrical resistance, is one of the key properties of BBB (Abbott et al. 2006). The absence of fenestrations, the existence of P-gp, which is implicated in drug efflux, and the presence of fewer endocytic vesicles, which reduces transcellular flow (Chang et al. 2009), are all factors that limit transcellular flux (Sanchez-Covarrubias et al. 2014). As a result, most medications are limited from crossing the BBB, and only tiny lipophilic molecules can pass the endothelium/BBB (Krol 2012).

Important nutrients are transported across the BBB by particular transportation mechanisms. However, it prohibits xenobiotic entrance, implying that each peptide molecule has a unique pathway for entry (Pardridge 1983). Endogenous and exogenous solutes are metabolized by mitochondria and metabolizing enzymes such as adenylate cyclase, guanylate cyclase, etc. (Brownlees and Williams 1993). As a result, these tight connections prevent ion or molecule exchange between the systemic circulation and the central nervous system. As a result, active drug transport into brain tissue is reduced, resulting in the failure of CNS therapies for diseases including Alzheimer's, schizophrenia, epilepsy, and depression (Löscher and Potschka 2005) (Fig. 19.1).

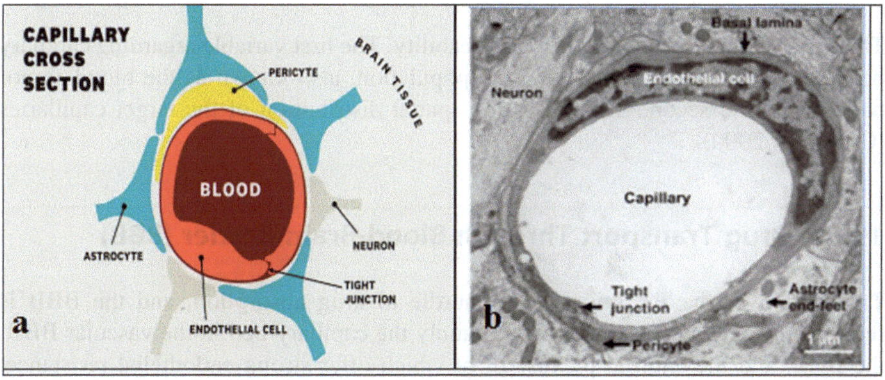

Fig. 19.1 (a) Representation and (b) Scanning Electron Microscope Image of the Capillary (Pardridge 2006)

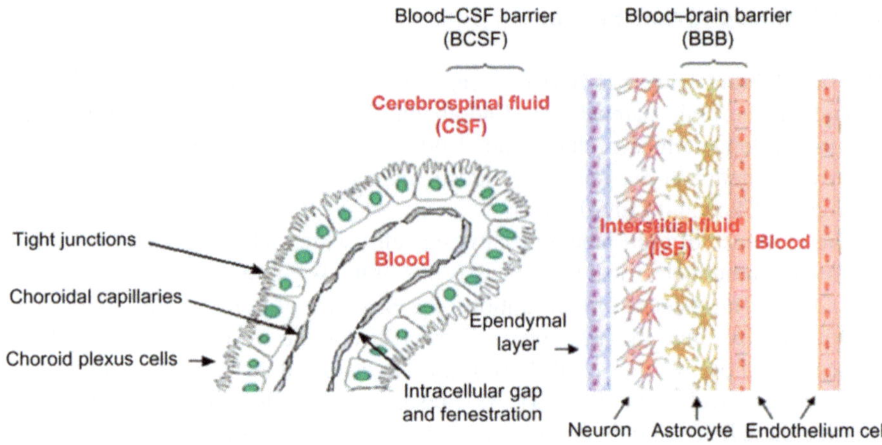

Fig. 19.2 Blood–cerebrospinal Fluid Barrier Schematic Diagram (BCB) (Alam et al. 2010)

19.2.2 Blood–Cerebrospinal Fluid Barrier

The blood–cerebrospinal fluid barrier (BCB) controls the passage of blood-carried chemicals into the CSF. CSF and brain parenchymal interstitial fluid can exchange chemicals. Because of their intimate connections, the arachnoid membrane and choroid plexus provide an important function as barriers between the blood and CSF (Misra et al. 2003). Figure 19.2 shows a diagram of the BCB (Alam et al. 2010).

19.2.3 Blood–Tumor Barrier

The major issue in brain tumors is permeability. The first variable regarding capillary permeability is the tumor microvessel population, also known as the blood–tumor barrier, and the second variable is the spatial distribution of the target capillaries (Groothuis 2000).

19.3 Drug Transport Through Blood-Brain-Barrier (BBB)

Transport over the BBB is a major hurdle to drug absorption, and the BBB is reinforced by many parallel barriers, mainly the capillary bed in the vascular BBB, the BCB, and the choroid plexus, all of which offer strong endothelial resistance. Essential solutes enter the CNS via one of the two methods listed below (Pardridge 2006). Figure 19.3 depicts the route of drug penetration through the BBB.

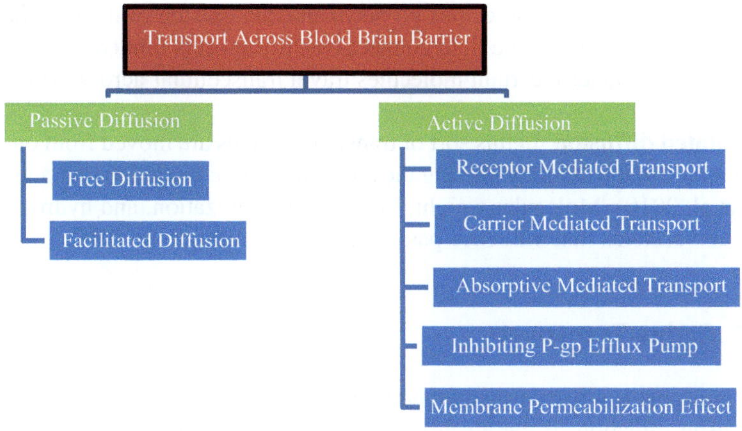

Fig. 19.3 Drug penetration route across the blood–brain barrier

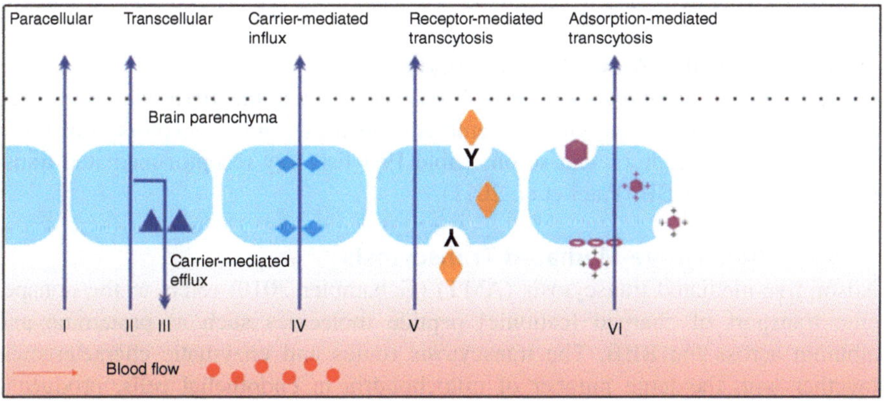

Fig. 19.4 (**a**) Efflux transport mechanisms prevent numerous chemicals from entering the brain. (**b**) Saturable transport is in charge of the transfer. (**c**) Transcellular lipophilic route for lipophilic agent diffusion. (**d**) A biological route that allows extremely tiny quantities of hydrophilic substances to diffuse. (**e**) Endocytosis of essential proteins such as insulin, transferrin, and cytokines is mediated by particular receptors. (**f**) Adsorptive endocytosis is caused by the cationization of certain molecules, such as albumin

19.3.1 Passive Diffusion

As illustrated in Fig. 19.4, highly lipophilic molecules of low molecular weight can extensively infiltrate the BBB via passive diffusion via multiple routes. Uncharged molecules partition easily between tissues and are absorbed in the brain due to their lipophilic nature (Jouyban and Soltani 2012). A molecule moving passively across the BBB has an average molecular mass of 357 Da, or less than 1 nm (Jain 2012).

There are two processes that control passive diffusion:

Free diffusion: Some molecules go between cells outside of the cell membrane (e.g., Sucrose). Because of the tight connections, transport is limited, particularly for hydrophobic substances, certain molecules travel transcellular across the cells (e.g., ethanol).

Facilitated diffusion: In this sort of transport, solutes are moved from one side of the membrane to the other by altering the conformation of certain membrane proteins (Alam et al. 2010). Molecular weight, lipophilicity, ionization, and hydrogen bonding are all variables that influence passive diffusion through the BBB.

19.3.2 Active Transport

Active transport is a sort of membrane transport that is energy driven and works in the opposite direction of the electrochemical gradient. This method is required to avoid any interruption in the passage of important substances such as glucose from the blood to the brain and vice versa for normal brain metabolism and activity (Nau et al. 2010; Begley 2004).

19.3.2.1 Receptor-Mediated Transcytosis

Large chemicals necessary for brain homeostasis enter the BBB via specialized receptors found in the endothelial membrane, such as insulin receptors, transferrin receptors, LDL receptors, and so on. Table 19.1 lists key receptor-mediated transporter systems at BBB (Patel et al. 2011).

19.3.2.2 Adsorptive-Mediated Transcytosis

Adsorptive-mediated transcytosis (AMT) (Gabathuler 2010) refers to the nonspecific transport of charged [cationic] peptide molecules such as protamine and albumin across the BBB. The transcytosis routes and enzymatic characteristics, together with the large number of mitochondria in endothelial cells, produce a shove for molecules to travel across the endothelial cytoplasm (Sauer et al. 2005; Xiao and Gan 2013).

Table 19.1 Key receptor-mediated transporter systems at BBB (Khosrow Tayebati et al. 2013)

Sr. No	RMT systems	Molecules transported
1	TfR	Transferrin
2	INSR	Insulin
3	LEPR	Leptin
4	SR-BI	LDL particles
5	FcGRT	Immunoglobulin
6	IGF-1R, IGF-2R	IGF

19.3.3 Carrier-Mediated Transport

Select membrane-confined transporters carry endogenous chemicals, including vital nutrients, from the blood to the brain parenchyma. Changing the drug/substrate to imitate a nutrient/endogenous substrate capable of transporting, or employing one of the current carrier systems, are two options for CMT.

19.3.3.1 Drug Delivery Strategies Across the BBB

In the case of CNS disorder, conventional drug delivery enters the brain through systemic circulation. To get the appropriate therapeutic concentration at the intended site, systemic drug levels must be increased by the repeated dosage or extended administration. This might lead to an increase in toxicity. Invasive, non-invasive, and miscellaneous techniques are the three types of strategies for crossing the BBB.

19.3.3.2 Intranasal Drug Delivery Systems

In CNS disorders, conventional medication delivery enters the brain via systemic circulation. Systemic drug levels must be raised by repeated dosing or longer administration to achieve the optimal therapeutic concentration at the target location. This might lead to an increase in toxicity. There are three types of strategies for crossing the BBB: invasive, non-invasive, and various.

19.4 The Nasal Cavity's Anatomy and Physiology

The human nasal cavity is around 12–14 cm long, with a total capacity of 15–20 mL. According to (Ali et al. 2010), the nasal cavity is divided into two parts by a vertical barrier called the nasal septum. In an anterior–posterior orientation, from the nostrils to the nasopharynx, the nasal cavity extends. Following delivery into the nasal canal, the drug might be deposited in one of three anatomically distinct sites. The drug must pass through the olfactory membrane as well as the arachnoid membrane, which covers the arachnoid space, which contains cerebrospinal fluid, before reaching the CNS through the nasal cavity (Vyas et al. 2005). Figure 19.5 depicts an overview of the nasal cavity.

19.4.1 Nasal Vestibule

The most anterior section of the nasal cavity is made up of a 0.6 cm^2 patch with sebaceous glands that is composed of stratified squamous and keratinized epithelium. Although the nasal vestibule provides considerable resistance to hazardous environmental chemicals, medication absorption becomes extremely difficult in this location (Djupesland 2013).

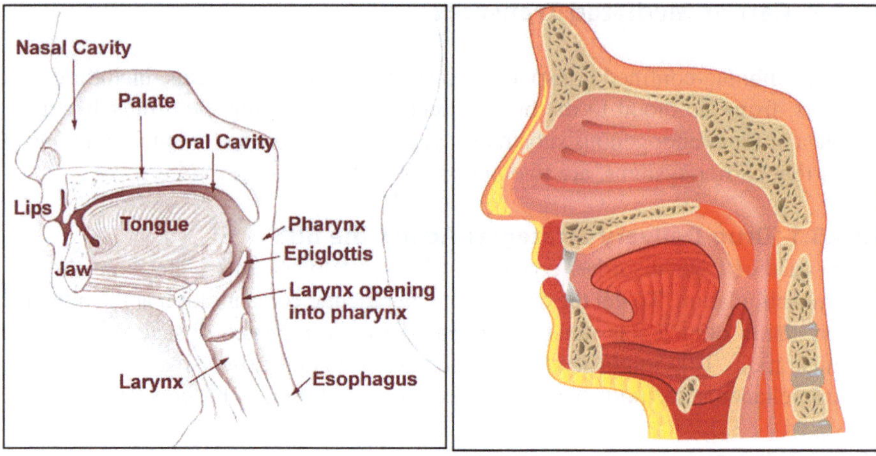

Fig. 19.5 A glimpse of the nasal cavity

19.4.2 Respiratory Region

In humans, the nasal respiratory area, also known as the conchae, is the biggest component, lining roughly 80–90% of the nasal cavity. It has a huge surface area and high vascularity, allowing for increased medication absorption via the respiratory epithelium. The cilia, which are hair-like projections on the surface of epithelial cells, are responsible for the epithelium's expanded surface area. Furthermore, these cilia aid in the transmission of mucus to the nasopharynx (Ozsoy et al. 2009). Mucociliary clearance is the activity of clearing mucous into the gastrointestinal channel (MCC). This protective effect of mucous results in the quick removal of medications from the body. This protective effect of mucous results in the quick clearance of medicines from the nasal cavity by reducing the drug's dwell time in the nasal cavity. Mucin, which is found in the mucous layer, is responsible for trapping big molecular weight drugs such as proteins and peptides (Charlton et al. 2007). Figure 19.6 displays an overview of respiratory mucosa with structural demonstration (Mandpe et al. 2013).

19.4.3 Olfactory Region

The olfactory region is made of three types of cells: olfactory receptor cells, supportive epithelial cells, and basal cells. The only portion of the CNS that is connected directly to the physical realm is the neuroepithelium (Lochhead and Thorne 2012; Talegaonkar and Mishra 2004). Figure 19.7 shows an overview of the organization of the human olfactory region (Desai et al. 2015).

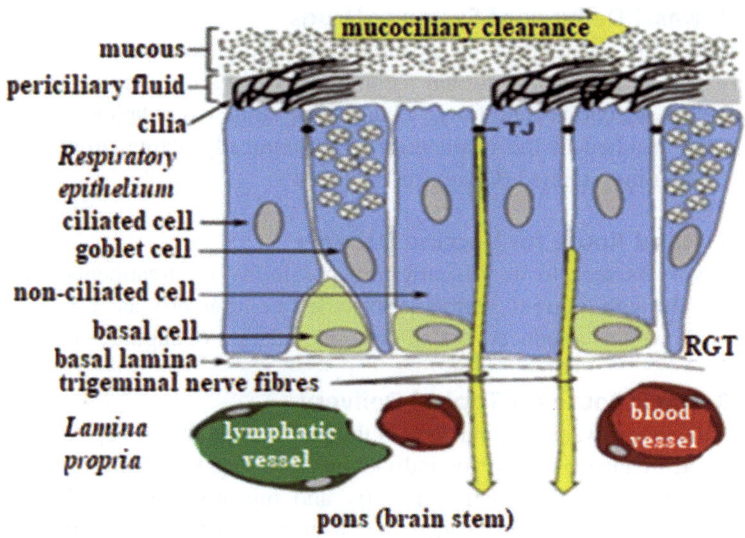

Fig. 19.6 Overview of respiratory mucosa with structural demonstration (Mandpe et al. 2013)

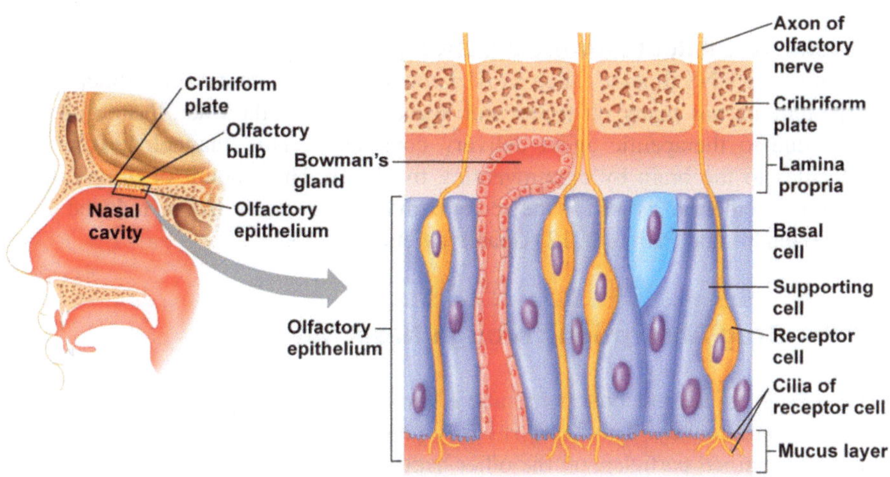

Fig. 19.7 Human olfactory region (Desai et al. 2015)

19.4.4 Applications of Nasal Route for Delivery and Therapeutic Opportunities

Many benefits of the nasal mucosal structure have been identified, including the capacity to deliver medications with a rapid beginning of the action, increased bioavailability, and patient compliance.

19.4.4.1 Nasal Delivery of Systemic Drugs

Nasal delivery is regarded to be an interesting option for drugs that are having difficulty with current conventional techniques. Needle-free administration as an alternative to injections and lower chances of drug degradation, among other benefits, has resulted in the availability of systemically acting drugs with many currently being investigated (Gangurde et al. 2019).

19.4.4.2 Nasal Route for Vaccine Delivery

Nasal delivery is seen to be an intriguing possibility for treatments that are not working well with current methods. Needle-free delivery as a substitute for injections, as well as a reduced risk of drug degradation (Gangurde et al. 2019).

19.4.4.3 Nasal Route for Topical Delivery

Allergies impact between 5% and 10% of the global population. Topical steroids are the preferred treatment option for individuals suffering from chronic allergic and non-allergic mucosal inflammation, rhinitis, and sinusitis (Illum 2012). Treatment with topical steroids remains inadequate in certain instances due to insufficient distribution to the nose and sinuses. This opens up the possibility of developing innovative topical medications with unique nasal delivery methods to treat diverse allergy diseases and increase patient compliance (Illum 2004).

19.4.4.4 Nasal Route for Brain Delivery

Treatment of CNS illnesses such as Alzheimer's disease, depression, epilepsy, and schizophrenia has many obstacles with the existing standard administration technique due to therapeutic agent delivery constraints. This channel delivers drug molecules to the brain by circumventing the BBB. The amount of drug entering systemic circulation, as well as the accompanying side effects, can be reduced in this way (Bahadur and Pathak 2012; Upadhyay et al. 2011).

19.5 Drug Absorption Through Nasal Mucosa

The passage of the mucous membrane is the initial stage in medication absorption from the nasal cavity. This layer is easily penetrated by small, unaltered particles. Large or charged particles, on the other hand, may have difficulties crossing them. Because of environmental changes, structural alterations in the mucous layer are conceivable (i.e., pH, temperature, etc.).

19.5.1 Nasal Drug Delivery Technique Benefits and Drawbacks

Advantages are as follows:

- Non-invasive,
- Bypass the BBB,

- Therapeutic agents do not need any modification,
- Does not require any trained person for medication,
- Avoidance of environmental conditions like chemical and enzymatic degradation of drugs,
- Rapid,
- Feasible and convenient route of delivery to CNS,
- Prevents first-pass metabolism,
- Allowing increased bioavailability (Jadhav et al. 2014).

Nasal delivery has a few disadvantages as follows:

- Including nasal irritation,
- Tissue toxicity,
- The risk of local side effects,
- Nasal congestion,
- Large molecular weight drugs cannot be given (Jadhav et al. 2014).

19.5.2 Factors Influencing Nasal Drug Absorption

To have a systemic influence or to penetrate the brain a medicine or formulation must pass through various membranes or barriers when administered intranasally. Furthermore, the rate and extent of nasal drug absorption are influenced by the drug's physicochemical properties and many more as shown in Fig. 19.8.

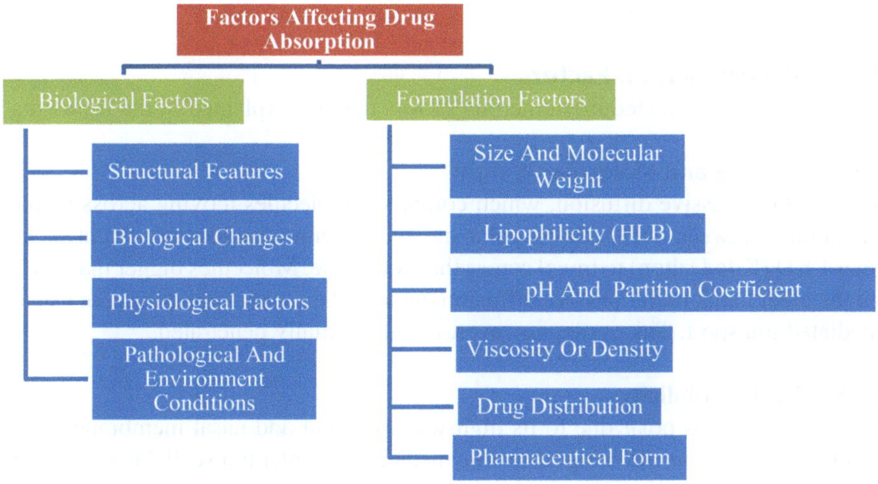

Fig. 19.8 Factors affecting drug absorption

19.5.2.1 Biological Factors

19.5.2.1.1 Structural Features

The nasal cavity is divided into five sections anatomically. The respiratory mucosa is the primary site of medication absorption via the nasal route. Columnar cells, vasculature, and other defense systems comprise it. The nasal cavity is well supplied with vasculature, which gets the greatest quantity of lachrymal secretion, which is responsible for the cavity's warmth and humidification.

19.5.2.1.2 Biochemical Changes

The nasal route was chosen over the oral route to increase bioavailability and prevent first-pass metabolism. Drugs are partially broken down by metabolic enzymes which are released into mucus. Enzymatic degradants include mono and di amino peptides, serine, and cysteine (Boddupalli et al. 2010).

19.5.2.2 Physiological Factors

Blood supply and neural regulation, nasal cavity pH, mucociliary clearance (MCC), and ciliary beat frequency are all physiological factors that influence nasal drug absorption (CBF).

19.5.2.3 Pathological and Environmental Conditions

Because all systems are interrelated, any change in the micro-environment in the nose, such as MCC, CBF, pH, or any slight alteration in the structure of the mucous layer, triggers a defensive reaction, reducing medicine bioavailability (Costantino et al. 2007).

Enzymatic degradation.
Transporters and efflux systems.
Effect of pathological conditions.

19.5.2.4 Formulation Factors

Drug transport is affected by numerous factors that are explained in the following:

19.5.2.4.1 Size and Molecular Weight

Transcellular passive diffusion, which comprises molecules moving across the cell membrane between cells, allows polar molecules with a molecular weight of less than 1 KD (Kilo Dalton) to travel across the membrane. Molecules bigger than 1 KD, on the other hand, prefer passive diffusion outside of the cell, transcytosis, or carrier-mediated transport. This route, however, is only partially penetrated.

19.5.2.4.2 Lipophilicity

Though mucous is polar due to its high-water content and nasal membrane underneath, it is lipophilic, allowing lipophilic medicines to enter transcellular and achieve fast bioavailability.

19.5.2.4.3 pH and Partition Coefficient

Drug absorption is also affected by pH and partition coefficient. HLB modification allows for the successful delivery of drugs with a wide pH range. Many researchers determined in the early 1980s that medications with unionized states have more absorption/penetration than those with ionized states. Unionized forms of drugs absorb better in a wide pH range (Kushwaha et al. 2011).

19.5.2.4.4 Viscosity/Density

Higher viscous liquids improve permeability by coating the CBF and MCC; drug retention duration is enhanced, and absorption is improved. For example, the incorporation of hydroxyl propyl methyl cellulose (HPMC) helped improve drug absorption by the nose by extending the therapeutic action of the formulation through the development of a high-viscosity gel (Chaturvedi et al. 2011).

19.5.2.4.5 Drug Distribution

A dosage form is essential in the field of application. When opposed to aerosols, nasal drops have a greater spreadability (coverage area) and a longer retention period on the mucosa. Typically, solutions produce dosage uniformity issues (Greimel et al. 2007).

19.5.2.4.6 Solubility

In rare cases, intranasal administration of drugs with limited water solubility and/or higher doses might be difficult (Greimel et al. 2007).

19.5.2.4.7 Pharmaceutical Form

Higher viscosity formulations may have difficulties in administration. Metered dosage pumps and actuators have recently become accessible, allowing nasal spray systems to provide precise doses ranging from 25 to 200 μL.

19.6 Nasal Drug Absorption Methodologies

As previously stated, nasally given drugs have a number of limitations, including low drug solubility, limited membrane penetration quick enzymatic degradation, and rapid MCC. Table 19.2 explains the challenges and possible solutions for bioavailability problems through the intranasal route.

19.7 Brain Disorders

19.7.1 Epilepsy

Seizures are neurological disorders caused by abnormally high neuronal activity in the brain. This aberrant brain activity can emerge in a variety of ways, ranging from convulsive activity to experiential events that are not easily observable by an

Table 19.2 Challenges and possible solutions for bioavailability problems through intranasal route

Problems	Challenges	Possible solution
Poor physical and chemical characteristics of the drug	Improve the drug's physicochemical characteristics and formulation	Cosolvents addition Use of cyclodextrins Pro-drugs Change pharmaceutical excipients Formulate novel carrier systems
The nasal membrane has a low permeability	Increase drug solubility and penetration across the lipid membrane in the bodily fluids	Co solvents Pro-drugs approach
	Modify the nasal membrane	Use of absorption enhancers
	Extend the length of the drug's stay in the nasal cavity	Mucosal adhesive systems Viscosity modifiers
Degradation by enzymes	Reduce the drug affinity to nasal enzymes	Pro-drugs approach
	Due to the presence of nasal enzymes, the drug is degraded	Use enzymatic inhibitors Co solvents Pro-drugs approach

observer, depending on the distribution of discharges. During a seizure, certain brain cells send out aberrant signals that prevent other cells from functioning normally. This anomaly may result in a brief shift in sensation, behavior, movement, or awareness (El-Zaafarany et al. 2018). Drugs used in epilepsy and their clinical status are given in Table 19.3.

19.7.2 Parkinson's Disease

Parkinson's disease is characterized by a loss of dopaminergic neurons in the brain's substantia nigra pars compacta. It is classified as a mobility disorder. Cerebral palsy, ataxia, and Tourette syndrome are examples of movement problems. They happen when a shift in the neurological system impairs a person's capacity to move or remain motionless. It affects the neurological system, and the symptoms worsen over time. It occurs when nerve cells in the brain do not create enough dopamine, a brain neurotransmitter. The bulk of unfavorable neuronal loss in Parkinson's disease is caused by complex 1 defects in the respiratory chain. The clinical status of drugs used in PD is given in Table 19.4.

Table 19.3 Clinical status of drugs used in epilepsy

S. No	Drug	Intranasal dose	Model	MOA	Outcome
1	Oxcarbazepine (OXC)	0.5 mg/kg	Rats	Blockage of voltage-sensitive sodium channels	In rats, PLGA NPs had a lower dosage frequency than the free drug and were able to control seizures (Alam et al. 2015)
2	Lamotrigine	–	–	Binding with sodium channels leads to stabilization of presynaptic neuronal membranes	In comparison to oral administration of drug solution, nano-formulation has a greater concentration and residence period of drug in the brain, resulting in a higher brain concentration (Patel et al. 2016a, b, c)
3	Carbamazepine (CMP)	0.18–0.2 mg/kg	Rats	Stabilization of hyper-excited neural membranes due to blockage of voltage-sensitive sodium channels	In comparison to IV CMPME, the IN CMPMME and CMPME/Brain/ Blood ratio rose 2–3 fold at all sample sites after IN administration of CMPMME, indicating a higher extent of CMP distribution in the brain (Czapp et al. 2008)
4	Phenobarbital	5.4 mg/kg	Rats	Acts on GABA receptors, increasing synaptic inhibition and elevating seizure threshold	IN SLN/2.4-fold rise in C_{max} values in cortical dialysates (Eskandari et al. 2011)
5	Valproic acid	20 mg/kg	Rat	Blockage of voltage-gated ion channels and inhibiting histone deacetylase	IN SLN/After intranasal delivery of VPA NLCs, the brain:plasma concentration ratio increased (Eskandari et al. 2011)

Table 19.4 Clinical status of drugs used in PD

S. No	Drug	Intranasal dose	Model	MOA	Outcome
1	Glutathione	100 mg GSH/mL	Human	Increase dopaminergic transporter activity	Increased glutathione level in CNS and improvement in motor movements (Mischley et al. 2016)
2	Bromocriptine	5 μL	Mice	Decreasing the amount of prolactin in the body	Stimulates the nerves that control movement (Md et al. 2014)
3	Rasagiline	0.00189/ day	Mice	Covalent binding to the flavin residue of the N5 nitrogen, resulting in irreversible inactivation of the enzyme	Following administration of RAS-loaded CG-NPs, there was a significant increase in bioavailability in the brain (Mittal et al. 2016)
4	Rotigotine	50 μL	Rat	It binds with D3 receptor with highest affinity. Inhibit dopamine uptake and prolactin secretion	Non-ergoline agonists of DOPA are found to be very effective for the treatment of PD (Bhattamisra et al. 2020)
5	Levodopa	16 mg/kg	Mice	Decarboxylates to doxylamine and stimulates the dopaminergic receptors	An increase in bioavailability, ultimately decreases the unfavorable side effects of levodopa (Arisoy et al. 2020)

19.7.3 Psychosis

Patients suffering from psychosis have a distorted worldview in which they are unable to discern between their personal, subjective experiences, and the reality of the outside world. They have hallucinations or delusions that they feel are real, and they may behave and converse inappropriately and incoherently. Delusions, hallucinations, and disorganized speech are some of the symptoms. The clinical status of drugs used in psychosis is given in Table 19.5.

19.7.4 Schizophrenia

Schizophrenia is a diverse condition characterized by disruptions in language, perception, thought, social behavior, emotion, and violation. Symptoms of

Table 19.5 Clinical status of drugs used in psychosis

Sr. No	Drug	Intranasal dose	Model	MOA	Outcome
1	Olanzapine (OZP)	1 mg/kg	Rats	Binding to histamine H-1, alpha-1, dopamine, muscarinic, and 5-HT2 receptors, among others	The brain/blood ratio increased by 5–6 times after OZP was injected into the brain. When compared to intravenous olanzapine microemulsion, it appeared to be greatest for intranasal olanzapine mucoadhesive microemulsion (Patel et al. 2016a, b, c)
2	Risperidone (RSP)	35 mg/kg	Rats	Decrease dopaminergic and serotonergic pathway activity in the brain	Risperidone mucoadhesive Nano Emulsion (I.N.) transfers RSP quicker and to a higher extent in the rat brain than Risperidone (I.V.) and Risperidone nanoemulsion (I.N.) and Risperidone nanoemulsion (I.V.) (Kumar et al. 2008)
3	Haloperidol	2 mg/kg	Rats	Inhibit the effects of dopamine and increase its turnover	The behavioral effects generated by intranasal administration of 6.7 times lower doses of the dendrimer haloperidol formulation were comparable to those caused by intraperitoneal injection of haloperidol formulations (Katare et al. 2015)
4	Remoxipride	4–16 mg/kg	Rats	Weaker binding to D_2 dopaminergic receptors that dopamine. This weaker binding is thought to account for the reduced incidence of parkinsonism	Bioavailability increased by 89% following intranasal dosing. When compared to IV delivery, IN resulted in longer brain ECF exposure (Stevens et al. 2011)

(continued)

Table 19.5 (continued)

Sr. No	Drug	Intranasal dose	Model	MOA	Outcome
5	Clozapine	0.05 mg/ 0.1 mL	Rat	Blockage of serotonin (5-HT_{2A}/5HT_{2C}) receptors and dopamine(D1–4) receptors	When compared to oral delivery, there is a 3.56-fold increase in brain/plasma drug concentration and a 5.28-fold lower drug dosage (Ravikrishna and Janapareddi 2019)

schizophrenia include hallucinations, confused ideas, and emotions of fear and paranoia. Psychiatrists evaluate symptoms, examinations, and medical history before prescribing drugs and psychotherapy. There are no pathogenomic characteristics. Table 19.6 depicts the clinical status of drugs used in schizophrenia.

19.7.5 Alzheimer Disease

Alzheimer's disease (AD) is a progressive decline of mental capacities that begins with moderate memory loss and progresses to loss of ability to keep employment, plan and execute normal tasks, reason, and exercise judgment. Skills in communication, emotions, and attitude are all harmed. Alzheimer's disease is the fourth leading cause of death in individuals, behind heart disease, cancer, and stroke. Alzheimer's Association (Alzheimer's Association 2017); Alzheimer's disease causes cognitive changes that follow a regular pattern, starting with memory loss and advancing to language and visuospatial problems. The symptom starts with memory loss, such as forgetfulness, and progresses to more distressing losses. The patient may forget the names of family members or what was stated at the start of a sentence. Table 19.7 explains the clinical status of drugs used in Alzheimer's disease.

19.8 Novel Formulation Approaches for Intranasal Drug Delivery

The pharmacological characteristics, delivery mechanism, and nasal physiology should all be researched in order to generate a nasal formulation with acceptable performance and commercial features. It is better to concentrate on extending the stay and ensuring optimal medicine absorption. Table 19.8 summarizes recent research on intranasal drug delivery devices in the treatment of brain diseases.

Table 19.6 Clinical status of drugs used in schizophrenia

S. No	Drug	Intranasal dose	Model	MOA	Outcome
1	Haloperidol	0.89 mg/kg	Rat	Blocks dopamine D_2 receptor in brain	Increase conc. of drug in brain, increase in bioavailability (Katare et al. 2017)
2	Apomorphine hydrochloride	2 mg/kg	Mice	It is a dopaminergic agonist that stimulates regions of brain involved in motor control	Enhances motor function, suppress prolactin release, cause vasodilatation and behavior effect (Patel et al. 2016a, b, c)
3	Paliperidone (PALI)	12 mg/kg	Rats	Antagonist activity at D2 receptor (α_1 and α_2), adrenergic receptor, and H_1 receptor	Increased brain activity approc 6–8 times after IN administration of PALI-Mucoadhesive microemulsion as compared to intravenous PALI-Microemulsion (Scarff and Casey 2011)
4	Zotepine (ZTP)	4.4 mg/kg	Rats	Serotonin receptors antagonism and inhibition of noradrenaline reuptake and serotonergic activity	In comparison to intravenous ZTP solution, intranally given Zotepine nano suspension resulted in an 8.6-fold rise in AUC0–24 h and a 10.79-fold increase in AUC0–24 h (Pailla et al. 2019)
5	Iloperidone	1 mg/kg	Rat and mice	Antagonism of certain receptors, such as dopamine D2 and serotonin 5-hydroxytryptamine (5HT) receptor 2A	Intranasal administration of iloperidone nanostructured lipid carriers resulted in a 3.05-fold increase in AUC0–∞ when compared to pure drug solution (Mandpe et al. 2013)

19.8.1 Mucoadhesive Solutions

Mucoadhesive solutions, such as chitosan, cellulose polymers, polycarbophil, poloxamers, and other mucoadhesive polymers, have been found to increase

Table 19.7 Clinical status of drugs used in Alzheimer's disease

S. No.	Drug	Intranasal dose	Model	MOA	Outcome
1	Insulin	160 IU	Human or animal	It increases brain insulin receptor sensitivity	Improved cognitive processes in both healthy and impaired humans due to increased brain signaling (Hallschmid et al. 2008)
2	Deferoxamine	200 mg / KG	Mice	Shifting APP processing to a nonamyloidogenic route reduced APP protein expression and phosphorylation	Intranasal iron chelation decreased iron-induced amyloidogenic APP processing and corrected behavioral changes (Fine et al. 2015)
3	Donepezil	0.5 mg/ mL	Rat	It binds reversibly to ach esterase enzyme and inhibits the hydrolysis of acetylcholine	Facilitate the cholinergic neurotransmission and regulate memory function (Noetzli and Eap 2013)
4	Rivastigmine	0.068 mg/ kg	Rat	It promotes cholinergic function by inhibiting both AChE and BChE enzymes in the brain	Increase conc. of Ach in brain which improve cognition and memory function (Fazil et al. 2012)
5	Galantamine	3 mg/kg	Rat	Reversibly inhibit AChE and enhance the intrinsic action of acetylcholine and nicotine	Increase cholinergic neurotransmission level in CNS (Hanafy et al. 2016)

intranasal drug penetration. Gel formulations with the appropriate rheological properties lengthen the time in contact with the mucosa at the absorption site. The polymer in the gel has mucoadhesive properties as well as the formulation's physicochemical characteristics, restricting removal through protective mechanisms in the nose and eyes (Ugwoke et al. 1999; Vidgren et al. 1992). Several studies have stated that mucoadhesive polymers such as cellulose derivatives, polyacrylates, starch, and so on can be used in the formulation of mucoadhesive solutions. (Gavini et al. 2012) and colleagues created a mucoadhesive formulation that demonstrated regulated dopamine release in the brain following intranasal application.

Table 19.8 Recent investigations on intranasal drug delivery systems used in brain disorders

Sr. No	Drug	Approach	Recent investigation	Composition	Application	Reference
1	Resveratrol + curcumin	Mucoadhesive nanoemulsion	Increased concentration of both the polyphenols when they were encapsulated in mucoadhesive nanoemulsion with antioxidant activity preserved	Hyaluronic acid, Labrafac PG, tween 80 Cremophor RH 40	Alzheimer	Nasr (2016)
2	Nimodipine	Mucoadhesive Nanoemulsion	Improved behavioral activities with nanoformulations like nanoemulsion showed high plasma and brain concentrations of nimodipine	Carbopol 934 P, chitosan, Capmul MCM, Labrasol, Transcutol P	Senile dementia	Pathak et al. (2014)
3	Selegiline	Mucoadhesive Nanoemulsion	Tremendous improvement in behavioral activities of rats when administered intranasal selegeline nanoemulsion as compared to oral administration	Tween 80, Grape seed oil, Sefsol 218	Parkinson	Fang et al. (2009)
4	Celecoxib	Liposomes	Celecoxib liposomes when administered intranasally had much high concentration as compared to free drug when determined by fluoroscence analysis	Phospholipid:Cholesterol	Alzheimer	Guo et al. (2017)
5	Donepezil	Liposomes	Pharmacokinetic parameters like higher C_{max}, longer half-life and higher AUC (doubled) was achieved with the liposomal drug as compared to free drug	DSPC:CHOL:PEG	AD	Al Asmari et al. (2016)
6	Rivastigmine	Liposomes	In a study comparing intranasally delivered solution with liposomes, it was discovered that intranasally administered solution had a greater drug level and a larger plasma AUC	EPC:CHOL Poloxamer 188	AD	Yang et al. (2013)

(continued)

Table 19.8 (continued)

Sr. No	Drug	Approach	Recent investigation	Composition	Application	Reference
7	Lurasidone-HCl	Polymeric micelles	When compared to pure drug, in vivo (single dose) pharmacokinetic investigations following intranasal delivery revealed a significant concentration of drug in the brain	Gelucire 44/14 (GL44), Pluronic F127 (PF127) Carbapol-940	Bipolar disorder	Pokharkar et al. (2020)
8	Zolmitriptan	Micelles	In vivo biodistribution tests of zolmitriptan showed that it was far superior than the drug's intravenous and nasal formulations when it came to brain targeting	TCP and BA. PF127, PEG-400 TvPGS	Migraine	Jain et al. (2010)
9	Insulin	Nanogel	With regard to free insulin administration, an increase in NG-in distribution to different brain locations and biological activity, as evaluated by Akt activation levels, was shown	PVP K60 and acrylic acid (AA), dimethyl sulfoxide (DMSO), carbodimide hydrochloride (EDC), N-hydroxysulfosuccinimide sodium salt (Sulfo-NHS), Acido 2-(N-morpholino) ethanesulfonic acid hydrate	Alzheimer	Picone et al. (2018)
10	Clonazepam	Microspheres	The clonazepam microspheres had a greater brain concentration and a high drug targeting index, indicating that they have a lot of promise as a brain-targeted drug delivery method	Sodium lauryl sulfate, paraffin oil, glutaraldehyde-saturated toluene, chitosan PEG 400	Seizures	Shaji et al. (2009)
11	Methotrexate (MTX)	Microspheres	MTX-loaded chitosan microspheres exhibited good swelling index, showed controlled release pattern and minor irritation to cilia. Relative bioavailability increased to 118%	Chitosan	Brain cancer	Sun et al. (2012)

12	Haloperidol	Dendrimers	A 6.7 times reduction in the dose of dedrimer base haloperidol formulation when administered intranasally gave same behavioral response when compared with the other haloperidol formulations given intraperitoneally	PAMAM dendrimer 1,4diaminobutane core, amine ended, gen 5.0, tween 20, ethanol, and ammonium acetate, PAMAM dendrimer 1,4 diaminobutane core, amine terminated	Anti-psychotic	Katare et al. (2015)
13	^{32}P-labeled siRNA	Dendriplexes	Increased radioactivity was observed with in the olfactory bulb when ^{32}PSiRNa was complexed with dendrimers and incorporated in the gel	Poly(amidoamine)G7, dendrimers, lipofectamine™ 2000		Perez et al. (2012)
14	Huperzine A	PLGA nanoparticles	The distribution of HUPA in the brain was improved after intranasal administration of PLGA nanoparticles, and the drug targeting index in the mouse was measured at various intervals	PLGA 50:50 2A,Chitosan Lf, 2-iminothiolan bicinchoninic acid	Alzheimer	Meng et al. (2018)
15	Oxcarbazepine	PLGA nanoparticles	Immunohistochemical investigations demonstrated that OXC PLGA NPs had a neuroprotective effect after 16 days of treatment, demonstrating the viability of developing a novel non-invasive nose-to-brain delivery technique	Methanol, acetonitrile, ethyl acetate, polymer Resomer® 502 H poly-OXC, tween® 80, 7-n-propylxanthine (7n-PX), and bovine serum albumin (BSA)	Epilepsy	Musumeci et al. (2018)

19.8.2 Microspheres

Microspheres are tiny spherical particles that range in size from 1 to 1000 mm in diameter. Microspheres are also called microparticles in some situations. Microspheres may be made from a variety of polymeric materials, both natural/ synthetic, as well as inorganic materials.

Microspheres, particularly mucoadhesive microspheres, are a prominent type of novel drug delivery technology. Microspheres may stay in touch with the nasal mucosa for extended periods, resulting in increased drug absorption rate and volume (Ugwoke et al. 2005). Microspheres for nasal applications are typically made from biocompatible materials such as starch, albumin, hyaluronic acid, chitosan, and gelatin, as well as HPMC, Carbopol 934P, dextran, and other polymer combinations (Patil and Sawant 2011; Takeuchi et al. 2003).

A team of researchers created a comparison report between Methotrexate (MTX) solution and MTX-chitosan loaded microspheres, which revealed that MTX-loaded chitosan microspheres exhibited rapid and controlled drug release after intranasal administration compared to MTX solution due to their ability to transport more drug to the brain. Furthermore, chitosan employed as a mucoadhesive polymer was found to be safe and aid in medication mucoadhesion in the olfactory area, resulting in enhanced drug bioavailability (Sun et al. 2012).

19.8.3 Liposomes

Liposomes are spherical vesicles that have a hydrophilic core within and an impenetrable lipophilic phospholipid bilayer on the outside, similar to cells. They may entrap drugs in both aqueous and lipid phases due to their unique structure. Lipophilic drugs are nearly completely enclosed in the lipid bilayers of liposomes, and because they are not water-soluble, problems like drug loss during storage are rare. Hydrophilic drugs can be located in the watery cores of liposomes or the surrounding water phase.

Liposome nasal clearance half-life was four times longer than usual human nasal clearance half-life, showing liposomal systems' mucoadhesion capabilities and nasal application potential (Jaafari et al. 2010; Andresen et al. 2005). Furthermore, liposomes have the benefit of being able to be changed to efficiently target a specific region of interest, such as the BBB. As a result, targeted liposomes have mostly been explored for the detection and therapy of brain illnesses (Mourtas et al. 2011).

Liposome research has yielded encouraging findings in drug delivery via the nasal transport system. Seok Hong and colleagues explored how surface modification of liposomes with targeting moieties, PEGylation methods, and the integration of various carrier systems with liposomes can boost drug bioavailability.

19.8.4 Nanoparticles

The drug substance is either encapsulated inside the matrix or conjugated or adsorbs to the particle surface in nanoparticles. To enable delayed, regulated drug therapy, nanoparticles are frequently comprised of polymers, lipids, or a mix of the two.

19.8.5 Exosomes

19.8.5.1 Natural Bioactivities of Exosomes

Exosomes are membrane nanovesicles that develop spontaneously and have a diameter of 50–150 nm (Simons and Raposo 2009). They came from the endosomes of immune cells such as dendritic cells (Théry et al. 2006), macrophages (Bhatnagar et al. 2007), B-cells (Clayton et al. 2005), and T-cells (Nolte't Hoen et al. 2009), as well as mesenchymal stem cells, cancer cells, and endothelial cells.

19.8.6 Nanoemulsions

Nanoemulsions (NEs) are surfactant-stabilized oil-in-water (O/W) or water-in-oil (W/O) colloidal carriers of two immiscible liquids with a mean droplet diameter of around 100 nm, but maximum size limitations of up to 300 nm have been recorded in the literature. Because the droplet size is much smaller than the wavelength of visible light, NEs seem translucent or milky-white (Rodrigues et al. 2005).

NEs can be made into a number of dosage forms, such as liquids (Sarker 2005), creams, gels, foams, sprays, and so on, and can be taken orally, parenterally, ocularly, or nasally. Much study has been done on nanoemulsions that target the brain via the nasal route.

19.8.7 Polymeric Micelles

Polymeric micelles are nano-shell structures generated by the co-polymerization of hydrophobic and hydrophilic block copolymers. These micelles have the ability to operate and change their shape for medication delivery (Wang et al. 2012). The medication, which is present inside the brain and targets the hydrophobic core, is stabilized from the external aqueous environment by the hydrophilic shell, which increases polymer water solubility. Karami et al. (2019) investigated magnetic brain targeting of naproxen-loaded polymeric micelles in recent work. The findings indicated that polymeric magnetic micelles with diameters less than 150 nm might be a suitable carrier for improving therapeutic drug accumulation in the brain for the treatment of CNS illnesses.

19.8.8 Nanogels

Nanogels are nano-sized polymeric systems having cross-linked swellable polymeric hydrophilic chains that can withstand significant amounts of water when immersed in an aqueous media without dissolving. They have several qualities, including the capacity to be ready physically or chemically manipulated, good swelling properties, high encapsulation efficiency, high permeability, stability, and so forth.

This polymeric system has received extensive investigation, highlighting its critical role in brain targeting. Because of their site-specificity, nanogels offer a wide range of applications in the treatment of cancers and neurological illnesses. Various studies show that nanogels may be used to deliver oligonucleotides, tiny therapies, proteins, as contrast agents, optical imaging, multimodal imaging, and diagnostic agents, among other things (Cherry 2006).

The effectiveness of cCHP-based nanoemulsion as a universal protein-based antigen-delivery vehicle for adjuvant-free intranasal vaccination has been demonstrated (Nochi et al. 2010). These nanogels are made by self-assembly in water and trap a variety of proteins via hydrophobic interactions, displaying chaperone-like activity due to protein hydration without aggregation. They may then be gradually released back to their original condition (Desale et al. 2013).

19.8.9 Dendrimers

Dendrimers are a new type of polymer with well-defined structures created by the sequential addition of layers to the branching groups around the core molecule. This results in structural regularity and minimal heterogeneity, making them an appealing candidate for next-generation nanomedicines (Harush-Frenkel et al. 2008). Their unusual design boosts their utility as radioligands, detecting agents, and targeting components, among other things. Recent improvements demonstrate their great utility in gene transfer, diagnostic agents, contrast agents, photodynamic treatment, and other fields. (Zaman et al. 2011) created a polyacrylate dendritic polymer-based completely synthesized peptide sub-unit vaccination candidate.

19.8.10 Niosomes

Niosomes are intriguing new nanocarriers in which the medication is encapsulated in multi-lamellar structures generated by the self-association of nonionic surfactants. Niosomes are biocompatible, biodegradable, non-immunogenic, non-toxic, and non-carcinogenic, among other properties (Mahale et al. 2012; Seleci et al. 2016). The capacity of a nonionic surfactant to generate bilayer vesicles is determined by the surfactant's HLB value. The functioning of niosomes is similar to that of liposomes in that they improve medication bioavailability by decreasing clearance.

19.9 Conclusion

Intranasal medicine delivery systems have several advantages over typical oral dosage methods. Several intranasal medicine delivery systems have been created, allowing medications to be administered directly to the brain through the nasal mucosa. Recent trends and clinical research suggest that novel carrier systems with improved bioavailability, efficiency, and fewer side effects should be used. The properties of the drug candidate, the nose-to-brain transport pathway, and transit to and within the brain must all be better understood.

Conflicts of Interest None of the authors have any conflict of interest.

References

Abbott NJ, Rönnbäck L, Hansson E et al (2006) Astrocyte–endothelial interactions at the blood–brain barrier. Nat Rev Neurosci 7:41 –53

Al Asmari AK, Ullah Z, Tariq M et al (2016) Preparation, characterization, and in vivo evaluation of intranasally administered liposomal formulation of donepezil. Drug Des Dev Ther 10:205

Alam MI, Beg S, Samad A et al (2010) Strategy for effective brain drug delivery. Eur J Pharm Sci 40:385 –403

Alam T, Pandit J, Vohora D et al (2015) Optimization of nanostructured lipid carriers of lamotrigine for brain delivery: in vitro characterization and in vivo efficacy in epilepsy. Expert Opin Drug Deliv 12:181 –194

Ali J, Ali M, Baboota S et al (2010) Potential of nanoparticulate drug delivery systems by intranasal administration. Curr Pharm Des 16:1644 –1653

Alzheimer's Association (2017) Alzheimer's disease facts and figures. Alzheimers Dement 13: 325 –373

Andresen TL, Jensen SS, Jørgensen K (2005) Advanced strategies in liposomal cancer therapy: problems and prospects of active and tumor specific drug release. Prog Lipid Res 44:68 –97

Arisoy S, Sayiner O, Comoglu T (2020) In vitro and in vivo evaluation of levodopa-loaded nanoparticles for nose to brain delivery. Pharm Dev Technol 25:735 –747

Bahadur S, Pathak K (2012) Physicochemical and physiological considerations for efficient nose-to-brain targeting. Expert Opin Drug Deliv 9:19 –31

Barchet TM, Amiji MM (2009) Challenges and opportunities in CNS delivery of therapeutics for neurodegenerative diseases. Expert Opin Drug Deliv 6:211 –225

Begley DJ (2004) Delivery of therapeutic agents to the central nervous system: the problems and the possibilities. Pharmacol Ther 104:29 –45

Bhatnagar S, Shinagawa K, Castellino FJ et al (2007) Exosomes released from macrophages infected with intracellular pathogens stimulate a pro-inflammatory response in vitro and in vivo. Blood 110:3234 –3244

Bhattamisra SK, Shak AT, Xi LW et al (2020) Nose to brain delivery of rotigotine loaded chitosan nanoparticles in human SH-SY5Y neuroblastoma cells and animal model of Parkinson's disease. Int J Pharm 579:119 –148

Boddupalli BM, Mohammed ZN, Nath RA et al (2010) Mucoadhesive drug delivery system: an overview. J Adv Pharm Technol Res 1:381

Bozdağ Pehlivan S (2013) Nanotechnology-based drug delivery systems for targeting, imaging and diagnosis of neurodegenerative diseases. Pharmaceutical Research. 30:2499–2511

Brownlees J, Williams CH (1993) Peptidases, peptides, and the mammalian blood–brain barrier. J Neurochem 60:793 –803

Chang J, Jallouli Y, Barras A et al (2009) Drug delivery to the brain using colloidal carriers. Prog Brain Res 180:2 –17

Charlton S, Jones NS, Davis SS et al (2007) Distribution and clearance of bio-adhesive formulations from the olfactory region in man: effect of polymer type and nasal delivery device. Eur J Pharm Sci 30:295 –302

Chaturvedi M, Kumar M, Pathak K et al (2011) A review on mucoadhesive polymer used in nasal drug delivery system. J Adv Pharm Technol Res 2:215

Chen Y, Liu L (2012) Modern methods for delivery of drugs across the blood–brain barrier. Adv Drug Deliv Rev 64:640 –665

Cherry SR (2006) Multimodality in vivo imaging systems: twice the power or double the trouble? Annu Rev Biomed Eng 8:35 –62

Clayton A, Turkes A, Navabi H et al (2005) Induction of heat shock proteins in B-cell exosomes. J Cell Sci 118:3631 –3638

Costantino HR, Illum L, Brandt G et al (2007) Intranasal delivery: physicochemical and therapeutic aspects. Int J Pharm 337:1 –24

Czapp M, Bankstahl JP, Zibell G et al (2008) Brain penetration and anticonvulsant efficacy of intranasal phenobarbital in rats. Epilepsia 49:1142 –1150

Desai P, Shete H, Adnaik R et al (2015) Therapeutic targets and delivery challenges for Alzheimer's disease. World J Pharmacol 4:236 –264

Desale SS, Cohen SM, Zhao Y et al (2013) Biodegradable hybrid polymer micelles for combination drug therapy in ovarian cancer. J Control Release 171:339 –348

Djupesland PG (2013) Nasal drug delivery devices: characteristics and performance in a clinical perspective—a review. Drug Deliv Transl Res 3:42 –62

El-Zaafarany GM, Soliman ME, Mansour S et al (2018) A tailored thermosensitive PLGA-PEG-PLGA/emulsomes composite for enhanced oxcarbazepine brain delivery via the nasal route. Pharmaceutics 10:217

Eskandari S, Varshosaz J, Minaiyan M et al (2011) Brain delivery of valproic acid via intranasal administration of nanostructured lipid carriers: in vivo pharmacodynamic studies using rat electroshock model. Int J Nanomedicine 6:363

Fang JY, Hung CF, Chi CH et al (2009) Transdermal permeation of selegiline from hydrogel-membrane drug delivery systems. Int J Pharm 380:33 –39

Fazil M, Md S, Haque S et al (2012) Development and evaluation of rivastigmine loaded chitosan nanoparticles for brain targeting. Eur J Pharm Sci 47:6 –15

Fine JM, Renner DB, Forsber AC et al (2015) Intranasal deferoxamine engages multiple pathways to decrease memory loss in the APP/PS1 model of amyloid accumulation. Neurosci Lett 584: 362 –367

Gabathuler R (2010) Approaches to transport therapeutic drugs across the blood–brain barrier to treat brain diseases. Neurobiol Dis 37:48 –57

Gangurde PK, Ajitkumar BN, Kumar L (2019) Lamotrigine lipid nanoparticles for effective treatment of epilepsy: a focus on brain targeting via nasal route. J Pharm Innov 14:91 –111

Gavini E, Rassu G, Ciarnelli V et al (2012) Mucoadhesive drug delivery systems for nose-to-brain targeting of dopamine. J Neurosci Res 2:47 –55

Greimel A, Bernkop-Schnürch A, Dorly Del Curto M et al (2007) Transport characteristics of a beta sheet breaker peptide across excised bovine nasal mucosa. Drug Dev Ind Pharm 33:71 –77

Groothuis DR (2000) The blood-brain and blood-tumor barriers: a review of strategies for increasing drug delivery. J Neuro-Oncol 2:45 –59

Guo JW, Guan PP, Ding WY et al (2017) Erythrocyte membrane-encapsulated celecoxib improves the cognitive decline of Alzheimer's disease by concurrently inducing neurogenesis and reducing apoptosis in APP/PS1 transgenic mice. Biomaterials 145:106 –127

Hallschmid M, Benedict C, Schultes B et al (2008) Obese men respond to cognitive but not to catabolic brain insulin signaling. Int J Obes 32:275 –282

Hanafy AS, Farid RM, Helmy MW (2016) Pharmacological, toxicological and neuronal localization assessment of galantamine/chitosan complex nanoparticles in rats: future potential contribution in Alzheimer's disease management. Drug Deliv 23:3111 –3122

Harush-Frenkel O, Rozentur E, Benita S et al (2008) Surface charge of nanoparticles determines their endocytic and transcytotic pathway in polarized MDCK cells. Biomacromolecules 9:435 – 443

Illum L (2004) Is nose-to-brain transport of drugs in man a reality? J Pharm Pharmcol 56:3 –17

Illum L (2012) Nasal drug delivery—recent developments and future prospects. J Control Release 161:254 –263

Jaafari MR, Tafaghodi M, Sa ST (2010) Evaluation of the clearance characteristics of liposomes in the human nose by gamma-scintigraphy. Iran J Pharm Res 20:3 –11

Jadhav AJ, Gondkar SB, Ravindra BS (2014) A review on nasal drug delivery system. WJPPS 3: 231 –254

Jain KK (2012) Nanobiotechnology-based strategies for crossing the blood–brain barrier. Nanomedicine 7:1225 –1233

Jain R, Nabar S, Dandekar P (2010) Micellar nanocarriers: potential nose-to-brain delivery of zolmitriptan as novel migraine therapy. Pharm Res 27:655 –664

Jouyban A, Soltani S (2012) Blood brain barrier permeation. Toxic Drug Test 10:1 –24

Karami Z, Sadighian S, Rostamizadeh K et al (2019) Magnetic brain targeting of naproxen-loaded polymeric micelles: pharmacokinetics and biodistribution study. Mater Sci Eng C 100:771 –780

Katare YK, Daya RP, Sookram GC et al (2015) Brain targeting of a water insoluble antipsychotic drug haloperidol via the intranasal route using PAMAM dendrimer. Mol Pharm 12:3380 –3388

Katare YK, Piazza JE, Bhandari J et al (2017) Intranasal delivery of antipsychotic drugs. Schizophr Res 184:2 –13

Khosrow Tayebati S, Ejike Nwankwo I, Amenta F (2013) Intranasal drug delivery to the central nervous system: present status and future outlook. Curr Pharm Des 19:510 –526

Krol S (2012) Challenges in drug delivery to the brain: nature is against us. J Control Release 164: 145 –155

Kumar M, Misra A, Babbar AK et al (2008) Intranasal nanoemulsion based brain targeting drug delivery system of risperidone. Int J Pharm 358:285 –291

Kushwaha SK, Keshari RK, Rai AK et al (2011) Advances in nasal trans-mucosal drug delivery. J Appl Pharm Sci 1:21

Lesniak MS, Brem H (2004) Targeted therapy for brain tumours. Nat Rev Drug Discov 3:499 –508

Lochhead JJ, Thorne RG (2012) Intranasal delivery of biologics to the central nervous system. Adv Drug Deliv Rev 64:614 –628

Löscher W, Potschka H (2005) Blood-brain barrier active efflux transporters: ATP-binding cassette gene family. NeuroRx 2:86 –98

MacDonald BK, Cockerell OC, Sander JW et al (2000) The incidence and lifetime prevalence of neurological disorders in a prospective community-based study in the UK. Brain 123:665 –676

Mahale NB, Thakkar PD, Mali RG et al (2012) Niosomes: novel sustained release nonionic stable vesicular systems—an overview. Adv Colloid Interf Sci 183:46 –54

Mandpe L, Kyadarkunte A, Pokharkar V (2013) Assessment of novel iloperidone- and idebenone-loaded nanostructured lipid carriers: brain targeting efficiency and neuroprotective potential. Ther Deliv 4:1365 –1383

Md S, Haque S, Fazil M et al (2014) Optimised nanoformulation of bromocriptine for direct nose-to-brain delivery: biodistribution, pharmacokinetic and dopamine estimation by ultra- HPLC/ mass spectrometry method. Expert Opin Drug Deliv 11:827 –842

Meng Q, Wang A, Hua H et al (2018) Intranasal delivery of Huperzine a to the brain using lactoferrin-conjugated N-trimethylated chitosan surface-modified PLGA nanoparticles for treatment of Alzheimer's disease. Int J Nanomedicine 13:705

Mischley LK, Conley KE, Shankland EG et al (2016) Central nervous system uptake of intranasal glutathione in Parkinson's disease. Parkinson's Dis 2:1

Misra A, Ganesh S, Shahiwala A et al (2003) Drug delivery to the central nervous system: a review. J Pharm Pharm Sci 6:252 –273

Mittal D, Md S, Hasan Q et al (2016) Brain targeted nanoparticulate drug delivery system of rasagiline via intranasal route. Drug Deliv 23:130 –139

Mourtas S, Canovi M, Zona C et al (2011) Curcumin-decorated nanoliposomes with very high affinity for amyloid-β1-42 peptide. Biomaterials 32:1635 –1645

Musumeci T, Serapide MF, Pellitteri R et al (2018) Oxcarbazepine free or loaded PLGA nanoparticles as effective intranasal approach to control epileptic seizures in rodents. Eur J Pharm Biopharm 133:309 –320

Nasr M (2016) Development of an optimized hyaluronic acid-based lipidic nanoemulsion co-encapsulating two polyphenols for nose to brain delivery. Drug Deliv 23:1444 –1452

Nau R, Sörgel F, Eiffert H (2010) Penetration of drugs through the blood-cerebrospinal fluid/blood-brain barrier for treatment of central nervous system infections. Clin Microbiol Rev 23:858 – 883

Nochi T, Yuki Y, Takahashi H et al (2010) Nanogel antigenic protein-delivery system for adjuvant-free intranasal vaccines. Nat Mater 9:572 –578

Noetzli M, Eap CB (2013) Pharmacodynamic, pharmacokinetic and pharmacogenetic aspects of drugs used in the treatment of Alzheimer's disease. Clin Pharmacokinet 52:225 –241

Nolte't Hoen EN, Buschow SI, Anderton SM et al (2009) Activated T cells recruit exosomes secreted by dendritic cells via LFA-1. Blood 113:1977 –1981

Ozsoy Y, Gungor S, Cevher E (2009) Nasal delivery of high molecular weight drugs. Molecules 14:3754 –3779

Pailla SR, Talluri S, Rangaraj N et al (2019) Intranasal zotepine nanosuspension: intended for improved brain distribution in rats. Daru 27:541 –556

Pardridge WM (1983) Brain metabolism: a perspective from the blood-brain barrier. Physiol Rev 63:1481 –1535

Pardridge WM (2006) Molecular trojan horses for blood–brain barrier drug delivery. Curr Opin Pharmacol 6:494 –500

Patel P, Acharya S, Acharya N (2011) Potential of surface functionalized nanoparticles for improved therapy of refractory central nervous system disorders. Pharm Res 4:1093 –1099

Patel MR, Patel RB, Bhatt KK et al (2016a) Paliperidone microemulsion for nose-to-brain targeted drug delivery system: pharmacodynamic and pharmacokinetic evaluation. Drug Deliv 23:346 – 354

Patel RB, Patel MR, Bhatt KK et al (2016b) Evaluation of brain targeting efficiency of intranasal microemulsion containing olanzapine: pharmacodynamic and pharmacokinetic consideration. Drug Deliv 23:307 –315

Patel RB, Patel MR, Bhatt KK et al (2016c) Micro-emulsion-based drug delivery system for transnasal delivery of carbamazepine: preliminary brain-targeting study. Drug Deliv 23:207 – 213

Pathak R, Dash RP, Misra M et al (2014) Role of mucoadhesive polymers in enhancing delivery of nimodipine microemulsion to brain via intranasal route. Acta Pharm Sin B 4:151 –160

Patil SB, Sawant KK (2011) Chitosan microspheres as a delivery system for nasal insufflation. Colloids Surf B Biointerfaces 84:384 –389

Perez AP, Mundiña-Weilenmann C, Romero EL et al (2012) Increased brain radioactivity by intranasal 32P-labeled siRNA dendriplexes within in situ-forming mucoadhesive gels. Int J Nanomedicine 7:1373

Picone P, Sabatino MA, Ditta LA, Amato A et al (2018) Nose-to-brain delivery of insulin enhanced by a nanogel carrier. J Control Release 270:23 –36

Pokharkar V, Suryawanshi S, Dhapte-Pawar V (2020) Exploring micellar-based polymeric systems for effective nose-to-brain drug delivery as potential neurotherapeutics. Drug Deliv Transl Res 10:1019 –1031

Ravikrishna V, Janapareddi K (2019) Development and evaluation of clozapine intranasal mucoadhesive in situ gels for brain targeting. JDDT 9:198 –207

Ricci M, Blasi P, Giovagnoli S et al (2006) Delivering drugs to the central nervous system: a medicinal chemistry or a pharmaceutical technology issue? Curr Med Chem 13:1757 –1575
Rodrigues RF, Costa IC, Almeida FB et al (2005) Development and characterization of evening primrose (Oenothera biennis) oil nanoemulsions. Rev Bras Farmacogn 25:422 –425
Sanchez-Covarrubias L, Slosky LM, Thompson BJ et al (2014) Transporters at CNS barrier sites: obstacles or opportunities for drug delivery? Curr Pharm Des 20:1422 –1449
Sarker DK (2005) Engineering of nanoemulsions for drug delivery. Curr Drug Deliv 2:297 –310
Sauer I, Dunay IR, Weisgraber K et al (2005) An apolipoprotein E-derived peptide mediates uptake of sterically stabilized liposomes into brain capillary endothelial cells. Biochemistry 44:2021 – 2029
Scarff JR, Casey DA (2011) Newer oral atypical anti-psychotic agents: a review. Pharm Ther 36: 832
Seleci M, Seleci DA, Joncyzk R et al (2016) Smart multifunctional nanoparticles in nanomedicine. Bionanomaterials 17:33 –41
Shaji J, Poddar A, Iyer S (2009) Brain-targeted nasal clonazepam microspheres. Indian J Pharm Sci 71:715
Simons M, Raposo G (2009) Exosomes–vesicular carriers for intercellular communication. Curr Opin Cell Biol 21:575 –581
Stevens J, Ploeger BA, Van der Graaf PH et al (2011) Systemic and direct nose-to-brain transport pharmacokinetic model for remoxipride after intravenous and intranasal administration. Drug Metab Dispos 9:2275 –2282
Sun Y, Shi K, Wan F (2012) Methotrexate-loaded microspheres for nose to brain delivery: in vitro/ in vivo evaluation. J Drug Deliv Sci Technol 22:167 –174
Takeuchi H, Matsui Y, Yamamoto H et al (2003) Mucoadhesive properties of carbopol or chitosan-coated liposomes and their effectiveness in the oral administration of calcitonin to rats. J Control Release 86:235 –242
Talegaonkar S, Mishra PR (2004) Intranasal delivery: an approach to bypass the blood brain barrier. Indian J Pharmacol 36:140
Théry C, Amigorena S, Raposo G et al (2006) Isolation and characterization of exosomes from cell culture supernatants and biological fluids. Curr Protoc Cell Biol 30:3 –22
Ugwoke MI, Exaud S, Van Den Mooter G et al (1999) Bioavailability of apomorphine following intranasal administration of mucoadhesive drug delivery systems in rabbits. Eur J Pharm Sci 9: 213 –219
Ugwoke MI, Agu RU, Verbeke N et al (2005) Nasal mucoadhesive drug delivery: background, applications, trends and future perspectives. Adv Drug Deliv Rev 57:1640 –1665
Upadhyay S, Parikh A, Joshi P et al (2011) Intranasal drug delivery system-a glimpse to become maestro. J Appl Pharm Sci 1:34 –44
Vidgren P, Vidgren M, Arppe J et al (1992) In vitro evaluation of spray-dried mucoadhesive micropheres for nasal administration. Drug Dev Ind Pharm 18:581 –597
Vyas TK, Shahiwala A, Marathe S et al (2005) Intranasal drug delivery for brain targeting. Curr Drug Deliv 2:165 –175
Wang Y, Wang C, Gong C et al (2012) Polysorbate 80 coated poly (ε-caprolactone)–poly (ethylene glycol)–poly (ε-caprolactone) micelles for paclitaxel delivery. Int J Pharm 434:1 –8
World Health Organization (2006) Neurological disorders: public health challenges. World Health Organization, Geneva
Xiao G, Gan LS (2013) Receptor-mediated endocytosis and brain delivery of therapeutic biologics. Int J Cell Biol 2013:703545 . https://doi.org/10.1155/2013/703545; Epub 2013 Jun 11. PMID: 23840214; PMCID: PMC3693099
Yang ZZ, Zhang YQ, Wang ZZ et al (2013) Enhanced brain distribution and pharmacodynamics of rivastigmine by liposomes following intranasal administration. Int J Pharm 452:344 –354
Zaman M, Skwarczynski M, Malcolm JM et al (2011) Self-adjuvanting polyacrylic nanoparticulate delivery system for group a streptococcus (GAS) vaccine. Nanomedicine 7:168 –173

Current Challenges and Nanotechnology-Based Pharmaceutical Approaches for the Treatment and Control of Visceral Leishmaniasis

20

Ganesh Yadagiri and Shyam Lal Mudavath

Abstract

Leishmaniasis, a neglected tropical disease (NTD), is caused by a kinetoplastid protozoan parasite *Leishmania*. Visceral leishmaniasis (VL) is a significant public health concern, with 50,000–90,000 new cases and high mortality and morbidity worldwide. The unavailability of vaccines and vector control measures poses significant challenges in treating and controlling VL. Additionally, extensive reliance on chemotherapy has led to severe life-threatening side effects, the emergence of drug-resistant parasites, high costs, and prolonged hospitalization. There is an immediate need for drug development to overcome the issues mentioned above, which is costly and time-consuming. Therefore, as an innovative intervention, pharmaceutical nanotechnology has emerged as a potential alternative for treating VL. Pharmaceutical nanoformulations could overcome the challenges of conventional chemotherapy, i.e., increased therapeutic index, minimal toxicity, cost-effectiveness, and improved patient compliance. Herein, we focused on the current antileishmanial therapeutic failures and nanotechnological approaches' role in treating VL. Several preclinical studies have demonstrated the enhanced efficacy of developed antileishmanial nanoformulations. Extensive research is crucial for developing cost-effective formulations for poverty-ridden neglected tropical diseases.

Keywords

Antileishmanials · Drug resistance · Efficacy · Nanotechnological approaches · Visceral leishmaniasis

G. Yadagiri · S. L. Mudavath (✉)
Infectious Disease Biology Laboratory, Chemical Biology Unit, Institute of Nano Science and Technology, Mohali, Punjab, India
e-mail: shyamlal@inst.ac.in

P. P. Singh (ed.), *Recent Advances in Pharmaceutical Innovation and Research*,
https://doi.org/10.1007/978-981-99-2302-1_20

20.1 Introduction

Leishmaniasis is caused by a kinetoplastid protozoan parasite, *Leishmania*, affecting approximately 12 million people in 98 countries, including developing and under-developed countries worldwide (Yadagiri and Singh 2018; Alvar et al. 2012). The disease is transmitted by a female *Phlebotomine* sandfly, a hemoflagellate vector to mammals (Ready 2013). Leishmaniasis gained considerable attention due to the higher incidence of morbidity; therefore, the London Declaration on Neglected Tropical Diseases was established to eradicate it as a public health issue by the year 2020 (de Vlas et al. 2016). Clinically, leishmaniasis exists in three forms: mucocutaneous (MCL), cutaneous (CL), and visceral leishmaniasis (VL). Globally, leishmaniasis (CL, MCL, and VL) carries one of the highest disease burdens. According to the World Health Organization (WHO), leishmaniasis is one of the most important tropical diseases. It has been implicated as a significant health issue with many clinical manifestations and possibly fatal outcomes (Yadagiri and Singh 2018; Torres-Guerrero et al. 2017). VL is the most critical form of leishmaniasis, characterized by irregular fever, progressive cachexia, hepatosplenomegaly, and pancytopenia, and is fatal if left untreated (Yadagiri and Mudavath 2020). *L. infantum* is the VL causative pathogen in Europe, Latin America, and North Africa. However, *L. donovani* is the causative agent in East Africa and India (Ready 2014). According to the WHO, around 50,000–90,000 new leishmaniasis cases are reported, with 95% of new cases reported in ten countries (https://www.who.int/news-room/fact-sheets/detail/leishmaniasis, 2019). CL is the most common form of the disease, which is caused by *L. major* and is characterized by skin lesions. The global incidence of CL is around 0.6–1 million cases per year. Most CL cases (>95%) are reported in America, the Mediterranean basin, Central Asia, and the Middle East, and the majority of new CL cases are reported in Colombia, Afghanistan, Iran, Brazil, Algeria, and the Syrian Arab Republic (https://www.who.int/news-room/fact-sheets/detail/leishmaniasis, 2019) (Iddawela et al. 2018). MCL is characterized by impaired mucus membranes, especially the nose, mouth, larynx, and pharynx. According to the WHO, around 90% of new MCL cases were reported in Ethiopia, Brazil, Bolivia, and Peru (Palumbo 2020). Post Kala-azar dermal leishmaniasis (PKDL) involves the presence of macular and nodular rashes on the body, with the majority of cases found in East Africa and India. Approximately 5–10% of VL patients develop PKDL, which serves as a potential VL reservoir in India (Zijlstra et al. 2003). The joint elimination program in 2005, constituted by the government of India, Bangladesh, and Nepal, aimed to minimize VL cases to <1/10,000 patients at health care facilities in endemic countries by 2015 (Singh et al. 2016).

20.2 Life Cycle of *Leishmania* Parasite

Leishmania parasite exists in two distinct morphological forms, i.e., an extracellular promastigote and an intracellular amastigote with the anthroponotic mode of transmission. The life cycle begins when an infected sandfly bites and transfers the

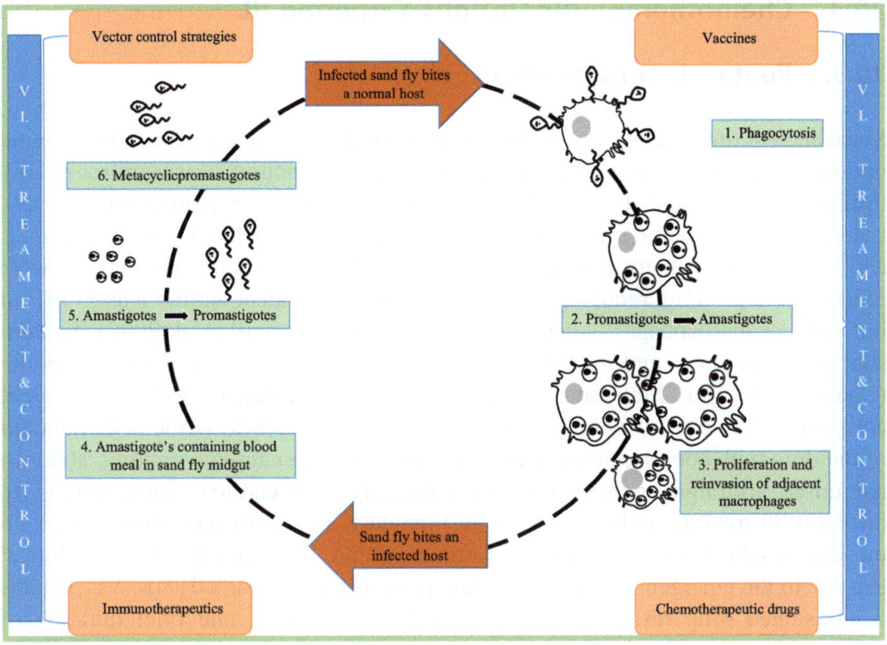

Fig. 20.1 The life cycle of the parasite and therapeutic interventions targeting specific stages

metacyclic promastigotes into the bloodstream, which are engulfed by the macrophages and are transformed into amastigotes. The amastigotes multiply and proliferate in the phagocytic vacuoles of macrophages. The infected macrophages burst and release free amastigotes which infect other cells. The female sandflies pick up the infected cells when they bite a host. In the midgut of the insect, the amastigotes are transformed into metacyclic form and multiply. These metacyclic forms are found in the anterior part of the midgut of sandflies and can be transmitted to another vertebrate host (Fig. 20.1) (Sunter and Gull 2017; Dostalova and Volf 2012).

Despite being the mainstay therapy for treating and controlling VL, chemotherapy involves severe toxic effects and prolonged hospitalization (Yadagiri and Singh 2018). Therefore, there is an immediate need for cost-effective and efficacious therapies for treating VL (Moore and Lockwood 2010). Several nanoformulations have demonstrated the augmented efficacy of drugs following approval by the US FDA (Patra et al. 2018). This chapter focused on the detailed analysis of conventional chemotherapy and nanotechnological strategies in developing novel therapies targeting VL.

20.3 Chemotherapeutic Options Targeting VL

20.3.1 Pentavalent Antimonials

Since ancient times, antimony (stibium) compounds have been used as a therapeutic option. Urea stibamine, an antimony-containing compound, was synthesized by Professor Upendranath Brahmachari, an Indian scientist, for treating VL in 1920. Later, pentavalent antimonials were introduced as a major therapy for treating CL and VL-infected patients. It is generally assumed that pentavalent antimonials (SbV) function as a prodrug, reducing into toxic and active trivalent antimonials (SbIII) in intracellular macrophages, where amastigotes reside (Frezard et al. 2009). Sodium stibogluconate binds and inhibits DNA topoisomerase-I of the *Leishmania* parasite (Chakraborty and Majumder 1988). Prolonged treatment with pentavalent antimonials via painful intramuscular administration leads to tissue accumulation of the drug with common toxic severe effects, including nausea, vomiting, abdominal upset, arthralgia, pancreatitis, nephrotoxicity, and cardiotoxicity, leading to difficulty in treating and controlling leishmaniasis (Yadagiri and Singh 2018). In the Indian subcontinent, pentavalent antimonials are not recommended for clinical use due to the emergence of drug-resistant parasites. In the early 1980s, VL patients demonstrated drug resistance to significantly lesser doses and brief duration of antimonials (10 mg/kg for 6–10 days). Furthermore, it exhibited treatment failures even at toxic doses (20 mg/kg) for a prolonged duration (Tiwari et al. 2017; Mohapatra 2014).

20.3.2 Amphotericin B

Amphotericin B (AmB) deoxycholate, an anti-fungal agent, was introduced as a major drug for treating VL. AmB effectively treated VL-infected patients in North Bihar, a VL hyper-endemic zone in India (Sundar et al. 2007). Amphotericin B can bind and sequester ergosterol, present in the membrane of the *Leishmania* parasite, leading to pore formation (Kumar Saha et al. 1986). AmB showed enhanced efficacy at 0.75–1 mg/kg dose, i.v. infusions for 15–20 times daily or on alternate days. Nonetheless, a maximum of VL-infected patients undergo infusion-related reactions, like fever, thrombophlebitis, chills, hypokalemia, myocarditis, and nephrotoxicity. These adverse effects require direct monitoring and hospitalization of patients, thereby increasing the cost of therapy (Sundar and Jaya 2010). To overcome the toxic side effects of AmB, liposomes AmB (AmBisome) was introduced with a unique safety profile (Stone et al. 2016). A single-dose AmBisome (10 mg/kg) demonstrated the highest efficacy and safety in South Asia. The WHO Committee on controlling leishmaniasis recommended using AmBisome as a major therapy for VL in South Asia (Balasegaram et al. 2012).

20.3.3 Aminosidine

Aminoglycoside antibiotics (paromomycin/aminosidine) exhibit a unique spectrum of antileishmanial activity. Aminosidine interacts with the ribosomes in mitochondria and leads to respiratory dysfunction in *Leishmania* parasites. Aminoglycoside antibiotics bind to the ribosomes (30S subunit) leading to protein synthesis inhibition (Jhingran et al. 2009). Paromomycin (11 mg/kg/day for 3 weeks, i.m. route) was approved as a VL therapy in endemic zones in India (Sundar et al. 2009). However, paromomycin suffers from severe ototoxicity and nephrotoxicity in VL-infected patients (Jamil et al. 2015).

20.3.4 Miltefosine

Miltefosine (hexadecylphosphocholine, HePC), an alkyl phospholipids compound, was developed as an oral therapy for treating VL. Initially, it was formulated as an anticancer drug for treating breast cancer and other tumors. Miltefosine was approved as an oral antileishmanial agent in India in 2002 (Sundar and Olliaro 2007). Miltefosine interferes with lipid metabolism by preventing phosphatidylcholine synthesis, thereby affecting parasite signaling and membrane synthesis. Miltefosine interacts with the acidocalcisome and activates the *Leishmania* parasite's sphingosine-dependent plasma membrane calcium channels (Pinto-Martinez et al. 2018). The administration of 2.5 mg/kg/day of miltefosine for 28 days effectively treated VL and CL patients (Dorlo et al. 2012). However, extensive research revealed that miltefosine was teratogenic in pregnant women (Sundar and Olliaro 2007).

20.3.5 Pentamidine

Pentamidine or 4-[5-(4-carbamimidoylphenoxy)pentoxy]benzenecarboximidamide, an orphan drug, has been approved for treating pneumonia caused by *Pneumocystis carinii* in the United States. The antiprotozoal effects of pentamidine were demonstrated in *Trypanosoma cruzi* in 1938. Soon after, reports of antimonial drug resistance were observed. It was later established as second-line antileishmanial therapy in VL-infected patients (Pearson and Hewlett 1985). Pentamidine isethionate is recommended as a prophylaxis to impede the occurrence in HIV-VL immunocompromised patients (Diro et al. 2015). Pentamidine isethionate binds to adenine-thymine base pairs in kinetoplastid DNA (No 2016). Intravenous administration of pentamidine causes multi-organ adverse effects, including gastrointestinal discomfort, nephrotoxicity, cardiotoxicity, and metabolic disorders (Yeung et al. 1996).

20.3.6 Sitamaquine

Sitamaquine was formulated by the Walter Reed Army Institute Research (WRAIR, USA) in collaboration with GlaxoSmithKline (GSK). Sitamaquine is an antileishmanial drug administered orally and has been proven effective against *L. donovani*-infected patients in Kenya (Singh et al. 2016). Sitamaquine enters the parasite through diffusion in a sterol-dependent and energy-independent electrical gradient fashion (Coimbra et al. 2010). Nephrotoxicity and methemoglobinemia are common side effects associated with higher doses of sitamaquine in VL-infected patients. Further, clinical development of sitamaquine has been abandoned due to safety issues (Table 20.1) (Fig. 20.2) (Loiseau et al. 2011).

20.4 Nano Pharmaceutical Approaches Targeting Visceral Leishmaniasis

Current antileishmanial therapy mainly relies on chemotherapy, involving severe side effects, drug-resistant parasites, and treatment failure (Uliana et al. 2018). To overcome the failures associated with conventional drug therapies, drug discovery and development are essential which are effective and time-consuming (Singh et al. 2016). Several researchers worldwide synthesized and evaluated the efficacy of novel compounds against the *Leishmania* parasite. However, the infective parasite resides and multiplies intracellularly in the phagolysosome of the macrophages, thereby limiting the entry of conventional antileishmanial chemotherapeutic agents into the macrophages to kill or suppress the growth of *Leishmania* parasite (Podinovskaia and Descoteaux 2015). The development of nano pharmaceuticals has gained significant potential for site-specific delivery with enhanced efficacy and bioavailability (Rizvi and Saleh 2018; Gujjari et al. 2022). Nanomedicine involves developing nanomaterials encapsulating drugs as a potential therapy to deliver drugs (Masri et al. 2019). Combining antileishmanial drugs with nanomaterials has evolved as a promising strategy for treating visceral leishmaniasis. The nanocarriers enter the phagolysosomes of macrophages, release the encapsulated chemotherapeutic agent, and eventually kill the protozoan parasite. Target-specific drug delivery against intracellular parasites by using nanocarriers can increase efficacy and bioavailability (Bruni et al. 2017). Additionally, using nanomaterials enhances efficacy, safety, and selectivity, augmenting the pharmacokinetics and solubility of the drug, preventing drug interaction, and endorsing uniform release of the drug at the specific site. Encapsulating chemotherapeutic agents in nanocarriers leads to a combinatorial synergistic effect (De Almeida et al. 2017). Surface modification of nanocarriers encapsulating specific drugs can enhance macrophage internalization and parasite selectivity (Qie et al. 2016). Liposomal amphotericin B (AmBisome) is a basic example of successful nanoformulation approved by the FDA for treating visceral leishmaniasis with increased efficacy in comparison with conventional amphotericin B (Meyerhoff 1999). Several nanoparticles, including metallic, emulsions, polymeric, lipid, and carbon based, have demonstrated potential applications in drug

Table 20.1 Clinically available drugs targeting leishmaniasis

S. No.	Treatment	Mechanism of action	Dose and route of administration	Advantage	Limitation
1.	Pentavalent antimonials	The toxic trivalent form induces oxidative stress in intracellular parasites	20 mg/kg/day for 20–30 days (i.m.)	Cost-effective and readily available	Drug resistance in endemic regions
2.	AmB	Sequesters ergosterol in the parasite membrane	0.75–1.0 mg/kg (i.v.) infusions daily or on alternate days (15–20 times)	Efficacious in the Indian subcontinent, with resistance reports	Infusion-related reactions, like fever, thrombophlebitis, chills, hypokalemia, myocarditis, and nephrotoxicity
3.	Liposomal amphotericin B	Target specific drug delivery against liver, spleen, and bone marrow macrophages	10 mg/kg (i.v.), single dose	High efficacy and safety	Expensive
4.	Aminosidine	Binds to ribosomal 30S subunit and inhibits protein synthesis	11 mg/kg/day (i.m.) for 3 weeks	Effective in combinatorial therapy	Ototoxicity and nephrotoxicity
5.	Miltefosine	Interferes with the lipid metabolism of the parasite	2.5 mg/kg/day (p.o.) for 28 days	Only orally active antileishmanial drug	Teratogenic effects in pregnant women
6.	Pentamidine	Binds to adenine-thymine base pairs of kinetoplastid DNA of the parasite	4 mg/kg/day thrice (i.m. or i.v.) weekly for 15–20 doses	Effective in combinatorial therapy	Nephrotoxicity and cardiotoxicity

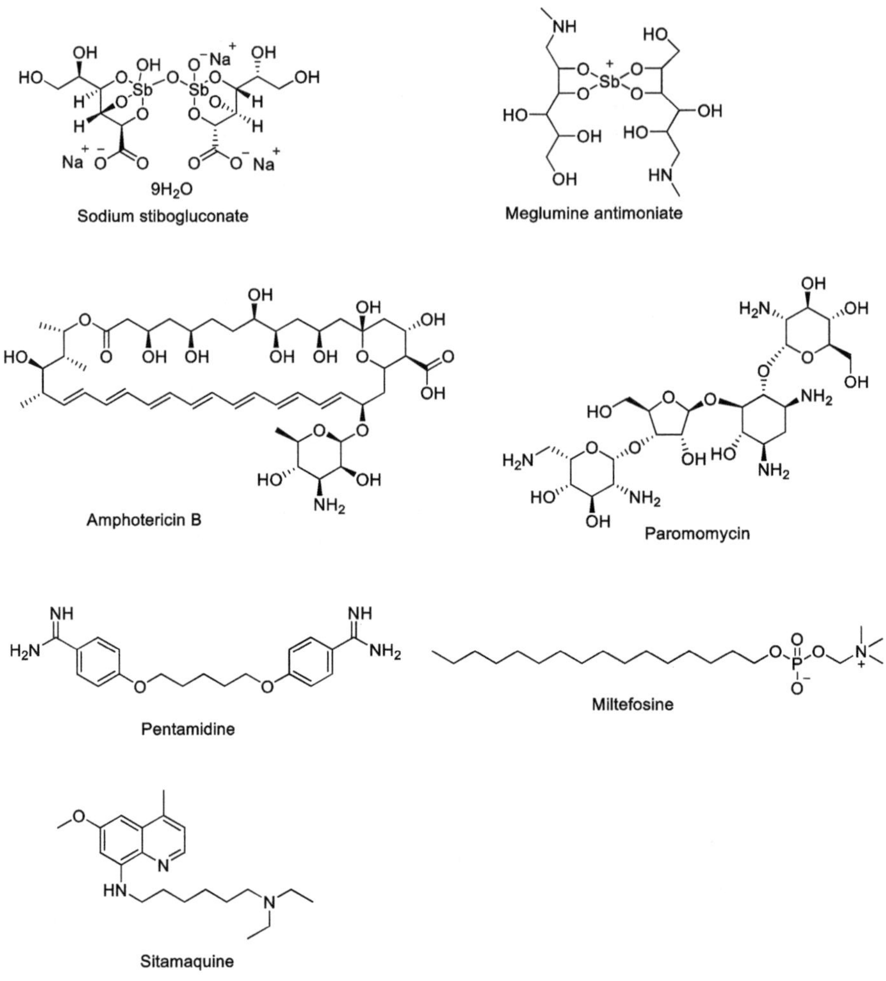

Fig. 20.2 The chemical structures of drugs for treating visceral leishmaniasis

therapy (Table 20.2) (Fig. 20.3) (Naseri et al. 2015). Nanoformulations exhibited enhanced efficacy, safety, and cost-effectiveness, thereby improving patient compliance (Chowdhury et al. 2017). Several studies have focused on developing nanoformulations encapsulating antileishmanial drugs and demonstrated enhanced efficacy in vivo, with potential applications in NTDs. Nonetheless, a greater attentiveness of the scientific community is crucial for poverty-ridden neglected tropical diseases.

Table 20.2 The advantages and disadvantages of nanotechnology-based carrier-based systems

S. No	Nanocarrier based-system	Advantages	Limitations
1.	Liposomes	Encapsulate hydrophilic and hydrophobic drugs, stable, biodegradable, biocompatible, flexible, administered parenterally and cutaneously augmented therapeutic index, surface functionalization	High production cost, shorter half-life, leakage into the systemic circulation
2.	Polymeric nanoparticles	Biocompatible, biodegradable, cost-effective, bypass RES system, flexibility, no leakage	Scale-up is challenging
3.	Solid lipid nanoparticles (SLNs)	Biocompatible, scale-up is easy, safe, and stable, drugs protected from various environmental conditions, used for oral, parenteral, and cutaneous routes, encapsulate hydrophilic and hydrophobic	Reduced drug loading, burst release, shorter half-life
4.	Nanostructured lipid carriers (NLCs)	Stable, higher drug loading, longer shelf life, scale-up is easy	Toxic side effects related to the type of surfactant
5.	Nanoemulsions	Easy formulation, longer shelf life, encapsulate hydrophilic and hydrophobic drugs, used for oral, parenteral, and cutaneous routes, thermodynamically stable	Toxic side effects related to the type of surfactants
6.	Metallic nanoparticles	Antibacterial, uniform, stable	Toxic side effects
7.	Nanotubes	Stable attachment with therapeutic molecules and target-specific drug delivery	Safety-related issues

20.4.1 Liposomes

Liposomes are vesicular structures made of cholesterol or nontoxic phospholipids. The idiosyncratic set of characteristics including biocompatibility, small size, and encapsulation of hydrophilic and hydrophobic drugs make them a drug delivery vehicle to reckon with. They demonstrate higher retention and half-life (Momeni et al. 2013; Lasic 1998). They were initially named smectic mesophases by Bangham and later renamed liposomes by G. Weismann (Bangham et al. 1965). The drugs encapsulated within the liposomes were efficiently internalized by the macrophages leading to enhanced intracellular activity (Ahsan et al. 2002). Liposomes demonstrated an augmented efficacy and pharmacokinetic parameters leading to a desired effect. Macrophages play a significant role in controlling several cellular functions via receptor-mediated pathways, and surface modification of liposomes with mannose and 4-sulfated acetyl galactosamine can selectively target the macrophages and enhance the antileishmanial effect (Singodia et al. 2012).

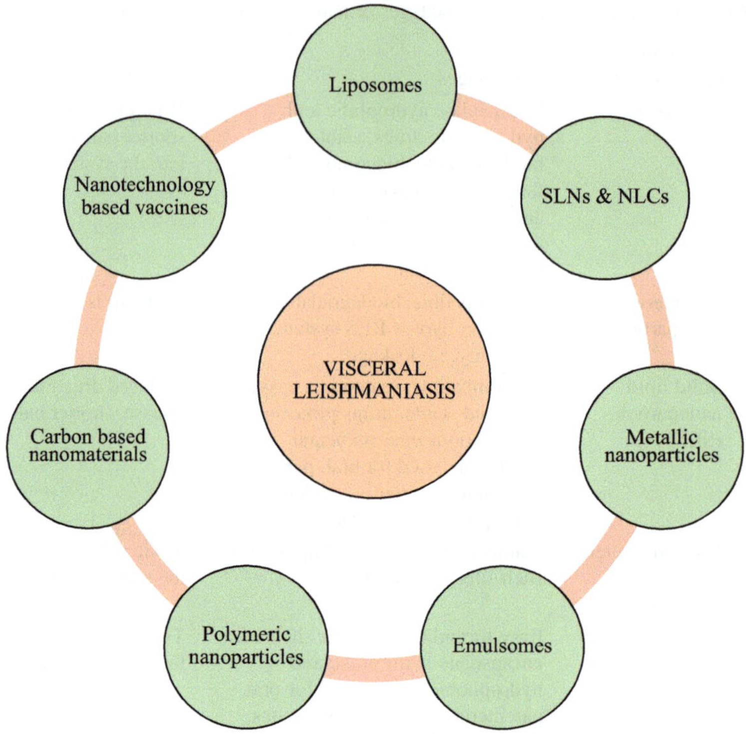

Fig. 20.3 Nanotechnological strategies for combating visceral leishmaniasis

Liposomal amphotericin B (AmBisome) alleviated disease severity in 95.7% of patients at a dose of 10 mg/kg, which showed similar results as obtained after treatment with conventional AmB (1 mg/kg, i.v) for one month (Sundar et al. 2010). AmBisome showed 90% cure rates with a prolonged half-life and prominent pharmacodynamics. AmBisome, compared to other antimonials, showed hepatic accumulation (Berman et al. 1986; Sundar et al. 2004). Carboxymethyl chitosan-coated liposomes encapsulating AmB showed enhanced activity and biocompatibility in preclinical antileishmanial models (Singh et al. 2022). Want et al. demonstrated that nanoliposomal artemisinin showed improved efficacy in *Leishmania*-infected BALB/c mice model with restored host immune response than with treatment with artemisinin alone (Want et al. 2017). Amphotericin B cholesterol dispersion (ABCD) eliminated the hepatic parasite burden and displayed fourfold higher efficacy than free amphotericin B (Berman et al. 1992).

20.4.2 Solid Lipid Nanoparticles (SLNs) and Nanostructured Lipid Carriers (NLCs)

Both SLNs and NLCs differ based on the properties of their core (Muller et al. 2000). They represent an alternative therapeutic approach in drug delivery and have a rigid core composed of a layer of phospholipids more stable than liposomes, microemulsions, and nanocapsules. SLNs were introduced in 1991. They are spherical-shaped structures composed of lipids stabilized by emulsifiers and remain solid at body temperatures. SLNs and NLCs are biocompatible and biodegradable, with a size of 100–1000 nm (Thukral et al. 2014). They can be prepared by high-pressure homogenization technique, which is cost-effective and safe compared to other colloidal carriers. SLNs have demonstrated a high therapeutic potential in cancer, malaria, tuberculosis, leishmaniasis, and human trypanosomiasis (Parvez et al. 2020a). Surface enhancement of SLNs enhances efficacy and safety by selective targeting (Singh et al. 2020b; Parvez et al. 2020c). Nanocarrier formulations of chitosan-coated SLNs loaded with amphotericin B treated leishmanial infections with higher efficacy than liposomal formulations, like AmBisome and Fungizone (Srivastava and Sundar 2011). SLNs-loaded paromomycin sulfate showed higher efficiency in suppressing intracellular parasites and upregulating Th1 type cytokines in *L. tropica*-infected mouse model as compared to paromomycin sulfate given alone (Heidari-Kharaji et al. 2016). Surface modification with chitosan further enhances the oral bioavailability of encapsulated antileishmanial drugs within the SLNs (Das et al. 2017; Parvez et al. 2020b). Paromomycin-encapsulated SLNs improved the drug interaction within the macrophages for treating CL (Kharaji et al. 2015). Chitosan-coated SLNs encapsulated with amphotericin B showed protection in infected cells via enhancement of the Th1 cytokines and nitric oxide (NO) (Jain et al. 2014). Oral administration of SLNs encapsulating melatonin and AmB enhanced the antileishmanial efficacy in infected BALB/c mice (Parvez et al. 2021). The major limitation of SLNs is the crystalline structure, which can lead to drug explosion during storage (Monteiro et al. 2017). NLCs are termed the second generation of SLNs. NLCs are not crystalline and have a better loading capacity than SLNs which can combat drug explosion and disintegration (Kar et al. 2017). Chitosan oligosaccharides coated NLCs loaded with ursolic acid showed improved leishmanicidal effects both in vitro and in vivo.

20.4.3 Metallic Nanoparticles

Targeted killing of *Leishmania* can be achieved by using metallic nanoparticle-mediated delivery of antileishmanial drugs, which are less toxic and highly effective. Metallic nanoparticles were introduced as therapeutics in the 1850s. Encapsulation of amphotericin B with glycine-coated iron oxide particles (10–15 nm) showed controlled release and enhanced killing of parasites in an infected spleen (Khatami et al. 2017). Zinc oxide nanoparticles demonstrated apoptosis in a concentration and time-dependent manner in the parasites (Delavari et al. 2014). Gold nanoparticles are potential carriers due to their enhanced gastric residence time in the systemic circulation. Gallic acid bioconjugated gold nanoparticles showed enhanced antileishmanial effects against sodium stibogluconate-sensitive and resistant strains of parasites (Das et al. 2018b). The enhanced antileishmanial activity was observed in quercetin-conjugated gold nanoparticles than other antileishmanial drugs in resistant strains (Das et al. 2013). Andrographolide-engineered gold nanoparticles exhibited strong antileishmanial effects on antimony/paromomycin sensitive and resistant strains of the parasite (Das et al. 2018a). These nanoparticles are small aggregates of noble metals. The physical characteristics of the nanoparticles can be managed by the capping agent and the reaction conditions (Tauran et al. 2013). Silver nanoparticles inhibited the metabolic activity of the parasite (Allahverdiyev et al. 2011).

20.4.4 Emulsomes

Emulsomes (nanoemulsions) are a new colloidal drug delivery system wherein the internal core consists of triglycerides and fats stabilized due to a high concentration of lecithin in oil in water emulsion. They are easy to synthesize and encapsulate water and fat-soluble drugs. Emulsomes are composed of an inner hydrophobic layer as in emulsions and stabilization with phospholipid bilayers as in liposomes (Gupta and Vyas 2007). Nanoemulsions loaded with amphotericin B showed higher encapsulation efficiency, extended stability, and improved efficacy and safety in *L. donovani*-infected macrophages compared to conventional amphotericin B (Caldeira et al. 2015). Nanoemulsions carrying *Copaifera paupera* oleoresins showed increased antileishmanial activity against *Leishmania infantum* (Rodrigues et al. 2018). Nanoemulsion template-based chitosan nanocapsules encapsulating AmB showed protection in *Leishmania*-infected macrophages and hamsters with increased internalization and Th1-type cytokines (Asthana et al. 2013). Nanoemulsions encapsulated sulfonamides showed increased selectivity and specificity against the β-carbonic anhydrase of the parasite. Drug delivery through nanoemulsion carrier systems is beneficial for enhancing the membrane penetration of hydrophilic drugs (da Silva Cardoso et al. 2018).

20.4.5 Polymeric Nanoparticles

Polymeric nanoparticles are essential for treating intracellular pathogen-intervened diseases because the drugs reach the targeted site. As polymeric nanoparticles cross the cell membrane easily, they enhance the subsequent uptake of therapeutic molecules, thereby enhancing the efficacy of the drug. Polymeric nanoparticles are constituted of synthetic and natural polymers with high biocompatibility and biodegradability (Hirenkumar and Steven 2012). Drugs are carried out by different methods, like adsorption, dissolution, encapsulation, and chemical interaction with polymeric nanoparticles. These nanoparticles can encapsulate all compounds with high biological stability and can withstand physiological stress. They are cleared by the reticuloendothelial system (Couvreur and Vauthier 2006). Slow and sustained release of hydrolyzed polymers (polylactide-co-glycolic acid, gelatin, albumin, polyisohexylcyanoacrylate, and polydiethylmethylidene malonate) is recommended for developing formulations because of their controlled release pattern. It can be easily modified to target intracellular pathogens (Chaurasia et al. 2016; Gupta et al. 2014). Intravenous administration of amphotericin B in conjugation with N-(2-hydroxypropyl)methacrylamide (HPMA) copolymer via Gly-Phe-Leu-Gly linker to *L. donovani*-infected mice at 3 and 1 mg/kg doses showed 99.6% and 93.8% inhibition in liver parasite burden, respectively (Nicoletti et al. 2020). Most polymers in nanoparticles are polyesters with polylactide–polyglycolide copolymers (PLGA) approved by the FDA (Jain 2000). Firstly, Gaspar and coworkers estimated the potential efficacy of polymeric nanoparticles for treating leishmaniasis. PLGA nanospheres are prepared with mannose, mannan, and mannosamine by nanoprecipitation. Co-culturing nanospheres with macrophages initiates immune-modulatory and pro-inflammatory responses killing the parasite (Barros et al. 2015). Van De Ven et al. reported the effective killing of the parasite with AmB-loaded PLGA nanoparticles. They were effective as AmBisome, as demonstrated by efficacy studies (Van De Ven et al. 2012).

20.4.6 Carbon-Based Nanomaterials

Carbon nanotubes have shown significant potential as the most encouraging nanomaterial in the twenty-first century for various biomedical applications. Carbon nanotubes are arranged in the form of tubes. Nanotubes are composed of large internal volumes with easy functionalization of external surfaces. Stable attachment of the nanotube backbone with therapeutic molecules could avoid the chances of macromolecular desorption (Srivastava and Sundar 2011; Prajapati et al. 2012). These carbon nanotubes have been proven excellent nanocarriers for targeting several parasitic infectious diseases. Recently, it has been shown that functionalized carbon nanotubes attached to amphotericin B have shown improved efficacy and safety in infected macrophage and hamster models (Srivastava and Sundar 2011). Furthermore, oral administration of AmB functionalized carbon nanotubes has been designed to treat VL, evaluated against infected hamsters, and exhibited 99% parasite suppression without showing any significant toxic effects (Prajapati et al. 2012). Functionalized carbon nanotubes attached to botulin formulation have shown enhanced leishmanicidal effects and safety in *L. donovani*-infected J774A.1 macrophages compared to botulin treatment alone (Saudagar and Dubey 2014). Nanoformulation of cisplatin bounded with carbon nanotubes effectively treated the infection compared to cisplatin alone (Akhtari et al. 2019). Amine functionalized carbon composite AmB nanoparticles showed protection in *L. donovani*-infected Syrian golden hamsters (Gedda et al. 2020). Mudavath et al. demonstrated the protective role of amine-modified graphene conjugated with amphotericin B. They significantly reduced parasite burden in vitro (Singh et al. 2020a) and in vivo with no significant hepatic and renal toxicities in mice (Mudavath et al. 2014).

20.5 Nanotechnology-Based Vaccines

Therapeutic options targeting VL are limited and exhibit severe side effects. Nanovaccines could overcome the limitations associated with conventional antileishmanial therapies and eliminate VL. Efforts are underway by researchers to develop a proper vaccine(s) in combating VL (Ghorbani and Farhoudi 2018). Broadly, the vaccines against leishmaniasis can be classified as first generation (whole-killed parasites) and second generation (recombinant proteins). The vaccines can be easily produced in developing countries at a minimal cost; however, registration of such vaccines derived from parasites is a challenge (Sundar and Singh 2014). On the contrary, recombinant proteins with adjuvants or expression in heterologous microbial vectors are used as a second-generation vaccine against leishmaniasis. Scalable and cost-effective production approaches can be achieved by recombinant technology and imply a more viable alternative for mass vaccination campaigns; however, stability should be evaluated (Duthie et al. 2012). Several vaccine candidates (Leish-F1, F2, and F3) are in clinical trials developed by Infectious Disease Research Laboratory (IDRI) against leishmaniasis. Leish-F1, a human recombinant vaccine, could confer some protective immunity against *Leishmania* (Moafi et al. 2019). Nanovaccines prepared by chitosan nanoparticles encapsulating recombinant superoxide dismutase of the *Leishmania* parasite resulted in an enhanced immune response. It increased IgG2a levels in the *Leishmania*-infected mice model (Danesh-Bahreini et al. 2011). Phage display technology was used to determine possible new vaccine candidates against leishmaniasis. Dendritic cells are specific antigen-presenting cells that activate innate and acquired immunity. Mature dendritic cell migrates from the infected site to the nearer lymph node to present antigens to naive T-cells to produce selective antibodies and offers protection against *Leishmania* (Janssen et al. 2003). *Leishmania*-infected mice immunized with kinetoplastid membrane protein 11 (KMP-11) encapsulated in PLGA nanoparticles, an 11-kDa *Leishmania* vaccine candidate, stimulated the innate immunity and induced parasite killing effectively (Santos et al. 2013). Also, vaccination with poly(lactide-co-glycolide) nanocarriers encapsulated with soluble antigens and TNF-α-mimicking peptide or monophosphoryl lipid conferred protection in infected BALB/c mice (Margaroni et al. 2017).

20.6 Conclusion

VL is a severe medical and public health concern affecting poverty-ridden populations across the globe. The archaic pharmacological drugs for treating VL are not deprived of severe toxic side effects. Further, to deter the antileishmanial manifestations from reaching the RES, there is a decisive obligation to administer the drugs near the required site. Nanotechnology-based therapy demonstrated significant potential in treating VL. Additionally, they have shown promising results in preclinical trials, exhibiting enhanced therapeutic efficacy. Various nanotechnology-based drug delivery systems have been successful in preclinical studies against VL. Further, clinical trials must confirm the results to corroborate the preclinical data.

Acknowledgments This work is supported by the Government of India for funding under DST-SERB (EEQ/2020/000563).

References

Ahsan F, Rivas IP, Khan MA et al (2002) Targeting to macrophages: role of physicochemical properties of particulate carriers-liposomes and microspheres - on the phagocytosis by macrophages. J Control Release 79:29–40

Akhtari J, Faridnia R, Kalani H et al (2019) Potent in vitro antileishmanial activity of a nanoformulation of cisplatin with carbon nanotubes against *Leishmania major*. J Glob Antimicrob Resist 16:11–16

Allahverdiyev AM, Abamor ES, Bagirova M et al (2011) Antileishmanial effect of silver nanoparticles and their enhanced antiparasitic activity under ultraviolet light. Int J Nanomedicine 6:2705–2714

Alvar J, Velez ID, Bern C et al (2012) Leishmaniasis worldwide and global estimates of its incidence. PLoS One 7:e35671

Asthana S, Jaiswal AK, Gupta PK et al (2013) Immunoadjuvant chemotherapy of visceral leishmaniasis in hamsters using amphotericin B-encapsulated nanoemulsion template-based chitosan nanocapsules. Antimicrob Agents Chemother 57:1714–1722

Balasegaram M, Ritmeijer K, Lima MA et al (2012) Liposomal amphotericin B as a treatment for human leishmaniasis. Expert Opin Emerg Drugs 17:493–510

Bangham AD, Standish MM, Watkins JC (1965) Diffusion of univalent ions across the lamellae of swollen phospholipids. J Mol Biol 13:238–252

Barros D, Lima SAC, Cordeiro-Da-Silva A (2015) Surface functionalization of polymeric nanospheres modulates macrophage activation: relevance in leishmaniasis therapy. Nanomed J 10:387–403

Berman JD, Hanson WL, Chapman WL et al (1986) Antileishmanial activity of liposome-encapsulated amphotericin B in hamsters and monkeys. Antimicrob Agents Chemother 30: 847–851

Berman JD, Ksionski G, Chapman WL et al (1992) Activity of amphotericin B cholesterol dispersion (Amphocil) in experimental visceral leishmaniasis. Antimicrob Agents Chemother 36:1978–1980

Bruni N, Stella B, Giraudo L et al (2017) Nanostructured delivery systems with improved leishmanicidal activity: a critical review. Int J Nanomedicine 12:5289–5311

Caldeira LR, Fernandes FR, Costa DF et al (2015) Nanoemulsions loaded with amphotericin B: a new approach for the treatment of leishmaniasis. Eur J Pharm Sci 70:125–131

Chakraborty AK, Majumder HK (1988) Mode of action of pentavalent antimonials: specific inhibition of type I DNA topoisomerase of *Leishmaniadonovani*. Biochem Biophys Res Commun 152:605–611

Chaurasia S, Chaubey P, Patel RR et al (2016) Curcumin-polymeric nanoparticles against colon-26 tumor-bearing mice: cytotoxicity, pharmacokinetic and anticancer efficacy studies. Drug Dev Ind Pharm 42:694–700

Chowdhury A, Kunjiappan S, Panneerselvam T et al (2017) Nanotechnology and nanocarrier-based approaches on treatment of degenerative diseases. Int Nano Lett 7:91–122

Coimbra ES, Libong D, Cojean S et al (2010) Mechanism of interaction of sitamaquine with *Leishmania donovani*. J Antimicrob Chemother 65:2548–2555

Couvreur P, Vauthier C (2006) Nanotechnology: intelligent design to treat complex disease. Pharm Res 23:1417–1450

da Silva Cardoso V, Vermelho AB, Ricci Junior E et al (2018) Antileishmanial activity of sulphonamide nanoemulsions targeting the β-carbonic anhydrase from *Leishmania* species. J Enzyme Inhib Med Chem 33:850–857

Danesh-Bahreini MA, Shokri J, Samiei A et al (2011) Nanovaccine for leishmaniasis: preparation of chitosan nanoparticles containing Leishmania superoxide dismutase and evaluation of its immunogenicity in BALB/c mice. Int J Nanomedicine 6:835–842

Das S, Roy P, Mondal S et al (2013) One pot synthesis of gold nanoparticles and application in chemotherapy of wild and resistant type visceral leishmaniasis. Colloids Surf B Biointerfaces 107:27–34

Das S, Ghosh S, De AK et al (2017) Oral delivery of ursolic acid-loaded nanostructured lipid carrier coated with chitosan oligosaccharides: development, characterization, in vitro and in vivo assessment for the therapy of leishmaniasis. Int J Biol Macromol 102:996–1008

Das S, Halder A, Mandal S et al (2018a) Andrographolide engineered gold nanoparticle to overcome drug resistant visceral leishmaniasis. Artif Cells Nanomed Biotechnol 46:751–762

Das S, Halder A, Roy P et al (2018b) Biogenic gold nanoparticles against wild and resistant type visceral leishmaniasis. Mater Today Proc 5:2912–2920

De Almeida L, Fujimura AT, Cistia MLD et al (2017) Nanotechnological strategies for treatment of leishmaniasis—a review. J Biomed Nanotechnol 13:117–133

de Vlas SJ, Stolk EA, Hontelez JA et al (2016) Concerted efforts to control or eliminate neglected tropical diseases: how much health will be gained? PLoS Negl Trop Dis 2:e0004386

Delavari M, Dalimi A, Ghaffarifar F et al (2014) In vitro study on cytotoxic effects of ZnO nanoparticles on promastigote and amastigote forms of *Leishmania major* (MRHO/IR/75/ER). Iran J Parasitol 9:6–13

Diro E, Ritmeijer K, Boelaert M et al (2015) Use of pentamidine as secondary prophylaxis to prevent visceral leishmaniasis relapse in HIV infected patients, the first twelve months of a prospective cohort study. PLoS Negl Trop Dis 9:1–15

Dorlo TPC, Balasegaram M, Beijnen JH et al (2012) Miltefosine: a review of its pharmacology and therapeutic efficacy in the treatment of leishmaniasis. Antimicrob Agents Chemother 67:2576–2597

Dostalova A, Volf P (2012) Leishmania development in sand flies: parasite-vector interactions overview. Parasit Vectors 5:1–12

Duthie MS, Raman VS, Piazza FM et al (2012) The development and clinical evaluation of second-generation leishmaniasis vaccines. Vaccine 30:134–141

Frezard F, Demicheli C, Ribeiro RR (2009) Pentavalent antimonials: new perspectives for old drugs. Molecules 14:2317–2336

Gedda MR, Madhukar P, Vishwakarma AK et al (2020) Evaluation of safety and antileishmanial efficacy of amine functionalized carbon-based composite nanoparticle appended with amphotericin B: an in vitro and preclinical study. Front Chem 8:510

Ghorbani M, Farhoudi R (2018) Leishmaniasis in humans: drug or vaccine therapy? Drug Des Devel Ther 12:25–40

Gujjari L, Kalani H, Pindiprolu SK et al (2022) Current challenges and nanotechnology-based pharmaceutical strategies for the treatment and control of malaria. Parasite Epidemiol Control 17:00244

Gupta S, Vyas SP (2007) Development and characterization of amphotericin B bearing emulsomes for passive and active macrophage targeting. J Drug Target 15:206–217

Gupta PK, Asthana S, Jaiswal AK et al (2014) Exploitation of lectinized lipo-polymerosome encapsulated amphotericin b to target macrophages for effective chemotherapy of visceral leishmaniasis. Bioconjug Chem 25:1091–1102

Heidari-Kharaji M, Taheri T, Doroud D et al (2016) Solid lipid nanoparticle loaded with paromomycin: in vivo efficacy against Leishmania tropica infection in BALB/c mice model. Appl Microbiol Biotechnol 100:7051–7060

Hirenkumar M, Steven S (2012) Poly lactic-co-glycolic acid (PLGA) as biodegradable controlled drug delivery carrier. Polymers 3:1–19

Iddawela D, Malinda S, Vithana P et al (2018) Clinical and epidemiological characteristics of cutaneous leishmaniasis in Sri Lanka. BMC Infect Dis 18:108

Jain RA (2000) The manufacturing techniques of various drug loaded biodegradable poly(lactide-co-glycolide) (PLGA) devices. Biomaterials 21:2475–2490

Jain V, Gupta A, Pawar VK (2014) Chitosan-assisted immunotherapy for intervention of experimental leishmaniasis via amphotericin B-loaded solid lipid nanoparticles. Appl Biochem Biotechnol 174:1300–1339

Jamil KM, Haque R, Rahman R et al (2015) Effectiveness study of paromomycin IM injection (PMIM) for the treatment of visceral leishmaniasis (VL) in Bangladesh. PLoS Negl Trop Dis 9: 1–11

Janssen EM, Lemmens EE, Wolfe T et al (2003) CD4+ T cells are required for secondary expansion and memory in CD8+ T lymphocytes. Nature 421:852–856

Jhingran A, Chawla B, Saxena S et al (2009) Paromomycin: uptake and resistance in *Leishmania donovani*. Mol Biochem Parasitol 142:111–117

Kar N, Chakraborty S, De AK et al (2017) Development and evaluation of a cedrol-loaded nanostructured lipid carrier system for in vitro and in vivo susceptibilities of wild and drug resistant *Leishmania donovani* amastigotes. Eur J Pharm Sci 104:196–211

Kharaji MH, Doroud D, Taheri T et al (2015) Drug targeting to macrophages with solid lipid nanoparticles harboring paromomycin: an in vitro evaluation against *L. major* and *L. tropica*. AAPS PharmSciTech 17:1110–1119

Khatami M, Alijani H, Sharifi I et al (2017) Leishmanicidal activity of biogenic Fe3O4 nanoparticles. Sci Pharm 85:1–36

Kumar Saha A, Mukherjee T, Bhaduri A (1986) Mechanism of action of amphotericin B on *Leishmania donovani* promastigotes. Mol Biochem Parasitol 19:195–200

Lasic DD (1998) Novel applications of liposomes. Trends Biotechnol 16:307–321

Loiseau PM, Cojean S, Schrevel J (2011) Sitamaquine as a putative antileishmanial drug candidate: from the mechanism of action to the risk of drug resistance. Parasite 18:115–119

Margaroni M, Agallou M, Athanasiou E et al (2017) Vaccination with poly(D,L-lactide-co-glycolide) nanoparticles loaded with soluble *leishmania* antigens and modified with a TNF-α-mimicking peptide or monophosphoryl lipid aconfers protection against experimental visceral leishmaniasis. Int J Nanomedicine 12:6169–6184

Masri A, Anwar A, Khan NA et al (2019) The use of nanomedicine for targeted therapy against bacterial infections. Antibiotics 8:1–12

Meyerhoff A (1999) US Food and Drug Administration approval of AmBisome (liposomal amphotericin B) for treatment of visceral leishmaniasis. Clin Infect Dis 28:42–48

Moafi M, Sherkat R, Taleban R et al (2019) *Leishmania* vaccines entered in clinical trials: a review of literature. Int J Prev Med 10:1–6

Mohapatra S (2014) Drug resistance in leishmaniasis: newer developments. Tropenmed Parasitol 4: 4–9

Momeni A, Rasoolian M, Momeni A et al (2013) Development of liposomes loaded with anti-leishmanial drugs for the treatment of cutaneous leishmaniasis. J Liposome Res 23:134–144

Monteiro LM, Lobenberg R, Cotrim PC et al (2017) Buparvaquone nanostructured lipid carrier: development of an affordable delivery system for the treatment of leishmaniases. Biomed Res Int 2017:9781603

Moore EM, Lockwood DN (2010) Treatment of visceral leishmaniasis. Int J Infect Dis 2:151–158

Mudavath SL, Talat M, Rai M et al (2014) Characterization and evaluation of amine-modified graphene amphotericin B for the treatment of visceral leishmaniasis: *in vivo* and *in vitro* studies. Drug Des Devel Ther 8:1235–1247

Muller RH, Mader K, Gohla S (2000) Solid lipid nanoparticles (SLN) for controlled drug delivery–a review of the state of the art. Eur J Pharm Biopharm 50:161–177

Naseri N, Valizadeh H, Zakeri-Milani P (2015) Solid lipid nanoparticles and nanostructured lipid carriers: structure preparation and application. Adv Pharm Bull 5:305–313

Nicoletti S, Seifert K, Gilbert IH (2020) N-(2-hydroxypropyl) methacrylamide – amphotericin B (HPMA – AmB) copolymer conjugates as antileishmanial agents. Int J Antimicrob Agents 33: 441–448

No JH (2016) Visceral leishmaniasis: revisiting current treatments and approaches for future discoveries. Acta Trop 155:113–123

Palumbo E (2020) Treatment strategies for mucocutaneous leishmaniasis. J Global Infect Dis 2: 147–150

Parvez S, Yadagiri G, Gedda MR et al (2020a) Modified solid lipid nanoparticles encapsulated with amphotericin B and paromomycin: an effective oral combination against experimental murine visceral leishmaniasis. Sci Rep 10:1–14

Parvez S, Yadagiri G, Karole A et al (2020b) Recuperating biopharmaceutical aspects of amphotericin B and paromomycin using a chitosan functionalized nanocarrier via oral route for enhanced anti-leishmanial activity. Front Cell Infect Microbiol 10:570573

Parvez S, Yadagiri G, Singh A et al (2020c) Improvising anti-leishmanial activity of amphotericin B and paromomycin using co-delivery in d-α-tocopheryl polyethylene glycol 1000 succinate (TPGS) tailored nano-lipid carrier system. Chem Phys Lipids 231:104946

Parvez S, Yadagiri G, Arora K et al (2021) Coalition of biological agent (Melatonin) with chemotherapeutic agent (Amphotericin B) for combating visceral leishmaniasis via oral administration of modified solid lipid nanoparticles. ACS Biomater Sci Eng. https://doi.org/10.1021/acsbiomaterials.1c00859

Patra JK, Das G, Fraceto LF et al (2018) Nano based drug delivery systems: recent developments and future prospects. J Nanobiotechnol 16:1–33

Pearson RD, Hewlett EL (1985) Pentamidine for the treatment of *Pneumocystis carinii* pneumonia and other protozoal diseases. Ann Intern Med 103:782–786

Pinto-Martinez AK, Rodriguez-Duran J, Serrano-Martin X et al (2018) Mechanism of action of miltefosine on *Leishmania donovani* involves the impairment of acidocalcisome function and the activation of the sphingosine-dependent plasma membrane Ca^{2+} channel. Antimicrob Agents Chemother 62:1–10

Podinovskaia M, Descoteaux A (2015) *Leishmania* and the macrophage: a multifaceted interaction. Future Microbiol 10:111–129

Prajapati VK, Awasthi K, Yadav TP (2012) An oral formulation of amphotericin B attached to functionalized carbon nanotubes is an effective reatment for experimental visceral leishmaniasis. J Infect Dis 205:333–336

Qie Y, Yuan H, Roemeling C et al (2016) Surface modification of nanoparticles enables selective evasion of phagocytic clearance by distinct macrophage phenotypes. Sci Rep 6:1–10

Ready PD (2013) Biology of phlebotomine sand flies as vectors of disease agents. Annu Rev Entomol 58:227–250

Ready PD (2014) Epidemiology of visceral leishmaniasis. Clin Epidemiol 6:147–154

Rizvi SAA, Saleh AM (2018) Applications of nanoparticle systems in drug delivery technology. Saudi Pharm J 26:64–70

Rodrigues IA, Ramos ADS, Falcao DQ et al (2018) Development of nanoemulsions to enhance the antileishmanial activity of copaifera paupera oleoresins. Biomed Res Int 2018:9781724

Santos DM, Carneiro MW, Moura TR et al (2013) PLGA nanoparticles loaded with KMP-11 stimulate innate immunity and induce the killing of *Leishmania*. Nanomedicine 9:985–995

Saudagar P, Dubey VK (2014) Carbon nanotube based betulin formulation shows better efficacy against *Leishmania* parasite. Parasitol Int 63:772–776

Singh OP, Singh B, Chakravarty J et al (2016) Current challenges in treatment options for visceral leishmaniasis in India: a public health perspective. Infect Dis Poverty 5:19

Singh A, Sharma S, Yadagiri G et al (2020a) Sensible graphene oxide differentiates macrophages and leishmania: a bio-nano interplay in attenuating intracellular parasite. RSC Adv 10:27502–27511

Singh A, Yadagiri G, Parvez S et al (2020b) Formulation, characterization and in vitro anti-leishmanial evaluation of amphotericin B loaded solid lipid nanoparticles coated with vitamin B12-stearic acid conjugate. Mater Sci Eng C 117:111279

Singh A, Yadagiri G, Negi M et al (2022) Carboxymethyl chitosan modified lipid nanoformulations as a highly efficacious and biocompatible oral anti-leishmanial drug carrier system. Int J Biol Macromol 204:373–385

Singodia D, Verma A, Verma RK et al (2012) Investigations into an alternate approach to target mannose receptors on macrophages using 4-sulfated N-acetyl galactosamine more efficiently in comparison with mannose-decorated liposomes: an application in drug delivery. Nanomedicine 8:468–477

Srivastava ON, Sundar S (2011) Targeted killing of *Leishmania donovani in vivo* and *in vitro* with amphotericin B attached to functionalized carbon nanotubes. J Antimicrob Chemother 66:874–879

Stone NRH, Bicanic T, Salim R et al (2016) Liposomal amphotericin B (AmBisome®): a review of the pharmacokinetics, pharmacodynamics, clinical experience and future directions. Drugs 76: 485–500

Sundar S, Jaya J (2010) Liposomal amphotericin B and leishmaniasis: dose and response. J Global Infect Dis 2:159

Sundar S, Olliaro PL (2007) Miltefosine in the treatment of leishmaniasis: clinical evidence for informed clinical risk management. Ther Clin Risk Manage 3:733–740

Sundar S, Singh B (2014) Identifying vaccine targets for anti-leishmanial vaccine development. Expert Rev Vaccines 13:489–505

Sundar S, Mehta H, Suresh AV et al (2004) Amphotericin B treatment for Indian visceral leishmaniasis: conventional versus lipid formulations. Clin Infect Dis 38:377–383

Sundar S, Chakravarty J, Rai VK et al (2007) Amphotericin B treatment for Indian visceral leishmaniasis: response to 15 daily versus alternate-day infusions. Clin Infect Dis 45:556–561

Sundar S, Agrawal N, Arora R et al (2009) Short-course paromomycin treatment of visceral leishmaniasis in India: 14-day vs 21-day treatment. Clin Infect Dis 49:914–918

Sundar S, Chakravarty J, Agarwal D et al (2010) Single-dose liposomal amphotericin B for visceral leishmaniasis in India. N Engl J Med 362:504–512

Sunter J, Gull K (2017) Shape, form, function and Leishmania pathogenicity: from textbook descriptions to biological understanding. Open Biol 7:170165

Tauran Y, Brioude A, Coleman AW et al (2013) Molecular recognition by gold, silver and copper nanoparticles. World J Biol Chem 4:35

Thukral D, Dumoga S, Mishra A (2014) Solid lipid nanoparticles: promising therapeutic nanocarriers for drug delivery. Curr Drug Deliv 11:771–791

Tiwari N, Gedda MR, Tiwari VK et al (2017) Limitations of current therapeutic options, possible drug targets and scope of natural products in control of leishmaniasis. Mini Rev Med Chem 18: 26–41

Torres-guerrero E, Quintanilla-cedillo MR, Ruiz-esmenjaud J et al (2017) Leishmaniasis: a review. F1000research 6:1–15

Uliana SRB, Trinconi CT, Coelho AC (2018) Chemotherapy of leishmaniasis: present challenges. Parasitology 145:464–480

Van De Ven H, Paulussen C, Feijens PB et al (2012) PLGA nanoparticles and nanosuspensions with amphotericin B: potent *in vitro* and *in vivo* alternatives to Fungizone and AmBisome. J Control Release 161:795–803

Want MY, Islammudin M, Chouhan G et al (2017) Nanoliposomal artemisinin for the treatment of murine visceral leishmaniasis. Int J Nanomedicine 12:2189–2204

Yadagiri G, Mudavath SL (2020) Enkephalins as a therapeutic intervention for visceral leishmaniasis. Med Hypotheses 144:109956

Yadagiri G, Singh PP (2018) Chemotherapy and experimental models of visceral leishmaniasis. In: Infectious diseases and your health. Springer, Berlin, pp 63–97

Yeung KT, Chan M, Chan CKN (1996) The safety of IV pentamidine administered in an ambulatory setting. Chest 110:136–140

Zijlstra EE, Musa AM, Khalil EAG et al (2003) Post-kala-azar dermal leishmaniasis. BMC Infect Dis 18:108

Polymeric Nanoparticles-Based Drug Delivery Systems for Anticancer Therapy

21

Neeta Solanki and Harish Dureja

Abstract

Cancer, a large group of diseases, remains the major cause of morbidity and mortality across the globe and is characterized by uncontrolled cell growth and abnormality in the cell division process. Chemotherapeutic treatment is the foremost treatment strategy for cancer. Non-specificity in targeting cancerous cells due to conventional chemotherapeutic agents leads to toxicity in normal healthy cells. Moreover, rapid excretion and non-specific distribution into targeted organs result in unwanted toxicity issues. To minimize major constraints associated with conventional cancer treatment, the need for advanced drug delivery technologies has become very vital. Among these technologies, nanotechnology has gained much more attention and popularity in the past few years. Nanoparticle-based drug delivery systems provide significant delivery benefits to effectively reduce these constraints related to drug delivery systems and are considered promising carriers to target the drug moieties and improve the treatment outcomes. Various types of nanoparticle systems are currently being explored but the majority of types including metallic and inorganic nanoparticles are utilized for the purpose of drug delivery and are characterized by toxic effects on normal cells. Among all, the nanotechnological polymeric nanoparticles have a variety of advantages over other delivery systems with minor toxicity issues due to biocompatibility and biodegradability of materials. A wide variety of biocompatible, biodegradable, natural, and synthetic polymers such as chitosan, poly (lactide-co-glycolide), polylactide, polyglycolide, and polycaprolactone can be utilized because these polymers can be easily modified to impart desired characteristics. In this chapter, an attempt has been made to highlight the

N. Solanki · H. Dureja (✉)
Faculty of Pharmaceutical Sciences, Maharshi Dayanand University, Rohtak, India
e-mail: harishdureja@mdurohtak.edu.in

P. P. Singh (ed.), *Recent Advances in Pharmaceutical Innovation and Research*,
https://doi.org/10.1007/978-981-99-2302-1_21

importance of nanoparticles for cancer treatment, the advantages and different preparation techniques, and the materials employed for the development of polymeric nanoparticles.

Keywords

Cancer · Nanoparticles · Polymers

21.1 Introduction

Cancer continues to be one of the most destructive groups of diseases for the human body and the foremost cause of mortality due to its complex and progressive nature, possessed by the uncontrolled growth of cells. Cancer is a group of diseases characterized by a lack of normal cell proliferation process within a normal tissue of the body (Murthy and Mathew 2004; Nguyen 2011; Huang et al. 2012). At the molecular and genetic level, the complexity of cancer disease is many-sided involving absences of early identification and detection of the disease and thereby decreasing the survival rate in comparison to other chronic diseases. Cancer affects millions of people across the world of all age groups. The American Cancer Society has stated that the prospects for the occurrence of cancer disease during one's lifetime are one in two for males and one in three for females. Cancer cells undergo rapid cell division and growth because of abnormalities in the DNA of the affected cells resulting in the development of an extra mass of tissue known as a tumor. Tumors are designated as benign or malignant and the formation of tumors occurs when body cells divide rapidly without control over cell division and follow an unregulated cell growth program (Zhang et al. 2006; Bharali et al. 2009; Anajwala et al. 2010; Zaid et al. 2010). The morbidity and mortality of cancer patients are mainly due to tumor cells' invasion of surrounding tissues and different parts of the body. A deep understanding of the biological processes or mechanisms through which normal healthy cells are transformed into malignant tumors has been the subject of continuous research in the biomedical field (Seyfried and Shelton 2010). A number of treatment approaches are being practiced for the management of cancer disease. Every treatment approach used for cancer is associated with significant merits and some sort of demerits along with side effects. Chemotherapy, radiation therapy, hormonal therapy, and surgical removal of tumors are the most frequently used modalities for cancer treatment (Natalie and Mandal 2007; Pucci et al. 2019).

21.2 Cancer Nanotechnology

Intensive research and continuous efforts made at biological, molecular, and cellular levels provide a detailed understanding of cancer and prospects of improved antitumor efficacy of chemotherapeutic agents with a reduction in side effects. Current cancer therapies are associated with some constraints that decrease therapeutic success rate including a lack of site-specific distribution of active drug moieties, and inappropriate drug concentration reaching to the target site (Kılıçay et al. 2011; Huang et al. 2012). Nanotechnology has given advanced insights into the avenue of science and technology, which is continuously growing with better approaches and signifies one of the most imperative directions in the technological advancements of the leading countries (Balaji and Parimala 2010). There are proven shreds of evidence for the application of nanotechnology in the medicinal field, diagnostic purpose, and treatment of various diseases (Peer et al. 2007). Implementation of nanotechnology in cancer disease has been significantly studied and investigated for cancer treatment, providing several major advancements in diagnosis, detection, and management concerns of the disease. Nanotechnology applications in the case of cancer disease offer wide benefits ranging from active and passive drug targeting, to enhanced aqueous solubility with improved bioavailability of poorly soluble chemotherapeutic drugs, and many other innovative therapies. The important aspect of nanotechnology is that nanoparticles (NPs) ranging between 1 and 1000 nm are a type of colloidal drug delivery system, mainly intended for the diagnosis, therapeutic benefits, and biomedical tools for research (Surendiran et al. 2009; Qiao et al. 2010; Agrawal et al. 2011; Plapied et al. 2011; Yu et al. 2021).

21.3 Polymeric Nanoparticles

Nanoparticles (NPs) are types of colloidal drug carriers, particulate dispersions, or solid particles with a size in the range of 1–1000 nm generally made up of polymers and these can be constructed from a variety of materials (Mohanraj and Chen 2006; Cho et al. 2008; Hamidi et al. 2008). Polymeric NPs are one of the most investigated colloidal formulations in the nanomedicine field and promising carriers for the controlled and sustained release of the active ingredients (Rajan and Raj 2013; Heera and Shanmugam 2015; Banik et al. 2016; Pandey et al. 2019; Solanki et al. 2020). Drug moieties can be incorporated into the matrix of NPs in several ways including encapsulation, dissolution, and attachment to the NP matrix. Based on the development technique used, polymeric NPs are categorized as nanospheres (the drug is uniformly dispersed) and nanocapsules (the polymeric membrane that surrounds the matrix core) (Gelperina et al. 2005; Jasmine and Prabhu 2014). Two approaches are mainly used for cancer cells targeted by nanoparticulate carriers—active and passive targeting. Active targeting is used for anticancer drugs to target the cancer cells based on recognition in such a way that they directly interact with the defected cells. To achieve active targeting, the surface-modified NPs are used for targeting the cancer cells. Ligand receptor interaction or antibody–antigen recognition approach is generally used for NPs that target cancer cells (Tang et al. 2010; Nguyen 2011; Sutradhar and Amin 2014). Tumor cells have leaky vascular nature and drugs are encapsulated in NPs targeted passively by passive diffusion through this leaky tumor capillary fenestration into the tumor cell. Selective accumulation of drug-loaded NPs then occurs by enhanced permeability retention (EPR) effect.

Anticancer drugs encapsulated within NPs can effectively increase drug concentration in cancer cells resulting in improved anticancer efficacy at the cellular level (Haley and Frenkel 2008; Cho et al. 2008; Kumari et al. 2010; Paus et al. 2021). Depending on the application (diagnosis, imaging, or therapy), different types of NPs have been prepared for different purposes, and a few of them are utilized for more than one purpose. Nanoparticles are classified into two major categories: organic and inorganic NPs. The first category of nanocarriers includes dendrimers, micelles, liposomes, hybrid, and polymeric NPs while the second category includes quantum dots, fullerenes, silver, and gold NPs. Several marketed products of NPs are available in the market revealing the significance of nano size in cancer management.

21.3.1 Advantages of Nanoparticles

In the arena of drug delivery systems, NPs possess various advantages in contrast to other drug delivery systems including increasing the stability of protein and other biomolecules and some unstable drugs by modulating their release properties. Several advantages linked with nanoparticulate drug delivery systems are discussed here (Mohanraj and Chen 2006; Rawat et al. 2006; Jahanshahi and Babaei 2008; Sanvicens and Marco 2008; Manmode et al. 2009; Joshi et al. 2020; Luo et al. 2021; Bhan and Solanki 2021).

- Particle size and surface characteristics of NPs significantly affect the biological response and these characteristics can be modified to accomplish both passive and active targeting of drugs administered by the parenteral route.
- Nanoparticles can effectively target active ingredients to specific areas within the body.
- The variations in particle size distribution and diameter of particles and surface properties can lead to control of the release behavior of drugs at the site of localization.
- Attachment of target ligands and other substrates to the NP surface provides site-specific drug localization to improve therapeutic efficacy.
- There is a wide possibility of drug administration by different routes including oral, parenteral, nasal, and intraocular, depending upon the desired therapeutic response.
- The choice to use a variety of polymeric materials and the facility to modulate drug release from NPs increase their applicability in drug delivery and make them ideal drug carriers for cancer treatment, vaccine administration, contraceptives, proteins, and peptide delivery.
- Nanoparticles offer several merits over conventional approaches to drug administration in terms of therapeutic effectiveness.
- Administration of anticancer drugs (poor water-soluble drugs) in nano size is an effective approach to enhance the aqueous solubility, which consequently leads to improving the oral bioavailability and increasing their efficacy in cancer therapy.

- Nanocarriers can reduce the resistance presented by the physiological systems because the particle size of the formulations greatly affects the effective delivery of drugs to different parts of the body.
- Targeted NPs provide effective drug distribution, which reduces the chances of drug toxicity.
- The small size of particles in the nano range and targeting applications make these NPs suitable to administer biotechnological drug products to different anatomically complex body systems such as the blood-brain barrier.

21.4 Methods of Nanoparticle Preparation

Different development techniques have been used for NPs' preparation, permitting wide modulation of their structural integrity, composition, and physicochemical characteristics as per the requirement of their usage (Rao and Geckeler 2011). The selection of an appropriate technique to prepare NP formulations mainly depends upon the physicochemical characteristics of the active drug molecule to be encapsulated along with the properties of polymeric materials used for encapsulation of drugs. The selection of polymers depends on various aspects including particle size required, intrinsic properties of the drugs (water solubility and stability aspects), permeability and surface charge on NPs, biocompatibility and biodegradability requirements, and drug release behavior (Krishna et al. 2006; Reis et al. 2006; Vauthier and Bouchemal 2009; Mahapatro and Singh 2011; Rao and Geckeler 2011; Hu and Zhang 2012; Zielinska et al. 2020; Mitchell et al. 2021). Most frequently used techniques for NPs' preparation are briefly discussed here.

21.4.1 Solvent Evaporation Method

This method is the most frequently employed method to formulate polymeric NPs based on the dispersion of preformed polymers. This method involves the utilization of volatile solvents to solubilize drugs and polymers. Dichloromethane and chloroform were the most frequently used solvents for NPs' preparation, but due to toxicity concerns nowadays ethyl acetate is preferred because of its better toxicity profile (Kumari et al. 2010; Mudshinge et al. 2011). Magnetic stirring at room temperature or ultrasonication or high-speed homogenizer is generally used for solvent evaporation from NPs' dispersion. Ultracentrifugation is a suitable step to collect solidified NPs. The resulting sediment of NPs is washed three to four times with distilled water to separate additives such as surfactants and then the product is freeze-dried to obtain NPs in powder form (Maruthi et al. 2011; Rao and Geckeler 2011; Zielinska et al. 2020; Subjakova et al. 2021).

To highlight the anticancer potential of polymeric NPs in cancer therapy, Kumari et al. (2021) investigated the anticancer potential of nanoparticle-based formulations of curcumin against breast cancer cells. The anticancer NPs of curcumin were prepared by using the solvent evaporation technique involving polyglycerol-malic acid-dodecanedioic acid polymer. Breast cancer cell lines, i.e., MCF-7 and

MDA-MB-231, were selected to investigate the anticancer potential of curcumin. Acridine orange/Ethidium bromide staining assays were performed and apoptotic abnormalities were revealed in the analysis. Further western blot analysis was done to confirm curcumin's role in apoptosis due to overexpression of caspase-9 (Kumari et al. 2021). Similarly, Ozturk et al. (2020) studied the anticancer effects of Telmisartan polymeric NPs of poly(D, L-lactide) on breast cancer cell lines (MCF-7 and MDA-MB-231 breast cancer cell lines). NPs were developed to improve the solubility of drug, using the solvent evaporation technique. Polymeric NPs of Telmisartan showed sustained release of drug from NPs and revealed a 45% decrease in viability of breast cancer cells (Ozturk et al. 2020).

21.4.2 Solvent Diffusion Method

When the solvent evaporation method is subjected to modification, the resulting method is known as the spontaneous emulsification method. In this technique, the water-miscible solvents (acetone and methanol), as well as a small amount of an organic solvent (water-immiscible), are employed as an oily phase. The solvents diffuse spontaneously due to which interfacial turbulence is generated between two solutions leading to the formation of NPs. The particle size of prepared NPs is strongly affected by the concentration of water-miscible solvent. If the concentration of water-miscible solvent increases, the particle size decreases. This technique is suitable for entrapment of both water-insoluble and water-soluble drugs. For encapsulation of water-soluble drugs, the formation of multiple emulsions (w/o/w) is a prime requirement (Manmode et al. 2009; Sailaja et al. 2011; Zielinska et al. 2020; Subjakova et al. 2021).

Hulangamuwa et al. (2021) developed nanoformulation of Garcinol using poly (lactic-co-glycolic acid) for targeting the breast cancer stem cells cultured under hypoxic conditions. Garcinol-loaded polylactic-co-glycolic acid (PLGA) NPs were developed by the emulsion solvent diffusion technique. WST-1 cell proliferation assay was used to investigate the anti-proliferative effects of polymeric NPs. Apoptotic effects of NPs were checked by the caspase-3/7 assay. Effects of prepared NPs on the expression of genes related to hypoxia-inducing factors were analyzed using real-time polymerase chain reaction (PCR). Overall findings of the study confirm that Hyaluronic acid-Garcinol-Nanoparticles (HA-GA-NPs) can be considered an effective nano delivery system to target breast cancer stem cells grown under the hypoxic microenvironment (Hulangamuwa et al. 2021).

21.4.3 Nanoprecipitation Method

The nanoprecipitation method is also recognized as a solvent displacement method (interfacial deposition) and allows the precipitation of polymeric material from an organic phase in the presence of surfactant. Diffusion of organic solvent in the aqueous phase leads to the formation of small size particles. Nanoprecipitation technique creates interfacial deposition of used polymeric material between organic

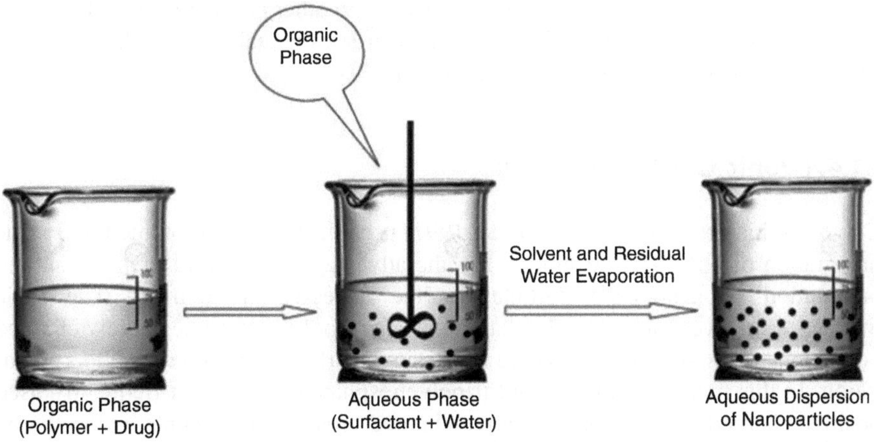

Fig. 21.1 Polymeric nanoparticle preparation by nanoprecipitation method

phase and aqueous phase and spontaneous diffusion of organic solvent allows the formation of NPs' suspension. The utility of this process is applicable to only water-miscible solvents having sufficient diffusion rate to undergo the spontaneous formation of an emulsion. Some water-miscible solvents show a certain type of instability after mixing with water and instant emulsification may not be observed. Water-miscible solvents such as acetone are frequently used solvents to entrap a variety of drugs and this method is useful for encapsulation of water-insoluble drugs (Hans and Lowman 2002; Reis et al. 2006; Sundar et al. 2010; Maravajhala et al. 2012; Ansari 2017; Solanki and Dureja 2018; Neerooa et al. 2021). The schematic representation of the nanoprecipitation method is shown in Fig. 21.1.

Sivanesan et al. (2021) developed NPs of the hydrophobic drug (5-fluorouracil) to improve solubility using biocompatible polymer polyglycerol sebacate using the nanoprecipitation technique. The prepared NPs were biocompatible and had minimal cytotoxicity against the MDA-MB-231 and A549 cell lines, even at a high concentration of 100 mg/mL. With the polyglycerol sebacate NPs loaded with 5-fluorouracil, the cell viability was decreased (Sivanesan et al. 2021). Sharma et al. (2020) determined the anticancer efficacy of diosgenin using polyglycerol malate co-dodecanedioate polymer by nanoprecipitation technique. In vitro, cytotoxic and morphological studies were performed to evaluate the toxicity of prepared drug-loaded nanoparticles in A549 lung cancer cell lines. The inhibitory concentration (IC_{50}) after 48 h was observed and found to be 27.14 μm, 15.15 μm, and 13.91 μm for free diosgenin, polyglycerol malate co-dodecanedioate (7:3), and polyglycerol malate co-dodecanedioate (6:4) NPs, respectively, when administered in A549 lung carcinoma cell lines (Sharma et al. 2020). Pieper et al. (2019) encapsulated doxorubicin into NPs' formulation by nanoprecipitation technique to bypass efflux-mediated drug resistance. Different polymers such as PLGA (polylactic-co-glycolic acid), polylactic acid, and PEGylated PLGA (PLGA-PEG) were selected for preparation of NPs and evaluated for anticancer activity in neuroblastoma cells, including ABCB1-expressing cell lines. PLGA NPs exhibited sustained release of doxorubicin in comparison with other polymeric NPs. The small size of doxorubicin-loaded PLGA-PEG nanoparticles revealed the highest anticancer effects in neuroblastoma cells (Pieper et al. 2019).

21.4.4 Ionic Gelation Method

Calvo and his co-workers (Calvo et al. 1997) used the ionic gelation method for the preparation of NPs. This technique has the advantages of ease of preparation and preparation of nano-sized particles in the aqueous environment. The ionic interactions between a positively charged group of polymer and negatively charged polyanions (cross-linking agent) are the basic underlying principle for NPs' preparation. In this method, chitosan and tripolyphosphate are the most frequently used positive-charged polymer and negative-charged cross-linking agents, respectively. This technique involves the preparation of two aqueous solutions; the first solution consists of polymer (chitosan) with a stabilizer such as a tween 80 and the other is an aqueous solution of polyanion or cross-linking agent. Dropwise addition of tripolyphosphate solution to the chitosan solution under magnetic stirring or homogenization leads to the formation of NPs. Surface characteristics and mean diameter of NPs can be modified by varying chitosan and tripolyphosphate ratios (Tiyaboonchai 2003; Sonia and Sharma 2011; Pandey et al. 2019; Solanki et al. 2020). The schematic representation of the ionic gelation method is represented in Fig. 21.2.

Ruman et al. (2021) formulated a nanocarrier-based drug delivery system of sorafenib-loaded chitosan and folate-coated sorafenib chitosan NPs for effective cellular uptake of sorafenib in human HepG2 (hepatocellular carcinoma) and HT29 (colorectal adenocarcinoma) cell lines. The above NPs were successfully prepared by ionic gelation technique using chitosan as a biocompatible and biodegradable polymer. In vitro cytotoxic studies showed enhanced anticancer activity against HepG2 and HT29 cancer cell lines in comparison with free drugs with the least toxicity toward normal human dermal fibroblast adult cells (Ruman et al. 2021).

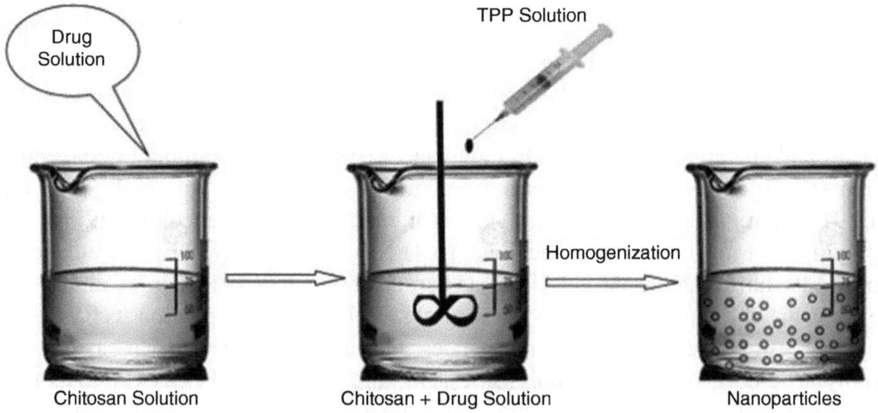

Fig. 21.2 Polymeric nanoparticle preparation by ionic gelation method

Solanki et al. (2020) prepared boswellic acids extracted from the oleo gum resin of plant *Boswellia serrata*-loaded chitosan NPs using the ionic gelation method. In vitro cytotoxic studies exhibited enhanced cytotoxicity of boswellic acid-loaded NPs in contrast to the free boswellic acids having an IC_{50} value of 17.29 and 29.59 μM, respectively. Flow cytometry confirmed that treatment of cells with 40 μg/mL had arrested 22.75 ± 0.3% at the $SubG_0$ phase of the cell cycle when compared with untreated A459 cells. The results obtained confirmed that the boswellic acid-loaded chitosan nanoparticles were effective due to greater cellular uptake.

21.4.5 Salting Out

This technique is based on the salting-out mechanism of strong electrolytes. Commonly used electrolytes for salting-out purposes include magnesium chloride, calcium chloride, and magnesium acetate. Drug and polymeric materials are dissolved in water-miscible solvents (acetone) while the aqueous phase consists of colloidal stabilizing agents like polyvinyl pyrrolidone with high concentrations of electrolytes. Then the organic phase is emulsified into the aqueous phase. The addition of a sufficient quantity of water to the above emulsion results in the diffusion of acetone into the water, which is the rate-limiting step for NPs' preparation. Cross-flow filtration is an effective approach used for the removal of salts and solvents (Krishna et al. 2006; Jawahar and Meyyanathan 2012; Patel et al. 2017).

Sailaja and Jyothi (2020) prepared NPs of 5-fluorouracil by simply salting out the method by utilizing Eudragit-100 as a synthetic polymeric material. The developed NPs were evaluated for particle size, zeta potential, in vitro drug release studies, and interaction studies. Drug-loaded NPs prepared by salting-out methods showed a higher dissolution rate for formulation and a high percentage of drug entrapment (Sailaja and Jyothi 2020).

21.4.6 Supercritical Fluid Technology

Various conventional techniques used for NPs' preparation involve the utilization of organic solvents that are harmful to the physiological systems of humans and the environment also. Therefore, considerable research efforts have been made for effective and safe encapsulation of therapeutically active ingredients into a variety of micro- and nano-sized drug carriers. If solvent impurities remain in the polymeric matrix, it may lead to the degradation of active ingredients and produce toxic changes in therapeutically active molecules. Supercritical fluids are environmentally safe solvents and nowadays have become promising alternatives due to the wide variety of advantages of these fluids on the biological system (lack of availability of traces of organic solvents), and they can be cost-effectively utilized to prepare pure products. Utilization of supercritical fluids results in the preparation of solvent-free precipitates. Supercritical CO_2 is a good example of a supercritical fluid with a variety of advantages such as non-toxicity, economically favorable, mild critical temperature conditions, and non-flammability. Unfortunately, the practical applicability of this technique is less due to little or no solubility of most of the polymers in supercritical fluids (Soppimath et al. 2001; Franco and Marco 2021).

Chen et al. (2020) designed nanoformulation of fisetin to circumvent hydrophobicity and improve its applicability in pharmaceutical dosage forms. Fisetin-loaded polyvinyl pyrrolidone-encapsulated NPs were prepared by using a supercritical anti-solvent technique to enhance its intrinsic solubility. In vitro drug release and cytotoxic studies indicated that the polymeric NPs exhibited improved performance efficiency in contrast to the free drug (Chen et al. 2020).

21.4.7 Dialysis

A dialysis method is an effective approach for the development of nanoformulations. For the development of polymeric NPs, polymers and drugs are dissolved in an organic solvent. The organic phase is then poured inside the dialysis tube and dialysis is performed against immiscible solvent. Inside the membrane, solvent displacement occurs, which leads to progressive polymer aggregation due to a decrease in solubility resulting in the formation of NPs' suspension (Tiruwa 2016; Zielinska et al. 2020). Ren et al. (2020) synthesized doxorubicin-loaded nanospheres of tea polyphenol as a biodegradable polymer that is biodegraded by an acidic microenvironment. In vitro cytotoxic studies were performed on polyphenol functionalized doxorubicin-loaded calcium phosphate nanospheres which confirmed the improved cell-killing outcomes on target cancer cell lines.

21.5 Purification and Freeze-Drying (Lyophilization) of Nanoparticles

The basic criteria used for the selection of the purification method depend upon the characteristics of the prepared NPs such as release properties of the NPs because an unsuitable purification method leads to loss of incorporated active ingredient. The purification of developed NPs is achieved by centrifugation, filtration, and size exclusion chromatography or a combination of these methods. Currently, freeze-drying or lyophilization is an effective approach used in drug delivery technology to enhance the long-term stability of NPs. The freeze-drying process is based on the principle of sublimation and desorption under reduced pressure. This process involves a series of steps for complete drying of suspension in which freezing is the initial step of freeze-drying. The freezing step is important in drying because during this step pure water present inside the suspension is converted into ice crystals. After sufficient freezing, maximum water present in the liquid suspension is frozen and leads to the concentration of suspension and increases the viscosity. A sufficient increase in viscosity of the suspension inhibits further crystallization. This highly viscous and concentrated liquid solidifies and an amorphous, crystalline, or

Table 21.1 Examples of polymeric nanoparticles prepared by using different techniques

Author	Development technique	Drug	Polymer used
Kalam et al. (2022)	Ionic gelation method	Tedizolid phosphate	Chitosan
Biswas et al. (2022)	Solvent evaporation	Eprosartan mesylate	PLGA
Wathoni et al. (2021)	Ionic gelation with spray drying	α-Mangostin	Chitosan
Solanki et al. (2020)	Ionic gelation	Boswellic acids	Chitosan
Pandey et al. (2019)	Ionic gelation with spray drying	Erlotinib	Chitosan
Solanki and Dureja (2018)	Nanoprecipitation	Boswellic acids	Pluronic F-127
Al-Nemrawi et al. (2018)	Solvent evaporation	Tobramycin	Chitosan, PLGA
Gazi and Sailaja (2018)	Salting out	Paracetamol	Eudragit-S100
Umerska et al. (2018)	Solvent evaporation	Curcumin	Eudragit® RLPO, PLGA, and PCL
Pattanayek and Puranik (2018)	Ionic gelation	Ketoprofen	Chitosan
Sukmawati et al. (2018)	Ionic gelation	Doxorubicin, curcumin	Chitosan
Wright et al. (2018)	Nanoprecipitation, solvent evaporation	Ramizol	PLGA
Arafa et al. (2018)	Ionic gelation	Metformin	Chitosan
Kumar et al. (2017)	Ionic gelation	Glimepiride	Gelatin
Salatin et al. (2017)	Nanoprecipitation	Rivastigmine hydrogen tartrate	Eudragit-RL100
Abdulaziz et al. (2017)	Ionic gelation	Hyaluronic acid, alginic acid	Chitosan, Alginate
Maciel et al. (2017)	Ionic gelation	Insulin	Chitosan, Pectin
Katuwavila et al. (2016)	Ionic gelation	Iron	Alginate
Popiolski et al. (2016)	Nanoprecipitation	Lavender oil	Poly(ethylene oxide)-*b*-poly(lactic acid)
Ru et al. (2016)	Solvent evaporation	9-Nitrocamptothecin	PLGA
Senthilnathan et al. (2016)	Solvent displacement	Pregabalin	Eudragit-S100
Colombo et al. (2016)	Emulsion polymerization	Paclitaxel	PCL
Ram et al. (2016)	Solvent displacement	Glipizide	PLGA

(continued)

Table 21.1 (continued)

Author	Development technique	Drug	Polymer used
Guedj et al. (2015)	Solvent evaporation	Bovine serum albumin	PLGA
Zhao et al. (2015)	Supercritical fluid	Paclitaxel	PEG
Ghaderi et al. (2015)	Solvent evaporation	Gamma oryzanol	Ethyl cellulose
Maaz et al. (2015)	Nanoprecipitation	Gatifloxacin	PCL
Jelvehgari et al. (2014)	Ionic gelation	Chlorpheniramine maleate	Alginate, chitosan
Korpe et al. (2014)	Ionic gelation	Bovine serum albumin	Chitosan, alginate
Khan and Schneider (2013)	Nanoprecipitation	Macromolecule	Gelatin
Sarei et al. (2013)	Ionic gelation	Diphtheria toxoid	Alginate
Nesalin and Smith (2013)	Ionic gelation	Stavudin	Chitosan
Akbarzadeh et al. (2012)	Polymerization	Doxorubicin	*N*-isopropylacrylamide and methacrylic acid
Santos et al. (2012)	Nanoprecipitation	3,5,3′-Triiodothyroacetic acid	Poly(methyl methacrylate)
Tan et al. (2012)	Solvent evaporation	Pyridostigmine bromide	PLA
Arora et al. (2011)	Ionic gelation	Amoxicillin	Chitosan, alginate
Hoa et al. (2011)	Solvent evaporation	Ketoprofen	Eudragit-E100, Eudragit-RS
Yadav and Sawant (2010)	Nanoprecipitation	Cytarabine	PLGA

PLGA poly lactic-co-glycolic acid, *RLPO* RL in powder form, *PCL* polycaprolactone, *PEG* polyethylene glycol

combined amorphous–crystalline product is obtained (Abdelwahed et al. 2006; Solanki and Dureja 2018; Solanki et al. 2020).

Wide-ranging research efforts have been undertaken worldwide in the area of polymeric nanotechnology. Various polymeric materials are used by several researchers for the development of NPs for targeting of drug moieties in different diseases. Some recently used examples of polymeric nanoparticles prepared by using different development techniques, materials, and active drugs are displayed in Table 21.1.

References

Abdelwahed W, Degobert G, Stainmesse S et al (2006) Freeze-drying of nanoparticles: formulation, process and storage considerations. Adv Drug Deliv Rev 58:1688–1713

Abdulaziz A, Alradwan I, Kalam MA et al (2017) Effect of cryoprotection on particle size stability and preservation of chitosan nanoparticles with and without hyaluronate or alginate coating. Saudi Pharm J 25:861–867

Agrawal A, Brahme R, Faria M et al (2011) Nanotechnology in cancer: a clinical review. J Appl Pharm Sci 3:25–29

Akbarzadeh A, Samiei M, Joo SW et al (2012) Synthesis, characterization and *in vitro* studies of doxorubicin-loaded magnetic nanoparticles grafted to smart copolymers on A549 lung cancer cell line. J Nanobiotechnol 10:1–7

Al-Nemrawi NK, Alshraiedeh NA, Zayed AL et al (2018) Low molecular weight chitosan-coated PLGA nanoparticles for pulmonary delivery of tobramycin for cystic fibrosis. Pharmaceuticals 11:1–28

Anajwala CC, Jani GK, Swamy SV (2010) Current trends of nanotechnology for cancer therapy. Int J Pharm Sci Nanotechnol 3:1043–1056

Ansari MJ (2017) Factors affecting preparation and properties of nanoparticles by nanoprecipitation method. Indo Am J Pharm Sci 4:4854–4858

Arafa K, Shamma RN, El-Gazayerly ON et al (2018) Facile development, characterization, and optimization of new metformin-loaded nanocarrier system for efficient colon cancer adjunct therapy. Drug Dev Ind Pharm 44:1158–1170

Arora S, Gupta S, Narang RK et al (2011) Amoxicillin loaded chitosan–alginate polyelectrolyte complex nanoparticles as muco penetrating delivery system for *H. pylori*. Sci Pharm 79:673–694

Balaji SU, Parimala B (2010) Nanotechnology and cancer- an overview. Int J Pharm Biopharm 1: 186–201

Banik BL, Fattahi P, Brown JL (2016) Polymeric nanoparticles: the future of nanomedicine. Wiley Interdiscip Rev Nanomed Nanobiotechnol 8:271–299

Bhan S, Solanki N (2021) Promising benefits of nanoparticles in drug delivery system. J Med Pharm Allied Sci 11:4105–4107

Bharali DJ, Khalil M, Gurbuz M et al (2009) Nanoparticles and cancer therapy: a concise review with emphasis on dendrimers. Int J Nanomedicine 4:1–7

Biswas E, Banerjee KK, Karmakar S et al (2022) Preparation and evaluation of eprosartan mesylate loaded PLGA nanostructures. Res J Pharm Technol 15:103–112

Calvo P, Remunan-Lopez C, Vila-Jato JL et al (1997) Novel hydrophilic chitosan-polyethylene oxide nanoparticles as protein carriers. J Appl Polym Sci 63:125–132

Chen LF, Xu PY, Fu CP et al (2020) Fabrication of supercritical antisolvent (SAS) process assisted fisetin encapsulated polyvinyl pyrrolidone nanocomposites for improved anticancer therapy. Nano 10:1–14

Cho K, Wang XU, Nie S et al (2008) Therapeutic nanoparticles for drug delivery in cancer. Clin Cancer Res 14:1310–1316

Colombo C, Morosi L, Bello E et al (2016) Pegylated nanoparticles obtained through emulsion polymerization as paclitaxel carriers. Mol Pharm 13:40–46

Franco P, Marco LD (2021) Nanoparticles and nanocrystals by supercritical CO_2-assisted techniques for pharmaceutical applications: a review. Appl Sci 11:1–27

Gazi AS, Sailaja AK (2018) Preparation and characterization of paracetamol loaded Eudragit S100 nanoparticles by salting out technique. J Dev Drugs 7:1–4

Gelperina S, Kisich K, Iseman MD et al (2005) The potential advantages of nanoparticle drug delivery systems in chemotherapy of tuberculosis. Am J Respir Crit Care Med 172:1487–1490

Ghaderi S, Ghanbarzadeh S, Hamishehkar H (2015) Evaluation of different methods for preparing nanoparticle containing gammaoryzanol for potential use in food fortification. Pharm Sci 20: 130–134

Guedj AS, Kell AJ, Barnes M et al (2015) Preparation, characterization and safety evaluation of poly(lactide-co-glycolide) nanoparticles for protein delivery into macrophages. Int J Nanomedicine 10:5965–5979

Haley B, Frenkel E (2008) Nanoparticles for drug delivery in cancer treatment. Urol Oncol Semin Orig Investig 26:57–64

Hamidi M, Azadi A, Rafiei P (2008) Hydrogel nanoparticles in drug delivery. Adv Drug Deliv Rev 60:1638–1649

Hans ML, Lowman AM (2002) Biodegradable nanoparticles for drug delivery and targeting. Curr Opin Solid State Mater Sci 6:319–327

Heera P, Shanmugam S (2015) Nanoparticle characterization and application: an overview. Int J Curr Microbiol App Sci 4:379–386

Hoa LTM, Chi NT, Triet NM et al (2011) Preparation and characterization of nanoparticles containing ketoprofen and acrylic polymers prepared by emulsion solvent evaporation method. J Exp Nanosci 7:189–197

Hu CM, Zhang L (2012) Nanoparticle-based combination therapy toward overcoming drug resistance in cancer. Biochem Pharmacol 8:1104-1111.

Huang ZW, Laurent V, Chetouani G et al (2012) New functional degradable and bio-compatible nanoparticles based on poly (malic acid) derivatives for site-specific anti-cancer drug delivery. Int J Pharm 423:84–92

Hulangamuwa AC, Ediriweera MK, Rajagopalan U et al (2021) Development of a new nanocarrier for dietary garcinol: characterization and in *vitro* efficacy evaluation using breast cancer stem cells grown in hypoxia. J Food Qual 2021:1–10

Jahanshahi M, Babaei Z (2008) Protein nanoparticle: a unique system as drug delivery vehicles. Afr J Biotechnol 7:4926–4934

Jasmine MJ, Prabhu VV (2014) Polymeric nanoparticles-the new face in drug delivery and cancer therapy. Malaya J Biosci 1:1–7

Jawahar N, Meyyanathan SN (2012) Polymeric nanoparticles for drug delivery and targeting: a comprehensive review. Int J Health Allied Sci 1:217–223

Jelvehgari M, Barghi L, Barghi F (2014) Preparation of chlorpheniramine maleate-loaded alginate/ chitosan particulate systems by the ionic gelation method for taste masking. Jundishapur J Nat Pharm Prod 9:39–48

Joshi G, Patel M, Chaudhary D et al (2020) Preparation and surface modification of polymeric nanoparticles for drug delivery: state of the art. Recent Patent Drug Deliv Formul 14:201–213

Kalam MA, Iqbal M, Alshememry A et al (2022) Development and evaluation of chitosan nanoparticles for ocular delivery of tedizolid phosphate. Molecules 27:2326

Katuwavila NP, Perera AD, Dahanayake D et al (2016) Alginate nanoparticles protect ferrous from oxidation: potential iron delivery system. Int J Pharm 513:404–409

Khan SA, Schneider M (2013) Improvement of nanoprecipitation technique for preparation of gelatin nanoparticles and potential macromolecular drug loading. Macromol Biosci 13:455–463

Kılıçay E, Demirbilek M, Turk M et al (2011) Preparation and characterization of poly (3-hydroxybutyrate-co-3-hydroxyhexanoate) (PHBHHX) based nanoparticles for targeted cancer therapy. Eur J Pharm Sci 44:310–320

Korpe DA, Malekghasemi S, Aydın U et al (2014) Fabrication of monodispersive nanoscale alginate-chitosan core-shell particulate systems for controlled release studies. J Nanopart Res 16:2754–2760

Krishna RS, Shivakumar HG, Gowda DV et al (2006) Nanoparticles-a novel colloidal drug delivery system. Indian J Pharm Educ Res 40:15–21

Kumar P, Pathak D, Singh P et al (2017) Formulation and evaluation of biodegradable nanoparticle of glimepiride by ionic gelation method. J Pharm Res 11:96–102

Kumari A, Yadav SK, Yadav SC (2010) Biodegradable polymeric nanoparticles-based drug delivery systems. Colloids Surf B Biointerfaces 75:1–8

Kumari M, Sharma N, Manchanda R et al (2021) PGMD/curcumin nanoparticles for the treatment of breast cancer. Sci Rep 11:1–11

Luo M, Hua S, Shang Q (2021) Application of nanotechnology in drug delivery systems for respiratory diseases (review). Mol Med Rep 23:1–17

Maaz A, Abdelwahed W, Tekko IA (2015) Influence of nanoprecipitation method parameters on nanoparticles loaded with gatifloxacin for ocular drug delivery. Int J Acad Sci Res 3:1–12

Maciel VB, Yoshida CM, Pereira SM et al (2017) Electrostatic self-assembled chitosan-pectin nano-and microparticles for insulin delivery. Molecules 22:1–21

Mahapatro A, Singh DK (2011) Biodegradable nanoparticles are excellent vehicle for site directed *in-vivo* delivery of drugs and vaccines. J Nanotechnol 9:55–62

Manmode AS, Sakarkar DM, Mahajan NM (2009) Nanoparticles-tremendous therapeutic potential: a review. Int J PharmTech Res 1:1020–1027

Maravajhala V, Papishetty S, Bandlapalli S (2012) Nanotechnology in development of drug delivery system. Int J Pharm Sci Res 3:84–96

Maruthi G, Smith AA, Manavalan R (2011) Nanoparticles a review. J Adv Sci Res 2:12–19

Mitchell MJ, Billingsley MM, Haley RM et al (2021) Engineering precision nanoparticles for drug delivery. Nat Rev Drug Discov 20:102–124

Mohanraj VJ, Chen Y (2006) Nanoparticles—a review. Trop J Pharm Res 5:561–573

Mudshinge SR, Deore AB, Patil S et al (2011) Nanoparticles: emerging carriers for drug delivery. Saudi Pharm J 19:129–141

Murthy NS, Mathew A (2004) Cancer epidemiology, prevention and control. Curr Sci 86:518–527

Natalie PP, Mandal TK (2007) Engineered nanoparticles in cancer therapy. Recent Patent Drug Deliv Formul 1:37–51

Neerooa BN, Ooi L, Shameli K et al (2021) Development of polymer assisted nanoparticles and nanogels for cancer therapy: an update. Gels 7:1–30

Nesalin JA, Smith AA (2013) Preparation and evaluation of stavudine loaded chitosan nanoparticles. J Pharm Res 6:268–274

Nguyen KT (2011) Targeted nanoparticles for cancer therapy: promises and challenges. J Nanomed Nanotechnol 2:1–2

Ozturk N, Kara A, Vural I (2020) Formulation and *in vitro* evaluation of telmisartan nanoparticles prepared by emulsion solvent evaporation technique. Turk J Pharm Sci 17:492–499

Pandey P, Dua K, Dureja H (2019) Erlotinib loaded chitosan nanoparticles: formulation, physicochemical characterization and cytotoxic potential. Int J Biol Macromol 139:1304–1316

Patel A, Khanna S, Xavier GK et al (2017) Polymeric nanoparticles for tumor targeting a review. Int J Drug Dev Res 9:50–59

Pattanayek S, Puranik SB (2018) Development and evaluation of ketoprofen loaded nanoparticles. Int J Pharma Pharm Res 11:225–237

Paus C, Voort RV, Cambi A (2021) Nanomedicine in cancer therapy: promises and hurdles of polymeric nanoparticles. Explor Med 2:167–185

Peer D, Karp JM, Hong S et al (2007) Nanocarriers as an emerging platform for cancer therapy. Nat Nanotechnol 2:751–760

Pieper S, Onafuye H, Mulac D et al (2019) Incorporation of doxorubicin in different polymer nanoparticles and their anticancer activity. Beilstein J Nanotechnol 10:2062–2072

Plapied L, Duhem N, Rieux A (2011) Fate of polymeric nanocarriers for oral drug delivery. Curr Opin Colloid Interface Sci 16:228–237

Popiolski TM, Otsuka I, Halila S et al (2016) Preparation of polymeric micelles of poly(ethylene oxide-b-lactic acid) and their encapsulation with lavender oil. Mater Res 19:1356–1365

Pucci C, Martinelli C, Ciofani G (2019) Innovative approaches for cancer treatment: current perspectives and new challenges. Cancer Med Sci 961:1–13

Qiao W, Wang B, Wang Y et al (2010) Cancer therapy based on nanomaterials and nanocarrier systems. J Nanomater 2010:1–9

Rajan M, Raj V (2013) Potential drug delivery applications of chitosan based nanomaterials. Int Rev Chem Eng 5:145–155

Ram AS, Gowda DV, Srivastava A et al (2016) Development and characterization of glipizide-loaded biodegradable nanoparticles by solvent displacement technique. ASEM 8:1001–1006

Rao JP, Geckeler KE (2011) Polymer nanoparticles: preparation techniques and size-control parameters. Prog Polym Sci 36:887–913

Rawat M, Singh D, Saraf S et al (2006) Nanocarriers: promising vehicle for bioactive drugs. Biol Pharm Bull 29:1790–1798

Reis CP, Neufeld RJ, Ribeiro AJ et al (2006) Nanoencapsulation I. methods for preparation of drug-loaded polymeric nanoparticles. Nanomed Nanotechnol Biol Med 2:8–21

Ren X, Zhang P, Chen Z (2020) Dialysis preparation of smart redox and acidity dual responsive tea polyphenol functionalized calcium phosphate nanospheres as anticancer drug carriers. Molecules 25:1–11

Ru G, Han L, Qing J et al (2016) Effects of borneol on the pharmacokinetics of 9-nitrocamptothecin encapsulated in PLGA nanoparticles with different size via oral administration. Drug Deliv 23: 3417–3423

Ruman U, Buskaran K, Pastorin G et al (2021) Synthesis and characterization of chitosan based nanodelivery systems to enhance the anticancer effect of sorafenib drug in hepatocellular carcinoma and colorectal adenocarcinoma cells. Nanomaterials 11:1–26

Sailaja PB, Jyothi JB (2020) Development and *in vitro* evaluation of 5-fluorouracil nanoparticles by salting out technique. Asian J Pharm Clin Res 13:166–171

Sailaja A, Amareshwar P, Chakravarty P (2011) Different techniques used for the preparation of nanoparticles using natural polymers and their application. Int J Pharm Pharm Sci 3:45–50

Salatin S, Barar J, Barzegar-Jalali M et al (2017) Development of a nanoprecipitation method for the entrapment of a very water soluble drug into eudragit RL nanoparticles. Res Pharm Sci 12:1–14

Santos KC, Silva MF, Pereira-Filho ER (2012) Polymeric nanoparticles loaded with the 3, 5, 3′-triiodothyroacetic acid (Triac), a thyroid hormone: factorial design, characterization and release kinetics. Nanotechnol Sci Appl 5:37–48

Sanvicens N, Marco MP (2008) Multifunctional nanoparticles–properties and prospects for their use in human medicine. Trends Biotechnol 26:425–433

Sarei F, Dounighi NM, Zolfagharian H et al (2013) Alginate nanoparticles as a promising adjuvant and vaccine delivery system. Indian J Pharm Sci 75:442–449

Senthilnathan B, Maheswaran A, Gopalasatheeskumar K et al (2016) Formulation and evaluation of pregabalin loaded eudragit S100 nanoparticles. Int J Eng Technol Sci 6:64–70

Seyfried TN, Shelton L (2010) Cancer as a metabolic disease. Nutr Metab 7:1–7

Sharma N, Singhal M, Kumari RM et al (2020) Diosgenin loaded polymeric nanoparticles with potential anticancer efficacy. Biomol Ther 10:1679–1685

Sivanesan D, Verma RS, Prasad E (2021) 5FU encapsulated polyglycerol sebacate nanoparticles as anti-cancer drug carriers. RSC Adv 11:18984–18993

Solanki N, Dureja H (2018) Impact of surfactants on formulation parameters and *in vitro* cytotoxicity of boswellic acids loaded nanoparticles on human colon cancer cell lines. Indian J Pharm Educ Res 52:S229–S236. https://doi.org/10.5530/ijper.52.4s.102

Solanki N, Mehta M, Chellappan DK et al (2020) Antiproliferative effects of boswellic acid-loaded chitosan nanoparticles on human lung cancer cell line A549. Future Med Chem 12:2019–2034

Sonia TA, Sharma CP (2011) Chitosan and its derivatives for drug delivery perspective. Adv Polym Sci 243:23–54

Soppimath KS, Aminabhavi TM, Kulkarni AR et al (2001) Biodegradable polymeric nanoparticles as drug delivery devices. J Control Release 70:1–20

Subjakova V, Oravczova V, Hianik T (2021) Polymer nanoparticles and nanomotors modified by DNA/RNA aptamers and antibodies in targeted therapy of cancer. Polymers 13:1–41

Sukmawati A, Utami W, Yuliani R et al (2018) Effect of tween 80 on nanoparticle preparation of modified chitosan for targeted delivery of combination doxorubicin and curcumin analogue. Mater Sci Eng A 311:1–5

Sundar S, Kundu J, Kundu SC (2010) Biopolymeric nanoparticles. Sci Technol Adv Mat 11:1–13

Surendiran A, Sandhiya S, Pradhan SC et al (2009) Novel applications of nanotechnology in medicine. Indian J Med Res 130:689–701

Sutradhar KB, Amin ML (2014) Nanotechnology in cancer drug delivery and selective targeting, vol 2014. ISRN Nanotechnol, pp 1–12
Tan QY, Xu ML, Wu JY et al (2012) Preparation and characterization of poly (lactic acid) nanoparticles for sustained release of pyridostigmine bromide. Die Pharmazie Int J Pharm Sci 67:311–318
Tang M, Lei L, Guo S et al (2010) Recent progress in nanotechnology for cancer therapy. Chin J Cancer 29:775–780
Tiruwa R (2016) A review on nanoparticles-preparation and evaluation parameters. Indian J Pharm Biol Res 4:27–31
Tiyaboonchai W (2003) Chitosan nanoparticles: a promising system for drug delivery. Naresuan Univ J 11:51–66
Umerska A, Gaucher C, Oyarzun-Ampuero F et al (2018) Polymeric nanoparticles for increasing oral bioavailability of curcumin. Antioxidants 4:1–18
Vauthier C, Bouchemal K (2009) Methods for the preparation and manufacture of polymeric nanoparticles. Pharm Res 26:1025–1058
Wathoni N, Meylina L, Rusdin A et al (2021) The potential cytotoxic activity enhancement of α-Mangostin in chitosan-kappa carrageenan-loaded nanoparticle against MCF-7 cell line. Polymers 13:1–13
Wright L, Rao S, Thomas N et al (2018) Ramizol® encapsulation into extended release PLGA micro-and nanoparticle systems for subcutaneous and intramuscular administration: *in vitro* and *in vivo* evaluation. Drug Dev Ind Pharm 44:1–7
Yadav KS, Sawant KK (2010) Modified nanoprecipitation method for preparation of cytarabine-loaded PLGA nanoparticles. AAPS PharmSciTech 11:1456–1465
Yu Z, Gao L, Chen K et al (2021) Nanoparticles: a new approach to upgrade cancer diagnosis and treatment. Nanoscale Res Lett 16:88
Zaid H, Rayan A, Said O et al (2010) Cancer treatment by Greco-Arab and Islamic herbal medicine. Open Nutraceut J 3:203–212
Zhang JL, Lan CQ, Post M et al (2006) Design of nanoparticles as drug carriers for cancer therapy. Cancer Genomics Proteomics 3:147–157
Zhao Z, Li Y, Zhang Y (2015) Preparation and characterization of paclitaxel loaded SF/PLLA-PEG-PLLA nanoparticles via solution-enhanced dispersion by supercritical CO_2. J Nanomater 201:1–7
Zielinska A, Carreiro F, Oliveira AM et al (2020) Polymeric nanoparticles: production, characterization, toxicology and ecotoxicology. Molecules 25:1–20

Carbon Nanotubes: A Review of Toxicity and Applicability in Biomedical Applications

22

Taranpreet Kaur and Raman Preet Singh

Abstract

The commercial production of carbon nanotubes (CNTs) and research in their biomedical applications have increased enormously in the past decade. Despite the augmentation in demand and a concomitant increase in commercial supply of CNTs, the feasibility of their biomedical applications has been compounded by reports of adverse effects. This chapter essentially reviews four aspects of CNTs from a biomedical application point of view. The first part focuses on in vitro and in vivo toxicity of carbon nanotubes, while the second part emphasizes the mechanisms of cellular uptake and toxicity. The third part highlights the experimental parameters that influence toxicity and efforts to reduce untoward effects by tweaking these parameters as well as CNTs. The last part concentrates on the regulatory aspects of CNT applications and tries to bring to the fore the regulatory challenges that CNTs are facing and will continue to face in the foreseeable future.

Keywords

Biomedical applications · Carbon nanotubes · Toxicity · Nanotoxicology

T. Kaur
Department of Biotechnology, Government Mohindra College, Patiala, Punjab, India

R. P. Singh (✉)
Department of Pharmacy, Government Polytechnic College, Bathinda, Punjab, India

Department of Pharmaceutical Sciences, Government Polytechnic College for Girls, Patiala, Punjab, India
e-mail: rpsingh@gpcgpatiala.edu.in

P. P. Singh (ed.), *Recent Advances in Pharmaceutical Innovation and Research*,
https://doi.org/10.1007/978-981-99-2302-1_22

22.1 Introduction

Carbon-based nanoparticles (NPs) are the most widely used form of NPs. These NPs are composed of carbon, which is a hollow sphere, ellipsoid or tube-shaped. The spherical, cage-like structures were the first to be discovered and are called fullerenes or buckyballs (Kroto et al. 1985; Kroto 1992). Originally, fullerenes were discovered to compose 60 carbon atoms but later fullerenes of other sizes were also discovered (Terrones and Terrones 2003). Later, cylindrical structures of carbon were also discovered and called carbon nanotubes (CNTs) (Iijima 1991). The CNTs resemble sheets of graphite, rolled to form hollow tubular structures. The wall of CNTs may either be one atom thick (single-walled CNTs; SWCNTs) or may consist of concentric tubes where each tube is one atom in thickness (multi-walled CNTs; MWCNTs).

CNTs possess excellent chemical and electronic properties and are found to be useful in electronics and other industries. However, during the past few years, the chemical and biological applications of CNTs have also been investigated. The CNTs were proposed to be used as drug carriers and biological carriers (Pantarotto et al. 2004a, b; Bianco et al. 2005a, b; Wu et al. 2005; Dumortier et al. 2006; Klumpp et al. 2006). As the commercial production of these materials increases, the safety concerns due to occupational exposure of workers increase (Kuempel 2011; Dahm et al. 2012; Ling et al. 2012; Castranova et al. 2013; Erdely et al. 2013; Hedmer et al. 2014). The present review focuses on the reported toxic potential of CNTs, mechanism of toxicity, factors modifying their toxicity and challenges in toxicity screening of these particles.

22.2 Toxicity of CNTs

In vitro and in vivo models have been extensively employed to understand the cytotoxicity of CNTs. Depending on the model system used (cultured cells or animal models), significant insights can be gained into different aspects of CNT toxicity as summarized in Fig. 22.1. In general, low doses of CNTs have been demonstrated to elicit toxic reactions under in vivo conditions. Contrastingly, relatively higher, generally supraphysiological, concentrations of CNTs are required to elicit toxic reactions under in vitro conditions. Based on reports from our laboratory, more than tenfold higher concentrations are required under in vitro conditions to elicit a toxic response than required under in vivo conditions (Singh et al. 2012).

22.2.1 Pulmonary Toxicity

CNTs are fluffy powders and can escape into the environment during manufacturing and handling procedures. The CNTs may disperse in the air and show pulmonary toxicity when inhaled. SWCNTs can be released into the air during handling and may also deposit on the gloves (Maynard et al. 2004). MWCNTs can also be released into the environment during manufacturing and handling (Maynard et al.

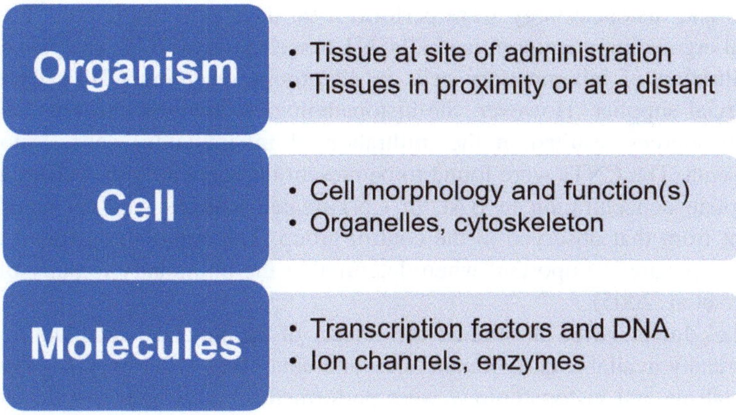

Fig. 22.1 Parameters studied at different levels of biological organization

2007). Murr and coworkers reported the presence of MWCNT aggregates in kitchens using natural gas or propane. These CNTs were found to be similar to a commercial MWCNT sample that exhibited cytotoxicity in a murine alveolar macrophage cell line (RAW 267.9 cells). The authors proposed that MWCNTs released from the combustion of gases may contribute to the development of asthma (Murr et al. 2005). In a later study, the group reported the presence of CNTs in indoor air (produced by gas combustion) and outdoors and correlated it to the incidence of asthma. The outdoor and indoor number concentrations of CNTs were found to be 10^2–10^3/m^3 and 10^3–10^5/m^3, respectively (Murr et al. 2006). The MWCNTs could also be detected during the combustion of various types of fuels and can induce toxicity in a mouse lung macrophage cell line (Murr et al. 2005). Similarly, combustion-derived soot containing MWCNTs showed cytotoxicity, cytokine release and free radical production (Garza et al. 2008).

22.2.1.1 Guinea-Pig Model

Huczko et al. were the first to determine the effects of CNTs in animals. In their study, soot containing CNTs was prepared by an arc-discharge method. Guinea pigs received 25 mg soot containing CNTs by intratracheal instillation. The control group received 25 mg of CNT-free soot. The respiratory parameters (tidal volume, breathing frequency and lung resistance) and bronchoalveolar lavage (BAL) were analysed 4 weeks post-instillation. Treatment with CNT-containing soot did not show a significant change in respiratory parameters or BAL parameters (total and differential counts and total protein concentration) as compared to the control group indicating the non-toxic nature of CNTs. However, the group did not report lung histopathology and type (SWCNT and MWCNT) and content of CNTs in soot (Huczko et al. 2001). In a later study, the group administered MWCNTs prepared by two methods: chemical vapor deposition and carbon arc discharge. Apart from these, commercially available CNTs from various manufacturers were also used. Guinea pigs received 15 mg CNTs intratracheally. The lung resistance, BAL analysis and histopathology were performed 90 days post-instillation. The group reported organizing pneumonitis in all CNT-treated groups. BAL analysis revealed the infiltration of inflammatory cells in all groups receiving CNTs, except one commercial supplier. However, the histopathological analysis showed that CNTs from all sources resulted in the infiltration of inflammatory cells, lesions and emphysema. The CNTs were found to be present intracellularly as well as in BAL. The protein concentration in BAL of CNT-treated animals was not significantly different from that observed in the control group. The authors concluded that the time of exposure is important when determining the pulmonary toxicity of CNTs (Huczko et al. 2005).

Grubek-Jaworska et al. studied the effects of intratracheal instillation of four commercially available carbon nanotubes and nanofibres (12.5 mg) in guinea pigs. BAL analysis and histopathology were performed 90 days after instillation. BAL analysis showed an increase in interleukin-8 (IL-8) after installation of MWCNTs in 3 out of 4 samples of CNTs studied, while there was no increase in protein concentration. However, pulmonary lesions and bronchiolitis obliterans organizing pneumonia were observed in all samples, though the intensity of lesions differed with samples (Grubek-Jaworska et al. 2006).

22.2.1.2 Mouse Model

Experimental studies in mice also showed the pulmonary inflammation and toxicity associated with CNTs. The experimental exposure of mice to CNTs was achieved by intratracheal instillation, intranasal instillation and pharyngeal aspiration or inhalation of CNTs.

Lam et al. reported that mice received 0.1 and 0.5 mg of SWCNTs by intratracheal instillation. In this study, three types of SWCNTs were used: raw HiPco CNTs containing high iron content (27%), purified HiPco CNTs with low iron content (2%) and commercially available SWCNTs containing high nickel content (26%). The lungs of mice were examined by histopathology on day 7 or day 90 post-instillation. The SWCNT instillation resulted in inflammation and lesions in mice lungs; the severity of inflammation increased with the dose of SWCNTs as well as time. However, there were no signs of inflammation observed in mice that received carbon black (negative control). On the other hand, SWCNTs appeared to have higher pro-inflammatory potential as compared to quartz (positive control) on a weight basis (Lam et al. 2004). Chou et al. demonstrated that intratracheal instillation of SWCNT (0.5 mg) in mice resulted in pulmonary inflammation and macrophage activation. Microarray analysis of macrophages revealed that SWCNT uptake by macrophages activates nuclear factor B (NF-κB) and activator protein 1 (AP-1). This leads to the production of cytokines, oxidative stress, infiltration and activation of inflammatory cells and expression of protective and antiapoptotic genes. These changes at the genetic level may account for the observed inflammatory response towards SWCNTs (Chou et al. 2008). Intranasal instillation of SWCNTs and MWCNTs in mice also resulted in worsening of the airway hyper-responsiveness and caused inflammatory cell infiltration in the lungs (Hamilton et al. 2007).

Shvedova et al. described an aspiration method for pulmonary delivery of CNTs. The mice were administered SWCNTs at doses of 10, 20 and 40 μg/mouse. The SWCNTs used for the study were >99% pure and contained 0.23% iron. The mice were sacrificed on days 1, 3, 7, 28 and 60 post-aspiration. The analysis of BAL revealed a dose-dependent increase in markers of cellular toxicity (lactate dehydrogenase; LDH and total protein levels), inflammation (transforming growth factor-beta; TGF-β1, tumour necrosis factor-alpha; TNF-α, interleukin 1beta; IL-1β) and oxidative stress (4-hydroxynonenal). This was accompanied by increased activity of γ-glutamyltranspeptidase activities and reduced glutathione levels in the lungs. Histopathological analysis revealed the presence of granulomas and fibrosis in the lungs of mice. Further, SWCNT resulted in a dose-dependent increase in expiratory time (Shvedova et al. 2005). In a later study, the group showed that mice fed with a vitamin-E deficient diet showed a 90-fold reduction in α-tocopherol levels in the lungs. This was accompanied by a reduction in glutathione and ascorbate and increased levels of malondialdehyde. The lowered antioxidant levels were associated with increased levels of cellular and biochemical markers of inflammation and cytotoxicity (Shvedova et al. 2007). The role of Nicotinamide adenine dinucleotide phosphate (NADPH) oxidase in SWCNT-induced pulmonary inflammation was studied in NAPDH knock-out mice. The knock-out mice showed an increase in inflammatory markers and apoptosis. This was accompanied by a decrease in anti-inflammatory cytokines, pro-fibrotic cytokines and collagen deposition. These results indicate that NAPDH-generated reactive oxygen species (ROS) may play a crucial role in switching from acute to chronic inflammation (Shvedova et al. 2008c). Saxena et al. reported the in vivo toxicity of SWCNTs functionalized by acid treatment. The acid-functionalized SWCNTs (AF-SWCNT) were negatively charged and were well dispersed in an aqueous medium. The functionalized SWCNTs were found to produce a strong inflammatory response in mice after pharyngeal aspiration of AF-SWCNT as compared to pristine SWCNT (Saxena et al. 2007). The pulmonary toxicity studies reported the presence of agglomerates in lungs surrounded by inflammatory cells. Mercer et al. determined the pulmonary toxicity of uniformly dispersed SWCNTs in mice. Mice were administered SWCNTs (10 μg/mouse) by pharyngeal aspiration. The SWCNTs were found to be dispersed 1 day after administration. The pulmonary toxicity was assessed 1 h, 1 day, 7 days and 1 month post-instillation. BAL and histopathology showed an initial increase followed by a fall in inflammatory parameters. There was no evidence of granulomatous lesions or intramacrophage SWCNTs. There was a time-dependent increase in collagen deposition (Mercer et al. 2008).

Inhalation of SWCNTs (5 mg/m^3, 5 h/day for 4 days) resulted in inflammation, granulomas, fibrosis and oxidative stress. When compared to pharyngeal aspiration, inhalation induced a higher degree of inflammation (Mercer et al. 2008). Similarly, inhalation of MWCNTs was also reported. Mice were exposed to CNTs (0.3, 1 or 5 mg/m^3) in a whole-body inhalation chamber 6 h/day for 7 or 14 days. No signs of inflammation or cytotoxicity were observed in BAL or by histopathology. There was no change in gene expression of cytokine (IL-6 and IL-10) and oxidative stress indicator (NAD(P)H oxidoreductase 1) genes in the lungs. The alveolar macrophages were found to contain CNTs in BAL as well as in lung sections. Further, immunosuppression was observed on day 14 but not on day 7. There was a reduction in T-cell-dependent antibody response to antigen (sheep erythrocytes), reduced mitogen-stimulated T-cell proliferation, reduction in NK cell activity and increased expression of IL-10 and NAD(P)H oxidoreductase 1 mRNA in the spleen (Mitchell et al. 2007).

22.2.1.3 Rat Model

The pulmonary inflammation and toxicity observed in mice are also shown in rats. In the study by Warheit et al. rats received intratracheal instillation of SWCNTs (1 and 5 mg/kg). The inflammatory reaction of SWCNTs was examined after 1, 7, 30 and 90 days after SWCNT instillation. The authors reported a transient increase in markers of inflammatory response and cell injury in BAL after 24 h. Histopathological investigation of lungs revealed the presence of multifocal granulomas. The granulomas were found to be independent of dose and were non-uniform and non-progressive after 30 days. The granulomas consisted of SWCNTs at the core with macrophages surrounding the CNTs (Warheit et al. 2004, 2005). The two prior studies showed different results with regard to the inflammatory activity of SWCNTs—the severity of inflammation increased with time in a study by Lam et al. (2004), while time-dependency of inflammation was not observed by Warheit et al. (2004).

Muller et al. reported the pulmonary toxicity of MWCNTs in rats following intratracheal instillation of 0.5, 2 and 5 mg doses. The group used 2 types of MWCNTs: unground MWCNTs and ground MWCNTs. The study measured the biopersistence, inflammation and lung pathology after the instillation of CNTs. Biopersistence of CNTs was measured up to 60 post-instillation and the group reported that nearly 80% of unground CNTs and 40% of ground CNTs remained in the lungs on day 60. BAL analysis on days 3 and 15 revealed inflammatory cell infiltration and increased levels of markers of cellular toxicity (LDH and total protein) and inflammation (TNF-α). Histopathological examination of lungs, on 60-day post-instillation, revealed the presence of MWCNTs in the lungs. This was accompanied by pulmonary inflammation and granulomas along with fibrosis due to the presence of large CNT aggregates. The unground CNTs were observed mainly in the bronchial lumen with few particles observed in the parenchymal region. The unground CNTs induced granulomas that protruded in the bronchial lumen along with inflammation in the surrounding regions. On the other hand, ground CNTs were better dispersed in the parenchymal and interstitium region and induced granulomas. Biochemical evaluation of the lungs revealed the production of markers of inflammation (TNF-α) and fibrosis (hydroxyproline and soluble collagen) (Muller et al. 2005).

Liu et al. exposed rats to MWCNTs (1, 3, 5 and 7 mg/kg) by intratracheal administration, and lung histopathology was performed on 1 day, 1 week, 1 month and 3 months after exposure. The rats showed an increased degree of inflammation with an increase in the dose of CNTs. Transmission electron microscopy (TEM) analysis revealed that after 3 months CNTs were present in capillary vessels and cytoplasmic vacuoles of pulmonary tissues indicating the capacity of CNTs to cross capillary vessels and enter cytoplasmic vacuoles (Liu et al. 2008a). Intratracheal administration was also accompanied by a dose-dependent increase in markers of cytotoxicity, inflammation and oxidative stress with a concomitant decrease in antioxidant defence mechanisms (Liu et al. 2008b).

Mangum et al. studied the fibrogenic potential of SWCNTs in rats. The rats were administered SWCNTs (produced by chemical vapour deposition) by oropharyngeal aspiration. Histopathological examination and BAL analysis did not show signs of inflammation on days 1 and 21 post-treatment, although small focal fibrotic lesions were observed on day 21. SWCNT treatment showed an increase in cell proliferation in these lesions. Further, 'carbon bridges' were observed between macrophages present in BAL as well as between macrophages present in lungs. An increase in mRNA expression of platelet-derived growth factor (PDGF)-A, PDGF-B and PDGF-C was observed on day 1. However, no change in the mRNA expression of PDGFs was observed on day 21 (Mangum et al. 2006). In recent years, pulmonary toxicity studies have resulted in the identification of the gene, protein and metabolic signatures contributing to inflammation and other effects induced by CNTs (Migliaccio et al. 2021; Ndika et al. 2021; Seidel et al. 2021; Wang et al. 2021; Wils et al. 2021a, b; Hieronimus et al. 2022; Hojo et al. 2022; Pei et al. 2022; Sager et al. 2022; Saleh et al. 2022; Salih and Ghobadi 2022). Additionally, the pulmonary toxicity of CNTs has also been linked to gut microflora (Bhattacharya et al. 2022).

22.2.1.4 In Vitro Models

When the CNTs enter airways, they come in contact with two types of cells: epithelial cells and alveolar macrophages. The pulmonary inflammation and cytotoxicity observed in experimental models of pulmonary inflammation have also been documented in vitro in lung cells and alveolar macrophages. Magrez et al. demonstrated in vitro toxicity of MWCNTs, carbon nanofibres and carbon nanoparticles in a lung tumour cell line. The group reported that the NPs were toxic and observed that the toxicity was size-dependent. Further, in vitro toxicity was higher with acid-functionalized NPs (Magrez et al. 2006). Similar results showing higher toxicity were also reported for LA4 murine epithelial cells. The functionalized CNTs were more toxic and inhibited cell cycling of the cells. However, neutralization of the negative charge of functionalized CNTs by prior treatment with poly-L-lysine abrogated the enhanced toxic effect. Pristine and functionalized CNTs showed induction of apoptosis to a similar extent (Saxena et al. 2007). This is by the higher inflammatory potential of acid-functionalized CNTs following pharyngeal aspiration in mice (Saxena et al. 2007).

The SWCNTs and MWCNTs have been shown to exert cytotoxicity towards lung cells (A549 cell line). SWCNTs were reported to show mild toxicity towards A549 cells. Further, rat alveolar macrophages (NR8383), when incubated with CNTs, showed no change in biochemical markers of inflammation. However, there was an increase in ROS production and a decrease in mitochondrial membrane potential with CNTs in A549 and NR8383 cells (Davoren et al. 2007; Pulskamp et al. 2007). Other studies have also demonstrated little or no toxicity of SWCNTs towards A549 cells (Worle-Knirsch et al. 2006; Belyanskaya et al. 2007; Pulskamp 2008). The SWCNTs were detected intracellularly, induced ROS production and activated the NF-κB pathway (Pulskamp 2008). Similarly, MWCNTs also showed mild toxicity in A549 cultures, released IL-6 and increased ROS production (Garza et al. 2008). MWCNTs were shown to be present intracellularly and the toxicity was found to be independent of CNT length and residual metal catalysts (Simon-Deckers et al. 2008).

SWCNTs stimulated ROS production in rat lung epithelial cells in a dose- and time-dependent manner. Further, the mitochondria did not appear to be a source of enhanced ROS production. The enhanced ROS production was associated with depletion of glutathione stores and a decrease in superoxide dismutase (Sarkar et al. 2007).

SWCNTs and MWCNTs also showed disruption of the airway barrier (Yacobi et al. 2007; Rotoli et al. 2008). SWCNT exposure showed no change in LDH levels in primary rat alveolar epithelial monolayers indicating a lack of toxicity. Transmonolayer resistance was also reduced after 24 h of SWCNT incubation (Yacobi et al. 2007). Similarly, transepithelial electrical resistance was reduced in human airway epithelial cell (Calu-3) monolayers along with an increase in mannitol permeability. The adverse effect of CNTs on airway barrier properties was more with MWCNT treatment as compared to SWCNT treatment. The viability of the cells remained unchanged during the study (Rotoli et al. 2008).

In real-life situations, pulmonary exposure to nanoparticles may be accompanied by exposure to environmental contaminants, pollutants and microbes. The effect of CNTs was evaluated in presence of *Listeria monocytogenes*. SWCNTs resulted in a dose-dependent decrease in *L. monocytogenes* clearance from the lungs (Shvedova et al. 2005). In a later study, the group showed that pre-exposure of SWCNTs, followed by *L. monocytogenes* infection, mounted a severe inflammatory response but was associated with a reduction in *L. monocytogenes* clearance as compared to PBS pre-exposed mice. The pre-exposure to SWCNT resulted in reduced phagocytosis and reduced nitric oxide production by macrophages. However, there was a persistent increase in biochemical and cellular markers of inflammation and toxicity in BAL (Shvedova et al. 2008a). However, intratracheal administration of CNTs with lipopolysaccharide (LPS; a component of the bacterial cell wall) resulted in enhanced production of inflammatory markers in mice; the effect was more pronounced in SWCNTs as compared to MWCNTs (Inoue et al. 2008). These effects were attributed to increased PDGF signalling (Cesta et al. 2010). Similarly, SWCNTs and MWCNTs also exacerbated ovalbumin-induced airway inflammation through activation of Th response, induction of oxidative stress and production of inflammatory markers (Inoue et al. 2009, 2010; Grecco et al. 2011; Yamaguchi et al. 2012; Nygaard et al. 2013; Sayers et al. 2013). Further, pulmonary administration of SWCNTs and MWCNTs resulted in worsening of airway allergy in ovalbumin-treated mice (Nygaard et al. 2009; Ryman-Rasmussen et al. 2009; Mizutani et al. 2012). Contrastingly, pharyngeal aspiration of SWCNTs followed by systemic infection with *Toxoplasma gondii* in mice did not affect the host immune response to the parasite (Swedin et al. 2012). Han et al. administered MWCNT by pharyngeal aspiration (20 μg/mouse) followed by exposure to ozone. The degree of cytotoxicity and inflammation was lower in mice that received CNTs and ozone as compared to CNTs alone (Han et al. 2008).

22.2.2 Cardiovascular Toxicity

SWCNTs exhibited toxicity in a rat heart cell line as evident by morphological changes, reduced cell viability and apoptosis induction (Garibaldi et al. 2006). On the other hand, vascular adventitial fibroblasts incubated with SWCNTs exhibited enhanced transformation into myofibroblasts with no significant effect on cell viability (Lin et al. 2012), while higher cell viability was observed in cardiac myocytes that were cultured on MWCNT supports (Martinelli et al. 2012). SWCNTs also enhanced oxidative stress in rat aorta endothelial cells and vascular adventitial fibroblasts with changes in the expression of several signalling proteins (Zhiqing et al. 2010; Cheng et al. 2011, 2012; Lin et al. 2012). SWCNTs showed no change in myofibrillar structure or ROS levels in neonatal rat ventricular cardiomyocytes. The effect on impulse conduction was also minimal (Helfenstein et al. 2008). SWCNTs were also found to inhibit the growth of rat aortic smooth muscle cells (Raja Pavan et al. 2007). SWCNTs and MWCNTs showed concentration-dependent cytotoxicity along with a reduction in tubule formation and disruption of actin filament and VE-cadherin in human aortic endothelial cells (Walker et al. 2009). Further, the SWCNTs can also interact with ion channels and may exert untoward effects on ion channel conductance (Park et al. 2003).

Intratracheal instillation of SWCNTs in rats was associated with a significant reduction in arterial baroreflex sequences suggesting pulmonary exposure to CNTs may influence cardiac autonomic regulation (Legramante et al. 2009). Further, pulmonary instillation of MWCNTs resulted in increased coronary vasoconstriction and exacerbation of cardiac injury, while inhalation of MWCNTs resulted in a partially reversible reduction in endothelium-dependent vasodilation (Stapleton et al. 2012; Thompson et al. 2014). Intravenously administered MWCNTs caused increased calcification of the aorta and damaged the aortic epithelium in rats (Xu et al. 2012). The pharyngeal aspiration of SWCNTs in mice was associated with mitochondrial DNA damage and protein carbonyl formation in the aorta after 7, 28 and 60 days post-aspiration. This was attributed to an alteration in the expression of MCP-1 and VCAM-1 in the heart (Bianco et al. 2005a). In a later study, ApoE−/− mice received 20 μg SWCNT on alternate weeks for 8 weeks by pharyngeal aspiration. After 8 weeks, histopathological examination revealed pulmonary inflammation and fibrotic response. This was associated with an increase in plaque area in aortas along with increased atherosclerotic lesion size in brachiocephalic arteries. The authors proposed that the cardiac effects of pulmonary administered CNTs may arise because of inflammatory mediators released from the lungs. Alternatively, the CNTs may translocate from the lungs into the systemic circulation (Li et al. 2006, 2007b). Similarly, ApoE−/− mice received 25.6 μg MWCNT/week for 5 weeks by intratracheal instillation. The exposure resulted in pulmonary inflammation, oxidative stress and vascular activation response genes. In vitro studies conducted in human endothelial cells and in monocyte-endothelial co-cultures exhibited changes characteristic of atherosclerosis progression (Cao et al. 2014). Pharyngeal aspiration of CNTs (SWCNTs and MWCNTs) also increased the expression of cell adhesion molecules, IL-6 and HO-1 in the aorta. Further, an increased expression of arginase II, MT-1 and Hif-3α was observed in the heart and aorta (Erdely et al. 2009). In a later study, the authors reported an increase in several markers of inflammation in blood following pharyngeal aspiration exposure in mice (Erdely et al. 2011). The oropharyngeal aspiration of MWCNTs also worsened cardiac ischaemia/reperfusion injury in mice (Urankar et al. 2012).

An increase in plasma angiotensin-I converting enzyme levels was observed in spontaneously hypertensive rats suggesting the ability of SWCNTs to induce peripheral vascular thrombosis following intratracheal instillation (Ge et al. 2012). Intratracheal administration of MWCNTs in mice resulted in platelet activation as evidenced by circulating platelet-leukocyte conjugates. Later, an increase in plasma procoagulant microvesicular tissue factor activity was observed along with increased peripheral thrombogenicity dependent on P-selectin (Nemmar et al. 2007). Radomski et al. demonstrated that CNTs can stimulate ferric chloride-induced platelet aggregation and enhance vascular thrombosis in rat carotid arteries. The SWCNTs were found to be more effective than MWCNTs. CNTs upregulated the expression of GPIIb/IIIa in platelets and affected platelet aggregation pathways dependent on thromboxane, ADP, matrix metalloproteinase and protein kinase C (Radomski et al. 2005).

22.2.3 Dermal Toxicity

In a study to determine dermal toxicity of CNTs, 40 human volunteers with susceptibilities to various allergic susceptibilities were subjected to a patch test and monitored for 96 h. The CNTs used in the study were prepared by arc discharge method and soot containing CNTs was used for the patch test. After 96 h, no signs of irritation or allergy were observed. The results were confirmed in rabbit which received suspension of CNT-containing soot in one eye while the other eye served as control. No signs of irritation were observed till 96 h. These results, in humans and rabbit, indicate the absence of dermal toxicity and irritancy of CNTs (Huczko and Lange 2001). Similarly, SWCNTs and MWCNTs were found to be non-irritant in rabbits and devoid of skin sensitization potential in guinea pigs (Kishore et al. 2009; Ema et al. 2011). Further, subcutaneous implantation of MWCNTs in mice showed lower carcinogenicity as compared to carbon black (Takanashi et al. 2012).

Contrary to the in vivo studies, in vitro studies suggest the dermal toxicity of CNTs. Manna et al. demonstrated the in vitro toxicity of SWCNTs towards human keratinocytes. The SWCNTs inhibited the growth of human keratinocytes and increased free radical production. The mechanism of toxicity appeared to be dependent on protein kinase activity leading to activation of the nuclear factor NF-κB (Manna et al. 2005). SWCNTs showed a reduction in cell viability along with enhanced production of free radicals in human keratinocytes and foreskin cells

(Shvedova et al. 2003; Sarkar et al. 2007). Similarly, SWCNTs also induced oxidative stress and inflammation in various in vitro models (Murray et al. 2009). Further, 6-aminohexanoic acid-derivatized SWCNTs were also found to show toxicity towards human epidermal keratinocytes (HEK) and showed an increase in IL-8 levels. The SWCNTs were also detected in cytoplasmic vacuoles (Zhang et al. 2007). Sayes et al. demonstrated the toxicity of functionalized SWCNTs towards human dermal fibroblasts and compared the toxicity of SWNT-phenyl-SO(3)H and SWNT-phenyl-SO(3)Na with pristine SWCNTs. The toxicity of SWCNTs was inversely related to the degree of functionalization. The functionalized SWCNTs interacted with cell membrane and deposited on membrane (Sayes et al. 2006).

MWCNTs have also been found to be toxic towards keratinocytes. The MWCNTs can induce the release of IL-8 from HEKs. Further, the MWCNTs were also observed in cytoplasmic vacuoles (Monteiro-Riviere et al. 2005b). The authors also reported the effect of surfactants on toxicity of MWCNTs and observed that Pluronic F127 was the least toxic surfactant. Further, the IL-8 release by MWCNTs was reduced in the presence of Pluronic F127 (Monteiro-Riviere et al. 2005a). MWCNTs were found to be toxic in human dermal and mouse fibroblasts as well (Zhang et al. 2011a).

Whole-genome expression studies showed that cytotoxic concentrations of MWCNOs and MWCNTs induced apoptotic or necrotic cell death. Both the NPs activated different gene sets. MWCNTs induced genes responsible for inflammatory response while MWCNOs did not induce these genes (Ding et al. 2005).

22.2.4 Immunotoxicity

Macrophages are the main cells responsible for the removal of foreign particles from the body. These cells, in conjunction with lymphocytes, are responsible for mounting a finely orchestrated immune response against foreign matter. The crosstalk between macrophages, lymphocytes and other cells leads to the inflammatory response. This inflammatory response is mediated and maintained by cytokines, chemokines and chemotactic factors released by immunocytes. In the context of NPs, macrophages are the main cells responsible for the uptake and clearance of NPs from the host body.

The in vitro effect of SWCNTs was studied in a murine macrophage cell line (RAW 264.7). The cells were incubated with 100 mg/mL SWCNTs for 4–6 h and cytokine release was studied. The cells produced TGF-b1 in levels comparable to those observed with zymosan. However, IL-10, IL-1b and TNF-α releases were less as compared to zymosan. Further, no evidence of superoxide or nitric oxide production, SWCNT phagocytosis or apoptosis was evident after 6 h of incubation (Shvedova et al. 2005). In the study by Muller et al. rat peritoneal macrophages were incubated with 20, 50 and 100 mg/well of unground and ground MWCNTs. The ground CNTs showed concentration-dependent toxicity (LDH release) and TNF-α production after 24 h of incubation. However, concentration-dependency was not established in unground CNTs. Macrophages incubated with CNTs for 6 h showed an increase in TNF-α mRNA with ground CNTs but not with unground CNTs. This is probably because unground CNTs show a higher tendency towards aggregation and formed large floating aggregates. This leads to a lower concentration of CNTs available for interaction with the cells and hence an apparent absence of concentration-dependency (Muller et al. 2005). In a similar study by Sato et al. the group incubated THP-1 cells, a human monocytic cell line, with 5, 50 and 500 ng/mL of 220-CNTs and 825-CNTs for 16 h. The CNTs resulted in increased release of TNF-α. However, there was no effect of length on TNF-α release (Sato et al. 2005b). Further, THP-1 cells cultured on SWCNTs showed a slight reduction in cellular metabolic activity (Kalbacova et al. 2006). In the study by Cherukuri et al. the authors showed that incubation of J774.1 murine macrophage cell line with low concentrations of SWCNT resulted in phagocytosis of the SWCNTs without affecting cell morphology and viability. The uptake was found to be concentration- and time-dependent. Further, incubation at a lower temperature (27 °C) resulted in lesser uptake as compared to the physiological temperature (37 °C) indicating the active nature of uptake (Cherukuri et al. 2004). In the study by Kagan et al. two types of HiPco SWCNTs were used: iron-rich containing 26wt% iron and iron-stripped containing 0.23wt% iron. The CNTs were incubated with RAW 264.7 cells for 1–2 h at 0.12–0.5 mg/mL concentrations. Flow cytometric and fluorescence microscopy of cells did not show stimulation of superoxide or nitric oxide production after incubation with either type of CNTs. However, iron-rich SWCNT produced higher concentrations of hydroxyl radical as determined by EPR spin-trapping as compared to iron-stripped CNTs. Further, incubation with iron-rich SWCNTs resulted in the loss of low molecular weight thiols (like glutathione) and increased concentration of lipid hydroperoxides which was partially abrogated by exogenously added catalase (Kagan et al. 2006). Similarly, the presence of iron also increased cytotoxic, genotoxic and oxidative stress in mouse alveolar macrophages (Aldieri et al. 2013).

Murr and coworkers studied the effects of SWCNTs and MWCNTs on RAW 267.9 cells, a murine macrophage cell line. The group studied the concentration-dependent (0.005–10 mg/mL) cytotoxicity of commercially available SWCNTs from a single supplier and MWCNTs from two suppliers. The in vitro tolerable concentrations, after 48 h of incubation, were found to be 2.5 mg/mL above which a reduction in cell viability was observed. The cytotoxicity was found to be similar to chrysotile asbestos nanotubes. No change in cytokine levels (IL-10 and IL-12) was observed at the concentrations tested (Murr et al. 2005). In a later study, the group reported cytotoxicity of CNTs after 2 days and 2 weeks of incubation with RAW 267.9 macrophage cells (Murr et al. 2006). Dutta et al. reported that SWCNTs can inhibit cyclooxygenase-2 (COX-2) induction by LPS in RAW 267.9 cells. Fucoidan, a scavenger receptor antagonist, inhibited the uptake of SWCNTs by RAW 267.9 cells and abrogated SWCNT-induced suppression of COX-2 in response to LPS (Dutta et al. 2007). However, in presence of SWCNTs or MWCNTs, an increase in cytokine production was observed in LPS-stimulated peripheral blood monocytes (Inoue et al. 2008). Further, macrophages grown on CNT implants showed a reduction in cell adhesion and proliferation (Sung et al. 2008). Concentration-dependent cytotoxicity, increased free radical production and induction of apoptosis were observed in RAW 264.7 cells incubated with MWCNTs (Liu et al. 2008c; Chen et al. 2011a; Zhang et al. 2012). Further, MWCNTs were found to induce cell death through activation of mitochondrial pathways (Chen et al. 2012; Wang et al. 2012b; Ye et al. 2012).

Jia et al. reported the toxicity of SWCNTs and MWCNTs towards guinea pig alveolar macrophages after 6 h of incubation. Both types of CNTs were found to be toxic; however, SWCNT showed a higher degree of toxicity as compared to MWCNT. Electron microscopy studies showed signs of necrosis as well as apoptosis. Further, SWCNTs suppressed the phagocytic activity of macrophages at lower concentrations as compared to MWCNTs (Jia et al. 2005). In a similar study, pre-incubation of naive alveolar macrophages with SWCNT also resulted in decreased nitric oxide generation and suppressed phagocytic activity towards *L. monocytogenes* (Shvedova et al. 2008a). However, in murine alveolar macrophages, exposure to SWCNTs and MWCNTs increased the antigen-presenting function of macrophages. The CNTs were found to accumulate in the plasma membrane and disrupted lipid rafts (Hamilton et al. 2007). Contrastingly, highly purified SWCNTs were weakly toxic to human macrophages and were taken up in small amounts by human macrophages. Further, these SWCNTs did stimulate NO release from murine macrophages (Fiorito et al. 2006). Similarly, the CNTs showed no toxicity towards monocytes but impaired their phagocytic activity. Further, straight fibres were more efficient in the induction of proinflammatory cytokines and ROS as compared to curved fibres (Brown et al. 2007). The MWCNTs were found to be more biocompatible as compared to SWCNTs (Schrand et al. 2007).

Incubation of primary B-cells with CNTs showed no effect on cell viability. Additionally, CNTs neither activated B-cells nor did they interfere with the activation of B-cells by stimulants (Cai et al. 2007). However, SWCNTs stimulated the proliferation of splenocytes at 25 and 50 mg/mL. SWCNTs inhibited Concanavalin A-stimulated T-cell proliferation, LPS-stimulated B-cell proliferation and NK cell activity (Zhang et al. 2008). SWCNTs showed a reduction in metabolic activity of human peripheral blood lymphocytes and did not induce DNA damage (Zeni et al. 2008). Dumortier et al. studied the effect of functionalization of CNT on macrophages and lymphocytes (B- and T- cells). The pristine SWCNTs were functionalized by the 1,3-dipolar cycloaddition reaction (f-CNTs 1) or the oxidation/amidation treatment (f-CNTs 3). The f-CNT1 was highly water-soluble while f-CNT 3 formed stable suspensions in an aqueous medium. Mouse spleen cells incubated with f-CNTs (10 mg/mL) for 4–24 h showed the intracellular presence of FITC-labelled f-CNT. Incubation of isolated macrophages, B- and T- cells with FITC-f-CNTs showed that CNTs were present in all three cell types. However, the intracellular concentration was more in macrophages as compared to lymphocytes. f-CNT 4 was found inside macrophages, while bundles of f-CNT 4 were present around lymphocytes. Further, incubation of purified B- and T- cells with f-CNTs (10 and 50 mg/mL for 48 h) showed no signs of toxicity or changes in functionality (lymphoproliferation). On the other hand, f-CNT 3 stimulated macrophages resulting in the release of TNF-α and IL-6, while f-CNT 1 showed no effect on cytokine levels after 24 h of incubation at 10 and 50 mg/mL CNT concentrations. Further, pre-incubation of lymphocytes and macrophages with f-CNT 1 showed no effect on LPS responsiveness, while pre-incubation with f-CNT 3 suppressed LPS responsiveness of macrophages while leaving the lymphocyte responsiveness intact (Dumortier et al. 2006). Incubation of naïve T-cells with SWCNTs resulted in Th1 polarization and reduction in T cytolytic response (Alam et al. 2013; Dong et al. 2013).

Bottini et al. reported the toxicity of pristine and oxidized (produced by acid treatment) MWCNTs towards human T-cells and Jurkat cells, a T-cell cell line. The incubation of Jurkat cells with MWCNTs resulted in a concentration- and time-dependent reduction in cell viability; the oxidized MWCNTs showed more toxicity than pristine CNTs. The oxidized MWCNTs induced a concentration-dependent increase in the number of apoptotic cell populations of Jurkat cells as well as human T cells. Further, the CNTs showed no effect on T-cell receptor signal transduction but slightly stimulated tyrosine phosphorylation (Bottini et al. 2006). Further, the toxicity of SWCNTs and MWCNTs was dependent on the size and impurities present in CNTs (Nicola et al. 2008). On the other hand, the toxicity of SWCNTs and MWCNTs did not significantly affect cell viability in U937 lymphoma cells (Nicola et al. 2007).

22.2.5 Neurotoxicity

In vitro toxicity of SWCNTs was reported in PC12 cells. SWCNTs induced cell death, oxidative stress and apoptosis in PC12 cells (Zhang et al. 2010; Wang et al. 2011). These effects were reduced in presence of vitamin E suggesting oxidative stress as a possible mechanism of toxicity (Wang et al. 2012a). Further, PEGylation of SWCNTs resulted in reduced toxicity and oxidative stress with concomitant changes in gene expression (Zhang et al. 2011b).

The MWCNTs, when injected into the cerebral cortex, did not induce toxicity or apoptosis in mice brains. In vitro, the MWCNTs protected primary cortical neurons from Pluronic F127-induced apoptosis (Bardi et al. 2009a). MWCNTs were detected intracellularly, mainly in macrophages, after 24 h of intracranial injection in mice without evident signs of toxicity (Vanhandel et al. 2009). Functionalized CNTs also induced chemoprotection in dopaminergic neurons after intracranial and intravenous administration (Kim et al. 2020).

The MWCNTs were ingested by microglia and glioma cells without evident signs of toxicity or alterations in cytokine levels (Kateb et al. 2007), while SWCNTs induced changes in the transcriptome of mouse microglial cells (Yang et al. 2020a). The MWCNTs were found to be more biocompatible as compared to SWCNTs in neuronal cells. The incubation with CNTs was accompanied by an increase in ROS production (Schrand et al. 2007). The SWCNTs were also shown to directly stimulate hippocampal neurons (Mazzatenta et al. 2007). Neurons cultured on CNT substrates also showed an increase in electrical activity and enhanced spreading (McKenzie et al. 2004; Lovat et al. 2005; Malarkey and Parpura 2007; Nguyen-

Vu et al. 2007; Dubin et al. 2008). On the other hand, functionalized SWCNTs can also be used as support for neuronal growth. Further, SWCNTs can also be electrically coupled to neuronal cells and show voltage-activated currents (Liopo et al. 2006). The neurite outgrowth was reduced on pristine SWCNT supports. However, the neurite outgrowth was dependent on the type of functional groups attached to SWCNTs (Ni et al. 2005; Liopo et al. 2006). PEG-functionalized SWCNTs blocked membrane endocytosis resulting in neurite elongation (Malarkey et al. 2008) and improved axonal growth and repair in rats (Roman et al. 2011). CNTs also promoted the differentiation of mouse embryonic stem cells into neurons (Yang et al. 2014) and amino-functionalized SWCNTs showed protection in the stroke model (Lee et al. 2011). Pulmonary exposure to CNTs has been shown to influence neurotransmitter release in the brain (Calzetta et al. 2021; Mostovenko et al. 2021).

22.2.6 Systemic Toxicity

Deng et al. demonstrated a lack of hepatic toxicity following intravenous administration of taurine-functionalized MWCNTs, while Wang et al. demonstrated an increase in biochemical markers of toxicity following intratracheal instillation of taurine-functionalized MWCNTs (Deng et al. 2007; Wang et al. 2007). Subcutaneous implantation of hat-stacked carbon nanofibres (H-CNFs) in rats showed signs of mild inflammation with infiltration of inflammatory cells at the implantation site. The microscopic evaluation also showed the presence of H-CNFs in lysosomal vacuoles of infiltrating macrophages (Yokoyama et al. 2005). On the other hand, water-soluble carboxylic functionalized H-CNFs and CHAPS-treated H-CNFs induced the production of TNF-α and activated NF-κB in THP-1 cells. The CHAPS-treated H-CNFs showed a higher degree of THP-1 cell activation (Sato et al. 2005a). The size-dependent effect of MWCNTs produced by the chemical vapor deposition method was also studied. The CNTs were cut to a smaller size by utilizing acid refluxing and ultrasonication and CNTs of 2 different lengths, 220 nm (220-CNTs) and 825 nm (825-CNTs), were used. The CNT samples were implanted subcutaneously in rats (0.1 mg/rat). Histopathological examination after 1 and 4 weeks post-implantation showed that smaller CNTs were predominantly present inside the macrophages, while longer 825-CNTs were present inside the macrophages as well as in intercellular spaces. The inflammatory response was higher in the 825-CNTs as compared to 220-CNTs after 1 week. Further, after 4 weeks, the inflammatory response persisted in the 825-CNTs group, while it subsided to very low levels in the 220-CNTs group. The smaller CNTs were found to be present intracellularly and could be detected in lysosomes. However, the longer CNTs were present in the cytoplasm of macrophages and were not enclosed in a membrane. Further, the inflammation in either group was minor with the absence of apoptosis, neutrophil infiltration and degeneration. In vitro, the CNTs led to the induction of human monocyte-macrophage cells (THP-1) as evidenced by stimulation of TNF-α release. However, the induction of THP-1 cells was independent of the size of CNTs (Sato et al. 2005b; Tohji and Sato 2006). Subcutaneous implantation of MWCNTs in rats showed signs of mild inflammation (Sato et al. 2008). After 1 year of implantation, the MWCNTs could be observed in macrophages and showed deteriorative structural changes in the CNT present intracellularly (Sakaguchi et al. 2008). Intraperitoneal injection of MWCNTs resulted in the development of mesothelioma in mice (Takagi et al. 2008). Further, MWCNTs can also induce inflammation and granulomas in the mesothelial lining (Poland et al. 2008). However, no signs of alteration in humoral immunity were observed following intraperitoneal injection of MWCNTs (Chiaretti et al. 2008). Intravenous administration of PEG-functionalized SWCNTs in rats increased complement activation. In vitro studies using human serum showed that MASP-2- activation, leading to C4 cleavage, leads to complement activation (Hamad et al. 2008).

22.2.7 Other Cells

Meng et al. demonstrated that SWCNTs enhanced adhesion and proliferation of fibroblasts along with the presence of organized cytoskeletal structures (Meng et al. 2006b). Similarly, MWCNTs were found to show good biocompatibility with fibroblasts (Chlopek et al. 2006). However, water-soluble MWCNTs induced apoptosis in fibroblasts (Meng et al. 2009). The CNTs were detected intracellularly in feline kidney fibroblasts without loss of cell viability (Pensabene et al. 2007). Contrastingly, Tian et al. demonstrated the toxicity of SWCNTs towards fibroblasts and, surprisingly, the toxicity of refined SWCNTs was more than that of unrefined SWCNTs (Tian et al. 2006). SWCNTs also demonstrate toxicity towards 3T3 cells (Nimmagadda et al. 2006) and mouse embryonic fibroblasts (Yang et al. 2007a) in a concentration-dependent manner. Further, the toxicity was dependent on the purity and functionalization state of the SWCNTs (Nimmagadda et al. 2006). The toxicity of CNTs towards chick embryo fibroblasts (Xiao-Feng et al. 2005) and normal human dermal fibroblasts (Bonafè et al. 2008) has also been reported.

Fibroblasts grown on CNT scaffold/supports have been demonstrated to show higher cell adhesion and proliferation. Cells grown on SWCNTs scaffolds displayed higher adhesion and proliferation without an adverse effect on the cytoskeleton (Meng et al. 2006a). Similarly, mouse fibroblasts (L929 cells) grown on vertically aligned MWCNTs showed variable effects on cell viability. The effect on cell viability was dependent on the substrate for MWCNT growth. MWCNTs grown on silica supports showed high cell viability, while low cell viability was observed when titanium was used as support (Lobo et al. 2007, 2008).

Prylutska et al. demonstrated that MWCNTs showed no effect on the viability of rat erythrocytes and thymocytes at concentrations up to 25 μg/mL. However, an increase in CNT concentration to 50 μg/mL resulted in haemolysis and a reduction in viable thymocytes. This was associated with the inhibition of the mitochondrial electron-transporting chain. The CNTs could not generate O_2 from H_2O_2 but showed free radical scavenging activity in presence of hydroxyl and superoxide radicals (Prylutska et al. 2008, 2009, 2013). SWCNTs can also inhibit the growth of HEK293 cells. Incubation with SWCNTs resulted in secretion of proteins, cell aggregation, cell cycle arrest at the G1 phase and apoptosis. SWCNTs can alter gene expression of several cell cycles, cell signalling, cell adhesion and adhesion-associated proteins (Cui et al. 2005).

The CNTs have been shown to exert cytotoxicity when incubated with HeLa (Mahmood et al. 2008b) and MCF-7 cells (Chiaretti et al. 2008). The SWCNTs were detected intracellularly in vacuoles without loss of cell viability or mitochondrial superoxide levels (Yehia Hadi et al. 2007). Several reports also exist demonstrating little or no toxicity of CNTs towards other cell types; viz, human umbilical vein endothelial cells (Flahaut et al. 2006), osteoblasts (Chlopek et al. 2006; Kalbacova et al. 2007a), human smooth muscle cells (Chiaretti et al. 2008), Caco-2 cells (Chiaretti et al. 2008), adipocytes (Bardi et al. 2009b) and lung epithelial cells (Sharma et al. 2007). The toxicity of SWCNTs towards human osteoblasts was dependent on CNT surface chemistry (Kalbacova et al. 2007b). The toxicity of CNTs has also been demonstrated in osteocytes (Mahmood et al. 2008a, b). The CNTs may also show effects on human mesenchymal stem cells and localize in the nucleus (Mooney et al. 2008). PEGylated SWCNTs were found to enter the nucleus without affecting cellular functions. The entry of SWCNTs into the nucleus and cells was found to be reversible (Cheng et al. 2008).

MWCNTs were intraperitoneally administered to pregnant mice on the day of mating and day 3 of gestation. No effect on reproductive performance of dams or neurobehavioural and physical parameters of pups was observed (Ivani et al. 2012). Similarly, oral administration of MWCNTs to pregnant mice showed no maternal toxicity up to 200 mg/kg/day and no embryo-foetal toxicity up to 1000 mg/kg/day (Lim et al. 2011). Similarly, intratracheal instillation of MWCNTs in female mice showed no evidence of embryo or foetal toxicity (Hougaard et al. 2013; Johansson et al. 2017). However, hydroxyl-functionalized CNTs induced foetal toxicity after oral administration of 10 mg/kg CNTs (single dose) during organogenesis (Philbrook et al. 2011). Further, repeated administration of water-soluble MWCNTs to male mice resulted in reversible testis damage without an observable effect on fertility parameters (Bai et al. 2010). CNTs have also been shown to inhibit steroidogenesis and damage sperm in rats and mice (Aminzadeh et al. 2017; Nirmal et al. 2017; Qu et al. 2017; Farshad et al. 2020). Long MWCNTs were found to cross the placental barrier and induce foetal toxicity, while SWCNTs and short MWCNTs could not cross the barrier and did not show foetal toxicity. Further, the foetal toxicity was found to be dependent on the genotype of mice receiving CNTs (Huang et al. 2014). Recent studies have established the reproductive toxicity of CNTs in the model nematode *Caenorhabditis elegans* (Lu et al. 2022; Yao et al. 2022; Zhao et al. 2022).

22.2.8 Genotoxicity

CNTs have been shown to possess genotoxic effects in both, in vitro and in vivo conditions. CNTs induced DNA damage, as assessed by Comet assay, in mouse embryo fibroblasts (Yang et al. 2008a). However, SWCNTs showed mild genotoxicity in mouse lung epithelial cell lines (Jacobsen et al. 2008). Similarly, SWCNTs also induced DNA damage (comet assay) along with slight induction of micronucleus formation in V79 lung fibroblasts. The SWCNTs did not induce mutations in *Salmonella typhimurium* as well (Kagan et al. 2007; Kisin et al. 2007). Later studies also identified a lack of genotoxic potential in in vitro bacterial reverse mutation, in vitro chromosomal aberration and in vivo micronucleus assay (Szendi and Varga 2008; Naya et al. 2011; Ema et al. 2013a; Kim et al. 2013). SWCNTs were found to induce DNA damage in human bronchial epithelial cells (Lindberg et al. 2008), human colon carcinoma cells (Pelka et al. 2013) and rat kidney epithelial cells (Nam et al. 2011). Further, SWCNTs also induced genotoxicity in rats following intratracheal instillation (Naya et al. 2012).

MWCNTs also did not show genotoxic potential in the Ames test and in *Escherichia coli*. Further, incubation of MWCNTs with liver S9 fraction showed no effect on bacterial colonies (Di Sotto et al. 2008). Further, MWCNTs also did not show signs of genotoxicity in in vitro assays (Wirnitzer et al. 2008; Kim et al. 2011; Ema et al. 2012; Ponti et al. 2013). MWCNTs were found to induce DNA damage (comet assay) in A549 cells (Karlsson et al. 2008). MWCNTs showed signs of DNA damage in mouse embryonic stem cells, activated p53 and increased mutation rate (Zhu et al. 2007). Intratracheal instillation of MWCNTs resulted in the induction of micronuclei in pneumocytes. Similarly, MWCNTs also induced micronuclei in rat lung epithelial cells (Muller et al. 2008a, b). Contrastingly, comet assay showed no evidence of genotoxicity in rat lung cells following intratracheal instillation of MWCNTs (Ema et al. 2013b). Further, MWCNTs induced genotoxicity following inhalation exposure in rats and mice (Kim et al. 2012, 2014).

Although several reports have demonstrated no or low level of genotoxicity of CNTs, several groups have investigated the possibility of oxidative DNA damage by SWCNTs and MWCNTs. Oxidative DNA damage has been demonstrated in mammalian cell cultures (Migliore et al. 2010; Di Giorgio et al. 2011; Kato et al. 2013) as well as in vivo (Folkmann et al. 2009; Moller et al. 2012).

22.3 Bio-Distribution

Following in vivo administration, molecules and particles are absorbed from the site of administration and enter the bloodstream. Once in the systemic circulation, the particles travel throughout the body and are preferentially accumulated in various tissues and organs. This is called translocation or biodistribution. Following biodistribution, the substances may be metabolized by enzymes, a process called biotransformation, in an attempt to detoxify the compounds. The compounds may be excreted unchanged or in the metabolized form in urine and faeces. Several methodologies for determining in vivo fate of small molecules have been developed over decades. The in vivo fate of molecules may be determined by use of fluorescent-tagged or radiolabelled compounds or by employing the intrinsic spectral properties of these compounds (fluorescence or absorption in ultraviolet, visible and infrared regions of the electromagnetic spectrum). Similar strategies have also been used to determine the in vivo fate of CNTs in experimental animals. CNTs can also be detected in vivo by tagging them with radiolabels, fluorophores or spectroscopy (infrared or Raman spectroscopy).

Cherukuri et al. administered 75 μg pristine HiPco SWCNTs intravenously in rabbits and monitored them for 24 h using near-infrared fluorescence. Spectral analysis revealed that surfactant molecules coating the CNTs were quickly displaced by blood proteins. The rate of disappearance of CNTs from blood followed first-order kinetics with a half-life of 1.0 ± 0.1 h. After 24 h of CNT administration, CNTs were found in the liver. Further, few or no CNTs were detected in other organs (Cherukuri et al. 2006). Similarly, Yang et al. studied the biodistribution of pristine ^{13}C-labelled SWNTs in mice. The SWNTs were found to be distributed mainly in the liver, lungs and spleen. However, the SWCNTs were detected in other organs also. Later, the group showed that intravenously administered SWCNTs persisted in the lungs, liver and spleen of mice for over 3 months. This was associated with an elevation in biochemical markers of toxicity in serum. Histological examination also showed toxicity and inflammation. Biochemical examination showed an increase in malondialdehyde levels and a decrease in glutathione levels (Yang et al. 2007b, 2008b, c; Chou et al. 2008; Yang and Liu 2012). Similarly, intravenously administered SWCNTs were concentrated in the kidney, liver and spleen of mice. The hepatic uptake was associated with histopathological changes and inflammation. The blood half-life of SWCNTs was 3–4 h (Kang et al. 2009).

McDevitt et al. reported biodistribution of SWCNT labelled with yttrium-86 (^{86}Y-CNT) and indium-111(^{111}In-CNT) in nude mice. The mice received ^{86}Y-CNT by the intravenous or intraperitoneal route and were imaged by positron emission tomography (PET) 1, 3 and 24 h after injection (McDevitt et al. 2007a, b). After 3 h of i.v. injection, ^{86}Y-CNT were mainly present in the renal cortex, liver, spleen and bone. The ^{86}Y-CNTs were not detected in blood after 3 h. Similarly, 3 h after i.p administration, the activity was detected in the kidneys, liver, spleen and bone. However, liver and spleen contained lesser CNT content as compared to i.v. injection. In both groups, renal clearance was observed after 24 h, while there was no significant change in liver and spleen content. Similar studies performed with ^{111}In-CNT showed a biodistribution profile similar to ^{86}Y-CNT (McDevitt et al. 2007a, b). Similarly, pristine and functionalized SWCNTs were found to accumulate in the liver and spleen after intravenous administration (Al Faraj et al. 2010, 2011; Doan et al. 2012; Tsyboulski et al. 2013). Singh et al. determined the clearance of intravenously administered functionalized SWCNTs in mice. SWNTs were functionalized with diethylentriaminepentaacetic (DTPA), labelled with radioactive indium (^{111}In) and administered intravenously (60 μg/mouse, 20 μCi). The fate of CNTs was determined by using gamma scintigraphy. The results of the study indicate that the CNTs are not retained by any organ studied, including the liver and spleen and are excreted from circulation by the renal route. The half-life of functionalized SWCNTs was found to be 3 h. The urinary excretion of functionalized SWCNTs and MWCNTs was determined. Mice were intravenously injected with 400 μg of SWCNTs or MWCNTs and urine was collected for 18 h. The

CNT morphology following excretion was determined by TEM and found to contain intact CNTs. These results indicate that CNTs are excreted as intact CNTs in urine (Singh et al. 2006).

Lacerda et al. studied the effect of functionalization on biodistribution of MWCNTs after 24 h. Purified MWCNTs were found to accumulate in the lungs and liver. The bioaccumulation of CNTs in organs had an inverse relation to the degree of ammonium functionalization without signs of toxicity. Functionalization with carboxylic groups also showed no signs of toxicity (Lacerda et al. 2008a, b). Jain et al. later reported functionalization density-dependent accumulation and excretion of Tc-99m-labelled MWCNTs intravenously injected in mice. Pristine MWCNTs were functionalized by acid refluxing followed by Tc-99m. Acid-functionalization resulted in the shortening of CNTs along with the introduction of carboxyl groups. Functionalized CNTs did not accumulate in organs of the reticulo-endothelial system and were excreted through the renal route. The renal excretion was found to be proportional to the degree of oxidation. The authors proposed that pristine MWCNTs as well as MWCNTs with a lower degree of functionalization might be excreted through the biliary route (Jain et al. 2011). A similar functionalization-dependent accumulation and renal clearance of MWCNTs was also reported in a later study (Al-Jamal et al. 2011). Similarly, several other groups have also reported uptake of pristine or functionalized MWCNTs by lung, liver and spleen and proposed excretion of CNTs through renal or faecal routes (Wu et al. 2011; Abe et al. 2012; Wei et al. 2012).

Deng et al. reported the biodistribution of MWCNTs in mice using [^{14}C]-taurine-MWCNTs (10 μg/mouse, 27×10^3 cpm). The MWCNTs were administered by three routes: oral, intravenous and intratracheal instillation. Following oral administration, the CNTs were detected in the stomach, small intestine, large intestine and faeces. The CNTs were mainly concentrated in the small intestine after 1 h and decreased thereafter. The concentration of CNTs increased in the large intestine after 3 h and decreased thereafter. There was a consistent increase in CNT content in faeces starting from 1 h and reached about 75% of the administered dose after 12 h. These results indicate that CNTs are not absorbed and do not enter systemic circulation after oral administration. On the other hand, intravenous administration resulted in the detection of CNTs in the heart, liver, lung and spleen but not in the brain, bone, intestine, muscle and stomach. More than 75% of the administered CNTs were found in the liver after 10 min of exposure and the levels remained unchanged up to 28 days and declined to about 20% after 90 days. During the duration of the experiment (90 days), around 5% of CNTs remained in the spleen. TEM analysis showed that CNTs were accumulated in Kupffer cells. Following intratracheal exposure, CNT distribution was restricted to lungs only. The CNTs were cleared from the lungs with concentrations dropping from 78% on day 1 to 20% after day 28. The CNTs were not detected in blood and other organs, even when radioactivity in the lungs was largely reduced. The authors proposed that the clearance of CNTs from the lungs may involve two mechanisms: the escalator pathway and lymphatic drainage. The CNTs showed no changes in LDH, TBIL, TBA, ALP and ALT at doses of 1–600 μg/mouse (i.v.) over a period of 28 days.

Similarly, no time-dependent changes in LDH, TBIL, TBA, ALP and ALT levels were observed at 100 μg/mouse dose (i.v.) measured 6 h, 1, 3, 7, 14 and 28 days post-administration. The lack of hepatic toxicity was confirmed by histopathology (Deng et al. 2007). In a later study, the group also used radioiodinated taurine-conjugated MWCNTs (^{125}I-tau-MWNTs). Alternatively, MWCNTs were non-covalently linked to Tween-80 and labelled with ^{125}I (^{125}I-Tween-MWNTs). No significant difference in tissue distribution was observed between [^{14}C]-taurine-MWCNTs and ^{125}I-tau-MWNTs. However, ^{125}I-Tween-MWNTs, unlike taurine functionalized CNTs, were distributed in the stomach, kidney, large intestine and small intestine apart from the major presence in the liver, spleen and lungs. The ^{125} I -Tween-MWNTs were present in lower concentrations in the liver and spleen, while blood concentrations were higher at all times as compared to tau-MWCNTs. This indicates that Tween prevents the MWCNT uptake by RES and increases their systemic availability (Deng et al. 2008).

Wang et al. also used radiolabelled (C-14 labelled) taurine-functionalized MWCNTs to study the translocation of MWCNTs in Kunming mice after intratracheal instillation. The taurine functionalized MWCNTs were administered at doses of 0.125, 0.25, 0.5 and 1 mg/kg and biochemical parameters were assessed on days 1, 7, 14 and 28 post-instillation. Blood biochemistry was altered after 7 days of CNT administration: the peak values of alkaline phosphatase (ALP) and lactate dehydrogenase (LDH) were observed on day 7. The ALP values were restored to normal after 28 days, while LDH levels remained significantly higher than control. Lung histopathology showed infiltration of inflammatory cells and characteristics of inflammation. The radiolabelled CNTs were detected predominantly in the lungs and decreased with time (Wang et al. 2007).

Wang et al. studied the biodistribution of radiolabelled (I-125) hydroxylated SWCNTs in mice. After intraperitoneal injection, the CNTs were found to be distributed throughout the body and accumulated in bone. Most of the CNTs were eliminated in faeces and urine 11 days post-injection. The authors reported that the CNTs could freely move in various body compartments like small molecules (Wang et al. 2004). Similarly, I-131 labelled hydroxylated SWCNTs were distributed throughout the body 2 min post-injection, except in the brain (Wang et al. 2008).

Guo et al. used chemical vapor deposition to produce MWCNTs which were derivatized with glucosamine (MWCNT-G) to impart water solubility. The solubilized CNTs were then labelled with ^{99m}Tc. Mice received intraperitoneal injection of CNTs and tissue distribution was determined by gamma imaging. The radioactivity was predominantly present in the intragastric area and stomach after 1 h of administration. Relatively small proportions of the administered dose were present in blood, coat, liver, lungs, spleen, kidneys, intestines, blood and muscles. After 6 h of injection, there was a decrease in CNT in blood, heart, muscles, lungs, liver and kidney. However, an increase in signal was observed in the stomach and intragastric area. The systemic half-life was found to be about 5.5 h and more than 70% of CNTs were excreted in urine and faeces within 24 h (Guo et al. 2007).

Liu et al. studied the in vivo fate of HiPco SWCNTs non-covalently functionalized with PEGylated phospholipids in mice. The group studied the effect of PEG chain length (molecular wt. 2000 and 5400) on the bio-distribution of CNTs. Radioactive copper (^{64}Cu) was attached to CNTs by covalently attaching a chelator to the PEG chain terminus. The ^{64}Cu served as a tracer of CNT distribution and was monitored using PET. The CNTs were found to accumulate in the liver. The liver uptake was less with long-chain PEG-derivatized CNTs. Further, long-chain PEG-derivatization resulted in higher blood clearance time as compared to small-chain PEG derivatization. The CNTs were detected in all the major organs after 24 h; however, a major proportion of injected CNTs were present in the liver and spleen suggesting uptake by RES (Liu et al. 2007). In a later study, mice were injected with 20 μg SWCNTs intravenously and their concentration in blood was monitored. Functionalization with branched-chain and long-chain PEGs resulted in a longer duration of blood circulation and reduced RES uptake. An increase in PEG chain length from 2 to 5 kDa resulted in an increase in circulation time from 1.2 to 5 h. Further increase in chain length to 7 and 12 kDa had little effect on circulation time. However, functionalization with 7 kDa branched chain PEG increased the circulation time to about 15 h, and CNTs were found up to 1-day post-injection. Mice were injected with 100 μg CNTs and tissue bio-distribution was determined on 1, 30, 60 and 90 days. The SWCNTs were detected predominantly in the liver and spleen and small quantities of SWNT were also detected in the bone, intestine, lung, kidney and stomach. The concentration of SWCNTs gradually decreased over 3 months in both the lungs and spleen. The SWCNTs functionalized with branched-chain PEG were found to be present in the least quantities. The SWCNTs functionalized with linear chain PEGs showed an inverse relationship between tissue concentration and chain length: low molecular weight PEG functionalization resulted in higher retention in lungs and spleen and vice versa. The CNTs were found in the intestine and faeces after 24 h indicating a biliary pathway of excretion. Similarly, CNTs were observed in the kidney and bladder indicating urinary excretion (Liu et al. 2008d). Yang et al. demonstrated that PEG-functionalized SWCNTs showed a half-life of 22.5 h and were detected in all major organs, including the brain (Yang et al. 2008b). Campagnolo et al. demonstrated that intravenously injected amino-functionalized SWCNTs were detected in the placenta and yolk sac of pregnant mice. Although CNTs were not detected in embryos, teratogenic effects were observed (Campagnolo et al. 2013).

Intratracheal instillation of MWCNTs in rats showed that MWCNTs could be detected in lungs even after 6 months post-installation. The MWCNT content was found to be reduced over 6 months although the CNTs had limited access to the systemic circulation. The CNTs were found to be chemically modified in the lungs (Elgrabli et al. 2008). MWCNTs induced extrapulmonary toxicity as well suggesting their translocation to lymphatics and further translocation to the liver and kidney (Reddy et al. 2010; Aiso et al. 2011; Mercer et al. 2013; Porter et al. 2013; Czarny et al. 2014). Matthews et al. suggested that 0.05% and 0.15% of SWCNTs may translocate to extra-pulmonary sites over 90 min and 14 days, respectively (Matthews et al. 2013). Intramuscular implantation of MWCNTs showed the presence of large MWCNT aggregates. On the other hand, SWCNTs administered intramuscularly were present in macrophages and translocated to lymph nodes (Fraczek et al. 2008).

The above studies on bio-distribution were further corroborated and supplemented by more recent studies using a wide range of non-invasive detection techniques (Huth et al. 2018; Shaik et al. 2019; Takeuchi et al. 2019; Galassi et al. 2020; Zhang et al. 2021a). In addition to pristine and functionalized CNTs, CNTs loaded with drugs have often been developed for monitoring bio-distribution and drug release in real-time (Das et al. 2013a, b; Zhang et al. 2018; Biagiotti et al. 2019; Cui et al. 2019; Gonzalez-Carter et al. 2019; Attri et al. 2021).

22.4 Challenges in CNT Toxicity Screening

The toxicity of CNTs could be altered by a variety of factors such as CNT morphology, CNT length and width, the suspending agent used for dispersion and aggregation status, surface chemistry, experimental systems employed and the route of administration. These factors are summarized in Fig. 22.2.

The factors include morphological characteristics (length, diameter, end-capping and defects), dispersion characteristics (surfactant and its concentration, degree of aggregation), surface characteristics (functionalization) and model systems used (in vitro, in vivo).

CNTs exist as elongated structures. The morphology of CNTs may be governed by the synthetic procedure and the processing after production. It was demonstrated that the introduction of structural defects by grinding CNTs of CNTs imparts free radical scavenging activity to the ground CNTs (Fenoglio et al. 2008). However, pulmonary toxicity and genotoxicity of MWCNTs were dependent on the presence of structural defects (Fenoglio et al. 2008). Similarly, the toxicity of MWCNTs towards mouse macrophages was also demonstrated to be dependent on structural defects (Bussy et al. 2012). The in vitro cytotoxicity in A549 and HepG2 cells and in vivo pulmonary toxicity in mice was also influenced by structural defects in CNTs (Requardt et al. 2019; Taylor-Just et al. 2020). Single-walled carbon nanohorns (SWCNHs) share structural similarities with SWCNTs. The SWCNHs did not

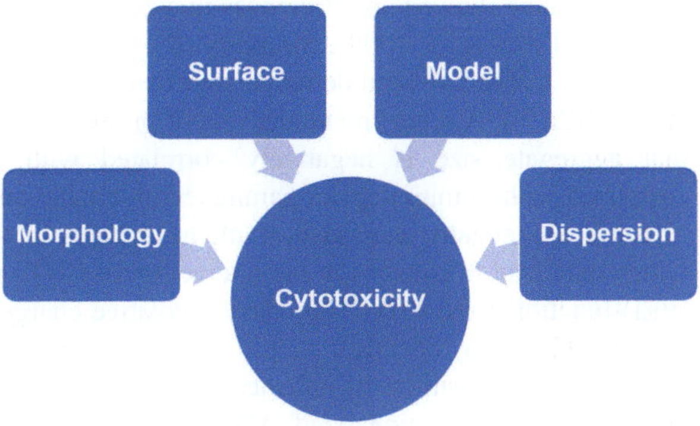

Fig. 22.2 Factors which influence cytotoxicity of CNTs

induce inflammation and cytotoxicity following pharyngeal aspiration in mice (Lynch et al. 2007). Similar results were reported in other cell lines as well (Garriga et al. 2020; Sano et al. 2020). This is in contrast to previous studies showing pulmonary inflammation and toxicity induced by SWCNTs. Closed-end SWCNTs and open-end SWCNTs differ in their ability to block ion channels (Park et al. 2003). The authors have also demonstrated length-dependent toxicity and excretion of MWCNTs following intravenous administration in mice and in vitro incubation with RAW 264.7 cells (Jain et al. 2011; Singh et al. 2012). In a recent meta-analysis, Gernand et al. demonstrated that CNT toxicity is positively related to CNT diameter and negatively correlated with CNT length (Gernand and Casman 2014). Diameter-dependent cytotoxicity of MWCNTs has also been demonstrated under in vitro conditions (Zhao et al. 2019; Yang et al. 2020b), while length-dependent effects have been studied in both SWCNTs and MWCNTs (Chatterjee and Choi 2020; Jiang et al. 2020).

CNTs do not form aqueous solutions and their aqueous dispersions are unstable. This leads to aggregation of CNTs in culture medium and biological fluids leading to heterogeneity in the concentration/dose of the CNTs in different regions. To obtain homogeneity in exposed concentrations, suspending agents are commonly used to suspend CNTs in aqueous media. However, the toxicity profile may be altered by different suspending agents. CNTs suspended in bovine serum albumin (BSA) showed better reproducibility in in vitro and in vivo results as compared to CNTs suspended in saline. The high reproducibility may be due to a lower degree of aggregation in BSA as compared to saline (Elgrabli et al. 2007). Further, SWCNTs suspended in phospholipids increased the toxicity of SWCNTs in A549 cells as compared to SWCNTs suspended in a culture medium. However, the toxicity profile of SWCNTs towards primary lung epithelial cells remained unaffected in either of the suspending media (Casey et al. 2008). Phosphatidylserine-coated SWCNTs were efficiently phagocytosed by phagocytic cells along with changes in cytokine release profile. The phosphatidylserine-coated SWCNTs were phagocytosed by alveolar macrophages following aspiration in mice (Konduru et al. 2009). The CNT toxicity may also be a result of the toxicity of the surfactant used to suspend the CNTs (Monteiro-Riviere et al. 2005a; Dong et al. 2008). On the other hand, Tabet et al. showed that MWCNTs dispersed in dipalmitoyl lecithin, ethanol and phosphate-buffered saline showed no differences in cytotoxicity towards A549 and mesothelial cell line (Tabet et al. 2009). Similarly, the influence of suspending agents on the toxicity of CNTs has been demonstrated by others as well (Chen et al. 2011b; Holt et al. 2012a, b). A recent meta-analysis of pulmonary toxicity studies demonstrated that aggregate size is negatively correlated with cytotoxicity and positively correlated with immunological parameters (Gernand and Casman 2014).

Surface chemistry may also result in changes in the surface charge of CNTs which may lead to changes in the interaction of CNTs with cells. Polyethylenimine (PEI)-functionalized MWCNTs carry a positive charge and were found to be toxic in vitro. However, chemical modifications aimed at reversing or neutralizing the positive charge resulted in enhanced biocompatibility of MWCNTs (Shen et al. 2009). Further, functionalized MWCNTs were found to be more genotoxic compared to pristine MWCNTs (Patlolla et al. 2010). Further, contrasting results have been published on the effect of functionalization on CNT toxicity in vitro and in vivo test systems (Saxena et al. 2007; Sohaebuddin et al. 2010; Luo et al. 2012; Montes-Fonseca et al. 2012; Singh et al. 2012; Yang et al. 2013; Yu et al. 2013; Chatterjee and Choi 2020; Jiang et al. 2020; Kyriakidou et al. 2020; Mohammadi et al. 2020; Nahle et al. 2020; Bhardwaj and Saxena 2021; Burdanova et al. 2021).

The choice of cell type and animal species determines the extent of toxicity of CNTs. The susceptibility of a cell type to CNT toxicity may depend on several factors like the basal levels of oxidative stress, protective mechanisms and other uncharacterized mechanisms. It has been shown that malignant mesothelial cells are more susceptible to MWCNT-induced cytotoxicity as compared to normal cells. On the other hand, the degree of activation of signalling molecules also depends on the cell type. For example, MWCNT-induced PARP activation is more pronounced in normal mesothelial cells, while higher levels of MAPK activation are observed in malignant mesothelial cells (Pacurari et al. 2008). Further, A549 cells may secrete lung surfactant that protects the cells from the cytotoxic effects of CNTs (Salvador-Morales et al. 2007). Similarly, the epithelial cells and mesothelial cells differ in their susceptibility to the toxic effects of SWCNTs (Kaiser et al. 2008). MWCNTs have been shown to induce pyroptosis in macrophages, while this was not observed in neutrophils (Keshavan et al. 2021). The in vivo effects of CNTs may be dependent on the species and strain of the animal used. In vivo studies show that the mouse model is susceptible to pulmonary effects of CNTs as compared to other animal species. Further, a comparison of mouse strains shows that ApoE knock-out mice are more susceptible to pulmonary effects of NPs (including SWCNTs) as compared to C57 mice (Jacobsen et al. 2009). Similarly, the genotype of female mice was found to influence foetal development following intravenous injection of MWCNTs (Huang et al. 2014).

Amongst the methods for pulmonary administration, intra-tracheal instillation of MWCNTs (50 μg/mouse) resulted in bronchial inflammation and alveolar destruction. On the other hand, inhalation of MWCNTs resulted in a lesser intensity of alveolar damage (Li et al. 2007a). Contrastingly, the pulmonary toxicity of SWCNTs was higher in the inhalation group as compared to the pharyngeal aspiration group (Shvedova et al. 2008b). Similar studies in mice also demonstrated route-dependent toxicity of MWCNTs (Carrero-Sanchez et al. 2006).

22.5 Biomedical Applications

CNTs have been investigated in drug delivery applications, generally as multifunctional materials bearing at least of the following: the target drug, a targeting agent, a stealth agent and a reporter probe. Amongst the drugs, anticancer drugs have been most commonly conjugated with CNTs (Choudhury et al. 2020; Kale et al. 2020; Mehta et al. 2020; Allemailem et al. 2021; Lyra et al. 2021; Pedziwiatr-Werbicka et al. 2021; Yadav et al. 2021; Alhalmi et al. 2022; Banthia et al. 2022; Boreggio et al. 2022), while other classes like antimicrobials (Chaudhari et al. 2019; De Mori et al. 2020; Gobi et al. 2021; Gong et al. 2021; Lagana et al. 2021; Teixeira-Santos et al. 2021; Zhang et al. 2021b) and other therapeutic classes (Negri et al. 2020; Sarecka-Hujar et al. 2020; Rabiee et al. 2021; Rahamathulla et al. 2021; Ramanunny et al. 2021; Nikeafshar et al. 2022) appear less frequently in literature. CNTs have also been reported for imaging and diagnostic purposes. Additionally, CNTs have been reported to induce in vitro tissue differentiation of cell lines and stem cells (Hindman and Ma 2018; Patel et al. 2018; Kim et al. 2019; Dominguez-Alfaro et al. 2020; Xiang et al. 2020; Oliveira et al. 2021; Ye et al. 2021; Zhang et al. 2021c; Prajapati et al. 2022).

References

Abe S, Itoh S, Hayashi D et al (2012) Biodistribution of aqueous suspensions of carbon nanotubes in mice and their biocompatibility. J Nanosci Nanotechnol 12:700–706

Aiso S, Kubota H, Umeda Y et al (2011) Translocation of intratracheally instilled multiwall carbon nanotubes to lung-associated lymph nodes in rats. Ind Health 49:215–220

Al Faraj A, Bessaad A, Cieslar K et al (2010) Long-term follow-up of lung biodistribution and effect of instilled SWCNTs using multiscale imaging techniques. Nanotechnology 21:175103

Al Faraj A, Fauvelle F, Luciani N et al (2011) In vivo biodistribution and biological impact of injected carbon nanotubes using magnetic resonance techniques. Int J Nanomedicine 6:351–361

Alam A, Sachar S, Puri N et al (2013) Interactions of polydispersed single-walled carbon nanotubes with T cells resulting in downregulation of allogeneic CTL responses in vitro and in vivo. Nanotoxicology 7:1351–1360

Aldieri E, Fenoglio I, Cesano F et al (2013) The role of iron impurities in the toxic effects exerted by short multiwalled carbon nanotubes (MWCNT) in murine alveolar macrophages. J Toxicol Environ Health A 76:1056–1071

Alhalmi A, Beg S, Almalki WH et al (2022) Recent advances in nanotechnology-based targeted therapeutics for breast cancer management. Curr Drug Metab 23(8):587–602

Al-Jamal KT, Gherardini L, Bardi G et al (2011) Functional motor recovery from brain ischemic insult by carbon nanotube-mediated siRNA silencing. Proc Natl Acad Sci U S A 108:10952–10957

Allemailem KS, Almatroudi A, Alsahli MA et al (2021) Novel strategies for disrupting cancer-cell functions with mitochondria-targeted antitumor drug-loaded nanoformulations. Int J Nanomedicine 16:3907–3936

Aminzadeh Z, Jamalan M, Chupani L et al (2017) In vitro reprotoxicity of carboxyl-functionalised single- and multi-walled carbon nanotubes on human spermatozoa. Andrologia 49(9)

Attri A, Thakur D, Kaur T et al (2021) Nanoparticles incorporating a fluorescence turn-on reporter for real-time drug release monitoring, a chemoenhancer and a stealth agent: Poseidon's trident against cancer? Mol Pharm 18:124–147

Bai Y, Zhang Y, Zhang J et al (2010) Repeated administrations of carbon nanotubes in male mice cause reversible testis damage without affecting fertility. Nat Nanotechnol 5:683–689

Banthia P, Gambhir L, Sharma A et al (2022) Nano to rescue: repository of nanocarriers for targeted drug delivery to curb breast cancer. 3 Biotech 12:70

Bardi G, Tognini P, Ciofani G et al (2009a) Pluronic-coated carbon nanotubes do not induce degeneration of cortical neurons in vivo and in vitro. Nanomedicine 5:96–104

Bardi G, Vittorio O, Maffei M et al (2009b) Adipocytes differentiation in the presence of Pluronic F127-coated carbon nanotubes. Nanomedicine 5:378–381

Belyanskaya L, Manser P, Spohn P et al (2007) The reliability and limits of the MTT reduction assay for carbon nanotubes-cell interaction. Carbon 45:2643–2648

Bhardwaj D, Saxena RK (2021) Poly-dispersed acid-functionalized single-walled carbon nanotubes (AF-SWCNTs) are potent inhibitor of BCG induced inflammatory response in macrophages. Inflammation 44:908–922

Bhattacharya SS, Yadav B, Rosen L et al (2022) Crosstalk between gut microbiota and lung inflammation in murine toxicity models of respiratory exposure or co-exposure to carbon nanotube particles and cigarette smoke extract. Toxicol Appl Pharmacol 447:116066

Biagiotti G, Pisaneschi F, Gammon ST et al (2019) Multiwalled carbon nanotubes for combination therapy: a biodistribution and efficacy pilot study. J Mater Chem B 7:2678–2687

Bianco A, Hoebeke J, Godefroy S et al (2005a) Cationic carbon nanotubes bind to CPG oligodeoxynucleotides and enhance their immunostimulatory properties. J Am Chem Soc 127:58–59

Bianco A, Kostarelos K, Prato M (2005b) Applications of carbon nanotubes in drug delivery. Curr Opin Chem Biol 9:674–679

Bonafè R, Bardini P, Bussi S et al (2008) Cytotoxic effects of carbon nanotubes (CNTs) and fullerenes on human fibroblasts: choosing the best suspension method. Toxicol Lett 180:S220–S221

Boreggio M, Rosini E, Gambarotti C et al (2022) Unveiling the bio-corona fingerprinting of potential anticancer carbon nanotubes coupled with d-amino acid oxidase. Mol Biotechnol 64: 1164

Bottini M, Bruckner S, Nika K et al (2006) Multi-walled carbon nanotubes induce T lymphocyte apoptosis. Toxicol Lett 160:121–126

Brown DM, Kinloch IA, Bangert U et al (2007) An in vitro study of the potential of carbon nanotubes and nanofibres to induce inflammatory mediators and frustrated phagocytosis. Carbon 45:1743–1756

Burdanova MG, Kharlamova MV, Kramberger C et al (2021) Applications of pristine and functionalized carbon nanotubes, graphene, and graphene nanoribbons in biomedicine. Nanomaterials 11:3020

Bussy C, Pinault M, Cambedouzou J et al (2012) Critical role of surface chemical modifications induced by length shortening on multi-walled carbon nanotubes-induced toxicity. Part Fibre Toxicol 9:46

Cai D, Doughty CA, Potocky TB et al (2007) Carbon nanotube-mediated delivery of nucleic acids does not result in non-specific activation of B lymphocytes. Nanotechnology 18:365101

Calzetta L, Pietroiusti A, Page C et al (2021) Multi-walled carbon nanotubes induce airway hyperresponsiveness in human bronchi by stimulating sensory C-fibers and increasing the release of neuronal acetylcholine. Expert Rev Respir Med 15:1473–1481

Campagnolo L, Massimiani M, Palmieri G et al (2013) Biodistribution and toxicity of pegylated single wall carbon nanotubes in pregnant mice. Part Fibre Toxicol 10:21

Cao Y, Jacobsen NR, Danielsen PH et al (2014) Vascular effects of multiwalled carbon nanotubes in dyslipidemic Apoe-/- mice and cultured endothelial cells. Toxicol Sci 138:104–116

Carrero-Sanchez JC, Elias AL, Mancilla R et al (2006) Biocompatibility and toxicological studies of carbon nanotubes doped with nitrogen. Nano Lett 6:1609–1616

Casey A, Herzog E, Lyng FM et al (2008) Single walled carbon nanotubes induce indirect cytotoxicity by medium depletion in A549 lung cells. Toxicol Lett 179:78–84

Castranova V, Schulte PA, Zumwalde RD (2013) Occupational nanosafety considerations for carbon nanotubes and carbon nanofibers. Acc Chem Res 46:642–649

Cesta MF, Ryman-Rasmussen JP, Wallace DG et al (2010) Bacterial lipopolysaccharide enhances PDGF signaling and pulmonary fibrosis in rats exposed to carbon nanotubes. Am J Respir Cell Mol Biol 43:142–151

Chatterjee N, Choi J (2020) Endoplasmic reticulum stress mediated apoptosis via JNK in MWCNT-exposed in vitro systems: size, surface functionalization and cell type specificity. J Toxicol Sci 45:305–317

Chaudhari AA, Joshi S, Vig K et al (2019) A three-dimensional human skin model to evaluate the inhibition of staphylococcus aureus by antimicrobial peptide-functionalized silver carbon nanotubes. J Biomater Appl 33:924–934

Chen B, Liu Y, Song WM et al (2011a) In vitro evaluation of cytotoxicity and oxidative stress induced by multiwalled carbon nanotubes in murine raw 264.7 macrophages and human A549 lung cells. Biomed Environ Sci 24:593–601

Chen H, Shen J, Longhua G et al (2011b) Cellular response of raw 264.7 to spray-coated multi-walled carbon nanotube films with various surfactants. J Biomed Mater Res A 96:413–421

Chen T, Zang J, Wang H et al (2012) Water-soluble taurine-functionalized multi-walled carbon nanotubes induce less damage to mitochondria of raw 264.7 cells. J Nanosci Nanotechnol 12: 8008–8016

Cheng J, Fernando KA, Veca LM et al (2008) Reversible accumulation of pegylated single-walled carbon nanotubes in the mammalian nucleus. ACS Nano 2:2085–2094

Cheng WW, Lin ZQ, Wei BF et al (2011) Single-walled carbon nanotube induction of rat aortic endothelial cell apoptosis: reactive oxygen species are involved in the mitochondrial pathway. Int J Biochem Cell Biol 43:564–572

Cheng WW, Lin ZQ, Ceng Q et al (2012) Single-wall carbon nanotubes induce oxidative stress in rat aortic endothelial cells. Toxicol Mech Methods 22:268–276

Cherukuri P, Bachilo SM, Litovsky SH et al (2004) Near-infrared fluorescence microscopy of single-walled carbon nanotubes in phagocytic cells. J Am Chem Soc 126:15638–15639

Cherukuri P, Gannon CJ, Leeuw TK et al (2006) Mammalian pharmacokinetics of carbon nanotubes using intrinsic near-infrared fluorescence. Proc Natl Acad Sci U S A 103:18882–18886

Chiaretti M, Mazzanti G, Bosco S et al (2008) Carbon nanotubes toxicology and effects on metabolism and immunological modification in vitro and in vivo. J Phys Condens Matter 20: 474203/474201–474203/474210

Chlopek J, Czajkowska B, Szaraniec B et al (2006) In vitro studies of carbon nanotubes biocompatibility. Carbon 44:1106–1111

Chou CC, Hsiao HY, Hong QS et al (2008) Single-walled carbon nanotubes can induce pulmonary injury in mouse model. Nano Lett 8:437–445

Choudhury H, Maheshwari R, Pandey M et al (2020) Advanced nanoscale carrier-based approaches to overcome biopharmaceutical issues associated with anticancer drug 'etoposide'. Mater Sci Eng C Mater Biol Appl 106:110275

Cui D, Tian F, Ozkan CS et al (2005) Effect of single wall carbon nanotubes on human HEK293 cells. Toxicol Lett 155:73–85

Cui X, Wan B, Yang Y et al (2019) Carbon nanomaterials stimulate HMGB1 release from macrophages and induce cell migration and invasion. Toxicol Sci 172:398–410

Czarny B, Georgin D, Berthon F et al (2014) Carbon nanotube translocation to distant organs after pulmonary exposure: insights from in situ (14)C-radiolabeling and tissue radioimaging. ACS Nano 8:5715–5724

Dahm MM, Evans DE, Schubauer-Berigan MK et al (2012) Occupational exposure assessment in carbon nanotube and nanofiber primary and secondary manufacturers. Ann Occup Hyg 56:542–556

Das M, Datir SR, Singh RP et al (2013a) Augmented anticancer activity of a targeted, intracellularly activatable, theranostic nanomedicine based on fluorescent and radiolabeled, methotrexate-folic acid-multiwalled carbon nanotube conjugate. Mol Pharm 10:2543–2557

Das M, Singh RP, Datir SR et al (2013b) Surface chemistry dependent "switch" regulates the trafficking and therapeutic performance of drug-loaded carbon nanotubes. Bioconjug Chem 24: 626–639

Davoren M, Herzog E, Casey A et al (2007) In vitro toxicity evaluation of single walled carbon nanotubes on human A549 lung cells. Toxicol In Vitro 21:438–448

De Mori A, Jones RS, Cretella M et al (2020) Evaluation of antibacterial and cytotoxicity properties of silver nanowires and their composites with carbon nanotubes for biomedical applications. Int J Mol Sci 21:2303

Deng X, Jia G, Wang H et al (2007) Translocation and fate of multi-walled carbon nanotubes in vivo. Carbon 45:1419–1424

Deng X, Yang S, Nie H et al (2008) A generally adoptable radiotracing method for tracking carbon nanotubes in animals. Nanotechnology 19:075101

Di Giorgio ML, Di Bucchianico S, Ragnelli AM et al (2011) Effects of single and multi walled carbon nanotubes on macrophages: cyto and genotoxicity and electron microscopy. Mutat Res 722:20–31

Di Sotto A, Chiaretti M, Carru A et al (2008) Multi-walled carbon nanotubes: lack of mutagenic activity in the bacterial reverse mutation assay. Toxicol Lett 184:192–197

Ding L, Stilwell J, Zhang T et al (2005) Molecular characterization of the cytotoxic mechanism of multiwall carbon nanotubes and nano-onions on human skin fibroblast. Nano Lett 5:2448–2464

Doan BT, Seguin J, Breton M et al (2012) Functionalized single-walled carbon nanotubes containing traces of iron as new negative MRI contrast agents for in vivo imaging. Contrast Media Mol Imaging 7:153–159

Dominguez-Alfaro A, Alegret N, Arnaiz B et al (2020) Toward spontaneous neuronal differentiation of SH-SY5Y cells using novel three-dimensional electropolymerized conductive scaffolds. ACS Appl Mater Interfaces 12:57330–57342

Dong L, Joseph KL, Witkowski CM et al (2008) Cytotoxicity of single-walled carbon nanotubes suspended in various surfactants. Nanotechnology 19:255702/255701–255702/255705

Dong PX, Wan B, Wang ZX et al (2013) Exposure of single-walled carbon nanotubes impairs the functions of primarily cultured murine peritoneal macrophages. Nanotoxicology 7:1028–1042

Dubin RA, Callegari G, Kohn J et al (2008) Carbon nanotube fibers are compatible with mammalian cells and neurons. IEEE Trans Nanobioscience 7:11–14

Dumortier H, Lacotte S, Pastorin G et al (2006) Functionalized carbon nanotubes are non-cytotoxic and preserve the functionality of primary immune cells. Nano Lett 6:1522–1528

Dutta D, Sundaram SK, Teeguarden JG et al (2007) Adsorbed proteins influence the biological activity and molecular targeting of nanomaterials. Toxicol Sci 100:303–315

Elgrabli D, Abella-Gallart S, Aguerre-Chariol O et al (2007) Effect of BSA on carbon nanotube dispersion for in vivo and in vitro studies. Nanotoxicology 1:266–278

Elgrabli D, Floriani M, Abella-Gallart S et al (2008) Biodistribution and clearance of instilled carbon nanotubes in rat lung. Part Fibre Toxicol 5:20

Ema M, Matsuda A, Kobayashi N et al (2011) Evaluation of dermal and eye irritation and skin sensitization due to carbon nanotubes. Regul Toxicol Pharmacol 61:276–281

Ema M, Imamura T, Suzuki H et al (2012) Evaluation of genotoxicity of multi-walled carbon nanotubes in a battery of in vitro and in vivo assays. Regul Toxicol Pharmacol 63:188–195

Ema M, Imamura T, Suzuki H et al (2013a) Genotoxicity evaluation for single-walled carbon nanotubes in a battery of in vitro and in vivo assays. J Appl Toxicol 33:933–939

Ema M, Masumori S, Kobayashi N et al (2013b) In vivo comet assay of multi-walled carbon nanotubes using lung cells of rats intratracheally instilled. J Appl Toxicol 33:1053–1060

Erdely A, Hulderman T, Salmen R et al (2009) Cross-talk between lung and systemic circulation during carbon nanotube respiratory exposure. potential biomarkers. Nano Lett 9:36–43

Erdely A, Liston A, Salmen-Muniz R et al (2011) Identification of systemic markers from a pulmonary carbon nanotube exposure. J Occup Environ Med 53:S80–S86

Erdely A, Dahm M, Chen BT et al (2013) Carbon nanotube dosimetry: From workplace exposure assessment to inhalation toxicology. Part Fibre Toxicol 10:53

Farshad O, Heidari R, Zamiri MJ et al (2020) Spermatotoxic effects of single-walled and multi-walled carbon nanotubes on male mice. Front Vet Sci 7:591558

Fenoglio I, Greco G, Tomatis M et al (2008) Structural defects play a major role in the acute lung toxicity of multiwall carbon nanotubes: physicochemical aspects. Chem Res Toxicol 21:1690–1697

Fiorito S, Serafino A, Andreola F et al (2006) Effects of fullerenes and single-wall carbon nanotubes on murine and human macrophages. Carbon 44:1100–1105

Flahaut E, Durrieu MC, Remy-Zolghadri M et al (2006) Investigation of the cytotoxicity of CCVD carbon nanotubes towards human umbilical vein endothelial cells. Carbon 44:1093–1099

Folkmann JK, Risom L, Jacobsen NR et al (2009) Oxidatively damaged DNA in rats exposed by oral gavage to C60 fullerenes and single-walled carbon nanotubes. Environ Health Perspect 117:703–708

Fraczek A, Menaszek E, Paluszkiewicz C et al (2008) Comparative in vivo biocompatibility study of single- and multi-wall carbon nanotubes. Acta Biomater 4:1593–1602

Galassi TV, Antman-Passig M, Yaari Z et al (2020) Long-term in vivo biocompatibility of single-walled carbon nanotubes. PloS One 15:e0226791

Garibaldi S, Brunelli C, Bavastrello V et al (2006) Carbon nanotube biocompatibility with cardiac muscle cells. Nanotechnology 17:391–397

Garriga R, Herrero-Continente T, Palos M et al (2020) Toxicity of carbon nanomaterials and their potential application as drug delivery systems: in vitro studies in CaCo-2 and MCF-7 cell lines. Nanomaterials (Basel) 10:1617

Garza KM, Murr LE, Soto KF (2008) Cytotoxicity and related inflammatory response for some manufactured metal oxide and carbon nanoparticulate material aggregates. In: Preprints of Extended Abstracts, ACS National Meetings, American Chemical Society Division of Environmental Chemistry, vol 48, pp 281–285

Ge C, Meng L, Xu L et al (2012) Acute pulmonary and moderate cardiovascular responses of spontaneously hypertensive rats after exposure to single-wall carbon nanotubes. Nanotoxicology 6:526–542

Gernand JM, Casman EA (2014) A meta-analysis of carbon nanotube pulmonary toxicity studies-how physical dimensions and impurities affect the toxicity of carbon nanotubes. Risk Anal 34: 583–597

Gobi R, Ravichandiran P, Babu RS et al (2021) Biopolymer and synthetic polymer-based nanocomposites in wound dressing applications: a review. Polymers (Basel) 13:1962

Gong X, Huang D, Liu Y et al (2021) Nanoscale zerovalent iron, carbon nanotubes and biochar facilitated the phytoremediation of cadmium contaminated sediments by changing cadmium fractions, sediments properties and bacterial community structure. Ecotoxicol Environ Saf 208: 111510

Gonzalez-Carter D, Goode AE, Kiryushko D et al (2019) Quantification of blood-brain barrier transport and neuronal toxicity of unlabelled multiwalled carbon nanotubes as a function of surface charge. Nanoscale 11:22054–22069

Grecco AC, Paula RF, Mizutani E et al (2011) Up-regulation of t lymphocyte and antibody production by inflammatory cytokines released by macrophage exposure to multi-walled carbon nanotubes. Nanotechnology 22:265103

Grubek-Jaworska H, Nejman P, Czuminska K et al (2006) Preliminary results on the pathogenic effects of intratracheal exposure to one-dimensional nanocarbons. Carbon 44:1057–1063

Guo J, Zhang X, Li Q et al (2007) Biodistribution of functionalized multiwall carbon nanotubes in mice. Nucl Med Biol 34:579–583

Hamad I, Christy Hunter A, Rutt KJ et al (2008) Complement activation by PEGylated single-walled carbon nanotubes is independent of C1q and alternative pathway turnover. Mol Immunol 45:3797–3803

Hamilton RF, Buford MC, Wood MB et al (2007) Engineered carbon nanoparticles alter macrophage immune function and initiate airway hyper-responsiveness in the BALB/c mouse model. Nanotoxicology 1:104–117

Han SG, Andrews R, Gairola CG et al (2008) Acute pulmonary effects of combined exposure to carbon nanotubes and ozone in mice. Inhal Toxicol 20:391–398

Hedmer M, Isaxon C, Nilsson PT et al (2014) Exposure and emission measurements during production, purification, and functionalization of arc-discharge-produced multi-walled carbon nanotubes. Ann Occup Hyg 58:355

Helfenstein M, Miragoli M, Rohr S et al (2008) Effects of combustion-derived ultrafine particles and manufactured nanoparticles on heart cells in vitro. Toxicology 253:70–78
Hieronimus L, Demazy R, Christiaens L et al (2022) Mouse innate-like B-1 lymphocytes promote inhaled particle-induced in vitro granuloma formation and inflammation in conjunction with macrophages. Arch Toxicol 96:585–599
Hindman B, Ma Q (2018) Carbon nanotubes and crystalline silica induce matrix remodeling and contraction by stimulating myofibroblast transformation in a three-dimensional culture of human pulmonary fibroblasts: role of dimension and rigidity. Arch Toxicol 92:3291–3305
Hojo M, Maeno A, Sakamoto Y et al (2022) Two-year intermittent exposure of a multiwalled carbon nanotube by intratracheal instillation induces lung tumors and pleural mesotheliomas in F344 rats. Part Fibre Toxicol 19:38
Holt BD, Dahl KN, Islam MF (2012a) Cells take up and recover from protein-stabilized single-wall carbon nanotubes with two distinct rates. ACS Nano 6:3481–3490
Holt BD, McCorry MC, Boyer PD et al (2012b) Not all protein-mediated single-wall carbon nanotube dispersions are equally bioactive. Nanoscale 4:7425–7434
Hougaard KS, Jackson P, Kyjovska ZO et al (2013) Effects of lung exposure to carbon nanotubes on female fertility and pregnancy. A study in mice. Reprod Toxicol 41:86–97
Huang X, Zhang F, Sun X et al (2014) The genotype-dependent influence of functionalized multiwalled carbon nanotubes on fetal development. Biomaterials 35:856–865
Huczko A, Lange H (2001) Carbon nanotubes: experimental evidence for a null risk of skin irritation and allergy. Fuller Sci Technol 9:247–250
Huczko A, Lange H, Calstrokko E et al (2001) Physiological testing of carbon nanotubes: are they asbestos-like? Fuller Nanotubes Carbon Nanostr 9:251–254
Huczko A, Lange H, Bystrzejewski M et al (2005) Pulmonary toxicity of 1-D nanocarbon materials. Fuller Nanotubes Carbon Nanostr 13:141–145
Huth K, Glaeske M, Achazi K et al (2018) Fluorescent polymer-single-walled carbon nanotube complexes with charged and noncharged dendronized perylene bisimides for bioimaging studies. Small 14:e1800796
Iijima S (1991) Helical microtubules of graphic carbon. Nature 354:56–58
Inoue K-i, Takano H, Koike E et al (2008) Effects of pulmonary exposure to carbon nanotubes on lung and systemic inflammation with coagulatory disturbance induced by lipopolysaccharide in mice. Exp Biol Med 233:1583–1590
Inoue K, Koike E, Yanagisawa R et al (2009) Effects of multi-walled carbon nanotubes on a murine allergic airway inflammation model. Toxicol Appl Pharmacol 237:306–316
Inoue K, Yanagisawa R, Koike E et al (2010) Repeated pulmonary exposure to single-walled carbon nanotubes exacerbates allergic inflammation of the airway: possible role of oxidative stress. Free Radic Biol Med 48:924–934
Ivani S, Karimi I, Tabatabaei SR (2012) Biosafety of multiwalled carbon nanotube in mice: a behavioral toxicological approach. J Toxicol Sci 37:1191–1205
Jacobsen NR, Pojana G, White P et al (2008) Genotoxicity, cytotoxicity, and reactive oxygen species induced by single-walled carbon nanotubes and C(60) fullerenes in the FE1-mutatrade markmouse lung epithelial cells. Environ Mol Mutagen 49:476–487
Jacobsen NR, Moller P, Jensen KA et al (2009) Lung inflammation and genotoxicity following pulmonary exposure to nanoparticles in Apoe-/- mice. Part Fibre Toxicol 6:2
Jain S, Thakare VS, Das M et al (2011) Toxicity of multiwalled carbon nanotubes with end defects critically depends on their functionalization density. Chem Res Toxicol 24:2028–2039
Jia G, Wang H, Yan L et al (2005) Cytotoxicity of carbon nanomaterials: single-wall nanotube, multi-wall nanotube, and fullerene. Environ Sci Technol 39:1378–1383
Jiang T, Amadei CA, Gou N et al (2020) Toxicity of single-walled carbon nanotubes (SWCNTs): effect of lengths, functional groups and electronic structures revealed by a quantitative toxicogenomics assay. Environ Sci Nano 7:1348–1364

Johansson HKL, Hansen JS, Elfving B et al (2017) Airway exposure to multi-walled carbon nanotubes disrupts the female reproductive cycle without affecting pregnancy outcomes in mice. Part Fibre Toxicol 14:17

Kagan VE, Tyurina YY, Tyurin VA et al (2006) Direct and indirect effects of single walled carbon nanotubes on raw 264.7 macrophages: role of iron. Toxicol Lett 165:88–100

Kagan V, Rao K, Kisin E et al (2007) Pulmonary effects of single-walled carbon nanotubes: inflammatory response, oxidative stress/signaling, and recognition by macrophages. In: Nanotechnology—toxicological issues and environmental safety and environmental safety, pp 65–75

Kaiser JP, Wick P, Manser P et al (2008) Single walled carbon nanotubes (SWCNT) affect cell physiology and cell architecture. J Mater Sci Mater Med 19:1523–1527

Kalbacova M, Kalbac M, Dunsch L et al (2006) The study of the interaction of human mesenchymal stem cells and monocytes/macrophages with single-walled carbon nanotube films. Phys Status Solidi B 243:3514–3518

Kalbacova M, Kalbac M, Dunsch L et al (2007a) Influence of single-walled carbon nanotube films on metabolic activity and adherence of human osteoblasts. Carbon 45:2266–2272

Kalbacova M, Kalbac M, Dunsch L et al (2007b) The effect of SWCNT and nano-diamond films on human osteoblast cells. Phys Status Solidi B 244:4356–4359

Kale N, Nandi S, Patil A et al (2020) Nanocarrier anticancer drug-conjugates cause higher cellular deformations: culpable for mischief. Biomater Sci 8:5729–5738

Kang B, Yu D, Dai Y et al (2009) Biodistribution and accumulation of intravenously administered carbon nanotubes in mice probed by Raman spectroscopy and fluorescent labeling. Carbon 47: 1189–1192

Karlsson HL, Cronholm P, Gustafsson J et al (2008) Copper oxide nanoparticles are highly toxic: a comparison between metal oxide nanoparticles and carbon nanotubes. Chem Res Toxicol 21: 1726–1732

Kateb B, Van Handel M, Zhang L et al (2007) Internalization of MWCNTs by microglia: possible application in immunotherapy of brain tumors. Neuroimage 37(Suppl 1):S9–S17

Kato T, Totsuka Y, Ishino K et al (2013) Genotoxicity of multi-walled carbon nanotubes in both in vitro and in vivo assay systems. Nanotoxicology 7:452–461

Keshavan S, Gupta G, Martin S et al (2021) Multi-walled carbon nanotubes trigger lysosome-dependent cell death (pyroptosis) in macrophages but not in neutrophils. Nanotoxicology 15: 1125–1150

Kim JS, Lee K, Lee YH et al (2011) Aspect ratio has no effect on genotoxicity of multi-wall carbon nanotubes. Arch Toxicol 85:775–786

Kim JS, Sung JH, Song KS et al (2012) Persistent DNA damage measured by comet assay of Sprague Dawley rat lung cells after five days of inhalation exposure and 1 month post-exposure to dispersed multi-wall carbon nanotubes (MWCNTs) generated by new MWCNT aerosol generation system. Toxicol Sci 128:439–448

Kim JS, Song KS, Yu IJ (2013) Evaluation of in vitro and in vivo genotoxicity of single-walled carbon nanotubes. Toxicol Ind Health 31:747

Kim JS, Sung JH, Choi BG et al (2014) In vivo genotoxicity evaluation of lung cells from Fischer 344 rats following 28 days of inhalation exposure to MWCNTs, plus 28 days and 90 days post-exposure. Inhal Toxicol 26:222–234

Kim HB, Jin B, Patel DK et al (2019) Enhanced osteogenesis of human mesenchymal stem cells in presence of single-walled carbon nanotubes. IEEE Trans Nanobioscience 18:463–468

Kim OH, Park JH, Son JI et al (2020) Both intracranial and intravenous administration of functionalized carbon nanotubes protect dopaminergic neuronal death from 6-hydroxydopamine. Int J Nanomedicine 15:7615–7626

Kishore AS, Surekha P, Murthy PB (2009) Assessment of the dermal and ocular irritation potential of multi-walled carbon nanotubes by using in vitro and in vivo methods. Toxicol Lett 191:268–274

Kisin ER, Murray AR, Keane MJ et al (2007) Single-walled carbon nanotubes: geno- and cytotoxic effects in lung fibroblast V79 cells. J Toxicol Environ Health A 70:2071–2079
Klumpp C, Kostarelos K, Prato M et al (2006) Functionalized carbon nanotubes as emerging nanovectors for the delivery of therapeutics. Biochim Biophys Acta 1758:404–412
Konduru NV, Tyurina YY, Feng W et al (2009) Phosphatidylserine targets single-walled carbon nanotubes to professional phagocytes in vitro and in vivo. PloS One 4:e4398
Kroto HW (1992) C60, buckminsterfullerene: the celestial sphere that fell to earth. Angew Chem 31:111–129
Kroto HW, Heath JR, O'Brien SC et al (1985) C60: Buckminsterfullerene. Nature 318:162–163
Kuempel ED (2011) Carbon nanotube risk assessment: implications for exposure and medical monitoring. J Occup Environ Med 53:S91–S97
Kyriakidou K, Brasinika D, Trompeta AFA et al (2020) In vitro cytotoxicity assessment of pristine and carboxyl-functionalized MWCNTs. Food Chem Toxicol 141:111374
Lacerda L, Ali-Boucetta H, Herrero MA et al (2008a) Tissue histology and physiology following intravenous administration of different types of functionalized multiwalled carbon nanotubes. Nanomedicine 3:149–161
Lacerda L, Herrero MA, Venner K et al (2008b) Carbon-nanotube shape and individualization critical for renal excretion. Small 4:1130–1132
Lagana P, Visalli G, Facciola A et al (2021) Is the antibacterial activity of multi-walled carbon nanotubes (MWCNTs) related to antibiotic resistance? An assessment in clinical isolates. Int J Environ Res Public Health 18:9310
Lam C-W, James JT, McCluskey R et al (2004) Pulmonary toxicity of single-wall carbon nanotubes in mice 7 and 90 days after intratracheal instillation. Toxicol Sci 77:126–134
Lee HJ, Park J, Yoon OJ et al (2011) Amine-modified single-walled carbon nanotubes protect neurons from injury in a rat stroke model. Nat Nanotechnol 6:121–125
Legramante JM, Valentini F, Magrini A et al (2009) Cardiac autonomic regulation after lung exposure to carbon nanotubes. Hum Exp Toxicol 28:369–375
Li ZJ, Chapman R, Hulderman T et al (2006) Relationship between pulmonary exposure to multiple doses of single wall carbon nanotubes and atherosclerosis in Apoe -/- mouse model. Toxicologist 90:318. (Abstract # 1555)
Li JG, Li WX, Xu JY et al (2007a) Comparative study of pathological lesions induced by multiwalled carbon nanotubes in lungs of mice by intratracheal instillation and inhalation. Environ Toxicol 22:415–421
Li Z, Hulderman T, Salmen R et al (2007b) Cardiovascular effects of pulmonary exposure to single-wall carbon nanotubes. Environ Health Perspect 115:377–382
Lim JH, Kim SH, Shin IS et al (2011) Maternal exposure to multi-wall carbon nanotubes does not induce embryo-fetal developmental toxicity in rats. Birth Defects Res B Dev Reprod Toxicol 92:69–76
Lin Z, Liu L, Xi Z et al (2012) Single-walled carbon nanotubes promote rat vascular adventitial fibroblasts to transform into myofibroblasts by SM22-alpha expression. Int J Nanomedicine 7: 4199–4206
Lindberg HK, Falck GCM, Suhonen S et al (2008) Genotoxicity of nanomaterials: DNA damage and micronuclei induced by carbon nanotubes and graphite nanofibres in human bronchial epithelial cells in vitro. Toxicol Lett 186:166. https://doi.org/10.1016/j.toxlet.2008.1011.1019
Ling MP, Lin WC, Liu CC et al (2012) Risk management strategy to increase the safety of workers in the nanomaterials industry. J Hazard Mater 229-230:83–93
Liopo AV, Stewart MP, Hudson J et al (2006) Biocompatibility of native and functionalized single-walled carbon nanotubes for neuronal interface. J Nanosci Nanotechnol 6:1365–1374
Liu Z, Cai W, He L et al (2007) In vivo biodistribution and highly efficient tumour targeting of carbon nanotubes in mice. Nat Nanotechnol 2:47–52
Liu A, Sun K, Yang J et al (2008a) Toxicological effects of multi-wall carbon nanotubes in rats. J Nanopart Res 10:1303–1307

Liu Y, Li W-h, Ichihara G et al (2008b) Acute pulmonary toxicity of multi-wall carbon nanotubes in rats. Huanjing Yu Zhiye Yixue 25:121–124

Liu Y, Song W, Li W et al (2008c) In vitro cytotoxicity and oxidative damage effects of multi-wall carbon nanotube on raw264.7 macrophages. Weisheng Yanjiu 37:281–284

Liu Z, Davis C, Cai W et al (2008d) Circulation and long-term fate of functionalized, biocompatible single-walled carbon nanotubes in mice probed by Raman spectroscopy. Proc Natl Acad Sci U S A 105:1410–1415

Lobo AO, Antunes EF, Palma MBS et al (2007) Cell viability and adhesion on as grown vertically aligned carbon nanotubes. In: Advanced Materials Materials Research Society Symposium Proceedings 950E:No pp given, Paper #: 0950-D0904-0916

Lobo AO, Antunes EF, Palma MBS et al (2008) Biocompatibility of multi-walled carbon nanotubes grown on titanium and silicon surfaces. Mater Sci Eng C 28:532–538

Lovat V, Pantarotto D, Lagostena L et al (2005) Carbon nanotube substrates boost neuronal electrical signaling. Nano Lett 5:1107–1110

Lu JH, Hou WC, Tsai MH et al (2022) The impact of background-level carboxylated single-walled carbon nanotubes (SWCNTs-COOH) on induced toxicity in Caenorhabditis elegans and human cells. Int J Environ Res Public Health 19:1218

Luo M, Deng X, Shen X et al (2012) Comparison of cytotoxicity of pristine and covalently functionalized multi-walled carbon nanotubes in raw 264.7 macrophages. J Nanosci Nanotechnol 12:274–283

Lynch RM, Voy BH, Glass DF et al (2007) Assessing the pulmonary toxicity of single-walled carbon nanohorns. Nanotoxicology 1:157–166

Lyra KM, Kaminari A, Panagiotaki KN et al (2021) Multi-walled carbon nanotubes decorated with guanidinylated dendritic molecular transporters: an efficient platform for the selective anticancer activity of doxorubicin. Pharmaceutics 13:858

Magrez A, Kasas S, Salicio V et al (2006) Cellular toxicity of carbon-based nanomaterials. Nano Lett 6:1121–1125

Mahmood M, Vyas K, Plotkin LI et al (2008a) Cytotoxicity of nanostructural materials in mlo-y4 bone cells. In: 235th ACS National Meeting, New Orleans, LA, United States

Mahmood MW, Xu Y, Whitlow M et al (2008b) Cytotoxicity and biological effects of functional nanomaterials delivered to various cell lines. In: 64th Southwest Regional Meeting of the American Chemical Society, Little Rock, AR, United States

Malarkey EB, Parpura V (2007) Applications of carbon nanotubes in neurobiology. Neurodegener Dis 4:292–299

Malarkey EB, Reyes RC, Zhao B et al (2008) Water soluble single-walled carbon nanotubes inhibit stimulated endocytosis in neurons. Nano Lett 8:3538–3542

Mangum JB, Turpin EA, Antao-Menezes A et al (2006) Single-walled carbon nanotube (SWCNT)-induced interstitial fibrosis in the lungs of rats is associated with increased levels of PDGF mRNA and the formation of unique intercellular carbon structures that bridge alveolar macrophages in situ. Part Fibre Toxicol 3:15

Manna SK, Sarkar S, Barr J et al (2005) Single-walled carbon nanotube induces oxidative stress and activates nuclear transcription factor-kappaB in human keratinocytes. Nano Lett 5:1676–1684

Martinelli V, Cellot G, Toma FM et al (2012) Carbon nanotubes promote growth and spontaneous electrical activity in cultured cardiac myocytes. Nano Lett 12:1831–1838

Matthews IP, Gregory CJ, Aljayyoussi G et al (2013) Maximal extent of translocation of single-walled carbon nanotubes from lung airways of the rat. Environ Toxicol Pharmacol 35:461–464

Maynard AD, Baron PA, Foley M et al (2004) Exposure to carbon nanotube material: aerosol release during the handling of unrefined single-walled carbon nanotube material. J Toxicol Environ Health A 67:87–107

Maynard A, Ku B, Emery M et al (2007) Measuring particle size-dependent physicochemical structure in airborne single walled carbon nanotube agglomerates. J Nanopart Res 9:85–92

Mazzatenta A, Giugliano M, Campidelli S et al (2007) Interfacing neurons with carbon nanotubes: electrical signal transfer and synaptic stimulation in cultured brain circuits. J Neurosci 27:6931–6936
McDevitt MR, Chattopadhyay D, Jaggi JS et al (2007a) Pet imaging of soluble yttrium-86-labeled carbon nanotubes in mice. PloS One 2:e907
McDevitt MR, Chattopadhyay D, Kappel BJ et al (2007b) Tumor targeting with antibody-functionalized, radiolabeled carbon nanotubes. J Nucl Med 48:1180–1189
McKenzie JL, Waid MC, Shi R et al (2004) Decreased functions of astrocytes on carbon nanofiber materials. Biomaterials 25:1309–1317
Mehta L, Kumari S, Singh RP (2020) Carbon nanotubes modulate activity of cytotoxic compounds via a trojan horse mechanism. Chem Res Toxicol 33:1206–1214
Meng J, Song L, Kong H et al (2006a) Using single-walled carbon nanotubes nonwoven films as scaffolds to enhance long-term cell proliferation in vitro. J Biomed Mater Res A 79:298–306
Meng J, Song L, Meng J et al (2006b) Using single-walled carbon nanotubes nonwoven films as scaffolds to enhance long-term cell proliferation in vitro. J Biomed Mater Res A 79:298–306
Meng J, Yang M, Song L et al (2009) Concentration control of carbon nanotubes in aqueous solution and its influence on the growth behavior of fibroblasts. Colloids Surf B Biointerfaces 71:148
Mercer RR, Scabilloni J, Wang L et al (2008) Alteration of deposition pattern and pulmonary response as a result of improved dispersion of aspirated single-walled carbon nanotubes in a mouse model. Am J Physiol Lung Cell Mol Physiol 294:L87–L97
Mercer RR, Scabilloni JF, Hubbs AF et al (2013) Extrapulmonary transport of MWCNT following inhalation exposure. Part Fibre Toxicol 10:38
Migliaccio CT, Hamilton RF Jr, Shaw PK et al (2021) Respiratory and systemic impacts following MWCNT inhalation in B6C3F1/N mice. Part Fibre Toxicol 18:16
Migliore L, Saracino D, Bonelli A et al (2010) Carbon nanotubes induce oxidative DNA damage in raw 264.7 cells. Environ Mol Mutagen 51:294–303
Mitchell LA, Gao J, Wal RV et al (2007) Pulmonary and systemic immune response to inhaled multiwalled carbon nanotubes. Toxicol Sci 100:203–214
Mizutani N, Nabe T, Yoshino S (2012) Exposure to multiwalled carbon nanotubes and allergen promotes early- and late-phase increases in airway resistance in mice. Biol Pharm Bull 35:2133–2140
Mohammadi E, Zeinali M, Mohammadi-Sardoo M et al (2020) The effects of functionalization of carbon nanotubes on toxicological parameters in mice. Hum Exp Toxicol 39:1147–1167
Moller P, Folkmann JK, Danielsen PH et al (2012) Oxidative stress generated damage to DNA by gastrointestinal exposure to insoluble particles. Curr Mol Med 12:732–745
Monteiro-Riviere NA, Inman AO, Wang YY et al (2005a) Surfactant effects on carbon nanotube interactions with human keratinocytes. Nanomedicine 1:293–299
Monteiro-Riviere NA, Nemanich RJ, Inman AO et al (2005b) Multi-walled carbon nanotube interactions with human epidermal keratinocytes. Toxicol Lett 155:377–384
Montes-Fonseca SL, Orrantia-Borunda E, Aguilar-Elguezabal A et al (2012) Cytotoxicity of functionalized carbon nanotubes in J774A macrophages. Nanomedicine 8:853–859
Mooney E, Dockery P, Greiser U et al (2008) Carbon nanotubes and mesenchymal stem cells: biocompatibility, proliferation and differentiation. Nano Lett 8:2137–2143
Mostovenko E, Saunders S, Muldoon PP et al (2021) Carbon nanotube exposure triggers a cerebral peptidomic response: barrier compromise, neuroinflammation, and a hyperexcited state. Toxicol Sci 182:107–119
Muller J, Huaux F, Moreau N et al (2005) Respiratory toxicity of multi-wall carbon nanotubes. Toxicol Appl Pharmacol 207:221–231
Muller J, Decordier I, Hoet PH et al (2008a) Clastogenic and aneugenic effects of multi-wall carbon nanotubes in epithelial cells. Carcinogenesis 29:427–433
Muller J, Huaux F, Fonseca A et al (2008b) Structural defects play a major role in the acute lung toxicity of multiwall carbon nanotubes: toxicological aspects. Chem Res Toxicol 21:1698–1705

Murr LE, Garza KM, Soto KF et al (2005) Cytotoxicity assessment of some carbon nanotubes and related carbon nanoparticle aggregates and the implications for anthropogenic carbon nanotube aggregates in the environment. Int J Environ Res Public Health 2:31–42

Murr LE, Soto KF, Garza KM et al (2006) Combustion-generated nanoparticulates in the El Paso, TX, USA/Juarez, Mexico Metroplex: their comparative characterization and potential for adverse health effects. Int J Environ Res Public Health 3:48–66

Murray AR, Kisin E, Leonard SS et al (2009) Oxidative stress and inflammatory response in dermal toxicity of single-walled carbon nanotubes. Toxicology 257:161–171

Nahle S, Cassidy H, Leroux MM et al (2020) Genes expression profiling of alveolar macrophages exposed to non-functionalized, anionic and cationic multi-walled carbon nanotubes shows three different mechanisms of toxicity. J Nanobiotechnol 18:36

Nam CW, Kang SJ, Kang YK et al (2011) Cell growth inhibition and apoptosis by SDS-solubilized single-walled carbon nanotubes in normal rat kidney epithelial cells. Arch Pharm Res 34:661–669

Naya M, Kobayashi N, Mizuno K et al (2011) Evaluation of the genotoxic potential of single-wall carbon nanotubes by using a battery of in vitro and in vivo genotoxicity assays. Regul Toxicol Pharmacol 61:192–198

Naya M, Kobayashi N, Endoh S et al (2012) In vivo genotoxicity study of single-wall carbon nanotubes using comet assay following intratracheal instillation in rats. Regul Toxicol Pharmacol 64:124–129

Ndika J, Karisola P, Kinaret P et al (2021) Profiling non-coding RNA changes associated with 16 different engineered nanomaterials in a mouse airway exposure model. Cell 10:1085

Negri V, Pacheco-Torres J, Calle D et al (2020) Carbon nanotubes in biomedicine. Top Curr Chem (Cham) 378:15

Nemmar A, Hoet PH, Vandervoort P et al (2007) Enhanced peripheral thrombogenicity after lung inflammation is mediated by platelet-leukocyte activation: Role of p-selectin. J Thromb Haemost 5:1217–1226

Nguyen-Vu TD, Chen H, Cassell AM et al (2007) Vertically aligned carbon nanofiber architecture as a multifunctional 3-D neural electrical interface. IEEE Trans Biomed Eng 54:1121–1128

Ni Y, Hu H, Malarkey EB et al (2005) Chemically functionalized water soluble single-walled carbon nanotubes modulate neurite outgrowth. J Nanosci Nanotechnol 5:1707–1712

Nicola MD, Gattia DM, Bellucci S et al (2007) Effect of different carbon nanotubes on cell viability and proliferation. J Phys Condens Matter 19:395013

Nicola MD, Bellucci S, Traversa E et al (2008) Carbon nanotubes on Jurkat cells: effects on cell viability and plasma membrane potential. J Phys Condens Matter 20:474204

Nikeafshar S, Khazaei A, Tahvilian R (2022) Inhibition of methamphetamine-induced cytotoxicity in the U87-cell line by atorvastatin-conjugated carbon nanotubes. Appl Biochem Biotechnol 194:479–503

Nimmagadda A, Thurston K, Nollert MU et al (2006) Chemical modification of SWNT alters in vitro cell-SWNT interactions. J Biomed Mater Res A 76A:614–625

Nirmal NK, Awasthi KK, John PJ (2017) Effects of hydroxyl-functionalized multiwalled carbon nanotubes on sperm health and testes of Wistar rats. Toxicol Ind Health 33:519–529

Nygaard UC, Hansen JS, Samuelsen M et al (2009) Single-walled and multi-walled carbon nanotubes promote allergic immune responses in mice. Toxicol Sci 109:113–123

Nygaard UC, Samuelsen M, Marioara CD et al (2013) Carbon nanofibers have IgE adjuvant capacity but are less potent than nanotubes in promoting allergic airway responses. Biomed Res Int 2013:476010

Oliveira FC, Carvalho JO, Magalhaes L et al (2021) Biomineralization inspired engineering of nanobiomaterials promoting bone repair. Mater Sci Eng C Mater Biol Appl 120:111776

Pacurari M, Yin XJ, Ding M et al (2008) Oxidative and molecular interactions of multi-wall carbon nanotubes (MWCNT) in normal and malignant human mesothelial cells. Nanotoxicology 2: 155–170

Pantarotto D, Briand JP, Prato M et al (2004a) Translocation of bioactive peptides across cell membranes by carbon nanotubes. Chem Commun:16–17

Pantarotto D, Singh R, McCarthy D et al (2004b) Functionalized carbon nanotubes for plasmid DNA gene delivery. Angew Chem 43:5242–5246

Park KH, Chhowalla M, Iqbal Z et al (2003) Single-walled carbon nanotubes are a new class of ion channel blockers. J Biol Chem 278:50212–50216

Patel SC, Alam O, Sitharaman B (2018) Osteogenic differentiation of human adipose derived stem cells on chemically crosslinked carbon nanomaterial coatings. J Biomed Mater Res A 106: 1189–1199

Patlolla AK, Hussain SM, Schlager JJ et al (2010) Comparative study of the clastogenicity of functionalized and nonfunctionalized multiwalled carbon nanotubes in bone marrow cells of Swiss-Webster mice. Environ Toxicol 25:608–621

Pedziwiatr-Werbicka E, Horodecka K, Shcharbin D et al (2021) Nanoparticles in combating cancer: opportunities and limitations. A brief review. Curr Med Chem 28:346–359

Pei L, Yang W, Cao Y (2022) Influences of unmodified and carboxylated carbon nanotubes on lipid profiles in THP-1 macrophages: a lipidomics study. Int J Toxicol 41:16–25

Pelka J, Gehrke H, Rechel A et al (2013) DNA damaging properties of single walled carbon nanotubes in human colon carcinoma cells. Nanotoxicology 7:2–20

Pensabene V, Vittorio O, Raffa V et al (2007) Investigation of CNTs interaction with fibroblast cells. Conf Proc IEEE Eng Med Biol Soc 2007:6621–6624

Philbrook NA, Walker VK, Afrooz AR et al (2011) Investigating the effects of functionalized carbon nanotubes on reproduction and development in drosophila melanogaster and CD-1 mice. Reprod Toxicol 32:442–448

Poland CA, Duffin R, Kinloch I et al (2008) Carbon nanotubes introduced into the abdominal cavity of mice show asbestos-like pathogenicity in a pilot study. Nat Nanotechnol 3:423–428

Ponti J, Broggi F, Mariani V et al (2013) Morphological transformation induced by multiwall carbon nanotubes on BALB/3T3 cell model as an in vitro end point of carcinogenic potential. Nanotoxicology 7:221–233

Porter DW, Hubbs AF, Chen BT et al (2013) Acute pulmonary dose-responses to inhaled multi-walled carbon nanotubes. Nanotoxicology 7:1179–1194

Prajapati SK, Malaiya A, Kesharwani P et al (2022) Biomedical applications and toxicities of carbon nanotubes. Drug Chem Toxicol 45:435–450

Prylutska SV, Grynyuk II, Matyshevska OP et al (2008) Estimation of multi-walled carbon nanotubes toxicity in vitro. Phys E Low Dimens Syst Nanostruct 40:2565–2569

Prylutska SV, Grynyuk II, Grebinyk SM et al (2009) Comparative study of biological action of fullerenes C60 and carbon nanotubes in thymus cells. Materwiss Werksttech 40:238–241

Prylutska S, Bilyy R, Shkandina T et al (2013) Comparative study of membranotropic action of single- and multi-walled carbon nanotubes. J Biosci Bioeng 115:674–679

Pulskamp K (2008) Influence of carbon nanoparticles on cell physiology—mechanistic studies on the toxic effects. Wissenschaftliche Berichte FZKA 7401:A-200

Pulskamp K, Diabate S, Krug HF (2007) Carbon nanotubes show no sign of acute toxicity but induce intracellular reactive oxygen species in dependence on contaminants. Toxicol Lett 168: 58–74

Qu Y, Yang B, Jiang X et al (2017) Multiwalled carbon nanotubes inhibit steroidogenesis by disrupting steroidogenic acute regulatory protein expression and redox status. J Nanosci Nanotechnol 17:914–925

Rabiee N, Bagherzadeh M, Ghadiri AM et al (2021) Turning toxic nanomaterials into a safe and bioactive nanocarrier for co-delivery of DOX/pCRISPR. ACS Appl Bio Mater 4:5336–5351

Radomski A, Jurasz P, Alonso-Escolano D et al (2005) Nanoparticle-induced platelet aggregation and vascular thrombosis. Br J Pharmacol 146:882–893

Rahamathulla M, Bhosale RR, Osmani RAM et al (2021) Carbon nanotubes: current perspectives on diverse applications in targeted drug delivery and therapies. Materials (Basel) 14:6707

Raja Pavan MV, Connolley J, Ganesan Gopal P et al (2007) Impact of carbon nanotube exposure, dosage and aggregation on smooth muscle cells. Toxicol Lett 169:51–63

Ramanunny AK, Wadhwa S, Gulati M et al (2021) Nanocarriers for treatment of dermatological diseases: principle, perspective and practices. Eur J Pharmacol 890:173691

Reddy AR, Krishna DR, Reddy YN et al (2010) Translocation and extra pulmonary toxicities of multi wall carbon nanotubes in rats. Toxicol Mech Methods 20:267–272

Requardt H, Braun A, Steinberg P et al (2019) Surface defects reduce carbon nanotube toxicity in vitro. Toxicol In Vitro 60:12–18

Roman JA, Niedzielko TL, Haddon RC et al (2011) Single-walled carbon nanotubes chemically functionalized with polyethylene glycol promote tissue repair in a rat model of spinal cord injury. J Neurotrauma 28:2349–2362

Rotoli BM, Bussolati O, Bianchi MG et al (2008) Non-functionalized multi-walled carbon nanotubes alter the paracellular permeability of human airway epithelial cells. Toxicol Lett 178:95–102

Ryman-Rasmussen JP, Tewksbury EW, Moss OR et al (2009) Inhaled multi-walled carbon nanotubes potentiate airway fibrosis in murine allergic asthma. Am J Respir Cell Mol Biol 40:349–358

Sager TM, Umbright CM, Mustafa GM et al (2022) Pulmonary toxicity and gene expression changes in response to whole-body inhalation exposure to multi-walled carbon nanotubes in rats. Inhal Toxicol 34:200–218

Sakaguchi N, Watari F, Yokoyama A et al (2008) High-resolution electron microscopy of multi-wall carbon nanotubes in the subcutaneous tissue of rats. J Electron Microsc (Tokyo) 57:159–164

Saleh DM, Luo S, Ahmed OHM et al (2022) Assessment of the toxicity and carcinogenicity of double-walled carbon nanotubes in the rat lung after intratracheal instillation: a two-year study. Part Fibre Toxicol 19:30

Salih SJ, Ghobadi MZ (2022) Evaluating the cytotoxicity and pathogenicity of multi-walled carbon nanotube through weighted gene co-expression network analysis: a nanotoxicogenomics study. BMC Genom Data 23:12

Salvador-Morales C, Townsend P, Flahaut E et al (2007) Binding of pulmonary surfactant proteins to carbon nanotubes; potential for damage to lung immune defense mechanisms. Carbon 45: 607–617

Sano M, Izumiya M, Haniu H et al (2020) Cellular responses of human lymphatic endothelial cells to carbon nanomaterials. Nanomaterials (Basel) 10:1374

Sarecka-Hujar B, Banys A, Ostrozka-Cieslik A et al (2020) Evaluation of the potential of nanoparticles containing active substances in selected chronic diseases. Adv Clin Exp Med 29:385–397

Sarkar S, Sharma C, Yog R et al (2007) Analysis of stress responsive genes induced by single-walled carbon nanotubes in BJ foreskin cells. J Nanosci Nanotechnol 7:584–592

Sato Y, Shibata K, Kataoka H et al (2005a) Strict preparation and evaluation of water-soluble hat-stacked carbon nanofibers for biomedical application and their high biocompatibility: influence of nanofiber-surface functional groups on cytotoxicity. Mol Biosyst 1:142–145

Sato Y, Yokoyama A, Shibata K et al (2005b) Influence of length on cytotoxicity of multi-walled carbon nanotubes against human acute monocytic leukemia cell line THP-1 in vitro and subcutaneous tissue of rats in vivo. Mol Biosyst 1:176–182

Sato Y, Yokoyama A, Kasai T et al (2008) In vivo rat subcutaneous tissue response of binder-free multi-walled carbon nanotube blocks cross-linked by de-fluorination. Carbon 46:1927–1934

Saxena RK, Williams W, McGee JK et al (2007) Enhanced *in vitro* and *in vivo* toxicity of poly-dispersed acid-functionalized single-wall carbon nanotubes. Nanotoxicology 1:291–300

Sayers BC, Taylor AJ, Glista-Baker EE et al (2013) Role of cyclooxygenase-2 in exacerbation of allergen-induced airway remodeling by multiwalled carbon nanotubes. Am J Respir Cell Mol Biol 49:525–535

Sayes CM, Liang F, Hudson JL et al (2006) Functionalization density dependence of single-walled carbon nanotubes cytotoxicity in vitro. Toxicol Lett 161:135–142

Schrand AM, Dai L, Schlager JJ et al (2007) Differential biocompatibility of carbon nanotubes and nanodiamonds. Diamond Relat Mater 16:2118–2123

Seidel C, Zhernovkov V, Cassidy H et al (2021) Inhaled multi-walled carbon nanotubes differently modulate global gene and protein expression in rat lungs. Nanotoxicology 15:238–256

Shaik AS, Shaik AP, Bammidi VK et al (2019) Effect of polyethylene glycol surface charge functionalization of SWCNT on the in vitro and in vivo nanotoxicity and biodistribution monitored noninvasively using MRI. Toxicol Mech Methods 29:233–243

Sharma CS, Sarkar S, Periyakaruppan A et al (2007) Single-walled carbon nanotubes induces oxidative stress in rat lung epithelial cells. J Nanosci Nanotechnol 7:2466–2472

Shen M, Wang SH, Shi X et al (2009) Polyethyleneimine-mediated functionalization of multiwalled carbon nanotubes: synthesis, characterization, and in vitro toxicity assay. J Phys Chem C 113: 3150–3156

Shvedova A, Castranova V, Kisin E et al (2003) Exposure to carbon nanotube material: assessment of nanotube cytotoxicity using human keratinocyte cells. J Toxicol Environ Health A 66:1909–1926

Shvedova AA, Kisin ER, Mercer R et al (2005) Unusual inflammatory and fibrogenic pulmonary responses to single-walled carbon nanotubes in mice. Am J Physiol Lung Cell Mol Physiol 289: L698–L708

Shvedova AA, Kisin ER, Murray AR et al (2007) Vitamin e deficiency enhances pulmonary inflammatory response and oxidative stress induced by single-walled carbon nanotubes in C57bl/6 mice. Toxicol Appl Pharmacol 221:339–348

Shvedova AA, Fabisiak JP, Kisin ER et al (2008a) Sequential exposure to carbon nanotubes and bacteria enhances pulmonary inflammation and infectivity. Am J Respir Cell Mol Biol 38:579–590

Shvedova AA, Kisin E, Murray AR et al (2008b) Inhalation vs. aspiration of single-walled carbon nanotubes in C57bl/6 mice: inflammation, fibrosis, oxidative stress, and mutagenesis. Am J Physiol Lung Cell Mol Physiol 295:L552–L565

Shvedova AA, Kisin ER, Murray AR et al (2008c) Increased accumulation of neutrophils and decreased fibrosis in the lung of NADPH oxidase-deficient c57bl/6 mice exposed to carbon nanotubes. Toxicol Appl Pharmacol 231:235–240

Simon-Deckers A, Gouget B, Mayne-L'Hermite M et al (2008) In vitro investigation of oxide toxicity and intracellular accumulation in A549 human pneumocytes. Toxicology 253:137–146

Singh R, Pantarotto D, Lacerda L et al (2006) Tissue biodistribution and blood clearance rates of intravenously administered carbon nanotube radiotracers. Proc Natl Acad Sci U S A 103:3357–3362

Singh RP, Das M, Thakare V et al (2012) Functionalization density dependent toxicity of oxidized multiwalled carbon nanotubes in a murine macrophage cell line. Chem Res Toxicol 25:2127–2137

Sohaebuddin SK, Thevenot PT, Baker D et al (2010) Nanomaterial cytotoxicity is composition, size, and cell type dependent. Part Fibre Toxicol 7:22

Stapleton PA, Minarchick VC, Cumpston AM et al (2012) Impairment of coronary arteriolar endothelium-dependent dilation after multi-walled carbon nanotube inhalation: a time-course study. Int J Mol Sci 13:13781–13803

Sung JH, Ji JH, Yoon JU et al (2008) Lung function changes in Sprague-Dawley rats after prolonged inhalation exposure to silver nanoparticles. Inhal Toxicol 20:567–574

Swedin L, Arrighi R, Andersson-Willman B et al (2012) Pulmonary exposure to single-walled carbon nanotubes does not affect the early immune response against toxoplasma gondii. Part Fibre Toxicol 9:16

Szendi K, Varga C (2008) Lack of genotoxicity of carbon nanotubes in a pilot study. Anticancer Res 28:349–352

Tabet L, Bussy C, Amara N et al (2009) Adverse effects of industrial multiwalled carbon nanotubes on human pulmonary cells. J Toxicol Environ Health A 72:60–73
Takagi A, Hirose A, Nishimura T et al (2008) Induction of mesothelioma in p53+/- mouse by intraperitoneal application of multi-wall carbon nanotube. J Toxicol Sci 33:105–116
Takanashi S, Hara K, Aoki K et al (2012) Carcinogenicity evaluation for the application of carbon nanotubes as biomaterials in rash2 mice. Sci Rep 2:498
Takeuchi T, Iizumi Y, Yudasaka M et al (2019) Characterization and biodistribution analysis of oxygen-doped single-walled carbon nanotubes used as in vivo fluorescence imaging probes. Bioconjug Chem 30:1323–1330
Taylor-Just AJ, Ihrie MD, Duke KS et al (2020) The pulmonary toxicity of carboxylated or aminated multi-walled carbon nanotubes in mice is determined by the prior purification method. Part Fibre Toxicol 17:60
Teixeira-Santos R, Gomes M, Gomes LC et al (2021) Antimicrobial and anti-adhesive properties of carbon nanotube-based surfaces for medical applications: a systematic review. iScience 24: 102001
Terrones M, Terrones H (2003) The carbon nanocosmos: novel materials for the twenty-first century. Phil Trans R Soc Lond A 361:2789–2806
Thompson LC, Frasier CR, Sloan RC et al (2014) Pulmonary instillation of multi-walled carbon nanotubes promotes coronary vasoconstriction and exacerbates injury in isolated hearts. Nanotoxicology 8:38–49
Tian F, Cui D, Schwarz H et al (2006) Cytotoxicity of single-wall carbon nanotubes on human fibroblasts. Toxicol In Vitro 20:1202–1212
Tohji K, Sato Y (2006) Influence of purity and morphology on tissue reactivity of carbon nanotubes. Baiomateriaru 24:333–344
Tsyboulski DA, Liopo AV, Su R et al (2013) Enabling in vivo measurements of nanoparticle concentrations with three-dimensional optoacoustic tomography. Biophotonics 7:581
Urankar RN, Lust RM, Mann E et al (2012) Expansion of cardiac ischemia/reperfusion injury after instillation of three forms of multi-walled carbon nanotubes. Part Fibre Toxicol 9:38
Vanhandel M, Alizadeh D, Zhang L et al (2009) Selective uptake of multi-walled carbon nanotubes by tumor macrophages in a murine glioma model. J Neuroimmunol 208:3–9
Walker VG, Li Z, Hulderman T et al (2009) Potential in vitro effects of carbon nanotubes on human aortic endothelial cells. Toxicol Appl Pharmacol 236:319
Wang H, Wang J, Deng X, Sun H, Shi Z, Gu Z et al (2004) Biodistribution of carbon single-wall carbon nanotubes in mice. J Nanosci Nanotechnol 4:1019–1024
Wang X, Deng X, Wang H et al (2007) Bio-effects of water soluble taurine multi-wall carbon nanotubes on lungs of mice. Zhonghua Yu Fang Yi Xue Za Zhi 41:85–90
Wang J, Deng X, Yang S et al (2008) Rapid translocation and pharmacokinetics of hydroxylated single-walled carbon nanotubes in mice. Nanotoxicology 2:28–32
Wang J, Sun P, Bao Y et al (2011) Cytotoxicity of single-walled carbon nanotubes on pc12 cells. Toxicol In Vitro 25:242–250
Wang J, Sun P, Bao Y et al (2012a) Vitamin e renders protection to pc12 cells against oxidative damage and apoptosis induced by single-walled carbon nanotubes. Toxicol In Vitro 26:32–41
Wang X, Guo J, Chen T et al (2012b) Multi-walled carbon nanotubes induce apoptosis via mitochondrial pathway and scavenger receptor. Toxicol In Vitro 26:799–806
Wang J, Tian X, Zhang J et al (2021) Postchronic single-walled carbon nanotube exposure causes irreversible malignant transformation of human bronchial epithelial cells through DNA methylation changes. ACS Nano 15:7094–7104
Warheit DB, Laurence BR, Reed KL et al (2004) Comparative pulmonary toxicity assessment of single-wall carbon nanotubes in rats. Toxicol Sci 77:117–125
Warheit DB, Laurence BR, Reed KL et al (2005) Lung toxicity bioassay study in rats with single-wall carbon nanotubes. ACS Symp Ser 890:67–90
Wei Q, Zhan L, Juanjuan B et al (2012) Biodistribution of co-exposure to multi-walled carbon nanotubes and nanodiamonds in mice. Nanoscale Res Lett 7:473

Wils RS, Jacobsen NR, Di Ianni E et al (2021a) Reactive oxygen species production, genotoxicity and telomere length in FE1-mutamouse lung epithelial cells exposed to carbon nanotubes. Nanotoxicology 15:661–672

Wils RS, Jacobsen NR, Vogel U et al (2021b) Inflammatory response, reactive oxygen species production and DNA damage in mice after intrapleural exposure to carbon nanotubes. Toxicol Sci 183:184–194

Wirnitzer U, Herbold B, Voetz M et al (2008) Studies on the in vitro genotoxicity of baytubes((r)), agglomerates of engineered multi-walled carbon-nanotubes (MWCNT). Toxicol Lett 186:160

Worle-Knirsch JM, Pulskamp K, Krug HF (2006) Oops they did it again! Carbon nanotubes hoax scientists in viability assays. Nano Lett 6:1261–1268

Wu W, Wieckowski S, Pastorin G et al (2005) Targeted delivery of amphotericin b to cells by using functionalized carbon nanotubes. Angew Chem 44:6358–6362

Wu H, Liu G, Zhuang Y et al (2011) The behavior after intravenous injection in mice of multiwalled carbon nanotube/Fe3O4 hybrid MRI contrast agents. Biomaterials 32:4867–4876

Xiang C, Zhang Y, Guo W et al (2020) Biomimetic carbon nanotubes for neurological disease therapeutics as inherent medication. Acta Pharm Sin B 10:239–248

Xiao-Feng P, Le-Wei L, Qiang Z et al (2005) Biological effects of the carbon nanotubes. Conf Proc IEEE Eng Med Biol Soc 2:1240–1243

Xu YY, Yang J, Shen T et al (2012) Intravenous administration of multi-walled carbon nanotubes affects the formation of atherosclerosis in Sprague-Dawley rats. J Occup Health 54:361–369

Yacobi NR, Phuleria HC, Demaio L et al (2007) Nanoparticle effects on rat alveolar epithelial cell monolayer barrier properties. Toxicol In Vitro 21:1373–1381

Yadav A, Singh S, Sohi H et al (2021) Advances in delivery of chemotherapeutic agents for cancer treatment. AAPS PharmSciTech 23:25

Yamaguchi A, Fujitani T, Ohyama K et al (2012) Effects of sustained stimulation with multi-wall carbon nanotubes on immune and inflammatory responses in mice. J Toxicol Sci 37:177–189

Yang K, Liu Z (2012) In vivo biodistribution, pharmacokinetics, and toxicology of carbon nanotubes. Curr Drug Metab 13:1057–1067

Yang H, Yang D-f, Zhang H-s et al (2007a) Study on cytotoxicity of four typical nanomaterials in mouse embryo fibroblasts. Shengtai Duli Xuebao 2:427–434

Yang ST, Guo W, Lin Y et al (2007b) Biodistribution of pristine single-walled carbon nanotubes in vivo. J Phys Chem C 111:17761–17764

Yang H, Liu C, Yang D et al (2008a) Comparative study of cytotoxicity, oxidative stress and genotoxicity induced by four typical nanomaterials: the role of particle size, shape and composition. J Appl Toxicol 29:69–78

Yang ST, Fernando KA, Liu JH et al (2008b) Covalently pegylated carbon nanotubes with stealth character in vivo. Small 4:940–944

Yang ST, Wang X, Jia G et al (2008c) Long-term accumulation and low toxicity of single-walled carbon nanotubes in intravenously exposed mice. Toxicol Lett 181:182–189

Yang M, Flavin K, Kopf I et al (2013) Functionalization of carbon nanoparticles modulates inflammatory cell recruitment and NLRP3 inflammasome activation. Small 9:4194–4206

Yang D, Li T, Xu M et al (2014) Graphene oxide promotes the differentiation of mouse embryonic stem cells to dopamine neurons. Nanomedicine 9:2445

Yang D, Yang SJ, Del Bonis-O'Donnell JT et al (2020a) Mitigation of carbon nanotube neurosensor induced transcriptomic and morphological changes in mouse microglia with surface passivation. ACS Nano 14:13794–13805

Yang T, Chen J, Gao L et al (2020b) Induction of lipid droplets in thp-1 macrophages by multi-walled carbon nanotubes in a diameter-dependent manner: a transcriptomic study. Toxicol Lett 332:65–73

Yao Y, Zhang T, Tang M (2022) A critical review of advances in reproductive toxicity of common nanomaterials to caenorhabditis elegans and influencing factors. Environ Pollut 306:119270

Ye S, Jiang Y, Zhang H et al (2012) Multi-walled carbon nanotubes induce apoptosis in raw 264.7 cell-derived osteoclasts through mitochondria-mediated death pathway. J Nanosci Nanotechnol 12:2101–2112

Ye L, Ji H, Liu J et al (2021) Carbon nanotube-hydrogel composites facilitate neuronal differentiation while maintaining homeostasis of network activity. Adv Mater 33:e2102981

Yehia Hadi N, Draper Rockford K, Mikoryak C et al (2007) Single-walled carbon nanotube interactions with hela cells. J Nanobiotechnol 5:8

Yokoyama A, Sato Y, Nodasaka Y et al (2005) Biological behavior of hat-stacked carbon nanofibers in the subcutaneous tissue in rats. Nano Lett 5:157–161

Yu KN, Kim JE, Seo HW et al (2013) Differential toxic responses between pristine and functionalized multiwall nanotubes involve induction of autophagy accumulation in murine lung. J Toxicol Environ Health A 76:1282–1292

Zeni O, Palumbo R, Bernini R et al (2008) Cytotoxicity investigation on cultured human blood cells treated with single-wall carbon nanotubes. Sensors 8:488–499

Zhang LW, Zeng L, Barron AR et al (2007) Biological interactions of functionalized single-wall carbon nanotubes in human epidermal keratinocytes. Int J Toxicol 26:103–113

Zhang J-C, Ji X-Y, Liu C-L et al (2008) Effect of single-walled carbon nanotubes on primary immune cells in vitro. Front Mater Sci China 2:228–232

Zhang Y, Ali SF, Dervishi E et al (2010) Cytotoxicity effects of graphene and single-wall carbon nanotubes in neural phaeochromocytoma-derived PC12 cells. ACS Nano 4:3181–3186

Zhang Y, Wang B, Meng X et al (2011a) Influences of acid-treated multiwalled carbon nanotubes on fibroblasts: proliferation, adhesion, migration, and wound healing. Ann Biomed Eng 39:414–426

Zhang Y, Xu Y, Li Z et al (2011b) Mechanistic toxicity evaluation of uncoated and pegylated single-walled carbon nanotubes in neuronal PC12 cells. ACS Nano 5:7020–7033

Zhang T, Tang M, Kong L et al (2012) Comparison of cytotoxic and inflammatory responses of pristine and functionalized multi-walled carbon nanotubes in raw 264.7 mouse macrophages. J Hazard Mater 219-220:203–212

Zhang M, Wang W, Cui Y et al (2018) Magnetofluorescent carbon quantum dot decorated multiwalled carbon nanotubes for dual-modal targeted imaging in chemo-photothermal synergistic therapy. ACS Biomater Sci Eng 4:151–162

Zhang M, Xu Y, Yang M et al (2021a) Comparative assessments of the biodistribution and toxicity of oxidized single-walled carbon nanotubes dispersed with two different reagents after intravenous injection. Nanotoxicology 15:798–811

Zhang W, Zhao L, Gao C et al (2021b) Highly resilient, biocompatible, and antibacterial carbon nanotube/hydroxybutyl chitosan sponge dressing for rapid and effective hemostasis. J Mater Chem B 9:9754–9763

Zhang X, Luo M, Zhang J et al (2021c) Carbon nanotubes promote alveolar macrophages toward M2 polarization mediated epithelial-mesenchymal transition and fibroblast-to-myofibroblast transdifferentiation. Nanotoxicology 15:588–604

Zhao X, Chang S, Long J et al (2019) The toxicity of multi-walled carbon nanotubes (MWCNTs) to human endothelial cells: the influence of diameters of MWCNTs. Food Chem Toxicol 126:169–177

Zhao Y, Xu R, Hua X et al (2022) Multi-walled carbon nanotubes induce transgenerational toxicity associated with activation of germline long non-coding RNA linc-7 in C. elegans. Chemosphere 301:134687

Zhiqing L, Zhuge X, Fuhuan C et al (2010) Icam-1 and vcam-1 expression in rat aortic endothelial cells after single-walled carbon nanotube exposure. J Nanosci Nanotechnol 10:8562–8574

Zhu L, Chang DW, Dai L et al (2007) DNA damage induced by multiwalled carbon nanotubes in mouse embryonic stem cells. Nano Lett 7:3592–3597

Recent Trends in Topical Herbal Nanotherapeutics for Psoriasis

23

Deepti Katiyar, Abhishek Kumar, Priya Bansal, Surya Prakash, and N. G. Raghavendra Rao

Abstract

Psoriasis, an autoimmune inflammatory skin ailment, is among the most prevalent immune-mediated illnesses afflicting people worldwide. At the moment, traditional psoriasis therapeutic approaches have a lot of shortcomings such as limited effectiveness, itchy skin, immunosuppression, and so forth. As a result, researchers must develop a much more effective and promising medication strategy for psoriasis therapies. Because of its restorative potential widely in terms of drug administration routes, natural medicine continues to be a significant source of potential drug development. Nevertheless, the typical composition of psoriasis conventional medicines is ineffective, limiting any usage of herbal remedies. Given the constraints noted previously, the utilization of nanocarriers in the manufacture of these therapeutic plant-based natural remedies might be extremely advantageous in improving the effectiveness of topical medications. Emerging data suggests the use of nanocarriers in a transdermal formulation as a promising strategy for improved efficacy and drug uptake directed to the target

D. Katiyar
Department of Pharmacognosy, KIET School of Pharmacy, KIET Group of Institutions, Ghaziabad, UP, India

A. Kumar · P. Bansal
Department of Pharmacology, KIET School of Pharmacy, KIET Group of Institutions, Ghaziabad, UP, India

S. Prakash
Department of Pharmaceutical Chemistry, KIET School of Pharmacy, KIET Group of Institutions, Ghaziabad, UP, India

N. G. R. Rao (✉)
Department of Pharmaceutics, KIET School of Pharmacy, KIET Group of Institutions, Delhi-NCR, Ghaziabad, UP, India
e-mail: raghavendra.rao@kiet.edu

P. P. Singh (ed.), *Recent Advances in Pharmaceutical Innovation and Research*,
https://doi.org/10.1007/978-981-99-2302-1_23

tissue with minimal adverse impact. This chapter emphasizes the pathogenic process of psoriasis, traditional medicinal formulation, bioactive constituents of medicinal herbs, and nano preparation utilized in the management of the psoriatic disorder.

Keywords

Herbal · Nanotherapeutics · Psoriasis · Topical administration

23.1 Introduction

Psoriasis is among the most prevalent immune-mediated illness disorders, in addition to autoimmune inflammatory skin ailment that affects people all over the world (Chen et al. 2019). It has been linked to a slew of issues, including higher incidence, comorbidities, and chronicity in patients' physical and mental health (Min et al. 2020). This disease is further distinguished by red-colored plaques that develop anywhere on the skin (especially on the extensor regions of the sacral regions, elbows, knees, and head) and are generally escorted with silvery or colorless multilayer flakes (Mabuchi et al. 2012). Psoriatic lesions were histopathologically observed to have neutrophil recruitment and hyperplastic epithelium and immune-cell infiltration in the corium, and enhanced angiogenesis (Na et al. 2019).

Transdermal medication administration is commonly regarded as the primary line of therapy approach used clinically for good patient compliance in any occurrence of skin-related disorders. However, topical medicine administration is fraught with difficulties. The skin is the most important organ for delivering medicine. The skin's primary function of resistance to infections and physical harm makes it difficult to permeate. The epidermal, dermal, and subcutaneous tissue layers comprise the skin. The epidermal layer is even further differentiated into strata, papillary layer, and reticular layer, which is the main barrier to percutaneous diffusion on topical administration of drugs (Ahmad et al. 2018). At the present, standard psoriasis treatment approaches (such as ointments, creams, and patches) have several problems such as limited effectiveness, immunosuppression, and so on. As a result, researchers must develop a more effective and versatile medication strategy for psoriasis therapies (Ma et al. 2019).

Natural medicine continues to be a significant source of novel drug development due to its therapeutic importance in a variety of illness situations. Nevertheless, the classic topical pharmaceutical preparations for psoriasis are ineffective, limiting the utilization of herbal drugs. Assuming the constraints stated above, the employment of nanocarriers in the formulation of these topical herbal preparations might be extremely advantageous in improving therapeutic efficiency. Furthermore, mounting data suggests the use of nanocarriers in transdermal formulations as a promising strategy for improved potency, targeted impregnation of the drug at the active site, and minimal or no toxic effects (Abdelgawad et al. 2017).

23.2 Epidemiology

Psoriasis is associated with numerous other health problems. This condition affects around 60 million individuals worldwide. Identifying the worldwide prevalence of psoriasis is crucial research precedence for the World Health Organization. The organization issued a declaration in 2014 identifying psoriasis as a "chronic, non-communicable, painful, disfiguring, and debilitating illness with no treatment." Psoriasis affects both males and females equally, with 33 years as the mean age of inception. It can manifest sooner in females, with a bimodal beginning at the ages of 16 to 22 years and 55 to 60 years, and is linked with two distinct subcategories depending upon the immunological and genetic traits: premature commencement (75% of instances) and delayed inception (beyond 40 years). Only 19% of nations across the globe have a documented incidence of psoriasis, which is unevenly distributed among ecological regions and is frequent in white people largely (Griffiths et al. 2021).

Most psoriasis patients have some reduction in their quality of life because of the condition, and many experience a significant, detrimental impact on their mental well-being. Furthermore, psoriasis-related avoidance coping is frequently the single most significant daily stressor in patients' life. The individuals suffering from extreme psoriasis experience a stronger psychological impact. If compared to the normal population, such patients exhibit a greater tendency of being sad (up to 20%) and have suicidal ideas (Griffiths et al. 2021).

23.3 Risk Factors

The risk factors associated with the onset and exacerbation of psoriasis can be categorized under two heads: extrinsic factors and intrinsic factors (Fig. 23.1). The various extrinsic factors include air pollution, mechanical stress, and drugs, vaccination, infection, smoking, and alcohol. However, metabolic syndrome, obesity, diabetes, dyslipidemia, hypertension, and mental stress are the intrinsic risk factors (Kamiya et al. 2019).

23.4 Immune Pathway Connected to Psoriasis Therapy

Numerous types of skin cells participate in the advancement of psoriasis. Cells of the immune system are assumed as the most important essential cell categories in psoriasis. The pathophysiology of this disease is centered on the differentiation of keratinocyte cells, as well as immune system dysfunction. Some keratinocyte-derived autoantigens (the LL37 nucleic acid composite and the recently formed lipid antigen) have been recognized as initiating T-cell stimulation, specifically the subgroup of T lymphocytes articulating interleukin-17A (IL-17A) (Th17 cells). Th17 cells generate intermediaries such as IL-17A, IL-17F, IL-22, and IL-26, in response to stimulation (Huang et al. 2019). Furthermore, IL-26 increases vascular endothelial cell proliferation and hose construction via the extracellular signal-

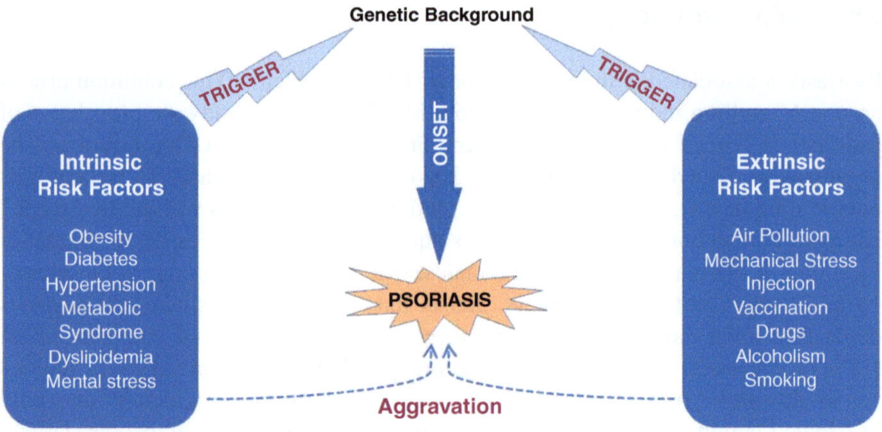

Fig. 23.1 Risk factors associated with the onset and aggravation of psoriasis

regulated kinase, protein kinase B, and nuclear factor kappa B (NF-B) pathways (Itoh et al. 2019). Keratinocyte proliferation, as well as the generation of inflammatory markers (tumor necrosis factor-alpha [TNF-α] and IL-23), is stimulated by IL-17F and IL-17A. Furthermore, the IL-23 fragment stimulates Th17 cells and produces inflammatory molecules that exaggerate and worsen psoriasis (Boehncke and Brembilla 2018). An increasing number of dendritic cells (DCs) in the epidermal and dermal layers (Michelle et al. 2014) also triggers psoriasis. IL-6, IL-23, Monocyte Chemoattractant Protein-1 (MCP-1), and TNF-α are inflammatory cytokines expressed by activated DCs. These cytokines promote keratinocyte proliferation, resulting in keratosis and keratosis pilaris (Fuentes-Duculan et al. 2010). The IL-23 fragment stimulates Th17 cells, causing them to generate a variety of chemokines (IL-17 and IL-22) that attract heterophil granulocytes resulting in psoriasis (Song et al. 2016).

Moreover, neutrophils are attracted into the epidermis in the premature phases of psoriasis to agglomerate known as Munro's microabscess. Inside skin inflammation, neutrocytes emit pro-inflammatory chemokines (IL-23, IL-17, IL-6, and IL-1), which promote keratinocyte hyper-propagation and stimulation. IL-1, for example, induces corneocytes to generate IL-19. TNF-α stimulates neutrocytes, allowing them to infiltrate deeper into the epidermis. TNF-α persuaded increase in Intercellular Adhesion Molecule-1 (ICAM-1) and Vascular Cell Adhesion Molecule-1 (VCAM-1) expression in microvascular endothelial cells may boost neutrophil migration (Zhu et al. 2020). Figure 23.2 depicts the immunological process of psoriasis.

23.5 Herbal Medicine

Natural medicine has received a lot of interest in the hunt for new therapeutics since it offers a lot of resources and possible active substances. The inclusion of penetration enhancers within herbal medications may also boost bioavailability. Many

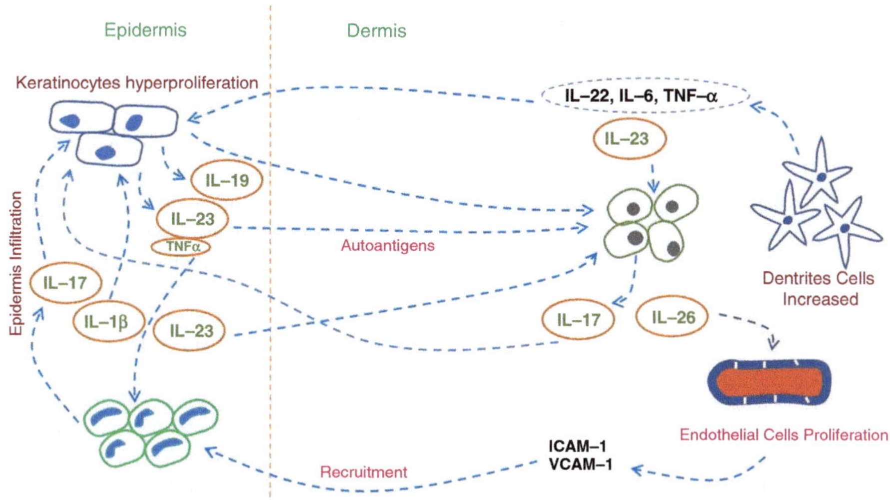

Fig. 23.2 Immune mechanism associated with the management of psoriasis

researches have been conducted to investigate the possibility of multifarious prescriptions and vigorous ingredients in herbal therapeutics to cure psoriasis. The "white mange combination" is an example of a natural pharmaceutical recipe that is useful in the treatment of psoriasis. Amomom, Figwort, Dong Quai, baical skullcap, Radix Arnebiae, Honeysuckle, Madder, Rhizoma imperatae, Botanpi, and Glycyrrhiza are among the 10 natural medicinal plants in it. Guo and Liu (Guo and Liu 2019) investigated the concoction model, dosage reactions, and serum levels of granulocyte colony-stimulating factor (GM-CSF). According to the findings, the level of GM-CSF in the model serum grew considerably and declined in the diverse cluster's dosages. Chen et al. discovered that "white mange combination" had an important role in psoriasis therapy by lowering TNF-α and IL-6 expression in mRNA and serum (Chen et al. 2015). Numerous herbs either single or in combination have exhibited their potency in the treatment of Psoriasis, few of which are enlisted in Table 23.1.

23.6 Herbal Nano-Medicine

The herbal nano-medicine helps in enhancing the safety, efficacy, and bioavailability. Further, they also improve the dose-response and targeting ability of the formulation. They have also exhibited their remarkable potential in the treatment of numerous skin diseases. Thus, they offer a trending approach to the management of psoriasis. Figure 23.3 displays the varieties of nanoformulations used in the treatment of a psoriatic disorder. Each category is further described and various examples of herbal nanoformulations are tabulated in Table 23.2.

Table 23.1 Herbal drugs used in treatment of psoriasis

S. no.	Herbal drugs	Target	Major findings
1	*Aloe vera*	Patients	The cream containing *A. vera* healed 25 patients out of 30
2	*Baphicacanthus cusia*	Patients	Psoriasis was significantly reduced as compared to the control group
3	*Capsicum frutescens*	Patients	Capsaicin preparation (cream) showed far better effects than a placebo
4	*Cassia tora*	Albino mice	When contrasted to tretinoin, cream containing *C. tora* extract showed a substantial decrease in percent of comparable dermal thickening (Singhal and Kansara 2012)
5	Celastrol	Cultured human keratinocyte (cells)	Suppression of NF-B pathway and downregulation of Bcl-2 protein appearance (Zhou et al. 2011)
6	Convollatoxin	Albino, immunodeficient laboratory-bred mice	Apoptotic cell death in HaCaT cells, which was driven by oxidative stress (Chen et al. 2020)
7	*Curcuma longa*	Patients	Progressive thickness diminution, accompanied by decreased redness and itching, culminating in modest to satisfactory remission; in rare instances, considerable clearance of psoriatic lesions
8	Cycloastragenol	C57black/6 lab strain mice	Controlled the decrease of infiltration of dermal macrophage and lowered the magnitude of pro-inflammatory cytokines in skin (Deng et al. 2019)
9	Danshensu	Cultured human keratinocyte (cells)	Retarded cell development and arrested cell cycle (Jia et al. 2020)
10	Diosgenin	HaCaT cell line	Diosgenin decreased HaCaT cell growth by arresting the cell cycle and inhibiting NF-B, while inducing apoptosis by modulating Caspase3, Bax, and Bcl-2 protein expression (Wu et al. 2019)
11	Epigallocatechin 3 gallate	Albino, immunodeficient laboratory-bred mice	By decreasing the appearance of epidermal PCNA, EGCG can enhance caspase-14 expression. It also reduces T cell infiltration and inflammatory factor expression (Zhang et al. 2016)
12	Geraniol	HEK-293 cells	Decreased the discharge of chemokines (e.g., IL-2 and TNF-α). It could also improve worsening histology and reduce the extent of splenomegaly (Ye et al. 2019)
13	Ginsenoside compound K	Wild-type C57BL/6	Inhibited the appearance of REG3A in keratinocytes (Fan et al. 2019)
14	*Hypericum perforatum*	Patients	When compared to the placebo group, *H. perforatum* ointment improved clinical ratings

(continued)

Table 23.1 (continued)

S. no.	Herbal drugs	Target	Major findings
15	*Indigo naturalis*	Patients	The drop in nail psoriasis severity index was greater than the control group
16	Khellactone	C57BL/6 mice	cis-khellactone decreased cutaneous macrophage infiltration and levels of chemokines (Feng et al. 2019)
17	*Kigelia africana*	Albino mice	Stem methanol extracts caused orthokeratosis in mouse tail parakeratotic regions, with substantial impacts on epidermal thickness (Oyedeji and Bankole-Ojo 2012)
18	*Mahonia aquifolium*	Patients	Noteworthy improvement in psoriasis area and severity index and quality of life index
19	Naringin	HPBMCs	The combination of naringin and sericin can considerably suppressed the expression of mRNA as well as the manufacture of pro-inflammatory cytokines such as IL-23, IL-6 and TNF-α (Deenonpoe et al. 2019)
20	*Nigella sativa*	Albino mice	The extract generated the same degree of orthokeratosis in the epidermis as tazarotene (Dwarampudi et al. 2012)
21	Oxymatrine	Patients	Oxymatrine can protect against experimental psoriasis by slowing epidermal cell proliferation, decreasing serum IL-2 and TNF-, and boosting serum IL-10 levels (Cheng et al. 2020)
22	*Persea americana*	Patients	In terms of Psoriasis Area and Severity Index (PASI) score, cream containing vitamin B12 and avocado oil outperformed calcipotriol cream
23	*Rubia cordifolia*	Albino, immunodeficient laboratory-bred mice	EA fraction enhanced granular layer and epidermal thickness in a dose-dependent manner; keratinocyte differentiation potency was 5% (Lin et al. 2010)
24	*Scutellaria baicalensis*	Albino, immunodeficient laboratory-bred mice	Creams containing baicalin block CHS response to a lesser extent than tacrolimus ointment (Wu et al. 2015)
25	*Smilax China*	Swiss albino mice	The flavonoid quercetin has substantial orthokeratosis, anti-inflammatory, and antiproliferative properties (Vijayalakshmi et al. 2012)
26	*Strobilanthes formosanus*	Patients	Psoriasis was cleared or almost cleared in 31 of 42 individuals after therapy with herbal ointment.
27	*Thespesia populne*	Wistar rats	TpF-2 and pet-ether extract enhanced the orthokeratotic area (Shrivastav et al. 2009)

(continued)

Table 23.1 (continued)

S. no.	Herbal drugs	Target	Major findings
28	Withanolides	C57 black 6 mice	Reduced the dysplasia in epidermal layer and infiltration of inflammatory cells (Li et al. 2019)
29	*Wrightia tinctoria*	Albino mice	When compared to isoretinoic acid, the extract caused substantial orthokeratosis and increased epidermal thickness (Raj et al. 2012)

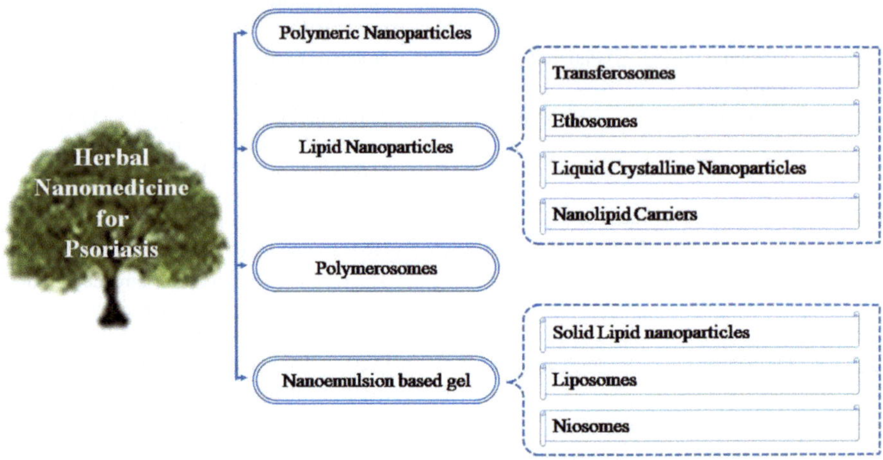

Fig. 23.3 Categories of herbal nano-medicine for psoriasis

23.6.1 Polymeric Nanoparticles

Polymeric nanoparticles are particle colloidal carriers or solidified materials in which the medicine can be encased, adsorbed, or disseminated in the appropriate polymer matrices. They may be either self-assembled or non-self-assembled. The latter encompass microspheres or nanospheres, which are array structures wherein substances are entombed, and microcapsules or nano-capsules, which are hub casing structures in which the center is a resource containing an active component in solid, liquid, or molecular dispersion. Hydrogels are cross-linked polymeric networks created by the self-arrangement of monomers, in self-assembled polymeric nanoparticles. Drug trapping in hydrogel can be accomplished either by loading during the process or at a later stage. Typically, the size of each particle ranges from 10 to 1000 nm (Saleem et al. 2020). The capability of hydrophobic medicines to be encapsulated in a nanocomposite is employed as a possible carter for psoriasis treatment over the derma. In the treatment of psoriasis, PCL (polycaprolactone) nanoparticles of hydrocortisone exhibit primary influence of drug discharge and greater absorption into epidermal strata, as well as a diminution in toxic effects of the

Table 23.2 Herbal nanoformulations for psoriasis

S. no.	Drug	Common name	Part used	Family	Anti-psoriatic phytomolecule	Disadvantages	Nano-formulation	Outcomes	Ref.
1.	*Tripterygium wilfordii*	Thunder God Wine	Roots	Celastraceae	Celestrol	Poor water solubility	Niosome hydrogel	Enhanced aqueous solubility and skin permeability. Scaling and erythema reduced in mouse models.	Meng et al. (2019)
2.	*Glycyrrhiza glabra*	Liquorice	Roots	Fabaceae	Ammonium glycyrrhizinate	Poor skin deposition	Niosome	Improved skin tolerability and increased anti-inflammatory potential	Marianecci et al. (2012)
3.	*Capsicum annum*	Capsicum	Fruits	Solanaceae	Capsaicin	Skin irritation	(SLNs) (NLCs)	Solubility of drug increased. The irritation reduced to minimum	Agrawal et al. (2014)
4.	*Taxus brevifolia*	Pacific yew tree	Bark	Taxaceae	Paclitaxel	Poor water solubility	TyroSpheres	Solubility of paclitaxel increased to about 4000 times. Provided	Kilfoyle et al. (2012)
5.	*Cornus officinalis*	Japanese cornelian cherry	Fruits	Cornaceae	Polyphenols	Lesser cell penetration	AuNPs AgNPs	Enhanced penetration and more	Crisan et al. (2018)

(continued)

Table 23.2 (continued)

S. no.	Drug	Common name	Part used	Family	Anti-psoriatic phytomolecule	Disadvantages	Nano-formulation	Outcomes	Ref.
								inhibition of macrophage activation and cytokine production	
6.	*Psoralea corylifolia*	Babchi	Seeds	Fabaceae	Babichi oil	Poor skin deposition, skin Irritation	Nanogel	Delayed the cell propagation, restricted synthesis of DNA	Kumar et al. (2019)
						Poor skin deposition, skin Irritation	Ethosomes liposomes	Improved solubility of drug and enhanced permeability	Fathalla et al. (2020)
7.	*Capsicum annuum*	Bell pepper	Fruits	Solanaceae	Capsaicin	Initiation of erythema accompanied by sting ness	NLC	Skin permeation improved	Agrawal et al. (2014)
							Emulsomal gel	Accretion of drug improved in various layers of skin	Gupta et al. (2016)
8.	*Nigella sativa*	Kalonji	Seeds	Ranunculaceae	Thymoquinone	Hydrophobic and photosensitive	Ethosomes	Enhanced solubility, entrapment and	Negi et al. (2019)

								penetration of drug	
					Kalonji oil	Less skin access and auxiliary side effects	Nano emulsion	Improved local concentration and less systemic exposure	Sahu et al. (2018)
9.	*Smilax China*	China root	Roots	Smilacaceae	Quercetin	Less bioavailability	NLC	Improved efficacy	Qadir et al. (2020)
10.	*Curcuma longa*	Turmeric	Rhizome (root)	Zingiberaceae	Curcumin	Poor water solubility	Ethosomes	Accrual of curcumin increased in skin	Zhang et al. (2019)

drug. Poly(lactic-co-glycolic acid) (PLGA) nanoparticles containing cyclosporine exhibit enhanced permeability into layers of skin, namely the dermal and epidermal layer, resulting in greater accretion in the skin and therefore reduced systemic absorption, as well as a general benefit in psoriasis. The ex vivo investigation revealed increased absorption of the medication into the dermal and epidermal layers when contrasted to free pharmaceuticals, indicating the improved effectiveness of polymeric nanoparticle preparation in drug delivery. Clobetasol 17-propionate packed with PLGA nanoparticles provides prolonged liberation of medicine into the skin in the management of psoriatic disease, enhancing medication steadiness and dramatically reducing adverse effects compared to standard preparations. According to one research, betamethasone-loaded PLGA/PLA nanoparticles containing zinc give a higher therapeutic effect due to the presence of systemic zinc, as well as superior encapsulation effectiveness with the better release of medication up to 8 days in the management of skin diseases (Saleem et al. 2020).

A research work developed and studied silibinin-loaded polymeric micelles for loading and liberation of drug molecules, as well as in vitro penetration in normal and psoriatic mouse skin. The improved lot was utilized to treat psoriasis lesions in a mouse model. In comparison to the control, polymeric micelles enhanced silibinin localization in psoriatic skin. There was no indication of psoriasis in sufferers after 7–10 days of therapy with silibinin-loaded polymeric micelle, and histological investigation revealed no trace of psoriasis (Chavoshy et al. 2020).

23.6.2 Polymerosomes

These are polymeric particles made up of amphiphilic polymers that can assemble by themselves in an aqueous surrounding and are bounded by a substantial membrane bilayer. On the interior, cationic lipids are present, while on the outside, fusogenic lipids are covered with polyethylene glycol. The electrostatic force connects the nucleic acid through a negative charge to the positive head group of cationic lipids, assisting in core encapsulation. They improve the stability of medicines as compared to traditional formulations. It comprises novel (F-NALP) fusogenic nucleic acid-lipid particles referred to as polymersomes. Anti-STAT3 siRNA and anti-TNF—siRNA are two different nucleic acids. The particles can transfer nucleic acid fragments to the epidermis and they demonstrated the symbiotic activity of both patterns in the therapy of psoriasis. When compared to standard formulations, novel-created F-NALP can dramatically suppress the production of STAT3, TNF-mRNAs, and Ki-67 proteins, resulting in a diminution in overall psoriasis effects (Saleem et al. 2020).

23.6.3 Lipid Nanoparticles

23.6.3.1 Transfersomes

Transfersomes are flexible malleable lipid packets with a dimension of less than 100 nm. These are composed of phospholipids mixed with periphery activators in a preferred 85:15 ratio. Vesicles are made up of phospholipids, which act as lipid nanoparticles and can self-assemble into bilayer vesicles. An edge activator is applied to the lipid membranes to increase the flexibility of the vesicles and to enable skin permeability (Sala et al. 2016). Transferosomes transport drugs through the skin using a motivating factor that may be a hydro tactic, and vesicles are transported in the shape of non-fragmented bilayers via a semi-permeable barrier. Transfer transit through the skin is not regulated by the transfer of some adherence to stratum corneum lipids or fluidification. The permeation, on the other hand, is fully independent of the partition coefficient and relative concentration. The research found that transfersomes created for the therapy of psoriasis and packed with betamethasone di-propionate have the capacity to realize the medicine deep into the epidermal cells while also providing superior acceptability to individuals with psoriasis plaque. Transferosomes improve anti-psoriatic action and durability when compared to traditional formulations (El et al. 2017). According to survey findings, transfersomes packed with tacrolimus (gel) accumulated more in the dermis and epidermis of skin than standard medicament. The deposited tacrolimus concentration in the epidermis and dermis was observed to be 3.8 times and 4.2 times higher, correspondingly, in contrast to the traditional ointment, resulting in greater effectiveness along with the increased topical distribution of tacrolimus in the treatment of psoriasis (Lei et al. 2013).

As compared to standard preparations, transfersomes carrying dexamethasone give a response in the duration of treatment, enhance the risk-benefit ratio, and decrease the frequency of treatment (Cevc and Blume 2004). Transferosomes loaded with triamcinolone demonstrated effectiveness as well as a tenfold decrease in dosage when compared to standard formulations. The use of transfersomes resulted in a significant decrease in skin erythema as well as a decrement of 12.1% in skin width (Fesq et al. 2003).

23.6.3.2 Ethosomes

Ethosomes are non-invasive transporters with rapid transdermal fluxes and considerable skin layer penetration. It is mostly composed of ethanol, phospholipids, and water. Because of their high ethanol concentration, ethosomes can infiltrate the stratum corneum and deepest stratum of the targeted skin. Zhang et al. created a hyaluronic acid-modified ethosome (HA-ES) transporter for curcumin target CD44 in the inflamed epidermis; the research found that the HA-ES cumulative transdermal quantity was around 1.6-fold that of ES, additionally 3.1-fold that of a curcumin propylene glycol solution, resulting in greater curcumin withholding in the skin. In vivo, the HA-ES faction had less inflammatory indications and poorer CCR6 protein appearance than the ES and PGS groups (Zhang et al. 2019). Fathalla et al.

discovered that psoralen-loaded ethosomes had 3.5- and 2.15-fold the transdermal flow and skin deposition of liposomes, respectively (Fathalla et al. 2020).

In vivo tests suggest that ethosomes and liposomes are nontoxic after 7 days of application to rat skin (Zhang et al. 2019). In vitro, skin permeability of resveratrol-ethosome preparation was 3.5-fold greater than firm liposomal formulation, according to Arora and Nanda. Furthermore, the absorption pace and permeability coefficient of resveratrol were 2.18 and 2.04 times greater, respectively, as compared to conventional emulsion. The dermal withholding of traditional cream formulation is 4.6 times lower than that of ethosomal formulation and ethosomal hydrogel (Arora and Nanda 2019). Ethosomes have been shown to have higher percutaneous permeability than traditional liposomes. Because ethosomes contain 20–45% ethanol, they may have a greater influence on skin penetration and retention than liposome preparations. As an excellent penetration enhancer, ethanol can help medicinal drugs penetrate deeper into the epidermal layer and into the circulatory system (Nainwal et al. 2018). Furthermore, altering the ethosomal formulations to precisely target receptor proteins appears to be a more optimistic strategy. Hyaluronic acid is a natural CD44 protein ligand that is extensively utilized to enhance targeting effectiveness and medication accumulation at injury sites, as well as to reduce the detrimental impacts of aimed nanodrug delivery methods. HA-modified Ethosomes offer a unique therapeutic tactic for psoriasis and could be a potential research focus.

23.6.3.3 Liquid Crystalline Nanoparticles

Liquid crystalline nanoparticles are associated with encapsulating the medications like cyclosporine and tacrolimus. They combine with Cell Permeating Peptides (CPPs), or cell permeating peptides, and thus this method is used to increase epidermal cell uptake through the plasma membrane. Liquid crystalline nanoparticles are created by combining the characteristics of solid crystal and isotropic liquid. The lyotropic LCs exhibit a phase change that is dependent on heat and concentration of the liquid crystal molecules, which are frequently amphiphilic in aqueous solution. Liquid crystalline nanoparticles are one possible method for delivery of cutaneous medication. These nanoparticles have been linked to the encapsulation of medications like tacrolimus and cyclosporine. They combine with CPPs, or cell permeating peptides, and thus this method is used to increase epidermal cell uptake through the plasma membrane. Liquid crystalline nanoparticles are created by combining the characteristics of solid crystal and isotropic liquid. According to one study, when tacrolimus packed liquid crystalline nanoparticle (LCNP) was implemented, the drug retention was 65%, whereas only 25% retention was noticed when the medication was administered as a mixture in propylene glycol. The enhanced permeability and retainment of tacrolimus are attributed to the bio-adhesiveness of the LCNPs and are capable of interrupting the lipid phase in skin, thereby increasing nanoparticle consumption (Thapa and Yoo 2014). Additionally, a research is targeted to create nanostructured lipid carriers (NLCs)-based topical gels containing curcumin (CUR) and caffeine for psoriasis therapy. NLCs loaded with CUR and caffeine were prepared using heat homogenization and ultrasonication methods before being integrated into topical gels. A completely randomized factorial design (32) was used to generate and assess all nine alternative experimental runs. Comprehensive assessment studies for NLC and NLC-based gels were carried out. The mouse model of imiquimod-induced psoriasis was used for in vivo animal experiments to improve the formulation (Iriventi and Gupta 2020).

Another study focused on the development of an extremely penetrating liquid crystalline nanoparticles (LCNPs) reservoir containing a solubility modified berberine oleate (Brb-OL) composite for successful psoriasis therapy. The hydrotrope approach was used to create Brb-OL-loaded LCNPs (Brb-OL-LCNPs). In vivo investigations demonstrated that topical treatment of Brb-OL-LCNPs hydrogel dramatically decreased psoriasis symptoms and psoriatic inflammatory cytokines. The formulation of Brb-OL in the LCNPs regulated the drug's liberation, retention, and penetration across skin layers, which is critical for psoriasis therapy (Freag et al. 2019).

Mild psoriasis is treated with vitamin D3 analogues, retinoids, anthralin, and corticosteroids. Topical therapy is chosen over systemic administration due to advantages such as site-specific activity, no systemic adverse effects, and ease of application.

23.6.3.4 Nanolipid Carriers

Dianthrol is an anthralin imitative that is utilized in treating psoriatic disease. Dianthrol is a very efficient anti-psoriatic compound with an antagonistic impact on keratinocyte growth. When dianthrol interacts with the DNA and Electron Transport Chain (ETC), the target organelle of dianthrol reduces ATP production. Dianthrol has an irritative effect when applied topically due to the production of free radicals due to auto-oxidation. Danthron and danthron dimer are formed when dithranol is broken down. Dianthrone has little effect on psoriasis; however, it has the capacity to stain. To address these issues, formulations like as liposomes, niosomes, and dendrimers are utilized. NLCs are often used in topical preparations. Because of the mixture of liquid and solid lipids, NLCs have the capacity to infiltrate cells and keep them hydrated. For ease of application, NLCs can be used as gels or creams. Gel is made up of two parts: a condensed bulk surrounded by a liquid. Because psoriatic skin is particularly dry, gel treatment will be useful. There are no issues with dianthrol NLC; the primary goal is the comparison between dianthrol NLCs and dianthrol ointment. In addition, to assess physio-chemical properties, the optimization characteristics of the dianthrol-loaded NLC gel were effective, i.e., encapsulation effectiveness was 100%, particle dimensions was less than 300 nm, and PDI was fewer than 0.3. NLC are particularly effective because they have delayed drug discharge, deep drug penetration, and slow drug discharge; hence, topical use of NLC in the IMQ produced psoriatic plaque replica exhibited improvement in psoriasis manifestations. Dianthrol-loaded NLC was shown to be more effective than ointment; PASI score, histological analysis, and enzyme-linked immunosorbent assay (ELISA) were all contrasted, and NLC yielded improved results.

Agrawal et al. formulated Solid Lipid nanoparticles (SLNs) and NLCs loaded with capsaicin that could accumulate Cellulose acetate phthalate (CAP) in the sub cutaneous (SC) and penetrate through the skin. The investigation found that NLCs had roughly five times the skin permeation of free CAP, whereas SLNs had around three times the skin permeation of free CAP. Skin irritation experiments revealed that SLNs and NLCs caused little or no exasperation after 72 h of application on comparing to the liquid formulation (Agrawal et al. 2014). Rapalli et al. used probing sonication to create curcumin NLC. The results indicated that the in vitro liberation of curcumin NLC was prolonged to 2 days, whereas free curcumin demonstrated altogether drug release after 4 h (Rapalli et al. 2020).

23.6.4 Nanoemulsion-Based Gel

Prospective naturally occurring phytomolecules possessing anti-psoriatic action include resveratrol, curcumin, and thymoquinone. However, these phytoconstituents have weak water solubility and restricted skin permeability which limit their efficient release and possible healing result. Thus, research focused on the development and improvement of nanoemulsion (NE) gel formulations for the simultaneous administration of all three medicines. Oleic acid constitutes the lipid phase while Tween 20 and PEG 200 act as the surfactant and co-surfactant respectively in the NE system. Carbopol 940 (0.5% w/v) was chosen as the gelling substance. The improved preparation inhibited growth more effectively on A-431 cells and revealed strong anti-angiogenic action in the Hen's Egg Test-Chorioallantoic Membrane (HET-CAM) test. Lastly, the research on animals in the Balb/c mouse model demonstrated enhanced anti-psoriatic circumstances, indicating that the amalgamation of three phytocompounds in the NE preparation is useful in the treatment of psoriasis. The gelling agent carbopol 940 (0.5% w/v) was employed to make the NE gel, which had a nice textural profile (Khatoon et al. 2021).

23.6.4.1 Solid Lipid Nanoparticles

The lipid-based vesicular system class, to which SLN belongs, provides a significant benefit since drug release is regulated and site-specific. Because of its capacity to produce lipid layers, SLN keeps the skin moisturized. Furthermore, deeper penetration is accredited to the incidence of lipid vesicles in the preparation. According to one study, Triamcinolone manufactured as SLN had greater penetration in the skin and a longer release profile when used topically to treat psoriasis. Keratinocyte development was slowed, and epidermal thickness was reduced. One benefit over typical creams was a reduction in skin irritation, as well as a limited distribution of the dermis and epidermis (Pradhan et al. 2016). Another study discovered that mometasone-loaded SLNs had enhanced skin permeation. Penetration was increased by 15.21-times and the drug liberation sequence was sustained for treating psoriasis which has a benefit over traditional formulations (Madan et al. 2014). Additional research claimed that encapsulating betamethasone into solid lipid nanoparticles caused the medicine to concentrate in the epidermal layer of the skin, forming a reservoir that is employed for targeting the skin's surface layers. They also asserted that they have an excellent medication release profile and that adverse reactions are decreased in the therapy of psoriatic disorder. The drug's systemic absorption was additionally reduced by using it topically (Nagy et al. 2012).

23.6.4.2 Liposomes

In the 1960s, Alec D Bangham invented liposomes as a possible pharmaceutical delivery system. They are composed of a single or saturated lipid bilayer wrapped in an aqueous compartment. Because of their great biocompatibility and biodegradability, liposomes enclose equally hydrophilic and hydrophobic medications, which have a substantial influence on the delivery of drugs (Abu and Ishida 2017). Liposome architecture is by far the most important consideration in the development of drug-to-skin preparations in formulation development for the delivery of drugs to the epidermis. The liposome bilayer may reside in a "gel phase," defined by a rigid and inflexible bilayer, or in a "liquid crystal" condition, described by a bilayer that is somewhat versatile and easy to modify (Mukul et al. 2016). Sinico et al. also looked at the impact of liposome charging on skin permeation and persistence. Because the epidermis is a negatively charged layer, medication persistence in the surface layer of the skin is enhanced by liposomes (Sinico et al. 2005).

Chen et al. investigated liposome gels containing turmeric and tretinoin. Turmeric and tretinoin may be integrated into liposomes with great encapsulation efficiency, significantly improving medication stability and solubility. Skin penetration and retention tests have revealed that liposomes significantly lengthen the lifetime of medications in hair follicles and keep more pharmaceuticals in mouse skin. Notably, the in vivo effectiveness of zedoary turmeric oil and tretinoin-loaded liposome gel is equivalent to the market-leading calcipotriol ointment for the management of psoriasis (Chen et al. 2020). Doppalapudi created cationic liposomes and anionic liposomes from psoralen. Psoralen is incorporated into liposomes, which significantly reduces skin irritation. According to in vitro release studies, the release profiles of cationic liposomes and anionic liposomes were identical. A skin penetration investigation revealed a fivefold augment in psoralen penetration with liposome gel carriers. The impact of the psoralen liposome group in vivo was shown to be comparable to that of Betamethasone Valerate (BMV) (reference chemical) (Doppalapudi et al. 2017).

Liposomes are amphiphilic lipid vesicles composed mostly of cholesterol and phospholipids. The architecture of drug-loaded liposome gel is substantially comparable to SC lipids in the skin, which may explain why it promotes skin penetration and retention. The administration of herbal medicines and liposomes has increased drug solubility, transdermal permeability, and effectiveness against psoriasis. In prospect, the use of the liposome as a carrier for herbals appears intriguing, with the potential to improve outcomes in psoriasis therapy (Cheng et al. 2020).

A study attempted to create a topical medication delivery system using zedoary turmeric oil (ZTO) and tretinoin (TRE)-loaded liposomal gel. To systematically improve the encapsulation process of compound liposomes, researchers employed a mixture of single-factor experiments and orthogonal experiments. The improved liposome preparations were stable even after studying for 30 days at (42)°C. An in vitro study found that liposome formulations might dramatically extend medication penetration into mouse tresses follicles and additional medicines in the skin when compared to traditional gel formulations. The liposomal gel was found to be more successful than traditional gel in healing psoriatic skin, with a substantial dose-dependent impact. In conclusion, the liposomal gel is considered a suitable carrier for ZTO and TRE topical drug delivery systems (Chen et al. 2021).

23.6.4.3 Niosomes

Because of the surfactants utilized, niosomes, a form of the vesicle with a bilayered crust mostly composed of non-ionic surfactant containing cholesterol, have superior chemical steadiness and a longer storage period than liposomes. They can improve medication availability and control by confining drug effects to target cells in designated carriers (Moghassemi and Hadjizadeh 2014). Celastrol-loaded niosomes, a novel topical medicinal formulation, were created by Meng et al. (Meng et al. 2019). This study found that when compared to the raw medication, penetration ability was enhanced in vitro and significantly relieved erythema and scaling on the dorsal skin of psoriasis mice models. Celastrol niosomes also decreased inflammatory factor expression, including IL-22, IL-23, and IL-17. Non-ionic surfactant vesicles filled with ammonium glycyrrhizinate considerably increase the drug's anti-inflammatory efficacy and skin irritation in mice. Furthermore, the vesicle enhanced chemically induced cutaneous erythema in people (Marianecci et al. 2012).

These studies demonstrated that non-ionic surfactant vesicles are a good carrier in psoriasis therapy due to their variable physicochemical properties. Niosomes are very compatible with biological systems due to their non-ionic nature and minimal toxicity. Natural medications combined with niosomes can minimize toxicity, and skin irritation, and improve the therapeutic benefits of psoriasis.

23.7 Conclusion

Owing to the high prevalence rate and disadvantages of topical formulations, it becomes very essential to develop highly effective medications for this disease. Drugs from nature continue to be a significant reservoir of novel drug development owing to their medicinal potential in a variety of illness situations. Assessing particular candidate subunits, pro-drugs, and drug components from conventional remedies, as well as drug combination therapy, might be a possible drug development method. Nanocarriers have the potential to alleviate the issues associated with conventional medication formulations, such as excessive injection rate, dosage dependency, and harmful impacts. Liposomes, ethosomes, NLCs, SLNs, AgNPs, AuNPs, and several other nanoparticles have been utilized effectively as medication transporters for psoriasis. Nano-medicines present an extensive variety of potential applications in the management and cure of other numerous skin disorders such as dermatitis and psoriasis. This chapter focuses on the topical herbal nanotherapeutics used in the management of psoriasis. Natural products offer very safe and effective therapeutics for chronic skin diseases. Nevertheless, sufficient clinical data is required to establish the destiny of topical herbal nanotherapeutics as potential anti-psoriatic medicines.

References

Abdelgawad R, Nasr M, Moftah NH et al (2017) Phospholipid membrane tubulation using ceramide doping "cerosomes": characterization and clinical application in psoriasis treatment. Eur J Pharm Sci 101:258–268

Abu LA, Ishida T (2017) Liposomal delivery systems: design optimization and current applications. Biol Pharm Bull 40:1–10

Agrawal U, Gupta M, Vyas SP (2014) Capsaicin delivery into the skin with lipidic nanoparticles for the treatment of psoriasis. Artif Cell Nanomed B 43:33–39

Ahmad U, Ahmad Z, Khan AA et al (2018) Strategies in development and delivery of nanotechnology based cosmetic products. Drug Res 68:545–552

Arora D, Nanda S (2019) Quality by design driven development of resveratrol loaded ethosomal hydrogel for improved dermatological benefits via enhanced skin permeation and retention. Int J Pharm 567:118448

Boehncke W, Brembilla NC (2018) Unmet needs in the field of psoriasis: pathogenesis and treatment. Clin Rev Allergy Immunol 55:295–311

Cevc G, Blume G (2004) Hydrocortisone and dexamethasone in very deformable drug carriers have increased biological potency, prolonged effect, and reduced therapeutic dosage. Biochim Biophys Acta Biomembr 1663:61–73

Chavoshy F, Zadeh BS, Tamaddon AM, Anbardar MH (2020) Delivery and anti-psoriatic effect of silibinin-loaded polymeric micelles: an experimental study in the psoriatic skin model. Curr Drug Deliv 17:787–798

Chen X, Lu YP, Li XH (2015) Effects of white mange mixture on the expression of proliferation and apoptosis of HaCaT cells in vitro. Zhonghua Zhongyiyao Xuekann 33:2961–2963

Chen ZX, Zhou DM, Wang Y et al (2019) Fire needle acupuncture or moxibustion for chronic plaque psoriasis: study protocol for a randomized controlled trial. Trials 20:674

Chen J, Ma Y, Tao Y et al (2020) Formulation and evaluation of a topical liposomal gel containing a combination of zedoary turmeric oil and tretinoin for psoriasis activity. J Liposome Res 31:1–15

Chen J, Ma Y, Tao Y et al (2021) Formulation and evaluation of a topical liposomal gel containing a combination of zedoary turmeric oil and tretinoin for psoriasis activity. J Liposome Res 31:130–144

Cheng YC, Li TS, Su HL et al (2020) Transdermal delivery systems of natural products applied to skin therapy and care. Molecules 25:5051

Crisan D, Scharffetter-Kochanek K, Crisan M (2018) Topical silver and gold nanoparticles complexed with Cornus mas suppress inflammation in human psoriasis plaques by inhibiting NF-κB activity. Exp Dermatol 27:1166–1169

Deenonpoe R, Prayong P, Thippamom N et al (2019) Anti-inflammatory effect of naringin and sericin combination on human peripheral blood mononuclear cells (hPBMCs) from patient with psoriasis. BMC Complement Altern Med 19:168

Deng GL, Chen WJ, Wang P et al (2019) Inhibition of NLRP3 inflammasome mediated pyroptosis in macrophage by cycloastragenol contributes to amelioration of imiquimod-induced psoriasis-like skin inflammation in mice. Int Immunopharmacol 74:105682

Doppalapudi S, Jain A et al (2017) Psoralen loaded liposomal nanocarriers for improved skin penetration and efficacy of topical PUVA in psoriasis. Eur J Pharm Sci 96:515–529

Dwarampudi LP, Palaniswamy D, Nithyanantham M et al (2012) Antipsoriatic activity and cytotoxicity of ethanolic extract of *Nigella sativa* seeds. Pharmacogn Mag 8:268–272

El GS, Fadel M, Mourad B (2017) Betamethasone dipropionate gel for treatment of localized plaque psoriasis. Int J Pharm Pharm Sci 9:173–182

Fan H, Wang Y, Zhang XL et al (2019) Ginsenoside compound K ameliorates imiquimod-induced psoriasis-like dermatitis through inhibiting REG3A/RegIIIγ expression in keratinocytes. Biochem Biophys Res Commun 515:665–671

Fathalla D, Youssef EMK, Soliman GM (2020) Liposomal and ethosomal gels for the topical delivery of anthralin: preparation, comparative evaluation and clinical assessment in psoriatic patients. Pharmaceutics 12:446

Feng L, Song PP, Xu F et al (2019) Cis-Khellactone inhibited the proinflammatory macrophages via promoting autophagy to ameliorate imiquimod-induced psoriasis. J Invest Dermatol 139:1946–1956

Fesq H, Lehmann J, Kontny A et al (2003) Improved risk–benefit ratio for topical triamcinolone acetonide in Transfersome® in comparison with equipotent cream and ointment: a randomized controlled trial Br. J Dermatol 3:611–619

Freag MS, Torky AS, Nasra MM et al (2019) Liquid crystalline nanoreservoir releasing a highly skin-penetrating berberine oleate complex for psoriasis management. Nanomedicine 14:931–954

Fuentes-Duculan J, Suarez-Farinas M, Zeba LC et al (2010) A subpopulation of CD163-positive macrophages is classically activated in psoriasis. J Invest Dermatol 130:2412–2422

Griffiths CE, Armstrong AW, Gudjonsson JE et al (2021) Lancet 10281:1301–1315

Guo J, Liu J (2019) Effect of white mange mixture in a murine model of psoriasis. Exp Ther Med 18:881–887

Gupta R, Gupta M, Mangal S et al (2016) Capsaicin-loaded vesicular systems designed for enhancing localized delivery for psoriasis therapy. Artif Cells Nanomed Biotechnol 44:825–834

Huang TM, Lin CF, Alalaiwe A et al (2019) Apoptotic or antiproliferative activity of natural products against keratinocytes for the treatment of psoriasis. Int J Mol Sci 20:2558

Iriventi P, Gupta NV (2020) Topical delivery of curcumin and caffeine mixture-loaded nanostructured lipid carriers for effective treatment of psoriasis. Phcog Mag 16(Suppl S1): 206–217

Itoh T, Hatano R, Komiya E et al (2019) Biological effects of IL-26 on T cell-mediated skin inflammation, including psoriasis. J Invest Dermatol 139:878–889

Jia JJ, Mo XM, Liu JF et al (2020) Mechanism of danshensu-induced inhibition of abnormal epidermal proliferation in psoriasis. Eur J Pharmacol 868:172881

Kamiya K, Kishimoto M, Sugai J et al (2019) Risk factors for the development of psoriasis. Int J Mol Sci 20:4347

Khatoon K, Ali A, Ahmad FJ et al (2021) Novel nanoemulsion gel containing triple natural bio-actives combination of curcumin, thymoquinone, and resveratrol improves psoriasis therapy: in vitro and in vivo studies. Drug Deliv Transl Res 11:1245–1260

Kilfoyle BE, Sheihet L, Zhang Z et al (2012) Development of paclitaxel-TyroSpheres for topical skin treatment. J Control Release 163:18–24

Kumar S, Singh KK, Rao R (2019) Enhanced anti-psoriatic efficacy and regulation of oxidative stress of a novel topical babchi oil *(Psoralea corylifolia)* cyclodextrin-based nanogel in a mouse tail model. J Microencapsul 2:140–155

Lei W, Yu C, Lin H et al (2013) Development of tacrolimus-loaded transfersomes for deeper skin penetration enhancement and therapeutic effect improvement in vivo. Asian J Pharm Sci 6:336–345

Li T, Wei Z, Sun Y et al (2019) Withanolides, extracted from datura Metel L. inhibit keratinocyte proliferation and imiquimod-induced psoriasis-like dermatitis via the STAT3/P38/ERK1/2 pathway. Molecules 24:2596

Lin ZX, Jiao BW, Che CT et al (2010) Ethyl acetate fraction of the root of *Rubia cordifolia* L. inhibits keratinocyte proliferation in vitro and promotes keratinocyte differentiation in vivo: potential application for psoriasis treatment. Phytother Res 24:1056–1064

Ma Z, Zhang B, Fan YQ et al (2019) Traditional Chinese medicine combined with hepatic targeted drug delivery systems: a new strategy for the treatment of liver diseases. Biomed Pharmacother 117:109128

Mabuchi T, Timothy W, Quinter S et al (2012) Chemokine receptors in the pathogenesis and therapy of psoriasis. J Dermatol Sci 65:4–11

Madan JR, Khude PA, Dua K (2014) Development and evaluation of solid lipid nanoparticles of mometasone furoate for topical delivery. Int J Pharm Investig 2:60–64

Marianecci C, Rinaldi F, Mastriota M et al (2012) Anti-inflammatory activity of novel ammonium glycyrrhizinate/niosomes delivery system: human and murine models. J Control Release 164: 17–25

Meng S, Sun L, Wang L et al (2019) Loading of water-insoluble celastrol into niosome hydrogels for improved topical permeation and anti-psoriasis activity. Colloid Surf B 182:110352

Michelle AL, Mayte SF, James GK (2014) Immunology of psoriasis. Annu Rev Immunol 32:227–255

Min C, Kim M, Oh DJ et al (2020) Bidirectional association between psoriasis and depression: two longitudinal follow-up studies using a national sample cohort. J Affect Disord 262:126–132

Moghassemi S, Hadjizadeh A (2014) Nano-niosomes as nanoscale drug delivery systems: an illustrated review. J Control Release 185:22–36

Mukul A, Kalpa N, Alfred F (2016) Transdermal delivery from liposomal formulations-evolution of the technology over the last three decades. J Control Release 242:126–140

Na CH, Chung J, Simpson EL (2019) Quality of life and disease impact of atopic dermatitis and psoriasis on children and their families. Children 6:133

Nagy ZK, Balogh A, Vajna B et al (2012) Comparison of electrospun and extruded Soluplus®--based solid dosage forms of improved dissolution. J Pharmacol Sci 1:322–332

Nainwal N, Jawla S, Singh R et al (2018) Transdermal applications of ethosomes—a detailed review. J Liposome Res 29:103–113

Negi P, Sharma I, Hemrajani C et al (2019) Thymoquinone-loaded lipid vesicles: a promising nanomedicine for psoriasis. BMC Complement Altern Med 1:1–9

Oyedeji FO, Bankole-Ojo OS (2012) Quantitative evaluation of the antipsoriatic activity of sausage tree (*Kigelia africana*). Afr J Pure Appl Chem 6:214–218

Pradhan M, Singh D, Singh MR (2016) Influence of selected variables on fabrication of triamcinolone acetonide loaded solid lipid nanoparticles for topical treatment of dermal disorders. Artif Cells Nanomed Biotechnol 44:392–400

Qadir A, Aqil M, Ali A, Warsi MH, Mujeeb M, Ahmad FJ, Ahmad S, Beg S (2020) Nanostructured lipidic carriers for dual drug delivery in the management of psoriasis: systematic optimization, dermatokinetic and preclinical evaluation. J Drug Del Sci Tech 57:101775

Raj BA, Muruganantham N, Praveen TK et al (2012) Screening of *Wrightiam tinctoria* leaves for anti psoriatic activity. Hygeia J Drug Med 4:73–78

Rapalli VK, Kaul V, Waghule T et al (2020) Curcumin loaded nanostructured lipid carriers for enhanced skin retained topical delivery: optimization, scale-up, in-vitro characterization and assessment of ex-vivo skin deposition. Eur J Pharm Sci 152:105438

Sahu S, Katiyar SS, Kushwah V et al (2018) Active natural oil-based nanoemulsion containing tacrolimus for synergistic antipsoriatic efficacy. Nanomedicine 16:1985–1998

Sala M, Elaissari A, Fessi H (2016) Advances in psoriasis physiopathology and treatments: up to date of mechanistic insights and perspectives of novel therapies based on innovative skin drug delivery systems (ISDDS). J Control Release 239:182–202

Saleem S, Iqubal MK, Garg S et al (2020) Trends in nanotechnology-based delivery systems for dermal targeting of drugs: an enticing approach to offset psoriasis. Expet Opin Drug Deliv 6: 817–838

Shrivastav S, Sindhu R, Kumar S et al (2009) Anti-psoriatic and phytochemical evaluation of *Thespesia populnea* bark extracts. Int J Pharm Sci 1:176–185

Singhal M, Kansara N (2012) *Cassia tora* Linn cream inhibits ultraviolet-B induced psoriasis in rats. ISRN Dermatol 2012:346510

Sinico C, Manconi M, Peppi M et al (2005) Liposomes as carriers for dermal delivery of tretinoin: in vitro evaluation of drug permeation and vesicle-kin interaction. J Control Release 103: 123136

Song HS, Kim SJ, Park TI et al (2016) Immunohistochemical comparison of IL-36 and the IL-23/Th17 axis of generalized pustular psoriasis and acute generalized exanthematous pustulosis. Ann Dermatol 28:451

Thapa RK, Yoo BK (2014) Evaluation of the effect of tacrolimus-loaded liquid crystalline nanoparticles on psoriasis-like skin inflammation. J Dermatol Treat 25:22–25

Vijayalakshmi A, Ravichandiran V, Velraj M et al (2012) Screening of flavonoid "quercetin" from the rhizome of *Smilax China* Linn. For anti-psoriatic activity. Asian Pac J Trop Biomed 2:269–275

Wu J, Li H, Li M (2015) Effects of baicalin cream in two mouse models: 2,4-dinitrofluorobenzene-induced contact hypersensitivity and mouse tail test for psoriasis. Int J Clin Exp Med 8:2128–2137

Wu S, Zhao MJ, Sun YH (2019) The potential of diosgenin in treating psoriasis: studies from HaCaT keratinocytes and imiquimod induced murine model. Life Sci:241

Ye CJ, Li SA, Zhang Y et al (2019) Geraniol targets K1.3 ion channel and exhibits anti-inflammatory activity in vitro and in vivo. Fitoterapia 139:104394

Zhang SS, Liu XD, Mei LH et al (2016) Epigallocatechin-3-gallate (EGCG) inhibits imiquimod-induced psoriasis-like inflammation of BALB/c mice. BMC Complement Altern Med 16:334

Zhang Y, Xia Q, Li Y et al (2019) CD44 assists the topical anti-psoriatic efficacy of curcumin-loaded hyaluronan-modified ethosomes: a new strategy for clustering drug in inflammatory skin. Theranostics 1:48–64

Zhou LL, Lin ZX, Fuang KP et al (2011) Celastrol-induced apoptosis in human HaCaT keratinocytes involves the inhibition of NF-κB activity. Eur J Pharmacol 670:399–408

Zhu Z, Chen JL, Lin YT et al (2020) Aryl hydrocarbon receptor in cutaneous vascular endothelial cells restricts psoriasis development by negatively regulating neutrophil recruitment. J Invest Dermatol 140:1233–1243

Part IV

Pharmaceutical Microbiology

Antibiotics: Past, Present, Future, and Clinical Pipeline

24

Raja Singh and Vibha Tandon

Abstract

A significant threat to global public health is antimicrobial resistance (AMR). The estimated cause of at least 700,000 deaths each year worldwide is medication-resistant bacterial infections (including tuberculosis). Estimates suggest that around ten million deaths are expected annually by 2050, due to drug-resistant bacteria. The World Health Organization (WHO) surveillance report concedes resistance in common human pathogens like *Escherichia coli*, *Pseudomonas aeruginosa*, *Klebsiella pneumonia*, and *Staphylococcus aureus* is one of the biggest threats to humankind. Antibiotic resistance is becoming globalized due to the significant high evolutionary pressure. Thus, developing new therapeutic agents against new targets or antibacterials with different approaches to target pathogenic bacteria has become imperative. This review listed out various antibiotics act with different mechanisms, and bacteria also acquire different mechanisms to dodge these antibiotics. The mechanism includes removing antibiotics out of the cell by increased efflux, not allowing the antibiotic to enter by decreased uptake, decreasing the antibiotic binding by modifying the target, degrading the antibiotics by enzyme, etc. So, the novel antibacterial agent uses four criteria: absence of known cross-resistance, new class, new target, and a new mode of action developed to treat AMR.

Keywords

Antibiotic · Antimicrobial resistance · Fluoroquinoles · β-lactam · Pathogens · Topoisomerases

R. Singh · V. Tandon (✉)
Special Centre for Molecular Medicine, Jawaharlal Nehru University, New Delhi, India
e-mail: vtandon@mail.jnu.ac.in

P. P. Singh (ed.), *Recent Advances in Pharmaceutical Innovation and Research*,
https://doi.org/10.1007/978-981-99-2302-1_24

24.1 Introduction

Bacteria are unicellular organisms divided by binary fission. The capability of a bacterium to cause disease is called pathogenicity. A bacteria that can cause disease is called a pathogen. A wide range of bacteria can cause severe harm to the human body and become lethal. The multi-dollar pharmaceutical industry and research groups expend time and money to combat these pathogenic bacteria through antibiotics. Antibiotics are microbial metabolites or synthetic analogs that inhibit microorganisms' survival and growth without severe toxicity to the host. Today's most commonly given medications are antibiotics, although bacterial tolerance comes from evolutionary pressures, and misuse threatens their continued efficacy. The clinical usefulness of regular antibiotics has been improved through chemical manipulations in the basic structure, leading to a wide-ranging antibacterial, lesser toxicity, better potency, appropriate administration, and surplus pharmacokinetic gains. There is a decrease in the pace of novel anti-infective discovery, increased regulatory constraints, and declined research prominence on antimicrobial agents. This corresponds with a drastic rise in bacterial resistance to chemotherapy that threatens an austere future in which humankind may once again experience bacterial diseases with few existing countermeasures.

24.2 Issues of Concern

In the past, novel antibiotics were improved to overcome the diverse types of antimicrobial resistance (AMR) to stay ahead of AMR. The recent development of bacterial resistance to vancomycin and different subtypes of methicillin-resistant *Staphylococcus aureus* (MRSA), vancomycin-resistant *Staphylococcus aureus* (VRSA) can only be minimized through stewardship (Watkins et al. 2019). The growing use of antibiotics by medical and community health specialists also creates AMR, and they have to come together to reduce the inappropriate usage of antibiotics. Physicians have to equilibrium the jeopardies of not treating or imperfectly treating against the risk of prescribing antibiotic use regarding adverse effects. It exposes the patients to unwelcome effects or drug-drug contacts, upsurges medical care costs, and contributes to AMR development. The are multiple approaches to managing the AMR informative, computer-assisted notifications, antimicrobial formulary restrictions, management guidelines, future audits and opinions, molecular testing technology, and interprofessional schemes. An antibiotic supervision program's core components are management assurance, responsibility, antibiotic knowledge, actions to antibiotic use, following antibiotic consumption, recording the use of antibiotics, and enlightening physicians on the right antibiotic use (Eggermont et al. 2018; Prasetyoputri et al. 2019).

Alternatively, it also needs new, novel, more effective antibiotics for our health systems to tackle AMR. During a meeting held in Geneva on January 25–27, 2017 by the coordinating group and the advisory board, the World Health Organization (WHO) reviewed the global priority pathogens list (PPL) of antibiotic-resistant

Table 24.1 List of WHO priority pathogens list (PPL) with their antibiotic resistance details

Priority category	Pathogens	Antibiotic resistance	Gram stain
Critical	*Pseudomonas aeruginosa*	Carbapenem-resistant	Gram-negative
	Acinetobacter baumannii		
	Pseudomonas aeruginosa		
	Enterobacteriaceae (*Escherichia coli, Proteus* spp., *Enterobacter* spp., *Klebsiella pneumonia Serratia* spp.)	Carbapenem-resistant, Third-generation Cephalosporin-resistant	
High	*Neisseria gonorrhoeae*	Third generation Cephalosporin-resistant Fluoroquinolone-resistant	Gram-negative
	Staphylococcus aureus	Methicillin-resistant Vancomycin intermediate and resistant	Gram-positive
	Salmonella spp.,	Fluoroquinolone-resistant	Gram-negative
	Campylobacter		
	Helicobacter pylori	Clarithromycin-resistant	
	Enterococcus faecium	Vancomycin-resistant	Gram-positive
Medium	*Streptococcus pneumoniae*	Penicillin-non-susceptible	Gram-positive
	Shigella spp.	Fluoroquinolone-resistant	Gram-negative
	Haemophilus influenzae	Ampicillin-resistant	

bacteria to prioritize the investigation and improvement of new and useful antibiotic treatments (Organización Mundial de la Salud 2016). The ranking exercise was performed by selecting antibiotic-resistant bacteria to be prioritized, selecting criteria for ranking, data mining and synthesis, scoring of alternatives, weighing criteria by experts, and finalizing pathogens' ranking. The experts approved prioritizing pathogens and gave a priority list (critical, high, and medium) of pathogens to reduce deaths due to resistant infections worldwide (Tacconelli and Magrini 2017; Rello et al. 2019). The list of WHO priority pathogens is given in Table 24.1.

24.2.1 *Acinetobacter baumannii*

An aerobic gram-negative bacteria *A. baumannii* has come under the category of critical in the WHO priority pathogen list (Ayobami et al. 2019). Infections from *A. baumannii* are associated with hospital-acquired and public health importance worldwide because of fast growth resistance to antibiotics by diverse mechanisms

(Almasaudi 2018; Ayobami et al. 2019) such as the β-lactamases deactivate β-lactams, overexpression of the multidrug-resistant (MDR) efflux pump. *A. baumannii* expresses adenyltransferases, transacetylase, and phosphotransferases (PTS), which synthetically change aminoglycosides. *A. baumannii* changes the envelope permeability, which causes permeability defects in porins. Porins play a vital role in the mechanism of resistance (Lee et al. 2017). CarO, Omp porins, decreasing in expression is associated with *A. baumannii* carbapenem resistance. *A. baumannii* shows resistance through altering the target site, such as DNA gyrase mutation, and overexpression of penicillin-binding proteins (PBPs), which makes *A. Baumannii* resistant against quinolone, tetracycline, and imipenem. *A. baumannii* have integrons, an exceptional ability to express and cluster drug resistance genes. Combination therapy of carbapenem, ampicillin, and sulbactam is useful for treating multidrug-resistant bacteria (MDR) *A. baumannii.* But *A. baumannii* has continuously developed resistance to combined therapy and it is required new strategies and novel targeted antibiotics to regulate *A. baumannii* infection successfully (Lin 2014; Lee et al. 2017; Almasaudi 2018).

24.2.2 *Pseudomonas aeruginosa*

Pseudomonas aeruginosa is classified as an ESKAPE gram-negative aerobic bacterium usually found in bowel flora. *P. aeruginosa* acquired various mechanisms that can contribute to its antibiotic resistance. The upregulation of the efflux genes and decreased external permeability help *P. aeruginosa* inhibit the antibiotic. It is typical for ventilator-related pneumonia, the bacteria for bloodstream infections, and the most universally isolated nosocomial pathogen. *P. aeruginosa* is commonly expresses classes A, B, C, and D of β-lactamases, which disrupt the β-lactam ring (amide bond), due to which third- and fourth-generation cephalosporins like cefpimizole and cefozopran became ineffective (Bassetti et al. 2018). It has transferable aminoglycoside altering enzymes (AMEs) like, aminoglycoside adenylyl transferases (AADs), aminoglycoside phosphoryl transferases (APHs), and aminoglycoside acetyltransferases (AACs); each of these three will discharge aminoglycosides by attached to the target site. AMEs attach the phosphate, adenyl or acetyl radical to the antibiotic, and decrease the affinity to bind with the target. An effective anti-pseudomonas agent combined with colistin, such as imipenem, piperacillin, aztreonam, ceftazidime, or ciprofloxacin, has been found to be effective against MDR *Pseudomonas* spp. (Pachori et al. 2019).

Escherichia coli, Enterobacter spp., *Klebsiella pneumonia, Serratia* spp., *Proteus* spp., and *Providencia* spp, *Morganella* spp come under the *Enterobacteriaceae* family generally cause blood-stream infections, healthcare-associated pneumonia, and urinary tract infections (UTIs). The *Enterobacteriaceae* family's resistance was found due to the production of extended-spectrum beta-lactamases (ESBLs), and other resistance mechanisms are also developed, which lead to MDR (Paterson 2006). The WHO list puts *Enterobacteriaceae*, third-generation cephalosporin-resistant, and carbapenem-resistant into the critical category. The resistance to aminopenicillins (amoxicillin, ampicillin) and primary cephalosporins occurs due to the production of SHV-1, TEM-2, TEM-1, class A β-lactamases, and third-generation cephalosporins arise with a mutation in TEM-1, TEM-2, or SHV-1 (Paterson 2006). There are two types of carbapenem-resistant *Enterobacteriaceae* (CRE), the first CRE-producing carbapenemase (CP-CRE) and the second no CRE-producing carbapenemase (non-CP-CRE). This makes *Enterobacteriaceae* isolate resistant to imipenem, ertapenem, meropenem, or any other carbapenem antibiotics (van Duin 2017).

24.2.3 *Enterococcus faecium*

Enterococcus species are usual residents of the surroundings and an indispensable component of healthy humans and animals' intestinal microbiota (Gilmore et al. 2013; Ahmed and Baptiste 2018). Urinary tract infections, sepsis (bloodstream infections), and endocarditis caused by Enterococci, an opportunistic pathogen became severe and life-threatening (Fisher and Phillips 2009; Gilmore et al. 2013; Lebreton et al. 2014; Neelakanta et al. 2015). *Enterococci* evolve to alter the peptidoglycan synthesis pathway, specifically the substitution to D-Alanine-D-Lactate (D-Ala-D-Lac) or D-Alanine-D-Serine (D-Ala-D-Ser) from D-Alanine-D-Alanine (D-Ala-D-Ala), this cause resistance to glycopeptide (e.g., vancomycin)(Ahmed and Baptiste 2018). These alterations of Enterococci can reduce the affinity of binding of glycopeptide antibiotics, ~1000-fold lessened binding affinity for D-Ala-D-lac, and ~sevenfold for D-Ala-D-Ser as compared to the regular cell wall precursors. *Enterococcus* spp. had vancomycin resistance (Van) operon is responsible for glycopeptides resistance. Example, the van operon contains *vanS-vanR*, a response regulator; *van (A, B, C, D, E, G, L, M, N,* and *H*) are nine different genes known as a variable ligase, *vanH* a D-lactate dehydrogenase gene, and *vanX* a D-Ala-D-Ala dipeptidase gene (Faron et al. 2016).

24.2.4 *Staphylococcus aureus*

MRSA infection is a threat to the worldwide population. For the treatment of MRSA infections, vancomycin is used as a first-line drug. Now in recent years, *S. aureus* emerged as utterly resistant to vancomycin. *S. aureus* encodes penicillin-binding protein 2A (PBP2A) or PBP2ALGA, responsible for crosslinking the peptidoglycans and *mecA*, the *mecC* gene responsible for methicillin resistance. The vancomycin treatment failure was caused by vancomycin-intermediate *S. aureus* (VISA), heterogeneous VISA (hVISA), and vancomycin-resistant *S. aureus* (VRSA). The mechanism of VISA, hVISA, and VRSA is not defined completely. The main reason for resistance against vancomycin is the gradual mutation accumulation. Vancomycin-resistant enterococci (VRE) was plasmid-mediated by transposons, which raised significant worry about the possibility of the distribution of vancomycin-resistant elements to commonly susceptible bacteria of medical importance, *S. aureus.* The van element transfer from *Enterococcus faecalis* to MRSA strain in mix-infection in mice confirmed the theory. Vancomycin resistance in *S. aureus* is controlled by *van* (*S, R, H, A*, and *X*) genes known as *vanA* gene cluster. In the presence of vancomycin, the expression of VanS and VanR forms a two-component system upregulated, which results in high-level vancomycin resistance (Harris et al. 2010; Cong et al. 2020; Shariati et al. 2020).

24.2.5 *Helicobacter pylori*

Helicobacter pylori mainly causes gastric cancer, peptic ulcers, and gastritis in humans. *H. pylori* infection was treated with clarithromycin, but it became resistant to clarithromycin due to extensive use. Resistance may occur because of the 23S rRNA gene mutation in Domain V in 2142 (A to C or G) or A2143G. The other cause of resistance is translation initiation factor IF-2, ribosomal protein L22, and efflux pumps activation or overexpression (Ghotaslou 2015; Marques et al. 2020).

24.2.6 *Salmonella* spp.

Salmonella spp. is the causative agent of Salmonellosis, common Salmonella gastroenteritis (fever, abdominal pains, and diarrhea), and enteric fevers or typhoid fever, life-threatening infections that required antibiotics for treatment (Giannella 1996). *Salmonella* is resistant to chloramphenicol, ampicillin, and co-trimoxazole directed to consistent use of ciprofloxacin and the third-generation cephalosporin ceftriaxone, and this causes *Salmonella* spp. to quickly managed to get resistance to these antibiotics. The leading cause for resistance is the mutation in the chromosomal *gyr* and *par* genes, and another reason for resistance is plasmid-mediated quinolone resistance (PMQR). The aac (60)-lb-cr gene reduces the function of fluoroquinolones, and *oqxAB* and *qepA* encode quinolone efflux pumps. Qnr genes are responsible for the physical protection of quinolones (Cuypers et al. 2018; Breijyeh et al. 2020).

24.2.7 *Neisseria gonorrhoeae*

A gram-negative bacteria *Neisseria gonorrhoeae* is an obligate humanoid pathogen that causes the sexually transmitted disease gonorrhea to include cervicitis in women and urethritis, pharyngitis, and proctitis in both sexes (Ng and Martin 2005). Extensive use of antibiotics causes *Neisseria gonorrhoeae* resistance to third-generation cephalosporins and fluoroquinolone. PBP2 altered by the *penA* gene, which reduces the penicillin-binding insertion of D-345a, is the explanation for resistance. The decrease in cytosolic antibacterial concentration by inhibiting its entry or pumping out by the efflux pump mechanism *MtrC-D-E*, a mutation in the porin gene *mtrR* and *penB,* results in efflux upregulation and reduced permeability to antimicrobial agents are other explanations for resistance (Barry and Klausner 2009). *Neisseria gonorrhoeae is* acquired resistance to fluoroquinolones through point mutation in *parC* genes, single or more mutation in the amino acid position of 91, 95, and 102 of *gyrA*. The topoisomerase IV (*parC* gene) and DNA gyrase (*gyrA* gene) mutant proteins are responsible for the increase in resistance to ciprofloxacin and do not allow the drug to bind to target enzymes. Overexpression of NorM efflux pumps is responsible for resistance to fluoroquinolones and decreases in the outer membrane porin protein expression cause a decline in drug permeability (Costa-Lourenço et al. 2017; Młynarczyk-Bonikowska et al. 2020).

24.2.8 *Streptococcus pneumoniae*

The infectious diseases meningitis and pneumonia are caused by *Streptococcus pneumoniae* or pneumococcus. It is typically found inhabited at the upper respiratory tract of humans, a place of entry for the development of invasive pneumococcal disease (IPD) and a reservoir (Pinto et al. 2019). PBPs were mainly affected by β-Lactam antibiotics, which interact with the PBPs. During the bacterial cell wall assembly, PBPs catalyze the polymerization and transpeptidation of the cell wall's glycan strands. The non-susceptibility of *Streptococcus pneumonia* evolves by altered PBPs, specifically PBP1a, PBP2x, and PBP2 (Diawara et al. 2017).

24.2.9 *Haemophilus influenza*

H. influenza infection is treated by cell wall synthesis inhibitors like ampicillin. There are several mechanisms for resistance, such as beta-lactamases acquired resistance or efflux activation, or PBP target modifications. In *Haemophilus influenza* TEM-1 and ROB-1, beta-lactamases are identified as responsible for a decline in the binding affinity of ampicillin to PBP (Heinz 2018).

24.2.10 *Shigella* spp.

Shigella infection is treated by fluoroquinolone antibiotics, but due to resistance, the effectiveness decreases. The main reason for resistance is a mutation at *gyrA, gyrB, parC,* and *parE*, multiple mutations in the quinolone resistance determining region (QRDR). Other reasons for resistance are *tolC, mdfA, marA, ydhE,* efflux pump mediators, and plasmid-mediated quinolone resistance due to Qnr genes (Nüesch-Inderbinen et al. 2016; Qin et al. 2017).

24.3 Clinical Significance of Prioritizing Pathogens

The theory of AMR in clinical practice is highly complex, and it is essential to recognize the complexity. The prevention and control of AMR involve dealing with several pathogens, each of which causes a definite threat to the population. It is necessary to prioritize the pathogens to rationally distribute the limited resources for research, investigation, and other activities. Prioritizing the bacterial pathogens based on infection and resistance profile had the following significance.

The WHO triage implementation recommends that antibiotic development and research policies emphasize the immediate requirement of novel antibiotics that are precisely active against PPL.

- Worldwide investigation and improvement approaches should also include antibiotics activity against more general community bacteria, like MDR *Helicobacter pylori.*, *Campylobacter* spp., and. *Salmonella* spp.
- Prioritization efforts will look at how oral formulations for high-fatality MDR infection, such as *Neisseria gonorrhoeae* MDR and *Enterobacteriaceae* immune to cephalosporin in the third generation, may be boosted.
- The results were backing prioritization in developing new antibacterial agents with novel targets and action mechanisms without cross-resistance to existing antibiotic classes. However, this prioritization work does not cover all the potential resistance and co-resistance patterns.
- Prioritization has been driving enduring plans for researching and developing new antibiotics in pharmaceutical and research centers. The WHO list must be aligned with increased political awareness in a global, multifaceted strategy to reduce the burden of resistant infections.
- The WHO PPL emphasizes the need for a novel antibacterial agent for the population and user-friendly formulations for infections caused by drug-resistant bacteria with high morbidities, such as multidrug- and extensively drug-resistant *Mycobacterium tuberculosis*, third-generation cephalosporin-resistant, fluoroquinolone-resistant *Neisseria gonorrhoeae*, fluoroquinolone-resistant *Campylobacter* spp., and *Salmonella* spp., *clarithromycin*-resistant *Helicobacter pylori*, and third-generation cephalosporin-resistant *Enterobacteriaceae*.
- Prioritizing the bacterial pathogens should be responsible for growing and regulating the implementation of control approaches.

24.4 Mechanism of Antibiotic Resistance

Bacteria develop to escape the effect of antibiotics through numerous diverse mechanisms and become antibiotic-resistant. The resistant mechanisms such as, active efflux, chemical inhibition of antibiotic, change in target site (e.g., mutation or structural modification) diminish the effect of antibiotic in the bacteria (Wilson et al. 2020). The mechanism of antibiotic resistance is categorized into the following:

- *Inherent resistance*: Through evolution, bacteria change the structure, and components of the antibiotic target site and survive against the antibiotic. For example, penicillin cannot kill bacteria that do not have a cell wall because penicillin affects the bacterial cell wall-building mechanism.
- *Acquired resistance*: Bacteria get resistant against a particular antibiotic to which bacteria was susceptible. The resistance was achieved through a new genetic mutation that allows the bacteria to doge the antibiotic and survive or receive DNA from resistant bacteria. An example is *Mycobacterium tuberculosis* resistance to rifamycin.
- *Genetic change*: Due to mutation in the bacterial DNA, the structure and production of protein change or alter, which leads to unrecognized by the antibacterial agents. The genomics of the bacteria would transform through intrinsic genetic determinants of resistance. An example is *E. coli* and *Haemophilus influenza* resistance to trimethoprim.
- *DNA transfer*: Bacteria can exchange genetic material and transfer resistant genes by horizontal gene transfer. An example is MRSA.

24.5 Drug Targets and Their Resistance Mechanism

The primary drug-target interaction of antibiotics induces bacterial cell death by inhibiting essential cellular functions. Antibiotics kill the bacteria (bactericidal drugs) or hinder cell growth (bacteriostatic drugs) depending on the system or cellular component they affect (Von Döhren 2009; Kohanski et al. 2010). Most clinically successful general antibiotic targets five bacterial metabolism areas, cell wall synthesis, DNA synthesis, protein synthesis, RNA synthesis and intermediary metabolism.

24.5.1 Cell Wall Synthesis Inhibitors

The cell wall is critical for a bacterium to stay alive in environmental conditions (Höltje 1998). The bacterial cell wall is layers of peptidoglycan (PG, or murein) (Bugg and Walsh 1992). The maintenance PG required transglycosylase and transpeptidase enzyme activity. To extend the glycan strands of PG these enzymes add disaccharide pentapeptides, and cross-link with the peptide strands of PG. The β-lactams and glycopeptides classes of antibiotics have interfered with specific homeostatic cell wall biosynthesis steps (Tomasz 1979). β-lactams classes antibiotics inhibits the PG synthesis by terminating the PG cross-linking (Tipper and Strominger 1965; Wise and Park 1965; Höltje 1998) (penicillins, carbapenems, and cephalosporins, etc.). Vancomycin, a glycopeptide antibiotic binding at the D-alanyl-

D-alanine dipeptide with PG units, inhibits PG synthesis by hindering transglycosylase activity and transpeptidase activity (Kahne et al. 2005).

24.5.1.1 Penicillins

These antibiotics have a nucleus of 6-animopenicillanic acid ring and additional ringside chains (Fig. 24.1). The group contains natural penicillins, agents immune to beta-lactamase, aminopenicillins, carboxypenicillins, and ureidopenicillins.

24.5.1.2 Cephalosporins

Cephalosporins have a nucleus of 7-aminocephalosporanic acid and a side-chain containing H-1,3-thiazane 3,6-dihydro-2 rings. Traditionally, cephalosporins are classified into five groups or generations (Fig. 24.2), but it is not universal to accept this terminology.

24.5.1.3 Carbapenems

Their distinguishing structure, a β-lactam ring, is coupled with carbapenem that provides protection against most beta-lactamases (Fig. 24.3). Resistance to these compounds, however, is a major problem and occurs mainly among gram-negative pathogens (e.g., *K. pneumoniae, P. aeruginosa, and A. baumannii*), which develop different groups of beta-lactamases called carbapenemase.

24.5.1.4 Monobactams

The β-lactam ring, standing alone, is not fused to another ring (Fig. 24.3).

24.5.1.5 Beta-lactamase Inhibitors

They primarily work by disabling serine β-lactamase, which hydrolyzes and inactivates the beta-lactam ring. These specialists incorporate the first-generation beta-lactamase inhibitors (clavulanic corrosive, sulbactam, and tazobactam) (Fig. 24.3) and the more current avibactam and vaborbactam active against carbapenemase such as *Klebsiella pneumoniae* carbapenemase (KPC).

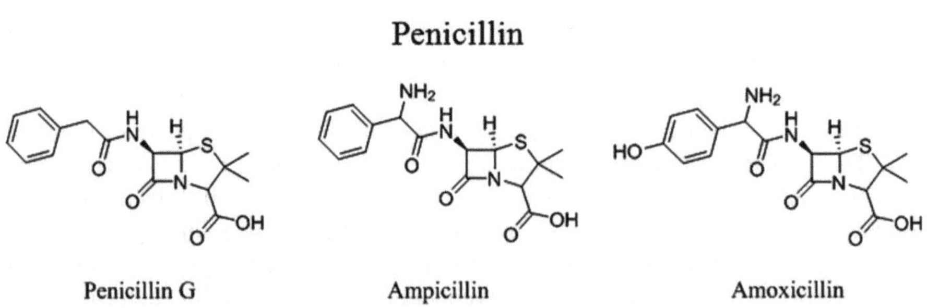

Fig. 24.1 Structure of penicillin derivatives

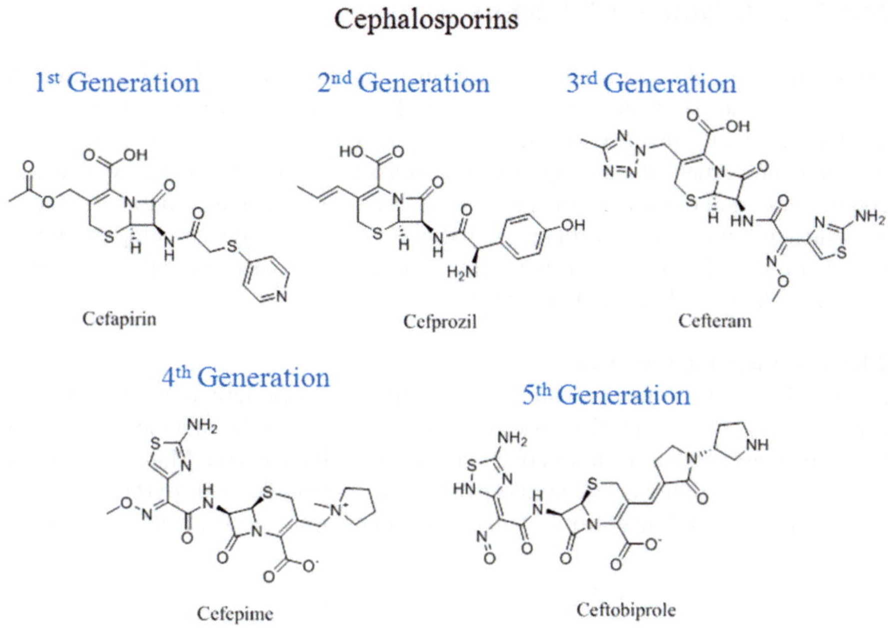

Fig. 24.2 Structure of five generations of cephalosporins antibiotics

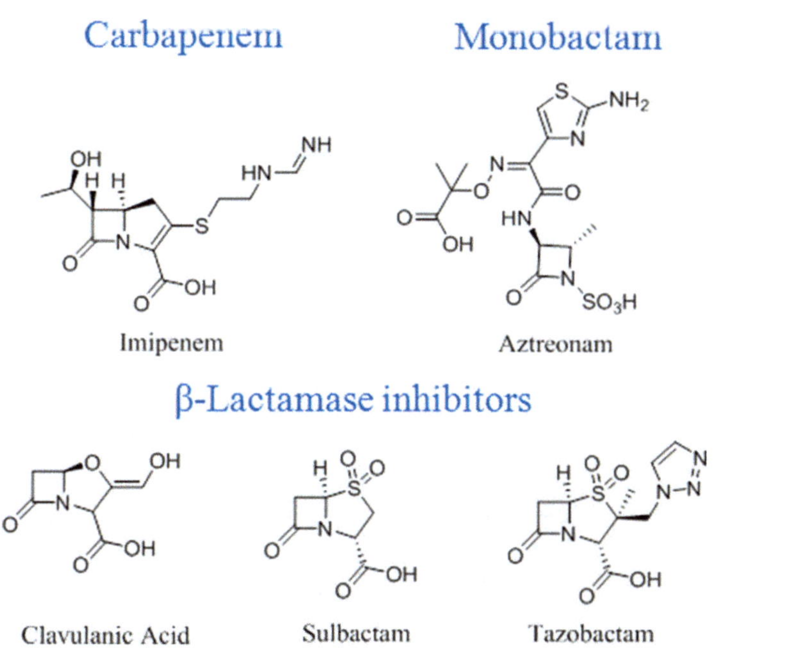

Fig. 24.3 Structure of generation of carbapenem, monobactam, and β-lactamase inhibitors

24.5.2 DNA Synthesis Inhibitors

Quinolones are a class of antibiotic agents that bind to gyrase and topoisomerase IV and prevent bacteria from replicating DNA (Naqvi et al. 2018). The discovery of nalidixic acid in 1962 led to a drastic rise in the development and comparable structural properties of other quinolones antibiotics (Andersson and MacGowan 2003). Fluoroquinolones are often used for bacterial treatment due to their outstanding efficacy against a wide range of bacteria. The DNA-gyrase complex during replication is stabilized by an interaction with the 3-oxo-4 carboxylic acid center (Zhang et al. 2018; Tsakou et al. 2020).

24.5.2.1 First Generation

In 1962, Lesher G Y and his colleagues identified the key derivative of quinolone, nalidixic acid (Fig. 24.4). Nalidixic acid has been used for the treatment of urinary tract infections and has a strong effect on gram-negative bacteria. Many derivatives, such as cinoline (cinoxacin), pyrido-pyrimidine (pipemidic acid, pyromidic acid), naphthyridine (nalidixic corrosive), and quinolones, were subsequently synthesized

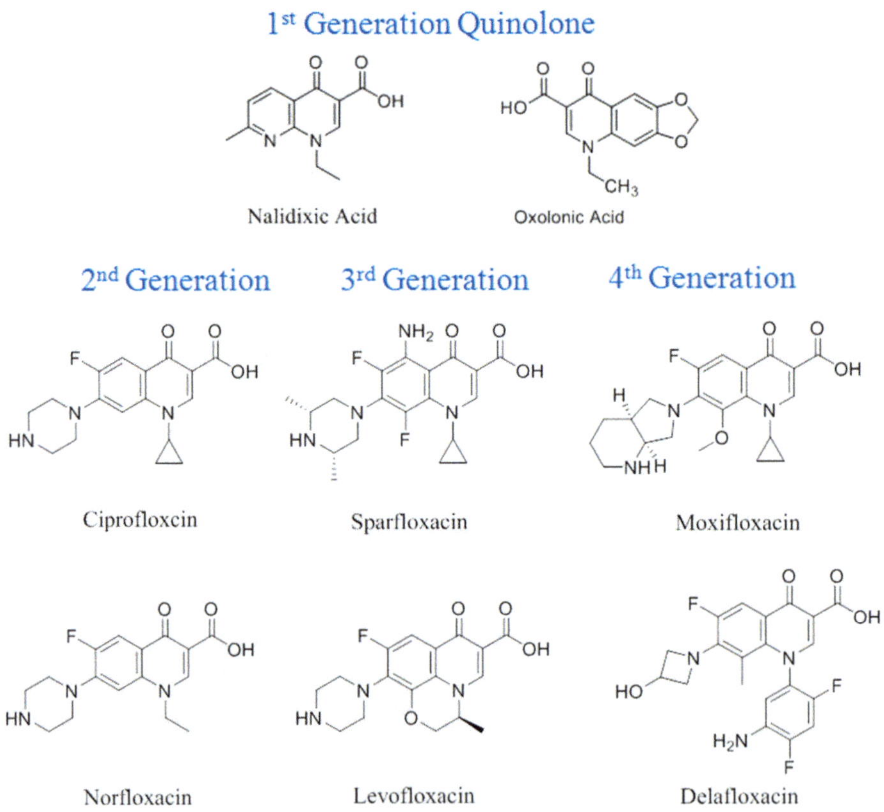

Fig. 24.4 Structure of generations of quinolone

(oxolinic acid, miloxacin, tioxacin, etc.). These derivatives have two common pharmacological properties with separate structures, an appropriate antibacterial range, extraordinarily planned for Enterobacteriaceae, and pharmacokinetics allowing for rapid disposal and interesting decreased tissue, as it allowed them to be used as urinary cleaning agents (Mimouni et al. 2019).

24.5.2.2 Second Generation

Norfloxacin, synthesized by Koga and his collaborators in 1980, is the main fluorine atom in quinolone, replaced by carbon-6 and carbon-7 of piperazine. The primary fluoroquinolones with extended antimicrobial activity were norfloxacin and ciprofloxacin (Fig. 24.4), active against a wide variety of bacteria (Mimouni et al. 2019).

24.5.2.3 Third Generation

Levofloxacin and sparfloxacin are the third-generation fluoroquinolones agents. Their extended action against gram-positive organisms is why these molecules are classified into a third class, counting the penicillin-resistant *Streptococcus pneumoniae* (Mimouni et al. 2019).

24.5.2.4 Fourth Generation

The fourth generation, such as moxifloxacin and delafloxacin (Fig. 24.4), has better activity than levofloxacin. The fourth generation is active against gram-positive bacteria also and shows less resistance. The fourth-generation fluoroquinolones show phototoxicity, neurotoxicity, and cartilage harm.

In addition, moxifloxacin and levofloxacin are used to treat *M. tuberculosis* and treat other infections with MDR (Lucena et al. 2000; Bush et al. 2020). Some very promising fluoroquinolones, including trovafloxacin and grepafloxacin, were removed from the clinic because of safety concerns, despite their effectiveness (Fig. 24.5) (Mandell and Tillotson 2002). However, many have remained in the

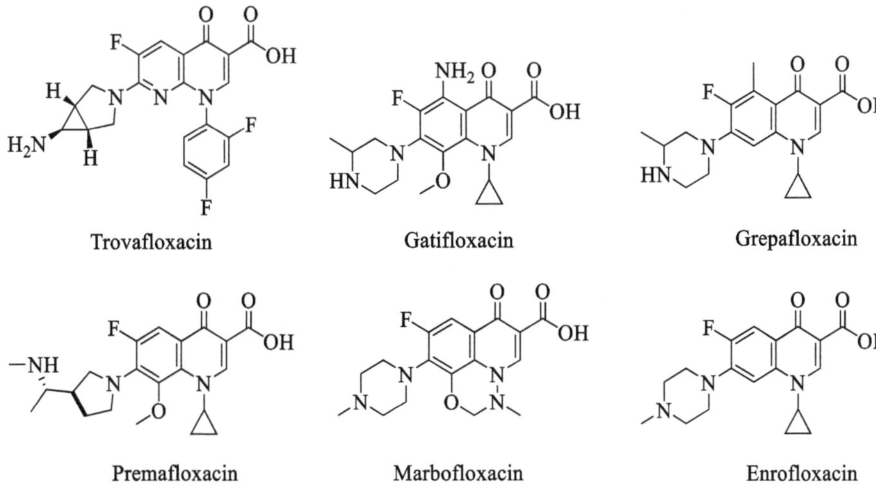

Fig. 24.5 Structure of fluoroquinolones inhibitors

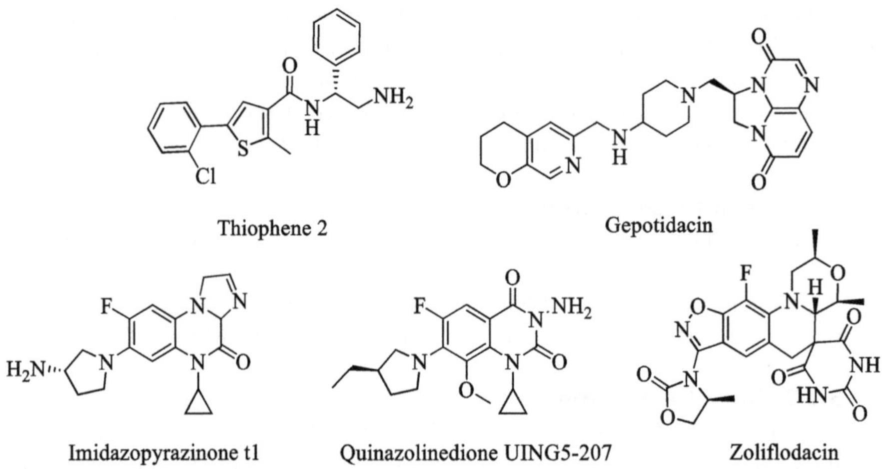

Fig. 24.6 Structures of new fluoroquinolones, gepotidacin, and zoliflodacin are in phase III clinical trials and structure of non-fluoroquinolone topoisomerase inhibitors, quinazolinedione UING5–207, imidazopyrazinone t1, and thiophene 2

clinic, with ciprofloxacin continuing to be one of the most clinically important antibiotics to date (Fig. 24.6).

24.5.3 Protein Synthesis Inhibition

Protein synthesis involved the translation of mRNA to its active biological form, which is the essential process involving ribosomes. Many enzymes and conformational alignments are required for protein synthesis because it is a complex, multi-step process. The 70S bacterial ribosome had two subunits, 30S, and 50S, most targeted by an inhibitor to block the protein synthesis such as tetracyclines, doxycycline (Fig. 24.7) and inhibit binding aminoacyl-tRNA by blocking the A (aminoacyl) site of the 30S ribosome (Witzky et al. 2019).

24.5.4 RNA Synthesis Inhibitors

RNA is essential for life because it can perform very miscellaneous and crucial functions and be used for drug development targets. One of the well-studied targets of RNA synthesis is rifampicin. Rifampicin (Fig. 24.8) has a catastrophic effect on bacterial cells and thus acts as bactericidal (Floss and Yu 2005; Goldstein 2014; Penchovsky 2018).

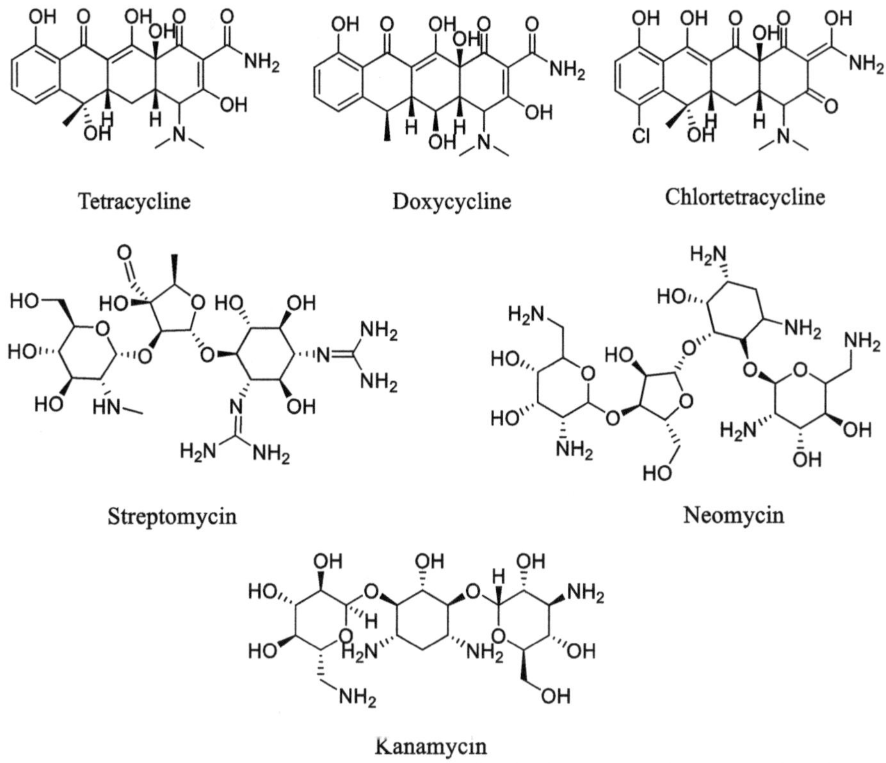

Fig. 24.7 Structure of protein synthesis inhibitors

24.5.5 Folic Acid Synthesis Inhibitors

Bacteria use PABA (para-aminobenzoic acid), pteridine, and glutamate to prepare folic acid. Folic acid is a vitamin for humans but is unable to synthesize; that is why these pathways are selectively targeted for antimicrobial agents like trimethoprim and sulfamethoxazole (Fig. 24.9) (Fernández-Villa et al. 2019).

24.5.6 Macrolides

The macrolides are bacteriostatic antibiotics with a varied range of action against several bacteria (gram-positive), such as azithromycin and erythromycin (Fig. 24.10). This suggests the urgent need to prevent cross-resistance by new and better-quality antimicrobial agents with a new target and a novel molecular structure.

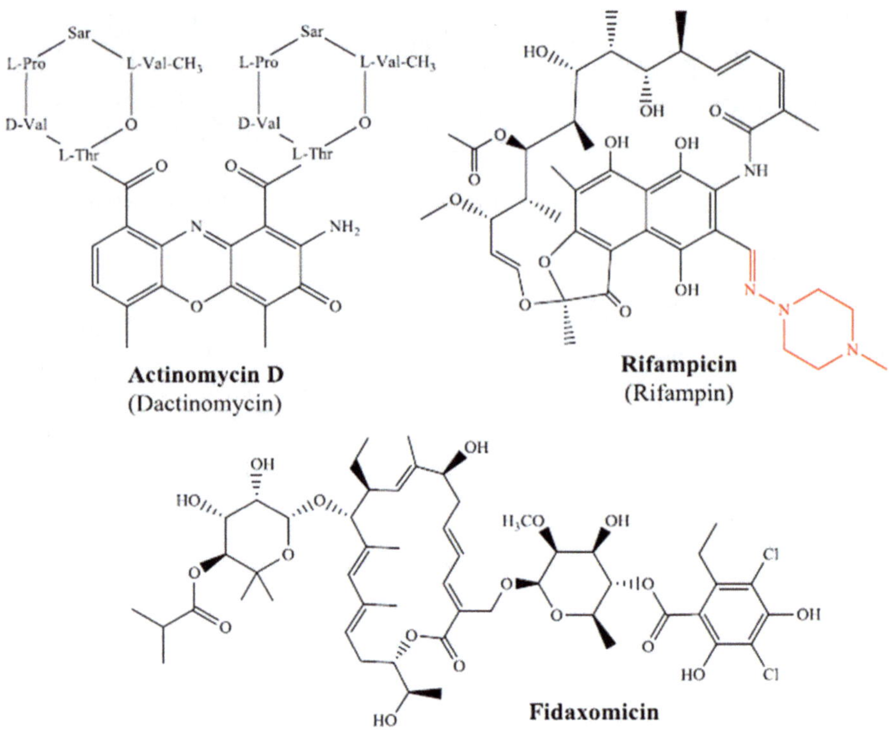

Fig. 24.8 Structure of RNA synthesis inhibitors

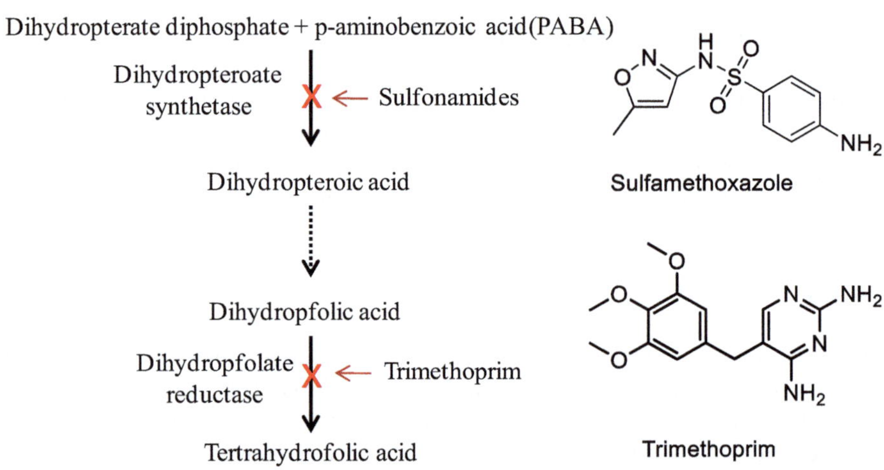

Fig. 24.9 Structure of folic acid synthesis inhibitors

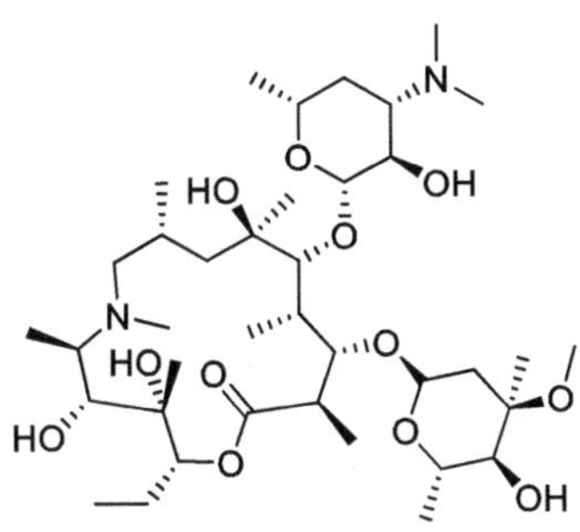

Azithremycin

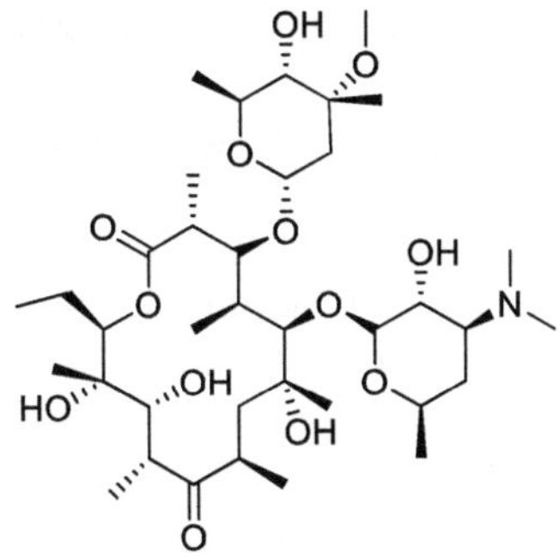

Eytheomycin

Fig. 24.10 Structure of macrolides

24.6 New Antibacterial Targets

The antibacterial resistance is increasing day by day, and the rate of antibacterial resistance is higher than the antibacterial agent development. Physicians are prescribing a drug in combination, and to deal with these resistant microbes, they have to turn to their second or even third choice of antibiotic (McDevitt et al. 2002). This indicates the urgent necessity for novel and better-quality antibacterial agents with a unique target and structure to avoid cross-resistance (Belete 2019).

24.6.1 Two-Component Signal Transduction (TCST)

TCST systems are ubiquitous in bacteria; use this system to monitor their external environment and adapt to survive. The TCST as a novel target was proposed by characterizing it. Bioinformatics analysis of genomic data reveals that many TCST systems are essential for infection, and at least one (*YycFG*) is necessary for the growth of *S. pneumoniae* and *S. aureus*.

24.6.2 Biofilm Formation Inhibitor

A large proportion of the drug-tolerant population of bacteria that synthesize matrix around the surface attached cell mass is known as a biofilm. Biofilm helps bacteria to survive in harsh condition and increase the resistance to 1000 times. Antibiotics failed to clear the biofilm; therefore, an antibiofilm inhibitor is required. Biofilm formation and virulence were increased with the increase of c-di-GMP in bacteria. To inhibit biofilm infection, first need to inhibition or modification of c-di-GMP is needed, and second is inhibition of the quorum sensing pathway.

24.6.3 Quorum Sensor Synthesis Inhibitor

An autoinducer, a signaling substance, is synthesized by bacteria to communicate and estimate their population. Bacteria proliferate to increase the number when the autoinducer level is too low. Quorum sensing regulates many mechanisms, including the production of biofilm (Kirmusaoglu 2016; Paluch et al. 2020), siderophores (Cornelis and Aendekerk 2004), exoenzymes (Pena et al. 2019), competence for horizontal gene transfer (Blokesch 2012; Shanker and Federle 2017), membrane vesicles (Toyofuku 2019), secondary metabolites with antimicrobial activity (Barnard et al. 2007; Kareb and Aïder 2020), swarming motility (Daniels et al. 2004), and microbial physiology. Inhibition of quorum sensing will decrease the pathogenicity and eradicate the system's microorganisms. These processes are associated with virulence, and many other functions (Krzyżek 2019); almost all the gram-positive and gram-negative bacteria produce signal substance acylhomoserine lactones (AHL), a quorum-sensing signaling substance. Inhibiting AHL and AHL pathways easily inhibit the quorum-sensing.

24.6.4 Teichoic Acids Inhibitors

Teichoic acid played an essential role in bacterial survival, cell division, growth, cell division, and growth, maintaining cell shape and resistance to antimicrobial agents. All these abilities make Teichoic acid a potential drug target, and human has no homolog of this.

24.6.5 Toxin Neutralization

All the antibacterial mAbs currently marketed work through toxin neutralization. Their effectiveness is based on their binding affinity to the toxin. By forming antibody-toxin complexes, which are mainly cleared by the reticulum endothelial system, antibacterial mAbs neutralize soluble exotoxins. All antibacterial mAbs currently on the market function through the neutralization of toxins. Their effectiveness is based on their binding affinity to the toxin.

24.6.6 Type IA Topoisomerase

At least one topoisomerase is present in the bacteria, and type IA topoisomerases are two types, topoisomerase I and topoisomerase III. This topoisomerase class splits a single strand of DNA at the active site with a tyrosine nucleophile forming a covalent complex through a 5′-phosphotyrosine bond with the split DNA.

DNA was rejoined by the movement of the DNA strands used to alter the topology of the DNA. For solving topological obstacles, an ssDNA transit in the topoisomerase type IA function is necessary. Inhibiting the type IA topoisomerase

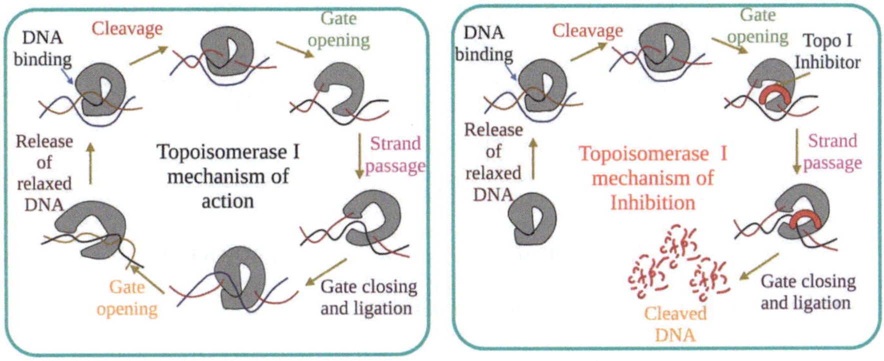

Fig. 24.11 Mechanisms of action and inhibition type IA topoisomerase

by catalytic inhibitor would have antibacterial efficacy (Fig. 24.11). *M. tuberculosis* (Ahmed et al. 2014; Ravishankar et al. 2015), *S. pneumonia* (Liu et al. 2017), *P. aeruginosa* (Yan et al. 2019), and *H. pylori* (Suerbaum et al. 1998) topoisomerase is essential for their viability, found in different studies (Seddek et al. 2021).

DNA intertwining is created by fundamental cellular processes, including recombination, replication, translation, transcription, and chromosome segregation (Nurse et al. 2003; Nicholls et al. 2018; Bizard et al. 2018; Lee et al. 2019; Rani and Nagaraja 2019). To maintain cell viability and genome stability, the intertwining of DNA needs to resolve. The solution for resolving the intertwining of DNA is type IA topoisomerases. It can relax the negatively supercoiled DNA, inhibit hyper negative supercoiling, catalyze DNA decatenation, and wind or unwind ssDNA (Liu et al. 1976; Brown and Cozzarelli 1981; Chen et al. 2013). The accumulation of hypernegative-superhelicity in DNA happened due to the Loss of topoisomerase's DNA relaxing activity, leading to damaged DNA accumulation, genomic instability, and cell death (Brochu et al. 2020). This indicated the inhibiting type IA topoisomerases lead to bacterial cell death. All topoisomerases control an active site tyrosine to form the covalent intermediate, a poison inhibitor can trap the covalent complex formed by type IA topoisomerases, but to date, there are no approved drugs against type IA topoisomerases (Seddek et al. 2021). Several bacterial topoisomerase I inhibitors have been described as follows.

24.6.7 Natural Products as Dual Inhibitors of Type IA & Type IIA Topoisomerases

Lichen Hypotrachyna sp. extract inhibits topoisomerase IA with an imp4213 mutation that improves penetrability for small molecules through antibacterial action against *E. coli* strain BAS3023 (Braun and Silhavy 2002; Tse-Dinh 2015). *Lichen Hypotrachyna* sp. (Lecanorales, Parmeliaceae) has fractionated and produced anziaic acid (Fig. 24.12) from Costa Rica, which acts as a poison inhibitor of

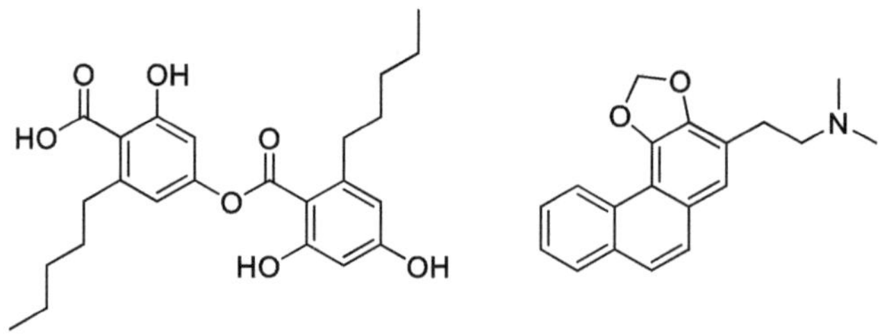

Anziaic acid Stephenanthrine

Fig. 24.12 Structure of *E. coli* topoisomerase I inhibitors

bacterial topoisomerase I. It can inhibit the growth of *Bacillus subtilis* but is not active against *E. coli*. It also inhibits type IIA topoisomerase activities of EcGyrase and HuTopo IIα but had less active against type IB HuTopo I. It could potentially affect the enzyme divalent ion interactions at the active site (Cheng et al. 2013).

24.6.8 Phenanthrene Compounds

Through HTS, based on the induction of SOS response in *E. coli,* the bacterial topoisomerase I poison inhibitor, aphenanthrene compound stephenanthrine (Fig. 24.12), was identified. Two phenanthrenes, seconeolitsin and N-methyl-seconeolitsin, semi-synthesis and bolden alkaloid evaluation, were found to inhibit the relaxation activity of *S. pneumoniae* topoisomerase I and cell growth at low levels.

24.6.9 Protein Inhibitors of Bacterial Topoisomerase I

The relaxation activity of bacterial topoisomerase I was also inhibited by specific proteins, like the *E. coli* Tn5 transposase and MazF homolog Rv1495 of *M. tuberculosis* (Tse-Dinh 2009; Huang and He 2010). DNA cleavage activity can also be inhibited by Rv1495. The cell development of *M. smegmatis* was restrained by the overexpression of Rv1495 or its N-terminal part. It is conceivable that the activity of protein inhibitors of bacterial topoisomerase I may be mimicked by the small compounds. There are monoclonal antibodies reported that can inhibit *M. tuberculosis* and *M. smegmatis* topoisomerase I's relaxation activity. These antibodies can stimulate DNA cleavage (Leelaram et al. 2013) and enhance the closing of the DNA gate after DNA cleavage (Leelaram et al. 2012). For the design of novel bacterial topoisomerase I inhibitors, these antibodies could be used (Tse-Dinh 2015).

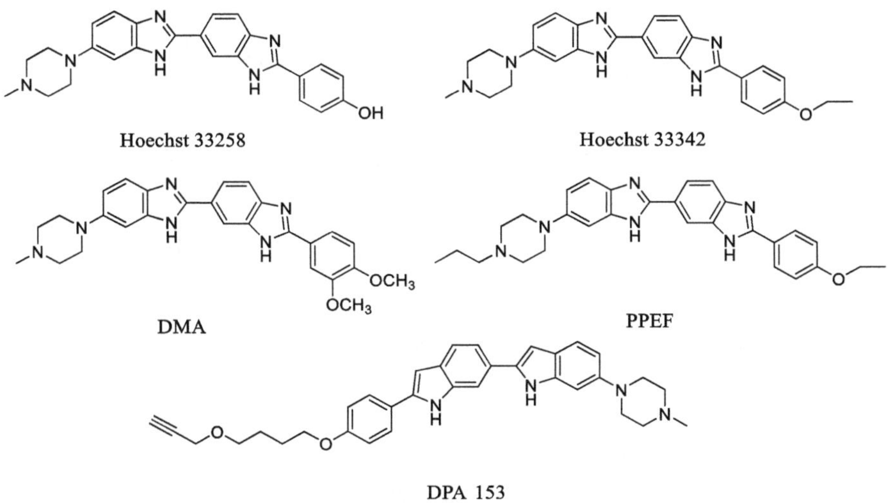

Fig. 24.13 Structure of *E. coli* topoisomerase I inhibitors benzimidazole derivatives

24.6.10 Bisbenzimidazoles

The Hoechst 33342 and Hoechst 33258 (Fig. 24.13) DNA minor groove binders serve as poison inhibitors for mammalian DNA topoisomerase I (Chen et al. 1993).

However, 3,4-Dimethoxyphenyl bisbenzimidazole (DMA) analogs of Hoechst can selectively target prokaryotic topoisomerase I DNA over HuTopoI (Bansal et al. 2010, 2012). With DMA therapy, the amount of DNA cleavage products increases and can act as a poison inhibitor for EcTopo IA. DMA kills bacterial cells by trapping the covalent complex of topoisomerase I. Another benzimidazole DPA 153 with a terminal alkyne substitution inhibits EcTopo IA much more intensely than EcGyrase, HuTopoI, and HuTopo Iα (Ranjan et al. 2014).

24.7 Fluoroquinophenoxazine Derivatives

Fluoroquinophenoxazine derivatives have antibacterial activity and selectivity in the inhibition of topoisomerases I. Fluoroquinophenoxazine NSC648059 (Fig. 24.14) inhibit the catalytic activity of *E. coli* topoisomerase I and also have antituberculosis activity (Yu et al. 2017). The structure of fluoroquinophenoxazine is similar to fluoroquinolones, but it shows greater effectiveness for inhibition of type IA bacterial topoisomerase than *E. coli* gyrase, human-type topoisomerase I, and II (Garcia et al. 2019).

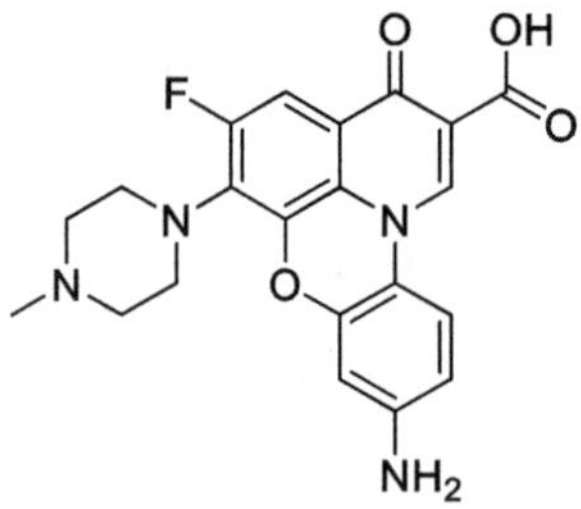

NSC648059 derivative 11a

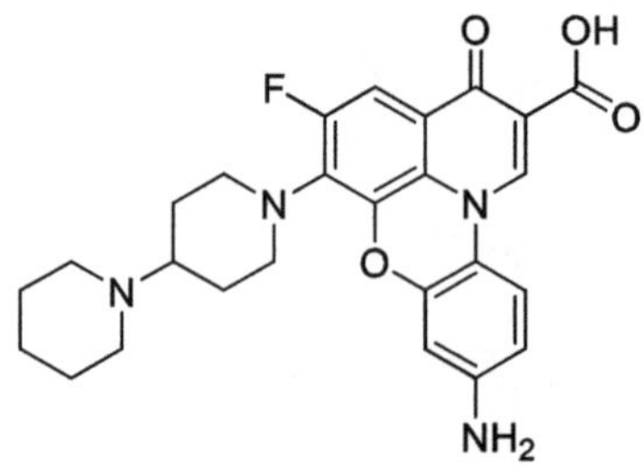

NSC648059 derivative11g

Fig. 24.14 Structure of *E. coli* topoisomerase I inhibitors fluoroquinophenoxazine NSC648059 derivatives

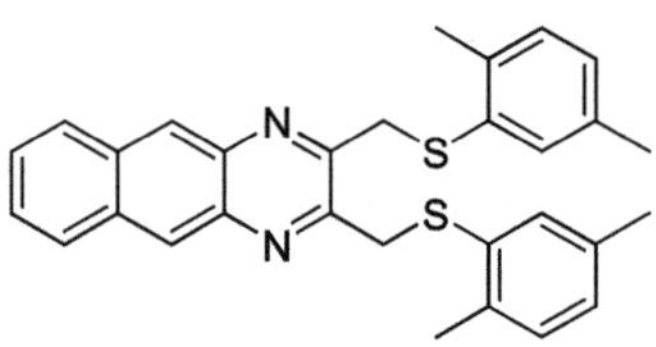

VCC891909

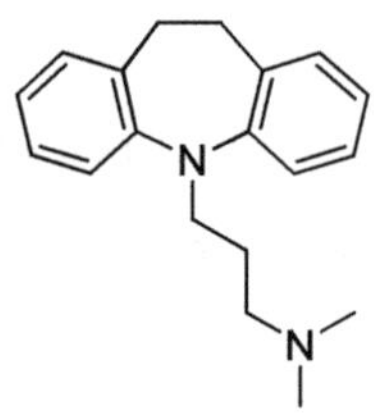

Imipramine

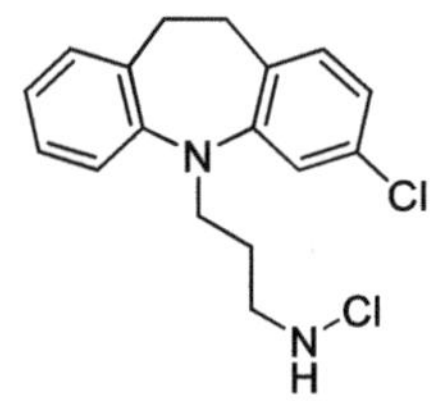

Norclomipramine

Fig. 24.15 Structures of benzo(g) quinoxalines compound and tricyclic antidepressants

24.8 Benzo(g)quinoxaline Compound

Vichem's Nested Chemical Libarary (NCL) have 17,000 molecules and out of which a selected set of seven Vichem NCL molecules with benzo(g)-quinoxaline, quinoxaline, or styryl-benzo(g)-quinazoline scaffolds were have the antibacterial activity. Further characterization revealed that it binds to specific sites on *M. tuberculosis* topoisomerase I through interactions with R167 and R114. Through this, VCC891909 (Fig. 24.15) was suggested to be an excellent antibacterial candidate with no cytotoxicity in all the human cell lines and efficient inhibition of topoisomerase I activity.

24.8.1 Tricyclic Assigned as *M.tuberculosis* Topoisomerase I Inhibitor

A library of FDA-approved drugs was screened through in silico docking for their potential activity against a homolog model of *M. tuberculosis* topoisomerase I. Imipramine and norclomipramine (Fig. 24.15) are clinically prescribed tricyclic antidepressants (clomipramine) that showed inhibition activity against *M. tuberculosis* topoisomerase I.

24.8.2 New FDA-Approved Antibacterial

Most antibiotics being advanced are hardly any improvement on present drugs. The limited research undertaken is mostly by small or medium-sized companies, with big pharmaceutical companies permanently exiting the field. Since 2017, eight new have been approved worldwide, with limited clinical benefits. The list is given below in Table 24.2.

24.8.2.1 Delafloxacin

Delafloxacin (Fig. 24.16) is currently the only antibiotic with in vitro efficacy against MRSA and *P. aeruginosa*, a new fluoroquinolone for the treatment of critical bacterial skin and skin organization infections, approved by the FDA.

24.8.2.2 Vabomere

Meropenem–vaborbactam has demonstrated in vitro activity and in vivo clinical efficacy against most isolates of *Enterobacter cloacae species* complex, *Escherichia coli*, and *Klebsiella pneumoniae*. In vitro data are available with unknown clinical significance for these gram-negative bacteria: *M. morgannii, C. koseri, Providencia spp., C. freundii, K.oxytoca, E. aerogenes, P. aeruginosa, P. mirabilis, and S. marcescens.*

24.8.2.3 Plazomicin

Asemisynthetic aminoglycoside bactericidal antibiotic drug is Plazomicin sulfate (Zemdri) (Fig. 24.16) acting similar to other aminoglycosides by suppressing the 30S bacterial ribosomal subunit. Vital, whereas other aminoglycosides can be inactivated by aminoglycoside-modifying proteins, plazomicin is safe for the activity of these enzymes.

24.8.2.4 Lefamulin

Lefamulin (Xenleta) (Fig. 24.16) could be a semisynthetic pleuromutilin antimicrobial that ties to the peptidyl transferase center of the 50S bacterial ribosomal subunit, hindering protein amalgamation inside bacteria. Comes about of these trials illustrated that lefamulin was as viable as moxifloxacin in the treatment of community-procured bacterial pneumonia.

24.8.2.5 Eravacycline

Eravacycline dihydrochloride (Xerava) (Fig. 24.16) could be a completely manufactured bacteriostatic fluorocycline and a tetracycline-class antibacterial operator that ties bacterial 30S ribosomal subunit. It has two structural substitutions compared to other tetracyclines that make the drug function on some gram-negative and gram-positive bacteria strains that typically have mechanisms of

Table 24.2 Antibacterial agents that have been FDA or EMA certified since 2017

S. no.	Name (trade name)	Antibiotic class	Approved by (date)	Indications	Mechanism
1	Delafloxacin (USA: Baxdela, EU: Quofenix)	Fluoro-quinolone	FDA (6/2017 ABSSSI, 10/2019 CAP) MAA	ABSSSI, CAP	Inhibit both bacterial topoisomerase IV and DNA gyrase (topoisomerase II)
2	Vaborbactam and meropenem (Vabomere)	Boronate BLI and carbapenem	FDA (8/2017) EMA (11/2018)	cUTI	Usage in conjunction with the carbapenem (meropenem) antibiotic Vaborbactam protects meropenem via a novel mechanism of enzyme inhibition against serine carbapenemase degradation
3	Plazomicin (Zemdri)	Aminoglyco-side	FDA (7/2018)	cUTI	Inhibit bacterial 30S ribosomal subunit
4	Relebactam and imipenem and cilastatin (Recarbrio)	FDA (7/2019) MAA	FDA (7/2019) MAA	cUTI, cIAI	Imipenem-cilastatin-relebactam (IMI-REL) is a novel β-lactam-β-lactamase inhibitor use in combination imipenem, a carbapenem antibacterial drug, cilastatin, a renal dehydropeptidase inhibitor, and relebactam, a diazabicyclooctane beta lactamase inhibitor
5	Lefamulin (Xenleta)	FDA (8/2019) MAA	FDA (8/2019) MAA	CAP	Inhibition of protein synthesis by binding the 50S bacterial ribosome to the peptidyltransferase core, thereby preventing the binding of transfer RNA for peptide transfer
6	Cefiderocol (Fetroja)	FDA (11/2019) MAA	FDA (11/2019) MAA	cUTI	Inhibit penicillin-binding proteins (PBPs), preventing cell wall synthesis

7	Eravacycline (Xerava)	Tetracycline	FDA (8/2018) EMA (9/2018)	cIAI	Disrupts the synthesis of bacterial proteins by binding to the 30S ribosomal subunit, preventing amino acid residues from being integrated into elongating peptide chains
8	Omadacycline (Nuzyra)	Tetracycline	FDA (10/2018)	CAP (intravenous), ABSSSI (intravenous, oral)	Inhibits the synthesis of bacterial proteins by binding the primary tetracycline binding site to the bacterial ribosome subunit 30S

ABSSSI acute bacterial skin and skin structure infections, *cIAI* complicated intra-abdominal infections, *CAP* community-acquired pneumonia, *CRAB* carbapenem-resistant *Acinetobacter baumannii*, *EMA* European Medicines Agency, *MAA* Marketing Authorization Application (EMA)

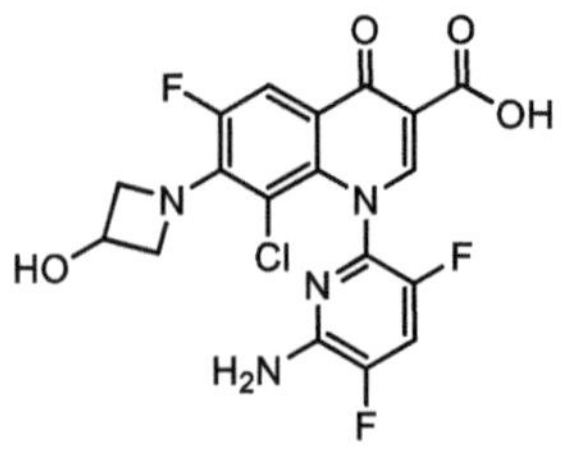

Delafloxacin

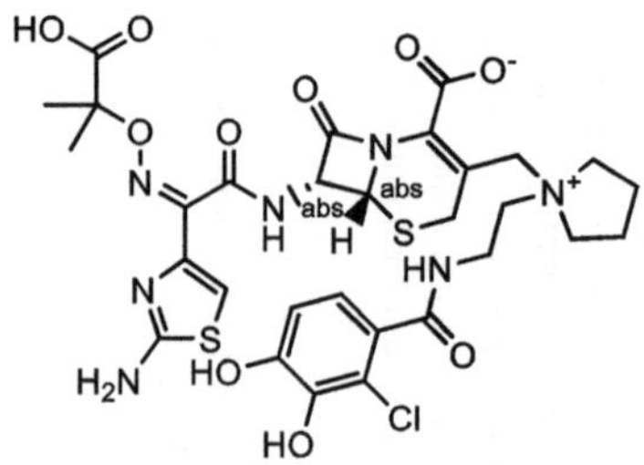

Cefiderocol

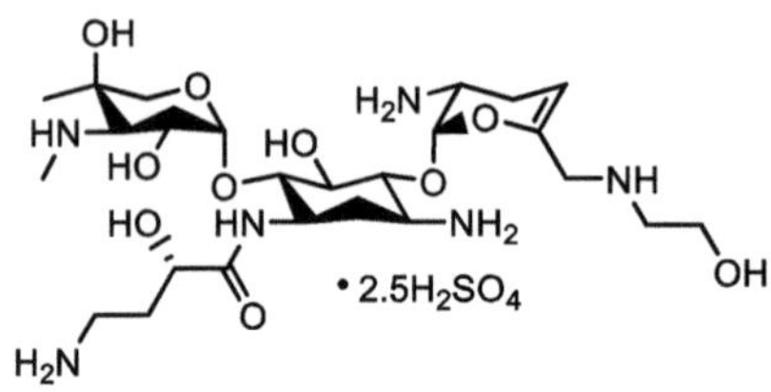

Plazomicin(Zemdri)

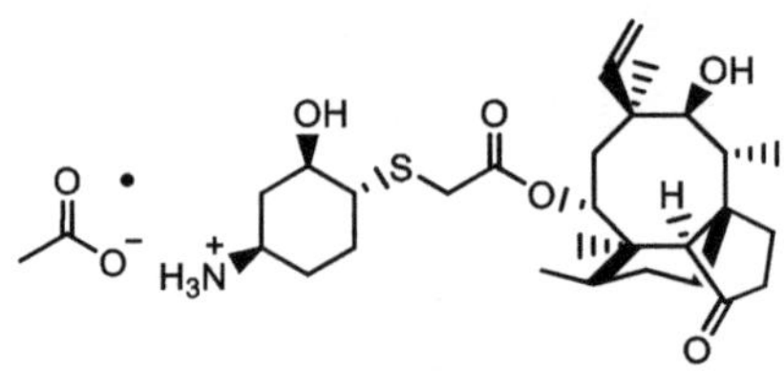

Xenleta

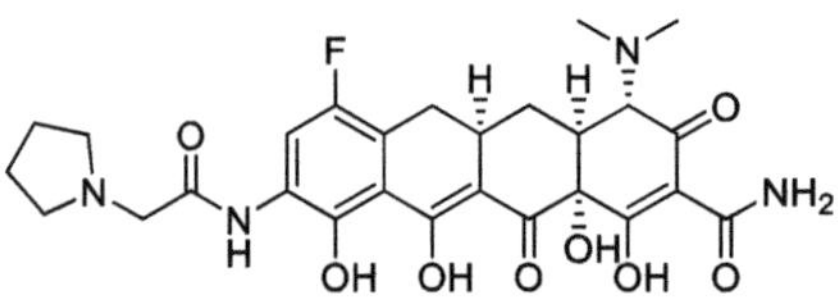

Eravacycline

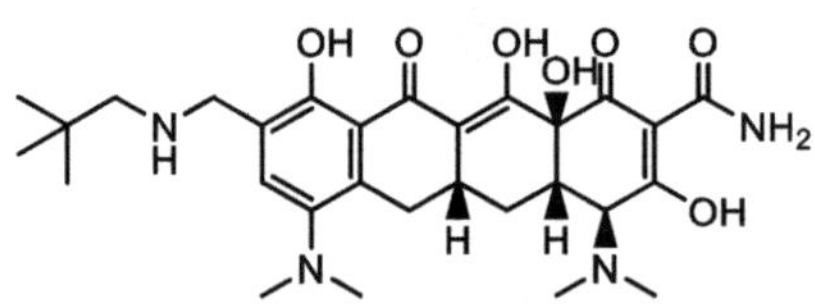

Omadacycline

Fig. 24.16 Structure of antibiotics approved since 2017

tetracycline-specific resistance. Remarkably, in the presence of some β-lactamaseses, eravacycline can be used (at least in cell culture) to attack Enterobacteriaceae.

24.8.2.6 Omadacycline

Omadacycline (Nuzyra) belongs to the tetracycline class and is an aminomethylcycline antibiotic. It inhibits bacterial ribosomal subunit 30S. Omadacycline has structural variations at the C9, and C7 positions of the core tetracycline ring relative to other tetracycline antibiotics, allowing ribosomal defense mechanisms and stability in the efflux pump relevant to tetracycline antibiotic resistance.

24.8.2.7 Imipenem, Cilastatin, and Relebactam (Recarbrio)

Recarbrio is a regimen consisting of imipenem, the antibacterial penem, cilastatin, the renal dehydropeptidase inhibitor, and relebactam, the inhibitor of β-lactamase.

Gram-negative microorganisms such as *K. pneumoniae*, *P. aeruginosa*, *E. cloacae*, *E. coli*, *K. oxytoca*, and *E. aerogenes*, some bacteroid strains, and other susceptible bacteria are susceptible to Recarbrio.

24.8.2.8 Cefiderocol

Cefiderocol (Fetroja) is an antibacterial siderophore cephalosporin drug intended to treat a wide variety of bacterial pathogens, such as βlactam-resistant and carbapenem-resistant species. Cefiderocol targets a wide variety of gram-negative bacteria of clinical interest, including but not limited to *Enterobacteriaceae* spp., such as non-fermenting species of bacteria such as *Acinetobacter* and *Pseudomonas*, *Enterobacter* spp., (*Shigella flexneri, Klebsiella* spp., *Vibrio* spp., *Salmonella* spp., *Yersinia* spp., *Serratia marcescens*, *Proteus* spp.).

24.9 The Compound in Clinical Phase Trials

According to the WHO report (2019), only 60 molecules are in the clinical phases for development against the WHO priority pathogens. Out of these 60 molecules, 32 molecules are active against the WHO priority pathogens, 12 against Tb, and six against *Clostridioides difficile.* All antibiotics and biological treatments currently being developed against the WHO priority list of pathogens are listed in the WHO clinical antibacterial pipeline review (Table 24.3). The future is bleak: in clinical phases 1 to 3, only 60 items are present. WHO experts group define Four criteria that should be followed to get the novel antibacterial agent: (1) absence of known cross-resistance, (2) new class, (3) new target, and (4) new mode of action. Only six molecules that inhibit the WHO priority pathogens fulfill at minimum one of the four criteria (Beyer and Paulin 2020).

24.10 Conclusion

The morbidity and mortality rates are high in humans and animals due to MDR bacterial strains-induced infectious diseases. New antibacterial agents and innovative antimicrobial approaches are required to combat increasing antibiotic resistance in different populations. It is vital to identify molecules with the novel target to defeat resistance or at least slow its emergence. DNA topoisomerase maintains DNA topology by cleaving and rejoining the DNA strand. Topoisomerases play an essential role in regulating physiological function and DNA processing (Wang 1996, 2002; Kathiravan et al. 2013), such as DNA synthesis, RNA synthesis, and DNA recombination (Lim et al. 1986; Shuman 1991; Merino et al. 1993; Kretzschmar et al. 1993; Pommier et al. 2006), chromosome decondensation, repair, and sister chromatids (Stevnsner and Bohr 1993; Nitiss 1994; Yeh et al. 1994; Kathiravan et al. 2013). Thus, topoisomerase is considered the druggable target for developing various antibacterial and anticancer agents. Topoisomerase II is the validated bacterial target for two types of antibiotic medications, with the first

Table 24.3 Antibiotics currently in global clinical phase III development

S. no	Drug name	Company	Drug class	Target	Expected activity against WHO critical threat pathogen?	Novelty
Antibiotics currently in global clinical phase III development						
1.	Cefepime + taniborbactam (VNRX-5133)	Venatorx Pharmaceuticals Inc.	β-Lactam (cephalosporin) + β-lactamase inhibitor (cyclic boronate)	PBP + β-lactamase	Yes (CRE), possibly (CRPA)	No
2.	Cefepime + enmetazobactam	Allecra Therapeutics GmbH	β-Lactam (cephalosporin) + β-lactamase inhibitor (penicillanic acid sulfone)	PBP + β-lactamase	Yes (ESBL), possibly (CRE)	No
3.	Cefilavancin	R-Pharm	Glycopeptide-β-lactam (cephalosporin) hybrid	PG chain elongation + PBP	No	No
4.	Ceftobiprole	BasileaPharmaceutica Ltd.	β-Lactam (cephalosporin)	PBP	No	No
5.	Contezolid (MRX-I) and contezolidacefosamil (MRX-4)	MicuRx Pharmaceuticals Inc.	Oxazolidinone	50S subunit of bacterial ribosome	No	No
6.	Gepotidacin (GSK2140944)	GlaxoSmithKline PLC	Triazaacenaphthylene	Bacterial type II topoisomerase (novel A subunit site)	Yes (drug-resistant *N. gonorrhoeae*), possibly (ESBL)	Yes
7.	Iclaprim	Motif Bio PLC	2,4-diaminopyrimidine	Dihydrofolatereductase	No	No
8.	Ridinilazole	Summit Therapeutics PLC	Bis-benzimidazole	Inhibition of cell division and reduction of toxin production	Yes (*C. difficile*)	No
9.	Sulbactam + durlobactam	Entasis Therapeutics Inc.	β-Lactam (sulbactam) + β-lactamase inhibitor (diazabicyclooctane)	PBP + β-lactamase	Yes (CRAB)	No
10.	Sulopenem/sulopenem-etzadroxil- probenecid	Iterum Therapeutics PLC	β-Lactam (carbapenem)	PBP	Yes (drug-resistant *N. gonorrhoeae*, ESBL)	No

11.	T-4288 (solithromycin)	Toyama Chemical Co. Ltd.	Macrolide	50S subunit of bacterial ribosome	Yes (drug-resistant *N. gonorrhoeae*)	No
12.	Tebipenem/ tebipenempivoxilhydrobromide	Spero Therapeutics Inc.	β-Lactam (carbapenem)	PBP	Yes (ESBL)	No
13.	Zoliflodacin (ETX0914)	Entasis Therapeutics Inc.	Spiropyrimidinetrione	Bacterial type II topoisomerase (gyrB)	Yes (drug-resistant *N. gonorrhoeae*)	Yes
Antibiotics currently in global clinical phase II development						
14.	Afabicin (Debio 1450)	Debiopharm International SA	Benzofurannaphthyridine	FabI	No	Yes
15.	ARV-1801 (fusidic acid)	Arrevus Inc.	Fusidane	Elongation factor G	No	No
16.	Benapenem	Sihuan Pharmaceutical Co. Ltd.	Carbapenem	PBP	No	No
17.	BOS-228	Boston Pharmaceuticals Inc.	β-Lactam (monobactam)	PBP	Yes (CRE)	No
18.	Brilacidin	Innovation Pharmaceuticals Inc.	Defensin mimetic	Cell membrane	No	Yes
19.	CG-549	CrystalGenomics Inc.	Benzyl pyridinone	FabI	No	Yes
20.	DNV3837/ DNV3681	Deinove SA	Oxazolidinone- quinolone hybrid	50S subunit of bacterial ribosome, bacterial type II topoisomerase	Yes (*C. difficile*)	No
21.	Finafloxacin	MerLion Pharmaceuticals GmbH	Fluoroquinolone	Bacterial type II topoisomerase	Possibly (ESBL)	No
22.	MGB-BP-3	MGB Biopharma Ltd.	Distamycin	DNA minor groove binder	Yes (*C. difficile*)	Yes
23.	Nafithromycin (WCK 4873)	Wockhardt Ltd.	Macrolide	50S subunit of bacterial ribosome	No	No
24.	Taigexyn (nemonoxacin)	TaiGen Biotechnology Co.	Quinolone	Bacterial type II topoisomerase	No	No

(continued)

Table 24.3 (continued)

S. no	Drug name	Company	Drug class	Target	Expected activity against WHO critical threat pathogen?	Novelty
25.	TNP-2092	TenNor Therapeutics Ltd.	Rifamycin-quinolone hybrid	RNA polymerase, DNA gyrase, DNA topoisomerase IV	Yes (*C. difficile*), possibly (drug-resistant *N. gonorrhoeae*)	No
Antibiotics currently in global clinical phase I development						
26.	Apramycin	Juvabis AG	Aminoglycoside	30S subunit of bacterial ribosome	Yes (CRE)	No
27.	CRS3123	Crestone Inc.	Diaryldiamine	Methionyl-tRNAsynthetase	Yes (*C. difficile*)	Yes
28.	Delpazolid (LCB01–0371)	LegoChem Biosciences Inc. (Shanghai HaiHe Pharmaceutical Co. Ltd./ CSPC Pharmaceutical Group Ltd. licensees)	Oxazolidinone	50S subunit of bacterial ribosome	No	No
29.	ETX0282CPDP/ ETX1317	Entasis Therapeutics Inc.	β-Lactam (cephalosporin) + β-lactamase inhibitor (diazabicyclooctane)	PBP + β-lactamase	Yes (CRE)	No
30.	Ibezapolstat (ACX-362E)	Acurx Pharmaceuticals LLC	DCBG ([dichlorobenzy] guanine)	*C. difficile* DNA polymerase IIIC	Yes (*C. difficile*)	No
31.	KBP-7072	KBP BioSciences Pharmaceutical Technical Co. Ltd.	Tetracycline	30S subunit of bacterial ribosome	Possibly (CRAB)	No
32.	Meropenem + nacubactam (OP0595/ RG6080)	Meiji Seika Pharma Co. Ltd./ Fedora Pharmaceuticals Inc.	β-Lactam (carbapenem) + β-lactamase inhibitor (diazabicyclooctane)	PBP + β-lactamase/ PBP2	Yes (CRE)	No
33.	Oravance (QPX-2015 + QPX-7728)	QpexBiopharma/Brii Biosciences	β-Lactam + β-lactamase inhibitor	PBP + β-lactamase	Possibly (CRE)	No

34.	SPR206	Spero Therapeutics Inc.	Polymyxin	Outer membrane	Yes (CRE, CRAB, CRPA)	No
35.	SPR741	Spero Therapeutics Inc.	Polymyxin	Outer membrane	Possibly (CRE, CRAB, CRPA)	No
36.	TP-271	Tetraphase Pharmaceuticals Inc.	Tetracycline	30S subunit of bacterial ribosome	Possibly (CRAB)	No
37.	TP-6076	Tetraphase Pharmaceuticals Inc.	Tetracycline	30S subunit of bacterial ribosome	Yes (CRE and CRAB)	No
38.	TXA709/TXA707	Taxis Pharmaceuticals Inc.	FtsZ inhibitor	Cell wall division	No	No
39.	VNRX-7145 + ceftibuten	Venatorx Pharmaceuticals Inc.	β-Lactam (cephalosporin) + β-lactamase inhibitor (cyclic boronate)	PBP + β-lactamase	Yes (CRE)	No
40.	WCK 5222 (cefepime + zidebactam)	Wockhardt Ltd.	β-Lactam (cephalosporin) + β-lactamase inhibitor (diazabicyclooctane)	PBP + β-lactamase	Yes (CRE), possibly (CRAB, CRPA)	No

class being aminocoumarin and the second being quinolones. But the evidence of target-specific mutation generated against these drugs has limited their use (Bush et al. 2020); topoisomerase II inhibitor's antibacterial resistance has emerged because of its widespread use as a medicine. Antibiotic inactivation, target modification, interference with nucleic acid synthesis, altered permeability, and bypass metabolic pathway are the main reasons for bacterial resistance development. Recent reports suggest topoisomerases IA can serve as a novel druggable target for developing antibacterial agents (Seddek et al. 2021). Thus, while developing an antibacterial agent, it is necessary to understand the biological processes that the bacteria may undergo to build resistance.

Acknowledgments V.T. is thank to AMR-DBT-BIRAC for funding this work. RS is thankful to ICMR for the research fellowship.

Funding AMR-DBT-BIRAC (Sanction No. BT/PR31944/MED/29/1408/2019) funded VT for this work.

Transparency Declarations V.T has received a DBT-BIRAC grant from the Department of Biotechnology. R.S. and V.T. have no conflicts of interest related to this work.

Author Contributions R.S. wrote the first draft of the review; R.S added parts of new texts and figures; V.T edited and analysed the data.

References

Ahmed MO, Baptiste KE (2018) Vancomycin-resistant enterococci: a review of antimicrobial resistance mechanisms and perspectives of human and animal health. Microb Drug Resist 24: 590–606

Ahmed W, Menon S, Godbole AA et al (2014) Conditional silencing of topoisomerase I gene of *mycobacterium tuberculosis* validates its essentiality for cell survival. FEMS Microbiol Lett 353:116–123

Almasaudi SB (2018) Acinetobacter spp. as nosocomial pathogens: epidemiology and resistance features. Saudi J Biol Sci 25:586–596

Andersson MI, MacGowan AP (2003) Development of the quinolones. J Antimicrob Chemother 51:1–11

Ayobami O, Willrich N, Harder T et al (2019) The incidence and prevalence of hospital-acquired (carbapenem-resistant) Acinetobacter baumannii in Europe, eastern Mediterranean and Africa: a systematic review and meta-analysis. Emerg Microbes Infect 8:1747–1759

Bansal S, Tawar U, Singh M et al (2010) Old class but new dimethoxy analogue of benzimidazole: a bacterial topoisomerase I inhibitor. Int J Antimicrob Agents 35:186–190

Bansal S, Sinha D, Singh M et al (2012) 3,4-dimethoxyphenyl bis-benzimidazole, a novel DNA topoisomerase inhibitor that preferentially targets Escherichia coli topoisomerase I. J Antimicrob Chemother 67:2882–2891

Barnard AML, Bowden SD, Burr T et al (2007) Quorum sensing, virulence and secondary metabolite production in plant soft-rotting bacteria. Philos Trans R Soc B Biol Sci 362:1165–1183

Barry PM, Klausner JD (2009) The use of cephalosporins for gonorrhea: the impending problem of resistance. Expert Opin Pharmacother 10:555–577

Bassetti M, Vena A, Croxatto A et al (2018) A continuous publication, open access, peer-reviewed journal citation. Drugs Context 7:212527
Belete TM (2019) Novel targets to develop new antibacterial agents and novel alternatives to antibacterial agents. Hum Microbiome J 11:100052
Beyer P, Paulin S (2020) Priority pathogens and the antibiotic pipeline: an update. Bull World Health Organ 98:151
Bizard AH, Yang X, Débat H et al (2018) TopA, the Sulfolobus solfataricus topoisomerase III, is a decatenase. Nucleic Acids Res 46:861–872
Blokesch M (2012) A quorum sensing-mediated switch contributes to natural transformation of vibrio cholerae. Mob Genet Elements 2:224–227
Braun M, Silhavy TJ (2002) Imp/OstA is required for cell envelope biogenesis in Escherichia coli. Mol Microbiol 45:1289–1302
Breijyeh Z, Jubeh B, Karaman R (2020) Resistance of gram-negative bacteria to current antibacterial agents and approaches to resolve it. Molecules 25:1340
Brochu J, Breton ÉV, Drolet M (2020) Supercoiling, R-loops, replication and the functions of bacterial type 1A topoisomerases. Genes (Basel) 11:249–267
Brown PO, Cozzarelli NR (1981) Catenation and knotting of duplex DNA by type 1 topoisomerases: a mechanistic parallel with type 2 topoisomerases. Proc Natl Acad Sci U S A 78:843–847
Bugg TDH, Walsh CT (1992) Intracellular steps of bacterial cell wall peptidoglycan biosynthesis: enzymology, antibiotics, and antibiotic resistance. Nat Prod Rep 9:199–215
Bush NG, Diez-Santos I, Abbott LR et al (2020) Quinolones: mechanism, lethality and their contributions to antibiotic resistance. Molecules 25:5562–5589
Chen AY, Yu C, Gatto B et al (1993) DNA minor groove-binding ligands: a different class of mammalian DNA topoisomerase I inhibitors. Proc Natl Acad Sci U S A 90:8131–8135
Chen SH, Chan NL, Hsieh TS (2013) New mechanistic and functional insights into DNA topoisomerases. Annu Rev Biochem 82:139–170
Cheng B, Cao S, Vasquez V et al (2013) Identification of anziaic acid, a lichen depside from hypotrachyna sp., as a new topoisomerase poison inhibitor. PLoS One 8:e60770
Cong Y, Yang S, Rao X (2020) Vancomycin resistant *Staphylococcus aureus* infections: a review of case updating and clinical features. J Adv Res 21:169–176
Cornelis P, Aendekerk S (2004) A new regulator linking quorum sensing and iron uptake in Pseudomonas aeruginosa. Microbiology 150:752–756
Costa-Lourenço APR, Barros dos Santos KT et al (2017) Antimicrobial resistance in Neisseria gonorrhoeae: history, molecular mechanisms and epidemiological aspects of an emerging global threat. Brazilian J Microbiol 48:617–628
Cuypers WL, Jacobs J, Wong V et al (2018) Fluoroquinolone resistance in Salmonella: insights by wholegenome sequencing. Microb Genomics 4:e000195
Daniels R, Vanderleyden J, Michiels J (2004) Quorum sensing and swarming migration in bacteria. FEMS Microbiol Rev 28:261–289
Diawara I, Barguigua A, Katfy K et al (2017) Molecular characterization of penicillin non-susceptible Streptococcus pneumoniae isolated before and after pneumococcal conjugate vaccine implementation in Casablanca. Morocco Ann Clin Microbiol Antimicrob 16:23
Eggermont D, Smit MAM, Kwestroo GA et al (2018) The influence of gender concordance between general practitioner and patient on antibiotic prescribing for sore throat symptoms: a retrospective study. BMC Fam Pract 19:175
Faron ML, Ledeboer NA, Buchan BW (2016) Resistance mechanisms, epidemiology, and approaches to screening for vancomycin-resistant enterococcus in the health care setting. J Clin Microbiol 54:2436–2447
Fernández-Villa D, Aguilar MR, Rojo L (2019) Folic acid antagonists: antimicrobial and immunomodulating mechanisms and applications. Int J Mol Sci 20:20
Fisher K, Phillips C (2009) The ecology, epidemiology and virulence of enterococcus. Microbiology 155:1749–1757

Floss HG, Yu TW (2005) Rifamycin—mode of action, resistance, and biosynthesis. Chem Rev 105: 621–632

Garcia PK, Annamalai T, Wang W et al (2019) Mechanism and resistance for antimycobacterial activity of a fluoroquinophenoxazine compound. PLoS One 14:e0207733

Ghotaslou R (2015) Prevalence of antibiotic resistance in helicobacter pylori: a recent literature review. World J Methodol 5:164

Giannella RA (1996) Salmonella. In: Baron S (ed) Medical microbiology, 4th edn. University of Texas Medical Branch, Galveston, TX

Gilmore MS, Lebreton F, van Schaik W (2013) Genomic transition of enterococci from gut commensals to leading causes of multidrug-resistant hospital infection in the antibiotic era. Curr Opin Microbiol 16:10–16

Goldstein BP (2014) Resistance to rifampicin: a review. J Antibiot (Tokyo) 67:625–630

Harris SR, Feil EJ, Holden MTG et al (2010) Evolution of MRSA during hospital transmission and intercontinental spread. Science 327:469–474

Heinz E (2018) The return of pfeiffer's bacillus: rising incidence of ampicillin resistance in haemophilus influenzae. Microb Genomics 4:e000214

Höltje J-V (1998) Growth of the stress-bearing and shape-maintaining murein sacculus of *Escherichia coli*. Microbiol Mol Biol Rev 62:181–203

Huang F, He ZG (2010) Characterization of an interplay between a Mycobacteriumtuberculosis MazF homolog, Rv1495 and its sole DNA topoisomerase i. Nucleic Acids Res 38:8219–8230

Kahne D, Leimkuhler C, Lu W et al (2005) Glycopeptide and lipoglycopeptide antibiotics. Chem Rev 105:425–448

Kareb O, Aïder M (2020) Quorum sensing circuits in the communicating mechanisms of bacteria and its implication in the biosynthesis of bacteriocins by lactic acid bacteria: a review. Probiotics Antimicrob Proteins 12:5–17

Kathiravan MK, Khilare MM, Nikoomanesh K et al (2013) Topoisomerase as target for antibacterial and anticancer drug discovery. J Enzyme Inhib Med Chem 28:419–435

Kirmusaoglu S (2016) Staphylococcal biofilms: pathogenicity, mechanism and regulation of biofilm formation by quorum-sensing system and antibiotic resistance mechanisms of biofilm-embedded microorganisms. In: Microbial biofilms—importance and applications. InTechOpen, London

Kohanski MA, Dwyer DJ, Collins JJ (2010) How antibiotics kill bacteria: from targets to networks. Nat Rev Microbiol 8:423–435

Kretzschmar M, Meisterernst M, Roeder RG (1993) Identification of human DNA topoisomerase I as a cofactor for activator-dependent transcription by RNA polymerase II. Proc Natl Acad Sci U S A 90:11508–11512

Krzyżek P (2019) Challenges and limitations of anti-quorum sensing therapies. Front Microbiol 10: 2473

Lebreton F, Willems RJL, Gilmore MS (2014) Enterococcus diversity, origins in nature, and gut colonization. Massachusetts Eye and Ear Infirmary, Boston, MA

Lee CR, Lee JH, Park M et al (2017) Biology of Acinetobacter baumannii: pathogenesis, antibiotic resistance mechanisms, and prospective treatment options. Front Cell Infect Microbiol 7:1–35

Lee CM, Wang G, Pertsinidis A et al (2019) Topoisomerase III acts at the replication fork to remove precatenanes. J Bacteriol 201:e00563

Leelaram MN, Bhat AG, Hegde SM et al (2012) Inhibition of type IA topoisomerase by a monoclonal antibody through perturbation of DNA cleavage-religation equilibrium. FEBS J 279:55–65

Leelaram MN, Bhat AG, Godbole AA et al (2013) Type IA topoisomerase inhibition by clamp closure. FASEB J 27:3030–3038

Lim M, Liu LF, Jacobson-Kram D et al (1986) Induction of sister chromatid exchanges by inhibitors of topoisomerases. Cell Biol Toxicol 2:485–494

Lin M-F (2014) Antimicrobial resistance in Acinetobacter baumannii : from bench to bedside. World J Clin Cases 2:787

Liu LF, Depew RE, Wang JC (1976) Knotted single-stranded DNA rings: a novel topological isomer of circular single-stranded DNA formed by treatment with Escherichia coli ω protein. J Mol Biol 106:439–452

Liu X, Gallay C, Kjos M et al (2017) High-throughput CRISPRi phenotyping identifies new essential genes in Streptococcus pneumoniae. Mol Syst Biol 13:931

Lucena MI, Andrade RJ, Rodrigo L et al (2000) Trovafloxacin-induced acute hepatitis. Clin Infect Dis 30:400–401

Mandell L, Tillotson G (2002) Safety of fluoroquinolones: an update. Can J Infect Dis 13:54–61

Marques AT, Vítor JMB, Santos A (2020) Trends in helicobacter pylori resistance to clarithromycin: from phenotypic to genomic approaches. Micro Genomics 6:e000344

McDevitt D, Payne DJ, Holmes DJ et al (2002) Novel targets for the future development of antibacterial agents. J Appl Microbiol 92:28–34

Merino A, Madden KR, Lane WS et al (1993) DNA topoisomerase I is involved in both repression and activation of transcription. Nature 365:227–232

Mimouni FZ, Belboukhari N, Sekkoum K (2019) Mini review: is fluoroquinolone drug or poison? J Complex Heal Sci 2:70–76

Młynarczyk-Bonikowska B, Majewska A et al (2020) Multiresistant Neisseria gonorrhoeae: a new threat in second decade of the XXI century. Med Microbiol Immunol 209:95–108

Naqvi SAR, Roohi S, Iqbal A et al (2018) Ciprofloxacin: from infection therapy to molecular imaging. Mol Biol Rep 45:1457–1468

Neelakanta A, Sharma S, Kesani VP et al (2015) Impact of changes in the NHSN catheter-associated urinary tract infection (CAUTI) surveillance criteria on the frequency and epidemiology of CAUTI in intensive care units (ICUS). Infect Control Hosp Epidemiol 36:346–349

Ng LK, Martin IE (2005) The laboratory diagnosis of Neisseria gonorrhoeae. Can J Infect Dis Med Microbiol 16:15–25

Nicholls TJ, Nadalutti CA, Motori E et al (2018) Topoisomerase 3α is required for decatenation and segregation of human mtDNA. Mol Cell 69:9–23.e6

Nitiss JL (1994) Roles of DNA topoisomerases in chromosomal replication and segregation. Adv Pharmacol 29A:103–134

Nüesch-Inderbinen M, Heini N, Zurfluh K et al (2016) Shigella antimicrobial drug resistance mechanisms, 2004–2014. Emerg Infect Dis 22:1083–1085

Nurse P, Levine C, Hassing H et al (2003) Topoisomerase III can serve as the cellular decatenase in Escherichia coli. J Biol Chem 278:8653–8660

Organización Mundial de la Salud (2016) WHO | global action plan on AMR. WHO, Geneva

Pachori P, Gothalwal R, Gandhi P (2019) Emergence of antibiotic resistance Pseudomonas aeruginosa in intensive care unit; a critical review. Genes Dis 6:109–119

Paluch E, Rewak-Soroczyńska J, Jędrusik I et al (2020) Prevention of biofilm formation by quorum quenching. Appl Microbiol Biotechnol 104:1871–1881

Paterson DL (2006) Resistance in gram-negative bacteria: enterobacteriaceae. Am J Med 119:20–70

Pena RT, Blasco L, Ambroa A et al (2019) Relationship between quorum sensing and secretion systems. Front Microbiol 10:1–14

Penchovsky R (2018) RNA as a potent target for antibacterial drug discovery. Biomed J Sci Tech Res 10:7752–7753

Pinto TCA, Neves FPG, Souza ARV et al (2019) Evolution of penicillin non-susceptibility among streptococcus pneumoniae isolates recovered from asymptomatic carriage and invasive disease over 25 years in Brazil, 1990–2014. Front Microbiol 10:1–10

Pommier Y, Barcelo JM, Rao VA et al (2006) Repair of topoisomerase i-mediated DNA damage. Prog Nucleic Acid Res Mol Biol 81:179–229

Prasetyoputri A, Jarrad AM, Cooper MA et al (2019) The eagle effect and antibiotic-induced persistence: two sides of the same coin? Trends Microbiol 27:339–354

Qin T, Qian H, Fan W et al (2017) Newest data on fluoroquinolone resistance mechanism of Shigella flexneri isolates in Jiangsu Province of China. Antimicrob Resist Infect Control 6:97

Rani P, Nagaraja V (2019) Genome-wide mapping of topoisomerase I activity sites reveal its role in chromosome segregation. Nucleic Acids Res 47:1416–1142

Ranjan N, Fulcrand G, King A et al (2014) Selective inhibition of bacterial topoisomerase i by alkynyl- bisbenzimidazoles. Medchemcomm 5:816–825

Ravishankar S, Ambady A, Awasthy D et al (2015) Genetic and chemical validation identifies mycobacterium tuberculosis topoisomerase i as an attractive anti-tubercular target. Tuberculosis 95:589–598

Rello J, Kalwaje Eshwara V, Lagunes L et al (2019) A global priority list of the TOp TEn resistant microorganisms (TOTEM) study at intensive care: a prioritization exercise based on multi-criteria decision analysis. Eur J Clin Microbiol Infect Dis 38:319–323

Seddek A, Annamalai T, Tse-Dinh Y-C (2021) Type IA topoisomerases as targets for infectious disease treatments. Microorganisms 9:86

Shanker E, Federle MJ (2017) Quorum sensing regulation of competence and bacteriocins in Streptococcus pneumoniae and mutans. Genes (Basel) 8:15

Shariati A, Dadashi M, Moghadam MT et al (2020) Global prevalence and distribution of vancomycin resistant, vancomycin intermediate and heterogeneously vancomycin intermediate Staphylococcus aureus clinical isolates: a systematic review and meta-analysis. Sci Rep 10:12689

Shuman S (1991) Recombination mediated by vaccinia virus DNA topoisomerase I in Escherichia coli is sequence specific. Proc Natl Acad Sci U S A 88:10104–10108

Stevnsner T, Bohr VA (1993) Studies on the role of topoisomerases in general, gene- and strand-specific DNA repair. Carcinogenesis 14:1841–1850

Suerbaum S, Brauer-Steppkes T, Labigne A et al (1998) Topoisomerase I of helicobacter pylori: juxtaposition with a flagellin gene (flaB) and functional requirement of a fourth zinc finger motif. Gene 210:151–161

Tacconelli E, Magrini N (2017) Global priority list of antibiotic-resistant bacteria to guide research, discovery, and development of new antibiotics. Organ Mund la Salud, pp 1–7

Tipper DJ, Strominger JL (1965) Mechanism of action of penicillins: a proposal based on their structural similarity to acyl-D-alanyl-D-alanine. Proc Natl Acad Sci U S A 54:1133–1141

Tomasz A (1979) The mechanism of the irreversible antimicrobial effects of penicillins: how the beta-lactam antibiotics kill and lyse bacteria. Annu Rev Microbiol 33:113–137

Toyofuku M (2019) Bacterial communication through membrane vesicles. Biosci Biotechnol Biochem 83:1599–1605

Tsakou F, Jersie-Christensen R, Jenssen H et al (2020) The role of proteomics in bacterial response to antibiotics. Pharmaceuticals 13:1–27

Tse-Dinh YC (2009) Bacterial topoisomerase I as a target for discovery of antibacterial compounds. Nucleic Acids Res 37:731–737

Tse-Dinh YC (2015) Targeting bacterial topoisomerase i to meet the challenge of finding new antibiotics. Future Med Chem 7:459–471

van Duin D (2017) Carbapenem-resistant Enterobacteriaceae: what we know and what we need to know. Virulence 8:379–382

Von Döhren H (2009) Antibiotics: actions, origins, resistance, by C. Walsh. 2003. Washington, DC: ASM press. 345 pp. $99.95 (hardcover). Protein Sci 13:3059–3060

Wang JC (1996) DNA topoisomerases. Annu Rev Biochem 65:635–692

Wang JC (2002) Cellular roles of DNA topoisomerases: a molecular perspective. Nat Rev Mol Cell Biol 3:430–440

Watkins RR, Holubar M, David MZ (2019) Antimicrobial resistance in methicillin-resistant staphylococcus aureus to newer antimicrobial agents. Antimicrob Agents Chemother 63: e01216–e01219

Wilson DN, Hauryliuk V, Atkinson GC et al (2020) Target protection as a key antibiotic resistance mechanism. Nat Rev Microbiol 18:637–648

Wise EM, Park JT (1965) Penicillin: its basic site of action as an inhibitor of a peptide cross-linking reaction in cell wall mucopeptide synthesis. Proc Natl Acad Sci U S A 54:75–81

Witzky A, Tollerson R, Ibba M (2019) Translational control of antibiotic resistance. Open Biol 9: 190051

Yan R, Hu S, Ma N et al (2019) Regulatory effect of DNA topoisomerase I on T3SS activity, antibiotic susceptibility and quorum-sensing-independent pyocyanin synthesis in Pseudomonas aeruginosa. Int J Mol Sci 20:1116

Yeh YC, Liu HF, Ellis CA et al (1994) Mammalian topoisomerase I has base mismatch nicking activity. J Biol Chem 269:15498–15504

Yu X, Zhang M, Annamalai T et al (2017) Synthesis, evaluation, and CoMFA study of fluoroquinophenoxazine derivatives as bacterial topoisomerase IA inhibitors. Eur J Med Chem 125:515–527

Zhang GF, Liu X, Zhang S et al (2018) Ciprofloxacin derivatives and their antibacterial activities. Eur J Med Chem 146:599–612

Innovative Strategies to Tackle Antimicrobial Resistance

25

Varsha Gupta and Nimitha K. Mohan

Abstract

Antimicrobial resistance (AMR) has been a dangerous threat to the effective treatment of the ever-increasing range of infections caused by bacteria, parasites, viruses, and fungi. It has resulted in decreased efficacy of antibacterial, anti-parasitic, antiviral, and antifungal drugs, making the treatment of patients difficult, costly, or even impossible. The impact of AMR results in prolonged illness and increased mortality in the vulnerable patient population. The extent of the situation worldwide and the influence of AMR on human health, and on costs for the healthcare sector, are still largely unknown. In this review, we would be stressing how AMR came into existence and how to deal with it, to prevent AMR globally.

Keywords

Antimicrobial resistance · Emerging infections

25.1 History

Several years before the introduction of penicillin as a therapeutic drug, a bacterial penicillinase enzyme was identified already by two members of the discovery team, after which Penicillin was discovered by Alexander Fleming in 1928 and 1940 (Abraham and Chain 1940). Interestingly, the identification of a bacterial

V. Gupta (✉)
Department of Microbiology, Government Medical College and Hospital, Chandigarh, India

N. K. Mohan
Aster Medcity, Kochi, India

P. P. Singh (ed.), *Recent Advances in Pharmaceutical Innovation and Research*,
https://doi.org/10.1007/978-981-99-2302-1_25

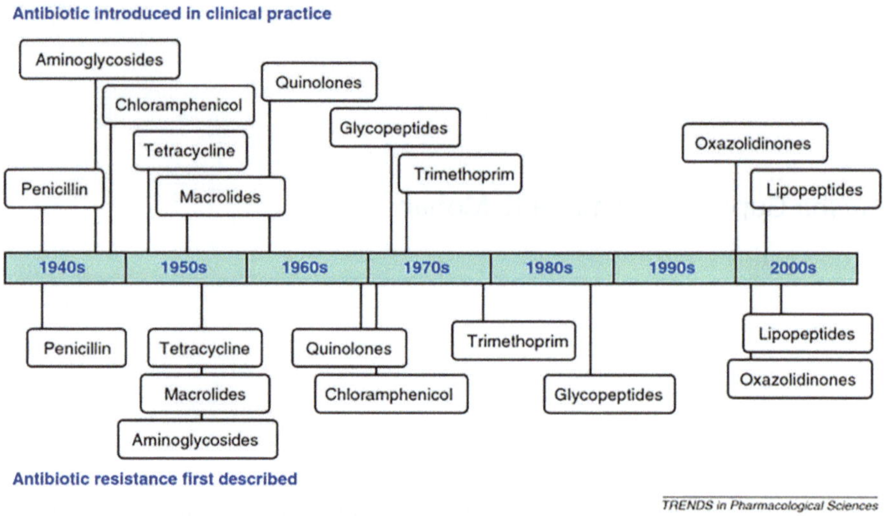

Fig. 25.1 The timeline of antibiotic discovery and development of antibiotic discovery and development of resistance to various antibiotics (reproduced from Google)

penicillinase before the use of the antibiotic can now be appreciated in the light of recent findings that a large number of antibiotic resistance genes are components of natural microbial populations (D'Costa et al. 2006). Which came first, the antibiotic or resistance? The time line of discovery of antibiotics and its resistance mechanisms which have developed over time are shown in Fig. 25.1. The discovery of transferable antibiotic resistance by means of various genes have changed the whole picture and it introduced the heretical genetic concept that the collections of antibiotic genes could be disseminated by bacterial conjugation throughout an entire population of bacterial pathogens (Davies 1995).

25.2 Mechanism of Drug Resistance

The evolutionary perspective in bacteria uses two major genetic strategies to adapt to the antibiotic attack, namely (a) mutations in genes and (b) acquisition of foreign DNA coding for resistance determinants through horizontal gene transfer. In the first case, mutational resistance, a subset of bacterial cells which is derived from a susceptible population, will develop mutations in genes that will affect the activity of the drug, which results in preserved cell survival in the presence of the antimicrobial molecule. Once this type of resistant mutant emerges, the antibiotic eliminates the susceptible population and thus the resistant bacteria will predominate. In the second case, horizontal gene transfer, in which acquisition of foreign DNA material occurs is the other most important cause associated with the development of antimicrobial resistance.

25.3 Molecular Mechanism of Drug Resistance

There are various biochemical mechanisms that result in the origin of antimicrobial resistance. A single bacterial cell can adopt various mechanisms to survive the effect of an antibiotic which are given in Fig. 25.2.

1. *Enzymatic alteration*: Bacteria produce various enzymes which are capable of bringing chemical changes to the antimicrobial structure which is a well-known mechanism of acquired antibiotic resistance in both Gram-negative and Gram-positive bacteria. The mechanism of resistance to the class of beta-lactam antibiotics occurs primarily through the hyper-production of β-lactamases, which are the enzymes that inactivate these antibiotics by breaking the amide bond of the β-lactam ring. Based on structure and arrangement of amino acids, β-lactamases are classified into various Ambler classes (molecular). The mechanism of substrate profile and the susceptibility to β-lactamase inhibitors are done according to Bush--Jacoby--Medeiros classes (functional). Various other enzymes are also responsible for this type of resistance mechanism. The alternate mechanism in which various modifying enzymes are the driving force for the resistance mechanism and their reaction have been described, namely (1) acetylation (aminoglycosides, chloramphenicol, streptogramins), (2) phosphorylation (aminoglycosides, chloramphenicol), and (3) adenylation (aminoglycosides, lincosamides). The best example which exhibits resistance via modification of the drug is the presence of aminoglycoside modifying enzymes (AMEs) that covalently modify the hydroxyl or amino groups of the aminoglycoside molecule.
2. *Decreased permeability to antibiotics*
   Outer membrane permeability

1. Enzymatic alteration
2. Decreased permeability to antibiotics
3. Promotion of efflux
4. Alteration of target site
5. Protection of target site
6. Overproduction of target
7. Bypass of inhibited process

Fig. 25.2 Various molecular mechanisms of antibiotic resistance

The rate of diffusion of antibiotics through the outer membrane depends on the number and nature of the porin channels. The loss of specific porin channels as a result of mutation determines the increased resistance to β-lactam antibiotics. Resistance mechanism in aminoglycosides and carbapenems emerges due to lack of production of outer membrane proteins. For example, imipenem resistance during therapy is observed in up to 25% of *Pseudomonas aeruginosa* infections, has been attributed to mutational loss of its OprD protein (also known as the D2 porin) (Nikaido and Vaara 1985).

Inner membrane permeability

The entry of certain antibiotics within the bacterial cells results in binding with an anionic transporter, they retain their positive charge and are later "pulled" across the cytoplasmic membrane by the internal negative charge of the cell (Bryan and Kwan 1983). This requires immense energy expenditure and thus a threshold minimum of the internal negative charge of the cell has to be present, which is the proton motive force. This type of action is noticed in aminoglycosides (Mates et al. 1982). The alteration in this mechanism occurs in mutants, which subsequently occurs in the course of long-term aminoglycoside therapy.

3. Promotion of efflux

   The commonest mechanisms of resistance in many clinically relevant pathogens are by the phenomenon of active efflux of antimicrobial agents. This type of resistance mechanism occurs in various groups of antibiotics like tetracyclines, streptogramins, macrolides, and fluoroquinolones. The resistant determinant is found on the bacterial chromosome or plasmid. In the case of tetracyclines, it is inducible by the sub-inhibitory concentration of tetracycline. The genes designated are tetA, tet B, tet 30, etc.

4. *Alteration of target site*

   The alteration of the target site can occur in various forms. This may occur in the following ways.

   (a) Alteration of ribosomal target site
   (b) Alteration of cell wall precursor targets
   (c) Alteration of target enzymes

   Alteration of ribosomal target site

   This method of resistance may result from the alteration of ribosomal binding target sites. It occurs in a wide variety of antimicrobial agents, including tetracyclines, macrolides, lincosamides, streptogramins, and aminoglycosides. Failure of the antibiotic to bind to its target site results in inhibition of protein synthesis and cell growth. The resistance among macrolide antibiotics, is mediated by the products of the erm (erythromycin ribosome methylation) gene, a variety of methylase enzymes (MLSB−determinant) that demethylate adenine residues on the 23S ribosomal RNA (rRNA) of the 50S subunit of the prokaryotic ribosome. This results in the inhibiting the binding of MLS to the ribosome. This can be explained by the demonstration of D test which results in clindamycin resistance.

   Alteration of cell wall precursor targets

The alteration in the cell wall targets is seen vancomycin and other glycopeptide antibiotics, which binds to d-alanine-d alanine (d-ala-d-ala) at the terminal end of the stem peptide in peptidoglycan precursors. This molecule prevents the incorporation of cell wall precursors. Resistance to vancomycin by the gram-positive bacteria has been attributed to this type of mechanism.

Alteration of target enzymes

The β-lactam antibiotics binds covalently to penicillin-binding proteins in the cytoplasmic membrane. They catalyze the synthesis of the peptidoglycan that forms the cell wall of bacteria. Any alterations in these proteins can lead to β-lactam antibiotic resistance. The development of Methicillin-resistant *Staphylococcus aureus* is also a result of this mechanism. *Staphylococcus aureus* exhibits resistance by the expression of the *mecA* gene which encodes for PBP2α. This mechanism has a very low affinity to β-lactam antibiotics.

5. *Protection of target site*
   Protection of target site is seen in the group of tetracyclines. Bacteria interfere with the ability of tetracycline to bind to the ribosome. The genes responsible for the protection of ribosomes from antibiotic action *are tetM* and others. The presence of tetracycline allows the *tetM* gene to generate a protein with elongation factor-like activity which stabilizes ribosome transfer RNA interactions.
6. *Overproduction of target*
   The competition between sulfonamides and para-aminobenzoic acid in the binding of the enzyme DHPS and halting the generation of pteridines and nucleic acids. The resistance mechanism to these drugs is mediated by the overproduction of the synthetic enzyme DHPS. The gene responsible for DHPS is felP, and strains of bacteria which secrete an excess amount of DHPS can overwhelm sulfa inhibition.
7. Bypass of inhibited process
   The development of auxotrophs also results in the acquisition of resistance to certain antibiotics. These auxotrophic mutant strains require various substrates that are normally synthesized by the target enzymes, and if the substrates are present in the environment, the organisms grow despite the inhibition of the synthetic enzyme. For example, *Enterococcus* spp. can be folate auxotrophs requiring the environmental acquisition of folic acid for growth.
8. Bind up antibiotic
   This type of resistance mechanism is shown by VISA strains. They express unusually thick peptidoglycan cell walls. The vancomycin gets absorbed on the excess false binding site, which prevents it from reaching its target and thus allows the peptidoglycan synthesis to continue (Fig. 25.3).

The polymyxins

Polymyxins are a group of drugs that include five different chemical compounds (polymyxins A, B, C, D, and E). The most common ones which are being used in clinical practice are colistin A (polymyxin E1) and colistin B (polymyxin E2). These two groups only differ in their fatty acid tails (Kasiakou et al. 2005; Li et al. 2002). The colistin acts rapidly on the bacterial cells and results in the prompt killing but the exact mechanism by which colistin can kill

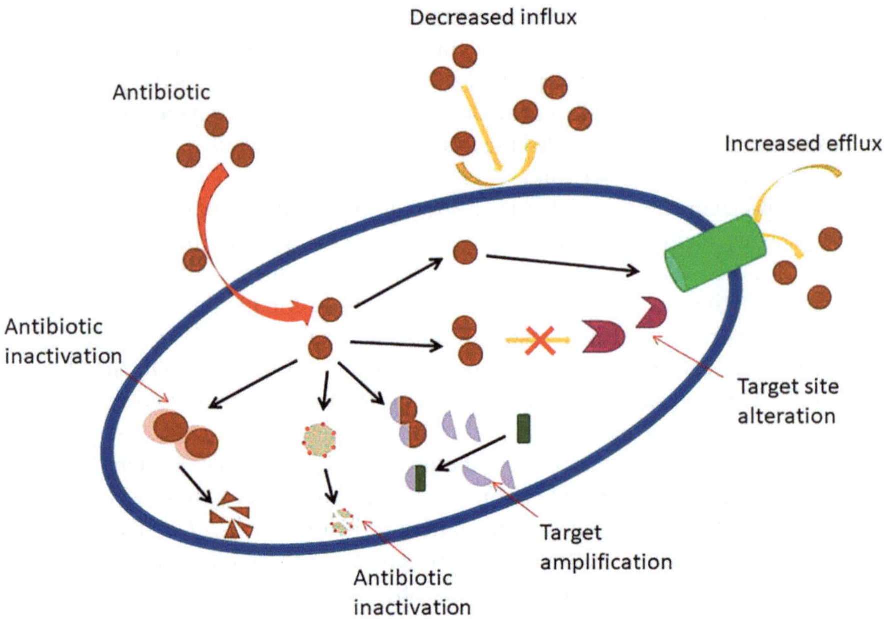

Fig. 25.3 Various antibiotic resistance mechanisms

bacterial cells is currently unclear. The action on Gram-negative bacteria has a detergent-like effect, and mediated through a two-step mechanism. Initially there occurs binding with electrostatic interactions between the polycationic ring of colistin to cell envelope components, causing the displacement, competitively, of the calcium (Ca2þ) and magnesium (Mg2þ) ions from the phosphate groups of LPS that act as membrane stabilizers, leading to disruption of the outer membrane and the loss of cellular contents, thus killing the bacterium (Fig. 25.4) (Mendes and Burdmann 2009; Moffatt et al. 2010; Evans et al. 1999).

25.4 Colistin Resistance

The incidence of colistin resistance is considered low globally, but it is seen in isolates such as *Acinetobacter baumannii, Klebsiella pneumoniae,* and *P. aeruginosa* (Park et al. 2011). The resistance to this antibiotic is due to LPS modification via diverse routes which include: (1) specific modification of outer membrane porins and reductions in the overall negative charge of the LPS, (2) overexpression of efflux pump systems, (3) overproduction of capsule polysaccharide (Kim et al. 2014), and (4) although enzymatic mechanisms of resistance haven't been reported so far, strains of *Bacillus polymyxa* are known to produce colistinase (Ito-Kagawa and Koyama 1980).

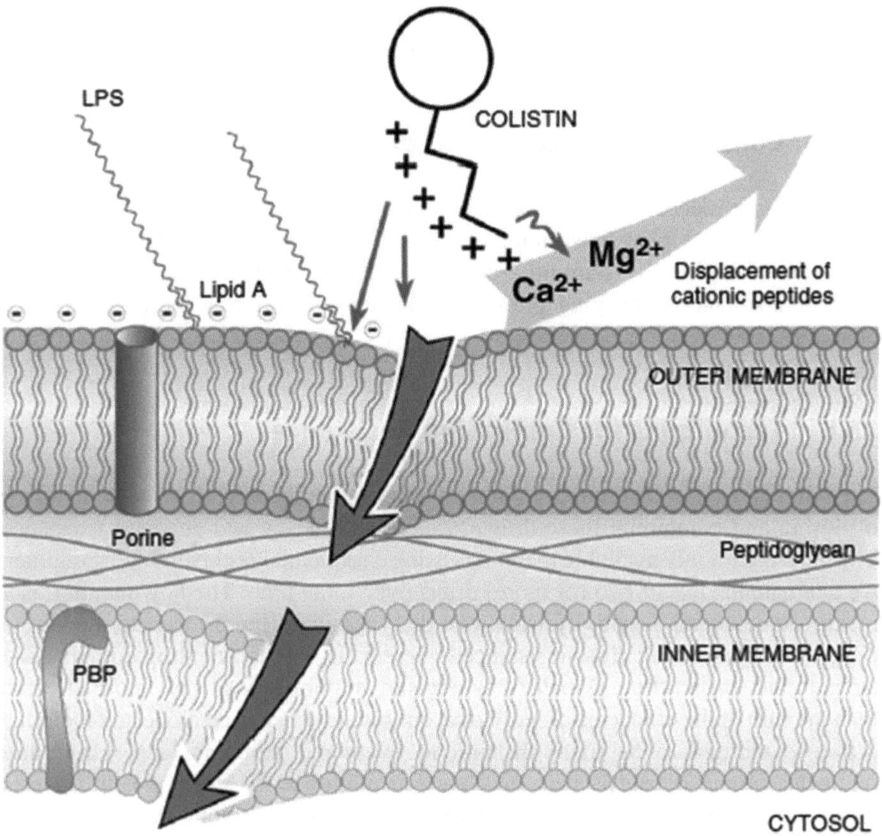

Fig. 25.4 The mechanism of action of polymyxins

25.5 Colistin Resistance in *A. baumannii*

Existing data show a relatively less prepondence of colistin resistance in *A. baumannii*, and little is known about its mechanism. The loss of LPS production in *A. baumannii* results in the inactivation of a lipid A biosynthesis gene, lpxA, lpxC, or lpxD, which leads to colistin resistance in these species (Moffatt et al. 2010).

25.6 Colistin Resistance in *P. aeruginosa*

The mechanism of resistance to colistin is mediated by the presence of *PmrAB* genes which directly control the pmr HIJKLM operon, the gene products of which are responsible for the synthesis of N4-aminoarabinose, later bind to lipid A moieties and neutralize the negatively charged phospholipids, causing a change in the negatively charged cell membrane which leads to colistin resistance (Ly et al. 2012).

25.7 Colistin Resistance in *Enterobacteriaceae*

The resistance mechanism existing in *Salmonella typhimurium* and *Escherichia coli* is due to having a higher substitution of the ester-linked phosphate group in the lipid. A portion of the LPS by 4-amino-4-deoxy-L-arabinose and showing larger amounts of 2-aminoethanol esterifying phosphates in the core oligosaccharide (Groisman et al. 1997).

25.8 Techniques for Combating Resistance

Even though we have a limited set of antibiotics in our armory, there are better ways of dosing and combining these drugs to increase efficacy and decrease resistance. The most commonly available method is using a sequential regimen. In this regimen, we alternate the use of two (or more) drugs over some time. The key intention is to maximize collateral sensitivity when one drug sensitizes the bacteria to the second drug while minimizing cross-resistance, where resistance to one drug confers resistance to the second drug. A pair of synergistic antibiotics is always more effective than the sum of the efficacies of individual antibiotics when used alone and their dual action is thought to be more difficult to overcome. Even in single-drug dose regimens, proper dosing is very necessary to minimize the development of resistance. A more proactive, diagnosis-driven approach for antibiotic-resistant bacteria allows the lifespans of current antibiotics to be maximized when it is used in combination with wider efforts such as antimicrobial stewardship and increased public awareness of antimicrobial resistance.

25.9 Molecular Strategies to Combat Antibiotic Resistance

The most important factor which keeps health and disease balance is maintained by interactions between host body and colonization of microbial flora. It is prudent to understand the mechanism of protein interactions which occurs during an infection. This understanding about the basics of protein interactions gives way to target non-essential mechanism like bacterial adhesion, communication, or host hijacking. The various ways of protein interactions are discussed below.

25.9.1 Bacterial Communication: Quorum Sensing

In bacteria, chemical communication involves the production, release, detection, and response to small hormone-like molecules called auto-inducers (Waters and Bassler

2005). The mechanism for the bacterial communication is by the phenomenon of Quorum sensing (QS). The bacteria can sense the environment and other microbial communities and can use this information to increase the concentration of auto-inducers to reach a threshold level which results them to sense a critical cell mass, which in turn activates or represses gene expression (Vasil 2003).

Enzymatic and non-enzymatic inactivation of these signals which helps in the inhibition of quorum sensing is under pipeline. The lactonases, acylases, or oxidoreductases help in the degradation of acyl-homoserine lactones (AHL) or its precursor (La Sarre and Federle 2013; Leadbetter and Greenberg 2000; Dong and Zhang 2000; Dong et al. 2001; Uroz et al. 2005). The development of various monoclonal antibodies and certain peptides are also being reviewed which are some other interesting facts (Rutherford and Bassler 2012; MDowell et al. 2001).

25.9.2 Host–Pathogen Interactions

Host–pathogen interactions can be defined as how microorganisms sustain themselves within host organisms on a molecular, cellular, organismal, or population level. This phenomenon can be explained by cell adhesion, virulence factors, and signaling and which can be targeted to avoid disease onset.

25.9.2.1 Targeting of Cell Adhesion and Colonization

For establishing an infection, the bacterial population initially adheres to the host cell, then colonizes and invades inside these cells. All these steps are mandatory for the infection to happen in the body. These secretion systems are present on the cell membranes of bacteria and are used to inject proteins to assist in the infection of host cells. They are essential for the manipulation of the host and to establish a replicative niche (Green and Mecsas 2016). For example, the pathogenicity in certain gram-negative bacteria can be explained by this mechanism. So, targeting this interaction between the proteins of the secretion systems can be used as a strategy to prevent different infections caused by species of *Salmonella, Shigella, Yersinia, or Pseudomonas* (Wilson et al. 2001).

25.9.2.2 Targeting of Virulence Factors and Toxins

Virulence factors are proteins secreted by the microbes, which enable them to colonize a host, creating an infection, and thus evading the immune system. Various studies have been under trial to inhibit these virulence factors used by bacteria. Monoclonal antibodies are used for such practices. So far, there exist only three monoclonal antibodies which have been licensed for the treatment of bacterial infections: palivizumab which is used for the prevention of respiratory syncytial virus in high-risk infants, and raxibacumab and obiltoxaximab for prophylaxis and treatment of anthrax (Sparrow et al. 2017). These molecules bind with the virulence factors giving the host a chance to clear the infection immunologically. Therapeutic antibodies against toxins, like Shiga toxin, produced by Shiga toxin-producing *E. coli*, are currently undergoing clinical trials (López et al. 2010; Bitzan et al. 2009).

25.9.2.3 Targeting of Signaling

After the adherence and colonization to the host cells, bacteria will hijack certain host cell pathways, thereby promoting it for replication and spread within the infected organism. Pathways like the NF-κB dependent pathway, mitochondrial pro-death pathway, etc. (Rudel et al. 2010) are inhibited during the infection thus permitting them to control the infection. So, targeting bacterial proteins which inhibit these pathways is an alternative for bacterial infection treatment.

25.10 Enhancing Host Cell Immune Response

Strengthening the host immune system or recruitment of more and more macrophages to the site of infection or by enabling more efficient bacterial opsonization is an eye opener to avoid the need of using antibiotics. Antimicrobial peptides and proteins (AMPs) are evolutionarily conserved molecules of less than 60 amino acids in length (Zasloff 2002; Ganz 2003). The antibacterial activity by enhancing host cell immune response is ascribed to the membrane lysis; in effect, the antimicrobial peptides and proteins (AMPs) have also been demonstrated to function in host immune modulation, often by enhancing protective immunity and suppressing inflammation. To give an instance, the AMP LL-37, which is mostly found in the granules of neutrophils, is an antiseptic agent which results in the overexpression of interleukin-8 secretion and leads to the recruitment of leukocytes to the site of infection (Scott et al. 2002; Tjabringa et al. 2003).

25.11 Summary

Conventional antibiotics function by interferes with cellular processes of the bacteria to kill them. The misuse of antibiotics and non-adherence to strict antimicrobial stewardship practices will lead to a situation where bacterial infections are projected to be the main cause of death. The above-mentioned remedies could minimize the appearance of resistance. In short, more focus on non-vital bacterial processes like host–pathogen interactions, cell attachment, or immunosuppression will give us the way to introduction of new alternatives for producing drugs that can inhibit the bacteria or empower the host to avoid disease onset. These alternatives help us in understanding the affected pathways which ease the development of new molecules targeting protein interactions. So, the introduction of these strategies into action in the future will help us to overcome antibiotic resistance.

References

Abraham EP, Chain E (1940) An enzyme from bacteria able to destroy penicillin. Rev Infect Dis 10: 677–678

Bitzan M et al (2009) Safety and pharmacokinetics of chimeric anti-Shiga toxin 1 and anti-Shiga toxin 2 monoclonal antibodies in healthy volunteers. Antimicrob Agents Chemother 53:3081–3087

Bryan LE, Kwan S (1983) Roles of ribosomal binding, membrane potential, and electron transport in bacterial uptake of streptomycin and gentamicin. Antimicrob Agents Chemother 23:835–845

D'Costa et al (2006) Sampling the antibiotic resistome. Science 311:374–377

Davies J (1995) Vicious circles looking back on resistance plasmids. Genetics 139:1465–1468

Davis BK et al (2011) The inflammasome NLRs in immunity, inflammation, and associated diseases. Annu Rev Immunol 29:707–735

Dong YH, Zhang LH (2000) An enzyme that inactivates the acylhomoserine lactone quorum sensing signal and attenuates the virulence of Erwinia carotovora. Proc Natl Acad Sci U S A 97:3526–3531

Dong YH et al (2001) Quenching quorum sensing dependent bacterial infection by an N-acyl homoserine lactonase. Nature 411:813–817

Evans ME et al (1999) Polymyxin B sulfate and colistin old antibiotics for emerging multiresistant gram-negative bacteria. Ann Pharmacother 33:960–967

Ganz T (2003) Defensins: antimicrobial peptides of innate immunity. Nat Rev Immunol 3:710–720

Green ER, Mecsas J (2016) Bacterial secretion systems—an overview. Microbiol Spect 4:1

Groisman EA et al (1997) Regulation of polymyxin resistance and adaptation to low-Mg2þ environments. J Bacteriol 179:7040–7045

Hospital Infection Control Practices Advisory Committee (HICPAC) (1995) Recommendations for preventing the spread of vancomycin resistance. Infect Control Hosp Epidemiol 16:105–113

Ito-Kagawa M, Koyama Y (1980) Selective cleavage of a peptide antibiotic, colistin by colistinase. J Antibiot 33:1551–1555

Kasiakou SK et al (2005) Combination therapy with intravenous colistin for management of infections due to multidrug-resistant Gram-negative bacteria in patients without cystic fibrosis. Antimicrob Agents Chemother 49:3136–3146

Kim Y et al (2014) In vivo emergence of colistin resistance in Acinetobacter baumannii clinical isolates of sequence type 357 during colistin treatment. Diagn Microbiol Infect Dis 79:362–366

La Sarre B, Federle MJ (2013) Exploiting quorum sensing to confuse bacterial pathogens. Microbiol Mol Biol Rev 77:73–111

Leadbetter JR, Greenberg EP (2000) Metabolism of acyl-homoserine lactone quorum sensing signals by Variovorax paradoxus. J Bacteriol 182:6921–6926

Li J, Milne RW, Nation RL (2002) Simple method for assaying colistin methane sulfonate in plasma and urine using high-performance liquid chromatography. Antimicrob Agents Chemother 46: 3304–3307

López EL et al (2010) Safety and pharmacokinetics of urtoxazumab, a humanized monoclonal antibody, against Shiga-like toxin 2 in healthy adults and in pediatric patients infected with Shiga-like toxin-producing Escherichia coli. Antimicrob Agents Chemother 54:239–243

Ly NS, Yang J et al (2012) Impact of two-component regulatory systems PhoP-PhoQ and PmrA-PmrB on colistin pharmacodynamics in Pseudomonas aeruginosa. Antimicrob Agents Chemother 56:3453–3456

Mates SM et al (1982) Membrane potential and gentamicin uptake in Staphylococcus aureus. Proc Natl Acad Sci U S A 79:6693–6697

MDowell P et al (2001) Structure, activity and evolution of the group I thiolactone peptide quorum sensing system of Staphylococcus aureus. Mol Microbiol 41:503–512

Mendes CA, Burdmann EA (2009) Polymyxins – review with emphasis on nephrotoxicity. Rev Assoc Med Bras 55:752–759

Moffatt JH et al (2010) Colistin resistance in Acinetobacter baumannii is mediated by complete loss of lipopolysaccharide production. Antimicrob Agents Chemother 54:4971–4977

Nikaido H, Vaara M (1985) Molecular basis of bacterial outer membrane permeability. Microbiol Rev 49:1–32

Park YK et al (2011) Correlation between overexpression and amino acid substitution of the PmrAB locus and colistin resistance in Acinetobacter baumannii. Int J Antimicrob Agents 37: 525–530

Reddy KVR et al (2004) Antimicrobial peptides: premises and promises. Int J Antimicrob Agents 24:536–547

Rudel T et al (2010) Interactions between bacterial pathogens and mitochondrial cell death pathways. Nat Rev Microbiol 8:693–705

Rutherford ST, Bassler BL (2012) Bacterial quorum sensing its role in virulence and possibilities for its control. Cold Spring Harb Perspect 2(11):a012427. https://doi.org/10.1101/cshperspect.a012427

Scott MG et al (2002) The human antimicrobial peptide LL-37 is a multifunctional modulator of innate immune responses. J Immunol 169:3883–3891

Sparrow E et al (2017) Therapeutic antibodies for infectious diseases. Bull World Health Organ 95: 235–237

Tjabringa GS et al (2003) The antimicrobial peptide LL-37 activates innate immunity at the airway epithelial surface by transactivation of the epidermal growth factor receptor. J Immunol 171: 6690–6696

Uroz S et al (2005) N-Acylhomoserine lactone quorum-sensing molecules are modified and degraded by Rhodococcuserythropolis W2 by both amidolytic and novel oxidoreductase activities. Microbiology 151:3313–3322

Vasil M (2003) DNA microarrays in analysis of quorum sensing: strengths and limitations. J Bacteriol 185:2061–2065

Waters CM, Bassler BL (2005) Quorum sensing: cell-to-cell communication in bacteria. Annu Rev Cell Dev Biol 21:319–346

Wilson RK et al (2001) Role of EscF, a putative needle complex protein, in the type III protein translocation system of enteropathogenic Escherichia coli. Cell Microbiol 3:753–762

Zasloff M (2002) Antimicrobial peptides of multicellular organisms. Nature 415:389–395

Zhang L, Gallo RL (2016) Antimicrobial peptides. Curr Biol 26:R14–R19

Molecular and Mechanistic Insights of Yeast Flocculation

26

Santhosh Kumar Sariki, Ramesh Kumawat, Ranu Singh, and Raghuvir Singh Tomar

Abstract

Flocculation is a survival strategy adopted by the yeast *Saccharomyces cerevisiae* to survive under unfavorable environmental conditions and depends on the expression of flocculation genes such as *FLO1*, *FLO5*, *FLO9*, *FLO10*, and *FLO11*. Yeast flocculation is crucial for efficient beer, wine, bread, bio-ethanol, and biopharmaceutical production. It is also used for the removal of heavy metal contaminants from wastewater as bioremediation. The transcriptional activity of the *FLO* genes is influenced by the nutritional status as well as pH, cations, agitation, and temperature. Flocculation is also affected by anomalies in cell wall integrity. This implies that, during industrial fermentation processes, flocculation is affected by numerous factors, as mentioned above. Tentatively, the wise use of these factors offers the possibility of attaining control over the flocculation process. To gain control over flocculation, it is ideal to have profound knowledge about the elements and genes that are involved and influence its regulation under extremely variable conditions. This chapter decisively discusses the various molecular and mechanical insights that influence flocculation in fermentation technology, such as cations, pH, temperature, cell wall integrity, and histones. Moreover, many of the conclusions will also be useful in different industrial processes where control over yeast flocculation is desirable.

S. K. Sariki
Laboratory of Chromatin Biology, Department of Biological Sciences, Indian Institute of Science Education and Research, Bhopal, India

Molecular and Integrative Physiology Department, University of Michigan, Ann Arbor, MI, USA

R. Kumawat · R. Singh · R. S. Tomar (✉)
Laboratory of Chromatin Biology, Department of Biological Sciences, Indian Institute of Science Education and Research, Bhopal, India
e-mail: rkumawat@iiserb.ac.in; ranu@iiserb.ac.in; rst@iiserb.ac.in

P. P. Singh (ed.), *Recent Advances in Pharmaceutical Innovation and Research*,
https://doi.org/10.1007/978-981-99-2302-1_26

Keywords

Yeast flocculation · CWI pathway · Slt2 · Tup1 · Histones · *FLO* genes

26.1 Introduction

The ability of microbial cells to form aggregates or to adhere to a substrate is a characteristic phenotype of different mechanisms like flocculation, invasive growth, biofilm formation, and sexual reproduction (Van Mulders et al. 2009). In yeast, adhesion phenotypes are mainly dependent on the properties of the cell wall. They are primarily regulated in the response to environmental stress like nitrogen availability, glucose depletion, changes in pH, and ethanol levels (Braus et al. 2003; Sampermans et al. 2005). Adhesion proteins play an essential role in fungal cell interactions that participate in events like mating, flocculation, filamentation, biofilm formation, colony morphology changes, and interaction with biotic and abiotic surfaces. Well-studied adhesin protein families are Flo (flocculation), Epa (epithelial adhesin) galectins, and Als (agglutinin-like sequence) protein families from *Saccharomyces cerevisiae*, *Candida glabrata,* and *Candida albicans* strains respectively. In *S. cerevisiae,* agglutinin proteins Aga1p, Aga2p, and Fig2p play a crucial role in the completion of mating and also in a mating-dependent invasion in agar (Cappellaro et al. 1994; Guo et al. 2000). The flocculation adhesin protein family includes different flocculin proteins such as Flo1p, Flo5p, Flo9p, Flo10p (Teunissen and Steensma 1995) and cell wall Epa (epithelial adhesins) proteins such as Epa1p, Epa6p, and Epa7p in opportunistic human pathogenic yeast *C. glabrata* which are involved in adhesion to epithelial cells (Linder and Gustafsson 2008; Maestre-Reyna et al. 2012; Lockhart et al. 2017). The peptide-binding Als proteins (Als1 to Als7 and Als9) regulate colony morphology and biofilm formation in *Candida albicans* (Fu et al. 1998). Epa and Als mediate interactions with their host involved commensal and pathogenic conditions (Gaur et al. 1999; Perlroth et al. 2007). Adhesive proteins Mam3p and Map4 have been identified in fission yeast *Schizosaccharomyces pombe* as mating type-specific agglutinins (Sharifmoghadam et al. 2006; Mata and Bähler 2006).

Flocculation, an adhesion-dependent phenotype, is also used as a model for studying the regulation of cell wall properties in *S. cerevisiae*. Yeast flocculation can be defined as the non-sexual homotypic and reversible aggregation of yeast cells into multicellular masses. The word floc derives from the Latin phrase floccus, which means a tuft of wool. Flocculation is a survival strategy of yeast cells to survive under unfavorable environmental conditions like nutritional conditions (Straver et al. 1993). Non-flocculating cells are dispersed all over the culture but flocculating cells are settled at bottom of the culture tube and also show clumps when placed in culture dishes (Fig. 26.1) (unpublished data). Like kin selection, inner cells of flocs die and autolyzed to maintain a nutritionally rich microenvironment for surrounding cells to keep the community alive in adverse conditions (Jin and Speers 2000). Apart from flocculation, *S. cerevisiae* forms clusters/aggregates for other

Fig. 26.1 Cell–cell adhesion of flocculating yeast strain: analysis of flocculation in wild-type (WT) and a flocculating strain. Flocculating strain shows flocculation behavior when grown in a rich YPD medium so that all cells clump together and sink to the bottom of the tube, while wild-type (WT) shows virtually no flocculation, leaving all cells in suspension. The same is observed in culture dishes as wild-type cells are dispersed, and the flocculating strain shows flocculation behavior by clumping together

processes like sexual aggregation, co-flocculation, pseudohyphae, and biofilm formation. In several industries, flocculation property has been utilized for rapid biomass recovery or clarification of fermented products.

Sexual aggregation occurs during complementary mating types (a and α) of haploid strains of *S. cerevisiae* after the exchange of pheromones a and α, respectively. The appearance of additional molecules (proteins) on the surface of cells facilitates the fusion of the haploid cells (Bilinski et al. 1986). *Co-flocculation* is a heterotypic aggregation process that occurs among non-flocculating or weakly flocculating strains. But when the two strains are mixed in the presence of Ca^{2+}, they flocculate rapidly (Stewart and Garrison 1972; Mortier and Soares 2007). Co-flocculation has also been known as mutual aggregation or mutual flocculation. In co-flocculation, a surface protein of one strain interacts with the surface carbohydrate component of another strain (Nishihara et al. 1982), which can be reversible upon treatment of monosaccharides (likely glucose, mannose, fructose, and galactose), urea and guanidine. The extent of co-flocculation can be calculated through flocculation assay for co-cultured *S. cerevisiae* and the non-Saccharomyces strain in (1:1) cells/cells ratio in the presence of Ca^{2+} ions at pH 4.5. And Co-flocculation percentage is calculated by subtracting the expected flocculation percentage (average flocculation percentage in pure cultures) from the observed flocculation percentage (experimentally determined flocculation percentage in co-cultured strains) (Rossouw et al. 2015).

$$\%\mathrm{Co-flocculation} = (O_{\%\mathrm{F}} - E_{\%\mathrm{F}})$$

where $O_{\%\mathrm{F}}$ = Observed % of flocculation and $E_{\%\mathrm{F}}$ = Expected % of flocculation.

Chain formation occurs when the young daughter buds failed to separate from the mother cell; this failure in separation makes the cells physically linked to each other and leads to an aggregate formation (Mamvura et al. 2017). General chain formation is also known as pseudohyphal growth, which contains around 30–100 cells. Upon the disruption of this aggregation employing physical disruption, sonication, or enzymatic treatment, cells will not be able to re-aggregate before a subsequent growth cycle (Stewart 2018). *Biofilm* formation is another aggregation phenomenon in *S. cerevisiae* that is mainly regulated by *FLO11* expression. *FLO11* is primarily recognized as the only *FLO* gene to be involved in the regulation of cell-surface adhesions, agar invasive growth, and pseudohyphae formation in many yeast strains, especially *S. cerevisiae* ∑1278b (Guo et al. 2000; Reynolds and Fink 2001; Zara et al. 2005; Verstrepen and Klis 2006). In *S. cerevisiae* S288c deletion of *FLO1*, *5*, and *9* shows decreased biofilm formation on plastic surfaces and loss of adhesion properties, suggesting that apart from *FLO11*, *FLO1*, *5*, and *9* also play a crucial role in biofilm formation (Fichtner et al. 2007; Di et al. 2017; Yang et al. 2018).

26.2 Genetic Basis of Flocculation

In *S. cerevisiae,* the flocculation mechanism initiates through interaction between cell surface receptor (lectin-like protein, flocculins) of flocculating cell and ligand (α-mannose residue of mannan molecule) of either flocculating or non-flocculating cells (Miki et al. 1982). In this process, Ca^{2+} ions act as cofactors to maintain the active conformation of surface proteins, thereby enhancing the ability of flocculins to interact with α-mannan carbohydrates, which leads to the formation of cellular

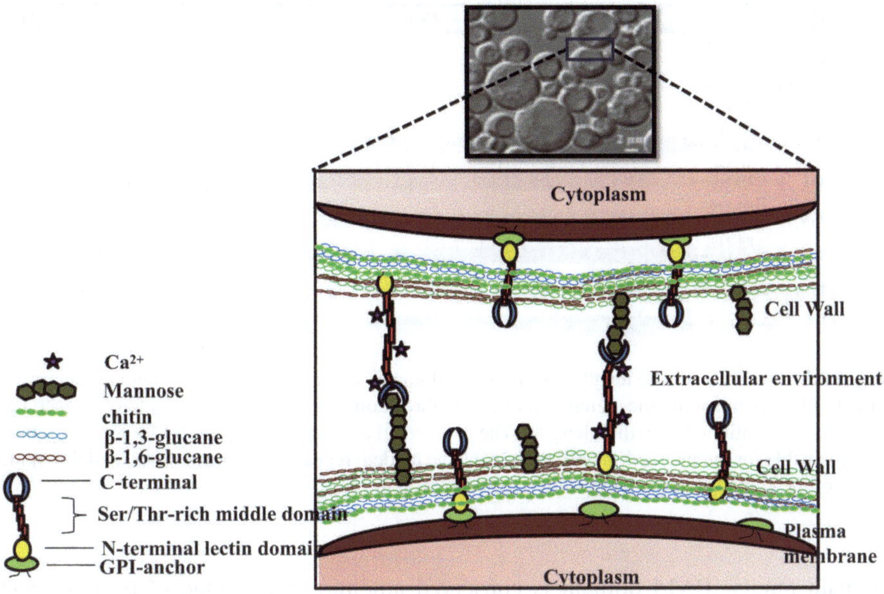

Fig. 26.2 Cell–surface anchoring of fungal adhesins: mature fungal adhesins consist of three distinct domains. During transport through the secretory pathway, adhesins undergo extensive posttranslational modifications. It is believed that the short *O*-linked oligosaccharide side chains enable the adhesions to obtain a long, semi-rigid rod-like structure that is stabilized by Ca^{2+} Ions. Upon arrival at the plasma membrane, the GPI anchor is cleaved off, and the adhesin is ultimately linked covalently through a GPI remnant to the beta-1,6 glucans that are an integral part of the fungal cell wall. The active conformation of flocculin bind to the mannose residue present on the cell wall of the neighboring cell, which leads to floc formation

aggregation and ultimately settling (Miki et al. 1982; Stratford 1989) (Fig. 26.2). A family of subtelomeric genes, *FLO1*, *FLO5*, *FLO9*, and *FLO10*, encode specific lectins that are responsible for flocculation (Teunissen and Steensma 1995). On the other hand, nonsub telomeric genes, including *FLO11/MUC1* (Lambrechts et al. 1996; Lo and Dranginis 1996) encode a protein that has been involved in flocculation, invasive growth, and substrate adhesion (Guo et al. 2000; Zara et al. 2005; Bayly et al. 2005).

All the Flo proteins are known to GPI anchored (glycosylphosphatidylinositol) glycoproteins. All Flo proteins share a common three-domain structure containing the N-terminal domain, a central domain, and a carboxyl-terminal domain (2) (Verstrepen and Klis 2006). The N-terminal domain is the lectin domain, which interacts with mannose residues. The N-terminal domain of Flo1p is heavily N- and O-glycosylated and contains β-sheets (37.7% β-sheets and 6.5% α-helices) (Goossens et al. 2011). The N-terminal domain of Flo1p was a variant of the PA14

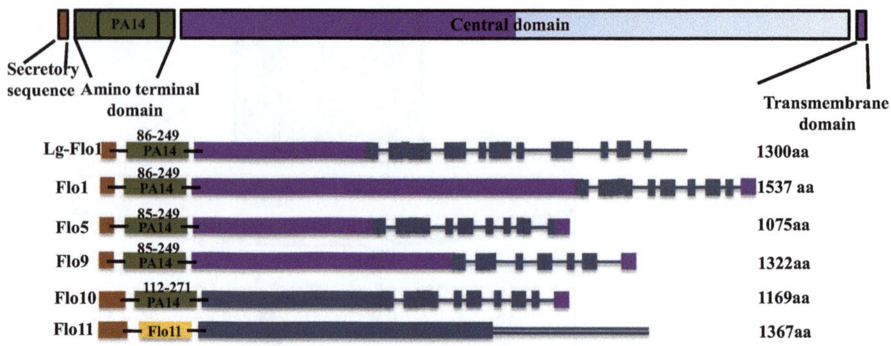

Fig. 26.3 Overview of the lengths of important domains of each Flo protein: schematic representation of Flo1p indicating the signal sequence and three domains. The domain structure for each Flo protein was mentioned with their lengths. The N-terminal domain contains the PA14 domain, which is responsible for the interaction with the sugar. The tandem repeats of the serine (Ser) and threonine (Thr) of the central domain were shown for each protein

domain family. PA14 domain is conserved among a wide range of bacterial and eukaryotic proteins, like adhesins, in some bacterial toxins such as anthrax protective antigen (PA). The crystal structure of PA toxin indicates that the PA14 domain consists of a series of antiparallel β-strands (Petosa et al. 1997). Ca^{2+}dependent carbohydrate-binding site is a common element present in the PA14 domain family (Yoshida et al. 2010). Ca^{2+} ions form a bridge between flocculating cells by crosslinking the carboxyl group of glycans with hydrogen bonding.

A central domain of Flo1p contains serine and threonine-rich residue (approx. 46%) and carries many tandem repeats; hence this region is a hot spot for recombination during DNA replicative events (Fig. 26.3). Extensive O-glycosylation and O-linked oligosaccharide side-chains enable the flocculins to achieve a long, semi-rigid rod-like structure, which is further stabilized by Ca^{2+} ions (Jentoft 1990; Verstrepen and Klis 2006). The presence of the proline residues in this region also prevents the central domain from forming any compact domains (Dranginis et al. 2007). The Carboxyl-terminal domain contains a GPI-attachment site. The adhesins are covalently linked to the β-1,6-glucans of the yeast cell wall through the GPI-remnant (de Nobel and Lipke 1994; de Groot et al. 2003). Hydrophobicity studies of the Flo1 protein showed that the N- and C-terminal domains of the protein were more hydrophobic than the central domain (Kyte and Doolittle 1982). The length of each domain varies for different *FLO* genes.

26.3 Genes Responsible for Flocculation

The genes that encode flocculating lectins are known as *FLO* genes. In the *S. cerevisiae* genome, the Flo adhesin protein family is subdivided into two groups. Four dominant *FLO* genes (*FLO1*, *FLO5*, *FLO9*, and *FLO10*) encode the members of the first group of proteins. The overexpression of these four genes induces flocculation by promoting cell–cell adhesion. *FLO2* and *FLO4* genes are the alleles of *FLO1*. *FLO5*, *FLO9*, and *FLO10* genes share significant sequence homology with the *FLO1* gene, viz., 96%, 94%, and 58% respectively (Teunissen and Steensma 1995). *FLO1*, *FLO9*, and *FLO8* genes are located on chromosome I. *FLO5* and *FLO10* genes are located on chromosomes VIII and XI, respectively. All the four dominant members of the *FLO* gene family are closest to their respective telomeres (Guo et al. 2000). In addition to the dominant genes, *FLO3*, *FLO6*, and *FLO7* genes have been described as recessive/semi-dominant genes. The second group of adhesion protein encoding genes are *FIG2*, *AGA1*, and *FLO11*. Fig2p and Aga1p are induced during the mating phenomenon (Erdman et al. 1998). Flo11p is also known as Muc1p and is involved in the regulation of haploid invasive growth and diploid pseudohyphal formation (Lambrechts et al. 1996; Lo and Dranginis 1996). *FLO1* is well known most dominant gene among other *FLO* genes. The expression of Flo1p leads to a strong flocculation phenotype (Kobayashi et al. 1998). All *FLO* genes exhibit various degrees of flocculation induced by their expression and sensitivity to/for sugars (Govender et al. 2008; Van Mulders et al. 2009). Flo10p is not only involved in flocculation but is also responsible for filamentation and the adhesion to agar and plastic surfaces (Guo et al. 2000). The Flo11p protein has the same domain structure as the other four dominant Flo proteins, but it has a different amino acid sequence. The Flo11p is induced in pseudohyphae formation, agar invasion, adhesion to substrates, as well as for filamentation (Lo and Dranginis 1996, 1998; Guo et al. 2000). *FLO8* was identified as a transcriptional activator for *FLO1* (Kobayashi et al. 1999), *FLO11* (Rupp et al. 1999), and *STA1* (Kim et al. 2003) and also essential for their expression. Interestingly, the mutated *FLO8* gene by a nonsense mutation in most laboratory strains like S288c and W303-1A causes transcriptional silencing in flocculation and invasive growth in these backgrounds (Fichtner et al. 2007).

26.4 Types of Flocculent Strains

Depending upon the growth phase or physiological conditions, one or more transcription factors bind to the promoter regions of the flocculation genes either temporarily or permanently to modulate *FLO* gene expression and flocculation in yeast cells. Based on hybridization studies (Teunissen and Steensma 1995; Teunissen et al. 1995), yeast strains are classified into four classes. First class includes strains that will never flocculate, as they lack intact *FLO* genes. The second class of strains that repressed through the *TUP1/SSN6* regulatory cascade. Upon disruption of this regulatory cascade, these strains show flocculation. The third class of strains is transcriptionally regulated and the fourth class of strains flocculates constitutively. Apart from this, the strains having New Flo phenotype (found in many brewers strains) or Flo phenotype (present in most laboratory strains) are also categorized separately based on the differential sugar specificity and sensitivity to physiological conditions such as pH and salt concentration (Stratford 1989; Stratford and Assinder 1991). Despite having these differences, the flocculation mechanism is fundamentally the same for both the New Flo and laboratory Flo flocculating phenotypes.

26.5 Measurement of Flocculation

Various methods have been described to measure flocculation. Flocculation in yeast cells can be measured based on four different criteria, such as bond strength, morphology, extent, and rate of sedimentation (Stratford and Keenan 1988). Bond strength can be estimated after deflocculation by the addition of mannose or by thermal deflocculation or by the critical cell density method (Miki et al. 1982; Bony et al. 1998). These tests are based on counts of free cells in a flocculating culture, which will be compared with the total number of cells before flocculation or after deflocculation. In many cases, the time course of flocculation is also monitored, so that both the extent of flocculation and the rate of flocculation will be measured in one test. Briefly, in these tests, yeast cells of flocculating culture were deflocculated using deflocculation buffer containing EDTA followed by two times washing with distilled water. After that, cells were resuspended in the flocculation buffer containing Ca^{2+} ions. Percentage flocculation ability (F) (Singh et al. 2015; Church et al. 2017) was determined by the following equation:

$$F = (1 - B/A) \times 100\%$$

where A = Absorbance at 600 nm (immediately before cells were shaken in flocculation buffer) and B = Absorbance at 600 nm (after flocculation).

26.6 Extrinsic and Intrinsic Factors that Determine Flocculation Phenotype of Yeast

26.6.1 Physiological Environment

Flocculation is an ideal and cost-effective procedure for removing yeast culture from a fermentor in the brewing process at the desired time required for the product recovery (Brückner and Mösch 2012). The flocculation phenomenon is influenced by a variety of factors. Apart from the genetic makeup of flocculating strains like *FLO* genes and associated modulators (suppressor and activator), there are many other factors that affect flocculation. These factors include pH, availability of nutrients, ions, presence of sugars, cell density, temperature, and age of cells.

26.6.1.1 Cations

Cations play a vital role in the flocculation of *S. cerevisiae*. In flocculating yeast strains, cells express cell-adhesion molecules having the ability to bind with mannose polymer located on the surface of an adjacent cell in presence of Ca^{2+}. Upon addition of $CaCl_2$ in growth media at pH 3.5–4.0, Ca^{2+} ion induces flocculation.

Hence, in culturing medium available Ca^{2+} concentration is more important than total Ca^{2+} concentration. Some other ions are also act as flocculation inducers like Cd^{2+}, Cu^{2+}, Co^{2+}, Ni^{2+}, Zn^{2+}, Fe^{2+}, Rb^{+}, Cs^{+}, Al^{3+}, and mainly Mg^{2+} and Mn^{2+}. Besides these inducers, some ions also play an inhibitory role competitively in yeast flocculation like Ba^{2+}, Sr^{2+}, and Pb^{2+} due to the similarity of their ionic ratio with Ca^{2+} (Soares 2011). Ability of flocculating strains to bind with divalent cation, make it more useful in bioremediation for the removal of heavy metal contamination from contaminated water (Wang and Chen 2006).

26.6.1.2 Impact of pH and Nutrients on Yeast Flocculation

Yeast cell wall carries a negative charge because of the presence of the ionized carboxylic group. This negatively charged surface repels adjacent cells and inhibits flocculation (Soares 2011). There is a positive correlation between cell surface hydrophobicity (CSH) and flocculation. The pH of the medium has a strong effect on yeast flocculation. Changes in pH affect flocculation by changing the ionization of the functional group of flocculins, which eventually changes their conformation. Depletion of glucose, maltose, and sucrose causes the onset of flocculation. In a rich medium (with yeast extract and peptone) or minimal medium (YNB), the onset of flocculation coincides with the attainment of a residual concentration of sugars (glucose, fructose, or maltose) (Sampermans et al. 2005). In fermenters, it is possible to trigger the flocculation under synergistic/additive response of ethanol stress and a deprived nutritional environment (Soares 2011).

26.6.1.3 Physical Environment Affecting Yeast Flocculation

Cell density and mechanical agitation play a crucial role in yeast cell flocculation (Speers et al. 1992). Dense/saturated cultures have less nutrient availability, which triggers flocculation. Every time cell sedimentation is not the ultimate result of flocculation, sometimes cells can overcome from agitation to maintain their suspension state. While during agitation, smaller flocs tend to settle faster than the bigger flocs because they face lesser friction and retardation by mixing (Speers et al. 1992). It is important to have proper agitation and flow rate, which can promote collision and exert less frictional retardation between cells.

26.7 Transcriptional Regulation of *FLO* Genes

26.7.1 Transcriptional Regulation of *FLO* Genes by Repressor Complex

Yeast flocculation phenotype is directly related to the abundance of cell wall-associated Flo1p protein (Goossens and Willaert 2010) which is encoded by the dominant and best-characterized flocculation gene, *FLO1* (Bony et al. 1998; Goossens et al. 2011). *FLO1* gene expression is regulated by its associated repressor–activator complex activity at the epigenomic level in the promoter and coding region (Fleming and Pennings 2001). Constitutively, *FLO1* gene expression can be regulated by co-repressor complex Tup1-Cyc8, which creates a regular array of the hypoacetylated nucleosome in promoter and intergenic regions to create hindrance between the interaction of RNA pol II and ORF (Fleming et al. 2014). Tup1-Cyc8 complex has a widespread function in the repression of mating specific genes, hypoxia genes, and glucose repression (Treitel and Carlson 1995). However, in some stresses such as oxidative stress, ER stress, and osmotic stress, this repressor complex also works as an activator and de-represses the expression of genes that are involved in the cellular homeostasis maintenance (Proft and Struhl 2002; Kumawat and Tomar 2022). This complex lacks the DNA-binding region, so for repressive or de-repressive activity at any promoter, it requires promoter sequence-specific protein for recruitment and binding. Mediator protein changes with genes under consideration. In the case of the *FLO1* gene, these mediator proteins are histone deacetylases (HDACs) Hda1 and Rpd3 (Davie et al. 2003; Fleming et al. 2014). These HDACs occupy the repressed *FLO1* promoter region in a Tup1-Cyc8 dependent manner and coordinate histone deacetylation and nucleosome stabilization on the promoter and intergenic region. Tup1 interacts with hypoacetylated or under-acetylated H3 and H4 histone tails of the promoter region's nucleosome, so any mutation on these tail residues has a synergistic effect on Tup1-Cyc8 mediated gene repression (Fig. 26.4).

For the *FLO1* de-repression ATP-dependent remodeling complex, Swi-snf acts as an antagonist for the Tup1-Cyc8 repressor complex and disrupts nucleosome positioning on *FLO1* promoter and upstream region up to 5 kb. The hypoacetylated state of promoter nucleosome interrupts the recruitment of co-activator complex Swi-snf. For *FLO1* expression, independent chromatin remodeling required in the promoter and coding region is required. Histone eviction, pol II, and Swi-snf co-activator complex recruitment are essential for transcription initiation, but acetylation of H3 (on Lysine 14) and H4 in the gene coding region by HATs Gcn5 and Sas2 is critically important for proper transcription elongation (Church and Fleming 2018). It is shown that in the absence of Gcn5 and Sas2, RNA polII was recruited on the Flo1 promoter but could not traverse through the coding region.

26.7.2 Transcriptional Regulation of *FLO* Genes by Histones

Histones are the sites for post-translational modification, which affects gene expression. Any mutation can alter the binding of nucleosome with gene sequence which will affect the RNA pol II interaction with gene and thus its transcription. Modifications at H3K14 affect the *FLO1* transcription, as this residue is essential for RNA pol II promoter eviction (Kruger et al. 1995). H4 sin (Swi-snf independent mutant) variant S47C, where 47th serine is mutated to a cysteine, which is adjacent to the sin domain, eliminates the requirement of Swi-snf co-activator complex in de-repression of *FLO1* gene (Flaus et al. 1996). This mutation is proposed to disrupt the DNA–histone interaction. H4S47C mutation allows the de-repression of the *FLO1* gene when the Tup1-Cyc8 complex is relieved without further requirement of Swi-snf (Fleming and Pennings 2001). With the reference of *SUC2* expression, H4T73C mutation, which lies in the sin domain adjacent to conserved tyrosine

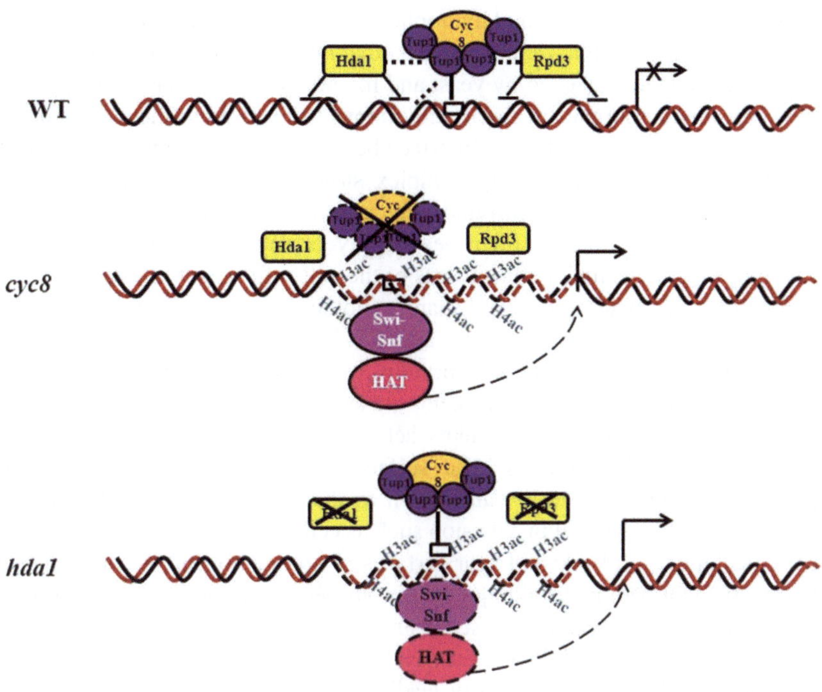

Fig. 26.4 Model for the role of Tup1-Cyc8 repressor complex in *FLO1* gene repression: Tup1-Cyc8 binds to a distinct site at the *FLO1* promoter in association with the histone deacetylases (HDACs), Rpd3p, and Hda1p and contributes to nucleosome positioning, histone deacetylation, and gene repression. In the *cyc8* mutant, there is a remodeling of the *FLO1* promoter through nucleosome acetylation, rearrangement, and eviction, which accompanies the *FLO1* gene de-repression. In the *rpd3 hda1* double mutant, Tup1p occupancy persists. Swi-Snf occupancy, nucleosome rearrangement, histone acetyltransferase (HAT) recruitment, and eviction occur at the *FLO1* promoter and upstream chromatin site that is previously occupied by Tup1-Cyc8, and *FLO1* expression is de-repressed

residue show Swi-snf independency but a similar effect is not observed in the case of the *FLO1* gene (Fleming and Pennings 2001; Santisteban et al. 1997; Makalowska et al. 1999).

26.8 The Role of the Cell Wall in Yeast Flocculation

Yeast flocculation is an asexual aggregation involving Ca^{2+} dependent interaction between lectins and cell wall polysaccharides, especially mannans and glucans. The yeast cell wall plays a major role in governing the rate and the extent of flocculation. Physical factors such as cell wall hydrophobicity, cell surface charge, cell surface topography, and age of the cells significantly contribute to the flocculating ability of yeast cells (Amory et al. 1988; Straver et al. 1993). Hydrophobicity of yeast cell surface shows a positive correlation with their flocculating ability (Straver and Kijne

1996). Zymolectin binding and hydrophobicity of the cells are the main biochemical "drivers," causing cells to attach to each other (Azeredo et al. 1997). The cell wall is an external envelope shared by yeast and filamentous fungi that define the interface between the microorganism and its environment. It surrounds the plasma membrane and is strategically placed at the interface between the cell and its environment. The yeast cell wall is an extremely complex structure consisting of polysaccharides (glucans and chitin) that surrounds the plasma membrane and contains a wide array of different proteins (often heavily glycosylated) anchored in the cell wall. The glucans are the major constituents of the cell wall (about 30–60%), involved in maintaining cell wall rigidity and shape. Another important component of the yeast cell wall is chitin, which constitutes about 1–3% of the cell wall. The outer layer of the cell wall is composed of mannoprotein, accounting for 25–50% of the cell wall. It has been shown that the mannoproteins are anchored to the cell wall through linkage with β-(1-6) glucan. Mannoproteins act as the receptors in the flocculation and sexual agglutination process. The surface receptor proteins are the lectin-like proteins present on the cell wall known as the cell wall proteins (CWPs), which are known to be involved in adhesion such as cell–cell attachment (flocculation) and cell attachment to plastic and agar surfaces in non-pathogenic yeast. In pathogenic yeasts, the cell wall plays a major role in virulence, morphogenesis, pathogenicity, antigenicity, and adhesion to host substrates.

Under extreme conditions, yeast cell wall stress sensors sense the stress and activate the CWI MAPK pathway leading to the up-regulation of genes required for cell wall maintenance (Arias et al. 2011). The Slt2 MAPK pathway (Fig. 26.5) is a linear cascade that comprises sensors, Wsc1/2/3, Mid2, GTPase Rho1, Pkc1, and a MAPKKK (Bck1), a pair of redundant MAPKK (Mkk1/2), and a MAPK (Slt2/Mpk1).

Under stress conditions, membrane proteins Mid2 and Wsc1/2/3 act as the major sensors of this pathway. The protein kinase C, Pkc1, is activated as an outcome of the activation of the small GTPase Rho1 by the guanine nucleotide exchange factor Rom2. Pkc1 triggers the sequential phosphorylation of the Bck1, Mkk1/2, and the MAPK Slt2 (also known as Mpk1). This pathway involves in regulation of the actin cytoskeleton organization (Pujol-Carrion et al. 2013), pexophagy (Manjithaya et al. 2010), and cell cycle progression (Soriano-Carot et al. 2012), and biosynthesis of cell wall polymers (Arias et al. 2011). In response to cell wall stressors, the Slt2 MAPK binds and activates its transcription factors; SBF and Rlm1 (Sanz et al. 2018). A negative-feedback regulation loop of MAPK signaling is activated by the transcriptional induction of many phosphatases that act upon MAPKs under the control of their target MAPK. Dual specificity protein phosphatases Msg5, Ptp2/3, or Sdp1 act as negative regulators of Slt2 (Liu and Levin 2018), and maintain the Slt2 at a low activity state in the absence of stress (Martín et al. 2000). The phosphorylated Slt2 activates its two transcription factors: SBF (SCB-binding factor) and Rlm1 (Jung et al. 2002). Phosphorylation of these transcription factors by Slt2 triggers the recruitment of transcriptional machinery at their target genes (Sanz et al. 2016). Phosphorylation of Slt2 is the major event in the activation of the CWI signaling pathway, leading to the recruitment of transcription factor Rlm1 at gene promoters of

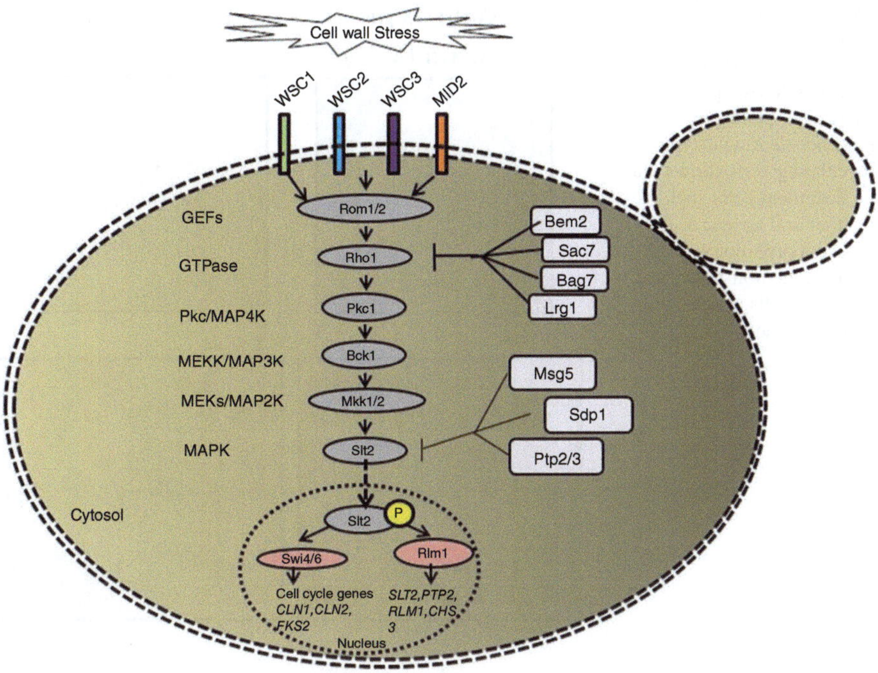

Fig. 26.5 Schematic diagram of the cell wall integrity pathway: cell surface sensors Wsc1/2/3 and Mid2 initiate the signals at the plasma membrane (PM). The sensors activateRom2, a GEF for the Rho1 GTPase. Rho1 activates the Pkc1-MAP kinase cascade, which is comprised of Bck1, Mkk1/2, and Mpk1(Slt2). Under cell wall stress conditions, Slt2 phosphorylates Rlm1, SBF complex (Swi4/Swi6), is targeted to the promoters of CWI-responsive genes via interaction with the activated transcription factor. Several MAP kinase phosphatases Msg5, Sdp1, and Ptp2/3, downregulate the Mpk1. The relative input of each sensor is indicated by the arrows

various genes, interdependently with SWI2/SNF2 to regulate the expression of the CWI responsive genes (Jung et al. 2002; Sanz et al. 2012, 2018). Deletion of *SLT2* and *RLM1* shows a reduction of flocculation indicates the importance of the CWI MAPK pathway in the regulation of flocculation (Sariki et al. 2019). The re-modeling of the yeast cell wall is vital in the brewing perspective as it has been demonstrated that the flocculation potential of brewing yeast strains is changed by growth in aerobic and anaerobic conditions (Lawrence and Smart 2007). The modified expression of the mannoproteins may vary cell wall composition and structure, favoring lectin–receptor interactions (Gibson et al. 2007). The differences in the cell wall architecture have been observed in flocculating strains aggravated strong mechanical changes of the cell wall, accompanied by significant topological changes such as rough surface morphology from SEM studies (Fig. 26.6) (Smukalla et al. 2008; Sariki et al. 2019).

It is essential to study the cell wall properties of industrial yeast strains to have a better understanding of the phenomenon of flocculation. This study aimed to attain greater insight into the fermentation performance of yeast strains and how their

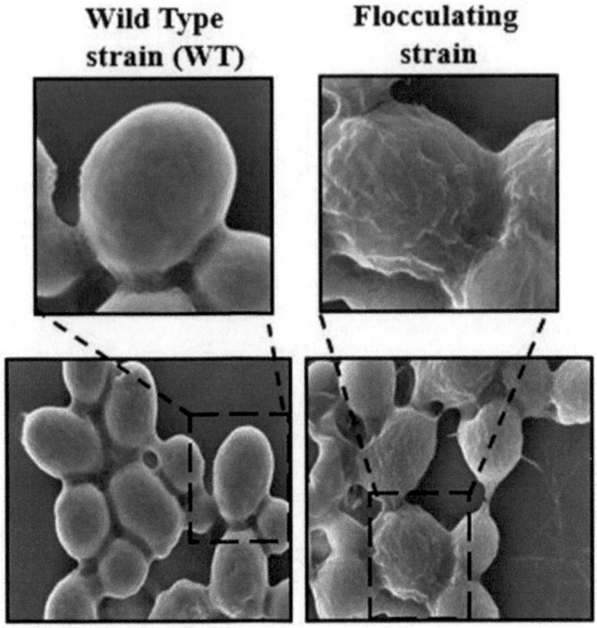

Fig. 26.6 Surface morphology image of flocculating yeast: scanning electron microscopy of centrifuged pellets of non-flocculent wild-type and flocculating cells shows that the flocculent cells stick together and appear rough surface morphology. By contrast, the non-flocculent wild-type cells appear with smooth surfaces

flocculation behavior is affected by changes in cell surface properties, which in turn are affected by changes in nutrient availability and physico-chemical conditions.

26.9 Future Perspectives

Although there is a lot of information available regarding the physiological condition which regulates flocculation, there is a lack of knowledge about how the signal travels through the cytoplasm and how it reaches inside the nucleus till the *FLO1* gene promoter? It is important to know the signaling pathways which identify different inductions and stress condition that leads to *FLO1* transcription. A better understanding of the cellular and molecular arrangement of flocculation will help in providing a better yield, and efficient strain removal from a fermentor in brewing industries. Adhesion to plastic surfaces is a way for the formation of biofilm. Biofilms shield yeast cells by conferring resistance to antifungal drugs. Yeast adhesion and infection to surgical implants cause severe post-surgical infections and are accountable for the increased mortality rate in hospitals (Wisplinghoff et al. 2004). Adhesion to surfaces such as catheters or prostheses can serve as a pool of pathogenic cells ready to gain access to the bloodstream of patients (Jabra-Rizk et al. 2004). A detailed study of flocculation and filamentous growth will help in designing improved antimycotics and muco-inhibitors for the safe packaging of surgical plastic equipment.

Acknowledgement Council of Scientific and Industrial Research (CSIR), India is acknowledged for fellowship support to SKS, RK, and RS. Members of the chromatin biology laboratory are recognized for their advice and helpful discussions.

References

Amory D, Rouxhet P, Dufour J (1988) Flocculence of brewery yeasts and their surface properties: chemical composition, electrostatic charge and hydrophobicity. J Inst Brew 94:79–84

Arias P, Díez-Muñiz S, García R, Nombela C, Rodríguez-Peña JM, Arroyo J (2011) Genome-wide survey of yeast mutations leading to activation of the yeast cell integrity MAPK pathway: novel insights into diverse MAPK outcomes. BMC Genomics 12:1–17

Azeredo J, Ramos I, Rodriguks L, Oliveira R, Teixeira J (1997) Yeast flocculation: a new method for characterising cell surface interactions. J Inst Brew 103:359–361

Bayly JC, Douglas LM, Pretorius IS, Bauer FF, Dranginis AM (2005) Characteristics of Flo11-dependent flocculation in Saccharomyces cerevisiae. FEMS Yeast Res 5:1151–1156

Bilinski C, Russell I, Stewart G (1986) Analysis of sporulation in Brewer's yeast: induction of tetrad formation. J Inst Brew 92:594–598

Bony M, Barre P, Blondin B (1998) Distribution of the flocculation protein, flop, at the cell surface during yeast growth: the availability of flop determines the flocculation level. Yeast 14:25–35

Braus GH, Grundmann O, Brückner S, Mösch H-U (2003) Amino acid starvation and Gcn4p regulate adhesive growth and FLO11 gene expression in Saccharomyces cerevisiae. Mol Biol Cell 14:4272–4284

Brückner S, Mösch H-U (2012) Choosing the right lifestyle: adhesion and development in Saccharomyces cerevisiae. FEMS Microbiol Rev 36:25–58

Cappellaro C, Baldermann C, Rachel R, Tanner W (1994) Mating type-specific cell-cell recognition of Saccharomyces cerevisiae: cell wall attachment and active sites of a- and alpha-agglutinin. EMBO J 13:4737–4744

Church MC, Fleming AB (2018) A role for histone acetylation in regulating transcription elongation. Transcription 9:225–232

Church M, Smith KC, Alhussain MM, Pennings S, Fleming AB (2017) Sas3 and Ada2(Gcn5)-dependent histone H3 acetylation is required for transcription elongation at the de-repressed FLO1 gene. Nucleic Acids Res 45:4413–4430

Davie JK, Edmondson DG, Coco CB, Dent SYR (2003) Tup1-Ssn6 interacts with multiple class I histone deacetylases in vivo. J Biol Chem 278:50158–50162

De Groot PWJ, Hellingwerf KJ, Klis FM (2003) Genome-wide identification of fungal GPI proteins. Yeast 20:781–796

De Nobel H, Lipke PN (1994) Is there a role for GPIs in yeast cell-wall assembly? Trends Cell Biol 4:42–45

Di PG, Tesnière C, Suzzi G, Blondin B, Tofalo R (2017) FLO5 gene controls flocculation phenotype and adhesive properties in a Saccharomyces cerevisiae sparkling wine strain. Sci Rep 7:10786–10786

Dranginis AM, Rauceo JM, Coronado JE, Lipke PN (2007) A biochemical guide to yeast adhesins: glycoproteins for social and antisocial occasions. Microbiol Mol Biol Rev 71:282–294

Erdman S, Lin L, Malczynski M, Snyder M (1998) Pheromone-regulated genes required for yeast mating differentiation. J Cell Biol 140:461–483

Fichtner L, Schulze F, Braus GH (2007) Differential Flo8p-dependent regulation of FLO1 and FLO11 for cell–cell and cell–substrate adherence of S. cerevisiae S288c. Mol Microbiol 66: 1276–1289

Flaus A, Luger K, Tan S, Richmond TJ (1996) Mapping nucleosome position at single base-pair resolution by using site-directed hydroxyl radicals. Proc Natl Acad Sci U S A 93:1370–1375

Fleming AB, Pennings S (2001) Antagonistic remodelling by Swi-Snf and Tup1-Ssn6 of an extensive chromatin region forms the background for FLO1 gene regulation. EMBO J 20: 5219–5231

Fleming AB, Beggs S, Church M, Tsukihashi Y, Pennings S (2014) The yeast Cyc8-Tup1 complex cooperates with Hda1p and Rpd3p histone deacetylases to robustly repress transcription of the subtelomeric FLO1 gene. Biochim Biophys Acta 1839:1242–1255

Fu Y, Rieg GN, Fonzi WA, Belanger PH, Edwards JE Jr, Filler SG (1998) Expression of the Candida albicans gene ALS1 in Saccharomyces cerevisiae induces adherence to endothelial and epithelial cells. Infect Immun 66:1783–1786

Gaur NK, Klotz SA, Henderson RL (1999) Overexpression of the Candida albicans ALA1 gene in Saccharomyces cerevisiae results in aggregation following attachment of yeast cells to extracellular matrix proteins, adherence properties similar to those of Candida albicans. Infect Immun 67:6040–6047

Gibson BR, Lawrence SJ, Leclaire JPR, Powell CD, Smart KA (2007) Yeast responses to stresses associated with industrial brewery handling. FEMS Microbiol Rev 31:535–569

Goossens K, Willaert R (2010) Flocculation protein structure and cell–cell adhesion mechanism in Saccharomyces cerevisiae. Biotechnol Lett 32:1571–1585

Goossens KVY, Stassen C, Stals I, Donohue DS, Devreese B, De Greve H, Willaert RG (2011) The N-terminal domain of the Flo1 flocculation protein from Saccharomyces cerevisiae binds specifically to mannose carbohydrates. Eukaryot Cell 10:110–117

Govender P, Domingo JL, Bester MC, Pretorius IS, Bauer FF (2008) Controlled expression of the dominant flocculation genes FLO1, FLO5, and FLO11 in Saccharomyces cerevisiae. Appl Environ Microbiol 74:6041–6052

Guo B, Styles CA, Feng Q, Fink GR (2000) A Saccharomyces gene family involved in invasive growth, cell–cell adhesion, and mating. Proc Natl Acad Sci U S A 97:12158–12163

Jabra-Rizk MA, Falker WA, Meiller TF (2004) Fungal biofilms and drug resistance. Emerg Infect Dis 10:14–20

Jentoft N (1990) Why are proteins O-glycosylated? Trends Biochem Sci 15:291–294

Jin Y-L, Speers RA (2000) Effect of environmental conditions on the flocculation of Saccharomyces cerevisiae. J Am Soc Brew Chem 58:108–116

Jung US, Sobering AK, Romeo MJ, Levin DE (2002) Regulation of the yeast Rlm1 transcription factor by the Mpk1 cell wall integrity MAP kinase. Mol Microbiol 46:781–789

Kim TS, Ahn JY, Yoon JH, Kang HS (2003) STA10 repression of STA gene expression is caused by a defective activator, flo8, in Saccharomyces cerevisiae. Curr Genet 44:261–267

Kobayashi O, Hayashi N, Kuroki R, Sone H (1998) Region of FLO1 proteins responsible for sugar recognition. J Bacteriol 180:6503–6510

Kobayashi O, Yoshimoto H, Sone H (1999) Analysis of the genes activated by the FLO8 gene in Saccharomyces cerevisiae. Curr Genet 36:256–261

Kruger W, Peterson CL, Sil A, Coburn C, Arents G, Moudrianakis EN, Herskowitz I (1995) Amino acid substitutions in the structured domains of histones H3 and H4 partially relieve the requirement of the yeast SWI/SNF complex for transcription. Genes Dev 9:2770–2779

Kumawat R, Tomar R (2022) Heavy metal exposure induces Yap1 and Hac1 mediated derepression of GSH1 and KAR2 by Tup1-Cyc8 complex. J Hazard Mater 429:128367

Kyte J, Doolittle RF (1982) A simple method for displaying the hydropathic character of a protein. J Mol Biol 157:105–132

Lambrechts MG, Bauer FF, Marmur J, Pretorius IS (1996) Muc1, a mucin-like protein that is regulated by Mss10, is critical for pseudohyphal differentiation in yeast. Proc Natl Acad Sci U S A 93:8419–8424

Lawrence SJ, Smart KA (2007) Impact of CO2-induced anaerobiosis on the assessment of brewing yeast flocculation. J Am Soc Brew Chem 65:208–213

Linder T, Gustafsson CM (2008) Molecular phylogenetics of ascomycotal adhesins—a novel family of putative cell-surface adhesive proteins in fission yeasts. Fungal Gent Biol 45:485–497

Liu L, Levin DE (2018) Intracellular mechanism by which genotoxic stress activates yeast SAPK Mpk1. Mol Biol Cell 29:2898–2909

Lo W-S, Dranginis A (1996) FLO11, a yeast gene related to the STA genes, encodes a novel cell surface flocculin. J Bacteriol 178:7144–7151

Lo W-S, Dranginis AM (1998) The cell surface flocculin Flo11 is required for pseudohyphae formation and invasion by Saccharomyces cerevisiae. Mol Biol Cell 9:161–171

Lockhart SR, Etienne KA, Vallabhaneni S, Farooqi J, Chowdhary A, Govender NP, Colombo AL, Calvo B, Cuomo CA, Desjardins CA (2017) Simultaneous emergence of multidrug-resistant Candida auris on 3 continents confirmed by whole-genome sequencing and epidemiological analyses. Clin Infect Dis 64:134–140

Maestre-Reyna M, Diderrich R, Veelders MS, Eulenburg G, Kalugin V, Brückner S, Keller P, Rupp S, Mösch H-U, Essen L-O (2012) Structural basis for promiscuity and specificity during Candida glabrata invasion of host epithelia. Proc Natl Acad Sci U S A 109:16864–16869

Makalowska I, Ferlanti ES, Baxevanis AD, Landsman D (1999) Histone sequence database: sequences, structures, post-translational modifications and genetic loci. Nucleic Acids Res 27: 323–324

Mamvura TA, Paterson AE, Fanucchi D (2017) The impact of pipe geometry variations on hygiene and success of orbital welding of brewing industry equipment. J Inst Brew 123:81–97

Manjithaya R, Jain S, Farré J-C, Subramani S (2010) A yeast MAPK cascade regulates pexophagy but not other autophagy pathways. J Cell Biol 189:303–310

Martín NH, Rodríguez-Pachón JM, Ruiz C, Nombela C, Molina MA (2000) Regulatory mechanisms for modulation of signaling through the cell integrity Slt2-mediated pathway in Saccharomyces cerevisiae. J Biol Chem 275:1511–1519

Mata J, Bähler J (2006) Global roles of Ste11p, cell type, and pheromone in the control of gene expression during early sexual differentiation in fission yeast. Proc Natl Acad Sci U S A 103: 15517–15522

Miki BL, Poon NH, James AP, Seligy VL (1982) Possible mechanism for flocculation interactions governed by gene FLO1 in Saccharomyces cerevisiae. J Bacteriol 150:878–889

Mortier A, Soares EV (2007) Separation of yeasts by addition of flocculent cells of Saccharomyces cerevisiae. World J Microbiol Biotechnol 23:1401–1407

Nishihara H, Toraya T, Fukui S (1982) Flocculation of cell walls of Brewer's yeast and effects of metal ions, protein-denaturants and enzyme treatments. Arch Microbiol 131:112–115

Perlroth J, Choi B, Spellberg B (2007) Nosocomial fungal infections: epidemiology, diagnosis, and treatment. Med Mycol 45:321–346

Petosa C, Collier RJ, Klimpel KR, Leppla SH, Liddington RC (1997) Crystal structure of the anthrax toxin protective antigen. Nature 385:833–838

Proft M, Struhl K (2002) Hog1 kinase converts the Sko1-Cyc8-Tup1 repressor complex into an activator that recruits SAGA and SWI/SNF in response to osmotic stress. Mol Cell 9:1307–1317

Pujol-Carrion N, Petkova MI, Serrano L, De La Torre-Ruiz MA (2013) The MAP kinase Slt2 is involved in vacuolar function and actin remodeling in Saccharomyces cerevisiae mutants affected by endogenous oxidative stress. Appl Environ Microbiol 79:6459–6471

Reynolds TB, Fink GR (2001) Bakers' yeast, a model for fungal biofilm formation. Science 291: 878–881

Rossouw D, Bagheri B, Setati M, Bauer F (2015) Co-flocculation of yeast species, a new mechanism to govern population dynamics in microbial ecosystems. PLoS One 10:e0136249

Rupp S, Summers E, Lo HJ, Madhani H, Fink G (1999) MAP kinase and cAMP filamentation signaling pathways converge on the unusually large promoter of the yeast FLO11 gene. EMBO J 18:1257–1269

Sampermans S, Mortier J, Soares EV (2005) Flocculation onset in Saccharomyces cerevisiae: the role of nutrients. J Appl Microbiol 98:525–531

Santisteban MS, Arents G, Moudrianakis EN, Smith MM (1997) Histone octamer function in vivo: mutations in the dimer-tetramer interfaces disrupt both gene activation and repression. EMBO J 16:2493–2506

Sanz AB, García R, Rodríguez-Peña JM, Díez-Muñiz S, Nombela C, Peterson CL, Arroyo J (2012) Chromatin remodeling by the SWI/SNF complex is essential for transcription mediated by the yeast cell wall integrity MAPK pathway. Mol Biol Cell 23:2805–2817
Sanz AB, García R, Rodríguez-Peña JM, Nombela C, Arroyo J (2016) Cooperation between SAGA and SWI/SNF complexes is required for efficient transcriptional responses regulated by the yeast MAPK Slt2. Nucleic Acids Res 44:7159–7172
Sanz AB, García R, Rodríguez-Peña JM, Nombela C, Arroyo J (2018) Slt2 MAPK association with chromatin is required for transcriptional activation of Rlm1 dependent genes upon cell wall stress. Biochim Biophys Acta 1861:1029–1039
Sariki SK, Kumawat R, Singh V, Tomar RS (2019) Flocculation of Saccharomyces cerevisiae is dependent on activation of Slt2 and Rlm1 regulated by the cell wall integrity pathway. Mol Microbiol 112:1350–1369
Sharifmoghadam MR, Bustos-Sanmamed P, Valdivieso M-H (2006) The fission yeast Map4 protein is a novel adhesin required for mating. FEBS Lett 580:4457–4462
Singh V, Azad GK, Sariki SK, Tomar RS (2015) Flocculation inSaccharomyces cerevisiae is regulated by RNA/DNA helicase Sen1p. FEBS Lett 589:3165–3174
Smukalla S, Caldara M, Pochet N, Beauvais A, Guadagnini S, Yan C, Vinces MD, Jansen A, Prevost MC, Latgé J-P, Fink GR, Foster KR, Verstrepen KJ (2008) FLO1 is a variable green beard gene that drives biofilm-like cooperation in budding yeast. Cell 135:726–737
Soares EV (2011) Flocculation in Saccharomyces cerevisiae: a review. J Appl Microbiol 110:1–18
Soriano-Carot M, Bañó MC, Igual JC (2012) The yeast mitogen-activated protein kinase Slt2 is involved in the cellular response to genotoxic stress. Cell Div 7:1–14
Speers RA, Tung MA, Durance TD, Stewart GG (1992) Colloidal aspects of yeast flocculation: a review. J Inst Brew 98:525–531
Stewart GG (2018) Yeast flocculation—sedimentation and flotation. Fermentation 4:28
Stewart G, Garrison I (1972) Some observations on co-flocculation in Saccharomyces cerevisiae. Am Soc Brew Chem 1972:118–131
Stratford M (1989) Evidence for two mechanisms of flocculation in Saccharomyces cerevisiae. Yeast 5:S441–S445
Stratford M, Assinder S (1991) Yeast flocculation: Flo1 and NewFlo phenotypes and receptor structure. Yeast 7:559–574
Stratford M, Keenan MHJ (1988) Yeast flocculation: quantification. Yeast 4:107–115
Straver MH, Kijne JW (1996) A rapid and selective assay for measuring cell surface hydrophobicity of Brewer's yeast cells. Yeast 12:207–213
Straver MH, Kijne JW, Smit G (1993) Cause and control of flocculation in yeast. Trends Biotechnol 11:228–232
Teunissen A, Steensma HY (1995) The dominant flocculation genes of Saccharomyces cerevisiae constitute a new subtelomeric gene family. Yeast 11:1001–1013
Teunissen AWRH, Van Den Berg JA, de Steensma H (1995) Transcriptional regulation of flocculation genes in Saccharomyces cerevisiae. Yeast 11:435–446
Treitel MA, Carlson M (1995) Repression by SSN6-TUP1 is directed by MIG1, a repressor/activator protein. Proc Natl Acad Sci U S A 92:3132–3136
Van Mulders SE, Christianen E, Saerens SM, Daenen L, Verbelen PJ, Willaert R, Verstrepen KJ, Delvaux FR (2009) Phenotypic diversity of Flo protein family-mediated adhesion in Saccharomyces cerevisiae. FEMS Yeast Res 9:178–190
Verstrepen KJ, Klis FM (2006) Flocculation, adhesion and biofilm formation in yeasts. Mol Microbiol 60:5–15
Wang J, Chen C (2006) Biosorption of heavy metals by Saccharomyces cerevisiae: a review. Biotechnol Adv 24:427–451
Wisplinghoff H, Bischoff T, Tallent SM, Seifert H, Wenzel RP, Edmond MB (2004) Nosocomial bloodstream infections in US hospitals: analysis of 24,179 cases from a prospective nationwide surveillance study. Clin Infect Dis 39:309–317

Yang L, Zheng C, Chen Y, Ying H (2018) FLO genes family and transcription factor MIG1 regulate saccharomyces cerevisiae biofilm formation during immobilized fermentation. Front Microbiol 9:1860–1860

Yoshida E, Hidaka M, Fushinobu S, Koyanagi T, Minami H, Tamaki H, Kitaoka M, Katayama T, Kumagai H (2010) Role of a PA14 domain in determining substrate specificity of a glycoside hydrolase family 3 β-glucosidase from Kluyveromyces marxianus. Biochem J 431:39–49

Zara S, Bakalinsky AT, Zara G, Pirino G, Demontis MA, Budroni M (2005) FLO11-based model for air-liquid interfacial biofilm formation by Saccharomyces cerevisiae. Appl Environ Microbiol 71:2934–2939

Management of Hepatitis C Virus: Current and Future Treatments

27

Disha Arora, Smriti Parashar, and Rupesh K. Gautam

Abstract

Globally, hepatitis C virus (HCV), the most important health issue, is a meticulous factor in liver disease. Its cure is now possible for most people affected by this chronic infection. A complete treatment of the same is challenging in a huge preponderance and nearly half of the cases do not achieve a detectable amount of HCV. Many drugs are in development phase targeting viral or host factors, and some will certainly be accepted in the coming years. At present, the current medication for hepatitis C virus is pegylated interferon (Peg-INF) with ribavirin. Pegylated interferon is expensive and in developing countries, the major treatment for the disease is normal interferon. Herbal medicines are another way of treating the disease although no scientific report is documented yet. New remedial measures and current scientific evaluation pay attention to HCV non-structural protein (NS)3 and NS5a RNA polymerase inhibitors. The present paper evaluates the present standard treatment, including herbals, for hepatitis C virus and provides the background of medicine expansion.

Keywords

Antivirals · Hepatitis C · HCV · Medicinal plants · Natural compounds

D. Arora
Chandigarh College of Pharmacy, Mohali, Punjab, India

S. Parashar
Vedic Institutes of Pharmaceutical Education and Research, Sagar, Madhya Pradesh, India

R. K. Gautam (✉)
Department of Pharmacology, Indore Institute of Pharmacy, IIST Campus, Indore, Madhya Pradesh, India
e-mail: rupesh.gautam@indoreinstitute.com

P. P. Singh (ed.), *Recent Advances in Pharmaceutical Innovation and Research*,
https://doi.org/10.1007/978-981-99-2302-1_27

27.1 Introduction

The hepatitis C virus (HCV) belongs to the RNA virus of the family Flaviviridae that replicates in the cytoplasm of liver cells (Lauer and Walker 2001). It brings about severe and persistent infections in persons, which can be lifelong. If hepatitis C (chronic) is not treated, it can cause carcinoma (liver) and cirrhosis (Alter and Seeff 2000). The HCV genetic material is most likely released into the cytoplasmic matrix, interpreted into viral proteins, and starts viral replication (Moradpour et al. 2007). Further, HCV stimulates the epithelial–mesenchymal transition (EMT) cells, a key procedure linked with tumor development (Bose et al. 2012).

27.2 Pathophysiology of Hepatitis C Virus

In 1989, Choo and Kuo for the first time demonstrated HCV and further acknowledged it as a reason for virus ailment, which previously was considered as post-transfusion non-A and non-B hepatitis virus infection (Kuo et al. 1989; Choo et al. 1989). This virus is non-cytopathic as it propagates in liver cells without killing their host cells. Further, it starts replication simultaneously in the liver cells followed by several mechanisms like hepatic steatosis, immune-mediated cytolysis, insulin resistance, and oxidative stress resulting in cell necrosis (Irshad and Dhar 2006). The proteins encrypted by HCV subgenomic regions and their quasi-species have a substantial part in the pathophysiology of HCV infectivity.

27.2.1 HCV Virus Entry

The hepatitis C virus enters the liver by circulation in the blood. For its entry, scavenger receptor on host cell surface class B type I, Occludin, CD81 (Cluster of Differentiation 81), CLDNI (Claudin-I), and host-derived factors are required (Masciopinto et al. 2002). The host cell surface has a CD81 molecule that acts as a viral receptor, binds to HCV, and helps its access to the hepatic cells (Zeisel et al. 2013). All cell surfaces contain CD81 and these complexes with other B-cell receptors like CD19 and CD21, further sending signals to liver host cells (Maecker et al. 1997). Due to a lack of RNA polymerase ability to read the virus, HCV highly mutates. Thus, HCV occurs in numerous other distinctive forms but is closely associated with real virus species within the infected person. These distinctive forms of the virus are well known as HCV quasi-species (Irshad et al. 2013).

27.2.2 HCV Mechanisms of Carcinogenesis

Major factors for hepatocellular carcinoma (HCC) development are chronic HCV and related liver cirrhosis because it is a multistep process involving progressive addition of different genetic mutations resulting in malignant transformation

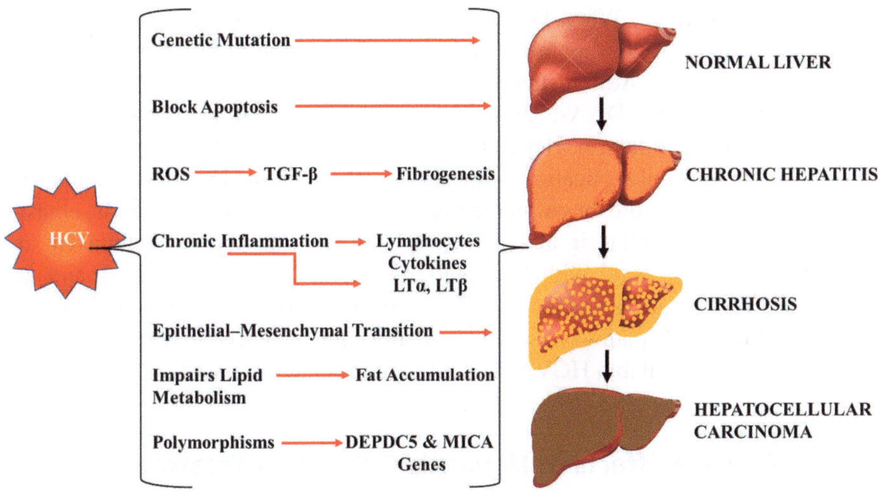

Fig. 27.1 Events involved in infectivity by hepatitis C virus

(Webster et al. 2015) (Fig. 27.1). Malignancy is induced by chronic liver injury and HCV causes inflammation and oxidative stress, which leads to regeneration. HCV protein induces a process that further enhances carcinogenesis or HCC development by blocking apoptosis, inducing reactive oxygen species (ROS) production, tenacious inflammation, building-up of white blood cells, and cytokines emergence, for example, LT-β (Lymphotoxin) and (Lymphotoxin) LT-α (Yang and Roberts 2010).

Prolonged inflammation due to HCV worsens ROS production resulting in mutations of genetic material. In addition, ROS persuade initiation of the (Transforming Growth Factor) TGF-β pathway, which causes activation of hepatic stellate cell and fibrogenesis. TGF-β and (Toll Like Receptor) TLR4 initiate losing the polarity of the epithelial cells and cell–cell adhesion, which enhances the transition of epithelial cells to mesenchymal cells (EMT) (El-Serag 2012). HCV impairs lipid metabolism, triggering liver fat accumulation. Polymorphisms, chiefly in *DEPDC5* and *MICA* genes, also increase the risk of emerging HCC (Vescovo et al. 2016).

27.3 Hepatitis C Treatment

HCV infection treatment involves the use of direct-acting antivirals (DAAs). Currently, in the HCV treatment regimen, various direct-acting antivirals (DAAs) are recommended. To target this virus, three possible mechanisms are required to work. Each of the DAA acts in one of three ways: first, as an inhibitor of non-structural protein (NS)3-4A serine protease (NS3/4A), second, by inhibiting non-structural protein 5A (NS5A) polymerase, and, third, by inhibiting non-structural protein 5B (NS5B) polymerase. NS3-4A serine protease is HCV protease whereas non-structural protein 5A (NS5A) and non-structural protein 5B (NS5B) are the viral proteins of HCV. When DAAs are given in combination they act with more than one mechanism, as shown in Table 27.1. One of the DAAs approved by the U.S. Food and Drug Administration (FDA) is ribavirin used in chronic HCV in combination with other DAAs. RNA polymerase, which is HCV RNA-dependent, is directly targeted by ribavirin. Hepatitis C RNA-dependent RNA polymerase (RdRp) is responsible for viral replication, which is directly inhibited by ribavirin. It is used in combination with sofosbuvir and ledipasvir, sofosbuvir and velpatasvir, elbasvir and grazoprevir, ombitasvir and paritaprevir/ritonavir/dasabuvir, ombitasvir and paritaprevir/ritonavir, and with Peg-interferon alpha-2a (Peg-INF-α) (Anand and Shalimar 2021). Pegasys® binds to human type 1 interferon receptors and helps to activate JAK/STAT pathway, which increases innate antiviral response. So, in Table 27.1 some available HCV treatment regimens are discussed.

27.4 Adverse Effects of Hepatitis C Antiviral Treatment

Hepatitis C therapy involves pan-genotypic direct-acting antivirals (DAAs) as recommended by the World Health Organization (WHO). Most HCV infection is cured by DAAs, and, usually, treatment takes about 12–24 weeks; time taken differs depending on the presence of the cirrhosis condition. Still, in many countries, pan-genotypic DAAs are expensive; that is why their generic versions are introduced, which are cost-effective. Reportedly, many adverse effects are associated with DAAs. Common adverse effects include fever, arthralgias, nausea, constitutional symptoms, headache, myalgias, depression, neuropsychiatric indications, rash, hematological warning signs, neutropenia, anemia, and thrombocytopenia. About 50–60% of treated patients show fatigue, headache, and fever as adverse effects whereas hematological symptoms are common with the use of ribavirin (Sung et al. 2011; Hayes Kaleen et al. 2021).

Globally about 60,000 adverse drug reaction (ADR) reports were studied and concluded unexpected ADRs related to DAA therapy. Severe ADRs include major pulmonary and cardiac events. Dyspnea, pneumonia, and respiratory failure as pulmonary complications were reported with ledipasvir/sofosbuvir use. Cardiac events such as myocardial infarction and cardiac arrest are visible in some ADR reports. In some cases, DAA treatment comprising sofosbuvir and amiodarone in combination results in bradyarrhythmia, a probable precursor to cardiac arrest. So, here is a try to study the alternative option of herbal plants, which can be a future prospective add on to therapy of HCV with more beneficiary effects and less complications (Sung et al. 2011; Hayes Kaleen et al. 2021).

27.5 Plants with Potent Anti-HCV Effects

Since ancient times, plant flora has been used for treating human disorders. As "natural" materials, their effectiveness and safety are sure (Stickel et al. 2005). The use of natural plants is more common in patients having persistent infections, so

Table 27.1 DAA treatment regimen for hepatitis C infection

DAA Generic name	DAA Trade name	DAA Treatment regimen	Ribavirin needed	Mechanism
Ribavirin (RBV) (Mathur et al. 2018)	Viramid® Ribavirin®	–	–	• A direct effect against the HCV RNA-dependent RNA polymerase (RdRp)
Ledipasvir/ sofosbuvir (LDV/SOF) (Sood et al. 2017)	Harvoni®	LDV/SOF ± RBV	Sometimes	• LDV inhibits non-structural protein 5A (NS5A) polymerase • SOF inhibits non-structural protein 5B (NS5B) nucleotide RdRp polymerase
Sofosbuvir (SOF) (Sood et al. 2017)	Sovaldi®	SOF ± RBV	Yes	• SOF inhibits nucleotide NS5B (RdRp) polymerase
Elbasvir/grazoprevir (EBR/GZR) (Cheng et al. 2021)	Zepatier®	EBR/GZR ± RBV	Sometimes	• EBR inhibits non-structural protein (NS5A) polymerase • GZR inhibits NS3/4A serine protease
Ombitasvir/ paritaprevir/ritonavir (OBV/PTV/RTV)	Viekirax®	OBV/ PTV/RTV ± RBV	Yes	• OBV inhibits non-structural protein (NS5A) polymerase • PTV inhibits NS3/4A serine protease • RTV is also a protease inhibitor; it inhibits CYP3A
Sofosbuvir/ Velpatasvir (SOF/VEL) (Jordi et al. 2019)	Epclusa®	SOF/VEL ± RBV	Sometimes	• SOF inhibits nucleotide NS5B (RdRp) polymerase • VEL inhibits non-structural protein (NS5A) polymerase
Ombitasvir/ dasabuvir/ paritaprevir/ritonavir OBV/DSV/PTV/ RTV	Viekira Pak	OBV/ PTV/ RTV/DSV	Yes	• OBV inhibits non-structural protein (NS5A) Polymerase • DSV inhibits non-nucleotide NS5B (RdRp) Polymerase • PTV inhibits NS3/4A serine protease • RTV inhibits CYP3A, it is also a protease inhibitor
Sofosbuvir/ voxilaprevir/ velpatasvir (SOF/VOX /VEL) (Jordi et al. 2019)	Vosevi™	SOF/VEL/ VOX	No	• SOF inhibits nucleotide NS5B (RdRp) polymerase • VEL inhibits non-structural protein (NS5A) polymerase

(continued)

Table 27.1 (continued)

DAA Generic name	DAA Trade name	DAA Treatment regimen	Ribavirin needed	Mechanism
Peg-interferon alpha-2a (peg-IFN-α) (Anand and Shalimar 2021)	Pegasys®	peg-IFN α ± RBV	Yes	• It binds human type 1 interferon receptor and activates JAK/STAT pathway, which increases the innate antiviral response
Glecaprevir/ pibrentasvir (GLE/PIB) (Lamb 2017)	Mavyret™	GLE/PIB	No	• GLE inhibits NS3/4A protease • PIB inhibits NS5A polymerase

supplementary plant therapy is not astonishing in chronic hepatic ailments (Strader et al. 2002; Stickel and Schuppan 2007). From molecules of biological origin, different common medications have come (Lee 2000; Balunas and Kinghorn 2005).

Few drugs (natural) revealed antiviral effects next to many viruses like the influenza virus (Amoros et al. 1987; Chang and Yeung 1988), human immunodeficiency virus (Cavallaro et al. 1995; Yao et al. 1992), hepatitis viruses (B and C), and herpes simplex virus (Kitazato et al. 2007; Hudson 1989). Table 27.2 shows the list of plants with potent effects.

The plants with potent anti-hepatitis C virus activity, investigated in vivo or on models (cellular), include:

1. *Cardus marianus:* Treating hepatic infection with this plant is a very old therapy. Silymarin or silibinin is the chemical constituent (flavanolignans) of the plant. Particular components are there in the plant that are accountable for a variety of anti-HCV effects (Polyak et al. 2007, 2010). Silymarin produces hindrance of HCV genotype 3a core expression (NS5B-independent), using a heterologous expression system (Ashfaq et al. 2011). Silymarin is ineffective in HCV due to its low bioavailability when given orally (Gordon et al. 2006; Hawke et al. 2010; Fried et al. 2012).
2. *Camellia sinensis*: Flavonoid (−)-epigallocatechin-3-gallate (EGCG) inhibits HCV entry (Calland et al. 2012). In primary human liver cells and hepatoma-derived cells, it repressed HCVpp entry (genotype-independent) (Ciesek et al. 2011; Chen et al. 2012). Intake of flavonoid (up to 800 mg) is harmless (Chow et al. 2003), suggesting its possible use in medical investigations (Fukazawa et al. 2012).
3. *Marrubium peregrinum:* Ladanein (genotype-independent) repressed HCV entry (Inhibitory Concentration, IC_{50} 2.5 μM). It acted synergistically on HCV infection when used with r HCV replication inhibitor (cyclosporin A). In mice, 0.25 mg/kg, p.o. was orally bioavailable (Haid et al. 2012).
4. *Citrus paradisi:* After naringenin treatment, a dose-related reduction of protein (core), contagious molecules, HCV-positive RNA, and ApoB was found in the

Table 27.2 List of plants with effective anti-hepatitis C virus activity

S. No.	Plant name	Chemical constituent/ plant extract	Family	Reference
1.	*Silybum marianum*	Silymarin or silibinin	Asteraceae	Polyak et al. (2007, 2010), Ashfaq et al. (2011), Hawke et al. (2010), Gordon et al. (2006), Fried et al. (2012)
2.	*Camellia sinensis*	(−)-Epigallocatechin-3-gallate	Theaceae	Ciesek et al. (2011), Calland et al. (2012), Chen et al. (2012), Chow et al. (2003), Fukazawa et al. (2012)
3.	*Marrubium peregrinum*	Ladanein	Lamiaceae	Haid et al. (2012)
4.	*Citrus paradisi*	Naringenin	Rutaceae	Nahmias et al. (2008), Goldwasser et al. (2011)
5.	*Embelia ribes*	Quercetin	Myrsinacae	Gonzalez et al. (2009), Bachmetov et al. (2012)
6.	*Luteolin and Apigenin*	Flavones	Flavonoid family	Liu et al. (2012), Luo et al. (2000)
7.	*Magnolia officinalis*	Honokiol	Magnoliaceae	Lan et al. (2012)
8.	*Swietenia macrophylla*	3-hydroxy caruilignan C	Meliaceae	Wu et al. (2012)
9.	*Aurantii nobilis*	Nobiletin	Rutaceae	Suzuki et al. (2005)
10.	*Stylogne cauliflora*	SCH 644343 and SCH 644342	Flaviviridae	Hegde et al. (2003)
11.	*Saxifraga melanocentra*	1,2,3,4,6-penta-*O*-galloyl-*β*-D-glucoside	Saxifragaceae	Zuo et al. (2005)
12.	*Galla Chinese*	1,2,6-tri-*O*-galloyl-*β*-D-glucose, 1,2,3,6-tetra-*O*-galloyl-*β*-D-glucose, and 1,2,3,4,6-penta-*O*-galloyl-*β*-D-glucose	Anacardiaceae	Duan et al. (2004)
13.	*Excoecaria agallocha*	Excoecariphenol D, corilagin	Euphorbiaceae	Li et al. (2012)
14.	*Acacia nilotica*	–	Fabaceae	Evers et al. (1987), Rehman et al. (2011)

(continued)

Table 27.2 (continued)

S. No.	Plant name	Chemical constituent/ plant extract	Family	Reference
15.	*Toona sureni*	Ethanol extract	Meliaceae	Wahyuni et al. (2013)
16.	*Melicope latifolia*	Ethanol extract	Rutaceae	Wahyuni et al. (2013)
17.	*Melanolepis multiglandulosa*	Ethanol extract	Euphorbiaceae	Wahyuni et al. (2013)
18.	*Ficus fistulosa*	Ethanol extract	Moraceae	Wahyuni et al. (2013)
19.	*Solanum nigrum*	Methanol, chloroform extracts	Solanaceae	Leitz and Muller (1987)
20.	*Viola yedoensis*	5,5′-bi-(6,7-dihydroxy-coumarin)	Violaceae	Zhang et al. (2013)
21.	*Diosgenin*	Sapogenin	Dioscoreaceae	Wang et al. (2011)
22.	*Lamium album*	Lamiridosins A/B (1/2), iridoids	Lamiaceae	Zhang et al. (2009)
23.	*Griffithsin* and *scytovirin*	Red and blue-green algae	Coronaviridae/ Scytonemataceae	Mori et al. (2005), Bokesch et al. (2003), McFeeters et al. (2007), Moulaei et al. (2007), Ziolkowska et al. (2006)
24.	*Limonium sinense*	Aqueous extract	Plumbaginaceae	Hsu et al. (2015)
25.	*Bupleurum kaoi*	Saikosaponins	Apiaceae	Lin et al. (2015)
26.	*Delphinidin*	Polyphenol	Polyphenol family	Calland et al. (2015)
27.	*Garcinia mangostana*	Ethanol extract	Guttiferae	Choi et al. (2014)
28.	*Quercetin*	Flavonoids	Flavonoid family	Pisonero-Vaquero et al. (2014)
29.	*Oxymatrine and matrine*	Alkaloids	Alkaloid family	Chen et al. (2001), Liu et al. (1994), Azzam et al. (2007)
30.	*Fructus schiznadrae*	*Schizandrin C*	Magnoliaceae	Liu (2009), Huang et al. (2019)
31.	*Phyllanthus urinaria*	Loliolide	Euphorbiaceae	Chung et al. (2016)
32.	*Rhodiola kirilowii*	Ethanol extract	Crassulaceae	Zuo et al. (2007)
33.	*Zingiber officinale*	Ethanol extract	Zingiberaceae	Abdel-Moneim et al. (2013)
34.	*Platycodon grandiflorum*	Triterpenoid saponins	Campanulaceae	Kim et al. (2012, 2013)

(continued)

Table 27.2 (continued)

S. No.	Plant name	Chemical constituent/ plant extract	Family	Reference
35.	*Ruta angustifolia*	–	Rutaceae	Wahyuni et al. (2019)
36	*Kalanchoe pinnata*	Methanol extract	Crassulaceae	Aoki et al. (2014)
37	*Pterogyne nitens*	Flavonoids	Fabaceae	Shimizu et al. (2017)
38	*Camel milk*	Lactoferrin	–	Gader and Alhaider (2016), Redwan et al. (2014)

lysate of contaminated Huh-7 cells (Nahmias et al. 2008). 200 μM naringenin shows maximal (74%) suppression of production of HCV RNA and ApoB together with an IC_{50} value of 109 μM (Goldwasser et al. 2011).

5. *Embelia ribes:* In cell culture, quercetin (flavonol) had a repressive (strong) outcome on HCV production at 50 μM (Gonzalez et al. 2009). *Embelia ribes* plant extracts were responsible for inhibiting NS3 protease activity. The authors found a weaker impact on replication but at a 10 μg/mL (≈33 μM) dose, HCV particle production was strongly inhibited (Bachmetov et al. 2012).
6. *Luteolin* and *Apigenin:* Anti-HCV agent flavones (luteolin and apigenin) were identified via pharmacophore search (Liu et al. 2012). Both displayed an anti-HCV activity with EC_{50} values of 4.3 μM and 7.9 μM, respectively, in a cell-based antiviral assay. A good inhibition of NS5B polymerase enzymatic function with an IC_{50} of 1.12 μM is exhibited by *Luteolin* (Luo et al. 2000).
7. *Magnolia officinalis:* Honokiol (lignin) in the plant, at non-toxic concentrations, strongly inhibited HCVcc infection. Its inhibition of HCV infection was shown from multiple effects on the HCV life cycle. It inhibited the expression levels of NS3, NS5A, and NS5B in a dose-dependent manner (Lan et al. 2012).
8. *Swietenia macrophylla*: 3-HCL-C in *Swietenia macrophylla* stem that is chemically 3-hydroxy caruilignan C was reported to be an anti-HCV agent. In the subgenomic replicon system, 3-HCL-C reduced both protein (NS3) and RNA levels of HCV with an EC_{50} value of 10.5 μg/mL. Combinations of 3-HCL-C and IFN-α, 2′-C-methylcytidine, or telaprevir increased the suppression of HCV RNA replication (Wu et al. 2012).
9. *Aurantii nobilis*: Nobiletin from Citrus unshiu peel was found to be liable for the anti-HCV activity. When examined in the (T Lymphoblast Cell Line) MOLT-4 cells infection assay, it showed an antiviral activity at a 10 μg/mL dose (Suzuki et al. 2005).
10. *Stylogne cauliflora*: SCH 644342 and SCH 644343 (oligophenolic compounds) in the plant were identified as inhibitors of HCV NS3 protease activity in vitro

with IC_{50} of 0.3 μM and 0.8 μM, respectively. SCH 644343 was also active in an NS3 binding assay ($IC_{50} = 2.8$ μM) (Hegde et al. 2003).

11. *Saxifraga melanocentra*: 1,2,3,4,6-penta-*O*-galloyl-*β*-D-glucoside was uncovered to be a strong repressor of NS3 protease (IC_{50} 0.68 μM). Up to 6 mg/mL, it has no adverse effects (Zuo et al. 2005).
12. *Galla Chinese*: Constituents obtained from ethyl acetate fraction of this plant, 1,2,6-tri-*O*-galloyl-*β*-D-glucose, help in inhibiting in vitro NS3 protease with IC_{50} values of 1.89 μM. Additionally, 1,2,3,6-tetra-*O*-galloyl-*β*-D-glucose from plants shows inhibition with IC_{50} values of 0.75 μM and 1,2,3,4,6-penta-*O*-galloyl-*β*-D-glucose with IC_{50} values 1.60 μM, respectively (Duan et al. 2004).
13. *Excoecaria agallocha*: Two polyphenolic compounds excoecariphenol D and corilagin were also found to repress NS3 protease (in vitro), and inhibit replicon assay (IC_{50} 12.6 and 13.5 μM, respectively) (Li et al. 2012).
14. *Acacia nilotica*: It shows in vitro repression of HCV titer established through (Polymerase Chain Reaction) PCR (Evers et al. 1987; Rehman et al. 2011).
15. Ethanol extracts of *Melicope latifolia* leaves (MLL) have IC_{50} of 3.5, *Toona sureni* leaves (TSL) exhibited an IC_{50} value of 13.9, *Melanolepis multiglandulosa* stem (MMS) exhibited IC_{50} of 17.1, and *Ficus fistulosa* leaves (FFL) showed IC_{50} of 15.0. Among all of these, MLL, TSL, FFL, and MMS exhibited antiviral activity against all genotypes of HCV, and thus it is suggested that these plants may prove good candidates to develop novel inhibitory drugs against HCV (Wahyuni et al. 2013).
16. *Solanum nigrum*: The plant and its extracts (methanol, chloroform) showed noteworthy suppression against HCV and HCV protease in infected hepatocytes (Leitz and Muller 1987).
17. *Viola yedoensis*: The plant has been tested for anti-HCV protease. By different chromatographic procedures, three coumarins have been isolated and characterized from *Viola yedoensis.* Among the isolated compounds, a dimeric coumarin 5,5′-bi (6,7-dihydroxycoumarin) has significantly inhibited NS3/4A protease with an IC_{50} value of 0.5 μg/mL. Thus, this dicoumarin can serve as an important molecular template to design novel anti-HCV drugs (Zhang et al. 2013).
18. *Diosgenin*: This sapogenin efficiently obstructs the duplication of the hepatitis C virus subgenomic replicon system at both the mRNA and the protein level. A decrease in the activator of transcription factor 3 and signal transducer has been observed. The (Effective Concentration) EC_{50} value of diosgenin was 3.8 μmol with no cellular toxicity. Another antiviral system, it showed inhibition of viral replication at 20 μmol concentrations (Wang et al. 2011).
19. *Lamium album*: From aqueous extract *lamina album*, lamiridosins A/B and iridoids are obtained, which decrease the entry of HCVpp by interrupting HCVE2 proteins with the CD81 receptor (Zhang et al. 2009).
20. *Griffithsin* and *Scytovirin*: The obstruction of HCV entry by proteins isolated from red (Mori et al. 2005) and blue-green algae (Bokesch et al. 2003) was reported by Takebe et al. (2013). By their unique structural characteristics, these proteins can bind to multiple carbohydrate moieties and demonstrated to inhibit

HCV in both cell culture and pseudoparticle assays by targeting HCV envelope glycoproteins E1 and E2 (McFeeters et al. 2007; Moulaei et al. 2007; Ziolkowska et al. 2006).

21. *Limonium sinense*: Without altering viral replication, cell–cell propagation, and translation, an aqueous extract (underground part) of the plant significantly repressed early HCV entry events by interrupting virus attachment and entry/fusion (Hsu et al. 2015).
22. *Bupleurum kaoi*: Initial repression of hepatitis C virus life cycle was detected by treating HuH 7.5 cells infected with Gaussia luciferase reporter viruses (Jc1-GLuc) with a saponin-rich methanol extract of *Bupleurum kaoi* roots (BK) or its associated saikosaponins SSa, SSb2, SSc, and SSd (Lin et al. 2015).
23. *Delphinidin*: Against all HCV genotypes, delphinidin was observed to be an entry blocker. This polyphenol acts on E1 and E2 glycoproteins inducing conformational changes in the viral particles, affecting the interaction between the viral particles and the cell surface (Calland et al. 2015).
24. *Garcinia mangostana*: In infectious replicon genotypes 1b and 2a, mangosteen ethanol extract (MG-EtOH) blocks the replication of the HCV genome by decreasing RNA and protein levels (Choi et al. 2014).
25. *Quercetin*: In Huh7 cells, quercetin repressed HCV replication expressing complete genotype 1b replicated HCV (dose-related) alone or with interferon-α and HCV non-structural 5A (NS5A) protein core articulation was also decreased (Pisonero-Vaquero et al. 2014).
26. Alkaloids *Oxymatrine* and *Matrine*: Oxymatrine from aqueous extract when tested in cell culture for HCV shows antiviral activity and when examined in animal models shows hepatoprotective activity (Chen et al. 2001; Liu et al. 1994; Azzam et al. 2007).
27. Derivative of *Schizandrin C, Bicyclol*: It is isolated from *Fructus schiznadrae*. Bicyclol significantly inhibited HCV replication in vitro and in hepatitis C patients. The activity is through the modulation of cytotoxic T-lymphocytes and by up-regulating the host restrictive factor for HCV replication and causing the spontaneous restriction of HCV replication (Liu 2009; Huang et al. 2019).
28. *Phyllanthus urinaria*: The plant blocked the early entry of HCV. Loliolide (monoterpene) interrupts virus addition with the host and therefore suppresses its entrance (Chung et al. 2016).
29. *Rhodiola kirilowii*: Ethanol extract of *Rhodiola kirilowii* (rhizomes) is effective in opposition to HCV NS3 serine protease (Zuo et al. 2007).
30. *Zingiber officinale*: Intake of ethanol extract (*Nigella sativa* and *Zingiber officinale*) in hepatitis C virus cases showed major healing interests by reducing viral burden along with mitigating distorted functions of the liver (Abdel-Moneim et al. 2013).
31. *Platycodon grandiflorum*: Saponins (triterpenoid) from the roots of the plant exert anti-HCV activity. Combinatorial treatment of the plant saponin mixture and IFN-α competently repressed colony configuration with noteworthy deduction in drug-resistant forms of hepatitis C virus (Kim et al. 2013). Hot water extract from the plant inhibits HCV RNA replication (Kim et al. 2012).

32. *Ruta angustifolia*: The extract of the plant possesses probable anti-HCV effect with no toxic effects (IC_{50} 3.0 μg/mL). The extract also inhibits the post-entry step along with the reduction in protein amount of HCV NS3 and NS5A (Wahyuni et al. 2019).
33. *Kalanchoe pinnata*: Methanol extract of the plant showed anti-hepatitis C effect (IC_{50} 17.2 μg/mL) (Aoki et al. 2014).
34. *Pterogyne nitens*: Flavonoids present in the plant repressed in vitro hepatitis C viral entry by hampering the early phase of the virus cycle (Shimizu et al. 2017).
35. *Camel milk lactoferrin*: It appreciably hinders the genotype for infection of hepatitis C via obstructing the viral entry into the cells (Gader and Alhaider 2016). In a cell culture medium, the antibody results of the same on human hepatoma cell line (Huh-7) contaminated with HCV were examined. Inhibition of virus growth along with the destruction of viral synthesized peptides was reported (Redwan et al. 2014).

27.6 Conclusion

Hepatitis C persistent infection determines the clinical approach and its management. By understanding various mechanisms of viral transmission, the prevention of HCV infection is possible. HCV is always the main reason for cirrhosis, loss of liver functions, and liver cell carcinoma regardless of current spectacular approaches to its treatment. The little amount of HCV RNA is infectious even in cases effectively managed for persistent hepatitis C. Liver disease or hepatocellular carcinoma may occur in some patients even after complete elimination of the virus. The prolonged effectiveness of therapy with the potential resistant virus and a new generation of direct-acting antiviral agents (interferon-free regimens) are not known. How HCV facilitates the development of the disease is under examination to reveal the virological actions and imbricate in rising suitable healing methods.

Newer compounds should ideally offer useful, more acceptable therapy to a greater section of patients with this disease. A variety of isolated plant components revealed efficiency next to HCV and their assay systems, which permit improved assessment of the antiviral action of components. We look ahead to the improvement of future agents that will hopefully interpret into better efficiency in clinical trials, and then consequently be translated into clinical practice, where the great preponderance of patients seeks management and treatment. The enduring new developments comprise the growth of pre-emptive jabs for universal applications as well as improved knowledge of actions of perseverance along with pathogenesis of enthralling disease.

References

Abdel-Moneim A, Morsy BM, Mahmoud AM et al (2013) Beneficial therapeutic effects of *Nigella sativa* and/or *Zingiber officinale* in HCV patients in Egypt. EXCLI J 12:943–955

Alter HJ, Seeff LB (2000) Recovery, persistence, and sequelae in hepatitis C virus infection: a perspective on long-term outcome. Semin Liver Dis 20:17–35

Amoros M, Fauconnier B, Girre RL (1987) *In vitro* antiviral activity of a saponin from *Anagallis arvensis*, Primulaceae, against herpes simplex virus and poliovirus. Antivir Res 8:13–25

Anand A, Shalimar (2021) Hepatitis C virus in India: challenges and successes. Clin Liver Dis 18: 150–154. https://doi.org/10.1002/cld.1137

Aoki C, Hartatib S, Santi MR et al (2014) Isolation and identification of substances with anti-hepatitis C virus activities from *Kalanchoe pinnata*. Int J Pharm Pharm Sci 6:211–215

Ashfaq UA, Javed T, Rehman S et al (2011) Inhibition of HCV 3a core gene through Silymarin and its fractions. Virol J 8:153

Azzam HS, Goertz C, Fritts M et al (2007) Natural products and chronic hepatitis C virus. Liver Int 27:17–25

Bachmetov L, Gal-Tanamy M, Shapira A et al (2012) Suppression of hepatitis C virus by the flavonoid quercetin is mediated by inhibition of NS3 protease activity. J Viral Hepat 19:e81–e88

Balunas MJ, Kinghorn AD (2005) Drug discovery from medicinal plants. Life Sci 78:431–441

Bokesch HR, O'Keefe BR, McKee TC et al (2003) A potent novel anti-HIV protein from the cultured cyanobacterium *Scytonema varium*. Biochemistry 42:2578–2584

Bose SK, Meyer K, Di Bisceglie AM et al (2012) Hepatitis C virus induces epithelial-mesenchymal transition in primary human hepatocytes. J Virol 86:13621–13628

Calland N, Albecka A, Belouzard S et al (2012) (−)-Epigallocatechin-*3*-gallate is a new inhibitor of hepatitis C virus entry. Hepatology 55:720–729

Calland N, Sahuc ME, Belouzard S et al (2015) Polyphenols inhibit hepatitis C virus entry by a new mechanism of action. J Virol 89:10053–10063

Cavallaro L, Garcia G, Broussalis A (1995) Antiherpetic *in vitro* activity of *Gamochaeta simplicicaulis* extract. Phytother Res 9:176–179

Chang RS, Yeung HW (1988) Inhibition of growth of human immunodeficiency virus *in vitro* by crude extracts of Chinese medicinal herbs. Antivir Res 9:163–175

Chen Y, Li J, Zeng M, Lu L, Qu D, Mao Y et al (2001) The inhibitory effect of oxymatrine on hepatitis C virus *in vitro*. Zhonghua Gan Zang Bing Za Zhi 9(Suppl):12–14

Chen C, Qiu H, Gong J et al (2012) (−)-Epigallocatechin-*3*-gallate inhibits the replication cycle of hepatitis C virus. Arch Virol 157:1301–1312

Cheng PN, Chen CY, Yu ML et al (2021) Elbasvir/grazoprevir is effective and tolerable for the treatment of HCV GT1-infected patients: a real world multicenter observatory study in Taiwan. J Microbiol Immunol Infect 54:588–595

Choi M, Kim YM, Lee S et al (2014) Mangosteen xanthones suppress hepatitis C virus genome replication. Virus Genes 49:208–222

Choo QL, Kuo G, Weiner AJ et al (1989) Isolation of a cDNA clone derived from a blood-borne non-A, non-B viral hepatitis genome. Science 244:359–362

Chow HHS, Cai Y, Hakim IA et al (2003) Pharmacokinetics and safety of green tea polyphenols after multiple-dose administration of epigallocatechin gallate and polyphenon E in healthy individuals. Clin Cancer Res 9:3312–3319

Chung CY, Liu CH, Burnouf T et al (2016) Activity-based and fraction-guided analysis of *Phyllanthus urinaria* identifies loliolide as a potent inhibitor of hepatitis C virus entry. Antivir Res 130:58–68

Ciesek S, von Hahn T, Colpitts CC et al (2011) The green tea polyphenol, epigallocatechin-*3*-gallate, inhibits hepatitis C virus entry. Hepatology 54:1947–1955

Duan D, Li Z, Luo H et al (2004) Antiviral compounds from traditional Chinese medicines *Galla Chinese* as inhibitors of HCV NS3 protease. Bioorg Med Chem Lett 14:6041–6044

El-Serag HB (2012) Epidemiology of viral hepatitis and hepatocellular carcinoma. Gastroenterology 142:1264–1273

Evers MP, Zelle B, Peeper DS et al (1987) Molecular cloning of a pair of human pepsinogen A genes which differ by a Glu-Lys mutation in the activation peptide. Hum Genet 77:182–187

Fried MW, Navarro VJ, Afdhal N et al (2012) Effect of silymarin (milk thistle) on liver disease in patients with chronic hepatitis C unsuccessfully treated with interferon therapy: a randomized controlled trial. JAMA 308:274–282

Fukazawa H, Suzuki T, Wakita TA (2012) Cell-based, microplate colorimetric screen identifies 7,8-benzoflavone and green tea gallate catechins as inhibitors of the hepatitis C virus. Biol Pharm Bull 35:1320–1327

Gader AG, Alhaider AA (2016) The unique medicinal properties of camel products: a review of the scientific evidence. J Taibah Univ Sci 11:98–103

Goldwasser J, Cohen PY, Lin W et al (2011) Naringenin inhibits the assembly and long-term production of infectious hepatitis C virus particles through a PPAR-mediated mechanism. J Hepatol 55:963–971

Gonzalez O, Fontanes V, Raychaudhuri S et al (2009) The heat shock protein inhibitor Quercetin attenuates hepatitis C virus production. Hepatology 50:1756–1764

Gordon A, Hobbs DA, Bowden DS et al (2006) Effects of *Silybum marianum* on serum hepatitis C virus RNA, alanine aminotransferase levels and well-being in patients with chronic hepatitis C. J Gastroenterol Hepatol 21:275–280

Haid S, Novodomska A, Gentzsch J et al (2012) A plant-derived flavonoid inhibits entry of all HCV genotypes into human hepatocytes. Gastroenterology 143:213–222

Hawke RL, Schrieber SJ, Soule TA et al (2010) Silymarin ascending multiple oral dosing phase I study in noncirrhotic patients with chronic hepatitis C. J Clin Pharmacol 50:434–449

Hayes Kaleen N, Burkard T, Weiler S, Tadrous M et al (2021) Global adverse events reported for direct-acting antiviral therapies for the treatment of hepatitis C: an analysis of the World Health Organization VigiBase. Eur J Gastroenterol Hepatol 33:e1017–e1021

Hegde VR, Pu H, Patel M et al (2003) Two antiviral compounds from the plant *Stylogne cauliflora* as inhibitors of HCV NS3 protease. Bioorg Med Chem Lett 13:2925–2928

Hsu WC, Chang SP, Lin LC et al (2015) *Limonium sinense* and gallic acid suppress hepatitis C virus infection by blocking early viral entry. Antivir Res 118:139–147

Huang MH, Li H, Xue R et al (2019) Up-regulation of glycolipid transfer protein by bicyclol causes spontaneous restriction of hepatitis C virus replication. Acta Pharm Sin B 9:769–781

Hudson JB (1989) Plant photosensitizers with antiviral properties. Antivir Res 12:55–74

Irshad M, Dhar I (2006) Hepatitis C virus core protein: an update on its molecular biology, cellular functions and clinical implications. Med Princ Pract 15:405–416

Irshad M, Mankotia DS, Irshad K (2013) An insight into the diagnosis and pathogenesis of hepatitis C virus infection. World J Gastroenterol 19:7896–7909. https://doi.org/10.3748/wjg.v19.i44.7896

Jordi L, Riveiro-Barciela M, Sabela L et al (2019) Effectiveness and safety of sofosbuvir/velpatasvir/voxilaprevir in patients with chronic hepatitis C previously treated with DAAs. J Hepatol 71:666–672

Kim TW, Lim JH, Song IB et al (2012) Hepatoprotective and anti-hepatitis C viral activity of *Platycodon grandiflorum* extract on carbon tetrachloride induced acute hepatic injury in mice. J Nutr Sci Vitaminol 58:187–194

Kim JW, Park SJ, Lim JH, Yang JW, Shin JC, Lee SW et al (2013) Triterpenoid saponins isolated from *Platycodon grandiflorum* inhibit hepatitis C virus replication. eCAM 2013:560417. https://doi.org/10.1155/2013/560417

Kitazato K, Wang Y, Kobayashi N (2007) Viral infectious disease and natural products with antiviral activity. Drug Discov Ther 1:14–22

Kuo G, Choo QL, Alter HJ et al (1989) An assay for circulating antibodies to a major etiologic virus of human non-A, non-B hepatitis. Science 244:362–364

Lamb YN (2017) Glecaprevir/pibrentasvir: first global approval. Drugs **77**:1797–1804. https://doi.org/10.1007/s40265-017-0817-y

Lan KH, Wang YW, Lee WP et al (2012) Multiple effects of Honokiol on the life cycle of hepatitis C virus. Liver Int 32:989–997

Lauer GM, Walker BD (2001) Hepatitis C virus infection. New Engl J Med 345:41–52

Lee KH (2000) Research and future trends in the pharmaceutical development of medicinal herbs from Chinese medicine. Public Health Nutr 3:515–522

Leitz T, Muller WA (1987) Evidence for the involvement of PI-signaling and diacylglycerol second messengers in the initiation of metamorphosis in the hydroid *Hydractinia echinata* Fleming. Dev Biol 121:82–89

Li Y, Yu S, Liu D et al (2012) Inhibitory effects of polyphenols toward HCV from the mangrove plant *Excoecaria agallocha* L. Bioorg Med Chem Lett 22:1099–1102

Lin LT, Chung CY, Hsu WC et al (2015) Saikosaponin b2 is a naturally occurring terpenoid that efficiently inhibits hepatitis C virus entry. J Hepatol 62:541–548

Liu G (2009) Bicyclol: a novel drug for treating chronic viral hepatitis B and C. Med Chem 5:29–43

Liu J, Liu Y, Klaassen CD (1994) The effect of Chinese hepatoprotective medicines on experimental liver injury in mice. J Ethnopharmacol 42:183–191

Liu MM, Zhou L, He PL et al (2012) Discovery of flavonoid derivatives as anti-HCV agents *via* pharmacophore search combining molecular docking strategy. Eur J Med Chem 52:33–43

Luo G, Hamatake RK, Mathis DM et al (2000) De novo initiation of RNA synthesis by the RNA-dependent RNA polymerase (NS5B) of hepatitis C virus. J Virol 74:851–863

Maecker HT, Todd SC, Levy S (1997) The tetraspanin superfamily: molecular facilitators. FASEB J 11:428–442

Mathur P, Kottilil S, Wilson E (2018) Use of ribavirin for hepatitis C treatment in the modern direct-acting antiviral era. J Clin Transl Hepatol 6:431–437. https://doi.org/10.14218/JCTH.2018.00007

Masciopinto F, Freer G, Burgio VL et al (2002) Expression of human CD81 in transgenic mice does not confer susceptibility to hepatitis C virus infection. Virology 304:187–196

McFeeters RL, Xiong C, O'Keefe BR et al (2007) The novel fold of scytovirin reveals a new twist for antiviral entry inhibitors. J Mol Biol 369:451–461

Moradpour D, Penin F, Rice CM (2007) Replication of hepatitis C virus. Nat Rev Microbiol 5:453–463

Mori T, O'Keefe BR, Sowder RC et al (2005) Isolation and characterization of griffithsin, a novel HIV-inactivating protein, from the red alga *Griffithsia* sp. J Biol Chem 280:9345–9353

Moulaei T, Botos I, Ziolkowska NE et al (2007) Atomic-resolution crystal structure of the antiviral lectin scytovirin. Protein Sci 16:2756–2760

Nahmias Y, Goldwasser J, Casali M (2008) Apolipoprotein B-dependent hepatitis C virus secretion is inhibited by the grapefruit flavonoid naringenin. Hepatology 47:1437–1445

Pisonero-Vaquero S, Garcia-Mediavilla MV, Jorquera F et al (2014) Modulation of PI3K-LXR-α-dependent lipogenesis mediated by oxidative/nitrosative stress contributes to inhibition of HCV replication by quercetin. Lab Investig 94:262–274

Polyak SJ, Morishima C, Shuhart MC et al (2007) Inhibition of T-cell inflammatory cytokines, hepatocyte NF-kappaB signaling, and HCV infection by standardized Silymarin. Gastroenterology 132:1925–1936

Polyak SJ, Morishima C, Lohmann V et al (2010) Identification of hepatoprotective flavonolignans from silymarin. Proc Natl Acad Sci U S A 107:5995–5999

Redwan EM, El-Fakharany EM, Uversky VN et al (2014) Screening the anti infectivity potentials of native N-and C-lobes derived from the camel lactoferrin against hepatitis C virus. BMC Complement Altern Med 14:219

Rehman S, Ashfaq UA, Riaz S et al (2011) Antiviral activity of *Acacia nilotica* against Hepatitis C virus in liver infected cells. Virol J 8:220

Shimizu JF, Lima CS, Pereira CM et al (2017) Flavonoids from *Pterogyne nitens* inhibit hepatitis C virus entry. Sci Rep 7:16127. https://doi.org/10.1038/s41598-017-16336-y

Sood A, Midha V, Mahajan R et al (2017) Results of sofosbuvir-based combination therapy for chronic hepatitis C cohort of Indian patients in real-life clinical practice. J Gastroenterol Hepatol 32:894–900

Stickel F, Schuppan D (2007) Herbal medicine in the treatment of liver diseases. Dig Liver Dis 39: 293–304

Stickel F, Patsenker E, Schuppan D (2005) Herbal hepatotoxicity. J Hepatol 43:901–910
Strader DB, Bacon BR, Lindsay KL et al (2002) Use of complementary and alternative medicine in patients with liver disease. Am J Gastroenterol 97:2391–2397
Sung H, Chang M, Saab S (2011) Management of hepatitis C antiviral therapy adverse effects. Curr Hepat Rep 10:33–40
Suzuki M, Sasaki K, Yoshizaki F et al (2005) Anti-hepatitis C virus effect of citrus unshiu peel and its active ingredient nobiletin. Am J Chin Med 33:87–94
Takebe Y, Saucedo CJ, Lund G et al (2013) Antiviral lectins from red and blue-green algae show potent *in vitro* and *in vivo* activity against hepatitis C virus. PLoS One 8:1–10
Vescovo T, Refolo G, Vitagliano G et al (2016) Molecular mechanisms of hepatitis C virus–induced hepatocellular carcinoma. Clin Microbiol Infect 22:853–861
Wahyuni TS, Tumewu L, Permanasari AA et al (2013) Antiviral activities of Indonesian medicinal plants in the East Java region against hepatitis C virus. Virol J 10:259
Wahyuni TS, Mahfud H, Permatasari AA et al (2019) Synergistic anti-hepatitis C virus activity of *Ruta angustifolia* extract with NS3 protein inhibitor. J Basic Clin Physiol Pharmacol 30(6):348. https://doi.org/10.1515/jbcpp-2019-0348
Wang YJ, Pan KL, Hsieh TC et al (2011) Diosgenin, a plant-derived sapogenin, exhibits antiviral activity *in vitro* against hepatitis C virus. J Nat Prod 74:580–584
Webster DP, Klenerman P, Dusheiko GM et al (2015) Lancet 385:1124–1135
Wu SF, Lin CK, Chuang YS et al (2012) Anti-hepatitis C virus activity of 3-hydroxy caruilignan C from *Swietenia macrophylla* stems. J Viral Hepat 19:364–370
Yang JD, Roberts LR (2010) Hepatocellular carcinoma: a global view. Nat Rev Gastroenterol Hepatol 7:448–458
Yao XJ, Wainberg MA, Parniak MA (1992) Mechanism of inhibition of HIV-1 infection *in vitro* by purified extract of *Prunella vulgaris*. Virology 187:56–62
Zeisel MB, Felmlee DJ, Baumert TF (2013) Hepatitis C virus entry. Curr Top Microbiol Immunol 369:87–112
Zhang H, Rothwang K, Mesecar AD et al (2009) Lamiridosins, hepatitis C virus entry inhibitors from *Lamium album*. J Nat Prod 72:2158–2162
Zhang L, Li MY, Wang LW et al (2013) Isolation, identification, quantification and inhibitory activity on HCV protease of coumarins from *Viola yedoensis*. Can Chem Trans 1:157–164
Ziolkowska NE, O'Keefe BR, Mori T et al (2006) Domain-swapped structure of the potent antiviral protein Griffiths in and its mode of carbohydrate binding. Structure 14:1127–1135
Zuo GY, Li ZQ, Chen LR et al (2005) *In vitro* anti-HCV activities of *Saxifraga melanocentra* and its related polyphenolic compounds. Antivir Chem Chemother 16:393–398
Zuo G, Li Z, Chen L et al (2007) Activity of compounds from Chinese herbal medicine *Rhodiola kirilowii* (Regel) Maxim against HCV NS3 serine protease. Antivir Res 76:86–92

Tuberculosis: Experimental Models, Innovations, and Challenges

28

Sarbjit Singh Jhamb, Raman Preet Singh, and Prati Pal Singh

Abstract

Tuberculosis (TB) is among the top ten killer diseases and remains the number one cause of death due to infection. A major bottleneck in TB research remains the availability of suitable animal models to understand the disease pathogenesis and progression, immune responses elicited by the pathogen, new molecule and vaccine testing, development and validation of diagnostics, and genetics of the pathogen about these myriad aspects of the infection. Although a broad range of organisms has been employed in TB research, most of the studies have been performed in mice due to cost-effectiveness, ease of handling, availability of immune reagents, and genetically-modified strains as well as ease of availability of strains with a relatively uniform genetic background. The commonly used mouse strains do not mimic human disease progression characteristics. More relevant models like guinea pig and macaque are not frequently employed due to high costs and/or lack of availability of immune reagents. Several models involving alternate, non-pathogenic mycobacteria have been evaluated in mammals and non-mammalian species like fish, frogs, nematodes, and protists. In vitro models such as macrophage infection and co-culture systems provide insights into drug activity and host cell-mycobacterial interactions. An even more straightforward approach relies on using mycobacterial cultures to evaluate drug sensitivity and drug activity. However, the in vitro models suffer from a

S. S. Jhamb (✉) · P. P. Singh
Department of Pharmacology and Toxicology, National Institute of Pharmaceutical Education and Research, Mohali, India
e-mail: sarabjit@niper.ac.in

R. P. Singh
Department of Pharmacy, Government Polytechnic College, Bathinda, India

Department of Pharmaceutical Sciences, Government Polytechnic College for Girls, Patiala, India

P. P. Singh (ed.), *Recent Advances in Pharmaceutical Innovation and Research*,
https://doi.org/10.1007/978-981-99-2302-1_28

shortcoming that compounds which require metabolic activation by enzymes, such as prodrugs and drug conjugates, could be falsely rejected as being inactive. This is because the cells/tissues employed for in vitro assays may not express the activating enzymes.

Keywords

Experimental models · In vitro · In vivo · Latent tuberculosis · Tuberculosis

28.1 Introduction

Tuberculosis (TB) continues to inflict mankind since time immemorial and has assumed even more significance in recent decades. TB is the number one killer in the world due to a bacterial infection. TB is a deadly disease that has killed more people than any other infectious disease. According to the World Health Organization (WHO), nearly 10 million people were infected and 1.5 million died in the year 2018 alone making TB one of the top 10 causes of death globally. Clinically, pulmonary TB caused by *Mycobacterium tuberculosis* is the most prevalent among non-HIV-positive patients, while *M. tuberculosis* and *M. avium* complex infection occurs in HIV-positive patients (Iacobino et al. 2020).

Despite efforts from various resources, the dream of TB elimination remains a distant reality. To accomplish it, sustainable and affordable programs are needed with anti-TB measures. To accomplish this task, three areas, vaccination, diagnosis, and treatment, need to be explored. Advancing these areas requires a deeper knowledge of host-pathogen interactions and better experimental models are needed. Animal models of TB are important tools for the assessment of the efficacy of vaccines and potential drug candidates as well as the identification and validation of disease biomarkers (Cardona and Williams 2017; Zhan et al. 2017; Bucsan et al. 2019; Gong et al. 2020).

Treatment of TB is compounded by the long duration of treatment (6 months to 2 years) and the side effects of anti-TB drugs. Both factors contribute to low patient compliance resulting in the re-emergence of infection after an initial recession as well as the emergence of multidrug-resistant strains. Therefore, the focus of new drug development has been to develop drugs that reduce treatment duration, have lesser side effects, and are active against multidrug-resistant strains. The long treatment duration is considered largely due to the continuous backflow of latent TB bacilli; hence, drugs active against latent TB bacilli are desirable (Defraine et al. 2018; Cohen et al. 2019). After a lull of over half a century, two new anti-TB drugs were approved—bedaquiline (2012) and delamanid (2013) followed by proteomanid (2019). Additionally, investigational molecules like diarylquinolines, fluoroquinolones, nitroimidazoles, and oxazolidinone are in clinical development with a large proportion being that of oxazolidinone (AZD5847, contezolid/MRX-1, sutezolid, delpazolid). Drug repurposing is yet another approach that has resulted in the identification of linezolid and auranofin as treatments for TB. Pretomanid is a novel compound developed by TB Alliance which has been granted authorization

(Khare et al. 2019). The activity of anti-TB drugs has been shown to depend on the immune status of the host which results in lower drug efficacy in immune-compromised patients. Hence, another approach for treatment has been to stimulate the host immune system for bacterial clearance, either as a standalone therapy or in combination with anti-mycobacterial agents (Ahmad et al. 2010, 2011; Gupta et al. 2012; Zhang et al. 2020). These approaches include immunotherapy with small molecules (Mourik et al. 2017; Bryk et al. 2020; Rao Muvva et al. 2021) or microbes and microbial products (Chaturvedi et al. 1999; Hernandez-Pando et al. 2008; Rodrigues et al. 2015). Drug repurposing of existing drugs is yet another viable alternative for the discovery of immune modulator compounds for TB (Mishra et al. 2018). We have found that low doses of morphine can protect infected mice from TB, the protection being comparable to standard anti-TB drugs (Singh et al. 2008). Bacillus Calmette-Guerin (BCG), the only approved TB vaccine, can prevent childhood TB but is ineffective in adults. Over a dozen vaccine candidates are in various stages of clinical trials but are years away from commercialization (Kaur et al. 2019; Li et al. 2020). Other targets of anti-TB drugs have focused on inducing autophagy in macrophages (Pelaez Coyotl et al. 2020; Rao Muvva et al. 2021), disruption of mycobacterial biofilms (Wang et al. 2019), and use of efflux pump inhibitors to overcome drug resistance (Grossman et al. 2015; Pieterman et al. 2018; Xu et al. 2018).

The mouse has been predominantly used as an experimental model of active and latent TB following intravenous inoculation or inhalation exposure to the mycobacterium. Guinea pig is considered a better model for pulmonary TB but is not frequently used due to the risk of aerosol transmission to persons handling these animals. Monkeys remain the most relevant models as they mimic several aspects of pulmonary and extrapulmonary TB which are not observed in other models. However, their high cost of maintenance, high inter-group variability, and limited availability of immune reagents are major obstacles to their application (Cardona and Williams 2017; Bucsan et al. 2019; Gong et al. 2020).

Despite the health impact of TB, research in TB has remained slow. This sluggish pace can be attributed to two factors: the pathogenicity of the organism and the slow growth rate of the organism. *Mycobacterium tuberculosis* (Mtb) infects macrophages (primarily alveolar macrophages) and adapts to the hostile intracellular milieu due to a variety of defense mechanisms. TB is primarily a disease of the respiratory system and the cycle of TB infection commences with the release of Mtb-carrying aerosols. A dose of 1–10 Mtb dispersed in the air is likely to cause a risk of transmission. Following their entry into the lung, Mtb are phagocytized by alveolar macrophage cells where they may either be completely cleared by the immune reactions or may reside and proliferate in macrophages. Under suitable conditions, Mtb may divide and invade the epithelial cells as well. Experimental studies in TB require biosafety level 3 laboratories, which are costly to develop and maintain making these unaffordable to most microbiology laboratories (Bucsan et al. 2019; Gong et al. 2020). The slow growth of *M. tuberculosis* makes it extremely difficult to perform studies since chances of microbial contamination are increased during the 1–2 months long incubation period required to observe colonies on agar

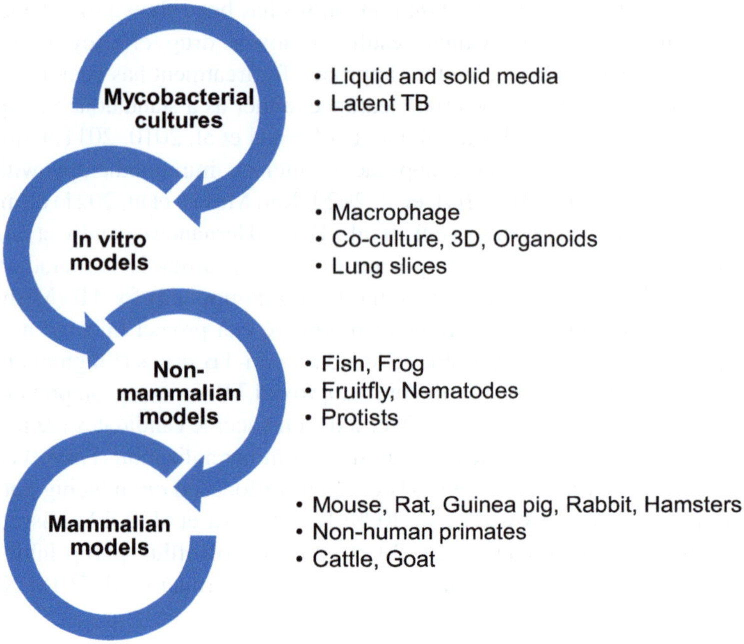

Fig. 28.1 Schematic diagram showing the different experimental models used currently in tuberculosis research

plates. In recent years, models using non-pathogenic and/or fast-growing alternates such as *M. smegmatis* (Jhamb and Singh 2009; Singh et al. 2009; Altaf et al. 2010; Costa et al. 2016; Arthur et al. 2019), *M. fortuitum* (Alim et al. 2017), *M. bovis* BCG (Altaf et al. 2010), *M. aurum* (Gupta et al. 2009; Gupta and Bhakta 2012), and *M. marinum* (Lienard and Carlsson 2017) have been developed to circumvent these issues. The models commonly employed in the laboratory are summarized in Fig. 28.1. Additionally, luminescence- and fluorescence-based methods employing genetically-engineered bacteria expressing luciferase (Zhang et al. 2012; Andreu et al. 2013) or fluorescent proteins (Zelmer et al. 2012; MacGilvary et al. 2019), as well as the use of fluorescent dyes (Amin et al. 2009), have hastened the screening process not only in experimental studies (Durkee et al. 2019) but also antimicrobial sensitivity of clinical isolates for drug therapy decisions (Amin et al. 2009; Cui et al. 2013). Additionally, polymerase chain reaction-based methods have also been evaluated as faster alternates compared to traditional methods based on the colony-forming unit (CFU) counts (Pathak et al. 2012; da Silva et al. 2017; de Knegt et al. 2017).

28.2 In Vivo Models of TB

28.2.1 Mouse Model

The mouse has remained a model of choice to study disseminated and pulmonary TB. Nearly a century ago, murine TB was experimentally induced in mice using bovine strains or BCG (Lewis and Margot 1914; Murphy and Ellis 1914; Grumbach et al. 1967; Collins et al. 1975; Forget et al. 1981), whereas Mtb infection models in mice appeared much later (Youmans and Mc 1945; Martin 1946; Mc et al. 1946; Youmans and Williston 1946). Several inbred strains of mice have been investigated resulting in their classification as susceptible (Balb/c, C57BL/6, B10.A, I/St, SWR) and resistant (C3H/HeCr, A/J, DBA/2, A/Sn) (Pierce et al. 1947; Kelley and Collins 1999; Nikonenko et al. 2000; Turner et al. 2003b); nevertheless, contrasting classification of mouse strains (C3H/HeJ as sensitive and C57BL/6 as resistant) has also appeared in the literature (Chackerian et al. 2001) highlighting the importance of Mtb strain, inoculum size, and route of administration on susceptibility to infection (Actor et al. 1999; Chackerian and Behar 2003). Apart from the inbred strains, outbred strains such as Swiss (Lynch et al. 1965; Lecoeur et al. 1989) and ICR mice (Shkurupy et al. 2020) have also been used which have not been equivocally classified as being susceptible or resistant. The dissemination model requires intravenous injection of millions of CFUs which results in a significant bacterial load in the lungs, liver, and spleen with a small number of bacilli also detectable in other organs and blood. The treatment with test compounds is typically initiated either on the day of inoculation or 24 h post-inoculation and CFU counts in target organs are determined after 1 month of treatment/inoculation (Singh et al. 2008). This model, although convenient for experimental screening of compounds, does not represent the actual pathology of the disease in humans since pulmonary TB is the major manifestation in humans (Cardona et al. 1999). Alternatively, a low-dose aerosol model has been employed which requires exposure of mice to a relatively lower number of bacteria (typically, 50 CFUs) via the inhalation route. The aerosol droplets, owing to their small size, deliver the bacilli in alveoli. After treatment with test compounds, lung CFU counts are determined after 3 or 4 weeks (Kelly et al. 1996; De Groote et al. 2011). A similar approach relies on intratracheal instillation of about a million CFUs in mice which results in the development of aspirating pneumonia but the pathology does not mimic pulmonary TB (Dormans et al. 2004; Eruslanov et al. 2004). Mouse model of pulmonary M. tuberculosis infection exhibits immune responses similar to that observed in humans but the disease characteristics differ significantly. Several pathological hallmarks of TB infection in humans such as caseous necrosis, granulomas, and lung cavitations are not observed in mouse strains (Bucsan et al. 2019). The pathogens traffic intracellularly in murine lungs of commonly used BALB/c and C57BL/6 strains in contrast to observations in DBA/2 and 129/Sv mice. This difference translates into differences in pathological outcomes whereby inflammation ensues in murine models but without development of necrotic lesions (Medina and North 1998; Guirado et al. 2006). On the other hand, the disease is progressive in nature in humans and other experimental models along with development of necrotic lesions with extracellular bacteria. The susceptibility of mice to infection is also impacted by the strain implying a role of genotype. For example, C57BL/6 mice are resistant compared to BALB/c mice, while C3HeB/FeJ mice exhibit development of necrotic granulomas similar to those observed in humans (Driver et al. 2012; Harper et al. 2012; Lee et al. 2018; Moreira-Teixeira et al. 2020). B6.C3H^{sst1} mice exhibit hypoxic lesions (Kramnik 2008), while CBA/J IL-10 knockout mice develop mature fibrotic granulomas (Cyktor et al. 2013; Bucsan et al. 2019). In recent years, humanized mice have been developed which not only mimic human pathology but also enable study of HIV/TB co-infection as well as anti-mycobacterial drug screening in mice (Calderon et al. 2013; Heuts et al. 2013; Nusbaum et al. 2016; Grover et al. 2017; Arrey et al. 2019; Corleis et al. 2019; Gong et al. 2020; Huante et al. 2020). Mouse strains also differ in their response to BCG and consequent protection from TB. Balb/C mice exhibit higher degree of immune response compared to C57BL/6 mice but afforded comparable protection from Mtb infection (Garcia-Pelayo et al. 2015). In another study, the effect of prior BCG vaccination on protection from Mtb aerosol infection has also been compared in susceptible (C3Heb/FeJ) and resistant (C3H/HeOuJ) mouse strains (Henao-Tamayo et al. 2015).

The commonly used mouse strains have been criticized in recent years as an oversimplification of human pathology since the effects of allelic variations on disease pathology as well as treatment and vaccination effects could not be studied. Recently, collaborative cross (CC) and diversity outbred (DO) models have been developed which could be used to study the effects of allelic variations on TB. CC model is a panel of recombinant inbred mouse strains derived from an eight-way cross. Five parental strains included two used in mouse genetics (C57BL/6J and 129S1/SvImJ) and three models of common diseases (A/J, NOD/ShiLtJ, and NZO/HiLtJ), while three founder strains included wild-inbred strains (CAST/EiJ, PWK/PhJand WSB/EiJ) (Churchill et al. 2004; Noll et al. 2019). CC mice have been shown to be susceptible to Mtb infection (Smith et al. 2016, 2019). The diversity outbred (DO) model was obtained from the same eight strains used to obtain the CC model. However, in contrast to the funnel breeding used in the CC model, the DO model was obtained by extensive inbreeding in these strains resulting in outbred DO strains (Churchill et al. 2012). DO model developed at Jackson Laboratories was obtained by using 160 CC mice as founder strains. Pulmonary infection of DO mice with Mtb resulted in super-susceptible, susceptible, and resistant phenotypes (Niazi et al. 2015; Tavolara et al. 2020). BCG vaccination of DO mice followed by aerosol exposure to Mtb also exhibited different intensities of TB infection (Kurtz et al. 2020).

Models of extrapulmonary TB have also been developed in mice representing brain infection. These models typically employ intravenous injection or intratracheal delivery of Mtb strains in Balb/C mice which disseminate to the brain and other organs (van Well et al. 2007; Be et al. 2008; Hernandez Pando et al. 2010; Gupta et al. 2016; Husain et al. 2017). These studies have revealed that Mtb dissemination to the brain is Mtb strain/genotype-dependent. Another model of TB meningitis relies on the intracerebral injection of Mtb which offers two advantages over other models of central nervous system (CNS) infection. First, the infection is localized to the brain unlike in other models where the infection is disseminated to other organs. Second, the Mtb strains which do not cause meningitis in other models can cause brain infection; thus, a very broad range of Mtb strains could be evaluated (van Well et al. 2007). The dissemination model has also been employed to model intraocular (Abhishek et al. 2019; Basu et al. 2020) and musculoskeletal TB in mice (Kager et al. 2014). Another model employs NOS_2−/− mice where intradermal injection of one thousand CFUs in the ear dermis resulted in hypoxia and granuloma formation in the lungs along with significant bacillary load in the spleen (Reece et al. 2010; Kupz et al. 2016).

Latent TB is yet another challenging area that suffers from a lack of suitable models. A mouse model of latent TB called the Cornell mouse model is the most commonly employed model. The original model required infection of mice with Mtb (by intravenous administration) followed by antimicrobial chemotherapy with two drugs for 12 weeks and a rest period of 90 days to obtain detectable CFUs in organs. Several modifications of this model have appeared in literature which vary in the inoculum size of infection, duration between inoculation of mice and commencement of treatment, dose of anti-TB drugs, duration of treatment, and duration of rest period (Scanga et al. 1999). Another model of latent TB requires low-dose aerosol infection in mice followed by a rest period of up to 3 months (Scanga et al. 1999). More recently, a model based on NOS2−/− mice has also been reported (Kupz et al. 2016). Studies in mice have revealed important insights into the persistence of TB suggesting that the microbe can persist in adipose tissue even after clearance from the lungs (Agarwal et al. 2014, 2016; Ayyappan et al. 2019) which is also corroborated by findings in humans (Neyrolles et al. 2006) and rabbits (Ayyappan et al. 2018). Mesenchymal stem cells have also been identified as a home to dormant Mtb in mice (Das et al. 2013; Beamer et al. 2014; Garhyan et al. 2015; Tornack et al. 2017; Fatima et al. 2020; Jain et al. 2020) as well as in humans (Garhyan et al. 2015; Tornack et al. 2017).

Apart from Mtb, several other species of mycobacteria have been used for infection in mice. These studies aimed at either developing short-term models of human TB to decrease the time required for screening of anti-TB compounds or using non-pathogenic strains/species for adoption in non-BSL3 facilities. *M. smegmatis* has been proposed for the screening of antimycobacterial agents in a mouse model (Jhamb and Singh 2009; Singh et al. 2009). *M. smegmatis* has also been employed to understand molecular mechanisms of Mtb pathogenesis (Sha et al. 2017, 2021; Sun et al. 2017; Yang et al. 2017; Li et al. 2019; Guo et al. 2021) as well as expression of proteins for vaccination purposes (Junqueira-Kipnis et al. 2013; Liu et al. 2015a; Kannan et al. 2020; Safar et al. 2020).

Mouse infection models have also been developed to mimic avian (Fujita et al. 2010; Haug et al. 2013; Andrejak et al. 2015; Cha et al. 2015; Bruffaerts et al. 2017; Dong et al. 2017; Babrak and Bermudez 2018) and bovine TB (Logan et al. 2008; Waters et al. 2014; Garcia-Pelayo et al. 2016; Garcia et al. 2020). Mouse models have also provided insights into the role of co-morbidities such as diabetes (Martens et al. 2007; Alim et al. 2017, 2019, 2020) and co-infections such as malaria (Mueller et al. 2012, 2014; Blank et al. 2016a, b), influenza (Florido et al. 2013; Redford et al. 2014; Ring et al. 2019), herpes (Miller et al. 2019), HIV (Nusbaum et al. 2016), and helminth infections (Monin et al. 2015; Rafi et al. 2015; McFarlane et al. 2017) on the progression of TB.

28.2.2 Guinea Pig Model

Guinea pigs were the preferred model for understanding TB pathogenesis and diagnosis as well as drug and vaccine screening (Negre and Bretey 1945; Steenken Jr. and Wagley 1945; Dessau et al. 1949; Steenken Jr. and Pratt 1949; Soltys and Jennings 1950; Wasz-Hockert and Backman 1954; Lithander 1957; Collymore et al. 2018; Williams et al. 2020). Their use in diagnosis has ceased since the introduction of culture medium and other diagnostic tests (Mitchison et al. 1973; Saxena and Sharma 1982; Martin et al. 1989; Smith et al. 1991). Nevertheless, guinea pigs are the second most employed model after mice for drug and vaccine efficacy studies and in understanding disease pathology (Morton 1916; Goyal 1938; Gharpure 1945; Kerr 1946). Guinea pigs exhibit several characteristic features of human TB pathology as observed in humans such as the development of granulomas, caseous necrosis, and secondary lesions after systemic dissemination (Wilkinson and White 1966; Narayanan et al. 1981; Shakila et al. 1999; McMurray 2003; Turner et al. 2003a; Basaraba et al. 2006; Ordway et al. 2007; Via et al. 2008). Although guinea pigs have been addressed as being highly susceptible to TB infection, high CFU counts need to be administered compared to mice. Guinea pigs also do not show any significant observable signs and symptoms of the disease even weeks after the Mtb challenge making this species unsuitable for studies where the death of the animal is a study parameter (Smith et al. 1991; Shakila et al. 1999; Williams et al. 2005). BCG vaccine has been shown to be more protective in guinea pigs compared to mice thereby raising concern that guinea pigs may not be a suitable model to screen vaccines better than BCG (Sugawara et al. 2007; Ly et al. 2008; Cardona and Williams 2017; Gong et al. 2020). Further, the lack of immunological reagents is also an impediment to employing guinea pigs in vaccine screening. Guinea pigs have also been employed to study non-pulmonary TB such as pleuritis (Phalen and McMurray 1993), ocular TB (Rao et al. 2009; Thayil et al. 2011), and central nervous system dissemination (Be et al. 2011) as well as TB in co-morbid conditions (Podell et al. 2014). Although highly virulent strains of Mtb such as H37Rv and Erdman strains (Palanisamy et al. 2008; Li et al. 2010) as well as clinical isolates (Shanley et al. 2013; Aiyaz et al. 2014; Pardieu et al. 2015) have been used, experimental models of *M. bovis* and BCG infection have also been reported in guinea pigs (Aygun et al. 2000; Chambers et al. 2001). Guinea pig model of latent TB has also been reported (Kashino et al. 2008; Klinkenberg et al. 2008; Rifat et al. 2009; Sugawara et al. 2009; Patel et al. 2011) and found to be suitable for study of latent TB.

28.2.3 Non-human Primates

Non-human primates are known to be susceptible to TB and reports have emerged showing the spontaneous spread of infection in wild and captive animals (Schroeder 1938). These include rhesus macaques (Lindsey and Melby Jr. 1966), stump-tailed macaques (Wolf et al. 1967; Indzhiia et al. 1977), squirrel monkeys (Chrisp et al. 1968; Hessler and Moreland 1968; da Silva et al. 2017), spider monkeys (Rocha et al. 2011), cebus monkeys or capuchins (Leathers and Hamm Jr. 1976; Broncyk and Kalter 1980; Ehlers et al. 2020), owl monkeys (Bone and Soave 1970; Snyder et al. 1970), pig-tailed monkeys (Sedgwick et al. 1970; Lau et al. 1972; Stockinger et al. 2011; Engel et al. 2012), lemur (Knezevic and McNulty 1967), langurs (Plesker et al. 2010), baboons (Heywood et al. 1970; Broncyk and Kalter 1980; Fourie and Odendaal 1983; Martino et al. 2007; Wolf et al. 2016), chimpanzees (Chaparas et al. 1970; Broncyk and Kalter 1980; Coscolla et al. 2013; Wolf et al. 2016), mandrills (Amado et al. 2006), grivet (Broncyk and Kalter 1980), marmosets (Broncyk and Kalter 1980; Via et al. 2013), and several species of New World monkeys (Alfonso et al. 2004; Rosenbaum et al. 2015). Non-human primates genetically resemble humans due to evolutionary proximity and hence, exhibit characteristic hallmarks of human TB. These characteristics include the development of caseous necrosis, granulomas, and dissemination of pulmonary TB to other organs (Via et al. 2008; Mattila et al. 2013, 2017; Pacheco et al. 2013; Dutta et al. 2014b; Marino et al. 2015; Esaulova et al. 2020; Wessler et al. 2020). Apart from the characteristic pulmonary pathology, several non-human primates have also been found to exhibit non-pulmonary manifestations of TB such as hepatic (Stockinger et al. 2011), spinal (Martin et al. 1968; Fox et al. 1974), cerebral (Machotka et al. 1975), cutaneous (Bellinger and Bullock 1988) and ocular (West et al. 1981) as well as infection by other species of mycobacteria, including non-tuberculous mycobacteria (Smith et al. 1973; Renner and Bartholomew 1974; Latt 1975; Sesline et al. 1975; Sapolsky and Else 1987; Brammer et al. 1995; Alfonso et al. 2004; Henrich et al. 2007; Chege et al. 2008; Parsons et al. 2010; Wachtman et al. 2011; Via et al. 2013; Rahim et al. 2017; Min et al. 2018). Further, immunological reagents targeted towards human proteins show reactivity with NHP proteins, and vice versa, due to the high degree of sequence and structural homology. Additionally, co-infection with simian immunodeficiency virus also mimics HIV/TB co-infection (Kuroda et al. 2018) and has been employed to study the effect of antiretroviral therapy on active and latent TB progression (Ganatra et al. 2020; Sterling and Lin 2020). However, the high cost of procurement and maintenance along with stringent ethical protocols restrict the use of NHPs to very few laboratories (Gong et al. 2020). Several species of NHP have been investigated as models for screening of anti-TB compounds as well as vaccines but the major species include cynomolgus macaques (*Macaca fascicularis*) (Marino et al. 2004; Dutta et al. 2014b; Tsujimura et al. 2020; Winchell et al. 2020) and rhesus macaques (*Macaca mulatta*) (Fremming et al. 1957; Pacheco et al. 2013; Rayner et al. 2013; Gong et al. 2020; Sterling and Lin 2020). Significant differences between the two species have been reported with regard to TB susceptibility and response to vaccination. Rhesus macaques have been found to be more susceptible to the development of active TB but BCG vaccination showed poor protection in this species compared to cynomolgus macaques (Langermans et al. 2001). The higher susceptibility of rhesus macaques to develop active TB, compared to cynomolgus macaques, has been attributed to differences in innate immune responses (Maiello et al. 2018; Dijkman et al. 2019) and monocyte: lymphocyte ratios in the two species (Sibley et al. 2019). Further, mutations in the natural resistance-associated macrophage protein 1 (NRAMP1) gene have been linked to differences in intraspecies susceptibility to TB in rhesus macaques (Deinard et al. 2002). The role of the route of administration on disease pathology has also been demonstrated: a uniform disease was obtained following aerosol exposure, while bronchoscopic instillation resulted in disease localized at the instillation site (Sibley et al. 2016). In contrast to rhesus macaques which develop active TB, cynomolgus macaques have been found to develop latent TB following low-dose pulmonary delivery of Mtb. These macaques remain asymptomatic, with no clinical manifestations in chest radiography, but show positive tuberculin tests after at least 6 months of Mtb administration (Walsh et al. 1996; Capuano et al. 2003; Lin et al. 2006; Flynn et al. 2015; Gideon et al. 2015; Sharpe et al. 2016). Nevertheless, a model of asymptomatic TB has also been described in rhesus macaques (Gormus et al. 2004; Lin et al. 2009). The macaque model has also been used to study the reactivation of latent TB in SIV-TB co-infection models (Diedrich et al. 2020; Ganatra et al. 2020) as well as identify biochemical and cellular markers in latent TB (Esaulova et al. 2020). More recent studies using PET-CT (Coleman et al. 2014a, b; Lin et al. 2016; Stammes et al. 2021), serial intravascular staining (Potter et al. 2021), in silico/mathematical models (Marino et al. 2016; Marino and Kirschner 2016; Pienaar et al. 2016; Sershen et al. 2016; Evans et al. 2020), omics studies (Mehra et al. 2010; Kunnath-Velayudhan et al. 2012; Luo et al. 2014; Gideon et al. 2016; Javed et al. 2016; Pienaar et al. 2016; Hudock et al. 2017; Martin et al. 2017; Thompson et al. 2018; Duffy et al. 2019; Ault et al. 2020), and other methods have been found to be useful in the study of TB pathogenesis in macaques (Lewinsohn et al. 2006; Lerche et al. 2008; Sharpe et al. 2009; Hudock et al. 2014; Pena and Ho 2016).

28.2.4 Other Mammalian Models

Rabbits have been employed in TB for a long time. The severity and nature of the infection have been attributed to the strain of the infecting organism as well as the rabbit strain employed (Dorman et al. 2004; Subbian et al. 2013a; Tsenova et al. 2020). Following aerosol challenge with *M. bovis*, rabbits exhibit several characteristics of human disease such as cavitation and granuloma formation (Via et al. 2008; Subbian et al. 2013b; Gong et al. 2020) as well as extrapulmonary dissemination (Nedeltchev et al. 2009). Notably, most of the rabbit strains are not susceptible to common human virulent strains (Gong et al. 2020); however, these strains could produce pulmonary lesions (Bishai et al. 1999; Manabe et al. 2003). The rabbit model has also been modified to study extrapulmonary TB such as meningitis (Tsenova et al. 2005, 2007; Tucker et al. 2016; O'Brien et al. 2020), spinal TB (Geng et al. 2015; Liu et al. 2015b) and bladder TB (Liu et al. 2015b). Imaging studies have demonstrated localization of administered drugs in pulmonary necrotic lesions thus providing a pharmacokinetic basis for the comparison of drug activity (Kjellsson et al. 2012; Via et al. 2012; Pienaar et al. 2017; Blanc et al. 2018a, b; Rifat et al. 2018; Tucker et al. 2018; Sarathy et al. 2019). A skin infection model has recently been reported in rabbits to assess the virulence of mycobacterial strains and liquefaction potential (Zhang et al. 2010; Sun et al. 2012). The rabbit model has also been investigated for the study of latent TB but the model has not been extensively studied (Manabe et al. 2008; Kesavan et al. 2009; Subbian et al. 2012, 2013b).

The earliest report on a study of TB in rats is over a century old (Bodkin 1918); however, their use in the assessment of drug effects on the course of TB was studied several years later (Smith et al. 1946a, b; Scheid and Mendheim 1949; Michael Jr. et al. 1950; Cummings et al. 1952; Grumbach 1960). Despite their early applications, rats were not extensively investigated as a model of TB. In recent years, several strains and species of rat have been employed such as Fischer rats (Sugawara et al. 2004a), Lewis rats (Sugawara et al. 2004b), Sprague-Dawley rats (Li et al. 1998), Wistar rats (Gaonkar et al. 2010; Singhal et al. 2011a, b), cotton rats (Daigeler 1952; Elwood et al. 2007; McFarland et al. 2010), vole rats (Jespersen 1974) and others (Sugawara et al. 2004c, 2006; Clarke et al. 2007; Sugawara and Mizuno 2008). Nevertheless, preliminary studies have demonstrated the formation of granulomas and pulmonary lesions in rats (McFarland et al. 2010; Heng et al. 2011). The application of rats in studying the effects of vaccines has been a recent development with preliminary studies indicating their utility in screening vaccines (McFarland et al. 2010; Singhal et al. 2011b; Cardona and Williams 2017; Gong et al. 2020). Rats offer additional advantages compared to mice such as the ability to collect multiple blood samples which makes them an attractive alternative to mice for pharmacokinetic studies (Kumar et al. 2014).

Apart from studying the effect of drugs and vaccines, rats have also been investigated for understanding the role of co-morbidities in TB such as diabetes (Sugawara and Mizuno 2008) and silicosis (Dong et al. 2014). Rats have also been investigated in the diagnosis of TB such as cotton rats (*Sigmodon hispidus hispidus*) (Daigeler 1952). African pouched rats have been studied for the olfactory detection of TB in clinical samples. Pouched rats were found to be more sensitive than smear microscopy in the detection of Mtb (Mahoney et al. 2012; Mgode et al. 2012; Ellis et al. 2017; Mulder et al. 2017; Webb et al. 2020).

Hamsters have also been investigated as a model for the study of human and bovine TB pathology (Steenken Jr. and Wagley 1945; Glover 1946; Dennis and Gaboe 1949; Rozenberg and Pisarenko 1965). Pulmonary infection of hamsters has been shown to exhibit tubercle formation and the pathological outcome was dependent on a diet (Ratcliffe and Palladino 1953; Merrick and Ratcliffe 1957). The cheek pouch has also been used as an inoculation site that exhibits granulomatous lesions (de Arruda and Montenegro 1995). Hamsters have also been used to study the antimycobacterial effects of compounds (Rozenberg and Pisarenko 1965; Gupta and Mathur 1969; Righi et al. 1999; Ugaz et al. 1999; Domingues-Junior et al.

2000; Palermo-Neto et al. 2001) as well as the effect of BCG on TB progression (Viallier and Cayre 1955; Rozenberg and Pisarenko 1965). Hamsters, like guinea pigs, have also been investigated in the diagnosis of TB but are not extensively used (Hussel 1951; Eskuchen 1952; Starck and Viehmann 1955). Minipigs have also been investigated as a model of pulmonary Mtb infection which exhibits characteristics of human pulmonary lesions such as granuloma formation (Gil et al. 2010; Ramos et al. 2017).

Several other mammalian models have been developed to model TB in wild animals and cattle (Palmer et al. 2012; Reis et al. 2020). These models typically rely on the induction of *M. bovis* infection in animals such as badgers, boars, deer, and possums (Palmer et al. 2012; Reis et al. 2020). These animals act as reservoirs of TB in the wild and play a key role in the spread of TB in wild animals and domesticated cattle (Fulford et al. 2002; Corner et al. 2003; Green et al. 2008; Fenwick 2012; Donnelly and Nouvellet 2013; Nugent et al. 2015; Sichewo et al. 2020). Experimental models of *M. bovis* infections have been developed in badger (Lesellier et al. 2008; Gormley and Corner 2017; Queiros and Vicente 2018), boar (Naranjo et al. 2006; Ballesteros et al. 2009; Gasso et al. 2016; Lopez et al. 2016), deer (Palmer et al. 1999; Mackintosh et al. 2000; Waters et al. 2003; Stringer et al. 2011) and possum (Dennis and Gaboe 1949; Skinner et al. 2002; Cooke et al. 2003; Nugent et al. 2013a, b; Rouco et al. 2016) which have provided insights into disease pathology progression, transmission as well as effects of vaccination on disease control. A model of aerosol infection has also been described in ferrets as a replacement for the badger model (McCallan et al. 2011). Apart from these reservoirs of infection, models of *M. bovis* infection have also been reported in goats (Schinkothe et al. 2016a, b), buffalo (De Klerk et al. 2006) and cattle (Kao et al. 1997, 2007; Joardar et al. 2002; Palmer et al. 2002; Griffin et al. 2006; Rodgers et al. 2007). Additionally, the *M. caprae* infection model has also been reported in goats (Bezos et al. 2010; de Val Perez et al. 2011). Cattle have been reported to be resistant to Mtb (Whelan et al. 2010) but *M. bovis* infection in cattle has been proposed as an alternate model for human TB for evaluating the effect of drugs and vaccines (Dean et al. 2008; Van Rhijn et al. 2008; Waters et al. 2014).

28.2.5 Fish and Other Models

Zebrafish infection with *M. marinum* has been a subject of considerable interest in recent years. *M. marinum* induces granuloma formation in zebrafish which resembles lung granulomas in humans (Prouty et al. 2003; Swaim et al. 2006; Davis and Ramakrishnan 2009; Carvalho et al. 2011; Cheng et al. 2020). The investigation of mechanisms of granuloma formation in zebrafish has provided important insights into the mechanisms operable in humans, including mechanisms operable in presence of co-morbidities (Benard et al. 2016; Kenyon et al. 2017; Bouz and Al Hasawi 2018; Johansen et al. 2018; Luukinen et al. 2018; Harjula et al. 2020; Oehlers et al. 2020; Hosseini et al. 2021). The optically transparent adult zebrafish and embryos allow easy visualization of disease progression while also allowing studies with large sample sizes due to the low cost of maintenance as well as the ability to conduct studies in BSL2 facilities (Myllymaki et al. 2016; Sommer and Cole 2019; Cheng et al. 2020; Gong et al. 2020; Hogset et al. 2020). The zebrafish model has been employed for the screening of anti-TB compounds and candidate vaccines (Oksanen et al. 2013; Lopez et al. 2018; Risalde et al. 2018; Sommer and Cole 2019; Commandeur et al. 2020; Nie et al. 2020; Saralahti et al. 2020; van Wijk et al. 2020). Genetically engineered zebrafish, expressing drug-metabolizing enzymes, has also been employed for studying the activity of anti-TB prodrugs (Ho et al. 2021). The zebrafish model has also been extended to study ocular (Takaki et al. 2018) and latent TB (Parikka et al. 2012) as well as tuberculous meningitis (van Leeuwen et al. 2014; Chen et al. 2018). In vivo models of *M. marinum* infection have also been described in goldfish (Ruley et al. 2002; Hodgkinson et al. 2012) and medaka (Broussard and Ennis 2007; Broussard et al. 2009) as well as in vitro models employing a carp cell line (El-Etr et al. 2001).

M. marinum infection model has also been proposed in frogs which results in granuloma formation (Ramakrishnan and Falkow 1994; Ramakrishnan et al. 1997; Cosma et al. 2006; Rhoo et al. 2019). Frog tadpoles are resistant to infection compared with adults (Rhoo et al. 2019) but tadpoles exhibit immune responses against mycobacteria similar to those observed in mammals (Hyoe and Robert 2019). In vitro studies using frog macrophages have demonstrated contrasting roles of cytokines in susceptibility to *M. marinum* infection (Popovic et al. 2019). Additionally, *M. marinum* infection model has also been described in the fruit fly (Dionne et al. 2003; Oh et al. 2013; Pushkaran et al. 2019), silkworm (Yagi et al. 2017, 2021), nematodes (Lopez Hernandez et al. 2015; Galbadage et al. 2016), and protists (Solomon et al. 2003; Andersson et al. 2006; Hagedorn and Soldati 2007; Arafah et al. 2013; Kolonko et al. 2014; Sanchez-Hidalgo et al. 2017; Trofimov et al. 2018). *Galleria mellonella* larvae have also been reported to be susceptible to a wide range of mycobacteria, including Mtb, (Asai et al. 2019b, 2020; Budell et al. 2020) and have been employed for screening antimycobacterial compounds (Entwistle and Coote 2018; Asai et al. 2019a). Models of avian TB have also been described in chick and quail (Chaudhuri et al. 1980; Tell et al. 2003).

28.3 In Vitro Models

Mycobacteria reside in macrophages and dendritic cells; hence, Mtb-infected macrophages have been frequently used as in vitro models to study drug activity and molecular aspects of pathology (Chingwaru et al. 2016; Keiser and Purdy 2017; Pi et al. 2019). Alveolar macrophages are the primary target for pulmonary TB (Cardona et al. 2003; Cohen et al. 2018) while hepatic and splenic macrophages are targets for systemic infection (Ozeki et al. 2006; Sivangala Thandi et al. 2020). Although primary macrophages obtained from the lungs, liver, and spleen appear to be obvious choices for in vitro studies, their application (particularly, alveolar and hepatic macrophages) is thwarted by low abundance and difficulty in the isolation of pure cell types. Splenic macrophages can be obtained in large quantities but exhibit much lower phagocytic activity compared to alveolar or hepatic macrophages (Guirado et al. 2013). These problems have resulted in a search for more convenient and representative sources of macrophages. Bone-marrow-derived macrophages have found particular interest in this regard as the precursor cells can be obtained in large amounts and can be differentiated into desired cell types using cytokines or conditioned culture media (Keiser and Purdy 2017). However, the high cost of cytokines could be a limiting factor. Peritoneal macrophages are yet another model which has been frequently employed for decades. The naïve/unelicited macrophages are obtained in relatively lower amounts but their numbers can be increased by eliciting the mice with chemicals. The yield of elicited macrophages is several folds higher compared to unelicited macrophages thereby reducing the number of animals required for experimentation. In recent years, several cell lines of murine alveolar macrophage origin have been developed. The cell lines offer several advantages over primary cells such as a virtually unlimited supply of cells with uniformity in genetic, biochemical, and physiological characteristics. Primary macrophages, as well as cell lines derived from a variety of cell lineages from mice (Chingwaru et al. 2016; Andreu et al. 2017), rat (Weikert et al. 2000; Hino et al. 2005; Markova et al. 2005; Hirota et al. 2010) and other animals, have been investigated as in vitro models for the study of Mtb-cell interactions (El-Etr et al. 2001; Hino et al. 2005; Keiser and Purdy 2017).

Human alveolar, hepatic and splenic macrophages are difficult to obtain due to ethical reasons; however, in recent years, these have become commercially available but their cost remains a major stumbling block (Henao et al. 2007). Human peripheral blood mononuclear cells (PBMCs) are relatively much easier to obtain, technically as well as ethically, and have also been widely used. These cells are differentiated into macrophages using cytokines or human serum and can then be used for infection with mycobacteria (Duque et al. 2014; Zhang et al. 2018). Several cell lines of human origin have also been used—the THP-1 monocytic cell line is the most frequently used. This cell line can be differentiated into macrophages by treatment with phorbol myristate acetate and then used for Mtb infection (Bai et al. 2010; Mendoza-Coronel and Castanon-Arreola 2016).

Under physiological conditions, macrophages phagocytose the mycobacteria while cytokines released by macrophages and T-cells contribute to macrophage activation and subsequent killing of the intracellular bacteria. The macrophage infection model is considered relevant for in vitro screening of anti-TB activity of test compounds since the ability of the test compound to cross biological membranes (plasma and phagosomal membranes of host and mycobacterial cell membrane) and exert activity in a biological relevant milieu can be determined (Clemens et al. 2019). However, this model is an oversimplification of the immune response and macrophage-T-cell co-cultures have been used to decipher the molecular basis of crosstalk between these cell types (Skinner et al. 1997; Lyadova et al. 1998; Gautam et al. 2018). As an alternative, Mtb has been incubated in whole blood to determine immune responses as well as study drug effects (Al-Attiyah et al. 2006; Newton et al. 2011; Raposo-Garcia et al. 2017; Cross et al. 2019; Kwan et al. 2020).

Three-dimensional culture and organoids have attracted immense interest in recent years since these methods are more closely related to in vivo conditions and have been successfully employed for drug and vaccine screening. A 3D model employing human PBMCs in an extracellular matrix has been shown to mimic human granulomas and found relevant as a model of latent TB (Crouser et al. 2017). Similar 3D models have been employed which either contain a single cell type or co-culture of macrophages with other cell types to mimic lung tissue or granuloma (Braian et al. 2015; Benmerzoug and Quesniaux 2017; Tezera et al. 2017a, b; Palucci et al. 2019; Thacker et al. 2020; Walter et al. 2020). Additionally, precision-cut lung slices and other models have also been reported to study disease pathology and drug effects (Carranza-Rosales et al. 2017; Carius et al. 2020). The hollow fiber system was developed over a decade ago and approved by European Medicines Agency to study as an in vitro model for pharmacokinetic/pharmacodynamic studies (Cavaleri and Manolis 2015). This method has been successfully employed for pharmacokinetic/pharmacodynamic studies involving single-drug or multi-drug regimens (Gumbo et al. 2015a, b; Pasipanodya et al. 2015; Srivastava et al. 2016; Kloprogge et al. 2019; Pieterman et al. 2021) including the ability to extend the results to children (Srivastava et al. 2020). Additionally, microfluidic systems have also been developed to study environmental milieu and signaling in granulomas as well as a study of drug resistance (Bielecka et al. 2017; Berry et al. 2020).

28.4 Mycobacterial Cultures

The field of mycobacterial culture has witnessed a steady improvement in terms of the development of culture media and detection methods. These methods are useful in detecting direct-acing anti-TB compounds but are irrelevant for indirect-acting compounds (such as immunomodulators) or those requiring metabolic activation (such as prodrugs) as well as vaccines.

In the case of compounds that act both directly and indirectly, the anti-TB activity determined by these methods is expected to be much lower compared to that observed under in vivo conditions. Currently, Middlebrook 7H9 broth is the liquid medium of choice while Middlebrook 7H10 medium and Middlebrook 7H11 medium are commonly employed solid media for experimental purposes. On the other hand, the Lowenstein-Jensen medium is the preferred solid medium for isolating Mtb from clinical samples. Mtb is a slow-growing bacterium with a doubling time of approximately 20 h which requires 1–2 months of incubation for CFU determination. Therefore, broth media are commonly employed to study the anti-TB activity of test compounds whereby mycobacterial growth is determined using turbidimetry or optical absorbance (Franzblau et al. 2012; Parish 2020). These methods, although widely used traditionally, suffer from low reliability in quantifying live bacteria due to possible interference by cell debris and have hence been superseded by dye-based methods. Colorimetric dyes such as Alamar blue or MTT are converted to fluorescent or colored products by viable bacteria. The metabolic conversion of dye, and resulting fluorescence/color intensity, is proportional to the number of viable bacteria (Amin et al. 2009; Cui et al. 2013). This provides the advantage that these methods could be adopted to high throughput formats, requires extremely small amounts of test compounds, and decreases operator exposure to pathogenic strains due to automation.

Commercially available radiometric systems such as BACTEC were employed which provided results relatively faster but have been replaced with fluorescence detection systems like MGIT in recent years due to concerns arising out of the use of radioactive media components (Franzblau et al. 2012; Jhamb et al. 2014). Additionally, luminescent methods based on the assay of ATP or luciferase-expressing Mtb have also been employed (Idh et al. 2017; Parish 2020).

A total of 80 amide derivatives had been tested for their anti-TB activity against metabolically active M. tuberculosis H37Rv. Out of this 34 compounds were found active at 6.25 µg/mL concentration and 11 of them were further tested for MIC determination. MIC values of these compounds ranged between ≤0.39 µg/mL and 6.25 µg/mL (unpublished data). Daily percent growth inhibitions values of all the compounds were evaluated in comparison to standard anti-TB drugs INH and rifampin. Further, 19 more compounds for MIC determination against Mycobacterium tuberculosis H37Rv were tested. All of these were tested at five different concentrations (0.39, 0.78, 1.56, 3.125, and 6.25 µg/mL) for MIC determination. Out of these, six compounds showed MIC values of ≤0.39 µg/mL (unpublished data). The rest of the compounds exhibited MIC values between 0.39 and 6.25 µg/mL. Day-wise percent growth inhibition by compounds was also studied to know the possible mode of action. MGIT 960 TB system was also established in our lab for anti-TB drug susceptibility testing. A total of 15 active compounds were tested at a concentration of 6.25 µg/mL by the MGIT- 960 method (unpublished data). All the 15 compounds were found to be active by MGIT as well as BACTEC 460 methods at a concentration of 6.25 µg/ml and the results correlated well with both the methods. A total of 16 Indian isolates were collected from different institutions in India. The isolates were tested by the BACTEC method against standard anti-TB drugs INH, rifampin, streptomycin, and ethambutol at critical drug concentrations of 0.1, 2.0, 2.0, and 2.5 µg/mL, respectively. All the isolates were sensitive to rifampin whereas some isolates were resistant to INH which is one of the two critical first-line anti-TB drugs. A total of 15 compounds were tested at 6.25 µg/mL concentration against one of the isolates which were resistant to INH. Interestingly three compounds were found to be inactive against this isolate whereas 12 compounds were active against this isolate which was resistant to INH. Intra-macrophage anti-TB activity determination against *M. tuberculosis* (in mouse non-activated peritoneal macrophages) was also established and MIC of standard anti-TB drugs was determined using this assay (our unpublished data).

In vitro mycobacterial cultures have also been investigated for the study of latent TB phenotypes. These models aim at mimicking the conditions observed in granulomas such as hypoxia (Aly et al. 2006; Harper et al. 2012; Dutta et al. 2014a) and nutrient starvation (Sarathy et al. 2018; Yuan and Sampson 2018) to induce a latent phenotype in Mtb (Via et al. 2008). The Wayne model was one of the earliest models described whereby bacteria are cultured in sealed containers. Cessation of aeration in the culture results in a decrease in dissolved oxygen concentration resulting in a shift towards hypoxia. After an extended duration of growth arrest, the bacteria could re-enter logarithmic growth if the cultures are aerated. The dormant stage under hypoxic conditions has been termed non-replicating persistence (NRP) and two distinct stages of NRP have been identified. NRP stage I, also described as microaerophilic, is reached when oxygen saturation decreases to 1% and is characterized by growth arrest, steady ATP levels, and increased glycine dehydrogenase production. As oxygen falls below 0.06% saturation, the bacteria enter anaerobic conditions, termed NRP stage II, which is characterized by a decrease in glycine dehydrogenase (Wayne and Hayes 1996; Wayne 2001). NRP stage II exhibits a reversal in the antimicrobial activity of metronidazole whereby the drug shows bactericidal activity in NRP stage II bacilli but is ineffective in aerobically growing bacilli. Based on the Wayne model, hypoxic resazurin reduction assay, as well as MTT assay, has been developed which enables high throughput screening of drugs against latent TB (Martin et al. 2006; Meinzen et al. 2016). A luciferase reporter has also been used to monitor bacterial growth using a protocol similar to the Wayne model. In this method, termed low oxygen recovery assay, the luminescence readout has been used to study the activity of drugs. A red fluorescent protein-expressing Mtb has also been used whereby the reporter protein expression could be monitored to determine the stage of bacterial growth (Sohaskey and Voskuil 2015; Gibson et al. 2018). In order to hasten and improve readout, the BACTEC method has also been employed to determine persisters (Kharatmal et al. 2009). Models based on hypoxia-induced dormancy have been most frequently employed; however, others in vitro models such as nutrient deprivation and selective carbon sources have also attracted attention in recent years (Patel et al. 2011; Gibson et al. 2018; Parish 2020). Additionally, nitric oxide and streptomycin have also been used as stressors to induce NRP-like conditions under in vitro conditions. A multi-stress model employing low oxygen and low pH has also been reported (Patel et al. 2011; Gibson et al. 2018; Parish 2020).

We evaluated the in vitro efficacy of satranidazole, a novel nitroimidazole-based bioreductive prodrug, against non-replicating persistent latent *M. tuberculosis* under oxygen depletion and nutrient starvation models/conditions. It exhibited a concentration-dependent effect (2–50 μg/mL) in both models; however, the maximum effect was observed at 50 μg/mL. Moreover, it showed statistically significant activity as compared to metronidazole. However, at lower concentrations (<10μg/mL), no significant difference was observed between satranidazole and metronidazole. In conclusion, the noteworthy activity of satranidazole against latent *M. tuberculosis* makes it an attractive drug candidate to target latent tuberculosis. Nevertheless, further detailed investigations, along these lines, using suitable animal models of latent tuberculosis are warranted.

Acknowledgments We thank the Department of Pharmaceuticals, Ministry of Chemical and Fertilizers, Govt. of India, New Delhi, and Indo-European Union, Seventh Framework Programme (FP-7) for the financial support.

References

Abhishek S, Ryndak MB, Choudhary A et al (2019) Transcriptional signatures of mycobacterium tuberculosis in mouse model of intraocular tuberculosis. Pathog Dis 77:45

Actor JK, Olsen M, Jagannath C, Hunter RL (1999) Relationship of survival, organism containment, and granuloma formation in acute murine tuberculosis. J Interf Cytokine Res 19:1183–1193

Agarwal P, Khan SR, Verma SC, Beg M, Singh K, Mitra K et al (2014) Mycobacterium tuberculosis persistence in various adipose depots of infected mice and the effect of anti-tubercular therapy. Microbes Infect 16:571–580

Agarwal P, Pandey P, Sarkar J, Krishnan MY (2016) Mycobacterium tuberculosis can gain access to adipose depots of mice infected via the intra-nasal route and to lungs of mice with an infected subcutaneous fat implant. Microb Pathog 93:32–37

Ahmad Z, Pinn ML, Nuermberger EL, Peloquin CA, Grosset JH, Karakousis PC (2010) The potent bactericidal activity of streptomycin in the guinea pig model of tuberculosis ceases due to the presence of persisters. J Antimicrob Chemother 65:2172–2175

Ahmad Z, Fraig MM, Pinn ML, Tyagi S, Nuermberger EL, Grosset JH et al (2011) Effectiveness of tuberculosis chemotherapy correlates with resistance to mycobacterium tuberculosis infection in animal models. J Antimicrob Chemother 66:1560–1566

Aiyaz M, Bipin C, Pantulwar V, Mugasimangalam R, Shanley CA, Ordway DJ et al (2014) Whole genome response in guinea pigs infected with the high virulence strain mycobacterium tuberculosis tt372. Tuberculosis 94:606–615

Al-Attiyah R, El-Shazly A, Mustafa AS (2006) Assessment of in vitro immunity to mycobacterium tuberculosis in a human peripheral blood infection model using a luciferase reporter construct of m. Tuberculosis h37rv. Clin Exp Immunol 145:520–527

Alfonso R, Romero RE, Diaz A, Calderon MN, Urdaneta G, Arce J et al (2004) Isolation and identification of mycobacteria in new world primates maintained in captivity. Vet Microbiol 98: 285–295

Alim MA, Sikder S, Bridson TL, Rush CM, Govan BL, Ketheesan N (2017) Anti-mycobacterial function of macrophages is impaired in a diet induced model of type 2 diabetes. Tuberculosis 102:47–54

Alim MA, Sikder S, Sathkumara H, Kupz A, Rush CM, Govan BL et al (2019) Dysregulation of key cytokines may contribute to increased susceptibility of diabetic mice to mycobacterium bovis bcg infection. Tuberculosis 115:113–120

Alim MA, Kupz A, Sikder S, Rush C, Govan B, Ketheesan N (2020) Increased susceptibility to mycobacterium tuberculosis infection in a diet-induced murine model of type 2 diabetes. Microbes Infect 22:303–311

Altaf M, Miller CH, Bellows DS, O'Toole R (2010) Evaluation of the mycobacterium smegmatis and bcg models for the discovery of mycobacterium tuberculosis inhibitors. Tuberculosis 90: 333–337

Aly S, Wagner K, Keller C, Malm S, Malzan A, Brandau S et al (2006) Oxygen status of lung granulomas in mycobacterium tuberculosis-infected mice. J Pathol 210:298–305

Amado A, Albuquerque T, Goncalves A, Duarte E, Botelho A, Fernandes T et al (2006) Tuberculosis in mandrills at the lisbon zoo. Vet Rec 159:643

Amin AG, Angala SK, Chatterjee D, Crick DC (2009) Rapid screening of inhibitors of mycobacterium tuberculosis growth using tetrazolium salts. Methods Mol Biol 465:187–201

Andersson JO, Hirt RP, Foster PG, Roger AJ (2006) Evolution of four gene families with patchy phylogenetic distributions: influx of genes into protist genomes. BMC Ecol Evol 6:27

Andrejak C, Almeida DV, Tyagi S, Converse PJ, Ammerman NC, Grosset JH (2015) Characterization of mouse models of mycobacterium avium complex infection and evaluation of drug combinations. Antimicrob Agents Chemother 59:2129–2135

Andreu N, Zelmer A, Sampson SL, Ikeh M, Bancroft GJ, Schaible UE et al (2013) Rapid in vivo assessment of drug efficacy against mycobacterium tuberculosis using an improved firefly luciferase. J Antimicrob Chemother 68:2118–2127

Andreu N, Phelan J, de Sessions PF, Cliff JM, Clark TG, Hibberd ML (2017) Primary macrophages and j774 cells respond differently to infection with mycobacterium tuberculosis. Sci Rep 7: 42225

Arafah S, Kicka S, Trofimov V, Hagedorn M, Andreu N, Wiles S et al (2013) Setting up and monitoring an infection of dictyostelium discoideum with mycobacteria. Methods Mol Biol 983:403–417

Arrey F, Lowe D, Kuhlmann S, Kaiser P, Moura-Alves P, Krishnamoorthy G et al (2019) Humanized mouse model mimicking pathology of human tuberculosis for in vivo evaluation of drug regimens. Front Immunol 10:89

Arthur PK, Amarh V, Cramer P, Arkaifie GB, Blessie EJS, Fuseini MS et al (2019) Characterization of two new multidrug-resistant strains of mycobacterium smegmatis: tools for routine in vitro screening of novel anti-mycobacterial agents. Antibiotics 8:4

Asai M, Li Y, Khara JS, Gladstone CA, Robertson BD, Langford PR et al (2019a) Use of the invertebrate galleria mellonella as an infection model to study the mycobacterium tuberculosis complex. J Vis Exp 2019:148

Asai M, Li Y, Khara JS, Robertson BD, Langford PR, Newton SM (2019b) Galleria mellonella: an infection model for screening compounds against the mycobacterium tuberculosis complex. Front Microbiol 10:2630

Asai M, Li Y, Spiropoulos J, Cooley W, Everest D, Robertson BD et al (2020) A novel biosafety level 2 compliant tuberculosis infection model using a deltaleuddeltapancd double auxotroph of mycobacterium tuberculosis h37rv and galleria mellonella. Virulence 11:811–824

Ault RC, Headley CA, Hare AE, Carruthers BJ, Mejias A, Turner J (2020) Blood rna signatures predict recent tuberculosis exposure in mice, macaques and humans. Sci Rep 10:16873

Aygun C, Ozen H, Kocagoz T, Saribas Z, Aki T, Tekin I (2000) Induction of mycobacteremia by intravesical bacillus calmette guerin instillation in an experimental animal model and detection with polymerase chain reaction. J Urol 163:1588–1590

Ayyappan JP, Vinnard C, Subbian S, Nagajyothi JF (2018) Effect of mycobacterium tuberculosis infection on adipocyte physiology. Microbes Infect 20:81–88

Ayyappan JP, Ganapathi U, Lizardo K, Vinnard C, Subbian S, Perlin DS et al (2019) Adipose tissue regulates pulmonary pathology during tb infection. MBio 10:e02771

Babrak L, Bermudez LE (2018) Response of the respiratory mucosal cells to mycobacterium avium subsp. Hominissuis microaggregate. Arch Microbiol 200:729–742

Bai X, Kim SH, Azam T, McGibney MT, Huang H, Dinarello CA et al (2010) Il-32 is a host protective cytokine against mycobacterium tuberculosis in differentiated thp-1 human macrophages. J Immunol 184:3830–3840

Ballesteros C, Garrido JM, Vicente J, Romero B, Galindo RC, Minguijon E et al (2009) First data on eurasian wild boar response to oral immunization with bcg and challenge with a mycobacterium bovis field strain. Vaccine 27:6662–6668

Basaraba RJ, Dailey DD, McFarland CT, Shanley CA, Smith EE, McMurray DN et al (2006) Lymphadenitis as a major element of disease in the guinea pig model of tuberculosis. Tuberculosis 86:386–394

Basu S, Rao N, Elkington P (2020) Animal models of ocular tuberculosis: Implications for diagnosis and treatment. Ocul Immunol Inflamm 2020:1–7

Be NA, Lamichhane G, Grosset J, Tyagi S, Cheng QJ, Kim KS et al (2008) Murine model to study the invasion and survival of mycobacterium tuberculosis in the central nervous system. J Infect Dis 198:1520–1528

Be NA, Klinkenberg LG, Bishai WR, Karakousis PC, Jain SK (2011) Strain-dependent cns dissemination in guinea pigs after mycobacterium tuberculosis aerosol challenge. Tuberculosis 91:386–389

Beamer G, Major S, Das B, Campos-Neto A (2014) Bone marrow mesenchymal stem cells provide an antibiotic-protective niche for persistent viable mycobacterium tuberculosis that survive antibiotic treatment. Am J Pathol 184:3170–3175

Bellinger DA, Bullock BC (1988) Cutaneous mycobacterium avium infection in a cynomolgus monkey. Lab Anim Sci 38:85–86

Benard EL, Rougeot J, Racz PI, Spaink HP, Meijer AH (2016) Transcriptomic approaches in the zebrafish model for tuberculosis-insights into host- and pathogen-specific determinants of the innate immune response. Adv Genet 95:217–251

Benmerzoug S, Quesniaux VFJ (2017) Bioengineered 3d models for studying human cell-tuberculosis interactions. Trends Microbiol 25:245–246

Berry SB, Gower MS, Su X, Seshadri C, Theberge AB (2020) A modular microscale granuloma model for immune-microenvironment signaling studies in vitro. Front Bioeng Biotechnol 8:931

Bezos J, de Juan L, Romero B, Alvarez J, Mazzucchelli F, Mateos A et al (2010) Experimental infection with mycobacterium caprae in goats and evaluation of immunological status in tuberculosis and paratuberculosis co-infected animals. Vet Immunol Immunopathol 133:269–275

Bielecka MK, Tezera LB, Zmijan R, Drobniewski F, Zhang X, Jayasinghe S et al (2017) A bioengineered three-dimensional cell culture platform integrated with microfluidics to address antimicrobial resistance in tuberculosis. MBio 8:e02073

Bishai WR, Dannenberg AM Jr, Parrish N, Ruiz R, Chen P, Zook BC et al (1999) Virulence of mycobacterium tuberculosis cdc1551 and h37rv in rabbits evaluated by lurie's pulmonary tubercle count method. Infect Immun 67:4931–4934

Blanc L, Daudelin IB, Podell BK, Chen PY, Zimmerman M, Martinot AJ et al (2018a) High-resolution mapping of fluoroquinolones in tb rabbit lesions reveals specific distribution in immune cell types. elife 7:e41115

Blanc L, Sarathy JP, Alvarez Cabrera N, O'Brien P, Dias-Freedman I, Mina M et al (2018b) Impact of immunopathology on the antituberculous activity of pyrazinamide. J Exp Med 215:1975–1986

Blank J, Behrends J, Jacobs T, Schneider BE (2016a) Mycobacterium tuberculosis coinfection has no impact on plasmodium berghei anka-induced experimental cerebral malaria in c57bl/6 mice. Infect Immun 84:502–510

Blank J, Eggers L, Behrends J, Jacobs T, Schneider BE (2016b) One episode of self-resolving plasmodium yoelii infection transiently exacerbates chronic mycobacterium tuberculosis infection. Front Microbiol 7:152

Bodkin GE (1918) A case of tuberculosis in a rat. J Hyg 17:10–12

Bone JF, Soave OA (1970) Experimental tuberculosis in owl monkeys (aotus trivirgatus). Lab Anim Care 20:946–948

Bouz G, Al Hasawi N (2018) The zebrafish model of tuberculosis - no lungs needed. Crit Rev Microbiol 44:779–792

Braian C, Svensson M, Brighenti S, Lerm M, Parasa VR (2015) A 3d human lung tissue model for functional studies on mycobacterium tuberculosis infection. J Vis Exp 2015:53084

Brammer DW, O'Rourke CM, Heath LA, Chrisp CE, Peter GK, Hofing GL (1995) Mycobacterium kansasii infection in squirrel monkeys (saimiri sciureus sciureus). J Med Primatol 24:231–235

Broncyk LH, Kalter SS (1980) Bacteriological findings in a nonhuman primate colony. Dev Biol Stand 45:23–28

Broussard GW, Ennis DG (2007) Mycobacterium marinum produces long-term chronic infections in medaka: a new animal model for studying human tuberculosis. Comp Biochem Physiol C Toxicol Pharmacol 145:45–54

Broussard GW, Norris MB, Schwindt AR, Fournie JW, Winn RN, Kent ML et al (2009) Chronic mycobacterium marinum infection acts as a tumor promoter in japanese medaka (oryzias latipes). Comp Biochem Physiol C Toxicol Pharmacol 149:152–160

Bruffaerts N, Vluggen C, Roupie V, Duytschaever L, Van den Poel C, Denoel J et al (2017) Virulence and immunogenicity of genetically defined human and porcine isolates of m. Avium subsp. Hominissuis in an experimental mouse infection. PLoS One 12:e0171895
Bryk R, Mundhra S, Jiang X, Wood M, Pfau D, Weber E et al (2020) Potentiation of rifampin activity in a mouse model of tuberculosis by activation of host transcription factor eb. PLoS Pathog 16:e1008567
Bucsan AN, Mehra S, Khader SA, Kaushal D (2019) The current state of animal models and genomic approaches towards identifying and validating molecular determinants of mycobacterium tuberculosis infection and tuberculosis disease. Pathog Dis 77:37
Budell WC, Germain GA, Janisch N, McKie-Krisberg Z, Jayaprakash AD, Resnick AE et al (2020) Transposon mutagenesis in mycobacterium kansasii links a small rna gene to colony morphology and biofilm formation and identifies 9,885 intragenic insertions that do not compromise colony outgrowth. Microbiol Open 9:e988
Calderon VE, Valbuena G, Goez Y, Judy BM, Huante MB, Sutjita P et al (2013) A humanized mouse model of tuberculosis. PLoS One 8:e63331
Capuano SV, Croix DA, Pawar S, Zinovik A, Myers A, Lin PL et al (2003) Experimental mycobacterium tuberculosis infection of cynomolgus macaques closely resembles the various manifestations of human m. Tuberculosis infection. Infect Immun 71:5831–5844
Cardona PJ, Williams A (2017) Experimental animal modelling for tb vaccine development. Int J Infect Dis 56:268–273
Cardona PJ, Cooper A, Luquin M, Ariza A, Filipo F, Orme IM et al (1999) The intravenous model of murine tuberculosis is less pathogenic than the aerogenic model owing to a more rapid induction of systemic immunity. Scand J Immunol 49:362–366
Cardona PJ, Gordillo S, Diaz J, Tapia G, Amat I, Pallares A et al (2003) Widespread bronchogenic dissemination makes dba/2 mice more susceptible than c57bl/6 mice to experimental aerosol infection with mycobacterium tuberculosis. Infect Immun 71:5845–5854
Carius P, Horstmann JC, de Souza Carvalho-Wodarz C, Lehr CM (2020) Disease models: Lung models for testing drugs against inflammation and infection. Handb Exp Pharmacol 265:157
Carranza-Rosales P, Carranza-Torres IE, Guzman-Delgado NE, Lozano-Garza G, Villarreal-Trevino L, Molina-Torres C et al (2017) Modeling tuberculosis pathogenesis through ex vivo lung tissue infection. Tuberculosis 107:126–132
Carvalho R, de Sonneville J, Stockhammer OW, Savage ND, Veneman WJ, Ottenhoff TH et al (2011) A high-throughput screen for tuberculosis progression. PLoS One 6:e16779
Cavaleri M, Manolis E (2015) Hollow fiber system model for tuberculosis: the European medicines agency experience. Clin Infect Dis 61(1):1–4
Cha SB, Jeon BY, Kim WS, Kim JS, Kim HM, Kwon KW et al (2015) Experimental reactivation of pulmonary mycobacterium avium complex infection in a modified cornell-like murine model. PLoS One 10:e0139251
Chackerian AA, Behar SM (2003) Susceptibility to mycobacterium tuberculosis: lessons from inbred strains of mice. Tuberculosis 83:279–285
Chackerian AA, Perera TV, Behar SM (2001) Gamma interferon-producing cd4+ t lymphocytes in the lung correlate with resistance to infection with mycobacterium tuberculosis. Infect Immun 69:2666–2674
Chambers MA, Williams A, Gavier-Widen D, Whelan A, Hughes C, Hall G et al (2001) A guinea pig model of low-dose mycobacterium bovis aerogenic infection. Vet Microbiol 80:213–226
Chaparas SD, Hedrick SR, Clark RG, Garman R (1970) Comparison of the lymphocyte transformation test with the tuberculin test in rhesus monkeys and chimpanzees. Am J Vet Res 31:1437–1441
Chaturvedi V, Jyoti D, Srivastava S, Gupta HP (1999) Secretory proteins of mycobacterium habana induce a protective immune response against experimental tuberculosis. FEMS Immunol Med Microbiol 26:143–151
Chaudhuri M, Squibb RL, Solotorovsky M (1980) Effects of glucose and fructose loading on glycogenesis in chicks infected with avian tuberculosis. Poult Sci 59:1736–1741

Chege GK, Warren RM, Gey NC, van Pittius WA, Burgers, Wilkinson RJ, Shephard EG et al (2008) Detection of natural infection with mycobacterium intracellulare in healthy wild-caught chacma baboons (papio ursinus) by esat-6 and cfp-10 ifn-gamma elispot tests following a tuberculosis outbreak. BMC Microbiol 8:27

Chen Z, Shao XY, Wang C, Hua MH, Wang CN, Wang X et al (2018) Mycobacterium marinum infection in zebrafish and microglia imitates the early stage of tuberculous meningitis. J Mol Neurosci 64:321–330

Cheng T, Kam JY, Johansen MD, Oehlers SH (2020) High content analysis of granuloma histology and neutrophilic inflammation in adult zebrafish infected with mycobacterium marinum. Micron 129:102782

Chingwaru W, Glashoff RH, Vidmar J, Kapewangolo P, Sampson SL (2016) Mammalian cell cultures as models for mycobacterium tuberculosis-human immunodeficiency virus (hiv) interaction studies: a review. Asian Pac J Trop Med 9:832–838

Chrisp CE, Cohen BJ, Ringler DH, Abrams GD (1968) Tuberculosis in a squirrel monkey (saimiri sciureus). J Am Vet Med Assoc 153:918–922

Churchill GA, Airey DC, Allayee H, Angel JM, Attie AD, Beatty J et al (2004) The collaborative cross, a community resource for the genetic analysis of complex traits. Nat Genet 36:1133–1137

Churchill GA, Gatti DM, Munger SC, Svenson KL (2012) The diversity outbred mouse population. Mamm Genome 23:713–718

Clarke KA, Fitzgerald SD, Zwick LS, Church SV, Kaneene JB, Wismer AR et al (2007) Experimental inoculation of meadow voles (microtus pennsylvanicus), house mice (mus musculus), and norway rats (rattus norvegicus) with mycobacterium bovis. J Wildl Dis 43:353–365

Clemens DL, Lee BY, Silva A, Dillon BJ, Maslesa-Galic S, Nava S et al (2019) Artificial intelligence enabled parabolic response surface platform identifies ultra-rapid near-universal tb drug treatment regimens comprising approved drugs. PLoS One 14:e0215607

Cohen SB, Gern BH, Delahaye JL, Adams KN, Plumlee CR, Winkler JK et al (2018) Alveolar macrophages provide an early mycobacterium tuberculosis niche and initiate dissemination. Cell Host Microbe 24:439–446

Cohen A, Mathiasen VD, Schon T, Wejse C (2019) The global prevalence of latent tuberculosis: a systematic review and meta-analysis. Eur Respir J 54:1900655

Coleman MT, Chen RY, Lee M, Lin PL, Dodd LE, Maiello P et al (2014a) Pet/ct imaging reveals a therapeutic response to oxazolidinones in macaques and humans with tuberculosis. Sci Transl Med 6:265

Coleman MT, Maiello P, Tomko J, Frye LJ, Fillmore D, Janssen C et al (2014b) Early changes by (18)fluorodeoxyglucose positron emission tomography coregistered with computed tomography predict outcome after mycobacterium tuberculosis infection in cynomolgus macaques. Infect Immun 82:2400–2404

Collins FM, Auclair L, Montalbine V (1975) Effect of t-cell depletion on the growth of bcg in the mouse footpad. Int Arch Allergy Appl Immunol 48:680–690

Collymore C, Kent L, Ahn SK, Xu W, Li M, Liu J et al (2018) Humane endpoints for guinea pigs used for mycobacterium tuberculosis vaccine research. Comp Med 68:41–47

Commandeur S, Iakobachvili N, Sparrius M, Nur MM, Mukamolova GV, Bitter W (2020) Zebrafish embryo model for assessment of drug efficacy on mycobacterial persisters. Antimicrob Agents Chemother 64:e00801

Cooke MM, Alley MR, Manktelow BW (2003) Experimental infection with BCG as a model of tuberculosis in the brushtail possum (trichosurus vulpecula). N Z Vet J 51:132–138

Corleis B, Bucsan AN, Deruaz M, Vrbanac VD, Lisanti-Park AC, Gates SJ et al (2019) Hiv-1 and siv infection are associated with early loss of lung interstitial cd4+ t cells and dissemination of pulmonary tuberculosis. Cell Rep 26:1409–1418

Corner LA, Pfeiffer DU, Morris RS (2003) Social-network analysis of mycobacterium bovis transmission among captive brushtail possums (trichosurus vulpecula). Prev Vet Med 59:147–167

Coscolla M, Lewin A, Metzger S, Maetz-Rennsing K, Calvignac-Spencer S, Nitsche A et al (2013) Novel mycobacterium tuberculosis complex isolate from a wild chimpanzee. Emerg Infect Dis 19:969–976

Cosma CL, Swaim LE, Volkman H, Ramakrishnan L, Davis JM (2006) Zebrafish and frog models of mycobacterium marinum infection. Curr Protoc Microbiol 10:10–12

Costa SS, Lopes E, Azzali E, Machado D, Coelho T, da Silva PE et al (2016) An experimental model for the rapid screening of compounds with potential use against mycobacteria. Assay Drug Dev Technol 14:524–534

Cross GB, Yeo BC, Hutchinson PE, Tan MC, Verma R, Lu Q et al (2019) Impact of selective immune-cell depletion on growth of mycobacterium tuberculosis (mtb) in a whole-blood bactericidal activity (wba) assay. PLoS One 14:e0216616

Crouser ED, White P, Caceres EG, Julian MW, Papp AC, Locke LW et al (2017) A novel in vitro human granuloma model of sarcoidosis and latent tuberculosis infection. Am J Respir Cell Mol Biol 57:487–498

Cui Z, Wang J, Lu J, Huang X, Zheng R, Hu Z (2013) Evaluation of methods for testing the susceptibility of clinical mycobacterium tuberculosis isolates to pyrazinamide. J Clin Microbiol 51:1374–1380

Cummings MM, Hudgins PC, Whorton MC, Sheldon WH (1952) The influence of cortisone and streptomycin on experimental tuberculosis in the albino rat. Am Rev Tuberc 65:596–602

Cyktor JC, Carruthers B, Kominsky RA, Beamer GL, Stromberg P, Turner J (2013) Il-10 inhibits mature fibrotic granuloma formation during mycobacterium tuberculosis infection. J Immunol 190:2778–2790

da Silva DA, Rego AM, Ferreira NV, de Andrade MAS, Campelo AR, Caldas PCS et al (2017) Detection of mycobacterial infection in non-human primates using the xpert mtb/rif molecular assay. Tuberculosis 107:59–62

Daigeler A (1952) The cotton rat (sigmodon hispidus hispidus) as an experimental animal in the diagnosis of tuberculosis. Z Hyg Infekt 135:588–591

Das B, Kashino SS, Pulu I, Kalita D, Swami V, Yeger H et al (2013) Cd271(+) bone marrow mesenchymal stem cells may provide a niche for dormant mycobacterium tuberculosis. Sci Transl Med 5:170ra113

Davis JM, Ramakrishnan L (2009) The role of the granuloma in expansion and dissemination of early tuberculous infection. Cell 136:37–49

de Arruda MS, Montenegro MR (1995) The hamster cheek pouch: an immunologically privileged site suitable to the study of granulomatous infections. Rev Inst Med Trop Sao Paulo 37:303–309

De Groote MA, Gilliland JC, Wells CL, Brooks EJ, Woolhiser LK, Gruppo V et al (2011) Comparative studies evaluating mouse models used for efficacy testing of experimental drugs against mycobacterium tuberculosis. Antimicrob Agents Chemother 55:1237–1247

De Klerk L, Michel AL, Grobler DG, Bengis RG, Bush M, Kriek NP et al (2006) An experimental intratonsilar infection model for bovine tuberculosis in african buffaloes, syncerus caffer. Onderstepoort J Vet Res 73:293–303

de Knegt GJ, Dickinson L, Pertinez H, Evangelopoulos D, McHugh TD, Bakker-Woudenberg I et al (2017) Assessment of treatment response by colony forming units, time to culture positivity and the molecular bacterial load assay compared in a mouse tuberculosis model. Tuberculosis 105:113–118

de Val Perez B, Lopez-Soria S, Nofrarias M, Martin M, Vordermeier HM, Villarreal-Ramos B et al (2011) Experimental model of tuberculosis in the domestic goat after endobronchial infection with mycobacterium caprae. Clin Vaccine Immunol 18:1872–1881

Dean GS, Rhodes SG, Coad M, Whelan AO, Wheeler P, Villareal-Ramos B et al (2008) Isoniazid treatment of mycobacterium bovis in cattle as a model for human tuberculosis. Tuberculosis 88: 586–594

Defraine V, Fauvart M, Michiels J (2018) Fighting bacterial persistence: current and emerging anti-persister strategies and therapeutics. Drug Resist Updat 38:12–26

Deinard AS, Lerche NW, Smith DG (2002) Polymorphism in the rhesus macaque (macaca mulatta) nramp1 gene: Lack of an allelic association to tuberculosis susceptibility. J Med Primatol 31:8–16

Dennis EW, Gaboe FC (1949) Experimental tuberculosis of the syrian hamster, cricetus auratus. Ann N Y Acad Sci 52:646–661

Dessau FI, Yeager RL, Kulish M (1949) A simplified guinea pig test for tuberculostatic agents. Am Rev Tuberc 60:223–227

Diedrich CR, Rutledge T, Maiello P, Baranowski TM, White AG, Borish HJ et al (2020) Siv and mycobacterium tuberculosis synergy within the granuloma accelerates the reactivation pattern of latent tuberculosis. PLoS Pathog 16:e1008413

Dijkman K, Vervenne RAW, Sombroek CC, Boot C, Hofman SO, van Meijgaarden KE et al (2019) Disparate tuberculosis disease development in macaque species is associated with innate immunity. Front Immunol 10:2479

Dionne MS, Ghori N, Schneider DS (2003) Drosophila melanogaster is a genetically tractable model host for mycobacterium marinum. Infect Immun 71:3540–3550

Domingues-Junior M, Pinheiro SR, Guerra JL, Palermo-Neto J (2000) Effects of treatment with amphetamine and diazepam on mycobacterium bovis-induced infection in hamsters. Immunopharmacol Immunotoxicol 22:555–574

Dong H, Jing W, Yabo Y, Xiaokang Y, Wan W, Min M et al (2014) Establishment of rat model of silicotuberculosis and its pathological characteristic. Pathog Glob Health 108:312–316

Dong H, Lv Y, Sreevatsan S, Zhao D, Zhou X (2017) Differences in pathogenicity of three animal isolates of mycobacterium species in a mouse model. PLoS One 12:e0183666

Donnelly CA, Nouvellet P (2013) The contribution of badgers to confirmed tuberculosis in cattle in high-incidence areas in england. PLoS Curr 5:776098

Dorman SE, Hatem CL, Tyagi S, Aird K, Lopez-Molina J, Pitt ML et al (2004) Susceptibility to tuberculosis: clues from studies with inbred and outbred New Zealand white rabbits. Infect Immun 72:1700–1705

Dormans J, Burger M, Aguilar D, Hernandez-Pando R, Kremer K, Roholl P et al (2004) Correlation of virulence, lung pathology, bacterial load and delayed type hypersensitivity responses after infection with different mycobacterium tuberculosis genotypes in a balb/c mouse model. Clin Exp Immunol 137:460–468

Driver ER, Ryan GJ, Hoff DR, Irwin SM, Basaraba RJ, Kramnik I et al (2012) Evaluation of a mouse model of necrotic granuloma formation using c3heb/fej mice for testing of drugs against mycobacterium tuberculosis. Antimicrob Agents Chemother 56:3181–3195

Duffy FJ, Weiner J, Hansen S, Tabb DL, Suliman S, Thompson E et al (2019) Immunometabolic signatures predict risk of progression to active tuberculosis and disease outcome. Front Immunol 10:527

Duque C, Arroyo L, Ortega H, Montufar F, Ortiz B, Rojas M et al (2014) Different responses of human mononuclear phagocyte populations to mycobacterium tuberculosis. Tuberculosis 94: 111–122

Durkee MS, Cirillo JD, Maitland KC (2019) Fluorescence modeling of in vivo optical detection of mycobacterium tuberculosis. Biomed Opt Express 10:5445–5460

Dutta NK, Illei PB, Jain SK, Karakousis PC (2014a) Characterization of a novel necrotic granuloma model of latent tuberculosis infection and reactivation in mice. Am J Pathol 184:2045–2055

Dutta NK, McLachlan J, Mehra S, Kaushal D (2014b) Humoral and lung immune responses to mycobacterium tuberculosis infection in a primate model of protection. Trials Vaccinol 3:47–51

Ehlers LP, Bianchi MV, Argenta FF, Lopes BC, Taunde PA, Wagner PGC et al (2020) Mycobacterium tuberculosis var. Tuberculosis infection in two captive black capuchins (sapajus nigritus) in southern brazil. Braz J Microbiol 51:2169–2173

El-Etr SH, Yan L, Cirillo JD (2001) Fish monocytes as a model for mycobacterial host-pathogen interactions. Infect Immun 69:7310–7317

Ellis H, Mulder C, Valverde E, Poling A, Edwards T (2017) Reproducibility of african giant pouched rats detecting mycobacterium tuberculosis. BMC Infect Dis 17:298

Elwood RL, Wilson S, Blanco JC, Yim K, Pletneva L, Nikonenko B et al (2007) The American cotton rat: a novel model for pulmonary tuberculosis. Tuberculosis 87:145–154

Engel GA, Wilbur AK, Westmark A, Horn D, Johnson J, Jones-Engel L (2012) Naturally acquired mycobacterium tuberculosis complex in laboratory pig-tailed macaques. Emerg Microbes Infect 1:e30

Entwistle FM, Coote PJ (2018) Evaluation of greater wax moth larvae, galleria mellonella, as a novel in vivo model for non-tuberculosis mycobacteria infections and antibiotic treatments. J Med Microbiol 67:585–597

Eruslanov EB, Majorov KB, Orlova MO, Mischenko VV, Kondratieva TK, Apt AS et al (2004) Lung cell responses to m. Tuberculosis in genetically susceptible and resistant mice following intratracheal challenge. Clin Exp Immunol 135:19–28

Esaulova E, Das S, Singh DK, Choreno-Parra JA, Swain A, Arthur L et al (2020) The immune landscape in tuberculosis reveals populations linked to disease and latency. Cell Host Microbe 29:165–178

Eskuchen (1952) Diagnosis of tuberculosis with guinea pigs and gold hamster. Tuberkulosearzt 6: 356–358

Evans S, Butler JR, Mattila JT, Kirschner DE (2020) Systems biology predicts that fibrosis in tuberculous granulomas may arise through macrophage-to-myofibroblast transformation. PLoS Comput Biol 16:e1008520

Fatima S, Kamble SS, Dwivedi VP, Bhattacharya D, Kumar S, Ranganathan A et al (2020) Mycobacterium tuberculosis programs mesenchymal stem cells to establish dormancy and persistence. J Clin Invest 130:655–661

Fenwick NI (2012) Modelled impacts of badger culling on cattle tb in a real area with geographic boundaries. Vet Rec 170:177

Florido M, Grima MA, Gillis CM, Xia Y, Turner SJ, Triccas JA et al (2013) Influenza a virus infection impairs mycobacteria-specific t cell responses and mycobacterial clearance in the lung during pulmonary coinfection. J Immunol 191:302–311

Flynn JL, Gideon HP, Mattila JT, Lin PL (2015) Immunology studies in non-human primate models of tuberculosis. Immunol Rev 264:60–73

Forget A, Skamene E, Gros P, Miailhe AC, Turcotte R (1981) Differences in response among inbred mouse strains to infection with small doses of mycobacterium bovis bcg. Infect Immun 32:42–47

Fourie PB, Odendaal MW (1983) Mycobacterium tuberculosis in a closed colony of baboons (papio ursinus). Lab Anim 17:125–128

Fox JG, Campbell LH, Snyder SB, Reed C, Soave OA (1974) Tuberculous spondylitis and pott's paraplegia in a rhesus monkey (macaca mulatta). Lab Anim Sci 24:335–339

Franzblau SG, DeGroote MA, Cho SH, Andries K, Nuermberger E, Orme IM et al (2012) Comprehensive analysis of methods used for the evaluation of compounds against mycobacterium tuberculosis. Tuberculosis 92:453–488

Fremming BD, Benson RE, Young RJ, Harris MD Jr (1957) Antituberculous therapy in macaca mulatta monkeys. Am Rev Tuberc 76:225–231

Fujita M, Harada E, Matsumoto T, Mizuta Y, Ikegame S, Ouchi H et al (2010) Impaired host defence against mycobacterium avium in mice with chronic granulomatous disease. Clin Exp Immunol 160:457–460

Fulford GR, Roberts MG, Heesterbeek JA (2002) The metapopulation dynamics of an infectious disease: tuberculosis in possums. Theor Popul Biol 61:15–29

Galbadage T, Shepherd TF, Cirillo SL, Gumienny TL, Cirillo JD (2016) The caenorhabditis elegans p38 mapk gene plays a key role in protection from mycobacteria. Microbiol Open 5:436–452

Ganatra SR, Bucsan AN, Alvarez X, Kumar S, Chatterjee A, Quezada M et al (2020) Antiretroviral therapy does not reduce tuberculosis reactivation in a tuberculosis-hiv coinfection model. J Clin Invest 130:5171–5179

Gaonkar S, Bharath S, Kumar N, Balasubramanian V, Shandil RK (2010) Aerosol infection model of tuberculosis in wistar rats. Int J Microbiol 2010:426035

Garcia EA, Blanco FC, Muniz XF, Eirin ME, Klepp LI, Bigi F (2020) Elimination of esat-6 and cfp-10 from a candidate vaccine against bovine tuberculosis impaired its protection efficacy in the balbc mouse model. Int J Mycobacteriol 9:417–421

Garcia-Pelayo MC, Bachy VS, Kaveh DA, Hogarth PJ (2015) Balb/c mice display more enhanced bcg vaccine induced th1 and th17 response than c57bl/6 mice but have equivalent protection. Tuberculosis 95:48–53

Garcia-Pelayo MC, Kaveh DA, Sibly L, Webb PR, Bull NC, Cutting SM et al (2016) Boosting bcg with inert spores improves immunogenicity and induces specific il-17 responses in a murine model of bovine tuberculosis. Tuberculosis 98:97–103

Garhyan J, Bhuyan S, Pulu I, Kalita D, Das B, Bhatnagar R (2015) Preclinical and clinical evidence of mycobacterium tuberculosis persistence in the hypoxic niche of bone marrow mesenchymal stem cells after therapy. Am J Pathol 185:1924–1934

Gasso D, Vicente J, Mentaberre G, Soriguer R, Jimenez Rodriguez R, Navarro-Gonzalez N et al (2016) Oxidative stress in wild boars naturally and experimentally infected with mycobacterium bovis. PLoS One 11:e0163971

Gautam US, Foreman TW, Bucsan AN, Veatch AV, Alvarez X, Adekambi T et al (2018) In vivo inhibition of tryptophan catabolism reorganizes the tuberculoma and augments immune-mediated control of mycobacterium tuberculosis. Proc Natl Acad Sci U S A 115:E62–E71

Geng G, Wang Q, Shi J, Yan J, Niu N, Wang Z (2015) Establishment of a new zealand rabbit model of spinal tuberculosis. J Spinal Disord Tech 28:140–145

Gharpure PV (1945) Guinea-pig inoculation in the diagnosis of tuberculosis. Ind Med Gaz 80:327

Gibson SER, Harrison J, Cox JAG (2018) Modelling a silent epidemic: a review of the in vitro models of latent tuberculosis. Pathogens 7:88

Gideon HP, Phuah J, Myers AJ, Bryson BD, Rodgers MA, Coleman MT et al (2015) Variability in tuberculosis granuloma t cell responses exists, but a balance of pro- and anti-inflammatory cytokines is associated with sterilization. PLoS Pathog 11:e1004603

Gideon HP, Skinner JA, Baldwin N, Flynn JL, Lin PL (2016) Early whole blood transcriptional signatures are associated with severity of lung inflammation in cynomolgus macaques with mycobacterium tuberculosis infection. J Immunol 197:4817–4828

Gil O, Diaz I, Vilaplana C, Tapia G, Diaz J, Fort M et al (2010) Granuloma encapsulation is a key factor for containing tuberculosis infection in minipigs. PLoS One 5:e10030

Glover RE (1946) Susceptibility of the golden hamster (cricetus auratus) to mycobacterium tuberculosis hominis and bovis. J Pathol Bacteriol 58:107–110

Gong W, Liang Y, Wu X (2020) Animal models of tuberculosis vaccine research: an important component in the fight against tuberculosis. Biomed Res Int 2020:4263079

Gormley E, Corner LAL (2017) Pathogenesis of mycobacterium bovis infection: the badger model as a paradigm for understanding tuberculosis in animals. Front Vet Sci 4:247

Gormus BJ, Blanchard JL, Alvarez XH, Didier PJ (2004) Evidence for a rhesus monkey model of asymptomatic tuberculosis. J Med Primatol 33:134–145

Goyal RK (1938) The guinea-pig in the laboratory diagnosis of tuberculosis. Ind Med Gaz 73:282–283

Green DM, Kiss IZ, Mitchell AP, Kao RR (2008) Estimates for local and movement-based transmission of bovine tuberculosis in british cattle. Proc Biol Sci 275:1001–1005

Griffin JF, Rodgers CR, Liggett S, Mackintosh CG (2006) Tuberculosis in ruminants: characteristics of intra-tonsilar mycobacterium bovis infection models in cattle and deer. Tuberculosis 86:404–418

Grossman TH, Shoen CM, Jones SM, Jones PL, Cynamon MH, Locher CP (2015) The efflux pump inhibitor timcodar improves the potency of antimycobacterial agents. Antimicrob Agents Chemother 59:1534–1541

Grover A, Troy A, Rowe J, Troudt JM, Creissen E, McLean J et al (2017) Humanized nog mice as a model for tuberculosis vaccine-induced immunity: a comparative analysis with the mouse and guinea pig models of tuberculosis. Immunology 152:150–162

Grumbach F (1960) Experimental antituberculous chemotherapy in the white rat. Ann Inst Pasteur 98:485–493

Grumbach F, Canetti G, Grosset J, le Lirzin M (1967) Late results of long-term intermittent chemotherapy of advanced, murine tuberculosis: limits of the murine model. Tubercle 48:11–26

Guirado E, Gordillo S, Gil O, Diaz J, Tapia G, Vilaplana C et al (2006) Intragranulomatous necrosis in pulmonary granulomas is not related to resistance against mycobacterium tuberculosis infection in experimental murine models induced by aerosol. Int J Exp Pathol 87:139–149

Guirado E, Schlesinger LS, Kaplan G (2013) Macrophages in tuberculosis: friend or foe. Semin Immunopathol 35:563–583

Gumbo T, Pasipanodya JG, Nuermberger E, Romero K, Hanna D (2015a) Correlations between the hollow fiber model of tuberculosis and therapeutic events in tuberculosis patients: learn and confirm. Clin Infect Dis 61(1):18–24

Gumbo T, Pasipanodya JG, Romero K, Hanna D, Nuermberger E (2015b) Forecasting accuracy of the hollow fiber model of tuberculosis for clinical therapeutic outcomes. Clin Infect Dis 61(1): 25–31

Guo Q, Bi J, Wang H, Zhang X (2021) Mycobacterium tuberculosis esx-1-secreted substrate protein espc promotes mycobacterial survival through endoplasmic reticulum stress-mediated apoptosis. Emerg Microbes Infect 10:19–36

Gupta A, Bhakta S (2012) An integrated surrogate model for screening of drugs against mycobacterium tuberculosis. J Antimicrob Chemother 67:1380–1391

Gupta SK, Mathur IS (1969) A cheap and quick method of screening potential antimycobacterial agents in the syrian or golden hamster (cricetus auratus). Experientia 25:782–783

Gupta A, Bhakta S, Kundu S, Gupta M, Srivastava BS, Srivastava R (2009) Fast-growing, non-infectious and intracellularly surviving drug-resistant mycobacterium aurum: a model for high-throughput antituberculosis drug screening. J Antimicrob Chemother 64:774–781

Gupta A, Ahmad FJ, Ahmad F, Gupta UD, Natarajan M, Katoch V et al (2012) Efficacy of mycobacterium indicus pranii immunotherapy as an adjunct to chemotherapy for tuberculosis and underlying immune responses in the lung. PLoS One 7:e39215

Gupta UD, Abbas A, Kashyap RP, Gupta P (2016) Murine model of tb meningitis. Int J Mycobacteriol 5(Suppl 1):S178

Hagedorn M, Soldati T (2007) Flotillin and rach modulate the intracellular immunity of dictyostelium to mycobacterium marinum infection. Cell Microbiol 9:2716–2733

Harjula SE, Saralahti AK, Ojanen MJT, Rantapero T, Uusi-Makela MIE, Nykter M et al (2020) Characterization of immune response against mycobacterium marinum infection in the main hematopoietic organ of adult zebrafish (danio rerio). Dev Comp Immunol 103:103523

Harper J, Skerry C, Davis SL, Tasneen R, Weir M, Kramnik I et al (2012) Mouse model of necrotic tuberculosis granulomas develops hypoxic lesions. J Infect Dis 205:595–602

Haug M, Awuh JA, Steigedal M, Frengen Kojen J, Marstad A, Nordrum IS et al (2013) Dynamics of immune effector mechanisms during infection with mycobacterium avium in c57bl/6 mice. Immunology 140:232–243

Henao J, Sanchez D, Munoz CH, Mejia N, Arias MA, Garcia LF et al (2007) Human splenic macrophages as a model for in vitro infection with mycobacterium tuberculosis. Tuberculosis 87:509–517

Henao-Tamayo M, Obregon-Henao A, Creissen E, Shanley C, Orme I, Ordway DJ (2015) Differential mycobacterium bovis bcg vaccine-derived efficacy in c3heb/fej and c3h/heouj mice exposed to a clinical strain of mycobacterium tuberculosis. Clin Vaccine Immunol 22:91–98

Heng Y, Seah PG, Siew JY, Tay HC, Singhal A, Mathys V et al (2011) Mycobacterium tuberculosis infection induces hypoxic lung lesions in the rat. Tuberculosis 91:339–341

Henrich M, Moser I, Weiss A, Reinacher M (2007) Multiple granulomas in three squirrel monkeys (saimiri sciureus) caused by mycobacterium microti. J Comp Pathol 137:245–248

Hernandez Pando R, Aguilar D, Cohen I, Guerrero M, Ribon W, Acosta P et al (2010) Specific bacterial genotypes of mycobacterium tuberculosis cause extensive dissemination and brain infection in an experimental model. Tuberculosis 90:268–277

Hernandez-Pando R, Aguilar D, Orozco H, Cortez Y, Brunet LR, Rook GA (2008) Orally administered mycobacterium vaccae modulates expression of immunoregulatory molecules in balb/c mice with pulmonary tuberculosis. Clin Vaccine Immunol 15:1730–1736

Hessler JR, Moreland AF (1968) Pulmonary tuberculosis in a squirrel monkey (saimiri sciureus). J Am Vet Med Assoc 153:923–927

Heuts F, Gavier-Widen D, Carow B, Juarez J, Wigzell H, Rottenberg ME (2013) Cd4+ cell-dependent granuloma formation in humanized mice infected with mycobacteria. Proc Natl Acad Sci U S A 110:6482–6487

Heywood R, Medd RK, Street AE (1970) The early clinical diagnosis of tuberculosis in baboons. Br Vet J 126:372–382

Hino M, Oda M, Yoshida A, Nakata K, Kohchi C, Nishizawa T et al (2005) Establishment of an in vitro model using nr8383 cells and mycobacterium bovis calmette-guerin that mimics a chronic infection of mycobacterium tuberculosis. In Vivo 19:821–830

Hirota K, Hasegawa T, Nakajima T, Inagawa H, Kohchi C, Soma G et al (2010) Delivery of rifampicin-plga microspheres into alveolar macrophages is promising for treatment of tuberculosis. J Control Release 142:339–346

Ho VQT, Verboom T, Rong MK, Habjan E, Bitter W, Speer A (2021) Heterologous expression of etha and katg in mycobacterium marinum enables the rapid identification of new prodrugs active against mycobacterium tuberculosis. Antimicrob Agents Chemother 65(4):e01445

Hodgkinson JW, Ge JQ, Grayfer L, Stafford J, Belosevic M (2012) Analysis of the immune response in infections of the goldfish (carassius auratus l.) with mycobacterium marinum. Dev Comp Immunol 38:456–465

Hogset H, Horgan CC, Armstrong JPK, Bergholt MS, Torraca V, Chen Q et al (2020) In vivo biomolecular imaging of zebrafish embryos using confocal raman spectroscopy. Nat Commun 11:6172

Hosseini R, Lamers GEM, Bos E, Hogendoorn PCW, Koster AJ, Meijer AH et al (2021) The adapter protein myd88 plays an important role in limiting mycobacterial growth in a zebrafish model for tuberculosis. Virchows Arch 479(2):265–275

Huante MB, Saito TB, Nusbaum RJ, Naqvi KF, Chauhan S, Hunter RL et al (2020) Small animal model of post-chemotherapy tuberculosis relapse in the setting of hiv co-infection. Front Cell Infect Microbiol 10:150

Hudock TA, Lackner AA, Kaushal D (2014) Microdissection approaches in tuberculosis research. J Med Primatol 43:294–297

Hudock TA, Foreman TW, Bandyopadhyay N, Gautam US, Veatch AV, LoBato DN et al (2017) Hypoxia sensing and persistence genes are expressed during the intragranulomatous survival of mycobacterium tuberculosis. Am J Respir Cell Mol Biol 56:637–647

Husain AA, Gupta UD, Gupta P, Nayak AR, Chandak NH, Daginawla HF et al (2017) Modelling of cerebral tuberculosis in balb/c mice using clinical strain from patients with cns tuberculosis infection. Indian J Med Res 145:833–839

Hussel L (1951) Suitability of the golden hamster as laboratory animal in tuberculosis diagnosis. Zentralbl Bakteriol Orig 156:445–450

Hyoe RK, Robert J (2019) A xenopus tadpole alternative model to study innate-like t cell-mediated anti-mycobacterial immunity. Dev Comp Immunol 92:253–259

Iacobino A, Fattorini L, Giannoni F (2020) Drug-resistant tuberculosis 2020: Where we stand. Appl Sci 10:2153

Idh J, Andersson B, Lerm M, Raffetseder J, Eklund D, Woksepp H et al (2017) Reduced susceptibility of clinical strains of mycobacterium tuberculosis to reactive nitrogen species promotes survival in activated macrophages. PLoS One 12:e0181221

Indzhiia LV, Yakovleva LA, Simovonjan VG, Dshikidze EK, Kovaljova VI, Popova VN (1977) The character and results of comparative experimental therapy of tuberculosis in macaca arctoides monkeys. Z Versuchstierkd 19:13–25

Jain N, Kalam H, Singh L, Sharma V, Kedia S, Das P et al (2020) Mesenchymal stem cells offer a drug-tolerant and immune-privileged niche to mycobacterium tuberculosis. Nat Commun 11: 3062

Javed S, Marsay L, Wareham A, Lewandowski KS, Williams A, Dennis MJ et al (2016) Temporal expression of peripheral blood leukocyte biomarkers in a macaca fascicularis infection model of tuberculosis; comparison with human datasets and analysis with parametric/non-parametric tools for improved diagnostic biomarker identification. PLoS One 11:e0154320

Jespersen A (1974) Infection of arvicola terrestris (vole rat) with m. Tuberculosis and m. Bovis. Acta Pathol Microbiol Scand B: Microbiol Immunol 82:667–675

Jhamb SS, Singh PP (2009) A short-term model for preliminary screening of potential anti-tubercular compounds. Scand J Infect Dis 41:886–889

Jhamb SS, Goyal A, Singh PP (2014) Determination of the activity of standard anti-tuberculosis drugs against intramacrophage mycobacterium tuberculosis, in vitro: Mgit 960 as a viable alternative for bactec 460. Braz J Infect Dis 18:336–340

Joardar SN, Ram GC, Goswami T (2002) Dynamic changes in cellular immune responses in experimental bovine tuberculosis. Med Sci Monit 8:471–480

Johansen MD, Kasparian JA, Hortle E, Britton WJ, Purdie AC, Oehlers SH (2018) Mycobacterium marinum infection drives foam cell differentiation in zebrafish infection models. Dev Comp Immunol 88:169–172

Junqueira-Kipnis AP, de Oliveira FM, Trentini MM, Tiwari S, Chen B, Resende DP et al (2013) Prime-boost with mycobacterium smegmatis recombinant vaccine improves protection in mice infected with mycobacterium tuberculosis. PLoS One 8:e78639

Kager LM, Runge JH, Nederveen AJ, Roelofs JJ, Stoker J, Maas M et al (2014) A new murine model to study musculoskeletal tuberculosis. Tuberculosis 94:306–310

Kannan N, Haug M, Steigedal M, Flo TH (2020) Mycobacterium smegmatis vaccine vector elicits cd4+ th17 and cd8+ tc17 t cells with therapeutic potential to infections with mycobacterium avium. Front Immunol 11:1116

Kao RR, Roberts MG, Ryan TJ (1997) A model of bovine tuberculosis control in domesticated cattle herds. Proc Biol Sci 264:1069–1076

Kao RR, Gravenor MB, Charleston B, Hope JC, Martin M, Howard CJ (2007) Mycobacterium bovis shedding patterns from experimentally infected calves and the effect of concurrent infection with bovine viral diarrhoea virus. J R Soc Interface 4:545–551

Kashino SS, Napolitano DR, Skobe Z, Campos-Neto A (2008) Guinea pig model of mycobacterium tuberculosis latent/dormant infection. Microbes Infect 10:1469–1476

Kaur G, Das DK, Singh S, Khan J, Sajid M, Bashir H et al (2019) Tuberculosis vaccine: past experiences and future prospects. In: Hasnain SE, Ehtesham NZ, Grover S (eds) Mycobacterium tuberculosis: molecular infection biology, pathogenesis, diagnostics and new interventions. Springer, Singapore, pp 463–495

Keiser TL, Purdy GE (2017) Killing mycobacterium tuberculosis in vitro: what model systems can teach us. Microbiol Spectr 5:28

Kelley CL, Collins FM (1999) Growth of a highly virulent strain of mycobacterium tuberculosis in mice of differing susceptibility to tuberculous challenge. Tuber Lung Dis 79:367–370

Kelly BP, Furney SK, Jessen MT, Orme IM (1996) Low-dose aerosol infection model for testing drugs for efficacy against mycobacterium tuberculosis. Antimicrob Agents Chemother 40: 2809–2812

Kenyon A, Gavriouchkina D, Zorman J, Napolitani G, Cerundolo V, Sauka-Spengler T (2017) Active nuclear transcriptome analysis reveals inflammasome-dependent mechanism for early neutrophil response to mycobacterium marinum. Sci Rep 7:6505

Kerr EG (1946) Survey of the efficiency of cultures and guinea pig inoculations in the diagnosis of tuberculosis. Bull Phila Pa Hosp Ayer Clin Lab 3:447–457

Kesavan AK, Brooks M, Tufariello J, Chan J, Manabe YC (2009) Tuberculosis genes expressed during persistence and reactivation in the resistant rabbit model. Tuberculosis 89:17–21

Kharatmal S, Jhamb SS, Singh PP (2009) Evaluation of bactec 460 tb system for rapid in vitro screening of drugs against latent state mycobacterium tuberculosis h37rv under hypoxia conditions. J Microbiol Methods 78:161–164

Khare G, Nangpal P, Tyagi AK (2019) Challenges and advances in tb drug discovery. In: Hasnain SE, Ehtesham NZ, Grover S (eds) Mycobacterium tuberculosis: molecular infection biology, pathogenesis, diagnostics and new interventions. Springer, Singapore, pp 463–495

Kjellsson MC, Via LE, Goh A, Weiner D, Low KM, Kern S et al (2012) Pharmacokinetic evaluation of the penetration of antituberculosis agents in rabbit pulmonary lesions. Antimicrob Agents Chemother 56:446–457

Klinkenberg LG, Sutherland LA, Bishai WR, Karakousis PC (2008) Metronidazole lacks activity against mycobacterium tuberculosis in an in vivo hypoxic granuloma model of latency. J Infect Dis 198:275–283

Kloprogge F, Hammond R, Kipper K, Gillespie SH, Della Pasqua O (2019) Mimicking in-vivo exposures to drug combinations in-vitro: anti-tuberculosis drugs in lung lesions and the hollow fiber model of infection. Sci Rep 9:13228

Knezevic AL, McNulty WP (1967) Tuberculosis in lemur mongoz. Folia Primatol 6:153–159

Kolonko M, Geffken AC, Blumer T, Hagens K, Schaible UE, Hagedorn M (2014) Wash-driven actin polymerization is required for efficient mycobacterial phagosome maturation arrest. Cell Microbiol 16:232–246

Kramnik I (2008) Genetic dissection of host resistance to mycobacterium tuberculosis: the sst1 locus and the ipr1 gene. Curr Top Microbiol Immunol 321:123–148

Kumar N, Vishwas KG, Kumar M, Reddy J, Parab M, Manikanth CL et al (2014) Pharmacokinetics and dose response of anti-tb drugs in rat infection model of tuberculosis. Tuberculosis 94:282–286

Kunnath-Velayudhan S, Davidow AL, Wang HY, Molina DM, Huynh VT, Salamon H et al (2012) Proteome-scale antibody responses and outcome of mycobacterium tuberculosis infection in nonhuman primates and in tuberculosis patients. J Infect Dis 206:697–705

Kupz A, Zedler U, Staber M, Kaufmann SH (2016) A mouse model of latent tuberculosis infection to study intervention strategies to prevent reactivation. PLoS One 11:e0158849

Kuroda MJ, Sugimoto C, Cai Y, Merino KM, Mehra S, Arainga M et al (2018) High turnover of tissue macrophages contributes to tuberculosis reactivation in simian immunodeficiency virus-infected rhesus macaques. J Infect Dis 217:1865–1874

Kurtz SL, Rossi AP, Beamer GL, Gatti DM, Kramnik I, Elkins KL (2020) The diversity outbred mouse population is an improved animal model of vaccination against tuberculosis that reflects heterogeneity of protection. mSphere 5:e00097

Kwan PKW, Lin W, Naim ANM, Periaswamy B, De Sessions PF, Hibberd ML et al (2020) Gene expression responses to anti-tuberculous drugs in a whole blood model. BMC Microbiol 20:81

Langermans JA, Andersen P, van Soolingen D, Vervenne RA, Frost PA, van der Laan T et al (2001) Divergent effect of bacillus calmette-guerin (bcg) vaccination on mycobacterium tuberculosis infection in highly related macaque species: implications for primate models in tuberculosis vaccine research. Proc Natl Acad Sci U S A 98:11497–11502

Latt RH (1975) Runyon group III atypical mycobacteria as a cause of tuberculosis in a rhesus monkey. Lab Anim Sci 25:206–209

Lau DT, Fuller JM, Sumner PE (1972) Tuberculosis in a pig-tailed macaque. J Am Vet Med Assoc 161:696–699

Leathers CW, Hamm TE Jr (1976) Naturally occurring tuberculosis in a squirrel monkey and a cebus monkey. J Am Vet Med Assoc 169:909–911

Lecoeur HF, Lagrange PH, Truffot-Pernot C, Gheorghiu M, Grosset J (1989) Relapses after stopping chemotherapy for experimental tuberculosis in genetically resistant and susceptible strains of mice. Clin Exp Immunol 76:458–462

Lee BY, Clemens DL, Silva A, Dillon BJ, Maslesa-Galic S, Nava S et al (2018) Ultra-rapid near universal tb drug regimen identified via parabolic response surface platform cures mice of both conventional and high susceptibility. PLoS One 13:e0207469

Lerche NW, Yee JL, Capuano SV, Flynn JL (2008) New approaches to tuberculosis surveillance in nonhuman primates. ILAR J 49:170–178

Lesellier S, Corner L, Costello E, Sleeman P, Lyashchenko K, Greenwald R et al (2008) Antigen specific immunological responses of badgers (meles meles) experimentally infected with mycobacterium bovis. Vet Immunol Immunopathol 122:35–45

Lewinsohn DM, Tydeman IS, Frieder M, Grotzke JE, Lines RA, Ahmed S et al (2006) High resolution radiographic and fine immunologic definition of tb disease progression in the rhesus macaque. Microbes Infect 8:2587–2598

Lewis PA, Margot AG (1914) The function of the spleen in the experimental infection of albino mice with bacillus tuberculosis. J Exp Med 19:187–194

Li X, Grossman CJ, Mendenhall CL, Hurtubise P, Rouster SD, Roselle GA et al (1998) Host response to mycobacterial infection in the alcoholic rat: male and female dimorphism. Alcohol 16:207–212

Li YL, Chen BW, Xu M, Luo YA, Wang GZ, Shen XB et al (2010) A guinea pig model of latent mycobacterium tuberculosis h37rv infection. Zhonghua Jie He He Hu Xi Za Zhi 33:684–687

Li Z, Liu H, Li H, Dang G, Cui Z, Song N et al (2019) Pe17 protein from mycobacterium tuberculosis enhances mycobacterium smegmatis survival in macrophages and pathogenicity in mice. Microb Pathog 126:63–73

Li J, Zhao A, Tang J, Wang G, Shi Y, Zhan L et al (2020) Tuberculosis vaccine development: From classic to clinical candidates. Eur J Clin Microbiol Infect Dis 39:1405–1425

Lienard J, Carlsson F (2017) Murine mycobacterium marinum infection as a model for tuberculosis. Methods Mol Biol 1535:301–315

Lin PL, Pawar S, Myers A, Pegu A, Fuhrman C, Reinhart TA et al (2006) Early events in mycobacterium tuberculosis infection in cynomolgus macaques. Infect Immun 74:3790–3803

Lin PL, Rodgers M, Smith L, Bigbee M, Myers A, Bigbee C et al (2009) Quantitative comparison of active and latent tuberculosis in the cynomolgus macaque model. Infect Immun 77:4631–4642

Lin PL, Maiello P, Gideon HP, Coleman MT, Cadena AM, Rodgers MA et al (2016) Pet ct identifies reactivation risk in cynomolgus macaques with latent M. tuberculosis. PLoS Pathog 12:e1005739

Lindsey JR, Melby EC Jr (1966) Naturally occurring primary cutaneous tuberculosis in the rhesus monkey. Lab Anim Care 16:369–385

Lithander A (1957) A comparison between concentrations and guinea-pig tests in the bacteriological diagnosis of tuberculosis in sputum. Acta Pathol Microbiol Scand 40:61–66

Liu L, Fu R, Yuan X, Shi C, Wang S, Lu X et al (2015a) Differential immune responses and protective effects in avirulent mycobacterial strains vaccinated balb/c mice. Curr Microbiol 71: 129–135

Liu X, Jia W, Wang H, Wang Y, Ma J, Wang H et al (2015b) Establishment of a rabbit model of spinal tuberculosis using mycobacterium tuberculosis strain h37rv. Jpn J Infect Dis 68:89–97

Logan KE, Gavier-Widen D, Hewinson RG, Hogarth PJ (2008) Development of a mycobacterium bovis intranasal challenge model in mice. Tuberculosis 88:437–443

Lopez Hernandez Y, Yero D, Pinos-Rodriguez JM, Gibert I (2015) Animals devoid of pulmonary system as infection models in the study of lung bacterial pathogens. Front Microbiol 6:38

Lopez V, Villar M, Queiros J, Vicente J, Mateos-Hernandez L, Diez-Delgado I et al (2016) Comparative proteomics identifies host immune system proteins affected by infection with mycobacterium bovis. PLoS Negl Trop Dis 10:e0004541

Lopez V, Risalde MA, Contreras M, Mateos-Hernandez L, Vicente J, Gortazar C et al (2018) Heat-inactivated mycobacterium bovis protects zebrafish against mycobacteriosis. J Fish Dis 41: 1515–1528

Luo Q, Mehra S, Golden NA, Kaushal D, Lacey MR (2014) Identification of biomarkers for tuberculosis susceptibility via integrated analysis of gene expression and longitudinal clinical data. Front Genet 5:240

Luukinen H, Hammaren MM, Vanha-Aho LM, Parikka M (2018) Modeling tuberculosis in mycobacterium marinum infected adult zebrafish. J Vis Exp 140:58299
Ly LH, Barhoumi R, Cho SH, Franzblau SG, McMurray DN (2008) Vaccination with bacille-calmette guerin promotes mycobacterial control in guinea pig macrophages infected in vivo. J Infect Dis 198:768–771
Lyadova I, Yeremeev V, Majorov K, Nikonenko B, Khaidukov S, Kondratieva T et al (1998) An ex vivo study of t lymphocytes recovered from the lungs of i/st mice infected with and susceptible to mycobacterium tuberculosis. Infect Immun 66:4981–4988
Lynch CJ, Pierce-Chase CH, Dubos R (1965) A genetic study of susceptibility to experimental tuberculosis in mice infected with mammalian tubercle bacilli. J Exp Med 121:1051–1070
MacGilvary NJ, Kevorkian YL, Tan S (2019) Potassium response and homeostasis in mycobacterium tuberculosis modulates environmental adaptation and is important for host colonization. PLoS Pathog 15:e1007591
Machotka SV, Chapple FE, Stookey JL (1975) Cerebral tuberculosis in a rhesus monkey. J Am Vet Med Assoc 167:648–650
Mackintosh CG, Qureshi T, Waldrup K, Labes RE, Dodds KG, Griffin JF (2000) Genetic resistance to experimental infection with mycobacterium bovis in red deer (cervus elaphus). Infect Immun 68:1620–1625
Mahoney A, Weetjens BJ, Cox C, Beyene N, Reither K, Makingi G et al (2012) Pouched rats' detection of tuberculosis in human sputum: comparison to culturing and polymerase chain reaction. Tuberc Res Treat 2012:716989
Maiello P, DiFazio RM, Cadena AM, Rodgers MA, Lin PL, Scanga CA et al (2018) Rhesus macaques are more susceptible to progressive tuberculosis than cynomolgus macaques: a quantitative comparison. Infect Immun 86:e00505
Manabe YC, Dannenberg AM Jr, Tyagi SK, Hatem CL, Yoder M, Woolwine SC et al (2003) Different strains of mycobacterium tuberculosis cause various spectrums of disease in the rabbit model of tuberculosis. Infect Immun 71:6004–6011
Manabe YC, Kesavan AK, Lopez-Molina J, Hatem CL, Brooks M, Fujiwara R et al (2008) The aerosol rabbit model of tb latency, reactivation and immune reconstitution inflammatory syndrome. Tuberculosis 88:187–196
Marino S, Kirschner DE (2016) A multi-compartment hybrid computational model predicts key roles for dendritic cells in tuberculosis infection. Computation 4:39
Marino S, Pawar S, Fuller CL, Reinhart TA, Flynn JL, Kirschner DE (2004) Dendritic cell trafficking and antigen presentation in the human immune response to mycobacterium tuberculosis. J Immunol 173:494–506
Marino S, Cilfone NA, Mattila JT, Linderman JJ, Flynn JL, Kirschner DE (2015) Macrophage polarization drives granuloma outcome during mycobacterium tuberculosis infection. Infect Immun 83:324–338
Marino S, Gideon HP, Gong C, Mankad S, McCrone JT, Lin PL et al (2016) Computational and empirical studies predict mycobacterium tuberculosis-specific t cells as a biomarker for infection outcome. PLoS Comput Biol 12:e1004804
Markova N, Michailova L, Kussovski V, Jourdanova M, Radoucheva T (2005) Intranasal application of lentinan enhances bactericidal activity of rat alveolar macrophages against mycobacterium tuberculosis. Pharmazie 60:42–48
Martens GW, Arikan MC, Lee J, Ren F, Greiner D, Kornfeld H (2007) Tuberculosis susceptibility of diabetic mice. Am J Respir Cell Mol Biol 37:518–524
Martin AR (1946) The use of mice in the examination of drugs for chemotherapeutic activity against mycobacterium tuberculosis. J Pathol Bacteriol 58:580–585
Martin JE, Cole WC, Whitney RA Jr (1968) Tuberculosis of the spine (pott's disease) in a rhesus monkey (macaca mulatta). J Am Vet Med Assoc 153:914–917
Martin T, Cheke D, Natyshak I (1989) Broth culture: The modern 'guinea-pig' for isolation of mycobacteria. Tubercle 70:53–56

Martin A, Takiff H, Vandamme P, Swings J, Palomino JC, Portaels F (2006) A new rapid and simple colorimetric method to detect pyrazinamide resistance in mycobacterium tuberculosis using nicotinamide. J Antimicrob Chemother 58:327–331

Martin CJ, Cadena AM, Leung VW, Lin PL, Maiello P, Hicks N et al (2017) Digitally barcoding mycobacterium tuberculosis reveals in vivo infection dynamics in the macaque model of tuberculosis. MBio 8:e00312

Martino M, Hubbard GB, Schlabritz-Loutsevitch N (2007) Tuberculosis (mycobacterium tuberculosis) in a pregnant baboon (papio cynocephalus). J Med Primatol 36:108–112

Mattila JT, Ojo OO, Kepka-Lenhart D, Marino S, Kim JH, Eum SY et al (2013) Microenvironments in tuberculous granulomas are delineated by distinct populations of macrophage subsets and expression of nitric oxide synthase and arginase isoforms. J Immunol 191:773–784

Mattila JT, Beaino W, Maiello P, Coleman MT, White AG, Scanga CA et al (2017) Positron emission tomography imaging of macaques with tuberculosis identifies temporal changes in granuloma glucose metabolism and integrin alpha4beta1-expressing immune cells. J Immunol 199:806–815

Mc CA, Katsampes CP, Clausen SW (1946) Effects of intranasal inoculation with tubercle bacilli on vitamin a stores and tissues of mice and rats. Am Rev Tuberc 54:84–91

McCallan L, Corbett D, Andersen PL, Aagaard C, McMurray D, Barry C et al (2011) A new experimental infection model in ferrets based on aerosolised mycobacterium bovis. Vet Med Int 2011:981410

McFarland CT, Ly L, Jeevan A, Yamamoto T, Weeks B, Izzo A et al (2010) Bcg vaccination in the cotton rat (sigmodon hispidus) infected by the pulmonary route with virulent mycobacterium tuberculosis. Tuberculosis 90:262–267

McFarlane AJ, McSorley HJ, Davidson DJ, Fitch PM, Errington C, Mackenzie KJ et al (2017) Enteric helminth-induced type i interferon signaling protects against pulmonary virus infection through interaction with the microbiota. J Allergy Clin Immunol 140:1068–1078

McMurray DN (2003) Hematogenous reseeding of the lung in low-dose, aerosol-infected guinea pigs. Unique features of the host-pathogen interface in secondary tubercles. Tuberculosis 83:131–134

Medina E, North RJ (1998) Resistance ranking of some common inbred mouse strains to mycobacterium tuberculosis and relationship to major histocompatibility complex haplotype and nramp1 genotype. Immunology 93:270–274

Mehra S, Pahar B, Dutta NK, Conerly CN, Philippi-Falkenstein K, Alvarez X et al (2010) Transcriptional reprogramming in nonhuman primate (rhesus macaque) tuberculosis granulomas. PLoS One 5:e12266

Meinzen C, Proano A, Gilman RH, Caviedes L, Coronel J, Zimic M et al (2016) A quantitative adaptation of the wayne test for pyrazinamide resistance. Tuberculosis 99:41–46

Mendoza-Coronel E, Castanon-Arreola M (2016) Comparative evaluation of in vitro human macrophage models for mycobacterial infection study. Pathog Dis 74:52

Merrick JV, Ratcliffe HL (1957) Tuberculosis induced by droplet nuclei infection; its developmental pattern in hamsters in relation to levels of dietary protein. Am J Pathol 33:107–129

Mgode GF, Weetjens BJ, Nawrath T, Cox C, Jubitana M, Machang'u RS et al (2012) Diagnosis of tuberculosis by trained african giant pouched rats and confounding impact of pathogens and microflora of the respiratory tract. J Clin Microbiol 50:274–280

Michael M Jr, Cummings MM, Bloom WL (1950) Course of experimental tuberculosis in the albino rat as influenced by cortisone. Proc Soc Exp Biol Med 75:613–616

Miller HE, Johnson KE, Tarakanova VL, Robinson RT (2019) Gamma-herpesvirus latency attenuates mycobacterium tuberculosis infection in mice. Tuberculosis 116:56–60

Min F, He L, Luo Y, Huang S, Pan J, Wang J et al (2018) Dynamics of immune responses during experimental mycobacterium kansasii infection of cynomolgus monkeys (macaca fascicularis). Mediat Inflamm 2018:8354902

Mishra AK, Yabaji SM, Dubey RK (2018) Evaluation of isoprinosine to be repurposed as an adjunct anti-tuberculosis chemotherapy. Med Hypotheses 115:77–80

Mitchison DA, Allen BW, Lambert RA (1973) Selective media in the isolation of tubercle bacilli from tissues. J Clin Pathol 26:250–252
Monin L, Griffiths KL, Lam WY, Gopal R, Kang DD, Ahmed M et al (2015) Helminth-induced arginase-1 exacerbates lung inflammation and disease severity in tuberculosis. J Clin Invest 125: 4699–4713
Moreira-Teixeira L, Stimpson PJ, Stavropoulos E, Hadebe S, Chakravarty P, Ioannou M et al (2020) Type i ifn exacerbates disease in tuberculosis-susceptible mice by inducing neutrophil-mediated lung inflammation and netosis. Nat Commun 11:5566
Morton JJ (1916) A rapid method for the diagnosis of renal tuberculosis by the use of the X-rayed guinea pig. J Exp Med 24:419–427
Mourik BC, Leenen PJ, de Knegt GJ, Huizinga R, van der Eerden BC, Wang J et al (2017) Immunotherapy added to antibiotic treatment reduces relapse of disease in a mouse model of tuberculosis. Am J Respir Cell Mol Biol 56:233–241
Mueller AK, Behrends J, Hagens K, Mahlo J, Schaible UE, Schneider BE (2012) Natural transmission of plasmodium berghei exacerbates chronic tuberculosis in an experimental co-infection model. PLoS One 7:e48110
Mueller AK, Behrends J, Blank J, Schaible UE, Schneider BE (2014) An experimental model to study tuberculosis-malaria coinfection upon natural transmission of mycobacterium tuberculosis and plasmodium berghei. J Vis Exp 84:e50829
Mulder C, Mgode GF, Ellis H, Valverde E, Beyene N, Cox C et al (2017) Accuracy of giant african pouched rats for diagnosing tuberculosis: comparison with culture and xpert((r)) mtb/rif. Int J Tuberc Lung Dis 21:1127–1133
Murphy JB, Ellis AW (1914) Experiments on the role of lymphoid tissue in the resistance to experimental tuberculosis in mice. J Exp Med 20:397–403
Myllymaki H, Bauerlein CA, Ramet M (2016) The zebrafish breathes new life into the study of tuberculosis. Front Immunol 7:196
Naranjo V, Ayoubi P, Vicente J, Ruiz-Fons F, Gortazar C, Kocan KM et al (2006) Characterization of selected genes upregulated in non-tuberculous european wild boar as possible correlates of resistance to mycobacterium bovis infection. Vet Microbiol 116:224–231
Narayanan RB, Badenoch-Jones P, Turk JL (1981) Experimental mycobacterial granulomas in guinea pig lymph nodes: ultrastructural observations. J Pathol 134:253–265
Nedeltchev GG, Raghunand TR, Jassal MS, Lun S, Cheng QJ, Bishai WR (2009) Extrapulmonary dissemination of mycobacterium bovis but not mycobacterium tuberculosis in a bronchoscopic rabbit model of cavitary tuberculosis. Infect Immun 77:598–603
Negre L, Bretey J (1945) Influence exerted on guinea pig tuberculosis by the bgg administered by cutaneous scarification. Ann Inst Pasteur 71:161–167
Newton S, Martineau A, Kampmann B (2011) A functional whole blood assay to measure viability of mycobacteria, using reporter-gene tagged bcg or m.Tb (bcglux/m.Tb lux). J Vis Exp 55:3332
Neyrolles O, Hernandez-Pando R, Pietri-Rouxel F, Fornes P, Tailleux L, Barrios Payan JA et al (2006) Is adipose tissue a place for mycobacterium tuberculosis persistence? PLoS One 1:e43
Niazi MK, Dhulekar N, Schmidt D, Major S, Cooper R, Abeijon C et al (2015) Lung necrosis and neutrophils reflect common pathways of susceptibility to mycobacterium tuberculosis in genetically diverse, immune-competent mice. Dis Model Mech 8:1141–1153
Nie WJ, Xie ZY, Gao S, Teng TL, Zhou WQ, Shang YY et al (2020) Efficacy of moxifloxacin against mycobacterium abscessus in zebrafish model in vivo. Biomed Environ Sci 33:350–358
Nikonenko BV, Averbakh MM, Lavebratt C, Schurr E, Apt AS (2000) Comparative analysis of mycobacterial infections in susceptible i/st and resistant a/sn inbred mice. Tuber Lung Dis 80: 15–25
Noll KE, Ferris MT, Heise MT (2019) The collaborative cross: a systems genetics resource for studying host-pathogen interactions. Cell Host Microbe 25:484–498
Nugent G, Whitford EJ, Yockney I, Perry M, Tompkins DM, Holtslag N et al (2013a) Percutaneous interdigital injection of mycobacterium bovis as a model for tuberculous lesion development in wild brushtail possums (trichosurus vulpecula). J Comp Pathol 148:33–42

Nugent G, Yockney I, Whitford J, Cross ML (2013b) Mortality rate and gross pathology due to tuberculosis in wild brushtail possums (trichosurus vulpecula) following low dose subcutaneous injection of mycobacterium bovis. Prev Vet Med 109:168–175

Nugent G, Buddle BM, Knowles G (2015) Epidemiology and control of mycobacterium bovis infection in brushtail possums (trichosurus vulpecula), the primary wildlife host of bovine tuberculosis in new zealand. N Z Vet J 63(Suppl 1):28–41

Nusbaum RJ, Calderon VE, Huante MB, Sutjita P, Vijayakumar S, Lancaster KL et al (2016) Pulmonary tuberculosis in humanized mice infected with hiv-1. Sci Rep 6:21522

O'Brien P, Vinnard C, Subbian S (2020) An improved protocol to establish experimental tuberculous meningitis in the rabbit. MethodsX 7:100832

Oehlers SH, Hortle E, Cook KM (2020) A zebrafish model of tuberculosis comorbidity and the effects of hif-activating intervention. FEBS J 287:3917–3920

Oh CT, Moon C, Choi TH, Kim BS, Jang J (2013) Mycobacterium marinum infection in drosophila melanogaster for antimycobacterial activity assessment. J Antimicrob Chemother 68:601–609

Oksanen KE, Halfpenny NJ, Sherwood E, Harjula SK, Hammaren MM, Ahava MJ et al (2013) An adult zebrafish model for preclinical tuberculosis vaccine development. Vaccine 31:5202–5209

Ordway D, Palanisamy G, Henao-Tamayo M, Smith EE, Shanley C, Orme IM et al (2007) The cellular immune response to mycobacterium tuberculosis infection in the guinea pig. J Immunol 179:2532–2541

Ozeki Y, Tsutsui H, Kawada N, Suzuki H, Kataoka M, Kodama T et al (2006) Macrophage scavenger receptor down-regulates mycobacterial cord factor-induced proinflammatory cytokine production by alveolar and hepatic macrophages. Microb Pathog 40:171–176

Pacheco SA, Powers KM, Engelmann F, Messaoudi I, Purdy GE (2013) Autophagic killing effects against mycobacterium tuberculosis by alveolar macrophages from young and aged rhesus macaques. PLoS One 8:e66985

Palanisamy GS, Smith EE, Shanley CA, Ordway DJ, Orme IM, Basaraba RJ (2008) Disseminated disease severity as a measure of virulence of mycobacterium tuberculosis in the guinea pig model. Tuberculosis 88:295–306

Palermo-Neto J, Santos FA, Guerra JL, Santos GO, Pinheiro SR (2001) Glue solvent inhalation impairs host resistance to mycobacterium bovis-induced infection in hamsters. Vet Hum Toxicol 43:1–5

Palmer MV, Whipple DL, Olsen SC (1999) Development of a model of natural infection with mycobacterium bovis in white-tailed deer. J Wildl Dis 35:450–457

Palmer MV, Waters WR, Whipple DL (2002) Aerosol delivery of virulent mycobacterium bovis to cattle. Tuberculosis 82:275–282

Palmer MV, Thacker TC, Waters WR, Gortazar C, Corner LA (2012) Mycobacterium bovis: a model pathogen at the interface of livestock, wildlife, and humans. Vet Med Int 2012:236205

Palucci I, Battah B, Salustri A, De Maio F, Petrone L, Ciccosanti F et al (2019) Ip-10 contributes to the inhibition of mycobacterial growth in an ex vivo whole blood assay. Int J Med Microbiol 309:299–306

Pardieu C, Casali N, Clark SO, Hooper R, Williams A, Velji P et al (2015) Correlates between models of virulence for mycobacterium tuberculosis among isolates of the central asian lineage: a case for lysozyme resistance testing? Infect Immun 83:2213–2223

Parikka M, Hammaren MM, Harjula SK, Halfpenny NJ, Oksanen KE, Lahtinen MJ et al (2012) Mycobacterium marinum causes a latent infection that can be reactivated by gamma irradiation in adult zebrafish. PLoS Pathog 8:e1002944

Parish T (2020) In vitro drug discovery models for mycobacterium tuberculosis relevant for host infection. Expert Opin Drug Discovery 15:349–358

Parsons SD, de Villiers C, Gey NC, van Pittius, Warren RM, van Helden PD (2010) Detection of mycobacterium kansasii infection in a rhesus macaque (macaca mulatta) using a modified quantiferon-tb gold assay. Vet Immunol Immunopathol 136:330–334

Pasipanodya JG, Nuermberger E, Romero K, Hanna D, Gumbo T (2015) Systematic analysis of hollow fiber model of tuberculosis experiments. Clin Infect Dis 61(Suppl 1):S10–S17

Patel K, Jhamb SS, Singh PP (2011) Models of latent tuberculosis: their salient features, limitations, and development. J Lab Phys 3:75–79

Pathak S, Awuh JA, Leversen NA, Flo TH, Asjo B (2012) Counting mycobacteria in infected human cells and mouse tissue: a comparison between qpcr and cfu. PLoS One 7:e34931

Pelaez Coyotl EA, Barrios Palacios J, Mucino G, Moreno-Blas D, Costas M, Montiel Montes T et al (2020) Antimicrobial peptide against mycobacterium tuberculosis that activates autophagy is an effective treatment for tuberculosis. Pharmaceutics 12:1071

Pena JC, Ho WZ (2016) Non-human primate models of tuberculosis. Microbiol Spectr 4:4

Phalen SW, McMurray DN (1993) T-lymphocyte response in a guinea pig model of tuberculous pleuritis. Infect Immun 61:142–145

Pi J, Shen L, Shen H, Yang E, Wang W, Wang R et al (2019) Mannosylated graphene oxide as macrophage-targeted delivery system for enhanced intracellular m.Tuberculosis killing efficiency. Mater Sci Eng C Mater Biol Appl 103:109777

Pienaar E, Matern WM, Linderman JJ, Bader JS, Kirschner DE (2016) Multiscale model of mycobacterium tuberculosis infection maps metabolite and gene perturbations to granuloma sterilization predictions. Infect Immun 84:1650–1669

Pienaar E, Sarathy J, Prideaux B, Dietzold J, Dartois V, Kirschner DE et al (2017) Comparing efficacies of moxifloxacin, levofloxacin and gatifloxacin in tuberculosis granulomas using a multi-scale systems pharmacology approach. PLoS Comput Biol 13:e1005650

Pierce C, Dubos RJ, Middlebrook G (1947) Infection of mice with mammalian tubercle bacilli grown in tween-albumin liquid medium. J Exp Med 86:159–174

Pieterman ED, Te Brake LHM, de Knegt GJ, van der Meijden A, Alffenaar JC, Bax HI et al (2018) Assessment of the additional value of verapamil to a moxifloxacin and linezolid combination regimen in a murine tuberculosis model. Antimicrob Agents Chemother 62:e00312

Pieterman ED, van den Berg S, van der Meijden A, Svensson EM, Bax HI, de Steenwinkel JEM (2021) Higher dosing of rifamycins does not increase activity against M. tuberculosis in the hollow fibre infection model. Antimicrob Agents Chemother 65(4):e02255

Plesker R, Teschner K, Behlert O, Prenger-Berninghoff E, Hillemann D (2010) Airborne mycobacterium avium infection in a group of red-shanked douc langurs (pygathrix nemaeus nemaeus). J Med Primatol 39:129–135

Podell BK, Ackart DF, Obregon-Henao A, Eck SP, Henao-Tamayo M, Richardson M et al (2014) Increased severity of tuberculosis in guinea pigs with type 2 diabetes: a model of diabetes-tuberculosis comorbidity. Am J Pathol 184:1104–1118

Popovic M, Yaparla A, Paquin-Proulx D, Koubourli DV, Webb R, Firmani M et al (2019) Colony-stimulating factor-1- and interleukin-34-derived macrophages differ in their susceptibility to mycobacterium marinum. J Leukoc Biol 106:1257–1269

Potter EL, Gideon HP, Tkachev V, Fabozzi G, Chassiakos A, Petrovas C et al (2021) Measurement of leukocyte trafficking kinetics in macaques by serial intravascular staining. Sci Transl Med 13: 4582

Prouty MG, Correa NE, Barker LP, Jagadeeswaran P, Klose KE (2003) Zebrafish-mycobacterium marinum model for mycobacterial pathogenesis. FEMS Microbiol Lett 225:177–182

Pushkaran AC, Vinod V, Vanuopadath M, Nair SS, Nair SV, Vasudevan AK et al (2019) Combination of repurposed drug diosmin with amoxicillin-clavulanic acid causes synergistic inhibition of mycobacterial growth. Sci Rep 9:6800

Queiros J, Vicente J (2018) Inbreeding shapes tuberculosis progression in female adult badgers (meles meles). J Anim Ecol 87:1497–1499

Rafi W, Bhatt K, Gause WC, Salgame P (2015) Neither primary nor memory immunity to mycobacterium tuberculosis infection is compromised in mice with chronic enteric helminth infection. Infect Immun 83:1217–1223

Rahim Z, Thapa J, Fukushima Y, van der Zanden AGM, Gordon SV, Suzuki Y et al (2017) Tuberculosis caused by mycobacterium orygis in dairy cattle and captured monkeys in bangladesh: A new scenario of tuberculosis in south asia. Transbound Emerg Dis 64:1965–1969

Ramakrishnan L, Falkow S (1994) Mycobacterium marinum persists in cultured mammalian cells in a temperature-restricted fashion. Infect Immun 62:3222–3229
Ramakrishnan L, Valdivia RH, McKerrow JH, Falkow S (1997) Mycobacterium marinum causes both long-term subclinical infection and acute disease in the leopard frog (rana pipiens). Infect Immun 65:767–773
Ramos L, Obregon-Henao A, Henao-Tamayo M, Bowen R, Lunney JK, Gonzalez-Juarrero M (2017) The minipig as an animal model to study mycobacterium tuberculosis infection and natural transmission. Tuberculosis 106:91–98
Rao Muvva J, Ahmed S, Rekha RS, Kalsum S, Groenheit R, Schon T et al (2021) Immunomodulatory agents combat multidrug-resistant tuberculosis by improving antimicrobial immunity. J Infect Dis 224:332–344
Rao NA, Albini TA, Kumaradas M, Pinn ML, Fraig MM, Karakousis PC (2009) Experimental ocular tuberculosis in guinea pigs. Arch Ophthalmol 127:1162–1166
Raposo-Garcia S, Guerra-Laso JM, Garcia-Garcia S, Juan-Garcia J, Lopez-Fidalgo E, Diez-Tascon C et al (2017) Immunological response to mycobacterium tuberculosis infection in blood from type 2 diabetes patients. Immunol Lett 186:41–45
Ratcliffe HL, Palladino VS (1953) Tuberculosis induced by droplet nuclei infection; initial homogeneous response of small mammals (rats, mice, guinea pigs, and hamsters) to human and to bovine bacilli, and the rate and pattern of tubercle development. J Exp Med 97:61–68
Rayner EL, Pearson GR, Hall GA, Basaraba RJ, Gleeson F, McIntyre A et al (2013) Early lesions following aerosol infection of rhesus macaques (macaca mulatta) with mycobacterium tuberculosis strain h37rv. J Comp Pathol 149:475–485
Redford PS, Mayer-Barber KD, McNab FW, Stavropoulos E, Wack A, Sher A et al (2014) Influenza a virus impairs control of mycobacterium tuberculosis coinfection through a type i interferon receptor-dependent pathway. J Infect Dis 209:270–274
Reece ST, Loddenkemper C, Askew DJ, Zedler U, Schommer-Leitner S, Stein M et al (2010) Serine protease activity contributes to control of mycobacterium tuberculosis in hypoxic lung granulomas in mice. J Clin Invest 120.3365–3376
Reis AC, Ramos B, Pereira AC, Cunha MV (2020) The hard numbers of tuberculosis epidemiology in wildlife: a meta-regression and systematic review. Transbound Emerg Dis 68(6):3257–3276
Renner M, Bartholomew WR (1974) Mycobacteriologic data from two outbreaks of bovine tuberculosis in nonhuman primates. Am Rev Respir Dis 109:11–16
Rhoo KH, Edholm ES, Forzan MJ, Khan A, Waddle AW, Pavelka MS Jr et al (2019) Distinct host-mycobacterial pathogen interactions between resistant adult and tolerant tadpole life stages of xenopus laevis. J Immunol 203:2679–2688
Rifat D, Bishai WR, Karakousis PC (2009) Phosphate depletion: a novel trigger for mycobacterium tuberculosis persistence. J Infect Dis 200:1126–1135
Rifat D, Prideaux B, Savic RM, Urbanowski ME, Parsons TL, Luna B et al (2018) Pharmacokinetics of rifapentine and rifampin in a rabbit model of tuberculosis and correlation with clinical trial data. Sci Transl Med 10:7786
Righi DA, Pinheiro SR, Guerra JL, Palermo-Neto J (1999) Effects of diazepam on mycobacterium bovis-induced infection in hamsters. Braz J Med Biol Res 32:1145–1153
Ring S, Eggers L, Behrends J, Wutkowski A, Schwudke D, Kroger A et al (2019) Blocking il-10 receptor signaling ameliorates mycobacterium tuberculosis infection during influenza-induced exacerbation. JCI Insight 5:e126533
Risalde MA, Lopez V, Contreras M, Mateos-Hernandez L, Gortazar C, de la Fuente J (2018) Control of mycobacteriosis in zebrafish (danio rerio) mucosally vaccinated with heat-inactivated mycobacterium bovis. Vaccine 36:4447–4453
Rocha VC, Ikuta CY, Gomes MS, Quaglia F, Matushima ER, Ferreira Neto JS (2011) Isolation of mycobacterium tuberculosis from captive ateles paniscus. Vector Borne Zoonotic Dis 11:593–594

Rodgers JD, Connery NL, McNair J, Welsh MD, Skuce RA, Bryson DG et al (2007) Experimental exposure of cattle to a precise aerosolised challenge of mycobacterium bovis: a novel model to study bovine tuberculosis. Tuberculosis 87:405–414

Rodrigues RF, Zarate-Blades CR, Rios WM, Soares LS, Souza PR, Brandao IT et al (2015) Synergy of chemotherapy and immunotherapy revealed by a genome-scale analysis of murine tuberculosis. J Antimicrob Chemother 70:1774–1783

Rosenbaum M, Mendoza P, Ghersi BM, Wilbur AK, Perez-Brumer A, Cavero Yong N et al (2015) Detection of mycobacterium tuberculosis complex in new world monkeys in peru. EcoHealth 12:288–297

Rouco C, Richardson KS, Buddle BM, French NP, Tompkins DM (2016) Sex difference in the survival rate of wild brushtail possums (trichosurus vulpecula) experimentally challenged with bovine tuberculosis. Res Vet Sci 107:102–105

Rozenberg AM, Pisarenko NN (1965) The gold hamster as an experimental model for the study of tuberculosis and antitubercular vaccination. Zh Mikrobiol Epidemiol Immunobiol 42:131–136

Ruley KM, Reimschuessel R, Trucksis M (2002) Goldfish as an animal model system for mycobacterial infection. Methods Enzymol 358:29–39

Safar HA, Mustafa AS, Amoudy HA, El-Hashim A (2020) The effect of adjuvants and delivery systems on th1, th2, th17 and treg cytokine responses in mice immunized with mycobacterium tuberculosis-specific proteins. PLoS One 15:e0228381

Sanchez-Hidalgo A, Obregon-Henao A, Wheat WH, Jackson M, Gonzalez-Juarrero M (2017) Mycobacterium bovis hosted by free-living-amoebae permits their long-term persistence survival outside of host mammalian cells and remain capable of transmitting disease to mice. Environ Microbiol 19:4010–4021

Sapolsky RM, Else JG (1987) Bovine tuberculosis in a wild baboon population: Epidemiological aspects. J Med Primatol 16:229–235

Saralahti AK, Uusi-Makela MIE, Niskanen MT, Ramet M (2020) Integrating fish models in tuberculosis vaccine development. Dis Model Mech 13:45716

Sarathy JP, Via LE, Weiner D, Blanc L, Boshoff H, Eugenin EA et al (2018) Extreme drug tolerance of mycobacterium tuberculosis in caseum. Antimicrob Agents Chemother 62:e02266

Sarathy J, Blanc L, Alvarez-Cabrera N, O'Brien P, Dias-Freedman I, Mina M et al (2019) Fluoroquinolone efficacy against tuberculosis is driven by penetration into lesions and activity against resident bacterial populations. Antimicrob Agents Chemother 63:e02516

Saxena PS, Sharma RK (1982) Value of histopathology, culture and guinea pig inoculation in osteoarticular tuberculosis. Int Surg 67:540–542

Scanga CA, Mohan VP, Joseph H, Yu K, Chan J, Flynn JL (1999) Reactivation of latent tuberculosis: variations on the cornell murine model. Infect Immun 67:4531–4538

Scheid G, Mendheim H (1949) About the simultaneous occurrence of spontaneous tuberculosis and malignant spontaneous tumors in the white laboratory rat. Tuberkulosearzt 3:88–91

Schinkothe J, Kohler H, Liebler-Tenorio EM (2016a) Characterization of tuberculous granulomas in different stages of progression and associated tertiary lymphoid tissue in goats experimentally infected with mycobacterium avium subsp. Hominissuis. Comp Immunol Microbiol Infect Dis 47:41–51

Schinkothe J, Mobius P, Kohler H, Liebler-Tenorio EM (2016b) Experimental infection of goats with mycobacterium avium subsp. Hominissuis: a model for comparative tuberculosis research. J Comp Pathol 155:218–230

Schroeder CR (1938) Acquired tuberculosis in the primate in laboratories and zoological collections. Am J Public Health Nations Health 28:469–475

Sedgwick C, Parcher J, Durham R (1970) Atypical mycobacterial infection in the pig-tailed macaque (macaca nemestrina). J Am Vet Med Assoc 157:724–725

Sershen CL, Plimpton SJ, May EE (2016) Oxygen modulates the effectiveness of granuloma mediated host response to mycobacterium tuberculosis: a multiscale computational biology approach. Front Cell Infect Microbiol 6:6

Sesline DH, Schwartz LW, Osburn BI, Thoen CO, Terrell T, Holmberg C et al (1975) Mycobacterium avium infection in three rhesus monkeys. J Am Vet Med Assoc 167:639–645
Sha S, Shi X, Deng G, Chen L, Xin Y, Ma Y (2017) Mycobacterium tuberculosis rv1987 induces th2 immune responses and enhances mycobacterium smegmatis survival in mice. Microbiol Res 197:74–80
Sha S, Shi Y, Tang Y, Jia L, Han X, Liu Y et al (2021) Mycobacterium tuberculosis rv1987 protein induces m2 polarization of macrophages through activating the pi3k/akt1/mtor signaling pathway. Immunol Cell Biol 99:570–585
Shakila H, Jayasankar K, Ramanathan VD (1999) The clearance of tubercle bacilli & mycobacterial antigen vis a vis the granuloma in different organs of guinea pigs. Indian J Med Res 110:4–10
Shanley CA, Streicher EM, Warren RM, Victor TC, Orme IM (2013) Characterization of w-beijing isolates of mycobacterium tuberculosis from the western cape. Vaccine 31:5934–5939
Sharpe SA, Eschelbach E, Basaraba RJ, Gleeson F, Hall GA, McIntyre A et al (2009) Determination of lesion volume by mri and stereology in a macaque model of tuberculosis. Tuberculosis 89:405–416
Sharpe S, White A, Gleeson F, McIntyre A, Smyth D, Clark S et al (2016) Ultra low dose aerosol challenge with mycobacterium tuberculosis leads to divergent outcomes in rhesus and cynomolgus macaques. Tuberculosis 96:1–12
Shkurupy VA, Cherdantseva LA, Kovner AV, Troitskiy AV, Bystrova AV, Starostenko AA (2020) Efficacy of inhalations of antituberculous compositions with different length of experimental therapy course in mice. Bull Exp Biol Med 168:743–747
Sibley L, Dennis M, Sarfas C, White A, Clark S, Gleeson F et al (2016) Route of delivery to the airway influences the distribution of pulmonary disease but not the outcome of mycobacterium tuberculosis infection in rhesus macaques. Tuberculosis 96:141–149
Sibley L, Gooch K, Wareham A, Gray S, Chancellor A, Dowall S et al (2019) Differences in monocyte: lymphocyte ratio and tuberculosis disease progression in genetically distinct populations of macaques. Sci Rep 9:3340
Sichewo PR, Etter EMC, Michel AL (2020) Wildlife-cattle interactions emerge as drivers of bovine tuberculosis in traditionally farmed cattle. Prev Vet Med 174:104847
Singh RP, Jhamb SS, Singh PP (2008) Effects of morphine during mycobacterium tuberculosis h37rv infection in mice. Life Sci 82:308–314
Singh RP, Jhamb SS, Singh PP (2009) Effect of morphine on mycobacterium smegmatis infection in mice and macrophages. Indian J Microbiol 49:276–282
Singhal A, Aliouat M, Herve M, Mathys V, Kiass M, Creusy C et al (2011a) Experimental tuberculosis in the wistar rat: a model for protective immunity and control of infection. PLoS One 6:e18632
Singhal A, Mathys V, Kiass M, Creusy C, Delaire B, Aliouat M et al (2011b) Bcg induces protection against mycobacterium tuberculosis infection in the wistar rat model. PLoS One 6: e28082
Sivangala Thandi R, Radhakrishnan RK, Tripathi D, Paidipally P, Azad AK, Schlesinger LS et al (2020) Ornithine-a urea cycle metabolite enhances autophagy and controls mycobacterium tuberculosis infection. Nat Commun 11:3535
Skinner MA, Yuan S, Prestidge R, Chuk D, Watson JD, Tan PL (1997) Immunization with heat-killed mycobacterium vaccae stimulates cd8+ cytotoxic t cells specific for macrophages infected with mycobacterium tuberculosis. Infect Immun 65:4525–4530
Skinner MA, Keen DL, Parlane NA, Yates GF, Buddle BM (2002) Increased protection against bovine tuberculosis in the brushtail possum (trichosurus vulpecula) when bcg is administered with killed mycobacterium vaccae. Tuberculosis 82:15–22
Smith MI, Mc CW, Emmart EW (1946a) Influence of streptomycin and promin on proliferation of tubercle bacilli in the tissues of albino rat. Proc Soc Exp Biol Med 62:157–162
Smith MI, Mc CW, Emmart EW (1946b) The influence of streptomycin and promin on the proliferation of tubercle bacilli in the tissues of the albino rat. Fed Proc 5:204

Smith EK, Hunt RD, Garcia FG, Fraser CE, Merkal RS, Karlson AG (1973) Avian tuberculosis in monkeys. A unique mycobacterial infection. Am Rev Respir Dis 107:469–471

Smith DW, Balasubramanian V, Wiegeshaus E (1991) A guinea pig model of experimental airborne tuberculosis for evaluation of the response to chemotherapy: the effect on bacilli in the initial phase of treatment. Tubercle 72:223–231

Smith CM, Proulx MK, Olive AJ, Laddy D, Mishra BB, Moss C et al (2016) Tuberculosis susceptibility and vaccine protection are independently controlled by host genotype. MBio 7: e01516

Smith CM, Proulx MK, Lai R, Kiritsy MC, Bell TA, Hock P et al (2019) Functionally overlapping variants control tuberculosis susceptibility in collaborative cross mice. MBio 10:e02791

Snyder S, Peace T, Soave O, Lund J (1970) Tuberculosis in an owl monkey (aotus trivirgatus). J Am Vet Med Assoc 157:712–713

Sohaskey CD, Voskuil MI (2015) In vitro models that utilize hypoxia to induce non-replicating persistence in mycobacteria. Methods Mol Biol 1285:201–213

Solomon JM, Leung GS, Isberg RR (2003) Intracellular replication of mycobacterium marinum within dictyostelium discoideum: efficient replication in the absence of host coronin. Infect Immun 71:3578–3586

Soltys MA, Jennings AR (1950) The dissemination of tubercle bacilli in experimental tuberculosis in the guinea pig. Am Rev Tuberc 61:399–406

Sommer R, Cole ST (2019) Monitoring tuberculosis drug activity in live animals by near-infrared fluorescence imaging. Antimicrob Agents Chemother 63(12):e01280

Srivastava S, Pasipanodya JG, Ramachandran G, Deshpande D, Shuford S, Crosswell HE et al (2016) A long-term co-perfused disseminated tuberculosis-3d liver hollow fiber model for both drug efficacy and hepatotoxicity in babies. EBioMedicine 6:126–138

Srivastava S, van Zyl J, Cirrincione K, Martin K, Thomas T, Deshpande D et al (2020) Evaluation of ceftriaxone plus avibactam in an intracellular hollow fiber model of tuberculosis: Implications for the treatment of disseminated and meningeal tuberculosis in children. Pediatr Infect Dis J 39: 1092–1100

Stammes MA, Bakker J, Vervenne RAW, Zijlmans DGM, van Geest L, Vierboom MPM et al (2021) Recommendations for standardizing thorax pet-ct in non-human primates by recent experience from macaque studies. Animals 11:204

Starck HJ, Viehmann P (1955) Usefulness of syrian gold hamster in diagnosis of tuberculosis. Zentralbl Bakteriol Orig 162:446–451

Steenken W Jr, Pratt PC (1949) Streptomycin in experimental tuberculosis; effect on the pathogenesis of early tuberculosis in the guinea pig infected with streptomycin-sensitive h37 rv tubercle bacilli. Am Rev Tuberc 59:664–673

Steenken W Jr, Wagley PF (1945) Comparison of the golden hamster with the guinea pig following inoculations of virulent tubercle bacilli. Proc Soc Exp Biol Med 60:255–257

Sterling TR, Lin PL (2020) Treatment of latent m. Tuberculosis infection and use of antiretroviral therapy to prevent tuberculosis. J Clin Invest 130:5102–5104

Stockinger DE, Roellich KM, Vogel KW, Eiffert KL, Torrence AE, Prentice JL et al (2011) Primary hepatic mycobacterium tuberculosis complex infection with terminal dissemination in a pig-tailed macaque (macaca nemestrina). J Am Assoc Lab Anim Sci 50:258–262

Stringer LA, Wilson PR, Heuer C, Hunnam JC, Mackintosh CG (2011) Effect of vaccination and natural infection with mycobacterium avium subsp. Paratuberculosis on specificity of diagnostic tests for bovine tuberculosis in farmed red deer (cervus elaphus). N Z Vet J 59:218–224

Subbian S, Tsenova L, O'Brien P, Yang G, Kushner NL, Parsons S et al (2012) Spontaneous latency in a rabbit model of pulmonary tuberculosis. Am J Pathol 181:1711–1724

Subbian S, Bandyopadhyay N, Tsenova L, O'Brien P, Khetani V, Kushner NL et al (2013a) Early innate immunity determines outcome of mycobacterium tuberculosis pulmonary infection in rabbits. Cell Commun Signal 11:60

Subbian S, O'Brien P, Kushner NL, Yang G, Tsenova L, Peixoto B et al (2013b) Molecular immunologic correlates of spontaneous latency in a rabbit model of pulmonary tuberculosis. Cell Commun Signal 11:16

Sugawara I, Mizuno S (2008) Higher susceptibility of type 1 diabetic rats to mycobacterium tuberculosis infection. Tohoku J Exp Med 216:363–370

Sugawara I, Udagawa T, Yamada H (2004a) Rat neutrophils prevent the development of tuberculosis. Infect Immun 72:1804–1806

Sugawara I, Yamada H, Mizuno S (2004b) Pathological and immunological profiles of rat tuberculosis. Int J Exp Pathol 85:125–134

Sugawara I, Yamada H, Mizuno S (2004c) Pulmonary tuberculosis in spontaneously diabetic goto kakizaki rats. Tohoku J Exp Med 204:135–145

Sugawara I, Yamada H, Mizuno S (2006) Nude rat (f344/n-rnu) tuberculosis. Cell Microbiol 8:661–667

Sugawara I, Yamada H, Mizuno S (2007) Bcg vaccination enhances resistance to m. tuberculosis infection in guinea pigs fed a low casein diet. Tohoku J Exp Med 211:259–268

Sugawara I, Udagawa T, Aoki T, Mizuno S (2009) Establishment of a guinea pig model of latent tuberculosis with gfp-introduced mycobacterium tuberculosis. Tohoku J Exp Med 219:257–262

Sun H, Ma X, Zhang G, Luo Y, Tang K, Lin X et al (2012) Effects of immunomodulators on liquefaction and ulceration in the rabbit skin model of tuberculosis. Tuberculosis 92:345–350

Sun C, Yang G, Yuan J, Peng X, Zhang C, Zhai X et al (2017) Mycobacterium tuberculosis hypoxic response protein 1 (hrp1) augments the pro-inflammatory response and enhances the survival of mycobacterium smegmatis in murine macrophages. J Med Microbiol 66:1033–1044

Swaim LE, Connolly LE, Volkman HE, Humbert O, Born DE, Ramakrishnan L (2006) Mycobacterium marinum infection of adult zebrafish causes caseating granulomatous tuberculosis and is moderated by adaptive immunity. Infect Immun 74:6108–6117

Takaki K, Ramakrishnan L, Basu S (2018) A zebrafish model for ocular tuberculosis. PLoS One 13: e0194982

Tavolara TE, Niazi MKK, Ginese M, Piedra-Mora C, Gatti DM, Beamer G et al (2020) Automatic discovery of clinically interpretable imaging biomarkers for mycobacterium tuberculosis supersusceptibility using deep learning. EBioMedicine 62:103094

Tell LA, Woods L, Foley J, Needham ML, Walker RL (2003) A model of avian mycobacteriosis: clinical and histopathologic findings in japanese quail (coturnix coturnix japonica) intravenously inoculated with mycobacterium avium. Avian Dis 47:433–443

Tezera LB, Bielecka MK, Chancellor A, Reichmann MT, Shammari BA, Brace P et al (2017a) Dissection of the host-pathogen interaction in human tuberculosis using a bioengineered 3-dimensional model. elife 6:e21283

Tezera LB, Bielecka MK, Elkington PT (2017b) Bioelectrospray methodology for dissection of the host-pathogen interaction in human tuberculosis. Bio Protoc 7:e00312

Thacker VV, Dhar N, Sharma K, Barrile R, Karalis K, McKinney JD (2020) A lung-on-chip model of early mycobacterium tuberculosis infection reveals an essential role for alveolar epithelial cells in controlling bacterial growth. elife 9:e59961

Thayil SM, Albini TA, Nazari H, Moshfeghi AA, Parel JM, Rao NA et al (2011) Local ischemia and increased expression of vascular endothelial growth factor following ocular dissemination of mycobacterium tuberculosis. PLoS One 6:e28383

Thompson EG, Shankar S, Gideon HP, Braun J, Valvo J, Skinner JA et al (2018) Prospective discrimination of controllers from progressors early after low-dose mycobacterium tuberculosis infection of cynomolgus macaques using blood rna signatures. J Infect Dis 217:1318–1322

Tornack J, Reece ST, Bauer WM, Vogelzang A, Bandermann S, Zedler U et al (2017) Human and mouse hematopoietic stem cells are a depot for dormant mycobacterium tuberculosis. PLoS One 12:e0169119

Trofimov V, Kicka S, Mucaria S, Hanna N, Ramon-Olayo F, Del Peral LV et al (2018) Antimycobacterial drug discovery using mycobacteria-infected amoebae identifies anti-infectives and new molecular targets. Sci Rep 8:3939

Tsenova L, Ellison E, Harbacheuski R, Moreira AL, Kurepina N, Reed MB et al (2005) Virulence of selected mycobacterium tuberculosis clinical isolates in the rabbit model of meningitis is dependent on phenolic glycolipid produced by the bacilli. J Infect Dis 192:98–106

Tsenova L, Harbacheuski R, Sung N, Ellison E, Fallows D, Kaplan G (2007) Bcg vaccination confers poor protection against m. Tuberculosis hn878-induced central nervous system disease. Vaccine 25:5126–5132

Tsenova L, Fallows D, Kolloli A, Singh P, O'Brien P, Kushner N et al (2020) Inoculum size and traits of the infecting clinical strain define the protection level against mycobacterium tuberculosis infection in a rabbit model. Eur J Immunol 50:858–872

Tsujimura Y, Shiogama Y, Soma S, Okamura T, Takano J, Urano E et al (2020) Vaccination with intradermal bacillus calmette-guerin provides robust protection against extrapulmonary tuberculosis but not pulmonary infection in cynomolgus macaques. J Immunol 205:3023–3036

Tucker EW, Pokkali S, Zhang Z, DeMarco VP, Klunk M, Smith ES et al (2016) Microglia activation in a pediatric rabbit model of tuberculous meningitis. Dis Model Mech 9:1497–1506

Tucker EW, Guglieri-Lopez B, Ordonez AA, Ritchie B, Klunk MH, Sharma R et al (2018) Noninvasive (11)c-rifampin positron emission tomography reveals drug biodistribution in tuberculous meningitis. Sci Transl Med 10:965

Turner OC, Basaraba RJ, Orme IM (2003a) Immunopathogenesis of pulmonary granulomas in the guinea pig after infection with mycobacterium tuberculosis. Infect Immun 71:864–871

Turner OC, Keefe RG, Sugawara I, Yamada H, Orme IM (2003b) Swr mice are highly susceptible to pulmonary infection with mycobacterium tuberculosis. Infect Immun 71:5266–5272

Ugaz EM, Pinheiro SR, Guerra JL, Palermo-Neto J (1999) Effects of prenatal diazepam treatment on mycobacterium bovis-induced infection in hamsters. Immunopharmacology 41:209–217

van Leeuwen LM, van der Kuip M, Youssef SA, de Bruin A, Bitter W, van Furth AM et al (2014) Modeling tuberculous meningitis in zebrafish using mycobacterium marinum. Dis Model Mech 7:1111–1122

Van Rhijn I, Godfroid J, Michel A, Rutten V (2008) Bovine tuberculosis as a model for human tuberculosis: advantages over small animal models. Microbes Infect 10:711–715

van Well GT, Wieland CW, Florquin S, Roord JJ, van der Poll T, van Furth AM (2007) A new murine model to study the pathogenesis of tuberculous meningitis. J Infect Dis 195:694–697

van Wijk RC, Hu W, Dijkema SM, van den Berg DJ, Liu J, Bahi R et al (2020) Anti-tuberculosis effect of isoniazid scales accurately from zebrafish to humans. Br J Pharmacol 177:5518–5533

Via LE, Lin PL, Ray SM, Carrillo J, Allen SS, Eum SY et al (2008) Tuberculous granulomas are hypoxic in guinea pigs, rabbits, and nonhuman primates. Infect Immun 76:2333–2340

Via LE, Schimel D, Weiner DM, Dartois V, Dayao E, Cai Y et al (2012) Infection dynamics and response to chemotherapy in a rabbit model of tuberculosis using [(1)(8)f]2-fluoro-deoxy-d-glucose positron emission tomography and computed tomography. Antimicrob Agents Chemother 56:4391–4402

Via LE, Weiner DM, Schimel D, Lin PL, Dayao E, Tankersley SL et al (2013) Differential virulence and disease progression following mycobacterium tuberculosis complex infection of the common marmoset (callithrix jacchus). Infect Immun 81:2909–2919

Viallier J, Cayre RM (1955) Behavior of the golden hamster inoculated with bcg vaccine and with strains of homologous tubercle bacilli. C R Seances Soc Biol Fil 149:1991–1993

Wachtman LM, Miller AD, Xia D, Curran EH, Mansfield KG (2011) Colonization with nontuberculous mycobacteria is associated with positive tuberculin skin test reactions in the common marmoset (callithrix jacchus). Comp Med 61:278–284

Walsh GP, Tan EV, dela Cruz EC, Abalos RM, Villahermosa LG, Young LJ et al (1996) The philippine cynomolgus monkey (macaca fasicularis) provides a new nonhuman primate model of tuberculosis that resembles human disease. Nat Med 2:430–436

Walter FR, Gilpin TE, Herbath M, Deli MA, Sandor M, Fabry Z (2020) A novel in vitro mouse model to study mycobacterium tuberculosis dissemination across brain vessels: A combination granuloma and blood-brain barrier mouse model. Curr Protoc Immunol 130:e101

Wang C, Zhang Q, Tang X, An Y, Li S, Xu H et al (2019) Effects of cwlm on autolysis and biofilm formation in mycobacterium tuberculosis and mycobacterium smegmatis. Int J Med Microbiol 309:73–83

Wasz-Hockert O, Backman A (1954) Effect of various vaccines on the course of experimental guinea pig tuberculosis. Ann Paediatr Fenn 1:91–98

Waters WR, Palmer MV, Olsen SC, Sacco RE, Whipple DL (2003) Immune responses of elk to mycobacterium bovis bacillus calmette guerin vaccination. Vaccine 21:1518–1526

Waters WR, Maggioli MF, McGill JL, Lyashchenko KP, Palmer MV (2014) Relevance of bovine tuberculosis research to the understanding of human disease: historical perspectives, approaches, and immunologic mechanisms. Vet Immunol Immunopathol 159:113–132

Wayne LG (2001) In vitro model of hypoxically induced nonreplicating persistence of mycobacterium tuberculosis. Methods Mol Med 54:247–269

Wayne LG, Hayes LG (1996) An in vitro model for sequential study of shiftdown of mycobacterium tuberculosis through two stages of nonreplicating persistence. Infect Immun 64:2062–2069

Webb EK, Saccardo CC, Poling A, Cox C, Fast CD (2020) Rapidly training african giant pouched rats (cricetomys ansorgei) with multiple targets for scent detection. Behav Process 174:104085

Weikert LF, Lopez JP, Abdolrasulnia R, Chroneos ZC, Shepherd VL (2000) Surfactant protein a enhances mycobacterial killing by rat macrophages through a nitric oxide-dependent pathway. Am J Physiol Lung Cell Mol Physiol 279:L216–L223

Wessler T, Joslyn LR, Borish HJ, Gideon HP, Flynn JL, Kirschner DE et al (2020) A computational model tracks whole-lung mycobacterium tuberculosis infection and predicts factors that inhibit dissemination. PLoS Comput Biol 16:e1007280

West CS, Vainisi SJ, Vygantas CM, Beluhan FZ (1981) Intraocular granulomas associated with tuberculosis in primates. J Am Vet Med Assoc 179:1240–1244

Whelan AO, Coad M, Cockle PJ, Hewinson G, Vordermeier M, Gordon SV (2010) Revisiting host preference in the mycobacterium tuberculosis complex: experimental infection shows m. Tuberculosis h37rv to be avirulent in cattle. PLoS One 5:e8527

Wilkinson PC, White RG (1966) The role of mycobacteria and silica in the immunological response of the guinea-pig. Immunology 11:229–241

Williams A, Hatch GJ, Clark SO, Gooch KE, Hatch KA, Hall GA et al (2005) Evaluation of vaccines in the eu tb vaccine cluster using a guinea pig aerosol infection model of tuberculosis. Tuberculosis 85:29–38

Williams WR, Troudt J, Creissen E, Bielefeldt-Ohmann H, Johnston MS, Kendall LV et al (2020) Evaluation of peripheral blood markers as early endpoint criteria in guinea pigs (cavia porcellus) when testing tuberculosis vaccine candidates. Comp Med 70:45–55

Winchell CG, Mishra BB, Phuah JY, Saqib M, Nelson SJ, Maiello P et al (2020) Evaluation of il-1 blockade as an adjunct to linezolid therapy for tuberculosis in mice and macaques. Front Immunol 11:891

Wolf RH, Bullock BC, Clarkson TB (1967) Tuberculosis in the stumptailed macaque (macaca speciosa). J Am Vet Med Assoc 151:914–917

Wolf TM, Sreevatsan S, Singer RS, Lipende I, Collins A, Gillespie TR et al (2016) Noninvasive tuberculosis screening in free-living primate populations in gombe national park, tanzania. EcoHealth 13:139–144

Xu J, Tasneen R, Peloquin CA, Almeida DV, Li SY, Barnes-Boyle K et al (2018) Verapamil increases the bioavailability and efficacy of bedaquiline but not clofazimine in a murine model of tuberculosis. Antimicrob Agents Chemother 62:e01692

Yagi A, Uchida R, Hamamoto H, Sekimizu K, Kimura KI, Tomoda H (2017) Anti-mycobacterium activity of microbial peptides in a silkworm infection model with mycobacterium smegmatis. J Antibiot 70:685–690

Yagi A, Yamazaki H, Terahara T, Yang T, Hamamoto H, Imada C et al (2021) Development of an in vivo-mimic silkworm infection model with mycobacterium avium complex. Drug Discov Ther 14:287–295

Yang G, Luo T, Sun C, Yuan J, Peng X, Zhang C et al (2017) Ppe27 in mycobacterium smegmatis enhances mycobacterial survival and manipulates cytokine secretion in mouse macrophages. J Interf Cytokine Res 37:421–431
Youmans GP, Mc CJ (1945) Streptomycin in experimental tuberculosis; its effect on tuberculous infections in mice produced by m. Tuberculosis var. Hominis. Am Rev Tuberc 52:432–439
Youmans GP, Williston EH (1946) Effect of streptomycin on experimental infections produced in mice with streptomycin resistant strains of m. Tuberculosis var. Hominis. Proc Soc Exp Biol Med 63:131–134
Yuan T, Sampson NS (2018) Hit generation in tb drug discovery: from genome to granuloma. Chem Rev 118:1887–1916
Zelmer A, Carroll P, Andreu N, Hagens K, Mahlo J, Redinger N et al (2012) A new in vivo model to test anti-tuberculosis drugs using fluorescence imaging. J Antimicrob Chemother 67:1948–1960
Zhan L, Tang J, Sun M, Qin C (2017) Animal models for tuberculosis in translational and precision medicine. Front Microbiol 8:717
Zhang G, Zhu B, Shi W, Wang M, Da Z, Zhang Y (2010) Evaluation of mycobacterial virulence using rabbit skin liquefaction model. Virulence 1:156–163
Zhang T, Li SY, Nuermberger EL (2012) Autoluminescent mycobacterium tuberculosis for rapid, real-time, non-invasive assessment of drug and vaccine efficacy. PLoS One 7:e29774
Zhang X, Mardinoglu A, Joosten LAB, Kuivenhoven JA, Li Y, Netea MG et al (2018) Identification of discriminating metabolic pathways and metabolites in human pbmcs stimulated by various pathogenic agents. Front Physiol 9:139
Zhang N, Strydom N, Tyagi S, Soni H, Tasneen R, Nuermberger EL et al (2020) Mechanistic modeling of mycobacterium tuberculosis infection in murine models for drug and vaccine efficacy studies. Antimicrob Agents Chemother 64:e01727

Targeting Outer Membrane to Tackle Multidrug-Resistant Bacterial Pathogens

29

Grace Kaul, Manjulika Shukla, Isha Soni, Vaishnavi Veerareddy, Ravi Akunuri Kumar, Arunava Dasgupta, Nanduri Srinivas, and Sidharth Chopra

Abstract

Infections caused due to Gram-negative bacterial pathogens, under community as well as nosocomial settings, are increasing affecting healthcare systems worldwide, leading to significant effects on mortality and morbidity. This, in conjunction, with increasing multidrug resistance being encountered has prompted increasing efforts directed towards identifying novel targets and drugs active against such pathogens. In this context, the outer membrane of Gram-negative bacteria represents a very promising drug target since it is responsible for disallowing entry to antibiotics and toxins as well as is directly responsible for maintaining cell shape and viability. Indeed, there have been several examples of molecules targeting the outer membrane of Gram-negative bacteria which have reached the clinical stage, although none has been approved yet. Citing examples, we focus on describing the various details of such inhibitors including several which are in the hit or preclinical stage, thus exemplifying the criticality of outer membrane as a drug target. This is the first comprehensive review to compile all

Grace Kaul, Manjulika Shukla and Isha Soni contributed equally with all other contributors.

G. Kaul · A. Dasgupta · S. Chopra (✉)
Division of Microbiology, CSIR-Central Drug Research Institute, Lucknow, UP, India

Academy of Scientific and Innovative Research, Ghaziabad, India
e-mail: a.dasgupta@cdri.res.in; Skchopra.007@cdri.res.in

M. Shukla · I. Soni
Division of Microbiology, CSIR-Central Drug Research Institute, Lucknow, UP, India

V. Veerareddy · R. A. Kumar · N. Srinivas (✉)
Department of Medicinal Chemistry, National Institute of Pharmaceutical Education and Research, Hyderabad, Telangana, India
e-mail: srinivasn@niperhyd.ac.in

P. P. Singh (ed.), *Recent Advances in Pharmaceutical Innovation and Research*,
https://doi.org/10.1007/978-981-99-2302-1_29

known outer membrane inhibitors from hit to clinical stage and to provide a detailed overview on this most intractable of all problems facing humanity.

Keywords

Gram-negative · LPS · LpxA · Multi drug resistance · Outer membrane

29.1 Introduction

World Health Organization (WHO), US Centers for Disease Control (CDC), and a host of other international agencies have declared antimicrobial resistance (AMR) as one of the biggest public health challenges of our time (Centers for Disease Control and Prevention (US), National Center for Emerging Zoonotic and Infectious Diseases (US), Division of Healthcare Quality Promotion, Antibiotic Resistance Coordination and Strategy Unit 2019; O'Neill 2016). Since the beginning of the so-called antibiotic era, AMR has been a daunting yet underappreciated problem in clinical medicine, so much so that given the current rates of AMR, we are expected to soon enter "Post antibiotic era" with frightening consequences affecting healthcare systems worldwide (Ventola 2015). Unfortunately, AMR disproportionally affects developing countries, which is often a result of multitude of factors including misuse and overuse of antimicrobials, lack of antimicrobial stewardship program in hospitals, slow rate of development of new antibiotics, and discovery of novel targets which are associated with increasing rates of morbidity and mortality (O'Neill 2016).

In order to rationally encourage drug discovery, WHO published a list of priority pathogens, majority among which are multidrug-resistant Gram-negative bacterial (MDR-GNB) pathogens such as *Acinetobacter baumannii*, *Pseudomonas aeruginosa,* and pathogens belonging to *Enterobacteriaceae* responsible for severe nosocomial and community acquired infections (World Health Organization 2017). One of the major AMR mechanisms employed by the MDR-GNB is a complex outer membrane (OM). The OM is an asymmetric lipid bilayer composed of phospholipids (PL), lipopolysaccharides (LPS), integral membrane proteins, and pore forming beta barrel proteins. Phospholipids are found in the inner leaflet while LPS is predominantly located in the outer leaflet. It is the presence of LPS which increases the rigidity, lowers the pore size of the OM as compared to normal phospholipid bilayer, and is the reason behind decreased diffusion of hydrophobic and hydrophilic compounds (Zgurskaya et al. 2015). Although highly variant in among GNB, LPS is made up of a basic tripartite structure: (1) Lipid A which anchors LPS to OM, (2) non-repeating core polysaccharides, and (3) O-antigen repeats (Raetz and Whitfield 2002).

The OM is a major permeability barrier to a number of molecules including antibiotics and innate immunity molecules such as cationic antimicrobial peptides, without compromising its ability to exchange vital substances such as nutrients (Delcour 2009). Thus, circumventing the permeability barrier represents a promising strategy to combat these pathogens. For example, Huang lab has recently reported that pre-weakening the OM of the pathogenic bacteria increases the effect of antibiotics which target other cell processes (Rojas et al. 2018). In addition, OM biogenesis pathway is unique to bacteria and absent in higher eukaryotes, thus making it a highly promising drug target. Indeed, LPS and Lipid A biosynthesis are critical steps in maintaining OM and inhibiting synthesis of lipid A leads to bactericidal effect by disrupting OM (Erwin 2016).

In context as a drug target, significant research to generate extensive understanding of biogenesis, regulation, and functioning of the OM has been carried out. Given the current state of lack of novel antibiotics against MDR-GNB and the concomitantly increasing AMR prompted us to shine a spotlight on this extremely important drug target. Here we will discuss recent advances in the studies on OM as a potential drug target as well as will enlist potential candidates which are in clinical and preclinical phases.

29.2 OM Structure and Synthesis

As described above, OM is an asymmetric lipid bilayer composed of PL, LPS, integral membrane proteins, and beta barrel proteins. The various potential drug targets in the inner and outer membrane are depicted in Fig. 29.1.

Structurally, LPS is composed of Lipid A, non-repeating core polysaccharides, and O-antigen repeats. Lipid A is synthesized by a set of nine conserved proteins (LpxA-LpxM) present in the cytoplasm and inner leaflet of the inner membrane and is then transported from the inner membrane to the outer membrane. The

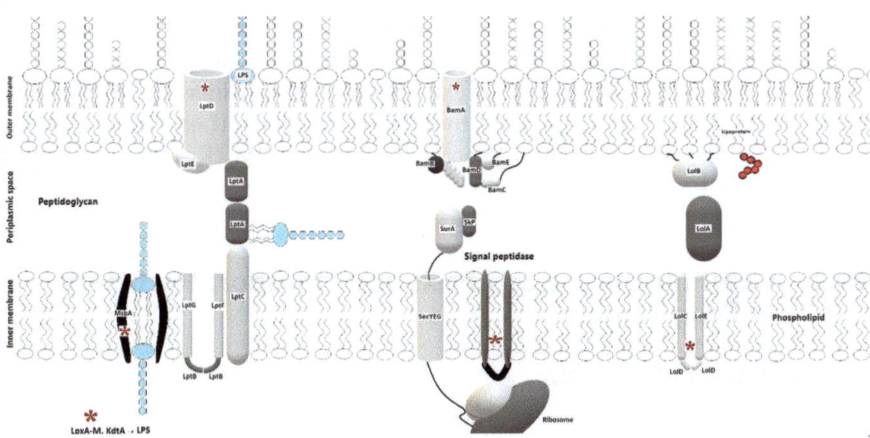

Fig. 29.1 Drug targets in the cell envelope of Gram-negative bacteria. The various inhibitors are depicted as red stars

biosynthesis of Lipid A is depicted in Fig. 29.2 but a detailed description is beyond the scope of this review (Raetz and Whitfield 2002).

For the transport of LPS to the OM, MsbA, an ABC transporter present in the inner membrane acts as a flippase and orients nascent LPS towards the periplasm. LPS is further transported to the outer membrane by LPS transport system (Lpt), which is a complex of an ABC protein, a chaperone, and a beta barrel protein (Simpson et al. 2015). The LPS transporter $LptB_2FGC$ complex transports LPS to the periplasmic space in an ATP driven manner where LptB, an ABC transporter, hydrolyzes ATP to release energy. LptA, a periplasmic chaperone, then takes over LPS and delivers it to LptD/LptE complex, where LptD assembles LPS in the OM (Dong et al. 2017) (Fig. 29.1). Ruiz et al. in 2015 have elaborated LPS transport to the OM in greater detail (Simpson et al. 2015; May et al. 2015).

Core polysaccharides and O-antigen, which are an essential part of LPS, are also synthesized on the inner leaflet of IM where the latter is synthesized on a polyisoprenoid carrier. This carrier flips O-antigen to the outer leaflet and finally it is ligated to the LPS core in the outer leaflet of the inner membrane. It is further transported and assembled along with LPS in the OM (Raetz and Whitfield 2002).

The outer membrane proteins are synthesized in the cytoplasm and are then translocated to the OM. The amino terminus leader sequence of the nascent OMP precursor interacts with the cytoplasmic chaperone and is then carried through the SecYEG complex to the periplasm. The signal peptidase interacts with the leader sequence and cleaves it, handing over the OMP to periplasmic chaperones SurA, Skp, and DegP (Bodelón et al. 2015). These chaperons perform via SurA pathway and Skp-DegP pathway. The assembly of OMPs in OM finally is facilitated by the β-barrel assembly machinery (BAM) including BamA, an integral membrane protein and accessory lipoproteins including BamB, BamC, and BamD (Knowles et al. 2009; Bodelón et al. 2015) (Fig. 29.1).

Like OMPs, lipoproteins also have amino-terminal signal sequence and are further translocated by the Sec machinery. A signal peptidase II removes the signal sequence and fatty acyl chain added at the cysteine amino group which functions as the amino terminus of lipoprotein. Most of the lipoproteins are then transported to the OM by Lol system. Lipoprotein is extracted and actively transported from inner membrane to periplasmic carrier LolA by an ABC transporter (LolCDE). It is further passed to LolB which assembles lipoproteins into the OM (Kovacs-Simon et al. 2011) (Fig. 29.1).

29.3 Outer Membrane Synthesis and Transport Machinery as Drug Targets

29.3.1 Essentiality of OM

Genetic essentiality to viability is a very important feature of any drug target. It has been recently reported that any deformity to the OM negatively affects the cell and is bactericidal (Rojas et al. 2018). These deformities could be chemical in nature such

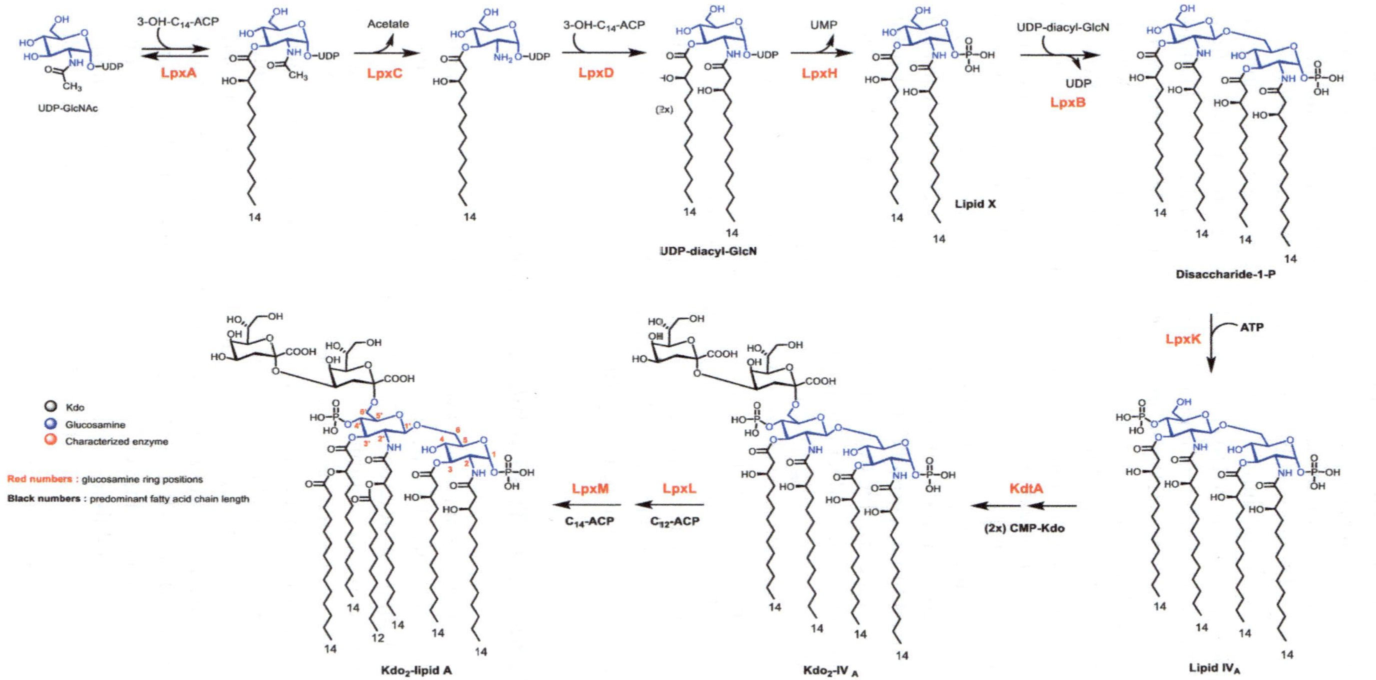

Fig. 29.2 A model lipopolysaccharide (LPS) biosynthetic pathway. Lipid A is assembled in the cytosol by a set of nine conserved enzymes present in the cytoplasm and inner leaflet of the inner membrane. The various enzymes such as LpxA (acyl-[acyl-carrier-protein])-UDP-N-acetylglucosamine O-acyltransferase (EC: 2.3.1.129), LpxC (UDP-3-O-acyl)-N-acetylglucosamine deacetylase (EC: 3.5.1.108), LpxD is UDP-3-O-(3-hydroxymyristoyl) glucosamine N-acyltransferase (EC: 2.3.1.191), LpxH is UDP-2,3-diacylglucosamine hydrolase (EC: 3.6.154), LpxB is lpxB which is lipid-A-disaccharide synthase (EC: 2.4.1.182), LpxK is Tetraacyldisaccharide 4′-kinase (EC: 2.7.1.130), KdtA is 3-deoxy-D-manno-octulosonic acid transferase, (EC: 2.4.99.12), LpxL is Lipid A biosynthesis lauroyltransferase (EC:2.3.1.241), LpxM is Lipid A biosynthesis myristoyltransferase (EC:2.3.1.243), Kdo is 3-deoxy-D-manno-oct-2-ulosonic acid

as treatment with lipid A biosynthesis inhibitors or genetic as mutation(s) or inhibitors of essential enzymes of lipid A biosynthesis. Among the various enzymes involved in its synthesis, the first committed step is catalyzed by LpxC, a member of zinc metalloenzyme family. The irreversible nature of this step makes LpxC a prime drug target. Indeed, extensive in vitro and in vivo studies have been done for generating LpxC inhibitors (Erwin 2016). Additionally, it has been demonstrated that *msbA* mutants cause accumulation of core LPS on inner membrane and not able to flip it, leading to growth cessation (Polissi and Georgopoulos 1996). Also, studies have revealed that bacterial mutants not bearing SurA, Skp or SurA and DegP are either nonviable or display defected OMP assembly (Sonntag et al. 1978; Bodelón et al. 2015). Hirota et al. reported that mutants with no lipoprotein make pores on the surface with periplasmic enzymes leaking into the growth medium, thus significantly affecting viability (Sonntag et al. 1978). Hypersensitivity to detergents and various antibiotics have been reported in *lptD* deletion mutants (Balibar and Grabowicz 2016). It has also been reported that mutations in biosynthesis of lipid A leads to hypersensitivity against antibiotics, possibly because the compromised OM structure allows entry of otherwise impermeable drugs (García-Quintanilla et al. 2016). Thus, targeting OM can have detrimental effects on the bacterial stability and viability.

One of the advantages of targeting OM is the potential lower propensity for resistance development since OM is such a conserved and essential target across GNBs. Indeed, the frequency of isolation of resistant mutants was $\sim 10^9$ in *E. coli* for CHIR-090, $\sim 10^{-10}$ in *P. aeruginosa* against PF-5081090, and $\sim 10^{-8}$ in *K. pneumoniae* against PF-5081090 (Tomaras et al. 2014; Erwin 2016). These frequencies are significantly higher than those reported for other drugs, thus reinforcing the OM as a preferred drug target (Erwin 2016).

29.3.2 Various Targets and Their Inhibitors in OM

As expected, owing to the importance of OM in the survival of GNB, there have been numerous studies on molecules targeting OM in drug development (Brown 2016). Despite establishing a relatively healthy preclinical pipeline, it was disheartening that only two OM inhibitors managed to reach the clinical trial stage, although none was approved as a therapeutic (Serio et al. 2013; Martin-Loeches et al. 2018). One of the major issues was limited spectrum of hits such as L-573,655, L-159,692, and L-161,240, reported against *E. coli*, were inactive against *P. aeruginosa* and other GNB (Onishi et al. 1996). This was explained on the basis of molecular differences between *E. coli* and *P. aeruginosa* LpxA-LpxM structures (Mdluli et al. 2006). Afterwards, researchers from the University of Washington and Chiron, in a medicinal chemistry program funded by Cystic Fibrosis Foundation, reported new inhibitors exhibiting activity against *P. aeruginosa* LpxC enzyme as well (Andersen et al. 2004; Andersen et al. 2011). Since then, numerous LpxC inhibitors have been identified. The search for novel LpxC inhibitors is focused on small molecules which share common structural features with enzyme like Zn^{2+}-chelating group, facilitating an efficient binding of the inhibitor to the enzyme's catalytic Zn^{2+}-ion, as well as a lipophilic side chain, addressing lipophilic tunnel of LpxC, etc. Among these, the *N*-aroyl-L-threonine- and methylsulfone-based hydroxamate LpxC inhibitors represent the most promising classes with respect to LpxC inhibitory and antibacterial activities (Brown et al. 2012; Montgomery et al. 2012; Tomaras et al. 2014; Erwin 2016; Lemaître et al. 2017; Piizzi et al. 2017).

The future development of new chemical entities against LPS biosynthesis is all the more exciting because LPS affects GNB growth and viability. Interestingly, LPS-deficient strains of *Neisseria meningitides*, *Moraxella catarrhalis,* and *Acinetobacter baumannii* have been reported whose growth is not affected and colistin resistance has been observed due to loss of LPS (Steeghs et al. 1998; Peng et al. 2005; Moffatt et al. 2010; Moffatt et al. 2011). Nevertheless, LpxC inhibitors such as PF-5081090 and LpxC-1 that have not been able to demonstrate good in vitro activity against *Acinetobacter baumannii* have successfully augmented its susceptibility to standard antibiotics and also protected mice from inflammation caused due to resistant *Acinetobacter* (Lin et al. 2012; García-Quintanilla et al. 2016). There is still plenty of scope for further studies in the area of affecting viability of LPS-deficient strains.

With respect to lipoprotein transport, despite multiple efforts to develop new antimicrobials affecting Lol system since 2000s, potent inhibitors with nanomolar MIC values have yet not been identified (Iwai et al. 2002; Ito et al. 2007; Pathania et al. 2009; Barker et al. 2013; McLeod et al. 2015; Nayar et al. 2015).

Another class of inhibitors are antimicrobial peptides (AMP) that are being increasingly used to bind and inhibit essential enzymes, validate intracellular targets, develop target-specific assays, and study biosynthetic pathways (Hancock and Sahl 2006). However due to their susceptibility to proteases and poor bioavailability, "peptidomimetics" which have greater proteolytic stability, bioavailability, and enhanced structural characteristics compared to natural peptides have gained prominence (Yin and Hamilton 2005). The peptidomimetic class inhibitors of OMPs are an extension of varied structural data available about proteins involved in LPS biosynthesis, LPS transport machinery, and outer membrane proteins (OMPs) assembly (Zerbe et al. 2017). Along with these, interest in synthesizing small molecules which are analogs of specific Outer Membrane Proteins (OMP) transport substrates is also gaining momentum (Eren et al. 2012).

In Table 29.1, we have summarized the candidates in clinical phases and others in advanced preclinical stages that have been extensively studied with at least in vivo activity and have scope for transition into clinical studies. Other than these, some recently studied molecules that have only been screened in vitro but have novel OM inhibition approaches and look promising are also described. All these inhibitors have targets in the synthesis, transport, and assembly of major components of the OM.

Table 29.1 Potential candidates against OM targets in GNB pathogens

Target	Drug candidate	Class	Function	GNB tested	MIC_{90} (μg/mL)	Stage of development	References
LPS biogenesis	JB-95	β-hairpin macrocyclic peptide	Interacting with *BamA&LptD*	*E. coli*	~0.25	Preclinical	Urfer et al. (2016)
	POL7001	Protein epitope mimetic (PEM)	Inhibiting LptD	*P. aeruginosa*	≤0.125	Preclinical	Srinivas et al. (2010), Cigana et al. (2016)
	Murepavadin (POL7080)	Protein epitope mimetic	Inhibiting LptD	*P. aeruginosa*	≤0.25	Phase 3 (IV)[a]; phase 1 (inhalation formulation)	Polyphor Ltd (2017, 2018a, b, 2019), Martin-Loeches et al. (2018), Spexis (2020)
Lipid A biosynthesis	ACHN-975	Hydroxamic acid	Inhibiting LpxC	*P. aeruginosa*/*E. coli*	≤1	Phase 1[b]	Serio et al. (2013), Castanheira et al. (2013), Achaogen Inc (2013)
	PF-5081090	Hydroxamic acid	Inhibiting LpxC	*P. aeruginosa*/*E. coli*	≤1	Preclinical	Tomaras et al. (2014), García-Quintanilla et al. (2016)
	LPC-069	Difluoromethyl-l-allo-threonyl-hydroxamate	Inhibiting LpxC	*E. coli*/*K. pneumoniae*/*P. aeruginosa*/*A. baumannii*/*Enterobacter* sp/*Salmonella*/*Y. pestis*	0.2–3.2	Preclinical	Liang et al. (2016), Lee et al. (2016), Lemaître et al. (2017)
	FG-944	Non-hydroxamate	Inhibiting LpxC	*E. coli* /*K. pneumoniae*/*Enterobacter* sp	0.06-8	Preclinical	Taganov et al. (2018), Forge Therapeutics (2018), Teng et al. (2018), Munguia et al. (2018)
	RJPXD33	Peptide	Inhibiting LpxA and LpxD	*E. coli*	LpxA (K_d = 22 μm) and LpxD (K_d = 6 μm)	Preclinical	Jenkins and Dotson (2012), Jenkins et al. (2014)

	CHIR090	*N*-aroyl-L-threonine hydroxamic acid	Inhibiting LpxC enzyme	*E. coli*/*P. aeruginosa*	0.2–16	Preclinical	Andersen et al. (2004), McClerren et al. (2005), Barb et al. (2007a, b), Bodewits et al. (2010), Liang et al. (2016), Lee et al. (2016), Tan et al. (2017)

[a] The study was stopped due to nephrotoxicity issues
[b] The study was stopped due to inflammation at injection sites

29.3.3 Outer Membrane Inhibitors in Clinical Stages

29.3.3.1 Murepavadin (POL7080)

Murepavadin (POL7080), developed by Polyphor, is a 14 amino acid cyclic peptide and first of class under outer membrane protein-targeting antibiotic (OMPTA) platform for treatment of hospital-acquired pneumonia (HAP) and ventilator-associated pneumonia (VAP) caused due to *P. aeruginosa* (Fig. 29.3). POL7080 was developed as a synthetic cyclic beta-hairpin peptidomimetic derived by utilizing Polyphor's proprietary chemistry platform and is based on protegrin I (PG-1). POL7080 targets LptD which is crucial for LPS biogenesis. POL7080 demonstrated bactericidal effects by binding to LptD and causing LPS modifications. In preclinical studies, it has demonstrated potent and specific activity against *P. aeruginosa* including MDR (in vitro MIC_{90} $\leq$0.25 μg/mL; >1500 strains tested) (Table 29.2). It demonstrated potent in vivo activity in neutropenic lung infection model against XDR *P. aeruginosa* isolates with the 5 log_{10} reduction in CFU as compared to Polymyxin B (Polyphor Ltd. 2017; Martin-Loeches et al. 2018).

POL7080 has been evaluated in 6 phase I and 2 phase II studies. Six phase I studies were performed in healthy volunteers with intravenous (IV) administration including amikacin and colistin. Pharmacokinetic studies suggested that it was metabolized and excreted primarily by the kidney and exhibited linear kinetics with $t_{1/2}$ of 6–8 h with good lung tissue penetration and a high distribution in epithelial lung fluid (Polyphor Ltd. 2017; Martin-Loeches et al. 2018).

In 2 phase II studies, POL7080 was evaluated to treat VAP and Non-Cystic Fibrosis Associated Bronchiectasis patients. POL7080 exhibited a low rate of

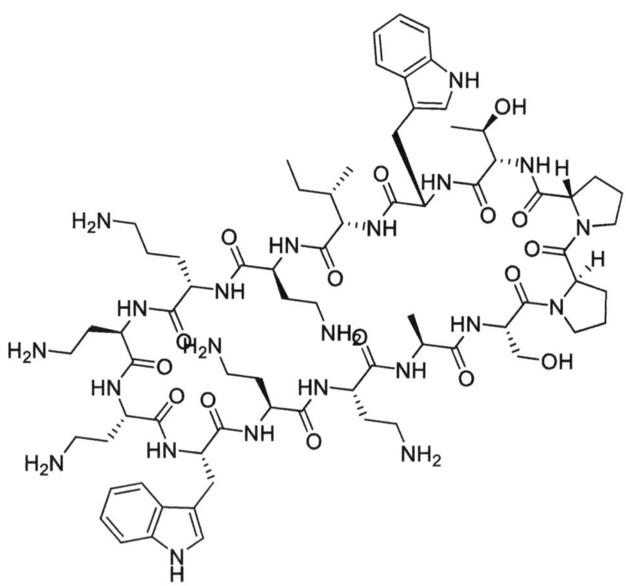

Fig. 29.3 Chemical structure of murepavadin (POL7080)

Table 29.2 MIC of murepavadin (POL7080) against various pathogenic microorganisms (Polyphor Ltd. 2017)

S. No.	Type strain	ATCC/DSM	MIC (μg/mL)
1	*Escherichia coli*	ATCC 25922	>64
2	*Klebsiella pneumonia*	ATCC 13883	>64
3	*Acinetobacter baumannii*	ATCC 19606	>64
4	*Burkholderia cepacia*	ATCC 25416	>64
5	*Stenotrophomonas maltophilia*	ATCC 13637	>64
6	*Staphylococcus aureus*	ATCC 29213	>64
7	*P. aeruginosa*	ATCC 27853	0.06
8	*P. aeruginosa*	PAO1	0.25
9	*Pseudomonas putida*	DSM 291	0.06
10	*Pseudomonas fluorescens*	DSM 6147	0.06
11	*Pseudomonas aureofaciens*	ATCC 15926	0.06
12	*Pseudomonas syringae*	ATCC 12271	0.008

resistance development (~10^{-9} by spontaneous mutations) with no cross-resistance to any FDA-approved antibiotics utilized for treatment of *Pseudomonas* infections because of the novel mechanism of action. POL7080 exhibited mild and reversible side effects of paresthesia and dizziness. It was considered safe and well tolerated in the studies (Polyphor Ltd. 2017; Martin-Loeches et al. 2018). However, Romano et al. (2019) isolated a series of POL7080-resistant strains with mutations in *pmrB* conferring high-level resistance to POL7080 and moderate cross-resistance to colistin in *P. aeruginosa*. The group raised concerns over the possibility of resistance generation to colistin and other polymyxins if POL7080 was successfully developed for market use (Romano et al. 2019).

Murepavadin as an IV formulation underwent two separate phase 3 trials—FDA-approved study (PRISM-UDR) and European Medicines Agency (EMA)-approved study (PRISM-MDR) to test its efficacy in HAP and VAP infections caused due to *P. aeruginosa* (Polyphor Ltd. 2018a, b). However, Polyphor terminated IV formulation development in 2019 due to higher-than-expected acute kidney injury incidences in murepavadin arm of PRISM-MDR trial (Polyphor Ltd. 2019). After this, Polyphor continued preclinical evaluation of murepavadin as an inhalation formulation and obtained the clinical trial authorization for development of murepavadin by MHRA in United Kingdom in December 2020. As of 2021, first-in-human clinical trial of inhaled murepavadin is underway following the successful completion of preclinical program (Spexis 2020).

Recent studies have also revealed that besides its direct activity against *P. aeruginosa*, murepavadin may aid in bacterial clearance and accelerate wound healing by activating human mast cells (MCs) via MRGPRX2 and murine MCs via MrgprB2 (Amponnawarat et al. 2021).

29.3.3.2 ACHN-975

ACHN-975 was developed by Achaogen Inc., USA, as a LpxC inhibitor designed against MDR-GNB including *P. aeruginosa* (Serio et al. 2013). Chemically, ACHN-975 is N-((S)-3-amino-1-(hydroxyamino)-3-methyl-1-oxobutan-2-yl)-4-(((1R,2R)-2-(hydroxymethyl)cyclopropyl) buta-1,3-diyn-1-yl) benzamide and its structure is given in Fig. 29.4.

ACHN-975 showed potent antibacterial activity against MDR *P. aeruginosa* including carbapenem-resistant *P. aeruginosa* carrying MβL enzyme and *Enterobacteriaceae* family carrying genes encoding KPC-2, KPC-3, NDM-1, and CTX-M-15 but was inactive against *A. baumannii* and Gram-positive bacteria as shown in Table 29.3. Being the first member of a novel class, no known pre-existing clinical resistance was found against it (Castanheira et al. 2013). ACHN-975 showed bactericidal activity including concentration-dependent killing against *P. aeruginosa*

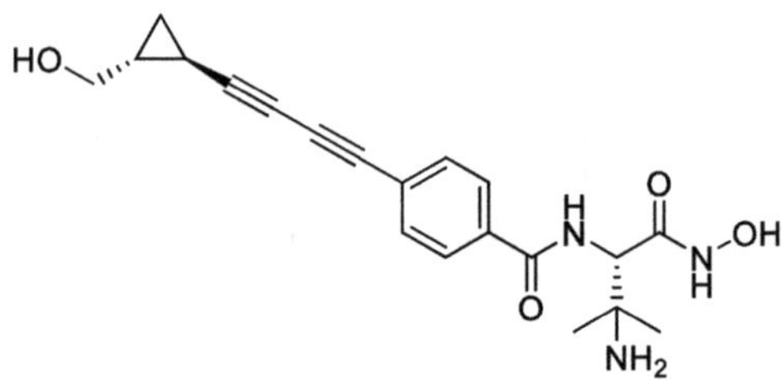

Fig. 29.4 Chemical structure for ACHN-975

Table 29.3 MIC of ACHN-975 against various pathogenic microorganisms (Serio et al. 2013)

S. No.	Family	Organism	*n* = number of clinical isolates	MIC (μg/mL)
1	Gram-negative (*n* = 454)	*E. coli*	*n* = 100	0.03–2
2		*Klebsiella*	*n* = 113	0.25–4
3		*Serratia*	*n* = 41	0.06–1
4		*Enterobacter*	*n* = 52	0.25–4
5		Indole positive *Proteus*	*n* = 50	0.06–0.5
6		*Salmonella* sp./ *Shigella*	*n* = 39	0.125–2
7		*Proteus mirabilis*	*n* = 42	0.25–2
8		*Citrobacter*	*n* = 28	0.125–1
9		*P. aeruginosa*	*n* = 100	0.008–0.5
10		*Acinetobacter baumannii*	*n* = 28	4 to >64
11	Gram-positive	*Staphylococcus aureus*	ATCC 29213	>128
12		*E. faecalis*	ATCC 29212	>128
13		*E. faecium*	[a]	>128
14		*S. pneumonia*	ATCC 49619	>128
15		*S. epidermidis*	[a]	>128

[a] Not available

and time-dependent killing against *E. coli* and *K. pneumonia* (Serio et al. 2013). In vitro studies demonstrated that altered regulation of RND efflux system affected ACHN-975's MIC. ACHN-975 MIC against *ΔacrAB E. coli* was 0.015 μg/mL, whereas it was 0.5 μg/mL against *E. coli* over-expressing AcrAB as compared to 0.125 μg/mL for wild-type *E. coli*. Similarly, against the ΔmexAB-oprM, ΔmexCD-oprJ, and ΔmexEF-oprN strains, MIC was 0.015 μg/mL and 1 μg/mL against the over-expressing MexAB system as compared to 0.06 μg/mL for wild type (Serio et al. 2013). At 4× MIC of ACHN-975, frequency of resistance ranged from 10^{-7} to 10^{-10} in *P. aeruginosa* clinical isolates (Krause et al. 2019) but it induced bradycardia in preclinical animal studies (Bornheim et al. 2013).

A phase 1 (NCT01597947) double blind, randomized, placebo-controlled, single ascending dose study was started and completed in 2012 for assessment of safety, tolerability, and pharmacokinetics in normal healthy volunteers with IV ACHN-975 (Achaogen Inc. 2012). The trial results revealed peak plasma concentration (C_{max})-driven dose-limiting toxicity (DLT) of transient hypotension without tachycardia, thus making the therapeutic window of ACHN-975 insufficient for the drug to advance in development (Krause et al. 2019). Unfortunately, another phase 1 multiple-dose study (NCT01870245) conducted in 2013 was terminated because of injection site problems (Achaogen Inc. 2013; Kalinin and Holl 2017).

Further studies revealed that a C_{max}-driven effect, achieved through less frequent doses, appeared to provide maximal efficacy against *P. aeruginosa* compared to time-dependent effect in *Enterobacteriaceae* (Krause et al. 2019). Hence, an optimization campaign was started to discover LpxC inhibitors related to ACHN-975 maintaining activity against *P. aeruginosa* specifically while reducing cardiovascular toxicity associated with the class (Cohen et al. 2019). However, further development of the lead molecules from this drug discovery program had to be halted as well due to the re-emergence of the cardiovascular toxicity in preclinical animal models (Cohen et al. 2019).

29.3.4 Inhibitors in Preclinical Stages

29.3.4.1 LPS Transport Inhibitors

29.3.4.1.1 POL7001

POL7001, a LPS transport inhibitor, was developed by Polyphor to target only susceptible and drug-resistant *P. aeruginosa* with nanomolar MIC (Srinivas et al. 2010). POL7001 is a macrocycle protein epitope mimetic (PEM) antibiotic inspired from Protegrin I (PG I) but with a D-proline-L-proline template, which not only helps to stabilize its β-hairpin conformation but also results in no notable hemolytic activity up to 100 μg/mL (Fig. 29.5) (Steinberg et al. 1997; Srinivas et al. 2010). POL7001 was being developed along with murepavadin (POL7080) and L27-11 among notable peptides out of which POL7001 and murepavadin displayed much improved plasma half-lives while retaining a potent and selective action against *P. aeruginosa*. For POL7001 and POL7080, MIC_{90} values were 0.13 and 0.25

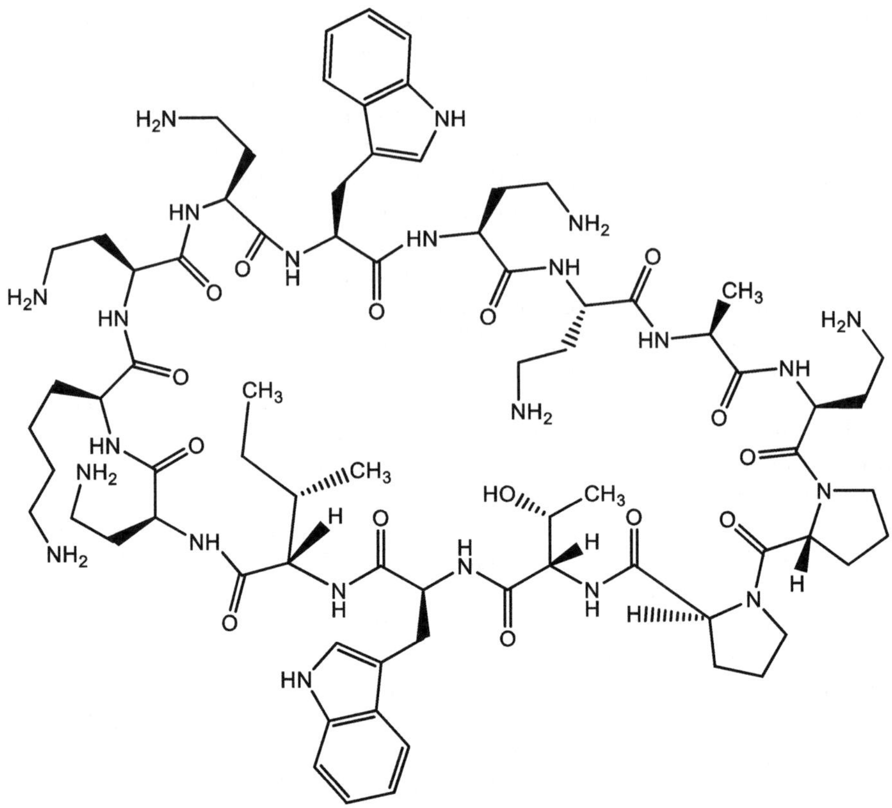

POL7001

Fig. 29.5 Chemical structure for POL7001

Table 29.4 MIC of POL7001 against various pathogenic microorganisms (Srinivas et al. 2010)

Strains	MIC (μg/mL)
P. aeruginosa ATCC 27853	0.008
P. aeruginosa PAO1	0.008
A. baumannii DSM3008	>64
K. pneumoniae ATCC 13883	>64
S. maltophilia ATCC 13637	>64
E. coli ATCC 25922	>64
E. faecalis DSM 12956	>64
S. aureus ATCC 29213	>64

mg/mL, respectively, against 90% of clinical isolates of *P. aeruginosa* tested with most of them resistant to standard drugs (Table 29.4) (Srinivas et al. 2010).

The rate of killing of *P. aeruginosa* cells by POL7001 was found to be slower than PG I and Polymyxin B when compared at 4× MIC. The *P. aeruginosa* cell membrane permeabilizing activity of POL7001, PG I, or Polymyxin was studied by measuring fluorescence increase in presence of SYTOX. While there was a quick increase in fluorescence in presence of PG-I or polymyxin B and SYTOX, there was no apparent increase in fluorescence over 120 min when exposed to POL7001 and SYTOX at the same concentrations. Investigations for its mechanism of action lead to no notable effects on protein, nucleic acid, or cell wall biosynthesis. The target for these peptidomimetic antibiotics was found to be the perturbation of LptD via photoaffinity labeling experiments (Srinivas et al. 2010).

The in vivo activity reported for POL7001 is in a mouse septicemia model at doses of 10, 3, 1, 0.3, and 0.1 mg/kg after infection with *P. aeruginosa* ATCC 9027 or *P. aeruginosa* ATCC 27853. The calculated median effective doses (ED_{50}) were reported to be in the range of 0.25–0.55 mg/kg, as compared to gentamicin (positive control), which showed an ED_{50} of 3.1 and 2.9 mg/kg, respectively, against these strains (Srinivas et al. 2010). POL7001 has also shown significant in vivo activity in chronic and acute pneumonia model of lung infection via pulmonary administration. 1 $\log_{10}$ reduction in CFUs has been reported as compared to pre-treatment indicating a bactericidal effect with the total daily dose of 15 mg/kg of POL7001. These results support the further evaluation of POL7001 as a novel therapeutic agent for the treatment of *P. aeruginosa* pulmonary infections (Cigana et al. 2016).

29.3.4.2 LPS Biosynthesis Inhibitors

29.3.4.2.1 RJPXD33

RJPXD33 was developed by Jenkins and Company and is an example of a dual targeting antimicrobial agent targeting LpxA and LpxD (Fig. 29.6). The sequence of RJPXD33 is TNLYMLPKWDIP-NH2 with affinity and inhibitory activity for both LpxA ($K_D = 20$ μM) and LpxD ($K_D = 6$ μM) enzymes. RJPXD33 was identified in a

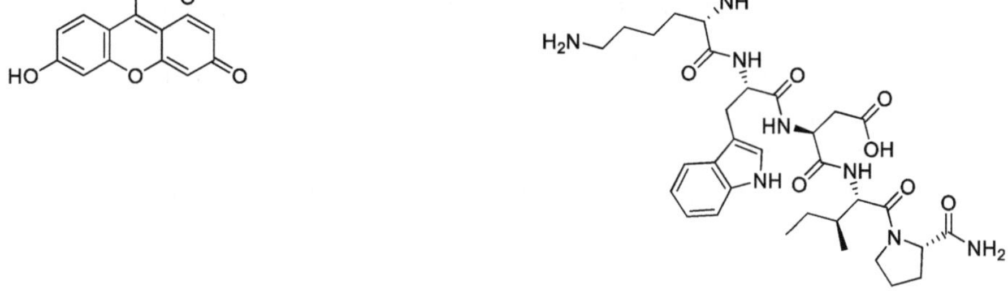

Fig. 29.6 Chemical structure of RJPXD33

screen from a random phage bound peptide library with subsequent ligand-competitive phage elution for identifying LpxD binding peptides. RJPXD33 displayed IC_{50} of 3.5 ± 0.08 μM and 19 ± 1.2 μM against LpxD and LpxA, respectively consistent with its binding affinity for both enzymes (Jenkins and Dotson 2012).

The crystal structure of *E. coli* LpxA in complex with RJPXD33 has been solved at 1.9 Å resolution via X-ray crystallography to understand the molecular basis of binding of peptide to LpxA and LpxD. The studies were done to further help in designing potent peptidomimetics and small-molecule antimicrobials with the same unique dual binding properties of the parent peptide. The *E. coli* LpxA-RJPXD33 superimposition with *L. interrogans* LpxA-(R)-β-Hydroxylauroylmethylphosphopantetheine co-crystal structure (PDB Code 3I3A) revealed that peptide binds in same way as (R)-β-hydroxyacylphosphopantetheine (acyl-PPan) moiety of the acyl-acyl carrier protein (acyl-ACP), natural substrate of LpxA, binds to LpxA. However, it has been suggested that the RJPXD33 binding is mutually exclusive from native substrate acyl-ACP (Jenkins and Dotson 2012).

Furthermore, binding studies of truncations of RJPXD33 to acyltransferases were also carried out to find out smaller peptidomimetics that could bind with similar affinity as RJPXD33 and also capable of crossing the cell membrane. In this respect, fluorescence polarization experiments demonstrated that FITC-RJPXD33Δ6 lacking six terminal amino acids of full-length FITC-RJPXD33 was found to have comparable affinity for LpxA with K_d value 12 ± 1 μM (K_d value for full-length FITC-RJPXD33 17 ± 1.6 μM) (Jenkins and Dotson 2012). Although attempts to co-crystallize RJPXD33 with *E. coli* LpxD were unsuccessful, an overlay of bound crystal structure of LpxARJPXD33 with *E. coli* LpxD (PDB code 3EH0) revealed a complementary peptide binding pocket within LpxD that can be used for future characterization of binding of RJPXD33 to LpxD. For further probe into this interaction, comparison of RJPXD33-LpxA complex with acyl-ACP-LpxD complex (PDB code 4IHF) resulted into alignment of RJPXD33 with R-3 hydroxymyristoylphosphopantetheine prosthetic group of acyl-ACP. This suggested that RJPXD33 would overlap with acyl-PPan arm of acyl-ACP, thereby inhibiting acyl-ACP from binding to LpxD (Jenkins et al. 2014). However, in a study published in 2019, researchers have revealed an unforeseen interaction mode of a FITC-labeled RJPXD33 with *E. coli* LpxD using the displacement soaking method for small-molecule co-structures (Ma et al. 2019a, b). The study demonstrated that RJPXD33 formed a beta-hairpin that is perpendicular to the LpxD threefold symmetry axis when bound to LpxD in contrast to the proposed binding model of RJPXD33-LpxD by Jenkins et al. (2014). Differences in trajectory and conformation of RJPXD33 bound to LpxA and LpxD were also observed affirming the fact that its conformational versatility is an important structural feature that allows the same peptide to interact with diversified acyltransferases LpxD and LpxA (Ma et al. 2019a, b). These studies have demonstrated that RJPXD33 can serve as template for further development of peptidomimetic molecules against acyltransferase enzyme targets in GNB.

29.3.4.2.2 CHIR-090

CHIR-090 is a *N*-aroyl-L-threonine hydroxamic acid compound developed by a collaboration of Chiron and University of Washington in 2004 (Fig. 29.7). It demonstrated excellent antibiotic activity against *E. coli* and *P. aeruginosa,* which was explained by its ability to inhibit LpxC orthologs (Andersen et al. 2004). CHIR-090 inhibited growth of *E. coli* and *P. aeruginosa* with potency comparable to tobramycin and ciprofloxacin in a bacterial disk diffusion assay. Since then, numerous derivatives have been discovered such as LPC-058, LPC-011, and LPC-069 with improved and broader spectrum of activity against GNB (Liang et al. 2016; Lee et al. 2016).

The inhibition of CHIR-090 comprises two steps: first step being fast and reversible and second step slow and irreversible where enzyme-inhibitor (EI) complex converts into a tightly bound form with a half-life of ~1 min (McClerren et al. 2005). CHIR-090 exhibits MIC of 0.2 and 1.6 μg/mL against *E. coli* and *P. aeruginosa* (Barb et al. 2007a, b). In a study, CHIR-090 exhibited activity against all representative strains of *B. multivorans*, *B. vietnamiensis*, *B. dolosa*, and *B. ambifaria.* The CHIR-090 MICs were strain dependent with values ranging from 0.1 to 100 μg/mL against *Burkholderia cepacia* complex (Bodewits et al. 2010). The solution structure of the *Aquifex aeolicus* AaLpxC-CHIR-090 was determined by NMR and demonstrated that CHIR-090 exploits the hydrophobic passage and essential active-site residues of LpxC which are its conserved features. CHIR-090 is adjacent to, but does not occupy, UDP-binding pocket of LpxC and its biphenyl acetylene moiety inserts into LpxC hydrophobic passage near the active site and protrudes from the opposite end of this passage. Additionally, key residues in hydrophobic passage modulate time-dependent inhibition and CHIR-090 resistance was also identified. The data attributed the shape, rigidity, and hydrophobicity of CHIR-090 for its high affinity to LpxC, apart from its chelation activity of the active-site zinc ion (Barb et al. 2007a, b). Recently in 2017, CHIR-090 was evaluated for its potential in inhibiting *P. aeruginosa* biofilms in synergy with colistin under in vitro and in vivo conditions. The study concluded that CHIR-090 works synergistically with colistin in inhibiting both colistin susceptible and resistant *P. aeruginosa* biofilms in vitro and in vivo (Tan et al. 2017). The compound is by far the most studied LpxC inhibitor but has still not reached the clinical stage, although it has driven the discovery of many important molecules in advanced preclinical stages.

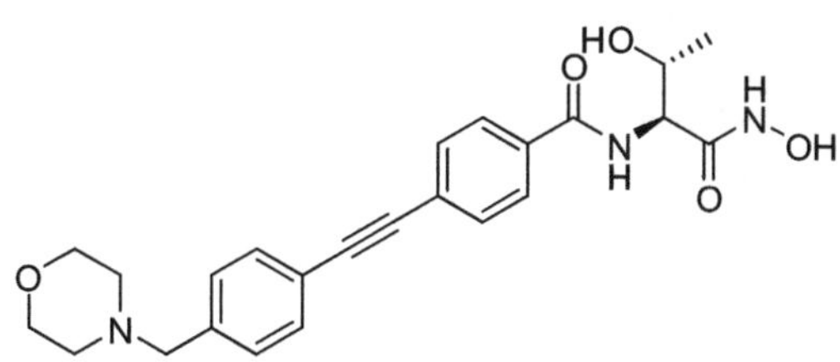

Fig. 29.7 Chemical structure of CHIR-090

29.3.4.2.3 PF-5081090

PF-5081090, also known as LpxC-4, is a Pyridone Methylsulfone Hydroxamate compound belonging to the Zn^{2+} binding class of hydroxamic acids and was developed by Pfizer (Fig. 29.8).

PF-5081090 effectively inhibits LpxCin, a broad range of GNB barring *Acinetobacter baumannii* (Table 29.5). It has shown more potent antibacterial activity with decreased MIC_{90} (μg/mL) as compared to lead molecules LpxC-2 and LpxC-3 described in previous studies especially against clinical isolates of *P. aeruginosa* (Brown et al. 2012; Montgomery et al. 2012). The enzyme inhibition activity of LpxC-4 was obtained via functional LpxC enzyme assays and MIC as well as Static Time kill studies (STK) were performed as per CLSI guidelines. It was observed in the STK studies that LpxC-4 had slower bacterial killing kinetics against the cystic fibrosis (CF) clinical isolate PA-1955, which constitutively expresses mexEF-oprN efflux pump, as compared to PA strain UC12120 where rapid killing was observed. However, re-growth was limited to drug concentration of ≤2X MICs against both the strains. Similarly, in KPC-producing clinical isolate of *K. pneumonia,* rapid bacterial killing was observed with re-growth limited to ~4× MIC drug concentration. The investigators performed standard frequency-of-resistance (FOR) analysis for assessing the resistance inducing potential of LpxC-4 with representative strains of *P. aeruginosa* and *K. pneumoniae.* The results indicated that while efflux pump and/or target upregulation in *P. aeruginosa* is the main reason for rare LpxC-4 resistance, mutation of fabZ is first step in LpxC-4 resistance

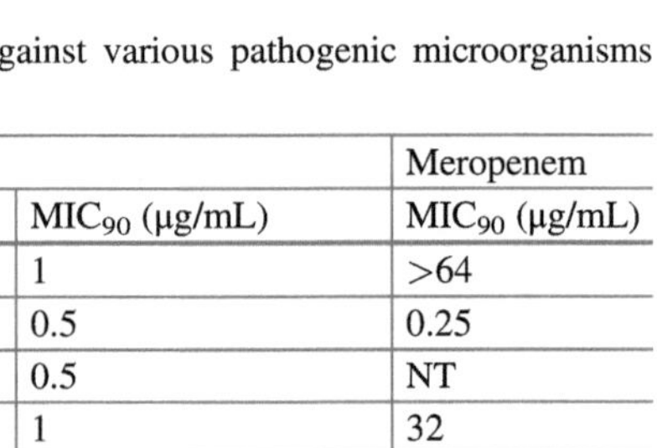

Fig. 29.8 Chemical structure of PF-5081090

Table 29.5 MIC of PF-5081090 and comparators against various pathogenic microorganisms (Tomaras et al. 2014)

	PF-5081090		Meropenem
Organism (no. of strains)	IC_{50} (nM)	MIC_{90} (μg/mL)	MIC_{90} (μg/mL)
P. aeruginosa (138)	1.1	1	>64
P. aeruginosa PAO1 WT	1.1	0.5	0.25
P. aeruginosa PAO1 M62R	2.1	0.5	NT
K. pneumoniae (98)	0.069	1	32
E. coli (79)	NT	0.25	NT
Enterobacter spp. (52)	NT	0.5	0.25
A. baumannii (31)	183	>64	32
B. cepacia (30)	NT	0.5	8
S. maltophilia (30)	NT	2	>64

NT not tested

mechanism in *Enterobacteriaceae* family especially in *E. coli* and *K. pneumoniae* (Tomaras et al. 2014).

It has been suggested via preliminary preclinical pharmacokinetic studies of LpxC-4 in mice that a clinical dose of ~1.2 g every 8 h would be sufficient to treat GNB strains with MICs $\leq$1 μg/mL. LpxC-4 demonstrated potent in vivo efficacy in acute septicemia, neutropenic thigh infection, and pulmonary infection model of activity against *P. aeruginosa* PA-1950, UC12120, PA-1955, and *K. pneumoniae* KP-1487 strains (Tomaras et al. 2014). The ED_{50} ranged from 7.4 mg/kg of body weight for PA-1950 (MIC 0.25 μg/mL) to 55.9 mg/kg for PA-1955 (MIC 2 μg/mL) in septicemia model. For pneumonia model, ED_{50} of <25 mg/kg was achieved in coherence with an ED_{50} of 16.8 mg/kg against PA-1950 in neutropenic thigh model. LpxC-4 also retains efficacy in vivo against mutants expressing different first-step resistance mechanisms in both *P. aeruginosa* and *K. pneumoniae*, demonstrating the potential to be moved further in their class. However, LpxC-4 failed to treat second-step *P. aeruginosa* small RNA (sRNA) mutants with increased LpxC expression in vivo but these mutations are exceedingly rare (Tomaras et al. 2014). In another study, LpxC inhibition in *Acinetobacter baumannii* by PF-5081090 (MIC 32 mg/L) increases its susceptibility to certain antibiotics by decreasing MIC of Rifampin by around 1000-folds and 8–16-folds for vancomycin (García-Quintanilla et al. 2016). This is postulated to happen by increasing the cell permeability.

29.3.4.2.4 LPC-069

LPC-069, jointly developed by Duke University and French National Institute of Health and Medical Research, is a morpholine-substituted biphenylacetylene-based inhibitor of LpxC with difluoromethyl-L-allo-threonyl-hydroxamate head group (Fig. 29.9). LPC-069 is active in vitro against a vast panel of GNB clinical isolates including several MDR and extremely drug-resistant (XDR) isolates including *Yersinia pestis*, causative agent of bubonic plague. Its activity was compared with a previously reported lead molecule LPC-058, whose in vitro MIC against *Y. pestis* CO92 experienced vast changes in the presence of serum albumin at 37 or 28 °C (Liang et al. 2016; Lee et al. 2016; Lemaître et al. 2017). It was observed that although LPC-069 activity (MIC 0.8 μg/mL at 28 °C and 0.1 μg/mL at 37 °C) was lesser than LPC-058 (MIC 0.05 μg/mL at 28 °C and 0.025 μg/mL at 37 °C) in the absence of albumin but in albumin presence, its MIC was fourfold lower (LPC-069 MIC 0.4 μg/mL, LPC-058 MIC 1.6 μg/mL) with lesser deviation from without albumin MIC values. Also, its MIC_{90} ranged from 0.2 to 0.8 μg/mL for MDR and

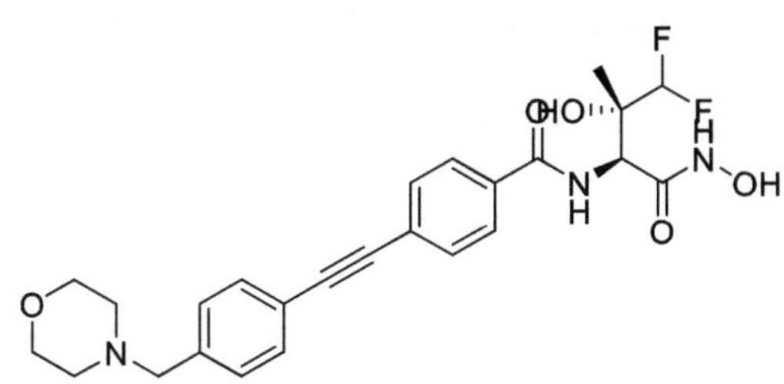

Fig. 29.9 Chemical structure of LPC-069

XDR strains of *Enterobacteriaceae* and 3.2 μg/mL for *P. aeruginosa* and *Acinetobacter baumannii* (Lemaître et al. 2017).

Its co-crystal structure with *Aquifex aeolicus* LpxC was determined for comparing it with the binding mechanism of previously reported LPC-058 to LpxC. It was deduced that LPC-069 binds in the same manner as LPC-058 with its curved tail group occupying the L-shaped hydrophobic substrate passage of LpxC and terminal morpholine group adopting a chair conformation (Lemaître et al. 2017).

The microsomal stability and pharmacokinetics parameters of LPC-069 along with LPC-058 in vivo were evaluated. The half-life and intrinsic clearance rates were 95 min and 24 μL/min/mg for LPC-058 and 44 min and 52 μL/min/mg for LPC-069, respectively, in hepatic microsomes. Whereas LPC-058 displayed a longer half-life of 53 min at single dose administration of 20 mg/kg weight of mouse, LPC-069 was found to have a shorter half-life (10 min) at a single IV dose of 40 mg/kg. The tissue distribution volume for LPC-058 was 1.1 L/Kg and LPC-069 had a larger distribution volume of 3.8 L/kg than LPC-058. The toxicity studies over a 5-day regimen however revealed that LPC-058 administration was associated with diarrhea, liver toxicity, and accumulation of polymorph nuclear cells in lung and large intestine cells. LPC-069, on the other hand, did not elicit any diarrhea or detectable toxicity in any of the regimens tested (including 200 mg/kg q8h). The compound has also shown potent in vivo activity in murine model of bubonic plague and potent efficacy at 200 mg/kg q8h, which was found to be as efficacious as the reference treatment with doxycycline (Lemaître et al. 2017).

29.3.4.2.5 FG-944

FG-944 is a non-hydroxamate compound being developed by Forge Therapeutics, San Diego, USA, that inhibits LpxC. Its in vitro and in vivo antibacterial activity data from preclinical studies against resistant GNB infections such as UTI and respiratory infections caused by *E. coli*, *E. cloacae,* and *K. pneumoniae* has been presented at American Society for Microbiology's Microbe (ASMM 2018) which showed potent activity of FG-944 (Taganov et al. 2018; Forge Therapeutics 2018; Teng et al. 2018). Its activity has also been observed against *Enterobacteriaceae* and *Acinetobacter baumannii* when combined with RNA synthesis inhibitors such as Rifampicin and the aryl myxopyronin investigational lead APY281. The FIC after checkerboard assay was determined as 0.3–0.4 for FG-944 and rifampin and 0.3–0.8 for FG-944 and APY281 (Munguia et al. 2018). Thus, there is plenty of scope for this molecule to enter clinical phase studies.

29.3.4.3 Lipoprotein Transport Inhibitors

29.3.4.3.1 G0507

G0507, developed by Genentech, USA, is a novel pyrrolopyrimidinedione compound that binds to LolCDE, which is an ABC transporter essential for bacterial survival (Fig. 29.10). G0507 was identified in a phenotypic screen of ~35,000 structurally diverse synthetic molecules that induced extracytoplasmic stress response in *E. coli.* Although inactive against wild-type *E. coli*, G0507 had the

Fig. 29.10 Chemical structure of G0507

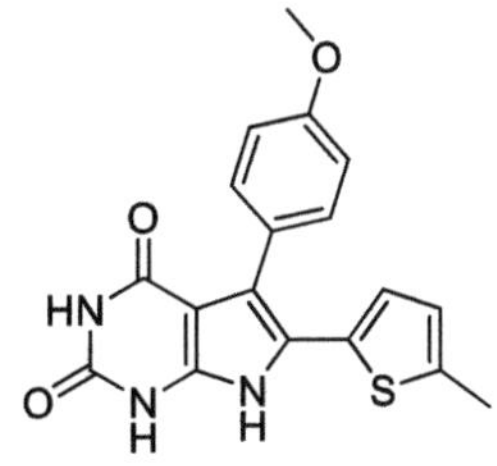

MIC of 0.5 and 1 μg/mL against efflux-deficient (*ΔtolC*) and an OM-compromised (*imp4213*) laboratory strain of *E. coli* MG1655, respectively. It also activates extracytoplasmic stress response in *E. coli* in a concentration- and time-dependent manner. When *lolC, lolD,* and *lolE* were mutated, resistance to G0507 was observed. It was also demonstrated that G0507 treatment results in accumulation of OM lipoprotein lpp in inner membrane. G0507 is believed to bind with LolCDE and stimulate its ATPase activity. If not as an OM inhibitor, this compound can surely be used in understanding the trafficking mechanisms of lipoproteins resulting in the discovery of novel molecules hitting Lol machinery (Nickerson et al. 2018).

29.3.4.4 Signal Peptidases Inhibitor

29.3.4.4.1 Arylomycin

Arylomycins are natural products derived from *Streptomyces* and are well-known inhibitors of type 1 signal peptidases (SPase I or LepB) that act as serine proteases (Fig. 29.11) (Smith and Romesberg 2012). Signal peptidases are an important component of lipoproteins translocation into the inner membrane and OM since they generate the signal peptide on lipoproteins necessary for their transfer. There are two classes of signal peptidases: type 1 and type 2 (SPaseII or LspA) (Paetzel et al. 2002). Due to a naturally occurring mutation in the active site, where a serine residue is exchanged with a proline residue, many bacterial species were intrinsically resistant to the class of arylomycins. Circumventing this problem, in a recent study, chemical optimization of arylomycins was reported. They reported a broad-spectrum molecule G0775, which inhibits essential bacterial type I signal peptidase through an unprecedented molecular mechanism. G0775 has significantly low MIC ranging from 0.125 to 2 μg/mL against a broad range of contemporary GNB namely *P. aeruginosa*, *K. pneumoniae,* and *Enterobacter.* The optimized moiety has worked excellently in in vivo infection models as well. In a neutropenic thigh infection model against *E. coli* strain ATCC 25922 and *K. pneumoniae* ATCC 43816, G0775 was required at just 1 mg/kg and 5 mg/kg for exhibiting potent activity, respectively. A co-crystal structure of LepB in complex with G0775 was solved at 2.4 Å resolution, thus deciphering the binding mechanism and possible mutation sites (Smith et al. 2018). In a 2019 study, it was demonstrated that even at sub-MIC levels, arylomycins inhibit a broad range of virulence and other extracytoplasmic processes in bacteria. This proposed that since they possess a broad spectrum of antibacterial activity, arylomycins are likely to inhibit virulence and horizontal gene transfer

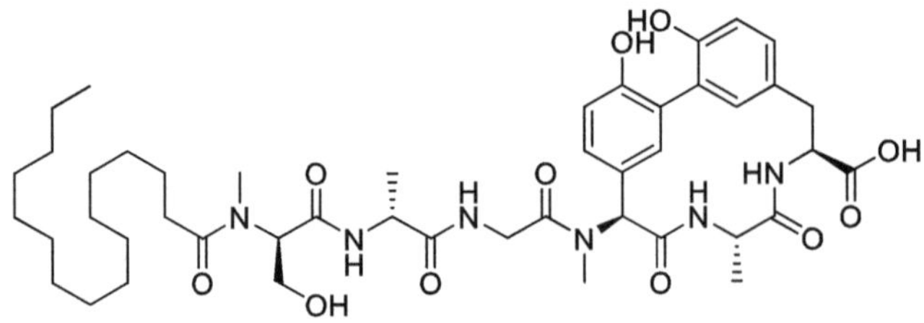

Arylomycin A-C_{16}

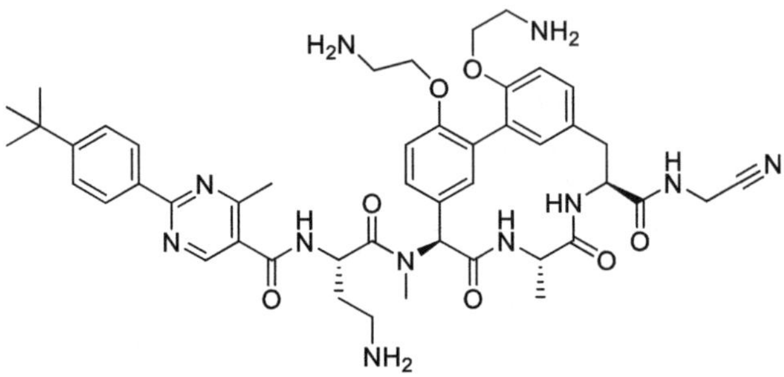

G0775

Fig. 29.11 Chemical structure of arylomycin A-C_{16} and G0775

(HGT) in a broad range of pathogens as well. The proteins whose proper localization was inhibited by arylomycins included components of the BAM complex LptD and LolA that themselves are promising drug targets (Walsh et al. 2019). Thus, arylomycins possess a great scope in being developed as broad-spectrum antibiotics by disarming the bacteria at many fronts.

29.3.4.5 Others (Hit to Preclinical)

29.3.4.5.1 L27-11

L27-11 is a peptidomimetic that interacts with LptD of LPS transport machinery. It was discovered along with POL7001 and murepavadin exhibiting in vitro activity specifically against susceptible and resistant *P. aeruginosa* with MIC 0.004–0.01 μg/mL (Table 29.6) (Srinivas et al. 2010). Furthermore, lptD mutant studies confirmed its mechanism of action to be interacting with lptD (Werneburg et al. 2012). Recently in 2018, it was shown that L27-11 cross-links to periplasmic segment of LptD, containing a β-jellyroll domain and an N-terminal insert domain characteristic of *Pseudomonas* sp. This binding of L27-11 to periplasmic segment of *Pseudomonas* blocks LPS transport, consistent with the proposed mode of action and observed specificity of these antibiotics for *Pseudomonas* as previously reported (Schmidt et al. 2013; Vetterli et al. 2016; Andolina et al. 2018). A study published in 2020 identified genes required for resistance to polymixin B, colistin and peptidomimetic antibiotics, including L27-11 by transposon sequencing (Vitale et al. 2020). It was

Table 29.6 MIC of L-27-11 against various pathogenic microorganisms (Srinivas et al. 2010)

Strains	MIC (μg/mL)
P. aeruginosa ATCC 27853	0.01
P. aeruginosa PAO1	0.004
A. baumannii DSM3008	>64
K. pneumoniae ATCC 13883	>64
S. maltophilia ATCC 13637	>64
E. coli ATCC 25922	64
E. faecalis DSM 12956	>64
S. aureus ATCC 29213	>64

revealed that 13 core genes affected resistance to all tested antibiotics, most of which encoded for enzymes involved in LPS modification or control their expression. But while assessing the possible cross-resistances, L27-11 showed fewer overlaps with the other antibiotics indicating its distinct mechanism of action. However, it was also described that spermidine caused a decrease in the antibacterial activity of L27-11 with an eightfold increase in MICs (Vitale et al. 2020).

29.3.4.6 MsbA Inhibitors

Zhang et al. devised a cell-based screen to identify inhibitors of MsbA, an inner membrane protein which is an ABC ATP-dependent transporter and helps in flipping the LPS onto the periplasmic side of inner membrane (Zhang et al. 2018). In this study, screens were done against *Acinetobacter baumannii* that lacked efflux pumps and LPS modifying enzymes and compared its effects with wild type. The best identified inhibitor compound **1** exhibited MIC 0.5 μM against *Acinetobacter baumannii* lacking efflux pumps out, while retaining activity in wild-type strain (Zhang et al. 2018). The probable mechanism of action is revealed to be the stimulation of ATPase activity of MsbA resulting in its decoupling with LPS abolishing LPS transport and subsequent elevated LPS levels in inner membrane (Zhang et al. 2018).

In another study, quinolone inhibitors were discovered to be inhibiting MsbA in a high-throughput in vitro assay with purified MsbA. Of the three active molecules identified G592, G913 and G332, G332 was found to be the most potent with MIC 2.8–6 μg/mL against *E. coli, Klebsiella pneumoniae,* and *Enterobacter cloacae* (Fig. 29.12). However due to high levels of plasma protein binding, these could not be progressed (Alexander et al. 2018). Crystal structure was also obtained at 2.9Å of *E. coli* MsbA with bound quinoline inhibitor G907, which showed the protein trapped in an inward-facing LPS-bound conformation blocking both ATP hydrolysis and LPS transport (Ho et al. 2018). The detailed chemical optimization campaign behind the G-inhibitors (Alexander et al. 2018) with activity in micromolar range and low frequency of resistance against three wild-type strains of *E. coli, K. pneumoniae,* and *E. cloacae* including MDR strains was published in 2022 (Verma et al. 2022). The study also suggested that for activity against inner membrane targets in wild-type bacteria, inhibitors with periplasmically accessible binding

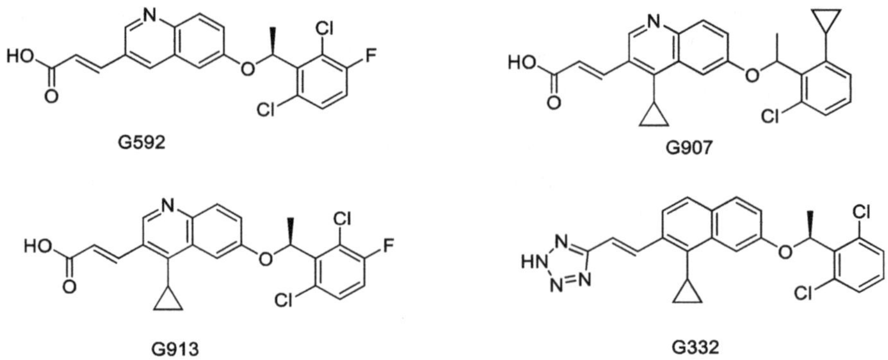

Fig. 29.12 Chemical structure of various MsbA inhibitors

sites should be encouraged over inhibitors that have binding regions fully encapsulated within the IM of GNBs (Verma et al. 2022).

A study published in 2021 delved into the mechanism of action of both the classes of MsbA inhibitors discovered by Zhang et al. and Alexander et al. (Thélot et al. 2021). The group demonstrated that both the classes of inhibitors had contrasting binding mechanisms with MsbA which provided important insights into ABC transporter pharmacology. Compounds discovered by Zhang et al. bound to MsbA differently than small-molecule modulators for other ABC transporters known. Two molecules of the inhibitor asymmetrically occupied the substrate-binding site, which lead to a collapsed inward-facing conformation with decreased distance between the nucleotide-binding domains (NBDs) (Thélot et al. 2021). In contrast, G-inhibitors discovered by Alexander et al. bound symmetrically to MsbA and increased NBD distance in a wide inward-open state of MsbA, which is similar to the ABC transporter inhibitors reported before (Thélot et al. 2021). These studies validate MsbA as an antibacterial target and have provided screening methods that can be used to discover its new inhibitors.

29.3.4.7 BamA Inhibitors

29.3.4.7.1 LlpA

Llp is a lectin-like bacteriocin that targets BamA, mediating the inhibition of selective GNB. However, the mechanism of inhibition is still not fully understood. In a recent study by Ghequire and group, BamA was identified as a receptor for LlpA-mediated killing in *Pseudomonas* sp. (Ghequire et al. 2013). Genetic analyses were performed to study the interaction between BamA and LlpA, which suggested that lectin-like domains of LlpA interact with specific carbohydrate moieties in LPS (Ghequire et al. 2018). Natural antimicrobial peptides are steadily gaining importance as antibacterial agents due to potent activity if their toxicity and stability can be controlled.

29.3.4.7.2 JB-95

JB-95 is a large conformationally constrained macrocycle β-hairpin peptidomimetic similar to POL group (Fig. 29.13). The most notable difference is the interchange of six lysine residues present in POL7001 with arginine residues in JB-95. JB-95 is active against a broad range of Gram-positive and GNB, *E. coli* in particular, and is not selective to *Pseudomonas* sp. as POL group. Its MIC against *E. coli* ATCC 25922, *A. baumannii* ATCC 17978, *P. aeruginosa* ATCC 27853, and *S. aureus* ATCC 29213 are 0.25, 1, 4, and 2 μg/mL respectively. JB-95 has also shown consistent antibacterial activity with MIC 0.25–0.5 μg/mL against a range of resistant *E. coli* clinical isolates. The time kill kinetics of JB-95 against *E. coli* ATCC 25922 revealed its bactericidal action at 2–8× MIC, although its bactericidal activity

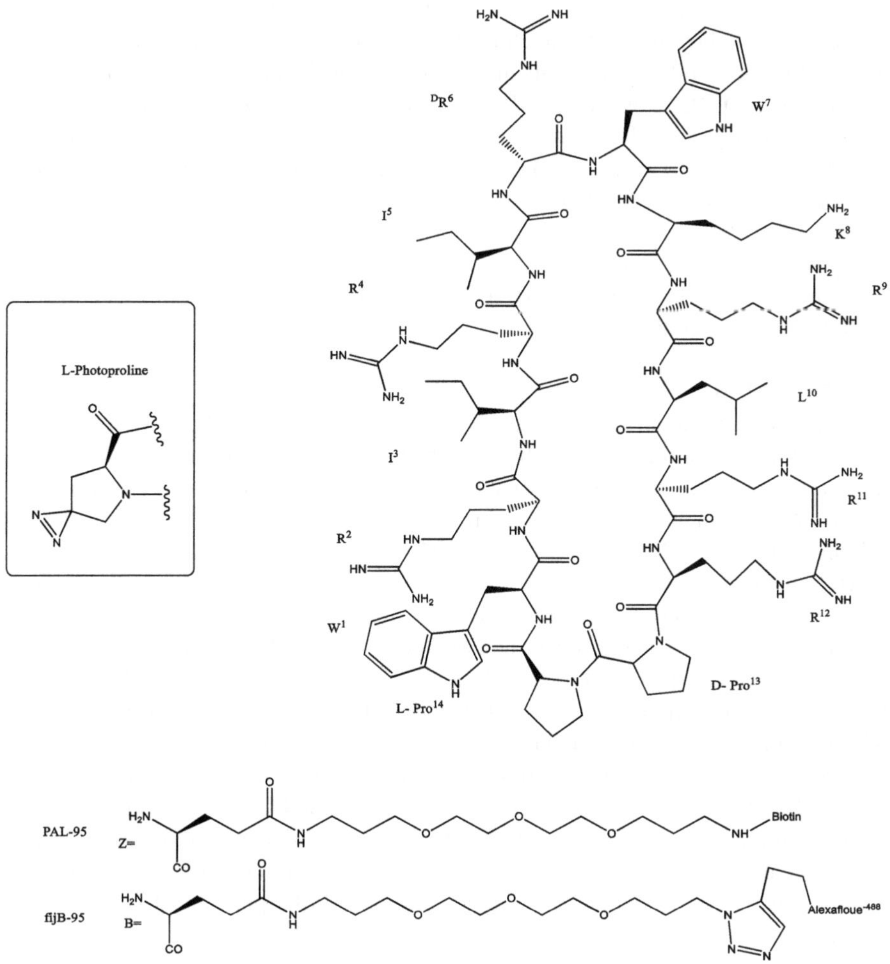

Fig. 29.13 Chemical structure of JB-95, PAL-95, fijB-95

is not as rapid as Protegrin I and took several hours, which is similar to POL peptidomimetics (Urfer et al. 2016).

NMR solution structure of JB-95 revealed that it adopts a stable β-hairpin conformation in aqueous solution. There was no hemolytic activity observed up to 100 μg/mL, which is the most common problem with peptide-based molecules. Fluorescence studies with SYTOX Green showed no rapid increase in fluorescence over 1 h indicating slow permeabilization of *E. coli* ATCC 25922 by JB-95 as opposed to polymyxin, which was able to permeabilize within 20 min. A fluorescently labeled form of JB-95 (flJB-95) was constructed with comparable antimicrobial activity against *E. coli* ATCC 25922 for high-resolution fluorescence microscopy of *E. coli* exposed to peptide. The fluorescence pattern showed that peptide was concentrated in a punctuated manner in OM rather than being distributed uniformly over OM. It was deduced that this was probably due to the binding of the peptide to OM protein patches such as BamA. This conclusion was further supported by the photo-labeling experiments which showed interaction of the peptide to several β-barrel proteins in OM. The photo-labeling experiments with the photoprobe PAL-95 revealed that JB-95 interacts with Bam A as well as LptD, both of which have β-barrel assembly complex in common being OM proteins. In PAL-95, L-Pro14 was replaced by photoreactive diazirine-containing L-photoproline and L-Arg9 by a Glu (PEG2-biotin) retaining the antimicrobial activity of JB-95. In summary, it was suggested that the interaction of peptide to OM proteins is the reason for the OM disruption activity of JB-95. Since LptD also helps in translocation of LPS to OM, so JB-95 has bigger potential to target OM from both ways (Urfer et al. 2016).

29.3.4.7.3 MAB1

Recently, a monoclonal antibody MAB1 was isolated from a library of anti-BamA IgG monoclonal antibodies (mAbs) which were screened against mutant *E. coli (ΔwaaD)* strains displaying a truncated LPS. MAB1 showed bactericidal activity against *E. coli* in nanomolar range. It was deduced that MAB1 inhibits BamA folding activity by lowered levels of OMPs in *E. coli* in antibody's presence. MAB1 induced σ^E stress response and caused permeabilization of cell envelope which was confirmed by enhanced uptake of ethidium bromide, a rapid loss of periplasmic mCherry, and a slower release of cytoplasmic Green Fluorescent Protein (GFP) from cells exposed to it (Storek et al. 2018). Further studies by Storek et al. revealed that OM environment is a critical determinant for the search for potential inhibitors of BamA and that an increase in the membrane fluidity through changes in the growth environment or alterations to LPS in the OM can provide resistance to MAB1 (Storek et al. 2019). Mutations in the periplasmic POTRA domain and within a transmembrane β-strand of BamA can also result in resistance to MAB1 inhibition (Storek et al. 2019). Future studies keeping optimal membrane fluidity as critical for Bam A function are needed to elucidate how extracellular portions of OMPs can be targeted via antibodies.

29.3.4.7.4 Darobactin

Darobactin was discovered by Imai and group while screening extracts of *Photorhabdus* and *Xenorhabdus* bacterial species residing in the gut of entomopathogenic nematodes as symbionts (Imai et al. 2019). The rationale behind the screening was to discover antimicrobial compounds produced by these nematophilic bacteria that have the same requirements for antibiotics as humans to fend off invading environmental microorganisms. Darobactin was identified as the active compound from a *Photorhabdus khanii* HGB1456 concentrated extract producing a small zone of *E. coli* growth inhibition on a petri dish. Mass spectroscopy and NMR studies revealed that darobactin is a modified heptapeptide with an amino acid sequence of WNWSKSF and characterized by a unique fused bicyclic peptide core expressed post-translationally. Its expression as a pro-peptide is from *darA* located within the *dar* operon which also encodes an ABC-type transenvelope exporter of darobactin (Imai et al. 2019).

Darobactin exhibited potent selective activity with MIC ranging from 2 to 16 μg/mL against GNB pathogens of WHO priority pathogen list (World Health Organization 2017). Against drug-resistant *E. coli* and *K. pneumoniae*, darobactin was bactericidal with MIC and MBC values of 2 and 8 μg/mL respectively (Imai et al. 2019). In a septicemia model of susceptible and polymyxin-resistant *E. coli*, carbapenemase producing *K. pneumoniae* and polymixin-resistant *P. aeruginosa* infection, a single dose of darobactin at 2.5, 50, and 50 mg/kg respectively resulted in 100% survival while the untreated animals died within 24 h of infection. In a murine thigh infection model with polymyxin-resistant *E. coli mcr-1*, darobactin given as a single injection of 50 mg/kg or as three injections of 25 mg/kg every 6 h significantly reduced bacterial burden (Imai et al. 2019).

Bam A was identified as target of darobactin through multiple experiments (Imai et al. 2019). Resistant mutants generated against darobactin had mutations in *bamA* shown to be solely responsible for resistance against it. Darobactin inhibited folding of protease OmpT in an in vitro folding assay, which is involved in BAM complex reconstitution. Also, it interacted with BamA with K_D 1.2 μM measured by isothermal titration colorimetry (Imai et al. 2019).

A study by Kaur et al. later validated through cryo-EM and crystal structure of Bam complex interacting with darobactin, that darobactin binds with high affinity to β-strand of Bam A lateral gate at the same position as β signal in turn competing with β signal of nascent OMPs; thereby blocking their entry into the lateral gate (Kaur et al. 2021). The study also suggested that interactions of darobactin with OM lipids stabilized its interaction with Bam A (Kaur et al. 2021). Thus, darobactin promises to be a potent GNB antibacterial to be explored further and studies revolving around its derivatives are already underway (Groß et al. 2021; Böhringer et al. 2021; Ritzmann et al. 2022).

29.3.4.7.5 Thanatin

Thanatin is a naturally occurring 21 residue peptide (GSKKPVPIIYCNRRTGKCQRM), including a disulfide bond between Cys^{11} and Cys^{18} and was isolated from *Podisusma culiventris*, a hemipetran insect (Fehlbaum et al. 1996; Ma et al. 2016). It is a cationic, cell agglutinating AMP demonstrating potent activity against *E. coli*, *Klebsiella pneumoniae*, *Salmonella typhimurium,* and *Enterobacter cloacae* with MIC $\leq$1.5 μM and higher MICs of 10–12 μM against *Erwinia carotovora* and *P. aeruginosa* while no activity was obtained against *S. aureus* (Hou et al. 2011). It showed weak permeabilizing effects on inner membrane, OM, and non-hemolytic activity against erythrocytes at 100× MIC and high plasma stability (Hou et al. 2011). Thanatin demonstrated nanomolar affinity for LptA and LptD, therefore inhibiting protein-protein interactions required for assembly of Lpt (Vetterli et al. 2018). Several mutations were reported in spontaneous generated mutants by whole genome sequencing but LptA was the common one. No mutations were detected in LptD (Vetterli et al. 2018). Hence, it showed low inherent ability to induce drug resistance. The structure of thanatin in complex with LPS revealed four stranded antiparallel β-sheet in a "head-tail" dimeric topology. By contrast, thanatin in free solution assumed an antiparallel β-hairpin conformation. Dimeric structure of thanatin displayed higher hydrophobicity and cationicity with sites of LPS interactions. Molecular dynamic simulations and biophysical interactions analyses provided a mode of LPS recognition and perturbation of LPS micelle structures (Vetterli et al. 2018).

Another study proposed a dual mechanism of action of thanatin where it disrupted the OM of NDM-1 producing bacteria by competitively displacing divalent cations on OM and inhibiting NDM-1 by displacing zinc ions from its active site (Ma et al. 2019a, b). Thanatin could successfully reverse carbapenem resistance in NDM-1 producing bacteria both in vitro and in vivo. However, OM disruption described in this study contrasted with findings of Vetterli et al. who stated that thanatin had weak permeabilizing effects on inner membrane and OM. The authors Ma et al. attributed this to high concentrations of Ca^{2+} (1 mM) and Mg^{2+} (0.5 mM) used in the membrane permeabilization assays by Vetterli et al.

A 2020 study revealed via bacterial adenylate cyclase two-hybrid (BACTH) system that thanatin impairs lipopolysaccharide transport complex assembly by targeting LptC–LptA interaction which was observed at sub-MIC concentrations of thanatin (Moura et al. 2020). It was also proposed that thanatin binds to LPS as a self-promoted mechanism of entry in periplasm and then targets LptA that undergoes degradation. This suggested that disassembly of Lpt machinery and subsequent blocking of LPS transport is the mechanism of action of thanatin against GNB pathogens (Moura et al. 2020).

29.3.4.7.6 Phe-Arg-β-naphthylamide (PAβN)

In a recent study, Phe-Arg-β-naphthylamide (PAβN), an efflux pump inhibitor (EPI) in GNB, with activity against efflux transporters such as *AcrB* in *E. coli* was reported. The study provided evidence of *LpxM* as a target for PAβN in LPS layer via in vitro random mutagenesis. The drug-sensitizing activity of PAβN with large lipophilic drugs was attributed to membrane activity caused by inhibiting LpxM, while its EPI property plays a separate role (Schuster et al. 2019).

29.4 Conclusions and Future Directions

The rapid emergence and spread of AMR particularly in GNB pathogens poses an immediate need for not only newer drugs but drugs that hit novel targets. As already established in numerous studies and being the crux of this review, OM in GNB acts as a remarkable barrier for antibiotics and is crucial in maintaining the bacterial fitness and virulence. Thus, the inhibitors of OM integrity can not only act as direct killing agents but can also facilitate the entry of drugs that are inactive against GNB due to lack of access to their intracellular targets. Needless to say, given its essentiality, OM components constitute prominent drug targets in antibacterial drug discovery against GNB. The second enzyme in the biosynthesis pathway LpxC has been extensively studied with one inhibitor even reaching the phase I trial, although it got terminated. However, metal-chelating moiety such as a hydroxamic acid in most of compounds that have been discovered to date against LpxC endows them with Zn^{++} chelating property posing a possible effect on human metalloenzymes as well, a significant concern for cytotoxicity. Hence, development of non-hydroxamate LpxC inhibitors is critically essential. Additionally, owing to vast biochemical and structural information available about other enzymes involved, i.e., LpxA-LpxM as discussed here, there is a huge scope of novel drug development against these targets with few inhibitors already reported.

Along with biosynthesis, transport of LPS to OM also has many target proteins other than LptD that are yet not fully explored. After murepavadin, new molecules targeting this fascinating transport machinery are warranted. Just like LPS transport inhibitors, high-throughput screens based on peptidomimetics might be helpful in discovering new Lol system inhibitors as well, since no highly effective inhibitors have yet been identified. The transport of phospholipids which form the inner leaflet of OM also has potential to be novel drug targets against GNB.

Since trafficking of phospholipids across IM and to OM is yet not fully understood, extensive studies are required in this field. Thus, along with intracellular enzyme targets for OM biogenesis, further studies on the potential of various cell transport channels across OM including both influx and efflux pumps as drug targets are also required. The targeting of specific components of OM will lead to narrow-spectrum antibiotics as opposed to conventional approach of developing broad-spectrum drugs; however, it might just be the need of the hour when coupled with companion diagnostics.

Acknowledgments GK thanks DST-Inspire for her fellowship, MS thanks UGC, and RKA and VVR thank NIPER for their fellowship. This study was supported by the DST grant from the Department of Science and Technology, Govt. of India, to S.C. and N.S. (SERB/F/2215/2017-18). This manuscript bears a CSIR-CDRI communication number.

Competing Financial Interests The authors declare no competing financial interests and conflict of interest.

References

Achaogen Inc. (2012) A study to assess the safety, tolerability, and pharmacokinetics of ACHN-975 in healthy volunteers. https://clinicaltrials.gov/ct2/show/study/NCT01597947?term=NCT01597947&rank=1. Accessed 15 May 2022

Achaogen Inc. (2013) A multiple dose study to assess the safety, tolerability, and pharmacokinetics of ACHN-975 in healthy volunteers. https://clinicaltrials.gov/ct2/show/NCT01870245?term=NCT01870245&rank=1. Accessed 15 May 2022

Alexander MK, Miu A, Oh A et al (2018) Disrupting gram-negative bacterial outer membrane biosynthesis through inhibition of the lipopolysaccharide transporter MsbA. Antimicrob Agents Chemother 62:e01142

Amponnawarat A, Chompunud Na Ayudhya C, Ali H (2021) Murepavadin, a small molecule host defense peptide mimetic, activates mast cells via MRGPRX2 and MrgprB2. Front Immunol 12: 2486

Andersen NH, Bowman J, Erwin A et al (2011) Antibacterial agents. US Patent 2011172174A1, 14 July 2011

Andersen NH, Bowman J, Erwin A et al (2004) Antibacterial agents. WIPO (PCT) WO2004062601A2, 29 July 2004

Andolina G, Bencze LC, Zerbe K et al (2018) A peptidomimetic antibiotic interacts with the periplasmic domain of LptD from *P. aeruginosa*. ACS Chem Biol 13:666–675

Balibar CJ, Grabowicz M (2016) Mutant alleles of lptD increase the permeability of *P. aeruginosa* and define determinants of intrinsic resistance to antibiotics. Antimicrob Agents Chemother 60: 845–854

Barb AW, Jiang L, Raetz CR et al (2007a) Structure of the deacetylase LpxC bound to the antibiotic CHIR-090: time-dependent inhibition and specificity in ligand binding. Proc Natl Acad Sci 104: 18433–18438

Barb AW, McClerren AL, Snehelatha K et al (2007b) Inhibition of lipid A biosynthesis as the primary mechanism of CHIR-090 antibiotic activity in *Escherichia coli*. Biochemistry 46:3793–3802

Barker CA, Allison SE, Zlitni S et al (2013) Degradation of MAC13243 and studies of the interaction of resulting thiourea compounds with the lipoprotein targeting chaperone LolA. Bioorg Med Chem Lett 23:2426–2431

Bodelón G, Marín E, Fernández LÁ (2015) Analyzing the role of periplasmic folding factors in the biogenesis of OMPs and members of the type V secretion system. Methods Mol Biol 1329:77–110

Bodewits K, Raetz CR, Govan JR et al (2010) Antimicrobial activity of CHIR-090, an inhibitor of lipopolysaccharide biosynthesis, against the Burkholderia cepacia complex. Antimicrob Agents Chemother 54:3531–3533

Böhringer N, Green R, Liu Y et al (2021) Mutasynthetic production and antimicrobial characterization of darobactin analogs. Microbiol Spectr 9:e01535

Bornheim L, McKinnell J, Fuchs-Knotts T et al (2013) Preclinical safety evaluation of the novel LpxC inhibitor ACHN-975 in rat and monkey. In: Poster Presented at the 53rd Annual Interscience Conference on Antimicrobial Agents and Chemotherapy, Denver, CO, 10–13 Sept 2013

Brown DG (2016) Drug discovery strategies to outer membrane targets in Gram-negative pathogens. Bioorg Med Chem 24:6320–6331

Brown MF, Reilly U, Abramite JA et al (2012) Potent inhibitors of LpxC for the treatment of Gram-negative infections. J Med Chem 55:914–923

Castanheira M, Rhomberg PR, Dietrich RR et al (2013) Activity of the first-in-class LpxC Inhibitor ACHN-975 and comparators tested against gram-negative organisms including isolates with characterized resistance mechanisms. In: Poster presented at the 53rd Interscience Conference on Antimicrobial Agents and Chemotherapy, Denver, CO, 10–13 Sept 2013

Centers for Disease Control and Prevention (US), National Center for Emerging Zoonotic and Infectious Diseases (US), Division of Healthcare Quality Promotion, Antibiotic Resistance Coordination and Strategy Unit (2019) Antibiotic resistance threats in the United States, 2019. https://stacks.cdc.gov/view/cdc/82532. Accessed 15 May 2022

Cigana C, Bernardini F, Facchini M et al (2016) Efficacy of the novel antibiotic POL7001 in preclinical models of *P. aeruginosa* pneumonia. Antimicrob Agents Chemother 60:4991–5000

Cohen F, Aggen JB, Andrews LD et al (2019) Optimization of LpxC inhibitors for antibacterial activity and cardiovascular safety. ChemMedChem 14:1560–1572

Delcour AH (2009) Outer membrane permeability and antibiotic resistance. Biochim Biophys Acta 1794:808–816

Dong H, Tang X, Zhang Z et al (2017) Structural insight into lipopolysaccharide transport from the Gram-negative bacterial inner membrane to the outer membrane. Biochim Biophys Acta Mol Cell Biol Lipids 1862:1461–1467

Eren E, Vijayaraghavan J, Liu J et al (2012) Substrate specificity within a family of outer membrane carboxylate channels. PLoS Biol 10:e1001242

Erwin AL (2016) Antibacterial drug discovery targeting the lipopolysaccharide biosynthetic enzyme LpxC. Cold Spring Harb Perspect Med 6:a025304

Fehlbaum P, Bulet P, Chernysh S et al (1996) Structure-activity analysis of thanatin, a 21-residue inducible insect defense peptide with sequence homology to frog skin antimicrobial peptides. Proc Natl Acad Sci 93:1221–1225

Forge Therapeutics (2018) Forging new therapies for multidrug resistant infections. Poster presented at the 28th European congress of clinical microbiology and infectious diseases, Madrid, Spain, 21–24 April 2018

García-Quintanilla M, Caro-Vega JM, Pulido MR et al (2016) Inhibition of LpxC increases antibiotic susceptibility in Acinetobacter baumannii. Antimicrob Agents Chemother 60:5076–5079

Ghequire MGK, Garcia-Pino A, Lebbe EKM et al (2013) Structural determinants for activity and specificity of the bacterial toxin LlpA. PLoS Pathog 9:e1003199

Ghequire MGK, Swings T, Michiels J et al (2018) Hitting with a BAM: selective killing by lectin-like bacteriocins. MBio 9:e02138

Groß S, Panter F, Pogorevc D et al (2021) Improved broad-spectrum antibiotics against Gram-negative pathogens via darobactin biosynthetic pathway engineering. Chem Sci 12:11882–11893

Hancock REW, Sahl H-G (2006) Antimicrobial and host-defense peptides as new anti-infective therapeutic strategies. Nat Biotechnol 24:1551–1557

Ho H, Miu A, Alexander MK et al (2018) Structural basis for dual-mode inhibition of the ABC transporter MsbA. Nature 557:196–201

Hou Z, Lu J, Fang C et al (2011) Underlying mechanism of in vivo and in vitro activity of C-terminal–amidated thanatin against clinical isolates of extended-spectrum β-lactamase–producing *Escherichia coli*. J Infect Dis 203:273–282

Imai Y, Meyer KJ, Iinishi A et al (2019) A new antibiotic selectively kills Gram-negative pathogens. Nature 576:459–464

Ito H, Ura A, Oyamada Y et al (2007) A new screening method to identify inhibitors of the Lol (localization of lipoproteins) system, a novel antibacterial target. Microbiol Immunol 51:263–270

Iwai N, Nagai K, Wachi M (2002) Novel S-benzylisothiourea compound that induces spherical cells in *Escherichia coli* probably by acting on a rod-shape-determining protein (s) other than penicillin-binding protein 2. Biosci Biotechnol Biochem 66:2658–2662

Jenkins RJ, Dotson GD (2012) Dual targeting antibacterial peptide inhibitor of early lipid A biosynthesis. ACS Chem Biol 7:1170–1177

Jenkins RJ, Heslip KA, Meagher JL et al (2014) Structural basis for the recognition of peptide RJPXD33 by acyltransferases in lipid A biosynthesis. J Biol Chem 289:15527–15535

Kalinin DV, Holl R (2017) LpxC inhibitors: a patent review (2010-2016). Expert Opin Ther Pat 27: 1227–1250

Kaur H, Jakob RP, Marzinek J et al (2021) The antibiotic darobactin mimics a β-strand to inhibit outer membrane insertase. Nature 593:125–129

Knowles TJ, Scott-Tucker A, Overduin M et al (2009) Membrane protein architects: the role of the BAM complex in outer membrane protein assembly. Nat Rev Microbiol 7:206–214

Kovacs-Simon A, Titball RW, Michell SL (2011) Lipoproteins of bacterial pathogens. Infect Immun 79:548–561

Krause KM, Haglund CM, Hebner C et al (2019) Potent LpxC inhibitors with in vitro activity against multidrug-resistant *P. aeruginosa*. Antimicrob Agents Chemother 63:e00977

Lee C-J, Liang X, Wu Q et al (2016) Drug design from the cryptic inhibitor envelope. Nat Commun 7:1–7

Lemaître N, Liang X, Najeeb J et al (2017) Curative treatment of severe Gram-negative bacterial infections by a new class of antibiotics targeting LpxC. MBio 8:e00674

Liang X, Gopalaswamy R, Navas F III et al (2016) A scalable synthesis of the difluoromethyl-allo-threonyl hydroxamate-based LpxC inhibitor LPC-058. J Org Chem 81:4393–4398

Lin L, Tan B, Pantapalangkoor P et al (2012) Inhibition of LpxC protects mice from resistant *Acinetobacter baumannii* by modulating inflammation and enhancing phagocytosis. MBio 3: e00312

Ma B, Fang C, Lu L et al (2019a) The antimicrobial peptide thanatin disrupts the bacterial outer membrane and inactivates the NDM-1 metallo-β-lactamase. Nat Commun 10:3517

Ma B, Niu C, Zhou Y et al (2016) The disulfide bond of the peptide thanatin is dispensible for its antimicrobial activity in vivo and in vitro. Antimicrob Agents Chemother 60:4283–4289

Ma X, Prathapam R, Wartchow C et al (2019b) Structural and biological basis of small molecule inhibition of Escherichia coli LpxD acyltransferase essential for lipopolysaccharide biosynthesis. ACS Infect Dis 6:1480–1489

Martin-Loeches I, Dale GE, Torres A (2018) Murepavadin: a new antibiotic class in the pipeline. Expert Rev Anti-Infect Ther 16:259–268

May JM, Sherman DJ, Simpson BW et al (2015) Lipopolysaccharide transport to the cell surface: periplasmic transport and assembly into the outer membrane. Philos Trans R Soc B Biol Sci 370: 20150027

McClerren AL, Endsley S, Bowman JL et al (2005) A slow, tight-binding inhibitor of the zinc-dependent deacetylase LpxC of lipid A biosynthesis with antibiotic activity comparable to ciprofloxacin. Biochemistry 44:16574–16583

McLeod SM, Fleming PR, MacCormack K et al (2015) Small-molecule inhibitors of Gram-negative lipoprotein trafficking discovered by phenotypic screening. J Bacteriol 197:1075–1082

Mdluli KE, Witte PR, Kline T et al (2006) Molecular validation of LpxC as an antibacterial drug target in *P. aeruginosa*. Antimicrob Agents Chemother 50:2178–2184

Moffatt JH, Harper M, Adler B et al (2011) Insertion sequence IS Aba11 is involved in colistin resistance and loss of lipopolysaccharide in *Acinetobacter baumannii*. Antimicrob Agents Chemother 55:3022–3024

Moffatt JH, Harper M, Harrison P et al (2010) Colistin resistance in *Acinetobacter baumannii* is mediated by complete loss of lipopolysaccharide production. Antimicrob Agents Chemother 54: 4971–4977

Montgomery JI, Brown MF, Reilly U et al (2012) Pyridonemethylsulfone hydroxamate LpxC inhibitors for the treatment of serious gram-negative infections. J Med Chem 55:1662–1670

Moura ECCM, Baeta T, Romanelli A et al (2020) Thanatin impairs lipopolysaccharide transport complex assembly by targeting LptC–LptA interaction and decreasing LptA stability. Front Microbiol 11:909

Munguia J, Degen D, Taganov K et al (2018) Improved gram-negative antibacterial activity upon combination of LpxC inhibitor FG-944 and RNA polymerase inhibitors (Rifampin or APY281). Poster presented at the American Society for Microbiology (ASM) Microbe 2018, Atlanta, GA, 7–11 June 2018

Nayar AS, Dougherty TJ, Ferguson KE et al (2015) Novel antibacterial targets and compounds revealed by a high-throughput cell wall reporter assay. J Bacteriol 197:1726–1734

Nickerson NN, Jao CC, Xu Y et al (2018) A novel inhibitor of the LolCDE ABC transporter essential for lipoprotein trafficking in Gram-negative bacteria. Antimicrob Agents Chemother 62:e02151

O'Neill J (2016) Tackling drug-resistant infections globally: final report and recommendations. https://amr-review.org/sites/default/files/160518_Final%20paper_with%20cover.pdf. Accessed 16 May 2022

Onishi HR, Pelak BA, Gerckens LS et al (1996) Antibacterial agents that inhibit lipid A biosynthesis. Science 274:980–982

Paetzel M, Karla A, Strynadka NCJ et al (2002) Signal peptidases. Chem Rev 102:4549–4580

Pathania R, Zlitni S, Barker C et al (2009) Chemical genomics in *Escherichia coli* identifies an inhibitor of bacterial lipoprotein targeting. Nat Chem Biol 5:849–856

Peng D, Hong W, Choudhury BP et al (2005) Moraxella catarrhalis bacterium without endotoxin, a potential vaccine candidate. Infect Immun 73:7569–7577

Piizzi G, Parker DT, Peng Y et al (2017) Design, synthesis, and properties of a potent inhibitor of *P. aeruginosa* deacetylase LpxC. J Med Chem 60:5002–5014

Polissi A, Georgopoulos C (1996) Mutational analysis and properties of the msbA gene of *Escherichia coli*, coding for an essential ABC family transporter. Mol Microbiol 20:1221–1233

Polyphor Ltd (2017) Murepavadin (POL7080). https://www.fda.gov/media/103043/download. Accessed 16 May 2022

Polyphor Ltd (2018a) Pivotal study in VAP suspected or confirmed to be due to *P. aeruginosa* (PRISM-MDR). https://clinicaltrials.gov/ct2/show/NCT03409679?term=NCT03409679&draw=2&rank=1. Accessed 16 May 2022

Polyphor Ltd (2018b) Pivotal study in nosocomial pneumonia suspected or confirmed to be due to Pseudomonas (PRISM-UDR). https://clinicaltrials.gov/ct2/show/NCT03582007. Accessed 18 May 2022

Polyphor Ltd (2019) Polyphor temporarily halts enrollment in the phase III studies of murepavadin for the treatment of patients with nosocomial pneumonia. https://www.polyphor.com/news/corporate-news-details/?newsid=1775911. Accessed 18 May 2022

Raetz CRH, Whitfield C (2002) Lipopolysaccharide endotoxins. Annu Rev Biochem 71:635–700

Rojas ER, Billings G, Odermatt PD et al (2018) The outer membrane is an essential load-bearing element in Gram-negative bacteria. Nature 559:617–621

Romano KP, Warrier T, Poulsen BE et al (2019) Mutations in pmrB confer cross-resistance between the LptD inhibitor POL7080 and colistin in *P. aeruginosa*. Antimicrob Agents Chemother 63:e00511

Ritzmann N, Manioglu S, Hiller S et al (2022) Monitoring the antibiotic darobactin modulating the β-barrel assembly factor BamA. Structure 30:350–359

Schmidt J, Patora-Komisarska K, Moehle K et al (2013) Structural studies of β-hairpin peptidomimetic antibiotics that target LptD in *Pseudomonas* sp. Bioorg Med Chem 21:5806–5810

Schuster S, Bohnert JA, Vavra M et al (2019) Proof of an outer membrane target of the efflux inhibitor Phe-Arg-β-naphthylamide from random mutagenesis. Molecules 24:470

Serio AW, Kubo A, Lopez S et al (2013) Structure, potency and bactericidal activity of ACHN-975, a first-in-class LpxC inhibitor. Poster presented at the 53rd interscience conference on antimicrobial agents and chemotherapy, Denver, CO, 10–13 Sept 2013

Simpson BW, May JM, Sherman DJ et al (2015) Lipopolysaccharide transport to the cell surface: biosynthesis and extraction from the inner membrane. Philos Trans R Soc B Biol Sci 370: 20150029

Smith PA, Koehler MFT, Girgis HS et al (2018) Optimized arylomycins are a new class of Gram-negative antibiotics. Nature 561:189–194

Smith PA, Romesberg FE (2012) Mechanism of action of the arylomycin antibiotics and effects of signal peptidase I inhibition. Antimicrob Agents Chemother 56:5054–5060

Sonntag I, Schwarz H, Hirota Y et al (1978) Cell envelope and shape of *Escherichia coli*: multiple mutants missing the outer membrane lipoprotein and other major outer membrane proteins. J Bacteriol 136:280–285

Spexis (2020) Inhaled murepavadin: first targeted antibiotic being developed for treating *P. aeruginosa* infections in people with cystic fibrosis. https://spexisbio.com/pol7080/#:~:text=Inhaled%20Murepavadin%20is%20highly%20potent,preclinical%20stage%20of%20its%20development. Accessed 18 May 2022

Srinivas N, Jetter P, Ueberbacher BJ et al (2010) Peptidomimetic antibiotics target outer-membrane biogenesis in *P. aeruginosa*. Science 327:1010–1013

Steeghs L, den Hartog R, den Boer A et al (1998) Meningitis bacterium is viable without endotoxin. Nature 392:449

Steinberg DA, Hurst MA, Fujii CA et al (1997) Protegrin-1: a broad-spectrum, rapidly microbicidal peptide with in vivo activity. Antimicrob Agents Chemother 41:1738–1742

Storek KM, Auerbach MR, Shi H et al (2018) Monoclonal antibody targeting the β-barrel assembly machine of *Escherichia coli* is bactericidal. Proc Natl Acad Sci 115:3692–3697

Storek KM, Vij R, Sun D et al (2019) The *Escherichia coli* β-barrel assembly machinery is sensitized to perturbations under high membrane fluidity. J Bacteriol 201:e00517

Taganov K, Corbett D, Stein D et al (2018) Antimicrobial activity of non-hydroxamate LpxC inhibitors. Poster presented at the American Society for Microbiology (ASM) Microbe 2018, Atlanta, GA, 7–11 June 2018

Tan JH, Vidaillac C, Yam JKH et al (2017) In vitro and in vivo efficacy of an LpxC inhibitor, CHIR-090, alone or combined with colistin against *P. aeruginosa* biofilm. Antimicrob Agents Chemother 61:e02223

Teng M, Vaddi S, Williamson B et al (2018) In vivo efficacy of non-hydroxamate LpxC inhibitors in murine infection models. Poster presented at the American Society for Microbiology (ASM) Microbe 2018, Atlanta, GA, 7–11 June 2018

Thélot FA, Zhang W, Song K et al (2021) Distinct allosteric mechanisms of first-generation MsbA inhibitors. Science 374:580–585

Tomaras AP, McPherson CJ, Kuhn M et al (2014) LpxC inhibitors as new antibacterial agents and tools for studying regulation of lipid A biosynthesis in Gram-negative pathogens. MBio 5: e01551

Urfer M, Bogdanovic J, Monte F et al (2016) A peptidomimetic antibiotic targets outer membrane proteins and disrupts selectively the outer membrane in *Escherichia coli*. J Biol Chem 291: 1921–1932

Ventola CL (2015) The antibiotic resistance crisis: part 1: causes and threats. Pharm Ther 40:277

Verma VA, Wang L, Labadie SS et al (2022) Discovery of inhibitors of the lipopolysaccharide transporter MsbA: from a screening hit to potent wild-type gram-negative activity. J Med Chem 65:4085–4120

Vetterli SU, Moehle K, Robinson JA (2016) Synthesis and antimicrobial activity against P. aeruginosa of macrocyclic β-hairpin peptidomimetic antibiotics containing N-methylated amino acids. Bioorg Med Chem 24:6332–6339

Vetterli SU, Zerbe K, Müller M et al (2018) Thanatin targets the intermembrane protein complex required for lipopolysaccharide transport in *Escherichia coli*. Sci Adv 4:2634

Vitale A, Pessi G, Urfer M et al (2020) Identification of genes required for resistance to peptidomimetic antibiotics by transposon sequencing. Front Microbiol 11:1681

Walsh SI, Peters DS, Smith PA et al (2019) Inhibition of protein secretion in *Escherichia coli* and sub-MIC effects of arylomycin antibiotics. Antimicrob Agents Chemother 63:e01253–e01218

Werneburg M, Zerbe K, Juhas M et al (2012) Inhibition of lipopolysaccharide transport to the outer membrane in *P. aeruginosa* by peptidomimetic antibiotics. Chembiochem 13:1767–1775

World Health Organization (2017) Prioritization of pathogens to guide discovery, research and development of new antibiotics for drug-resistant bacterial infections, including tuberculosis. https://apps.who.int/iris/handle/10665/311820. Accessed 16 May 2022

Yin H, Hamilton AD (2005) Strategies for targeting protein–protein interactions with synthetic agents. Angew Chem Int Ed 44:4130–4163

Zerbe K, Moehle K, Robinson JA (2017) Protein epitope mimetics: from new antibiotics to supramolecular synthetic vaccines. Acc Chem Res 50:1323–1331

Zgurskaya HI, López CA, Gnanakaran S (2015) Permeability barrier of Gram-negative cell envelopes and approaches to bypass it. ACS Infect Dis 1:512–522

Zhang G, Baidin V, Pahil KS et al (2018) Cell-based screen for discovering lipopolysaccharide biogenesis inhibitors. Proc Natl Acad Sci 115:6834–6839

30 Diphtheria: A Paradigmatic Vaccine-Preventable Toxigenic Disease with Changing Epidemiology

Ivneet Kour, Lipika Singhal, and Varsha Gupta

Abstract

Diphtheria is a paradigmatic example of a toxigenic infectious disease. It was Klebs who first identified *Corynebacterium diphtheriae* as the causative agent for diphtheria in 1883. It is an acute respiratory infection characterized by pseudo-membrane formation in the throat but can also cause cutaneous infections. Systemic effects are a result of the production of diphtheria toxin, which is an exotoxin that inhibits protein synthesis and leads to cell death. The toxin can commonly cause myocarditis and neuropathy, which are associated with increased mortality. Clinical diagnosis is of utmost importance and timely diagnosis and management are lifesaving. An attempt to confirm the diagnosis by isolating and identifying *Corynebacterium diphtheriae* by microbiological culture should be made. Enzymatic and toxin detection tests should confirm the isolate. Treatment consists of the administration of diphtheria antitoxin and antimicrobial therapy. Mainly a vaccine-preventable childhood disease, this disease has re-emerged in countries where the recommended vaccination programs are not sustained, and not only children but also adults are becoming prey to the disease. In the South East Asia region, thousands of diphtheria cases are reported annually. Globally, small pockets of outbreaks still occur in developed countries. There has been a change in the epidemiological trend of diphtheria around the world. In order to prevent the spread of such toxigenic strains in communities, clinical and epidemiological investigations are necessary along with strict public health measures. Recent outbreaks have highlighted the importance of vaccination in reducing the incidence in children and its re-emergence in adults.

I. Kour · L. Singhal · V. Gupta (✉)
Department of Microbiology, Government Medical College and Hospital, Chandigarh, India

P. P. Singh (ed.), *Recent Advances in Pharmaceutical Innovation and Research*,
https://doi.org/10.1007/978-981-99-2302-1_30

Keywords

Adult diphtheria · Changing epidemiology · Pseudo-membrane · Vaccine-preventable

30.1 History

The origin of *diphtheriae* is from the Greek word for "leather" or "hide," which describes the coating that appears in the throat that is called the pseudo-membrane. Diphtheria gained its official name from French physician Pierre Bretonneau (1778–1862), who called the disease diphtérite (Bretonneau 1826).

Bretonneau also distinguished diphtheria from scarlet fever. It was Klebs who was the first to identify the organism in 1883 (Klebs 1940), and Loeffler was the first to cultivate the organism in 1884 (Loeffler 1884). The bacterium came to be called the "Klebs-Loeffler" bacillus. Roux and Yersin demonstrated that the organism could produce a toxin in 1888, and von Behring and Kitasato in 1890 demonstrated that after administering an "anti-serum" or "antitoxin," the antibody-containing serum produced in animals infected with an attenuated strain would prevent mortality (Behring 2013). The first successful treatment of a child with diphtheria using the antitoxin occurred in Germany in 1891. Antitoxin was invented in the late nineteenth century but the toxoid was developed further in 1920 (Smith 1909).

30.2 Epidemiology

In 1921, the United States recorded 206,000 cases of diphtheria, resulting in 15,520 deaths. But the disease rate dropped quickly due to the widespread use of vaccines, which resulted in two cases in 2004 and 2017. Since 1980, globally, there has been a decline in the number of diphtheria cases. However, between the mid-1990s, there occurred a re-emergence of diphtheria cases in the former Union of the Soviet Socialist Republic. From 1980 to 2015, the number of cases reported to the World Health Organization (WHO) has declined to 2500 from nearly 100,000. In 2016, globally, 7100 cases of diphtheria were reported to the World Health Organization (WHO 2016).

A recent evaluation of the literature suggests several outbreaks from sub-Saharan Africa (e.g., Nigeria and Madagascar) since 2000. In 2014, for example, 22 cases of confirmed diphtheria were reported in the European Union, and about half of these cases were in Latvia. According to the Global Epidemiology of Diphtheria 2000–2017, the final dataset consisted of 15,380 cases of diphtheria (15,068 including age data and 7242 including vaccination status data) from 34 countries. Recently, an outbreak has been experienced in Bangladesh among the Rohingya's refugee camps in the year 2017 (Rahman and Islam 2019). So, in recent years, Venezuela, Yemen, and Bangladesh are the three countries where diphtheria has been in the headlines. Although cases of diphtheria have been reported from across the globe, it is an endemic primarily in developing regions of Africa, Asia, and South America (Clarke et al. 2019).

30.2.1 Indian Scenario of Diphtheria

Over the last 25 years, diphtheria has continued to persist in India without much decline, according to the Government of India's data on vaccine-preventable diseases. During 2001–2015, India accounted for nearly half of all diphtheria cases reported worldwide (Dabbagh et al. 2007). From 2005 to 2014, the Central Bureau of Health Intelligence (CBHI) reported 41,672 cases with 897 deaths with case fatality ratio of about 2.2% whereas Joint Reporting Form on Immunisation (JRF) data since 2000 reported 64% (CBHI 2005). However, on reviewing the literature, >50% of cases were from India. According to the database, 84% cases have been reported from 10 states namely, Andhra Pradesh, Assam, Delhi, Gujarat, Haryana, Karnataka, Nagaland, Maharashtra, Rajasthan, and West Bengal. Collectively, in short, India, Nepal, and Indonesia reported 96–99% of the cases from the South-East Asia region since 2000 (Singh et al. 1999). According to the authors, there have been few published reports on adult diphtheria outbreaks in India in the past few years, suggesting a shift in the incidence of diphtheria from children to adolescents, as shown in Table 30.1.

Murheka et al. reported that the median age of diphtheria cases in most of the published studies was 5 years. In some Indian states, Muslim children were affected more. Most of the diphtheria cases were unvaccinated. The coverage of the primary diphtheria vaccine in the country is around 80%, whereas the coverage of diphtheria boosters, although not available, is expected to be low (Murhekar and Bitragunta 2011).

The most probable reasons for the resurgence in cases are lack of awareness among parents in following up on immunization, and ignoring the booster doses. To prevent this infection, the World Health Organization (WHO) introduced Expanded Program of Immunization (EPI), which recommends three doses of DPT—diphtheria, pertussis, and tetanus—vaccine starting at 6 weeks of age with booster doses of diphtheria vaccine in countries with adequate resources. Besides the routine doses given at 6, 10, and 14 weeks of age, the Universal Immunization Program (UIP) in India offers two booster doses at 18 months and 56–72 months of age (World Health Organization 2016). Historically, children have been the major victims of diphtheria, but recent trends have shown a rise in adult diphtheria also. Several recent series also reported that the majority of cases (60–85%) occurred in children aged 0–10 years. In contrast, an outbreak during the 1990s in the New Independent States of the former Soviet Union included many adults, with 70% of cases occurring in patients older than 15 years of age (Vitek and Wharton 1998).

Table 30.1 Outbreaks of adult diphtheria

Sl. no	Study	Case/ outbreaks	Region/state	Age (years)	Sex	Immunization status	Treatment/outcome
1.	Loganathan and Bin Yusof (2018)	Case	Malaysia	29	M	Partially immunized	IV erythromycin; outcome: recovered
							IV crystalline penicillin 20 lac International unit (IU); IV metrogyl 500 mg
2.	Swarna et al. (2020)	Case	No available data (NAD)	28	F	Fully immunized	Tab erythromycin for 3 days; outcome: recovered
3.	Sangal et al. (2017)	Outbreak	Kerala	18	F	Partially immunized	Treatment not specified; outcome: recovered
4.	Das et al. (2016)	Case	Assam	18	M	No record of vaccination	Tab erythromycin; outcome: recovered
5.	Rahman and Islam (2019)	Outbreak	Malappuram	16 and 20	F	Partially immunized	IV azithromycin, ceftriaxone, nafcillin, and steroids; outcome: died
6.	Lurie (2004)	Case	Pennsylvania	63	M	Non-immunized	IV azithromycin, ceftriaxone, nafcillin, and steroids; outcome: died
7.	Singhal et al. (2021)	Case	Chandigarh	24	M	No record of vaccination	Treatment: Inj. amoxicillin and clavulanic acid 1 g bis in die/ twice a day (BD) for 5 days; outcome: recovered

BD, *F* female, *IU*, *IV* intravenous, *M* male, *NAD*

30.3 Pathogenesis

The only known reservoir for *Corynebacterium diphtheriae* is humans, and the disease is transmitted by respiratory droplets, exudate from skin lesions, or fomites. Pathogenesis of diphtheria is based on two determinants: one that is associated with colonization of the host and is encoded by the bacteria, and the other determinant that is associated with virulence and is encoded by corynebacteriophages:

- The first ability of *Corynebacterium diphtheriae* is to colonize the nasopharyngeal cavity and/or the skin and the second ability is to produce toxins. Thus, in a susceptible host, it colonizes on the mucosa of the respiratory tract and within the next few days, the toxigenic strains produce a potent exotoxin that causes inflammation and necrosis of the local tissue resulting in the formation of pseudo-membrane that consists of exudate, inflammatory cells, necrotic tissue, and the organism itself. Epidemiologic studies have demonstrated that a given isotype may be supplanted by another isotype and the persistence and the emergence of a new isotype is presumably due to its ability to colonize.
- Toxin when gets absorbed into systemic circulation causes dysfunction of various organs (heart, nervous system, kidneys). The exotoxin has two protein fragments, A and B, of which the B fragment binds to cell receptors enabling fragment A to enter the cell cytosol and inhibit protein synthesis. Diphtheria toxin specifically cleaves its protease-sensitive loop into two polypeptide fragments, A and B. Fragment A is the N-terminal 21 kDa component of the toxin and contains the catalytic center for the Adenosine Diphosphate (ADP)-ribosylation of elongation factor 2 (EF-2), whereas fragment B carries the transmembrane and the receptor-binding domains of the toxin. Non-toxigenic strains usually cause mild to moderate pharyngitis and do not form the typical pseudo-membrane and by lysogenic conversion, these strains can become toxigenic with corynebacteriophage-ß (Tao et al. 1994).

30.4 Diphtheria Toxin Production

As is well-known, diphtheria is a vaccine-preventable disease that causes systemic organ damage due to the production of toxins, which leads to the development of both an effective antitoxin-based therapy and a highly successful toxoid. The tox gene, which is the structural gene for the diphtheria toxin, is carried by a family of closely related corynebacteriophages, among which the phage has been the subject of the most research about diphtheria. A DtxR-like iron-activated repressor is responsible for controlling gene expression. Iron dissociates from DtxR in circumstances when it becomes the growth-rate limiting substrate, depressing tox; hence the expression of the gene is dependent on the physiological state of *C. diphtheriae*. Because of its extreme potency, diphtheria toxin can be fatal in doses as little as 100–150 mg/kg of body weight. A 535 amino acid long polypeptide chain makes up the diphtheria toxin (Parveen et al. 2019). Additionally, X-ray crystallography and biochemical genetic studies reveal that the toxin is made up of three structural/functional domains, namely:

1. A catalytic domain, i.e., N-terminal ADP-ribosyl transferase.
2. A transmembrane domain, which is a region that facilitates the delivery of the catalytic domain across the cell membrane.
3. The receptor-binding domain.

The following mechanism is involved in intoxication. First, the toxin is internalized by receptor-mediated endocytosis as a result of binding to its cell surface receptor, which causes charged receptors to adhere to the coated pits. Here, an Adenosine Triphosphate (ATP)-driven proton pump causes the endocytic vesicle to become acidic, the transmembrane domain to be inserted into the membrane, and the catalytic domain to be more easily delivered into the cytosol. Finally, the ADP-ribosylation of EF-2 causes the irreversible inhibition of protein synthesis (Greenfield et al. 1983).

30.5 Treatment

Diphtheria is highly contagious but a vaccine-preventable disease. The patient may remain contagious for about 48–72 h even after starting antibiotic therapy and for up to 2–6 weeks in case of untreated or partially treated patients (Truelove et al. 2020).

Patients should be evaluated promptly including evaluation for severity of symptoms, overall clinical status, vital signs, and development of any stridor or other systemic complication. Most importantly, diphtheria antitoxin (DAT) must be obtained on an emergency basis and administered as soon as possible. Treatment with diphtheria antitoxin and antibiotics should be started immediately in cases that are clinically compatible with diphtheria, without waiting for culture results. All patients should be monitored closely for the development of any respiratory or cardiac complications. Severe cases having a toxic appearance require continuous monitoring with special attention to airway patency and any cardiovascular instability. Sometimes it may require resuscitative measures like tracheostomy or rarely temporary cardiac pacing. In cases with airway compromise, immediate action should be taken (e.g., tracheostomy, intubation, clearing of pseudo-membrane endoscopically by an otolaryngologist).

All suspected diphtheria cases should be kept in isolation immediately in the emergency department of the healthcare facility. The patient should be treated following all droplet precautions and contact precautions. Of note, the Centre for Disease Control and Prevention (CDC) recommends only droplet precautions for respiratory diphtheria, although it notes that fomite transmission is also possible. Therefore, it seems prudent to use both droplet and contact precautions until the patient has two negative cultures obtained 24 h apart (Atkinson et al. 2007).

30.6 Antitoxin

Diphtheria antitoxin (DAT) was first produced in the 1890s and is still being produced using serum from horses that are hyperimmunized with diphtheria toxoid. The evidence for the efficacy of DAT for the treatment of diphtheria is found in studies done several decades ago. Mortality rates exceeded more than 50% in the pre-antitoxin era for clinical diphtheria. Almost as soon as antitoxin was available, a dramatic decline in mortality was noticed after the introduction of DAT. In a controlled trial when patients at a hospital were allocated with antitoxin treatment or no antitoxin treatment on an alternating day schedule, it was noticed that mortality in treated patients was 3.3% as compared to 12.2% among the untreated patients. So, it was concluded that mortality increased progressively from the onset of illness to treatment, with a pointy increase from 4% in those treated with antitoxin to 16.1% in those treated on their third day of illness. With prolonged intervals, mortality also continued to increase reaching 29.9% in those treated on day 7 or more days after the onset of illness. Current thinking is that toxin fixes to susceptible cells early in the disease and the fixed toxin is not neutralized by antitoxin, so it was proved that early treatment with DAT is inversely associated with the duration of clinical illness preceding its administration (Sevigny et al. 2013).

30.7 Vaccination

Immunity in an individual can be acquired either naturally or by vaccination, but it does not prevent asymptomatic carriage in an individual. Diphtheria, a vaccine-preventable disease, remains a global health threat, more so in developing countries. The reason for it being endemic is incomplete childhood vaccinations and thereafter inadequate boosters. Toxoid-containing diphtheria vaccines stay the oldest vaccines and continue to be in use. The United States was the first to start with active immunization against diphtheria based on a mixture of toxin and antitoxin both in 1914. In 1923, the diphtheria toxoid vaccine was developed by formaldehyde detoxification. In 1926, a more immunogenic alum-precipitated diphtheria toxoid was developed and it was in the 1940s that diphtheria toxoid was combined with tetanus toxoid and pertussis antigens and labeled as DPT/DTP, which was widely used throughout the world. Diphtheria toxoid is usually included with other antigens in combination with vaccines of varying strengths. For the description of diphtheria vaccines, uppercase letters like *D*, *P*, or *T* denote the full strength of diphtheria, pertussis, and tetanus whereas lowercase letters *d* and *p* indicate the reduced concentrations of diphtheria. Acellular components of a vaccine are represented by the lowercase "*a*" (DTaP). A combined vaccine of DTP, which contained whole-cell pertussis (DTwP), was introduced in 1948 whereas in 1990, acellular pertussis equivalent (DTaP) became available (Chitkara et al. 2019).

Co-administration of vaccines containing DTaP or DTwP does not interfere with antibody response, so such vaccines can be combined with Bacille Calmette-Guerin

(BCG), conjugate pneumococcal vaccine (PCV), inactivated polio vaccine (IPV), oral polio vaccine (OPV), and measles and rubella vaccines (Skibinski et al. 2011)

The advantage of conjugate vaccines is that it contains diphtheria toxoid or a protein carrier that induces a booster response in previously immunized individuals, so they will be safe when co-administered with other conjugate vaccines. According to the WHO recommendations, the higher potency of the diphtheria vaccine (D) is used for the immunization of children up to 6 years of age. Tetanus-diphtheria (Td), low-dose diphtheria toxoid formulations, and tetanus-diphtheria-acellular pertussis (Tdap) formulations are licensed for use from 5 years of age and 3 years of age, respectively. This reduction in diphtheria toxoid potency is sufficient to provoke an antibody response in older children and adults. DTP vaccines have been combined with other vaccine antigens of *Haemophilus influenzae* type b (Hib), poliomyelitis, and hepatitis B (HepB) virus (HBV) to make it a pentavalent vaccine and labeled as DPT-HBV-Hib and DTaP-IPV/Hib. Antigens manufactured against all six diseases offer the general benefits of higher valent combination vaccines for children, parents, and to healthcare providers and these are available as hexavalent vaccines. Three IPV-containing hexavalent vaccines are available in India: DTwP-Hib/HepB-IPV (Panacea Biotec), DTaP-IPV-HB-Polyribosylribitol phosphate (PRP)~T (Sanofi Pasteur), and DTaP-HBV-IPV/Hib (GSK). The main difference in their composition is that DTwP-Hib/HepB-IPV contains a wP component, DTaP-IPV-HB-PRP~T contains two aP components, and DTaP-HBV-IPV/Hib contains three aP components. We will therefore refer to them as wP-hexa, 2aP-hexa, and 3aP-hexa, respectively. wP-hexa has been available since 2017 and is only available in India. 2aP-hexa was launched in 2013 and has been available in India since 2016. 3aP-hexa has been available in India since 2018. All hexavalent vaccines are well tolerated, although whole-cell pertussis-containing vaccines may result in more solicited local reactions and fever than those with acellular pertussis components (Madhi et al. 2011).

Dosage is scheduled in India during the first 2 years of life. The Universal Immunization Program (UIP) recommends vaccination against the six diseases covered by the hexavalent vaccines: oral poliovirus (OPV) and HBV vaccines are given at birth, the pentavalent vaccines like DPT-HBV-Hib plus OPV are given at the age of 6, 10, and 14 weeks, while fractional doses, i.e., 1/5 full dose, via the intradermal route of inactivated polio vaccine (IPV) are given at 6 and 14 weeks; and the boosters of OPV and DTP are provided at 16–24 months (Mohanty et al. 2018). Most diphtheria toxoid-containing vaccines are administered as a 0.5 mL dose, by intramuscular injection only (Lalwani et al. 2017).

However, the Indian Academy of Pediatrics (IAP) recommends OPV and HBV vaccines at birth; DTP, HBV, and Hib (or pentavalent vaccine), and intramuscular IPV at age 6–10–14 weeks; and DTP, Hib, and IPV at 16–18 months. The most common solicited local adverse events (AEs) were pain/tenderness and pain while the most common solicited systemic AEs were fever (wP-hexa), irritability (2aP-hexa), and temperature (3aP-hexa). Serious adverse events were rare ($<2\%$ in each study) and none were judged to be related to vaccination. All these studies have reported that the hexavalent vaccines were well tolerated (Madhi et al. 2011).

A finding on the epidemiology of diphtheria reported by six representative studies from different countries and decades documented that not only children but also many adults remain susceptible to diphtheria, primarily due to waning immunity and failure to get the recommended booster doses and pertussis in the vaccine. DTaP and DT vaccines are used through 6 years of age. The DTaP is also formulated with inactivated polio–hepatitis B (Pediarix). Adult Td can vaccinate individuals either 7 years or older than that, Tdap vaccines are available for young people aged 10 and 18 years, and Adacel is approved for individuals aged 11–64 years (Kulkarni et al. 2011).

In a study by Grasse et al., it was stated that in multivalent tetanus or diphtheria vaccines, diphtheria is more frequently left unprotected than tetanus, with disastrous results. According to a recent outbreak in Yemen where more than 333 cases were reported, the mortality rate was over 10%, and 39% in suspected cases who were vaccinated against diphtheria, indicating a need to improve the current vaccination strategy and/or the vaccine. Grasse et al. found a correlation between antibody levels against diphtheria in both young and old persons and a number of immunological markers. The study came to the conclusion that the immune responsiveness to the diphtheria component in multivalent vaccinations would be impacted by the in vivo administration of Granulocyte-Macrophage Colony Stimulating Factor (GM-CSF). Recombinant GM-CSF will therefore enhance the humoral and cellular immune responses to diphtheria while leaving the response to tetanus unaffected when administered in vivo. Additionally, GM-CSF raised antibody titers against human immunodeficiency virus (HIV), *Chlamydia trachomatis*, and H5N1 influenza virus and functions as an adjuvant for diphtheria, but not tetanus toxoid (Grasse et al. 2018).

Revised recommendations for diphtheria vaccination were introduced in August 2017 by the WHO, which stated that in addition to the three-dose primary series in infancy, three diphtheria toxoid-containing booster doses were to be given at 12–23 months of age, 4–7 years of age, and 9–15 years of age. These recommendations, which harmonize with the updated recommendations for tetanus boosters released in February 2017, emphasize the need for a life course vaccination approach and present new opportunities for synergies with other vaccines. According to the Global Epidemiology of Diphtheria, it was noticed that vaccination coverage in a country has increased and the percentage of case patients of more than 15 years has also increased. In India, the national-level health surveys depicted that coverage of three doses of the diphtheria vaccine was 80% in the year 2015–2016 (World Health Organization 2016).

Immunization programs also ensure providing booster doses of diphtheria toxoid-containing vaccine during childhood and adolescence, which will protect adolescence and adulthood. The diphtheria booster should be given in combination with tetanus toxoid at the same schedule at 12–23 months of age, 4–7 years of age, and 9–15 years of age (Michel and Lang 2011). However, it remains a patch whether a booster dose later in life will be necessary to ensure life-long protection or not.

30.8 Summary

To summarize, *Corynebacterium diphtheriae* is an aerobic Gram-positive bacillus. It has three biotypes, namely gravis, intermedius, and mitis. All strains may produce toxins and only toxigenic strains can cause severe disease. Toxigenicity occurs when the bacillus is lysogenic by a specific virus bacteriophage carrying the tox gene. The toxin, if absorbed systemically, can affect organs and tissues distant from the site of invasion resulting in myocarditis and neuritis. The global incidence of diphtheria has decreased by >90% during 1980–2016 after the initiation of the WHO Expanded Program on Immunization in 1974. Recently, there is a changing trend regarding the age distribution of cases, from children to adolescents and adults. This reminds us of non-vaccination and failure to receive boosters among specific subpopulations, which even today remain vulnerable to this severe vaccine-preventable disease. Clinicians and public health practitioners should remember that primary prevention is through proper vaccination coverage and completing boosters as well.

References

Atkinson W, Hamborsky J, McIntyre L (2007) Epidemiology and prevention of vaccine-preventable disease, vol 10. Public Health Foundation, Washington, DC, pp 59–70

Behring EV (2013) Ueber das zustandekommen der diphtherie-immunität und der tetanus-immunität bei thieren. Drucke 16:1113–1114

Bretonneau P (1826) Des inflammations spéciales du tissu muqueux, et en particular de la diphthérite, on inflammation pelliculaire. Chez Crevot, excerpt in Major. Classic Descriptions, Paris, pp 159–161

CBHI (2005) Onwards. http://www.cbhidghs.nic.in/index1.asp?linkid=267. Accessed 15 Apr 2017

Chitkara AJ, Parikh R, Mihalyi A et al (2019) Hexavalent vaccines in India: current status. Indian Pediatr 56:939–950

Clarke KE, MacNeil A, Hadler S et al (2019) Global epidemiology of diphtheria, 2000–2017. Emerg Infect Dis 25:1834–1842

Dabbagh A, Eggers R, Cochi S et al (2007) A new global framework for immunization monitoring and surveillance. Bull WHO 85:904–905

Das PP, Patgiri SJ, Saikia L (2016) Recent outbreaks of diphtheria in Dibrugarh district, Assam, India. J Clin Diagn Res 10:1–3

Grasse M, Meryk A, Miggitsch C et al (2018) GM-CSF improves the immune response to the diphtheria-component in a multivalent vaccine. Vaccine 36:4672–4680

Greenfield L, Bjorn MJ, Horn G (1983) Nucleotide sequence of the structural gene for diphtheria toxin carried by corynebacteriophage beta. Proc Natl Acad Sci U S A 80:6853–6857

Klebs E (1940) Ueber Diphtherie. Bull Hist Med 8:509

Kulkarni PS, Raut SK, Dhorje SP (2011) Diphtheria, tetanus, and pertussis immunity in Indian adults and immunogenicity of Td vaccine. Int Sch Res Notices 28:1–4

Lalwani SK, Agarkhedkar S, Sundaram B (2017) Immunogenicity and safety of 3-dose primary vaccination with combined DTPa-HBV-IPV/Hib in Indian infants. Hum Vaccin 13:120–127

Loeffler F (1884) Utersuchung uber die Bedeutung der Mikroorganismen fir die Entstehung der Diptherie beim Menschen, bei der taube und beim Kalbe. Mitth Adkaiserl 2:421–499

Loganathan T, Bin Yusof MP (2018) Adult diphtheria in Malaysia: a case report. Med J Malaysia 73:340–341

Lurie P (2004) Fatal respiratory diphtheria in a U.S. traveler to Haiti Pennsylvania. JAMA 291:937–938

Madhi SA, Mitha I, Cutland C et al (2011) Immunogenicity and safety of an investigational fully liquid hexavalent combination vaccine versus licensed combination vaccines at 6, 10, and 14 weeks of age in healthy South African infants. Pediatr Infect Dis J 30:68–74

Michel JP, Lang PO (2011) Promoting life course vaccination. Rejuvenation Res 14:75–81

Mohanty L, Sharma S, Behera B (2018) A randomized, open label trial to evaluate and compare the immunogenicity and safety of a novel liquid hexavalent DTwP-Hib/Hep B-IPV (EasySix™) to licensed combination vaccines in healthy infants. Vaccine 36:2378–2384

Murhekar M, Bitragunta S (2011) Persistence of diphtheria in India. Indian J Community Med 36: 164–166

Parveen S, Bishai WR, Murphy JR (2019) Corynebacterium diphtheriae: diphtheria toxin, the tox Operon, and its regulation by Fe2+ activation of apo-DtxR. Microbiol Spectr 7:7–4

Rahman M, Islam K (2019) Massive diphtheria outbreak among Rohingya refugees: lessons learnt. J Travel Med 26:1–3

Sangal L, Joshi S, Anandan S, Balaji V et al (2017) Resurgence of diphtheria in North Kerala, India, 2016: laboratory supported case-based surveillance outcomes. Front Public Health 5:218

Sevigny LM, Booth BJ, Rowley KJ (2013) Identification of a human monoclonal antibody to replace equine diphtheria antitoxin for treatment of diphtheria intoxication. Infect Immun 81: 3992–4000

Singh J, Harit AK, Jain DC et al (1999) Diphtheria is declining but continues to kill many children: analysis of data from a sentinel centre in Delhi, 1997. Epidemiol Infect 123:209–215

Singhal L, Kour I, Gupta V et al (2021) Diphtheria in an adult: a paradigm of waning immunity. J Clin Diagn Res 15:1–3

Skibinski DA, Baudner BC, Singh M et al (2011) Combination vaccines. J Glob Infect 3:63–72

Smith T (1909) Active immunity produced by so called balanced or neutral mixtures of diphtheria toxin and antitoxin. J Exp Med 11:241–256

Swarna S, Sivagurunathan S, Bharathi S (2020) Diphtheria in an adult – a case report. Ind J Case Rep 6:160–162

Tao X, Schiering N, Zeng HY et al (1994) Iron, DtxR, and the regulation of diphtheria toxin expression. Mol Microbiol 14:191–197

Truelove SA, Keegan LT, Moss WJ (2020) Clinical and epidemiological aspects of diphtheria: a systematic review and pooled analysis. Clin Infect Dis 71:89–97

Vitek CR, Wharton M (1998) Diphtheria in the former Soviet Union: reemergence of a pandemic disease. Emerg Infect Dis 4:539–550

WHO (2016) Immunization, vaccines and biologicals. reported cases of selected vaccine preventable diseases (VPDs). http://www.who.int/immunization/monitoring_surveillance/data/en/

World Health Organization (2016) Summary of WHO position papers: recommendations for routine immunization, Geneva. http://www.who.int/immunization/documents/positionpapers/en/

Emerging and Re-Emerging Infections: An Overview

31

Varsha Gupta and Ritu Garg

Abstract

The incidence of emerging infectious diseases in humans has increased in the recent past or threatens to increase shortly. Over 30 new infectious agents have been detected worldwide in the last three decades. Thirty-seven percent of emerging and re-emerging pathogens are viruses and prions followed by protozoa. The majority of emerging and re-emerging human pathogens are known to be zoonotic. Transmissibility between humans might be the risk factor for the emergence of zoonotic pathogens. It is very difficult to afford the tremendous economic, social and public health burden of these diseases. National commitment and comprehensive efforts are necessary at all levels of health services to meet the threat of emerging and re-emerging infections.

Keywords

Emerging · Re-emerging · Zoonotic pathogens

31.1 Introduction

Emerging and re-emerging pathogens present a huge challenge to human and veterinary medicine. The incidence of emerging infectious diseases in humans has increased in the recent past or threatens to increase shortly. Over 30 new infectious agents have been detected worldwide in the last three decades (Dikid et al. 2013).

V. Gupta (✉)
Department of Microbiology, Government Medical College and Hospital, Chandigarh, India

R. Garg
Department of Microbiology, Dr. B. R Ambedkar State Institute of Medical Sciences, Mohali, India

P. P. Singh (ed.), *Recent Advances in Pharmaceutical Innovation and Research*,
https://doi.org/10.1007/978-981-99-2302-1_31

Thirty-seven percent of emerging and re-emerging pathogens are viruses and prions followed by protozoa (25%). This indicates that emerging and re-emerging pathogens are disproportionately viruses (Woolhouse 2002). The majority of emerging and re-emerging human pathogens are known to be zoonotic (Dikid et al. 2013). Transmissibility between humans might be the risk factor for the emergence of zoonotic pathogens. Two major categories of emerging infections—newly emerging and re-emerging infectious diseases—can be defined, respectively, as diseases that are recognised in the human host for the first time; and diseases that historically have infected humans, but continue to appear in new locations or drug-resistant forms, or that reappear after apparent control or elimination. Emerging/re-emerging infections may exhibit successive stages of emergence. These stages include adaptation to a new host, an epidemic/pathogenic stage, an endemic stage and a fully adapted stage in which the organism may become non-pathogenic and potentially even beneficial to the new host (e.g. the human gut microbiome) or stably integrated into the host genome (e.g. as endogenous retroviruses). These successive stages characterise the evolution of certain microbial agents more than others. For example, we describe as 're-emerging' new or more severe diseases associated with the acquisition of new genes by an existing microbe, e.g. antibiotic-resistant genes, even when mutations cause entirely new diseases with unique clinical epidemiologic features, e.g. Brazilian purpuric fever. Similarly, we refer to Severe Acute Respiratory Syndrome (SARS) as an emerging disease a decade after it disappeared, and apply the same term to the related Middle East Respiratory Syndrome (MERS) β coronavirus that appeared in Saudi Arabia in late 2012 (Morens and Fauci 2013).

Definition of Emerging Infectious Disease: These are infectious diseases whose incidence in humans have increased in the past two decades or threaten to increase shortly. These diseases include (Shastri and Bhat 2020):

- New infections resulting from changes or evolution of existing organisms
- Known infections spreading to new geographic areas or existing pathogens
- Previously unrecognised infections appearing in areas undergoing ecological transformations

Definition of Re-Emerging Infectious Diseases: They are old infections that are clinically silent or reduced in incidence but have again re-emerged in the community, as a result of:

1. Antimicrobial resistance in known agents
2. Breakdown in public health measure

The re-emerging infections that have increased in frequency in the last decade as a result of the development of antimicrobial resistance include:

- Multidrug-resistant tuberculosis (MDRTB)
- Extensively drug-resistant tuberculosis (XDRTB)
- Methicillin-resistant *Staphylococcus aureus* (MRSA)

- Vancomycin-resistant Enterococci (VRE)
- Vancomycin-resistant *Staphylococcus aureus* (VRSA)
- Beta lactamase producers
- Extended spectrum beta lactamase producers (ESBL)
- Carbapenemases producers
- AmpC beta lactamase producers

31.2 Factors Responsible for the Emerging and Re-Emerging Infectious Diseases

Emergence is most commonly associated with factors given in Table 31.1. Epidemics or pandemics caused by these emerging and re-emerging infections rapidly spreading across borders are responsible for much concern and panic (Morens and Fauci 2013).

31.3 Emerging and Re-Emerging Pathogens in India

Developing countries such as India suffer disproportionately from the burden of infectious diseases. Because of favourable environment and demographic and socio-economic factors, there are risks of epidemics of emerging infections in India. Over the years, the increase in cases of drug-resistant malaria, tuberculosis (TB), human immunodeficiency virus (HIV)-TB co-infections and epidemics of avian influenza has demonstrated the vulnerability of India to the threat of evolving microbes. In the recent past, India has seen outbreaks of eight organisms of emerging and re-emerging diseases in various parts of the country. Six of these infections are zoonotic, and five are of viral origin (Dikid et al. 2013). These are listed in

Table 31.1 Major factors that underlie disease emergence and re-emergence

The microbial agent	The human host	The human environment
Genetic adaption and change	Human susceptibility to infection	Climate and weather
Polymicrobial disease	Human demographics and behaviour	Changing ecosystems
	International trade and travel	Economic developments and land use
	Intent to harm	Technology and industry
	Occupational exposure	Poverty and social inequality
	Inappropriate use of antibiotics	Lack of public health services
		Animal populations
		War and famine
		Lack of political will

Source: Morens and Fauci (2013)

Table 31.2 Re-emerging infectious diseases in India

Organism	Year and location
Vibrio cholerae O139	1992, Chennai
Plague	1994, Surat 2002, Shimla 2004, Uttarakhand
Diphtheria Nipah	2000, Delhi 2001, Siliguri 2018, Calicut (Kerala)
Chandipura	2003, Andhra Pradesh 2004, Gujarat 2007, Maharashtra
Chikungunya	2005, Hyderabad
H1N1	Almost all states
Crimean-Congo haemorrhagic fever (CCHF)	2011, Gujarat

Source: Shastri and Bhat (2020)

Table 31.2, and the emerging infectious diseases in the world, of public health importance, are given in Table 31.3.

The recent outbreak of a novel (new) coronavirus termed '2019-nCoV' was initially noticed in a seafood market in Wuhan City in the Hubei Province of China on 12 December 2019 and had spread across China and many countries. On 30 January 2020, the International Health Regulations Emergency Committee of the World Health Organization (WHO) declared the outbreak a 'public health emergency of international concern' (PHEIC). On 11 February 2020, the World Health Organization named the disease 'Coronavirus Disease-2019' (abbreviated 'COVID-19'). This is an emerging, rapidly evolving situation. The outbreak of the respiratory disease was caused by a novel (new) coronavirus that was first detected in Wuhan City, Hubei Province, China. The WHO raised the global risk to the highest level of alert on 29 February 2020, and on 10 March 2020 declared it as a near uneven controllable pandemic.

Coronaviruses are a large family of viruses that are common in many different species of animals, including camels, cattle, cats and bats. Rarely, animal coronaviruses can infect people and then spread between people such as with MERS, SARS and now with this new virus (named SARS-Coronavirus 2 [SARS-COV2]). The SARS-COV2 virus is a beta coronavirus, like MERS and SARs, both of which have their origins in bats. (Cdc.gov/coronavirus/2019-nCoV/Summary.html 2019, WHO Novel Coronavirus (2019-nCoV) Situation Report-22.)

Since 13 May 2022, cases of Monkeypox have been reported to WHO from ten Member States that are not endemic to the Monkeypox virus. According to the WHO, as of 21 May, 92 confirmed cases and 28 suspected cases of Monkeypox have been reported from 12 Member States that are not endemic for the Monkeypox virus. It was first discovered in 1958 from monkeys kept for research. The first human case was recorded in 1970 in the Democratic Republic of the Congo (DRC). It primarily

Table 31.3 Emerging infectious diseases in the world of public health importance

Year	Microbes	Disease
1975	Parvovirus-B-19	Fifth disease
1976	*Cryptosporidium parvum*	Cryptosporidiosis
1977	Ebola virus *Legionella pneumophila* Hantavirus *Campylobacter jejuni*	Ebola haemorrhagic fever Legionnaire's disease Korean haemorrhagic fever Gastroenteritis
1980	Human T-lymphotropic virus (HTLV-1)	T-cell leukaemia/lymphoma
1981	Toxin-producing strains of *Staphylococcus aureus*	Various infections
1982	*Escherichia coli* O157:H7 Human T-lymphotropic virus II (HTLV-II) *Borrelia burgdorferi*	Food poisoning Lymphoma Lyme disease
1983	Human immunodeficiency virus (HIV) *Helicobacter pylori*	Acquired immunodeficiency syndrome (AIDS) Duodenal and gastric ulcer, stomach cancer
1985	*Enterocytozoon bieneusi*	Microsporidiosis diarrhoea
1986	*Cyclospora cayetanensis*	Diarrhoea
1988	Hepatitis E virus	Hepatitis
1989	*Ehrlichia chaffeensis* Hepatitis C virus	Human monocytic Ehrlichiosis Hepatocellular carcinoma
1991	Guanarito virus *Encephalitozoon hellem* New species of *Babesia*	Venezuelan haemorrhagic fever – Babesiosis haemolytic disease
1992	*Vibrio cholerae* O139 *Bartonella henselae*	Cholera Bacteraemia, endocarditis, bacillary angiomatosis and peliosis hepatis
1993	Sin Nombre virus *Encephalitozoon cuniculi*	Hantavirus cardiopulmonary syndrome –
1994	Sabia virus	–
1995	Human herpes virus 8 (HHV-8)	Kaposi's sarcoma
1999	Nipah virus	–
2002	SARS corona virus	Severe acute respiratory syndrome
2003	Influenza A (H5N1)	Avian influenza
2009	Influenza A (H1N1)	Swine flu
2012	Novel coronavirus or MERS-CoV	Severe respiratory syndrome
2013	Severe fever with thrombocytopenia syndrome (SFTS) virus	Severe fever with thrombocytopenia syndrome
2019	SARS-COV2	COVID-19
2022	Monkey pox	

COVID-19 Coronavirus Disease-2019, *MERS-CoV* Middle East Respiratory Syndrome-Coronavirus, *SARS-COV2* Severe Acute Respiratory Syndrome-Coronavirus 2
Source: Shastri and Bhat (2020)

occurs in tropical rainforest areas of Central and West Africa. The cases that have been reported worldwide are both due to local transmission and due to travel to African countries (World Health Organization, 4 June 2022: Disease Outbreak News; Multi-country monkeypox outbreak in non-endemic countries).

WHO Top Eight Emerging Diseases: In 2015, the WHO published the top eight emerging diseases that are likely to cause severe outbreaks and major epidemics in the near future. These diseases include:

1. Crimean Congo haemorrhagic fever
2. Ebola virus disease
3. Marburg virus disease
4. Lassa fever
5. MERS coronavirus disease
6. SARS coronavirus disease
7. Nipah virus disease
8. Rift Valley fever

This list provides a basis for working on research and development (R&D) preparedness to try and control potential future outbreaks.

Three other diseases were designated as 'serious' requiring action by the WHO to promote R&D as soon as possible; these were:

- Chikungunya
- Severe fever with thrombocytopenia syndrome
- Zika virus disease

Ebola virus disease (EVD), formerly known as Ebola haemorrhagic fever, is a rare but severe, often fatal illness in humans. The virus is transmitted to people from wild animals and spreads in the human population through human-to-human transmission. The Ebola virus causes an acute, serious illness that is often fatal if untreated. EVD first appeared in 1976 in two simultaneous outbreaks, one in what is now Nzara, South Sudan, and the other in Yambuku, DRC. The latter occurred in a village near the Ebola River, from which the disease takes its name. The 2014–2016 outbreak in West Africa was the largest Ebola outbreak since the virus was first discovered in 1976. The current 2018–2019 outbreak in eastern DRC is highly complex, with insecurity adversely affecting public health response activities (WHO, Ebola Virus Disease, 10 February 2020).

31.4 Controlling Emerging and Re-Emerging Diseases

A strategic vision and an effective plan of action are developed to combat emerging and re-emerging infections. There is the application of sophisticated epidemiological and molecular technologies, with changes in human behaviour and a national perspective. The WHO has made several recommendations for national strategies including the need to strengthen epidemic preparedness and rapid response, public health infrastructure, risk communication, research and its utilisation, and advocacy for political commitment and partnership building (World Health Organization; accessed on June 3, 2012).

31.4.1 Initiatives Underway in the Country

31.4.1.1 Strengthening Surveillance and Rapid Response Mechanisms

Surveillance is an important keystone for the control of emerging and re-emerging infections. After the outbreak of plaque in 1994, the Government constituted a technical committee to suggest measures to prevent the recurrence of such outbreaks. In 1995, it was asserted that there is a need to strengthen the public health system to effectively implement and evaluate national health programmes and to prevent outbreaks that can have national and international consequences. In 1997, the National Surveillance Programme on Communicable Diseases (NSPCD) was established. In 2004, the Integrated Disease Surveillance Project (IDSP) was established in 101 districts and expanded to cover all states and districts in the country with a rapid response team to quickly manage the disease outbreak in any part of the country. Internet connectivity has been established with all the States, districts and medical colleges for rapid data transfer, video conferencing and distance learning activities (WHO Global Consultation on strengthening national capacities for surveillance and control of communicable disease 2003).

31.4.1.2 Building Capacity in Epidemiology

Short- and long-term Field Epidemiology Training Programmes (FETP) have been started to augment epidemiological capacities at the local, state and national levels. National Centre for Disease Control (NCDC), National Institute of Epidemiology in Chennai and Epic Intelligence Services run by the Centers for Disease Control and Prevention (CDC) Atlanta, with the collaboration of the Government of India are conducting training programmes to augment epidemiological capacity by creating a cadre of highly skilled field epidemiologists at local, state and national levels (Dhariwal and Jain 2010; MOHFW; EIS training programmes in India 2012).

31.5 Importance of Laboratory Diagnosis for Emerging and Re-Emerging Infectious Diseases

Accurate laboratory diagnosis is useful for clinical or infection control management and is needed for public health surveillance purposes (IDSP, accessed on April 26, 2012).

Recommended laboratory testing in emerging infections:

- Molecular testing: highly sensitive and specific
- Rapid tests need to confirm with molecular tests
- Dengue NS1 antigen

- Antimicrobial resistance: susceptibility testing, gene detection by polymerase chain reaction (PCR)

So, there is a need of strengthening laboratory networks in surveillance through a collaboration of international centres like Centers for Disease Control and Prevention (CDC) Atlanta, National Reference laboratories like National Institute of Communicable Diseases (NICD) Delhi, National Institute of Virology (NIV) Pune and National Institute of Cholera and Enteric Diseases (NICED) Kolkata, State laboratories, intermediate laboratories (district/medical colleges) and peripheral laboratories (Primary Health Centers [PHC], Community Health Centers [CHC]).

31.5.1 Research and Development

The Government of India created a new Department of Health Research in the Ministry of Health and Family Welfare in 2007. The Department has specific programmes focusing on emerging and re-emerging diseases such as: the identification of agents; the development of diagnostic tests; formulation of case management modules and preventive strategies; and establishment of laboratories to handle new, exotic and dangerous organisms. In the late 1990s, the Indian Council of Medical Research (ICMR) stepped up its funding in communicable diseases, which increased the speed of research inputs for emerging infections (MOHFW; Annual report to people on health 2010; Kant 2008).

31.5.2 Information Sharing and Partnerships

Under International Health Regulations (IHR), national focal points are required to work closely with relevant ministries in the timely identification of extraordinary public health events. As the national focal point for International Health Regulations (IHR) in India, the National Centre for Disease Control (NCDC) is in the process of identifying and partnering with other relevant ministries in the identification of public health emergencies of international concern (PHEIC) (Dikid et al. 2013).

31.5.3 Key Tasks in Dealing with Emerging Diseases

- Surveillance at national, regional and global levels
  - Epidemiological
  - Laboratory
  - Ecological
  - Anthropological
- Investigation and early control measures

- Implement preventive measures
  - Behavioural
  - Political
  - Environmental
- Monitoring and evaluation

31.6 Conclusions

As India is a country of extreme geo-climatic diversity, there is a continuous threat of emerging and re-emerging infections of public health importance. National commitment and comprehensive efforts are necessary at all levels of health services to meet the threat of emerging and re-emerging infections. Sensitive rapid response mechanisms at various levels of health service are the cornerstone to detecting public health threats and responding quickly enough to protect valuable human lives. There is also a persistent need to gain detailed insights into disease bionics, including vector biology and environmental factors influencing the diseases. It is also important to strengthen the emergency preparedness for these diseases and response by focusing on a 'one health' approach.

31.7 Summary

Emerging and re-emerging pathogens present a huge challenge to human and veterinary medicine. The majority of emerging and re-emerging human pathogens are known to be zoonotic. Transmissibility between humans might be the risk factor for the emergence of zoonotic pathogens. Because of the favourable environment and demographic and socio-economic factors, there is a risk of epidemics of emerging infections in India. Horizontal gene transfer plays a principal part in the molecular evolution of novel bacterial pathogens. Emerging infections are defined as those whose incidence in humans has increased over the past two decades or poses a threat of increasing in the near future. These diseases transcend national boundaries (Sastry and Bhat 2023). The WHO defines 're-emerging infectious diseases' as those that are 'due to the reappearance of, and an increase in the number of infections from a disease, which is known, but which had formerly caused so few infections that it had no longer been considered a public health problem' (Baveja 2021). Factors responsible for emerging and re-emerging infectious diseases are related to microbial agents, human hosts and the environment. India has seen outbreaks of eight organisms of emerging and re-emerging diseases in various parts of the country. Six of these infections are zoonotic, and five are of viral origin. In 2015, the WHO published the top eight emerging diseases that are likely to cause severe outbreaks and major epidemics shortly. A strategic vision and an effective plan of action are developed to combat emerging and re-emerging infections. Accurate laboratory diagnosis is useful for clinical or infection control management and is needed for public health surveillance purposes. National commitment and comprehensive efforts are necessary at all levels of health services to meet the threat of emerging and re-emerging infections.

References

Baveja CP (2021) Chapter 34 Emerging and re-emerging infections. In, Complete microbiology for MBBS, p 383

Cdc.gov (2019). Cdc.gov/coronavirus/2019-nCoV/Summary.html

Dhariwal AK, Jain SK (2010) Proceedings of South East Asia Regional Conference on Epidemiology, March 8–10. World Health Organizations, Regional Office for South-East Asia; Field epidemiology training programme in India: ten years of experience, New Delhi, pp 285–288

Dikid T, Jain SK, Sharma A et al (2013) Emerging & re-emerging infections in India: an overview. Indian J Med Res 138:19–31

Government of India, Ministry of Health and Family Welfare (MOHFW) (2010) Annual Report to People on Health. MOHFW, New Delhi

Government of India, Ministry of Health and Family Welfare (MOHFW) (2012) Compendium of operational guidelines for epidemic intelligence service (EIS) like training programme in India. MOHFW, New Delhi

IDSP (2012) Integrated disease surveillance project. Accessed 26 Apr 2012

Kant L (2008) Combating emerging infectious diseases in India: orchestrating a symphony. J Biosci 33:425–427

Morens DM, Fauci SA (2013) Emerging infectious diseases: threats to human health and global stability. PLoS Pathog 9:1–3

Shastri A, Bhat S (2020) Essentials of medical microbiology. Annexure 1, 2nd edn. Jaypee Brothers Medical, New Delhi

Sastry AS, Bhat S (2023) Essentials of medical microbiology, 4th edn. New Delhi, Jaypee Brothers, p 683

WHO (2020) Ebola Virus Disease. Accessed 10 Feb 2020

WHO Novel (2019) Coronavirus (2019-nCoV) Situation Report-22

Woolhouse MEJ (2002) Population biology of emerging and re-emerging pathogens. Trends Microbiol 10(10):S3–S7

World Health Organization (2003) Global Consultation on strengthening national capacities for surveillance and control of communicable disease. WHO/CDS/CSR/LYO/2005.18

World Health Organization (2012) Asia-pacific strategy for emerging diseases. World Health Organization, Geneva. Accessed 3 Jun 2012

World Health Organization (2022) Disease outbreak news; Multi-country monkeypox outbreak in non-endemic countries: update. https://www.who.int/emergencies/disease-outbreak-news/item/2022-DON390

MIX
Papier aus verantwortungsvollen Quellen
Paper from responsible sources
FSC® C105338

If you have any concerns about our products,
you can contact us on
ProductSafety@springernature.com

In case Publisher is established outside the EU,
the EU authorized representative is:
Springer Nature Customer Service Center GmbH
Europaplatz 3, 69115 Heidelberg, Germany

Printed by Libri Plureos GmbH
in Hamburg, Germany